Extracellular Matrix: Pathobiology and Signaling

Edited by Nikos K. Karamanos

Extracellular Matrix: Pathobiology and Signaling

Edited by Nikos K. Karamanos

DE GRUYTER

Editor
Professor Nikos K. Karamanos
Biochemistry, Biochemical Analysis & Matrix Pathobiology Research Group
Laboratory of Biochemistry
Department of Chemistry
University of Patras
261 10 Patras
Greece
n.k.karamanos@upatras.gr

Front cover image: Uptake of heparin by melanoma cells. FITC-labelled heparin was visualized by confocal microscopy (40 x magnification). Kindly provided by Dr. D. Nikitovic-Tzanakakis.

ISBN 978-3-11-025876-9
e-ISBN 978-3-11-025877-6

Library of Congress Cataloging-in-Publication Data

A CIP catalog record for this book has been applied for at the Library of Congress.

Bibliographic information published by the Deutsche Nationalbibliothek

The Deutsche Nationalbibliothek lists this publication in the Deutsche Nationalbibliografie; detailed bibliographic data are available in the internet at http://dnb.d-nb.de.

Typesetting: Apex CoVantage, LLC
Printing and binding: Hubert & Co. GmbH & Co. KG, Göttingen

♾ Printed on acid-free paper.
Printed in Germany
www.degruyter.com

Contents

Israel Vlodavsky, Michael Elkin, [illegible], Ji-Ping Li, Ralph D. Sanderson, and Neta Ilan

Preface

Over the past few decades cell biology and biological chemistry have increasingly turned their attention to the space among cells (extracellular matrix, ECM) and revealed elaborate networks of macromolecules essential for structural support, cell migration, adhesion, and signaling. Macromolecules, such as proteoglycans and glycoproteins, collagens, laminin, fibronectin, matrix proteases, enzymes regulating glycan synthesis, and degradation are all components of ECM. Upon interactions in the extracellular space they affect ECM dynamics and via cell surface receptors and co-receptors (e.g., integrins and syndecans) communicate with the cellular compartment.

Progress in biochemical tools and the advancement of existing and the introduction of new analytical and biotechnological techniques have enabled the determination of the structural characteristics of various classes of matrix macromolecules in unprecedented detail. Studies of their biological roles using approaches from the field of cell biology helped the growth and expansion of the field of ECM biology.

ECM constituents contribute to tissue organization, development, and remodeling as well as its properties, such as elasticity and strength. Due to their unique structural features, matrix components interact with each other, with cell surface receptors, and with the various growth factors, thus regulating cell signaling and function. It is well established that they are all partners of a dynamic elaborative network that regulates cell functions and properties; that is, proliferation and apoptosis, differentiation, adhesion, and migration.

Matrix-mediated regulation of the various cell functions has been demonstrated during the past decade based on *in vitro* and *in vivo* models. Therefore, ECM molecules are essential players in physiological and pathological conditions. Mutations and structural alterations of matrix effectors are closely related to the development and progression of common diseases, such as cancer, including invasion and metastasis, cardiovascular disease, musculoskeletal disorders, inflammatory, and rarer conditions.

The number of research articles published year by year about ECM is increasing so fast that it would not be possible to cover all ECM biology within the limited space of one book. Therefore, the presentation of key aspects of ECM organization and regulation in health and disease was a great challenge for this book's editor and section editors. Our ambition was to present a comprehensive handbook of ECM, and to that direction 53 chapters were organized into an introductory chapter and seven thematic sections, giving an overview of the current state of knowledge of matrix components (structure and function), their roles in health and disease (matrix pathobiology and signaling) as well as new concepts of pharmacological targeting.

Thanks to the enthusiastic contribution of the section editors and chapter authors, *Extracellular Matrix: Pathobiology and Signaling* covers major areas of importance. Each section commences with an introduction written by the section editor, providing the reader with a short historical background, where necessary, as well as the importance of the ECM molecules and their interacting networks on cell properties and function.

Each chapter is organized as a timely review including concluding remarks/future perspectives, and a take-home message.

The section edited by Evi Heldin focuses on the field of *glycobiology* and covers within six chapters key molecules and areas, among them hyaluronan and sulfated glycosaminoglycans in cancer, inflammations, and metabolic control. In addition to giving insights into the function of glycans, this section also highlights striking differences in their biosynthetic machinery. Furthermore, their role in the diagnosis and therapeutics of various diseases is discussed.

The section edited by Liliana Schaefer comprises seven chapters that cover the major families of *proteoglycans* (aggrecan, versican, perlecan, small leucine-rich proteoglycans, syndecans, glypicans, and serglycin). It summarizes our rapidly expanding knowledge in the field of proteoglycan biology with emphasis on proteoglycans' intricate composition and multiple functions. This section particularly aims to recapitulate the newly described "dual function" of proteoglycans acting both as structural components and signaling molecules, depending on the biological context. It describes the implications of these signaling events in various diseases. Furthermore, it promotes the idea to translate our knowledge of proteoglycan signaling into novel therapeutic strategies for the treatment of malignant, inflammatory, and fibrotic diseases.

The section edited by Jan-Olof Winberg has six chapters that review various aspects of some of the most well-studied families of the main *proteases* found in the extracellular space. This section emphasizes the role of proteases and their endogenous inhibitors in tissue remodeling, wound healing, as well as their regulatory role in health and disease. Complex formation of individual matrix metalloproteinases (MMPs) with other MMPs, ECM macromolecules, non-ECM molecules, and cell surface receptors as well as the biological role of these complexes are also described. Furthermore, new techniques used to study the degradome, that is, the complete set of proteinases produced by an organism, are also described.

The section edited by Donald Gullberg is divided into five chapters dedicated to *cell surface receptors* covering integrins, syndecans, CD44, DDR, and their roles in cell signaling and pathophysiology as well as mechanical strain and TGFβ signaling. The matrix receptor section predicts an exciting future in the field of ECM receptor research, especially with regard to the contribution of these receptors to different disease mechanisms. Equipped with new methodological tools and refined analysis techniques, it is suggested that in the years to come new liaisons and mechanisms of action will be identified where matrix receptors will play both supporting as well as leading roles.

The section edited by Ruggero Tenni encompasses six chapters that describe up-to-date information on *collagens*. Emphasis is on the expanding knowledge of some basic issues on collagens themselves, their folding, maturation and morphology, and mutations in collagen alpha chains. The issues discussed are: the characteristics of trimerization domains in different collagen types; a class of proteinases, originally identified as components of procollagen processing, involved, for example, in morphogenesis and repair; a deeper understanding the effect of several factors on the phenotype in human pathologies caused by missense mutations in the triple helix of collagens, such as Osteogenesis Imperfecta; collagen I fibrils' morphology itself, which still deserve a discussion of the state of the art and of the open questions. The high degree of connectivity of collagen with "countless" ligands is also emphasized, in two chapters, by discussing the interactome of several collagen types and by the detailed knowledge of the interaction for the few collagen ligands whose binding site in collagen is known.

The section edited by Achilleas Theocharis focuses, in its seven chapters, on the *emerging aspects in ECM pathobiology*, where the implication of certain ECM molecules, among them hyaluronidases, EMMPRIN, syndecan-1, and serglycin, in the development and progression of malignancies is discussed. A proteomic approach for studying ECM involvement in cancer as well as the utility of 3D matrix systems and new quantitative methods in studying cancer are also presented. It also exhibits the role of bone sialoprotein in skeletal infections.

The last section, which I edited, groups together a series of eight chapters dealing with novel approaches in respect to the *pharmacological targeting* at the ECM level in cancer. In particular, the importance of tumor microenvironment, targeting epithelial to mesenchymal transition at the level of protein-glycan interactions, the importance of growth factor effects in the expression of ECM molecules with possible pharmaceutical interventions, targeting shedding syndecans in cancer, fresh ideas on the proteoglycan receptors with phosphatase activity, and the importance of heparanase, a multifaceted protein in inflammation and cancer, are presented. Aspects regarding delivery systems as well as glycan analogues as therapeutic agents are also presented within the chapters.

The text and the contents of this book have been organized in such a way so as to be easily understood by researchers in distinct scientific fields; that is, biochemistry, cell and molecular biology, biotechnology, pharmacology, and medicine, in both academic and industrial sectors. We hope that this book will put forward the emerging concepts and future prospects of the field of *ECM Pathobiology and Signaling* and will also stimulate new studies in this very interesting scientific area that will provide an in-depth understanding of disease progression and treatment, and will attract new investigators in the field.

This book has been developed through the efforts of 124 well-known scientists for their achievements in the field of ECM, representing 16 countries and 4 continents. I wish to express my gratitude to all authors and section editors for their contributions, and the harmonic collaboration during all preparation steps of this book. Moreover, the contribution of my colleagues in this endeavor and particularly of Achilleas Theocharis and Chrisostomi Gialeli is greatly acknowledged. From my personal view, I feel amply rewarded learning a lot during reading and editing the chapters of this book.

I would also like to thank the editorial director, Dr. Stephanie Dawson, and the production team at de Gruyter for their contribution to the successful realization of this multisection and multiauthor book, both in a printed and electronic version (e-book).

Nikos K. Karamanos
Editor

Comments on the book *Extracellular Matrix: Pathobiology & Signaling*

by Dick Heinegård, MD, PhD
Department of Clinical Sciences Lund, Section Rheumatology, Lund University, Lund, Sweden

Many of the very common diseases will affect the extracellular matrix and the cells of this matrix leading to tissue malfunction.

Via the characterization of the genome from a number of species, we have learnt of the protein primary structure and in many instances obtained information that helps us to deduce higher orders of structure. The next challenge in understanding biology and pathology is to learn about the numerous posttranslational modifications that are such an integral part of the proteins. This is particularly the case outside the cell, where with few exceptions proteins are extensively modified. These modifications will change structure and function of the proteins and are often variable for a given gene product—protein. Examples of such modifications range from glycosylations exemplified by simple monosaccharides, in e.g. the collagen triple helix, variably complex *O*- or *N*-glycosidically linked oligosaccharides, much larger glycosaminoglycans with a potential for variability of similar order to the proteins, phosphorylations, tyrosine sulfations and importantly cleavage to yield fragments with a new activity or to release interactions or bound other molecules such as growth factors.

Despite the largely unknown regulation of these processes, the increasing understanding of players in the form of enzymes catalyzing modifications and others changing their character by cleaving them leads to new options to learn about the regulation. At the same time there are other modifications that do not depend on enzyme catalysis, but still can modify properties of target molecules exemplified by glucose reacting with amino groups in proteins.

The current text of *Extracellular Matrix: Pathobiology & Signaling* provides an extensive coverage of components in the extracellular matrix, interactions with cells and signals elicited. It contains a very comprehensive presentation of posttranslational modifications and their roles in modulating and/or accomplishing signals and cellular effects.

Book chapters cover a number of examples of mechanisms of the synthesis of glycosaminoglycans and the variability of the ensuing polysaccharide chains resulting in fine-tuning of functions. It is further described how enzymes can degrade the glycosaminoglycan chains to further modulate their structure and function. The role of the glycosaminoglycans in presentation of e.g. growth factors to cellular receptors is also discussed.

At the next level with glycosaminoglycans bound to variable protein cores proteoglycans, a number of biological roles of these molecules are highlighted, such as cell surface syndecans in associating with integrins and influencing cellular responses including migration. Interesting modulations of such interactions by glucose and metabolites, e.g. in diabetes, result in functional consequences. Other functions of the proteoglycans include regulating assembly of matrix and providing interactions stabilizing the matrix. Fragments, e.g. such of perlecan may have roles in vascularization. Important roles of the leucine rich proteoglycans in innate immune response and inflammation are also presented and discussed.

Interestingly, there is a number of examples provided of interactions between cells and the extracellular matrix and how these will change cellular behavior and activity. The interactions may be modulated by a variety of modifications in particular involving proteolysis. Among the many cell surface molecules discussed the discoidin receptors provide for collagen binding and the interactions modulate inflammation and cell migration.

Hyaluronan and its receptors are steadily becoming more in focus in a number of conditions. The understanding of the regulation of the synthetic enzymes as well as of the hyaluronidases accomplishing degradation resulting in variable polymer dimensions and the roles in inflammation and cancer is rapidly developing.

Extracellular matrix is dynamic with extensive turnover and modifications of molecules by cleavages resulting in constituents with modified functions as well as in tissue turnover, also such accelerated in a number of diseases. *Extracellular Matrix: Pathobiology & Signaling* covers also a number of proteolytic enzyme families and their roles, ranging from such at the cell surface to the interplay between the protease and its inhibitors regulating effects on targets. Fragments formed often have different activities and growth factors bound may be released to regulate cellular activities. Importantly, such activities may be very relevant in the communication between cancer cells and the surrounding tissue cells in metastasis and cancer growth. A number of the mechanisms and interactions described are potential future targets for intervention in disease. Examples of novel means for interference by the use of siRNA, shRNA and antisense are also discussed.

The book *Extracellular Matrix: Pathobiology & Signaling* provides a comprehensive and up to date collection of very relevant topics for understanding the various facets of extracellular matrix and its interactions with cells in normal tissue as well as in disease. It represents the current front-line and will serve as a reference for extracellular matrix and posttranslational modifications.

About the Editor/Section Editors

Editor

Nikos Karamanos

Nikos Karamanos, Ph.D., is a professor of biochemistry at the University of Patras and a collaborating member of ICE-TH/FORTH. He obtained his diploma (chemistry) in 1984 and Ph.D. (biochemistry) in 1988 from the University of Patras. He has carried out pre- and postdoctoral research at Karolinska Institute (School of Medicine, Sweden). His main area of research involves matrix pathobiology, cell signaling and gene expression, pharmacological targeting, drug preclinical evaluation, mimetic cell culture models, and development of bioanalytical assays for structure analysis, diagnostic, and pharmacokinetic purposes. Studies are focused on the implication of biological matrix effectors (proteoglycans, glycosaminoglycans, metalloproteinases, and glycoproteins) in tissue organization, pathogenesis and progression of various disorders, such as cancer growth, invasiveness, and metastatic potential. Currently he is on the editorial boards of *The Journal of Biological Chemistry, FEBS Journal, Current Medicinal Chemistry*, and so on, and is an academic editor of *PLoS ONE*. Dr. Karamanos has organized several scientific meetings, among them the FEBS Advanced Lecture Courses on *"Matrix Pathobiology, Signaling & Molecular Targets"*.

Section Editors

Donald Gullberg

Donald Gullberg, Ph.D., has been trained as a biochemist at the University of Uppsala, Sweden. After obtaining a Ph.D. at Uppsala University in 1990 on a thesis dealing with collagen-binding integrins, he did a postdoc at UCLA in the laboratory of John and Lisa Fessler on a Drosophila-related project. Upon returning to Uppsala University in 1993 he became an independent researcher in the Department of Animal Physiology headed by Peter Ekblom. Since 2004 he has been professor in the Department of Biomedicine in Bergen, Norway. He heads a group of matrix biology and has a long-standing interest in collagen-binding integrins and their role in health and disease. Currently Dr. Gullberg is on the editorial board of Matrix Biology and is an academic editor of *PLoS ONE*.

Paraskevi Heldin

Paraskevi Heldin, Ph.D., received her B. Sci. exam in 1979 at Uppsala University and her Ph.D. in 1987 in medical and physiological chemistry for studies on regulatory phosphorylation of proteins, also at Uppsala University. After her dissertation she changed her research direction and focused on the biology of hyaluronan at the Department of Medical and Physiological Chemistry, Uppsala University. Since 2001, she has been adjunct associate professor in the Department of Medical Biochemistry and Microbiology, and head of the Matrix Biology Group at the Ludwig Institute for Cancer Research, Uppsala Univerity, Sweden. Dr Heldin's interests are focused on understanding the mechanisms of how hyaluronan-CD44 complexes interact with the receptors for the growth factors PDGF and TGFβ for modulation of the proliferative and invasive behavior of malignant cells.

Liliana Schaefer

Liliana Schaefer, M.D., has been trained in internal medicine and nephrology at the University of Poznan, School of Medicine, Poland. She graduated as a medical doctor from the University of Wuerzburg, Germany. After working as a postdoc at the University of Wuerzburg, she became principal investigator in the Department of Medicine at the University of Muenster, Germany. Since 2006 she has been a professor of nephropharmacology at the Institute of Pharmacology and Toxicology, University of Frankfurt/Main with an area of expertise in matrix biology. Her laboratory is addressing the role(s) of the two TGF-beta-binding, small leucine-rich repeat proteoglycans decorin and biglycan in inflammation and fibrosis. Her work gave rise to the novel concept that, under certain conditions, matrix components in their soluble form may act as endogenous "danger" signals. These molecules are recognized by innate immunity receptors and are capable of triggering inflammatory response reactions. Currently, Dr. Schaefer is a member of the editorial board of *The Journal of Biological Chemistry*, associated editor of the *Journal of Histochemistry & Cytochemistry*, and president of the German Connective Tissue Society.

Ruggero Tenni

Ruggero Tenni, Ph.D., is a full professor at the Faculty of Medicine and Surgery, University of Pavia, Italy. His current teaching duties in the medical school include chemistry and biochemistry courses in the English language. His scientific research of the last 15 years concerns fibrillar collagen properties, structural features, and interactions with other components of the ECM. By using the CNBr peptides from collagens I and II able to trimerize in triple helical conformation, it was possible to ascertain that these collagen types have multiple binding sites for three small leucine-rich proteoglycans (decorin, fibromulin, and biglycan), and to discover a binding site for SPARC in collagen I. Derivatization in mild conditions of the two collagens and their CNBr peptides showed that some lysine/hydroxylysine residues are involved in collagen stability and interactions.

Achilleas Theocharis

Achilleas Theocharis, Ph.D., is assistant professor of biochemistry and molecular biology at the University of Patras. He received his diploma in chemistry in 1994 and his Ph.D. in biochemistry in 2000 from the University of Patras. He has carried out postdoctoral research at Karolinska Institute in Stockholm, Sweden. His research interests are focused in the areas of matrix pathobiology, cell signaling, and molecular targeting. He investigates the role of proteoglycans, glycosaminoglycans, and matrix metalloproteinases in tissue organization and their implication in the pathogenesis of various diseases, such as cancer and atherosclerosis.

Jan-Olof Winberg

Jan-Olof Winberg, Dr. philos., received his Cand. real. in 1982 and Dr. philos. in 1990 from Biochemical Institute, Faculty of Science, University of Oslo, Norway. Since 1994 he has been a professor in the Department of Medical Biology, Faculty of Health Sciences, University of Tromsø, Norway. He is a member of the editorial advisory board of the *Biochemical Journal* and consulted as a reviewer for various journals. His research interests include biochemical and kinetic characterization of enzymes belonging to the short-chain dehydrogenase/reductase (SDR) superfamily and the matrix metalloproteinase (MMP) family, the role of individual enzymes from these two families in diseases like cancer, the role of extracellular matrix on cells' expression of these enzymes, and on activation of MMPs.

List of contributing authors

Nikos Afratis
Laboratory of Biochemistry
Department of Chemistry
University of Patras
Patras, Greece

Alexios J. Aletras
Laboratory of Biochemistry
Department of Chemistry
University of Patras
Patras, Greece

Hans Peter Bächinger
Research Department
Shriners Hospital for Children
Portland, OR, USA
and
Department of Biochemistry and Molecular Biology
Oregon Health and Science University
Portland, OR, USA
hpb@shcc.org

Rebecca Bell
Rush University Medical Center
Chicago, IL, USA
bellrebecca@gmail.com

Michelle P. Bendeck
Department of Laboratory Medicine and Pathobiology
University of Toronto
Toronto, ON, Canada

Aikaterini Berdiaki
Department of Histology—Embryology
Medical School
University of Crete
Heraklion, Greece

Sergei P. Boudko
Research Department
Shriners Hospital for Children
Portland, OR, USA
and
Department of Biochemistry and Molecular Biology
Oregon Health and Science University
Portland, OR, USA

Helen Bouga
Department of Chemistry
University of Patras
Patras, Greece

Cord Brakebusch
Department of Biomedical Sciences
University of Copenhagen
Biocenter
Copenhagen, Denmark

Barbara Brodsky
Science & Technology Center
Department of Biomedical Engineering
Tufts University
Medford, MA, USA
Barbara.Brodsky@Tufts.edu

Mariana Capurro
Department of Molecular and Cellular Biology Sunnybrook Research Institute
Toronto, ON, Canada
and
Department of Medical Biophysics
University of Toronto,
Toronto, ON, Canada

Benito Casu
G. Ronzoni Institute for Chemical and Biochemical Research
Milan, Italy

John R. Couchman
Department of Biomedical Sciences
University of Copenhagen
Biocenter
Copenhagen, Denmark
john.couchman@bric.ku.dk

Anders Dagälv
Department of Medical Biochemistry and Microbiology
Uppsala University
Uppsala, Sweden

Audrey Deligny
Department of Medical Biochemistry and Microbiology
Uppsala University
Uppsala, Sweden

Giancarlo De Luca
Department of Surgery and Morphological Sciences
University of Insubria
Varese, Italy

J. Des Parkin
Dept. Medicine
Univ. Melbourne
Northern Health
Epping, Melbourne, Australia

Anh-Tri Do
Department of Medical Biochemistry and Microbiology
Uppsala University
Uppsala, Sweden

Katalin Dobra
Division of Pathology
Department of Laboratory Medicine
Karolinska Institutet
Stockholm, Sweden

Antoine Dufour
Department of Biochemistry and Molecular Biology
University of British Columbia
Vancouver, BC, Canada
and
Department of Oral Biological and Medical Sciences
University of British Columbia
Vancouver, BC, Canada

Michael Elkin
Sharett Oncology Institute
Hadassah-Hebrew University Medical Center
Jerusalem, Israel

Bodil Fadnes
Department of Medical Biology
Faculty of Health Sciences
University of Tromsø
Tromsø, Norway

Jorge Filmus
Department of Molecular and Cellular Biology Sunnybrook Research Institute
Toronto, ON, Canada
and
Department of Medical Biophysics
University of Toronto,
Toronto, ON, Canada
Jorge.filmus@sri.utoronto.ca

Serafoula Filou
Department of Chemistry
University of Patras
Patras, Greece

Eric E. Gabison
Laboratoire CRRET
Université Paris Est
Créteil, France

Congxiao Gao
RIKEN Alliance Laboratory
The Institute of Scientific and Industrial Research
Osaka University
Japan
and
Disease Glycomics Team
RIKEN Advanced Science Institute
Saitama, Japan

Shibnath Ghatak
Regenerative Medicine and Cell Biology
Medical University of South Carolina
Charleston, SC, USA

Chrisostomi Gialeli
Laboratory of Biochemistry
Department of Chemistry
University of Patras
Patras, Greece

Daniel J. Gorski
Rush University Medical Center
Chicago, IL, USA
Daniel_J_Gorski@rush.edu

Jianguo Gu
Division of Regulatory Glycobiology
Institute of Molecular Biomembrane and Glycobiology
Tohoku Pharmaceutical University
Japan

Donald Gullberg
Department of Biomedicine
University of Bergen
Bergen, Norway
donald.gullberg@biomed.uib.no

Elin Hadler-Olsen
Department of Medical Biology
Faculty of Health Sciences
University of Tromsø
Tromsø, Norway

Vincent Hascall
Department of Biomedical Engineering
Cleveland Clinic Lerner Research Institute
Cleveland, OH, USA

Paraskevi Heldin
Ludwig Institute for Cancer Research
Uppsala University
Biomedical Center
Uppsala, Sweden
Evi.Heldin@LICR.uu.se

Boris Hinz
Matrix Dynamics Group
Faculty of Dentistry
University of Toronto
Toronto, ON, Canada

Anders Hjerpe
Division of Pathology
Department of Laboratory Medicine
Karolinska Institutet
Stockholm, Sweden

Eric Huet
Laboratoire CRRET
Université Paris Est
Créteil, France

David J. S. Hulmes
Institut de Biologie et Chimie des Protéines
Université de Lyon
Lyon, France
d.hulmes@ibcp.fr

Neta Ilan
Cancer and Vascular Biology Research
Center The Rappaport Faculty of Medicine
Technion
Haifa, Israel

Renato V. Iozzo
Department of Pathology,
Anatomy and Cell Biology
and the Cancer Cell Biology and Signaling
Program, Kimmel Cancer Center
Thomas Jefferson University
Philadelphia, PA, USA
iozzo@kimmelcancercenter.org

Olena Jacenko
Dept. Animal Biology
School Veterinary Medicine
University of Pennsylvania
Philadelphia, PA, USA

Tiina A. Jokela
Institute of Biomedicine/Anatomy
School of Medicine
University of Eastern Finland
Kuopio, Finland

Veli-Matti Kähäri
Department of Dermatology
University of Turku and Turku University Hospital
Turku, Finland
and
MediCity Research Laboratory
University of Turku
Turku, Finland
veli-matti.kahari@utu.fi

Nikos K. Karamanos
Laboratory of Biochemistry
Department of Chemistry
University of Patras
Patras, Greece
n.k.karamanos@upatras.gr

Evgenia Karousou
Department of Surgery and Morphological Sciences
University of Insubria
Varese, Italy

Lena Kjellén
Department of Medical Biochemistry and Microbiology
Uppsala University
Uppsala, Sweden
lena.kjellen@imbim.uu.se

Dimitris Kletsas
Laboratory of Cell Proliferation & Ageing
Institute of Biology
National Centre for Scientific Research 'Demokritos'
Athens, Greece

Constantine Kolliopoulos
Department of Chemistry
University of Patras
Patras, Greece

Angeliki Korpetinou
Laboratory of Biochemistry
Department of Chemistry
University of Patras
Patras, Greece

Marina Koutsioumpa
Laboratory of Molecular Pharmacology
Department of Pharmacy
University of Patras
Patras, Greece

Ilona Kovalszky
1st Institute of Pathology and Experimental Cancer Research
Semmelweis University
Budapest, Hungary

Eliene O. Kozlowski
Laboratório de Bioquímica e Biologia Celular de Glicoconjugados
Programa de Glicobiologia
Instituto de Bioquímica Médica and Hospital Universitário Clementino Fraga Filho
Universidade Federal do Rio de Janeiro
Rio de Janeiro, RJ, Brasil

Vassiliki Labropoulou
Department of Internal Medicine
University Hospital of Patras
Patras, Greece

Ning Lu
Department of Biomedicine
University of Bergen
Bergen, Norway

Joseph S. Maffei
Department of Biomedical Engineering
Boston University
Boston, MA, USA
muffy2521@gmail.com

Synnøve Magnussen
Department of Medical Biology
Faculty of Health Sciences
University of Tromsø
Tromsø, Norway

Roger R. Markwald
Regenerative Medicine and Cell Biology
Medical University of South Carolina
Charleston, SC, USA

Christopher A. McCulloch
Matrix Dynamics Group
Faculty of Dentistry
University of Toronto
Toronto, ON, Canada
Christopher.Mcculloch@utoronto.ca

Colton McNicols
Rush University Medical Center
Chicago, IL, USA
mcnichols.md@gmail.com

Suzanne Menashi
Laboratoire CRRET
Université Paris Est
Créteil, France

Rosetta Merline
Pharmazentrum Frankfurt
Institut für Allgemeine Pharmakologie und Toxikologie
Klinikum der Goethe-Universität Frankfurt am Main
Frankfurt, Germany

Eleni Milia-Argeiti
Laboratory of Biochemistry
Department of Chemistry
University of Patras
Patras, Greece

Suniti Misra
Regenerative Medicine and Cell Biology
Medical University of South Carolina
Charleston, SC, USA
misra@musc.edu

Shuji Mizumoto
Laboratory of Proteoglycan Signaling and Therapeutics
Frontier Research Center for Post-Genomic Science and Technology
Hokkaido University Graduate School of Life Science
Sapporo, Hokkaido, Japan

Catherine Moali
Institut de Biologie et Chimie des Protéines
Université de Lyon
Lyon, France

Samia Mourah
Inserm
UMR-S 940 Laboratoire de Pharmacologie,
Paris, France

Hideaki Nagase
Kennedy Institute of Rheumatology
Nuffield Department of Orthopaedics
Rheumatology and Musculoskeletal Sciences
University of Oxford
London, United Kingdom
hideaki.nagase@kennedy.ox.ac.uk

Madalina V. Nastase
Pharmazentrum Frankfurt
Institut für Allgemeine Pharmakologie und Toxikologie
Klinikum der Goethe-Universität Frankfurt am Main
Frankfurt, Germany

Dragana Nikitovic
Department of Histology-Embryology
Medical School
University of Crete
Heraklion, Greece

Kazuaki Ohtsubo
RIKEN Alliance Laboratory
The Institute of Scientific and Industrial Research
Osaka University
Japan
and
Disease Glycomics Team
RIKEN Advanced Science Institute
Saitama, Japan

Sanna Oikari
Institute of Biomedicine/Anatomy
School of Medicine
University of Eastern Finland
Kuopio, Finland

Joseph P.R.O. Orgel
Center Synchrotron Radiation/
Instrumentation Dept. Biology, Chemistry, and Physical Sciences
Illinois Institutes of Technology
Chicago, IL, USA

Christopher M. Overall
Department of Biochemistry and Molecular Biology
University of British Columbia
Vancouver, BC, Canada
and
Department of Oral Biological and Medical Sciences
University of British Columbia
Vancouver, BC, Canada
chris.overall@ubc.ca

Evangelia Papadimitriou
Laboratory of Molecular Pharmacology
Department of Pharmacy
University of Patras
Patras, Greece
epapad@upatras.gr

Alberto Passi
Department of Surgery and Morphological Sciences
University of Insubria
Varese, Italy
alberto.passi@uninsubria.it

Csilla Pataki
Department of Biomedical Sciences
University of Copenhagen
Biocenter
Copenhagen, Denmark

Mauro S. G. Pavao
Laboratório de Bioquímica e Biologia
Celular de Glicoconjugados
Programa de Glicobiologia
Instituto de Bioquímica Médica and
Hospital Universitário Clementino Fraga
Filho
Universidade Federal do Rio de Janeiro
Rio de Janeiro, RJ, Brasil
mpavao@hucff.ufrj.br

Anton V. Persikov
Lewis-Sigler Institute for Integrative
Genomics Princeton University
Princeton, NJ, USA

Dagmar S. Pikas
Department of Medical Biochemistry and
Microbiology
Uppsala University
Uppsala, Sweden

Jin-Ping Li
Dept. of Medical Biochemistry and
Microbiology Uppsala University Uppsala,
Sweden

Anna Plaas
Rush University Medical Center
Chicago, IL, USA

Harris Pratsinis
Laboratory of Cell Proliferation & Ageing
Institute of Biology
National Centre for Scientific Research
'Demokritos'
Athens, Greece

Ida Pucci-Minafra
Centro di Oncobiologia Sperimentale
University of Palermo
Oncology Department *La Maddalena*
Palermo, Italy
ida.pucci@unipa.it

Ellen Puré
The Wistar Institute
Philadelphia, PA, USA
pure@Wistar.org

Mario Raspanti
Department of Surgical & Morphological
Sciences
Insubria University
Varese, Italy
mario.raspanti@uninsubria.it

Antonio S. Rocca
Division of Cardiology
Department of Medicine
University of Toronto
Toronto, ON, Canada

Cecilia Rydén
Dept. of Infectious diseases
Dept. of Medical Biochemistry
Microbiology and Immunology
Uppsala University
Akademiska Hospital
Uppsala, Sweden

James D. San Antonio
Stryker Orthobiologics
Malvern, PA, USA
james.sanantonio@stryker.com

Ralph D. Sanderson
Department of Pathology
University of Alabama at Birmingham
Birmingham, AL, USA

John D. Sandy
Rush University Medical Center
Chicago, IL, USA
jsandy44@gmail.com

Takako Sasaki
Department of Experimental Medicine I
Nikolaus-Fiebiger Center for Molecular
Medicine
University of Erlangen-Nürnberg
Erlangen, Germany
tsasaki@molmed.uni-erlangen.de

Judy Savige
Dept. Medicine
Univ. Melbourne
Northern Health
Epping, Melbourne, Australia

Liliana Schaefer
Pharmazentrum Frankfurt
Institut für Allgemeine Pharmakologie und Toxikologie
Klinikum der Goethe-Universität Frankfurt am Main
Frankfurt, Germany
schaefer@med.uni-frankfurt.de

Barbara P. Schick
Cardeza Foundation for Hematologic Research
Thomas Jefferson University
Philadelphia, PA, USA
barbara.schick@jefferson.edu

Spyros S. Skandalis
Laboratory of Biochemistry
Department of Chemistry
University of Patras
Patras, Greece

Rigmor Solberg
Department of Pharmaceutical Biosciences
School of Pharmacy
University of Oslo
Oslo, Norway
rigmor.solberg@farmasi.uio.no

Kazuyuki Sugahara
Laboratory of Proteoglycan Signaling and Therapeutics
Frontier Research Center for Post-Genomic Science and Technology
Hokkaido University Graduate School of Life Science
Sapporo, Hokkaido, Japan
k-sugar@sci.hokudai.ac.jp

Gunbjørg Svineng
Department of Medical Biology
Faculty of Health Sciences
University of Tromsø
Tromsø, Norway
gunbjorg.svineng@uit.no

Ingebrigt Sylte
Department of Medical Biology
Faculty of Health Sciences
University of Tromsø
Tromsø, Norway

Markku I. Tammi
Institute of Biomedicine/Anatomy
School of Medicine
University of Eastern Finland
Kuopio, Finland
tammi@uef.fi

Raija H. Tammi
Institute of Biomedicine/Anatomy
School of Medicine
University of Eastern Finland
Kuopio, Finland

Naoyuki Taniguchi
RIKEN Alliance Laboratory
The Institute of Scientific and Industrial Research
Osaka University
Japan
and
Disease Glycomics Team
RIKEN Advanced Science Institute
Saitama, Japan
tani52@wd5.so-net.ne.jp

Felipe C. O. B. Teixeira
Laboratório de Bioquímica e Biologia Celular de Glicoconjugados
Programa de Glicobiologia
Instituto de Bioquímica Médica and Hospital Universitário Clementino Fraga Filho
Universidade Federal do Rio de Janeiro
Rio de Janeiro, RJ, Brasil

Ruggero Tenni
Dipartimento di Biochimica "A. Castellani"
University of Pavia
Pavia, Italy
rtenni@unipv.it

Achilleas Theocharis
Laboratory of Biochemistry
Department of Chemistry
University of Patras
Patras, Greece
atheoch@upatras.gr

Harald Thidemann Johansen
Department of Pharmaceutical Biosciences
School of Pharmacy
University of Oslo
Oslo, Norway

Mervi Toriseva
Department of Dermatology
University of Turku and Turku University Hospital
Turku, Finland
and
MediCity Research Laboratory
University of Turku
Turku, Finland

Irene-Eva Triantaphyllidou
Department of Chemistry
University of Patras
Patras, Greece

Theodore Tsegenidis
Laboratory of Biochemistry
Department of Chemistry
University of Patras
Patras, Greece

George Tzanakakis
Department of Histology-Embryology
Medical School
University of Crete
Heraklion, Greece

Lars Uhlin-Hansen
Department of Medical Biology
Faculty of Health Sciences
University of Tromsø
Tromsø, Norway

Jennifer Velasco
Rush University Medical Center
Chicago, IL, USA
jennifer.velasco1@gmail.com

Davide Vigetti
Department of Surgery and Morphological Sciences
University of Insubria
Varese, Italy

Manuela Viola
Department of Surgery and Morphological Sciences
University of Insubria
Varese, Italy

Israel Vlodavsky
Cancer and Vascular Biology Research Center The Rappaport Faculty of Medicine
Technion
Haifa, Israel
vlodavsk@cc.huji.ac.il

Demitrios H. Vynios
Department of Chemistry
University of Patras
Patras, Greece
vynios@chemistry.upatras.gr

Vincent Wang
Rush University Medical Center
Chicago, IL, USA
vincent_wang@rush.edu

James R. Whiteford
Centre for Microvascular Research
William Harvey Research Institute
Barts and the London School of Medicine & Dentistry
Queen Mary University of London
London, United Kingdom

Thomas N. Wight
The Hope Heart Matrix Biology Program
Benaroya Research Institute at Virginia Mason
Seattle, WA, USA
twight@benaroyaresearch.org

Chris D. Willis
Department of Pathology,
Anatomy and Cell Biology
Thomas Jefferson University
Philadelphia, PA, USA

Jan-Olof Winberg
Department of Medical Biology
Faculty of Health Sciences
University of Tromsø
Tromsø, Norway
jan.o.winberg@uit.no

Xiaojie Xian
Department of Biomedical Sciences
University of Copenhagen
Biocenter
Copenhagen, Denmark

Muhammad H. Zaman
Department of Biomedical Engineering
Boston University
Boston, MA, USA
zaman@bu.edu

Fang Zong
Division of Pathology
Department of Laboratory Medicine
Karolinska Institutet
Stockholm, Sweden

Abbreviations and acronyms used

4-MU	4-methylumbelliferone
A2ar	Adenosine 2a receptor
ABCB5	ATP-binding cassette sub-family B member 5
ABCG2	ATP-binding cassette sub-family G member 2
ADAM	A disintegrin and metalloproteinase
ADAMTS	A disintegrin and metalloproteinase with thrombospondin motifs
ADCC	Antibody-dependent cellular cytotoxicity
ADME	Absorption, distribution, metabolism and excretion
ADP	Adenosine diphosphate
AFM	Atomic force microscopy
AFP	α-Fetoprotein
AGE	Advanced glycation end product
AICAR	5-aminoimidazole-4-carboxamide 1-β-D-ribofuranoside
AKT	Protein kinase B (PKB), a serine/threonine-specific kinase
ALDH	Aldehyde dehydrogenase
ALK	Activin-like receptor kinase
ALK	Anaplastic lymphoma kinase
ALL	Acute lymphocytic leukemia
AML	Acute myeloid leukemia
AMPK	AMP-activated protein kinase
ANXA2	Annexin A2
AP1	Activator protein 1
AP–2	Activating enhancer binding protein 2
APC	Adenomatus polyposis coli
APCE	Antiplasmin-cleaving enzyme
APMA	p-aminophenylmercum acetate
ApoE	Apolipoprotein E
APS	Adenosine-5′-phosphosulfate
AR	Androgen receptor
ARDS	Adult respiratory distress syndrome
AREG	Amphiregulin
ARMD1	Age-related macular degeneration 1
ASC	Apoptosis-associated speck-like protein containing CARD
aSMA	a-Smooth muscle actin
ASMCs	Arterial smooth muscle cells
ATCS	Adducted thrumb-club foot syndrome
ATP	Adenosine-5′-triphosphate
AXPC1	Posterior column ataxia with renitis pigmentosa
Aβ	Amyloid-beta protein

Bbp	Bone sialoprotein-binding protein
BCE	Bovine capillary endothelial cells
Bcl2	B-cell lymphoma 2
B-CLL	B-cell chronic lymphocytic leukemia
BclxL	B-cell lymphoma-extra large
bFGF	basic Fibroblast growth factor, also known as FGF-2
BID	BH3 interacting-domain death agonist
BL	Basal lamina
BM(s)	Basement membrane(s)
BMMC	Bone-marrow-derived mast cells
BMP(s)	Bone morphogenic protein(s)
BP(s)	Bisphosphonate(s)
BSP	Bone sialoprotein
BTC	Betacellulin
C1q	C1QA complement component 1, q subcomponent
C4ST/D4ST	Chondroitin/Dermatan 4-O-sulfotransferase
C6ST/D6ST	Chondroitin/Dermatan 6-O-sulfotransferase
CAA	Cerebral amyloid angiopathy
CAF	Cancer-associated fibroblast
CASK	Calcium/Calmodulin associated serine/threonine kinase
CAT	Catalytic domain
CBP/p300	CREB binding protein 300
c-Cbl	Casitas B-lineage lymphoma proto-oncogene
Ccl9s0	Chemokine (C-C motif) ligand(s)
CD14	Cluster of differentiation 14
CD44	Cluster of differentiation 44, a major receptor for hyaluronan
CD44s	Standard CD44
CD44v	CD44 variant
CDC37	Cell division cycle control protein
Cdk or CDK	Cyclin-dependent kinase
cFN	Cell form of fibronectin
CGD	Congenital glycosylation disorders
CHAPS	3[(3-cholamidopropyl)dimethylammonio] propanesulfonic acid
ChIP	Chromation immunoprecipitation
ChPF	Chondroitin polymerizing factor
ChSy	Chondroitin synthase
CMAP	Cystatin-like metastasis-associated protein
c-Met	Cellular MNNG HOS transforming gene that encodes a protein known as hepatocyte growth factor receptor
CMTs	Chemically modified tetracyclines
c-Myc	Cellular myelocytomatosis
CNS	Central nervous system
COB	Calvaria osteoblasts
COFRADIC	Combined fractional diagonal chromatography
COL	Collagenous domain
COMP	Cartilage oligomeric matrix protein

COPD	Chrononic obstructive pulmonary disease
COX2	Cyclooxygenase-2
CRC	Colon rectal cancer
CREB	cAMP responsive element binding protein
CRM197	A mutated diphtheria toxin
CRMO	Chronic recurrent multifocal osteomyelitis
CRP	C-reactive protein
CRTAP	Cartilage-associated protein
CS polymerase	GlcUA/GalNAc transferase
CS	Chondroitin sulfate
CS-A	CS rich in A units (units sulfated at C-4 position of GalNAc)
CS-C	CS rich in C units (units sulfated at C-6 position of GalNAc)
CSC(s)	Cancer stem cell(s)
CSF	Cerebrospinal fluid
CSF2	Colony stimulating factor 2
CS-K	CS containing K units
CT	Computerized tomography
CTGF	Connective tissue growth factor
CTL(s)	Cytotoxic T-lymphocyte(s)
CTX1	C-terminal telopeptides of type I collagen
CUB domain	Complement C1r/C1s, Uegf, Bmp1 protein domain
CXCE(s)	Chemokine (C-X-C motifs) receptor(s)
CXCL(s)	Chemokine (C-X-C motif) ligand(s) or CXC chemokine ligand(s)
CyPB	Cyclophilin B
DAPK	Death-associated protein kinase
DCM	Diabetic cardiomyopathy
DDR(s)	Discoidin domain receptor(s)
DDS(s)	Drug delivery system(s)
DED-1	Density-enhanced phosphatase 1
DN	Diabetic nephropathy
DR-1	Down-regulator of transcription 1
DS	Dermatan sulfate
DSE	DS-glucuronyl C5 epimerase
DSE(s)	DS C5-epimerase(s)
DSP(s)	Dual specific phosphatase(s)
DSPP	Dentin sialophosphoprotein
DSS	Dextran sodium sulfate
EBI 1/2	Epstein-Barr virus-induced genes 1 and 2
EBV	Epstein-Barr virus
EC	Endothelial cells
ECD	Extracellular domain
ECM(s)	Extracellular matrix (matrices)
ECM1/2	ECM protein 1 or 2
ECMX	ECM-like protein, X chromosome
ED	Embryonic day
ED-A	Extradomain A

EDS	Ehlers-Danlos syndrome
EGF	Epidermal growth factor
EGFR	Epidermal growth factor receptor
Egr–1	Early growth response protein 1
ELISA	Enzyme linked immunosorbent assay
EMMPRIN	Extracellular matrix metalloproteinase inducer
EMT	Epithelial to mesenchymal transition
EnMT	Endothelial mesenchymal transition
e-NOS	Endothelial nitric oxide synthase
EPGN	Epigen
EPR	Extended permeability and retention
ER	Endoplasmatic reticulum
ErbB2	Epidermal growth factor receptor 2, also known as HER2/new
EREG	Epiregulin
ERK	Extracellular signal-regulated kinase
ERM	Ezrin, radixin and moesin
ERα/β	Estrogen receptor α or β
ESR	Erythrocyte sedimentation rate
EXT	Exostosin
EXTL	EXT-like
FA(s)	Focal adhesion(s)
FACIT	Fibril-associated collagen with interrupted triple helices
FADD	Fas associated death domain
FAK	Focal adhesion kinase
FAP	Fibroblast activation protein
Fas	TNF receptor superfamily member 6
FDA	Food and drug administration
FDL	Flexor digitorum longus
FGF	Fibroblast growth factor
FGF-2	Fibroblast growth factor-2 (or basic)
FGFBP	FGF binding protein
FGFR	Fibroblast growth factor receptor
FLI1	Friend leukemia integration 1 transcription factor
FN	Fibronectin
FOX N1	Forkhead box protein 1
FS	Follistatin-like
FSP-1	Fibroblast specific protein-1
GABP	GA-binding protein
GAG(s)	Glycosaminoglycan(s)
GAIP	Ga-interacting protein
Gal	Galactose
GalNAc	N-acetyl galactosamine
GalNAc4S, 6S	4-O- and 6-O- sulfated GalNAc
GalNAc4S-6ST	N-acetyl-D-galactosamine-4-sulfate-6-O-sulfotransferase
GalNAcT-I	GalNAc tranferase I
GalNAcT-II	GalNAc tranferase II

GalT(s)	Galactosyltransferase(s)
GalT-I	β1,4-galactosyltranferase-I
GalT-II	β1,3-galactosyltranferase-II
GAPs	GTPase activating proteins
GBM	Glomerular basement membrane
GDIs	Guanine dissociation inhibitors
GDNF	glial cellulinecellulite-derived neurotrophic factor
GEFs	Guanine nucleotide exchange factors
GF(s)	Growth factor(s)
GFP	Green fluorescent protein
GFR(s)	Growth factor receptor(s)
GIPC	GAIP-interacting protein C terminus
GLCP/GPLC	Glypican
GlcA/GlcUA	Glucuronic acid
GlcA3S	3-O- sulfated GlcA
GlcAT(s)	Glucuronyltransferase(s)
GlcN	Glucosamine
GLUT(s)	Glucose transporter(s)
GMB	Glomerular basement membrane
GM-CSF	Granulocyte-macrophage colony-stimulating factor
GNP	Gross national product
GnT(s)	N-acetylglucosaminyltransferase(s)
GnT-IX	β1,6-N-acetylglucosaminyl-transferase IX
GnT-VI	β1,4-N-acetylglucosaminyl-transferase VI
GPI	Glycosyl phosphatidyl–inositol
Grb2	Growth factor receptor-bound protein 2
GROα/CXCL1	Growth-related oncogene α
GRP94	Glucose-regualated protein, 94 kDa
HA	Hyaluronan
HABR	hyaluronan-binding region
HARE/STAB	hyaluronic acid receptor for endocytosis
HARP	Heparin affinity regulatory peptide
HAS	Hyaluronan synthase
HAT	Histone acetyltransferase
HB-EGF	Heparin binding EGF
HC-4 or 4–HC	4–hydroperoxycyclophosphamide
HCC	Hepatocellular carcinoma
HCCAA	hereditary cystatin-C amyloid angiopathy
HCs	Heavy chains
HDL	High density lipoprotein
HGF	Hepatocyte growth factor
Hh	Hedgehog
HIF-1	Hypoxia inducible factor-1
HIV	Human immunodeficiency virus
hK	Human kallikrein
HMM	high molecular mass

HMW	High molecular weight
HNSCC	Head and neck squamous cell carcinoma
Hp	Heparin
HP1γ	Heterochromatic protein 1γ
Hpa-tg	Heparanase transgenic
HPV	Human papilloma virus
HPX	Hemopexin-like domain
HRG	Histidine-rich glycoprotein
HS	Heparan sulfate
HSP	Heat shock protein
HSPG(s)	Heparan sulfate proteoglycan(s)
HSV-1	Herpes simplex virus type 1
HUVECs	Human umbilical vein endothelial cells
HYAL(s)	Hyaluronidase(s)
IaI	Inter-a-trypsin inhibitor
IBD	Inflammatory bowel disease
IC_{50}	The half maximal inhibitory concentration
ICAM	Intracellular adhesion molecule
ICAT	Isotope-coded affinity tag
ICD	Intracellular domain
ICDS	Inactive catalytic-domain capture
IdoA	L-Iduronic acid
IdoA2S	2-O- sulfated IdoA
IFN-γ	Interferon gamma
IFP	Interstitial fluid pressure
Ig	Immunoglobulin
IGD	Interglobular domain
IGF-1	Insulin-like growth factor-1
IGFBP	IGF binding protein
IGF-IR	Insulin-like growth factor I receptor
IHC	Immunohistochemistry
IL(s)	Interleukin(s)
ILK	Integrin-linked kinase
INF	Interferon Inhibitors
IP	Intraperitoneal
IR	Ionizing radiation
ISEMF(s)	Intestinal subepithelial myofibroblast(s)
iTRAQ	Isotopic dimethylation, isobaric tags for relative and absolute quantification
JAK	Janus kinase
JNK(s)	c–jun N–terminal kinase(s)
KC	Keratinocyte chemoattractant
KGF	Keratinocytes growth factor
KS	Keratan sulfate

L1CAM	L1-cell adhesion molecule
LAIR	Leukocyte-associated immunoglobulin-like receptor
LAP	Latency-associated peptide
LAR	Leukocyte common antigen-related protein
LDL	Low-density lipoprotein
LDLR	LDL receptor
LEF	Lymphoid enhancing factor
LG3/4/5	Laminin-like globular 3, 4 or 5
Lgr5	Leucine-rich repeat-containing G-protein coupled receptor 5
LLC	Large latent complex
LLC-LM	Lewis lung carcinoma
LMW	Low molecular weight
LN	Laminin
LOXL2	Lysyl oxidase-like 2
LPS	Lipopolysaccharide
LRP	Low-density lipoprotein receptor-related protein
LRRs	Leucine rich repeats
LTBP	Latent TGF-β binding protein
LYVE-1	Lymphatic vessel endothelial hyaluronan receptor 1
MAC1	Macrophage antigen complex-1, αMβ2 integrin
MACIT	Membrane-associated collagen with interrupted triple helices
MadCAM-1	Mucosal addressin cell adhesison molecule-1
MAGUK	Membrane-associated guanylate kinase
MAL	Megakaryocytic acute leukemia protein
MALDI-TOF	Matrix-assisted laser desorption-ionization time-of-flight
MAPK	Mitogen-activated protein kinase
MAPKK	MAPK kinase
MARCKS	Myristylated alanine-rich protein kinase C substrate
MBL	Mannose binding lectin
MCP	Monocyte chemotactic protein
MCP	Mast cell protease, monocyte chemotactic protein
MD	Mineral density
MDCK	Madin Darby kanine kidney cell line
mDia	Mammalian Diaphanous-related formin protein
MDM2	Murine double minute 2
MDR	Multidrug resistance
MEFs	Mouse embryo fibroblasts
MEK/ERK	Mitogen-activated protein kinase pathway
MEROPS	http://merops.sanger.asc.uk/, the definitive database of proteases and inhibitors
MET	Mesenchymal to epithelial transition
MHCII	Monocyte histocompatibity complex class II
MIDAS	Metal ion-dependent adhesion site
MIG	Monocyte-induced by gamma interferon
MIP-1α/β/2	Macrophage inflammatory protein-1α, β or 2
miRNA	microRNA

miRNA/miR	micro RNA
MK	Megakaryocytes
MM	Multiple myeloma
MME	Membrane metallo-endopeptidase
MMP(s)	Matrix metalloproteinase(s)
MMP2/9	Gelatinase A/B
MMP26	Matrilysin-2/ endometase
MMP28	Epilysin
MMP7	Matrilysin
MMP8	Collagenase-2
MMPI(s)	MMP inhibitor(s)
MRI	Magnetic resonance tomography imaging
mRNA	Messenger RNA
MRSA	Methicillin resistant *S. aureus*
MRSE	Methicillin resistant *S. epidermidis*
MRTF-A	Myocardin-related transcription factor A
MSCRAMM	Microbial surface components recognizing adhesive matrix molecules
MT-MMP(s)	Membrane type MMP(s)
mTOR	Mammalian target of rapamycin
Multiplexin	Multiple triple collagens with interrupted triple helices
MyD88	Myeloid differentiation primary response protein
NAD	Nicotinamide adenine dinucleotide
NC	Non-collagenous domain
N-CAM	Neural cell adhesion molecule
NCoR1	Nuclear receptor corepressor-1
NDST(s)	N-deacetylase/N-sulfotransferase(s)
NeuroD	Neurogenic differentiation 1
NF-κB	Nuclear factor kappa-light-chain-enhancer of activated B cells
NG2	Neuron-glial antigen 2, Neuron/Glia type 2 antigen
Ng-CAM	Neuron-glia cell adhesion molecule
NHE1	Na^+ / H^+ exchanger 1
NK cells	Natural killer cells
NLRP3	NLR family, pyrin domain containing 3
NLS	Nuclear localization signal
NMR	Nuclear magnetic resonance
NSAIDs	Non-steroidal anti-inflammatory drugs
NSCCL	Non small-cell lung cancer
OA	Osteoarthritis
OCD	Osteochondritis dissecans
OF/LB	Oncofetal/Laminin-binding
OGT	O-GlcNAc transferase
OI	Osteogenesis imperfecta
OST(s)	*O*-sulafation by *O*-sulfotranferases
Ox-LDL	Oxidized LDL

P2X7	Purinergic receptor P2X, Ligand-gated ion channel, 7
p38	p38 Mitogen-activated protein kinase
P3H1	Propyl 3-hydroxylase 1
PA	Plasminogen activator
PAI	PA inhibitor
PAPS	3′-phosphoadenosine-5′-phosphosphosulfate
PAPSS	PAPS synthase
PBMC	Peripheral blood mononuclear cell
PBS	Phosphate buffered saline
PCAM	Platelet endothelial cell adhesion molecule
PCP	Planar cell polarity
PCP	Procollagen C-proteinase
PCPE(s)	Procollagen C-proteinase enhancer(s)
PCR	Polymerase chain reaction
PCT	Procalcitonin
PDB ID	Protein data bank identification number
PDCD	Programmed cell death
PDGF	Platelet-derived growth factor
PDGFR	PDGF receptor
PDK1	Phospholipid-dependent kinase 1
PDZ	Post synaptic density protein, Drosophila disc large tumor suppressor, Zonula occludens-1 protein
PEDF	Pigment epithelium-derived factor
PEG	Polyethylene glycol
PEI	Polyethylene imine
PF4	Platelet factor 4
PG(s)	Proteoglycan(s)
PGE2	Prostaglandin F2
PI3K	Phosphatidylinositol 3–kinase
PICS	Proteomic identification of proteome cleavageprotease cleavage site
PKA/B/C	Protein kinase A, B or C
PLAU	Human uPA
PLG	Plasminogen
PLOD2	Pro-collagen-lysine, 2-oxoglutarate 5-dioxygenase
PMA	Phorbol 12-myristate-13-acetate
Poly I:C	Polyinosinic-polycytidylic acid
PPARγ	Peroxisome proliferator-activated receptor γ
PRELP	Proline/arginine-rich end leucine-rich repeat protein
proLOX	Pro-lysyl oxidase
Pro-MMP	proform MMP, inactive MMP
PRPT (RPTP)	PTP Rreceptor type tyrosine protein phosphate
PRR(s)	Pattern recognition receptor(s)
PSA	Prostate-specific antigen
PSD	Post synaptic density protein
Ptc1	Patched-1, Sonic Hedgehog cell surface receptor
PTEN	Phosphatase and tensin homolog

PTK(s)	Protein tyrosine kinase(s)
PTN	Pleiotrophin
PTP(s)	Protein tyrosine phosphatase(s)
Pyk2	Proline-rich tyrosine kinase 2
RA	Rheumatoid arthritis
Rac1	Ras-related C3 botulinum toxin substrate-1, a subfamily of the Rho family of GTPases
Raf	Rapidly accelerated fibrosarcoma
RAGE	AGE receptor
RANK	Receptor activator of NF-kB
RANKL	RANK ligand
RANTES	Regulated upon Activation, Normal T-cell Expressed, and Secreted or CCL5 [Chemokine (C-C motif) ligand 5]
Ras	Rat sarcoma, a protein of small GTPases involved in cellular signal transduction
RECK	Reversion-inducing cysteine-rich protein with Kazal motifs
RGD	Arginine-glycine-aspartate or aspartic acid
RHAMM	Hyaluronan-mediated motility receptor
RhoA	Ras homolog gene family member A, a GTPase protein
ROCK	Rho-associated protein kinase
ROS	Reactive oxygen species
RPTP(s)	Receptor-type of PTP(s)
RTK	Receptor tyrosine kinase
RT–PCR	Reverse transcriptase polymerase chain reaction
S100A11	Calcium binding S10AB protein
S100A4	Calcium binding protein A4
SCC	Squamous cell carcinoma
SCF	Stem cell factor
SCID	Severe combined immunodeficiency
SD	Standard deviation
SDC1SDC/SND	Syndecan-1
SDF1	Stromal cell-derived factor 1
Sdr	Serine-aspartate repeat
SDS	Sodium dodecyl sulfate
SDS-PAGE	SDS-polyacrylamide gel electrophoresis
SEA	Sperm protein, enterokinase and agrin
SED(s)	Spondyloepiphyseal dysplasis(dyspalsias)
SEM	Scanning electron microscopy
SEMD	Spondyloepimetaphyseal dysplasia
SENPSs	Sentrin-specific proteases
SERPINF2	Serpin α2-antiplasmin
SF	Scatter factor
sFRPs	Secreted frizzled-related proteins
SG	Serglycin
SGBS	Simpson-Golabi-Behmel syndrome

sGLP3	Soluble glypican-3
Shc	SH2 homology 2 domain containing
SHH	Sonic hedgehog
SHP1	Src homology phosphatase-1
shRNA	Small hairpin RNA
SIBLING	Small integrin-binding ligand N-linked glycoprotein
siRNA	Small interfering RNA
SJS	Schwartz-Jampel syndrome
SLRPs	Small leucine rich proteoglycans
Smad(s)	Small and mothers agains decapentaplegic homology
SMB	Somatomedin-B domain
SMC	Smooth muscle cell
Soc(s)	Suppressor(s) of cytokine signaling
SP1	Specificity protein 1
SPARC	Secreted protein acidic and rich in cysteine
SPCs	Subtilisin-like pro-protein convertases specificity
SRF	Serum response factor
SSC	Squamous cell carcinoma
STAT	Signal transducer and activator of transcription
STZ	Streptozotocin
Sulfas	Sulfatases
SUMO	Small ubiquitin-like modifier
TACE	TNF-α converting enzyme
TAECs	Tumor-associated endothelial cells
TAILS	Terminal amine isotopic labeling of substrates
TALs	Tumor-associated lymphocytes
TAMs	Tumor-associated macrophages
TCF	T-cell factor
TEB	Terminal end buds
TEM	Transmission electron microscopy
TFIID	Transcription factor II D
TFPI	Tissue factor pathway inhibitor
Tf-R	Transferrin receptor
TGFα	Transforming growth factor alpha
TGFβ	Transforming growth factor beta
THBS	Thrombospondin
THP	Triple helical peptide
TIM	Triosephosphate isomerase
TIMP(s)	Tissue inhibitor(s) of matrix metalloproteinases
TKI(s)	Tyrosine kinase inhibitor(s)
TLR(s)	Toll like receptor(s)
TM	Trabecular meshwork
TMD	Transmembrane domain
TME	Tumor microenvironment
TNF	Tumor necrosis factor
TPA	Phorbolester 12-*O*-tetradecanoylphorbol-13-acetate

tPA	Tissue-type plasminogen activator
TR	Transferrin
TRAPP1	Transport protein particle 1
TS	Thrombospondin motifs type I
TTP	Thrombotic thrombocytopenic purpura
UC	Ulcerative colitis
UDP	Uuridine diphosphate
UFH	Unfractionated heparin
UGDH	UDP-α-D-glucose 6 dehydrogenase
uPA	Urokinase-type plasminogen activator
uPAR	uPA receptor
UTP	Uridine 5′-triphosphate
VAP1	Vascular apoptosis-inducing protein-1
VCAM	Vascular cell adhesion molecule-1
VEGF	Vascular endothelial growth factor
VEGFR	VEGF receptor
VSM	Vascular muscle cells
VWF	Von Willebrand factor
WBC	White blood cells
WGA	Wheat germ agglutinin
WISP	Wnt1-inducible signaling pathway
WISP-1	WNT1Wnt1-inducible induced -signaling pathway secreted protein 1
WOX1	WW domain containing oxidoreductase
WT	Wild type
WT1	Wilms tumor 1
XylT	Xylosyltransferase
Zfra	Zinc finger-like peptide that regulates apoptosis

1 An introduction to the extracellular matrix molecules and their importance in pathobiology and signaling

1.1 Extracellular matrix: a functional scaffold

Achilleas Theocharis, Chrisostomi Gialeli, Vincent Hascall, and Nikos K. Karamanos

Tissues in mammals are made by cells but also contain significant quantities of supporting matrices (i.e. extracellular mixtures of substances known as extracellular matrices [ECMs]). Cells must be properly supported and have contacts with neighboring cells and/or the ECM in order to function. ECMs are composed of a large variety of matrix macromolecules, including collagens, elastin, fibronectin (FN), laminins, tenascin, vitronectin, thrombospondin, secreted protein acidic and rich in cysteine (SPARC), various proteoglycans (PGs), and hyaluronan (HA). They are synthesized and secreted mainly by stromal cells, such as fibroblasts and osteoblasts, as well as immune cells and epithelial and endothelial cells. The ECMs of most tissues contain a set of related molecules interacting to form organized networks, adapted to the functional requirements of the particular tissue (Frantz et al., 2010). Variability is to a large extent an effect of different assembly, partly regulated by a limited number of molecules more unique to a specific tissue.

The major protein in the majority of ECMs is a fibril-forming collagen (i.e. collagen type I in most tissues and collagen type II in cartilage-related tissues). These collagens form fibrillar structures, and the process is tuned by other fibrillar collagens (e.g. collagen type V with type I and collagen type XI with type II) that are present in small amounts in the fibers. Fibril formation is further regulated by a number of other ECM proteins, such as members of the small leucine-rich repeat (SLRP) PG family (decorin, biglycan, fibromodulin, and lumican) as well as the thrombospondins, the matrilins, and many others. Many of these molecules remain bound at the surface of the completed fibril, thereby providing for interactions with other components of the matrix (Heinegård, 2009). Therefore, ECMs form a complex three-dimensional, multimolecular network in an organ-specific manner.

There are two major types of ECMs, the interstitial and pericellular matrices. The interstitial matrix is present in connective tissues and consisted of a variety of collagen types, elastins, FN, and tenascins among others, as well as PGs and HA. Pericellular matrices are in close contact with cells and significantly vary in composition from the surrounding interstitial ECMs. Basement membrane is the prototype of pericellular matrix. It is formed at the interface between parenchymal cells and connective tissue and serves as a supporting layer that organizes and holds them in place and prevents them from ripping apart. Basement membranes are especially tough sheetlike structures, primarily composed of laminins, collagen type IV, nidogen, fibulin, dystroglycan, and heparan sulfate (HS) PGs, and they separate these cell types from the adjacent connective tissue (LeBleu et al., 2007).

In the past, ECMs were considered a space-filling material between cells that provide tissues and organs with mechanical support and integrity without specific functions. During the last decades, ECMs have been recognized as dynamic structures with multiple functional properties that regulate cell behavior through various chemical and mechanical signals. Cells bind to ECMs through many specific cell surface receptors and molecules, including integrins, cell surface PGs (syndecans and glypicans), HA receptor CD44, discoidin domain receptors, and others.

ECMs also contain a variable proportion of growth factors, cytokines, and/or immune soluble mediators, which can be retained by macromolecules; this makes ECMs reservoirs for these important mediators that can be presented to specific receptors on cell surfaces at developmentally and/or physiologically relevant times (Kirkpatrick and Selleck, 2007; Rozario and DeSimone, 2010). In this way, the ECMs can contribute to the establishment of gradients of secreted signaling molecules or can concentrate these factors around certain cell types and facilitate their binding to their receptors (Kirkpatrick and Selleck, 2007; Rozario and DeSimone, 2010). Cell surface receptors interact with various ECM components, growth factors, and cytokines with different affinities, and these fine-tuned interactions transmit various signals within cells that affect gene expression and such diverse cellular responses as proliferation, adhesion, migration, polarization, differentiation, survival, and apoptosis (Frantz et al., 2010).

ECM remodeling occurs in physiological (e.g. during development) and pathological situations and involves several proteolytic enzymes, including matrix metalloproteinases (MMPs), a disintegrin and metalloproteinases (ADAMs), a disintegrin and metalloproteinase with thrombospondin motifs (ADAMTSs) proteinases, the system of plasmin and plasminogen activators, and various cysteine proteases that are released in ECMs in various situations. Glycosaminoglycan (GAG)–degrading enzymes, such as hyaluronidases and heparanase may also modify ECM in some cases. Cell-controlled degradation of ECMs can allow the free movement of the cells or initiate processing and deposition of new matrix. The action of these enzymes can either release growth factors bound to fragments of matrix molecules and facilitate their binding to signaling receptors on cell surfaces, or they can liberate ECM fragments (e.g. endostatin) that can regulate cell functions in a different way than the intact molecules (Whitelock et al., 2008; Theocharis et al., 2010).

In this introductory chapter, the structural and functional components of the main types of ECMs as well as the cell surface receptors that interact with them to regulate cell function are presented in a manner to give to the reader a short overview of their importance in the organization and biological roles of ECMs. A schematic presentation summarizing the main ECM components and their molecular interactions is given in ▶Figure 1.1.

1.1.1 ECM components: structural and functional properties

Proteoglycans-glycosaminoglycans

PGs are macromolecules composed of a specific core protein carrying covalently linked GAG chains named chondroitin sulfate (CS), dermatan sulfate (DS), keratan sulfate, heparin, and HS. GAGs are linear, negatively charged polysaccharides composed

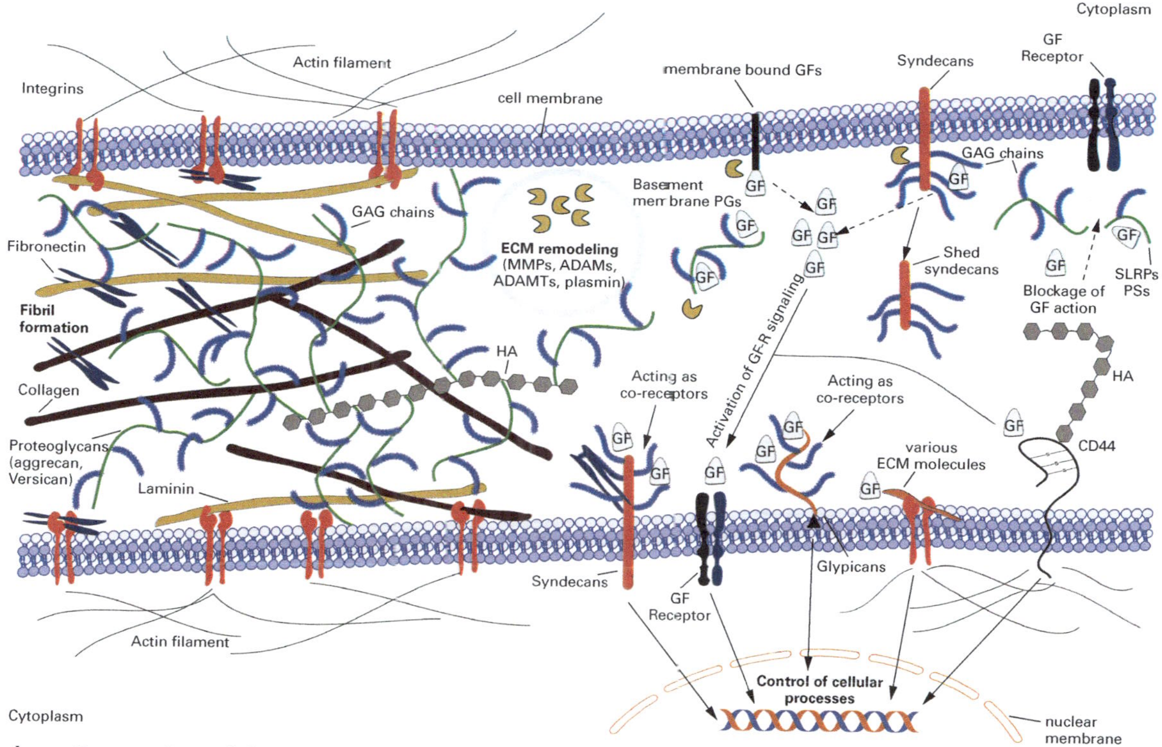

Figure 1.1 A schematic overview of the major extracellular matrix molecules and cell surface receptors contributing to cell-cell and cell-matrix interactions. Extracellularly secreted macromolecules, such as matrix proteoglycans, hyaluronan, collagen, and fibronectin, and cell-bound receptors, such as integrins, syndecans, and CD44, form a dynamic interacting network that regulates cell behavior. Growth factors and cytokines interacting with proteoglycans are stored in the ECM and protected from proteolytic degradation. Proteoglycans also modulate their signaling because they act as co-receptors presenting them to high-affinity receptors. On the other hand, proteolytic enzymes, such as MMPs, ADAMTSs, and plasmin play critical roles in normal remodeling and disease progression.

of repeating disaccharides of acetylated hexosamines (*N*-acetyl-galactosamine or *N*-acetyl-glucosamine) and mainly hexuronic acids (D-glucuronic acid or L-iduronic acid). GAGs are O-sulfated at various positions, and N-sulfation occurs in heparin and HS. Keratan sulfate is the only GAG that lacks a hexuronic acid and is composed of repeating disaccharides containing *N*-acetyl-glucosamine and galactose. GAG chains exhibit a large heterogeneity in their structure due to the variation in chain length and the extent and pattern of sulfation (Karamanos and Tzanakakis, 2012).

PGs can interact with almost all of the structural proteins in the ECM, with growth factors and cytokines, and with growth factor receptors. PGs exhibit numerous biological functions acting as structural components in tissue organization. They are implicated in cell signaling and affect several cellular parameters, such as cell proliferation, adhesion, migration, and differentiation. PG expression is markedly modified during the ECM remodeling that occurs in physiological and pathological situations, such as during tissue differentiation and in malignant development and metastasis. Tumor microenvironments are characterized by altered composition and structure of PGs. The type and fine structure of GAG chains attached to PGs are markedly affected in the context of malignant transformation as a result of the altered expression of GAG-synthesizing enzymes. Structural modifications of GAGs may facilitate tumorigenesis in various ways, modulating the functions of PGs. Furthermore, altered expression of PGs on tumor and stromal cell membranes affects cancer cell signaling, growth and survival, cell adhesion, migration, and angiogenesis.

PGs according to their localization can be classified into three main groups: ECM secreted, those localized at the cell surface, and intracellular. Each main group is further classified into subfamilies according to its gene homology, core protein properties, size, and modular composition (Theocharis et al., 2010).

Secreted ECM PGs are divided into families of large aggregating PGs, named hyalectans, SLRPs, and basement membrane PGs.

The subfamily of hyalectans includes versican, aggrecan, neurocan, and brevican. Hyalectans have the ability to bind HA through their N-terminal globular domains, the central domains carry most of the GAG chains, and the C-terminal globular domains can exhibit lectin-like activity. For example, versican interacts with HA through the N-terminal domain and with various other molecules via specific protein motifs in the C-terminal domain, which can regulate many cellular processes including adhesion, proliferation, apoptosis, migration, invasion, and ECM assembly. Elevated levels of versican have been reported in most malignancies to date and have been associated with cancer relapse and poor patient outcome in breast cancer. Versican is accumulated in the preclinical phase of breast cancer and is associated with malignancy, and its presence in mammographic findings is highly suggestive for malignancy in nonpalpable breast carcinomas (Skandalis et al., 2011). The N-terminal globular domain of versican stimulates proliferation by destabilizing cell adhesion, whereas the C-terminal globular domain induces proliferation, at least in part, by activating the epidermal growth factor receptor (EGFR) via the action of epidermal growth factor (EGF)–like motifs. Several proteases, including ADAMTSs, MMPs, and plasmin, can cleave versican, thereby generating fragments containing the globular domains that promote cancer cell motility and metastasis.

Aggrecan is a major structural component of cartilage ECM through formation of large aggregates with HA and link proteins, which provide cartilages with their ability to resist the compressive forces that are applied to skeletal structures. Aggrecan is also involved in the regulation of cartilage development, growth, and homeostasis. In addition to its essential structural role in the ECM of skeletal elements during development, aggrecan has important roles in reparative responses following tissue injury that involves regulation of transforming growth factor-beta 1 (TGF-β1) signaling.

The family of SLRPs includes 18 genes, which are divided into five subgroups (Iozzo and Schaefer, 2010). SLRPs are characterized by tandem leucine-rich repeats within the protein cores. Although SLRPs were initially considered structural components, they participate in a wide range of matrix-matrix and matrix-cell interactions. They bind and modulate the biological functions of various cytokines and growth factors participating in the development of proliferative, inflammatory, and fibrotic disorders. Decorin is the prototype of SLRPs and represents a powerful inhibitor of tumor cell growth and migration. Decorin is generally accumulated in the tumor stroma in various types of cancer as a consequence of increased biosynthesis by stromal fibroblasts. Reduced amounts of decorin were associated with poor prognosis in node-negative invasive breast cancer. Decorin expression is switched off in many tumor cells by proteasomal degradation and epigenetic control. Decorin binds directly to growth factor receptors (c-Met and EGFR/ErbB family) and downregulates their activity. For example, decorin binds to EGFR, which then dimerizes and is subsequently internalized and degraded in the lysosomes via caveolin-mediated internalization. Decorin inhibits tumor cell proliferation by evoking a signaling cascade that is different from the one evoked by EGF, possibly by inducing a different EGFR conformation and selectively activating phosphotyrosines in the receptor autophosphorylation domain. Decorin also suppresses the activity of ErbB2 and ErbB4 receptors via degradation.

The subfamily of basement membrane PGs has three main, well-characterized members: perlecan, collagen type XVIII, and agrin, which are almost universally decorated with HS side chains. These PGs are elongated molecules with a variety of domains that share structural and functional homology with numerous ECM proteins, growth factors, and surface receptors. They possess disparate biological functions for the maintenance of basement membrane homeostasis and modulation of growth factor activity and angiogenesis (Theocharis et al., 2010).

Cell-surface-associated PGs are classified into two main subfamilies, the transmembrane syndecans (four members: syndecan-1 to -4) and glycosylphosphatidylinositol (GPI)–anchored glypicans (six members: glypican-1 to -6). They can be cleaved from the cell surface and released into ECMs by the action of proteases and phospolipases, respectively (Manon-Jensen et al., 2010).

Glypicans show cell-type and developmental stage-specific expression. They are involved in fundamental biological processes such as cell-ECM interactions and the control of cell proliferation, differentiation, and morphogenesis. They are involved in the regulation of various signaling pathways (Filmus et al., 2008). Glypicans influence tumor development and progression, and their expression is abnormal in various human tumors. Glypican-1 is highly expressed in breast cancer and pancreatic cancer, where it may have a key role in promoting growth factor signaling in cancer cells. In contrast, glypican-3 negatively regulates Hedgehog signaling and has a negative role in cell proliferation and induces apoptosis in mesothelioma and breast cancer. Loss-of-function mutation of glypican-3 induces overgrowth in patients with Simpson-Golabi-Behmel syndrome, deregulating the balance between cell proliferation and apoptosis, which increases the risk for embryonic tumor development. In other tumor types, such as in hepatocellular carcinoma, overexpression of glypican-3 promotes tumor growth via promotion of Wnt signaling. Targeting glypican-3 is a promising therapeutic approach for the clinical management of hepatocellular carcinoma and selected other tumors that express this marker (Allegretta and Filmus, 2011).

Syndecan-1 is present in early stages of development and on epithelial and cancer cells in adults, whereas syndecan-2 is distributed in mesenchymal tissues, liver, and neuronal cells. Syndecan-3 is mainly associated with neural tissues, and syndecan-4 is ubiquitously distributed. Syndecans serve as matrix and cell surface receptors, coreceptors for growth factor signaling, internalization receptors, and soluble paracrine effectors. They are capable of interacting mostly through their HS/CS chains, with numerous ECM components including growth factors, chemokines, collagens, FN, and laminins (Couchman, 2010). Syndecans possess a cytoplasmic domain with various motifs that interact with other proteins like PDZ-domain-containing proteins, which can link syndecans with cytoskeleton.

Syndecans act as coreceptors for growth factors and work in concert with high-affinity growth factor receptors to properly transmit intracellular signals that regulate cell proliferation, differentiation, adhesion, and migration. In the case of other ECM molecules, the syndecans can act as coreceptors with integrins, which also interact with a great variety of ECM molecules (Lambaerts et al., 2009). Syndecan-ECM interactions can induce intracellular signals, which mostly synchronize with integrin signaling, thereby regulating cytoskeleton rearrangement, cell adhesion, and migration. Cross talk of syndecans with integrins induces the transcription of MMPs and regulates matrix remodeling. Syndecans may also transmit intracellular signals by themselves, perhaps independently, as described for syndecan-4, where elements of a signaling pathway have been elucidated (Couchman, 2010).

Syndecan-1 and syndecan-2 are upregulated in various malignancies and correlate with poor prognosis. By contrast, syndecan-1 is downregulated in other types of tumors. Syndecan-1 expression by tumor-associated stromal fibroblasts may promote tumorigenesis by regulating tumor cell spreading and adhesion, proliferation, and angiogenesis (Fears and Woods, 2006; Theocharis et al., 2010). Notably, syndecan-1 also mediates cell adhesion in cooperation with integrins. Syndecans promote proangiogenic signaling by binding fibroblast growth factor-2 (FGF2) and vascular endothelial growth factor (VEGF) and presenting them to their high-affinity receptors and also by protecting them from inactivation. Various growth factors and enzymes that are upregulated in solid tumors *in vivo* accelerate shedding of syndecans *in vitro*. Soluble syndecan-1 accelerates tumor growth *in vivo* and increases cell invasion. Soluble syndecan-1 may activate proangiogenic/protumor factors, like FGF2, and can also bind to mitogens, effectively creating a chemotactic gradient, and/or interact with proteases to protect them from contact with endogenous inhibitors (Fears and Woods, 2006; Theocharis et al., 2010). Syndecan-4 is a focal adhesion component in a range of cell types and mediates breast cancer cell adhesion and spreading. The directed homeostasis in syndecan-4 levels supports the optimal migration of tumor cells since cells that overexpress syndecan-4 exhibit decreased cell migration, whereas lack of syndecan-4 engagement promotes an amotile cell phenotype in which focal adhesion kinase (FAK) and Rho signaling are downregulated (Fears and Woods, 2006).

Serglycin is the only characterized intracellular PG. It has important functions related to the formation of several types of storage granules and to the biosynthesis of bioactive molecules in hematopoietic cells (Kolset and Pejler, 2011). It has been shown that serglycin can regulate immune system functions through inhibition of the classical and lectin pathways of the complement system (Skliris et al., 2011). A recent study also demonstrated that elevated expression of serglycin by tumor cells regulates nasopharygeal carcinoma metastasis via autocrine and paracrine routes and correlates with adverse outcomes for patients (Li et al., 2011).

The GAG HA is a large unbranched polymer that contains disaccharide units of D-glucuronic acid/*N*-acetylglucosamine with molecular mass ranging from 10^6 to 10^7 Da. HA is a GAG that does not have a core protein, and its synthesis is regulated by the action of three HA synthases (HAS1, 2, and 3), which differ in tissue distribution, rate of HA synthesis, and size of HA produced (Misra et al., 2011). The HA contents of ECMs and pericellular matrices are regulated by the action of hyaluronidases (Hyal1, 2, and 3 and PH-20), which degrade HA producing fragments of HA with diverse biological functions.

HA is found in pericellular matrices attached to HA-synthesizing enzymes, which are in plasma membranes and extrude the growing HA molecule into the extracellular space. HA also interacts with its cell surface receptors, such as the ubiquitous CD44. Pericellular HA is normally metabolically dynamic with a catabolic process that includes GPI-anchored Hyal2, which produces HA fragments that are internalized and transported to lysosomes. A variety of cell stresses, including endoplasmic reticulum stress, viral infection, and hyperglycemic cell division, induce formation of monocyte-adhesive HA matrices that are central to many pathologies (Wang et al., 2011). HA accumulates in sites of cell division and rapid matrix remodeling that occurs during embryonic morphogenesis, inflammation, and tumorigenesis. The regulation of transient interactions of HA with its HA-binding proteins is crucial for fundamental physiological processes and pathological conditions in which HA affects cell proliferation, migration, and differentiation (Misra et al., 2011). *HAS2* expression is essential for epithelial-to-mesenchymal transition during development. The overproduction of HA in cancer cells transfected with *HAS* genes triggered intracellular signaling pathways that promoted anchorage-independent growth and invasiveness, which correlated with increased expression of CD44 variants. Invasive and/or metastatic breast cancer cells deprived of HAS2 lost their aggressive phenotype.

Collagens

Collagens are the most abundant proteins of ECMs. Collagens provide both structural integrity and functional diversity within tissues and also provide signals to cells overlaying the collagen scaffold that can alter their behavior. So far, 28 different types of collagen have been identified in vertebrates. Collagen type I is the most extensively studied member of this family and the major structural ECM component of most tissues (Heino, 2007). Collagens are characterized by the assembly of three left-handed α-helical polypeptide chains. The α chains are encoded by 46 different genes and vary in the number of collagenous and noncollagenous regions present within their α chains. They are packaged in either homo- or heterotrimeric helical bundles to form the procollagen triple helix molecule. The α chains are twisted together into a right-handed coiled coil, producing a right-handed homotrimer or heterotrimer triple helix that has a ropelike structure. Procollagen is converted to the mature form by proteolytic removal of the

N- and C-propeptides, thereby triggering the process of collagen fibrillogenesis. These undergo self-assembly and are packaged to form collagen microfibrils. Continued cross-linking leads to the formation of large collagen fibers.

Polypeptide α chains contain a variable number of Gly-X-Y repeats, where X and Y can be any amino acid but often are proline and hydroxyproline, respectively. The presence of glycine with its small hydrogen atom side chain in every third residue within α chains allows them to pack tightly together in a triple helix with this residue in the interior (axis) of the helix and the rings of the proline and hydroxyproline pointing outward. The presence of the relatively high content of proline and hydroxyproline rings adds to the rigidity and stability of helical structures since their geometrically constrained carboxyl and secondary amino groups, along with the rich abundance of glycine, accounts for the tendency of the individual polypeptide strands to form left-handed helices spontaneously, without any intrachain hydrogen bonding.

Collagen molecules can be grouped according to their structural and functional properties into seven subfamilies (Heino, 2007):

1. Fibrillar collagens (collagen types I, II, III, V, XI, XXIV, and XXVII) are abundantly present in connective tissues. They provide tissues and organs with structural integrity and confer mechanical tensile strength.
2. Network-forming collagens (collagen types IV, VI, VIII, and X) have many interruptions in their triple helical structures. This gives them their molecular flexibility, and they can associate with each other forming extensive networks. Furthermore, they can interact with various other ECM components creating multimolecular complexes.
3. Fibril-associated collagens with interrupted triple helices (FACIT) (collagen types IX, XII, XIV, XVI, XIX, XX, XXI, XXII, and XXVI), have noncollagenous domains that interrupt the triple helical collagenous domains providing these molecules with flexibility. FACIT collagens do not wind together to form fibrils in ECMs. Rather, they interact with fibrillar collagens on their surface and link collagen fibers to one another and with other ECM molecules.
4. Anchoring fibrils are composed of collagen type VII. Collagen type VII is a homotrimer [α1(VII)]3 with a central collagenous triple helical structure. It is interrupted by a short noncollagenous domain and is flanked by N-terminal and C-terminal noncollagenous domains. Two such molecules are dimerized, and dimers are assembled together to form anchoring fibrils that connect the basement membranes to underlying stroma in the ECM.
5. Multiplexin collagens (collagen types XV and XVIII) have a central triple helical collagen domain that is interrupted by several noncollagenous domains and have long N-terminal and C-terminal noncollagenous regions. Collagen type XVIII carries HS GAG chains, whereas collagen type XV is covalently linked to CS PGs through disulfide bonds. Both multiplexin collagens are found in the basement membrane zones. Proteolytic degradation of the C-terminal domain of multiplexin collagens liberates the polypeptides endostatin (collagen type XVIII) and restin (collagen type XV) with similar biological properties. Endostatin is a potent antiangiogenic molecule that can reduce tumor growth, choroidal neovascularization, and wound healing (Iozzo et al., 2009).
6. Transmembrane collagens (collagen types XIII, XVII, XXIII, and XXV) have a single hydrophobic transmembrane domain with an extracellular C- terminal domain composed of collagenous parts separated by noncollagenous sequences and a cytoplasmic

N-terminal domain. They are expressed in several cells and tissues and act as cell surface receptors. Proteolytic enzymes can shed transmembrane collagens from the cell surface, thereby liberating a soluble ectodomain form of these molecules.

7. Collagen type XXVIII has its triple helix flanked by a von Willebrand factor A structure and is expressed in dorsal root ganglia and peripheral nerves.

Several mutations have been identified in collagens that may affect trimerization, formation of collagen networks, and cleavage of the C-propeptide. Mutations have been associated with various clinical pathologies, among them the Ehlers-Danlos syndrome, Stickler syndrome, Alport syndrome, and osteogenesis imperfecta. Collagen levels and structure in the ECMs are altered in various physiological and pathological conditions under the regulation of various stimuli. These alterations regulate cell attachment, which is important for the support and the functions of resident cells. For example, collagen is the major component of ECMs in solid tumors, and its density and stiffness within tumor stroma are important for tumor cell biology contributing to tumor initiation, invasion, and metastasis (Levental et al., 2009; Gao et al., 2010). Collagen deposition and enhanced collagen cross-linking mediated by lysyl oxidase promote growth and invasion by premalignant human mammary epithelial cells and hypoxia-induced metastasis of lung cancer through induction of integrin signaling, activation of FAK, phosphatidylinositol 3-kinase (PI3K), and Akt. The interaction of various cell types with collagen type I is mediated by several β1 integrins, and this interaction is crucial for endothelial cell migration and angiogenesis (Kim et al., 2011). Collagen type I remodeling in perivascular stroma by various collagen-degrading enzymes also represents an important step for endothelial cell reorganization into tubular structures during angiogenesis.

Elastin

Elastin is the major component of elastic fibers present in large elastic blood vessels such as the aorta and in the lungs, heart, elastic ligaments, skin, bladder, and the elastic cartilage. Elastin is secreted as a soluble precursor tropoelastin, which is rich in hydrophobic amino acids presenting hydrophobic regions alternating with cross-linking domains that contain lysine residues. In the ECM, tropoelastin molecules are highly cross-linked making a massive insoluble rubberlike network, in a reaction catalyzed by lysyl oxidase. The hydrophobic domains of elastin are responsible for the elastic properties of the network (Muiznieks et al., 2010). Nonelastin molecules, such as fibrillins and fibulins are also present in elastic fibers in association with elastin. Elastin is synthesized early in life and is very stable with very low or absent turnover. Elastin is degraded by various proteolytic enzymes, such as MMPs, in pathological processes with consequent failure of the tissue/organ (Duca et al., 2004).

Fibronectin

FN is a high-molecular-weight glycoprotein present in ECMs as a protein dimer, consisting of two nearly identical polypeptide chains linked by a pair of C-terminal disulfide bonds. Each FN monomer contains three types of modules: types I, II, and III. Type I and type II domains are stabilized by intrachain disulfide bonds, while the type III domain does not have any disulfide bonds. It contains the Arg-Gly-Asp (RGD) amino acid sequence (RGD motif) responsible for binding to integrins, domains for binding to other FN molecules, and heparin-binding domains. FN type I and II modules exhibit domains for binding to other FN molecules, collagen, and fibrin. FN modules are present in multiple copies within each protein (e.g. FN exhibits 15 FN III domains), which allows multiple interactions, thus providing a protein-binding platform (Pankov and Yamada, 2002). FN is assembled into an insoluble fibrillar matrix in a complex cell-mediated process facilitated by integrin receptors on the cell surface. The ability of FN to interact with numerous matrix macromolecules and cell surface receptors gives this molecule a major role in cell adhesion, growth, migration, and differentiation. FN is important for wound healing and development, whereas alterations in its expression, degradation, and organization have been associated with pathological situations, including cancer and fibrosis.

Laminins

Laminins are large heterotrimeric glycoproteins secreted and incorporated into cell-associated ECMs. They are present in basement membranes but also found in ECMs in animal tissues contributing to the tensile strength of the tissues. Laminins are composed of a combination of three chains (α, β, and γ), which exist in five, four, and three genetic variants, respectively, forming at least 16 different laminins (Durbeej, 2010). The laminin molecules are named according to their chain composition. Thus, laminin-221 contains α2, β2, and γ1 chains. Each chain has a number of globular, rodlike, and coiled coil regions. The long arm of laminin forms a helical coiled coil region by linking together all three chains by disulfide bonds. The long arm can bind to cells, which helps anchor organized tissue cells to the laminin matrices. The long-arm α chain is extended at the C terminus forming five homologous globular LG domains, which are involved in binding with several ECM molecules, including PGs, fibulin-1, and dystroglycans. The three shorter arms can also bind ECM molecules and different integrin receptors as well as other laminin molecules, which allow their polymerization and the formation of laminin sheets (Patarroyo et al., 2002). Different laminins are found at different stages of development and in different tissues where they influence cell differentiation, migration, and adhesion, and they are vital for the maintenance and survival of tissues. Defective laminins result in improperly formed muscles, leading to a form of muscular dystrophy, lethal skin blistering disease (junctional epidermolysis bullosa), and defects of kidney function (nephrotic syndrome) (Durbeej, 2010).

1.1.2 Matrix remodeling is accomplished by proteolytic enzymes

Proteolysis occurs in all biological processes, including cell proliferation, migration and invasion, apoptosis, angiogenesis, and embryogenesis, and is mediated by numerous classes of enzymes. The most important classes involved in the degradation of ECMs are MMP, ADAM, ADAMTS proteinases, the system of plasmin and plasminogen activators, and cysteine proteases.

Matrix metalloproteinases

MMPs produced by various cell types (fibroblasts, epithelial, endothelial, inflammatory, and tumor cells) are key enzymes in the ECM remodeling that occurs in physiological and pathological situations (Gialeli et al., 2011). MMPs belong to a zinc-dependent family of endopeptidases composed of 24 currently known enzymes in humans. They share several functional domains and are divided into matrix-secreted soluble MMPs and membrane-type MMPs (MT-MMPs) that are associated with the cell membrane via GPI anchors (MT4-MMP and MT6-MMP) or by protein transmembrane domains (MT1-MMP, MT2-MMP, MT3-MMP, and MT5-MMP).

MMPs secreted into ECMs can be associated with cell membranes via interactions with cell surface molecules, such as CD44, PGs, and integrins, and contribute to the degradation of ECM in cell microenvironments. Since excessive action of these enzymes may compromise the integrity of the tissue, their activity is tightly regulated both at the transcription and the activation levels. MMPs are synthesized as inactive enzymes that require proteolytic cleavage to be activated. For example, pro-MMP2 is activated at the cell surface by MT1-MMP through the formation of a trimolecular complex with tissue inhibitor of metalloproteinase-2 (TIMP2). The activities of MMPs are also regulated by their binding to a family of inhibitors of MMPs, including tissue inhibitor of metalloproteinases (TIMPs) (Brew and Nagase, 2010). MMPs are generally broad spectrum proteases that degrade various ECM components.

MMPs mediate matrix remodeling and regulate tissue homeostasis and development. MMPs are major mediators of the alterations observed in the ECM in pathologies including rheumatoid arthritis, inflammatory diseases, and cancer progression. The actions of MMPs in tumor microenvironments generate space for the cells to migrate, produce cryptic peptides with novel biological activity, and release ECM-stored growth factors that can stimulate malignant and endothelial cells to promote tumor growth, angiogenesis, and tumor spread (Kessenbrock et al., 2010).

ADAM/ADAMTS

A zinc-dependent family of proteinases related to the MMPs is represented by ADAM, which includes two subgroups: membrane-bound ADAMs and ADAMTSs. Recent studies show that ADAMs and ADAMTSs present altered expressions in diverse tumor types, suggesting that these proteases are involved in different steps of cancer progression, including carcinogenesis (Murphy, 2008; Rocks et al., 2008). ADAM molecules are implicated in tumor cell proliferation/apoptosis, cell adhesion/migration, and cell signaling. In particular, they exhibit proteolytic activity like MMPs, although their main roles focus on ectodomain shedding and nonproteolytic functions, such as binding to adhesion molecules, integrins, and interactions with phosphorylation sites for serine/threonine and tyrosine kinases, thus contributing to cancer development (Rocks et al., 2008).

Plasmin and plasminogen activator system

The plasmin and plasminogen activator system has important roles in normal physiology and in disease. Plasminogen has two main physiological activators, urokinase plasminogen activator (uPA) and tissue-type plasminogen activator (tPA). uPA is synthesized by endothelial cells, epithelial cells, leukocytes, monocytes, fibroblasts, and cancer cells (Hildenbrand et al., 2010), whereas tPA is synthesized mainly by vascular endothelial cells, keratinocytes, melanocytes, and neurons (Gebbink, 2011). uPA and tPA are serine proteases that act on their primary physiological substrate, plasminogen, the zymogen form of the serine protease plasmin. Activation of plasmin triggers a proteolysis cascade that, depending on the physiological environment, participates in the dissolution of fibrin clots and the regulation of pericellular proteolysis. Plasmin can also degrade various ECM components, including FN, thrombospondin, and laminin, and can also activate collagenases.

Elevated expression levels of urokinase and several other components of the plasminogen activation system correlate with tumorigenesis because plasminogen activation facilitates tumor cell invasion and, thus, contributes to metastasis. The most important inhibitors of the uPA/tPA system are the serpins, plasminogen activator inhibitor-1 (PAI1) and plasminogen activator inhibitor-2 (PAI2), which inhibit plasmin activity irreversibly. Plasmin is also inhibited by its main physiological inhibitor, α2-antiplasmin, and by the general protease inhibitor α2-macroglobulin. The uPA receptor (uPAR) is expressed in several tissues, and strong expression is often found in tissues that undergo remodeling. uPAR expression is often upregulated in various cancers (Ulisse et al., 2009). uPAR binds uPA and thereby restricts plasminogen activation to the immediate vicinity of the cell membrane (Blasi and Sidenius, 2010). The interaction of uPA with uPAR is involved in the regulation of several other aspects of cancer biology, such as cell adhesion, migration, and cellular mitotic pathways (Smith and Marshall, 2010).

Cysteine proteases

Cysteine proteases are proteolytic enzymes with a cysteine residue in the catalytic site and are classified in several families according to similarities in their secondary and/or tertiary structures. Some cysteine proteases, such as cathepsins and legumain, are primarily localized in lysosomes. However, several reports have demonstrated their localization and action in ECMs in diseases such as cancer, atherosclerosis, and osteoporosis. In order to be activated, these enzymes generally require an acidic environment, which is often present in pathological situations. Various ECM components are substrates for cysteine proteases. Therefore, they are implicated in ECM degradation and remodeling and may regulate important cellular processes, such as cell signaling, transformation and differentiation, motility, adhesion, invasion, angiogenesis, and metastasis (Obermajer et al., 2008).

1.1.3 Cell surface receptors mediate cell-cell and cell-matrix interactions

Integrins

Integrins are heterodimeric transmembrane receptors composed of 18 α subunits and 8 β subunits that can be noncovalently assembled into 24 combinations that help cells to respond to mechanical and biochemical changes in ECMs (Barczyk et al., 2010). Both α and β subunits of integrins bind to the RGD motif, which is present in many ECM proteins. The evolutionarily conserved three-residue RGD motif efficiently serves as the attachment site for integrin-mediated cell adhesion. The specificity of integrin binding to different matrix proteins is partly determined by other amino acids surrounding the RGD sequence (Ruoslahti, 1996). High density of RGD motifs is required to allow a precise spatial distribution pattern of integrins in order to initiate an optimal cellular response (Chen et al., 1997).

The integrin dimers bind to a great variety of different ECM molecules with overlapping binding affinities (Alam et al., 2007). Therefore, the specific integrin expression patterns by a cell can dictate the binding to a certain ECM substrate, and the composition of integrin molecules determines the downstream signaling events, and therefore the eventual cell behavior and fate. Integrins have the unique ability to respond to the molecular composition and physical properties of ECMs, and they integrate both mechanical and chemical signals through direct association with the cytoskeleton, which also determines the selection of specific integrin species to be involved. Integrins have been shown to regulate tumor cell growth and migration, matrix synthesis, turnover and fibrosis, angiogenesis, and epithelial-to-mesenchymal transition.

Integrins on the cell surface are allosteric proteins that exist in low-affinity, primed, and high-affinity states (Askari et al., 2009). Their transition between those states is influenced by a variety of factors, such as ligand engagement and the binding of intracellular proteins to their cytoplasmic domains. For example, the binding of integrin to an extracellular ligand can result in conformational changes in the integrin that alter the affinity of its cytoplasmic domain for proteins involved in numerous signaling pathways and cytoskeleton organization. On the other hand, the interaction of cytoplasmic proteins with the intracellular domain of integrins can modulate their ligand affinity (Askari et al., 2009). When inactive, integrins are unable to bind to ECM molecules or to other receptors necessary to form focal adhesions that are required for cell migration. The initial assembly of the nascent adhesion is mediated by anchoring proteins called talins, which interact both with the cytoplasmic tail of the β integrin subunit and actin. Binding of kindlin to β integrin is also required to achieve maximal integrin activation (Moser et al., 2009). The synergistic effect of talin and kindlin on the activation and the assembly of adhesion structures is enhanced by the binding of vinculin to talin, which triggers the clustering of activated integrins/actin complexes, further strengthening the focal adhesion interaction. The formation of larger focal adhesion assembly results in the conformational transition of integrin dimers to their active state, which is capable of high-affinity interactions with ECM ligands. This eventually leads to maturation of fully activated streak-like fibrillar focal adhesions that can transmit signals (Tadokoro et al., 2003).

The cytoplasmic domains of integrins have no catalytic activity of their own but are able to recruit accessory molecules that contribute to the actin cytoskeletal reorganization and endow catalytic activity to the focal adhesion. The integrin-associated proteins include FAK; Src, a nonreceptor tyrosine kinase; receptor-type tyrosine protein phosphatase a (RPTPa); and SH2-domain-containing protein tyrosine phosphatase-2 (SHP2). FAK is a nonreceptor tyrosine kinase that is activated upon integrin binding and subsequently binds to Src resulting in the formation of a stable and activated Src-FAK complex. Src further phosphorylates FAK and several downstream binding partners. FAK can activate mitogen-activated protein (MAP) kinase, and extracellular signal-regulated kinase (ERK) and is a regulator of Rho GTPases, thereby affecting cell motility. FAK activation leads to the recruitment of PI3K to focal adhesion, leading to activation of AKT, which regulates integrin-mediated cell survival. Moreover, FAK represents a cross-talk point for growth factor receptor pathways, since the signals from integrin-Src-FAK complexes can be integrated with those of growth factors and be transmitted through the same Ras-MEK-MAPK pathway, which modulates both focal adhesion dynamics and cellular functions (Kim and Kim, 2008). Therefore, ECM can augment signaling downstream of growth factor receptors by concentrating signaling substrates in close proximity of these receptors, and integrin-mediated adhesion to ECMs can even cluster and activate growth factor receptors in the absence of their ligand (Comoglio et al., 2003). In this way, integrin signaling has been shown to modulate proliferation, polarization, migration, differentiation, survival, and apoptosis.

Discoidin domain receptors

Apart from integrins, discoidin domain receptors (DDRs) have been recognized as novel collagen receptors. DDRs are tyrosine kinase receptors that have been highly conserved throughout evolution, and they possess an extracellular discoidin domain and a longer juxtamembrane region. They bind to collagens independently of the integrins and regulate cell proliferation, migration, and matrix remodeling (Vogel et al., 2006). DDRs are implicated in various pathologies including atherosclerosis, fibrotic diseases, and cancer.

CD44

CD44 is a type I transmembrane cell surface receptor involved in cell-cell interactions, adhesion, and migration. CD44 is encoded by a single gene, but, as a result of alternate splicing, multiple forms of CD44 are generated that are further modified by *N*- and *O*-linked glycosylations. Although CD44 is the major cell surface receptor for HA, its ectodomain can have CS and/or HS chains that enable CD44 to bind growth factors. The rather short cytoplasmic tail of CD44 binds to ankyrin, ezrin-radixin-moesin proteins, and merlin, thereby providing a link to the cytoskeleton (Misra et al., 2011).

The smallest CD44 isoform that lacks variant exons, designated standard CD44 (CD44s), is abundantly expressed by both normal and cancer cells, whereas the variant CD44 isoforms (CD44v) that contain a variable number of exon insertions (v1–v10) are expressed mostly by cancer cells. The modified CD44 structures in cancer due to alternate splicing and posttranslational modifications provide CD44 molecules with enhanced HA binding, which can lead to increased tumorigenicity (Misra et al., 2011). The effect of HA on cells depends on the size of this molecule. High-molecular-weight forms of HA predominate in physiological ECM, whereas lower-molecular-weight forms of HA accumulate at sites of inflammation and in tumors either as result of altered HA biosynthesis or due to increased activity of hyaluronidases. Binding of high-molecular-weight HA to CD44 can lead to transduction of antimitogenic and anti-inflammatory signals, while in contrast, low-molecular-weight HA binding can activate mitogenic and proinflammatory pathways. CD44 can also react with other molecules, including MMPs, growth factors, and growth factor receptors, thereby participating in signaling events that regulate various cellular functions. CD44 is of great importance in the onset of malignant transformation and has been recognized as a marker of cancer stem cells in breast, pancreas, and colorectal cancers (Zöller, 2011). This small population of tumor cells, referred to as cancer stem cells or cancer-initiating cells, exhibit stem cell properties and are responsible for maintaining the tumor and, possibly, for the formation of new tumors at metastatic loci.

1.1.4 Take-home message

ECMs, apart from their significance in providing tissues with structural integrity and mechanical properties, have been highlighted as a biological active scaffold that regulates tissue homeostasis and cellular functions. Loss-of-function mutations in matrix molecules and modifications of ECMs are associated with various pathologies. All recent data support the notion that cross talk between matrix molecules in cell microenvironments and tissue cells is crucial for the development and progression of diseases. The deeper understanding of these biological processes will help scientists to develop novel strategies for disease diagnosis and successful treatment of various diseases.

References

Alam, N., Goel, H. L., Zarif, M. J., et al. (2007). The integrin-growth factor receptor duet. J Cell Physiol *213*, 649–653.

Allegretta, M., and Filmus, J. (2011). Therapeutic potential of targeting glypican-3 in hepatocellular carcinoma. Anticancer Agents Med Chem *11*, 543–548.

Askari, J. A., Buckley, P. A., Mould, A. P., and Humphries, M. J. (2009). Linking integrin conformation to function. J Cell Sci *122*, 165–170.

Barczyk, M., Carracedo, S., and Gullberg, D. (2010). Integrins. Cell Tissue Res *339*, 269–280.

Blasi, F., and Sidenius, N. (2010). The urokinase receptor: focused cell surface proteolysis, cell adhesion and signaling. FEBS Lett *584*, 1923–1930.

Brew, K., and Nagase, H. (2010). The tissue inhibitors of metalloproteinases (TIMPs): an ancient family with structural and functional diversity. Biochim Biophys Acta *1803*, 55–71.

Chen, C. S., Mrksich, M., Huang, S., Whitesides, G. M., and Ingber, D. E. (1997). Geometric control of cell life and death. Science *276*, 1425–1428.

Comoglio, P. M., Boccaccio, C., and Trusolino, L. (2003). Interactions between growth factor receptors and adhesion molecules: breaking the rules. Curr Opin Cell Biol *15*, 565–571.

Couchman, J. R. (2010). Transmembrane signaling proteoglycans. Annu Rev Cell Dev Biol *26*, 89–114.

Duca, L., Floquet, N., Alix, A. J., Haye, B., and Debelle, L. (2004). Elastin as a matrikine. Crit Rev Oncol Hematol *49*, 235–244.

Durbeej, M. (2010). Laminins. Cell Tissue Res *339*, 259–268.

Fears, C. Y., and Woods, A. (2006). The role of syndecans in disease and wound healing. Matrix Biol *25*, 443–456.

Filmus, J., Capurro, M., and Rast, J. (2008). Glypicans. Genome Biol *9*, 224.

Frantz, C., Stewart, K. M., and Weaver, V. M. (2010). The extracellular matrix at a glance. J Cell Sci *123*, 4195–4200.

Gao, Y., Xiao, Q., Ma, H., et al. (2010). LKB1 inhibits lung cancer progression through lysyl oxidase and extracellular matrix remodeling. Proc Natl Acad Sci U S A *107*, 18892–18897.

Gebbink, M. F. (2011). Tissue-type plasminogen activator-mediated plasminogen activation and contact activation, implications in and beyond haemostasis. J Thromb Haemost *9* (Suppl. 1), 174–181.

Gialeli, C., Theocharis, A. D., and Karamanos, N. K. (2011). Roles of matrix metalloproteinases in cancer progression and their pharmacological targeting. FEBS J *278*, 16–27.
Heinegård, D. (2009). Proteoglycans and more – from molecules to biology. Int J Exp Pathol. *90*, 575–586.
Heino, J. (2007). The collagen family members as cell adhesion proteins. Bioessays *29*, 1001–1010.
Hildenbrand, R., Allgayer, H., Marx, A., and Stroebel, P. (2010). Modulators of the urokinase-type plasminogen activation system for cancer. Expert Opin Investig Drugs *19*, 641–652.
Iozzo, R. V., and Schaefer, L. (2010). Proteoglycans in health and disease: novel regulatory signaling mechanisms evoked by the small leucine-rich proteoglycans. FEBS J *277*, 3864–3875.
Iozzo, R. V., Zoeller, J. J., and Nyström, A. (2009). Basement membrane proteoglycans: modulators par excellence of cancer growth and angiogenesis. Mol Cells *27*, 503–513.
Karamanos, N. K., and Tzanakakis, G. N. (2012). Glycosaminoglycans: from "cellular glue" to novel therapeutical agents. Curr Opin Pharmacol *12*, 220–222.
Kessenbrock, K., Plaks, V., and Werb, Z. (2010). Matrix metalloproteinases: regulators of the tumor microenvironment. Cell *141*, 52–67.
Kim, S. H., and Kim, S. H. (2008). Antagonistic effect of EGF on FAK phosphorylation/dephosphorylation in a cell. Cell Biochem Funct *26*, 539–547.
Kim, S. H., Turnbull, J., and Guimond, S. (2011). Extracellular matrix and cell signalling: the dynamic cooperation of integrin, proteoglycan and growth factor receptor. J Endocrinol *209*, 139–151.
Kirkpatrick, C. A., and Selleck, S. B. (2007). Heparan sulfate proteoglycans at a glance. J Cell Sci *120*, 1829–1832.
Kolset, S. O., and Pejler, G. (2011). Serglycin: a structural and functional chameleon with wide impact on immune cells. J Immunol *187*, 4927–4933.
Lambaerts, K., Wilcox-Adelman, S. A., and Zimmermann, P. (2009). The signaling mechanisms of syndecan heparan sulfate proteoglycans. Curr Opin Cell Biol *21*, 662–669.
LeBleu, V. S., MacDonald, B., and Kalluri, R. (2007). Structure and function of basement membranes. Exp Biol Med *232*, 1121–1129.
Levental, K. R., Yu, H., Kass, L., et al. (2009). Matrix crosslinking forces tumor progression by enhancing integrin signaling. Cell *139*, 891–906.
Li, X. J., Ong, C. K., Cao, Y., et al. (2011). Serglycin is a theranostic target in nasopharyngeal carcinoma that promotes metastasis. Cancer Res *71*, 3162–3172.
Manon-Jensen, T., Itoh, Y., and Couchman, J. R. (2010). Proteoglycans in health and disease: the multiple roles of syndecan shedding. FEBS J *277*, 3876–3889.
Misra, S., Heldin, P., Hascall, V. C., et al. (2011). Hyaluronan-CD44 interactions as potential targets for cancer therapy. FEBS J *278*, 1429–1443.
Moser, M., Legate, K. R., Zent, R., and Fässler, R. (2009). The tail of integrins, talin, and kindlins. Science *324*, 895–899.
Muiznieks, L. D., Weiss, A. S., and Keeley, F. W. (2010). Structural disorder and dynamics of elastin. Biochem Cell Biol *88*, 239–250.
Murphy, G. (2008). The ADAMs: signalling scissors in the tumour microenvironment. Nat Rev Cancer *8*, 932–941.
Obermajer, N., Jevnikar, Z., Doljak, B., and Kos, J. (2008). Role of cysteine cathepsins in matrix degradation and cell signalling. Connect Tissue Res *49*, 193–196.
Pankov, R., and Yamada, K. M. (2002). Fibronectin at a glance. J Cell Sci *115*, 3861–3863.
Patarroyo, M., Tryggvason, K., and Virtanen, I. (2002). Laminin isoforms in tumor invasion, angiogenesis and metastasis. Semin Cancer Biol *12*, 197–207.
Rocks, N., Paulissen, G., El Hour, M., et al. (2008). Emerging roles of ADAM and ADAMTS metalloproteinases in cancer. Biochimie *90*, 369–379.

Rozario, T., and DeSimone, D. W. (2010). The extracellular matrix in development and morphogenesis: a dynamic view. Dev Biol *341*, 126–140.

Ruoslahti, E. (1996). RGD and other recognition sequences for integrins. Annu Rev Cell Dev Biol *12*, 697–715.

Skandalis, S. S., Labropoulou, V. T., Ravazoula, P., et al. (2011). Versican but not decorin accumulation is related to malignancy in mammographically detected high density and malignant-appearing microcalcifications in non-palpable breast carcinomas. BMC Cancer *11*, 314.

Skliris, A., Happonen, K. E., Terpos, E., et al. (2011). Serglycin inhibits the classical and lectin pathways of complement via its glycosaminoglycan chains: implications for multiple myeloma. Eur J Immunol *41*, 437–449.

Smith, H. W., and Marshall, C. J. (2010). Regulation of cell signalling by uPAR. Nat Rev Mol Cell Biol *11*, 23–36.

Tadokoro, S., Shattil, S. J., Eto, K., et al. (2003). Talin binding to integrin beta tails: a final common step in integrin activation. Science *302*, 103–106.

Theocharis, A. D., Skandalis, S. S., Tzanakakis, G. N., and Karamanos, N. K. (2010). Proteoglycans in health and disease: novel roles for proteoglycans in malignancy and their pharmacological targeting. FEBS J *277*, 3904–3923.

Ulisse, S., Baldini, E., Sorrenti, S., and D'Armiento, M. (2009). The urokinase plasminogen activator system: a target for anti-cancer therapy. Curr Cancer Drug Targets *9*, 32–71.

Vogel, W. F., Abdulhussein, R., and Ford, C.E . (2006). Sensing extracellular matrix: an update on discoidin domain receptor function. Cell Signal *18*, 1108–1116.

Wang, A., de la Motte, C., Lauer, M., and Hascall, V. (2011). Hyaluronan matrices in pathobiological processes. FEBS J *278*, 1412–1418.

Whitelock, J. M., Melrose, J., and Iozzo, R. V. (2008). Diverse cell signaling events modulated by perlecan. Biochemistry *47*, 11174–11183.

Zöller, M. (2011). CD44: can a cancer-initiating cell profit from an abundantly expressed molecule? Nat Rev Cancer *11*, 254–267.

2 Insights into the function of glycans

2.1 Introduction

Paraskevi Heldin

The majority of proteins are glycosylated, making glycans important components of the cell and its environment. Cell surface glycans mediate cellular interactions with other cells, bacteria, viruses, antibodies, hormones, and growth factors and thereby have crucial roles in cellular adhesion, recognition, and signaling. Abnormal glycan synthesis is associated with severe diseases, including chronic inflammation and cancer (▶Figure 2.1). The structures of the glycans vary extensively with regard to sugar number, linkage pattern, and modifications. Glycans can be classified as glycosaminoglycans (GAGs), *O*- and *N*-linked glycoproteins, as well as glycosylphosphatidylinositol (GPI)–anchored proteins and glycolipids. The focus of this section is on the structure and function of biosynthetic enzymes responsible for the synthesis, assembly, and modifications of nonsulfated (hyaluronan) and sulfated (heparin; heparin sulfate, HS, chondroitin sulfate, CS; dermatan sulfate, DS) GAGs, as well as of *N*-glycans.

Chapters 2.2, 2.3, and 2.4 focus on the metabolic control and biological functions of hyaluronan. Hyaluronan is unique among the GAGs in that it is not sulfated and is not covalently attached to proteins as the other sulfated GAGs. It possesses both structural (due to its physicochemical properties) and cell signaling (due to its interactions with specific cell surface receptors) functions and thereby affects cellular behavior and tissue homeostasis. A large body of research suggests that hyaluronan and hyaluronan-synthesizing or hyaluronan-degrading enzymes are possible targets for the development of drugs to suppress malignant growth and inflammation (Li and Heldin, 2001; Jacobson et al., 2002; Lokeshwar et al., 2010). *In vivo* studies revealed that the gene for one of the hyaluronan-synthesizing enzymes – namely, hyaluronan synthase-2 (*HAS2*) – undergoes amplification and/or rearrangements in 28% of grade 3 breast cancers (Unger et al., 2010). In addition, the high expression of *HAS2* predisposes Chinese shar-pei dogs to a wrinkled skin phenotype, as well as to inflammation and periodic fever syndrome

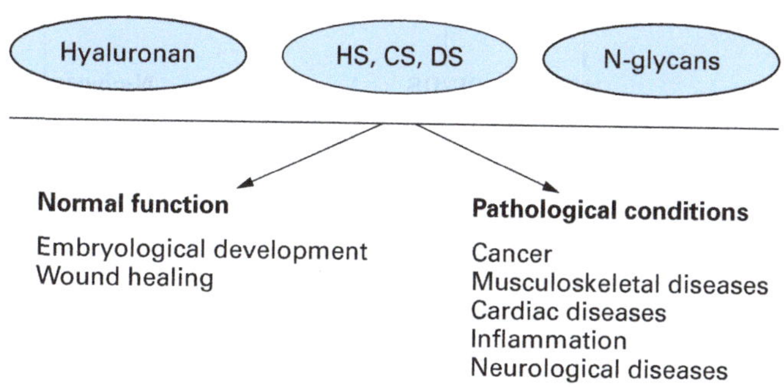

Figure 2.1 Glycans and glycan-binding proteins affect normal and pathological processes.

(Olsson et al., 2011). Interestingly, *HAS2* expression and, subsequently, hyaluronan production are crucial during normal embryonic and cardiac cushion morphogenesis because *HAS2* knockdown suppresses cushion cell transformation to a motile phenotype (Camenisch et al., 2000). Moreover, overexpression and aberrant variants of hyaluronan synthase-1 (*HAS1*) have been demonstrated in multiple myeloma patients, but not in healthy individuals (Adamia et al., 2005).

The major types of linear sulfated GAGs, HS and CS/DS (described in Chapters 2.4, 2.5, and 2.6), are synthesized and modified by a large number of enzymes (responsible for GAG building saccharide units, linkage region, copolymerization, modification, and epimerization) encoded by more than 40 genes. The expression and activity of these enzymes result in different sulfation patterns for the GAGs in different cell types (Zhang, 2010). Notably, their turnover is high, with half-lives of hours (Yanagishita and Hascall, 1992), rapidly changing in response to microenvironmental cues. Mutations on exostosin (EXT) family polymerases (glycosyltransferases that synthesize HS/heparin) results in exostoses (i.e. cartilage benign tumors) (Jennes et al., 2009). Furthermore, mutations of sulfotransferases that mediate the sulfonylations of proteoglycans can lead to dwarfism and osteochondrodysplasias (Dawson, 2011). In addition, mutations in the biosynthetic machinery of CS and DS also result in pathological conditions that are extensively described in ▶Table 2.5 in Chapter 2.6.

Chapter 2.7 covers the biosynthesis of branched *N*-glycans that are covalently linked to asparagine residues of proteins. Deficiencies in glycosyltransferases and glycosidases involved in *N*-glycan biosynthesis result in neurological and vascularization dysfunctions.

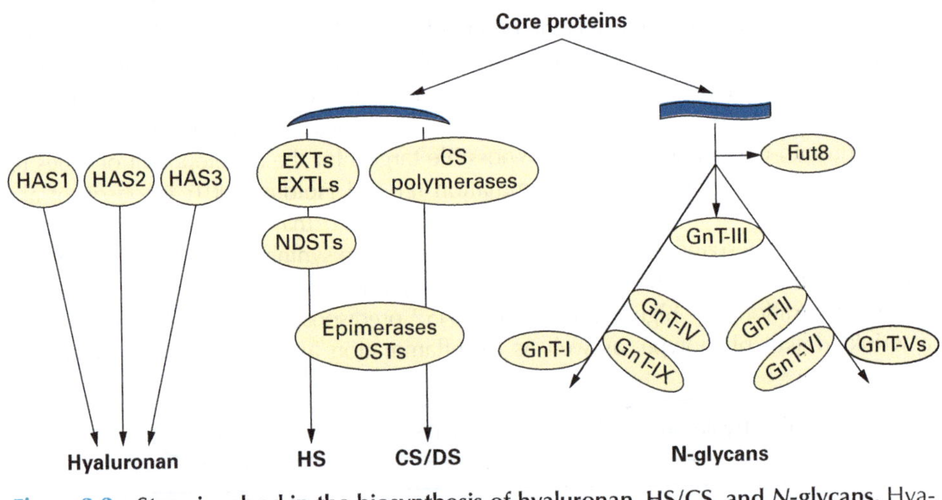

Figure 2.2 Steps involved in the biosynthesis of hyaluronan, HS/CS, and *N*-glycans. Hyaluronan is synthesized by three HASs. HS and CS chains are synthesized while attached to core proteins. Thereafter, HS is polymerized by the action of EXT/EXT-like (EXTL) polymerases, modified by *N*-deacetylation/*N*-sulfation by the action of sulfotransferases (NDSTs), and further modified by C-5 epimerase followed by *O*-sulfation by *O*-sulfotransferases (OSTs). Biosynthesis of CS is initiated by polymerization of *N*-acetylgalactosamine and glucuronic acid by CS polymerases followed by sulfation and epimerization. Schematic depiction of *N*-glycans biosynthesis involves TI, TII, TIII, TIV, TIX, and TV β-*N*-acetylglucosaminyltransferases and α1,6-fucosyltransferase (Fut8).

The major concept emerging from the chapters in this part is that the biological roles of GAGs and *N*-glycans are quite varied as might be expected from their ubiquitous expression combined with their specific and complex structures and functions. Glycan biosynthesis cannot be predicted directly from the DNA sequence because of the complex structural modifications (▶Figure 2.2). Specific changes at the levels of glycans are associated with inflammation, transformation and malignant progression. Thus, increased knowledge about the molecular mechanisms involved in the regulation of the activity of the biosynthetic enzymes might contribute to the development of tools for diagnosis and therapeutics of diseases, like cancer, inflammation and neurodegenerative diseases.

References

Adamia, S., Reiman, T., Crainie, M., Mant, M. J., Belch, A. R., and Pilarski, L. M. (2005). Intronic splicing of hyaluronan synthase 1 (HAS1): a biologically relevant indicator of poor outcome in multiple myeloma. Blood *105*, 4836–4844.

Camenisch, T. D., Spicer, A. P., Brehm-Gibson, T., et al. (2000). Disruption of hyaluronan synthase-2 abrogates normal cardiac morphogenesis and hyaluronan-mediated transformation of epithelium to mesenchyme. J Clin Invest *106*, 349–360.

Dawson, P. A. (2011). Sulfate in fetal development. Semin Cell Dev Biol *22*, 653–659.

Jacobson, A., Rahmanian, M., Rubin, K., and Heldin, P. (2002). Expression of hyaluronan synthase 2 or hyaluronidase 1 differentially affect the growth rate of transplantable colon carcinoma cell tumors. Int J Cancer *102*, 212–219.

Jennes, I., Pedrini, E., Zuntini, M., et al. (2009). Multiple osteochondromas: mutation update and description of the multiple osteochondromas mutation database (MOdb). Hum Mutat *30*, 1620–1627.

Li, Y., and Heldin, P. (2001). Hyaluronan production increases the malignant properties of mesothelioma cells. Br J Cancer *85*, 600–607.

Lokeshwar, V. B., Lopez, L. E., Munoz, D., et al. (2010). Antitumor activity of hyaluronic acid synthesis inhibitor 4-methylumbelliferone in prostate cancer cells. Cancer Res *70*, 2613–2623.

Olsson, M., Meadows, J. R., Truve, K., et al. (2011). A novel unstable duplication upstream of HAS2 predisposes to a breed-defining skin phenotype and a periodic fever syndrome in Chinese Shar-Pei dogs. PLoS Genet *7*, e1001332.

Unger, K., Wienberg, J., Riches, A., et al. (2010). Novel gene rearrangements in transformed breast cells identified by high-resolution breakpoint analysis of chromosomal aberrations. Endocr Relat Cancer *17*, 87–98.

Yanagishita, M., and Hascall, V. C. (1992). Cell surface heparan sulfate proteoglycans. J Biol Chem *267*, 9451–9454.

Zhang, L. (2010). Glycosaminoglycan (GAG) biosynthesis and GAG-binding proteins. Prog Mol Biol Transl Sci *93*, 1–17.

2.2 Metabolic control of hyaluronan synthesis

Alberto Passi, Giancarlo De Luca, Evgenia Karousou, Davide Vigetti, and Manuela Viola

2.2.1 Introduction

Hyaluronan (HA) is a polymer of increasing interest in the biomedical field due to its crucial role in several biological functions in cells. Even though HA was described for the first time in 1934 (Meyer and Palmer, 1934), until now several aspects of its metabolism were missing, and an increasing body of literature is shedding light on the secrets of this molecule.

From a structural point of view, HA often reaches a molecular mass of 10^7 Da with an extended length of more than 20 μm. Despite its simple structure, HA has appeared recently in life, and as a late molecule, it is present only in complex organisms and presents properties that are extraordinary in many ways. HA appeared only about 500 million years ago in chordates (DeAngelis, 2002). Nevertheless, the synthesis of HA can also take place in selected bacteria and archaea, including gram-positive streptococci (*S. pyogenes, S. equisimilis, S. uberis,* and *S. zooepidemicus*) and gram-negative *P. multocida* increasing their virulence. In mammals, HA can be synthesized by three isoenzymes (HA synthases, HAS1, 2, and 3), transferring both the glucuronic acid (GlcUA) and *N*-acetyl-glucosamine (GlcNAc) moieties alternating in the linear polysaccharide chain (Weigel and DeAngelis, 2007). These isoenzymes probably stemmed from an ancestral gene; in fact, they show structural identity of about 55%–70% (Spicer and McDonald, 1998). HAS2 is the more common isoenzyme in mammal tissues and seems to play a critical role for fetal survival in animals (Camenisch et al., 2000). HASs are integral membrane proteins active in the plasma membrane. Interestingly, the HASs show a peculiar structure, possessing two distinct binding domains for uridine diphosphate (UDP) sugars, whereas most described glycosyl transferases have only one catalytic domain. Moreover, it was suggested that HA remains attached to HAS while it is synthesized and simultaneously extruded into the extracellular space (Weigel and DeAngelis, 2007). Actually, a large proportion of cell surface HA can be bound to HAS, rather than HA receptors like CD44 or HA-mediated motility receptor (RHAMM) (Rilla et al. 2008). HA can bind membrane receptors such as CD44 and RHAMM, and recently, it was suggested that it can bind Toll-like receptors (TLRs) (Jiang et al., 2011). Its interaction with these receptors may trigger complex signaling pathways. The molecular mass of the bound HA and the specific plasma membrane locations of HASs are crucial.

The HA synthesis is therefore regulated by several factors, including among them gene expression and covalent modifications.

The expression of the HAS enzymes is the first and perhaps the most important determinant of HA synthesis. In different cells, the mRNA level of HASs may be very different. Studies at the protein level are scant, and this may be justified as the structural analysis of these enzymes is missing, which may be due to the membranous nature of the proteins. At the moment, no HASs have been crystallized, and no X-ray structural analysis has been performed. Only in silico was it possible to resolve the intracellular structure of HAS where it was evident that there was a portion of high flexibility close to the catalytic domains, confirming the peculiar characteristics of the catalytic domains of the enzyme.

The posttranslational modifications of HAS enzymes and the traffic of HASs to the plasma membrane are able to influence the enzymatic activity. Recently, the phosphorylation, ubiquitination, and *O*-GlcNAcylation of HAS, posttranslational modifications that strongly modify its enzymatic activity, were described. The traffic of HAS to the plasma membrane may be coupled to its posttranslational modifications, but information about this aspect is scant.

HA synthesis consumes large quantities of UDP-GlcUA and UDP-GlcNAc, which serve as substrates for the HAS enzymes. The cellular concentration of either of the UDP-sugars can become limiting in HA synthesis (Vigetti et al. 2006; Jokela et al., 2008). The content of the UDP-sugars varies between cell types and metabolic requirements, suggesting that the contribution of the UDP-GlcUA and UDP-GlcNAc content to the overall HA synthesis needs to be assessed.

2.2.2 Transcription of hyaluronan synthases

Embryonic development and *HAS* expression

The distribution of HAS during development could actually address the question of why three different isoenzymes have developed for the synthesis of a simple polymer like HA. The first HAS described in vertebrates was discovered in *Xenopus laevis* (XHas1) (Varki, 1996). It was demonstrated that three synthases are expressed in *Xenopus* at different stages of development (Vigetti et al., 2003). *XHas1* and *2* are widely expressed in the embryo, showing that *XHas1* has a very early onset. On the other hand, the transcription of *XHas3* is restricted to certain areas of the embryo, including the inner ear and the cement gland (Vigetti et al., 2003; Nardini et al., 2004). The three synthases are expressed in mice in different temporal patterns during development, *Has2* being the major contributor to HA synthesis in mice during development (Tien and Spicer, 2005). The importance of Has2 in mice development is demonstrated by the knockout phenotype of this enzyme; in fact, homozygous deletion of the *Has2* gene manifests severe cardiac and vascular abnormalities leading to death at midgestation (E9.5–10) (Camenisch et al., 2000). HA is abundant in fetal human tissues (Mack et al., 2003; Raio et al., 2005), but its amount decreases during development. It is replaced by collagen and proteoglycans, possibly as a matrix adaptation to more severe mechanical requirements (Mack et al., 2003). As Has1 and 2 are able to produce large-size HA and Has3 produces smaller-size HA, it is evident that the differential expression of these enzymes is coupled with different biological roles of HA in the tissues. The size in fact plays an important function, as oligosaccharide of HA have different effects from those exerted by large-size HA.

Comparative gene expression analysis of HASs suggests that the different roles HA plays during development are fulfilled by the spatiotemporally regulated transcription of the three different synthases.

Multiple growth factors and cytokines influence *Has* expression

As previously described, the expression of *Has* genes undergoes rapid and dramatic changes during embryonic development, corresponding to migration of cells to their final sites in organs (Camenisch et al., 2000; Mack et al., 2003). In adult tissues, HA synthesis is stimulated by during wound healing, inflammation, and neoplastic tumors. A number of cytokines and growth factors – such as platelet-derived growth factor (PDGF) (Jacobson et al., 2000; Evanko et al., 2001), fibroblast growth factor-2 (FGF2) (Shimabukuro et al., 2011), keratinocytes growth factor (KGF) (Karvinen et al., 2003), epidermal growth factor (EGF) (Pienimäki et al., 2001), transforming growth factor-beta (TGF-β) (Pasonen-Seppänen et al., 2003), interleukin-1β (IL-1β) (Suzuki et al., 2003; Vigetti et al., 2010), tumor necrosis factor-alpha (TNF-α) (Oguchi and Ishiguro, 2004), and interferon-gamma (IFN-γ) (Sayo, 2002) – are released from local cells, including platelets and leukocytes, and increase *Has* expression. *Has* expression and HA synthesis are also sensitive to local mediators like prostaglandins (Sussmann et al., 2004) and hormone-type effectors like corticosteroids (Zhang et al., 2000), which downregulate *Has2*, reducing HA synthesis. Conversely, retinoids induce *Has2* expression (Saavalainen et al., 2005; Pasonen-Seppanen et al., 2008). It is noteworthy to mention that the cell reactivity to these effectors is quite different. Moreover, some of the effectors modulate the expression of all *Has* genes, whereas some modulate just one or two of them. For example, *Has2* in epidermal keratinocytes is particularly sensitive to epidermal growth factor receptor (EGFR) ligands, while in MCF7 breast cancer cells this pathway responds weakly, if at all, to EGF. TGF-β downregulates *Has2* and *Has3* in keratinocytes but enhances the expression of *Has1* in fibroblasts (Sugiyama et al., 1998) and synoviocytes (Stuhlmeier and Pollaschek, 2004). Taken together, these findings indicate that there is a common ancestral gene, suggesting that the three *Has* genes have promoters that react to common transcriptional signals.

Response elements and transcription factors in *Has2* promotor

The *Has2* gene is the most active in tissues and therefore is often involved in regulating the synthesis rate of HA (Fülop et al., 1997; Sussmann et al., 2004; Vigetti et al., 2010). Therefore, the promoter area of *Has2* has been most actively studied for response elements that bind transcription factors.

In epidermal keratinocytes, the EGF-family growth factors, especially heparin-binding epidermal growth factor (HB-EGF) (Monslow et al., 2009), acting in an autocrine and paracrine fashion, is a major inducer of *Has2* expression. Inhibitors of EGFR (ErbB1) and phosphatidylinositol 3-kinase (PI3K) prevent this stimulation, which is mediated by the transcription factor signal transducer and activator of transcription (STAT).

Stimulants of HA synthesis mediated by G-protein-associated receptors act via cAMP-dependent protein kinase A (PKA) and the cAMP responsive element binding protein 1 (CREB1) pathway on the *Has2* promoter, while all-*trans*-retinoic acid acts through its nuclear receptor with a functional response element 1,300 bp upstream of the *Has2* transcription start site in keratinocytes (Makkonen et al., 2009).

Among the transcription factors, nuclear factor kappa B (NF-kB) has a central role in the induction and resolution of inflammation. In this latter process, HA has multiple effects depending on its molecular mass, and it has been shown that HA can participate, with other adhesive molecules (i.e. selectins and integrins), in the recruitment of circulating immune cells to the site of inflammation. For example, under physiological conditions endothelial cells do not synthesize HA, but after stimulation with TNF-α and IL-1β, such cells induce *Has2* transcription via NF-kB (Saavalainen et al. 2007), and the newly synthesized HA mediates monocyte adhesion through CD44 (Vigetti et al., 2010).

2.2.3 UDP-sugar substrates as limiting factors in hyaluronan synthesis

UDP-GlcUA and UDP-GlcNAc contents control hyaluronan synthesis

HA is produced using UDP precursors, which derive from glucose through the metabolic steps shown in ▶Figure 2.3. The synthesis of HA presents several peculiar aspects that underline the capacity of the cells to act in a very "economic" way in the

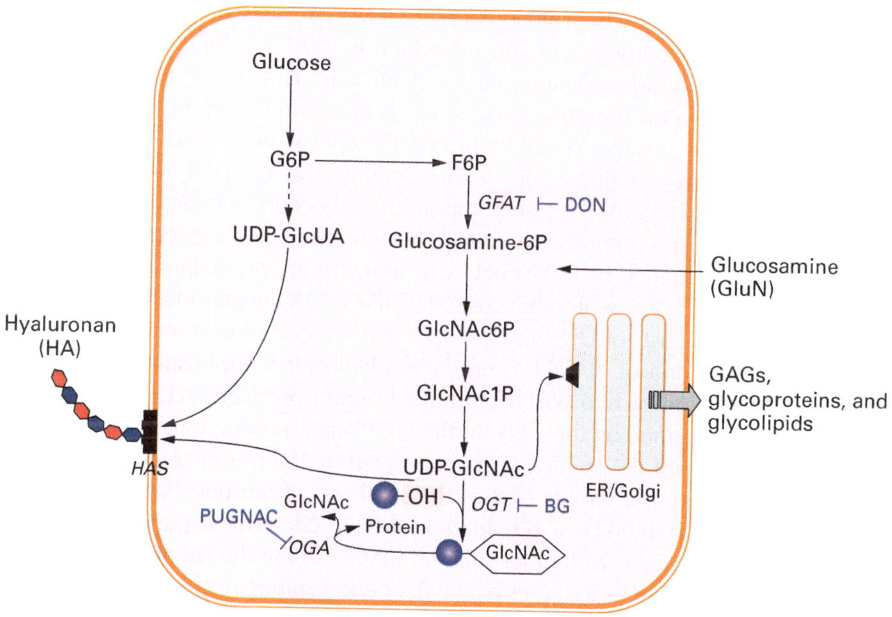

Figure 2.3 Intracellular biochemical pathway of HA precursors and intracellular biochemical pathway of glucosamine biosynthesis. G6P, glucose-6-phosphate; G1P, glucose-1-phosphate; UDP-G, UDP-glucose; UDP-GlcUA, UDP-glucuronic acid; Fru-6-P, fructose-6-phosphate; GlcNAc6P, *N*-acetyl-glucosamine-6-phosphate; GlcNAc1P, *N*-acetyl-glucosamine-1-phosphate; UDP-GlcNAc, UDP-*N*-acetyl-glucosamine; GLN, glutamine; GLU glutamate; Ac-CoA, acetyl-Coenzyme A; PPi, pyrophosphate; GFAT, glutamine:fructose-6-phosphate amidotransferase; DON, diazo-oxo-norleucine; BG, benzylguanine; OGT, UDP-*N*-acetylglucosamine–peptide *N*-acetylglucosaminyltransferase; OGA, *O*-GlcNAcase; PUGNAC, *O*-(2-acetamido-2-deoxy-D-glucopyranosylidene)amino *N*-phenyl carbamate. Each biochemical reaction is catalyzed by a specific enzyme reported as a number.

production of this polymer. The synthesis of HA can appear very expensive as it uses glucose, but all the energy requirements for its synthesis are completely balanced by the oxidation of UDP-glucose, a crucial step in the synthesis of UDP-GlcUA, the precursor of all glycosaminoglycans (GAGs). HA assembly strongly depends on the availability of UDP sugars in the cells (Vigetti et al., 2006). This aspect of cell biosynthesis implies that cell metabolism supports the anabolic pathways, and therefore the cells producing all GAGs must have a positive energy balance. In fact, UDP-sugar precursors represent molecules with a high cost in terms of energy, as UDP sugars compete with glycolysis for their synthesis. For this reason, it is evident that HA synthesis is easier in healthy tissues with a good oxygen supply, as these tissues downregulate glycolysis, while the presence of oxygen allows the oxidative reaction necessary for UDP-GlcUA synthesis (▶Figure 2.3).

The synthesis of UDP-GlcUA is a critical step for all GAGs, except keratan sulfate (KS), which does not contain uronic acid. Furthermore, KS is present primarily in tissues without a vascular system and consequently with a poor oxygen supply, such as the cornea and the inner portion of articular cartilage (De Luca et al., 1975; Balduini et al., 1992). It is not surprising that KS is almost absent in fetal tissue, whereas it increases in tissues characterized by low oxygen supply and during tissue aging. The synthesis of UDP-GlcUA requires the activity of the UDP-glucose dehydrogenase (UGDH), which produces UDP-GlcUA from precursor UDP-Glc. This reaction is possible in the presence of nicotinamide adenine dinucleotide (NAD), which is converted into NADH during the twofold oxidation of C-6 of UDP-Glc. This reaction is quite unusual.

A double oxidation on the same C-6, which is converted from an alcoholic to carboxylic group producing two NADH/mol of UDP-Glc, is very uncommon in nature. The NAD availability is critical for UGDH activity and for UDP-GlcUA synthesis; the NAD recycling from NADH is very important and occurs in cells with efficient mitochondrial activity and a good oxygen supply. Moreover, this aspect of HA synthesis introduces the importance of UDP-GlcUA in terms of energy balance. When all the steps of the anabolic pathway from glucose to UDP-GlcUA are considered (▶Figure 2.3), it is evident that the costs of UDP-GlcUA synthesis are completely repaid by the reoxidation in mitochondria of NADH to produce adenosine-5′-triphosphate (ATP) molecules. The reoxidation of two NADH in mitochondria produces ATP, which refunds the energy costs sustained by the cells for the UDP-sugar production. In fact, considering the stoichometry of HA, which is characterized by the presence of UA and hexosamine in a 1:1 ratio, the energy cost of the synthesis of the unsulfated backbone of HA is completely sustained by C-6 oxidation of UDP-Glc. This is particularly remarkable for HA in comparison with other sulfated GAGs, where the sulfation increases the energy costs. In HA synthesis, the energy balance to synthesize this molecule is completely compensated by the reoxidation of the NADH produced in UDP-GlcUA synthesis. One can consider the difference between the synthesis of glycogen and HA using UDP sugars in the cells. For glycogen, the synthesis of this intracellular polymer costs energy that is repaid only during glycolysis during glycogen degradation, whereas in the HA synthesis the energy costs are paid during the synthesis of the polymer before its extrusion from the cells.

On this basis, the availability of UDP-sugar precursors influences HA synthesis, and it has been demonstrated that the altered UDP-GlcUA concentration in the cells influences HA synthesis and also modifies the gene expression pattern of enzymes involved in its synthesis (Vigetti et al., 2006).

Confirming this mechanism is the use of a specific drug to reduce UDP-GlcUA inside the cells. In fact, a relatively specific inhibitor of HA synthesis is 4-methylumbelliferone (4-MU) (Nakamura et al., 1997; Kakizaki et al., 2004), a derivative of coumarin. The 4-MU inhibits HA synthesis by depleting the HAS substrate UDP-GlcUA, which is consumed by 4-MU glucuronidation (Kultti et al., 2009; Vigetti et al., 2009b), a reaction common in detoxification of foreign substances. The 4-MU treatment revealed the crucial role of UDP-GlcUA availability specifically for HA synthesis. Moreover, recent data suggest that the recovery after 4-MU with HA has an antiapoptotic effect (Vigetti et al., 2011b). Interestingly, other GAGs, such as chondroitin and heparan sulfates, were less sensitive to UDP-GlcUA deficiency because they are synthesized in the Golgi apparatus, a privileged compartment due to the high-affinity transporters that pump UDP sugars from the cytosol into the Golgi apparatus. Increasing the content of UDP-GlcUA by overexpression of UGDH enhances HA production (Vigetti et al., 2006) confirming the importance of the cellular level of UDP-GlcUA in HA production.

As far as the cellular content of UDP-GlcNAc is concerned, the concentration of this precursor in the cells is usually 3–7 times higher than of that of UDP-GlcUA and was previously not considered a limiting factor in HA synthesis. However, the K_m for UDP-GlcNAc is also up to ~10 times higher as compared to UDP-GlcUA in all HAS enzymes (Itano et al., 1999). From this point of view, it is not surprising that cells readily respond to mannose-induced UDP-GlcNAc depletion by an inhibition of HA synthesis (Kultti et al., 2006). The inhibition by mannose is likely due to reduced formation or enhanced catabolism of GlcN6P (▶Figure 2.3). Again, the synthesis of other GAGs, operating in the presumably higher UDP-GlcNAc concentration of the Golgi apparatus, are not affected (Jokela et al., 2008). As a confirming experiment, raising the content of UDP-GlcNAc by feeding glucosamine increases HA synthesis, indicating that the cellular content of UDP-GlcNAc is as important as UDP-GlcUA in controlling the level of HA synthesis, not only influencing the UDP-sugar availability but inducing covalent modification of key enzymes throughout glycosylation of specific residues.

UDP sugars control HAS access to plasma membrane

The activity of HAS enzymes is strongly influenced by the migration of these molecules from internal compartments to the plasma membranes. Therefore any mechanism altering this process influences the HA synthesis. Experiments based on transfection of cells with green fluorescent protein (GFP)–labeled *Has* constructs have enabled the examination of HAS localization within live cells and movements between the cellular compartments. When HA synthesis is inhibited with 4-MU depletion of UDP-GlcUA, GFP-Has3 gradually disappears from its regular location in plasma membrane protrusions, and the protrusions also wither (Kultti et al., 2006). In the same way, mannose-induced reduction of the UDP-GlcNAc pool, while it inhibits HA synthesis, also prevents HAS3 localization in the plasma membrane (Jokela et al., 2008). The same occurs with glucose deprivation, and the cells show GFP-HAS3 mainly in the Golgi area, while the same cell, after restoring glucose in the medium, presents HAS in the plasma membrane, typically in microvillous projections. Thus, synthesis of HA is coupled to HAS localization in the plasma membrane, suggesting that either the availability of UDP sugars stabilizes HAS in the plasma membrane or that sufficient amounts of the UDP sugars are required for triggering the transport of HAS into the plasma membrane. These findings may offer clues for future research on the enigma of why HA synthesis normally takes place only in the plasma membrane, while HAS inserted in endoplasmic reticulum (ER) and Golgi membranes normally remains inactive.

Nevertheless, it has been reported that in *in vitro* systems, the HASs are able to produce HA when microsomal preparations of different cell lines are incubated in the presence of substrates (Vigetti et al., 2009a). These findings shed light on the possibility of HA synthesis inside the cells in specific situations such as ER stress and inflammation (Hascall et al., 2004).

Cellular UDP-GlcNAc content controls *Has2* expression

A strong relationship between UDP availability and *HAS* expression was described both for altered availability of UDP-GlcUA (Vigetti et al., 2006) and UDP-GlcNAc. In the case of UDP-GlcNAc cytoplasmic concentration, the situation is even more complex. In fact, when UDP-GlcNAc was increased in keratinocytes by adding glucosamine in the medium, the expression of *Has2* was reduced by ~50%, while reducing cellular UDP-GlcNAc by feeding mannose increased *Has2* mRNA. By checking glucosamine-induced changes of transcription factors bound to the *Has2* promoter by using chromatin immunoprecipitation (ChIP) techniques, an accumulation of YY1 was revealed. Treatment with mannose reduced promoter binding of specificity protein 1 (SP1), another transcription factor. Following siRNA-mediated reduction of YY1 and SP1 levels, *Has2* mRNA was increased, confirming the suppressive role of YY1 and SP1 on *Has2* expression (Jokela et al., 2008).

The changes in *Has2* promoter binding of YY1 and SP1 were associated with increased *O*-GlcNAc modifications on these proteins. This is a recently discovered signaling pathway in which GlcNAc from UDP-GlcNAc is transferred to certain Ser and Thr residues of cytosolic and nuclear proteins by a specific enzyme (*O*-linked *N*-acetylglucosamine [*O*-GlcNAc] transferase [OGT]). This dynamic addition and removal of *O*-GlcNAc through a specific hydrolase (OGA) has been shown to control protein functions, sometimes by supporting the effects of phosphorylation or counteracting the influence of phosphorylation, often even competing with phosphorylation for certain Ser and Thr residues (Hart et al., 2007). The supply of UDP-GlcNAc to OGT influences the extent of *O*-GlcNAc modification on specific molecular targets. Accordingly, the levels of *O*-GlcNAcylation of YY1 and SP1 were increased and decreased, respectively, by glucosamine and mannose, and this explained the alterations in promoter binding and *Has2* expression. *Has*2 activity is influenced by OGT activity, and it was demonstrated that *O*-glucosamination occurs in the enzyme, increasing its activity (Vigetti, unpublished data).

2.2.4 Posttranslational processing of HAS

Posttranslational modulation of HAS activity

Rapid changes in HA synthesis are due to posttranslational covalent modifications, and a recent *in vitro* technique was described to address this aspect of HASregulation (Vigetti et al., 2009a). ▶Figure 2.4 summarizes these posttranslational modifications as reported in recent literature. Plasma membrane and cytoplasmic fractions (the latter

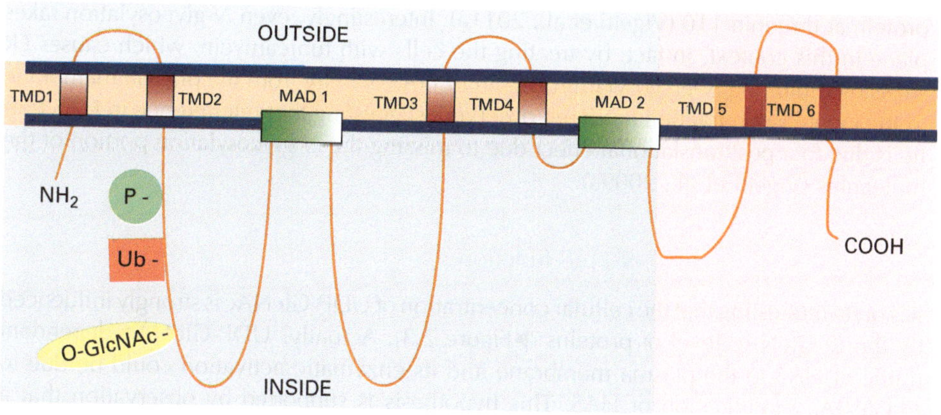

Figure 2.4 **Schematic representation of posttranslational modification of HAS.** TMD (1–6), transmembrane domain (1–6); MAD (1 and 2), membrane domain (1 and 2); P, site of phosphorylation; Ub, site of monoubiquitination; *O*-GlcNAc, site of *O*-GlcNAcylation.

including ER and Golgi membranes) derived from cell homogenates incubated with UDP precursors are able to synthesize HA in a controlled fashion. In this system, short (5–10 min.) treatment of cells with phorbol 12-myristate-13-acetate (PMA), IL-1β, and platelet derived growth factor-BB (PDGF-BB) induced a several-fold increase in HAS activity, both in the plasma membrane and cytoplasmic fractions (Vigetti et al., 2009a). Multiple receptors types, like those of cytokines, growth factors, and protein kinase C (PKC) activators, thus have immediate influences on the activity of HAS, in addition to their stimulation of *Has* transcription, as discussed previously.

Phosphorylation influences HAS activity

The receptor-mediated signaling mostly works through kinases, and HASs are possible targets (Goentzel et al., 2006). Pretreatment of the membrane preparations with alkaline phosphatase reduce the HAS activity induced by PMA and IL-1β, but not that by PDGF-BB (Vigetti et al., 2009a), demonstrating the diversity of the activation pathways. All three HAS isoenzyme can be phosphorylated by the extracellular signal-regulated kinase ERK/ErbB2, leading to increased synthetic activity (Bourguignon et al., 2007). By transfecting constructs of HASs in cells, it was possible to demonstrate that the phosphorylation could be removed by alkaline phosphatase, and the HAS activity increased several fold as compared to control membranes (Vigetti et al., 2011a). A recent study was able to demonstrate that phosphorylation of HAS is due to the AMP-activated protein kinase (AMPK) activity. In fact, this enzyme is involved in energy charge regulation, and its activity is high when the cells show a low energy charge level. This hypothesis was demonstrated by treating cells with inducers of energetic stress (i.e. 5-aminoimidazole-4-carboxamide 1-β-D-ribofuranoside [AICAR] or 2 deoxyglucose) leading to the specific HAS2 phosphorylation, coupled with a reduced HA synthetic activity (Vigetti et al., 2011a).

Experiments carried out using constructs with site-specific modified residues in the enzyme HAS2 demonstrated that phosphorylation modification is related to the cytoplasmatic portion of the molecule in particular in a very conserved sequence of the protein at threonin 110 (Vigetti et al., 2011a). Interestingly, even *N*-glycosylation takes place in this context. In fact, by treating the cells with tunicamycin, which causes ER stress by inhibiting *N*-glycosylation, a shift of HAS activity from the plasma membrane to the cytoplasmic fraction has been observed, suggesting that interference in HAS traffic is due to a posttranslational effect due to missing the *N*-glycosylation portion of the molecules (Vigetti et al., 2009a).

O-GlcNAc as a modifier of HAS function

It is quite interesting that the cellular concentration of UDP-GlcNAc is strongly influenced by the *O*-GlcNAc level of proteins (▶Figure 2.3). Actually, UDP-GlcNAc–dependent traffic of HAS to the plasma membrane and its enzymatic activation could be due to *O*-GlcNAc modification of HAS. This hypothesis is supported by observation that a lectin able to recognize *O*-GlcNAc epitopes is able to capture a membrane fraction with HAS activity and this interaction is abolished in modified HAS where the specific cytoplasmic consensus motif for OGT was mutated (D. Vigetti et al., unpublished).

HAS2 and HAS3 can dimerize

It was long assumed that HAS proteins worked without functional interaction with other proteins and as monomers on plasma membranes (Weigel and DeAngelis, 2007). However, it was recently demonstrated that HAS2 tagged with 6myc and Flag was coimmunoprecipitated with antibodies against both tags (Karousou et al., 2010), suggesting a dimeric complex. Furthermore, an antibody against 6myc-HAS2 tag also coimmunoprecipitated Flag-HAS3, suggesting heterodimerization on HAS2 and HAS3. Dimerization (or multimerization) explains how a single 65 kDa polypeptide can form a membrane pore for the extrusion of the bulky HA chain, in addition to its two different transferase activities and an ability to hold the growing chain (Weigel and DeAngelis, 2007). Actually, most of the glycosyl transferases, as well as pore-forming proteins in the plasma membrane, are multimers.

Monoubiquitination is required for HAS2 activity

Polyubiquitin chains built on cytoplasmic proteins form a label that leads the proteins into proteasomal degradation. In contrast, monoubiquitin is a signal that has been shown to change the function of proteins or their traffic to certain subcellular compartments. A part of 6myc- and Flag-HAS2 was found to carry (mono)ubiquitin on Lys190 of HAS2 (Karousou et al., 2010). The enzymatic activity of Has2 is lost with site-directed Lys190Arg mutation. The 190Arg construct was able to dimerize with the wild-type form of HAS2 but inhibited HA synthesis by the latter in a dominant negative fashion. These data strongly suggest that monoubiquitination is required for HAS2 activity and that HAS2 is only active as a dimer.

Cell stress changes the structure of hyaluronan on cell surface

It has been shown in several cell types that the compact layer of HA on the cell surface can be shattered into a loose, hairy orientation, extending far away from the cell.

Upon fixation with methanol, these HA chains are prone to lateral aggregation into structures called cables. Functionally, the formation of cables is associated with enhanced binding of leucocytes. Binding of leukocytes to the HA cables changes their behavior and presumably the course of the inflammation. This altered surface structure of HA may depend on the traffic and localization of HAS and has widespread influence on cell behavior as described in Wang et al. (2011).

2.2.5 Challenges and future prospects

It is important to shed light on the metabolism of HA as an increasing body of literature is showing the involvement of this polymer in several biological functions, including pathology. For instance, the biology of the microenvironment is strongly influenced by the presence of HA and its fragments, which play a critical role in inflammation. Novel data are now available on HA interaction with several receptors involved in innate immunity, such as TLR. Despite its increasing involvement in human pathology, basic information, such as structure resolved by crystallization, is still lacking. The recent findings on the role of HA in poor outcomes for patients with cancers and degenerative chronic inflammatory processes introduce the importance of the HASs as future targets for therapeutic approaches.

2.2.6 Take-home message

HASs are extremely efficient enzymes that undergo to several forms of regulation, both in terms of gene expression and covalent posttranslational modification. It is, therefore, evident that there is a fast regulation involving the covalent modification of enzymes and their migration to plasma membranes, an aspect that suggests a reservoir of "ready-to-use" enzymes in the cells, and a second slower regulation that involves the gene expression.

References

Balduini, C., De Luca, G., Passi, A., Rindi, S., Salvini, R., and Scott, J. E. (1992). Effect of oxygen tension and lactate concentration on keratan sulphate and chondroitin sulphate biosynthesis in bovine cornea. Biochim Biophys Acta *1115* (3), 187–191.

Bourguignon, L. Y., Gilad, E., and Peyrollier, K. (2007). Heregulin-mediated ErbB2-ERK signaling activates hyaluronan synthases leading to CD44-dependent ovarian tumor cell growth and migration. J Biol Chem *282*, 19426–19441.

Camenisch, T. D., Spicer, A. P., Brehm-Gibson, T., et al. (2000). Disruption of hyaluronan synthase-2 abrogates normal cardiac morphogenesis and hyaluronan-mediated transformation of epithelium to mesenchyme. J Clin Invest *106*, 349–360.

DeAngelis, P. L. (2002). Evolution of glycosaminoglycans and their glycosyltransferases: implications for the extracellular matrices of animals and the capsules of pathogenic bacteria. Anat Rec *268*, 317–326.

De Luca, G., Speziale, P., Balduini, C., and Castellani, A. A. (1975). Biosynthesis of glycosaminoglycans: uridine diphosphate glucose 4′-epimerase from cornea and epiphysial-plate cartilage. Connect Tissue Res *3* (1), 39–47.

Evanko, S. P., Johnson, P. Y., Braun, K. R., Underhill, C. B., Dudhia, J., and Wight, T. N. (2001). Platelet-derived growth factor stimulates the formation of versican-hyaluronan aggregates and pericellular matrix expansion in arterial smooth muscle cells. Arch Biochem Biophys *394*, 29–38.

Fülop, C., Kamath, R. V., Li, Y., et al. (1997). Coding sequence, exon-intron structure and chromosomal localization of murine TNF-stimulated gene 6 that is specifically expressed by expanding cumulus cell-oocyte complexes. Gene *202*, 95–102.

Goentzel, B. J., Weigel, P. H., and Steinberg, R. A. (2006). Recombinant human hyaluronan synthase 3 is phosphorylated in mammalian cells. Biochem J *396* (2), 347–354.

Hart, G. W., Housley, M. P., and Slawson C. (2007). Cycling of O-linked beta-N-acetylglucosamine on nucleocytoplasmic proteins. Nature *446* (7139), 1017–1022.

Hascall, V. C., Majors, A. K., De La Motte, C. A., et al. (2004). Intracellular hyaluronan: a new frontier for inflammation? Biochim Biophys Acta *1673*, 3–12.

Itano, N., Sawai, T., Yoshida, M., et al. (1999). Three isoforms of mammalian hyaluronan synthases have distinct enzymatic properties. *J Biol Chem 274*, 25085–25092.

Jacobson, A., Brinck, J., Briskin, M. J., Spicer, A. P., and Heldin, P. (2000). Expression of human hyaluronan synthases in response to external stimuli. Biochem J *348* (Pt 1), 29–35.

Jiang, D., Liang, J., and Noble, P. W. (2011). Hyaluronan as an immune regulator in human diseases. Physiol Rev *91* (1), 221–64.

Jokela, T. A., Jauhiainen, M., Auriola, S., et al. (2008). Mannose inhibits hyaluronan synthesis by down-regulation of the cellular pool of UDP-N-acetylhexosamines. J Biol Chem *283*, 7666–7673.

Kakizaki, I., Kojima, K., Takagaki, K., et al. (2004). A novel mechanism for the inhibition of hyaluronan biosynthesis by 4-methylumbelliferone. J Biol Chem *279*, 33281–33289.

Karousou, E., Kamiryo, M., Skandalis, S. S., et al. (2010). The activity of hyaluronan synthase 2 is regulated by dimerization and ubiquitination. J Biol Chem *285*, 23647–23654.

Karvinen, S., Pasonen-Seppänen, S., Hyttinen, J. M., et al. (2003). Keratinocyte growth factor stimulates migration and hyaluronan synthesis in the epidermis by activation of keratinocyte hyaluronan synthases 2 and 3. J Biol Chem *278*, 49495–49504.

Kultti, A., Pasonen-Seppanen, S., Jauhiainen, M., et al. (2009). 4-Methylumbelliferone inhibits hyaluronan synthesis by depletion of cellular UDP-glucuronic acid and downregulation of hyaluronan synthase 2 and 3. Exp Cell Res *315* (11), 1914–1923.

Kultti, A., Rilla, K., Tiihonen, R., Spicer, A. P., Tammi, R. H., and Tammi, M. I. (2006). Hyaluronan synthesis induces microvillus-like cell surface protrusions. J Biol Chem *281*, 15821–15828.

Mack, J. A., Abramson, S. R., Ben, Y., et al. (2003). Hoxb13 knockout adult skin exhibits high levels of hyaluronan and enhanced wound healing. FASEB J *17*, 1352–1354.

Makkonen, K. M., Pasonen-Seppänen, S., Törrönen, K., Tammi, M. I., and Carlberg, C. (2009). Regulation of the hyaluronan synthase 2 gene by convergence in cyclic AMP response element-binding protein and retinoid acid receptor signaling. J Biol Chem *284* (27), 18270–18281.

Meyer, K., and Palmer, J. W. (1934). The polysaccharide of the vitreous humor. J Biol Chem *107*, 629–634.

Monslow, J., Sato, N., Mack, J. A., and Maytin, E. V. (2009). Wounding-induced synthesis of hyaluronic acid in organotypic epidermal cultures requires the release of heparin-binding egf and activation of the EGFR. J Invest Dermatol *129*, 2046–2058.

Nakamura, T., Funahashi, M., Takagaki, K., et al. (1997). Effect of 4-methylumbelliferone on cell-free synthesis of hyaluronic acid. Biochem Mol Biol Int *43*, 263–268.

Nardini, M., Ori, M., Vigetti, D., Gornati, R., Nardi, I., and Perris, R. (2004). Regulated gene expression of hyaluronan synthases during Xenopus laevis development. Gene Expr Patterns *4*, 303–308.

Oguchi, T., and Ishiguro, N. (2004). Differential stimulation of three forms of hyaluronan synthase by TGF-beta, IL-1beta, and TNF-alpha. Connect Tissue Res *45*, 197–205.

Pasonen-Seppänen, S., Karvinen, S., Törrönen, K., et al. (2003). EGF upregulates, whereas TGF-beta downregulates, the hyaluronan synthases Has2 and Has3 in organotypic keratinocyte cultures: correlations with epidermal proliferation and differentiation. J Invest Dermatol *120*, 1038–1044.

Pasonen-Seppanen, S. M., Maytin, E. V., Torronen, K. J., et al. (2008). All-trans retinoic acid-induced hyaluronan production and hyperplasia are partly mediated by EGFR signaling in epidermal keratinocytes. J Invest Dermatol *128*, 797–807.

Pienimäki, J. P., Rilla, K., Fulop, C., et al. (2001). Epidermal growth factor activates hyaluronan synthase 2 in epidermal keratinocytes and increases pericellular and intracellular hyaluronan. J Biol Chem *276*, 20428–20435.

Raio, L., Cromi, A., Ghezzi, F., et al. (2005). Hyaluronan content of Wharton's jelly in healthy and Down syndrome fetuses. Matrix Biol *24*, 166–174.

Rilla, K., Tiihonen, R., Kultti, A., Tammi, M., and Tammi, R. (2008). Pericellular hyaluronan coat visualized in live cells with a fluorescent probe is scaffolded by plasma membrane protrusions. J Histochem Cytochem *56*, 901–910.

Saavalainen, K., Pasonen-Seppanen, S., Dunlop, T. W., Tammi, R., Tammi, M. I., and Carlberg, C. (2005). The human hyaluronan synthase 2 gene is a primary retinoic acid and epidermal growth factor responding gene. J Biol Chem *280*, 14636–14644.

Saavalainen, K., Tammi, M. I., Bowen, T., Schmitz, M. L., and Carlberg, C. (2007). Integration of the activation of the human hyaluronan synthase 2 gene promoter by common cofactors of the transcription factors retinoic acid receptor and nuclear factor kappaB. J Biol Chem *282* (15), 11530–11539.

Sayo, T., Sugiyama, Y., Takahashi, Y., et al. (2002). Hyaluronan synthase 3 regulates hyaluronan synthesis in cultured human keratinocytes. J Invest Dermatol *118*, 43–48.

Shimabukuro, Y., Terashima, H., Takedachi, M., et al. (2011). Fibroblast growth factor-2 stimulates directed migration of periodontal ligament cells via PI3K/AKT signaling and CD44/hyaluronan interaction. J Cell Physiol *226* (3), 809–821.

Spicer, A.P., and McDonald, J. A. (1998). Characterization and molecular evolution of a vertebrate hyaluronan synthase gene family. J Biol Chem *273*, 1923–1932.

Stuhlmeier, K. M., and Pollaschek, C. (2004). Differential effect of transforming growth factor beta (TGF-beta) on the genes encoding hyaluronan synthases and utilization of the p38 MAPK pathway in TGF-beta-induced hyaluronan synthase 1 activation. J Biol Chem *279*, 8753–8760.

Sugiyama, Y., Shimada, A., Sayo, T., Sakai, S., and Inoue, S. (1998). Putative hyaluronan synthase mRNA are expressed in mouse skin and TGF-beta upregulates their expression in cultured human skin cells. J Invest Dermatol *110*, 116–121.

Sussmann, M., Sarbia, M., Meyer-Kirchrath, J., Nusing, R. M., Schror, K., and Fischer, J. W. (2004). Induction of hyaluronic acid synthase 2 (HAS2) in human vascular smooth muscle cells by vasodilatory prostaglandins. Circ Res *94*, 592–600.

Suzuki, K., Yamamoto, T., Usui, T., Heldin, P., and Yamashita, H. (2003). Expression of hyaluronan synthase in intraocular proliferative diseases: regulation of expression in human vascular endothelial cells by transforming growth factor-beta. Jpn J Ophthalmol *47*, 557–564.

Tien, J. Y., and Spicer, A. P. (2005). Three vertebrate hyaluronan synthases are expressed during mouse development in distinct spatial and temporal patterns. Dev Dyn *233*, 130–141.

Varki, A. (1996). Does DG42 synthesize hyaluronan or chitin? A controversy about oligosaccharides in vertebrate development. Proc Natl Acad Sci U S A *93*, 4523–4525.

Vigetti, D., Clerici, M., Deleonibus, S., et al. (2011a). Hyaluronan synthesis is inhibited by adenosine monophosphate-activated protein kinase through the regulation of HAS2 activity in human aortic smooth muscle cells. J Biol Chem *286* (10), 7917–7924.

Vigetti, D., Genasetti, A., Karousou, E., et al. (2009a). Modulation of hyaluronan synthase activity in cellular membrane fractions. J Biol Chem *284*, 30684–30694.

Vigetti, D., Genasetti, A., Karousou, E., et al. (2010). Proinflammatory cytokines induce hyaluronan synthesis and monocyte adhesion in human endothelial cells through hyaluronan synthase 2 (HAS2) and the nuclear factor-kappaB (NF-kappaB) pathway. J Biol Chem *285*, 24639–24645.

Vigetti, D., Ori, M., Viola, M., et al. (2006). Molecular cloning and characterization of UDP-glucose dehydrogenase from the amphibian Xenopus laevis and its involvement in hyaluronan synthesis. J Biol Chem *281*, 8254–8263.

Vigetti, D., Rizzi, M., Moretto, P., et al. (2011b). Glycosaminoglycans and glucose prevent apoptosis in 4-methylumbelliferone treated human aortic smooth muscle cells. J Biol Chem *286*, 34497–34503.

Vigetti, D., Rizzi, M., Viola, M., et al. (2009b). The effects of 4-methylumbelliferone on hyaluronan synthesis, MMP2 activity, proliferation, and motility of human aortic smooth muscle cells. Glycobiology *19*, 537–546.

Vigetti, D., Viola, M., Gornati, R., et al. (2003). Molecular cloning, genomic organization and developmental expression of the Xenopus laevis hyaluronan synthase 3. Matrix Biol *22*, 511–517.

Wang, A., de la Motte, C., Lauer, M., and Hascall, V. (2011). Hyaluronan matrices in pathobiological processes. FEBS J *278*, 1412–1418.

Weigel, P. H., and DeAngelis, P. L. (2007). Hyaluronan synthases: a decade-plus of novel glycosyltransferases. J Biol Chem *282*, 36777–36781.

Zhang, W., Watson, C. E., Liu, C., Williams, K. J., and Werth, V. P. (2000). Glucocorticoids induce a near-total suppression of hyaluronan synthase mRNA in dermal fibroblasts and in osteoblasts: a molecular mechanism contributing to organ atrophy. Biochem J *349*, 91–97.

2.3 Multiple roles of hyaluronan as a target and modifier of the inflammatory response

Sanna Oikari, Tiina A. Jokela, Raija H. Tammi, and Markku I. Tammi

2.3.1 Introduction

Hyaluronan (HA; hyaluronic acid, hyaluronate) was isolated from the vitreous of the eye and was found to be responsible for the jellylike consistency of this tissue (Meyer and Palmer, 1934). The highly hydrated property of this extracellular polysaccharide was explained by its acidic nature, the carboxylates of its glucuronic acid moieties binding Na^+ counterions with large water shells. Later characterization of the physicochemical properties of this glycosaminoglycan stressed its tendency even at relatively small concentrations to swell and fill large volumes of water, excluding other large molecules and particles (Laurent, 1964). The softness, elasticity, and high water content of embryonic tissues were soon attributed to their high concentration of HA. In adults, it was soon realized that the swelling pressure created by the rapid increase of HA in injured tissues contributed to the first of the major signs of inflammation (i.e. swelling, redness, heat, and pain). For instance, the HA accumulation was found to explain periarticular swelling and morning stiffness, typical for joints affected by rheumatoid arthritis (Engström Laurent and Hällgren, 1987).

The "unspecific" role of HA just as a tissue space filler started to change first by the demonstration of its bonding with high specificity and affinity to cartilage link protein and the core protein of aggrecan, the major proteoglycan of cartilage (Hardingham and Muir, 1972; Hascall and Heinegård, 1974). HA and these bonds are required for the formation of proteoglycan aggregates, which are trapped in a collagen network, and crucial for proper functioning of cartilage. HA-dependent proteoglycan aggregates were also found in other tissues (Gardell et al., 1980), and specific bonds of HA to cell surface receptors (Underhill et al., 1987) and other matrix proteins (Yoneda et al., 1990) were discovered soon after.

While an increase of HA has been noted whenever some sort of tissue remodeling takes place, whether it be embryonic development, bacterial or viral infection, autoimmune inflammation, metabolic derangement, malignant tumor, injury, or wounding, the exact role of the cellular and molecular interactions of HA during these processes has only recently received some attention. It is obviously premature to attempt painting a complete picture on all the contributions of HA in inflammation, but subsequently, we highlight some important aspects that have recently emerged.

2.3.2 Endothelial permeability

One of the earliest signs of inflammation is increased permeability of the endothelium and flux of fluid and leukocytes from circulation into the interstitium. Blood vascular endothelial cells are covered with a layer of glycosaminoglycans, proteoglycans, and glycoproteins, collectively called glycocalyx (Gao and Lipowsky, 2010). The molecular architecture of endothelial glycocalyx is unknown, but it is also dependent on the presence of plasma proteins (Ebong et al., 2011). Specific fixation techniques with transmission electron microscopy revealed an endothelial glycocalyx thickness of 11 μm on cultured bovine aortic endothelial cells (Ebong et al., 2011), while fluorescent probes for wheat germ agglutinin (WGA) lectin, heparan sulfate, and HA suggested 2–5 μm thickness in mouse arteries (Reitsma et al., 2011). The endothelial glycocalyx is destroyed by hyaluronidase, which, like inhibition of global HA synthesis (Nagy et al., 2010), enhances the penetration of LDL, eventually promoting the development of arterial disease, now understood as an inflammatory process. Thus, intact HA is an important constituent of endothelial glycocalyx, the primary endothelial barrier, and shield against vascular leakage and edema that takes place in inflammation (Singleton et al., 2010). However, the role of endothelial HA in leukocyte rolling, adhesion, and extravasation is not settled because leukocyte and endothelial CD44 with attached HA can also be adhesive (DeGrendele et al., 1997; McDonald et al., 2008), depending on the experimental setup and biological state of the tissue. Research on the structure of endothelial glycocalyx and its interactions with blood cells is in a very intensive phase, and our insight into the role of HA will undoubtedly yield new details soon.

2.3.3 Angiogenesis

HA has an important role in both maintaining normal vascular integrity and in the growth of new blood vessels. Many studies have shown that high-molecular-mass (HMM) HA suppresses angiogenesis by inhibition of endothelial cell proliferation, migration, and capillary formation in three-dimensional matrices (review by Slevin et al., 2007). A recent study by Singleton et al. (2010) demonstrates that HMM HA can enhance endothelial cell barrier function and therefore inhibit vascular leakiness. They propose a model where binding of HMM HA to CD44 induces a complex between CD44, annexin A2, and protein S100-A10. The protein complex translocates to caveolin-enriched microdomains and further interacts with filamin-A and filamin-B. This results in actin cytoskeleton reorganization and increased endothelial cell barrier function. In contrast, HA oligosaccharides are pro-angiogenetic, as it has been reported that they induce endothelial cell proliferation, migration, and tube formation in collagen gels and that they have similar effects *in vivo* (review by Slevin et al., 2007). The influence of HA oligosaccharides seems to be mediated through CD44 and HA-mediated motility receptor (RHAMM), each having different effects, but the exact mechanism remains unclear. A recent paper describing the effect of siRNA against CD44 and RHAMM shows that RHAMM functions mainly through the extracellular regulated kinase 1/2 (ERK1/2) pathway, indicated earlier (Lokeshwar and Selzer, 2000), while CD44 affects protein kinase C and γ-adducin phosphorylation. In addition, both receptors can influence matrix metalloproteinase-9 activity (Matou-Nasri et al., 2009). Besides its direct effects on endothelial cells, HA and its fragments can block or promote angiogenesis, respectively, by modulating the functions of inflammatory cells and their cytokine production, as discussed subsequently. HA has a particularly important role in the angiogenesis during wound healing and tumor neovascularization.

2.3.4 Mechanisms of hyaluronan degradation

It is not known what determines termination of an HA chain (i.e. how the chain length is regulated). It is likely, however, that different sizes of HA are produced de novo. The HA produced by HAS3 is smaller than that of HAS1 and HAS2 in certain culture conditions, while sometimes a difference was not seen (Brinck and Heldin, 1999).

However, newly synthesized HMM HA can get fragmented immediately after, or even during, its synthesis. This can occur by the activity of hyaluronidase-2 (Hyal2) tethered to the cell surface by glycosylphosphatidylinositol or by reactive oxygen species (ROS) created during inflammation. Both Hyal2 and ROS probably cleave at random and are likely to produce relatively large fragments that can be further degraded. However, the subsequent degradation pathways are complex, since the fragments can either diffuse to lymph (Sabaratnam et al., 2005) or become endocytosed and catabolized locally in lysosomes. The fragments are therefore supposed to have a short half-life in the tissue, and very few studies have attempted to characterize them or measure their content (Bracke et al., 2010). Nevertheless, *in vitro* tests have shown that HA fragments have profound biological effects, influencing the expression of a number of genes (Takahashi et al., 2005).

At the moment, a large body of evidence indicates that ROS are involved in the inflammatory response and HA degradation. A role for Hyal2 was also suggested by a recent study (de la Motte et al., 2009), but mice without a functional *Hyal2* gene present a phenotype that has not yet been probed for its ability to induce or clear inflammation (Jadin et al., 2008).

2.3.5 Consequences of hyaluronan fragmentation

Hyaluronidase treatment and HA fragments activate ovulated cumulus cells in a Toll-like receptor-2 (TLR2)/Toll-like receptor-4 (TLR4)–dependent manner (Shimada et al., 2008). Low-molecular-mass HA increases keratinocyte production of β-defensin 2 in a TLR2- and TRL4-dependent manner (Gariboldi et al., 2008).

ROS created, for instance, by ozone exposure (Garantziotis et al., 2009) make HA fragments that increase airway mucus secretion by epidermal growth factor receptor (EGFR)/CD44 activation. The activation involves CD44 and kallikrein interaction, leading to epidermal growth factor (EGF) release from its membrane-bound precursor (Yu et al., 2011). HA fragments alone can also induce CD44/EGFR interaction and the activation of the MUC5B mucin (Casalino-Matsuda et al., 2009). Extracellular superoxide dismutase produced in the lungs attaches to HA and shields it against ROS attack, thus preventing HA fragmentation and inhibiting inflammation (Gao et al., 2008). HA has been suggested to activate macrophages with a mechanism distinct from that of lipopolysaccharide, but it involves TLR4 and MD2 proteins of the LPS receptor complex (Taylor et al., 2007). On the other hand, HMM HA applied intravenously may also counteract the inflammation induced in the lungs by lipopolysaccharide, as estimated by the tissue accumulation of neutrophils and monocytes (Liu et al., 2008). T-helper (Th1) lymphocytes stimulate dendritic cell production of HA, which in turn mediates binding between lymphocytes and dendritic cells (Bollyky et al., 2010). HA fragments as small as a hexamer stimulate inflammatory cytokine expression via CD44 and TLR4 (Campo et al., 2010a, 2010b), while HMM HA can suppress the activation of monocytes and inhibit their tumoricidal inflammatory response (del Fresno et al., 2005) and prostaglandin E2 production (Yasuda, 2010). Intact HA also promotes resolution of inflammation by induction of regulatory (TR1) lymphocytes from their precursors (Bollyky et al., 2011). Breast tumor stromal cells with conditionally blocked HAS2 expression showed reduced tumor angiogenesis, lymphangiogenesis, and macrophage recruitment, perhaps by reducing the HA that can be converted to active fragments (Kobayashi et al., 2010).

2.3.6 Hyaluronan cross-talk with leukocytes

During inflammation, several types of leukocytes participitate in the healing procress. HA has been shown to interact with almost all of these cells. Intraepithelial T cells (γδT cells) represent the first line of defense through their recognition of self-antigens expressed on damaged cells. Epidermal γδT cells induce keratinocyte HA synthesis, and after activation, they also synthesize HA. An elevated HA level enhances macrophage migration into the inflamed tissue (Jameson et al., 2005). CD44-HA interaction on the vessel wall is also essential for neutrophil and lymphocyte extravasation into the inflammation site (DeGrendele et al., 1997; McDonald et al., 2008). The CD44 on resting T cells is functionally inactive, while after activation CD44 splice variants V6 and V9 are expressed and HA binding is enhanced (Bollyky et al., 2007). CD44-HA interaction induces proliferation, adhesion, and spreading of T cells (DeGrendele et al., 1997; Lefebvre et al., 2010; Maeshima et al., 2011). Elevated HA binding has also been noticed in a rapidly proliferating memory-type T-cell population (Maeshima et al., 2011). HA-CD44 interaction also induces B-cell proliferation and differentiation (Rafi et al., 1997). HA increases the stability and secretion of granulocyte macrophage colony – stimulating factor (GM-CSF) mRNA, which prolongs survival of cultured eosinophiles (Esnault and Malter, 2003).

Several inflammatory agents secreted by leukocytes, connective tissue cells, and epithelial cells regulate HA synthesis or induce the adhesive forms of HA (▶Table 2.1). On the other hand, HA interaction with leukocyte surface receptors like CD44, TRL2, and TLR4 regulate the expression of many inflammatory agents (▶Table 2.2). Since HA synthesis is modified by inflammatory mediators and HA modifies the secretion of the mediators, an intricate network of feedback loops is created.

The molecular size of HA has an important role considering the interaction with leukocytes. High-molecular-weight (HMW) HA interaction with CD44 has been shown to suppress T-cell functions, while an opposite effect is seen with low-molecular-weight (LMW) HA (Bollyky et al., 2007). LMW HA is more effective in inducing eosinophile survival than HMW HA (Ohkawara et al., 2000). HMW HA generally upregulates

Table 2.1 **Inflammatory agents that regulate hyaluronan (HA) expression.** Polypeptide growth factors and cytokines known to be released from cells in inflamed tissues.

Agent	*Cell/ tissue*	*HA*	*Cables*	*Reference*
EGF	keratinocyte, primary skin fibroblast, oral mucosa fibroblast	↑	–	(Pasonen-Seppänen et al., 2003; Yamada et al., 2004)
IFN-γ	keratinocyte, lung fibroblast	↑	–	(Sampson et al., 1992; Sayo et al., 2002)
IGF	fibroblast, mesothelial	↑		(Honda et al., 1991; Kuroda et al., 2001)
IL-1β	skin fibroblast, oral mucosa fibroblast, synoviocyte from reumatoid arthritis, fetal skin fibroblast, endothelial cell	↑	+	(Kennedy et al., 2000; Oguchi and Ishiguro, 2004; Yamada et al., 2004; Vigetti et al., 2010)
IL-4	synovial membrane	↑		(Hyc et al., 2009)
IL-6	skin fibroblast	↑		(Duncan and Berman, 1991)
IL-15	endothelial cell	↑		(Estess et al., 1999)
KGF	keratinocytes	↑		(Karvinen et al., 2003)
PDGF	fibroblast, mesothelium, vascular endothelium, vascular *SMC	↑		(Heldin et al., 1989, 1992; Jacobson et al., 2000; Evanko et al., 2001; Suzuki et al., 2003)
PolyI:C	primary mucosal SMC, lung fibroblast	↑ –	+	(de la Motte et al., 2003; Evanko et al., 2009)
TGF-β	keratinocyte, synoviocyte	↓		(Kawakami et al., 1998; Sayo et al., 2002; Pasonen-Seppänen et al., 2003)
TGF-β	synoviocyte from reumatoid arthritis, fibroblast, keratinocyte, vascular endothelial cell	↑		(Heldin et al., 1989; Sugiyama et al., 1998; Suzuki et al., 2003; Oguchi and Ishiguro, 2004)
TNF-α	fetal skin fibroblast, lung fibroblast, endothelial cell, synovial membrane	↑	+	(Sampson et al., 1992; Kennedy et al., 2000; Hyc et al., 2009; Vigetti et al., 2010)
TNF-β	vascular endothelium	↑	+	(Vigetti et al., 2010)

*SMC = smooth muscle cell.
EGF = epidermal growth factor; IFN = interferon gamma; IGF = insulin-like growth factor; IL-1b = interleukin 1b; IL-4, 6, 15 = Interleukins 4, 6, and 15; KGF = keratinocyte growth factor; PDGF = platelet derived growth factor; poly I:C = polyinosinic-polycytidylic acid; TGF-b = transforming growth factor b; TNF a, b = tumor necrosis factors a and b.

anti-inflammatory cytokines like interleukin-10 (IL-10) and downregulates proinflammatory agents like interleukin-1β (IL-1β), interleukin-8 (IL-8), and matrix metalloproteinase-1 (MMP1), while LMW HA induces the expression of several proinflammatory agents (▶Table 2.2). In conclusion, HMW HA enhances anti-inflammatory actions of leukocytes while LMW HA induces and maintains inflammation.

Table 2.2 Inflammatory agents regulated by hyaluronan (HA).

Upregulated inflammatory agent	*Leukocyte type/ soluble phase*	*HA*	*Receptor*	*Reference*
Ccl2/ MCP-1	*PBMC, macrophage	LMW	TLR4, CD44	(McKee et al., 1996; Yamawaki et al., 2009)
Ccl3/ MIP-1α	monocyte	MIX		(Wallet et al., 2010)
Ccl3/ MIP-1α	macrophage	LMW	CD44	(McKee et al., 1996)
Ccl3/ MIP-1α	macrophage	LMW	TLR2, TLR4	(Scheibner et al., 2009)
Ccl4/ MIP-1β	macrophage	LMW	TLR2, TLR4	(Jiang et al., 2005)
Ccl4/ MIP-1β	macrophage	LMW	CD44	(McKee et al., 1996)
Ccl4/ MIP-1β	monocyte	MIX		(Wallet et al., 2010)
Ccl5 / Rantes	macrophage	LMW	CD44	(McKee et al., 1996)
CSF2/ GM-CSF	monocyte, eosinophil	LMW, MIX	CD44	(Esnault and Malter 2003; Wallet et al., 2010)
Cxcl1/KC	macrophage	LMW	CD44	(McKee et al., 1996)
Cxcl1/KC	bronchoalveolar lavage fluid	LMW	Non-TLR4	(Zhao et al., 2010)
Cxcl2/ MIP-2α	macrophage	LMW		(Bai et al., 2005)
Cxcl9/MIG	macrophage	LMW		(Horton et al., 1998)
Cxcl10/ crg-2/IP-10	macrophage, monocyte	LMW	CD44	(McKee et al., 1996; Horton et al., 1998; Wallet et al., 2010)
I-CAM	eosinophi	LMW		(Ohkawara et al., 2000)
IFN-γ	T lymphocyte	LMW		(Blass et al., 2001)
IGF-1	macrophage	LMW	Cd44	(Noble et al., 1993)
IL-1α	monocyte	MIX		(Wallet et al., 2010)
IL-1β	monocyte	MIX		(Wallet et al., 2010)
IL-1β	bronchoalveolar lavage fluid	LMW	Non-TLR4	(Zhao et al., 2010)
IL-6	monocyte, B lymphocyte	MIX	TLR4	(Iwata et al., 2009; Wallet et al., 2010)
IL-6	bronchoalveolar lavage fluid	LMW	Non-TLR4	(Zhao et al., 2010)
IL-6	PBMC	LMW	TLR4, CD44	(Yamawaki et al., 2009)
IL-8	macrophages monocyte	LMW	TLR4, CD44, MD-2	(McKee et al., 1996; Taylor et al., 2007)

(*Continued*)

Table 2.2 **Inflammatory agents regulated by hyaluronan (HA). (*Continued*)**

Upregulated inflammatory agent	*Leukocyte type/ soluble phase*	*HA*	*Receptor*	*Reference*
IL-10	monocyte, serum, T lymphocyte, B lymphocyte	HMW, MIX	TLR4	(Bollyky et al., 2007; Iwata et al., 2009; Asari et al., 2010; Wallet et al., 2010)
IL-12	macrophage	LMW	CD44	(Hodge-Dufour et al., 1997)
IL-12p40	monocyte	HMW		(Wallet et al., 2010)
MIP-2	bronchoalveolar lavage fluid	LMW	Non-TLR4	(Zhao et al., 2010)
MIP-2	macrophage, monocyte	LMW	TLR4, CD44, MD-2	(Taylor et al., 2007; Zheng et al., 2009)
MME	macrophage	LMW		(Horton et al., 1999)
MMP3	macrophage	LMW		(Taylor et al., 2007)
MMP12	macrophage	LMW		(Horton et al., 1999)
PAI1	macrophage	LMW		(Horton et al., 2000)
Socs3	macrophage	MIX		(Taylor et al., 2007)
TGF-β	eosinophil	LMW	CD44	(Ohkawara et al., 2000)
TGF-β	B lymphocyte	MIX	TLR4	(Iwata et al., 2009)
TGF-β2	Monocyte	LMW	CD44, MD-2, TLR4	(Taylor et al., 2007)
TNF	monocyte	MIX		(Wallet et al., 2010)
TNF-α	B lymphocyte	MIX	TLR4	(Iwata et al., 2009)
TNF-α	bronchoalveolar lavage fluid	LMW	Non-TLR4	(Zhao et al., 2010)
TNF-α	macrophage	LMW	TLR4	(Zheng et al., 2009)
Downregulated inflammatory agent	*Leukocyte type/ soluble phase*	*HA*	*Receptor*	*Reference*
A2ar	macrophage	LMW	CD44	(Collins et al., 2011)
MCP-5	serum	HMW		(Asari et al., 2010)
MIP-2	serum	HMW		(Asari et al., 2010)
p-selectin	serum	HMW		(Asari et al., 2010)
Rantes	serum	HMW		(Asari et al., 2010)
SCF	serum	HMW		(Asari et al., 2010)

(*Continued*)

Table 2.2 Inflammatory agents regulated by hyaluronan (HA). (*Continued*)

Downregulated inflammatory agent	*Leukocyte type/ soluble phase*	*HA*	*Receptor*	*Reference*
Socs7	macrophage	MIX		(Taylor et al., 2007)
uPA	macrophage	LMW		(Horton et al., 2000)
VEGF	serum	HMW		(Asari et al., 2010)

*PBMC = peripheral blood mononuclear cell.
Ccl2,3,4,5 = Chemokine (C-C motif) ligand 2,3,4,5; MCP-1 = monocyte chemotactic protein-1; MIP-1a,b,2 = macrophage inflammatory protein 1a and b and 2; Rantes = regulated upon activation, normal T-cell expressed, and secreted; CSF2 = colony-stimulating factor 2; GM-CSF = granulocyte-macrophage colony-stimulating factor; Cxcl1/KC = Cxcl1,2,10 = chemokine (C-X-C motif) ligand 1, 2, 9 and 10; KC = keratinocyte chemoattractant; MIG = monocyte induced by gamma interferon; I-CAM = intercellular adhesion molecule; IFN-g = interferon g; IL-1a,b,6,8,10,12, 12p40 = interleukins 1a,b, 6, 8, 10, 12, 12p40; MME = membrane metallo-endopeptidase; MMP3,12 = matrix metalloproteinase 3 and 12; PAI1 = plasminogen activator inhibitor-1; Socs3, 7 = suppressor of cytokine signaling 3 and 7; TGF-β = transforming growth factor β; TNF a = tumor necrosis factor a; A2ar = adenosine A2a receptor; MCP-5 = monocyte chemotactic protein 5; SCF = stem cell factor; uPA = urokinase-type plasminogen activator; VEGF = vascular endothelial growth factor.

2.3.7 Adhesion of leukocytes to hyaluronan

Special forms of HA coats, called HA cables, appear in inflammation, endoplasmic reticulum (ER) stress, and hyperglycemic conditions; after α1-adrenergig receptor simulation and viral infection; or by double-stranded RNA mimicked by polyinosinic: polycytidylic acid (Poly I:C) (de la Motte et al., 1999; Majors et al., 2003; Wang and Hascall 2004; Shi et al., 2006). Stressed cells reshape their HA into these elongated structures that adhere leukocytes and platelets (de la Motte et al., 2003, 2009). *In vitro*, treatments with IL-1β and tumor necrosis factor-alpha (TNF-α) stimulate HA cable formation (Jokela et al., 2008b; Vigetti et al., 2010). TNF-α induces cables in endothelial and epithelial cell cultures but not in smooth muscle cell cultures (de la Motte et al., 1999; Jokela et al., 2008b; Vigetti et al., 2010). HA cables have been reported in cultured smooth muscle cells, fibroblasts, mesangial cells, and epithelial and endothelial cells (de la Motte et al., 1999; Wang and Hascall, 2004;Jokela et al., 2008b; Evanko et al., 2009; Vigetti et al., 2010). Similar HA structures are present *in vivo* in the intestine of patients with inflammatory bowel disease (IBD) (Kessler et al., 2008). (See ▸Figure 2.5.)

HA cables can span several cells, up to millimeters in total length (Evanko et al., 2009). In endothelial cells, HAS2 and CD44 are essential to cable formation, probably acting as cell surface anchors for the HA (Vigetti et al., 2010). These massive structures of parallel HA chains probably need some supporting molecules, but complete molecular formula of the cables is still unknown. There seems to be a lot of divergence between cables expressed by different cell types. In tissue sections from IBD patients, HA cables associate with inter α-trypsin inhibitor (IαI) staining (de la Motte et al., 2003). Human primary colon mucosal smooth muscle cells and immortalized human

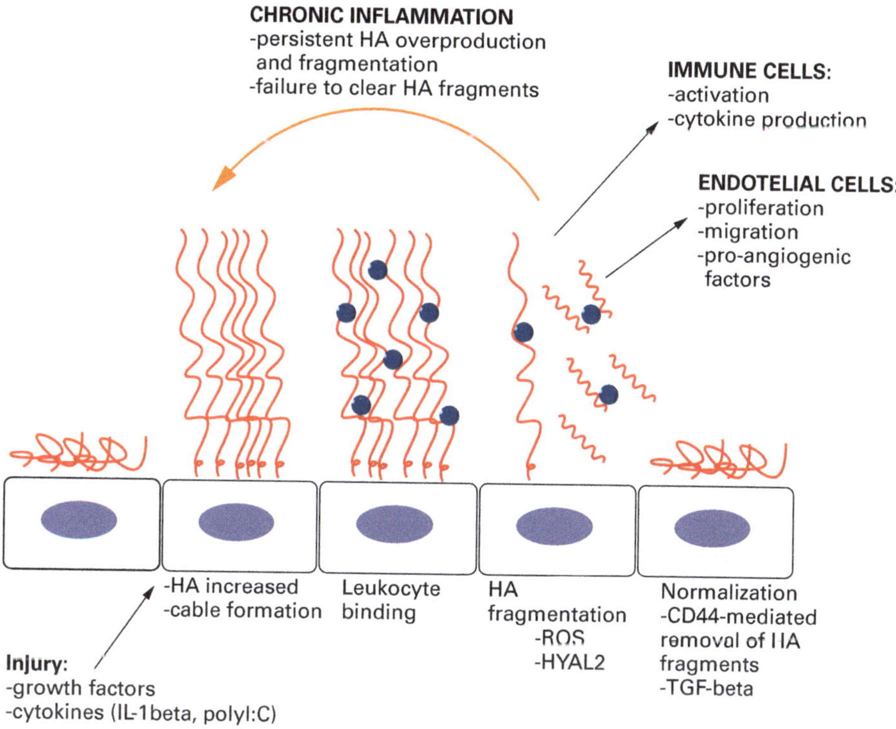

Figure 2.5 **Hyaluronan changes during the course of inflammation.** Small amounts of tightly attached hyaluronan (red) is normally present on the cell surface (left cell). Following injury, a set of signals from injured cells induces hyaluronan synthesis and its release into structures that cause leukocyte adhesion. Binding to hyaluronan influences gene expression and phenotype of the leukocytes. At the same time, at least macrophages can degrade the hyaluronan and create fragments that induce a gene expression profile and phenotype different from those of high-molecular-mass hyaluronan. Factors that eventually quench the vicious cycle of enhanced synthesis and fragmentation of hyaluronan are not known, but CD44 has been suggested to contribute to the elimination of the fragments. Image modified from de la Motte et al. (2009).

proximal tubular epithelial cells show HA cables colocalized with IαI and versican, that perhaps cross-link HA chains (de la Motte et al., 2003; Selbi et al., 2006). Tumor necrosis factor stimulated gene 6 (TSG6) remains the only documented enzyme capable of the covalent transfer of IαI-derived heavy chains (HCs) to HA, and oligomerized TSG-6 alone may cross-link HA (Baranova et al., 2011). However, it has been shown that cable formation is independent of TSG-6 (Selbi et al., 2006) and cables in murine and human airway smooth muscle cells are formed without IαI and versican (Lauer et al., 2009). Formation of cables does not require *de novo* protein synthesis and they can be built up in a serum-free medium (de la Motte et al., 2003). Cable structures can also form in the absence of an intact cytoskeleton and membrane protrusions (Evanko et al., 2009). Poly I:C stimulates HA cable formation in a signal transducer and activator of transcription 1 (STAT1) and phosphatidylinositol 3-kinase (PI3K)/protein kinase B (Akt)-dependent manner, independently of Toll-like receptor-3 (TLR3) (Bandyopadhyay et al., 2010).

Monocytes bind to HA cables via their CD44 receptors (de la Motte et al., 1999), and platelets have also the capability to bind HA organized in this manner. Interestingly, platelets have membrane-bound Hyal2 on their surface. In neutral pH, Hyal2 mediates HA fragmentation, and the fragments induce monocyte activation (de la Motte et al., 2009). HA cables can obscure leukocyte access to other cell surface adhesion molecules, like intracellular adhesion molecule (ICAM), and therefore inhibit signaling through these receptors (Zhang et al., 2005).

2.3.8 Hyaluronan removal in the late phase of inflammation

Removal of HA fragments is crucial for turning off inflammation, and defects in the removal result in chronic inflammation (Teder et al., 2002). In normal tissues, HA turnover is rapid; in human skin organ cultures, the half-life of epidermal HA is about 1 day (Tammi et al., 1991; Ågren et al., 1995), whereas in cartilage it is 2–3 weeks (Fraser et al., 1997). Most of the HA is probably degraded in the tissue in which it is synthesized (Armstrong and Bell, 2002c), and the rest drains into lymph to be catabolized in lymph nodes (Fraser et al., 1988), liver endothelial cells, and spleen (Fraser et al., 1985; McGary et al., 1989; Harris et al., 2008) that take up HA by a protein named hyaluronic acid receptor for endocytosis (HARE/STAB-2), which facilitates rapid endocytosis of HA and other glycosaminoglycans (McGary et al., 1989; Harris et al., 2007). In addition, the endothelium of lymphatic capillaries also has a second receptor proposed to function in HA clearance named lymphatic vessel endothelial HA receptor (LYVE-1) (Banerji et al., 1999).

About 85% of the HA leaving tissues is estimated to be degraded in lymph nodes, and the rest enters blood circulation and is rapidly catabolized in liver. The half-life of circulating HA is only 2–5 minutes (Fraser et al., 1981). The amount of HA is elevated in lymph during tissue injury or inflammation (reviewed by Laurent et al., 1996), perhaps due to increased synthesis and enhanced degradation, the latter facilitating its escape from tissue. Interestingly, it has been shown that postnodal lymph contains mainly low-molecular-mass or fragmented HA (<800,000 Da), while that in tissue and prenodal lymph is of high molecular mass (Armstrong and Bell, 2002a, 2002c). Despite effective catabolism before entrance into circulation, injury and inflammation increase HA concentration also in plasma (Engström-Laurent and Hellgren, 1987).

2.3.9 Local clearance of hyaluronan

In vivo, local degradation has been demonstrated following injection of labeled HA in skin and joints (Laurent et al., 1991, 1992). *In vitro*, many cell lines like keratinocytes (Tammi et al., 2001), chondrocytes (Aguiar et al., 1999), fibroblasts, smooth muscle cells (McGuire et al., 1987), macrophages (Culty et al., 1994a), and some breast cancer cells (Culty et al., 1994b) internalize and degrade HA. In most of the cells studied, the rate of HA degradation is correlated with their CD44 expression, and deletion of CD44 increases tissue HA in inflamed lung tissue (Teder et al., 2002). Therefore, CD44 is considered a major receptor for HA internalization in many cell types. In keratinocytes, CD44-mediated HA internalization is not associated with clathrin-coated vesicles or caveolae (Tammi et al., 2001). The C-terminal cytoplasmic domain of CD44 is important for internalization in chondrocytes, as selective blocking of an inhibitory CD44 variant missing this domain enhances HA uptake in chondrocytes (Jiang et al., 2002). Another factor regulating CD44-mediated HA degradation by chondrocytes is the posttranslational modification of the standard form of CD44 with fatty acids. Dual acylation of cysteines 286 and 295 is required for HA uptake, but not for binding (Thankamony and Knudson, 2006). Many posttranslational modifications of CD44 influence the affinity of CD44 to HA, and indirectly endocytosis. In addition, complex formation between CD44 and Hyal2 on the plasma membrane has been demonstrated to influence HA shedding (Duterme et al., 2009) or degradation (Bourguignon et al., 2004). This is relevant for inflammation as the endocytosis rate of HA fragments is faster than that of intact, anti-inflammatory HA (McGuire et al., 1987). It is believed that following internalization CD44 is recycled back to the plasma membrane and HA is further catabolized in lysosomes by Hyal1 and β-exoglycosidases (Tammi et al., 2001; Stern, 2004; Thankamony and Knudson, 2006). Concerning extinguishment of inflammation, macrophages have an important role in local HA clearance (Underhill et al., 1993; Teder et al., 2002), and their ability to bind HA is influenced by both inflammatory and anti-inflammatory agents (Ruffell et al., 2011).

In addition to CD44-dependent uptake of HA, some cell types also show CD44-independent mechanisms of HA internalization. The melanoma cell line B16-F10 utilizes macropinocytosis to internalize HA despite the presence of CD44. This endocytosis mechanism of HA is time and concentration dependent, and HA itself induces the process (Greyner et al., 2010).

2.3.10 Chronic inflammation

Chronic inflammation is not a trivial thing as many major health concerns of Western societies, like obesity, type 2 diabetes, and atherosclerosis are associated with an inflammatory component. Involvement of improper HA processing is evident in many of these diseases. In obesity, both glucose and saturated free fatty acids increase HA production of adipocytes and enhance monocyte adhesion. In adipose tissue, monocytes become macrophages. Accumulation of macrophages in adipose tissue is characteristic of obesity and causes systemic inflammation and insulin resistance (Han et al., 2007; Yeop Han et al., 2010). HA levels that are upregulated in both early and late atherosclerotic lesions contribute to the development of atherosclerosis, (Levesque et al., 1994; Bot et al., 2008). Increased HA levels can have multiple effects depending on the cell type and the form of HA. In aortic smooth muscle cells, HMW HA enhances motility (Vigetti et al., 2008), while fragmented HA functions as a proinflammatory agent and stimulates incorporation of oxidized low-density lipoprotein into macrophages (Seike et al., 2006). HA thus contributes to migration of smooth muscle cells into the tunica intima layer of blood vessels, sustainment of local inflammation, and uptake of low-density lipoproteins into macrophages, resulting in foam cells, all important factors in the development atherosclerosis. Yet another disease associated with disturbed HA metabolism is asthma, which is a chronic inflammatory disease of the airways. Increased levels of HA have been measured from bronchoalveolar lavage fluid of patients with asthma, and HA levels correlate with the severity of the disease (Bousquet et al., 1991). A recent study shows that fibroblasts isolated from asthma patients produce more HA than cells isolated from controls. Furthermore, macrophages isolated from the same patients have reduced expression of CD44 and increased levels of TLR2/4 compared to healthy controls (Liang et al., 2011). The results show how increased HA production combined with the reduced uptake caused by low CD44 levels and increased signaling through TLR2/4 can lead to chronic airway inflammation manifested as asthma. In addition, disturbed HA processing is involved in the pathophysiology of many other diseases associated with chronic inflammation (review by Jiang et al., 2011).

2.3.11 Hyaluronan increase in wounds

The increased HA content of the dermal compartment after wounding or inflammation is well known. However, recent experiments indicate that HA synthesis is also activated in epithelial cells undergoing wound reaction (Tammi et al., 2005) and in skin disorders with an inflammatory component (Tammi et al., 1994; Ohtani et al., 2009). In the dermal wound response, HA accumulation coincides with the formation of the fibrin clot and its invasion by the inflammatory cells, although a second HA peak may occur somewhat later (Dechert et al., 2006). In epidermis, an elevated HA level is detected on the first postwounding day and peaks on day 3 (Maytin et al., 2004; Tammi et al., 2005). The inflammatory cells in the clot are a rich source of various growth factors and cytokines, probably the major source of the HA stimulatory activity *in vivo* (▶Table 2.1). Nevertheless, experiments in an artificial skin model indicate that also local cells like keratinocytes alone can rapidly activate HA synthase (HAS) mRNA expression by the release of heparin-binding epidermal growth factor (HB-EGF), and the immediate response may also involve activation of existing HAS proteins (Monslow et al., 2010). Expression of the HA receptors CD44 and RHAMM are increased by wounding both in keratinocytes and fibroblasts (Lovvorn et al., 1998; Tammi et al., 2005) and have been suggested to correlate with fibroplasia (Lovvorn et al., 1998). Interestingly, it was recently shown that the reduction of HA content in the dermal wounds of *HAS1+3* knockout mice was associated with faster wound closure (Mack et al., 2012), while in fetal wounds the high level of HA has been associated with scarless wound healing (Mast et al., 1991; Estes et al., 1993). Supporting this view, Mack and coworkers showed that the adult *Hoxb13* knockout mice with similar wound healing characteristics as fetal wounds contained more epidermal and dermal HA than wild-type animals (Mack et al., 2003).

In addition to an increase in quantity, HA is believed to undergo qualitative changes rapidly after the onset of the inflammation, from a very HMM polysaccharide found in most normal resting tissues (Tammi et al., 1991; Armstrong and Bell, 2002b) into shorter fragments (Noble, 2002; Dechert et al., 2006; Stern et al., 2006). The accumulation of HMM HA may modulate the speed of formation and the structure of the fibrin clot in the wound bed (Weigel et al., 1989). It may also attract and entrap leukocytes and modulate their function (▶Table 2.1), while the fragmentation of HA may induce secretion of chemokines, cytokines, and growth factors in macrophages, monocytes, dendritic cells, and endothelial cells (▶Table 2.2) (Noble et al., 1996; reviewed by Jiang et al., 2011 and Noble, 2002). HA fragmentation may result from direct action by inflammation-induced reactive oxygen radicals, or the free radicals may induce the expression of Hyal2 (Monzon et al., 2010).

2.3.12 Support of migration and proliferation

HA production generally correlates with the rates of proliferation and migration in several cell types, activated by wounding or growth factors (reviewed by Evanko et al., 2007). However, experimental evidence on the direct effects of HA on these processes is far from congruent. Addition of exogenous HA has been reported to stimulate, inhibit, or have no effect on proliferation (Goldberg and Toole, 1987; Yoneda et al., 1988; West and Kumar, 1989; Reijnen et al., 2000; Ahrens et al., 2001; Bourguignon et al., 2001; Dube et al., 2001; Haider et al., 2003; Gomes et al., 2004). The same variability concerns the effects of overexpression of *HAS* genes (Dube et al., 2001; Liu et al., 2001; Wilkinson et al., 2006), and the effects on cell migration show the same variability (reviewed by Evanko et al., 2007). Reasons for this may come from the different synthesis levels of HA and growth factor receptor expressions in various cell types, since both are involved in the regulation of cell migration in different ways. For example, HA inhibits platelet-derived growth factor (PDGF)–induced dermal fibroblast migration through recruiting a CD44-associated tyrosine phosphatase to the PDGF receptor (Li et al., 2006), while it activates epidermal growth factor receptor 2 (ErbB2) signaling in colon carcinoma cells (Ghatak et al., 2005) and melanoma cells (Kim et al., 2008). The presence of matrix hyaladherins like different isoforms of versican may also modulate the proliferative and migratory responses to HA (Hernandez et al., 2011a, 2011b). The more universal inhibition of cell migration induced by down-modulation of *HAS2* and/or *HAS3* (Bakkers et al., 2004; Nishida et al., 2005; Udabage et al., 2005; Li et al., 2007) suggests that although HA is needed for cell motility, the level of endogenous HA in migrating cells often fulfills the need for this extracellular matrix component.

After tissue injury, keratinocytes start to migrate to cover the wound area, and fibroblasts migrate from the collagen-rich wound edge into the fibrin-laden matrix to build the granulation tissue essential for the healing. These processes involve activation of both cell migration and proliferation, which are classically regarded to be modulated by HA (Toole, 1997). HA may open extracellular spaces for migration and help the dividing cells to detach, or it may activate migration-associated signaling pathways through HA receptors or growth factor receptors, or it may attract the cells (reviewed by Thorne et al., 2004). Large pericellular HA coats are seen around mitotic cells, and HA accumulates in the cleavage furrow between daughter cells in cytokinesis (Tammi et al., 1991) and around the trailing edge of migrating cells, suggesting that HA contributes to the detachment of cells from substratum and from each other (Evanko et al., 1999). HA also accumulates in the intracellular compartment in dividing and migrating cells (Evanko and Wight, 1999; Tammi et al., 2001). Although the significance of this event is not fully understood at the moment, it may help the daughter chromatids to separate (Evanko and Wight, 1999), or it may be used to transport HA to cell surface locations where it is needed.

Manipulation of keratinocyte HA synthesis by growth factors (EGF, transforming growth factor-beta [TGF-β]) and hormones like hydrocortisone, retinoic acid, and the UDP-sugar-depleting agents 4-methylumbelliferone (4-MU) and mannose has indicated a positive correlation between keratinocyte HA synthesis rate and cell proliferation and migration (Ågren et al., 1997; Pasonen-Seppänen et al., 2003, 2008; Rilla et al., 2004; Jokela et al., 2008a). However, little difference in keratinocyte migration rate was achieved by adding exogenous HA or by overexpression of inducible HAS2 or HAS3 (Bart, unpublished). Furthermore, HA is not restricted to the proliferative cells in epidermis (Tammi et al., 1988), and keratinocyte proliferation rate in epidermis does not correlate with HA synthesis rate (Wang et al., 1992). In wounded skin, activated keratinocytes are strongly HA positive throughout the hyperplastic epidermis at the wound edge, and not only those in the proliferative cell pool (Maytin et al., 2004; Tammi et al., 2005; Monslow et al., 2009). Interestingly, keratinocytes migrating into the wound show less intense and uneven HA staining pattern than those in adjacent hyperplastic area (Tammi et al., 2005), suggesting that cell migration is associated with rapid HA turnover. Knockout of *HAS1* and *HAS3* genes does not influence epidermal thickness after wounding despite marked reduction in epidermal HA (Mack et al., 2012), indicating that at least the HA produced by these two enzymes is not necessary for the proliferative response. It remains to be shown whether the activation of HAS2-dependent HA synthesis in these animals is specifically associated with the hyperplastic reaction in wound healing. HAS2 is the main isoform present in human and rat keratinocytes cultured in monolayer cultures (Pienimäki et al., 2001; Jokela et al., 2008a), while the levels of *HAS2* and *HAS3* mRNA are approximately the same in differentiating, organotypic cultures (Bart, unpublished). In an *in vivo* wound model, epidermal *HAS2* showed more substantial upregulation than *HAS1* or *HAS3* (Tammi et al., 2005).

2.3.13 TGF-β and myofibroblasts

After tissue injury, fibroblasts proliferate, migrate to the wound, and differentiate to myofibroblasts. TGF-β is a major player in these processes, and its influence is modulated by the presence of HA in an SMAD-independent way (Meran et al., 2007; Webber et al., 2009). Dermal fibroblasts respond to TGF-β with cell proliferation, while oral fibroblasts treated with TGF-β show an antiproliferative response. Inhibition of HA synthesis blocks the proliferative response in dermal fibroblasts (Meran et al., 2007). Conversely, overexpression of HAS2 in oral fibroblasts results in a proliferative response to TGF-β (Meran et al., 2011). The effect of HA on TGF-β signaling depends on CD44-EGFR interaction, leading to increased activation of signaling pathways controlling cell proliferation (Meran et al., 2011). On the other hand, HA may provide mechanical support essential for myofibroblast differentiation by stabilizing focal adhesions by a RHAMM-dependent mechanism (Hall et al., 1994; Twarock et al., 2010). Despite the beneficial effect on the differentiation of myofibroblasts, HA may counteract their collagen contraction and wound closure (Sainio et al., 2010). In normal wound healing, myofibroblasts are eventually removed by apoptosis; however, continuous TGF-β–HA signaling may lead to their persistence and fibrosis (Webber et al., 2009; Guo et al., 2010).

2.3.14 Therapeutic applications

The fact that HA resides in the interface between endothelial, mesenchymal, epithelial, and leukocytic cells, and contributes to a number of processes in different phases of inflammation, complicates attempts to modulate inflammation by targeting HA.

HA synthesis within arterial wall by smooth muscle cells is associated with progression of atherosclerotic disease, suggesting that HA depletion would be beneficial (reviewed by Wight, 2008). However, long-term systemic administration of 4-MU, a relatively specific inhibitor of HA synthesis, disturbs arterial endothelial function, probably interfering with the formation of glycocalyx, and therefore promotes arterial disease (Nagy et al., 2010). Very few attempts have been done with HA-blocking agents in inflammation, and it is quite possible that those will find applications in specific inflammatory diseases (e.g. local treatments of skin disorders) (Rilla et al., 2004). Mannose is a novel HA synthesis inhibitor (Jokela et al. 2008b) that prevents HA-dependent leukocyte adhesion to keratinocytes (Jokela et al., unpublished) and influx of leukocytes into wound granulation tissue (Kossi et al., 1999), suggesting that it should also be tested for suppression of excessive or deleterious inflammation.

Cancer is a wound that never heals, and inflammation is always involved. Recent studies have demonstrated that instead of fighting against the neoplastic cells, leukocytes often promote malignant growth (reviewed by Allavena et al., 2008). A number of studies *in vitro* and on animal models have indicated that inhibition of HA synthesis by 4-MU blocks cancer growth and enhances the therapeutic efficiency of cytostatic drugs (Kultti et al., 2009). The same effect has been obtained with systemic administration of recombinant human PH-20 hyaluronidase (Thompson et al., 2010). It relieves swelling pressure in the tumor, suppresses its growth by mechanisms not known at present, and augments penetration of anticancer agents. Initial trials on humans are ongoing and will determine the value of this therapy and its possible adverse side effects.

While many of the inflammation-promoting effects of HA have been specifically assigned to HA fragments, intact HMM HA may actually suppress lung inflammation (Liu et al., 2008). Healing of chronic tympanic perforations is enhanced by application of HMM HA (Laurent et al., 1991). Intra-articular HA injections often show anti-inflammatory effects; relief of pain in particular has been reported (Gigante and Callegari, 2011).

2.3.15 Future perspectives

The level of tissue HA depends on the synthesis, executed by three individual *HAS* genes, and the corresponding HAS proteins. The catabolism of HA is mediated by Hyal enzymes, ROS, and cell surface receptors that retain HA and also contribute to its endocytosis. The balance between the activity of these factors determines the content and molecular mass distribution of HA in the different phases of inflammation.

An obvious target of future research is to further define the roles of the three HAS enzymes. This includes transcriptional and posttranslational regulation by the inflamed environment of each HAS in the different cells types that contribute to tissue HA. The mechanisms that control cell surface presentation of HA, determining its ability to bind and modify the function of leukocytes, deserves more elaborate studies.

The catabolism of HA in peripheral tissues, in particular the roles of Hyal1, Hyal2, and Hyal3 enzymes, has remained somewhat obscure in normal conditions, and even more so during inflammation when HA turnover is stimulated (Jadin et al. 2008; de la Motte et al. 2009; Duterme et al. 2009). Targeted disruption of Hyal functions in specific cell types (and this also concerns the HAS enzymes) should provide important information about the processes that govern HA turnover in inflammation.

Experiments using inhibitors of HA synthesis in experimental animals, and perhaps in clinical settings, will undoubtedly tell us not only about therapeutic opportunities but also about the general roles of HA in inflammation.

2.3.16 Take-home message

HA is rapidly increased in all kinds of inflammatory responses, including bacterial or viral infection, autoimmune inflammation, metabolic derangement, malignant tumor, injury, or wounding, presumably by cytokines and other mediators of inflammation that induce its synthesis. At the same time, fragmentation of HA is a signal that further increases the production of inflammatory mediators. Cutting this vicious circle by HA clearance is essential for healing, and drugs targeting HA are expected. The HA matrix at the interface between endothelial, mesenchymal, and epithelial cells, and the invading leukocytes, is involved in multiple specific processes that take place in inflamed tissue.

Acknowledgments

The research of the authors has been financially supported by the Academy of Finland (M. I. T.), Sigrid Juselius Foundation (R. H. T., M. I. T.), Cancer Center of the University of Eastern Finland, and EVO funds of the Kuopio University Hospital (M. I. T.).

References

Ågren, U. M., Tammi, M., and Tammi, R. (1995). Hydrocortisone regulation of hyaluronan metabolism in human skin organ culture. J Cell Physiol *164*, 240–248.

Ågren, U. M., Tammi, R. H., and Tammi, M. I. (1997). Reactive oxygen species contribute to epidermal hyaluronan catabolism in human skin organ culture. Free Radic Biol Med *23*, 996–1001.

Aguiar, D. J., Knudson, W., and Knudson, C. B. (1999). Internalization of the hyaluronan receptor CD44 by chondrocytes. Exp Cell Res *252*, 292–302.

Ahrens, T., Assmann, V., Fieber, C., et al. (2001). CD44 is the principal mediator of hyaluronic-acid-induced melanoma cell proliferation. J Invest Dermatol *116*, 93–101.

Allavena, P., Sica, A., Solinas, G., Porta, C., and Mantovani, A. (2008). The inflammatory micro-environment in tumor progression: the role of tumor-associated macrophages. Crit Rev Oncol Hematol *66*, 1–9.

Armstrong, S. E., and Bell, D. R. (2002a). Ischemia-reperfusion does not cause significant hyaluronan depolymerization in skeletal muscle. Microvasc Res *64*, 353–362.

Armstrong, S. E., and Bell, D. R. (2002b). Measurement of high-molecular-weight hyaluronan in solid tissue using agarose gel electrophoresis. Anal Biochem *308*, 255–264.

Armstrong, S. E., and Bell, D.R . (2002c). Relationship between lymph and tissue hyaluronan in skin and skeletal muscle. Am J Physiol Heart Circ Physiol *283*, H2485–H2494.

Asari, A., Kanemitsu, T., and Kurihara, H. (2010). Oral administration of high molecular weight hyaluronan (900 kDa) controls immune system via Toll-like receptor 4 in the intestinal epithelium. J Biol Chem *285*, 24751–24758.

Bai, K. J., Spicer, A. P., Mascarenhas, M. M., et al. (2005). The role of hyaluronan synthase 3 in ventilator-induced lung injury. Am J Respir Crit Care Med *172*, 92–98.

Bakkers, J., Kramer, C., Pothof, J., Quaedvlieg, N. E., Spaink, H. P., and Hammerschmidt, M. (2004). Has2 is required upstream of Rac1 to govern dorsal migration of lateral cells during zebrafish gastrulation. Development *131*, 525–537.

Bandyopadhyay, S. K., de la Motte, C. A., Majors, A. K., and Strong, S. A. (2010). Inhibition of the phosphatidylinositol-3-kinase pathway abrogates polyinosinic:polycytidylic acid-stimulated hyaluronan-mediated human mucosal smooth muscle cell binding of U937 monocytic cells. J Interferon Cytokine Res *30*, 809–816.

Banerji, S., Ni, J., Wang, S. X., et al. (1999). LYVE-1, a new homologue of the CD44 glycoprotein, is a lymph-specific receptor for hyaluronan. J Cell Biol *144*, 789–801.

Baranova, N. S., Nileback, E., Haller, F. M., et al. (2011). The inflammation-associated protein TSG-6 cross-links hyaluronan via hyaluronan-induced TSG-6 oligomers. J Biol Chem *286* (29), 25675–25686.

Blass, S. L., Pure, E., and Hunter, C. A. (2001). A role for CD44 in the production of IFN-gamma and immunopathology during infection with Toxoplasma gondii. J Immunol *166*, 5726–5732.

Bollyky, P. L., Evanko, S. P., Wu, R. P., et al. (2010). Th1 cytokines promote T-cell binding to antigen-presenting cells via enhanced hyaluronan production and accumulation at the immune synapse. Cell Mol Immunol *7*, 211–220.

Bollyky, P. L., Lord, J. D., Masewicz, S. A., et al. (2007). Cutting edge: high molecular weight hyaluronan promotes the suppressive effects of CD4+CD25+ regulatory T cells. J Immunol *179*, 744–747.

Bollyky, P. L., Wu, R. P., Falk, B. A., et al. (2011). ECM components guide IL-10 producing regulatory T-cell (TR1) induction from effector memory T-cell precursors. Proc Natl Acad Sci U S A *108*, 7938–7943.

Bot, P. T., Hoefer, I. E., Piek, J. J., and Pasterkamp, G. (2008). Hyaluronic acid: targeting immune modulatory components of the extracellular matrix in atherosclerosis. Curr Med Chem *15*, 786–791.

Bourguignon, L. Y., Singleton, P. A., Diedrich, F., Stern, R., and Gilad, E. (2004). CD44 interaction with Na+-H+ exchanger (NHE1) creates acidic microenvironments leading to hyaluronidase-2 and cathepsin B activation and breast tumor cell invasion. J Biol Chem *279*, 26991–27007.

Bourguignon, L. Y., Zhu, H., Zhou, B., Diedrich, F., Singleton, P. A., and Hung, M. C. (2001). Hyaluronan promotes CD44v3-Vav2 interaction with Grb2-p185(HER2) and induces Rac1 and ras signaling during ovarian tumor cell migration and growth. J Biol Chem *276*, 48679–48692.

Bousquet, J., Chanez, P., Lacoste, J. Y., et al. (1991). Indirect evidence of bronchial inflammation assessed by titration of inflammatory mediators in BAL fluid of patients with asthma. J Allergy Clin Immunol *88*, 649–660.

Bracke, K. R., Dentener, M. A., Papakonstantinou, E., et al. (2010). Enhanced deposition of low-molecular-weight hyaluronan in lungs of cigarette smoke-exposed mice. Am J Respir Cell Mol Biol *42*, 753–761.

Brinck, J., and Heldin, P. (1999). Expression of recombinant hyaluronan synthase (HAS) isoforms in CHO cells reduces cell migration and cell surface CD44. Exp Cell Res *252*, 342–351.

Campo, G. M., Avenoso, A., Campo, S., D'Ascola, A., Nastasi, G., and Calatroni, A. (2010a). Small hyaluronan oligosaccharides induce inflammation by engaging both toll-like-4 and CD44 receptors in human chondrocytes. Biochem Pharmacol *80*, 480–490.

Campo, G. M., Avenoso, A., Campo, S., et al. (2010b). Differential effect of molecular mass hyaluronan on lipopolysaccharide-induced damage in chondrocytes. Innate Immun *16*, 48–63.

Casalino-Matsuda, S. M., Monzon, M. E., Day, A. J., and Forteza, R. M. (2009). Hyaluronan fragments/CD44 mediate oxidative stress-induced MUC5B up-regulation in airway epithelium. Am J Respir Cell Mol Biol *40*, 277–285.

Collins, S. L., Black, K. E., Chan-Li, Y., et al. (2011). Hyaluronan fragments promote inflammation by down-regulating the anti-inflammatory A2a receptor. Am J Respir Cell Mol Biol *45* (4), 675–683.

Culty, M., O'Mara, T. E., Underhill, C. B., Yeager, H., Jr., and Swartz, R. P. (1994a). Hyaluronan receptor (CD44) expression and function in human peripheral blood monocytes and alveolar macrophages. J Leukoc Biol *56*, 605–611.

Culty, M., Shizari, M., Thompson, E. W., and Underhill, C. B. (1994b). Binding and degradation of hyaluronan by human breast cancer cell lines expressing different forms of CD44: correlation with invasive potential. J Cell Physiol *160*, 275–286.

Dechert, T. A., Ducale, A. E., Ward, S. I., and Yager, D. R. (2006). Hyaluronan in human acute and chronic dermal wounds. Wound Repair Regen *14*, 252–258.

DeGrendele, H. C., Estess, P., and Siegelman, M. H. (1997). Requirement for CD44 in activated T cell extravasation into an inflammatory site. Science *278*, 672–675.

de la Motte, C., Nigro, J., Vasanji, A., et al. (2009). Platelet-derived hyaluronidase 2 cleaves hyaluronan into fragments that trigger monocyte-mediated production of proinflammatory cytokines. Am J Pathol *174*, 2254–2264.

de la Motte, C. A., Hascall, V. C., Calabro, A., Yen-Lieberman, B., and Strong, S. A. (1999). Mononuclear leukocytes preferentially bind via CD44 to hyaluronan on human intestinal mucosal smooth muscle cells after virus infection or treatment with poly(I.C). J Biol Chem *274*, 30747–30755.

de la Motte, C. A., Hascall, V. C., Drazba, J., Bandyopadhyay, S. K., and Strong, S. A. (2003). Mononuclear leukocytes bind to specific hyaluronan structures on colon mucosal smooth muscle cells treated with polyinosinic acid:polycytidylic acid: inter-alpha-trypsin inhibitor is crucial to structure and function. Am J Pathol *163*, 121–133.

del Fresno, C., Otero, K., Gomez-Garcia, L., et al. (2005). Tumor cells deactivate human monocytes by up-regulating IL-1 receptor associated kinase-M expression via CD44 and TLR4. J Immunol *174*, 3032–3040.

Dube, B., Luke, H. J., Aumailley, M., and Prehm, P. (2001). Hyaluronan reduces migration and proliferation in CHO cells. Biochim Biophys Acta *1538*, 283–289.

Duncan, M. R., and Berman, B. (1991). Stimulation of collagen and glycosaminoglycan production in cultured human adult dermal fibroblasts by recombinant human interleukin 6. J Invest Dermatol *97*, 686–692.

Duterme, C., Mertens-Strijthagen, J., Tammi, M., and Flamion, B. (2009). Two novel functions of hyaluronidase-2 (Hyal2) are formation of the glycocalyx and control of CD44-ERM interactions. J Biol Chem *284*, 33495–33508.

Ebong, E. E., Macaluso, F. P., Spray, D. C., and Tarbell, J. M. (2011). Imaging the endothelial glycocalyx in vitro by rapid freezing/freeze substitution transmission electron microscopy. Arterioscler Thromb Vasc Biol *31* (8), 1908–1915.

Engström-Laurent , A., and Hällgren, R. (1987). Circulating hyaluronic acid levels vary with physical activity in healthy subjects and in rheumatoid arthritis patients. Relationship to synovitis mass and morning stiffness. Arthritis Rheum *30*, 1333–1338.

Esnault, S., and Malter, J. S. (2003). Hyaluronic acid or TNF-alpha plus fibronectin triggers granulocyte macrophage-colony-stimulating factor mRNA stabilization in eosinophils yet engages differential intracellular pathways and mRNA binding proteins. J Immunol *171*, 6780–6787.

Estes, J. M., Adzick, N. S., Harrison, M. R., Longaker, M. T., and Stern, R. (1993). Hyaluronate metabolism undergoes an ontogenic transition during fetal development: implications for scar-free wound healing. J Pediatr Surg *28*, 1227–1231.

Estess, P., Nandi, A., Mohamadzadeh, M., and Siegelman, M. H. (1999). Interleukin 15 induces endothelial hyaluronan expression in vitro and promotes activated T cell extravasation through a CD44-dependent pathway in vivo. J Exp Med *190*, 9–19.

Evanko, S. P., Angello, J. C., and Wight, T. N. (1999). Formation of hyaluronan- and versican-rich pericellular matrix is required for proliferation and migration of vascular smooth muscle cells. Arterioscler Thromb Vasc Biol *19*, 1004–1013.

Evanko, S. P., Johnson, P. Y., Braun, K. R., Underhill, C. B., Dudhia, J., and Wight, T. N. (2001). Platelet-derived growth factor stimulates the formation of versican-hyaluronan aggregates and pericellular matrix expansion in arterial smooth muscle cells. Arch Biochem Biophys *394*, 29–38.

Evanko, S. P., Potter-Perigo, S., Johnson, P. Y., and Wight, T. N. (2009). Organization of hyaluronan and versican in the extracellular matrix of human fibroblasts treated with the viral mimetic poly I:C. J Histochem Cytochem *57*, 1041–1060.

Evanko, S. P., Tammi, M. I., Tammi, R. H., and Wight, T. N. (2007). Hyaluronan-dependent pericellular matrix. Adv Drug Deliv Rev *59*, 1351–1365.

Evanko, S. P., and Wight, T. N. (1999). Intracellular localization of hyaluronan in proliferating cells. J Histochem Cytochem *47*, 1331–1342.

Fraser, J. R., Alcorn, D., Laurent, T. C., Robinson, A. D., and Ryan, G. B. (1985). Uptake of circulating hyaluronic acid by the rat liver. Cellular localization in situ. Cell Tissue Res *242*, 505–510.

Fraser, J. R., Kimpton, W. G., Laurent, T. C., Cahill, R. N., and Vakakis, N. (1988). Uptake and degradation of hyaluronan in lymphatic tissue. Biochem J *256*, 153–158.

Fraser, J. R., Laurent, T. C., and Laurent, U. B. (1997). Hyaluronan: its nature, distribution, functions and turnover. J Intern Med *242*, 27–33.

Fraser, J. R., Laurent, T. C., Pertoft, H., and Baxter, E. (1981). Plasma clearance, tissue distribution and metabolism of hyaluronic acid injected intravenously in the rabbit. Biochem J *200*, 415–424.

Gao, F., Koenitzer, J. R., Tobolewski, J. M., et al. (2008). Extracellular superoxide dismutase inhibits inflammation by preventing oxidative fragmentation of hyaluronan. J Biol Chem *283*, 6058–6066.

Gao, L., and Lipowsky, H. H. (2010). Composition of the endothelial glycocalyx and its relation to its thickness and diffusion of small solutes. Microvasc Res *80*, 394–401.

Garantziotis, S., Li, Z., Potts, E. N., et al. (2009). Hyaluronan mediates ozone-induced airway hyperresponsiveness in mice. J Biol Chem *284*, 11309–11317.

Gardell, S., Baker, J., Caterson, B., Heinegard, D., and Roden, L. (1980). Link protein and a hyaluronic acid-binding region as components of aorta proteoglycan. Biochem Biophys Res Commun *95*, 1823–1831.

Gariboldi, S., Palazzo, M., Zanobbio, L., et al. (2008). Low molecular weight hyaluronic acid increases the self-defense of skin epithelium by induction of beta-defensin 2 via TLR2 and TLR4. J Immunol *181*, 2103–2110.

Ghatak, S., Misra, S., and Toole, B. P. (2005). Hyaluronan constitutively regulates ErbB2 phosphorylation and signaling complex formation in carcinoma cells. J Biol Chem *280*, 8875–8883.

Gigante, A., and Callegari, L. (2011). The role of intra-articular hyaluronan (sinovial) in the treatment of osteoarthritis. Rheumatol Int *31*, 427–444.
Goldberg, R. L., and Toole, B. P. (1987). Hyaluronate inhibition of cell proliferation. Arthritis Rheum *30*, 769–778.
Gomes, J. A., Amankwah, R., Powell-Richards, A., and Dua, H. S. (2004). Sodium hyaluronate (hyaluronic acid) promotes migration of human corneal epithelial cells in vitro. Br J Ophthalmol *88*, 821–825.
Greyner, H. J., Wiraszka, T., Zhang, L. S., Petroll, W. M., and Mummert, M. E. (2010). Inducible macropinocytosis of hyaluronan in B16-F10 melanoma cells. Matrix Biol *29*, 503–510.
Guo, N., Li, X., Mann, M. M., Funderburgh, M. L., Du, Y., and Funderburgh, J. L. (2010). Hyaluronan synthesis mediates the fibrotic response of keratocytes to transforming growth factor beta. J Biol Chem *285*, 32012–32019.
Haider, A. S., Grabarek, J., Eng, B., et al. (2003). In vitro model of "wound healing" analyzed by laser scanning cytometry: accelerated healing of epithelial cell monolayers in the presence of hyaluronate. Cytometry A *53*, 1–8.
Hall, C. L., Wang, C., Lange, L. A., and Turley, E. A. (1994). Hyaluronan and the hyaluronan receptor RHAMM promote focal adhesion turnover and transient tyrosine kinase activity. J Cell Biol *126*, 575–588.
Han, C. Y., Subramanian, S., Chan, C. K., et al. (2007). Adipocyte-derived serum amyloid A3 and hyaluronan play a role in monocyte recruitment and adhesion. Diabetes *56*, 2260–2273.
Hardingham, T. E., and Muir, H. (1972). The specific interaction of hyaluronic acid with cartillage proteoglycans. Biochim Biophys Acta *279*, 401–405.
Harris, E. N., Kyosseva, S. V., Weigel, J. A., and Weigel, P. H. (2007). Expression, processing, and glycosaminoglycan binding activity of the recombinant human 315-kDa hyaluronic acid receptor for endocytosis (HARE). J Biol Chem *282*, 2785–2797.
Harris, E. N., Weigel, J. A., and Weigel, P. H. (2008). The human hyaluronan receptor for endocytosis (HARE/Stabilin-2) is a systemic clearance receptor for heparin. J Biol Chem *283*, 17341–17350.
Hascall, V. C., and Heinegård, D. (1974). Aggregation of cartilage proteoglycans. I. the role of hyaluronic acid. J Biol Chem *249*, 4232–4241.
Heldin, P., Asplund, T., Ytterberg, D., Thelin, S., and Laurent, T. C. (1992). Characterization of the molecular mechanism involved in the activation of hyaluronan synthetase by platelet-derived growth factor in human mesothelial cells. Biochem J *283* (Pt 1), 165–170.
Heldin, P., Laurent, T. C., and Heldin, C. H. (1989). Effect of growth factors on hyaluronan synthesis in cultured human fibroblasts. Biochem J *258*, 919–922.
Hernandez, D., Miquel-Serra, L., Docampo, M. J., Marco-Ramell, A., and Bassols, A. (2011a). Role of versican V0/V1 and CD44 in the regulation of human melanoma cell behavior. Int J Mol Med *27*, 269–275.
Hernandez, D., Miquel-Serra, L., Docampo, M. J., et al. (2011b). V3 versican isoform alters the behavior of human melanoma cells by interfering with CD44/ErbB-dependent signaling. J Biol Chem *286*, 1475–1485.
Hodge-Dufour, J., Noble, P. W., Horton, M. R., et al. (1997). Induction of IL-12 and chemokines by hyaluronan requires adhesion-dependent priming of resident but not elicited macrophages. J Immunol *159*, 2492–2500.
Honda, A., Noguchi, N., Takehara, H., Ohashi, Y., Asuwa, N., and Mori, Y. (1991). Cooperative enhancement of hyaluronic acid synthesis by combined use of IGF-I and EGF, and inhibition by tyrosine kinase inhibitor genistein, in cultured mesothelial cells from rabbit pericardial cavity. J Cell Sci *98* (Pt 1), 91–98.

Horton, M. R., McKee, C. M., Bao, C., et al. (1998). Hyaluronan fragments synergize with interferon-gamma to induce the C-X-C chemokines mig and interferon-inducible protein-10 in mouse macrophages. J Biol Chem *273*, 35088–35094.

Horton, M. R., Olman, M. A., Bao, C., et al. (2000). Regulation of plasminogen activator inhibitor-1 and urokinase by hyaluronan fragments in mouse macrophages. Am J Physiol Lung Cell Mol Physiol *279*, L707–L715.

Horton, M. R., Shapiro, S., Bao, C., Lowenstein, C. J., and Noble, P. W. (1999). Induction and regulation of macrophage metalloelastase by hyaluronan fragments in mouse macrophages. J Immunol *162*, 4171–4176.

Hyc, A., Osiecka-Iwan, A., Niderla-Bielinska, J., Jankowska-Steifer, E., and Moskalewski, S. (2009). Pro- and anti-inflammatory cytokines increase hyaluronan production by rat synovial membrane in vitro. Int J Mol Med *24*, 579–585.

Iwata, Y., Yoshizaki, A., Komura, K., et al. (2009). CD19, a response regulator of B lymphocytes, regulates wound healing through hyaluronan-induced TLR4 signaling. Am J Pathol *175*, 649–660.

Jacobson, A., Brinck, J., Briskin, M. J., Spicer, A. P., and Heldin, P. (2000). Expression of human hyaluronan synthases in response to external stimuli. Biochem J *348* (Pt 1), 29–35.

Jadin, L., Wu, X., Ding, H., et al. (2008). Skeletal and hematological anomalies in HYAL2-deficient mice: a second type of mucopolysaccharidosis IX? FASEB J *22*, 4316–4326.

Jameson, J. M., Cauvi, G., Sharp, L. L., Witherden, D. A., and Havran, W. L. (2005). Gammadelta T cell-induced hyaluronan production by epithelial cells regulates inflammation. J Exp Med *201*, 1269–1279.

Jiang, D., Liang, J., Fan, J., et al. (2005). Regulation of lung injury and repair by Toll-like receptors and hyaluronan. Nat Med *11*, 1173–1179.

Jiang, D., Liang, J., and Noble, P. W. (2011). Hyaluronan as an immune regulator in human diseases. Physiol Rev *91*, 221–264.

Jiang, H., Peterson, R. S., Wang, W., Bartnik, E., Knudson, C. B., and Knudson, W. (2002). A requirement for the CD44 cytoplasmic domain for hyaluronan binding, pericellular matrix assembly, and receptor-mediated endocytosis in COS-7 cells. J Biol Chem *277*, 10531–10538.

Jokela, T. A., Jauhiainen, M., Auriola, S., et al. (2008a). Mannose inhibits hyaluronan synthesis by down-regulation of the cellular pool of UDP-N-acetylhexosamines. J Biol Chem *283*, 7666–7673.

Jokela, T. A., Lindgren, A., Rilla, K., et al. (2008b). Induction of hyaluronan cables and monocyte adherence in epidermal keratinocytes. Connect Tissue Res *49* (3), 115–119.

Karvinen, S., Pasonen-Seppänen, S., Hyttinen, J. M., et al. (2003). Keratinocyte growth factor stimulates migration and hyaluronan synthesis in the epidermis by activation of keratinocyte hyaluronan synthases 2 and 3. J Biol Chem *278*, 49495–49504.

Kawakami, M., Suzuki, K., Matsuki, Y., et al. (1998). Hyaluronan production in human rheumatoid fibroblastic synovial lining cells is increased by interleukin 1 beta but inhibited by transforming growth factor beta 1. Ann Rheum Dis *57*, 602–605.

Kennedy, C. I., Diegelmann, R. F., Haynes, J. H., and Yager, D. R. (2000). Proinflammatory cytokines differentially regulate hyaluronan synthase isoforms in fetal and adult fibroblasts. J Pediatr Surg *35*, 874–879.

Kessler, S., Rho, H., West, G., Fiocchi, C., Drazba, J., and de la Motte, C. (2008). Hyaluronan (HA) deposition precedes and promotes leukocyte recruitment in intestinal inflammation. Clin Transl Sci *1*, 57–61.

Kim, Y., Lee, Y. S., Choe, J., Lee, H., Kim, Y. M., and Jeoung, D. (2008). CD44-epidermal growth factor receptor interaction mediates hyaluronic acid-promoted cell motility by activating protein kinase C signaling involving akt, Rac1, phox, reactive oxygen species, focal adhesion kinase, and MMP-2. J Biol Chem *283*, 22513–22528.

Kobayashi, N., Miyoshi, S., Mikami, T., et al. (2010). Hyaluronan deficiency in tumor stroma impairs macrophage trafficking and tumor neovascularization. Cancer Res *70*, 7073–7083.

Kössi, J., Peltonen, J., Ekfors, T., Niinikoski, J., and Laato, M. (1999). Effects of hexose sugars: glucose, fructose, galactose and mannose on wound healing in the rat. Eur Surg Res *31*, 74–82.

Kultti, A. H., Pasonen-Seppänen, S., Jauhiainen, M., Rilla, K. J., Kärnä, R., Pyöriä, E., Tammi, R. H., and Tammi, M. I. (2009). 4-methylumbelliferone inhibits hyaluronan synthesis by depletion of cellular UDP-glucuronic acid and downregulation of hyaluronan synthase 2 and 3. Exp Cell Res *315*, 1914–1923.

Kuroda, K., Utani, A., Hamasaki, Y., and Shinkai, H. (2001). Up-regulation of putative hyaluronan synthase mRNA by basic fibroblast growth factor and insulin-like growth factor-1 in human skin fibroblasts. J Dermatol Sci *26*, 156–160.

Lauer, M. E., Fulop, C., Mukhopadhyay, D., Comhair, S., Erzurum, S. C., and Hascall, V. C. (2009). Airway smooth muscle cells synthesize hyaluronan cable structures independent of inter-alpha-inhibitor heavy chain attachment. J Biol Chem *284*, 5313–5323.

Laurent, C., Soderberg, O., Anniko, M., and Hartwig, S. (1991). Repair of chronic tympanic membrane perforations using applications of hyaluronan or rice paper prostheses. ORL J Otorhinolaryngol Relat Spec *53*, 37–40.

Laurent, T. C. (1964). The interaction between polysaccharides and other macromolecules. 9. The exclusion of molecules from hyaluronic acid gels and solutions. Biochem J *93*, 106–112.

Laurent, T. C., Laurent, U. B., and Fraser, J. R. (1996). Serum hyaluronan as a disease marker. Ann Med *28*, 241–253.

Laurent, U. B., Dahl, L. B., and Reed, R. K. (1991). Catabolism of hyaluronan in rabbit skin takes place locally, in lymph nodes and liver. Exp Physiol *76*, 695–703.

Laurent, U. B., Fraser, J. R., Engstrom-Laurent, A., Reed, R. K., Dahl, L. B., and Laurent, T. C. (1992). Catabolism of hyaluronan in the knee joint of the rabbit. Matrix *12*, 130–136.

Lefebvre, D. C., Lai, J. C., Maeshima, N., et al. (2010). CD44 interacts directly with Lck in a zinc-dependent manner. Mol Immunol *47*, 1882–1889.

Levesque, H., Girard, N., Maingonnat, C., et al. (1994). Localization and solubilization of hyaluronan and of the hyaluronan-binding protein hyaluronectin in human normal and arteriosclerotic arterial walls. Atherosclerosis *105*, 51–62.

Li, L., Heldin, C. H., and Heldin, P. (2006). Inhibition of platelet-derived growth factor-BB-induced receptor activation and fibroblast migration by hyaluronan activation of CD44. J Biol Chem *281*, 26512–26519.

Li, Y., Li, L., Brown, T. J., and Heldin, P. (2007). Silencing of hyaluronan synthase 2 suppresses the malignant phenotype of invasive breast cancer cells. Int J Cancer *120*, 2557–2567.

Liang, J., Jiang, D., Jung, Y., et al. (2011). Role of hyaluronan and hyaluronan-binding proteins in human asthma. J Allergy Clin Immunol *128* (2), 403–411.

Liu, N., Gao, F., Han, Z., Xu, X., Underhill, C. B., and Zhang, L. (2001). Hyaluronan synthase 3 overexpression promotes the growth of TSU prostate cancer cells. Cancer Res *61*, 5207–5214.

Liu, Y. Y., Lee, C. H., Dedaj, R., et al. (2008). High-molecular-weight hyaluronan – a possible new treatment for sepsis-induced lung injury: a preclinical study in mechanically ventilated rats. Crit Care *12*, R102.

Lokeshwar, V. B., and Selzer, M. G. (2000). Differences in hyaluronic acid-mediated functions and signaling in arterial, microvessel, and vein-derived human endothelial cells. J Biol Chem *275*, 27641–27649.

Lovvorn, H. N., III, Cass, D. L., Sylvester, K. G., et al. (1998). Hyaluronan receptor expression increases in fetal excisional skin wounds and correlates with fibroplasia. J Pediatr Surg *33*, 1062–1070.

Mack, J. A., Abramson, S. R., Ben, Y., et al. (2003). Hoxb13 knockout adult skin exhibits high levels of hyaluronan and enhanced wound healing. FASEB J *17*, 1352–1354.

Mack, J. A., Feldman, R. J., Itano. N., et al. (2012). Enhanced inflammation and accelerated wound closure following tetraphorbol ester application or full-thickness wounding in mice lacking hyaluronan synthases Has1 and Has3. J Invest Dermatol. *132*, 198–207.

Maeshima, N., Poon, G. F., Dosanjh, M., et al. (2011). Hyaluronan binding identifies the most proliferative activated and memory T cells. Eur J Immunol *41*, 1108–1119.

Majors, A. K., Austin, R. C., de la Motte, C. A., et al. (2003). Endoplasmic reticulum stress induces hyaluronan deposition and leukocyte adhesion. J Biol Chem *278*, 47223–47231.

Mast, B. A., Flood, L. C., Haynes, J. H., et al. (1991). Hyaluronic acid is a major component of the matrix of fetal rabbit skin and wounds: implications for healing by regeneration. Matrix *11*, 63–68.

Matou-Nasri, S., Gaffney, J., Kumar, S., and Slevin, M. (2009). Oligosaccharides of hyaluronan induce angiogenesis through distinct CD44 and RHAMM-mediated signalling pathways involving Cdc2 and gamma-adducin. Int J Oncol *35*, 761–773.

Maytin, E. V., Chung, H. H., and Seetharaman, V. M. (2004). Hyaluronan participates in the epidermal response to disruption of the permeability barrier in vivo. Am J Pathol *165*, 1331–1341.

McDonald, B., McAvoy, E. F., Lam, F., et al. (2008). Interaction of CD44 and hyaluronan is the dominant mechanism for neutrophil sequestration in inflamed liver sinusoids. J Exp Med *205*, 915–927.

McGary, C. T., Raja, R. H., and Weigel, P. H. (1989). Endocytosis of hyaluronic acid by rat liver endothelial cells. Evidence for receptor recycling. Biochem J *257*, 875–884.

McGuire, P. G., Castellot, J. J., Jr., and Orkin, R. W. (1987). Size-dependent hyaluronate degradation by cultured cells. J Cell Physiol *133*, 267–276.

McKee, C. M., Penno, M. B., Cowman, M., et al. (1996). Hyaluronan (HA) fragments induce chemokine gene expression in alveolar macrophages. The role of HA size and CD44. J Clin Invest *98*, 2403–2413.

Meran, S., Luo, D. D., Simpson, R., et al. (2011). Hyaluronan facilitates transforming growth factor-{beta}1-dependent proliferation via CD44 and epidermal growth factor receptor interaction. J Biol Chem *286*, 17618–17630.

Meran, S., Thomas, D., Stephens, P., et al. (2007). Involvement of hyaluronan in regulation of fibroblast phenotype. J Biol Chem *282*, 25687–25697.

Meyer, K., and Palmer, J. W. (1934). The polysaccharide of the vitreous humor. J Biol Chem *107*, 629–634.

Mitsui, Y., Gotoh, M., Nakama, K., Yamada, T., Higuchi, F., and Nagata, K. (2008). Hyaluronic acid inhibits mRNA expression of proinflammatory cytokines and cyclooxygenase-2/prostaglandin E(2) production via CD44 in interleukin-1-stimulated subacromial synovial fibroblasts from patients with rotator cuff disease. J Orthop Res *26*, 1032–1037.

Monslow, J., Sato, N., Mack, J. A., and Maytin, E. V. (2009). Wounding-induced synthesis of hyaluronic acid in organotypic epidermal cultures requires the release of heparin-binding egf and activation of the EGFR. J Invest Dermatol *129*, 2046–2058.

Monzon, M. E., Fregien, N., Schmid, N., et al. (2010). Reactive oxygen species and hyaluronidase 2 regulate airway epithelial hyaluronan fragmentation. J Biol Chem *285*, 26126–26134.

Nagy, N., Freudenberger, T., Melchior-Becker, A., et al. (2010). Inhibition of hyaluronan synthesis accelerates murine atherosclerosis: novel insights into the role of hyaluronan synthesis. Circulation *122*, 2313–2322.

Nishida, Y., Knudson, W., Knudson, C. B., and Ishiguro, N. (2005). Antisense inhibition of hyaluronan synthase-2 in human osteosarcoma cells inhibits hyaluronan retention and tumorigenicity. Exp Cell Res *307*, 194–203.

Noble, P. W. (2002). Hyaluronan and its catabolic products in tissue injury and repair. Matrix Biol *21*, 25–29.

Noble, P. W., Lake, F. R., Henson, P. M., and Riches, D. W. (1993). Hyaluronate activation of CD44 induces insulin-like growth factor-1 expression by a tumor necrosis factor-alpha-dependent mechanism in murine macrophages. J Clin Invest *91*, 2368–2377.

Noble, P. W., McKee, C. M., Cowman, M., and Shin, H. S. (1996). Hyaluronan fragments activate an NF-kappa B/I-kappa B alpha autoregulatory loop in murine macrophages. J Exp Med *183*, 2373–2378.

Oguchi, T., and Ishiguro, N. (2004). Differential stimulation of three forms of hyaluronan synthase by TGF-beta, IL-1beta, and TNF-alpha. Connect Tissue Res *45*, 197–205.

Ohkawara, Y., Tamura, G., Iwasaki, T., Tanaka, A., Kikuchi, T., and Shirato, K. (2000). Activation and transforming growth factor-beta production in eosinophils by hyaluronan. Am J Respir Cell Mol Biol *23*, 444–451.

Ohtani, T., Memezawa, A., Okuyama, R., et al. (2009). Increased hyaluronan production and decreased E-cadherin expression by cytokine-stimulated keratinocytes lead to spongiosis formation. J Invest Dermatol *129*, 1412–1420.

Pasonen-Seppänen, S., Karvinen, S., Törrönen, K., et al. (2003). EGF upregulates, whereas TGF-beta downregulates, the hyaluronan synthases Has2 and Has3 in organotypic keratinocyte cultures: correlations with epidermal proliferation and differentiation. J Invest Dermatol *120*, 1038–1044.

Pasonen- Seppänen, S. M., Maytin, E. V., Törrönen, K. J., et al. (2008). All-trans retinoic acid-induced hyaluronan production and hyperplasia are partly mediated by EGFR signaling in epidermal keratinocytes. J Invest Dermatol *128*, 797–807.

Pienimäki, J. P., Rilla, K., Fulop, C., et al. (2001). Epidermal growth factor activates hyaluronan synthase 2 in epidermal keratinocytes and increases pericellular and intracellular hyaluronan. J Biol Chem *276*, 20428–20435.

Rafi, A., Nagarkatti, M., and Nagarkatti, P. S. (1997). Hyaluronate-CD44 interactions can induce murine B-cell activation. Blood *89*, 2901–2908.

Reijnen, M. M., Falk, P., van Goor, H., and Holmdahl, L. (2000). The antiadhesive agent sodium hyaluronate increases the proliferation rate of human peritoneal mesothelial cells. Fertil Steril *74*, 146–151.

Reitsma, S., Oude Egbrink, M. G., Vink, H., et al. (2011). Endothelial glycocalyx structure in the intact carotid artery: a two-photon laser scanning microscopy study. J Vasc Res *48*, 297–306.

Rilla, K., Pasonen-Seppänen, S., Rieppo, J., Tammi, M., and Tammi, R. (2004). The hyaluronan synthesis inhibitor 4-methylumbelliferone prevents keratinocyte activation and epidermal hyperproliferation induced by epidermal growth factor. J Invest Dermatol *123*, 708–714.

Ruffell, B., Poon, G. F., Lee, S. S., et al. (2011). Differential use of chondroitin sulfate to regulate hyaluronan binding by CD44 in inflammatory and interleukin 4 (IL-4) activated macrophages. J Biol Chem *286* (22), 19179–90.

Sabaratnam, S., Arunan, V., Coleman, P. J., Mason, R. M., and Levick, J. R. (2005). Size selectivity of hyaluronan molecular sieving by extracellular matrix in rabbit synovial joints. J Physiol *567*, 569–581.

Sainio, A., Jokela, T., Tammi, M. I., and Jarvelainen, H. (2010). Hyperglycemic conditions modulate connective tissue reorganization by human vascular smooth muscle cells through stimulation of hyaluronan synthesis. Glycobiology *20*, 1117–1126.

Sampson, P. M., Rochester, C. L., Freundlich, B., and Elias, J. A. (1992). Cytokine regulation of human lung fibroblast hyaluronan (hyaluronic acid) production. Evidence for cytokine-regulated hyaluronan (hyaluronic acid) degradation and human lung fibroblast-derived hyaluronidase. J Clin Invest *90*, 1492–1503.

Sayo, T., Sugiyama, Y., Takahashi, Y., et al. (2002). Hyaluronan synthase 3 regulates hyaluronan synthesis in cultured human keratinocytes. J Invest Dermatol *118*, 43–48.

Scheibner, K.A ., Boodoo, S., Collins, S., et al. (2009). The adenosine a2a receptor inhibits matrix-induced inflammation in a novel fashion. Am J Respir Cell Mol Biol *40*, 251–259.

Schimizzi, A. L., Massie, J. B., Murphy, M., et al. (2006). High-molecular-weight hyaluronan inhibits macrophage proliferation and cytokine release in the early wound of a preclinical postlaminectomy rat model. Spine J *6*, 550–556.

Seike, M., Ikeda, M., Matsumoto, M., Hamada, R., Takeya, M., and Kodama, H. (2006). Hyaluronan forms complexes with low density lipoprotein while also inducing foam cell infiltration in the dermis. J Dermatol Sci *41*, 197–204.

Selbi, W., de la Motte, C. A., Hascall, V. C., Day, A. J., Bowen, T., and Phillips, A. O. (2006). Characterization of hyaluronan cable structure and function in renal proximal tubular epithelial cells. Kidney Int *70*, 1287–1295.

Shi, T., Duan, Z. H., Papay, R., et al. (2006). Novel alpha1-adrenergic receptor signaling pathways: secreted factors and interactions with the extracellular matrix. Mol Pharmacol *70*, 129–142.

Shimada, M., Yanai, Y., Okazaki, T., et al. (2008). Hyaluronan fragments generated by sperm-secreted hyaluronidase stimulate cytokine/chemokine production via the TLR2 and TLR4 pathway in cumulus cells of ovulated COCs, which may enhance fertilization. Development *135*, 2001–2011.

Singleton, P. A., Mirzapoiazova, T., Guo, Y., et al. (2010). High-molecular-weight hyaluronan is a novel inhibitor of pulmonary vascular leakiness. Am J Physiol Lung Cell Mol Physiol *299*, L639–L651.

Slevin, M., Krupinski, J., Gaffney, J., et al. (2007). Hyaluronan-mediated angiogenesis in vascular disease: uncovering RHAMM and CD44 receptor signaling pathways. Matrix Biol *26*, 58–68.

Stern, R. (2004). Hyaluronan catabolism: a new metabolic pathway. Eur J Cell Biol *83*, 317–325.

Stern, R., Asari, A. A., and Sugahara, K. N. (2006). Hyaluronan fragments: an information-rich system. Eur J Cell Biol *85*, 699–715.

Sugiyama, Y., Shimada, A., Sayo, T., Sakai, S., and Inoue, S. (1998). Putative hyaluronan synthase mRNA are expressed in mouse skin and TGF-beta upregulates their expression in cultured human skin cells. J Invest Dermatol *110*, 116–121.

Suzuki, K., Yamamoto, T., Usui, T., Suzuki, K., Heldin, P., and Yamashita, H. (2003). Expression of hyaluronan synthase in intraocular proliferative diseases: regulation of expression in human vascular endothelial cells by transforming growth factor-beta. Jpn J Ophthalmol *47*, 557–564.

Takahashi, Y., Li, L., Kamiryo, M., et al. (2005). Hyaluronan fragments induce endothelial cell differentiation in a CD44- and CXCL1/GRO1-dependent manner. J Biol Chem *280*, 24195–24204.

Tammi, R., Pasonen-Seppänen, S., Kolehmainen, E., and Tammi, M. (2005). Hyaluronan synthase induction and hyaluronan accumulation in mouse epidermis following skin injury. J Invest Dermatol *124*, 898–905.

Tammi, R., Paukkonen, K., Wang, C., Horsmanheimo, M., and Tammi, M. (1994). Hyaluronan and CD44 in psoriatic skin. intense staining for hyaluronan on dermal capillary loops and reduced expression of CD44 and hyaluronan in keratinocyte-leukocyte interfaces. Arch Dermatol Res *286*, 21–29.

Tammi, R., Rilla, K., Pienimaki, J. P., et al. (2001). Hyaluronan enters keratinocytes by a novel endocytic route for catabolism. J Biol Chem *276*, 35111–35122.

Tammi, R., Ripellino, J. A., Margolis, R. U., and Tammi, M. (1988). Localization of epidermal hyaluronic acid using the hyaluronate binding region of cartilage proteoglycan as a specific probe. J Invest Dermatol *90*, 412–414.

Tammi, R., Saamanen, A. M., Maibach, H. I., and Tammi, M. (1991). Degradation of newly synthesized high molecular mass hyaluronan in the epidermal and dermal compartments of human skin in organ culture. J Invest Dermatol *97*, 126–130.

Taylor, K.R ., Yamasaki, K., Radek, K. A., et al. (2007). Recognition of hyaluronan released in sterile injury involves a unique receptor complex dependent on Toll-like receptor 4, CD44, and MD-2. J Biol Chem *282*, 18265–18275.

Teder, P., Vandivier, R. W., Jiang, D., et al. (2002). Resolution of lung inflammation by CD44. Science *296*, 155–158.

Thankamony, S. P., and Knudson, W. (2006). Acylation of CD44 and its association with lipid rafts are required for receptor and hyaluronan endocytosis. J Biol Chem *281*, 34601–34609.

Thompson, C. B., Shepard, H. M., O'Connor, P. M., et al. (2010). Enzymatic depletion of tumor hyaluronan induces antitumor responses in preclinical animal models. Mol Cancer Ther *9*, 3052–3064.

Thorne, R. F., Legg, J. W., and Isacke, C. M. (2004). The role of the CD44 transmembrane and cytoplasmic domains in co-ordinating adhesive and signalling events. J Cell Sci *117*, 373–380.

Toole, B. P. (1997). Hyaluronan in morphogenesis. J Intern Med *242*, 35–40.

Twarock, S., Tammi, M. I., Savani, R. C., and Fischer, J. W. (2010). Hyaluronan stabilizes focal adhesions, filopodia, and the proliferative phenotype in esophageal squamous carcinoma cells. J Biol Chem *285*, 23276–23284.

Udabage, L., Brownlee, G. R., Waltham, M., et al. (2005). Antisense-mediated suppression of hyaluronan synthase 2 inhibits the tumorigenesis and progression of breast cancer. Cancer Res *65*, 6139–6150.

Underhill, C. B., Green, S. J., Comoglio, P. M., and Tarone, G. (1987). The hyaluronate receptor is identical to a glycoprotein of mr 85,000 (gp85) as shown by a monoclonal antibody that interferes with binding activity. J Biol Chem *262*, 13142–13146.

Underhill, C. B., Nguyen, H. A., Shizari, M., and Culty, M. (1993). CD44 positive macrophages take up hyaluronan during lung development. Dev Biol *155*, 324–336.

Vigetti, D., Genasetti, A., Karousou, E., et al. (2010). Proinflammatory cytokines induce hyaluronan synthesis and monocyte adhesion in human endothelial cells through hyaluronan synthase 2 (HAS2) and the nuclear factor-kappaB (NF-kappaB) pathway. J Biol Chem *285*, 24639–24645.

Vigetti, D., Viola, M., Karousou, E., et al. (2008). Hyaluronan-CD44-ERK1/2 regulate human aortic smooth muscle cell motility during aging. J Biol Chem *283*, 4448–4458.

Wallet, M. A., Wallet, S. M., Guiulfo, G., Sleasman, J. W., and Goodenow, M. M. (2010). IFNgamma primes macrophages for inflammatory activation by high molecular weight hyaluronan. Cell Immunol *262*, 84–88.

Wang, A., and Hascall, V. C. (2004). Hyaluronan structures synthesized by rat mesangial cells in response to hyperglycemia induce monocyte adhesion. J Biol Chem *279*, 10279–10285.

Wang, C., Tammi, M., and Tammi, R. (1992). Distribution of hyaluronan and its CD44 receptor in the epithelia of human skin appendages. Histochemistry *98*, 105–112.

Webber, J., Meran, S., Steadman, R., and Phillips, A. (2009). Hyaluronan orchestrates transforming growth factor-beta1-dependent maintenance of myofibroblast phenotype. J Biol Chem *284*, 9083–9092.

Weigel, P. H., Frost, S. J., LeBoeuf, R. D., and McGary, C. T. (1989). The specific interaction between fibrin(ogen) and hyaluronan: possible consequences in haemostasis, inflammation and wound healing. Ciba Found Symp *143*, 248–264, 281–285.

West, D. C., and Kumar, S. (1989). The effect of hyaluronate and its oligosaccharides on endothelial cell proliferation and monolayer integrity. Exp Cell Res *183*, 179–196.

Wight, T. N. (2008). Arterial remodeling in vascular disease: a key role for hyaluronan and versican. Front Biosci *13*, 4933–4937.

Wilkinson, T. S., Bressler, S. L., Evanko, S. P., Braun, K. R., and Wight, T. N. (2006). Overexpression of hyaluronan synthases alters vascular smooth muscle cell phenotype and promotes monocyte adhesion. J Cell Physiol *206*, 378–385.

Yamada, Y., Itano, N., Hata, K., Ueda, M., and Kimata, K. (2004). Differential regulation by IL-1beta and EGF of expression of three different hyaluronan synthases in oral mucosal epithelial cells and fibroblasts and dermal fibroblasts: quantitative analysis using real-time RT-PCR. J Invest Dermatol *122*, 631–639.

Yamawaki, H., Hirohata, S., Miyoshi, T., et al. (2009). Hyaluronan receptors involved in cytokine induction in monocytes. Glycobiology *19*, 83–92.

Yasuda, T. (2010). Hyaluronan inhibits prostaglandin E2 production via CD44 in U937 human macrophages. Tohoku J Exp Med *220*, 229–235.

Yeop Han, C., Kargi, A. Y., Omer, M., et al. (2010). Differential effect of saturated and unsaturated free fatty acids on the generation of monocyte adhesion and chemotactic factors by adipocytes: dissociation of adipocyte hypertrophy from inflammation. Diabetes *59*, 386–396.

Yoneda, M., Suzuki, S., and Kimata, K. (1990). Hyaluronic acid associated with the surfaces of cultured fibroblasts is linked to a serum-derived 85-kDa protein. J Biol Chem *265*, 5247–5257.

Yoneda, M., Yamagata, M., Suzuki, S., and Kimata, K. (1988). Hyaluronic acid modulates proliferation of mouse dermal fibroblasts in culture. J Cell Sci *90* (Pt 2), 265–273.

Yu, H., Li, Q., Zhou, X., Kolosov, V. P., and Perelman, J. M. (2011). Role of hyaluronan and CD44 in reactive oxygen species-induced mucus hypersecretion. Mol Cell Biochem *352*, 65–75.

Zhang, X. L., Selbi, W., de la Motte, C., Hascall, V., and Phillips, A. O. (2005). Bone morphogenic protein-7 inhibits monocyte-stimulated TGF-beta1 generation in renal proximal tubular epithelial cells. J Am Soc Nephrol *16*, 79–89.

Zhao, H., Leu, S. W., Shi, L., et al. (2010). TLR4 is a negative regulator in noninfectious lung inflammation. J Immunol *184*, 5308–5314.

Zheng, L., Riehl, T. E., and Stenson, W. F. (2009). Regulation of colonic epithelial repair in mice by Toll-like receptors and hyaluronic acid. Gastroenterology *137*, 2041–2051.

2.4 Roles of sulfated and nonsulfated glycosaminoglycans in cancer growth and progression-therapeutic implications

Dragana Nikitovic and George N. Tzanakakis

2.4.1 Introduction

Cancer is a leading cause of mortality worldwide, with up to 90% of cancer-related deaths being due to metastatic dissemination (Fidler and Kripke, 2003). The consecutive steps of tumor growth, local invasion, intravasation, extravasation, and invasion of anatomically distant sites are obligatorily perpetrated through specific interactions of the tumor cells with their microenvironment (Chaffer and Weinberg, 2011; Hanahan and Weinberg, 2011; Karamanos and Tzanakakis, 2012).

Glycosaminoglycans (GAGs) comprise a class of linear, negatively charged polysaccharides composed of repeating disaccharide units of acetylated hexosamines (*N*-acetyl-galactosamine in the case of chondroitin sulfate [CS] and dermatan sulfate [DS]or *N*-acetyl-glucosamine [GlcNAc] in the case of heparin sulfate [HS]and heparin) and mainly of uronic acids (D-glucuronic acid or L-iduronic acid) being sulfated at various positions. The exception constitutes keratan sulfate (KS), in which galactose instead of uronic acid is present. Based on the epimeric form of uronic acid and the type of hexosamine in their repeating disaccharide units, GAGs are classified into four major types: hyaluronan (HA), CS and DS, heparin and HS, and KS. HA is synthesized in the absence of a protein core at the inner face of the plasma membrane and is consequently found in the form of free chains, whereas other GAG types are covalently bound into protein cores to form proteoglycans (PGs). With the exception of HA, all GAG types are variably sulfated, which contributes to the intricate complexity of their structures. Free GAG chains are secreted to the extracellular space and distributed both in the pericellular matrix and extracellular matrix (ECM) proper. GAGs bound into PGs are located to the ECM, basal membrane, and cell surface (Antonio and Iozzo, 2001).

Free GAGs and proteoglycan-containing GAGs, key effectors of cell surface, pericellular, and extracellular microenvironments, perform multiple functions in cancer by virtue of their coded structure and their ability to interact with both ligands and receptors that regulate neoplastic growth and dissemination (Iozzo and Sanderson, 2011; Theocharis et al., 2011). Cell type and tissue-specific alterations in fine GAG structure, which are strictly predetermined (Syrokou et al., 1999; Aguiar et al., 2005; Nikitovic et al., 2005), allow these molecules to modulate with high specificity different cellular processes (Antonio and Iozzo, 2001). Cell-associated and tumor microenvironment GAG content and distribution is markedly altered during tumor pathogenesis and progression (Pupa et al., 2002; Aguiar et al., 2005). Consecutively, GAGs have been proposed as potential prognostic (Lv et al., 2007; Muramatsu and Muramatsu, 2008; Ueno et al., 2009) or diagnostic markers for cancer type or differentiation stage (Suzuki, 1992; Laurent and Guinnepain, 1996; Ohyama, 2008; Pan et al., 2011; Tátrai et al., 2011). GAGs have also been promoted as therapeutic agents for cancer-related pathologies. The specific involvement of GAGs/PGs in the regulation of key tumor cell functions will be discussed.

2.4.2 Heparin and heparan sulfate affect key tumor cell functions

A large body of work has now highlighted a number of important roles that HS plays in the formation, vascularization, and spread of solid tumors. HS chains in the biological environment are covalently attached into a protein core to form proteoglycans (HSPGs) (Bernfield et al., 1999; Esko and Lindahl, 2001; Bishop et al., 2007) present abundantly on the cell surface and in the ECM, where they interact with numerous growth factors, growth factor receptors, extracellular proteases, protease inhibitors, cytokines, chemokines, and adhesive proteins (Bernfield et al., 1999; Esko and Lindahl, 2001; Forsberg and Kjellen, 2001; Bishop et al., 2007; Iozzo and Sanderson, 2011; Theocharis et al., 2011). Most of the interactions of HSPGs with the ligands are mediated directly by the HS moieties (Bernfield et al., 1999, Esko and Lindahl, 2001; Forsberg and Kjellen, 2001). Intriguingly, the distribution, length, and modification level of such domains appear to be tightly regulated in a cell/tissue-specific fashion, indicating that the regulatory role of HS is perpetrated in a spatiotemporal manner, by interacting with unique combinations of protein ligands in different tissues and at different pathological stages (Allen et al., 2001; Ledin et al., 2004). These complex interactions regulate the activity, gradient formation, and stability of specific ligand-receptor interactions, thus consecutively affecting key tumor cell functions (Ledin et al., 2004).

HS distribution and extent of sulfation in cancer

HSPGs have been delegated the role of the essential cellular machinery that senses, integrates, and responds to the physical and chemical environmental information either by directly connecting with the local adhesion sites or by regulating global cellular processes through growth factor receptor signaling pathways, leading to coordinated proliferation, cell migration/adhesion, or differentiation (Kim et al., 2011). The specificity of these interactions is well illustrated in the case of basic fibroblast growth factor-2/fibroblast growth factor receptor (FGF2/FGFR)–mediated signaling. To bind FGF2, HS chains require *N*-sulfated glucosamine units and 2-*O*-sulfated iduronic acid units (Guimond et al., 1993; Lundin et al., 2000; Jia et al., 2009). Concurrently, for HS chains to bind FGFR, they require 6-*O*-sulfated glucosamine residues in addition to 2-*O*-sulfated iduronic acid and *N*-sulfated glucosamine residues (Guimond et al., 1993; Sugaya et al., 2008; Jia et al., 2009). HS also tethers a multitude of growth factors, chemokines, cytokines, and enzymes to the ECM and cell surface, providing a low-affinity storage depot (Vlodavsky et al., 1987; Folkman et al., 1988; Bernfield et al., 1999). Up-to-date, overwhelming evidence correlates signaling by heparin-binding growth factors and sulfation state of HS chains to cancer progression. Thus, the extent of *N*-sulfation of the HS chains is directly correlated to increased cell migration in various melanoma cell lines (Baljinnyam et al., 2011), whereas the expression of highly sulfated endogenous HS is required for both basal and FGF2-stimulated proliferation of human melanoma cells (Nikitovic et al., 2008).

HS metabolism and cancer

Cancer cells have inherently altered the GAG fine structure that allows them to invade, metastasize, and grow uncontrollably (Pauli et al., 1983; Nakato and Kimata, 2002). These fine structural changes are due to errors in the regulation of enzymes involved in HS metabolism. HSulf1, a heparin-degrading endosulfatase, encodes a cell surface-associated enzyme that diminishes sulfation of cell surface HSPG (Lai et al., 2003). HSulf1 is widely expressed in normal tissue, but ovarian, pancreatic, and hepatocellular cell lines and ovarian tumor specimens had either a complete loss or markedly diminished expression of HSulf1 (Lai et al., 2003). The loss of HSulf-1 in breast cancer cells, under a hypoxic microenvironment, results in increased growth factor signaling, cell migration, and invasion (Khurana et al., 2011). Evaluation of HSulf-1 expression levels in primary invasive breast tumors by RNA in situ hybridization indicated that HSulf-1 is downregulated in the majority (60%) of tumors, with a predominant association with lobular histology (Narita et al., 2007). The prosurvival, antiapoptotic effect of extracellular heparan sulfate 6-*O*-endosulfatase (HSulf2) in hepatocellular carcinoma is mediated through activation of the phosphatidylinositol 3-kinase (PI3K)/serine–threonine-specific protein kinase (Akt) pathway (Lai et al., 2010). Lymphatic endothelial deficiency in *N*-deacetylase/*N*-sulfotransferase-1 (Ndst1), a key enzyme involved in sulfating nascent HS chains, resulted in altered lymph node metastasis in tumor-bearing gene targeted mice (Yin et al., 2011). HS secreted by lymphatic endothelium is required for chemokine (C-C motif) ligand 21 (CCL21)-dependent directional migration of murine as well as human lung carcinoma cells toward the targeted lymphatic endothelium demonstrating the genetic importance of host lymphatic HS in mediating chemokine-dependent tumor cell traffic in the lymphatic microenvironment (Yin et al., 2011). Furthermore, since natural killer (NK) cells recognize particular HS fine structural patterns, cancer cells modify their HS patterns to evade NK cells and the immune system (Damiens et al., 1998).

Heparanase is an endoglycosidase that specifically cleaves HS side chains of HSPGs. Cleavage of HS side chains not only alters the integrity of the ECM but also releases HS-bound biological mediators (Vlodavsky et al., 1990). Importantly, free HS fragments are also capable of modulating the activity of growth factors, such as FGF2 and vascular endothelial growth factor (VEGF), and enzymes, such as thrombin and lipoprotein lipase (Vlodavsky et al., 1993). Heparanase activity is strongly implicated in cell migration and invasion associated with tumor metastasis, angiogenesis, and inflammation; whereas heparanase upregulation was documented in an increasing number of human carcinomas, correlating with reduced postoperative survival rate and enhanced tumor angiogenesis (Ilan et al., 2006). On the other hand, it has been reported that reduced expression of HS in the epithelial component was associated with higher histological grades of gastric cancer as well as the presence of more extensive tumor infiltration (Lo et al., 2011). Since NK cells can only recognize cancer cells exhibiting certain 6-*O*-sulfation and *N*-acetylation patterns, pancreatic cancer cells that express increased extracellular heparanase and aberrant HSulf activity exhibit much lower NK cell recognition (Bloushtain et al., 2004).

Roles of HS moieties of HSPGs in cancer progression

HS chains bound to specific cell surface syndecan (SND) and glypican (GPC) members have been attributed discriminate roles in cancer pathogenesis. Glypican-1 (GPC1) and syndecan-1 (SND1) are greatly upregulated in late-stage breast cancer tissue (Matsuda et al., 2001), allowing increased growth factor signaling dependent on specific sulfated residue sequence in HS (Nurcombe et al., 2000). Glypican-3 (GPC3) is overexpressed in hepatocellular carcinoma, interacts with wingless (Wnt) and Hedgehog (Hh) signaling proteins, and is also able to bind basic growth factors such as FGF2 through its HS chains (Ho and Kim, 2011). Another glypican member, glypican-5 (GPC-5), stimulates Hh signaling by facilitating/stabilizing the interaction between Hh and patched 1 with its HS moieties to stimulate rhabdomyosarcoma proliferation (Li et al., 2011). The expression levels of syndecan-4 (SND4), in a manner dependent on its HS chains, regulate focal adhesion kinase (FAK) Y397 phosphorylation during fibronectin (FN)–mediated M5 melanoma cell adhesion, promoting their migration (Chalkiadaki et al., 2009).

Heparin and cancer

Heparin is a complex mixture of natural polyanionic, polydisperse, linear, highly sulfated GAG chains, which perform a variety of critical biological functions and have been extensively employed as anticoagulative agents (Linhardt, 2003; Malavaki et al., 2011). Biochemical modifications of heparin produce a wide range of heparin derivatives, including unfractionated heparin (UFH), low-molecular-weight heparin (LMWH), and modified heparins with different anticoagulant activities (Smorenburg and Van Noorden, 2001; Borsig et al., 2011). There is accumulating evidence that heparin and its derivatives, apart from their anticoagulant activity, have significant anti-metastatic capabilities. Implicated pathways of UFH action are the blocking of P- and L-selectin binding (Engelberg, 1999; Maraveyas et al., 2010), modulation of growth factors and cytokines activities, inhibition of heparanase and angiogenesis (Folkman and Shing, 1992; Vlodavsky and Friedmann, 2001), and the alteration of protease activity resulting in changes of ECM composition (Tyagi et al., 1997). Recent advances have demonstrated an inhibitory effect of UFH on melanoma cell attachment and migration, steps important in early metastatic cascade, through a novel p53/FAK/signaling pathway (Chalkiadaki et al., 2011).

Therapeutic implications of HS/heparin

HS mimetics, which have been developed to target the inhibition of both angiogenesis and heparanase activity, have been selected as the lead clinical candidates for oncology and are currently undergoing formal preclinical development as a novel treatment for advanced cancer (Chen and Sanderson, 2009). Expression of the heparin-degrading endosulfatase-2 (HSulf2) activates mitogen activated protein kinase (MAPK) (e.g. extracellular signal-regulated kinase [Erk], c-Jun N-terminal kinase [JNK], p38 mitogen-activated protein kinase [p38]) and Akt pathways, promotes hepatocellular cancer cell (HC) growth in vitro and in vivo, and is associated with decreased survival of patients. Targeting HSulf2 or the interaction between HSulf2 and HSulf1 may lead to novel therapeutics for the treatment of HCCs (Lai et al., 2010). Heparin and its derivatives are established anticoagulative agents (Abad Rico et al., 2010). An association between cancer and thrombosis has been recognized since 1865, and supplementary anticoagulation therapy is common in cancer patients (Akl et al., 2011). The antimetastaic effect of heparin is not based on its anticoagulative properties, as demonstrated with the utilization of borohydride-reduced heparin (RO-heparin) with low anticoagulant activities, which blocks L-selectin-mediated cell adhesion and prevents tumor metastasis (Chen et al., 2009). Recent clinical studies have demonstrated significant survival advantages when LMWH has been added to chemotherapy in cancer patients (Trujillo-Santos and Monreal, 2011).

2.4.3 Chondroitin sulfate participates in cancer cell, tumor stroma, and tumor microenvironement interactions to affect cancer progression

PGs containing CS/DS chain(s) are ubiquitous components of extracellular matrices and cell surfaces (Baggio et al., 1990). Chondroitin sulfate PGs (CSPGs) are key mediators of ECM organization, but they also contribute to cell signaling (Fthenou et al., 2009). CSPGs regulate the growth, differentiation, and migration of multipotent neural precursor cells through the integrin signaling pathway (Gu et al., 2009). Sulfation of CSPGs is necessary for proper Indian hedgehog signaling in the developing growth plate (Cortes et al., 2009) and, importantly, human diseases (Prabhakar and Sasisekharan, 2006). Indeed, tumor stroma and tumor fibrotic tissues (Ioachim, 1976) contain abnormally high concentrations of CSPGs (Theocharis and Theocharis, 2002). Even though, a clear correlation between the accumulation of CS/DS-PGs and cancer progression has been demonstrated, the exact biological role of CS/DS-PGs in the progression has not been determined.

CS distribution and extent of sulfation in cancer

The type and sulfation pattern of galactosaminoglycans is suggested to participate in cancer progression (Theocharis et al., 2006). The enrichment of the tumor stroma with galactosaminoglycan chains is accompanied by structural modifications as to their type and sulfation pattern (Wegrowski and Maquart, 2004). Several studies have identified potential roles of specific CS forms (determined as to the site of sulfation) in tumor progression, including chondroitin-4-sulfate (CS-A) and chondroitin-4,6-sulfate (CS-E). For example, higher amounts of CS-E were found on a highly metastatic mouse osteosarcoma cell line, when compared to the nonmetastatic parental tumor line (Basappa et al., 2009). Tissue colonization experiments demonstrated that preincubation of these metastatic tumor cells with an antibody against endogenous CS-E, or administration of exogenous CS-E together with tumor cells, could interfere with colonization of the liver (Bassapa et al., 2009). In ovarian adenocarcinomas, a strong expression of a specific CS-E epitope (rich in glucuronyl-N acetylgalactosamine4S6S [GlcA-GalNAc4S6S] units) was detected in the ECM and, occasionally, on tumor cells. No expression, however, was found in normal ovary and cystadenomas (Ten Dam et al., 2007). On the other hand, CS-A is expressed on the melanoma cell surface, but not CS-E, which specifically facilitates the formation of a tripartite complex between CS, membrane-type 3 matrix metalloproteinase (MT3-MMP), and pro–matrix metalloproteinase-2 (pro-MMP2), and the subsequent activation of MMP2. Activation of MMP2 can facilitate invasion of melanoma cells through basement membranes (Iida et al., 2007). In addition, CS-A has been shown to regulate fibrosarcoma cell adhesion, motility, and migration through JNK and tyrosine kinase signaling pathways (Fthenou et al., 2008). CS-E has also been shown to bind P-selectin on mouse breast cancer cells in vitro, suggestive of a potential role of CSPGs in tumor cell adhesion to the endothelium during extravasation of the circulating tumor cells to target organs (Monzavi-Karbassi et al., 2007). Other studies have shown an involvement of CS-A chains of cell surface PGs in FGF ligand-receptor interactions, thus potentially enhancing a tumor-promoting signaling pathway (Deepa et al., 2002; Nikitovic et al., 2008; Asada et al., 2009). Likewise, CS-A chains were found to enhance platelet-derived growth factor (PDGF)–mediated signaling in fibrosarcoma cells (Fthenou et al., 2006).

CS/DS metabolism in cancer

The functions of enzymes involved in the biosynthesis of CS in tumor biology have been targeted in a number of studies, and it is widely accepted that the aberration and alterations of these functions correlate to cancer progression. In colorectal carcinomas, it was shown that chondroitin polymerizing factor and glucuronyltransferase gradually increased with cancer stage and were expressed at much higher levels in adenomas compared to adjacent normal tissue (Kalathas et al., 2011b). The opposite profile was obtained for chondroitin synthase I (CS-I); whereas, the low basal levels of chondroitin synthase III (CS-III) were significantly higher in cancer samples (Kalathas et al., 2011b). DS epimerase was expressed exclusively in the laryngeal cancer tissue and to a minor extent in the adjoining healthy tissues, but it was not detected in the macroscopically normal samples (Kalathas et al., 2011a). Chondroitin-4-sulfotransferase-1 (C4ST-1) overexpression was detected in myeloma cells (Bret et al., 2009) and in microarray studies of human breast carcinoma samples (Potapenko et al., 2010). On the other hand, C4ST-1 expression seems to decrease with the increased stage of colorectal cancer (Kalathas et al., 2009). Together, these studies suggest that misregulation of C4ST-1 expression might play a role in progression in a number of different tumors; however, a functional analysis of any involvement of C4ST-1 in tumorigenesis is not available at present.

Roles of CS moieties of CSPGs in cancer progression

A specific CSPG detected by the mAbB5 antibody has a significantly higher expression in metastatic uveal melanoma as compared to primary uveal melanoma and has been proposed as a potential prognostic factor (Kiewe et al., 2006). Versican and the melanoma-associated proteoglycan (mel-CSPG) (Harper and Reisfeld, 1987) are two major proteoglycans expressed by malignant melanoma, and they have a role in the regulation of cell adhesion, migration, and differentiation (Burg et al., 1998; Eisenmann et al., 1999). The expression and function of a 250 kDa CSPG (NG2) in malignant gliomas has been reviewed by Gladson (1999), whereas a role in glioma progression has been proposed by Stallcup and Huang (2008). Moreover, perlecan and bamacan are two basement membrane CS/DS-PGs expressed in the Engelbreth-Holm-Swarm tumor matrix (Couchman et al., 1996). In a review by Ricciardelli et al. (2009), increased expression of the PG versican is reported to be strongly associated with poor outcome for many different cancers. This hypothesis is supported by functional studies demonstrating that versican is able to increase cancer cell motility (Ang et al., 1999; Cattaruzza et al., 2004), proliferation (LaPierre et al., 2007), and metastasis (Yee et al., 2007). Transforming growth factor-beta (TGF-β) has been shown to cause a dramatic upregulation of expression of the V0/V1 versican isoforms, which have antiadhesive properties and stimulate cell migration, thus potentiating the malignant phenotype of human osteosarcoma cells (Nikitovic et al., 2006). On the other hand, decorin, a CS/DS small leucine-rich proteoglycan (SLRP) is suggested to be an antitumor agent, as described in the review by Iozzo and Schaefer (2011), with the exception of human osteosarcoma cells (Zafiropoulos and Tzanakakis, 2008).

CS/DS as target and potent therapeutic agent

The chondroitin sulfate proteoglycan 4 (CSPG4) has been proposed as a potential target for immunotherapy to control malignant cell growth (Campoli et al., 2010; Wang et al., 2011). Taking into account the observation that CS is overexpressed in several highly metastatic tumors, CS may well be used as a target for the selective delivery of anticancer drugs by polyethylene glycol–coated liposomes (Lee et al., 2002). These liposomes contain a new cationic lipid, 3,5-dipentadecycloxybenzamidine hydrochloride, bound preferentially to CS and avidly internalized by highly metastatic tumor cells. When these liposomes are loaded with cisplatin, they effectively kill CS-expressing tumor cells (Lee et al., 2002); whereas better cellular uptake of doxorubicin by tumor cells was obtained with the utilization of CS nanogels (Huang et al., 2009). Pumphrey et al. (Pumphrey et al., 2002), showed that the injection of a carbodiimide-modified CS directly into breast tumors growing in nude mice reduced or abolished cancer-cell growth without causing apparent toxicity to the adjacent normal tissue. Patients with head and neck tumors excrete to urine a CS with a low degree of sulfation, which may be exploited as a new tool for diagnosis and follow-up of cancer therapy (Martins et al., 2000). Until now, the application of CS/DS and their derivatives to therapeutic strategies has been limited due to fact that in most cases the precisely sulfated distinct structures of CS/DS, which specifically bind to functional proteins, have not been elucidated. Therefore, structural elucidation of the functional domains of CS/DS is desired to obtain information for the development of new therapeutics.

2.4.4 HA synthesis is correlated to cancer progression

HA (hyaluronic acid, hyaluronate) is a free, unbranched GAG composed of repeating disaccharides of GlcNAc and glucuronic acid (GlcA) units, with molecular weights ranging from 10^5 to 10^7 Da (Antonio and Iozzo, 2001). It is synthesized in a unique manner by a family of HA synthases (HASs), degraded by hyaluronidases, and exerts a biological effect by binding to families of cellular receptors, the hyaladhedrins. Interactions of HA with CD44 and receptor for HA-mediated motility (RHAMM) have been linked to the pathophysiological conditions of cancer (Toole and Hascall, 2002; Twarock et al., 2011; Kouvidi et al., 2011). The biological roles of HA are also dependent on its hydrophilic and hydrodynamic properties, which allow it to retain water and play a structural role in tissue organization (Toole and Slomiany, 2008). HA forms loose, hydrated matrices with a pseudo random coil configuration creating space to facilitate cell division and migration in the embryo (Day and Prestwich, 2002) and increase cell motility and decrease cell-cell contact in postembryonic development (Stern, 2003) and cancer invasion (Sironen et al., 2011). HA is one of the major matrix molecules in human malignancies, and the amount of HA in the tumor stroma or in the neoplastic cell compartment impacts the overall outcome (Sironen et al., 2011). There is abundant accumulation of HA in a majority of human tumors, especially in various adenocarcinomas arising from the breast (Auvinen et al., 2000), lung (Pirinen et al., 2001), or ovary (Anttila et al., 2000). On the other hand, the poorly differentiated squamous cell carcinomas (SCCs) have a low HA content (Kosunen et al., 2004). This reduced HA content correlates with increased lymph node metastases in SCC (Hirvikoski et al., 1999). Malignant melanomas, arising from the melanocytes, also show a decreased amount of HA positive tumor cells like the SCCs (Karjalainen et al., 2000).

In the epithelial-mesenchymal transition (EMT), epithelial cells downregulate epithelial markers (e.g. E-cadherin) and upregulate mesenchymal markers (e.g. vimentin and smooth muscle actin), favoring the development of a promigratory phenotype that is essential for embryonic development but also for metastasis. HA overproduction induces EMT-like changes of cancer cells so that they obtain a migratory fibroblastic phenotype correlated to more aggressive phenotype (Chow et al., 2010).

HA metabolism and cancer

HA can be produced in mammals by three HAS isoenzymes: HAS1, HAS2, and HAS3. The expression of HAS enzymes is the first and perhaps most important determinant of the HA synthesis rate in a given cell type under specific circumstances (Weigel and DeAngelis, 2007). HAS mRNA levels are known to influence the content of HA in various tumor cells (Nikitovic et al., 2006; Berdiaki et al., 2009). Therefore, upregulation of HAS expression can contribute to the HA accumulation in tissues and promote tumor growth and metastasis, particularly when coexpressed with hyaluronidase (Itano et al., 2002; Simpson et al., 2002). On the other hand, levels of hyaluronidases, the enzymes that degrade HA, are variable in cancer – in some cases elevated and in others suppressed, relative to normal tissues (Stern, 2003). Thus, tumors of the male genitourinary tract and prostate (Lokeshwar et al., 1996) and urinary bladder cancers show increased hyaluronidase-1 (Hyal1) expression (Lokeshwar et al., 2005). Aggressiveness of other human cancers also correlates with hyaluronidase, including breast (Madan et al., 1999) and laryngeal cancer (Godin et al., 2000). On the other hand, overexpression of Hyal1 suppresses tumorigenicity in a model for colon carcinoma (Jacobson et al., 2002); whereas hyaluronidase administration delays the appearance of carcinogen-induced tumors (Pawlowski et al., 1979). This anomaly can be explained in several ways. HA exists in a high-molecular-weight (HMW) form as well as in a myriad of lower-molecular-weight (LMW) intra- and extracellular HAs. Both the actual amount and the mass of the HA polymer determine its physiological function. While HMW HA is a reflection of intact, healthy tissues, the fragmented forms, indicators of distress signals, occur in abundance in malignancies. They promote angiogenesis, stimulate production of inflammatory cytokines, and activate signaling pathways that are critical for cancer progression. These fragments may be truncated products of the synthetic reaction, but may also be the result of hyaluronidase activities (Stern, 2003, 2008).

Therapeutic implications of hyaluronan metabolism

Inhibition of HA synthesis in general (Simpson et al., 2002), and by restriction of the building block uridine diphosphate-α-D-glucuronic acid (UDP-GlcA) in particular (Huh et al., 2005), has been shown to slow down tumor growth. Biosynthesis of UDP-GlcA occurs through a twofold oxidation of UDP-α-D-glucose that is catalyzed by UDP-α-D-glucose 6-dehydrogenase (UGDH). Limitation of UDP-GlcA in tumor cells by UGDH antagonism is a novel therapeutic target in combating cancer (reviewed by Egger et al., 2010). The development of UGDH antagonists, however, is still in the early stages. Previously described irreversible inhibitors, such as UDP chloroacetol, which are directed against the active-site cysteine residue, are probably not useful. However, synthesized UDP-7-deoxy-α-D-gluco-hept-6-ulopyranose as a UDP-Glc mimic showed reversible inhibition of UDP-glucose-dehydrogenase from the bacterium *Streptococcus pyogenes* (SpUGDH) by this substrate analogue (Campbell and Tanner, 1999). The polyphenols gallic acid and quercetin were recently shown to inhibit human UGDH, which resulted in reduced proliferation of MCF7 human breast cancer cells in vitro (Hwang et al., 2008). In vivo preclinical data on a new bioconjugate of HA with 7-ethyl-10-hydroxy-camptothecin (SN-38) and irinotecan (CPT11 active metabolite) support the validity of the drug delivery strategy, implying the use of HA as a macromolecular carrier of antineoplastic drugs, an approach based on the overexpression of its target (CD44) in a wide variety of cancers (Serafino et al., 2011).

2.4.5 Challenges and future prospects

Both sulfated and nonsulfated GAGs play an important role in cancer biology as they are involved in various interactions with tumor cells, stromal cells, and ECM and through their specific chemical properties bind effective molecules to affect multiple signaling pathways (▶Figure 2.6). Numerous studies report the role of GAGs in promoting tumor growth and mediating events that enhance tumor invasion and metastasis, effects that can be abolished by modulating GAG expression, chemically modified GAGs, or the use of GAGs with altered sulfation patterns. Encouraging studies have demonstrated the potential use of GAGs as a potent anticancer agent. However, the goal of developing efficient GAG-related anticancer agents has still not been reached. The administration of GAG-related drugs should be made after thorough consideration of the specific cancer types and disease progression because the time frame of targeting GAGs differs depending on the stage of cancer and the expression profile, as well as the GAG function, differs in the early stage compared to advanced cancer. Therefore, a fine characterization of both GAG expression and sulfation patterns at all stages of cancer progression is imperative. The case of CS/DS is illustrative as the application of CS/DS and their derivatives to therapeutic strategies has been limited due to fact that in most cases the precisely sulfated distinct structures of CS/DS, which specifically

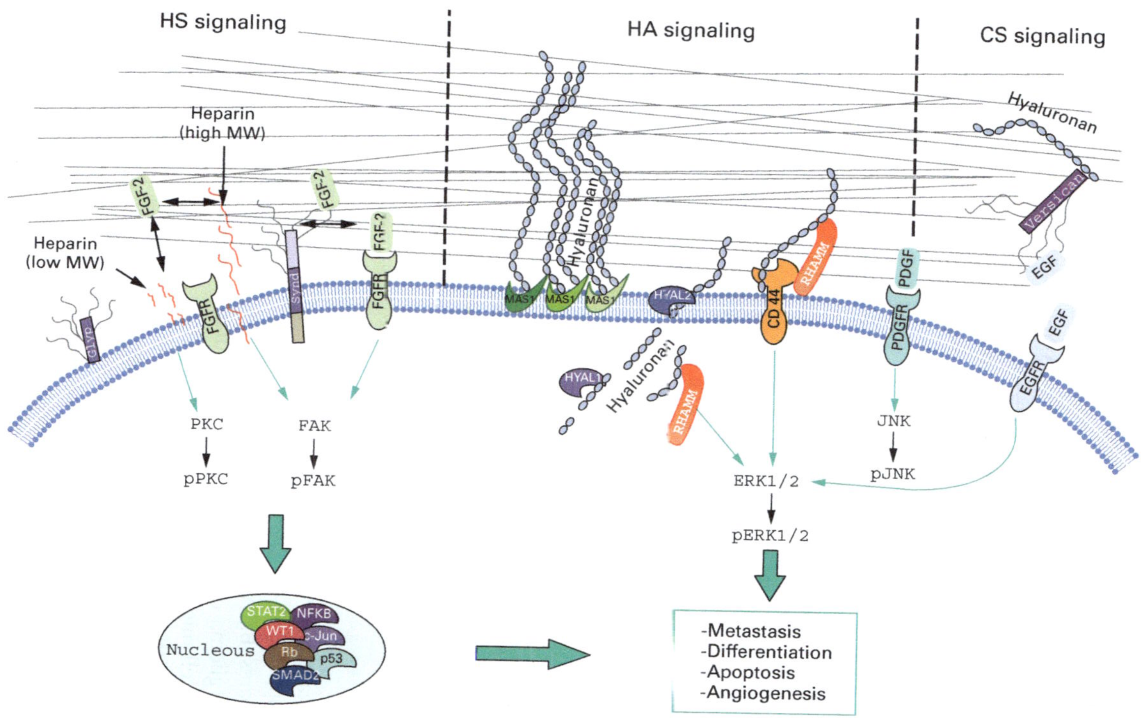

Figure 2.6 **The consecutive steps of tumor growth, local invasion, intravasation, extravasation, and invasion of anatomically distant sites are obligatorily perpetrated through specific interactions of the tumor cells with their microenvironment.** Free glycosaminoglycans (GAGs) and proteoglycan (PG)-containing GAGs, key effectors of cell surface, pericellular, and extracellular microenvironments, perform multiple functions in cancer by virtue of their coded structure and their ability to interact with growth factors, growth factor receptors, cytokines, chemokines, and fibrillar proteins, as well as cell-associated macromolecules that regulate cancer growth and dissemination. Cell type and tissue-specific alterations in fine GAG structure, which are strictly predetermined, allow these molecules to modulate different cellular processes with high specificity (HS, heparan sulfate; HA, hyaluronan; CS, chondroitin sulfate).

bind to functional proteins, have not been elucidated. On the other hand, the well-characterized HS mimetics that have been developed to target the inhibition of both angiogenesis and heparanase activity are currently undergoing formal preclinical development as a novel treatment for advanced cancer (Chen and Sanderson, 2009). Inhibition of HA synthesis by restriction of the building block UDP-GlcA in particular (Huh et al., 2005) has been shown to slow down tumor growth. Limitation of UDP-GlcA in tumor cells by UGDH antagonism is a novel therapeutic target in combating cancer (reviewed by Egger et al., 2010). The development of UGDH antagonists, however, is still in the early stages. Previously described irreversible inhibitors, such as UDP chloroacetol, which are directed against the active-site cysteine residue, are probably not useful due to toxicity issues; however, reversible inhibition of SpUGDH by UDP-7-deoxy-α-D-gluco-hept-6-ulopyranose as a substrate analogue showed promising in vitro results (Campbell and Tanner, 1999).

2.4.6 Take-home message

Encouraging studies, outlined in this chapter, have proved the potential use of GAGs as a potent anticancer agent. In conclusion, it is imperative to further develop GAG-dependent anticancer-related drugs and estimate their preclinical and clinical performance in this new and promising field of targeted anticancer therapy.

Acknowledgments

The kind help of Drs. Ai. Berdiaki and G. Chalkiadaki in the creation and design of the figure is gratefully acknowledged.

References

Abad Rico, J. I., Llau Pitarch, J. V., and Rocha, E. (2010). Overview of venous thromboembolism. Drugs *70* (Suppl. 2), 3–10.

Aguiar, C. B., Lobao-Soares, B., Alvarez-Silva, M., and Trentin, A. G. (2005). Glycosaminoglycans modulate C6 glioma cell adhesion to extracellular matrix components and alter cell proliferation and cell migration. BMC Cell Biol 6, 31.

Akl, E. A., Gunukula, S., Barba, M., et al. (2011). Parenteral anticoagulation in patients with cancer who have no therapeutic or prophylactic indication for anticoagulation. Cochrane Database Syst Rev CD006652.

Allen, B. L., Filla, M. S., and Rapraeger, A. C. (2001). Role of heparan sulfate as a tissue-specific regulator of FGF-4 and FGF receptor recognition. J Cell Biol *155*, 845–858.

Ang, L. C., Zhang, Y., Cao, L., et al. (1999). Versican enhances locomotion of astrocytoma cells and reduces cell adhesion through its G1 domain. J Neuropathol Exp Neurol *58*, 597–605.

Antonio, J. D. S., and Iozzo, R. V. (2001). Glycosaminoglycans: structure and biological functions. eLS doi: 10.1038/npg.els.0000704.

Anttila, M. A., Tammi, R. H., Tammi, M. I., Syrjanen, K. J., Saarikoski, S. V., and Kosma, V. M. (2000). High levels of stromal hyaluronan predict poor disease outcome in epithelial ovarian cancer. Cancer Res *60*, 150–155.

Asada, M., Shinomiya, M., Suzuki, M., et al. (2009). Glycosaminoglycan affinity of the complete fibroblast growth factor family. Biochim Biophys Acta *1790*, 40–48.

Auvinen, P., Tammi, R., Parkkinen, J., et al. (2000). Hyaluronan in peritumoral stroma and malignant cells associates with breast cancer spreading and predicts survival. Am J Pathol *156*, 529–536.

Baggio, B., Marzaro, G., Gambaro, G., Marchini, F., Williams, H. E., and Borsatti, A. (1990). Glycosaminoglycan content, oxalate self-exchange and protein phosphorylation in erythrocytes of patients with 'idiopathic' calcium oxalate nephrolithiasis. Clin Sci (Lond) *79*, 113–116.

Baljinnyam, E., Umemura, M., De Lorenzo, M. S., et al. (2011). Epac1 promotes melanoma metastasis via modification of heparan sulfate. Pigment Cell Melanoma Res *24* (4), 680–687.

Basappa, Murugan, S., Sugahara, K. N., Lee, C. M., et al. (2009). Involvement of chondroitin sulfate E in the liver tumor focal formation of murine osteosarcoma cells. Glycobiology *19*, 735–742.

Berdiaki, A., Nikitovic, D., Tsatsakis, A., Katonis, P., Karamanos, N. K., and Tzanakakis, G. N. (2009). bFGF induces changes in hyaluronan synthase and hyaluronidase isoform expression and modulates the migration capacity of fibrosarcoma cells. Biochim Biophys Acta *1790*, 1258–1265.

Bernfield, M., Gotte, M., Park, P. W., et al. (1999). Functions of cell surface heparan sulfate proteoglycans. Annu Rev Biochem *68*, 729–777.

Bishop, J. M., O'Ryan, C., and Jarvis, J. U. (2007). Social common mole-rats enhance outbreeding via extra-pair mating. Biol Lett *3*, 176–179.

Bloushtain, N., Qimron, U., Bar-Ilan, A., et al. (2004). Membrane-associated heparan sulfate proteoglycans are involved in the recognition of cellular targets by NKp30 and NKp46. J Immunol *173*, 2392–2401.

Borsig, L., Vlodavsky, I., Ishai-Michaeli, R., Torri, G., and Vismara, E. (2011). Sulfated hexasaccharides attenuate metastasis by inhibition of P-selectin and heparanase. Neoplasia *13*, 445–452.

Bret, C., Hose, D., Reme, T., et al. (2009). Expression of genes encoding for proteins involved in heparan sulphate and chondroitin sulphate chain synthesis and modification in normal and malignant plasma cells. Br J Haematol *145*, 350–368.

Burg, M., Heinemann, U., and Schmitz, D. (1998). Neuroactive steroids induce GABA(A) receptor-mediated depolarizing postsynaptic potentials in hippocampal CA1 pyramidal cells of the rat. Eur J Neurosci *10*, 2880–2886.

Campbell, R. E., and Tanner, M. E. (1999). UDP-glucose analogues as inhibitors and mechanistic probes of UDP-glucose dehydrogenase. J Org Chem *64*, 9487–9492.

Campoli, M., Ferrone, S., and Wang, X. (2010). Functional and clinical relevance of chondroitin sulfate proteoglycan 4. Adv Cancer Res *109*, 73–121.

Cattaruzza, S., Schiappacassi, M., Kimata, K., Colombatti, A., and Perris, R. (2004). The globular domains of PG-M/versican modulate the proliferation-apoptosis equilibrium and invasive capabilities of tumor cells. FASEB J *18*, 779–781.

Chaffer, C. L., and Weinberg, R. A. (2011). A perspective on cancer cell metastasis. Science *331*, 1559–1564.

Chalkiadaki, G., Nikitovic, D., Berdiaki, A., Katonis, P., Karamanos, N. K., and Tzanakakis, G. N. (2011). Heparin plays a key regulatory role via a p53/FAK-dependent signaling in melanoma cell adhesion and migration. IUBMB Life *63*, 109–119.

Chalkiadaki, G., Nikitovic, D., Berdiaki, A., et al. (2009). Fibroblast growth factor-2 modulates melanoma adhesion and migration through a syndecan-4-dependent mechanism. Int J Biochem Cell Biol *41*, 1323–1331.

Chen, L., and Sanderson, R. D. (2009). Heparanase regulates levels of syndecan-1 in the nucleus. PLoS One *4*, e4947.

Chen, Z., Jing, Y., Song, B., Han, Y., and Chu, Y. (2009). Chemically modified heparin inhibits in vitro L-selectin-mediated human ovarian carcinoma cell adhesion. Int J Gynecol Cancer. *19*, 540–546.

Chow, G., Tauler, J., and Mulshine, J. L. (2010). Cytokines and growth factors stimulate hyaluronan production: role of hyaluronan in epithelial to mesenchymal-like transition in non-small cell lung cancer. J Biomed Biotechnol 485468.

Cortes, M., Baria, A. T., and Schwartz, N. B. (2009). Sulfation of chondroitin sulfate proteoglycans is necessary for proper Indian hedgehog signaling in the developing growth plate. Development *136*, 1697–1706.

Couchman, J. R., Kapoor, R., Sthanam, M., and Wu, R. R. (1996). Perlecan and basement membrane-chondroitin sulfate proteoglycan (bamacan) are two basement membrane chondroitin/dermatan sulfate proteoglycans in the Engelbreth-Holm-Swarm tumor matrix. J Biol Chem *271*, 9595–9602.

Damiens, E., El Yazidi, I., Mazurier, J., et al. (1998). Role of heparan sulphate proteoglycans in the regulation of human lactoferrin binding and activity in the MDA-MB-231 breast cancer cell line. Eur J Cell Biol *77*, 344–351.

Day, A. J., and Prestwich, G. D. (2002). Hyaluronan-binding proteins: tying up the giant. J Biol Chem *277*, 4585–4588.

Deepa, S. S., Umehara, Y., Higashiyama, S., Itoh, N., and Sugahara, K. (2002). Specific molecular interactions of oversulfated chondroitin sulfate E with various heparin-binding growth factors. Implications as a physiological binding partner in the brain and other tissues. J Biol Chem *277*, 43707–43716.

Egger, S., Chaikuad, A., Kavanagh, K. L., Oppermann, U., and Nidetzky, B. (2010). UDP-glucose dehydrogenase: structure and function of a potential drug target. Biochem Soc Trans *38*, 1378–1385.

Eisenmann, K. M., McCarthy, J. B., Simpson, M. A., et al. (1999). Melanoma chondroitin sulphate proteoglycan regulates cell spreading through Cdc42, Ack-1 and p130cas. Nat Cell Biol *1*, 507–513.

Engelberg, H. (1999). Actions of heparin that may affect the malignant process. Cancer *85*, 257–272.

Esko, J. D., and Lindahl, U. (2001). Molecular diversity of heparan sulfate. J Clin Invest *108*, 169–173.

Fidler, I. J., and Kripke, M. L. (2003). Genomic analysis of primary tumors does not address the prevalence of metastatic cells in the population. Nat Genet *34*, 23; author reply 25.

Folkman, J., Klagsbrun, M., Sasse, J., Wadzinski, M., Ingber, D., and Vlodavsky, I. (1988). A heparin-binding angiogenic protein – basic fibroblast growth factor – is stored within basement membrane. Am J Pathol *130*, 393–400.

Folkman, J., and Shing, Y. (1992). Angiogenesis. J Biol Chem *267*, 10931–10934.

Forsberg, E., and Kjellen, L. (2001). Heparan sulfate: lessons from knockout mice. J Clin Invest *108*, 175–180.

Fthenou, E., Zafiropoulos, A., Katonis, P., Tsatsakis, A., Karamanos, N. K., and Tzanakakis, G. N. (2008). Chondroitin sulfate prevents platelet derived growth factor-mediated phosphorylation of PDGF-Rbeta in normal human fibroblasts severely impairing mitogenic responses. J Cell Biochem *103*, 1866–1876.

Fthenou, E., Zafiropoulos, A., Tsatsakis, A., Stathopoulos, A., Karamanos, N. K., and Tzanakakis, G. N. (2006). Chondroitin sulfate A chains enhance platelet derived growth factor-mediated signalling in fibrosarcoma cells. Int J Biochem Cell Biol *38*, 2141–2150.

Fthenou, E., Zong, F., Zafiropoulos, A., Dobra, K., Hjerpe, A., and Tzanakakis, G. N. (2009). Chondroitin sulfate A regulates fibrosarcoma cell adhesion, motility and migration through JNK and tyrosine kinase signaling pathways. In Vivo *23*, 69–76.

Gladson, C. L. (1999). The extracellular matrix of gliomas: modulation of cell function. J Neuropathol Exp Neurol *58*, 1029–1040.

Godin, D. A., Fitzpatrick, P. C., Scandurro, A. B., et al. (2000). PH20: a novel tumor marker for laryngeal cancer. Arch Otolaryngol Head Neck Surg *126*, 402–404.

Gu, W. L., Fu, S. L., Wang, Y. X., et al. (2009). Chondroitin sulfate proteoglycans regulate the growth, differentiation and migration of multipotent neural precursor cells through the integrin signaling pathway. BMC Neurosci *10*, 128.

Guimond, S., Maccarana, M., Olwin, B. B., Lindahl, U., and Rapraeger, A. C. (1993). Activating and inhibitory heparin sequences for FGF-2 (basic FGF). Distinct requirements for FGF-1, FGF-2, and FGF-4. J Biol Chem *268*, 23906–23914.

Hanahan, D., and Weinberg, R. A. (2011). Hallmarks of cancer: the next generation. Cell *144*, 646–674.

Harper, J. R., and Reisfeld, R. A. (1987). Cell-associated proteoglycans in human malignant melanoma. In Biology of Proteoglycans, T. N. Wight and R. P. Mecham, eds. Orlando, FL: Academic Press, pp. 345–366.

Hirvikoski, P., Tammi, R., Kumpulainen, E., et al. (1999). Irregular expression of hyaluronan and its CD44 receptor is associated with metastatic phenotype in laryngeal squamous cell carcinoma. Virchows Arch *434*, 37–44.

Ho, M., and Kim, H. (2011). Glypican-3: a new target for cancer immunotherapy. Eur J Cancer *47*, 333–338.

Huang, X., Kim, J. M., Kong, T. H., et al. (2009). GM-CSF inhibits glial scar formation and shows long-term protective effect after spinal cord injury. J Neurol Sci *277*, 87–97.

Huh, J. W., Choi, M. M., Yang, S. J., Yoon, S. Y., Choi, S. Y., and Cho, S. W. (2005). Inhibition of human UDP-glucose dehydrogenase expression using siRNA expression vector in breast cancer cells. Biotechnol Lett *27*, 1229–1232.

Hwang, E. Y., Huh, J. W., Choi, M. M., Choi, S. Y., Hong, H. N., and Cho, S. W. (2008). Inhibitory effects of gallic acid and quercetin on UDP-glucose dehydrogenase activity. FEBS Lett *582*, 3793–3797.

Iida, J., Wilhelmson, K. L., Ng, J., et al. (2007). Cell surface chondroitin sulfate glycosaminoglycan in melanoma: role in the activation of pro-MMP-2 (pro-gelatinase A). Biochem J *403*, 553–563.

Ilan, N., Elkin, M., and Vlodavsky, I. (2006). Regulation, function and clinical significance of heparanase in cancer metastasis and angiogenesis. Int J Biochem Cell Biol *38*, 2018–2039.

Ioachim, H. L. (1976). The stromal reaction of tumors: an expression of immune surveillance. J Natl Cancer Inst *57*, 465–475.

Iozzo, R. V., and Sanderson, R. D. (2011). Proteoglycans in cancer biology, tumour microenvironment and angiogenesis. J Cell Mol Med *15*, 1013–1031.

Iozzo, R. V., and Schaefer, L. (2010). Proteoglycans in health and disease: novel regulatory signaling mechanisms evoked by the small leucine-rich proteoglycans. FEBS J *277* (19), 3864–3875.

Itano, N., Atsumi, F., Sawai, T., et al. (2002). Abnormal accumulation of hyaluronan matrix diminishes contact inhibition of cell growth and promotes cell migration. Proc Natl Acad Sci U S A *99*, 3609–3614.

Jacobson, A., Rahmanian, M., Rubin, K., and Heldin, P. (2002). Expression of hyaluronan synthase 2 or hyaluronidase 1 differentially affect the growth rate of transplantable colon carcinoma cell tumors. Int J Cancer *102*, 212–219.

Jia, J., Maccarana, M., Zhang, X., Bespalov, M., Lindahl, U., and Li, J. P. (2009). Lack of L-iduronic acid in heparan sulfate affects interaction with growth factors and cell signaling. J Biol Chem *284*, 15942–15950.

Kalathas, D., Theocharis, D. A., Bounias, D., et al. (2009). Alterations of glycosaminoglycan disaccharide content and composition in colorectal cancer: structural and expressional studies. Oncol Rep *22*, 369–375.

Kalathas, D., Theocharis, D. A., Bounias, D., et al. (2011a). Chondroitin synthases I, II, III and chondroitin sulfate glucuronyltransferase expression in colorectal cancer. Mol Med Report *4*, 363–368.

Kalathas, D., Triantaphyllidou, I. E., Mastronikolis, N. S., et al. (2011b). The chondroitin/dermatan sulfate synthesizing and modifying enzymes in laryngeal cancer: expressional and epigenetic studies. Head Neck Oncol *2*, 27.

Karamanos, N. K., and Tzanakakis, G. N. (2012). Glycosaminoglycans: from "cellular glue" to novel therapeutical agents. Curr Opin Pharmacol *12*, 220–222.

Karjalainen, J. M., Tammi, R. H., Tammi, M. I., et al. (2000). Reduced level of CD44 and hyaluronan associated with unfavorable prognosis in clinical stage I cutaneous melanoma. Am J Pathol *157*, 957–965.

Khurana, A., Liu, P., Mellone, P., et al. (2011). HSulf-1 modulates FGF2and hypoxia-mediated migration and invasion of breast cancer cells. Cancer Res *71*, 2152–2161.

Kiewe, P., Bechrakis, N. E., Schmittel, A., et al. (2006). Increased chondroitin sulphate proteoglycan expression (B5 immunoreactivity) in metastases of uveal melanoma. Ann Oncol *17*, 1830–1834.

Kim, S. H., Turnbull, J., and Guimond, S. (2011). Extracellular matrix and cell signalling: the dynamic cooperation of integrin, proteoglycan and growth factor receptor. J Endocrinol *209*, 139–151.

Kosunen, A., Ropponen, K., Kellokoski, J., et al. (2004). Reduced expression of hyaluronan is a strong indicator of poor survival in oral squamous cell carcinoma. Oral Oncol *40*, 257–263.

Kouvidi, K., Berdiaki, A., Nikitovic, D., et al. (2011). Role of receptor for hyaluronic acid-mediated motility (RHAMM) in low molecular weight hyaluronan (LMWHA)-mediated fibrosarcoma cell adhesion. J Biol Chem *286*, 38509–38520.

Lai, J., Chien, J., Staub, J., et al. (2003). Loss of HSulf-1 up-regulates heparin-binding growth factor signaling in cancer. J Biol Chem *278*, 23107–23117.

Lai, J. P., Sandhu, D. S., Yu, C., et al. (2010). Sulfatase 2 protects hepatocellular carcinoma cells against apoptosis induced by the PI3K inhibitor LY294002 and ERK and JNK kinase inhibitors. Liver Int *30*, 1522–1528.

LaPierre, D. P., Lee, D. Y., Li, S. Z., et al. (2007). The ability of versican to simultaneously cause apoptotic resistance and sensitivity. Cancer Res *67*, 4742–4750.

Laurent, J., and Guinnepain, M. T. (1996). Hereditary angioedema with ascites. J Allergy Clin Immunol *98*, 999.

Ledin, J., Staatz, W., Li, J. P., et al. (2004). Heparan sulfate structure in mice with genetically modified heparan sulfate production. J Biol Chem *279*, 42732–42741.

Lee, H., Jeong, J. H., and Park, T. G. (2002). PEG grafted polylysine with fusogenic peptide for gene delivery: high transfection efficiency with low cytotoxicity. J Control Release *79*, 283–291.

Li, F., Shi, W., Capurro, M., and Filmus, J. (2011). Glypican-5 stimulates rhabdomyosarcoma cell proliferation by activating Hedgehog signaling. J Cell Biol *192*, 691–704.

Linhardt, R. J. (2003). 2003 Claude S. Hudson Award address in carbohydrate chemistry. Heparin: structure and activity. J Med Chem *46*, 2551–2564.

Lo, S. L., Thike, A. A., Tan, S. Y., et al. (2011). Expression of heparan sulfate in gastric carcinoma and its correlation with clinicopathological features and patient survival. J Clin Pathol *64*, 153–158.

Lokeshwar, V. B., Habuchi, T., Grossman, H. B., et al. (2005). Bladder tumor markers beyond cytology: International Consensus Panel on bladder tumor markers. Urology *66*, 35–63.

Lokeshwar, V. B., Lokeshwar, B. L., Pham, H. T., and Block, N. L. (1996). Association of elevated levels of hyaluronidase, a matrix-degrading enzyme, with prostate cancer progression. Cancer Res *56*, 651–657.

Lundin, L., Larsson, H., Kreuger, J., et al. (2000). Selectively desulfated heparin inhibits fibroblast growth factor-induced mitogenicity and angiogenesis. J Biol Chem *275*, 24653–24660.

Lv, H., Yu, G., Sun, L., Zhang, Z., Zhao, X., and Chai, W. (2007). Elevate level of glycosaminoglycans and altered sulfation pattern of chondroitin sulfate are associated with differentiation status and histological type of human primary hepatic carcinoma. Oncology *72*, 347–356.

Madan, A. K., Yu, K., Dhurandhar, N., Cullinane, C., Pang, Y., and Beech, D. J. (1999). Association of hyaluronidase and breast adenocarcinoma invasiveness. Oncol Rep *6*, 607–609.

Malavaki, C. J., Theocharis, A. D., Lamari, F. N., et al. (2011). Heparan sulfate: biological significance, tools for biochemical analysis and structural characterization. Biomed Chromatogr *25*, 11–20.

Maraveyas, A., Johnson, M. J., Xiao, Y. P., and Noble, S. (2010). Malignant melanoma as a target malignancy for the study of the anti-metastatic properties of the heparins. Cancer Metastasis Rev *29*, 777–784.

Martins, R. C., Lima, F. R., Werneck, C. C., Neto, V. M., and Silva, L. C. (2000). Patterns of synthesis and secretion of sulfated glycosaminoglycans in primary cortical and cerebellar astrocytes in vitro. Biol Cell *92*, 421–427.

Matsuda, K., Maruyama, H., Guo, F., et al. (2001). Glypican-1 is overexpressed in human breast cancer and modulates the mitogenic effects of multiple heparin-binding growth factors in breast cancer cells. Cancer Res *61*, 5562–5569.

Monzavi-Karbassi, B., Stanley, J. S., Hennings, L., et al. (2007). Chondroitin sulfate glycosaminoglycans as major P-selectin ligands on metastatic breast cancer cell lines. Int J Cancer *120*, 1179–1191.

Muramatsu, T., and Muramatsu, H. (2008). Glycosaminoglycan-binding cytokines as tumor markers. Proteomics *8*, 3350–3359.

Nakato, H., and Kimata, K. (2002). Heparan sulfate fine structure and specificity of proteoglycan functions. Biochim Biophys Acta *1573*, 312–318.

Narita, K., Chien, J., Mullany, S. A., et al. (2007). Loss of HSulf-1 expression enhances autocrine signaling mediated by amphiregulin in breast cancer. J Biol Chem *282*, 14413–14420.

Nikitovic, D., Assouti, M., Sifaki, M., et al. (2008). Chondroitin sulfate and heparan sulfate-containing proteoglycans are both partners and targets of basic fibroblast growth factor-mediated proliferation in human metastatic melanoma cell lines. Int J Biochem Cell Biol *40*, 72–83.

Nikitovic, D., Zafiropoulos, A., Katonis, P., et al. (2006). Transforming growth factor-beta as a key molecule triggering the expression of versican isoforms v0 and v1, hyaluronan synthase-2 and synthesis of hyaluronan in malignant osteosarcoma cells. IUBMB Life *58*, 47–53.

Nikitovic, D., Zafiropoulos, A., Tzanakakis, G. N., Karamanos, N. K., and Tsatsakis, A. M. (2005). Effects of glycosaminoglycans on cell proliferation of normal osteoblasts and human osteosarcoma cells depend on their type and fine chemical compositions. Anticancer Res *25*, 2851–2856.

Nurcombe, V., Smart, C. E., Chipperfield, H., Cool, S. M., Boilly, B., and Hondermarck, H. (2000). The proliferative and migratory activities of breast cancer cells can be differentially regulated by heparan sulfates. J Biol Chem *275*, 30009–30018.

Ohyama, C. (2008). Glycosylation in bladder cancer. Int J Clin Oncol *13*, 308–313.

Pan, J., Qian, Y., Weiser, P., et al. (2011). Glycosaminoglycans and activated contact system in cancer patient plasmas. Prog Mol Biol Transl Sci *93*, 473–495.

Pauli, B. U., Schwartz, D. E., Thonar, E. J., and Kuettner, K. E. (1983). Tumor invasion and host extracellular matrix. Cancer Metastasis Rev *2*, 129–152.

Pawlowski, A., Haberman, H. F., and Menon, I. A. (1979). The effects of hyalurodinase upon tumor formation in BALB/c mice painted with 7,12-dimethylbenz-(a)anthracene. Int J Cancer *23*, 105–109.

Pirinen, R., Tammi, R., Tammi, M., et al. (2001). Prognostic value of hyaluronan expression in non-small-cell lung cancer: increased stromal expression indicates unfavorable outcome in patients with adenocarcinoma. Int J Cancer *95*, 12–17.

Potapenko, I. O., Haakensen, V. D., Luders, T., et al. (2010). Glycan gene expression signatures in normal and malignant breast tissue; possible role in diagnosis and progression. Mol Oncol *4*, 98–118.

Prabhakar, V., and Sasisekharan, R. (2006). The biosynthesis and catabolism of galactosaminoglycans. Adv Pharmacol *53*, 69–115.

Pumphrey, C. Y., Theus, A. M., Li, S., Parrish, R. S., and Sanderson, R. D. (2002). Neoglycans, carbodiimide-modified glycosaminoglycans: a new class of anticancer agents that inhibit cancer cell proliferation and induce apoptosis. Cancer Res *62*, 3722–3728.

Pupa, S. M., Menard, S., Forti, S., and Tagliabue, E. (2002). New insights into the role of extracellular matrix during tumor onset and progression. J Cell Physiol *192*, 259–267.

Ricciardelli, C., Sakko, A. J., Stahl, J., Tilley, W. D., Marshall, V. R., and Horsfall, D. J. (2009). Prostatic chondroitin sulfate is increased in patients with metastatic disease but does not predict survival outcome. Prostate *69*, 761–769.

Serafino, A., Zonfrillo, M., Andreola, F., et al. (2011). CD44-targeting for antitumor drug delivery: a new SN-38-hyaluronan bioconjugate for locoregional treatment of peritoneal carcinomatosis. Curr Cancer Drug Targets *11*, 572–585.

Simpson, M. A., Wilson, C. M., Furcht, L. T., Spicer, A. P., Oegema, T. R., Jr., and McCarthy, J. B. (2002). Manipulation of hyaluronan synthase expression in prostate adenocarcinoma cells alters pericellular matrix retention and adhesion to bone marrow endothelial cells. J Biol Chem *277*, 10050–10057.

Sironen, R. K., Tammi, M., Tammi, R., Auvinen, P. K., Anttila, M., and Kosma, V. M. (2011). Hyaluronan in human malignancies. Exp Cell Res *317*, 383–391.

Smorenburg, S. M., and Van Noorden, C. J. (2001). The complex effects of heparins on cancer progression and metastasis in experimental studies. Pharmacol Rev *53*, 93–105.

Stallcup, W. B., and Huang, F. J. (2008). A role for the NG2 proteoglycan in glioma progression. Cell Adh Migr *2*, 192–201.

Stern, R. (2003). Devising a pathway for hyaluronan catabolism: are we there yet? Glycobiology *13*, 105R–115R.

Stern, R. (2008). Hyaluronidases in cancer biology. Semin Cancer Biol *18*, 275–280.

Sugaya, N., Habuchi, H., Nagai, N., Ashikari-Hada, S., and Kimata, K. (2008). 6-O-sulfation of heparan sulfate differentially regulates various fibroblast growth factor-dependent signalings in culture. J Biol Chem *283*, 10366–10376.

Suzuki, Y. (1992). Diagnostic criteria for human diffuse malignant mesothelioma. Acta Pathol Jpn *42*, 767–786.

Syrokou, A., Tzanakakis, G., Tsegenidis, T., Hjerpe, A., and Karamanos, N. K. (1999). Effects of glycosaminoglycans on proliferation of epithelial and fibroblast human malignant mesothelioma cells: a structure-function relationship. Cell Prolif *32*, 85–99.

Tátrai, P., Egedi, K., Somoracz, A., et al. (2011). Quantitative and qualitative alterations of heparan sulfate in fibrogenic liver diseases and hepatocellular cancer. J Histochem Cytochem *58*, 429–441.

Ten Dam, G. B., van de Westerlo, E. M., Purushothaman, A., et al. (2007). Antibody GD3G7 selected against embryonic glycosaminoglycans defines chondroitin sulfate-E domains highly up-regulated in ovarian cancer and involved in vascular endothelial growth factor binding. Am J Pathol *171*, 1324–1333.

Theocharis, A. D., Skandalis, S. S., Tzanakakis, G. N., and Karamanos, N. K. (2011). Proteoglycans in health and disease: novel roles for proteoglycans in malignancy and their pharmacological targeting. FEBS J *278*, 16–27.

Theocharis, A. D., and Theocharis, D. A. (2002). High-performance capillary electrophoretic analysis of hyaluronan and galactosaminoglycan disaccharides in gastrointestinal carcinomas. Differential disaccharide composition as a possible tool-indicator for malignancies. Biomed Chromatogr *16*, 157–161.

Theocharis, A. D., Tsolakis, I., Tzanakakis, G. N., and Karamanos, N. K. (2006). Chondroitin sulfate as a key molecule in the development of atherosclerosis and cancer progression. Adv Pharmacol *53*, 281–295.

Toole, B. P., and Hascall, V. C. (2002). Hyaluronan and tumor growth. Am J Pathol *161*, 745–747.

Toole, B. P., and Slomiany, M. G. (2008). Hyaluronan: a constitutive regulator of chemoresistance and malignancy in cancer cells. Semin Cancer Biol *18*, 244–250.

Trujillo-Santos, J., and Monreal, M. (2011). Anticoagulant use in cancer patients. Expert Opin Pharmacother *12*, 351–362.

Twarock, S., Tammi, M. I., Savani, R. C., and Fischer, J. W. (2011). Hyaluronan stabilizes focal adhesions, filopodia, and the proliferative phenotype in esophageal squamous carcinoma cells. J Biol Chem *285*, 23276–23284.

Tyagi, S. C., Kumar, S., and Katwa, L. (1997). Differential regulation of extracellular matrix metalloproteinase and tissue inhibitor by heparin and cholesterol in fibroblast cells. J Mol Cell Cardiol *29*, 391–404.

Ueno, M., Uchiyama, K., Ozawa, S., Hayami, S., Kiriyama, S., and Yamaue, H. (2009). A new prediction model of postoperative complications after major hepatectomy for hepatocellular carcinoma. Dig Surg *26*, 392–399.

Vlodavsky I., Bar-Shavit, R., Korner, G., and Fuks Z. (1993). Extracellular matrix-boundgrowth factors, enzymes and plasma proteins. In Basement membranes: Cellular and Molecular aspects, D. H.Rohrbach and R. Timpl, eds. Orlando, FL: Academic Press, pp. 327–343.

Vlodavsky, I., Folkman, J., Sullivan, R., et al. (1987). Endothelial cell-derived basic fibroblast growth factor: synthesis and deposition into subendothelial extracellular matrix. Proc Natl Acad Sci U S A *84*, 2292–2296.

Vlodavsky, I., and Friedmann, Y. (2001). Molecular properties and involvement of heparanase in cancer metastasis and angiogenesis. J Clin Invest *108*, 341–347.

Vlodavsky, I., Korner, G., Ishai-Michaeli, R., Bashkin, P., Bar-Shavit, R., and Fuks, Z. (1990). Extracellular matrix-resident growth factors and enzymes: possible involvement in tumor metastasis and angiogenesis. Cancer Metastasis Rev *9*, 203–226.

Wang, X., Wang, Y., Yu, L., et al. (2011). CSPG4 in cancer: multiple roles. Curr Mol Med *10*, 419–429.

Wegrowski, Y., and Maquart, F. X. (2004). Involvement of stromal proteoglycans in tumour progression. Crit Rev Oncol Hematol *49*, 259–268.

Weigel, P. H., and DeAngelis, P. L. (2007). Hyaluronan synthases: a decade-plus of novel glycosyltransferases. J Biol Chem *282*, 36777–36781.

Yee, A. J., Akens, M., Yang, B. L., et al. (2007). The effect of versican G3 domain on local breast cancer invasiveness and bony metastasis. Breast Cancer Res *9*, R47.

Yin, X., Truty, J., Lawrence, R., et al. (2011). A critical role for lymphatic endothelial heparan sulfate in lymph node metastasis. Mol Cancer *9*, 316.

Zafiropoulos, A., and Tzanakakis, G. N. (2008). Decorin-mediated effects in cancer cell biology. Connect Tissue Res *49*, 244–248.

2.5 Heparan sulfate design: regulation of biosynthesis

Dagmar S. Pikas, Anh-Tri Do, Audrey Deligny, Anders Dagälv, and Lena Kjellén

2.5.1 Heparan sulfate – an extracellular component with variable structure

Heparan sulfate (HS) is a linear polysaccharide of the glycosaminoglycan (GAG) type that interacts with a number of different proteins, such as growth factors, morphogens, enzymes, and extracellular matrix (ECM) components (Lindahl and Li, 2009). Being the glycan part of proteoglycans, HS is found covalently attached to core proteins at the cell surface and in the ECM, predominantly in basement membranes. The HS proteoglycans are named based on the type of core protein and include the syndecans and glypicans (membrane bound), perlecan, agrin and collagen XVIII (ECM components), and serglycin (intracellularly located to mast cell granules) (▶Figure 2.7). Through their HS chains, the proteoglycans can influence both embryonic development and adult pathophysiology (Bulow and Hobert, 2006; Bishop et al., 2007). The local distribution of HS proteoglycans enables growth factors to bind and form concentration gradients through the matrix. At the cell surface, HS proteoglycans can act as coreceptors (e.g. in fibroblast growth factor [FGF] signaling).

The building block of HS consists of repeating disaccharide units of glucosamine (GlcN) residues and negatively charged hexuronic acid residues (glucuronic acid [GlcA] or iduronic acid [IdoA]). The GlcN residues are essentially of two types, *N*-acetyl-D-glucosamine (GlcNAc) and *N*-sulfo-D-glucosamine ($GlcNSO_3^-$), but unsubstituted GlcN residues ($GlcNH_2$) have also been identified (Westling and Lindahl, 2002). The *N*-sulfate groups together with additional *O*-sulfate groups are distributed along the polysaccharide chain in such a way that contiguous sulfated stretches (NS domains) are formed with nonmodified regions (NA domains) in between. The overall organization of these domains is thought to be crucial for protein binding (Kreuger et al., 2006).

The fine structure, depending on the exact distribution of sulfate groups, varies along as well as between individual HS chains. However, the amount of HS produced by different tissues and the HS structure (analyzed as disaccharide composition after cleavage of the polysaccharide) show specific characteristics. These features seem to be strictly dependent on the cell type and developmental stage, while very few structural differences can be seen when the same tissues from different individuals are compared (Ledin et al., 2004).

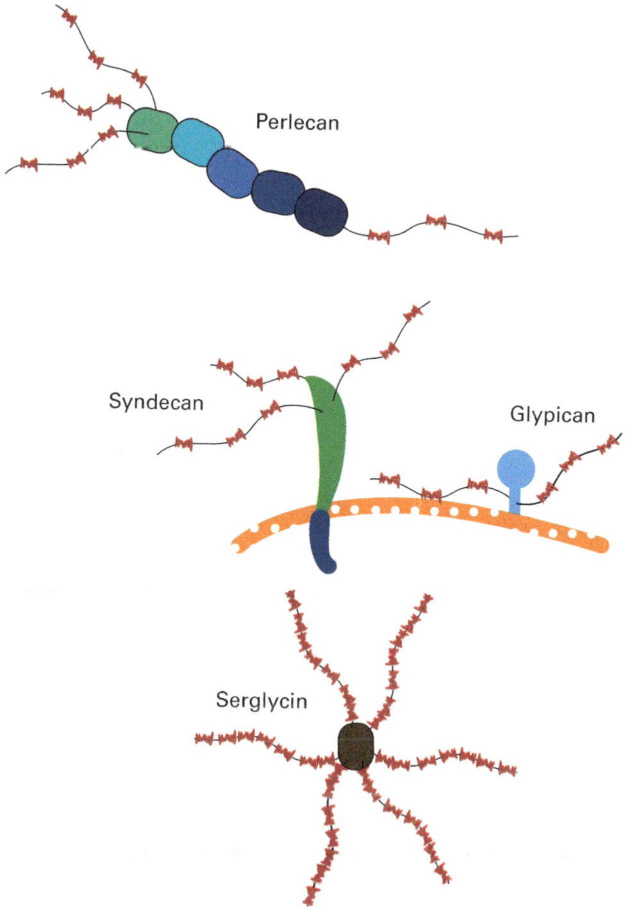

Figure 2.7 Heparan sulfate proteoglycans. Most HS proteoglycans are found in the extracellular space and at the cell surface. However, the proteoglycan serglycin, substituted with heparin (a heavily sulfated HS variant), is found intracellularly in mast cell granules. The main cell surface HS proteoglycans are the syndecans with transmembrane core proteins (four members in vertebrates) and the glypicans (six members), linked to the plasma membrane through a glycosylphospatidylinositol (GPI) anchor. Extracellular HS proteoglycans include perlecan (shown in the figure), collagen XVIII, and agrin. Sulfated NS domains are shown in red, while the unmodifed NA regions are black.

2.5.2 How is heparan sulfate synthesized and which enzymes contribute?

A newly synthesized core protein is substituted with GAG chains in the endoplasmic reticulum (ER)/Golgi compartments. The process (see Esko and Lindahl, 2001; illustrated in ▶Figure 2.8) starts with the synthesis of a tetrasaccharide linkage region (common for HS and chondroitin sulfate [CS]) to a serine residue in the protein by four different enzymes, resulting in a glucuronic acid-galactose-galactose-xylose-serine structure. Specific transferases, a GlcNAc-transferase in HS biosynthesis and an

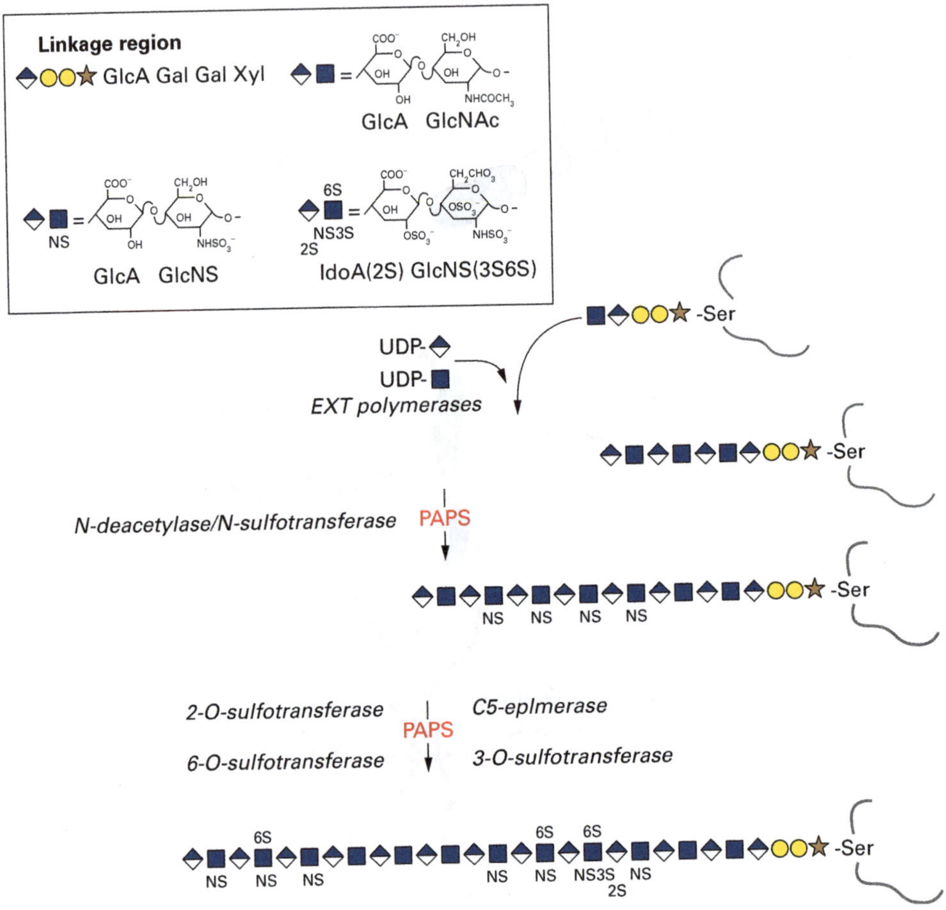

Figure 2.8 **Heparan sulfate biosynthesis.** The reducing end of the growing HS chain is attached to a serine residue in the core protein via a linkage region. Specific enzymes add the sugars in the linkage region, followed by exostosin (EXT) polymerases that elongate the chain. Symbols used for individual monosaccharides are explained in the box in the upper left. The extending HS chain is modified, first through *N*-deacetylation/*N*-sulfation, then through epimerization, 2-*O*-sulfation, 6-*O*-sulfation, and 3-*O*-sulfation by specific enzymes.

N-acetyl-D-galactosamine-transferase (GalNAc-transferase) in CS biosynthesis, will thereafter direct the biosynthesis toward HS or CS biosynthesis by adding the first sugar residue to the linkage region.

Thereafter, the chain is elongated and, at the same time, modified through epimerization and sulfation. This is carried out by a number of different enzymes, some of which exist in several isoforms. In the case of HS, the major enzymes involved are the GlcA/GlcNAc-polymerase complex (EXT1/EXT2), the *N*-deacetylase/*N*-sulfotransferase (NDST; four isoforms in mammals), a GlcA C5-epimerase, a 2-*O*-sulfotransferase, 6-*O*-sulfotransferases (6-OSTs; three isoforms), and 3-*O*-sulfotransferases (seven isoforms). The different sufotransferases all depend on the nucleotide 3′-phosphoadenosine 5′-phosphosulfate (PAPS) as the sulfate donor. Thus, the enzymes involved in the synthesis and transport of PAPS and sulfate are also essential for HS biosynthesis, as will be discussed in this chapter.

Polymerization of the chain

The polymerase complex, consisting of the EXT1 and EXT2 proteins, adds alternating GlcA and GlcNAc residues from the corresponding uridine diphosphate (UDP) sugars to the nonreducing end of the polysaccharide chain. *EXT1* and *EXT2* belong to a gene family of glycosyltransferases that also includes the EXT-like proteins 1–3 (EXTL1–3), all of which are involved in initiation and/or elongation of the HS chain (Kim et al., 2001; Busse et al., 2007). Chain elongation is promoted by concomitant sulfation and continues until the number of sugar residues added by the polymerase is in the range of 40–200. The final products are extended polysaccharides, for HS normally 22,000–45,000 Da (Lyon et al., 1994), but in mast cells that synthesize heparin, a highly sulfated variant of HS, the chains can be even longer (Robinson et al., 1978).

N-sulfation determines sulfation pattern

The modification reactions start with the exchange of selected GlcN-acetyl groups for GlcN-sulfate groups. This *N*-deacetylation/*N*-sulfation reaction is carried out by NDST enzymes (NDST1–4). NDSTs hold a key role in HS biosynthesis since further epimerization and *O*-sulfation will take place in the vicinity of these *N*-sulfated GlcN residues. Between the *N*-sulfated regions, unmodified stretches of alternating GlcA and GlcNAc residues will remain, resulting in the typical domain structure of HS.

What determines the *N*-sulfation pattern? Studies of NDSTs indicate that these enzymes act in a processive manner along the polysaccharide chain (Carlsson et al., 2008). The direction of *N*-sulfation has been suggested to be from the nonreducing end toward the reducing end (Sheng et al., 2011). Notably, this direction is toward the core protein and opposite to the direction of the growing polysaccharide chain. Since NDST1 shows affinity for EXT2 (Presto et al., 2008; see also Section 2.5.3), it is possible that this interaction physically restricts NDST1 from forming extended *N*-sulfated domains.

Epimerization and *O*-sulfation of *N*-sulfated domains

After *N*-sulfation, the C5-epimerase converts GlcA residues linked to these *N*-sulfated units (Jacobsson et al., 1984) into IdoA by epimerization of the C5 carboxyl group. The presence of IdoA residues will result in an increased flexibility of the polysaccharide chain. The 2-*O*-sulfation is closely associated with epimerization, so that most IdoA residues are 2-*O*-sulfated, while 2-*O*-sulfated GlcA residues are rare. The C5-epimerase and the 2-*O*-sulfotransferase both exist in single isoforms and are able to colocalize in the Golgi apparatus, supporting the idea that the HS biosynthetic enzymes can form complexes (Pinhal et al., 2001).

The 6- and 3-*O*-sulfotransferases carry out further *O*-sulfation of the polysaccharide chains. The three 6-OST isoforms differ slightly regarding their expression pattern and substrate preferences but can all sulfate GlcNAc and $GlcNSO_3$ residues (Jemth et al., 2003; Smeds et al., 2003). The 3-*O*-sulfotransferase exists in seven isoforms, which differ in terms of distribution pattern. The presence of 3-*O*-sulfate groups is sparse compared to sulfation in other positions of HS. Notably, a 3-*O*-sulfate group in a specific position is crucial for the formation of the antithrombin-binding pentasaccharide sequence in heparin and HS, a feature that makes heparin fragments useful as an anticoagulant drug (Petitou et al., 2003).

Another example of an HS motif with an unusual modification is the sequence permitting the binding of glycoprotein gD of herpes simplex virus type 1 (HSV-1) leading to viral infection of target cells (Shukla et al., 1999). Glycoprotein gD recognizes an octasaccharide containing the disaccharide unit (IdoAA2S-GlcNH$_2$3S±6S; see ▶Figure 2.8 for an explanation of abbreviations) (Liu et al., 2002). Most recently, it has been demonstrated that the binding of cyclophilin B (CyPB), a chemotactic factor that depend on interactions with HS chains for activity, involves *N*-unsubstituted 3-*O*-sulfated glucosamine residues (Vanpouille et al., 2007). This modification is rarely found in HS, suggesting that the specific HS sequence required for binding and activities of glycoprotein gD and CyPB could possess similar characteristics. In this way, HSV-1 might hijack 3-*O*-sulfated HS proteoglycans recognized by CyPB and take them out of their original functions to facilitate its entry within some target cells.

However, not all protein ligands recognize specific fine structure in HS. Instead, the interactions are influenced by the extension and/or distribution of the NS domains along the polysaccharide chain (Kreuger et al., 2006).

2.5.3 Fine-tuning of heparan sulfate structure in the right place, at the right time

HS biosynthesis may be regulated in many different ways. Expression levels of the different enzymes, modulation of their activities, and the availability of precursor molecules will obviously influence biosynthesis (▶Figure 2.9). The activity of an enzyme in an HS-synthesizing Golgi compartment may thus depend on both transcriptional and translational control of expression as well as posttranslational modification of the enzyme. Interactions between biosynthesis enzymes have been demonstrated, and it is possible that both elongation and modification of the HS chain takes place in "GAGosomes," molecular complexes of enzymes and other (so far unidentified) regulatory molecules. The completed HS chain may also be further modified after biosynthesis by heparanase, causing fragmentation of the chain, or by the extracellular sulfatases (Sulfs), removing 6-*O*-sulfate groups.

Regulation of enzyme activity

DNA level

Much more work is needed in this area where only limited information is available. The *NDST2* gene has been shown to be under regulation of the GA-binding protein, a transcription factor that is able to bind to a GGAA motif in the 5′-untranslated region of the *NDST2* gene and increase the amount of NDST2 transcript (Morii et al., 2001). The GA-binding protein may regulate the expression either by binding to the DNA or by binding the transcribed RNA and thereby stabilizing it. Some information about transcriptional

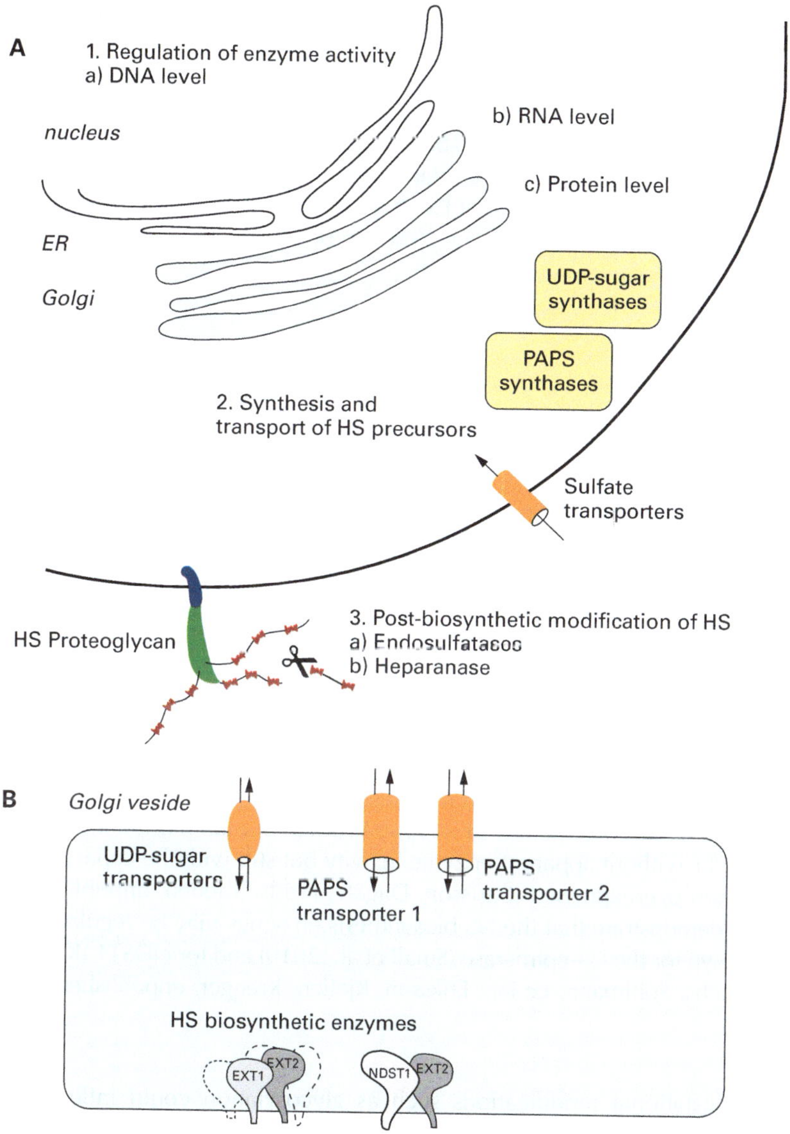

Figure 2.9 Regulation of heparan sulfate structure. (A) HS structure is regulated through (1) control of the activity of the biosynthetic enzymes on DNA, RNA, and protein level; (2) the synthesis and transport of precursors such as sulfate, PAPS, and UDP sugars; or (3) by postbiosynthetic modification of HS carried out by Sulfs and heparanase. (B) Transporters for UDP-sugars and PAPS (PAPST1 and PAPST2) are located in the Golgi membrane. The HS biosynthetic enzymes are transmembrane proteins spanning the ER/Golgi membranes, and it has been suggested that they interact with each other in molecular complexes (GAGosomes).

regulation of the PAPS synthases (Shimizu et al., 2001, 2002) as well as the Sulfs (Langsdorf et al., 2011) is also available. Recently, the Runt-related transcription factor 2 (Runx2) was shown to increase the expression of EXT1 and heparanase, and it may also alter the expression of several of the sulfotransferases (Teplyuk et al., 2009). EXT1 function has also been shown to be abrogated in human cancer cells by transcriptional silencing associated with CpG island promoter hypermethylation (Ropero et al., 2004). This epigenetic inactivation of EXT1 resulted in a loss of HS synthesis. The 3-*O*-sulfotransferase genes have also been shown to be subjected to epigenetic regulation (Bui et al., 2010; Tokuyama et al., 2010).

RNA level

The translational regulation of the NDSTs may be of specific interest because these enzymes are responsible for the overall design of the domain structure of the polysaccharide chain. Studies of murine and human NDST cDNAs revealed long, structured and AUG-rich 5′-untranslated regions (Grobe and Esko, 2002), in contrast to what is typical for most cellular mRNAs. The mRNA will fold into secondary structures that block conventional cap-dependent translation. Instead, translation is initiated by so called internal ribosome entry sites (IRES), allowing another level of control of protein expression. This type of translational control was also found for the HS biosynthetic enzymes in *Drosophila melanogaster* (Bornemann et al., 2008). Interestingly, translation of growth factors, some of which bind to HS, is regulated in a similar way (Prats and Prats, 2002). This may allow for a precise regulation of cell-specific expression levels of growth factors and their HS coreceptors during differentiation.

In mammals, there are three *6-OST* genes. The transcript of one of these (Hs6st2) is alternatively spliced giving rise to a long form and a short form with slightly different substrate specificities (Habuchi et al., 2003). Work in our lab has identified a short version of NDST1 without apparent enzyme activity but still with the capacity to alter HS structure when overexpressed (Carlsson, Dagälv, Presto, Kjellén, unpublished). Recent results also demonstrate that the HS biosynthesis enzymes may be regulated by microRNA, as shown for the C5-epimerase (Small et al., 2010) and for NDST1 (Kasza, Larsson, Fredlund Fuchs, Spillmann, Le Jan, Eriksson, Kjellén, Kreuger, unpublished).

Protein level

That posttranslational modifications such as glycosylation could influence enzyme activity has been shown for one of the CS sulfotransferases (Yusa et al., 2005). In addition, glycosylation of NDST1 affects its enzyme activity (Carlsson and Kjellén, unpublished).

Several of the enzymes have also been found to physically interact, offering an additional possibility for regulation of HS biosynthesis. The functional HS polymerase is a complex between EXT1 and EXT2 (Kobayashi et al., 2000; McCormick et al., 2000; Senay et al., 2000). Pairwise interactions have also been demonstrated for a xylosyltransferase and a galactosyltransferase (Schwartz, 1975), for the C5-epimerase and the 2-*O*-sulfotransferase (Pinhal et al., 2001), and for NDST1 and EXT2 (Presto et al., 2008). The ability of the individual enzymes (or isoforms of enzymes) to interact with others in a multienzyme complex (the GAGosome; Esko and Selleck, 2002) may be a way to regulate the HS structure formed (Ledin et al., 2006). In our study of the NDST1-EXT2 complex, we could demonstrate that the expression levels of EXT1 and EXT2 influenced the amount of active NDST1 enzyme in cells, in turn influencing HS structure (Presto et al., 2008). We hypothesized that NDST1 and EXT1 competed for binding to EXT2 in the GAGosome, resulting in lowered sulfation in the presence of a high concentration of EXT1, while sulfation increased when EXT2 was overexpressed.

Synthesis and transport of precursors

Another obvious way of regulating biosynthesis is by altering the availability of substrates. HS biosynthesis is dependent on transport of UDP sugars and PAPS into the Golgi compartment by specific transporter proteins. The UDP sugars are synthesized in the cytosol by coupling of a monosaccharide to uridine 5′-triphosphate (UTP) and are then translocated through energy-independent antiporters that simultaneously transport nucleotide monophosphates back to the cytosol (Berninsone and Hirschberg, 2000; Caffaro and Hirschberg, 2006). Both the synthesis of UDP sugars and the transport may be points of regulation of HS biosynthesis.

Any cellular sulfation reactions are dependent on the transport of inorganic sulfate into the cell by transporters in the plasma membrane. Recessively inherited disorders affecting bone and cartilage development have been related to mutations of genes encoding such sulfate transporters (Superti-Furga et al., 1996). The sulfate donor PAPS is synthesized in the cytosol from inorganic sulfate and adenosine-5′-triphosphate (ATP) by PAPS synthase (PAPSS). Two isoforms exist in vertebrates, PAPSS1 and PAPSS2 (Fuda et al., 2002; reviewed by Venkatachalam, 2003). While PAPSS2 is more restricted in its expression, PAPSS1 is ubiquitously distributed. Interestingly, PAPSS is not only a cytosolic protein, but has also been found in the nucleus (Besset et al., 2000), although its function there is not known. PAPS is translocated by transporter proteins (PAPST1 and 2) from the cytosol to Golgi compartments where HS biosynthesis takes place (Kamiyama et al., 2003, 2006).

The PAPS concentration may be a critical factor for regulating the length of NS domains. *In vitro* experiments with recombinant NDST enzyme have shown that in the absence of PAPS, the enzyme removes acetyl groups from the substrate in a more or less random fashion. With PAPS present, the NDST enzyme instead works in a processive manner along the chain to create NS-domains, the length of which depends on the PAPS concentration (Carlsson et al., 2008).

Postbiosynthetic modification of heparan sulfate

Endosulfatases

The Sulfs are to date the only known extracellular endosulfatases that can remodel sulfation after biosynthesis. Two member of the Sulf family are known, Sulf1 and Sulf2. The structural regions in Sulfs required for the active site formation are homologous with the lysosomal GlcNAc sulfatases, enzymes that catalyze the hydrolysis of 6-*O*-sulfate groups from GlcNAc residues of HS chains during degradation. In addition, the Sulfs have a distinctive hydrophilic domain and an N-terminal secretory signal domain (Dhoot et al., 2001; Ai et al., 2006) that direct them to their location in the plasma membrane.

Sulf1 and Sulf2 show similar enzymatic activities toward a specific subset of 6-*O*-sulfate groups on the trisulfated disaccharide unit, IdoA2S-GlcNS6S, within the highly sulfated domains of HS chains (Ai et al. 2003, 2006). By regulating HS-ligand interactions, Sulf1 has been shown to promote Wnt, glial cell line–derived neurotrophic factor (GDNF), and bone morphogenetic protein (BMP) signaling (Ai et al., 2003, 2007; Viviano et al., 2004; Wang et al., 2004). In contrast, Sulf2 inhibits FGF and vascular endothelial growth factor (VEGF) signaling by disrupting the formation of ternary complexes (Wang et al., 2004; Uchimura et al., 2006).

Expression and distribution pattern studies and the generation of Sulf-deficient mice in the past 10 years indicate that Sulf1 and Sulf2 have complementary but not completely overlapping roles in cell-signaling events during development and wound repair (Morimoto-Tomita et al., 2002; Ai et al., 2007; Holst et al., 2007; Langsdorf et al., 2007; Lum et al., 2007; Ratzka et al., 2008).

Heparanase

Heparanase is an enzyme with endoglucuronidase activity that takes part in HS degradation within the cell. It is also present extracellularly where it cleaves HS chains into fragments of M_r ~ 5,000–10,000. It cleaves HS at GlcA residues surrounded by sulfated sugar units (Pikas et al., 1998; Okada et al., 2002) and thus tends to excise heavily modified NS domains. Fragmentation of HS will lead to the release of bioactive HS fragments and HS-bound growth factors and may result in disassembly of extracellular barriers. Interestingly, upregulation of heparanase will increase sulfation of HS apparently through regulation of biosynthesis (Escobar Galvis et al., 2007).

2.5.4 Disturbed heparan sulfate biosynthesis in human pathobiology

Disturbed biosynthesis of HS resulting in altered HS structure will affect HS-protein interactions that may cause or contribute to human pathological conditions. Some pathological conditions are known that can be directly related to HS (or CS) biosynthesis, as exemplified subsequently.

Heterozygous mutations in the EXT polymerases result in hereditary multiple exostoses, characterized by the formation of cartilage-capped benign tumors (exostoses) that develop from the growth plate of endochondral bone (reviewed by Jennes et al., 2009). The disturbed HS biosynthesis increases the proliferation of chondrocytes and results in excessive bone growth during the first decades of life, often leading to short stature and deformities causing functional problems.

Due to mutations in one of the proteins (SLC26A2) that transport inorganic sulfate from the extracellular milieu into the cytoplasm, undersulfated proteoglycans are synthesized. This results in different types of human osteochondrodysplasias (Dawson, 2011), such as diastrophic dysplasia, multiple epiphyseal dysplasia, atelosteogenesis type II, and achondogenesis type IB. The clinical features are mild to severe and include joint pain and deformity, dwarfism, or respiratory insufficiency and death in the neonatal period, and so forth. Mutations in PAPSS2, which synthesizes the sulfate donor PAPS for sulfotransferase reactions, have also been linked to developmental dwarfism disorders (Faiyaz ul Haque et al., 1998). However, since all sulfation reactions will be affected by altered PAPS metabolism, it is not possible to conclude that HS (or CS) sulfation alone are responsible for the pathological changes seen.

Some families with idiopathic hypogonadotropic hypogonadism (IHH; absent or incomplete puberty by the age of 18 years in the absence of other causes) have mutations in 6-OST1 (Tornberg et al., 2011). These mutations result in defects in neuronal development associated with the olfactory nerve and the migration of the neurons secreting gonadotropin-releasing hormone that regulates sexual development. A connection between HS and IHH-associated genes, including *kal-1* (coding for the neural cell adhesion protein anosmin-1), *FGF*, and the FGF receptor, could be confirmed using *Caenorhabditis elegans* as the model.

An elevated concentration of heparanase and increased cleavage of extracellular HS will affect pathological processes such as inflammation and tumorigenesis (reviewed by Li and Vlodavsky, 2009 and Vlodavsky et al., 2011). The Sulfs have also been reported to have important roles in the pathogenesis of a number of different cancers. Sulf1 has generally been implicated as a tumor suppressor, as its expression is downregulated in a variety of tumors (Narita et al., 2007; Lai et al., 2009). However, the role of Sulf1 as a tumor suppressor has recently been questioned (Rosen and Lemjabbar-Alaoui, 2010). Sulf2 expression is upregulated in a wide range of human tumors where it may promote angiogenesis and cancer cell growth through activation of Wnt and receptor tyrosine kinases and their downstream pathways (Nawroth et al., 2007; Lai et al., 2008). Therefore, Sulf2 has been designated a tumorigenic role correlated with the aggressiveness and severity of the disease (Morimoto-Tomita et al., 2005; Lai et al., 2008).

Through studies in model organisms (e.g. mice, zebrafish, *Drosophila melanogaster*, and *Caenorhabditis elegans*) much has been learned about the requirement of correct HS biosynthesis for normal development. In addition, the phenotypes acquired by the animal mutants predict that regulation of HS biosynthesis is also important for normal tissue homeostasis in the adult organism (Bishop et al., 2007).

2.5.5 Take-home message

The biological functions of HS depend on the structure of the HS chain, which is largely determined during biosynthesis. Postbiosynthetic modifications by heparanase and/or endosulfatases may also modify the structure.

Regulation of biosynthesis involves control of both enzyme transcription and translation as well as posttranslational modifications of the enzymes. Synthesis and transport of UDP sugars and the sulfate donor PAPS will also affect the outcome of biosynthesis.

References

Ai, X., Do, A. T., Kusche-Gullberg, M., Lindahl, U., Lu, K., and Emerson, C. P., Jr. (2006). Substrate specificity and domain functions of extracellular heparan sulfate 6-O-endosulfatases, QSulf1 and QSulf2. J Biol Chem *281* (8), 4969–4976.

Ai, X., Do, A. T., Lozynska, O., Kusche-Gullberg, M., Lindahl, U., and Emerson, C. P., Jr. (2003). QSulf1 remodels the 6-O sulfation states of cell surface heparan sulfate proteoglycans to promote Wnt signaling. J Cell Biol *162* (2), 341–351.

Ai, X., Kitazawa, T., Do, A. T., Kusche-Gullberg, M., Labosky, P. A., and Emerson, C. P., Jr. (2007). SULF1 and SULF2 regulate heparan sulfate-mediated GDNF signaling for esophageal innervation. Development *134* (18), 3327–3338.

Berninsone, P. M., and Hirschberg, C. B. (2000). Nucleotide sugar transporters of the Golgi apparatus. Curr Opin Struct Biol *10* (5), 542–547.

Besset, S., Vincourt, J. B., Amalric, F., and Girard, J. P. (2000). Nuclear localization of PAPS synthetase 1: a sulfate activation pathway in the nucleus of eukaryotic cells. FASEB J *14* (2), 345–354.

Bishop, J. R., Schuksz, M., and Esko, J. D. (2007). Heparan sulphate proteoglycans fine-tune mammalian physiology. Nature *446* (7139), 1030–1037.

Bornemann, D. J., Park, S., Phin, S., and Warrior, R. (2008). A translational block to HSPG synthesis permits BMP signaling in the early Drosophila embryo. Development *135* (6), 1039–1047.

Bui, C., Ouzzine, M., Talhaoui, I., et al. (2010). Epigenetics: methylation-associated repression of heparan sulfate 3-O-sulfotransferase gene expression contributes to the invasive phenotype of H-EMC-SS chondrosarcoma cells. FASEB J *24* (2), 436–450.

Bulow, H. E., and Hobert, O. (2006). The molecular diversity of glycosaminoglycans shapes animal development. Annu Rev Cell Dev Biol *22*, 375–407.

Busse, M., Feta, A., Presto, J., et al. (2007). Contribution of EXT1, EXT2, and EXTL3 to heparan sulfate chain elongation. J Biol Chem *282* (45), 32802–32810.

Caffaro, C. E., and Hirschberg, C. B. (2006). Nucleotide sugar transporters of the Golgi apparatus: from basic science to diseases. Acc Chem Res *39* (11), 805–812.

Carlsson, P., Presto, J., Spillmann, D., Lindahl, U., and Kjellen, L. (2008). Heparin/heparan sulfate biosynthesis: processive formation of N-sulfated domains. J Biol Chem *283* (29), 20008–20014.

Dawson, P. A. (2011). Sulfate in fetal development. Semin Cell Dev Biol *22* (6), 653–659.

Dhoot, G. K., Gustafsson, M. K., Ai, X., Sun, W., Standiford, D. M., and Emerson, C. P., Jr. (2001). Regulation of Wnt signaling and embryo patterning by an extracellular sulfatase. Science *293* (5535), 1663–1666.

Escobar Galvis, M. L., Jia, J., Zhang, X., et al. (2007). Transgenic or tumor-induced expression of heparanase upregulates sulfation of heparan sulfate. Nat Chem Biol *3* (12), 773–778.

Esko, J. D., and Lindahl, U. (2001). Molecular diversity of heparan sulfate. J Clin Invest *108* (2), 169–173.

Esko, J. D., and Selleck, S. B. (2002). Order out of chaos: assembly of ligand binding sites in heparan sulfate. Annu Rev Biochem *71*, 435–471.

Faiyaz ul Haque, M., King, L. M., Krakow, D., et al. (1998). Mutations in orthologous genes in human spondyloepimetaphyseal dysplasia and the brachymorphic mouse. Nat Genet *20* (2), 157–162.

Fuda, H., Shimizu, C., Lee, Y. C., Akita, H., and Strott, C. A. (2002). Characterization and expression of human bifunctional 3′-phosphoadenosine 5′-phosphosulphate synthase isoforms. Biochem J *365* (Pt 2), 497–504.

Grobe, K., and Esko, J. D. (2002). Regulated translation of heparan sulfate N-acetylglucosamine N-deacetylase/n-sulfotransferase isozymes by structured 5′-untranslated regions and internal ribosome entry sites. J Biol Chem *277* (34), 30699–30706.

Habuchi, H., Miyake, G., Nogami, K., et al. (2003). Biosynthesis of heparan sulphate with diverse structures and functions: two alternatively spliced forms of human heparan sulphate 6-O-sulphotransferase-2 having different expression patterns and properties. Biochem J *371* (Pt 1), 131–142.

Holst, C. R., Bou-Reslan, H., Gore, B. B., et al. (2007). Secreted sulfatases Sulf1 and Sulf2 have overlapping yet essential roles in mouse neonatal survival. PLoS One *2* (6), e575.

Jacobsson, I., Lindahl, U., Jensen, J. W., Roden, L., Prihar, H., and Feingold, D. S. (1984). Biosynthesis of heparin. Substrate specificity of heparosan N-sulfate D-glucuronosyl 5-epimerase. J Biol Chem *259* (2), 1056–1063.

Jemth, P., Smeds, E., Do, A. T., et al. (2003). Oligosaccharide library-based assessment of heparan sulfate 6-O-sulfotransferase substrate specificity. J Biol Chem *278* (27), 24371–24376.

Jennes, I., Pedrini, E., Zuntini, M., et al. (2009). Multiple osteochondromas: mutation update and description of the multiple osteochondromas mutation database (MOdb). Hum Mutat *30* (12), 1620–1627.

Kamiyama, S., Sasaki, N., Goda, E., et al. (2006). Molecular cloning and characterization of a novel 3′-phosphoadenosine 5′-phosphosulfate transporter, PAPST2. J Biol Chem *281* (16), 10945–10953.

Kamiyama, S., Suda, T., Ueda, R., et al. (2003). Molecular cloning and identification of 3′-phosphoadenosine 5′-phosphosulfate transporter. J Biol Chem *278* (28), 25958–25963.

Kim, B. T., Kitagawa, H., Tamura, J., et al. (2001). Human tumor suppressor EXT gene family members EXTL1 and EXTL3 encode alpha 1,4- N-acetylglucosaminyltransferases that likely are involved in heparan sulfate/ heparin biosynthesis. Proc Natl Acad Sci U S A *98* (13), 7176–7181.

Kobayashi, S., Morimoto, K., Shimizu, T., Takahashi, M., Kurosawa, H., and Shirasawa, T. (2000). Association of EXT1 and EXT2, hereditary multiple exostoses gene products, in Golgi apparatus. Biochem Biophys Res Commun *268* (3), 860–867.

Kreuger, J., Spillmann, D., Li, J. P., and Lindahl, U. (2006). Interactions between heparan sulfate and proteins: the concept of specificity. J Cell Biol *174* (3), 323–327.

Lai, J. P., Sandhu, D. S., Moser, C. D., et al. (2009). Additive effect of apicidin and doxorubicin in sulfatase 1 expressing hepatocellular carcinoma in vitro and in vivo. J Hepatol *50* (6), 1112–1121.

Lai, J. P., Sandhu, D. S., Yu, C., et al. (2008). Sulfatase 2 up-regulates glypican 3, promotes fibroblast growth factor signaling, and decreases survival in hepatocellular carcinoma. Hepatology *47* (4), 1211–1222.

Langsdorf, A., Do, A. T., Kusche-Gullberg, M., Emerson, C. P., Jr., and Ai, X. (2007). Sulfs are regulators of growth factor signaling for satellite cell differentiation and muscle regeneration. Dev Biol *311* (2), 464–477.

Langsdorf, A., Schumacher, V., Shi, X., et al. (2011). Expression regulation and function of heparan sulfate 6-O-endosulfatases in the spermatogonial stem cell niche. Glycobiology *21* (2), 152–161.

Ledin, J., Ringvall, M., Thuveson, M., et al. (2006). Enzymatically active N-deacetylase/N-sulfotransferase-2 is present in liver but does not contribute to heparan sulfate N-sulfation. J Biol Chem *281* (47), 35727–35734.

Ledin, J., Staatz, W., Li, J. P., et al. (2004). Heparan sulfate structure in mice with genetically modified heparan sulfate production. J Biol Chem *279* (41), 42732–42741.

Li, J. P., and Vlodavsky, I. (2009). Heparin, heparan sulfate and heparanase in inflammatory reactions. Thromb Haemost *102* (5), 823–828.

Lindahl, U., and Li, J. P. (2009). Interactions between heparan sulfate and proteins-design and functional implications. Int Rev Cell Mol Biol *276*, 105–159.

Liu, J., Shriver, Z., Pope, R. M., et al. (2002). Characterization of a heparan sulfate octasaccharide that binds to herpes simplex virus type 1 glycoprotein D. J Biol Chem *277* (36), 33456–33467.

Lum, D. H., Tan, J., Rosen, S. D. and Werb, Z. (2007). Gene trap disruption of the mouse heparan sulfate 6-O-endosulfatase gene, Sulf2. Mol Cell Biol *27* (2), 678–688.

Lyon, M., Deakin, J. A., and Gallagher, J. T. (1994). Liver heparan sulfate structure. A novel molecular design. J Biol Chem *269* (15), 11208–11215.

McCormick, C., Duncan, G., Goutsos, K. T., and Tufaro, F. (2000). The putative tumor suppressors EXT1 and EXT2 form a stable complex that accumulates in the Golgi apparatus and catalyzes the synthesis of heparan sulfate. Proc Natl Acad Sci U S A *97* (2), 668–673.
Morii, E., Ogihara, H., Oboki, K., et al. (2001). Inhibitory effect of the mi transcription factor encoded by the mutant mi allele on GA binding protein-mediated transcript expression in mouse mast cells. Blood *97* (10), 3032–3039.
Morimoto-Tomita, M., Uchimura, K., Bistrup, A., et al. (2005). Sulf-2, a proangiogenic heparan sulfate endosulfatase, is upregulated in breast cancer. Neoplasia *7* (11), 1001–1010.
Morimoto-Tomita, M., Uchimura, K., Werb, Z., Hemmerich, S., and Rosen, S. D. (2002). Cloning and characterization of two extracellular heparin-degrading endosulfatases in mice and humans. J Biol Chem *277* (51), 49175–49185.
Narita, K., Chien, J., Mullany, S. A., et al. (2007). Loss of HSulf-1 expression enhances autocrine signaling mediated by amphiregulin in breast cancer. J Biol Chem *282* (19), 14413–14420.
Nawroth, R., van Zante, A., Cervantes, S., McManus, M., Hebrok, M., and Rosen, S. D. (2007). Extracellular sulfatases, elements of the Wnt signaling pathway, positively regulate growth and tumorigenicity of human pancreatic cancer cells. PLoS One *2* (4), e392.
Okada, Y., Yamada, S., Toyoshima, M., Dong, J., Nakajima, M., and Sugahara, K. (2002). Structural recognition by recombinant human heparanase that plays critical roles in tumor metastasis. Hierarchical sulfate groups with different effects and the essential target disulfated trisaccharide sequence. J Biol Chem *277* (45), 42488–42495.
Petitou, M., Casu, B., and Lindahl, U. (2003). 1976–1983, a critical period in the history of heparin: the discovery of the antithrombin binding site. Biochimie *85* (1–2), 83–89.
Pikas, D. S., Li, J. P., Vlodavsky, I., and Lindahl, U. (1998). Substrate specificity of heparanases from human hepatoma and platelets. J Biol Chem *273* (30), 18770–18777.
Pinhal, M. A., Smith, B., Olson, S., Aikawa, J., Kimata, K. and Esko, J. D. (2001). Enzyme interactions in heparan sulfate biosynthesis: uronosyl 5-epimerase and 2-O-sulfotransferase interact in vivo. Proc Natl Acad Sci U S A *98* (23), 12984–12989.
Prats, A. C., and Prats, H. (2002). Translational control of gene expression: role of IRESs and consequences for cell transformation and angiogenesis. Prog Nucleic Acid Res Mol Biol *72*, 367–413.
Presto, J., Thuveson, M., Carlsson, P., et al. (2008). Heparan sulfate biosynthesis enzymes EXT1 and EXT2 affect NDST1 expression and heparan sulfate sulfation. Proc Natl Acad Sci U S A *105* (12), 4751–4756.
Ratzka, A., Kalus, I., Moser, M., Dierks, T., Mundlos, S., and Vortkamp, A. (2008). Redundant function of the heparan sulfate 6-O-endosulfatases Sulf1 and Sulf2 during skeletal development. Dev Dyn *237* (2), 339–353.
Robinson, H. C., Horner, A. A., Hook, M., Ogren, S., and Lindahl, U. (1978). A proteoglycan form of heparin and its degradation to single-chain molecules. J Biol Chem *253* (19), 6687–6693.
Ropero, S., Setien, F., Espada, J., et al. (2004). Epigenetic loss of the familial tumor-suppressor gene exostosin-1 (EXT1) disrupts heparan sulfate synthesis in cancer cells. Hum Mol Genet *13* (22), 2753–2765.
Rosen, S. D., and Lemjabbar-Alaoui, H. (2010). Sulf-2: an extracellular modulator of cell signaling and a cancer target candidate. Expert Opin Ther Targets *14* (9), 935–949.
Schwartz, N. B. (1975). Biosynthesis of chondroitin sulfate: immunoprecipitation of interacting xylosyltransferase and galactosyltransferase. FEBS Lett *49* (3), 342–345.
Senay, C., Lind, T., Muguruma, K., et al. (2000). The EXT1/EXT2 tumor suppressors: catalytic activities and role in heparan sulfate biosynthesis. EMBO Rep *1* (3), 282–286.
Sheng, J., Liu, R., Xu, Y., and Liu, J. (2011). The dominating role of N-deacetylase/N-sulfotransferase 1 in forming domain structures in heparan sulfate. J Biol Chem *286* (22), 19768–19776.

Shimizu, C., Fuda, H., Lee, Y. C., and Strott, C. A. (2001). Transcriptional regulation of human 3′-phosphoadenosine 5′-phosphosulfate synthase 1. Biochem Biophys Res Commun *284* (3), 763–770.

Shimizu, C., Fuda, H., Lee, Y. C., and Strott, C. A. (2002). Transcriptional regulation of human 3′-phosphoadenosine 5′-phosphosulphate synthase 2. Biochem J *363* (Pt 2), 263–271.

Shukla, D., Liu, J., Blaiklock, P., et al. (1999). A novel role for 3-O-sulfated heparan sulfate in herpes simplex virus 1 entry. Cell *99* (1), 13–22.

Small, E. M., Sutherland, L. B., Rajagopalan, K. N., Wang, S., and Olson, E. N. (2010). MicroRNA-218 regulates vascular patterning by modulation of Slit-Robo signaling. Circ Res *107* (11), 1336–1344.

Smeds, E., Habuchi, H., Do, A. T., et al. (2003). Substrate specificities of mouse heparan sulphate glucosaminyl 6-O-sulphotransferases. Biochem J *372* (Pt 2), 371–380.

Superti-Furga, A., Hastbacka, J., Rossi, A., et al. (1996). A family of chondrodysplasias caused by mutations in the diastrophic dysplasia sulfate transporter gene and associated with impaired sulfation of proteoglycans. Ann N Y Acad Sci *785*, 195–201.

Teplyuk, N. M., Haupt, L. M., Ling, L., et al. (2009). The osteogenic transcription factor Runx2 regulates components of the fibroblast growth factor/proteoglycan signaling axis in osteoblasts. J Cell Biochem *107* (1), 144–154.

Tokuyama, Y., Takahashi, T., Okumura, N., et al. (2010). Aberrant methylation of heparan sulfate glucosamine 3-O-sulfotransferase 2 genes as a biomarker in colorectal cancer. Anticancer Res *30* (12), 4811–4818.

Tornberg, J., Sykiotis, G. P., Keefe, K., et al. (2011). Heparan sulfate 6-O-sulfotransferase 1, a gene involved in extracellular sugar modifications, is mutated in patients with idiopathic hypogonadotrophic hypogonadism. Proc Natl Acad Sci U S A *108* (28), 11524–11529.

Uchimura, K., Morimoto-Tomita, M., Bistrup, A., et al. (2006). HSulf-2, an extracellular endoglucosamine-6-sulfatase, selectively mobilizes heparin-bound growth factors and chemokines: effects on VEGF, FGF-1, and SDF-1. BMC Biochem *7*, 2.

Vanpouille, C., Deligny, A., Delehedde, M., et al. (2007). The heparin/heparan sulfate sequence that interacts with cyclophilin B contains a 3-O-sulfated N-unsubstituted glucosamine residue. J Biol Chem *282* (33), 24416–24429.

Venkatachalam, K. V. (2003). Human 3′-phosphoadenosine 5′-phosphosulfate (PAPS) synthase: biochemistry, molecular biology and genetic deficiency. IUBMB Life *55* (1), 1–11.

Viviano, B. L., Paine-Saunders, S., Gasiunas, N., Gallagher, J., and Saunders, S. (2004). Domain-specific modification of heparan sulfate by Qsulf1 modulates the binding of the bone morphogenetic protein antagonist Noggin. J Biol Chem *279* (7), 5604–5611.

Vlodavsky, I., Beckhove, P., Lerner, I., et al. (2011). Significance of heparanase in cancer and inflammation. Cancer Microenviron doi: 10.1007/s12307-011-0082-7.

Wang, S., Ai, X., Freeman, S. D., et al. (2004). QSulf1, a heparan sulfate 6-O-endosulfatase, inhibits fibroblast growth factor signaling in mesoderm induction and angiogenesis. Proc Natl Acad Sci U S A *101* (14), 4833–4838.

Westling, C., and Lindahl, U. (2002). Location of N-unsubstituted glucosamine residues in heparan sulfate. J Biol Chem *277* (51), 49247–49255.

Yusa, A., Kitajima, K., and Habuchi, O. (2005). N-linked oligosaccharides are required to produce and stabilize the active form of chondroitin 4-sulphotransferase-1. Biochem J *388* (Pt 1), 115–121.

2.6 Bone and skin disorders caused by a disturbance in the biosynthesis of chondroitin sulfate and dermatan sulfate

Shuji Mizumoto and Kazuyuki Sugahara

2.6.1 Introduction

Chondroitin sulfate (CS) and dermatan sulfate (DS) chains are classified as glycosaminoglycans (GAGs) and covalently attached to specific core proteins, forming proteoglycans (PGs). They are ubiquitously distributed at cell surfaces and in extracellular matrices (ECMs) (▶Figure 2.10). PGs often play roles through the GAG side chains in a variety of biological processes such as cell adhesion, proliferation, tissue morphogenesis, neurite outgrowth, infections of viruses/bacteria, and the regulation of various growth factors/cytokine effectors (▶Figure 2.10) (Sugahara et al., 2003; Bishop et al., 2007; Sugahara and Mikami, 2007). Aggrecan and decorin are representative chondroitin sulfate proteoglycans (CSPGs) and dermatan sulfate proteoglycans (DSPGs) and are involved in bone morphogenesis and the formation of collagen fibers/bundles, respectively (▶Figure 2.10). On the other hand, RPTP-β (receptor-type protein tyrosine phosphatase beta), which is a type I membrane protein CSPG with a transmembrane domain and intracellular tyrosine phosphatase domains, regulates the neuron–glia cell interactions, neuronal migration, adhesion, and neurite outgrowth and pathfinding in the brain (▶Figure 2.10) (Sugahara and Mikami, 2007).

The sugar back bones of CS and DS chains are linear polysaccharides consisting of disaccharide units, [-4GlcUAβ1–3GalNAcβ1-]n and [-4IdoUAα1–3GalNAcβ1-]n, where GlcUA, GalNAc, and IdoUA stand for D-glucuronic acid, *N*-acetyl-D-galactosamine, and L-iduronic acid, respectively (▶Figure 2.11). Both types of polysaccharides are often found as CS/DS hybrid chains in mammalian tissues (Sugahara and Mikami, 2007). CS/DS are modified by specific sulfotransferases, which catalyze the transfer of a sulfate group from 3′-phosphoadenosine 5′-phosphosulfate (PAPS) to C-2 of GlcUA and/or IdoUA, and C-4 and/or C-6 of GalNAc, yielding enormous structural diversity. Accordingly, the sulfated CS, DS, and CS/DS hybrid chains display a diverse structure with distinct combinations of characteristic disaccharide units as shown in ▶Figure 2.11. Monosulfated GlcUA/IdoUA-GalNAc(4-*O*-sulfate) or GlcUA/IdoUA-GalNAc(6-*O*-sulfate) disaccharides are abbreviated as A/iA and C/iC units. Representative disulfated disaccharide units are GlcUA(2-*O*-sulfate)/IdoUA(2-*O*-sulfate)-GalNAc(6-*O*-sulfate) (D/iD unit), GlcUA/IdoUA-GalNAc(4-*O*-, 6-*O*-sulfate) (E/iE unit), and IdoUA(2-*O*-sulfate)-GalNAc(4-*O*-sulfate) (iB unit) (▶Figure 2.11). A and C units are common and major components of CS chains in mammalian tissues, whereas the disulfated disaccharides such as D and E units are rare, but their proportions as well as those of A and C units change

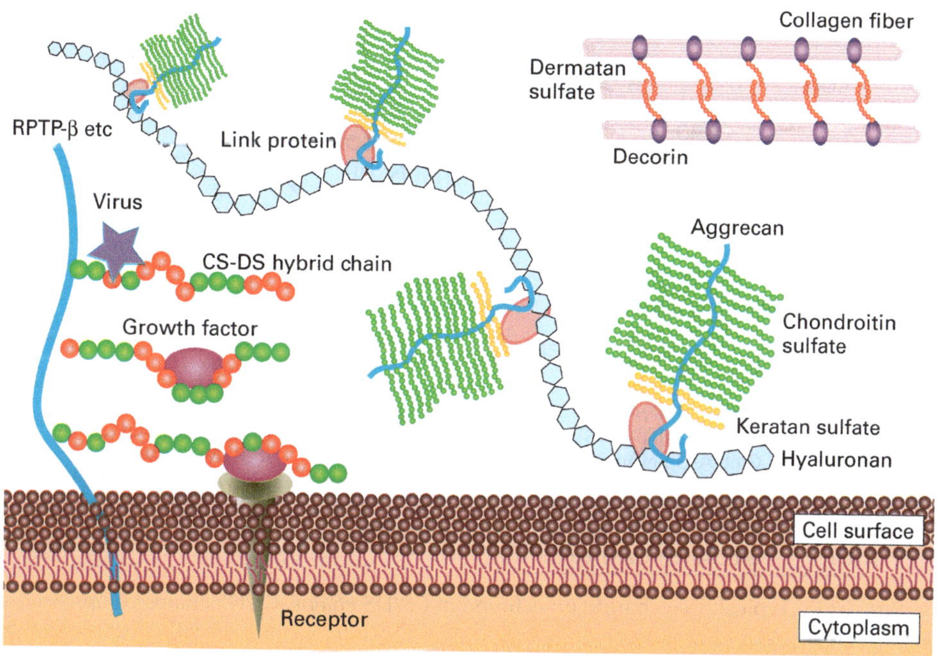

Figure 2.10 Various functions of CS/DS-PGs at cell surfaces and in extracellular matrices. CS and DS are widely distributed as sulfated GAG side chains of PGs in extracellular matrices and at cell surfaces. Aggrecan is a macromolecule bearing ~100 CS side chains and ~60 keratan sulfate side chains on a single core protein as one of the main components of cartilage. In addition, aggrecan binds to hyaluronan with the aid of a hyaluronan-link protein and contributes to elasticity, water retention, and the construction of the matrix in cartilage. On the other hand, decorin is abundantly synthesized in skin and contributes to the construction of collagen fibers and bundles. Furthermore, CS/DS-PGs at cell surfaces interact with a large number of proteins such as viral envelope proteins and various growth factors, resulting in infection and signal transduction by presenting such proteins to specific receptors, respectively. RPTP-β is a type I membrane protein CSPG with a transmembrane domain and intracellular tyrosine phosphatase domains and regulates brain development through the signaling.

during development and tissue morphogenesis (Sugahara and Mikami, 2007; Akatsu et al., 2011), under pathological conditions such as tumorigenesis, and with the metastatic capacity of tumor cells (Li et al., 2008). Moreover, CS structures containing C, D, and E units play distinct roles in bone development (Thiele et al., 2004; van Roij et al., 2008), neuritogenesis (Nadanaka et al., 1998; Bao et al., 2005), and viral infections (Uyama et al., 2006), respectively. Decorin PG bearing a DS structure containing iA units modulates the formation of collagen bundles in skin (Miyake et al., 2010).

Thus, the identification of altered functional CS/DS structures constructed by different combinations and sequential arrangements of distinct disaccharide units aid in the development of diagnostic and therapeutic methods for human diseases. Furthermore, recent studies involving genetic analyses using knockout mice, zebrafish, fruit flies, and nematodes have led to unanticipated findings of interesting biological phenotypes (Perrimon and Bernfield, 2000; Sugahara et al., 2003; Bishop et al., 2007).

CS DS

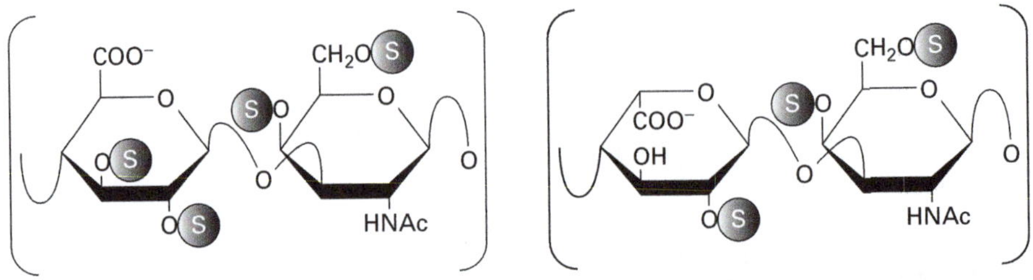

[-4GlcUAβ1-3GalNAcβ1-] [-4IdoUAα1-3GalNAcβ1-]

Unit	Sequence
O unit	GlcUA-GalNAc
A unit	GlcUA-GalNAc(4S)
C unit	GlcUA-GalNAc(6S)
B unit	GlcUA(2S)-GalNAc(4S)
D unit	GlcUA(2S)-GalNAc(6S)
E unit	GlcUA-GalNAc(4S,6S)
K unit	GlcUA(3S)-GalNAc(4S)

Unit	Sequence
iO unit	IdoUA-GalNAc
iA unit	IdoUA-GalNAc(4S)
iC unit	IdoUA-GalNAc(6S)
iB unit	IdoUA(2S)-GalNAc(4S)
iD unit	IdoUA(2S)-GalNAc(6S)
iE(H) unit	IdoUA-GalNAc(4S,6S)

Figure 2.11 Typical disaccharide units found in CS/DS chains. CS/DS chains comprise uronic acid (GlcUA or IdoUA) and GalNAc residues. DS is a stereoisomer of CS including IdoUA instead of or in addition to GlcUA. These sugar residues can be esterified by sulfate at various positions as indicated by "S" enclosed by a circle. Sulfation at the C-3 position of a GlcUA residue is occasionally found in CS-E and CS-K derived from the cartilage of squid and king crab, respectively. CS-K was named after the king crab and contains K units (Seno and Murakami, 1982). So far, no such sulfate group has been reported for vertebrates. The disaccharide units of CS chains are largely classified into O, A, C, D, B, and E units on the basis of their sulfation patterns. iA, iC, iD, iB, and iE units are the isomeric counterparts found in DS chains. 2S, 4S, and 6S represent the 2-*O*-. 4-*O*-, and 6-*O*-sulfate groups, respectively. The abbreviation "i" in iA–iE stands for IdoUA.

In this chapter, we will focus on recent advances in the study of biosynthetic mechanisms of CS and DS chains and hereditary disorders involving a disturbance of their structure.

2.6.2 Biosynthetic pathways of CS and DS chains

Uridine diphosphate sugars

All glycosyltransferases responsible for the biosynthesis of GAGs use uridine diphosphate (UDP) sugars as sugar donor substrates and contain at least one DXD motif, where D and X stand for aspartate and any amino acid, respectively, for the binding of a UDP sugar, which is characteristic of Golgi-residing glycosyltransferases (Wiggins and Munro, 1998; Pedersen et al., 2000, 2003). UDP sugars are synthesized in the cytosol by a panel of enzymes, except UDP-xylose, which is produced from UDP-GlcUA by UDP-GlcUA decarboxylase in the Golgi lumen (Moriarity et al., 2002). UDP-GlcUA is formed by the action of the UDP-Glc dehydrogenase on UDP-Glc in the cytosol (Spicer et al., 1998). The UDP sugars are transported into the lumen of the Golgi and endoplasmic reticulum by the corresponding UDP-sugar transporters, where a number of glycosyltransferases utilize UDP sugars to build up various glycans including GAGs. These transporters take various UDP sugars from the cytosol into the lumen of the Golgi and endoplasmic reticulum.

3′-Phosphoadenosine 5′-phosphosulfate

The sulfation of GAGs is crucial to their physiological functions. Sulfotransferases catalyze the transfer of a sulfate group from the donor substrate, PAPS, to respective acceptor substrates. PAPS is synthesized from adenosine triphosphate (ATP) and inorganic sulfate, which is transported into the cytosol by the sulfate transporter. The conjugation reaction occurs in two sequential steps: (1) inorganic sulfate reacts with ATP to form adenosine 5′-phosphosulfate (APS) and pyrophosphate by the actions of ATP sulfurylase (Sugahara and Schwartz, 1979); (2) the reaction of APS with ATP is catalyzed by APS kinase, resulting in the formation of PAPS and adenosine diphosphate (ADP). ATP sulfurylase and APS kinase have been cloned from bacteria, fungi, yeast, and plants, whereas in animals these enzymes are combined to form a bifunctional single enzyme and referred to as PAPS synthase (Schwartz et al., 1998).

Biosynthesis of the GAG-protein linkage region

The transfer of a D-xylose residue to specific serine (Ser) residues in core proteins by β-xylosyltransferase (XylT) takes place in the endoplasmic reticulum and *cis*-Golgi compartments (▶Figure 2.12, ▶Table 2.3) (Götting et al., 2000; Pönighaus et al., 2007). Next, β1,4-galactosyltransferase-I (GalT-I) and β1,3-galactosyltransferase-II (GalT-II) successively transfer two D-galactose residues to the xylose residue attached to the Ser residue, respectively (Almeida et al., 1999; Okajima et al., 1999b; Bai et al., 2001). β1,3-Glucuronyltransferase-I (GlcAT-I) completes the synthesis of the linkage region tetrasaccharide by transferring a GlcUA residue to the galactose-galactose-

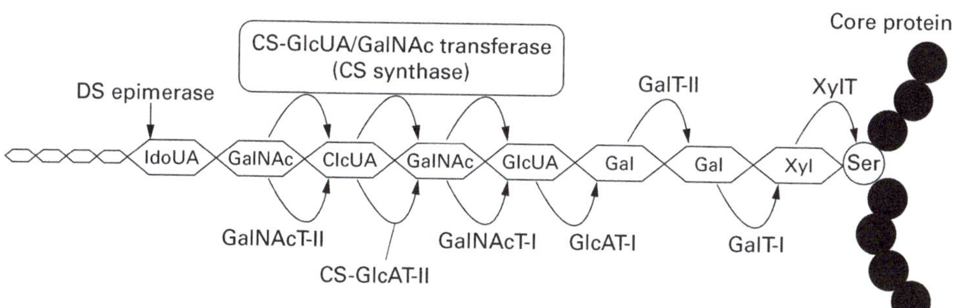

Figure 2.12 Schematic presentation of the biosynthetic assembly of the CS/DS backbones by various glycosyltransferases. A number of glycosyltransferases are required for the synthesis of the backbones of GAGs. XylT, xylosyltransferase; GalT-I, β1,4-galactosyltransferase-I; GalT-II, β1,3-galactosyltransferase-II, GlcAT-I; β1,3-glucuronyltransferase-I; GalNAcT-I, GalNAc transferase-I; CS GlcAT-II, CS GlcUA transferase-II; GalNAcT-II, GalNAc transferase-II; CS polymerase, GlcUA/GalNAc transferase.

Table 2.3 Human glycosyltransferases involved in the biosynthesis of CS and DS.

Enzymes (activity)	*Coding genes (synonym)*	*Chromosomal location*	*Amino acid*	*mRNA accession*
Linkage region glycosyltransferases				
Xylosyltransferase (XylT)	*XYLT1*	16p12.3	959	NM_022166
	XYLT2	17q21.3-q22	865	NM_022167
β1,4-Galactosyltransferase-I (GalT-I)	*B4GALT7*	5q35.2-q35.3	327	NM_007255
β1,3-Galactosyltransferase-II (GalT-II)	*B3GALT6*	1p36.33	329	NM_080605
β1,3-Glucuronyltransferase-I (GlcAT-I)	*B3GAT3*	11q12.3	335	NM_012200
CS/DS glycosyltransferase				
Chondroitin synthase (GalNAcT-II, CS-GlcAT-II)	*CHSY1*	15q26.3	802	NM_014918
	CHSY2 (*CSS3*)	5q23.3	882	NM_175856
	CHSY3 (*CHPF2, CSGLCA-T*)	7q36.1	772	NM_019015
Chondroitin polymerizing factor (GalNAcT-II, CS-GlcAT-II)	*CHPF* (*CSS2*)	2q35	775	NM_024536
Chondroitin *N*-acetylgalactosaminyltransferase (GalNAcT-I, GalNAcT-II)	*CSGALNACT1*	8p21.3	532	NM_018371
	CSGALNACT2	10q11.21	542	NM_018590

xylose trisaccharide sequence in the medial- and *trans*-Golgi compartments (Kitagawa et al., 1998).

Modifications of the GAG linkage region sugar residues include 2-*O*-phosphorylation of the xylose residue and sulfations at C-6 of the first galactose and C-4 or C-6 of the second galactose residue in the linkage region of CS/DS/heparan sulfate (HS) chains (Sugahara and Kitagawa, 2000). Moreover, the biological roles of these modifications, which are often but not always found, remain to be investigated, although the phosphorylation of xylose has been demonstrated to stimulate the synthesis of the linkage region (Moses et al., 1997), and the linkage trisaccharide with the phosphorylated xylose is one of the best acceptors for recombinant GlcAT-I *in vitro* but not for GalT-I (Gulberti et al., 2005; Tone et al., 2008). Recently, the enzymes responsible for the phosphorylation and sulfation in xylose and galactose at the reducing end have been identified as FAM20B and chondroitin 6-*O*-sulfotransferase-1 (C6ST-1), respectively (Kitagawa et al., 2008; Koike et al., 2009) (▶Table 2.4).

Table 2.4 Human kinase, epimerase, and sulfotransferases involved in the biosynthesis of CS and DS.

Enzymes	*Coding genes (synonym)*	*Chromosomal location*	*Amino acid*	*mRNA accession*
Linkage region kinase and sulfotransferase				
Xylose 2-*O*-kinase	*FAM20B* (*gxk1*)	1q25	409	NM_014864
Galactose 6-*O*-sulfotransferase	*CHST3* (*C6ST-1*)	10q22.1	479	NM_004273
CS/DS sulfotransferases and epimerases				
Uronyl 2-*O*-sulfotransferase	*UST*	6q25.1	406	NM_005715
Chondroitin 4-*O*-sulfotransferase	*CHST11* (*C4ST-1*)	12q	352	NM_018413
	CHST12 (*C4ST-2*)	7p22	414	NM_018641
	CHST13 (*C4ST-3*)	3q21.3	341	NM_152889
Dermatan 4-*O*-sulfotransferase	*CHST14* (*D4ST-1*)	15q15.1	376	NM_130468
Chondroitin 6-*O*-sulfotransferase	*CHST3* (*C6ST-1*)	10q22.1	479	NM_004273
N-Acetylgalactosamine-4-sulfate-6-*O*-sulfotransferase	*CHST15* (*GalNAc4S-6ST*)	10q26	561	NM_015892
Dermatan sulfate epimerases	*DSE*	6q22-q23	958	NM_013352
	DSEL	18q22.1	1,222	NM_032160

Biosynthesis of the chondroitin and dermatan backbone

Various glycosyltransferases responsible for the biosynthesis of CS/DS chains have been cloned (▶Table 2.3) (Kitagawa et al., 2001, 2003; Gotoh et al., 2002a, 2002b; Uyama et al., 2002, 2003; Sato et al., 2003; Yada et al., 2003a, 2003b; Izumikawa et al., 2007, 2008). Following completion of the synthesis of the linkage region tetrasaccharide, the first GalNAc residue is transferred to the linkage region built on specific Ser residues in the core protein by GalNAcT-I, which initiates the synthesis of CS (▶Figure 2.12) (Sugahara and Kitagawa, 2000; Uyama et al., 2002, 2003). Thereafter, polymerization of the chondroitin backbone occurs to build up the repeating disaccharide region composed of GalNAcβ1–4GlcUA. The enzymatic activities for the addition of GlcUA and GalNAc have been named CS GlcAT-II and GalNAcT-II activities, respectively (▶Figure 2.12). The formation of disaccharide units is catalyzed by the enzyme complex CS polymerase, consisting of chondroitin synthase (ChSy) and chondroitin polymerizing factor (ChPF) (Kitagawa et al., 2001, 2003; Izumikawa et al., 2007, 2008). It has also been demonstrated that complexes composed of two other enzymes belonging to the same gene family as ChSy/ChPF can also synthesize chondroitin with different chain lengths (Izumikawa et al., 2008). After the formation of the repeating disaccharide units, sulfation reactions at various positions and epimerization of a carboxy group on a GlcUA residue in the chondroitin backbone are catalyzed by a number of sulfotransferases (see the next section "Sulfation of Chondroitin and Dermatan") and by two distinct DS epimerases, respectively.

The conversion of GlcUA to IdoUA is catalyzed by DS C5-epimerases (DSEs), where the C-5 carboxy group of GlcUA is epimerized accompanied by an anomeric change of the glycosidic linkage of GlcUA from β to α. DSE catalyzes the interconversion between GlcUA and IdoUA but favors the IdoUA configuration in chondroitin and dermatan precursor chains (Malmström and Fransson, 1975; Malmström, 1984). Furthermore, nonsulfated chondroitin and dermatan are better acceptor substrates than DS, chondroitin 4-sulfate, and chondroitin 6-sulfate, which are mainly sulfated at the C-4 and C-6 positions of hydroxy groups of GalNAc residues, respectively (Malmström, 1984). Together, these observations suggest that the sulfate groups on CS and DS may prevent reverse reactions by DSE, resulting in the fixation of GlcUA and/or IdoUA configurations.

The gene encoding DSE was identified after purification of the enzyme from bovine spleen and was identical to *SART2*, which encodes squamous cell carcinoma antigen recognized by T-cell 2 (Maccarana et al., 2006). In addition, a second epimerase, DSE2, which is encoded by *DSE-like* (*DSEL*), was identified (Pacheco et al., 2009). Interestingly, the C-terminal portion of DSE2 is predicted to be an *O*-sulfotransferase domain, and hence DSE2 may be a bifunctional protein. However, no sulfotransferase activity has been reported so far. Furthermore, it has been reported that *DSE2* was a candidate for a causative gene for type II bipolar disorder, characterized by alternating episodes of depression and hypomania (Goossens et al., 2003).

Sulfation of chondroitin and dermatan

PAPS serves as the universal sulfate donor for all sulfotransferase reactions in the animal kingdom and is synthesized from 2 moles of ATP and 1 mole of inorganic sulfate in the cytosol (Schwartz et al., 1998). Then, PAPS synthesized in the cytosol is transported to the Golgi apparatus by the PAPS transporter (Kamiyama et al., 2003).

Following the formation of the repeating disaccharide region of chondroitin or dermatan, these backbones are modified by a number of sulfotransferases as shown in ▶Figure 2.13 and listed in ▶Table 2.4. C6ST transfers sulfate to position 6 of the GalNAc residue to produce C and D units (Fukuta et al., 1998). Chondroitin and dermatan 4-*O*-sulfotransferases (C4ST/D4ST) are responsible for 4-*O*-sulfation of a GalNAc residue next to GlcUA in chondroitin and IdoUA in dermatan, yielding A and iA units, respectively (Hiraoka et al., 2000; Yamauchi et al., 2000; Evers et al., 2001; Kang et al., 2002; Mikami et al., 2003). The 2-*O*-sulfation of both GlcUA and IdoUA is catalyzed by a single enzyme, uronyl 2-*O*-sulfotransferase (Kobayashi et al., 1999). This sulfotransferase is essential for the formation of the D and iB units, both of which are formed by two-step reactions. C6ST or D4ST catalyzes 6-*O*- or

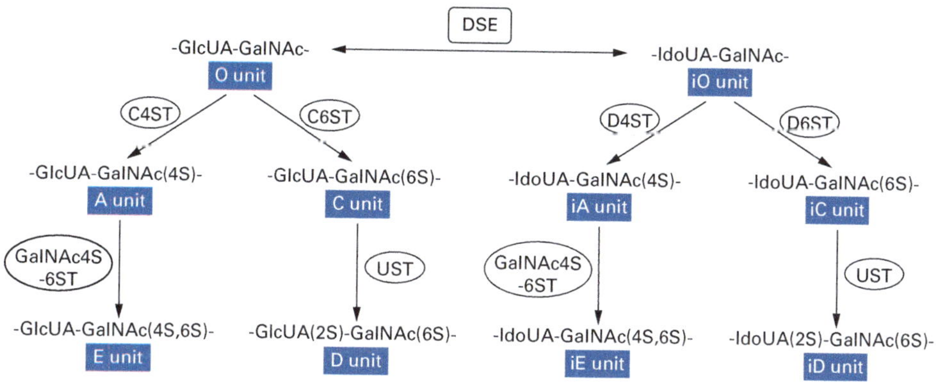

Figure 2.13 A schematic diagram of the biosynthetic modifications of CS/DS chains. DSE, DS C5-epimerase; C4ST, chondroitin 4-*O*-sulfotransferase; C6ST, chondroitin 6-*O*-sulfotransferase; D4ST, dermatan 4-*O*-sulfotransferase; D6ST, dermatan 6-*O*-sulfotransferase; UST, uronyl 2-*O*-sulfotransferase; GalNAc4S-6ST, GalNAc 4-sulfate 6-*O*-sulfotransferase.

4-*O*-sulfation of the GalNAc residue first. The 2-*O*-sulfation of the GlcUA or IdoUA residue is then catalyzed by uronyl 2-*O*-sulfotransferase (▶Figure 2.13).

GalNAc 4-sulfate 6-*O*-sulfotransferase (GalNAc4S-6ST) transfers sulfate to position 6 of GalNAc(4-*O*-sulfate) formed by C4ST and is responsible for the formation of E units (▶Figure 2.13) (Ohtake et al., 2001). Thus, the sulfated modifications and the amount of IdoUA in CS/DS chains are governed by the expression of the *sulfotransferases* and *DSE* genes depending on developmental and pathological conditions.

2.6.3 Human congenital disorders caused by mutations of the enzymes involved in the biosynthesis of CS and DS

Deficiency of β1,4-galactosyltransferase-7 (B4GALT7; GalT-I)

Mutations of the *B4GALT7* gene encoding GalT-I (▶Figure 2.12, ▶Table 2.3) cause a progeroidal appearance and symptoms of Ehlers-Danlos syndrome (EDS) (Quentin et al., 1990; Okajima et al., 1999a; Seidler et al., 2006). EDS is a group of inherited connective tissue disorders, classified into more than 10 types and caused mainly by a defect in the synthesis of collagen. In the skin, muscles, ligaments, blood vessels, and visceral organs, collagen plays a significant role with increased elasticity (Mao and Bristow, 2001). The features of the progeroid form of EDS include an aged appearance, developmental delay, short stature, craniofacial disproportion, generalized osteopenia, defective wound healing, hypermobile joints, hypotonic muscles, and loose but elastic skin (Quentin et al., 1990; Okajima et al., 1999a; Seidler et al., 2006). Fibroblasts, derived from a patient bearing a point mutation in *B4GALT7*, synthesize PGs devoid of GAGs, the core proteins of which are substituted with only a xylose residue. This mutant enzyme GalT-I is aberrantly thermolabile; incubation of fibroblasts at 41°C leads to a reduction in the production of mature PGs (Quentin et al., 1990). In addition, the mutant GalT-I shows a diffuse expression pattern in the cytoplasm, whereas the normal GalT-I shows localization in the Golgi apparatus (Okajima et al., 1999a).

Deficiency of β1,3-glucuronyltransferase-3 (B3GAT3; GlcAT-I)

Very recently, a family with recessive inheritance was found, exhibiting joint dislocations mainly affecting the elbow and congenital heart defects such as bicuspid aortic valves and aortic root dilatation (Baasanjav et al., 2011). A mutation (p.R227Q) in *B3GAT3*, the gene coding for GlcAT-I (▶Figure 2.12, ▶Table 2.3), was detected in these patients designated as having Larsen-like syndrome, B3GAT3 type. Patients' cells as well as the recombinant mutant protein showed reduced GlcAT-I activity. GlcAT-I is located in the *cis-*, *cis*-medial-, and *trans*-Golgi compartments, in mutant cells, and the protein level was dramatically reduced, whereas the level of *B3GAT3* mRNA was not significantly reduced. These observations imply the mutant GlcAT-I to be degraded or produced to a lesser extent than in control subjects. A mutant Chinese hamster ovary cell line, pgsG, prepared by introducing missense mutations, showed reduced enzymatic activity, decreased stability of the mutant, and subsequent proteasomal decay (Wei et al., 2004). Furthermore, CS, DS, and HS side chains of PGs were decreased in patient fibroblasts compared to the wild type. Taken together, a reduction in the enzymatic activity and/or amount of GlcAT-I impairs skeletal as well as heart development and results in variable combinations of heart malformations, including mitral valve prolapse, ventricular septal defects, and bicuspid aortic valves.

Deficiency of chondroitin sulfate β1,4-*N*-acetylgalactosaminyltransferase-1 (CSGALNACT1)

A mutation in the *CSGALNACT1* gene causes neuropathies including Guillain-Barré syndrome, chronic inflammatory demyelinating polyneuropathy, and hereditary motor and sensory neuropathy (Saigoh et al., 2011). Missense mutations were found in *CSGALNACT1* encoding a protein with GalNAcT-I and -II activities in two patients with neuropathy (▶Figure 2.12, ▶Table 2.3). The recombinant mutant proteins, p.H234R and M509R, showed no GalNAcT-II activity, indicating these mutations and/or CS/DS-PGs to be associated with the pathogenetic mechanisms of the peripheral neuropathies.

Deficiency of chondroitin sulfate synthase-1 (CHSY1)

A loss of human *CHSY1* function causes autosomal-recessive Temtamy preaxial brachydactyly syndrome, mostly characterized by limb malformations, short stature, and hearing loss (Li et al., 2010; Tian et al., 2010). Causative mutations of the syndrome have been mapped to 15q26-qterm and detected in *CHSY1* in five consanguineous families (Li et al., 2010). The notion that *CHSY1* is a candidate gene was also supported by results of antisense-mediated gene knockdown using zebrafish. Knockdown of *chsy1* using an antisense morpholino oligonucleotide caused defects in multiple developmental processes, some of which are presumed to also be causative of Temtamy preaxial brachydactyly syndrome (Li et al., 2010). In the inner ears of zebrafish larvae, *chsy1* is expressed similarly to the bone morphogenetic protein inhibitor *dan* and in a complementary fashion to *bmp2b*, indicating that bone morphogenetic protein signaling might affect ear development by supressing *chsy1* (Li et al., 2010).

Furthermore, Tian et al. (2010) also reported a syndromic recessive preaxial brachydactyly with partial duplication of proximal phalanges caused by the mutations in *CHSY1*. Based on the results from analyses *in vitro* and *in vivo*, they speculated that CHSY1 functions as a secreted FRINGE enzyme required for the adjustment of NOTCH signaling throughout embryogenesis, including limb patterning (Tian et al., 2010).

Deficiency of carbohydrate sulfotransferase-3 (CHST3; C6ST-1)

A missense mutation (p.R304Q) in C6ST-1 was found in patients with spondyloepiphyseal dysplasia (SED), Omani type, a recessively inherited skeletal dysplasia, which results in short stature, severe kyphoscoliosis, arthritic joints (in elbows, wrists, and knees), secondary large joint dislocations, rhizomelia, fusion of carpal bones, and mild brachydactyly (Thiele et al., 2004). Furthermore, two additional mutations (p.T141M and L286P) of C6ST-1 in SED patients were also identified in Turkish families with skeletal dysplasia phenotypes similar to but not identical with those of the SED, Omani type (van Roij et al., 2008; Tuysuz et al., 2009). The clinical phenotypes of the new patients consist of ventricular septal, mitral, and/or tricuspid defects, aortic regurgitations, deafness, and metacarpal shortening, and have not been reported for the SED, Omani type, described previously. The C6ST activity of the fibroblasts from these patients was dramatically reduced compared with that in a normal subject (van Roij et al., 2008). In addition, negligible levels of 6-*O*-sulfated GalNAc residues in CS chains were detected in fibroblasts and urine obtained from the patients (van Roij et al., 2008). Moreover, Superti-Furga and his colleagues have reported Larsen syndrome, humerospinal dysostosis (HSD), chondrodysplasia with multiple dislocations (CDMD, "Megarbane type"), Desbuquois syndrome, and SED caused by a deficiency in *C6ST-1* (Hermanns et al., 2008; Unger et al., 2010). These findings together suggest that there is wide diversity in the clinical consequences of a deficiency in 6-*O*-sulfation of CS chains and that the 6-*O*-sulfation in CS side chains on PGs plays crucial roles in the development of not only bone but also heart.

Deficiency of carbohydrate sulfotransferase-14 (CHST14; D4ST-1)

EDS is a heterogeneous connective tissue disorder involving skin and joint laxity with tissue fragility as described previously (Mao and Bristow, 2001). A new type of EDS, similar to the kyphoscoliosis type but without a deficiency of lysyl hydroxylase, has been identified, with homozygous mutations in *CHST14* encoding D4ST-1 (▶Figure 2.13, ▶Table 2.4) found in two familial cases from Japan and compound heterozygous mutations found in four sporadic cases (EDS type VIB) (Miyake et al., 2010). Mutations in D4ST-1 include one nonsense (p.K69X) and three missense (p.P281L, p.C289S, and p.Y293C) mutations but were absent in 376 normal Japanese subjects. Recombinant D4ST-1 with these mutations and lysates of fibroblasts from the patients show a loss of or considerable reduction in D4ST activity compared with that of control subjects. Furthermore, a DS side chain of decorin PG, a key regulator of the assembly of collagen fibrils (Reed and Iozzo, 2002), is completely lost and replaced by CS in the patients' fibroblasts. The replacement of the DS side chain by a CS chain on the decorin PG due to D4ST-1 mutations may prevent the formation or maintenance of adequate collagen bundles, whereas the sizes and shapes of collagen fibrils are unchanged as observed in the patients' dermal tissues, indicating the important role of decorin DS in the extracellular matrix and a novel pathogenic mechanism of EDS.

CHST14 has recently been identified as the causative gene for adducted thumb–clubfoot syndrome (ATCS), another autosomal recessive disorder (Dündar et al., 2009; Malfait et al., 2010). ATCS has been categorized as a new type of arthrogryposis, based on characteristic clinical features from birth to early childhood, including adducted thumbs and clubfoot as well as craniofacial dysmorphism, arachnodactyly with tapering fingers, cryptorchidism, inguinal hernia, atrial septal defect, kidney defects, cranial ventricular enlargement, and psychomotor retardation (Dundar et al., 1997; Sonoda and Kouno, 2000). A recent study revealed that ATCS could again be categorized as a connective tissue disorder, based on additional clinical features from childhood to adolescence, including skin fragility, bruisability, and translucency; joint laxity; and osteopenia (Dündar et al., 2009). Of 11 ATCS patients, 5 died in early infancy or childhood. Taken together, ATCS patients may have wider and more severe manifestations than patients with EDS type VIB, implying roles for DS and/or DSPGs, including decorin not only in connective tissue maintenance but also in embryonic development.

Due to limited space, other genetic diseases caused by variations of amino acids in a PAPS transporter and core proteins of various CS/DS-PGs are listed in ▶Table 2.5.

2.6.4 Challenges and future prospects

cDNA cloning of enzymes involved in the biosynthesis of CS and DS has led to a better understanding of not only the biosynthetic mechanisms but also the *in vivo* functions of CS and DS, especially through the discovery of a considerable number of human genetic disorders caused by a disturbance to the construction of both types of chains. Although analytical data for the *in vivo* functions of CS/DS chains, which have been obtained using model organisms, could not be discussed in this chapter due to limitations of space, functional analyses of the genes encoding biosynthetic enzymes for CS and DS using knockout mice in addition to mutants of fruit flies and nematodes have also revealed important roles of CS/DS chains (see Sugahara et al., 2003; Mizumoto et al., 2005; Bishop et al., 2007). Recent studies described previously have demonstrated that mutations of the biosynthetic enzymes for CS and DS mainly cause bone, dermal, and heart dysplasias including brachydactyly, kyphoscoliosis, cardiac valve defects, and cutis laxa. These findings would be helpful in establishing better diagnostic and therapeutic methods for these disorders. Although RPTP-β–deficient mice showed no obvious abnormalities in the central nervous system, they exhibited an impairment of spacial learning and enhancement of long-term potentiation (LTP) in the hippocampus due to anomalous activation of Rho-associated kinase (Harroch et al., 2000; Tamura et al., 2006). Most recently, it has been reported that disturbance of the biosynthesis of CS causes neuropathies (Saigoh et al., 2011). It is of interest how RPTP-β, the CS side chain, and/or the downsteam signalings are involved in the neuropathies and regulation of specific developmental processes in the central nervous system. Further investigations of the pathogenic mechanisms for these disorders are required. The identification of core proteins carrying CS/DS side chains and characterization of defective fine sugar structures including the sulfation pattern, chain length, and epimerization of

Table 2.5 Human genetic disorders caused by mutations involved in the biosynthesis of the CS/DS side chains and core protein of PGs.

Name of disorder	*Mendelian Inheritance in Man (MIM) no.*	*Gene (coding enzyme or protein)*	*Clinical condition*	*References*
Achondrogenesis type IB Atelosteogenesis type II Diastrophic dysplasia (DTD) Multiple epiphyseal dysplasia (MED)	600972 256050 222600 226900	*SLC26A2* (DTDST)	Autosomal recessive, lethal chondrodysplasia with severe underdevelopment of skeleton, extreme micromelia, and death before or immediately after birth	Rossi et al. (1996), Superti-Furga et al. (1996)
Spondyloepimetaphyseal dysplasia (SEMD)	612847	*PAPSS2* (PAPS synthase-2)	Short, bowed lower limbs, enlarged knee joints, kyphoscoliosis, a mild generalized brachydactyly	ul Haque et al. (1998)
Schneckenbecken dysplasia	269250	*SLC35D1* (UDP-GlcUA/ UDP-GalNAc transporter)	Neonatal lethal chondrodysplasi, platyspondyly with oval-shaped vertebral bodies, extremely short long bones with dumbbell-like appearance, and small ilia with snail-like appearance	Hiraoka et al. (2007)
Ehlers-Danlos syndrome (EDS) progeroid form	130070	*B4GALT7* (GalT-I)	Aged appearance, developmental delay, dwarfism, craniofacial disproportion, and generalized osteopenia	Quentin et al. (1990), Okajima et al. (1999a), Seidler et al. (2006)
Larsen-like syndrome, B3GAT3 type	245600	*B3GAT3* (GlcAT-I)	Joint dislocations mainly affecting the elbow and congenital heart defects such as bicuspid aortic valve and aortic root dilatation	Baasanjav et al. (2011)
Guillain-Barré syndrome Chronic inflammatory	–	*CSGALNACT1*	Guillain-Barré syndrome, chronic inflammatory demyelinating	Saigoh et al. (2011)

(*Continued*)

Table 2.5 **Human genetic disorders caused by mutations involved in the biosynthesis of the CS/DS side chains and core protein of PGs. (*Continued*)**

Name of disorder	*Mendelian Inheritance in Man (MIM) no.*	*Gene (coding enzyme or protein)*	*Clinical condition*	*References*
demyelinating polyneuropathy			polyneuropathy, and hereditary motor and sensory neuropathy	
Temtamy preaxial brachydactyly syndrome Syndromic recessive preaxial brachydactyly	605282	*CHSY1*	Limb malformations, short stature, and hearing loss	Li et al. (2010), Tian et al. (2010)
Spondyloepiphyseal dysplasia (SED), Omani type Chondrodysplasia with multiple dislocations (CDMD) Humerospinal dysostosis (HSD)	143095	*CHST3* (C6ST-1)	Short stature; severe kyphoscoliosis; arthritic joints (elbows, wrists, and knees); secondary large joint dislocations; rhizomelia; fusion of carpal bones and mild brachydactyly; ventricular septal, mitral, and/or tricuspid defects; aortic regurgitations; deafness; and metacarpal shortening	Thiele et al. (2004), Hermanns et al. (2008), van Roij et al. (2008), Tuysuz et al. (2009), Unger et al. (2010)
Bipolar disorder Depressive disorder Diaphragmatic hernia Microphthalmia evaluation	611125	*DSEL* (DS epimerase-2)	Alternating episodes of depression and mania or hypomania, and congenital malformation (birth defect) of the diaphragm	Goossens et al. (2003), Zayed et al. (2010), Shi et al. (2011)
EDS, Kosho type, musculocontractural type Adducted thumb–clubfoot syndrome (ATCS)	601776	*CHST14* (D4ST-1)	Distinct craniofacial characteristics; multiple congenital contractures; progressive joint and skin laxity and multisystem fragility-related manifestations; congenital	Dündar et al. (2009), Malfait et al. (2010), Miyake et al. (2010)

			contractures of thumbs and feet; heart, kidney, or intestinal defects	
Chondrodysplasia with joint dislocations, gPAPP type (gPAPP deficiency)	614078	*IMPAD1* (Golgi-resident 3′-phosphoadenosine 5′-phosphate 3′-phosphatase [gPAPP])	Short stature, chondrodysplasia with brachydactyly, congenital joint dislocations, cleft palate, and facial dysmorphism	Vissers et al. (2011)
SED, Kimberley type SEMD, aggrecan type Familial osteochondritis dissecans	608361 612813 165800	*ACAN* (aggrecan)	Short stature, stocky build, early onset osteoarthritis of the weight-bearing joints, and multiple osteochondritic lesions in knees and/or hips and/or elbows	Gleghorn et al. (2005), Tompson et al. (2009), Stattin et al. (2010)
Bipolar disorder	600826	*NCAN* (neurocan)	Recurrent episodes of mania and depression	Cichon et al. (2011)
Wagner syndrome 1 (Wagner vitreoretinal degeneration)	143200	*VCAN* (versican)	An empty vitreous with fibrillary condensations, avascular membrane, perivascular sheathing, and progressive chorioretinal dystrophy	Miyamoto et al. (2005)
Congenital stromal corneal dystrophy	610048	*DCN* (decorin)	A large number of flakes and spots throughout all layers of the stroma	Bredrup et al. (2005)
Camptodactyly-arthropathy-coxa vara-pericarditis syndrome	208250	*PRG4*	Congenital or early onset camptodactyly and noninflammatory arthropathy with synovial hyperplasia	Marcelino et al. (1999)
Osteogenesis imperfecta	610915	*LEPRE1* (leprecan)	Severe osteoporosis, shortened long bones, and a soft skull with wide open fontanel	Cabral et al. (2007)
Retinitis pigmentosa-56 (RP56)	613581	*IMPG2* (interphotoreceptor matrix proteoglycan-2)	Night blindness, visual field loss, optic disk pallor, attenuated vessels, and bone-spicule-like pigmentation	Bandah-Rozenfeld et al. (2010)

uronic acid would help to gain a deeper understanding of the biosynthetic mechanisms of CS/DS chains. These studies would provide insights into possible therapeutics.

2.6.5 Take-home message

The cDNA cloning during the last 10 years or so of the enzymes involved in the biosynthesis of GAG chains has led not only to an understanding of the *in vivo* functions of CS, DS, and HS chains revealed using model organisms, but also to discoveries of a variety of hereditary diseases caused by the mutations of these genes. It is hoped that further investigation of the molecular mechanisms of the pathogenesis will provide useful information for developing therapeutics for these diseases.

2.6.6 Abbreviations

CS, Chondroitin sulfate; DS, dermatan sulfate; GAG, glycosaminoglycans; PGs, proteoglycans; HS, heparan sulfate; GlcUA, D-glucuronic acid; IdoUA, L-iduronic acid; GalNAc, *N*-acetyl-D-galactosamine; ECM, extracellular matrix; DSE, DS-glucuronyl C5-epimerase; PAPS, 3′-phosphoadenosine 5′-phosphosulfate; C4ST, chondroitin 4-*O*-sulfotransferases; D4ST, dermatan 4-*O*-sulfotransferase; C6ST/D6ST, chondroitin/dermatan 6-*O*-sulfotransferase; GalNAc4S-6ST, *N*-acetyl-D-galactosamine 4-sulfate 6-*O*-sulfotransferase; VEGF, vascular endothelial growth factor; O, GlcUAβ1–3GalNAc; A, GlcUAβ1–3GalNAc(4S); iA, IdoUAα1–3GalNAc(4S); C, GlcUAβ1–3GalNAc(6S); iC, IdoUAα1–3GalNAc(6S); D, GlcUA(2S)β1–3GalNAc(6S); iD, IdoUA(2S)α1–3GalNAc(6S); B, GlcUA(2S)β1–3GalNAc(4S); iB, IdoUA(2S)α1–3GalNAc(4S); E, GlcUAβ1–3-GalNAc(4S,6S); iE, IdoUAα1–3GalNAc(4S,6S); 2S, 2-*O*-sulfate; 4S, 4-*O*-sulfate; 6S, 6-*O*-sulfate.

Acknowledgments

This work was supported in part by the Matching Program for Innovations in Future Drug Discovery and Medical Care (to K. S.) from the Ministry of Education, Culture, Sports, Science and Technology, Japan (MEXT), and by Grant-in-aid for Young Scientists (B) 23790066 (to S. M.) from the Japan Society for the Promotion of Science, Japan.

References

Akatsu, C., Mizumoto, S., Kaneiwa, T., et al. (2011). Dermatan sulfate epimerase 2 is the predominant isozyme in the formation of the chondroitin sulfate/dermatan sulfate hybrid structure in postnatal developing mouse brain. Glycobiology *21*, 565–574.

Almeida, R., Levery, S. B., Mandel, U., et al. (1999). Cloning and expression of a proteoglycan UDP-galactose: β-xylose β1,4-galactosyltransferase I: aseventh member of the human β4-galactosyltransferase gene family. J Biol Chem *274*, 26165–26171.

Baasanjav, S., Al-Gazali, L., Hashiguchi, T., et al. (2011). Faulty initiation of proteoglycan synthesis causes cardiac and joint defects. Am J Hum Genet *89*, 15–27.

Bai, X., Zhou, D., Brown, J. R., Crawford, B. E., Hennet, T., and Esko, J. D. (2001). Biosynthesis of the linkage region of glycosaminoglycans: cloning and activity of galactosyltransferase II, the sixth member of the β1,3-galactosyltransferase family (β3GalT6). J Biol Chem *276*, 48189–48195.

Bandah-Rozenfeld, D., Collin, R. W., Banin, F., et al. (2010). Mutations in *IMPG2*, encoding interphotoreceptor matrix proteoglycan 2, cause autosomal-recessive retinitis pigmentosa. Am J Hum Genet *87*, 199–208.

Bao, X., Muramatsu, T., and Sugahara, K. (2005). Demonstration of the pleiotrophin-binding oligosaccharide sequences isolated from chondroitin sulfate/dermatan sulfate hybrid chains of embryonic pig brains. J Biol Chem *280*, 35318–35328.

Bishop, J. R., Schuksz, M., and Esko, J. D. (2007). Heparan sulphate proteoglycans fine-tune mammalian physiology. Nature *446*, 1030–1037.

Bredrup, C., Knappskog, P. M., Majewski, J., Rødahl, E., and Boman, H. (2005). Congenital stromal dystrophy of the cornea caused by a mutation in the decorin gene. Invest Ophthalmol Vis Sci *46*, 420–426.

Cabral, W. A., Chang, W., Barnes, A. M., et al. (2007). Prolyl 3-hydroxylase 1 deficiency causes a recessive metabolic bone disorder resembling lethal/severe osteogenesis imperfecta. Nat Genet *39*, 359–365.

Cichon, S., Mühleisen, T. W., Degenhardt, F. A., et al. (2011). Genome-wide association study identifies genetic variation in neurocan as a susceptibility factor for bipolar disorder. Am J Hum Genet *88*, 372–381.

Dundar, M., Demiryilmaz, F., Demiryilmaz, I., et al. (1997). An autosomal recessive adducted thumb club foot syndrome observed in Turkish cousins. Clin Genet *51*, 61–64.

Dündar, M., Müller, T., Zhang, Q., et al. (2009). Loss of dermatan-4-sulfotransferase 1 function results in adducted thumb-clubfoot syndrome. Am J Hum Genet *85*, 873–882.

Evers, M. R., Xia, G., Kang, H. G., Schachner, M., and Baenziger, J. U. (2001). Molecular cloning and characterization of a dermatan-specific *N*-acetylgalactosamine 4-*O*-sulfotransferase. J Biol Chem *276*, 36344–36353.

Fukuta, M., Kobayashi, Y., Uchimura, K., Kimata, K., and Habuchi, O. (1998) Molecular cloning and expression of human chondroitin 6-sulfotransferase. Biochim Biophys Acta *1399*, 57–61.

Gleghorn, L., Ramesar, R., Beighton, P., and Wallis, G. (2005). A mutation in the variable repeat region of the aggrecan gene (*AGC1*) causes a form of spondyloepiphyseal dysplasia associated with severe, premature osteoarthritis. Am J Hum Genet *77*, 484–490.

Goossens, D., Van Gestel, S., Claes, S., et al. (2003). A novel CpG-associated brain-expressed candidate gene for chromosome 18q-linked bipolar disorder. Mol Psychiatry *8*, 83–89.

Gotoh, M., Sato, T., Akashima, T., et al. (2002a). Enzymatic synthesis of chondroitin with a novel chondroitin sulfate *N*-acetylgalactosaminyltransferase that transfers *N*-acetylgalactosamine to glucuronic acid in initiation and elongation of chondroitin sulfate synthesis. J Biol Chem *277*, 38189–38196.

Gotoh, M., Yada, T., Sato, T., et al. (2002b). Molecular cloning and characterization of a novel chondroitin sulfate glucuronyltransferase that transfers glucuronic acid to *N*-acetylgalactosamine. J Biol Chem *277*, 38179–38188.

Götting, C., Kuhn, J., Zahn, R., Brinkmann, T., and Kleesiek, K. (2000). Molecular cloning and expression of Human UDP-D-xylose: proteoglycan core protein β-D-xylosyltransferase and its first isoform XT-II. J Mol Biol *304*, 517–528.

Gulberti, S., Lattard, V., Fondeur, M., et al. (2005). Phosphorylation and sulfation of oligosaccharide substrates critically influence the activity of human β1,4-galactosyltransferase 7 (GalT-I) and β1,3-glucuronosyltransferase I (GlcAT-I) involved in the biosynthesis of the glycosaminoglycan-protein linkage region of proteoglycans. J Biol Chem *280*, 1417–1425.

Harroch, S., Palmeri, M., Rosenbluth, J., et al. (2000). No obvious abnormality in mice deficient in receptor protein tyrosine phosphatase β. Mol Cell Biol *20*, 7706–7715.

Hermanns, P., Unger, S., Rossi, A., et al. (2008). Congenital joint dislocations caused by carbohydrate sulfotransferase 3 deficiency in recessive Larsen syndrome and humero-spinal dysostosis. Am J Hum Genet *82*, 1368–1374.

Hiraoka, N., Nakagawa, H., Ong, E., Akama, T. O., Fukuda, M. N., and Fukuda, M. (2000). Molecular cloning and expression of two distinct human chondroitin 4-*O*-sulfotransferases that belong to the HNK-1 sulfotransferase gene family. J Biol Chem *275*, 20188–20196.

Hiraoka, S., Furuichi, T., Nishimura, G., et al. (2007). Nucleotide-sugar transporter SLC35D1 is critical to chondroitin sulfate synthesis in cartilage and skeletal development in mouse and human. Nat Med *13*, 1363–1367.

Izumikawa, T., Koike, T., Shiozawa, S., Sugahara, K., Tamura, J., and Kitagawa, H. (2008). Identification of chondroitin sulfate glucuronyltransferase as chondroitin synthase-3 involved in chondroitin polymerization: chondroitin polymerization is achieved by multiple enzyme complexes consisting of chondroitin synthase family members. J Biol Chem *283*, 11396–11406.

Izumikawa, T., Uyama, T., Okuura, Y., Sugahara, K., and Kitagawa, H. (2007). Involvement of chondroitin sulfate synthase-3 (chondroitin synthase-2) in chondroitin polymerization through its interaction with chondroitin synthase-1 or chondroitin polymerizing factor. Biochem J *403*, 545–552.

Kamiyama, S., Suda, T., Ueda, R., et al. (2003). Molecular cloning and identification of 3′-phosphoadenosine 5′-phosphosulfate transporter. J Biol Chem *278*, 25958–25963.

Kang, H. G., Evers, M. R., Xia, G., Baenziger, J. U., and Schachner, M. (2002). Molecular cloning and characterization of chondroitin-4-*O*-sulfotransferase-3: a novel member of the HNK-1 family of sulfotransferases. J Biol Chem *277*, 34766–34772.

Kitagawa, H., Izumikawa, T., Uyama, T., and Sugahara, K. (2003). Molecular cloning of a chondroitin polymerizing factor that cooperates with chondroitin synthase for chondroitin polymerization. J Biol Chem *278*, 23666–23671.

Kitagawa, H., Tone, Y., Tamura, J., et al. (1998). Molecular cloning and expression of glucuronyltransferase I involved in the biosynthesis of the glycosaminoglycan-protein linkage region of proteoglycans. J Biol Chem *273*, 6615–6618.

Kitagawa, H., Tsutsumi, K., Ikegami-Kuzuhara, A., et al. (2008). Sulfation of the galactose residues in the glycosaminoglycan-protein linkage region by recombinant human chondroitin 6-*O*-sulfotransferase-1. J Biol Chem *283*, 27438–27443.

Kitagawa, H., Uyama, T., and Sugahara, K. (2001). Molecular cloning and expression of a human chondroitin synthase. J Biol Chem *276*, 38721–38726.

Kobayashi, M., Sugumaran, G., Liu, J., Shworak, N. W., Silbert, J. E., and Rosenberg, R. D. (1999). Molecular cloning and characterization of a human uronyl 2-sulfotransferase that sulfates iduronyl and glucuronyl residues in dermatan/chondroitin sulfate. J Biol Chem *274*, 10474–10480.

Koike, T., Izumikawa, T., Tamura, J., and Kitagawa, H. (2009). FAM20B is a kinase that phosphorylates xylose in the glycosaminoglycan-protein linkage region. Biochem J *421*, 157–162.

Li, F., ten Dam, G. B., Murugan, S., et al. (2008). Involvement of highly sulfated chondroitin sulfate in the metastasis of the Lewis lung carcinoma cells. J Biol Chem *283*, 34294–34304.

Li, Y., Laue, K., Temtamy, S., et al. (2010). Temtamy preaxial brachydactyly syndrome is caused by loss-of-function mutations in chondroitin synthase 1, a potential target of BMP signaling. Am J Hum Genet *87*, 757–767.

Maccarana, M., Olander, B., Malmström, J., et al. (2006). Biosynthesis of dermatan sulfate: chondroitin-glucuronate C5-epimerase is identical to SART2. J Biol Chem *281*, 11560–11568.

Malfait, F., Syx, D., Vlummens, P., et al. (2010). Musculocontractural Ehlers-Danlos Syndrome (former EDS type VIB) and adducted thumb clubfoot syndrome (ATCS) represent a single clinical entity caused by mutations in the dermatan-4-sulfotransferase 1 encoding *CHST14* gene. Hum Mutat *31*, 1233–1239.

Malmström, A. (1984). Biosynthesis of dermatan sulfate. II. Substrate specificity of the C-5 uronosyl epimerase. J Biol Chem *259*, 161–165.

Malmström, A., and Fransson, L. A. (1975). Biosynthesis of dermatan sulfate. I. Formation of L-iduronic acid residues. J Biol Chem *250*, 3419–3425.

Mao, J. R., and Bristow, J. (2001). The Ehlers-Danlos syndrome: on beyond collagens. J Clin Invest *107*, 1063–1069.

Marcelino, J., Carpten, J. D., Suwairi, W. M., et al. (1999). *CACP*, encoding a secreted proteoglycan, is mutated in camptodactyly-arthropathy-coxa vara-pericarditis syndrome. Nat Genet *23*, 319–322.

Mikami, T., Mizumoto, S., Kago, N., Kitagawa, H., and Sugahara, K. (2003). Specificities of three distinct human chondroitin/dermatan *N*-acetylgalactosamine 4-*O*-sulfotransferases demonstrated using partially desulfated dermatan sulfate as an acceptor: implication of differential roles in dermatan sulfate biosynthesis. J Biol Chem *278*, 36115–36127.

Miyake, N., Kosho, T., Mizumoto, S., et al. (2010). Loss-of-function mutations of CHST14 in a new type of Ehlers-Danlos syndrome. Hum Mutat *31*, 966–974.

Miyamoto, T., Inoue, H., Sakamoto, Y., et al. (2005). Identification of a novel splice site mutation of the CSPG2 gene in a Japanese family with Wagner syndrome. Invest Ophthalmol Vis Sci *46*, 2726–2735.

Mizumoto, S., Uyama, T., Mikami, T., Kitagawa, H., and Sugahara, K. (2005). Biosynthetic pathways for differential expression of functional chondroitin sulfate and heparan sulfate. In Handbook of Carbohydrate Engineering, Kevin J. Yarema, ed. Boca Raton, FL: CRC Press, Taylor & Francis Group, pp. 289–324.

Moriarity, J. L., Hurt, K. J., Resnick, A. C., et al. (2002). UDP-glucuronate decarboxylase, a key enzyme in proteoglycan synthesis. J Biol Chem *277*, 16968–16975.

Moses, J., Oldberg, Å., Cheng, F., and Fransson, L. Å. (1997). Biosynthesis of the proteoglycan decorin: transient 2-phosphorylation of xylose during formation of the trisaccharide linkage region. Eur J Biochem *248*, 521–526.

Nadanaka, S., Clement, A., Masayama, K., Faissner, A., and Sugahara, K. (1998). Characteristic hexasaccharide sequences in octasaccharides derived from shark cartilage chondroitin sulfate D with neurite outgrowth promoting activity. J Biol Chem *273*, 3296–3307.

Ohtake, S., Ito, Y., Fukuta, M., and Habuchi, O. (2001). Human *N*-acetylgalactosamine 4-sulfate 6-*O*-sulfotransferase cDNA is related to human B cell recombination activating gene-associated gene. J Biol Chem *276*, 43894–43900.

Okajima, T., Fukumoto, S., Furukawa, K., Urano, T., and Furukawa, K. (1999a). Molecular basis for the progeroid variant of Ehlers-Danlos syndrome: identification and characterization of two mutations in galactosyltransferase I gene. J Biol Chem *274*, 28841–28844.

Okajima, T., Yoshida, K., Kondo, T., and Furukawa, K. (1999b). Human homolog of *Caenorhabditis elegans sqv-3* gene is galactosyltransferase I involved in the biosynthesis of the glycosaminoglycan-protein linkage region of proteoglycans. J Biol Chem *274*, 22915–22918.

Pacheco, B., Malmström, A., and Maccarana, M. (2009) Two dermatan sulfate epimerases form iduronic acid domains in dermatan sulfate. J Biol Chem *284*, 9788–9795.

Pedersen, L. C., Dong, J., Taniguchi, F., et al. (2003). Crystal structure of an α1, 4-*N*-acetylhexosaminyltransferase (EXTL2), a member of the exostosin gene family involved in heparan sulfate biosynthesis. J Biol Chem *278*, 14420–14428.

Pedersen, L. C., Tsuchida, K., Kitagawa, H., Sugahara, K., and Negishi, M. (2000). Heparan/chondroitin sulfate biosynthesis: structure and mechanism of human glucuronyltransferase I. J Biol Chem *275*, 34580–34585.

Perrimon, N., and Bernfield, M. (2000). Specificities of heparan sulphate proteoglycans in developmental processes. Nature *404*, 725–728.

Pönighaus, C., Ambrosius, M., Casanova, J. C., et al. (2007). Human xylosyltransferase II is involved in the biosynthesis of the uniform tetrasaccharide linkage region in chondroitin sulfate and heparan sulfate proteoglycans. J Biol Chem *282*, 5201–5206.

Quentin, E., Gladen, A., Rodén, L., and Kresse, H. (1990). A genetic defect in the biosynthesis of dermatan sulfate proteoglycan: galactosyltransferase I deficiency in fibroblasts from a patient with a progeroid syndrome. Proc Natl Acad Sci U S A *87*, 1342–1346.

Reed, C. C., and Iozzo, R. V. (2002). The role of decorin in collagen fibrillogenesis and skin homeostasis. Glycoconj J *19*, 249–255.

Rossi, A., van der Harten, H. J., Beemer, F. A., et al. (1996). Phenotypic and genotypic overlap between atelosteogenesis type 2 and diastrophic dysplasia. Hum Genet *98*, 657–661.

Saigoh, K., Izumikawa, T., Koike, T., Shimizu, J., Kitagawa, H., and Kusunoki, S. (2011). Chondroitin beta-1,4-*N*-acetylgalactosaminyltransferase-1 missense mutations are associated with neuropathies. J Hum Genet *56*, 143–146.

Sato, T., Gotoh, M., Kiyohara, K., et al. (2003). Differential roles of two *N*-acetylgalactosaminyltransferases, CSGalNAcT-1, and a novel enzyme, CSGalNAcT-2: initiation and elongation in synthesis of chondroitin sulfate. J Biol Chem *278*, 3063–3071.

Schwartz, N. B., Lyle, S., Ozeran, J. D., et al. (1998). Sulfate activation and transport in mammals: system components and mechanisms. Chem Biol Interact *109*, 143–151.

Seidler, D. G., Faiyaz-Ul-Haque, M., Hansen, U., et al. (2006). Defective glycosylation of decorin and biglycan, altered collagen structure, and abnormal phenotype of the skin fibroblasts of an Ehlers-Danlos syndrome patient carrying the novel Arg270Cys substitution in galactosyltransferase I (beta4GalT-7). J Mol Med *84*, 583–594.

Seno, N., and Murakami, K. (1982). Structure of disulfated disaccharides from chondroitin polysulfates, chondroitin sulfate D and K. Carbohydr Res *103*, 190–194.

Shi, J., Potash, J. B., Knowles, J. A., et al. (2011). Genome-wide association study of recurrent early-onset major depressive disorder. Mol Psychiatry *16*, 193–201.

Sonoda, T., and Kouno, K. (2000). Two brothers with distal arthrogryposis, peculiar facial appearance, cleft palate, short stature, hydronephrosis, retentio testis, and normal intelligence: a new type of distal arthrogryposis? Am J Med Genet *91*, 280–285.

Spicer, A. P., Kaback, L. A., Smith, T. J., and Seldin, M. F. (1998). Molecular cloning and characterization of the human and mouse UDP-glucose dehydrogenase genes. J Biol Chem *273*, 25117–25124.

Stattin, E.-L., Wiklund, F., Lindblom, K., et al. (2010). A missense mutation in the aggrecan C-type lectin domain disrupts extracellular matrix interactions and causes dominant familial osteochondritis dissecans. Am J Hum Genet *86*, 126–137.

Sugahara, K., and Kitagawa, H. (2000). Recent advances in the study of the biosynthesis and functions of sulfated glycosaminoglycans. Curr Opin Struct Biol *10*, 518–527.

Sugahara, K., and Mikami, T. (2007). Chondroitin/dermatan sulfate in the central nervous system. Curr Opin Struct Biol *17*, 536–545.

Sugahara, K., Mikami, T., Uyama, T., Mizuguchi, S., Nomura, K., and Kitagawa, H. (2003). Recent advances in the structural biology of chondroitin sulfate and dermatan sulfate. Curr Opin Struct Biol *13*, 612–620.

Sugahara, K., and Schwartz, N. B. (1979). Defect in 3′-phosphoadenosine 5′-phosphosulfate formation in brachymorphic mice. Proc Natl Acad Sci U S A *76*, 6615–6618.

Superti-Furga, A., Hastbacka, J., Wilcox, W. R., et al. (1996). Achondrogenesis type IB is caused by mutations in the diastrophic dysplasia sulphate transporter gene. Nat Genet *12*, 100–102.

Tamura, H., Fukada, M., Fujikawa, A., and Noda, M. (2006). Protein tyrosine phosphatase receptor type Z is involved in hippocampus-dependent memory formation through dephosphorylation at Y1105 on p190 RhoGAP. Neurosci Lett *399*, 33–38.

Thiele, H., Sakano, M., Kitagawa, H., et al. (2004). Loss of chondroitin 6-*O*-sulfotransferase-1 function results in severe human chondrodysplasia with progressive spinal involvement. Proc Natl Acad Sci U S A *101*, 10155–10160.

Tian, J., Ling, L., Shboul, M., et al. (2010). Loss of CHSY1, a secreted FRINGE enzyme, causes syndromic brachydactyly in humans via increased NOTCH signaling. Am J Hum Genet *87*, 768–78.

Tompson, S. W., Merriman, B., Funari, V. A., et al. (2009). A recessive skeletal dysplasia, SEMD aggrecan type, results from a missense mutation affecting the C-type lectin domain of aggrecan. Am J Hum Genet *84*, 72–79.

Tone, Y., Pedersen, L. C., Yamamoto, T., et al. (2008). 2-*O*-phosphorylation of xylose and 6-*O*-sulfation of galactose in the protein linkage region of glycosaminoglycans influence the glucuronyltransferase-I activity involved in the linkage region synthesis. J Biol Chem *283*, 16801–16807.

Tuysuz, B., Mizumoto, S., Sugahara, K., Celebi, A., Mundlos, S., and Turkmen, S. (2009). Omani-type spondyloepiphyseal dysplasia with cardiac involvement caused by a missense mutation in *CHST3*. Clin Genet *75*, 375–383.

ul Haque, M. F., King, L. M., Krakow, D., et al. (1998). Mutations in orthologous genes in human spondyloepimetaphyseal dysplasia and the brachymorphic mouse. Nat Genet *20*, 157–162.

Unger, S., Lausch, E., Rossi, A., et al. (2010). Phenotypic features of carbohydrate sulfotransferase 3 (CHST3) deficiency in 24 patients: congenital dislocations and vertebral changes as principal diagnostic features. Am J Med Genet A *152A*, 2543–2549.

Uyama, T., Ishida, M., Izumikawa, T., et al. (2006). Chondroitin 4-*O*-sulfotransferase-1 regulates E disaccharide expression of chondroitin sulfate required for herpes simplex virus infectivity. J Biol Chem *281*, 38668–38674.

Uyama, T., Kitagawa, H., Tamura, J., and Sugahara, K. (2002). Molecular cloning and expression of human chondroitin *N*-acetylgalactosaminyltransferase: the key enzyme for chain initiation and elongation of chondroitin/dermatan sulfate on the protein linkage region tetrasaccharide shared by heparin/heparan sulfate. J Biol Chem *277*, 8841–8846.

Uyama, T., Kitagawa, H., Tanaka, J., Tamura, J., Ogawa, T., and Sugahara, K. (2003). Molecular cloning and expression of a second chondroitin *N*-acetylgalactosaminyltransferase involved in the initiation and elongation of chondroitin/dermatan sulfate. J Biol Chem *278*, 3072–3078.

van Roij, M. H. H., Mizumoto, S., Yamada, S., et al. (2008). Spondyloepiphyseal dysplasia, Omani type: further definition of the phenotype. Am J Med Genet *146A*, 2376–2384.

Vissers, L. E., Lausch, E., Unger, S., et al. (2011). Chondrodysplasia and abnormal joint development associated with mutations in *IMPAD1*, encoding the Golgi-resident nucleotide phosphatase, gPAPP. Am J Hum Genet *88*, 608–615.

Wei, G., Bai, X., and Esko, J. D. (2004). Temperature-sensitive glycosaminoglycan biosynthesis in a Chinese hamster ovary cell mutant containing a point mutation in glucuronyltransferase I. J Biol Chem *279*, 5693–5698.

Wiggins, C. A. R., and Munro, S. (1998). Activity of the yeast MNN1 *a*-1,3-mannosyltransferase requires a motif conserved in many other families of glycosyltransferases. Proc Natl Acad Sci U S A *95*, 7945–7950.

Yada, T., Gotoh, M., Sato, T., et al. (2003a). Chondroitin sulfate synthase-2: molecular cloning and characterization of a novel human glycosyltransferase homologous to chondroitin sulfate glucuronyltransferase, which has dual enzymatic activities. J Biol Chem *278*, 30235–30247.

Yada, T., Sato, T., Kaseyama, H., et al. (2003b). Chondroitin sulfate synthase-3: molecular cloning and characterization. J Biol Chem *278*, 39711–39725.

Yamauchi, S., Mita, S., Matsubara, T., et al. (2000). Molecular cloning and expression of chondroitin 4-sulfotransferase. J Biol Chem *275*, 8975–8981.

Zayed, H., Chao, R., Moshrefi, A., et al. (2010). A maternally inherited chromosome 18q22.1 deletion in a male with late-presenting diaphragmatic hernia and microphthalmia-evaluation of DSEL as a candidate gene for the diaphragmatic defect. Am J Med Genet A *152A*, 916–923.

2.7 Biological functions of branched *N*-glycans related to physiology and pathology of extracellular matrix

Congxiao Gao, Kazuaki Ohtsubo, Jianguo Gu, and Naoyuki Taniguchi

2.7.1 Introduction

The extracellular matrix (ECM) is composed of some major biomolecules produced by cells and excreted to the extracellular space within the tissues that include collagen, elastin, fibronectin (FN), laminin, proteoglycans such as aggrecan, and so forth (Ruoslahti and Yamaguchi, 1991; Lodish et al., 2000; Alberts et al., 2002). The constituent materials in the ECM can vary widely depending on the types of cells with which ECMs are associated. Due to its diverse nature and composition, the ECM can serve many special functions besides mechanical support; for example, it helps to regulate communication between cells, regulates products of certain substances in the body, and provides a framework for cell adhesion, encouraging growth and stable healing (Joladarashi et al., 2011; Lefort et al., 2011; Menon and Beningo, 2011).

Being located on the outermost layer of the cell, carbohydrates may be the first molecules to be encountered and recognized by the other cells, antibodies, invading viruses, and bacteria (Brooks et al., 2010; Dasgupta et al., 2011; Mathias and Corthésy, 2011). Many molecules secreted into the ECM, such as hormones and toxins, have also been reported to bind to carbohydrate receptors on the cell surface (Gringhuis and Geijtenbeek, 2010; Murai et al., 2011). Among diverse carbohydrate structures, branched *N*-glycans, such as α1,6-fucose (core fucose); β1,4-GlcNAc (bisecting GlcNAc) and β1,6-GlcNAc have received attention because of their unique structures and special biological functions (Taniguchi, 2007). It will be easy to assume that aberrant glycosylation patterns can affect the cell-cell and cell-ECM interactions. This chapter briefly summarizes the special biological functions of major branched *N*-glycan structures related to the pathology of ECM.

2.7.2 Synthesis of branched *N*-glycans

β-1,4-*N*-acetylglucosaminyltransferase-III

β-1,4-*N*-acetylglucosaminyltranferase III (GnT-III; EC2.4.1.144) catalyzes the addition of GlcNAc to mannose that is β1,4-linked to an underlying *N*-acetylglucosamine, producing what is known as a "bisecting" GlcNAc linkage (▶Figure 2.14). GnT-III is ubiquitous in all tissues, although relatively higher GnT-III activity is found in the kidney

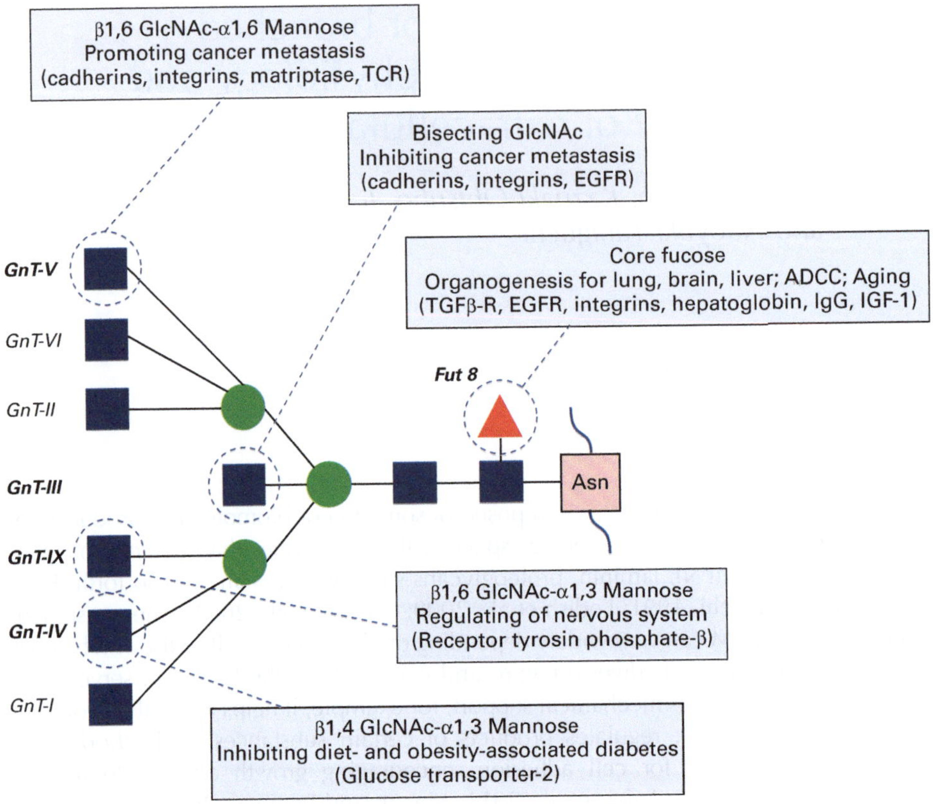

Figure 2.14 Glycosylation reactions catalyzed by the glycosyltransferases α1,6-fucosyltransferase (Fut8), GnT-III, GnT-IV, GnT-V, and β-1,6-*N*-acetylglucosaminyltransferase IX (GnT-IX). The target proteins and the biological functions are indicated.

and brain (Nishikawa et al., 1990). As the introduction of a bisecting GlcNAc suppresses additional processing and elongation of *N*-glycans, GnT-III is generally regarded as a key glycosyltransferase in *N*-glycan biosynthetic pathways (▶Figure 2.14). Those reactions, which are catalyzed *in vitro* by other glycosyltransferases, such as β-1,4-*N*-acetylglucosaminyltransferase-IV (GnT-IV), β-1,6-*N*-acetylglucosaminyltransferase-V (GnT-V), and β-1,4-*N*-acetylglucosaminyltranferase VI (GnT-VI), do not proceed, because the enzymes cannot utilize the bisected oligosaccharide as a substrate (Schachter, 1986).

β-1,4-*N*-acetylglucosaminyltransferase-IV

GnT-IV (EC2.4.1.145) catalyzes the addition of β1,4-linked GlcNAc residue to the Manα1–3 arm of the *N*-glycan core and initiates the synthesis of the GlcNAcβ1–4 branch structure for increasing branch complexity of *N*-glycans (▶Figure 2.14). GnT-IV has two isoenzymes, a and b, which have identical acceptor substrate specificities and comparable GnT-IV enzymatic activities, in vitro. However, *in vivo*, GnT-IVa predominantly exerts GnT-IV activities, whereas GnT-IVb contributes subtly. GnT-IVa is expressed in a highly restricted manner in gastrointestinal tissues and leukocytes, whereas GnT-IVb is ubiquitously and constitutively expressed among tissues in human and rodents (Yoshida et al., 1998, 1999; Ohtsubo et al., 2005; Takamatsu et al., 2010). Attenuation of GnT-IVa due to genetic disruption or induced by a high-fat diet evoked metabolic disorders such as type 2 diabetes, which was attributed to the deficiency of glucose-stimulated insulin secretion of pancreatic β cells that was a consequence of the dysglycosylation of the glucose transporter and its severely reduced cell surface residency (Ohtsubo et al., 2005). These results illuminate the physiological significance of glycan complexities on the regulation of cellular functions.

β-1,6-*N*-acetylglucosaminyltransferase: GnT-V and GnT-IX (GnT-Vb)

GnT-V (EC 2.4.1.155) catalyzes the addition of β1,6-linked GlcNAc units to the α1,6-linked mannose of the trimannosyl core to form tri- or tetra-antennary branches (▶Figure 2.14) (Gu et al., 1993). This branched structure can be further extended to poly-*N*-acetylactosamine chains that can be terminally modified with sialylated structure. The enhanced β1,6-GlcNAc branching structures, a major contributor to *N*-glycan heterogeneity, play important roles in tumor metastasis (Cummings et al., 1982; Shoreibah et al., 1993). On the other hand, GnT-V knockout mice differed in their response to various extrinsic conditions, including altered susceptibility to autoimmune diseases, enhanced delayed-type hypersensitivity, lowered T-cell activation thresholds, and cancer progression (Granovsky et al., 2000; Demetriou et al., 2001; Lagana et al., 2006).

GnT-IX, also known as GnT-Vb, is a brain-specific glycosyltransferase. This tissue-specific expression is found to be regulated by an epigenetic mechanism dependent on neurogenic differentiation 1 (NeuroD1) and CCCTC-binding factor (CTCF) (Kizuka et al., 2011). Receptor tyrosine phosphatase beta (RPTP-β) is a well-known molecule to be modified by GnT-IX (Abbott et al., 2008), and the attached glycan subsequently attenuated cell surface galectin-1 binding. These observations indicate that GnT-IX is an important regulator of cell-ECM interactions in the developing nervous system.

GDP-L-Fuc:*N*-acetyl-β-D-glucosaminide α1,6-fucosyltransferase

GDP-L-Fuc:*N*-acetyl-β-D-glucosaminide α1,6-fucosyltransferase (Fut8; EC 2.4.1.152) catalyzes the transfer of a fucose residue from GDP-fucose to position 6 of the innermost GlcNAc residue of hybrid and complex types of *N*-linked oligosaccharides (Uozumi et al., 1996a, 1996b; Yanagidani et al., 1997). The content of the formed core fucose structure (▶Figure 2.14) may be altered under pathological conditions, such as hepatocellular carcinoma (HCC), liver cirrhosis, and so forth (Miyoshi et al., 1999; Noda et al., 2003). Core fucosylated α-fetoprotein (AFP), which is barely detectable in benign liver disease, is a highly specific tumor marker for HCC (Taketa et al., 1990). Fucosylated hepatoglobin is also one of the biomarkers for pancreatic cancer (Okuyama et al., 2006).

On the other hand, the reduction of core fuscose on immunoglobulin G (IgG), a molecule secreted into the ECM, was reported to improve its binding to FcγRs and enhanced the antibody-dependent cellular cytotoxicity (ADCC) activity by as much as 50- to 100-fold.

2.7.3 Effect of *N*-glycosylation on ECM formation

Effect of *N*-glycosylation of laminin-332 on tumor progression

Basement membranes, which provide mechanical support, cell polarity, and so forth, are thin sheets of specialized ECM (Adams and Watt, 1993; Gustafsson and Fassler, 2000). As major components of all basement membranes, laminins are large extracellular glycoproteins and heterotrimers composed of α and β chains. α1β1, α3β1, α6β1, α6β4, and α7β1 integrins are generally regarded as the main receptors for laminins (Carter et al., 1991).

In the skin, laminin-332, associated with integrin α6β4, forms hemidesmosomes (Dajee et al., 2003) to maintain stable adhesion between the intracellular keratin cytoskeleton and the basement membrane. Laminin-332 undergoes terminal glycosylation within the Golgi apparatus, and then *N*-glycosylated laminin is secreted out of the cell. More recently, it was reported that an increase in bisecting GlcNAc structure on laminin-332 suppressed its cell adhesion and migration activities (Kariya et al., 2010). This effect is likely to be derived from impaired α3β1 integrin clustering and resultant focal adhesion formation by laminin-332.

Lack of *Fut8* destroys the ECM balance

The deficiency of protein core fucosylation by genetic disruption of *Fut8* leads to lethal abnormalities and the development of emphysematous lesions in the lung by attenuation of transforming growth factor-beta 1 (TGF-β1) receptor signaling (Wang et al., 2005), along with a marked overexpression of matrix metalloproteinases (MMPs) and lung-destructive phenotypes. TGF-β1 signaling is a pleiotropic factor that controls ECM homeostasis by regulating MMP expression and inducing ECM protein components. Loss of core fucosylation of TGF-β1 receptor attenuated the binding with TGF-β1 and phosphorylation of Smad2. In *Fut8*$^{-/-}$ mice, the epidermal growth factor receptor (EGFR)–trypsin–proteinase-activated receptor-2 (PAR2) signaling pathway was also downregulated (Li et al., 2006; Wang et al., 2006). The changes on both the TGF-β1 receptors and EGFR receptors are plausible explanations for the emphysematous changes and growth retardation observed in *Fut8*$^{-/-}$ mice.

Moreover, *Fut8* heterozygous-knockout (*Fut8*$^{+/-}$) mice developed emphysematous changes, including airspace enlargement, alveolar wall disruption, and dilated alveolar dusts, in the lungs after a 3-month period of cigarette smoke exposure, which is half the time required for the wild-type mice (Wright et al., 2008). The onset of emphysema in *Fut8*$^{+/-}$ mice is associated with the accumulation of macrophages in the ECM, the induction of MMP gene expression, and activation of MMP enzyme activities (Gao et al., 2012; Kamio et al., 2012).

MMP is one of the most important families participating in the remodeling of ECM. Under normal conditions, the actions of MMP are carefully regulated by specific enzyme inhibitors, and the expression level is controlled by variety of growth factors and cytokines (e.g. TGF-β). The previous results suggest that core fucose modification of functional proteins existing in the ECM may affect important physiological functions and finally the homeostasis of the ECM.

Modification and stabilization of active matriptase by GnT-V

Matriptase is known as an epithelial-derived, type II transmembrane serine protease, which acts as an upstream activator in metastasis and cancer invasion. The function of matriptase seemed to be affected by *N*-glycosylation (Ihara et al., 2002). Addition of α1,6-GlcNAc branch upregulated the protein expression level of matriptase in the conditioned media and on the cell surface without any changes in mRNA expression. Moreover, such increase in matriptase expression seemed to be directly linked to tumor metastasis because the overexpression of matriptase in gastric cancer cells led to lymph node metastasis in athymic mice.

The expression of GnT-V was correlated with tumor metastasis or poor prognosis in colon cancers (Fernandes et al., 1991; Seelentag et al., 1998; Murata et al., 2000) and the mammary gland (Seberger and Chaney, 1999), which show a high expression of matriptase. Therefore, the acquired resistance to degradation, as the result of the addition of α1,6-GlcNAc branching in their cancer cells, is a more important factor in the upregulation of matriptase, leading to tumor metastasis and malignant transformation.

2.7.4 Complexity of *N*-glycan branch modulates cellular functions via clustering cell surface proteins

GnT-IV and -V cooperate to increase the *N*-glycan branch complexity, which provides a variety of carbohydrate epitopes. Galectins are extracellular lectins that preferentially bind to β-galactoside in *N*-glycan branches generated by GnT-IV and -V. Galectins form multivalent complexes and are considered to organize lattice structures on the cell surface. It has been reported that the cross-linking of glycoproteins with galectins prevents endocytosis and increases cell surface residency of glycoproteins that contributes to sustain cellular responses to dynamically changing cellular microenvironments in a normal context. GnT-IVa deficiency attenuates the *N*-glycan complexity of glucose transporter-2 (GLUT2) and thereby abolishes the formation of the GLUT2–galectin-9 complex that consequently diminishes cell surface residency of GLUT2 and impairs glucose-stimulated insulin secretion of pancreatic β cells (Ohtsubo et al., 2005). The pathophysiological pathway to diet- and obesity-associated diabetes has been recently revealed, in which a high-fat diet leading to diabetes recapitulated the free fatty acid–induced oxidative stress in human pancreatic β cells that attenuated GnT-IVa expression by transcriptional impairments, which was coincident with nuclear exclusion of FOXA2 and HNF1α (Ohtsubo et al., 2011). GnT-V deficiency also impairs galectin-mediated clustering and residency of glycoproteins on the cell surface that elicits a variety of disorders (Demetriou et al., 2001; Partridge et al., 2004; Ohtsubo et al., 2005). These findings indicate that GnT-IV/V–mediated organization of cell surface protein clusters is a fundamental process to modulate cellular responses to extrinsic signal inputs, the paradigm of which can be practically applied to better understand the true nature of biological systems in general.

2.7.5 Branched *N*-glycans regulate the biological functions of integrins

Many of the protein molecules reported to participate in the metastatic process have *N*- and/or *O*-linked sugar chains, and alteration of these sugar chains could modify the biological function of these molecules (Guo et al., 2001). Among them, cadherins, integrins, and tyrosine kinase receptor were well studied.

Structure of integrins and their *N*-linked carbohydrates

Integrins belong to a family of heterodimeric transmembrane receptors of the ECM (Hynes, 2002). Integrins contain α and β subunits, both of which have a large extracellular region, a single transmembrane domain, and a short cytoplasmic domain (except for β4). The N-terminal domains of the α and β subunits contain the ECM binding site, while the C-terminal domains mediate the interactions with the cytoplasmic membrane, cytoskeleton, and signaling molecules. It is reported that both α3β1 and α5β1 contain 14 and 12 potential *N*-linked glycosylation sites, respectively. To date, 35 different oligosaccharides structures have been identified in α5β1 integrin purified from human placenta (Nakagawa et al., 1996), of which the most predominant structure was the biantennary di-α(2,3)-sialyl fucosyl sugar chain. In the case of integrin α3β1 from human ureter epithelium cells (HUV29), most of the oligosaccharides were found to be complex types with a wide heterogeneity, including bi-, tri-, and tetra-antennary sugar chains. With rich potential *N*-glycosylation sites, it is easy to deduce that the functions of integrins can be greatly remodeled by different sugar structures.

Integrin functions regulated by different branched *N*-glycans

A growing body of evidence indicates that the presence of the appropriate oligosaccharide can modulate integrin activation (Isaji et al., 2004). Culturing human fibroblasts with α-mannosidase-II led to the appearance of immature α5β1 on the cell surface and reduction of FN-dependent adhesion (Akiyama et al., 1989). Treatment of purified α5β1 with *N*-glycosidase F blocked the binding between α5β1and FN and prevented the inherent association between subunits (Zheng et al., 1994). Thus, altered expression of *N*-glycans on the α5β1might contribute to the adhesive property of tumor cells and to tumor formation.

Overexpression of GnT-III inhibited α5β1-mediated cell spreading and migration, and phosphorylation of focal adhesion kinase (FAK) (Gu et al., 2004). The binding affinity of α5β1 for FN was significantly reduced after introduction of a bisecting GlcNAc into the α5 subunit.

The expression of GnT-V or β1,6-branched *N*-glycans is reported to be strongly associated with cancer metastasis. It is found that cell migration toward FN and invasion through the matrigel were both substantially stimulated in GnT-V cDNA–transfected HT1080 cells (Guo et al., 2002). GnT-V modification appears to be selective since the greatly increased expression of β1,6-GlcNAc could only be found on the β1 subunit. Many studies support that the β1,6-GlcNAc structure leads to cancer invasion and metastasis properties. Similarly, α3β1 integrins contain many more β1,6-GlcNAc branches in metastatic cells than in nonmetastatic cells (Pocheć et al., 2003). It is well known that the reactions catalyzed by GnT-V are inhibited by GnT-III (Schachter, 1986). A recent study suggested that the priority of GnT-III for modification of the α3 subunit might explain inhibition of GnT-V–induced cell migration by GnT-III (Zhao et al., 2006b).

α3β1-mediated cell migration on laminin-5 and FAK phosphorylation induced by adhesion to laminin-5 are significantly reduced in *Fut8*$^{-/-}$ embryonic fibroblasts (Zhao et al., 2006a). Reintroduction of *Fut8* potentially reversed the impairment of migration and signaling.

Integrins have multiple potential *N*-glycosylation sites, but only a certain part of these sites – for example, the sites located on the β propeller of α5 or the I-like domain of β1 and β3 (Sato et al., 2009) – could regulate the conformation and biological function of integrins. Remodeling of glycosyltransferase-modified *N*-glycan structures either positively or negatively modulates cell adhesion and migration.

Role of *N*-glycosylation in integrin-ECM complex formation

Besides the branched N-glycans function in cell migration and cell signaling, integrin surface expression and activation appear to be dependent on branched *N*-glycans, which is important for galectin binding (Lagana et al., 2006). It is worth noting that the binding of galectin-3 to branched *N*-glycan ligands stimulates focal adhesion remodeling, FAK and phosphatidylinositol 3-kinase (PI3K) activation, local F-actin instability, and α5β1 translocation to fibrillar adhesion; therefore, FN polymerization and tumor cell motility are regulated. Conversely, deletion of GnT-V modification in mouse embryonic fibroblasts resulted in an enhanced integrin clustering and activation of α5β1 transcription by protein kinase C signaling, which in turn upregulated cell surface expression of α5β1 resulting in increased matrix adhesion and decreased migration (Guo et al., 2005).

It is known that integrins are associated with various members of the tetraspanin family, such as CD9 and CD82 (Hakomori, 2010). Integrin α3 or α5 subunit associates with tetraspanin CD9 or CD82, and this complex formation is affected by the *N*-glycosylation of both integrin and tetraspanins, as well as by gangliosides in the microdomain (Kawakami et al., 2002). CD82 with complete *N*-glycosylation reduces its association with α3 or α5 integrin (Ono et al., 2000); conversely, the association of CD9 with the α3 or α5 integrin subunits is not influenced by *N*-glycosylation since CD9 contains no *N*-glycosylation sites.

2.7.6 The mutual regulation of *N*-glycosylation and cadherins

N-glycosylation regulates E-cadherin functions

Cadherins comprise an important family of adhesion molecules that function in cell recognition, tissue morphogenesis, and tumor progression (Yagi and Takeichi, 2000). In epithelial-derived tumors, loss of cell-cell adhesion is correlated with downregulation of E-cadherin as well as increased proliferation and tumor invasiveness (Hazan et al., 2004). E-cadherin is the core transmembrane protein of the adherent junction and is required for binding and localization of a number of important cytoplasmic proteins, termed *catenins,* that connect the cadherin complex to the actin cytoskeleton and several signaling pathways.

E-cadherin can be posttranslationally modified by phosphorylation, *O*-glycosylation, and *N*-glycosylation, and the *N*-glycosylation at Asn-633 is essential for its expression, folding, and trafficking (Zhou et al., 2008). Overexpression of GnT-III increased the retention of E-cadherin at the cell border, which resulted in an enhancement of E-cadherin–mediated homotypic adhesion, whereas addition of β1,6-GlcNAc branches led to the acquisition of the malignant phenotype (Pinho et al., 2011). Pierce and co-workers reported that the increased β1,6-GlcNAc branch level decreased the rate of homophilic cell-cell adhesion depending on neural cadherin (N-cadherin), by promoting the phosphorylation of catenins through the EGFR and Src signaling pathways, as well as a stimulation of a cadherin-mediated cell-cell migration (Guo et al., 2009).

Mutual regulation of E-cadherin and GnT-III

GnT-III overexpression was reported to increase E-cadherin–mediated homotypic adhesion and suppressed phosphorylation of the E-cadherin–β-catenin complex during cell-cell adhesion. Conversely, GnT-III is regulated by E-cadherin–mediated cell-cell adhesion. GnT-III activity was increased under dense culture conditions compared with sparse culture conditions in an E-cadherin–dependent manner (Iijima et al., 2006). Recent studies suggested that shRNA knockdown of β-catenin, an essential molecule in both cadherin-mediated cell adhesion and canonical Wnt signaling, resulted in a dramatic increase in GnT-III expression and the bisected GlcNAc, while stimulation of the Wnt signaling pathway consistently and significantly inhibited GnT-III expression (Xu et al., 2011).

2.7.7 Challenges and future prospects

To date, more than 180 human glycogenes, which encode glycosyltransferases and related genes, have been cloned and characterized. Approximately 45 different kinds of congenital disorders of glycosylation (CGD) have been reported, and in addition, various human diseases such as type 2 diabetes and chronic obstructive pulmonary disease (COPD) have been associated with changes in glycan structures. The clinical relevance of glycans is currently a subject of intense interest because the glycans play roles in diverse biological phenomena, including development, aging, and disease susceptibility. On the other hand, the glycan structures, even on the same protein, are so heterogeneous, and their expression patterns so species-, organ-, and disease-specific, as to make research difficult.

Recent studies using the functional glycomics approach, which uses gene manipulation, such as gene knockout and transgenesis, as well as small interfering RNA, and characterization of glycan structures using mass spectra analysis, open new avenues toward the study of the physiological relevance of glycosylation. It would be a big challenge to construct easy-to-use databases of glycogene and glycan structure libraries. It is also important to have technologies to detect the subtle changes of glycan structures. In this postgenomic era, it is crucial to identify target proteins and to determine the functions of specific glycan structures in these different proteins.

Furthermore, another major challenge for the future is to offer glycan-scientific knowledge for medical application – for example, glycan imaging for diagnosis using glycobiomarkers, a drug delivery system depending on the organ-specificity or disease-specificity of glycans, and also glycan inhibitors due to their specific binding to target proteins. Prior to such applications, it is important to determine whether the phenomena observed *in vitro* are also observed *in vivo*. Use of animal models, for example *Fut8* knockout mice, will elucidate the relationships between Fut8-catalyzed core fucosylation and ECM homeostasis, which might ultimately lead to the development of novel COPD therapies.

In order to better understand the significant functions of glycans under pathophysiological conditions, it is crucial to combine glycoscience with other areas of research such as genomics, proteomics, and chemical biology.

2.7.8 Take-home message

Various biological functions of *N*-glycan in disease processes are summarized in this article. Identifying the target molecules involved in the *N*-glycosylation and analyzing the glycan structures are essential for understanding the function of glycan under pathophysiological conditions. A systems glycobiology approach using biochemical analyses, structural biology, and chemical biology is requisite for understanding the biochemical and molecular mechanisms of diseases.

Moreover, exploring the underlying mechanisms by which glycans are expressed in organ-specific and disease-specific manners is a promising challenge for the development of novel glycodrugs.

Acknowledgments

This work was partly supported by the Program for Promotion of Fundamental Studies in Health Science of the National Institute of Biomedical Innovation (NIBIO), and the Global Center of Excellent (GCOE) Program.

References

Abbott, K. L., Matthews, R. T., and Pierce, M. (2008). Receptor tyrosine phosphatase beta (RPTP-β) activity and signaling are attenuated by glycosylation and subsequent cell surface galectin-1 binding. J Biol Chem *283*, 33026–33035.

Adams, J. C., and Watt, F. M. (1993). Regulation of development and differentiation by the extracellular matrix. Development *117*, 1183–1198.

Akiyama, S. K., Yamada, S. S., and Yamada, K. M. (1989). Analysis of the role of glycosylation of the human fibronectin receptor. J Biol Chem *264*, 18011–18018.

Alberts, B., Johnson, A., Lewis, J., Raff, M., Roberts, K., and Walter, P. (2002). Cell junctions, cell adhesions, and the extracellular matrix. In Molecular Biology of the Cell. New York: Garland Science, pp. 1065–1125.

Brooks, C. L., Schietinger, A., Borisova, S. N., et al. (2010). Antibody recognition of a unique tumor-specific glycopeptide antigen. Proc Natl Acad Sci U S A *107*, 10056–10061.

Carter, W. G., Ryan, M. C., and Gahr, P. J. (1991). Epiligrin, a new cell adhesion ligand for integrin α3β1 in epithelial basement membranes. Cell *65*, 599–610.

Cummings, R. D., Trowbridge, I. S., and Kornfeld, S. (1982). A mouse lymphoma cell line resistant to the leukoagglutinating lectin from Phaseolus vulgaris is deficient in UDP-GlcNAc: alpha-D-mannoside beta 1,6 N-acetylglucosaminyltransferase. J Biol Chem *257*, 13421–13427.

Dajee, M., Lazarov, M., Zhang, J. Y., et al. (2003). NF-κB blockade and oncogenic Ras trigger invasive human epidermal neoplasia. Nature *421*, 639–643.

Dasgupta, J., Bienkowska-Haba, M., Ortega, M. E., et al. (2011). Structural basis of oligosaccharide receptor recognition by human papillomavirus. J Biol Chem *286*, 2617–2624.

Demetriou, M., Granovsky, M., Quaggin, S., and Dennis, J. W. (2001). Negative regulation of T-cell activation and autoimmunity by Mgat5 N-glycosylation. Nature *409*, 733–739.

Fernandes, B., Sagman, U., Auger, M., Demetrio, M., and Dennis, J. W. (1991). Beta 1–6 branched oligosaccharides as a marker of tumor progression in human breast and colon neoplasia. Cancer Res *51*, 718–723.

Gao, C., Maeno, T., Ota, F., et al. (2012). Sensitivity of heterozygous α1,6-fucosyltransferase knock-out mice to cigarette smoke-induced emphysema: implication of aberrant transforming growth factor-β signaling and matrix metalloproteinase gene expression. J Biol Chem *287*, 16699–16708.

Granovsky, M., Fata, J., Pawling, J., Muller, W. J., Khokha, R., and Dennis, J. W. (2000). Suppression of tumor growth and metastasis in Mgat5-deficient mice. Nat Med *6*, 306–312.

Gringhuis, S. I., and Geijtenbeek, T. B. (2010). Carbohydrate signaling by C-type lectin DC-SIGN affects NF-kappaB activity. Methods Enzymol *480*, 151–164.

Gu, J., Nishikawa, A., Tsuruoka, N., et al. (1993). Purification and characterization of UDP-N-acetylglucosamine: alpha-6-D-mannoside beta 1–6N-acetylglucosaminyltransferase (N-acetylglucosaminyltransferase V) from a human lung cancer cell line. J Biochem *113*, 614–619.

Gu, J., Zhao, Y., Isaji, T, et al. (2004). Beta1,4-N-acetylglucosaminyltransferase III down-regulates neurite outgrowth induced by costimulation of epidermal growth factor and integrins through the Ras/ERK signaling pathway in PC12 cells. Glycobiology *14*, 177–186.

Guo, H. B., Johnson, H., Randolph, M., and Pierce, M. (2009). Regulation of homotypic cell-cell adhesion by branched N-glycosylation of N-cadherin extracellular EC2 and EC3 domains. J Biol Chem *284*, 34986–34997.

Guo, H. B., Lee, I., Bryan, B. T., and Pierce, M. (2005). Deletion of mouse embryo fibroblast N-acetylglucosaminyltransferase V stimulates alpha5beta1 integrin expression mediated by the protein kinase C signaling pathway. J Biol Chem *280*, 8332–8342.

Guo, H. B., Lee, I., Kamar, M., Akiyama, S. K., and Pierce, M. (2002). Aberrant N-glycosylation of beta1 integrin causes reduced alpha5beta1 integrin clustering and stimulates cell migration. Cancer Res *62*, 6837–6845.

Guo, H. B., Zhang, Y., and Chen, H. L. (2001). Relationship between metastasis-associated phenotypes and N-glycan structure of surface glycoproteins in human hepatocarcinoma cells. J Cancer Res Clin Oncol *127*, 231–236.

Gustafsson, E., and Fassler, R. (2000). Insights into extracellular matrix functions from mutant mouse models. Exp Cell Res *261*, 52–68.

Hakomori, S. I. (2010). Glycosynaptic microdomains controlling tumor cell phenotype through alteration of cell growth, adhesion, and motility. FEBS Lett *584*, 1901–1906.

Hazan, R. B., Qiao, R., Keren, R., Badano, I., and Suyama K. (2004). Cadherin switch in tumor progression. Ann N Y Acad Sci *1014*, 155–163.

Hynes, R. O. (2002). Integrins: bidirectional, allosteric signaling machines. Cell *110*, 673–687.

Ihara, S., Miyoshi, E., Ko, J. H., et al. (2002). Prometastatic effect of N-acetylglucosaminyltransferase V is due to modification and stabilization of active matriptase by adding beta 1–6 GlcNAc branching. J Biol Chem *277*, 16960–16967.

Iijima, J., Zhao, Y., Isaji ,T., et al. (2006). Cell-cell interaction-dependent regulation of N-acetylglucosaminyltransferase III and the bisected N-glycans in GE11 epithelial cells. Involvement of E-cadherin-mediated cell adhesion. J Biol Chem *281*, 13038–13046.

Isaji, T., Gu, J., Nishiuchi, R., et al. (2004). Introduction of bisecting GlcNAc into integrin alpha5beta1 reduces ligand binding and down-regulates cell adhesion and cell migration. J Biol Chem *279*, 19747–19754.

Joladarashi, D., Salimath, P. V., and Chilkunda, N. D. (2011). Diabetes results in structural alteration of chondroitin sulfate/dermatan sulfate in the rat kidney: effects on the binding to extracellular matrix components. Glycobiology *7*, 960–972.

Kamio, K., Yoshida, T., Gao, C., et al. (2012). α1,6-Fucosyltransferase (Fut8) is implicated in vulnerability to elastase-induced emphysema in mice and a possible non-invasive predictive marker for disease progression and exacerbation in chronic obstructive pulmonary disease (COPD). Biochem Biophys Res Commun in press.

Kariya, Y., Kawamura, C., Tabei, T., and Gu, J. (2010). Bisecting GlcNAc residues on laminin-332 down-regulate galectin-3-dependent keratinocyte motility. J Biol Chem *285*, 3330–3340.

Kawakami, Y., Kawakami, K., Steelant, W. F., et al. (2002). Tetraspanin CD9 is a "proteolipid," and its interaction with alpha 3 integrin in microdomain is promoted by GM3 ganglioside, leading to inhibition of laminin-5-dependent cell motility. J Biol Chem *277*, 34349–34358.

Kizuka, Y., Kitazume, S., Yoshida, M., and Taniguchi, N. (2011). Brain-specific expression of N-aAcetylglucosaminyltransferase IX (GnT-IX) is regulated by epigenetic histone modifications. J Biol Chem *286*, 31875–31884.

Lagana, A., Goetz, J. G., Cheung, P., Raz, A., Dennis, J. W., and Nabi, I. R. (2006). Galectin binding to Mgat5-modified N-glycans regulates fibronectin matrix remodeling in tumor cells. Mol Cell Biol *26*, 3181–3193.

Lefort, C. T., Wojciechowski, K., and Hocking, D. C. (2011). N-cadherin cell-cell adhesion complexes are regulated by fibronectin matrix assembly. J Biol Chem *286*, 3149–3160.

Li, W., Nakagawa, T., Koyama, N., et al. (2006). Down-regulation of trypsinogen expression is associated with growth retardation in alpha1,6-fucosyltransferase-deficient mice: attenuation of proteinase-activated receptor 2 activity. Glycobiology *16*, 1007–1019.

Lodish, H., Berk, A., Zipursky, S. L., Matsudaira, P., Baltimore, D., and Darnell, J. (2000). Integrating cells into tissues. In Molecular Cell Biology. New York: W. H. Freeman & Co., pp. 968–1002.

Mathias, A., and Corthésy, B. (2011). Recognition of gram-positive intestinal bacteria by hybridoma- and colostrum-derived secretory immunoglobulin A is mediated by carbohydrates. J Biol Chem *286*, 17239–17247.

Menon, S., and Beningo, K.A. (2011). Cancer cell invasion is enhanced by applied mechanical stimulation. *PLoS One*. 6, e17277.

Miyoshi, E., Noda, K., Yamaguchi, Y., et al. (1999). The alpha1–6-fucosyltransferase gene and its biological significance. Biochim Biophys Acta *1473*, 9–20.

Murai, T., Maruyama, Y., Mio, K., Nishiyama, H., Suga, M., and Sato, C. (2011). Low cholesterol triggers membrane microdomain-dependent CD44 shedding and suppresses tumor cell migration. J Biol Chem *286*, 1999–2007.

Murata, K., Miyoshi, E., Kameyama, M., et al. (2000). Expression of N-acetylglucosaminyltransferase V in colorectal cancer correlates with metastasis and poor prognosis. Clin Cancer Res *6*, 1772–1777.

Nakagawa, H., Zheng, M., Hakomori, S., Tsukamoto, Y., Kawamura, Y., and Takahashi, N. (1996). Detailed oligosaccharide structures of human integrin alpha 5 beta 1 analyzed by a three-dimensional mapping technique. Eur J Biochem *237*, 76–85.

Nishikawa, A., Gu, J., Fujii, S., and Taniguchi, N. (1990). Determination of N-acetylglucosaminyltransferases III, IV and V in normal and hepatoma tissues of rats. Biochim Biophys Acta *1035*, 313–318.

Noda, K., Miyoshi, E., Gu, J., et al. (2003). Relationship between elevated FX expression and increased production of GDP-L-fucose, a common donor substrate for fucosylation in human hepatocellular carcinoma and hepatoma cell lines. Cancer Res *63*, 6282–6289.

Ohtsubo, K., Mark, Z. C., Olefsky, J. M., and Marth, J. D. (2011). Pathway to diet- and obesity-associated diabetes through attenuation of pancreatic beta cell glycosylation and glucose transport. Nat Med *17*, 1067–1075.

Ohtsubo, K., Takamatsu, S., Minowa, M. T., Yoshida, A., Takeuchi, M., and Marth, J. D. (2005). Dietary and genetic control of glucose transporter-2 glycosylation promotes insulin secretion in suppressing diabetes. Cell *123*, 1307–1321.

Okuyama, N., Ide, Y., Nakano, M., et al. (2006). Fucosylated haptoglobin is a novel marker for pancreatic cancer: a detailed analysis of the oligosaccharide structure and a possible mechanism for fucosylation. Int J Cancer *118*, 2803–2808.

Ono, M., Handa, K., Withers, D. A., and Hakomori, S. (2000). Glycosylation effect on membrane domain (GEM) involved in cell adhesion and motility: a preliminary note on functional alpha3, alpha5-CD82 glycosylation complex in ldlD 14 cells. Biochem Biophys Res Commun *279*, 744–750.

Partridge, E. A., Le Roy, C., Di Guglielmo, G. M., et al. (2004). Regulation of cytokine receptors by Golgi N-glycan processing and endocytosis. Science *306*, 120–124.

Pinho, S. S., Seruca, R., Gärtner, F., et al. (2011). Modulation of E-cadherin function and dysfunction by N-glycosylation. Cell Mol Life Sci *68*, 1011–1020.

Pocheć, E., Lityńska, A., Amoresano, A., and Casbarra, A. (2003). Glycosylation profile of integrin alpha 3 beta 1 changes with melanoma progression. Biochim Biophys Acta *1643*, 113–123.

Ruoslahti, E., and Yamaguchi, Y. (1991). Proteoglycans as modulators of growth factor activities. Cell *64*, 867–869.

Sato, Y., Isaji, T., Tajiri, M., et al. (2009). An N-glycosylation site on the beta-propeller domain of the integrin alpha5 subunit plays key roles in both its function and site-specific modification by beta1,4-N-acetylglucosaminyltransferase III. J Biol Chem *284*, 11873–11881.

Schachter, H. (1986). Biosynthetic controls that determine the branching and microheterogeneity of protein-bound oligosaccharides. Biochem Cell Biol *64*, 163–181.

Seberger, P. J., and Chaney, W. G. (1999). Control of metastasis by Asn-linked, beta1–6 branched oligosaccharides in mouse mammary cancer cells. Glycobiology *9*, 235–241.

Seelentag, W. K., Li, W. P., Schmitz, S. F., et al. (1998). Prognostic value of beta1,6-branched oligosaccharides in human colorectal carcinoma. Cancer Res *58*, 5559–5564.

Shoreibah, M., Perng, G. S., Adler, B., et al. (1993). Isolation, characterization, and expression of a cDNA encoding N-acetylglucosaminyltransferase V. J Biol Chem *268*, 15381–15385.

Takamatsu, S., Antonopoulos, A., Ohtsubo, K., et al. (2010). Physiological and glycomic characterization of *N*-acetylglucosaminyltransferase -IVα and -IVβ double deficient mice. Glycobiology *20*, 485–497.

Taketa, K., Sekiya, C., Namiki, M., et al. (1990). Lectin-reactive profiles of alpha-fetoprotein characterizing hepatocellular carcinoma and related conditions. Gastroenterology *99*, 508–518.

Taniguchi, N. (2007). A sugar-coated switch for cellular growth and arrest. Nat Chem Biol *3*, 307–309.

Uozumi, N., Teshima, T., Yamamoto, T., et al. (1996a). A fluorescent assay method for GDP-L-Fuc:N-acetyl-beta-D-glucosaminide alpha 1–6fucosyltransferase activity, involving high performance liquid chromatography. J Biochem *120*, 385–392.

Uozumi, N., Yanagidani, S., Miyoshi, E., et al. (1996b). Purification and cDNA cloning of porcine brain GDP-L-Fuc:N-acetyl-beta-D-glucosaminide alpha1->6fucosyltransferase. J Biol Chem *271*, 27810–27817.

Wang, X., Inoue, S., Gu, J., et al. (2005). Dysregulation of TGF-beta1 receptor activation leads to abnormal lung development and emphysema-like phenotype in core fucose-deficient mice. Proc Natl Acad Sci U S A *102*, 15791–15796.

Wang, X., Gu, J., Ihara, H., Miyoshi, E., Honke, K., and Taniguchi, N. (2006). Core fucosylation regulates epidermal growth factor receptor-mediated intracellular signaling. J Biol Chem *281*, 2572–2577.

Wright, J. L., Cosio, M., and Churg, A. (2008). Animal models of chronic obstructive pulmonary disease. Am J Physiol Lung Cell Mol Physiol *295*, L1–L5.

Xu, Q., Akama, R., Isaji, T., et al. (2011). Wnt/beta-catenin signaling down-regulates N-acetylglucosaminyltransferase III expression: the implications of two mutually exclusive pathways for regulation. J Biol Chem *286*, 4310–4318.

Yagi, T., and Takeichi, M. (2000). Cadherin superfamily genes: functions, genomic organization, and neurologic diversity. Genes Dev *14*, 1169–1180.

Yanagidani, S., Uozumi, N., Ihara, Y., Miyoshi, E., Yamaguchi, N., and Taniguchi, N. (1997). Purification and cDNA cloning of GDP-L-Fuc:N-acetyl-beta-D-glucosaminide:alpha1–6 fucosyltransferase (alpha1–6 FucT) from human gastric cancer MKN45 cells. J Biochem *121*, 626–632.

Yoshida, A., Minowa, M. T., Takamatsu, S., Hara, T., Ikenaga, H., and Takeuchi, M. (1998). A novel second isoenzyme of the human UDP-N-acetylgulcosamine: α1,3-D-mannoside β1,4-N-acetylglucosaminyltransferase family: cDNA cloning, expression, and chromosomal assignment. Glycoconj J *15*, 1115–1123.

Yoshida, A., Minowa, M. T., Takamatsu, S., et al. (1999). Tissue specific expression and chromosomal mapping of a human UDP-N-acetylgulcosamine: α1,3-D-mannoside β1,4-N-acetylglucosaminyltransferase. Glycobiology *9*, 303–310.

Zhao, Y., Itoh, S., Wang, X., et al. (2006a). Deletion of core fucosylation on alpha3beta1 integrin down-regulates its functions. J Biol Chem *281*, 38343–38350.

Zhao, Y., Nakagawa, T., Itoh, S., et al. (2006b). N-acetylglucosaminyltransferase III antagonizes the effect of N-acetylglucosaminyltransferase V on alpha3beta1 integrin-mediated cell migration. J Biol Chem *281*, 32122–32130.

Zheng, M., Fang, H., and Hakomori, S. (1994). Functional role of N-glycosylation in alpha 5 beta 1 integrin receptor. De-N-glycosylation induces dissociation or altered association of alpha 5 and beta 1 subunits and concomitant loss of fibronectin binding activity. J Biol Chem *269*, 12325–12331.

Zhou, F., Su, J., Fu, L., et al. (2008). Unglycosylation at Asn-633 made extracellular domain of E-cadherin folded incorrectly and arrested in endoplasmic reticulum, then sequentially degraded by ERAD. Glycoconj J *25*, 727–740.

3 Proteoglycans: structure, pathobiology, and signaling

3.1 Introduction

Liliana Schaefer

Proteoglycans (PGs) are a heterogeneous group of glycoconjugates consisting of a specific core protein posttranslationally modified and covalently linked with linear glycosaminoglycans (GAGs) consisting of repeating disaccharides (Ly et al., 2010; Schaefer and Schaefer, 2010). Hyaluronan (HA) is an exception to this definition, as it lacks a covalently bound protein core. Its unique structure and complex function is described in Section 2 of this book.

GAG chains are long, unbranched polysaccharides, negatively charged due to the presence of carboxyl and/or sulfate groups on their sugar moieties. Depending on the degree of sulfation, they can be classified into two categories: (1) sulfated GAGs, composed of chondroitin sulfate (CS), dermatan sulfate (DS), keratan sulfate (KS), heparin (Hp), and heparan sulfate (HS); and (2) nonsulfated GAGs such as HA. The sulfated GAGs CS, DS, and HS are linked to their respective protein cores via serine residues. KS can be assembled on *N*-linked oligosaccharides as well as on serine/threonine *O*-linked oligosaccharides, which are distinct from the linkage regions of all other GAGs. HS and Hp also contain *N*-sulfate residues (Oldberg et al., 1990; Kjellen and Lindahl, 1991; Iozzo, 1998; Lander and Selleck, 2000; Esko and Lindahl, 2001; Perrimon and Bernfield, 2001; Bulow and Hobert, 2006; Gandhi and Mancera, 2008; Seidler and Dreier, 2008). Due to their structural similarities, there is often confusion between HS and Hp. However, the most striking difference between the two structures is the fact that Hp displays a higher proportion of sulfation compared to HS (Mulloy and Forster, 2000).

Besides the core protein, GAG chains also contribute to the structural and functional diversity of PGs, being involved in cell growth, differentiation, cell migration, cancer, the coagulation cascade, or microbial pathogenesis.

Following their synthesis, which has been dealt with in extensive reviews (Kusche-Gullberg and Kjellen, 2003; Malavaki et al., 2008), PGs are directed to different intra- and extracelullar locations. Some PGs, such as the small leucine-rich proteoglycans (SLRPs) (Schaefer and Iozzo, 2008) and the larger-molecular-weight PGs versican (CS/DS-PG) and aggrecan (CS/KS-PG), are secreted into the pericellular environment or are incorporated into basement membranes (e.g. perlecan and agrin, both heparan sulfate proteoglycans [HSPGs]) (Bi et al., 2005; Bix and Iozzo, 2008). Syndecans have a core protein containing a transmembrane domain and are associated with the cell membrane (Couchman, 2003; Alexopoulou et al., 2007; Filmus et al., 2008), while glypicans are linked to the cell membrane by a glycophosphatidylinositol (GPI) anchor. Some PGs, such as serglycin (CS/HS-PG), which is found in all nucleated hematopoietic cells and platelets, blood vessels, and chondrocytes, also have intracellular functions (Prydz and Dalen, 2000; Perrimon and Bernfield, 2001; Kolset and Tveit, 2008;

Table 3.1 Glycosaminoglycan chains: structure and classification

GAG chain	*Disaccharide components*	*Disaccharide representation*	*Proteoglycans*
CS/DS	GlcA/IdoA → GalNAc(β1→3 linkage)	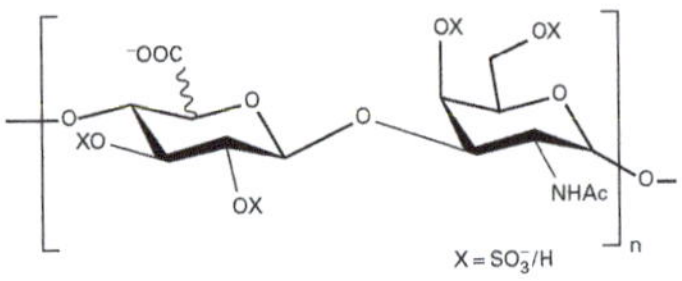	Decorin, Biglycan, Epiphycan, Versican, Aggrecan, Neurocan, Brevican, Perlecan, Leprecan, Testican, Serglycin
KS	GlcNAc → Gal (β1→3 linkage)	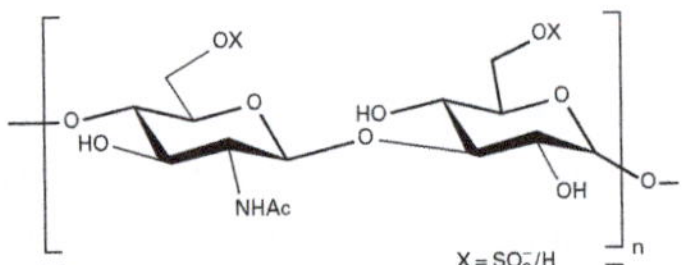	Fibromodulin, Lumican, Keratocan, Mimecan, Agreccan (human)
HS/Hp	GlcA or IdoA → GlcNAc(β/α1→4 linkage)	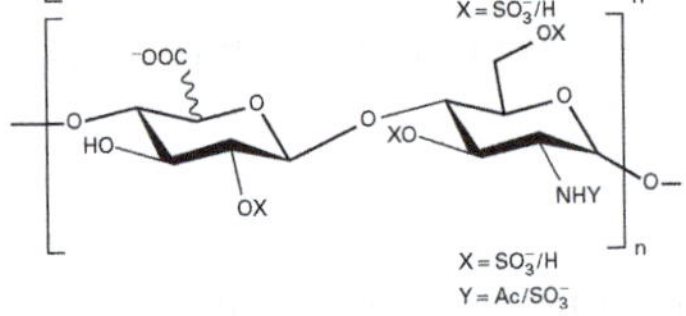	Perlecan, Agrin, Testican, Collagen type XVIII, Syndecans 1-4, Glypicans 1-6, Serglycin
HA	GlcA → GlcNAc (β1→3 linkage)	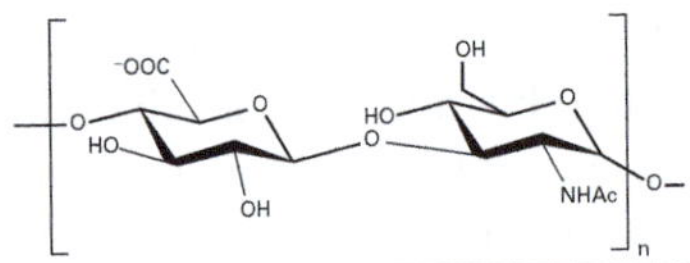	Binds non-covalently hyalectans

Abbreviations used: CS: chondroitin sulfate; DS: dermatan sulfate; KS: keratan sulfate; HS: heparan sulfate; Hp: heparin; HA: hyaluronan; GlcA: D-glucuronic acid; IdoA: L-iduronic acid; GalNAc: *N*-acetyl galactosamine; GlcNAc: *N*-acetyl glucosamine; Gal: galactose.

Pejler et al., 2009). A number of PGs, such as decorin and versican, are widely expressed in many tissues, while others appear to be tissue specific. The latter is the case for brevican, neurocan, neuroglycan C (all chondroitin sulfate proteoglycans [CSPGs]) and phosphacan (CS/KS-PG, a splice variant of the receptor-like protein tyrosine phosphatase-β), found mainly in the central nervous system (CNS), which is particularly rich in CSPGs (Haddock et al., 2007; Sugahara and Mikami, 2007; Kwok et al., 2008). (See ▶Figure 3.1.)

Based on the properties of the protein core, size, localization, type of GAGs, and modular composition, PGs can be classified into three groups: (1) modular PGs divided into two subfamilies of hyalectans (HA- and lectin-binding PGs) and the non-HA-binding PGs (Iozzo, 1998), (2) SLRPs, and (3) cell surface PGs (Iozzo and Murdoch, 1996). (See ▶Figure 3.1.)

Modular PGs can be defined as an assembly of protein modules and are divided into hyalectans (hyaluronan- and lectin-binding PGs) and non-hyaluronan-binding PGs. Hyalectans are represented by aggrecan, versican, brevican, and neurocan. Their main structural features are the N-terminal domain, which is capable of binding hyaluronan, the central part with the covalently attached GAG chains, and the C-terminal domain containing a C-type lectin domain (Iozzo and Murdoch, 1996). Four members are considered at present to be part of the nonhyaluronan-binding PGs class, also called

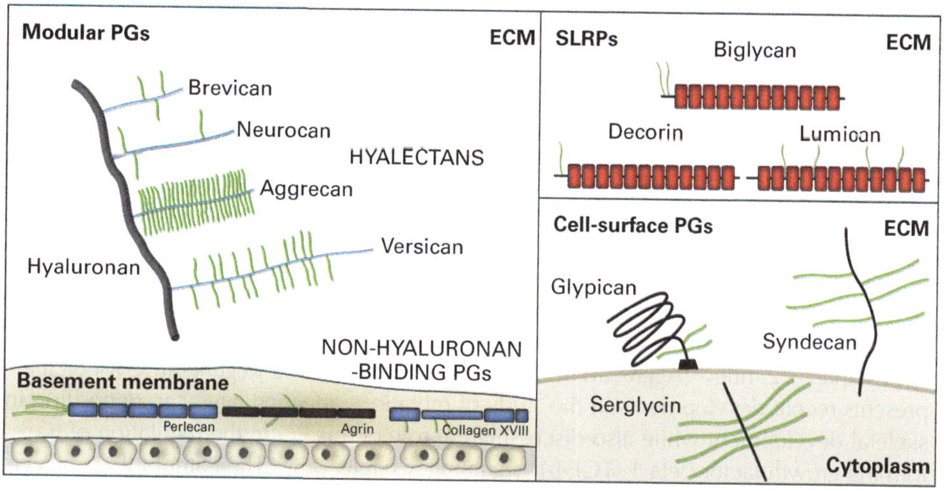

Figure 3.1 Main classes of proteoglycans and selected members.

basement membrane PGs: perlecan, agrin, collagen type XVIII, and leprecan. The first three carry HS chains, while leprecan contains CS chains.

SLRPs are characterized by the presence of leucine-rich repeats (LRRs) within their protein core and typically have cysteine-rich clusters at the N terminus and "ear repeats" at the C terminus (classes I–III). SLRPs are grouped into five classes depending on the characteristic N-terminal cysteine-rich clusters, C-terminal ear repeats, homologies at the genomic and protein levels, and their chromosomal organization (Schaefer and Iozzo, 2008).

Cell surface PGs are divided principally into two groups – namely, syndecans and glypicans. The former are characterized by a transmembrane domain, whereas the latter present a GPI anchor at the C-terminal part. Their particularity is the constant presence of an HS chain, which confers to them the ability to bind different functionally relevant ligands with high affinity (Bernfield et al., 1999). Besides these two groups, there are also other surface PGs, such as CSPG neuron-glial antigen 2 (NG2), cluster of differentiation 44 (CD44), or phosphacan. Serglycin, being an intracellular PG, cannot be included in any of these categories.

Recent advances in the research on PGs has provided solid evidence that the function of these molecules goes far beyond their role as structural components of the extracellular matrix (ECM). The specificity of the protein core, together with the number and various types of GAGs (in addition, differently modified), provides the basis for multifunctional cell-matrix and matrix-matrix interactions. Being mostly extracellular, they are upstream of many signaling cascades and are capable of affecting intracellular phosphorylation events and modulate directly or indirectly distinct pathways, including those driven by bone morphogenic protein/transforming growth factor superfamily members, receptor tyrosine kinases, Toll-like receptors, fibroblast growth factors, wingless and Int (Wnt), and the Hedgehog homologues, among others. The specific function of a particular PG depends not only on its structure but also on the availability of growth factors and receptors expressed by the target cell. Despite the structural similarity of PGs within one class or family, distinct members of the class/family have molecular and cell-specific affinities to different receptors, thereby exerting unique downstream signaling events. Futhermore, it is increasingly clear that the function of PGs varies between their sequestered and "unsequestered" soluble forms (e.g. SLRPs) or between intact and shed forms (e.g. cell surface PGs), defining the availability of PGs to interact with their receptors. The ability of a PG to interact with various receptors appears to be determined by the cellular expression and density of a given receptor. In addition, interaction is dependent on the affinity constant of the PG to each receptor, thereby permitting "hierarchical" receptor binding and activation (Iozzo and Schaefer, 2010).

The following chapters describe recent developments concerning the function and pathobiology of selected PGs and underline the importance of advancing our knowledge in this field. It begins with two chapters describing modular PGs of the hyalectan subfamily. Chapter 3.2, titled "Aggrecan in Skeletal Development and Regenerative Medicine," presents recent developments in the study of mutations affecting aggrecan deposition in skeletal development while also discussing the role of this PG in the regulation of transforming growth factor-beta 1 (TGF-β1) signaling. Chapter 3.3, "The Pathobiology of Versican," discusses the involvement of this PG in several cardiovascular, pulmonary, and ocular diseases and its role in malignancy. Chapter 3.4, "The Biology of Perlecan and Its Bioactive Modules," is a review of new developments in perlecan research. It contains a discussion on the structure, known interactions, and role of this basement membrane PG in cardiovascular disease and cancer, and it also presents some consequences of mutations in the *HSPG2* gene, which encodes for perlecan. Chapter 3.5, titled "Small Leucine-Rich Proteoglycans: Multifunctional Signaling Effectors," focuses on the signaling-dependent pathobiology of selected members of the first three classes of SLRPs: the class I SLRPs decorin and biglycan, the class II SLRPs fibromodulin and lumican, and epiphycan and opticin, which belong to class III. The cell surface PGs are addressed in Chapters 3.6 and 3.7. Chapter 3.6 describes the "Structure and Function of Syndecans." Additional information regarding the signaling effects of syndecans is included in Section 5. Chapter 3.7, "The Glypican Family," discusses the important role of glypicans in signaling and their role in pathology, with a particular focus on cancer. The final chapter, Chapter 3.8, titled "Serglycin Proteoglycan: Implications for Thrombosis, Inflammation, Atherosclerosis, and Metastasis," is a review focusing mainly on the high-affinity binding of serglycin to biologically active proteins in several cell types and the consequences of serglycin knockout in mice.

Despite the fact that none of the following chapters are dedicated to the other members of the PG family, these PGs are also of significant interest. Neurocan and brevican are two hyalectans localized in the CNS, particularly at the nodes of Ranvier. Neurocan is known to play an important role in brain development, tissue remodeling, and neuronal plasticity (Bekku et al., 2009; Bekku and Oohashi, 2010). The non-HA-binding PGs, agrin and collagen XVIII, were identified as constituents of the glomerular basement membrane (GBM) contributing to its negative charge. Agrin is expressed in podocytes, and by binding to laminin, dystroglycan, and integrins, it becomes part of the protein complex, which allows podocytes to attach to the GBM. Collagen XVIII is localized in the mesangial matrix and in the basement membrane of Bowman's capsule contributing to glomerular structure and function (Schaefer and Schaefer, 2010). Chondroadherin is a class IV SLRP found in the bone and is particularly abundant in cartilage. It has been shown in osteoblasts that by binding to integrin, chondroadherin is able to decrease interleukin (IL)-1, IL-6, and receptor activator of NF-κB (RANK) ligand. In addition, both the integrin-binding domain and the HS-binding domain of chondroadherin determine the formation of focal adhesion complexes and play a role in cell migration (Heinegard, 2009). Nyctalopin, part of the same PG class, is attached to the plasma membrane either via a GPI anchor in humans or through a transmembrane domain in mice. It has been shown that mutations of the gene encoding nyctalopin are responsible for the development of the retinal disorder called congenital stationary night blindness (O'Connor et al., 2005). Podocan is another SLRP member included in class V, which is able, similar to other SLRPs, to bind collagen I and inhibit cell growth via p21 (Schaefer and Iozzo, 2008). The latest SLRP to be discovered is podocan-like protein-1 (Podn1), which was found in bone matrix. It is classified as a class V SLRP member and is considered to be involved in matrix mineralization, but its role in this process is not completely understood and needs to be studied in more detail (Mochida et al., 2011).

Our knowledge of PG biology has significantly expanded over the past two decades with the discovery of a host of new members of this multifunctional family. Apart from being structural proteins, PGs also play a major role in signal transduction with regulatory functions in various cellular processes. Mechanistic insights into the molecular and cellular functions of PGs have revealed both the sophistication of these regulatory proteins and the challenges that remain in uncovering the entirety of their biological functions. This section on PGs aims to summarize the multiple functions of PGs with particular emphasis on their intricate composition and the newly described signaling events, in which these molecules play a key role.

References

Alexopoulou, A. N., Multhaupt, H. A., and Couchman, J. R. (2007). Syndecans in wound healing, inflammation and vascular biology. Int J Biochem Cell Biol *39*, 505–528.

Bekku, Y., and Oohashi, T. (2010). Neurocan contributes to the molecular heterogeneity of the perinodal ECM. Arch Histol Cytol *73*, 95–102.

Bekku, Y., Rauch, U., Ninomiya, Y., and Oohashi, T. (2009). Brevican distinctively assembles extracellular components at the large diameter nodes of Ranvier in the CNS. J Neurochem *108*, 1266–1276.

Bernfield, M., Gotte, M., Park, P. W., et al. (1999). Functions of cell surface heparan sulfate proteoglycans. Annu Rev Biochem *68*, 729–777.

Bi, Y., Stuelten, C. H., Kilts, T., et al. (2005). Extracellular matrix proteoglycans control the fate of bone marrow stromal cells. J Biol Chem *280*, 30481–30489.

Bix, G., and Iozzo, R. V. (2008). Novel interactions of perlecan: unraveling perlecan's role in angiogenesis. Microsc Res Tech *71*, 339–348.

Bulow, H. E., and Hobert, O. (2006). The molecular diversity of glycosaminoglycans shapes animal development. Annu Rev Cell Dev Biol *22*, 375–407.

Couchman, J. R. (2003). Syndecans: proteoglycan regulators of cell-surface microdomains? Nat Rev Mol Cell Biol *4*, 926–937.

Esko, J. D., and Lindahl, U. (2001). Molecular diversity of heparan sulfate. J Clin Invest *108*, 169–173.

Filmus, J., Capurro, M., and Rast, J. (2008). Glypicans. Genome Biol *9*, 224.

Gandhi, N. S., and Mancera, R. L. (2008). The structure of glycosaminoglycans and their interactions with proteins. Chem Biol Drug Des *72*, 455–482.

Haddock, G., Cross, A. K., Allan, S., et al. (2007). Brevican and phosphacan expression and localization following transient middle cerebral artery occlusion in the rat. Biochem Soc Trans *35*, 692–694.

Heinegard, D. (2009). Proteoglycans and more–from molecules to biology. Int J Exp Pathol *90*, 575–586.
Iozzo, R. V. (1998). Matrix proteoglycans: from molecular design to cellular function. Annu Rev Biochem *67*, 609–652.
Iozzo, R. V., and Murdoch, A. D. (1996). Proteoglycans of the extracellular environment: clues from the gene and protein side offer novel perspectives in molecular diversity and function. FASEB J *10*, 598–614.
Iozzo, R. V., and Schaefer, L. (2010). Proteoglycans in health and disease: novel regulatory signaling mechanisms evoked by the small leucine-rich proteoglycans. FEBS J *277*, 3864–3875.
Kjellen, L., and Lindahl, U. (1991). Proteoglycans: structures and interactions. Annu Rev Biochem *60*, 443–475.
Kolset, S. O., and Tveit, H. (2008). Serglycin–structure and biology. Cell Mol Life Sci *65*, 1073–1085.
Kusche-Gullberg, M., and Kjellen, L. (2003). Sulfotransferases in glycosaminoglycan biosynthesis. Curr Opin Struct Biol *13*, 605–611.
Kwok, J. C., Afshari, F., Garcia-Alias, G., and Fawcett, J. W. (2008). Proteoglycans in the central nervous system: plasticity, regeneration and their stimulation with chondroitinase ABC. Restor Neurol Neurosci *26*, 131–145.
Lander, A. D., and Selleck, S. B. (2000). The elusive functions of proteoglycans: in vivo veritas. J Cell Biol *148*, 227–232.
Ly, M., Laremore, T. N., and Linhardt, R. J. (2010). Proteoglycomics: recent progress and future challenges. OMICS *14*, 389–399.
Malavaki, C., Mizumoto, S., Karamanos, N., and Sugahara, K. (2008). Recent advances in the structural study of functional chondroitin sulfate and dermatan sulfate in health and disease. Connect Tissue Res *49*, 133–139.
Mochida, Y., Kaku, M., Yoshida, K., Katafuchi, M., Atsawasuwan, P., and Yamauchi, M. (2011). Podocan-like protein: a novel small leucine-rich repeat matrix protein in bone. Biochem Biophys Res Commun *410*, 333–338.
Mulloy, B., and Forster, M. J. (2000). Conformation and dynamics of heparin and heparan sulfate. Glycobiology *10*, 1147–1156.
O'Connor, E., Eisenhaber, B., Dalley, J., et al. (2005). Species specific membrane anchoring of nyctalopin, a small leucine-rich repeat protein. Hum Mol Genet *14*, 1877–1887.
Oldberg, A., Antonsson, P., Hedbom, E., and Heinegard, D. (1990). Structure and function of extracellular matrix proteoglycans. Biochem Soc Trans *18*, 789–792.
Pejler, G., Abrink, M., and Wernersson, S. (2009). Serglycin proteoglycan: regulating the storage and activities of hematopoietic proteases. Biofactors *35*, 61–68.
Perrimon, N., and Bernfield, M. (2001). Cellular functions of proteoglycans – an overview. Semin Cell Dev Biol *12*, 65–67.
Prydz, K., and Dalen, K. T. (2000). Synthesis and sorting of proteoglycans. J Cell Sci *113* (Pt 2), 193–205.
Schaefer, L., and Iozzo, R. V. (2008). Biological functions of the small leucine-rich proteoglycans: from genetics to signal transduction. J Biol Chem *283*, 21305–21309.
Schaefer, L., and Schaefer, R. M. (2010). Proteoglycans: from structural compounds to signaling molecules. Cell Tissue Res *339*, 237–246.
Seidler, D. G., and Dreier, R. (2008). Decorin and its galactosaminoglycan chain: extracellular regulator of cellular function? IUBMB Life *60*, 729–733.
Sugahara, K., and Mikami, T. (2007). Chondroitin/dermatan sulfate in the central nervous system. Curr Opin Struct Biol *17*, 536–545.

3.2 Aggrecan in skeletal development and regenerative medicine

Anna Plaas, Daniel J. Gorski, Jennifer Velasco, Colton McNicols, Rebecca Bell, Vincent Wang and John Sandy

3.2.1 Introduction

Aggrecan was the first proteoglycan isolated and studied in a purified form. Extensive biochemical and cell biological studies on the structure and biosynthesis of cartilage aggrecan were carried out in the period from 1970 to 1990, and much of this information can be found and referenced in some of the seminal papers from this period (Hascall and Heinegard, 1974; Hardingham et al. 1976; Roughley, 1977; Kimura et al., 1980; Buckwalter and Rosenberg, 1982; Plaas et al., 1983). These foundational investigations, along with histochemical analyses of chick and murine embryos showed that aggrecan is the major space-filling matrix component of embryonic skeletal structures. In this location, it confers on the tissue the capacity to expand (as in growth) and to compress reversibly under load (as in joint articulation). Further, it was established that these molecular functions of aggrecan derive from its osmotic properties, due to heavy substitution with polyanionic chondroitin sulfate (CS) chains, and also from its capacity to form supramolecular structures with hyaluronan (HA) and link protein, which are immobilized by the collagen network (Maroudas, 1976). In the current article, we discuss recent discoveries on mutations that affect aggrecan biosynthesis (core protein, glycosylation, and sulfation) and the abundance and structure, and thereby the function, of aggrecan in the extracellular matrix of the developing skeleton. In addition, we discuss recent studies that implicate a cell surface complex composed of aggrecan, HA, and CD44 in the phosphorylation events that follow binding of transforming growth factor-beta 1 (TGF-β1) to its primary receptor, transforming growth factor beta1 receptor II (TGFβ1RII). This recent work suggests that, in addition to its well-known structural role, aggrecan is also a major player in the modulation of cell signaling that follows soft tissue injury.

3.2.2 Aggrecan in skeletal development

Aggrecan core mutations and developmental anomalies

The cloning of the rat, human, and murine aggrecan genes (Doege et al., 1994; Walcz et al., 1994; Valhmu et al., 1995) provided the tools required for a detailed analysis of the aggrecan core protein domains necessary for normal skeletal development. Homozygous mutations resulting in the near total absence of aggrecan in skeletal elements during development are nearly always lethal, whereas in heterozygotes, which often survive, the relationships between aggrecan abundance, structure, and function and the resulting growth phenotypes can be examined. Histological analysis of the heterozygous *cmd* (cartilage matrix deficiency) mouse showed that its marked disproportionate dwarfism resulted from a reduction in CS-aggrecan content due to a mutant product containing only the N-terminal portion of the G1 domain of aggrecan and therefore no CS-attachment sites (Watanabe et al., 1994). A similar disproportionate phenotype in cattle, termed *bulldog dwarfism* (Cavanagh et al., 2007), results from a 4 bp insertion in exon 11 of *Acan*, which leads to decay of the mutant mRNA and, again, a major CS-aggrecan deficiency. A major role for aggrecan in development and expansion of the spine is illustrated by the description of aggrecan mutations, which can lead to one of the many subtypes of spondyloepiphyseal dysplasias (SEDs). In SED type Kimberley (Gleghorn et al., 2005), the mutation causes a frameshift and premature termination within the CS-1 domain. This site is similar to that found in the chick nanomelia mutant (Li et al., 1993), which results in a fatal skeletal deformity. In a severe and essentially proportionate human dwarfism, termed *spondyloepimetaphyseal dysplasia* (SEMD) (Tompson et al., 2009), the disturbed growth of all skeletal structures was due to a missense mutation resulting in a D2267N substitution in the aggrecan G3 domain. This single aspartic acid residue appears to control essential functions since it is highly conserved in all known aggrecan orthologues throughout evolution. The severity of the effect was attributed to the loss of aspartic acid–mediated calcium ion chelation at this site, leading to a G3 conformational change, which was shown to inactivate the normal binding of aggrecan to tenascin-C. Such binding may be essential for aggrecan accumulation since it has been shown (Chimal-Monroy and Diaz de Leon, 1999) that precartilage condensations in TGF-β1–treated murine mesenchymal cells synthesize tenascin-C before aggrecan begins to accumulate. Whether the interaction is required simply for entrapment of aggrecan, or whether pericellular aggrecan has secondary effects, such as differentiation of chondroprogenitor cells, is a topic discussed later in this chapter.

To underline the rather complex skeletal effects that can result from mutations in the aggrecan G3 domain, it has also been found that a M2303V substitution results in a mild and disproportionate short stature, and also a form of osteochondritis dissecans (OCD) characterized by the separation of cartilage and sclerotic subchondral bone from the surrounding tissue in confined areas (Stattin et al., 2010). In this case, the mutation causes a loss of aggrecan binding to fibulin-1, fibulin-2, and tenascin-R. This loss of function might explain the pathology since aggrecan-HA complexes can be networked by fibulins, which might be needed to stabilize the aggrecan matrix in the condensation phase of limb development (Olin et al., 2001) and also help to strengthen the cartilage-bone interface, a structure that fails in OCD (Ochs et al., 2011).

Aggrecan glycosylation and abnormal development

It is generally accepted that any skeletal dysplasia caused by a mutated aggrecan core protein results primarily from the inability of the CS-aggrecan-deficient cartilage limb template to grow by cell proliferation and matrix expansion. It is therefore not surprising that growth abnormalities, which are similar to those seen with protein core mutations, can result from inactivation of genes that are directly responsible for CS synthesis itself. A review of mutations that result in aberrant CS synthesis per se shows that except for a single example of a mutant nucleotide sugar transporter (Hiraoka et al., 2007), all others are in the sulfate transporters responsible for maintaining the intracellular sulfate concentration. This concentration must be maintained at a level that meets K_m requirements for reaction with adenosine triphosphate (ATP) in order to provide sufficient phosphoadenosine phosphosulfate (PAPS) for appropriate sulfation of the *N*-acetyl glucosamine residues of the growing chondroitin chain. Since each aggrecan core protein can carry up to 100 CS chains, and each CS chain can carry up to 30 sulfate moieties, it is not surprising that CS sulfation represents the major demand on chondrocyte intracellular sulfate, and that a decreased efficiency of uptake can markedly reduce the degree of sulfation of the secreted product.

The major sulfate uptake transporter in chondrocytes is SLC26A2, a 739-residue protein with at least 11 transmembrane domains that, in chondrocytes, facilitates sulfate exchange with chloride (Meredith et al., 2007). A functional analysis of SLC26A2 (Satoh et al., 1998) revealed the presence of two transcripts in chondrocytes, one with five exons and a splice variant lacking exon 3. A large number of mutations of SLC26A2, including premature stop codons, reduced mRNA levels, splice site mutations, and single amino-acid substitutions or deletions have been described, and many of these have been linked to specific skeletal phenotypes (Borochowitz et al., 1988; Rossi et al., 1996, 2001; Czarny-Ratajczak et al., 2010). In a study of a quite common genotype, ACG1B (Corsi et al., 2001), it was shown that retarded fetal growth was accompanied by not only a lack of total stainable aggrecan, but also an abnormal organization of the epiphyseal cartilage. Specifically, by staining with antibodies to biglycan and decorin, it was shown that the pericellular matrix was not affected in the mutant cartilage, as shown by normal biglycan staining, whereas the interterritorial matrix was essentially absent in the mutant, as shown by the lack of decorin staining. Pericellular deposition in the mutant might be due to a low rate of aggrecan synthesis per se, or it might be that deficient aggrecan sulfation interferes with deposition in the intercellular matrix. Biosynthetic studies with chondrocytes carrying this mutation might uncover novel aspects of the modulation of aggrecan biosynthesis and/or extracellular deposition.

From the previous discussion, it is clear that a partial lack of sulfated aggrecan in heterozygotes can have a severe effect on many skeletal structures. To examine whether aggrecan depletion due to a sulfation deficiency has any lasting effects on chondrocyte function, a mouse SLC26A2 mutant was generated (Forlino et al., 2005), in which undersulfation was found to have a growth-retarding effect by blocking Indian hedgehog–induced chondrocyte proliferation in the growth zone (Cornaglia et al., 2009). The multiple possible effects of aggrecan undersulfation on skeletal development has also been illustrated by studies (Sohaskey et al., 2008) of a chondrodysplastic mutant mouse, which exhibits severe alterations in growth, accompanied by a disturbance in the orientation and consequent positioning of the interphalangeal joints.

Proteolytic removal of aggrecan in development

Aggrecan is a suitable substrate for many peptidases, including those from all major superfamilies (Roughley, 1977). Endopeptidases that degrade aggrecan efficiently are the aspartic peptidase cathepsin D (Handley et al., 2001); the serine peptidases plasmin and elastase (McDonnell et al., 1993); the cysteine peptidases calpain (Oshita et al.,

2004) and cathepsins K and B (McDonnell et al., 1993); and the metalloproteinases matrix metalloproteinase-3 (MMP3) (Flannery et al., 1992), a disintegrin and metalloproteinase with thrombospondin motifs 4 (ADAMTS4), and a disintegrin and metalloproteinase with thrombospondin motifs 5 (ADAMTS5) (Durigova et al., 2011). Any one or more of these would be capable of degrading aggrecan, if that was required *in vivo* for limb expansion and enchondral ossification at the growth plate. Indeed there is direct evidence, from neoepitope antibodies, for the cleavage of aggrecan by both matrix metalloproteinases (MMPs) (Lee et al., 1998) and ADAMTS-aggrecanases (Lee et al., 2001) at the cartilage-bone interface. However, it appears that these specific cleavages are not required for bone elongation, since mice created with aggrecan, which is resistant to destructive proteolysis by MMPs or ADAMTS-aggrecanases, develop normally with no skeletal deformities (Little et al., 2005a, 2007). In addition, these mice do not accumulate excessive cartilage aggrecan during development, suggesting that metalloproteinase-mediated aggrecan degradation is not required at any stage of skeletal growth. Supportive evidence for this conclusion comes from studies with mice that are deficient in matrix metalloproteinase-13 (MMP13) or ADAMTS1, 4, or 5, none of which exhibit any marked growth abnormality or growth plate changes (Little et al., 2005b; Majumdar et al., 2007). This conclusion is also supported by studies of *Adamts5* expression and protein levels in murine development, which showed no evidence for *Adamts5* (*ts5*) message or ADAMTS5 protein in developing cartilages, but high levels in neuronal and collagenous tissues, such as tendon (McCulloch et al. 2009). The finding that neither MMPs nor ADAMTS-aggrecanases appear to be required for aggrecan removal from the hypertrophic zone of the growth plate suggests that either aggrecan is not removed (Mwale et al., 2002) or that nonspecific proteolysis, perhaps involving lysosomal peptidases released from apoptotic hypertrophic cells, or endocytosis of the aggrecan-rich matrix by macrophage-like cells (Lee et al., 2001), is sufficient for calcification and vascular invasion to proceed.

3.2.3 Aggrecan in regenerative medicine

Aggrecan in cartilage regeneration

The depletion of aggrecan from articular cartilages during progression of osteoarthritis (OA) has been widely considered an imbalance between synthesis and degradation of aggrecan by the resident population of chondrocytes. This model has been derived largely from studies with cartilage explants treated with different combinations of growth factors (such as TGF-β1and fibroblast growth factor-2 [FGF2]) and/or cytokines (such as interleukin-1 [IL-1] and tumor necrosis factor-alpha [TNF-α]) (Ma et al., 2007; Cawston and Young, 2010) predicted to be involved in cartilage degradation *in vivo*. However, not surprisingly, this model of the process is an oversimplification, which does not accommodate many observations made on OA cartilage *in vivo*. Thus, the loss of cartilage aggrecan in human OA, which appears to be primarily a result of ADAMTS-aggrecanase activity (Sandy, 2006), does not occur spatially as seen in intact or injured explants (Sui et al., 2009). Rather, the loss in OA occurs most rapidly from the articular surface layers and progresses downward as the cartilage fibrillates and loses its layered structure. In addition, histologically, the loss *in vivo* occurs preferentially from the pericellular matrix of single cells and cell clusters (Plaas et al., 2007) and appears to extend gradually into the intercellular matrix. Significantly, in human OA cartilages (Yuan et al., 2004; Yoshihara et al., 2008), many of the cells at the surface exhibit a fibrogenic (Kim and Spector, 2000; Lahm et al., 2010) rather than chondrogenic phenotype, and others express genes that are characteristic of those at the cartilage-bone interface in mature animals (Fukui et al., 2008).The origin of these phenotypically altered cells is not necessarily from the resident chondrocytes, but they may be partly, or largely, from progenitor populations in the cartilage (Grogan et al., 2009; Koelling et al., 2009), synovial fluid (Jones et al., 2008; Kurose et al., 2010), and surrounding fibrous tissues (Arufe et al., 2010; Giannoudis et al., 2010). The implication of this view is that the loss of aggrecan from OA cartilage is not due to the chondrocytic release of high levels of ADAMTS-aggrecanase(s), since this would be expected to degrade aggrecan uniformly throughout the tissue. Instead, the available data suggests that *in vivo* ADAMTS-mediated degradation occurs primarily within the pericellular pool of aggrecan, and that if this process is uncontrolled, it results in a cellular milieu in which neither resident nor progenitor cells can maintain a stable articular chondrocyte phenotype (Plaas et al., 2011b). It should be noted here that the existence of such a fast-turnover pool of pericellular aggrecan is also consistent with the kinetics of release of aggrecan in 35 sulfate-labeled cartilage explants treated with inflammatory cytokines (Carney et al., 1985; Quinn et al., 1998, 1999; Handley et al., 2002).The importance of this aggrecan pool to the phenotypic modulation of cells in collagenous tissues has been further illustrated by studies on dermal repair with the *Adamts5*$^{-/-}$mouse, in which it has been shown that removal of this aggrecan by ADAMTS5 is an essential step in the differentiation of progenitors to reparative dermal fibroblasts (Velasco et al., 2011; see ▶Figure 3.2). This chondrogenic/fibrogenic model has also been supported by our studies with murine OA models (Li et al., 2011), wherein cartilage "protection" in the *Adamts5*$^{-/-}$mouse is not due to elimination of cartilage aggrecanase per se, but due to specific protection of the pericellular pool of aggrecan. In this regard, it appears relevant that the chondrogenic growth factor, bone morphogenetic protein 7 (BMP7) , which has been found to form a robust pericellular matrix in the presence of TGF-β1 (Miyamoto et al., 2007), has also been shown to exhibit potent antifibrotic activity in a range of pathologies (Phillips and Fraser, 2010). In addition, consistent with this model is the finding that normal human chondrocytes in alginate beads assemble a robust aggrecan-rich pericellular matrix, whereas OA chondrocytes have lost this capacity (Wang et al., 2003).

With respect to regenerative medicine, there have been many different approaches taken to reestablish normal homeostatic aggrecan turnover in the cartilage of animals and humans with OA. These include combinations of biomechanical stimulation, pharmacological interventions, and also implantation of cell-laden matrix constructs. The goals of achieving effective tissue integration and of reconstructing the biomechanical properties of the layered nature of the articular tissue may never be totally achieved. However, an alternative approach to cartilage regeneration in OA has been suggested by the studies described previously on what appears to be a loss of chondrogenic potential in resident OA chondrocytes and/or reparative progenitors (Li et al., 2011; Velasco et al., 2011). Simply stated, treatments to establish a stable pericellular pool of aggrecan will be required to maintain the chondrogenic behavior of reparative cells. Studies with *Adamts5*$^{-/-}$mice (Plaas et al., 2011b) suggest that targeted and

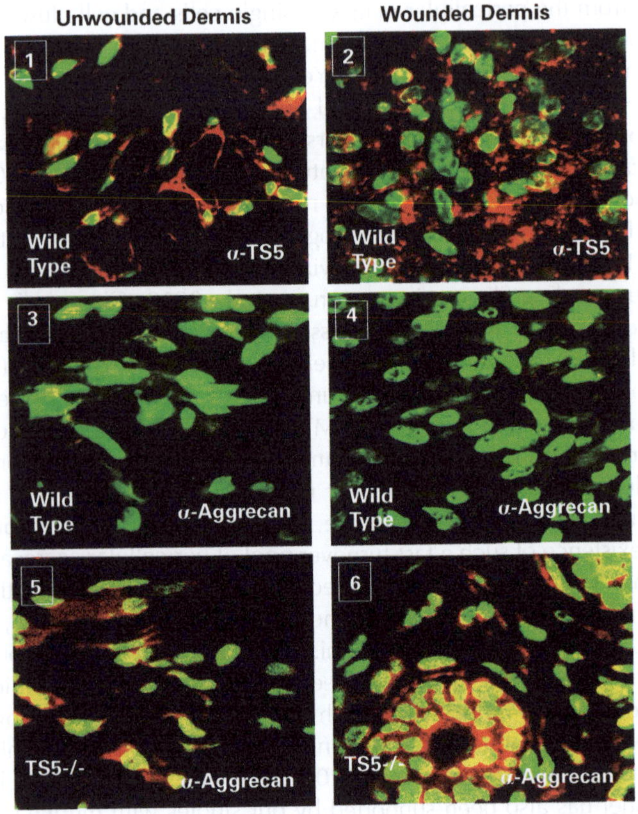

Figure 3.2 Immunolocalization of ADAMTS5 and aggrecan in murine dermal wounds illustrates the role of aggrecan in preventing dermal healing in *Adamts5*$^{-/-}$ mice.
ADAMTS5 is abundant in murine dermis where it is found near dermal fibroblasts in unwounded *wild-type* mice (panel 1) and near granulation tissue cells at the injury site of wounded *wild-type* mice, 8 days postwounding (panel 2). Aggrecan is essentially undetectable in the dermis of unwounded (panel 3) or wounded (panel 4) *wild-type* mice, which achieve effective dermal healing. However, aggrecan is readily detected in the dermis of unwounded *Adamts5*$^{-/-}$ mice (panel 5) and is particularly abundant in association with cell clusters, which are spread throughout the granulation tissue at the nonhealing injury site of wounded *Adamts5*$^{-/-}$ mice (panel 6). Antibodies and protocols used were anti-KNG for ADAMTS5 and anti-DLS for aggrecan (Plass et al., 2007). Also, the antibody notations are defined by the antigens they detect. (Green= nuclei; red= ADAMTS5 or aggrecan.)

effective inhibition of ADAMTS5 in this pericellular location will achieve this goal and directly enhance the potential for repair of adult cartilages in situ.

Aggrecan turnover in tendon and ligament homeostasis and repair

In tendons, aggrecan is concentrated in compressed fibrocartilaginous regions (at wrap-around pulley sites or bone insertion sites) but can also be found at low concentrations in normal tensional regions (Vogel et al., 1994; Vogel and Meyers, 1999; Samiric et al.,

2004; Ilic et al., 2005). Notably, excessive deposition of aggrecan in the tensile midsubstance of tendons is a feature of human tendinopathies (de Mos et al., 2009; Samiric et al. 2009). Moreover, in heritable equine suspensory ligament desmitis, the midsubstance of the ligament body and both branches accumulate two forms of high-molecular-weight aggrecan core protein, and this was accompanied by increased immunohistochemical (IHC) staining for aggrecan in the matrix surrounding groups of rounded cells that did not strictly align with collagen fibrils, as do normal tenocytes (Plaas et al., 2011b).

A possible framework for studying the role of aggrecan in the tensile regions of pathological tendon has come from our recent studies on the flexor digitorum longus (FDL) in *Adamts5*$^{-/-}$mice (Wang et al., 2012). These tendons accumulate aggrecan to about threefold normal levels, whereas the abundance of versican and decorin core proteins (and collagen 1 alpha chains) is not markedly altered. In addition, some of the accumulated aggrecan was present around cells with a normal fibroblastic morphology; however, much was found around others that resembled the loosely organized tendon progenitor cells seen during embryonic development (Brent et al., 2003; Pryce et al., 2009; Watson et al., 2009). Biomechanical analyses showed that the *Adamts5*$^{-/-}$tendons had inferior mechanical properties to the *wild type*, including a significantly lower maximum stress and tensile modulus (P=0.019 and 0.032, respectively). In addition, the mean cross-sectional area of *Adamts5*$^{-/-}$tendons was approximately 33% greater than that of the *wild type* (P=0.014), and the *Adamts5*$^{-/-}$tendons exhibited a greater fibril area fraction of cross-sectional images (P=0.027). The data suggest that aggrecan accumulation due to ablation of *ts5* is detrimental to the biomechanical properties of the FDL, which in turn, indicates that TS5-mediated removal of pericellular aggrecan in tendon is essential to the differentiation of progenitors into a stable fibrogenic phenotype (Wang et al., 2012).

Aggrecan synthesis and degradation in dermal wound healing

Healing of the wounded dermis involves a cascade of cellular responses, which together close the epidermal layer to generate a loosely organized granulation tissue in the dermis. This matrix serves as a molecular framework for the activity of dermal fibroblast progenitors, which synthesize and remodel the collagen and proteoglycan structures necessary for normal tissue properties. Specific functions for aggrecan in skin biology have not been established, although IHC has suggested a role in the maturation of fibroblast progenitors, which arise between the cell aggregates of the dermal papillae of hair follicles (Malgouries et al., 2008). This may be related to the finding that aggrecan expression is highly upregulated in skin fibroblasts obtained from Hutchinson-Gilford progeria syndrome patients (Lemire et al., 2006), an inherited condition characterized by premature skin aging.

We have recently shown (Velasco et al., 2011) that TS5-mediated cleavage of aggrecan is required for effective dermal repair in a model of murine excisional dermal wound healing. In *Adamts5*$^{-/-}$mice, but not in *wild type,* there is a pericellular accumulation of aggrecan bound to HA-CD44 complexes on fibroblast progenitor cells, and this promotes their assembly into epithelioid cell aggregates that disrupt the reforming dermis (▶Figure 3.2). In addition, we found that *Acan* (aggrecan) transcripts, but not *col2* (type 2 collagen) or *halpn1* (link protein-1), became abundant during the early phases of normal dermal regeneration, suggesting that aggrecan has an essential role in wound healing, and that this role is independent of *col2* and *halpn1* expression. We concluded that the ability of wild-type wounds to repair effectively is due to the pericellular deposition of aggrecan and its subsequent ADAMTS5-mediated removal, a process that is required for robust wound contraction and collagen deposition (Velasco et al., 2011).

Pericellular aggrecan in the regulation of TGF-β1 signaling

A previously unappreciated role for aggrecan in the regulation of TGF-β1 signaling was first suggested by mechanistic studies on TGF-β1 signaling with newborn fibroblasts from *wild-type, Adamts5$^{-/-}$,cd44$^{-/-}$*,and *Adamts5$^{-/-}$/cd44$^{-/-}$* double-knockout mice (Velasco et al., 2011). This revealed that the CD44-dependent accumulation of pericellular aggrecan around *Adamts5$^{-/-}$*cells was sufficient to switch TGF-β1 signaling away from the normal Smad2/3 profibrogenic response (Mori et al., 2004) to signaling through Smad1/5/8 (Otsuki et al., 2010). Elimination of aggrecan from the pericellular matrix of *Adamts5$^{-/-}$*fibroblasts by Strep.hyase treatment, or by ablation of *cd44*, restored Smad2/3 signaling and normal fibrogenesis. In other words, both the *in vivo* and isolated cell data indicated that the presence of CD44-bound pericellular HA-aggrecan weakens or prevents the association of TGFβ1RII with ALK1 and blocks profibrogenic Smad2/3 signaling (Chen et al., 2006).This was found to be accompanied by an upregulation of TGFβ1RII/activin-like receptor kinase 5 (ALK5) signaling through Smad1/5/8. The mechanism by which pericellular aggrecan alters TGF-β1 signaling remains to be fully elucidated. However, quantitative gene expression studies of *Acan, hapln1*,and *has2* (hyaluronan synthase-2), in murine newborn fibroblasts treated with TGF-β1, showed strong expression of *Acan* and *has2* but no detectable *halpn1* (Plaas, McNicols, Velasco, unpublished). In contrast, the same experiments with murine chondrocytes (2-week pups) showed strong expression of all three genes. Taken together, it appears that the pericellular aggrecan that modifies TGF-β1 signaling in fibroblasts is organized into an aggrecan/HA complex with CD44 at the cell surface, but without the participation of link protein. It may be this novel assembly at the cell surface that allows aggrecan to interfere with TGF-β1 receptor associations by steric hindrance or alterations in the ionic microenvironment.

3.2.4 Take-home message

In this chapter, we have very briefly reviewed the extensive literature on aggrecan, which was the first described of the many proteoglycans that are now under study. We have discussed how aggrecan, in link-protein-stabilized aggregates (Kimura et al., 1980; Plaas et al., 1983), plays a pivotal structural role in the extracellular matrix of skeletal elements during development, and also in providing the property of reversible compressibility to mature articular cartilages. This is followed by a discussion of some novel findings on the role of pericellular aggrecan, specifically its turnover by ADAMTS5, in the reparative responses of soft tissues (such as cartilage, skin, and tendon) to injury. Lastly, we have introduced a new area on the regulation of TGF-β1 signaling by the pericellular aggrecan of fibroblasts, where it is apparently complexed with HA and CD44, but not link protein. Elucidation of the mechanisms by which aggrecan achieves this effect on signaling should open new and interesting avenues for studies on a specific cell biological role for this proteoglycan. This involvement in TGF-β1 signaling may result in a renewed interest in aggrecan, a molecule that for many years has been considered to function only as a structural component of cartilage matrix.

References

Arufe, M. C., Fuente, A. D., Fuentes, I., de Toro, F. J., and Blanco, F. J., (2010). Chondrogenic potential of subpopulations of cells expressing mesenchymal stem cell markers derived from human synovial membranes. J Cell Biochem *111* (4), 834–845.

Borochowitz, Z., Lachman, R., Adomian, G. E., Spear, G., Jones, K., and Rimoin, D. L. (1988). Achondrogenesis type I: delineation of further heterogeneity and identification of two distinct subgroups. J Pediatr *112*, 23–31.

Brent, A. E., Schweitzer, R., and Tabin, C. J. (2003). A somitic compartment of tendon progenitors. Cell *113*, 235–248.

Buckwalter, J. A., and Rosenberg, L. C. (1982). Electron microscopic studies of cartilage proteoglycans. Direct evidence for the variable length of the chondroitin sulfate-rich region of proteoglycan subunit core protein. J Biol Chem *257*, 9830–9839.

Carney, S. L., Billingham, M. E., Muir, H., and Sandy, J. D. (1985). Structure of newly synthesised (35S)-proteoglycans and (35S)-proteoglycan turnover products of cartilage explant cultures from dogs with experimental osteoarthritis. J Orthop Res *3*, 140–147.

Cavanagh, J. A., Tammen, I., Windsor, P. A., et al. (2007). Bulldog dwarfism in Dexter cattle is caused by mutations in ACAN. Mamm Genome *18*, 808–814.

Cawston, T. E., and Young, D. A. (2010). Proteinases involved in matrix turnover during cartilage and bone breakdown. Cell Tissue Res *339*, 221–235.

Chen, Y., Shi-wen, X., Eastwood, M., et al. (2006). Contribution of activin receptor-like kinase 5 (transforming growth factor beta receptor type I) signaling to the fibrotic phenotype of scleroderma fibroblasts. Arthritis Rheum *54*, 1309–1316.

Chimal-Monroy, J., and Diaz de Leon, L. (1999). Expression of N-cadherin, N-CAM, fibronectin and tenascin is stimulated by TGF-beta1, beta2, beta3 and beta5 during the formation of precartilage condensations. Int J Dev Biol *43*, 59–67.

Cornaglia, A. I., Casasco, A., Casasco, M., Riva, F., and Necchi, V. (2009). Dysplastic histogenesis of cartilage growth plate by alteration of sulphation pathway: a transgenic model. Connect Tissue Res *50*, 232–242.

Corsi, A., Riminucci, M., Fisher, L. W., and Bianco, P. (2001). Achondrogenesis type IB: agenesis of cartilage interterritorial matrix as the link between gene defect and pathological skeletal phenotype. Arch Pathol Lab Med *125*, 1375–1378.

Czarny-Ratajczak, M., Bieganski, T., Rogala, P., Glowacki, M., Trzeciak, T., and Kozlowski, K. (2010). New intermediate phenotype between MED and DD caused by compound heterozygous mutations in the DTDST gene. Am J Med Genet A *152A*, 3036–3042.

deMos, M., Koevoet, W., van Schie, H. T., et al. (2009). In vitro model to study chondrogenic differentiation in tendinopathy. Am J Sports Med *37*, 1214–1222.

Doege, K. J., Garrison, K., Coulter, S. N., and Yamada, Y. (1994). The structure of the rat aggrecan gene and preliminary characterization of its promoter. J Biol Chem *269*, 29232–29240.

Durigova, M., Nagase, H., Mort, J. S., and Roughley, P. J. (2011). MMPs are less efficient than ADAMTS5 in cleaving aggrecan core protein. Matrix Biol *30*, 145–153.

Flannery, C. R., Lark, M. W., and Sandy, J. D. (1992). Identification of a stromelysin cleavage site within the interglobular domain of human aggrecan. Evidence for proteolysis at this site in vivo in human articular cartilage. J Biol Chem *267*, 1008–1014.

Forlino, A., Piazza, R., Tiveron, C., et al. (2005). A diastrophic dysplasia sulfate transporter (SLC26A2) mutant mouse: morphological and biochemical characterization of the resulting chondrodysplasia phenotype. Hum Mol Genet *14*, 859–871.

Fukui, N., Miyamoto, Y., Nakajima, M., et al. (2008). Zonal gene expression of chondrocytes in osteoarthritic cartilage. Arthritis Rheum *58*, 3843–3853.

Giannoudis, P. V., Goff, T., Roshdy, T., Jones, E., and McGonagle, D. (2010). Does mobilisation and transmigration of mesenchymal stem cells occur after trauma? Injury *41*, 1099–1102.

Gleghorn, L., Ramesar, R., Beighton, P., and Wallis, G. (2005). A mutation in the variable repeat region of the aggrecan gene (AGC1) causes a form of spondyloepiphyseal dysplasia associated with severe, premature osteoarthritis. Am J Hum Genet *77*, 484–490.

Grogan, S. P., Miyaki, S., Asahara, H., D'Lima, D. D., and Lotz, M. K. (2009). Mesenchymal progenitor cell markers in human articular cartilage:normal distribution and changes in OA. Arthritis Res Ther *11*, R85.

Handley, C. J., Mok, M. T., Ilic, M. Z., Adcocks, C., Buttle, D. J., and Robinson, H. C. (2001). Cathepsin D cleaves aggrecan at unique sites within the interglobular domain and chondroitin sulfate attachment regions that are also cleaved when cartilage is maintained at acid pH. Matrix Biol *20*, 543–553.

Handley, C. J., Winter, G. M., Ilic, M. Z., Ross, J. M., Anthony Poole, C., and Clem Robinson, H. (2002). Distribution of newly synthesized aggrecan in explant cultures of bovine cartilage treated with retinoic acid. Matrix Biol *21*, 579–592.

Hardingham, T. E., Ewins, R. J., and Muir, H. (1976). Cartilage proteoglycans. Structure and heterogeneity of the protein core and the effects of specific protein modifications on the binding to hyaluronate. Biochem J *157*, 127–143.

Hascall, V. C., and Heinegard, D. (1974). Aggregation of cartilage proteoglycans. I. The role of hyaluronic acid. J Biol Chem *249*, 4232–4241.

Hiraoka, S., Furuichi, T., Nishimura, G., et al. (2007). Nucleotide-sugar transporter SLC35D1 is critical to chondroitin sulfate synthesis in cartilage and skeletal development in mouse and human. Nat Med *13*, 1363–1367.

Ilic, M. Z., Carter, P., Tyndall, A., Dudhia, J., and Handley, C. J. (2005). Proteoglycans and catabolic products of proteoglycans present in ligament. Biochem J *385*, 381–388.

Jones, E. A., Crawford, A., English, A., et al. (2008). Synovial fluid mesenchymal stem cells in health and early osteoarthritis: detection and functional evaluation at the single-cell level. Arthritis Rheum *58*, 1731–1740.

Kim, A. C., and Spector, M. (2000). Distribution of chondrocytes containing alpha-smooth muscle actin in human articular cartilage. J Orthop Res *18*, 749–755.

Kimura, J. H., Hardingham, T. E., and Hascall, V. C. (1980). Assembly of newly synthesized proteoglycan and link protein into aggregates in cultures of chondrosarcoma chondrocytes. J Biol Chem *255*, 7134–7143.

Koelling, S., Kruegel, J., Irmer, M., et al. (2009). Migratory chondrogenic progenitor cells from repair tissue during the later stages of human osteoarthritis. Cell Stem Cell *4*, 324–335.

Kurose, R., Ichinohe, S., Tajima, G., et al. (2010). Characterization of human synovial fluid cells of 26 patients with osteoarthritis knee for cartilage repair therapy. Int J Rheum Dis *13*, 68–74.

Lahm, A., Mrosek, E., Spank, H., et al. (2010). Changes in content and synthesis of collagen types and proteoglycans in osteoarthritis of the knee joint and comparison of quantitative analysis with Photoshop-based image analysis. Arch Orthop Trauma Surg *130*, 557–564.

Lee, E. R., Lamplugh, L., Davoli, M. A., et al. (2001). Enzymes active in the areas undergoing cartilage resorption during the development of the secondary ossification center in the tibiae of rats ages 0–21 days: I. Two groups of proteinases cleave the core protein of aggrecan. Dev Dyn *222*, 52–70.

Lee, E. R., Lamplugh, L., Leblond, C. P., Mordier, S., Magny, M. C., and Mort, J. S. (1998). Immunolocalization of the cleavage of the aggrecan core protein at the Asn341-Phe342 bond, as an indicator of the location of the metalloproteinases active in the lysis of the rat growth plate. Anat Rec *252*, 117–132.

Lemire, J. M., Patis, C., Gordon, L. B., Sandy, J. D., Toole, B. P., and Weiss, A. S. (2006). Aggrecan expression is substantially and abnormally upregulated in Hutchinson-Gilford progeria syndrome dermal fibroblasts. Mech Ageing Dev *127*, 660–669.

Li, H., Schwartz, N. B., and Vertel, B. M. (1993). cDNA cloning of chick cartilage chondroitin sulfate (aggrecan) core protein and identification of a stop codon in the aggrecan gene associated with the chondrodystrophy, nanomelia. J Biol Chem *268*, 23504–23511.

Li, J., Anemaet, W., Diaz, M. A., et al. (2011). Knockout of ADAMTS5 does not eliminate cartilage aggrecanase activity but abrogates joint fibrosis and promotes cartilage aggrecan deposition in murine osteoarthritis models. J Orthop Res. *4*, 516–522.

Little, C. B., Meeker, C. T., Golub, S. B., et al. (2007). Blocking aggrecanase cleavage in the aggrecan interglobular domain abrogates cartilage erosion and promotes cartilage repair. J Clin Invest *117*, 1627–1636.

Little, C. B., Meeker, C. T., Hembry, R. M., et al. (2005a). Matrix metalloproteinases are not essential for aggrecan turnover during normal skeletal growth and development. Mol Cell Biol *25*, 3388–3399.

Little, C. B., Mittaz, L., Belluoccio, D., et al. (2005b). ADAMTS-1-knockout mice do not exhibit abnormalities in aggrecan turnover in vitro or in vivo. Arthritis Rheum *52*, 1461–1472.

Ma, H. L., Blanchet, T. J., Peluso, D., Hopkins, B., Morris, E. A., and Glasson, S. S. (2007). Osteoarthritis severity is sex dependent in a surgical mouse model. Osteoarthritis Cartilage *15*, 695–700.

Majumdar, M. K., Askew, R., Schelling, S., et al. (2007). Double-knockout of ADAMTS-4 and ADAMTS-5 in mice results in physiologically normal animals and prevents the progression of osteoarthritis. Arthritis Rheum *56*, 3670–3674.

Malgouries, S., Thibaut, S., and Bernard, B. A. (2008). Proteoglycan expression patterns in human hair follicle. Br J Dermatol *158*, 234–242.

Maroudas, A. I. 1976. Balance between swelling pressure and collagen tension in normal and degenerate cartilage. Nature *260*, 808–809.

McCulloch, D. R., Le Goff, C., Bhatt, S., Dixon, L. J., Sandy, J. D., and Apte, S. S. (2009). Adamts5, the gene encoding a proteoglycan-degrading metalloprotease, is expressed by specific cell lineages during mouse embryonic development and in adult tissues. Gene Expr Patterns *9*, 314–323.

McDonnell, J., Lobner, J. M., Knight, W. B., et al. (1993). Comparison of the proteoglycanolytic activities of human leukocyte elastase and human cathepsin G in vitro and in vivo. Connect Tissue Res *30*, 1–9.

Meredith, D., Gehl, K. A., Seymour, J., Ellory, J. C., and Wilkins, R. J. (2007). Characterization of sulphate transporters in isolated bovine articular chondrocytes. J Orthop Res *25*, 1145–1153.

Miyamoto, C., Matsumoto, T., Sakimura, K., and Shindo, H. (2007). Osteogenic protein-1 with transforming growth factor-beta1: potent inducer of chondrogenesis of synovial mesenchymal stem cells in vitro. J Orthop Sci *12*, 555–561.

Mori, Y., Ishida, W., Bhattacharyya, S., Li, Y., Platanias, L. C., and Varga, J. (2004). Selective inhibition of activin receptor-like kinase 5 signaling blocks profibrotic transforming growth factor beta responses in skin fibroblasts. Arthritis Rheum *50*, 4008–4021.

Mwale, F., Tchetina, E., Wu, C. W., and Poole, A. R. (2002). The assembly and remodeling of the extracellular matrix in the growth plate in relationship to mineral deposition and cellular hypertrophy: an in situ study of collagens II and IX and proteoglycan. J Bone Miner Res *17*, 275–283.

Ochs, B. G., Muller-Horvat, C., Albrecht, D., et al. (2011). Remodeling of articular cartilage and subchondral bone after bone grafting and matrix-associated autologous chondrocyte implantation for osteochondritis dissecans of the knee. Am J Sports Med *39*, 764–773.

Olin, A. I., Morgelin, M., Sasaki, T., Timpl, R., Heinegard, D., and Aspberg, A. (2001). The proteoglycans aggrecan and Versican form networks with fibulin-2 through their lectin domain binding. J Biol Chem *276*, 1253–1261.

Oshita, H., Sandy, J. D., Suzuki, K., et al. (2004). Mature bovine articular cartilage contains abundant aggrecan that is C-terminally truncated at Ala719-Ala720, a site which is readily cleaved by m-calpain. Biochem J *382*, 253–259.

Otsuki, S., Hanson, S. R., Miyaki, S., et al. (2010). Extracellular sulfatases support cartilage homeostasis by regulating BMP and FGF signaling pathways. Proc Natl Acad Sci U S A *107*, 10202–10207.

Phillips, A. O., and Fraser, D. J. (2010). BMP-7 stops TGF-{beta} in peritoneal fibrosis. Nephrol Dial Transplant *25*, 1036–1038.

Plaas, A., Osborn, B., Yoshihara, Y., et al. (2007). Aggrecanolysis in human osteoarthritis: confocal localization and biochemical characterization of ADAMTS5-hyaluronan complexes in articular cartilages. Osteoarthritis Cartilage *15*, 719–734.

Plaas, A., Sandy, J. D., Liu, H., et al. (2011a). Biochemical identification and immunolocalizaton of aggrecan, ADAMTS5 and inter-alpha-trypsin-inhibitor in equine degenerative suspensory ligament desmitis. J Orthop Res *29* (6), 900–906.

Plaas, A., Velasco, J., Gorski, D. J., et al. (2011b). The relationship between fibrogenic TGFbeta1 signaling in the joint and cartilage degradation in post-injury osteoarthritis. Osteoarthritis and Cartilage *19* (9), 1081–1090.

Plaas, A. H., Sandy, J. D., and Muir, H. (1983). Proteoglycan aggregate formation by articular chondrocytes. Decrease in link-protein synthesis during culture. Biochem J *214*, 855–864.

Pryce, B. A., Watson, S. S., Murchison, N. D., Staverosky, J. A., Dunker, N., and Schweitzer, R. (2009). Recruitment and maintenance of tendon progenitors by TGFbeta signaling are essential for tendon formation. Development *136*, 1351–1361.

Quinn, T. M., Grodzinsky, A. J., Hunziker, E. B., and Sandy, J. D. (1998). Effects of injurious compression on matrix turnover around individual cells in calf articular cartilage explants. J Orthop Res *16*, 490–499.

Quinn, T. M., Maung, A. A., Grodzinsky, A. J., Hunziker, E. B., and Sandy, J. D. (1999). Physical and biological regulation of proteoglycan turnover around chondrocytes in cartilage explants. Implications for tissue degradation and repair. Ann N Y Acad Sci *878*, 420–441.

Rossi, A., Bonaventure, J., Delezoide, A. L., Cetta, G., and Superti-Furga, A. (1996). Undersulfation of proteoglycans synthesized by chondrocytes from a patient with achondrogenesis type 1B homozygous for an L483P substitution in the diastrophic dysplasia sulfate transporter. J Biol Chem *271*, 18456–18464.

Rossi, A., and Superti-Furga, A. (2001). Mutations in the diastrophic dysplasia sulfate transporter (DTDST) gene (SLC26A2): 22 novel mutations, mutation review, associated skeletal phenotypes, and diagnostic relevance. Hum Mutat *17*, 159–171.

Roughley, P. J. (1977). The degradation of cartilage proteoglycans by tissue proteinases. Proteoglycan heterogeneity and the pathway of proteolytic degradation. Biochem J *167*, 639–646.

Samiric, T., Ilic, M. Z., and Handley, C. J. (2004). Characterisation of proteoglycans and their catabolic products in tendon and explant cultures of tendon. Matrix Biol *23*, 127–140.

Samiric, T., Parkinson, J., Ilic, M. Z., Cook, J., Feller, J. A., and Handley, C. J. (2009). Changes in the composition of the extracellular matrix in patellar tendinopathy. Matrix Biol *28*, 230–236.

Sandy, J. D. (2006). A contentious issue finds some clarity: on the independent and complementary roles of aggrecanase activity and MMP activity in human joint aggrecanolysis. Osteoarthritis Cartilage *14*, 95–100.

Satoh, H., Susaki, M., Shukunami, C., Iyama, K., Negoro, T., and Hiraki, Y. (1998). Functional analysis of diastrophic dysplasia sulfate transporter. Its involvement in growth regulation of chondrocytes mediated by sulfated proteoglycans. J Biol Chem *273*, 12307–12315.

Sohaskey, M. L., Yu, J., Diaz, M. A., Plaas, A. H., and Harland, R. M. (2008). JAWS coordinates chondrogenesis and synovial joint positioning. Development *135*, 2215–2220.

Stattin, E. L., Wiklund, F., Lindblom, K., et al. (2010). A missense mutation in the aggrecan C-type lectin domain disrupts extracellular matrix interactions and causes dominant familial osteochondritis dissecans. Am J Hum Genet *86*, 126–137.

Sui, Y., Lee, J. H., DiMicco, M. A., et al. (2009). Mechanical injury potentiates proteoglycan catabolism induced by interleukin-6 with soluble interleukin-6 receptor and tumor necrosis factor alpha in immature bovine and adult human articular cartilage. Arthritis Rheum *60*, 2985–2996.

Tompson, S. W., Merriman, B., Funari, V. A., et al. (2009). A recessive skeletal dysplasia, SEMD aggrecan type, results from a missense mutation affecting the C-type lectin domain of aggrecan. Am J Hum Genet *84*, 72–79.

Valhmu, W. B., Palmer, G. D., Rivers, P. A., et al. (1995). Structure of the human aggrecan gene: exon-intron organization and association with the protein domains. Biochem J *309* (Pt 2), 535–542.

Velasco, J., Li, J., Dipietro, L., Stepp, M. A., Sandy, J. D., and Plaas, A. (2011). ADAMTS5 ablation blocks murine dermal repair through CD44-mediated aggrecan accumulation and modulation of TGFbeta1 signaling. J Biol Chem, *286*, 26016–26027.

Vogel, K. G., and Meyers, A. B. (1999). Proteins in the tensile region of adult bovine deep flexor tendon. Clin Orthop Relat Res *367* (Suppl.), S344–355.

Vogel, K. G., Sandy, J. D., Pogany, G., and Robbins, J. R. (1994). Aggrecan in bovine tendon. Matrix Biol *14*, 171–179.

Walcz, E., Deak, F., Erhardt, P., et al. (1994). Complete coding sequence, deduced primary structure, chromosomal localization, and structural analysis of murine aggrecan. Genomics *22*, 364–371.

Wang, J., Verdonk, P., Elewaut, D., Veys, E. M., and Verbruggen, G. (2003). Homeostasis of the extracellular matrix of normal and osteoarthritic human articular cartilage chondrocytes in vitro. Osteoarthritis Cartilage *11*, 801–809.

Wang, V. M., Bell, R., Thakore, R., Li, J., Sandy, J. D., and Plaas, A. (2012). Knockout of adamts5 adversely affects the biomechanical properties of murine tendons. J Orthop Res *30*, 620–626.

Watanabe, H., Kimata, K., Line, S., et al. (1994). Mouse cartilage matrix deficiency (cmd) caused by a 7 bp deletion in the aggrecan gene. Nat Genet *7*, 154–157.

Watson, S. S., Riordan, T. J., Pryce, B. A., and Schweitzer, R. (2009). Tendons and muscles of the mouse forelimb during embryonic development. Dev Dyn *238*, 693–700.

Yoshihara, Y., Plaas, A., Osborn, B., et al. (2008). Superficial zone chondrocytes in normal and osteoarthritic human articular cartilages synthesize novel truncated forms of inter-alpha-trypsin inhibitor heavy chains which are attached to a chondroitin sulfate proteoglycan other than bikunin. Osteoarthritis Cartilage *16*, 1343–1355.

Yuan, G. H., Tanaka, M., Masuko-Hongo, K., et al. (2004). Characterization of cells from pannus-like tissue over articular cartilage of advanced osteoarthritis. Osteoarthritis Cartilage *12*, 38–45.

3.3 The pathobiology of versican

Thomas N. Wight

3.3.1 Introduction

Versican is a chondroitin sulfate proteoglycan (CSPG) that is found in the interstitial space of most soft tissues (reviewed in Wight, 2002; Wight et al., 2011). Due to its size, negative charge density, and ability to form high-molecular-weight aggregates with hyaluronan, versican can trap water, promoting a swelling pressure in the interstitial space that is offset by other extracellular matrix (ECM) components such as the collagens and elastic fibers. Versican can exist at least as four isoforms due to the alternative splicing of the major exons that code for the glycosaminoglycan (GAG) attachment regions in the core protein (▶Figure 3.3). Since versican interacts with a number of other ECM proteins, it is often considered a bridging molecule or "linker" involved in regulating the spatial organization and assembly of the ECM into higher-ordered structures contributing to tissue viscoelasticity (Wight, 2002; Wu et al., 2005; Wight et al., 2011). However, several studies over the years have shown that versican interacts with cells to influence their phenotype (Wight, 2002; Wight et al., 2011). In most tissues, the concentration of versican is low, occupying a small percentage of total ECM protein. However, in many diseases, the concentration of versican increases and contributes to expansion of tissue space and alterations in cell phenotypes facilitating events involved in tissue repair and remodeling (Wight, 2002).

3.3.2 Cardiovascular disease

Versican is present in the ECM of normal blood vessels (Yao et al., 1994) and increases dramatically in all forms of vascular disease (reviewed in Wight and Merrilees, 2004; Theocharis, 2008). Versican is prominent in the intima and adventitia of most arteries and veins (Merrilees et al., 2001) and increases mainly in the intima in most forms of vascular disease (▶Figure 3.4).

Restenosis

Accumulation of versican typifies early lesion development in response to vascular injury. Versican increases in the intima in the early phases of response to balloon injury and in vein graft repair and contributes to intimal expansion following vascular trauma (reviewed in Wight and Merrilees, 2004). Interference with versican expression either by antibodies to transforming growth factor-beta (TGF-β) (Wolf et al., 1994) or TGF-β antisense (Merrilees et al., 2000) or antisense to versican (Huang et al., 2006) blocks intimal expansion following injury. In addition, forced expression of the V3 variant form of versican in the vascular injury model mimics the versican antisense response and creates

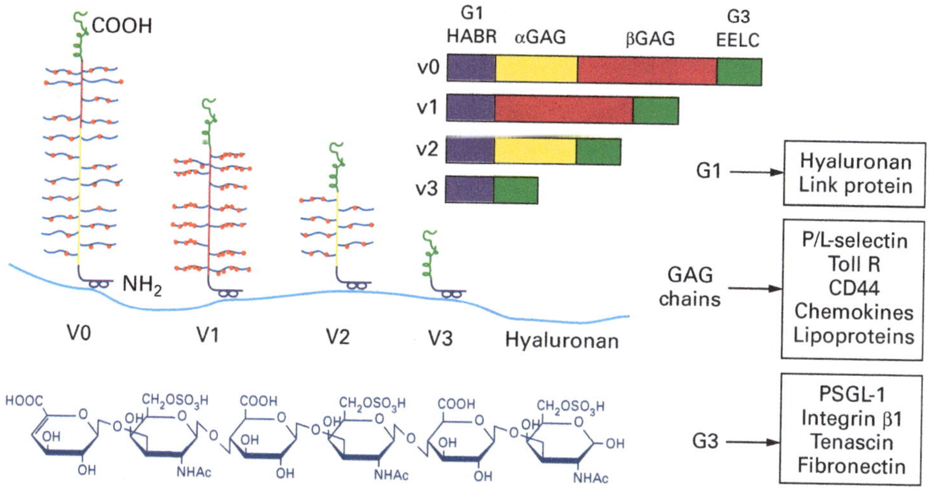

Figure 3.3 A model of the different isoforms generated by alternative splicing of the mRNA transcript for versican. All isoforms interact with hyaluronan and thus are capable of forming different-sized versican-hyaluronan aggregates, which in turn determines, in part, tissue volume. Different colors denote specific domains in the gene and in the protein product. Purple = hyaluronan-binding region (HABR); yellow = the α-GAG exon and protein product, red = β-GAG exon and the protein product; green = two epidermal growth factor repeats (EE), a lectin-binding domain (L) and a complement regulatory region. The GAG chains are shown in blue. Reproduced with modifications from Wight et al. (2011). Used with permission.

an intima that resists thickening and inflammatory cell infiltration (Merrilees et al., 2011). Furthermore, overexpression of miRNA-143 and -145, which decreases versican expression in arterial smooth muscle cells (ASMCs) (Wang et al., 2010), decreases neointimal formation in a rat model of acute vascular injury (Elia et al., 2009). Such results indicate that reducing or eliminating versican may be advantageous in preventing intimal hyperplasia, which is thought to be a precursor of the atherosclerotic lesion.

Versican is markedly increased in both stented and nonstented restenotic lesions in humans (reviewed in Wight and Merrilees, 2004) (▶Figure 3.4). Accumulation occurs primarily in the ASMC hypercellular myxoid regions of the lesions that have a very low content of collagen and elastic fibers. Hyaluronan, which associates with versican, also accumulates in the versican-rich regions of these lesions (Riessen et al., 1996) and forms high-molecular-weight hydrophilic complexes that trap water and cause tissue swelling. The rapid expansion of restenotic lesions could be due, in part, to the swelling pressure brought on by accumulation of hyaluronan-versican complexes. Conversely, loss or breakdown of the hyaluronan-versican complex could lead to expulsion of water and tissue shrinkage with reduction in arterial circumference. Interestingly, collagen gels impregnated with hyaluronan show CD44-dependent contraction when populated by ASMCs (Travis et al., 2001). The involvement of versican in regulating the ability of ASMCs to contract their matrices has not been investigated.

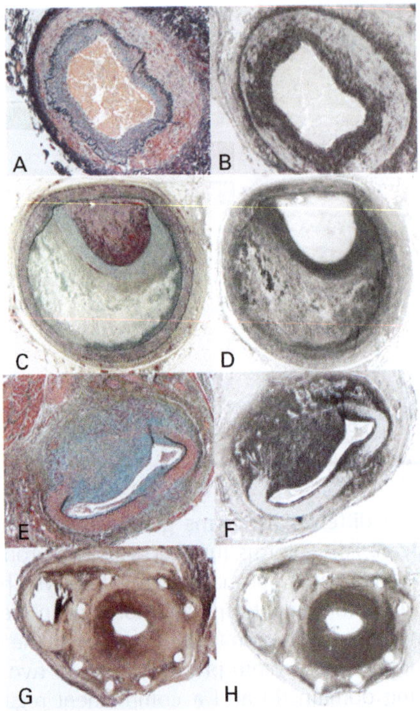

Figure 3.4 (A) Section from a human artery with intimal thickening stained with a modified Movat's stain that shows proteoglycans as blue, collagens as yellow, and elastic fibers as dark purple; x20. Note enrichment for proteoglycans in the thickened intima. (B) Adjacent section to that shown in (A), immunostained for versican. Note that the thickened intima is enriched in versican; x20. (C) Section from an eroded human coronary atherosclerotic plaque stained with a modified Movat's stain showing a proteoglycan-rich layer adjacent to an occluding thrombus; x20. Reproduced from Wight and Merrilees (2004). Used with permission. (D) Adjacent section to that shown in (C), immunostained for versican. Note the proteoglycan-enriched layer adjacent to the thrombus is enriched in versican; x20. (E) Section from a human temporal artery pseudoaneurysm stained with a modified Movat's stain showing proteoglycans in blue, collagens in yellow, and elastic fibers in dark purple. Note interruption of the elastic fibers and accumulation of proteoglycans in the developing lesion; x20. Sections kindly provided by Drs Alan Burke and Renu Virmani of the Armed Forces Institute of Pathology, Washington, DC. (F) Adjacent section to that shown in (E), immunostained for versican. Note the significant accumulation of versican in those regions where the lesions are developing; x20. Section kindly provided by Drs Alan Burke and Renu Virmani. (G) Section from an in-stent human restenotic coronary artery stained for hyaluronan (red) using a biotinylated hyaluronan-binding probe. Note the stent wires separate the old and new lesion and significant accumulation of hyaluronan in the new lesion; x20. Section kindly provided by Drs Andrew Farb and Renu Virmani of the Armed Forces Institute of Pathology, Washington, DC. (H) Adjacent section as shown in (G), immunostained for versican. Note excessive accumulation of versican in the new lesion on the luminal side of the stent wires. Immunostaining for versican parallels to a large extent the staining for hyaluronan, indicating the likelihood of large hyaluronan-versican complexes filling the ECM of these lesions; x20. Section kindly provided by Drs Andrew Farb and Renu Virmani.

The accumulation of versican in the neointima may be the result of a combination of elevated synthesis and/or decreased degradation. For example, areas high in expression of a disintegrin and metalloproteinase with thrombospondin motifs-1 and -4 (ADAMTS1 and ADAMTS4) in the vascular wall neointima correlate with greater versican degradation and production of the amino terminal DPEAAE-containing versican fragment (Sandy et al., 2001; Kenagy et al., 2006). Enhancing blood flow in experimental vascular grafts leads to increased generation of DPEAAE versican fragments and regression of intimal thickenings in an animal model of graft repair (Kenagy et al., 2005). ADAMTS4 is also elevated in vascular grafts and correlates with cell death in the regressed lesions (Kenagy et al., 2009, 2011). These findings are also of interest because recent studies indicate that cleaved versican regulates apoptosis during mammalian interdigital web regression (McCulloch et al., 2009). ADAMTS1 mRNA transcript is also abundant in human aorta and increases as ASMCs migrate and proliferate *in vitro* (Jonsson-Rylander et al., 2005). A polymorphism in the *ADAMTS1* gene has been associated with an increase in cardiovascular disease in two separate studies (Morrison et al., 2007; Sabatine et al., 2008). Whether the generation of versican fragments by ADAMTS enzymes directly contributes to events leading to vascular atrophy and/or vascular disease awaits further experimentation.

Atherosclerosis

Versican accumulates during all stages of human vascular disease (▶Figure 3.4). It is found in specific locations, such as in early pathological intimal thickenings, associated with lipid and macrophage accumulation, and at the plaque-thrombus interface (reviewed in Wight and Merrilees, 2004). Consistent with these findings in human tissues, versican has been identified as a major ECM component in atherosclerotic lesions in most animal models of vascular disease including mice (Jonsson-Rylander et al., 2005; Karra et al., 2005; Strom et al., 2006; Seidelmann et al., 2008). Mechanistically, versican plays multiple roles in promoting atherogenesis. It influences lipid retention within developing vascular lesions (Williams and Tabas, 1995), regulates ASMC proliferation and migration (Wight and Merrilees, 2004), interacts with proinflammatory leukocytes such as monocytes/macrophages (Evanko et al., 2009; Potter-Perigo et al., 2010), and influences coagulation and thrombosis (Mazzucato et al., 2002; McGee and Wagner, 2003; Zheng et al., 2006).

Lipids build up in atherosclerotic plaques. The "response-to-retention" hypothesis of atherosclerosis (Williams and Tabas, 1995), based on pioneering work from several laboratories (Srinivasan et al., 1970; Hollander, 1976; Camejo et al., 1980), invokes a critical role for proteoglycans in binding and trapping lipoproteins within the vascular wall during the progression of atherosclerosis. Versican, along with other proteoglycans, such as biglycan, has been found in association with lipoproteins within atherosclerotic lesions and in lipid-induced lesions in experimental animals. Conditions that promote chondroitin sulfate (CS) chain elongation in ASMCs, such as cell proliferation (Camejo et al., 1993), treatment of the cells with oxidized low-density lipoprotein (LDL) (Chang et al., 2000), and TGF-β (Little et al., 2002) also cause increased binding of versican to LDL. The composition of the CS chains of versican may affect LDL interaction as well. CS enriched in the 6-*O*-sulfate isomer binds LDL more avidly than that enriched in the 4-*O*-sulfate isomer (Cardoso and Mourão, 1994). Furthermore, the charge density on the GAG influences LDL interactions in that increases in the degree of sulfation of CS increase lipoprotein interaction with these proteoglycans (Sambandam et al., 1991).

While the precise structure(s) of the proteoglycans responsible for LDL binding have yet to be determined, proteoglycan binding sites in human apoprotein B of LDL have been located at residue 3,363 (Borén et al., 1998). Furthermore, transgenic mice overexpressing a defective proteoglycan-binding apoprotein, when fed a high-fat diet, failed to develop atherosclerosis after 20 weeks, even though serum cholesterols were significantly elevated (Skalen et al., 2002). These results indicate that subendothelial retention of apoprotein B-100–containing lipoproteins by proteoglycans is a critical step in the early stages of atherosclerosis (Nakashima et al., 2007, 2008). The precise role of versican in these early phases remains to be determined. Furthermore, lipoproteins bound to proteoglycans are more sensitive to modifications such as oxidation and enzymatic hydrolysis, thereby affecting their potential atherogenicity (Camejo et al., 1998; Hurt-Camejo et al., 1992, 2001).

In addition to regulating the extracellular retention of lipoproteins, versican may play a role in intracellular lipid accumulation as well. CSPG-LDL complexes are taken up rapidly by macrophages (Hurt-Camejo et al., 1992; Vijayagopal, 1994) and ASMCs (Ismail et al., 1994), and these cells internalize versican-LDL complexes through both LDL receptor and LDL receptor–related protein pathways (Llorente-Cortes et al., 2002). This uptake leads to accumulation of lipid and formation of foam cells, which characterize lipid-filled atherosclerotic plaques. Thus, versican plays fundamental roles in both the extracellular and intracellular retention of lipoproteins in atherogenesis.

ECM remodeling takes place throughout the different phases of atherosclerosis as part of an injury and inflammatory response. The sequence of changes is not unlike what is seen during wound repair in which the early ECM changes are characterized by ECM deposits that create a loose, open, and watery matrix (referred to as a "provisional ECM") (Clark et al., 1982; Clark and Henson, 1988), which allows cellular invasion and repair. This provisional matrix is then replaced by a more fibrous ECM enriched in collagens and assorted glycoproteins as wound healing progresses. A similar sequence is seen in restenotic lesions in human stented arteries (Chung et al., 2002). One ECM component that fails to reassemble is the elastic fiber. In fact, elastic fibers are conspicuously absent from restenotic and atherosclerotic lesions. The importance of elastic fibers in regulating intimal hyperplasia is highlighted by studies of the elastin knockout mouse. Disruption of elastin synthesis by targeting the promoter and first exon of the tropoelastin gene (Li et al., 1998) leads to subendothelial proliferation of ASMCs and obstruction and closure of the aorta in the elastin knockout mouse (Li et al., 1998). Thus, factors regulating elastic fiber formation may be critical to controlling vascular lesion formation. One factor that appears to inhibit elastic fiber assembly is CS, which is part of the versican molecule (Hinek et al., 1991). We have found that overexpression of the versican variant that lacks CS, V3, in ASMCs leads to changes in tropoelastin expression and accumulation of elastic fibers in long-term ASMC cultures (Merrilees et al., 2002). When these V3-transduced ASMCs are seeded into balloon-injured rat carotid arteries, a compact and highly structured neointima, enriched in elastic lamellae, develops (Merrilees et al., 2002). Furthermore, in a recent study, Merrilees and colleagues demonstrated that injecting rabbit V3-transduced ASMCs into injured rabbit carotid arteries in animals placed on a lipid-rich diet prevented lipid buildup and monocyte ingress over an 8-week period (Merrilees et al., 2011). Previous studies had shown that monocytes do not adhere well to elastin but adhere avidly to collagen (Liu et al., 2005), suggesting that elastin is a poor adhesive substrate for monocytes.

Versican and hyaluronan are present at the plaque-thrombus interface in advanced human atherosclerotic lesions (Kolodgie et al., 2002), suggesting a possible role in thrombosis (▶Figure 3.4). For example, versican promotes platelet adhesion at low shear rates and cooperates with collagens to promote platelet aggregation (Mazzucato et al., 2002). In addition, versican at the plaque-thrombus interface may, in part, regulate the water content of this region of the lesion and, in turn, promote coagulation, since water transfer is linked to the anticoagulation activity of molecules such as the CS in versican (McGee and Wagner, 2003). In addition, the G3 domain of versican can bind to tissue factor pathway inhibitor-1 (TFPI1), promoting blood coagulation through the extrinsic pathway (Zheng et al., 2006). Thus, multiple pathways exist for versican to have a role in thrombosis as part of vascular disease.

Aneurysms

Aortic aneurysms are a frequent complication of severe atherosclerosis and constitute an important clinical problem. Versican V0 mRNA is decreased by 40% in human abdominal aortic aneurysms, with significantly reduced immunostaining for versican (Theocharis et al., 2001) and with evidence of versican degradation and CS composition modifications (Theocharis et al., 2003). Recently, a proteomic approach has confirmed the loss of versican in abdominal aortic aneurysms along with a number of other ECM proteins (Didangelos et al., 2011). Some of this loss could be due to selective versican degradation as versican degradative enzymes, such as plasmin (Kenagy et al., 2002, 2006), increase in association with aneurysmal expansion (Lijnen, 2001; Lindholt et al., 2001). Thus, late-phase aneurysmal development may involve selective loss of versican. Psuedoaneurysms of the temporal artery show increased immunostaining for versican in areas where there is elastic fiber interruption (Burke et al., 2004), indicating a possible early involvement of versican in remodeling involving elastogenesis. In addition, a gene-mapping study of familial thoracic aortic aneurysms and dissections identified a gene locus for this disease at 5q13–14 (Guo et al., 2001), which is the chromosomal segment that contains the versican gene (Iozzo et al., 1992). However, sequencing of versican DNA from affected individuals failed to reveal any variation that could account for the phenotype. On the other hand, single nucleotide polymorphisms (SNPs) and haplotype analyses of the versican gene in intracranial aneurysms revealed a strong association, suggesting a potential role for versican in this type of aneurysm (Ruigrok et al., 2006). Of interest, the versican intracranial aneurysm susceptibility gene does not appear to be associated with abdominal aortic aneurysms (Baas et al., 2010).

3.3.3 Cancer

Overproduction of versican is a common feature of several tumor types, and a number of research groups have correlated levels of versican accumulation with tumor growth, metastatic potential, and poor prognosis (reviewed in Ricciardelli et al., 2009). The expression of versican in such a wide variety of cancers suggests an active role for these molecules in tumor development. The source of versican in tumors can be from stromal connective tissue cells surrounding the tumor or from the tumor cells themselves. In many cancers of epithelial origin, versican has been identified as a component of the tumor stroma (reviewed in Ricciardelli et al., 2009). This versican is thought to originate from the stromal connective tissue cells in response to the surrounding tumor cells (Dvorak, 1986; Iozzo, 1995; West and van de Rijn, 2007). On the other hand, versican can originate from the tumor cells themselves, such as seen in lung carcinoma (Kim et al., 2009; Wang et al., 2009), melanoma (Touab et al., 2002), leiomyosarcoma (Sobue et al., 1987; Cattaruzza et al., 2002), and leiomyoma (as a benign neoplasm) (Tsibris et al., 2002; Arslan et al., 2005; Malik and Catherino, 2007; Norian et al., 2009).

While the exact role of versican in tumorogenesis is not known, no doubt the accumulation of versican in the ECM of the tumor modulates the capacity of the tumor cells to adhere, proliferate, migrate, and survive (Ricciardelli et al., 2009). For example, an increase in versican expression in the ECM facilitates prostate tumor invasion and metastasis by decreasing cell-ECM adhesion (Sakko et al., 2003). In addition, reducing versican levels increased osteosarcoma cell adhesion in culture (Yamagata and Kimata, 1994), while increased expression of versican was correlated with low adhesion levels and metastatic spread of cancer cells (Yamagata et al., 1993; Yamagata and Kimata, 1994). In addition, versican isolated from Lewis lung carcinomas is capable of stimulating inflammatory cytokine production by bone marrow mononuclear cells, which is thought to be critical for metastasis in this tumor model (Kim et al., 2009). This study showed that versican can activate tumor-infiltrating myeloid cells through Toll-like receptor-2 (TLR2) and its coreceptors, Toll-like receptor-6 (TLR6) and CD14, which then can elicit the production of proinflammatory cytokines including tumor necrosis factor-alpha (TNF-α)–enhancing tumor metastasis (Kim et al., 2009; Wang et al., 2009). Furthermore, CS isolated from highly metastatic Lewis lung carcinoma cells is more highly sulfated than that from weakly metastatic lung carcinoma cells (Li et al., 2008), suggesting that versican may carry different proportions of highly sulfated CS chains capable of interacting with TLR2 on myeloid cells promoting inflammation and driving metastasis. This hypothesis awaits further testing.

Malignant melanoma tumor cells also express elevated levels of versican, which is thought to contribute to the highly metastatic and invasive nature of this tumor (Touab et al., 2002, 2003). The synthesis of versican by these tumor cells is regulated by several transcription factors such as AP1, Sp1, AP2, and two TCF4 sites. Promoter activation requires ERK/mitogen-activated protein kinase (MAPK) and JNK signaling pathways acting on the AP-1 site (Domenzain-Reyna et al., 2009). These results may indicate a link between the superactivation of ERK and elevated levels of versican in malignant melanoma. This study also demonstrated that there was cross talk between ERK and β-catenin involvement in versican upregulation in the highly metastatic melanoma cells, indicating once again a role for the Wnt/β-catenin pathway in regulating versican expression in this epithelial cancer (Domenzain-Reyna et al., 2009). Furthermore, expression of the versican variant V3 alters the behavior of human melanoma cells by interfering with CD44/ErbB-dependent signaling (Hernandez et al., 2011). In these studies, CD44 silencing had the same effect as V3 on cell proliferation and migration, suggesting that V3 may be acting through a CD44-mediated mechanism. In addition, these studies indicate that an epidermal growth factor receptor (EGFR)/ErbB2 complex, which interacts with CD44, and which also decreased in V3-overexpressing cells, may be the pathway that controls the proliferative and migratory phenotype of these cancer cells (Hernandez et al., 2011).

3.3.4 Lung

While versican content is low in normal lung, versican accumulates in lung disease. Versican increases in the airways of humans with pulmonary fibrosis (Huang et al., 1999; Roberts, 2003; Araujo et al., 2008), adult respiratory distress syndrome (ARDS) (Bensadoun et al., 1996; Morales et al., 2011), lymphangioleiomyomatosis (Merrilees et al., 2004), chrononic obstructive pulmonary disease (COPD) (Merrilees et al., 2008), and asthma (Roberts, 1995; Huang et al., 1999; Johnson, 2001), and in animal models of asthma (Zhu et al., 2002; Lowry et al., 2008). A significant increase in versican is noticed in the interstitial space of small and large airways in asthmatics (de Medeiros Matsushita et al., 2005). The increase in versican in the lungs of patients with lymphangioleiomyomatosis is of interest since these lungs are characterized by abnormal proliferation of cells in the interstitial spaces of the lungs and a loss of elastic fibers in regions of versican accumulation (Merrilees et al., 2004). This reciprocal relationship between elastic fibers and versican is not unlike the remodeling of the ECM seen in elastic tissues, such as blood vessels and skin, in which versican accumulation has been modified (Merrilees et al., 2002, 2011; Hinek et al., 2004; Huang et al., 2006) (see previous discussion). Lungs with COPD are also characterized by decreased elastin and increased versican in the small airways and alveoli. These changes have been correlated with increased airflow obstruction (Black et al., 2008; Merrilees et al., 2008). These results indicate that versican may be an important regulator in controlling elastogenesis in lung disease.

Cells isolated from diseased lungs also exhibit altered versican production. For example, bronchial fibroblasts cultured from human asthmatics exhibited elevated production of versican, along with other proteoglycans such as perlecan and biglycan (Malmstrom et al., 2002; Westergren-Thorsson et al., 2002; Ludwig et al., 2004). Similar increases in versican production have been shown for fibroblasts isolated from COPD patients compared to normal controls (Hallgren et al., 2010). Versican degradation has been associated with edematous lung disease in rabbits (Passi et al., 1998) and versican increases in noninfectious bleomycin-induced lung injury in rats (Venkatesan et al., 2000). Bleomycin-exposed lung fibroblasts exhibit increased production of versican, which appears to be TGF-β mediated (Venkatesan et al., 2002). Surprisingly, there have been few studies that address versican's involvement in the lung in response to bacterial and viral infectious agents.

3.3.5 Eye

In the eye, versican is found in the vitreous, the retina, the ciliary body, and the trabecular meshwork. Versican is thought to maintain the structure of the vitreous by maintaining a separation between the collagen molecules (Bishop, 2000). Wagner syndrome is a hereditary vitroretinopathy that maps to chromosome 5q13-q14 and is associated with mutations in the GAG acceptor site preceding exon 8, which codes the β-GAG domain (Miyamoto et al., 2005; Zhao and Russell, 2005; Kloeckener-Gruissem et al.,

2006; Mukhopadhyay et al., 2006; Ronan et al., 2009). Mutations have also been found in introns 7 and 8 that cause skipping of the V1 exon, leading to reduced expression of the V0 and V1 forms of versican and increases in the V2 and V3 forms or the appearance of an aberrently spliced versican transcript (Mukhopadhyay et al., 2006; Brezin et al. 2011). This imbalance in the variants of versican is believed to lead to the pathology, but the actual mechanism by which altered versican causes this retinopathy is not known.

Versican is a normal ECM component of the trabecular meshwork (TM) and ciliary muscle of the human eye (Miyamoto et al., 2005; Zhao and Russell, 2005; Ohno-Jinno et al., 2008). Recently, versican was found to be enriched in regions of high resistance to outflow in the anterior segments of the TM, with low levels in regions of low resistance and high levels in regions of high resistance (Keller et al., 2011). Furthermore, introducing shRNA versican into TM cells increased outflow in porcine eyes. Thus, versican appears to be a central component in outflow resistance in the eye, controlling open flow channels in the TM. Mechanical stretch modulates versican production in the eye. For example, when TM cells cultured from porcine eyes are subjected to mechanical stretch, overall mRNA for versican decreases during a 48-hour time interval. However, when mRNA for the different variants was analyzed, V1 mRNA increased nearly fourfold with mechanical stretch at 48 hours (Keller et al., 2007). Such results indicate that expression of the different splice variants of versican are under different regulatory control. In addition, the enzymes that degrade versican are also affected by mechanical stretch. For example, human TM subjected to mechanical stretch increases the expression and accumulation of ADAMTS4, but not ADAMTS1 or ADAMTS5 (Keller et al., 2009). Addition of recombinant ADAMTS4 to TM anterior segments in a perfusion culture promotes outflow consistent with a loss of versican (Keller et al., 2007, 2009, 2011), while addition of recombinant ADAMTS1 or ADAMTS5 had minimal effect on outflow facility (Keller et al., 2009). While all three of these enzymes exhibit versicanase activity and cleave versican in the G1 domain at Glu441 Ala 442 (Kenagy et al., 2006; Sandy, 2006; Apte, 2009), these results suggest that each enzyme may behave differently depending on location and presence of other ECM components (Keller et al., 2009).

3.3.6 Concluding remarks

This review has focused on those diseases in which versican has been found to play a significant role. The list of diseases or disease conditions in which versican contributes significantly to pathogenesis is growing. For example, versican is the most highly expressed gene in severe endometriosis, suggesting a possible role in the invasive capacity of the endometrial cells (Aghajanova and Giudice, 2011). Versican is also critical in cartilage development and joint morphogenesis, since knocking out versican in developing mice alters chondrogenesis and joint morphogenesis (Choocheep et al., 2010). Versican is a gene that exhibits abnormal exon splicing in amyotrophic lateral sclerosis (ALS), or "Lou Gherig's disease," suggesting that specific proteoglycans known to play a role in cell adhesion, such as versican, may be critical to preventing motor neuron degeneration (Rabin et al., 2010). Finally, versican is one of several CSPGs that increase following spinal cord injury and serve as a barrier to prevent nerve regeneration (Asher et al., 2002).

Clearly, versican is not only an ECM component that helps maintain a structural framework for most tissues, but also a bioactive molecule that interacts with other molecules to modulate their activity, whether they be enzymes, receptors, or structural proteins. While versican is one of many ECM macromolecules that changes during different pathological conditions, its unique structure and interactive nature provide it with unusual properties for regulating events fundamental to the pathogenesis of many diseases. Versican's multiple roles make it an attractive target for therapeutic intervention in the treatment of several different diseases (Jarvelainen et al., 2009; Ricciardelli et al., 2009; Theocharis et al., 2010)

3.3.7 Take-home message

Versican is a large proteoglycan that increases in the several diseases and in part regulates many of the events common to the pathogenesis of these diseases.

Acknowledgments

NIH grants HL018645 and HL098067 supported this work. Special thanks to Drs Virginia Green and Susan Potter-Perigo for their helpful suggestions and editing in the preparation of this manuscript. Due to space limitations, the author apologizes to those researchers whose original work on this topic was not cited directly, but rather in other reviews on this topic.

References

Aghajanova, L., and Giudice, L. C. (2011). Molecular evidence for differences in endometrium in severe versus mild endometriosis. Reprod Sci *18*, 229–251.

Apte, S. S. (2009). A disintegrin-like and metalloprotease (reprolysin-type) with thrombospondin type 1 motif (ADAMTS) superfamily: functions and mechanisms. J Biol Chem *284*, 31493–31497.

Araujo, B. B., Dolhnikoff, M., Silva, L. F., et al. (2008). Extracellular matrix components and regulators in the airway smooth muscle in asthma. Eur Respir J *32*, 61–69.

Arslan, A. A., Gold, L. I., Mittal, K., et al. (2005). Gene expression studies provide clues to the pathogenesis of uterine leiomyoma: new evidence and a systematic review. Hum Reprod *20*, 852–863.

Asher, R. A., Morgenstern, D. A., Shearer, M. C., Adcock, K. H., Pesheva, P., and Fawcett, J. W. (2002). Versican is upregulated in CNS injury and is a product of oligodendrocyte lineage cells. J Neurosci *22*, 2225–2236.

Baas, A. F., Medic, J., van't Slot, R., et al. (2010). The intracranial aneurysm susceptibility genes HSPG2 and CSPG2 are not associated with abdominal aortic aneurysm. Angiology *61*, 238–242.

Bensadoun, E. S., Burke, A. K., Hogg, J. C., and Roberts, C. R. (1996). Proteoglycan deposition in pulmonary fibrosis. Am J Respir Crit Care Med *154*, 1819–1828.

Bishop, P. N. (2000). Structural macromolecules and supramolecular organisation of the vitreous gel. Prog Retin Eye Res *19*, 323–344.

Black, P. N., Ching, P. S., Beaumont, B., Ranasinghe, S., Taylor, G., and Merrilees, M. J. (2008). Changes in elastic fibres in the small airways and alveoli in COPD. Eur Respir J *31*, 998–1004.

Borén, J., Olin, K., Lee, I., Chait, A., Wight, T. N., and Innerarity, T. L. (1998). Identification of a principal proteoglycan binding site in LDL: a single point mutation in apo B-100 severely affects proteoglycan interaction without affecting LDL receptor binding. J Clin Invest *101*, 2658–2664.

Brezin, A. P., Nedelec, B., Barjol, A., Rothschild, P. R., Delpech, M., and Valleix, S. (2011). A new VCAN/versican splice acceptor site mutation in a French Wagner family associated with vascular and inflammatory ocular features. Mol Vis *17*, 1669–1678.

Burke, A. P., Jarvelainen, H., Kolodgie, F. D., Goel, A., Wight, T. N., and Virmani, R. (2004). Superficial pseudoaneurysms: clinicopathologic aspects and involvement of extracellular matrix proteoglycans. Mod Pathol *17*, 482–488.

Camejo, G., Fager, G., Rosengren, B., Hurt-Camejo, E., and Bondjers, G. (1993). Binding of low density lipoproteins by proteoglycans synthesized by proliferating and quiescent human arterial smooth muscle cells. J Biol Chem *268*, 14131–14137.

Camejo, G., Hurt-Camejo, E., Wiklund, O., and Bondjers, G. (1998). Association of apo B lipoproteins with arterial proteoglycans: pathological significance and molecular basis. Atherosclerosis *139*, 205–222.

Camejo, G., Lalaguna, F., Lopez, F., and Starosta, R. (1980). Characterization and properties of a lipoprotein-complexing proteoglycan from human aorta. Atherosclerosis *35*, 307–320.

Cardoso, L. E. M., and Mourão, P. A. S. (1994). Glycosaminoglycan fractions from human arteries presenting diverse susceptibilities to atherosclerosis have different binding affinities to plasma LDL. Arterioscler Thromb *14*, 115–124.

Cattaruzza, S., Schiappacassi, M., Ljungberg-Rose, A., et al. (2002). Distribution of PG-M/versican variants in human tissues and de novo expression of isoform V3 upon endothelial cell activation, migration, and neoangiogenesis in vitro. J Biol Chem *277*, 47626–47635.

Chang, M. Y., Potter-Perigo, S., Tsoi, C., Chait, A., and Wight, T. N. (2000). Oxidized low density lipoproteins regulate synthesis of monkey aortic smooth muscle cell proteoglycans that have enhanced native low density lipoprotein binding properties. J Biol Chem *275*, 4766–4773.

Choocheep, K., Hatano, S., Takagi, H., Watanabe, H., Kimata, K., and Kongtawelert, P. (2010). Versican facilitates chondrocyte differentiation and regulates joint morphogenesis. J Biol Chem *285*, 21114–21125.

Chung, I. M., Gold, H. K., Schwartz, S. M., Ikari, Y., Reidy, M. A., and Wight, T. N. (2002). Enhanced extracellular matrix accumulation in restenosis of coronary arteries after stent deployment. J Am Coll Cardiol *40*, 2072–2081.

Clark, R. A. F., and Henson, P. M. (1988). The Molecular and Cellular Biology of Wound Repair. New York: Plenum Press.

Clark, R. A. F., Quinn, J. H., Winn, H. J., Lanigan, J. M., Dellepella, P., and Colvin, R. B. (1982). Fibronectin is produced by blood vessels in response to injury. J Exp Med *156*, 646–651.

de Medeiros Matsushita, M., da Silva, L. F., dos Santos, M. A., et al. (2005). Airway proteoglycans are differentially altered in fatal asthma. J Pathol *207*, 102–110.

Didangelos, A., Yin, X., Mandal, K., et al. (2011). Extracellular matrix composition and remodeling in human abdominal aortic aneurysms: a proteomics approach. Mol Cell Proteomics, *10*, m.111.008128.

Domenzain-Reyna, C., Hernandez, D., Miquel-Serra, L., et al. (2009). Structure and regulation of the versican promoter: the versican promoter is regulated by AP-1 and TCF transcription factors in invasive human melanoma cells. J Biol Chem *284*, 12306–12317.

Dvorak, H. F. (1986). Tumors: wounds that do not heal. Similarities between tumor stroma generation and wound healing. N Engl J Med *315*, 1650–1659.

Elia, L., Quintavalle, M., Zhang, J., et al. (2009). The knockout of miR-143 and -145 alters smooth muscle cell maintenance and vascular homeostasis in mice: correlates with human disease. Cell Death Differ *16*, 1590–1598.

Evanko, S. P., Potter-Perigo, S., Johnson, P. Y., and Wight, T. N. (2009). Organization of hyaluronan and versican in the extracellular matrix of human fibroblasts treated with the viral mimetic poly I:C. J Histochem Cytochem *57*, 1041–1060.

Guo, D., Hasham, S., Kuang, S. Q., et al. (2001). Familial thoracic aortic aneurysms and dissections: genetic heterogeneity with a major locus mapping to 5q13–14. Circulation *103*, 2461–2468.

Hallgren, O., Nihlberg, K., Dahlback, M., et al. (2010). Altered fibroblast proteoglycan production in COPD. Respir Res *11*, 55.

Hernandez, D., Miquel-Serra, L., Docampo, M. J., et al. (2011). V3 versican isoform alters the behavior of human melanoma cells by interfering with CD44/ErbB-dependent signaling. J Biol Chem *286*, 1475–1485.

Hinek, A., Braun, K. R., Liu, K., Wang, Y., and Wight, T. N. (2004). Retrovirally mediated overexpression of versican v3 reverses impaired elastogenesis and heightened proliferation exhibited by fibroblasts from Costello syndrome and Hurler disease patients. Am J Pathol *164*, 119–131.

Hinek, A., Mecham, R. P., Keeley, F., and Rabinovitch, M. (1991). Impaired elastin fiber assembly related to reduced 67-kD elastin-binding protein in fetal lamb ductus arteriosus and in cultured aortic smooth muscle cells treated with chondroitin sulfate. J Clin Invest *88*, 2083–2094.

Hollander, W. (1976). Unified concept on the role of acid mucopolysaccharides and connective tissue proteins in the accumulation of lipids, lipoproteins, and calcium in the atherosclerotic plaque. Exp Mol Pathol *25*, 106–120.

Huang, J., Olivenstein, R., Taha, R., Hamid, Q., and Ludwig, M. (1999). Enhanced proteoglycan deposition in the airway wall of atopic asthmatics. Am J Respir Crit Care Med *160*, 725–729.

Huang, R., Merrilees, M. J., Braun, K., et al. (2006). Inhibition of versican synthesis by antisense alters smooth muscle cell phenotype and induces elastic fiber formation in vitro and in neointima after vessel injury. Circ Res *98*, 370–377.

Hurt-Camejo, E., Camejo, G., Peilot, H., Oorni, K., and Kovanen, P. (2001). Phospholipase A(2) in vascular disease. Circ Res *89*, 298–304.

Hurt-Camejo, E., Camejo, G., Rosengren, B., et al. (1992). Effect of arterial proteoglycans and glycosaminoglycans on low density lipoprotein oxidation and its uptake by human macrophages and arterial smooth muscle cells. Arterioscler Thromb *12*, 569–583.

Iozzo, R. V. (1995). Tumor stroma as a regulator of neoplastic behavior. Agonistic and antagonistic elements embedded in the same connective tissue. Lab Invest *73*, 157–160.

Iozzo, R. V., Naso, M. F., Cannizzaro, L. A., Wasmuth, J. J., and McPherson, J. D. (1992). Mapping of the versican proteoglycan gene (CSPG2) to the long arm of human chromosome 5 (5q12–5q14). Genomics *14*, 845–851.

Ismail, N. A., Alavi, M. Z., and Moore, S. (1994). Lipoprotein-proteoglycan complexes from injured rabbit aortas accelerate lipoprotein uptake by arterial smooth muscle cells. Atherosclerosis *105*, 79–87.

Jarvelainen, H., Sainio, A., Koulu, M., Wight, T. N., and Penttinen, R. (2009). Extracellular matrix molecules: potential targets in pharmacotherapy. Pharmacol Rev *61*, 198–223.

Johnson, P. R. (2001). Role of human airway smooth muscle in altered extracellular matrix production in asthma. Clin Exp Pharmacol Physiol *28*, 233–236.

Jonsson-Rylander, A. C., Nilsson, T., Fritsche-Danielson, R., et al. (2005). Role of ADAMTS-1 in atherosclerosis: remodeling of carotid artery, immunohistochemistry, and proteolysis of versican. Arterioscler Thromb Vasc Biol *25*, 180–185.

Karra, R., Vemullapalli, S., Dong, C., et al. (2005). Molecular evidence for arterial repair in atherosclerosis. Proc Natl Acad Sci U S A *102*, 16789–16794.

Keller, K. E., Bradley, J. M., and Acott, T. S. (2009). Differential effects of ADAMTS-1, -4, and -5 in the trabecular meshwork. Invest Ophthalmol Vis Sci *50*, 5769–5777.

Keller, K. E., Bradley, J. M., Vranka, J. A., and Acott, T. S. (2011). Segmental versican expression in the trabecular meshwork and involvement in outflow facility. Invest Ophthalmol Vis Sci, *52*, 5049–5057.

Keller, K. E., Kelley, M. J., and Acott, T. S. (2007). Extracellular matrix gene alternative splicing by trabecular meshwork cells in response to mechanical stretching. Invest Ophthalmol Vis Sci *48*, 1164–1172.

Kenagy, R. D., Fischer, J. W., Davies, M. G., et al. (2002). Increased plasmin and serine proteinase activity during flow-induced intimal atrophy in baboon PTFE grafts. Arterioscler Thromb Vasc Biol *22*, 400–404.

Kenagy, R. D., Fischer, J. W., Lara, S., Sandy, J. D., Clowes, A. W., and Wight, T. N. (2005). Accumulation and loss of extracellular matrix during shear stress-mediated intimal growth and regression in baboon vascular grafts. J Histochem Cytochem *53*, 131–140.

Kenagy, R. D., Min, S. K., Clowes, A. W., and Sandy, J. D. (2009). Cell death-associated ADAMTS4 and versican degradation in vascular tissue. J Histochem Cytochem *57*, 889–897.

Kenagy, R. D., Min, S. K., Mulvihill, E., and Clowes, A. W. (2011). A link between smooth muscle cell death and extracellular matrix degradation during vascular atrophy. J Vasc Surg, *54*, 182–191.

Kenagy, R. D., Plaas, A. H., and Wight, T. N. (2006). Versican degradation and vascular disease. Trends Cardiovasc Med *16*, 209–215.

Kim, S., Takahashi, H., Lin, W. W., et al. (2009). Carcinoma-produced factors activate myeloid cells through TLR2 to stimulate metastasis. Nature *457*, 102–106.

Kloeckener-Gruissem, B., Bartholdi, D., Abdou, M. T., Zimmermann, D. R., and Berger, W. (2006). Identification of the genetic defect in the original Wagner syndrome family. Mol Vis *12*, 350–355.

Kolodgie, F. D., Burke, A. P., Farb, A., et al. (2002). Differential accumulation of proteoglycans and hyaluronan in culprit lesions: insights into plaque erosion. Arterioscler Thromb Vasc Biol *22*, 1642–1648.

Li, D. Y., Brooke, B., Davis, E. C., et al. (1998). Elastin is an essential determinant of arterial morphogenesis. Nature *393*, 276–280.

Li, F., Ten Dam, G. B., Murugan, S., et al. (2008). Involvement of highly sulfated chondroitin sulfate in the metastasis of the Lewis lung carcinoma cells. J Biol Chem *283*, 34294–34304.

Lijnen, H. R. (2001). Plasmin and matrix metalloproteinases in vascular remodeling. Thromb Haemost *86*, 324–333.

Lindholt, J. S., Jorgensen, B., Fasting, H., and Henneberg, E. W. (2001). Plasma levels of plasmin-antiplasmin-complexes are predictive for small abdominal aortic aneurysms expanding to operation-recommendable sizes. J Vasc Surg *34*, 611–615.

Little, P. J., Tannock, L., Olin, K. L., Chait, A., and Wight, T. N. (2002). Proteoglycans synthesized by arterial smooth muscle cells in the presence of transforming growth factor-beta1 exhibit increased binding to LDLs. Arterioscler Thromb Vasc Biol *22*, 55–60.

Liu, S. Q., Alkema, P. K., Tieche, C., et al. (2005). Negative regulation of monocyte adhesion to arterial elastic laminae by signal regulatory protein alpha and Src homology 2 domain-containing protein-tyrosine phosphatase-1. J Biol Chem *280*, 39294–39301.

Llorente-Cortes, V., Otero-Vinas, M., Hurt-Camejo, E., Martinez-Gonzalez, J., and Badimon, L. (2002). Human coronary smooth muscle cells internalize versican-modified LDL through LDL receptor-related protein and LDL receptors. Arterioscler Thromb Vasc Biol *22*, 387–393.

Lowry, M. H., McAllister, B. P., Jean, J. C., et al. (2008). Lung lining fluid glutathione attenuates IL-13-induced asthma. Am J Respir Cell Mol Biol *38*, 509–516.

Ludwig, M. S., Ftouhi-Paquin, N., Huang, W., Page, N., Chakir, J., and Hamid, Q. (2004). Mechanical strain enhances proteoglycan message in fibroblasts from asthmatic subjects. Clin Exp Allergy *34*, 926–930.

Malik, M., and Catherino, W. H. (2007). Novel method to characterize primary cultures of leiomyoma and myometrium with the use of confirmatory biomarker gene arrays. Fertil Steril *87*, 1166–1172.

Malmstrom, J., Larsen, K., Hansson, L., et al. (2002). Proteoglycan and proteome profiling of central human pulmonary fibrotic tissue utilizing miniaturized sample preparation: a feasibility study. Proteomics *2*, 394–404.

Mazzucato, M., Cozzi, M. R., Pradella, P., et al. (2002). Vascular PG-M/versican variants promote platelet adhesion at low shear rates and cooperate with collagens to induce aggregation. FASEB J *16*, 1903–1916.

McCulloch, D. R., Nelson, C. M., Dixon, L. J., et al. (2009). ADAMTS metalloproteases generate active versican fragments that regulate interdigital web regression. Dev Cell *17*, 687–698.

McGee, M., and Wagner, W. D. (2003). Chondroitin sulfate anticoagulant activity is linked to water transfer: relevance to proteoglycan structure in atherosclerosis. Arterioscler Thromb Vasc Biol *23*, 1921–1927.

Merrilees, M., Beaumont, B., Scott, L., Hermanutz, V., and Fennessy, P. (2000). Effect of TGF-beta(1) antisense S-oligonucleotide on synthesis and accumulation of matrix proteoglycans in balloon catheter-injured neointima of rabbit carotid arteries. J Vasc Res *37*, 50–60.

Merrilees, M., Ching, P., Beaumont, B., Hinek, A., Wight, T., and Black, P. (2008). Changes in elastin, elastin binding protein and versican in alveoli in chronic obstructive pulmonary disease. Respir Res *18*, 41–50.

Merrilees, M. J., Beaumont, B., and Scott, L. J. (2001). Comparison of deposits of versican, biglycan and decorin in saphenous vein and internal thoracic, radial and coronary arteries: correlation to patency. Coron Artery Dis *12*, 7–16.

Merrilees, M. J., Beaumont, B. W., Braun, K. R., et al. (2011). Neointima formed by arterial smooth muscle cells expressing versican variant v3 is resistant to lipid and macrophage accumulation. Arterioscler Thromb Vasc Biol *31*, 1309–1316.

Merrilees, M. J., Hankin, E. J., Black, J. L., and Beaumont, B. (2004). Matrix proteoglycans and remodelling of interstitial lung tissue in lymphangioleiomyomatosis. J Pathol *203*, 653–660.

Merrilees, M. J., Lemire, J. M., Fischer, J. W., et al. (2002). Retrovirally mediated overexpression of versican v3 by arterial smooth muscle cells induces tropoelastin synthesis and elastic fiber formation in vitro and in neointima after vascular injury. Circ Res *90*, 481–487.

Miyamoto, T., Inoue, H., Sakamoto, Y., et al. (2005). Identification of a novel splice site mutation of the CSPG2 gene in a Japanese family with Wagner syndrome. Invest Ophthalmol Vis Sci *46*, 2726–2735.

Morales, M. M., Pires-Neto, R. C., Inforsato, N., et al. (2011). Small airway remodeling in acute respiratory distress syndrome: a study in autopsy lung tissue. Crit Care *15*, R4.

Morrison, A. C., Bare, L. A., Chambless, L. E., et al. (2007). Prediction of coronary heart disease risk using a genetic risk score: the Atherosclerosis Risk in Communities Study. Am J Epidemiol *166*, 28–35.

Mukhopadhyay, A., Nikopoulos, K., Maugeri, A., et al. (2006). Erosive vitreoretinopathy and Wagner disease are caused by intronic mutations in CSPG2/Versican that result in an imbalance of splice variants. Invest Ophthalmol Vis Sci *47*, 3565–3572.

Nakashima, Y., Fujii, H., Sumiyoshi, S., Wight, T. N., and Sueishi, K. (2007). Early human atherosclerosis: accumulation of lipid and proteoglycans in intimal thickenings followed by macrophage infiltration. Arterioscler Thromb Vasc Biol *27*, 1159–1165.

Nakashima, Y., Wight, T. N., and Sueishi, K. (2008). Early atherosclerosis in humans: role of diffuse intimal thickening and extracellular matrix proteoglycans. Cardiovasc Res *79*, 14–23.

Norian, J. M., Malik, M., Parker, C. Y., et al. (2009). Transforming growth factor β3 regulates the versican variants in the extracellular matrix-rich uterine leiomyomas. Reprod Sci *16*, 1153–1164.

Ohno-Jinno, A., Isogai, Z., Yoneda, M., et al. (2008). Versican and fibrillin-1 form a major hyaluronan-binding complex in the ciliary body. Invest Ophthalmol Vis Sci *49*, 2870–2877.
Passi, A., Negrini, D., Albertini, R., De Luca, G., and Miserocchi, G. (1998). Involvement of lung interstitial proteoglycans in development of hydraulic- and elastase-induced edema. Am J Physiol *275*, L631–635.
Potter-Perigo, S., Johnson, P. Y., Evanko, S. P., et al. (2010). Polyinosine-polycytidylic acid stimulates versican accumulation in the extracellular matrix promoting monocyte adhesion. Am J Respir Cell Mol Biol *43*, 109–120.
Rabin, S. J., Kim, J. M., Baughn, M., et al. (2010). Sporadic ALS has compartment-specific aberrant exon splicing and altered cell-matrix adhesion biology. Hum Mol Genet *19*, 313–328.
Ricciardelli, C., Sakko, A. J., Ween, M. P., Russell, D. L., and Horsfall, D. J. (2009). The biological role and regulation of versican levels in cancer. Cancer Metastasis Rev *28*, 233–245.
Riessen, R., Wight, T. N., Pastore, C., Henley, C., and Isner, J. M. (1996). Distribution of hyaluronan during extracellular matrix remodeling in human restenotic arteries and balloon-injured rat carotid arteries. Circulation *93*, 1141–1147.
Roberts, C. (2003). Versican in the cell biology of pulmonary fibrosis. In Proteoglycans and Lung Disease, H. Garg, P. Roughley, and C. Hales, eds. New York: Marcel Dekker, pp. 191–212.
Roberts, C. R. (1995). Is asthma a fibrotic disease? Chest *107*, 111S–117S.
Ronan, S. M., Tran-Viet, K. N., Burner, E. L., Metlapally, R., Toth, C. A., and Young, T. L. (2009). Mutational hot spot potential of a novel base pair mutation of the CSPG2 gene in a family with Wagner syndrome. Arch Ophthalmol *127*, 1511–1519.
Ruigrok, Y. M., Rinkel, G. J., and Wijmenga, C. (2006). The versican gene and the risk of intracranial aneurysms. Stroke *37*, 2372–2374.
Sabatine, M. S., Ploughman, L., Simonsen, K. L., et al. (2008). Association between ADAMTS1 matrix metalloproteinase gene variation, coronary heart disease, and benefit of statin therapy. Arterioscler Thromb Vasc Biol *28*, 562–567.
Sakko, A. J., Ricciardelli, C., Mayne, K., et al. (2003). Modulation of prostate cancer cell attachment to matrix by versican. Cancer Res *63*, 4786–4791.
Sambandam, T., Baker, J. R., Christner, J. E., and Ekborg, S. L. (1991). Specificity of the low density lipoprotein-glycosaminoglycan interaction. Arterioscler Thromb *11*, 561–568.
Sandy, J. D. (2006). A contentious issue finds some clarity: on the independent and complementary roles of aggrecanase activity and MMP activity in human joint aggrecanolysis. Osteoarthritis Cartilage *14*, 95–100.
Sandy, J. D., Westling, J., Kenagy, R. D., et al. (2001). Versican V1 proteolysis in human aorta in vivo occurs at the Glu441-Ala442 bond, a site that is cleaved by recombinant ADAMTS-1 and ADAMTS-4. J Biol Chem *276*, 13372–13378.
Seidelmann, S. B., Kuo, C., Pleskac, N., et al. (2008). Athsq1 is an atherosclerosis modifier locus with dramatic effects on lesion area and prominent accumulation of versican. Arterioscler Thromb Vasc Biol *28*, 2180–2186.
Skalen, K., Gustafsson, M., Rydberg, E. K., et al. (2002). Subendothelial retention of atherogenic lipoproteins in early atherosclerosis. Nature *417*, 750–754.
Sobue, M., Takeuchi, J., Yoshida, K., et al. (1987). Isolation and characterization of proteoglycans from human nonepithelial tumors. Cancer Res *47*, 160–168.
Srinivasan, S. R., Lopez, A., Radhakrishnamurthy, B., and Berenson, G. S. (1970). Complexing of serum pre-beta and beta-lipoproteins and acid mucopolysaccharides. Atherosclerosis *12*, 321–334.
Strom, A., Olin, A. I., Aspberg, A., and Hultgardh-Nilsson, A. (2006). Fibulin-2 is present in murine vascular lesions and is important for smooth muscle cell migration. Cardiovasc Res *69*, 755–763.

Theocharis, A. D. (2008). Versican in health and disease. Connect Tissue Res *49*, 230–234.

Theocharis, A. D., Skandalis, S. S., Tzanakakis, G. N., and Karamanos, N. K. (2010). Proteoglycans in health and disease: novel roles for proteoglycans in malignancy and their pharmacological targeting. FEBS J *277*, 3904–3923.

Theocharis, A. D., Tsolakis, I., Hjerpe, A., and Karamanos, N. K. (2001). Human abdominal aortic aneurysm is characterized by decreased versican concentration and specific downregulation of versican isoform V(0). Atherosclerosis *154*, 367–376.

Theocharis, A. D., Tsolakis, I., Hjerpe, A., and Karamanos, N. K. (2003). Versican undergoes specific alterations in the fine molecular structure and organization in human aneurysmal abdominal aortas. Biomed Chromatogr *17*, 411–416.

Touab, M., Arumi-Uria, M., Barranco, C., and Bassols, A. (2003). Expression of the proteoglycans versican and mel-CSPG in dysplastic nevi. Am J Clin Pathol *119*, 587–593.

Touab, M., Villena, J., Barranco, C., Arumi-Uria, M., and Bassols, A. (2002). Versican is differentially expressed in human melanoma and may play a role in tumor development. Am J Pathol *160*, 549–557.

Travis, J. A., Hughes, M. G., Wong, J. M., Wagner, W. D., and Geary, R. L. (2001). Hyaluronan enhances contraction of collagen by smooth muscle cells and adventitial fibroblasts: role of CD44 and implications for constrictive remodeling. Circ Res *88*, 77–83.

Tsibris, J. C., Segars, J., Coppola, D., et al. (2002). Insights from gene arrays on the development and growth regulation of uterine leiomyomata. Fertil Steril *78*, 114–121.

Venkatesan, N., Ebihara, T., Roughley, P. J., and Ludwig, M. S. (2000). Alterations in large and small proteoglycans in bleomycin-induced pulmonary fibrosis in rats. Am J Respir Crit Care Med *161*, 2066–2073.

Venkatesan, N., Roughley, P. J., and Ludwig, M. S. (2002). Proteoglycan expression in bleomycin lung fibroblasts: role of transforming growth factor-beta(1) and interferon-gamma. Am J Physiol Lung Cell Mol Physiol *283*, L806–814.

Vijayagopal, P. (1994). Regulation of the metabolism of lipoprotein-proteoglycan complexes in human monocyte-derived macrophages. Biochem J *301*, 675–681.

Wang, W., Xu, G. L., Jia, W. D., et al. (2009). Ligation of TLR2 by versican: a link between inflammation and metastasis. Arch Med Res *40*, 321–323.

Wang, X., Hu, G., and Zhou, J. (2010). Repression of versican expression by microRNA-143. J Biol Chem *285*, 23241–23250.

West, R. B., and van de Rijn, M. (2007). Experimental approaches to the study of cancer-stroma interactions: recent findings suggest a pivotal role for stroma in carcinogenesis. Lab Invest *87*, 967–970.

Westergren-Thorsson, G., Chakir, J., Lafreniere-Allard, M. J., Boulet, L. P., and Tremblay, G. M. (2002). Correlation between airway responsiveness and proteoglycan production by bronchial fibroblasts from normal and asthmatic subjects. Int J Biochem Cell Biol *34*, 1256–1267.

Wight, T., Toole, B., and Hascall, V. (2011). Hyaluronan and the aggregating proteoglycans. In The Extracellular Matrix: An Overview, R. Mecham, ed. Heidelberg: Springer, pp. 147–195.

Wight, T. N. (2002). Versican: a versatile extracellular matrix proteoglycan in cell biology. Curr Opin Cell Biol *14*, 617–623.

Wight, T. N., and Merrilees, M. J. (2004). Proteoglycans in atherosclerosis and restenosis: key roles for versican. Circ Res *94*, 1158–1167.

Williams, K. J., and Tabas, I. (1995). The response-to-retention hypothesis of early atherogenesis. Arterioscler Thromb Vasc Biol *15*, 551–561.

Wolf, Y. G., Rasmussen, L. M., and Ruoslahti, E. (1994). Antibodies against transforming growth factor-beta1 suppress intimal hyperplasia in a rat model. J Clin Invest *93*, 1172–1178.

Wu, Y. J., La Pierre, D. P., Wu, J., Yee, A. J., and Yang, B. B. (2005). The interaction of versican with its binding partners. Cell Res *15*, 483–494.

Yamagata, M., and Kimata, K. (1994). Repression of a malignant cell-substratum adhesion phenotype by inhibiting the production of the anti-adhesive proteoglycan, PG-M/versican. J Cell Sci *107*, 2581–2590.

Yamagata, M., Saga, S., Kato, M., Bernfield, M., and Kimata, K. (1993). Selective distributions of proteoglycans and their ligands in pericellular matrix of cultured fibroblasts. Implications for their roles in cell-substratum adhesion. J Cell Sci *106*, 55–65.

Yao, L. Y., Moody, C., Schönherr, E., Wight, T. N., and Sandell, L. J. (1994). Identification of the proteoglycan versican in aorta and smooth muscle cells by DNA sequence analysis, in situ hybridization and immunohistochemistry. Matrix Biol *14*, 213–225.

Zhao, X., and Russell, P. (2005). Versican splice variants in human trabecular meshwork and ciliary muscle. Mol Vis *11*, 603–608.

Zheng, P. S., Reis, M., Sparling, C., et al. (2006). Versican G3 domain promotes blood coagulation through suppressing the activity of tissue factor pathway inhibitor-1. J Biol Chem *281*, 8175–8182.

Zhu, Z., Ma, B., Zheng, T., et al. (2002). IL-13-induced chemokine responses in the lung: role of CCR2 in the pathogenesis of IL-13-induced inflammation and remodeling. J Immunol *168*, 2953–2962.

3.4 The biology of perlecan and its bioactive modules

Chris D. Willis, Liliana Schaefer, and Renato V. Iozzo

3.4.1 Introduction

Perlecan is a modular heparan sulfate proteoglycan (HSPG) that represents a family in itself, not just because of its colossal size but, more importantly, because of its vast number of molecular interactions governing both structural and signaling events. The five modules of perlecan could be construed as five diverse structural units, and each individual domain does indeed regulate many biological processes involved in vascular development, muscle integrity, angiogenesis, and cancer. This chapter will review recent developments in perlecan research in the context of animal models of perlecan deficiency and human genetic diseases related to the expression of dysfunctional perlecan. Moreover, it will cover recent progress in deciphering the receptors for perlecan and how these interactions could lead to angiogenesis control.

3.4.2 Discovery

Since being identified more than 30 years ago, the proteoglycan perlecan has been implicated as a protein involved in a multitude of functions that extend across phyla and cell types. Perlecan protein was originally isolated from the basement membrane of the murine Engelbreth-Holm-Swarm sarcoma and early on was thought to act as an anionic filter. The name *perlecan* was coined from the observation that molecules visualized by rotary shadowing electron microscopy adopt an ultrastructure that resembles "beads on a string." This appearance of a series of globules separated by rods is well representative of the modular domain structure of the 4,391 amino acids that make up the human perlecan protein product. In addition, early on it was shown that perlecan self-assembles into dimers and oligomers through the C terminus of the protein, making larger multimeric structures.

3.4.3 Expression and localization

The mRNA levels of perlecan have been studied in a number of mouse models. Perlecan mRNA expression levels sharply increase in day-4 blastocysts, peaking at embryonic day (ED) 4.5 when blastocytes are attachment competent, suggesting that during development perlecan mRNA expression levels directly correlate with the cellular acquisition of attachment competence. Other experiments with mice have shown perlecan is expressed in tissues undergoing vasculogenesis as early as ED 10. A few days after, further expression is observed in developing cartilage. As ED 13–17.5 is reached, perlecan expression correlates with the development of other organs. As one would expect with the large number of total amino acids in conjunction with the elaborate posttranslational modifications, perlecan mRNA levels remain low in adult stages, likely as a result of minimal protein turnover.

Within epithelial and endothelial cells, perlecan localizes to the extracellular matrix (ECM) and, most commonly, within the basement membrane zone, although incorporation at the cell surface has also been shown. It is commonly associated with the microvasculature where growing neoplastic cells are supplied with essential nutrients and oxygen (Whitelock et al., 2008). The abundant expression of perlecan within the basement membrane is observed ubiquitously in mammals across all tissue types. Perlecan is expressed in mesenchymal organs and connective tissue in addition to avascular examples such as connective tissue stromas, articular cartilage, and intervertebral disks (Iozzo and Murdoch, 1996). Immunohistochemical staining has shown strong localization of perlecan to cartilaginous vertebral body rudiments, with weaker staining at the fetal intervertebral disk of human spines (Smith et al., 2009). Perlecan represents an example of a protein expressed in every organ of the body and has been examined in a number of model systems.

The preferential localization of perlecan to vascular basement membrane zones has been suggested to be involved in proper formation of this region. At very early stages of mouse development, embryo extracellular matrices begin perlecan incorporation. These studies also showed that within 12 hours of blastocyst activation, perlecan protein expression sharply increases. Together, this suggests that perlecan expression correlates with the acquisition of cell attachment competence. On the other hand, perlecan-null mice develop normal basement membranes. This would contradict the notion that multimeric macromolecules self-assembling into dynamic structures are indicative of early events in basement membrane formation.

3.4.4 Protein family

Perlecan is a secreted HSPG. Within the basement membrane zone, perlecan along with agrin and collagen XVIII compose the major HSPGs that help organize this region of the ECM (Iozzo et al., 2009). There are a total of four potential sites within the human protein that allow for the attachment of glycosaminoglycan (GAG) chains (Iozzo, 1998). Potential carbohydrate chain attachment is observed at three sites at the N terminus and results in observed molecular weights greater than 800 kDa. The possible GAGs include heparan sulfate (HS) and/or chondroitin sulfate (CS) chains resulting in unique amino sugars (Whitelock and Iozzo, 2005). Covalent chain attachment of the unbranched polysaccharide chains takes place preferentially on serine (Ser) residues and is initiated by a serine–glycine–aspartic acid (SGD) amino acid motif. Preferential incorporation of these chains is under the control of a variety of regulatory signals during posttranslational modification. The mechanisms of specific HS or CS addition are not well understood.

3.4.5 The *HSPG2* gene

Perlecan is encoded by the *HSPG2* gene and maps to the short arm of chromosome 1 at 1p36 in humans and a syntenic region of chromosome 4 in mice. The sequence is highly conserved among mammals with orthologs also present in *Caenorhabditis elegans* (*unc-52* gene) and *Drosphila melanogaster* (*trol* gene). The identification and sequencing of the *HSPG2* gene has indicated that perlecan developed from ancient ancestors through gene duplication and exon shuffling (Cohen et al., 1993). *HSPG2* is a single-copy gene whose amount of reported exons has increased through the years, but it is now accepted as possessing 97 exons based on the most recent working draft of human chromosome 1.

The *HSPG2* gene comprises ~120 kb of genomic DNA resulting in a ~15.5 kb mRNA (Iozzo et al., 1997). The promoter region of perlecan is equally complex with a significant amount of GC content, particularly within the 500 bp upstream of exon 1 (Iozzo et al., 1994). The *HSPG2* promoter does not contain canonical TATA or CATT boxes, which are characteristic of housekeeping genes. Instead, it contains two viral-enhanced activator protein 2 (AP2) motifs and three short palindromic direct and indirect repeats, suggesting the presence of multiple transcriptional start sites. Perlecan's promoter sequence also possesses four GC boxes and three GGGCGG nucleotide sequences suggesting specificity protein 1 (SP1) is another active transcription factor for the perlecan gene. Active response elements exist, allowing perlecan to be under both positive and negative control by transforming growth factor-beta (TGF-β) (Dodge et al., 1990) and interferon-γ (Sharma and Iozzo, 1998), respectively.

3.4.6 Domain structure and known interactions

Perlecan is a modular proteoglycan with sequences unique to perlecan, as well as conserved motifs seen in other signaling proteins such as growth factors (▶Figure 3.5). Domain I harbors a unique sperm protein, enterokinase, and agrin (SEA) module, characteristic of extracellular domains known to enhance *O*-glycosylation. This highly acidic region contains three consecutive SGD sequences, which are sites responsible for covalent GAG attachment. The three GAG substitutions are primarily responsible for domain I having the largest ligand-binding partners despite being composed of only 172 amino acids.

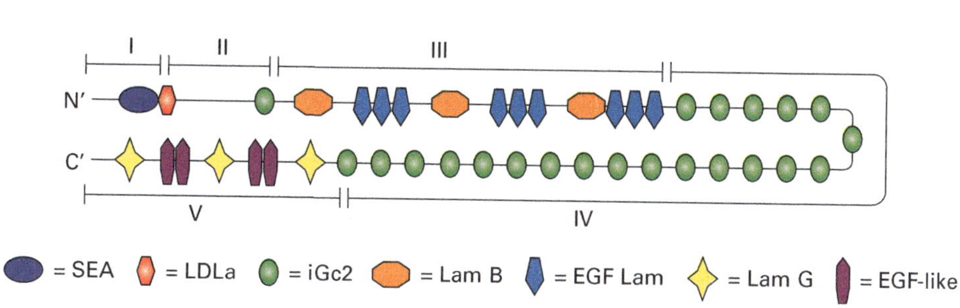

Figure 3.5 Schematic representation of the five modular domains of perlecan. The 4,391 amino acids of the human perlecan protein make up the functional domains as predicted by the Simple Modular Architecture Research Tool. Each domain is drawn to scale in terms of number of amino acids, beginning with the SEA module of domain I at the N terminus. The C-terminal domain V, also known as endorepellin, is an independently functional fragment upon proteolytic cleavage.

Domain I binds laminin, collagen type IV, nidogen, and fibronectin, interactions thought to provide stability within the ECM through protein-protein interactions. A host of growth factors have also been identified as binding to perlecan domain I, including fibroblast growth factor-2 (FGF2), platelet-derived growth factor (PDGF), vascular endothelial growth factor (VEGF), hepatocyte growth factor (HGF), bone morphogenetic protein-2 (BMP2), granulocyte-macrophage colony-stimulating factor (GM-CSF), angiopoietin-3, and activin A. The harboring and delivery of these growth factors to their native cell surface receptors helps describe the mechanism of perlecan's proangiogenic role in tumor angiogenesis. Other signaling proteins that bind domain I include thrombospondin-1, sonic hedgehog (SHH), and interleukin (IL)-2 and IL-8 (▶Figure 3.6). Sites within domain I have also been mapped for ECM scaffolding and anchoring proteins such as von Willebrand factor A domain-related protein (WARP), fibrillin-1, and proline/arginine-rich end leucine-rich repeat protein (PRELP) .

Domain II spans 311 amino acids and is composed of four low-density lipoprotein (LDL) receptor class A domains and one immunoglobulin c-2 type (IGc2) domain. Rotary shadowing electron microscopy images show that domain II adopts a globular structure connected to a short rodlike segment. LDL domains are characterized by six disulfide-bound cysteine residues and a highly conserved cluster of negatively charged amino acids that participate in calcium binding while modulating Wnt/calcium signaling. Perlecan has been proposed to mediate a distinct pathway for internalization of atherogenic lipoproteins enriched in lipoprotein lipase (Fuki et al., 2000).

Domain III is composed of three similar tandem laminin-B domains followed by three ~60-amino-acid laminin-type epidermal growth factor (EGF) domains. There are 27 exons that encode the second largest domain, and interestingly, there is no correlation between the exon arrangement and the protein subdomains. EGF domains contain four conserved disulfide bonds, which have been linked to diseases. Like domain II, the overall structure of this region forms repeating inflexible rodlike structures. There are overlapping binding partners for this region, including the previously mentioned PDGF and WARP proteins. Other binding partners include fibroblast growth factor-7 (FGF7), fibroblast growth factor-18 (FGF18), and fibroblast growth factor binding protein (FGFBP) (Mongiat et al., 2000, 2001). The interaction with FGF7 has been studied in an engineered human skin model, which showed that the absence of perlecan in keratinocytes results in a disrupted epidermis but can be partially rescued by exogenous FGF7 (Sher et al., 2006).

Domain IV is exclusively composed of IGc2 domains. There are 21 consecutive IGc2 repeats in humans but only 14 in mice, making it the largest of the five domains with a total of 2,010 amino acids in humans. A high degree of similarity exists between this matrix-binding region encoded by 40 exons and the immunoglobulin (Ig) superfamily member, neural cell adhesion molecule. All types of Ig domains possess a core Greek-key beta-sandwich structure allowing for the modular nature of this region. Studies using recombinant domain IV identified a number of extracellular proteins that bind specifically and with high affinity. The identified ligands include fibronectin, nidogen-1 and -2, fibulin-2, collagen type IV, and the laminin-nidogen ternary complex. The binding of these ligands provides further support for domain IV being involved in basement membrane stability via protein-protein interactions. PDGF once again also binds to this domain to promote growth factor availability. Later studies pinpointed the specific binding IGc2 domain as the ligand for PDGF, while other studies showed that four arginine residues of IGc2–5 bind heparin/sulfatides.

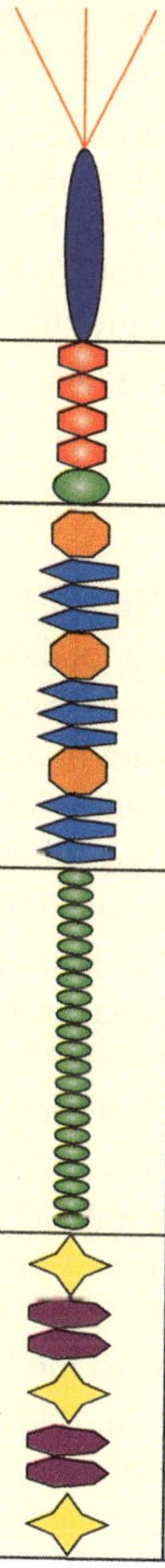

Perlecan Structural Domains	Binding Partners	Proposed Biological Function
I	Laminin, collagen IV, nidogen-1, FGF2, Fibronectin, PDGF, VEGF, HGF, BMP-2, GM-CSF, angiopoietin-3, activin A, thrombospondin-1, SHH, IL-2, IL-8, WARP, fibrillin-1, PRELP	Basement membrane formation and stability, scaffolding, growth factor harboring and signaling, proangiogenic activity, cell motility and adhesion, chondrogenesis, extracellular matrix organization
II	LDL, VLDL, DTGF, fibrillin-1, Wnt	Lipid metabolism, extracellular matrix scaffolding, chondrogenesis, calcium signaling
III	PDGF, WARP, FGF7, FGF18, FGFBP	Growth factor harboring and signaling, pro-angiogenic activity, extracellula matrix organization
IV	Fibronectin, nidogen-1/2, fibulin-2, collagen IV, PDGF, heparin, sulfatides	Growth factor harboring and signaling, basement membrane stability, cell motility and adhesion
V	Endostatin, fibulin-2, α2β1 integrin, VEGFR1/2, nidogen-1, acetylcholinesterase, heparin, PDGF, FGF7, α-dystroglycan, PRELP, ECM-1, progranulin	Neuromuscular junction scaffolding, pro and anti-angiogenic activity, receptor antagonization, growth factor harboring and signaling, cell adhesion, basement membrane stability, oligomerization

Figure 3.6 The structure-function relationship of perlecan. The five structural domains of perlecan are represented on the left, with the GAG chains attached to domain I shown as red lines at the N terminus. All reported binding partners that have been mapped to distinct domains are listed in the middle column. The proposed biological functions resulting from the protein-protein interactions are shown in the right column and discussed in more detail within the text.

Domain V, also referred to as endorepellin, contains three laminin-G (LG) domains that are separated by two pairs of EGF repeats (Mongiat et al., 2003b). Endorepellin consists of 705 amino acids and can exist on its own protein as a result of proteolytic processing (Bix and Iozzo, 2005). The enzymatic activity of cathepsin L is responsible for liberating endorepellin (Cailhier et al., 2008), whereas Tolloid-like metalloproteases further cleave the

25 kDa laminin-like globular 3 (LG3) C-terminal fragment, which possesses antiangiogenic activity on its own (Gonzalez et al., 2005). The opposing role of perlecan and endorepellin in angiogenesis is an emerging theme for similarly proteolytically processed proteins.

Overlapping binding partners of this domain include PDGF, FGF7, and PRELP. Unique binding partners for endorepellin involve the proangiogenic glycoprotein extracellular matrix protein 1 (ECM1) (Mongiat et al., 2003a) and progranulin (Gonzalez et al., 2003). Domain V also plays a role at the neuromuscular junction through forming a complex with the linker protein dystroglycan and the collagen-tailed form of the acetylcholinesterase enzyme. Lastly, endorepellin binds the I domain within the α2 region of the α2β1-integrin receptor (Bix et al., 2007). Along with vascular endothelial growth factor receptors 1 and 2 (VEGFR1/2), endorepellin is the only portion of perlecan that can concurrently bind two cell surface receptors and directly influence intracellular signaling (Goyal et al., 2011).

3.4.7 Genetic links to diseases

Given its large size and ubiquitous distribution, it is not surprising that *HSPG2* mutations have been causatively associated with disease phenotypes. There are a total of 37 reported unique mutations that map to domains II–V in Schwartz-Jampel syndrome (SJS), a rare autosomal recessive condition with myotonia and chondrodysplasia (Stum et al., 2006). The predicted effects of the mutations include faulty splicing, exon deletion/insertion, exon fusion, and frameshifts, as well as missense and nonsense mutations. These mutations lead to unstable mRNA and intracellular retention of mutant perlecan, which does not allow for proper trafficking to the basement membrane compartment (Stum et al., 2006). The resulting phenotype and severity vary, and multiple skeletal dysplasias have recently been reclassified as SJS. A similar phenotype has been observed in *Hspg2*$^{-/-}$mice, which exhibit muscle hyperexcitability and myotonia resulting from the lack of acetylcholinesterase at the neuromuscular junction. A hypomorphic mutation secondary to a missense substitution, corresponding to a human familial SJS mutation, exhibits a similar myotonia in a knockin mouse model (Echaniz-Laguna et al., 2009). Cartilage perlecan can be substituted with CS chains instead of HS chains, and this posttranslational modification enhances collage type II fibril formation (Kvist et al., 2006). This could contribute to the genesis of the perlecan-null chondrodysplasia. Finally, knockdown of perlecan in zebra fish causes severe myopathy characterized by abnormal actin filament orientations and disorganized sarcomeres in addition to a vascular phenotype (Zoeller et al., 2008).

Another disorder characterized by *HSPG2* mutations is dyssegmental dysplasia, Silverman-Handmaker type. Only three mutations have been reported, and all consist of frameshift mutations within domain III or IV, which lead to a truncated perlecan protein core lacking domain V (Arikawa-Hirasawa et al., 2001b). As a result of protein instability, the mutant perlecan is not properly secreted into the extracellular milieu (Arikawa-Hirasawa et al., 2001a). Cartilage matrix staining confirmed this observation and demonstrates the significance of native perlecan in the ECM during cartilage development.

3.4.8 Genetic models

The importance of perlecan within other species has also been illustrated through genetic manipulations. Approximately 40% of *Hspg2*$^{-/-}$ mice die at ED 10.5 from intrapericardial hemorrhage and lethal chondrodysplasia. Those embryos that survive display brain defects as well as malfunction of the cardiac outflow tract with complete transposition of great vessels, implicating a role for perlecan in cardiovascular development. This phenotype can be accompanied by malformations of semilunar valves (Costell et al., 2002), which ultimately results in perinetal death due to respiratory failure. The underlying biology of cartilage development is impeded by the lack of perlecan as immunohistochemical studies have shown disorganized growth plates with reduced chondrocyte proliferation and differentiation for *Hspg2*-null chondrocytes at ED 16.5 (Arikawa-Hirasawa et al., 1999).

Genetic studies of the *C. elegans* ortholog of perlecan, *unc-52*, also demonstrate adverse phenotypes upon gene disruption. Domain I of *unc-52* does not contain the SEA domain with GAG attachments but instead possesses a substituted arginine-rich domain. The mature protein is specific to muscle tissue in *C. elegans*. Mutational analysis revealed that insertions in domain IV result in a disorganized muscle phenotype characterized by disrupted sarcomeres and detachment of body-wall muscle (Rogalski et al., 1995). Further studies showed that a null *unc-52* gene is embryonic lethal, and other various mutations result in paralysis. Therefore, unc-52 plays a vital role in myofilament assembly in body-wall muscle during embryonic development. In skeletal muscle, perlecan deficiency results in muscle hypertrophy through a decrease in the TGF-β family member myostatin (Xu et al., 2010). The decreased signaling by myostatin leads to shifts in muscle fiber composition, suggesting that perlecan is critical for the makeup of skeletal muscle fibers.

Mutations in *trol*, the ortholog of perlecan in *Drosophila*, also gives rise to lethal alleles as well as abnormal larval brain lobes and small imaginal disk phenotypes (Datta and Kankel, 1992). *Trol* gene is developmentally expressed around the visceral mesoderm, the hindgut, and the central nervous system. Neuroblast proliferation is one role identified for *trol*, which involves modulating signaling of the previously mentioned binding partners, FGF and Hedgehog. Notably, diverse mutations in *trol* allow developmental progression to varying extents, suggesting that *trol* is directly involved in multiple cell-fate and patterning decisions (Lindner et al., 2007). Perlecan has been studied using a zebra fish model in which morpholino technology was used to inhibit perlecan mRNA translation (Zoeller et al., 2008, 2009). The cardiovascular defects seen in the transgenic mice correlate with the perlecan-morphant zebra fish, which lack sprouting in the intersegmental and subintestinal vessels, both of which derive from angiogenesis.

3.4.9 Perlecan role in cancer

Abnormal expression of proteoglycans in the stroma of various cancers has been linked to tumor progression and metastasis (Iozzo and Cohen, 1993; Iozzo, 1995). In addition, perlecan has been directly involved in regulating cancer growth and angiogenesis primarily because of its ability to act as a depot for powerful growth and proangiogenic factors (Iozzo and San Antonio, 2001). The direct angiogenic effects of perlecan result from the numerous functional roles of the protein within the basement membrane. Perlecan sequesters a plethora of growth factors, which contributes to the proangiogenic role through matrix signaling (▶Figure 3.6). *In vivo* tumor sample analysis has shown up to a 15-fold increase of perlecan mRNA levels in metastatic melanomas (Cohen et al., 1994). Upregulation was shown to occur within 10 minutes of neurotrophin stimulation, which suggests that perlecan is an early response gene upon tumor invasion. This is consistent with early studies showing that the perlecan protein core is found in abundance in blood vessel walls of many human primary tumors.

The abundance of perlecan within tumor cells is linked to the levels of growth factors in the basement membrane. For example, in human melanoma cells, antisense perlecan cDNA causes a suppression of the autocrine and paracrine functions for FGF2 (Aviezer et al., 1997). The human melanoma cells also show reduced proliferation and invasion resulting from the absence of HS chain binding. The other domains are responsible for similar binding events that protect the proangiogenic growth factors from degradation and misfolding.

Prostate cancer screens also have yielded high Gleason scores for correlation of perlecan expression and aggressive tumors (Datta et al., 2006). Increased perlecan levels result in rapid cell proliferation through SHH signaling, independent of androgen signaling. Suppression of perlecan reverses the phenotype and has also been demonstrated in mouse models. Gene expression and protein deposition of major basement membrane components and TGF-β in human breast cancer and antibodies specific for perlecan domain III show abundant deposits of perlecan in the stroma of malignant breast cancer tissue. On the contrary, decreased circulating levels of the C-terminal LG3 region also has been proposed as a biomarker for breast cancer (Chang et al., 2008). This paradigm of dual angiogenic roles for the different termini of perlecan is the following Section 3.4.10.

Staining of malignant colon cancer tissue also show numerous deposits of perlecan (Murdoch et al., 1994). Once again, mouse models have confirmed that suppression of perlecan results in substantial inhibition of tumor growth and neovascularization in colon carcinoma cells. An analogous growth-factor-suppressing mechanism for FGF7 was also elucidated in these studies (Sharma et al., 1998).The increases in perlecan levels observed in multiple cancers correspond to an enhanced metastatic potential. By localizing at the interface between proliferating tumor and migrating endothelial cells, perlecan is an ideal candidate to moderate angiogenesis (Whitelock et al., 2008). Vascular basement membranes repeatedly undergo remodeling during tumor formation, and this process likely involves growth factor release from perlecan by heparanases and MMPs (Whitelock et al., 1996).

The biological role of perlecan in angiogenesis is dependent on the cellular context. For example, in fibrosarcoma cells and xenografts, perlecan acts as a negative regulator of growth and invasion (Mathiak et al., 1997). In addition, human Kaposi's sarcoma cells with reduced perlecan expression display reduced tumor cell proliferation *in vitro* but increased tumor growth *in vivo* (Marchisone et al., 2000). These opposing observations suggest that the specific cellular circumstances or pathological processes might define the role of perlecan.

3.4.10 Perlecan role in vascular biology and angiogenesis

Endothelial cells expressing an antisense vector targeting domain III of perlecan show a reduced ability to inhibit the binding and mitogenic activity of FGF2 in vascular smooth muscle cells. Using these cells, Nugent and coworkers showed that perlecan is absolutely required for the prevention of thrombosis after deep vascular injury (Nugent et al., 2000). These results could be partially explained by the observation that perlecan immunopurified from various endothelial cell sources has different adhesive properties (Whitelock et al., 1999). Thus, it is possible that perlecan synthesized by injured endothelial cells has unique distinguishing features that could influence vascular biology. Interestingly, mechanical regulation of perlecan expression in endothelial cells is regulated by a mechanotransduction pathway that requires autocrine TGF-β signaling (Baker et al., 2008), a pathway that transcriptionally activates perlecan (Dodge et al., 1990).

The C-terminal domain of perlecan is proteolytically processed, resulting in a bioactive fragment of ~85 kDa. Upon the serendipitous discovery using a yeast two-hybrid screening, it was shown endorepellin can repress the chemotactic migration in response to VEGF treatment of endothelial cells (Mongiat et al., 2003b). Chorioallantoic membrane assays showed that endorepellin reduces the angiogenic activity of VEGF, counteracting the N-terminal proangiogenic role of perlecan. A similar antiangiogenic mechanism was previously observed for endostatin, suggesting that a family of C-terminal fragments that alone possess unique signaling properties exists.

The role of endorepellin as an antiangiogenic protein also derives from the identification of α2β1 integrin as the functional receptor that binds LG3. The signaling pathway through the α2β1-integrin receptor involves the disassembly of the actin cytoskeleton and focal adhesions in endothelial cells, leading to the inhibition of endothelial cell movement (Bix et al., 2004). Animals lacking the α2β1 receptor have confirmed the role of endorepellin as a key ligand by showing that endorepellin targets the tumor xenograft vasculature in an α2β1-integrin-dependent manner (Woodall et al., 2008). Reduced levels of α2β1 integrin result in a vascular phenotype, similar to the observation of blunted intersegmental vessels in α2β1-integrin-morphant zebra fish (San Antonio et al., 2009). The resulting intracellular signaling cascade induced by endorepellin binding to VEGFR2 has yet to be determined, although it has been shown to lead to activation of the tyrosine phosphatase Src homology phosphatase-1 (SHP1) while attenuating VEGF transcription (Goyal et al., 2011). This is consistent with previous observation that SHP1 is actively phosphorylated upon endorepellin stimulation in endothelial cells (Nyström et al., 2009).

The 25 kDa LG3 fragment requires correct Ca^{2+} coordination for its function and can bind the α2β1-integrin receptor to signal for the disassembly of actin stress fibers and focal adhesions (Bix et al., 2004; Gonzalez et al., 2005). LG3 can also be liberated from apoptotic endothelial cells and can induce an α2β1-integrin-dependent antiapoptotic pathway in fibroblasts (Laplante et al., 2006). The LG3 domain inhibits capillary morphogenesis and is specific for endothelial cells. Proteomic profiling has provided some clues about potential proteins involved in the antiangiogenic signaling of endorepellin in endothelial cells (Zoeller and Iozzo, 2008).

The potential for endorepellin as a protein therapeutic to combat tumor angiogenesis has been explored *in vivo* for mouse models bearing orthotopic squamous carcinoma xenografts or syngeneic Lewis lung carcinoma tumors (Bix et al., 2006). The results showed inhibition of tumor angiogenesis, enhanced tumor hypoxia, and a decrease in tumor metabolism rate. Therefore, endorepellin as a therapeutic has the potential to a block endothelial cell migration, survival, and proliferation for vascularized tumors.

3.4.11 Conclusions and future directions

With a proteoglycan as ubiquitous as perlecan, the future is bright to identify novel roles for this key cell-associated and basement membrane constituent. The possible relevance of perlecan as a therapeutic agent has yet to be explored in the clinical setting. Recently, there has been an emphasis on proteoglycans as novel therapeutics that approach bone healing and neuromuscular disorders with this class of proteins. Although the large number of amino acid residues coupled to the potential substitution with 3–4 HS chains and numerous oligosaccharides makes the potential of a protein therapeutic difficult, it is possible that smaller regions linked to specific pathways could be explored as treatments. Future research should be directed toward understanding the regulation of individual domain processing during pathophysiological events. It has been suggested that endorepellin and LG3 are liberated from perlecan during tissue remodeling and cancer growth. This mechanism has yet to be confirmed *in vivo*, so many questions remain before it can be considered as a potential treatment. Recently, it has also been reported that soluble forms of domain I of perlecan are biologically active and increase VEGF signaling through VEGFR2 in human bone marrow endothelial cells. It is possible that similar mechanisms for the individual domains are released during basement membrane turnover.

The therapeutic potential demonstrated in mouse tumor models is promising, and minimal immune response would be expected with use of a native recombinant protein. However, before this potential can be tested, a better understanding of how endorepellin affects the cell cycle, cell survival, and cancer growth must be understood. Recent findings of endorepellin as a dual receptor antagonist for the $\alpha 2\beta 1$ integrin and VEGFR2 could address these mechanisms and provide a better understanding of how tumor cells modulate their microenvironment.

Current trends to answer complex questions should also provide direction to propose hypotheses for understanding the role of perlecan in basement membranes. Just as the human genome project was indispensable in understanding the link between mutations and diseases, new missions such as the Protein Structure Initiative (PSI) should lead to novel structural information about perlecan. To date, no solved structures have been reported for any of the domains of perlecan. Since the goal of the PSI is to solve the three-dimensional structure of all proteins, once the HSPG family is addressed, this will provide insight into the structure-function relationship of perlecan. Homology modeling has provided some clues that endorepellin is likely to adopt a structure similar to agrin, which also contains three LG domains interspersed with three EGF domains. In addition, agrin belongs to this class of C-terminal cleavage fragments as neurotrypsin is responsible for liberating a 90 kDa fragment and an LG3 module.

The size of perlecan also makes it likely that functional noncoding DNA exists. There has yet to be an identified microRNA among the >120 kb of perlecan genomic DNA. To date, no microRNAs exist that have been identified that regulate perlecan mRNA either. In addition, there have yet to be reports of alternative splicing within the human gene despite clues that other isoforms exist. Alternatively spliced variants have been reported in mouse and nematode species. Specific amino acid sequences have been identified that suggest further proteolytic cleavage to give rise to other functional proteins, but it is possible that these are inaccessible cryptic sites. Further structural studies should aid in understanding how the folding of perlecan regulates its function.

3.4.12 Take-home message

The multiple functions of perlecan in development, cancer growth, and angiogenesis, outlined in this chapter, transcend its structural role and pave the way for new therapeutic modalities to combat angiogenesis.

References

Arikawa-Hirasawa, E., Watanabe, E., Takami, H., Hassell, J. R., and Yamada, Y. (1999). Perlecan is essential for cartilage and cephalic development. Nat Genet *23*, 354–358.

Arikawa-Hirasawa, E., Wilcox, W. R., Le, A. H., et al. (2001a). Dyssegmental dysplasia, Silverman-Handmaker type, is caused by functional null mutations of the perlecan gene. Nat Genet *27*, 431–434.

Arikawa-Hirasawa, E., Wilcox, W. R., and Yamada, Y. (2001b). Dyssegmental dysplasia, Silverman-Handmaker type: unexpected role of perlecan in cartilage development. Am J Med Genet *106*, 254–257.

Aviezer, D., Iozzo, R. V., Noonan, D. M., and Yayon, A. (1997). Suppression of autocrine and paracrine functions of basic fibroblast growth factor by stable expression of perlecan antisense cDNA. Mol Cell Biol *17*, 1938–1946.

Baker, A. B., Ettenson, D. S., Jonas, M., Nugent, M. A., Iozzo, R. V., and Edelman, E. R. (2008). Endothelial cells provide feedback control for vascular remodeling through a mechanosensitive autocrine TGF-b signaling pathway. Circ Res *103*, 289–297.

Bix, G., Castello, R., Burrows, M., et al. (2006). Endorepellin in vivo: targeting the tumor vasculature and retarding cancer growth and metabolism. J Natl Cancer Inst *98*, 1634–1646.

Bix, G., Fu, J., Gonzalez, E., et al. (2004). Endorepellin causes endothelial cell disassembly of actin cytoskeleton and focal adhesions through the alpha2beta1 integrin. J Cell Biol *166*, 97–109.

Bix, G., Iozzo, R. A., Woodall, B., et al. (2007). Endorepellin, the C-terminal angiostatic module of perlecan, enhances collagen-platelet responses via the alpha2beta1-integrin receptor. Blood *109*, 3745–3748.

Bix, G., and Iozzo, R. V. (2005). Matrix revolutions: "tails" of basement-membrane components with angiostatic functions. Trends Cell Biol *15*, 52–60.

Cailhier, J. F., Sirois, I., Raymond, M.-A., et al. (2008). Caspase-3 activation triggers extracellular release of cathepsin L and endorepellin proteolysis. J Biol Chem *283*, 27220–27229.

Chang, J. W., Kang, U.-B., Kim, D. H., et al. (2008). Identification of circulating endorepellin LG3 fragment: potential use as a serological biomarker for breast cancer. Proteomics Clin Appl *2*, 23–32.

Cohen, I. R., Grässel, S., Murdoch, A. D., and Iozzo, R. V. (1993). Structural characterization of the complete human perlecan gene and its promoter. Proc Natl Acad Sci U S A *90*, 10404–10408.

Cohen, I. R., Murdoch, A. D., Naso, M. F., Marchetti, D., Berd, D., and Iozzo, R. V. (1994). Abnormal expression of perlecan proteoglycan in metastatic melanomas. Cancer Res *54*, 5771–5774.

Costell, M., Carmona, R., Gustafsson, E., González-Iriarte, M., Fässler, R., and Munoz-Chápuli, R. (2002). Hyperplastic conotruncal endocardial cushions and transposition of great arteries in perlecan-null mice. Circ Res *91*, 158–164.

Datta, M. W., Hernandez, A. M., Schlicht, M. J., et al. (2006). Perlecan, a candidate gene for the CAPB locus, regulates prostate cancer cell growth via the Sonic Hedgehog pathway. Mol Cancer *5*, 9.

Datta, S., and Kankel, D. R. (1992). l(1)trol and l(1)devl, loci affecting the development of the adult central nervous system in Drosophila melanogaster. Genetics *130*, 523–537.

Dodge, G. R., Kovalszky, I., Hassell, J. R., and Iozzo, R. V. (1990). Transforming growth factor beta alters the expression of heparan sulfate proteoglycan in human colon carcinoma cells. J Biol Chem *265*, 18023–18029.

Echaniz-Laguna, A., Rene, F., Marcel, C., et al. (2009). Electrophysiological studies in a mouse model of Schwartz-Jampel syndrome demonstrate muscle fiber hyperactivity of peripheral nerve origin. Muscle Nerve *40*, 55–61.

Fuki, I., Iozzo, R. V., and Williams, K. J. (2000). Perlecan heparan sulfate proteoglycan. A novel receptor that mediates a distinct pathway for ligand catabolism. J Biol Chem *275*, 25742–25750.

Gonzalez, E. M., Mongiat, M., Slater, S. J., Baffa, R., and Iozzo, R. V. (2003). A novel interaction between perlecan protein core and progranulin: potential effects on tumor growth. J Biol Chem *278*, 38113–38116.

Gonzalez, E. M., Reed, C. C., Bix, G., et al. (2005). BMP-1/Tolloid-like metalloproteases process endorepellin, the angiostatic C-terminal fragment of perlecan. J Biol Chem *280*, 7080–7087.

Goyal, A., Pal, N., Concannon, M., et al. (2011). Endorepellin, the angiostatic module of perlecan, interacts with both the alpha2beta1 integrin and vascular endothelial growth factor receptor 2 (VEGFR2). J Biol Chem *286*, 25947–25962.

Iozzo, R. V. (1995). Tumor stroma as a regulator of neoplastic behavior. Agonistic and antagonistic elements embedded in the same connective tissue. Lab Invest *73*, 157–160.

Iozzo, R. V. (1998). Matrix proteoglycans: from molecular design to cellular function. Annu Rev Biochem *67*, 609–652.

Iozzo, R. V., and Cohen, I. (1993). Altered proteoglycan gene expression and the tumor stroma. Cell Mol Life Sci *49*, 447–455.

Iozzo, R. V., Cohen, I. R., Grässel, S., and Murdoch, A. D. (1994). The biology of perlecan: the multifaceted heparan sulphate proteoglycan of basement membranes and pericellular matrices. Biochem J *302*, 625–639.

Iozzo, R. V., and Murdoch, A. D. (1996). Proteoglycans of the extracellular environment: clues from the gene and protein side offer novel perspectives in molecular diversity and function. FASEB J *10*, 598–614.

Iozzo, R. V., Pillarisetti, J., Sharma, B., et al. (1997). Structural and functional characterization of the human perlecan gene promoter. Transcriptional activation by transforming factor-beta via a nuclear factor 1-binding element. J Biol Chem *272*, 5219–5228.

Iozzo, R. V., and San Antonio, J. D. (2001). Heparan sulfate proteoglycans: heavy hitters in the angiogenesis arena. J Clin Invest *108*, 349–355.

Iozzo, R. V., Zoeller, J. J., and Nyström, A. (2009). Basement membrane proteoglycans: modulators *par excellence* of cancer growth and angiogenesis. Mol Cells *27*, 503–513.

Kvist, A. J., Johnson, A. E., Mörgelin, M., et al. (2006). Chondroitin sulfate perlecan enhances collagen fibril formation. Implications for perlecan chondrodysplasias. J Biol Chem *281*, 33127–33139.

Laplante, P., Raymond, M.-A., Labelle, A., Abe, J.-I., Iozzo, R. V., and Hebért, M.-J. (2006). Perlecan proteolysis induces alpha2beta1 integrin and src-family kinases dependent anti-apoptotic pathway in fibroblasts in the absence of focal adhesion kinase activation. J Biol Chem *281*, 30383–30392.

Lindner, J. R., Hillman, P. R., Barrett, A. L., et al. (2007). The Drosophila perlecan gene trol regulates multiple signaling pathways in different developmental contexts. BMC Dev Biol *7*, 121.

Marchisone, C., Del Grosso, F., Masiello, L., Prat, M., Santi, L., and Noonan, D. M. (2000). Phenotypic alterations in Kaposi's sarcoma cells by antisense reduction of perlecan. Pathol Oncol Res *6*, 10–17.

Mathiak, M., Yenisey, C., Grant, D. S., Sharma, B., and Iozzo, R. V. (1997). A role for perlecan in the suppression of growth and invasion in fibrosarcoma cells. Cancer Res *57*, 2130–2136.

Mongiat, M., Fu, J., Oldershaw, R., Greenhalgh, R., Gown, A., and Iozzo, R. V. (2003a). Perlecan protein core interacts with extracellular matrix protein 1 (ECM1), a glycoprotein involved in bone formation and angiogenesis. J Biol Chem *278*, 17491–17499.

Mongiat, M., Otto, J., Oldershaw, R., Ferrer, F., Sato, J. D., and Iozzo, R. V. (2001). Fibroblast growth factor-binding protein is a novel partner for perlecan protein core. J Biol Chem *276*, 10263–10271.

Mongiat, M., Sweeney, S., San Antonio, J. D., Fu, J., and Iozzo, R. V. (2003b). Endorepellin, a novel inhibitor of angiogenesis derived from the C terminus of perlecan. J Biol Chem *278*, 4238–4249.

Mongiat, M., Taylor, K., Otto, J., et al. (2000). The protein core of the proteoglycan perlecan binds specifically to fibroblast growth factor-7. J Biol Chem *275*, 7095–7100.

Murdoch, A. D., Liu, B., Schwarting, R., Tuan, R. S., and Iozzo, R. V. (1994). Widespread expression of perlecan proteoglycan in basement membranes and extracellular matrices of human tissues as detected by a novel monoclonal antibody against domain III and by in situ hybridization. J Histochem Cytochem *42*, 239–249.

Nugent, M. A., Nugent, H. M., Iozzo, R. V., Sanchack, K., and Edelman, E. R. (2000). Perlecan is required to inhibit thrombosis after deep vascular injury and contributes to endothelial cell-mediated inhibition of intimal hyperplasia. Proc Natl Acad Sci U S A *97*, 6722–6727.

Nyström, A., Shaik, Z. P., Gullberg, D., et al. (2009). Role of tyrosine phosphatase SHP-1 in the mechanism of endorepellin angiostatic activity. Blood *114*, 4897–4906.

Rogalski, T. M., Gilchrist, E. J., Mullen, G. P., and Moerman, D. G. (1995). Mutations in the unc-52 gene responsible for body wall muscle defects in adult Caenorhabditis elegans are located in alternatively spliced exons. Genetics *139*, 159–169.

San Antonio, J. D., Zoeller, J. J., Habursky, K., et al. (2009). A key role for the integrin a2b1 in experimental and developmental angiogenesis. Am J Pathol *175*, 1338–1347.

Sharma, B., Handler, M., Eichstetter, I., Whitelock, J., Nugent, M. A., and Iozzo, R. V. (1998). Antisense targeting of perlecan blocks tumor growth and angiogenesis in vivo. J Clin Invest *102*, 1599–1608.

Sharma, B., and Iozzo, R. V. (1998). Transcriptional silencing of perlecan gene expression by interferon-gamma. J Biol Chem *273*, 4642–4646.

Sher, I., Zisman-Rozen, S., Eliahu, L., et al. (2006). Targeting perlecan in human keratinocytes reveals novel roles for perlecan in epidermal formation. J Biol Chem *281*, 5178–5187.

Smith, S. M., Whitelock, J. M., Iozzo, R. V., Little, C. B., and Melrose, J. (2009). Topographical variation in the distribution of versican, aggrecan and perlecan in the foetal human spine reflects their diverse functional roles in spinal development. Histochem Cell Biol *132*, 491–503.

Stum, M., Davoine, C. S., Vicart, S., et al. (2006). Spectrum of HSPG2 (perlecan) mutations in patients with Schwartz-Jampel syndrome. Hum Mutat *27*, 1082–1091.

Whitelock, J. M., Graham, L. D., Melrose, J., Murdoch, A. D., Iozzo, R. V., and Underwood, P. A. (1999). Human perlecan immunopurified from different endothelial cell sources has different adhesive properties for vascular cells. Matrix Biol *18*, 163–178.

Whitelock, J. M., and Iozzo, R. V. (2005). Heparan sulfate: a complex polymer charged with biological activity. Chem Rev *105*, 2745–2764.

Whitelock, J. M., Melrose, J., and Iozzo, R. V. (2008). Diverse cell signaling events modulated by perlecan. Biochemistry *47*, 11174–11183.

Whitelock, J. M., Murdoch, A. D., Iozzo, R. V., and Underwood, P. A. (1996). The degradation of human endothelial cell-derived perlecan and release of bound basic fibroblast growth factor by stromelysin, collagenase, plasmin and heparanases. J Biol Chem *271*, 10079–10086.

Woodall, B. P., Nyström, A., Iozzo, R. A., et al. (2008). Integrin a2b1 is the required receptor for endorepellin angiostatic activity. J Biol Chem *283*, 2335–2343.

Xu, Z., Ichikawa, N., Kosaki, K., et al. (2010). Perlecan deficiency causes muscle hypertrophy, a decrease in myostatin expression, and changes in muscle fiber composition. Matrix Biol *29*, 461–470.

Zoeller, J. J., and Iozzo, R. V. (2008). Proteomic profiling of endorepellin angiostatic activity on human endothelial cells. Proteome Sci 6, 7.

Zoeller, J. J., McQuillan, A., Whitelock, J., Ho, S.-Y., and Iozzo, R. V. (2008). A central function for perlecan in skeletal muscle and cardiovascular development. J Cell Biol *181*, 381–394.

Zoeller, J. J., Whitelock, J., and Iozzo, R. V. (2009). Perlecan regulates developmental angiogenesis by modulating the VEGF-VEGFR2 axis. Matrix Biol *28*, 284–291.

3.5 Small leucine-rich proteoglycans: multifunctional signaling effectors

Rosetta Merline, Madalina V. Nastase, Renato V. Iozzo, and Liliana Schaefer

3.5.1 Introduction

Small leucine-rich proteoglycans (SLRPs), named for their relatively small sizes, when compared to the cartilage proteoglycans, and for their pathognomonic tandem leucine-rich repeats (LRRs) within the protein cores, are a group of biologically active molecules of the extracellular matrix (ECM). Structurally, they are composed of a conserved protein core and covalently linked glycosaminoglycan (GAG) side chains of varied numbers and types – namely, chondroitin sulfate (CS), dermatan sulfate (DS), or keratan sulfate (KS). These side chains are negatively charged and contain sulfated, linear disaccharide repeating units of acetylated amino sugar moieties and uronic acid (CS and DS). In the KS GAG side chains, the disaccharide-repeating units consist of galactose (-4*N*-acetyl-glucosamine-β1, 3-galactose-β1). The vast majority of the SLRPs are secreted into the pericellular space with a ubiquitous tissue distribution (Iozzo and Schaefer, 2010).

The SLRP family includes 18 genes (not all are classical proteoglycans), which are divided into five classes based on the characteristic N-terminal cysteine-rich clusters, C-terminal ear repeats, homologies at the genomic and protein levels, chromosomal organization, and shared biological activity (Schaefer and Iozzo, 2008) (▶Table 3.2).

3.5.2 Physiological functions

The structural diversity of the SLRPs allows for a wide range of matrix-matrix and matrix-cell interactions. Initially, SLRPs were mainly considered structural components of the ECM. They play an important role in fibrillogenesis based on their ability to bind various types of collagens and, accordingly, regulate collagen fibril organization in skin, tendons, and the cornea. Additionally, SLRPs bind and modulate the biological functions of various cytokines, including transforming growth factor-beta (TGF-β), connective tissue growth factor (CTGF), tumor necrosis factor-alpha (TNF-α), and proteins such as the bone morphogenetic protein-4 (BMP4), Wnt1-induced secreted protein-1 (WISP1), von Willebrand factor (VWF), insulin-like growth factor-I (IGF-I), and platelet-derived growth factor (PDGF). These interactions have important therapeutic implications for the treatment of proliferative, inflammatory, and fibrotic disorders. Moreover, data from the last two decades indicate that soluble SLRPs (not bound to the ECM) act as signaling molecules. They bind to and signal via the insulin-like growth factor-I receptor (IGF-IR), epidermal growth factor receptor (EGFR), Met, Toll-like receptors (TLRs), low-density lipoprotein receptor-related protein-1 (LRP1), and integrin α2β1, thereby regulating

Table 3.2 Classification of SLRPs.

Class I	*Class II*	*Class III*	*Class IV*	*Class V*
Biglycan (CS/DS)	Fibromodulin (KS)	Epiphycan (CS/DS)	Chondroadherin	Podocan
Decorin (CS/DS)	Lumican (KS)	Opticin	Nyctalopin	Podocan-like protein-1
Asporin	PRELP (KS)	Osteoglycan (KS)	Tsukushi	
ECM2	Keratocan (KS)			
ECMX	Osteoadherin (KS)			

Classification of SLRPs into five distinct classes based on the characteristic N-terminal cysteine-rich clusters, C-terminal ear repeats, homologies at the genomic and protein levels, and chromosomal organization (Schaefer and Iozzo, 2008).
PRELEP = proline/arginine-rich end leucine-rich repeat protein; ECM2 = extracellular matrix protein 2; ECMX = ECM2-like protein, X chromosome.

cellular proliferation, differentiation, survival, adhesion, and migration (Schaefer and Iozzo, 2008). In this chapter, we will discuss selected SLRP members of classes I–III with particular attention to their signaling-dependent pathobiology.

3.5.3 Pathobiology of class I SLRPs

Decorin

Decorin, the most studied SLRP, plays a crucial role in collagen fibrillogenesis and assembly ("decorates" fibrils). Therefore, mice lacking decorin develop fragile skin and abnormal tendon phenotypes (Danielson et al., 1997; Reed and Iozzo, 2002).

Two decades of attention have focused on the antifibrotic properties of decorin as a TGF-β1–neutralizing factor in various organs, based on its physical interaction with TGF-β, interference with TGF-β signaling, either directly or indirectly by regulating other modulators of TGF-β activity (e.g. fibrillin-1, myostatin), and formation of decorin/TGF-β complexes (Yamaguchi et al., 1990; Border et al., 1992; Kolb et al., 2001; Schaefer et al., 2002; Brandan et al., 2008; Schaefer, 2011). Furthermore, the leucine-rich repeat-12 (LRR12) of the decorin protein core also interacts with CTGF and inhibits its biological activity (Vial et al., 2011).

The multifunctional effects of decorin on cell growth and differentiation, proliferation, apoptosis, angiogenesis, and synthesis of other ECM components became increasingly better understood by describing decorin as a signaling modulator. The interaction of decorin with four different receptor tyrosine kinases (RTKs) mediates direct and indirect regulation of several signaling pathways. In normal cells, decorin is a ligand for IGF-IR and LRP1, whereas in carcinoma cells decorin signals via EGFR and Met (Brandan et al., 2008; Schaefer and Iozzo, 2008; Iozzo and Schaefer, 2010). Via binding to and phosphorylation of the IGF-IR, decorin activates phosphatidylinositol 3-kinase (PI3K), protein kinase B (PKB)/Akt, thereby protecting endothelial (Schonherr et al., 2005) and renal tubular epithelial cells from apoptosis and exerting antifibrotic effects (Schaefer et al., 2007; Merline et al., 2009). In renal fibroblasts, decorin activates mammalian target of rapamycin (mTOR) and p70 S6 kinase downstream of Akt, thereby enhancing translation and synthesis of fibrillin-1 (Schaefer et al., 2007). Thus, decorin/IGF-IR signaling results in distinct, cell-type-specific biological outcomes: either protecting cells against apoptosis or directly regulating the synthesis of other ECM proteins.

In transformed cells, decorin induces apoptosis and inhibits proliferation by interaction with the EGF receptors and Met, thereby suppressing tumor growth and metastasis mediated through various signaling pathways (Goldoni and Iozzo, 2008; Iozzo and Sanderson, 2011). Decorin binds to EGFR and induces mitogen-activated protein kinase (MAPK), Ca^{2+} influx, and the cyclin-dependent kinase (CDK) inhibitor p21 with subsequent downregulation of this receptor due to internalization and degradation by caveolin-mediated endocytosis (De Luca et al., 1996; Moscatello et al., 1998; Iozzo et al., 1999; Csordas et al., 2000; Santra et al., 2002; Zhu et al., 2005). Decorin also binds to Met, leading to transient activation followed by E3 ubiquitin ligase casitas B-lineage lymphoma proto-oncogene (c-Cbl) recruitment and intracellular degradation of this RTK (Goldoni et al., 2009). Consequently, the downstream signaling of β-catenin and Myc is shut down, causing inhibition of cell scatter, evasion, and migration required for tumor progression (Buraschi et al., 2010). These mechanisms together with decorin-mediated induction of caspase-3 and apoptosis cause *in vivo* inhibition of tumor growth and metastatic spreading in various tumor xenograft models treated with decorin (Tralhao et al., 2003; Reed et al., 2005; Seidler et al., 2006; Hu et al., 2009).

Recently it was shown that the soluble form of decorin gives rise to proinflammatory signaling, linking innate immunity, inflammation, and tumor development (Merline et al., 2011; Moreth et al., 2012; Schaefer and Iozzo, 2012). Decorin was identified as a novel endogenous ligand of Toll-like receptor 2 and -4 (TLR2/4). Binding of decorin to TLR2 and TLR4 in macrophages results in rapid activation of the mitogen-activated protein kinase p38, extracellular signal-regulated kinase (ERK), and nuclear factor kappa-light-chain-enhancer of activated B-cells (NF-κB) pathways, and enhances synthesis of the proinflammatory cytokines tumor necrosis factor (TNF)α and interleukin (IL)-12. Furthermore, by signaling through TLR2/4, decorin acts as a transcriptional inducer of the tumor suppressor Programmed Cell Death 4 (PDCD4), a unique regulator of both tumorigenesis and inflammation. By a reduction in mature microRNA (miR)-21, an oncogene and a posttranscriptional repressor of PDCD4, decorin further contributes to the enhancement of PDCD4 protein abundance. This occurs independently of TLR2/4 and is based on decorin-mediated inactivation of TGF-β1, which normally enhances the levels of precursor and mature miR-21. The subsequent increase in PDCD4, a specific translational suppressor of IL-10, finally results in lower levels of antiinflammatory IL-10. Thus, decorin creates a proinflammatory environment by the stimulation of proinflammatory PDCD4, TNFα, and IL-12 as well as by inhibition of immunosuppressive TGF-β1 and antiinflammatory IL-10. This along with enhancement of the tumor suppressive PDCD4 and a reduction in the oncogene miR-21 might represent an attractive approach for cancer therapy. In fact, in a model of tumor xenograft growth adenoviral-mediated overexpression of decorin induced TLR2/4-driven synthesis of PDCD4, TNFα, IL-12, and TGFβ1/miR-21-mediated inhibition of PDCD4 suppression.

In consequence, the immune reaction is shifted to a more apoptotic and inflammatory response with strong antitumorigenic effects resulting in a marked retardation of tumor growth (Merline et al., 2011).

Thus, taking into account the diverse signaling pathways modulated by decorin in a cell-type-specific manner, this secreted proteoglycan possesses considerable therapeutic potential for the treatment of a wide range of disorders, including cancer and fibrosis.

Biglycan

The lessons from biglycan-deficient mice, which develop age-dependent osteopenia due to a reduced response of bone marrow stromal cells to TGF-β, show that biglycan plays a crucial role in the network of secreted proteins regulating BMP signaling in the musculoskeletal ECM (Ameye and Young, 2002; ; Chen et al., 2002; Bi et al., 2007). Furthermore, biglycan interacts with WISP1, thereby controlling differentiation and proliferation of osteogenic cells (Inkson et al., 2009). Together with fibromodulin, biglycan regulates chondrogenesis and ECM turnover in joint osteoarthritis (Embree et al., 2010). Biglycan and decorin are important for skeletal muscle formation and development of skeletal muscular dystrophies (Brandan et al., 2008). Recently, biglycan has been proposed as a therapeutic molecule in experimental Duchenne muscular dystrophy, based on its ability to recruit utrophin (a homolog of dystrophin) to the sarcolemma (Amenta et al., 2011).

A number of studies indicate that biglycan plays a critical role in cardiovascular diseases (Heegaard et al., 2007; Bereczki and Santha, 2008; Westermann et al., 2008; Csont et al., 2010; Melchior-Becker et al., 2011), stressing the importance of biglycan in stabilizing the collagen network of the heart to preserve cardiac hemodynamic function. Furthermore, biglycan has been suggested to promote atherosclerosis due to retention of atherogenic lipids and promotion of inflammation within atherosclerotic lesions (Little et al., 2008; Derbali et al., 2010; Lam et al., 2011).

There is growing evidence that soluble biglycan acts as a danger signal (Schaefer et al., 2005). Under homeostatic conditions, biglycan is sequestered in the ECM and is therefore immunologically inactive. However, upon tissue stress or injury, resident cells (e.g. fibroblasts) release proteases that cleave the ECM resulting in the release of soluble biglycan, which then may act as a signaling molecule (▶Figure 3.7). Soluble biglycan turns into an endogenous ligand of the innate immunity receptors TLR2 and TLR4, leading to the activation of p38, ERK, and NF-κB in a myeloid differentiation primary response protein (MyD88)-dependent manner and resulting in the synthesis of proinflammatory cytokines (TNF-α and pro–interleukin [IL]-1β) and chemoattractants for macrophages, T cells, and B cells (macrophage inflammatory protein-1α [MIP-1α], monocyte chemoattractant protein-1 [MCP1], regulated upon activation, normal T-cell expressed, and secreted [RANTES], chemokine [C-X-C motif] ligand-13 [CXCL13]) (Schaefer et al., 2005; Moreth et al., 2010). In addition, macrophages stimulated with proinflammatory factors are able to synthesize biglycan themselves, thereby creating a feed-forward boost to the inflammatory reaction. By clustering TLR2/4 and purinergic receptor P2X, ligand-gated ion channel, 7 ($P2X_7$) purinergic receptors, biglycan induces the NLR family, pyrin domain containing 3 (NLRP3)/apoptosis-associated speck-like protein containing CARD (ASC) inflammasome in an reactive oxygen species (ROS) and heat shock protein 70 (HSP70)-dependent manner, leading to activation of caspase-1 and to the release of mature IL-1β

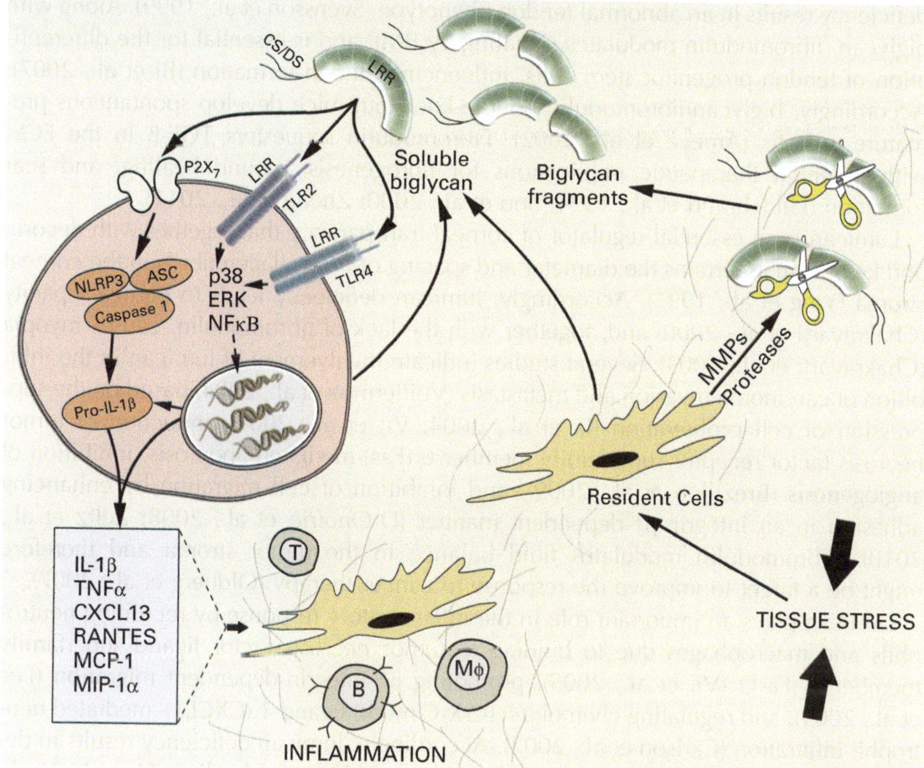

Figure 3.7 Soluble biglycan acts as a "danger" signal bridging innate and adaptive immunity.

(Babelova et al., 2009). In pathogen-mediated inflammation (lipopolysaccharide [LPS]- or zymosan-induced sepsis) biglycan potentiates the inflammatory response via a second TLR, which is not involved in pathogen sensing. In sterile inflammation (kidney and lungs), biglycan acts as an autonomous trigger of the inflammatory response (Schaefer, 2011). In fact, soluble biglycan is present in plasma during inflammation (e.g. in patients with systemic lupus erythematosus and in lupus-prone mice). Overexpression of biglycan in lupus mice triggers enhanced expression and synthesis of CXCL13 and B1 cell infiltration in the kidney, thereby enhancing albuminuria and renal damage. Conversely, deficiency of biglycan improves the renal outcome in lupus mice (Moreth et al., 2010). Thus, soluble biglycan induces cooperativity of innate immunity receptors, thereby bridging the innate and the adaptive immunity response.

3.5.4 Pathobiology of class II SLRPs

Fibromodulin and lumican

Fibromodulin and lumican are class II KS SLRPs involved in collagen assembly and are therefore important in the pathophysiology of the skeletal ECM (Roughley, 2006). Limited data are available regarding fibromodulin and lumican signaling. Fibromodulin deficiency results in an abnormal tendon phenotype (Svensson et al., 1999). Along with biglycan, fibromodulin modulates signaling by BMP and is essential for the differentiation of tendon progenitor stem cells, influencing tendon formation (Bi et al., 2007). Accordingly, biglycan/fibromodulin double-knockout mice develop spontaneous premature arthritis (Ameye et al., 2002). Fibromodulin sequesters TGF-β in the ECM with potential therapeutic implications for fibrogenesis, wound healing, and scar formation (Hildebrand et al., 1994; Soo et al., 2000; Zheng et al., 2011).

Lumican is an essential regulator of corneal transparency that together with decorin and keratocan ascertains the diameter and spacing of the collagen fibrils in the corneal stroma (Ying et al., 1997). Accordingly, lumican deficiency leads to corneal opacity (Chakravarti et al., 2000) and, together with the lack of fibromodulin, causes myopia (Chakravarti et al., 2003). Several studies indicate involvement of lumican in the inhibition of carcinoma invasion and metastasis (Vuillermoz et al., 2004) based on the suppression of cell proliferation (Li et al., 2004; Vij et al., 2004), induction of tumor necrosis factor receptor superfamily member 6 (Fas)-mediated apoptosis, inhibition of angiogenesis (Brezillon et al., 2009), and inhibition of cell migration by enhancing adhesion in an integrin-β1-dependent manner (D'Onofrio et al., 2008; Zeltz et al., 2010). Fibromodulin modulates fluid balance in the tumor stroma and therefore might be a target to improve the response to cancer therapy (Oldberg et al., 2007).

Lumican plays an important role in the inflammatory response by recruiting neutrophils and macrophages due to binding to tumor necrosis factor ligand superfamily member 6 (FasL) (Vij et al., 2005), promoting β2-integrin-dependent migration (Lee et al., 2009), and regulating chemokine (C-X-C motif) ligand-1 (CXCL1)–mediated neutrophil infiltration (Carlson et al., 2007). Accordingly, lumican deficiency results in decreased neutrophil recruitment and impaired corneal wound healing (Hayashi et al., 2010). In a model of colitis, *Lum*$^{-/-}$ mice have decreased levels of CXCL1, TNF-α, and less neutrophil infiltration and NF-κB activity, thereby impairing intestinal homeostasis. The latter is required to enhance the innate inflammatory response necessary in the early stages of this disease (Lohr et al., 2011). Moreover, lumican stimulates the TLR4 signaling pathway as the core protein of lumican binds LPS and presents it to cluster of differentiation 14 (CD14) (Wu et al., 2007).

There are some indications of the involvement of fibromodulin in inflammation and immunity. Fibromodulin binds complement C1QA complement component 1, q subcomponent (C1q) and activates the classical complement pathway (Sjoberg et al., 2009). In chronic lymphocytic leukemia, fibromodulin is overexpressed and serves as a tumor-associated antigen promoting the expansion of autologous tumor-specific T cells (Mayr et al., 2005). Enhanced fibromodulin expression was reported in chronic B-cell leukemia and in patients suffering from mantle-cell lymphoma (Mikaelsson et al., 2005). Further details have been discussed in recent reviews (Hassell and Birk, 2010; Iozzo and Schaefer, 2010; Iozzo and Sanderson, 2011).

3.5.5 Pathobiology of class III SLRPs

Epiphycan and opticin

Epiphycan was first isolated from bovine epiphyseal cartilage (Johnson et al., 1997). During cartilage development, epiphycan gene expression is seen at an earlier time point in the growth plate. In zebra fish, the gene was localized in the developing notochord and cartilage. The protein is found in the ECM surrounding the resting, proliferating, and hypertrophic chondrocytes, supporting a functional role for growth plate development and chondrogenesis (Johnson et al., 1999; Knudson and Knudson, 2001). Epiphycan-deficient mice develop osteoarthritis with age, indicating its role in the maintenance of joint integrity (Nuka et al., 2010).

Opticin was initially isolated from bovine vitreous collagen fibrils. Opticin mRNA expression has also been detected in ligaments, skin, and eye (Reardon et al., 2000). In the eye, several sequence variations in the opticin gene have been identified. Some of them appear to be a risk factor for developing high myopia (Majava et al., 2007), glaucoma (Bouhenni et al., 2011), proliferative retinopathies (Pattwell et al., 2010), retinal detachment, macular hole formation (Bishop, 2000; Bishop et al., 2002), and the inherited eye diseases age-related macular degeneration 1 (ARMD1) and posterior column ataxia with retinitis pigmentosa (AXPC1) (Ramesh et al., 2004). Lack of opticin is associated with neoplastic transformation as seen in adenoma and adenocarcinoma of the ciliary body epithelium (Assheton et al., 2007). Apart from the eye, recent reports indicate that opticin is also expressed in the synovial membrane and cartilage. In the osteoarthritic cartilage, opticin is preferentially degraded by MMP13 leading to progressive degeneration of the cartilage matrix (Monfort et al., 2008).

3.5.6 Take-home message

Two decades of investigations have provided strong evidence for the following: (1) SLRPs in their soluble form (not matrix bound) are capable of signaling. (2) Despite the structural similarity of SLRPs (e.g. decorin and biglycan), distinct members of the family have molecular and cell-specific affinities to different receptors, thereby exerting unique downstream signaling events. (3) The ability of a distinct SLRP to interact with various receptors appears to be determined by the cellular expression and density of a given receptor and by different affinity constants of decorin to each of these receptors, thereby permitting "hierarchical" receptor binding and activation (Iozzo and Schaefer, 2010). (4) Signaling triggered by the same ligand and receptor (e.g. decorin/IGF-IR) on different cells may result in distinct, cell-type-specific biological outcomes.

Taking into account the plethora of biological functions modulated by SLRPs and their abundance in tissues throughout the body, the roles these molecules play in the pathobiology of various diseases should be better appreciated. Despite the large amount of available data, the field is still in its infancy, and future research should explore the various receptors, adaptor molecules, signaling cascades, cross-talk possibilities, and autoregulatory mechanisms. A better understanding of these biological interactions mediated by SLRP family members is warranted and could lead to the development of novel therapeutic strategies for the treatment of malignant, inflammatory, and fibrotic diseases.

References

Amenta, A. R., Yilmaz, A., Bogdanovich, S., et al. (2011). Biglycan recruits utrophin to the sarcolemma and counters dystrophic pathology in mdx mice. Proc Natl Acad Sci U S A *108*, 762–767.

Ameye, L., Aria, D., Jepsen, K., Oldberg, A., Xu, T., and Young, M. F. (2002). Abnormal collagen fibrils in tendons of biglycan/fibromodulin-deficient mice lead to gait impairment, ectopic ossification, and osteoarthritis. FASEB J *16*, 673–680.

Ameye, L., and Young, M. F. (2002). Mice deficient in small leucine-rich proteoglycans: novel in vivo models for osteoporosis, osteoarthritis, Ehlers-Danlos syndrome, muscular dystrophy, and corneal diseases. Glycobiology *12*, 107R–116R.

Assheton, D. C., Guerin, E. P., Sheridan, C. M., Bishop, P. N., and Hiscott, P. S. (2007). Neoplastic transformation of ciliary body epithelium is associated with loss of opticin expression. Br J Ophthalmol *91*, 230–232.

Babelova, A., Moreth, K., Tsalastra-Greul, W., et al. (2009). Biglycan, a danger signal that activates the NLRP3 inflammasome via toll-like and P2X receptors. J Biol Chem *284*, 24035–24048.

Bereczki, E., and Santha, M. (2008). The role of biglycan in the heart. Connect Tissue Res *49*, 129–132.

Bi, Y., Ehirchiou, D., Kilts, T. M., et al. (2007). Identification of tendon stem/progenitor cells and the role of the extracellular matrix in their niche. Nat Med *13*, 1219–1227.

Bishop, P. N. (2000). Structural macromolecules and supramolecular organisation of the vitreous gel. Prog Retin Eye Res *19*, 323–344.

Bishop, P. N., Takanosu, M., Le Goff, M., and Mayne, R. (2002). The role of the posterior ciliary body in the biosynthesis of vitreous humour. Eye (Lond) *16*, 454–460.

Border, W. A., Noble, N. A., Yamamoto, T., et al. (1992). Natural inhibitor of transforming growth factor-beta protects against scarring in experimental kidney disease. Nature *360*, 361–364.

Bouhenni, R. A., Al Shahwan, S., Morales, J., et al. (2011). Identification of differentially expressed proteins in the aqueous humor of primary congenital glaucoma. Exp Eye Res *92*, 67–75.

Brandan, E., Cabello-Verrugio, C., and Vial, C. (2008). Novel regulatory mechanisms for the proteoglycans decorin and biglycan during muscle formation and muscular dystrophy. Matrix Biol *27*, 700–708.

Brezillon, S., Zeltz, C., Schneider, L., et al. (2009). Lumican inhibits B16F1 melanoma cell lung metastasis. J Physiol Pharmacol *60* (Suppl. 4), 15–22.

Buraschi, S., Pal, N., Tyler-Rubinstein, N., Owens, R. T., Neill, T., and Iozzo, R. V. (2010). Decorin antagonizes Met receptor activity and down-regulates {beta}-catenin and Myc levels. J Biol Chem *285*, 42075–42085.

Carlson, E. C., Lin, M., Liu, C. Y., Kao, W. W., Perez, V. L., and Pearlman, E. (2007). Keratocan and lumican regulate neutrophil infiltration and corneal clarity in lipopolysaccharide-induced keratitis by direct interaction with CXCL1. J Biol Chem *282*, 35502–35509.

Chakravarti, S., Paul, J., Roberts, L., Chervoneva, I., Oldberg, A., and Birk, D. E. (2003). Ocular and scleral alterations in gene-targeted lumican-fibromodulin double-null mice. Invest Ophthalmol Vis Sci *44*, 2422–2432.

Chakravarti, S., Petroll, W. M., Hassell, J. R., et al. (2000). Corneal opacity in lumican-null mice: defects in collagen fibril structure and packing in the posterior stroma. Invest Ophthalmol Vis Sci *41*, 3365–3373.

Chen, X. D., Shi, S., Xu, T., Robey, P. G., and Young, M. F. (2002). Age-related osteoporosis in biglycan-deficient mice is related to defects in bone marrow stromal cells. J Bone Miner Res *17*, 331–340.

Csont, T., Gorbe, A., Bereczki, E., et al. (2010). Biglycan protects cardiomyocytes against hypoxia/reoxygenation injury: role of nitric oxide. J Mol Cell Cardiol *48*, 649–652.

Csordas, G., Santra, M., Reed, C. C., et al. (2000). Sustained down-regulation of the epidermal growth factor receptor by decorin. A mechanism for controlling tumor growth in vivo. J Biol Chem *275*, 32879–32887.

Danielson, K. G., Baribault, H., Holmes, D. F., Graham, H., Kadler, K. E., and Iozzo, R. V. (1997). Targeted disruption of decorin leads to abnormal collagen fibril morphology and skin fragility. J Cell Biol *136*, 729–743.

De Luca, A., Santra, M., Baldi, A., Giordano, A., and Iozzo, R. V. (1996). Decorin-induced growth suppression is associated with up-regulation of p21, an inhibitor of cyclin-dependent kinases. J Biol Chem *271*, 18961–18965.

Derbali, H., Bosse, Y., Cote, N., et al. (2010). Increased biglycan in aortic valve stenosis leads to the overexpression of phospholipid transfer protein via Toll-like receptor 2. Am J Pathol *176*, 2638–2645.

D'Onofrio, M. F., Brezillon, S., Baranek, T., et al. (2008). Identification of beta1 integrin as mediator of melanoma cell adhesion to lumican. Biochem Biophys Res Commun *365*, 266–272.

Embree, M. C., Kilts, T. M., Ono, M., et al. (2010). Biglycan and fibromodulin have essential roles in regulating chondrogenesis and extracellular matrix turnover in temporomandibular joint osteoarthritis. Am J Pathol *176*, 812–826.

Goldoni, S., Humphries, A., Nystrom, A., et al. (2009). Decorin is a novel antagonistic ligand of the Met receptor. J Cell Biol *185*, 743–754.

Goldoni, S., and Iozzo, R. V. (2008). Tumor microenvironment: modulation by decorin and related molecules harboring leucine-rich tandem motifs. Int J Cancer *123*, 2473–2479.

Hassell, J. R., and Birk, D. E. (2010). The molecular basis of corneal transparency. Exp Eye Res *91*, 326–335.

Hayashi, Y., Call, M. K., Chikama, T., et al. (2010). Lumican is required for neutrophil extravasation following corneal injury and wound healing. J Cell Sci *123*, 2987–2995.

Heegaard, A. M., Corsi, A., Danielsen, C. C., et al. (2007). Biglycan deficiency causes spontaneous aortic dissection and rupture in mice. Circulation *115*, 2731–2738.

Hildebrand, A., Romaris, M., Rasmussen, L. M., et al. (1994). Interaction of the small interstitial proteoglycans biglycan, decorin and fibromodulin with transforming growth factor beta. Biochem J *302* (Pt 2), 527–534.

Hu, Y., Sun, H., Owens, R. T., et al. (2009). Decorin suppresses prostate tumor growth through inhibition of epidermal growth factor and androgen receptor pathways. Neoplasia *11*, 1042–1053.

Inkson, C. A., Ono, M., Bi, Y., Kuznetsov, S. A., Fisher, L. W., and Young, M. F. (2009). The potential functional interaction of biglycan and WISP-1 in controlling differentiation and proliferation of osteogenic cells. Cells Tissues Organs *189*, 153–157.

Iozzo, R. V., Moscatello, D. K., McQuillan, D. J., and Eichstetter, I. (1999). Decorin is a biological ligand for the epidermal growth factor receptor. J Biol Chem *274*, 4489–4492.

Iozzo, R. V., and Sanderson, R. D. (2011). Proteoglycans in cancer biology, tumour microenvironment and angiogenesis. J Cell Mol Med *15*, 1013–1031.

Iozzo, R. V., and Schaefer, L. (2010). Proteoglycans in health and disease: novel regulatory signaling mechanisms evoked by the small leucine-rich proteoglycans. FEBS J *277*, 3864–3875.

Johnson, H. J., Rosenberg, L., Choi, H. U., Garza, S., Hook, M., and Neame, P. J. (1997). Characterization of epiphycan, a small proteoglycan with a leucine-rich repeat core protein. J Biol Chem *272*, 18709–18717.

Johnson, J., Shinomura, T., Eberspaecher, H., Pinero, G., Decrombrugghe, B., and Hook, M. (1999). Expression and localization of PG-Lb/epiphycan during mouse development. Dev Dyn *216*, 499–510.

Knudson, C. B., and Knudson, W. (2001). Cartilage proteoglycans. Semin Cell Dev Biol *12*, 69–78.

Kolb, M., Margetts, P. J., Sime, P. J., and Gauldie, J. (2001). Proteoglycans decorin and biglycan differentially modulate TGF-beta-mediated fibrotic responses in the lung. Am J Physiol Lung Cell Mol Physiol *280*, L1327–1334.

Lam, V., Takechi, R., Pallebage-Gamarallage, M. M., Galloway, S., and Mamo, J. C. (2011). Colocalisation of plasma derived apo B lipoproteins with cerebral proteoglycans in a transgenic-amyloid model of Alzheimer's disease. Neurosci Lett *492*, 160–164.

Lee, S., Bowrin, K., Hamad, A. R., and Chakravarti, S. (2009). Extracellular matrix lumican deposited on the surface of neutrophils promotes migration by binding to beta2 integrin. J Biol Chem *284*, 23662–23669.

Li, Y., Aoki, T., Mori, Y., et al. (2004). Cleavage of lumican by membrane-type matrix metalloproteinase-1 abrogates this proteoglycan-mediated suppression of tumor cell colony formation in soft agar. Cancer Res *64*, 7058–7064.

Little, P. J., Osman, N., and O'Brien, K. D. (2008). Hyperelongated biglycan: the surreptitious initiator of atherosclerosis. Curr Opin Lipidol *19*, 448–454.

Lohr, K., Sardana, H., Lee, S., et al. (2011). Extracellular matrix protein lumican regulates inflammation in a mouse model of colitis. Inflamm Bowel Dis *18* (1), 143–151.

Majava, M., Bishop, P. N., Hagg, P., et al. (2007). Novel mutations in the small leucine-rich repeat protein/proteoglycan (SLRP) genes in high myopia. Hum Mutat *28*, 336–344.

Mayr, C., Bund, D., Schlee, M., et al. (2005). Fibromodulin as a novel tumor-associated antigen (TAA) in chronic lymphocytic leukemia (CLL), which allows expansion of specific CD8+ autologous T lymphocytes. Blood *105*, 1566–1573.

Melchior-Becker, A., Dai, G., Ding, Z., et al. (2011). Deficiency of biglycan causes cardiac fibroblasts to differentiate into a myofibroblast phenotype. J Biol Chem *286*, 17365–17375.

Merline, R., Lazaroski, S., Babelova, A., et al. (2009). Decorin deficiency in diabetic mice: aggravation of nephropathy due to overexpression of profibrotic factors, enhanced apoptosis and mononuclear cell infiltration. J Physiol Pharmacol *60* (Suppl. 4), 5–13.

Merline, R., Moreth, K., Beckmann, J., et al. (2011). Signaling by the matrix proteoglycan decorin controls inflammation and cancer through PDCD4 and MicroRNA-21. Sci Signal, *4*, ra75.

Mikaelsson, E., Danesh-Manesh, A. H., Luppert, A., et al. (2005). Fibromodulin, an extracellular matrix protein: characterization of its unique gene and protein expression in B-cell chronic lymphocytic leukemia and mantle cell lymphoma. Blood *105*, 4828–4835.

Monfort, J., Tardif, G., Roughley, P., et al. (2008). Identification of opticin, a member of the small leucine-rich repeat proteoglycan family, in human articular tissues: a novel target for MMP-13 in osteoarthritis. Osteoarthritis Cartilage *16*, 749–755.

Moreth, K., Brodbeck, R., Babelova, A., et al. (2010). The proteoglycan biglycan regulates expression of the B cell chemoattractant CXCL13 and aggravates murine lupus nephritis. J Clin Invest *120*, 4251–4272.

Moreth, K., Iozzo, R. V., and Schaefer, L. (2012). Small leucine-rich proteoglycans orchestrate receptor crosstalk during inflammation. Cell Cycle *11*, 2084–2091.

Moscatello, D. K., Santra, M., Mann, D. M., McQuillan, D. J., Wong, A. J., and Iozzo, R. V. (1998). Decorin suppresses tumor cell growth by activating the epidermal growth factor receptor. J Clin Invest *101*, 406–412.

Nuka, S., Zhou, W., Henry, S. P., et al. (2010). Phenotypic characterization of epiphycan-deficient and epiphycan/biglycan double-deficient mice. Osteoarthritis Cartilage *18*, 88–96.

Oldberg, A., Kalamajski, S., Salnikov, A. V., et al. (2007). Collagen-binding proteoglycan fibromodulin can determine stroma matrix structure and fluid balance in experimental carcinoma. Proc Natl Acad Sci U S A *104*, 13966–13971.

Pattwell, D. M., Sheridan, C. M., Le Goff, M., Bishop, P. N., and Hiscott, P. (2010). Localisation of opticin in human proliferative retinal disease. Exp Eye Res *90*, 461–464.

Ramesh, S., Bonshek, R. E., and Bishop, P. N. (2004). Immunolocalisation of opticin in the human eye. Br J Ophthalmol *88*, 697–702.

Reardon, A. J., Le Goff, M., Briggs, M. D., et al. (2000). Identification in vitreous and molecular cloning of opticin, a novel member of the family of leucine-rich repeat proteins of the extracellular matrix. J Biol Chem *275*, 2123–2129.

Reed, C. C., and Iozzo, R. V. (2002). The role of decorin in collagen fibrillogenesis and skin homeostasis. Glycoconj J *19*, 249–255.

Reed, C. C., Waterhouse, A., Kirby, S., et al. (2005). Decorin prevents metastatic spreading of breast cancer. Oncogene *24*, 1104–1110.

Roughley, P. J. (2006). The structure and function of cartilage proteoglycans. Eur Cell Mater *12*, 92–101.

Santra, M., Reed, C. C., and Iozzo, R. V. (2002). Decorin binds to a narrow region of the epidermal growth factor (EGF) receptor, partially overlapping but distinct from the EGF-binding epitope. J Biol Chem *277*, 35671–35681.

Schaefer, L. (2011). Small leucine-rich proteoglycans in kidney disease. J Am Soc Nephrol *22*, 1200–1207.

Schaefer, L., Babelova, A., Kiss, E., et al. (2005). The matrix component biglycan is proinflammatory and signals through Toll-like receptors 4 and 2 in macrophages. J Clin Invest *115*, 2223–2233.

Schaefer, L., and Iozzo, R. V. (2008). Biological functions of the small leucine-rich proteoglycans: from genetics to signal transduction. J Biol Chem *283*, 21305–21309.

Schaefer, L., and Iozzo, R. V. (2012). Small leucine-rich proteoglycans, at the crossroad of cancer growth and inflammation. Curr Opin Genet Dev *22*, 56–57.

Schaefer, L., Macakova, K., Raslik, I., et al. (2002). Absence of decorin adversely influences tubulointerstitial fibrosis of the obstructed kidney by enhanced apoptosis and increased inflammatory reaction. Am J Pathol *160*, 1181–1191.

Schaefer, L., Tsalastra, W., Babelova, A., et al. (2007). Decorin-mediated regulation of fibrillin-1 in the kidney involves the insulin-like growth factor-I receptor and mammalian target of rapamycin. Am J Pathol *170*, 301–315.

Schonherr, E., Sunderkotter, C., Iozzo, R. V., and Schaefer, L. (2005). Decorin, a novel player in the insulin-like growth factor system. J Biol Chem *280*, 15767–15772.

Seidler, D. G., Goldoni, S., Agnew, C., et al. (2006). Decorin protein core inhibits in vivo cancer growth and metabolism by hindering epidermal growth factor receptor function and triggering apoptosis via caspase-3 activation. J Biol Chem *281*, 26408–26418.

Sjoberg, A. P., Manderson, G. A., Morgelin, M., Day, A. J., Heinegard, D., and Blom, A. M. (2009). Short leucine-rich glycoproteins of the extracellular matrix display diverse patterns of complement interaction and activation. Mol Immunol *46*, 830–839.

Soo, C., Hu, F. Y., Zhang, X., et al. (2000). Differential expression of fibromodulin, a transforming growth factor-beta modulator, in fetal skin development and scarless repair. Am J Pathol *157*, 423–433.

Svensson, L., Aszodi, A., Reinholt, F. P., Fassler, R., Heinegard, D., and Oldberg, A. (1999). Fibromodulin-null mice have abnormal collagen fibrils, tissue organization, and altered lumican deposition in tendon. J Biol Chem *274*, 9636–9647.

Tralhao, J. G., Schaefer, L., Micegova, M., et al. (2003). In vivo selective and distant killing of cancer cells using adenovirus-mediated decorin gene transfer. FASEB J *17*, 464–466.

Vial, C., Gutierrez, J., Santander, C., Cabrera, D., and Brandan, E. (2011). Decorin interacts with connective tissue growth factor (CTGF)/CCN2 by LRR12 inhibiting its biological activity. J Biol Chem *286*, 24242–24252.

Vij, N., Roberts, L., Joyce, S., and Chakravarti, S. (2004). Lumican suppresses cell proliferation and aids Fas-Fas ligand mediated apoptosis: implications in the cornea. Exp Eye Res *78*, 957–971.

Vij, N., Roberts, L., Joyce, S., and Chakravarti, S. (2005). Lumican regulates corneal inflammatory responses by modulating Fas-Fas ligand signaling. Invest Ophthalmol Vis Sci *46*, 88–95.

Vuillermoz, B., Khoruzhenko, A., D'Onofrio, M. F., et al. (2004). The small leucine-rich proteoglycan lumican inhibits melanoma progression. Exp Cell Res *296*, 294–306.

Westermann, D., Mersmann, J., Melchior, A., et al. (2008). Biglycan is required for adaptive remodeling after myocardial infarction. Circulation *117*, 1269–1276.

Wu, F., Vij, N., Roberts, L., Lopez-Briones, S., Joyce, S., and Chakravarti, S. (2007). A novel role of the lumican core protein in bacterial lipopolysaccharide-induced innate immune response. J Biol Chem *282*, 26409–26417.

Yamaguchi, Y., Mann, D. M., and Ruoslahti, E. (1990). Negative regulation of transforming growth factor-beta by the proteoglycan decorin. Nature *346*, 281–284.

Ying, S., Shiraishi, A., Kao, C. W., et al. (1997). Characterization and expression of the mouse lumican gene. J Biol Chem *272*, 30306–30313.

Zeltz, C., Brezillon, S., Kapyla, J., et al. (2010). Lumican inhibits cell migration through alpha2beta1 integrin. Exp Cell Res *316*, 2922–2931.

Zheng, Z., Nguyen, C., Zhang, X., et al. (2011). Delayed wound closure in fibromodulin-deficient mice is associated with increased TGF-beta3 signaling. J Invest Dermatol *131*, 769–778.

Zhu, J. X., Goldoni, S., Bix, G., et al. (2005). Decorin evokes protracted internalization and degradation of the epidermal growth factor receptor via caveolar endocytosis. J Biol Chem *280*, 32468–32479.

3.6 Structure and function of syndecans

Csilla Pataki and John R. Couchman

3.6.1 Syndecan stucture

Syndecans are a family of type 1 transmembrane glycoproteins. They are cell surface heparan sulfate proteoglycans (HSPG) with a long evolutionary history, being expressed in Cnidaria and throughout the Bilateria (Chakravarti and Adams, 2006). Invertebrates and primitive chordates possess a single syndecan gene, while four syndecan family members (1–4) occur in higher vertebrates (Couchman, 2003, 2010). Most cell types, with the exception of erythrocytes, express at least one syndecan family member, and a few may even express all four. Syndecan-1 (CD138) is common to many epithelia, condensing mesenchyme in development, and to some leucocytes, such as plasma cells. Syndecan-2 (fibroglycan) is present predominantly in mesenchymal cells, fibroblasts and developing neural tissue. Syndecan-3 (N-syndecan) is enriched in neural tissue but is also present in the developing musculoskeletal system. In contrast, syndecan-4 (amphiglycan, ryudocan) is widely expressed in many cell types (Couchman, 2010; Xian et al., 2010). The most dramatic changes in syndecan expression occur during development and are associated with morphological transitions, cell differentiation, or changes in tissue organization (Yoneda and Couchman, 2003). Marked changes can also occur in a range of diseases (Götte, 2003; Tkachenko et al., 2005; Alexopoulou et al., 2007; and this volume).

Syndecans have three major structural domains: ectodomain with an N-terminal signal peptide and several glycosaminoglycan (GAG) attachment sites, a single transmembrane domain, and a short C-terminal cytoplasmic domain (Couchman, 2003) (▶Figure 3.8A). The ectodomains of syndecans display low sequence homology, whereas their transmembrane and cytoplasmic domains are highly conserved (Yoneda and Couchman, 2003) (▶Figure 3.8B).

Originally, it was believed that the 20–40 kDa core syndecan protein ectodomains had the sole function of being substituted by GAG chains. While there is not much sequence conservation between syndecans and even between, for example, syndecan-4 from different species, it is now known that there is a functional activity in the extracellular core protein (Rapraeger, 2001; Kramer and Yost, 2003; Beauvais et al., 2009; Xian et al., 2010). Heparan sulfate (HS) is the principal GAG present in all four syndecans, although syndecan-1 and -3 may also be substituted with chondroitin sulfate (CS) or dermatan sulfate (DS) (Zhang et al., 1995) (▶Figure 3.8A). The HS chains are usually close to the N terminus and attach to the serine-glycine (Ser-Gly) attachment sites on the core protein of which there are at least three in each syndecan (Multhaupt et al., 2009). Syndecan-3 has a relatively extended threonine-serine-proline (Thr-Ser-Pro)–rich area in the center of the extracellular domain, whose function is unknown but

A

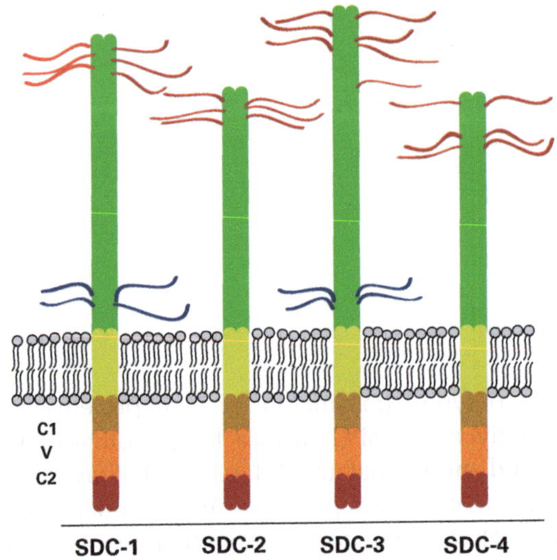

B

```
RMKKKDEGSYSLEEPKQANGGA-YQK-PTKQEEFYA
RMKKKDEGSYTLEEPKQA-SVT-YQK-PDKQEEFYA
RMRKKDEGSYDLGERK-PSSAA-YQKAPTK--EFYA
RMKKKDEGSYDLG-KK-PI----YKKAPT--NEFYA
     C1              V            C2
```

Figure 3.8 The syndecan family of transmembrane heparan sulphate proteoglycans. (A) The four syndecans are membrane spanning full-time proteoglycans with a short cytoplasmic domain, a transmembrane domain and a longer ectodomain. Toward the N termini the ectodomain is substituted with heparan sulphate chains (all syndecans in red) and sometimes membrane proximal chondroitin sulphate or dermatan sulphate chains (syndecan-1 and -3 in blue). The number and position of GAG chains varies between family members. (B) Cytoplasmic domain sequence alignment of the human syndecan family members.

may be rich in *O*-linked oligosaccharides. Syndecan-2 and -4 core proteins are smaller and belong to one subfamily, while syndecan-1 and -3 are members of a second subfamily (Choi et al., 2010), based on sequence homology.

HS is one of the most complex of all carbohydrates, synthesized in a Golgi apparatus–localized biosynthetic pathway requiring multiple enzymes (Bishop et al., 2007) and described elsewhere in this volume. These chains are a significant feature of syndecans and may dominate their physicochemical properties, interactions and functions. In contrast to the ectodomain, the transmembrane region of the syndecans is highly conserved. In its outer part, there is a GXXXG motif that promotes the formation of homodimers that are substantially SDS (sodium dodecyl sulfate) resistant in solution (Choi et al., 2005; Dews and Mackenzie, 2007). No evidence for naturally occurring heterodimers exists currently, perhaps because of differences in syndecan extracellular domains and the varying length of their cytoplasmic domains (Couchman, 2010). The small cytoplasmic domain of each syndecan has two highly conserved regions, C1 and C2, proximal and distal to the membrane, flanking a variable (V) region. The V region is unique to each syndecan but shows across-species sequence conservation (Couchman, 2003). Avian syndecan-2, for example, has a V-region sequence almost identical to that of mammals. Invertebrate V regions are longer, while the shortest is that of syndecan-4. Even where there are changes in the sequences, the structure remains conserved (Whiteford et al., 2008). Syndecan-4 V region binds to the membrane lipid phosphatidylinositol 4, 5-bisphosphate (PtdIns4,5P_2) and induces a conformational change revealed by nuclear magnetic resonance (NMR) spectroscopy (Shin et al., 2001).

3.6.2 Function of syndecans

One way to define the role of different molecules during development is through genetic ablation in different model organisms. Mice in which syndecan-1, -3, or -4 have been deleted all develop normally, are fertile, and show no apparent pathologies. This suggests redundancy or that individual syndecans have no critical roles during development (Ishiguro et al., 2000, 2001; Echtemeyer et al., 2001; Kaksonen et al., 2002; Stepp et al., 2002; Bellin et al., 2003). However, closer investigation shows that syndecan-1–null mice have epithelial cell migration defects (Alexander et al., 2000; Stepp et al., 2002). Syndecan-3 knockout mice exhibit impaired neural migration in development and partial resistence to obesity (Kaksonen et al., 2002; Bellin et al., 2003; Hienola et al., 2006). Syndecan-3 is also a component of the skeletal muscle satellite cell niche, where it plays a role in satellite cell maintenance, proliferation, and differentiation (Fuentealba et al., 1999). Syndecan-3–null muscles regenerate after injury but exhibit aberrant phenotypes (Cornelison et al., 2004). It has been shown recently that syndecan-3 cooperates with Notch signaling in regulating homeostasis of satellite cell populations and myofiber size (Pisconti et al., 2010).

The syndecan-4–null mouse has vascular defects in the fetal labyrinth and capillaries of the kidney (Ishiguro et al., 2000) and has a poor angiogenic response in postnatal wound healing (Echtermeyer et al., 2001). At present, there is no syndecan-2 knockout mouse; however, it has been shown that *Xenopus* syndecan-2 is important in left-right symmetry in embryonic development (e.g. the heart looping) and is also required for angiogenic sprouting in zebra fish (Chen et al., 2002). The use of morpholinos in zebra fish and *Xenopus* has also demonstrated a role for syndecan-4 proteoglycan in neural crest migration and convergent extension movements (Munoz et al., 2006; Matthews et al., 2008). Therefore, in lower vertebrates, syndecans may have important roles not apparent in mammals.

The *Drosophila* syndecan is encoded by a single gene and has a structure analogous to its vertebrate homologues with important roles in development. In the late-stage-16 embryo, syndecan is expressed in the lymph glands, the peripheral and central nervous system, and the basal surfaces of the gut epithelium (Bellin et al., 2003). Adult *Drosophila* syndecan mutants show axon guidance and visual-system defects suggesting aberrant cell migration (Rawson et al., 2005). Moreover, it has been suggested that the *Caenorhabditis elegans* syndecan is also involved in axon guidance and cell migration (Rhiner et al., 2005; Schwabiuk et al., 2009). An appropriate level of syndecan expression is also required for normal cell adhesion and morphogenesis of ascidian embryos (Satou et al., 1999) and in cell adhesion, migration, and division during postoral arm formation in the sea urchin embryo (Tomita et al., 2002).

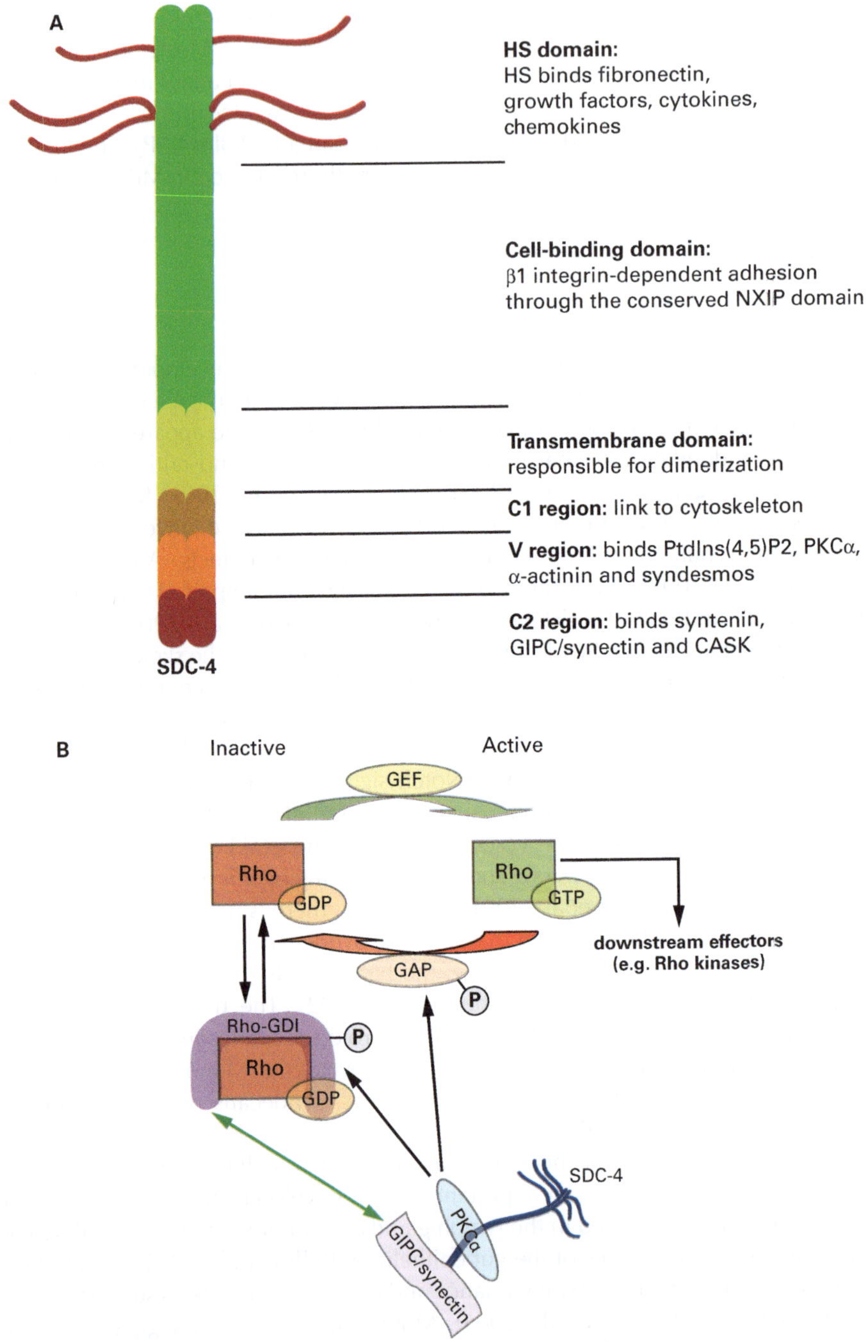

Figure 3.9 (A) Syndecan-4 protein domain structure and major domain-specific interactions. The HS chains are known to interact with matrix collagens and glycoproteins, growth factors, cytokines, chemokines and morphogens while the membrane proximal part of ectodomain mediates the β1 integrin-dependent cell adhesion through an NXIP motif. There is apparently

3.6.3 Syndecan domains and their roles

The syndecan ectodomains

Syndecans localize to the plasma membrane and likely function as cell surface receptors in cell adhesion, migration, cytoskeleton organization, and cell differentiation. It has been shown that a wide array of extracellular molecules interact with the HS chains, including growth factors, morphogens, chemokines, lipid-regulating enzymes, and collagen and glycoprotein components of the extracellular matrix (ECM), molecules responsible for cell-cell adhesion and blood coagulation (▶Figure 3.9A). Syndecans are often described as coreceptors since they can collaborate with other receptors like the vertebrate fibroblast growth factor receptors, the Wnt signaling Frizzled receptors (Luyten et al., 2008), Hedgehog family members (Shimo et al., 2004), and transforming growth factor-beta (TGF-β) receptors (Chen and Meng, 2004). However, these functions may not be syndecan specific, because glypicans lacking cytoplasmic domains have similar roles. The interactions of syndecan ectodomain and GAG chains with extracellular ligands and the consequences of these interactions are discussed in detail elsewhere in this volume.

The transmembrane domain and syndecan dimerization

The transmembrane domain of syndecans consists of 25 hydrophobic amino acid residues and is sufficient to induce dimerization/oligomerization of core proteins (Choi et al., 2005; Dews and Mackenzie, 2007). This oligomerization appears crucial for syndecans to transduce signals from the extracellular matrix to cytosol and in localizing syndecans to distinct membrane compartments (Rapraeger, 2001) (▶Figure 3.9A). All syndecans containing the transmembrane domain formed SDS-resistant oligomers, but not those lacking it. In chimeric syndecans where the transmembrane domain has been replaced with that of the platelet-derived growth factor receptor, SDS-resistant oligomerization was largely absent and was accompanied by loss of signaling function (Choi et al., 2005). Therefore, the syndecan homodimer may be an essential functional unit.

Figure 3.9 (***Continued***)

no direct syndecan-4/integrin interaction, while direct interactions have been shown for syndecan-1. The transmembrane domain is responsible for dimerization. On the cytoplasmic side, the C1 region links the molecule to the actin cytoskeleton, the V region binds PtdIns(4,5)P2, and interacts with PKCα, α-actinin and syndesmos. The last four amino acids of the C2 region (EFYA) interact with PDZ domain containing proteins for example syntenin, GIPC/synectin and CASK. (B) The Rho-GTPase activation cycle and syndecan-4. The Rho-GDP dissociation inhibitors (Rho-GDIs) sequester the inactive, GDP-bound form of Rho-GTPases (in this case Rho) in the cytoplasm. Following release by Rho-GDI, Rho-GDP is targeted to the plasma membrane, where its activation cycle is regulated by guanine nucleotide exchange factors (GEFs) that promote GTP loading and activation of Rho. Inactivation of Rho-GTP is mediated by GTPase-activating proteins (GAPs) that promote GTP hydrolysis to GDP. PKCα directly or indirectly phosphorylates Rho-GAP and Rho-GDI upon syndecan-4 dimerization, leading to focal adhesion formation. GIPC/synectin interacts both with syndecan-4 and Rho-GDI (green arrow).

Functions of the cytoplasmic domain

The cytoplasmic domains of syndecans have received much attention, and although lacking enzymatic activity, they are now known to undertake multiple interactions. Despite their small size, experiments have shown interactions with a variety of proteins, including kinases. All four syndecans are able to interact with the intracellular actin cytoskeleton (Yoneda and Couchman, 2003; Okina et al., 2009) via their conserved C1 and C2 or the highly variable V region.

The membrane-proximal C1 region

The C1 region contains a cationic sequence, common to many transmembrane molecules, and it has been shown that the syndecan-2 C1 region interacts with ezrin, one of a group of ezrin, radixin, moesin (ERM) proteins that regulate the submembranous actin cytoskeleton (Granes et al., 2003) (▶Figure 3.9A). The corresponding region of syndecan-3 was reported to bind c-Src (cellular-Src), and its substrate cortactin, in response to the heparin-binding ligand pleiotrophin (also known as heparin-binding growth-associated molecule -HB-GAM) (Kinnunen et al., 1998). Presumably, similar interactions can take place in the case of other vertebrate or invertebrate syndecans because this region is highly conserved.

The isoform-specific V region

The V region is unique for each syndecan but is highly conserved across species. Only with syndecan-4 is there substantial information regarding the V-region interactions. NMR spectroscopy identified the 12-amino-acid sequence within the V region as forming twisted clamp dimers (Ser183-Lys194 in rat syndecan-4). Syndecan-4 binds to PtdIns4,5P_2 via a KKXXXKK motif in this region, which promotes the formation and stabilization of syndecan-4 dimers (Oh et al., 1998; Whiteford et al., 2008) and provides a platform for binding of protein kinase C-alpha (PKC-α) leading to the formation of a ternary complex where the kinase is persistently activated. This interaction is a major basis of syndecan-4 signaling (Lim et al., 2003; Dovas et al., 2006; ▶Figure 3.9A and B).

Events downstream of PKC-α are becoming clearer. In combination with integrins such as α5β1, syndecan-4 promotes cell adhesion, in particular the assembly of actin stress fibers terminating in focal adhesions (Dovas et al., 2006; Bass et al., 2007). Several reports suggest that the small guanosine triphosphatases (GTPases) Rac and RhoA are downstream of PKC-α, suggesting that PKC-α directly or indirectly affects regulators such as guanine nucleotide exchange factors (GEFs), GTPase activating proteins (GAPs), or guanine dissociation inhibitors (GDIs) (Morgan et al., 2007; Couchman, 2010). For example, serine-34 of RhoGDI is a target for PKC-α leading specifically to RhoA activation (Dovas et al., 2010) (▶Figure 3.9B). Other reports suggest that RhoGDI when phosphorylated on serine-96, in endothelial cells, releases RhoA or RhoG, a close relative of Rac (Knezevic et al., 2007; Elfenbein et al., 2009). Syndecan-4, PKC-α, and RhoA may be in a linear pathway leading to focal adhesion formation (Saoncella et al., 1999; Dovas et al., 2010). However, a connection between syndecan-4 and Rac has also been demonstrated because the absence of syndecan-4 also leads to impaired migration and altered migrational persistence characterized by deregulated GTP-Rac levels (Bass et al., 2007). This might explain the defects in granulation tissue angiogenesis in wounded syndecan-4–null mice (Echtermeyer et al., 2001). Cell-matrix adhesion experiments suggest that integrin ligation leads to transient downregulation of GTP-RhoA levels through Src-mediated p190RhoGAP phosphorylation, whereas syndecan-4 targets p190RhoGAP to specific membrane domains (Bass et al., 2008). Syndecan-4 signaling to PKC-α is evolutionarily conserved because neural crest migration in *Xenopus* embryos requires syndecan-4 and an fibroblast growth factor (FGF)/extracellular signal-regulated kinase (ERK) pathway but also PKC-α, which inhibits RacGTPase by an unknown mechanism (Munoz et al., 2006; Kuriyama and Mayor, 2009).

High Rac activity at the leading edge of migrating cells is accompanied by a RhoA-driven contractile activity at the rear end in the same cell (Petrie et al., 2009). This polarized response links syndecan-4 to planar polarity. It has been demonstrated that syndecan-4 plays a role in convergent extension movements during *Xenopus* embryo development (Munoz et al., 2006), and in both zebra fish and *Xenopus* embryos, neural crest migration, but not induction, requires syndecan-4 that inhibits Rac activity (Matthews et al., 2008).

In addition to PKC-α, syndesmos (a 40 kDa protein) binds to the C1 and V regions of syndecan-4 (Denhez et al., 2002). Its function remains unknown besides an interaction with the focal adhesion component paxillin and the related Hic5. Hic5 is involved in nuclear RNA decapping, a property not shared with syndesmos. However, syndesmos has RNA-binding activity, but whether this is a functional attribute *in vivo* is unclear (Multhaupt et al., 2009).

Finally, syndecan-4 directly interacts with α-actinin, another cytoskeletal protein. This occurs via the V region of syndecan-4 (Greene et al., 2003; Choi et al., 2008). Because α-actinin has other binding partners, including the focal adhesion components vinculin and zyxin, this may provide a link from syndecan-4 to the focal adhesion and cytoskeleton.

Binding partners of the V region of all syndecans except syndecan-4 have been difficult to come by, perhaps in part because a dimeric structure is required. In *Xenopus*, syndecan-2 regulates heart looping, and in this case, the role of protein kinase C-gamma (PKC-γ) was shown. However, it is not clear whether this kinase associates with the syndecan-2 V region (Kramer et al., 2002). Recent evidence suggests that the syndecan-2 V region directly binds to Sarm1 protein, a negative regulator of Toll-like receptor-3 (TLR3) controlling dendritic arborization in mice (Chen et al., 2011). No interacting partners of the syndecan-1 and -3 V regions have been identified, and similarly, information is lacking from invertebrates.

The PDZ protein-binding C2 region

The C terminus of all four vertebrate syndecans is a highly conserved hydrophobic motif terminating at EFYA (glutamic acid-phenylalanine-tyrosine-alanine) or very similar residues. Through this motif, syndecans bind PDZ-domain-containing proteins that form submembranous complexes and have a role in regulating targeting and trafficking of cell surface molecules (Hung and Sheng, 2002). The PDZ motif is named after post-synaptic density protein (PSD-95), *Drosophila* discs large tumor suppressor (Dlg1) and zonula occludens-1 (zo-1) proteins. Several hundred mammalian proteins are predicted to contain PDZ motifs (Couchman, 2010). One of them is syntenin (also called *mda-9*, melanoma differentiation-associated gene-9) (Beekman and Coffer, 2008), which is known to promote tumor cell migration. Syntenin contains two PDZ domains; PDZ1 has higher affinity for PtdIns4,5P_2, while PDZ2 preferentially binds syndecans (Zimmermann et al., 2002). The EFYA motif of syndecans fits into the PDZ2 pocket and has been visualized by NMR spectroscopy and crystallography (Grembecka et al., 2006). However, many cell surface molecules are able to bind syntenin. Therefore, syntenin may have a scaffolding function at the cell membrane (Beekman and Coffer, 2008). Its lipid-binding function is essential for Arf6 (ADP ribosylation factor 6)-dependent trafficking of syntenin/syndecan complexes to the membrane (Zimmermann et al., 2005). Syntenin may play a scaffolding role in *Xenopus* noncanonical Wnt signaling by binding the Wnt receptor Frizzled 7 at its PDZ1 domain and syndecan-4 or -2 at its PDZ2 domain (Luyten et al., 2008).

Syndecan-2 and -4 interact with GIPC (Gα-interacting protein [GAIP]–interacting protein C terminus), also known as semcap-1 (semaphorin F-binding protein) and synectin (Tkachenko et al., 2005). Like syntenin, GIPC/synectin also interacts with other receptors, including neuropilin-1 (Cai and Reed, 1999) that promotes α5β1-integrin-mediated cell adhesion. When overexpressed in endothelial cells, GIPC/synectin suppresses cell migration in a syndecan-4-dependent manner (Gao et al., 2000). However, GIPC/synectin has also been implicated in syndecan removal from the cell surface by interacting with myosin VI and endocytic vesicles (Naccache et al., 2006).

Synbindin (also known as trs23) interacts with the syndecan-2 C-terminal EFYA motif and appears to be involved in postsynaptic vesicle trafficking and dendritic spine maturation in neurons (Ethell et al., 2001). This suggests that these two molecules act together in recruiting vesicles to the postsynaptic membrane. Synbindin is a component of the transport protein particle-1 (TRAPP1), involved in transport from the endoplasmic reticulum to the Golgi apparatus (Fan et al., 2009). However, recent structural studies suggest that due to its atypical PDZ domain, synbindin may not interact directly with the syndecan C terminus (Fan et al., 2009).

Calcium/calmodulin-associated serine/threonine kinase (CASK) is a membrane-associated guanylate kinase (MAGUK) that in rat brain was identified as a protein that binds all syndecans and neurexin (Hsueh, 2006). Its C-terminal guanylate kinase domain is a pseudokinase involved in targeting to the nucleus of neurons where it interacts with the transcription factor TBR1 (T-box, brain, 1) (Wang et al., 2004). Human mutations of CASK relate this protein to X-linked brain malformation (Najm et al., 2008). The balance of CASK distribution seems to be regulated by its binding partners (Multhaupt et al., 2009). Overexpression of syndecan-3 leads to a predominantly cytoplasmic distribution, while the increased Thr1 concentrates CASK in the nucleus (Ojeh et al., 2008). It has been shown that the syndecan-3 cytoplasmic domain (after cleavage by persenilin/γ-secretase) is released to the cytoplasm, which leads to a reduction in membrane targeting of CASK (Schulz et al., 2003). Therefore, a dual role of CASK at the cell surface and as a nuclear protein is apparent, the former perhaps related to interactions with syndecans (Multhaupt et al., 2009) (▶Figure 3.9A).

3.6.4 Take-home message

Syndecans are an ancient family of cell surface HSPGs. Recent advances in syndecan research implicate these molecules in many signaling functions that translate information between extra- and intracellular environments, leading to regulation of cell adhesion and migration. Syndecans have multiple extracellular and intracellular binding partners and may signal independently or in conjunction with other receptors. Syndecans presumably share a set of generic functions mediated by the conserved C1 and C2 cytoplasmic domains, but the individual functions of the intracellular variable region and the ectodomain need further investigation because knowledge is fragmentary. Further *in vitro* and *in vivo* genetic studies could provide significant information about the yet uncharacterized roles of invertebrate and vertebrate syndecans.

References

Alexander, C. M., Reichsman, F., Hinkes, M. T., et al. (2000). Syndecan-1 is required for Wnt-1 induced mammary tumourigenesis in mice. Nat Genet *25*, 329–332.

Alexopoulou, A. N., Multhaupt, H. A. B., and Couchman, J. R. (2007). Syndecans in wound healing, inflammation and vascular biology. Int J Biochem Cell Biol *39*, 5055–5528.

Bass, M. D., Morgan, M. R., Roach, K. A., Settleman, J., Goryachev, A. B., and Humphries, M. J. (2008). p190RhoGAP is the convergence point of adhesion signals from α5β1 integrin and syndecan-4. J Cell Biol *181*, 1013–1026.

Bass, M. D., Roach, K. A., Morgan, M. R., et al. (2007). Syndecan-4-dependent Rac1 regulation determines directional migration in response to the extracellular matrix. J Cell Biol *177*, 527–538.

Beauvais, D. M., Ell, B. J., McWhorter, A. R., and Rapraeger, A. C. (2009). Syndecan-1 regulates alphavbeta3 and alphavbeta5 integrin activation during angiogenesis and is blocked by synstatin a novel peptide inhibitor. J Exp Med *206*, 691–705.

Beekman, J. M., and Coffer, P. J. (2008). The ins and outs of syntenin, a multifunctional intracellular adaptor protein. J Cell Sci *121*, 1349–1355.

Bellin, R., Capila, I., Lincecum, J., Park, P. W., Reizes, O., and Bernfield, M. R. (2003). Unlocking the secrets of syndecans: transgenic organisms as a potential key. Glycoconj J *19*, 293–304.

Bishop, J. R., Schuksz, M., and Esco, J. D. (2007). Heparan sulphate proteoglycans fine-tune mammalian physiology. Nature *446*, 1030–1037.

Cai, H., and Reed, R. R. (1999). Cloning and characterization of neuropilin-1-interacting protein: a PSD-95/Dlg/ZO-1 domain-containing protein that interacts with the cytoplasmic domain of neuropilin-1. J Neurosci *19*, 6519–6527.

Chakravarti, R., and Adams, J. C. (2006). Comparative genomics of the syndecans defines an ancestral genomic context associated with matrilins in vertebrates. BMC Genomics *7*, 1–16.

Chen, C. Y., Lin C. W., Chang, C. Y., Jiang, S. T., and Hsueh, Y. P. (2011). Sarm1, a negative regulator of innate immunity, interacts with syndecan-2 and regulates neuronal morphology. J Cell Biol *193*, 769–784.

Chen, E., Hermanson, S., and Ekker, S. C. (2002). Syndecan-2 is essential for angiogenic sprouting during zebrafish development. Blood *103*, 1710–1719.

Chen, Y. G., and Meng, A. M. (2004). Negative regulation of TGF-beta signaling in development. Cell Res 14, 441–449.

Choi, S., Lee, E., Kwon, S., et al. (2005). Transmembrane domain-induced oligomerization is crucial for the functions of syndecan-2 and syndecan-4. J Biol Chem *280*, 42573–42579.

Choi, Y., Chung, H., Jung, H., Couchman, J. R., and Oh, E. S. (2010). Syndecans as cell surface receptors: unique structure equates with functional diversity. Matrix Biol *30*, 93–99.

Choi, Y., Kim, S., Lee, J., Han, I. O., Woods, A., and Oh, E. S. (2008). The oligomeric status of syndecan-4 regulates syndecan-4 interaction with alpha actinin. Eur J Cell Biol *87*, 807–815.

Cornelison, D. D., Wilcox-Adelman, S. A., Geotinck, P. F., Rauvala, H., Rapraeger, A. C., and Olwin, B. B. (2004). Essential and separable roles for Syndecan-3 and Syndecan-4 in skeletal muscle development and regeneration. Genes Dev *18*, 2231–2236.

Couchman, J. R. (2003). Syndecans: proteoglycan regulators of cell surface microdomains? Nat Rev Mol Cell Biol *4*, 926–937.

Couchman, J. R. (2010). Transmembrane signalling proteoglycans. Annu Rev Cell Dev Biol *26*, 89–114.

Denhez, F., Wilcox-Adelman, S. A., Baciu, P. C., et al. (2002). Syndesmos, a syndecan-4 cytoplasmic domain interactor, binds to the focal adhesion adaptor proteins paxillin and Hic-5. J Biol Chem *277*, 12270–12277.

Dews, I. C., and Mackenzie, K. (2007). Transmembrane domains of the syndecan family of growth factor coreceptors display a hierarchy of homotypic and heterotypic interactions. Proc Natl Acad Sci U S A *104*, 20782–20787.

Dovas, A., Choi, Y., Yoneda, A., et al. (2010). Serine 34 phosphorylation of Rho guanine dissociation inhibitor (RhoGDIα) links signalling from conventional protein kinase C to RhoGTPase in cell adhesion. J Biol Chem *285*, 23296–23308.

Dovas, A., Yoneda, A., and Couchman, J. R. (2006). PKC alpha-dependent activation of RhoA by syndecan-4 during focal adhesion formation. J Cell Sci *119*, 2837–2846.

Echtermeyer, F., Streit, M., Wilcox-Adelman, S., et al. (2001). Delayed wound repair and impaired angiogenesis in mice lacking syndecan-4. J Clin Invest *107*, R9–R14.

Elfenbein, A., Rhodes, J. M., Meller, J., Schwartz, M. A., Matsuda, M., and Simons, M. (2009). Suppression of RhoG activity is mediated by a syndecan-4-synectin-RhoGDI1 complex and is reversed by PKCα in a Rac1 activation pathway. J Cell Biol *186*, 75–83.

Ethell, I. M., Irie, F., Kalo, M. S.,Couchman, J. R., Pasquale, E. B., and Yamaguchi, Y. (2001). EphB/syndecan-2 signalling in dendritic spine morphogenesis. Neuron *31*, 1001–1013.

Fan, S., Wei, Z., Xu, H., and Gong, W. (2009). Crystal structure of human synbindin reveals two conformations of longin domain. Biochem Biophys Res Commun *378*, 338–343.

Fuentealba, L., Carey, D. J., and Brandan, E. (1999). Antisense inhibition of syndecan-3 expression during skeletal muscle differentiation accelerates myogenesis through a basic fibroblast growth factor-dependent mechanism. J Biol Chem *274*, 37876–37884.

Gao, Y., Li, M., Chen, W., and Simons, M. (2000). Synectin, syndecan-4 cytoplasmic domain binding PDZ protein, inhibits cell migration. J Cell Physiol *184*, 373–379.

Götte, M. (2003). Syndecans in inflammation. FASEB J *17*, 575–591.

Granes, F., Berndt, C., Roy, C., Mangeat, P., Reina, M., and Vilaro, S. (2003). Identification of a novel ezrin-binding site in syndecan-2 cytoplasmic domain. FEBS Lett *547*, 212–216.

Greene, D. K., Tumova, S., Couchman, J. R., and Woods, A. (2003). Syndecan-4 associates with alpha-actinin. J Biol Chem *278*, 7617–7623.

Grembecka, J.,Cierpicki, T., Devedjiev, Y., et al. (2006). The binding of the PDZ tandem of syntenin to target proteins. Biochemistry *45*, 3674–3683.

Hienola, A., Tumova, S., Kulesskiy, E., and Rauvala, H. (2006). N-syndecan deficiency impairs neural migration in brain. J Cell Biol *174*, 569–580.

Hsueh, Y. P. (2006). The role of the MAGUK protein CASK in neural development and synaptic function. Curr Med Chem *13*, 1915–1927.

Hung, A. Y., and Sheng, M. (2002). PDZ domains: structural modules for protein complex assembly. J Biol Chem *277*, 5699–5702.

Ishiguro, K., Kadamatsu, K., Kojima, K., et al. (2001). Syndecan-4 deficiency leads to high mortality of lipopolysaccharide-injected mice. J Biol Chem *276*, 47483–47488.

Ishiguro, K., Kadamatsu, K., Kojima, K., et al. (2000). Syndecan-4 deficiency impairs focal adhesion formation only under restricted conditions. J Biol Chem *275*, 5249–5252.

Kaksonen, M., Pavlov, I., Voikar, V., et al. (2002). Syndecan-3-deficient mice exhibit enhanced LTP and impaired hippocampus-dependent memory. Mol Cell Neurosci *21*, 158–172.

Kinnunen, T., Kaksonen, M., Saarinen, J., Kalkkinen, N., Peng, H. B., and Rauvala, H. (1998). Cortactin-Src kinase signalling pathway is involved in N-syndecan-dependent neurite outgrowth. J Biol Chem *273*, 10702–10708.

Knezevic, N., Roy, A., Timblin, B., et al. (2007). RhoGDI-1 phosphorylation switch at serine 96 induces RhoA activation and increased endothelial permeability. Mol Cell Biol *27*, 6323–6333.

Kramer, K. L., Barnette, J. E., and Yost, H. J. (2002). PKCγ regulates syndecan-2 inside-out signalling during Xenopus left-right development. Cell *111*, 981–990.

Kramer, K. L., and Yost, H. J. (2003). Heparan sulphate core proteins in cell-cell signalling. Annu Rev Genet *37*, 461–484.

Kuriyama, S., and Mayor, R. (2009). A role for syndecan-4 in neural induction involving ERK- and PKC-dependent pathways. Development *136*, 575–584.

Lim, S. T., Longley, R. L., Couchman, J. R., and Woods, A. (2003). Direct binding of syndecan-4 cytoplasmic domain to the catalytic domain of protein kinase C alpha (PKCα) increases focal adhesion localization of PKCα. J Biol Chem *278*, 13795–13802.

Luyten, A., Mortier, E., Van Campenhout, C., et al. (2008). The postsynaptic density 95/disc-large/zona occludens protein syntenin directly interacts with frizzled 7 and supports non-canonical Wnt signalling. Mol Biol Cell *19*, 1594–1604.

Matthews, H. K., Marchant, L., Carmona-Fontaine, C., et al. (2008). Directional migration of neural crest cells in vivo is regulated by syndecan-4/Rac1 and noncanonical Wnt signalling/RhoA. Development *135*, 1771–1780.

Morgan, M. R., Humphries, M. J., and Bass, M. D. (2007). Synergistic control of cell adhesion by integrins and syndecans. Nat Rev Mol Cell Biol *8*, 957–969.

Multhaupt, H. A. B., Yoneda, A., Whiteford, J., Oh, E. S., Lee, W., and Couchman, J. R. (2009). Syndecan signalling: when, where and why? J Physiol Pharmacol *60*, 31–38.

Munoz, R., Moreno, M., Oliva, C., Orbenes, C., and Larrain, J. (2006). Syndecan-4 regulates non-canonical Wnt signalling and is essential for convergent and extension movements in Xenopus embryos. Nat Cell Biol *8*, 492–500.

Naccache, S. N., Hasson, T., and Horowitz, A. (2006). Binding of internalized receptors to the PDZ domain of GIPC/synectin recruits myosin VI to endocytic vesicles. Proc Natl Acad Sci U S A *103*, 12735–12740.

Najm, J., Horn, D., Wimplinger, I., et al. (2008). Mutations of CASK cause an X-linked brain malformation phenotype with microcephaly and hypoplasia of the brainstem and cerebellum. Nat Genet *40*, 1065–1067.

Oh, E. S., Woods, A., Lim, S. T., Theibert, A., and Couchman, J. R. (1998). Syndecan-4 proteoglycan cytoplasmic domain and phosphatidylinositol 4,5-bisphosphate co-ordinately regulate protein kinase C activity. J Biol Chem *273*, 10624–10629.

Ojeh, N., Pekovic, V., Jahoda, C., and Maatta, A. (2008). The MAGUK family protein CASK is targeted to nuclei of the basal epidermis and controls keratinocyte proliferation. J Cell Sci *121*, 2705–2717.

Okina, E., Manon-Jensen, T., Whiteford, J., and Couchman, J. R. (2009). Syndecan proteoglycan contributions to cytoskeletal organization and contractility. Scand J Med Sci Sports *19*, 479–489.

Petrie, R. J., Doyle, A. D., and Yamada, K. M. (2009). Random versus directionally persistent cell migration. Nat Rev Mol Cell Biol *8*, 538–549.

Pisconti, A., Cornelison, D. D. W., Olguin, H. C., Antwine, L. T., and Olwin, B. B. (2010). Syndecan-3 and Notch cooperate in regulating adult myogenesis. J Cell Biol *190*, 427–441.

Rapraeger, A. C. (2001). Molecular interactions of syndecans during development. Cell Dev Biol *12*, 107–116.

Rawson, J. M., Dimitroff, B., Johnson, K. G., et al. (2005). The heparan sulphate proteoglycans Dally-like and syndecan have distinct functions in axon guidance and visual-system assembly in Drosophila. Curr Biol *15*, 833–838.

Rhiner, C., Gysi, S., Fröhli, E., Hengartner, M. O., and Hajnal, A. (2005). Syndecan regulates cell migration and axon guidance in C. elegans. Development *132*, 4621–4633.

Saoncella, S., Echtermeyer, F., Denhez, F., et al. (1999). Syndecan-4 signals cooperatively with integrins in a Rho-dependent manner in the assembly of focal adhesions and actin stress fibers. Proc Natl Acad Sci U S A *96*, 2805–2810.

Satou, Y., Chiba, S., and Satoh, N. (1999). Expression cloning of an ascidian syndecan suggests its role in embryonic cell adhesion and morphogenesis. Dev Biol *211*, 198–207.

Schulz, J. G., Annaert, W., Vandekerckhove, J., Zimmermann, P., De Strooper, B., and David, G. (2003). Syndecan 3 intramembrane proteolysis is presenilin/γsecretase-dependent and modulates cytosolic signalling. J Biol Chem *278*, 48651–48657.

Schwabiuk, M., Coudiere, L., and Merz, D. C. (2009). SDN-1/syndecan regulates growth factor signalling in distal tip cell migration in C. elegans. Dev Biol *334*, 235–242.

Shimo, T., Gentili, C., Iwamoto, M., Wu, C., Koyama, E., and Pacifici, M. (2004). Indian hedgehog and syndecan-3 coregulate chondrocyte proliferation and function during chick limb skeletogenesis. Dev Dyn *229*, 607–617.

Shin, J., Lee, W., Lee, D., et al. (2001). Solution structure of the dimeric cytoplasmic domain of syndecan-4. Biochemistry *40*, 8471–8478.

Stepp, M. A., Gibson, H. E., Gala, P. H., et al. (2002). Defects in keratinocyte activation during wound healing in the syndecan-1-deficient mouse. J Cell Sci *115*, 4517–4531.

Tkachenko, E., Rhodes, J. M., and Simons, M. (2005). Syndecans: new kids on the signalling block. Circ Res *96*, 488–500.

Tomita, K., Yamasu, K., and Suyemitsu, T. (2002). Role of syndecan in the elongation of postoral arms in sea urchin larvae. Dev Growth Differ *44*, 45–53.

Wang, T. F., Ding, C. N., Wang, G. S., et al. (2004). Identification of Tbr-1/CASK complex target genes in neurons. J Neurochem *91*, 1483–1492.

Whiteford, J. R., Ko, S., Lee, W., and Couchman, J. R. (2008). Structural and cell adhesion properties of zebrafish syndecan-4 are shared with higher vertebrates. J Biol Chem *283*, 29322–29330.

Xian, X., Gopal, S., and Couchman, J. R. (2010). Syndecans as receptors and organisers of the extracellular matrix. Cell Tissue Res *339*, 331–346.

Yoneda, A., and Couchman, J. R. (2003). Regulation of cytoskeletal organization by syndecan transmembrane proteoglycans. Matrix Biol *22*, 25–33.

Zhang, L., David, G., and Esko, J. D. (1995). Repetitive Ser-Gly sequences enhance heparan sulfate assembly in proteoglycans. J Biol Chem 270, 27127–12135.

Zimmermann, P., Meerschaert, K., Reekmans, G., et al. (2002). PIP(2)-PDZ domain binding controls the association of syntenin with the plasma membrane. Mol Cell *9*, 1215–1225.

Zimmermann, P., Zhang, Z., Degeest, G., et al. (2005). Syndecan recycling is controlled by syntenin-PIP_2 interaction and Arf6. Cell *9*, 377–388.

3.7 The glypican family

Jorge Filmus and Mariana Capurro

3.7.1 The structure of glypicans

The name *glypican* identifies a family of extracellular proteoglycans that are attached to the cell surface by a glycosylphosphatidylinositol anchor. Six members of the glypican family have been identified in mammals (glypican-1 [GPC1] to glypican-6 [GPC6]), and proteins with high homology to mammalian glypicans have been found in lower organisms, including two in *Drosophila* (Dally and Dlp) (Filmus et al., 2008). Structural features that are conserved across the family include the localization of 14 cysteine residues and of the insertion sites for the glycosaminoglycan (GAG) chains. All these insertion sites are close to the C terminus, placing the GAG chains in proximity to the cell surface and suggesting that these chains could mediate the interaction of glypicans with other cell surface proteins (Song and Filmus, 2002). The size of the core proteins is also highly conserved across the family.

The crystal structure of GPC1 lacking the C-terminal heparan sulfate attachment domain has been recently reported (Svensson et al., 2012). The structure indicates that GPC1 is a densely packed one-domain protein of cylindrical shape. The crystallized protein consists of 14 α-helices and three major loops. The overall structure of glypicans seems to be conserved across the family, since the structure of GPC1 is similar to that of Dlp (Kim et al., 2011). Interestingly, glypicans do not display domains with obvious homology to characterized domains found in other proteins, suggesting that glypicans have highly unique functions. The existence of a domain that extends from residue 200 to residue 300, which has weak homology to the cysteine-rich domain (CRD) of Frizzled proteins, has been reported (Topczewsky et al., 2001). However, the functional relevance of this finding has not been established.

Most glypicans are cleaved by furin-like endoproteases (Watanabe et al., 1995; De Cat et al., 2003). The cleavage site is located at the carboxy-terminal end of the presumptive CRD domain and generates two subunits that remain attached to each other by one or more disulfide linkages (De Cat et al., 2003). It should be noted, however, that this cleavage does not seem to be required for all glypican functions (Capurro et al., 2005b).

Glypicans display several GAG insertion sites (from two in glypican-3 [GPC3] to five in glypican-5 [GPC5]). The functional implications of this variation in the number of GAG chains are still unknown. Most glypicans have been shown to carry heparan sulfate (HS) chains. However, GPC5 produced in transfected cells also displays chondroitin sulfate (CS) chains (Saunders et al., 1997). Moreover, CS chains have also been found in endogenous GPC5 expressed by rhabdomyosarcoma (RMS) cells (Li et al., 2011). However, because GPC5 is overexpressed in these cells, it remains to be seen whether GPC5 produced in normal cells also displays CS chains. Interestingly, it has recently been reported that the HS chains of GPC3 and GPC5 produced in mouse embryo fibroblasts display a different degree of sulfation (Li et al., 2011). Because the binding specificity of HS chains is, at least in part, determined by the sulfation levels, it is highly likely that this difference in the degree of sulfation of the HS chains of GPC3 and GPC5 will have an impact on their functions.

3.7.2 The functions of glypicans

Regulation of growth factor/morphogen activity

A large number of genetic and biochemical studies have demonstrated that glypicans regulate growth factor/morphogen activity at the level of ligand-receptor interaction. This regulatory activity is based on the ability of glypicans to stimulate or inhibit the binding of the ligand to the receptor. Several signaling pathways have been shown to be regulated by glypicans, including those triggered by Hedgehogs (Hhs) (Capurro et al., 2008; Williams et al., 2010), Wnts (Capurro et al., 2005a; Yan et al., 2009), fibroblast growth factors (FGFs)(Gutierrez and Brandan, 2010), and bone morphogenetic proteins (BMPs)(Jackson et al., 1997; Akiyama et al., 2008; Dejima et al., 2011). Accumulated evidence suggests that the specific function of a particular glypican depends on the structural features of that glypican and on which growth factors and growth factor receptors are expressed by a specific cellular system.

Several studies in the *Drosophila* wing imaginal disk have demonstrated that, in addition to their ability to regulate cell autonomous growth factor/morphogen activity, glypicans also play a role in the secretion and transport of these peptides from the secreting cells to the receiving cells (Fujise et al., 2003; Belenkaya et al., 2004; Kreuger et al., 2004; Franch-Marro et al., 2005). This function of glypicans seems to be based on at least two mechanisms. One is the release of glypicans from the cell surface by the lipase Notum (Ayers et al., 2010). The second mechanism consists of apical-basolateral trancytosis (Gallet et al., 2008).

Hedgehog signaling

Glypicans have been shown to have both stimulatory and inhibitory functions in Hh signaling. For example, GPC5 promotes Hh signaling in RMS cells by increasing the binding of Hh to its signaling receptor, Patched (Li et al., 2011). Consistent with this function, it has been shown that GPC5 binds to Hh and Patched. Both interactions are mediated by the GAG chains, but the protein core is required for GPC5 activity because exogenous heparin does not stimulate Hh signaling (Li et al., 2011). It is well established that the interaction of Hh with Patched occurs at the primary cilia. As expected, therefore, GPC5 has also been shown to localize in the cilia (Li et al., 2011) (▶Figure 3.10).

Unlike GPC5, GPC3 displays Hh-inhibitory activity (Capurro et al., 2008, 2009). This activity is based on the ability of GPC3 to bind to Hh but not to Patched, and, consequently, to compete with Patched for Hh binding. Upon binding to Hh, the GPC3-Hh complex is endocytosed and degraded (Capurro et al., 2008). The endocytocis of this complex is mediated by the low-density-lipoprotein receptor-related protein-1 (Capurro et al., in press). Contrary to GPC5, GPC3 is located outside of the cilia, and

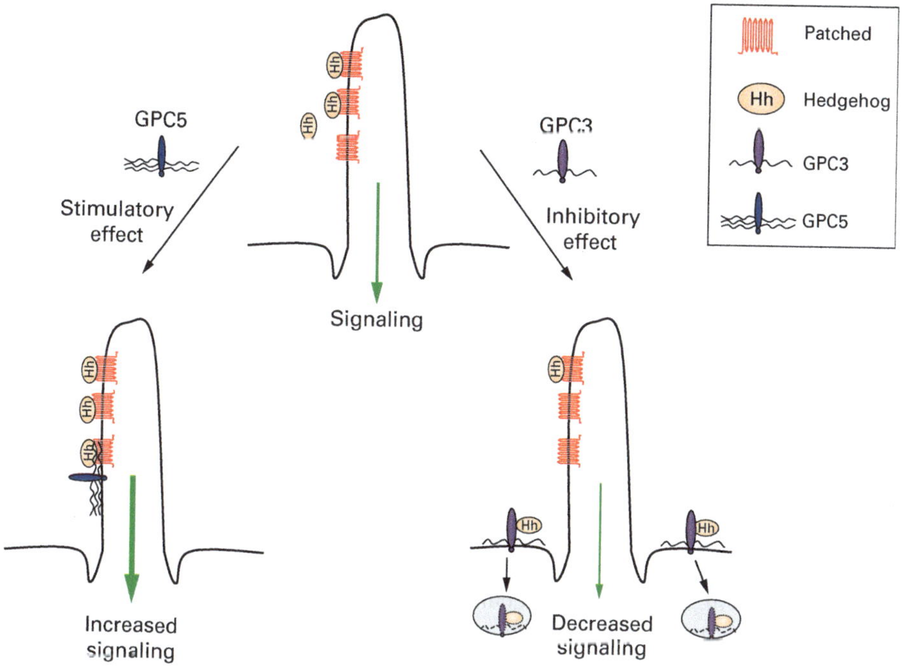

Figure 3.10 Opposite activity of GPC5 and GPC3 in Hedgehog (Hh) signaling. This diagram depicts the molecular basis for the opposite activity of GPC5 and GPC3 in Hh signaling. The interaction between Hh and its receptor Patched occurs at the cilium. Both GPC3 and GPC5 interact with Hh. Because GPC3 is located outside the cilium, it competes with Patched for Hh binding. The interaction of GPC3 with Hh triggers the endocytosis and degradation of the GPC3/Hh complex and inhibits signaling. GPC5 is located at the cilium, and it also interacts with Patched. Thus, this glypican stimulates the binding of Hh to Patched and increases Hh signaling.

the high-affinity binding of GPC3 to Hh is mediated by its core protein (Li et al., 2011) (▶Figure 3.10). Consistent with the Hh-inhibitory activity of GPC3, mouse embryos that lack this glypican display higher levels of Hh activity than their normal littermates and significant developmental overgrowth (Capurro et al., 2008, 2009). A cell autonomous Hh-regulatory activity has also been shown for Dlp in *Drosophila* (Lum et al., 2003). Like GPC5 in mammalian cells, Dlp stimulates Hh signaling (Lum et al., 2003). However, the molecular basis of this Dlp-induced stimulation is still controversial. One study reported that Dlp binds to both Hh and Patched (Yan et al., 2010). However, another study, which only investigated the interaction with Hh, could not detect any significant binding between Dlp and Hh (Williams et al., 2010).

A role for glypicans in the transport of Hh through the *Drosophila* wing imaginal disk has been extensively documented (Desbordes and Sanson, 2003; Ayers et al., 2010; Yan et al., 2010). In particular, the cleavage of Dally by Notum has been shown to play a critical role in apical long-range signaling (Ayers et al., 2010). In addition, glypicans play a role in large-scale Hh oligomerization in the secreting cells, which seems to be essential for long-range signaling (Vyas et al., 2008). Interestingly, Eugster et al.

(2007) have shown that Dally can bind to lipophorin, the *Drosophila* lipoprotein particle that plays a role in long-range Hh signaling. It has been proposed that Dally might increase Hh-signaling activity in lipophorin-targeted cells (Eugster et al., 2007).

Wnt signaling

Numerous studies in mammals and other organisms have demonstrated that glypicans stimulate canonical and noncanonical Wnt signaling (Topczewsky et al., 2001; Gallet et al., 2008; Shiba et al., 2010). For example, GPC3-null mice display reduced noncanonical Wnt signaling (Song et al., 2005), and loss of glypican-4 (GPC4) in *Xenopus* and zebra fish induces defective convergent extension movements during gastrulation as a result of a reduction in noncanonical Wnt signaling (Topczewsky et al., 2001; Ohkawara et al., 2003). On the other hand, a role of glypicans in canonical Wnt signaling has been shown in the context of *Drosophila* embryos (Lin and Perrimon, 1999; Tsuda et al., 1999; Franch-Marro et al., 2005) and in hepatocellular carcinoma cells (Capurro et al., 2005a). In the case of GPC4, it was reported that this glypican can form complexes with Wnt and its signaling receptor, Frizzled (Ohkawara et al., 2003). Based on this finding, it has been proposed that GPC4 stimulates Wnt signaling by presenting the ligand to the receptor. In other words, GPC4 would act as a nonsignaling coreceptor (Ohkawara et al., 2003). This mechanism is also consistent with the observation that high overexpression of GPC4 inhibits noncanonical Wnt signaling (Ohkawara et al., 2003). Presumably, if in a given cell there is more GPC4 than Frizzled, the excess of GPC4 could act as an inhibitor of signaling by competing with Frizzled for Wnt binding. Recent results by Yan et al. (2009) suggest a similar mechanism of action for Dlp in *Drosophila* cells. These authors showed that at low Dlp/Frizzled ratios, Dlp stimulates canonical Wnt signaling, whereas at high Dlp/Frizzled ratios, the glypican has an inhibitory effect (Yan et al., 2009). Certainly, glypicans could stimulate Wnt signaling by other mechanisms. For example, it has been recently proposed that GPC4 is required for the localization of Frizzled-7 on the cell membrane in *Xenopus* embryos (Shao et al., 2009).

As in the case of Hh, it was reported that Wnt can interact with the core protein of glypicans (Capurro et al., 2005a; Yan et al., 2009). However, we still don't know whether this is a high-affinity interaction. It should also be noted that it has been reported that the interaction between GPC1 and Wnt8 is mediated by the HS chains (Ai et al., 2003).

Dlp has been implicated in the transport of Wnt in the *Drosophila* wing imaginal disk (Franch-Marro et al., 2005; Gallet et al., 2008). The mechanism remains controversial. One report has proposed that Wnt spreading along the basolateral compartment requires Dlp-mediated apicobasal trancytosis (Gallet et al., 2008). Others have proposed that cleavage of Dlp by Notum is required for proper Wnt transport (Kreuger et al., 2004), but evidence against this model has been recently reported (Yan et al., 2009). Both *Drosophila* glypicans also play a role in Wnt cell autonomous signaling reception (Franch-Marro et al., 2005; Yan et al., 2009).

BMP signaling

Genetic evidence that glypicans can regulate BMP signaling has been provided by the study of *Gpc3*-mutant mice (Paine-Saunders et al., 2000; Hartwig et al., 2005). For example, a significant downregulation of BMP signaling was observed in the kidneys of *Gpc3*$^{+/-}$ mice, and the analysis of the abnormalities of *Gpc3*$^{+/-}$/*Bmp2*$^{+/-}$ double-mutant mice demonstrated a very strong genetic interaction (Hartwig et al., 2005). Similarly, a significant genetic interaction was observed in the analysis of *Gpc3*$^{-}$/*Bmp4*$^{+/-}$ double-mutant mice. These mice displayed polydactyly and rib malformations with high penetrance. This was not observed in each single-mutant mouse (Paine-Saunders et al., 2000).

A cell autonomous function for Dally in BMP signaling during the development of the *Drosophila* wing has been demonstrated in several genetic studies (Jackson et al., 1997; Fujise et al., 2003; Akiyama et al., 2008). Furthermore, one of these studies provided evidence suggesting that Dally stimulates BMP signaling by stabilizing Dpp (the *Drosophila* BMP) on the cell surface and inhibiting its endocytosis and degradation (Akiyama et al., 2008). In addition, it has been recently reported that Dally is required for Dpp-dependent maintenance of the germline stem-cell niche (Guo and Wang, 2009; Hayashi et al., 2009). Additional *in vitro* studies suggest that in this niche Dally regulates Dpp in *trans*, by stabilizing Dpp on the surface of the cells receiving the Dpp signal (Dejima et al., 2011). The HS chains of Dally were shown to be essential for its activity, but Dally cannot be replaced by heparin (Dejima et al., 2011).

Both Dally and Dlp have been implicated in the extracellular transport of Dpp across the *Drosophila* wing (Fujise et al., 2003; Belenkaya et al., 2004). Genetic studies showed that Dpp fails to move across cell clones that lack Dally and Dlp (Belenkaya et al., 2004). Based on these results, it was proposed that glypicans are involved in the movement of Dpp by restricted extracellular diffusion (Belenkaya et al., 2004).

FGF signaling

It is very well established that various families of heparan sulfate proteoglycans (HSPGs) can play a stimulatory role in FGF signaling (Bishop et al., 2007). Thus, results of genetic mammalian studies on the role of glypicans in FGF signaling could be difficult to interpret due to potential compensatory effects by various kinds of HSPGs. However, there are some genetic studies that strongly suggest that glypicans regulate FGF signaling in specific tissues.

GPC1-null mice display a highly significant reduction in brain size (Jen et al., 2009). Study of various signaling pathways in the mutant mice suggests that this GPC1-null phenotype is most likely due to a reduction in FGF signaling. Furthermore, analysis of compound mutants strongly suggests that GPC1 regulates brain size through fibroblast growth factor-17 (FGF17) (Jen et al., 2009).

In addition to embryonic overgrowth, GPC3-null mice show various developmental abnormalities, including cardiac malformations (Ng et al., 2009). Analysis of various markers of cardiac development in the GPC3-null mice suggests that fibroblast growth factor-9 (FGF9) signaling is deficient in the hearts of these mice (Ng et al., 2009).

Glypicans have also been shown to play a role in FGF signaling by a genetic study in *Drosophila*. Yan and Lin (2007) demonstrated that Dlp is required in FGF-receiving cells for FGF signaling during tracheal morphogenesis. Interestingly, this study shows that Dally, syndecan, or perlecan cannot rescue the Dlp phenotype, suggesting that the core protein of Dlp plays a critical role in this signaling system (Yan and Lin, 2007).

Other functions

Studies in *Droshopila* have demonstrated that Dlp is required for axon guidance (Rawson et al., 2005; Smart et al., 2011). Based on results from genetic interaction experiments and *in vitro* studies, it has been proposed that Dlp facilitates the interaction between the secreted growth cone repellent Slit and its receptor, Robo (Zhang et al., 2004; Smart et al., 2011). Analyses of Dlp mutants have also shown that Dlp is required for proper visual-system assembly, but the molecular interactions that mediate this Dlp function have not been elucidated yet (Rawson et al., 2005).

In addition to playing a role in axon guidance, Dlp has been shown to regulate synaptic morphogenesis (Johnson et al., 2006). Specifically, genetic studies have demonstrated that Dlp contributes to active zone stabilization, and that this contribution is based on its ability to inhibit the receptor phosphatase leukocyte antigen-related (LAR). Binding studies have shown that Dlp displays high-affinity binding to LAR (Johnson et al., 2006).

3.7.3 Pathobiology of glypicans

GPC3 and hepatocellular carcinoma

In 1997, Hsu et al. (1997) reported that GPC3 mRNA levels were significantly elevated in most hepatocellular carcinomas (HCCs) compared with normal liver and nonmalignant liver lesions. Later on, a study by Capurro et al. (2003) showed that GPC3 is also upregulated in most HCCs at the protein level.These authors detected GPC3 in tissue sections of 72% of HCC patients, whereas no GPC3 could be detected in normal liver or benign liver disease (Capurro et al., 2003). These results have been confirmed by many studies (Patil et al., 2005; Libbrecht et al., 2006; Luo et al., 2006; Di Tommaso et al., 2009). Consequently, the American Association for the Study of Liver Diseases has recently recommended clinical pathologists to immunostain for GPC3, together with heat shock protein 70 (HSP70) and glutamine synthase, for confirmation of HCC diagnosis (Bruix and Sherman, 2011). GPC3 has also been found in the blood of a large proportion of HCC patients (Capurro et al., 2003; Nakatsura et al., 2003; Hippo et al., 2004). Because normal individuals do not have detectable GPC3 levels in their blood, it has been suggested that a blood GPC3 test could be used for the early detection of HCC (Capurro et al., 2003). However, additional studies with a large number of patients will be required to verify this hypothesis.

In addition to being a diagnostic marker for HCC, GPC3 promotes the growth of this malignancy by stimulating canonical Wnt signaling (Capurro et al., 2005a). Significantly, a soluble GPC3 inhibits the *in vitro* and *in vivo* growth of HCC cells (Zittermann et al., 2010; Feng et al., 2011). This inhibition correlates with a reduction in canonical Wnt signaling, providing additional support to the hypothesis that membrane-bound GPC3 facilitates the interaction of Wnt with Frizzled (Zittermann et al., 2010).

Based on the fact that GPC3 is expressed by most HCCs but not by normal liver or benign liver lesions, several groups have investigated whether this glypican could be targeted by immunotherapy (Komori et al., 2006; Ho and Kim, 2010). Ishiguro et al. (2008), for example, have shown that an anti-GPC3 antibody induces antibody-dependent cellular cytotoxicity against GPC3-positive HCC xenografts. This antibody is currently being tested in phase 1 clinical trials (Zhu et al., 2011).

Other glypicans and cancer

It has been recently reported that pancreatic tumors subcutaneously injected in GPC1-null mice display significantly less angiogenesis than those injected in normal mice. Consistent with this observation, the metastatic growth of pancreatic cancer cells was dramatically inhibited in mice lacking GPC1 (Aikawa et al., 2008). This result suggests that stromal GPC1 plays a critical role in pancreatic tumor progression. Significantly, it has been observed that GPC1 levels are elevated in fibroblasts around human pancreatic tumors (Kleeff et al., 1998).

GPC1 has also been shown to play a role in human gliomas (Su et al., 2006). Immunohistochemical analysis revealed that GPC1 is overexpressed in human astrocytomas and oligodendrogliomas compared with nonneoplastic gliosis. Furthermore, ectopic GPC1 enhances the response of glioma cells to fibroblast growth factor-2 (FGF2) (Su et al., 2006).

As described above, another glypican involved in cancer progression is GPC5, which is overexpressed in RMSs (Li et al., 2011). Li et al. have recently reported that GPC5 stimulates RMS cell proliferation by activating Hh signaling. RNAi-induced downregulation of GPC5 in RMS cell lines significantly reduces their cell proliferation rate (Williamson et al., 2007; Li et al., 2011).

GPC3 and the Simpson-Golabi-Behmel syndrome

Loss-of-function mutations in the *GPC3* gene cause the Simpson-Golabi-Behmel syndrome (SGBS) (Pilia et al., 1996). This is an X-linked disorder characterized by developmental overgrowth and a broad spectrum of clinical features, including enlarged tongue, cleft palate, polydactyly, syndactyly, supernumerary nipples, cystic and dysplastic kidneys, congenital heart defects, rib and vertebral fusions, and umbilical and inguinal hernias (Neri et al., 1998; DeBaun et al., 2001). GPC3-null mice display many of the abnormalities seen in the SGBS patients, including developmental overgrowth (Cano-Gauci et al., 1999). As discussed above, embryos lacking GPC3 display increased Hh-signaling activity. This is due to elevated levels of Sonic Hh and Indian Hh and is consistent with the finding that GPC3 induces endocytosis and degradation of Hh (Capurro et al., 2008, 2009). Genetic studies have clearly demonstrated that elevated Hh signaling plays a causative role in the overgrowth of GPC3-null mice (Capurro et al., 2009). Given the broad spectrum of developmental abnormalities observed in SBGS patients, it is highly likely that at least some of these abnormalities are caused by alterations in signaling pathways other than the one triggered by Hh. For example, it has already been demonstrated that Wnt signaling is altered in GPC3-null mice (Song et al., 2005). Given the important role of Wnt signaling in developmental morphogenesis, it is reasonable to speculate that some of the clinical features observed in SGBS patients are due to deregulated Wnt signaling.

Glypicans and other diseases

Recessive omodysplasia is a genetic condition characterized by short stature, facial dysmorphism, and proximal limb shortening. Some patients also display cryptorchidism, hernias, congenital heart defects, and cognitive delay (Campos-Xavier et al., 2009). It has been recently reported that this genetic disease is caused by point mutations or by larger genomic rearrangements in *GPC6* (Campos-Xavier et al., 2009). Because all mutations cause the truncation of the GPC6 protein, it is highly likely that this disease is due to loss of GPC6 function. The clinical features of recessive omodysplasia indicate that GPC6 plays a role in endochondral ossification and skeletal growth. It remains to be investigated which signaling pathway(s) mediate the role of GPC6 in bone development.

In prion diseases, the cellular form of the prion protein, PrP^{C}, is converted into the infectious form scrapie (PrP^{Sc}). Taylor et al. (2009) have shown that GPC1 binds to PrP^{C} on the cell surface and promotes its association with lipid rafts. Furthermore, these authors showed that depletion of GPC1 in scrapie-infected cells significantly reduced PrP^{Sc} formation. Because GPC1 coimmunoprecipitates with both prion forms, the authors proposed that GPC1 acts as a scaffold facilitating their interaction and the conversion of PrP^{C} into PrP^{Sc} (Taylor et al., 2009). The interaction of GPC1 with the prion proteins was shown to be mediated by the HS chains. Further work will be required to determine whether other glypicans can also promote PrP^{Sc} formation, and whether GPC1 plays a similar function *in vivo*.

Nephrotic syndrome is characterized by severe proteinuria with concurrent decrease of serum protein. This syndrome can be associated with various glomerular diseases, including focal segmental glomerulosclerosis, membranous nephropathy, diabetes mellitus, amyloidosis, and systemic lupus erythematosus. Podocyte injury is thought to make a significant contribution to this syndrome. Recently, Okamoto et al. (2011) identified GPC5 as a susceptibility gene for acquired nephrotic syndrome by genome-wide association studies. They showed that GPC5 is expressed in podocytes, and that downregulation of GPC5 in mouse podocytes causes resistance to nephrotic syndrome induced by treatment with FGF2 and puromycin aminonucleoside. They proposed that GPC5 stimulates FGF2-induced pathological damage in the podocytes (Okamoto et al., 2011).

3.7.4 Future research

In the past few years, significant advances have been made in our understanding of the functions of glypicans. It is now clear that glypicans can regulate the activity of several signaling pathways, including those triggered by Wnts, Hhs, FGFs, and BMPs. However, very little is known about the structural features that determine the specific roles of each glypican in these pathways. Moreover, because some of these structural features of glypicans are most likely related to the type of modifications of their GAG chains, another important topic in future research will be the factors that determine core-protein-specific and cell-type-specific GAG modifications.

3.7.5 Take-home message

Glypicans are proteoglycans bound to the cell membrane by a lipid anchor. They regulate various signaling pathways including those triggered by Wnts, Hhs, FGFs, and BMPs. Their regulatory activity occurs mainly at the level of signal reception, and it could be stimulatory or inhibitory in nature, according to the cell context. The specific function of a particular glypican depends on its structure and on which growth factors and growth factor receptors are expressed by the target cell.

References

Ai, X., Do, A. T., Lozynska, O., Kusche-Gullberg, M., Lindhal, U., and Emerson, C. P. (2003). QSulf1 remodels the 6-O sulfation states of cell surface proteoglycans to promote Wnt signaling. J Cell Biol *162*, 341–351.

Aikawa, T., Whipple, C. A., Lopez, M. E., et al. (2008). Glypican-1 modulates the angiogenic and metastatic potential of human and mouse cancer cells. J Clin Invest *118*, 89–99.

Akiyama, T., Kamimura, K., Firkus, C., Takeo, S., Shimmi, O., and Nakato, H. (2008). Dally regulates Dpp morphogen gradient formation by stabilizing Dpp on the cell surface. Dev Biol *313*, 408–419.

Ayers, K. L., Gallet, A., Staccini-Lavenant, L., and Therond, P. P. (2010). The long-range activity of hedgehog is regulated in the apical extracellular space by the glypican Dally and the hydrolase Notum. Dev Cell *18*, 605–620.

Belenkaya, T. Y., Han, C., Yan, D., et al. (2004). Drosophila Dpp morphogen movement is independent of dynamin-mediated endocytosis but regulated by the glypican members of heparan sulfate proteoglycans. Cell *119*, 231–244.

Bishop, J. R., Schuksz, M., and Esko, J. D. (2007). Heparan sulphate proteoglycans fine-tune mammalian physiology. Nature *446*, 1030–1036.

Bruix, J., and Sherman, M. (2011). Management of hepatocellular carcinoma: an update. Hepatology *53*, 1020–1022.

Campos-Xavier, A. B., Martinet, D., Bateman, J., et al. (2009). Mutations in the heparan-sulfate proteoglycan glypican-6 (GPC6) impair endochodral ossification and cause recessive omodysplasia. Am J Hum Genet *84*, 760–770.

Cano-Gauci, D. F., Song, H. H., Yang, H., et al. (1999). Glypican-3-deficient mice exhibit the overgrowth and renal abnormalities typical of the Simpson-Golabi-Behmel syndrome. J Cell Biol *146*, 255–264.

Capurro, M., Wanless, I. R., Sherman, M., et al. (2003). Glypican-3: a novel serum and histochemical marker for hepatocellular carcinoma. Gastroenterology *125*, 81–90.

Capurro, M., Xiang, Y. Y., Lobe, C., and Filmus, J. (2005a). Glypican-3 promotes the growth of hepatocellular carcinoma by stimulating canonical Wnt signaling. Cancer Res *65*, 6245–6254.

Capurro, M. I., Li, F., and Filmus, J. (2009). Overgrowth of a mouse model of Simpson-Golabi-Behmel syndrome is partly mediated by Indian Hedgehog. EMBO Rep *10*, 901–907.

Capurro, M. I., Shi, W., Sandal, S., and Filmus, J. (2005b). Processing by convertases is not required for glypican-3-induced stimulation of hepatocellular carcinoma. J Biol Chem *280*, 41201–41206.

Capurro, M. I., Xu, P., Shi, W., Li, F., Jia, A., and Filmus, J. (2008). Glypican-3 inhibits hedgehog signaling during development by competing with Patched for Hedgehog binding. Dev Cell *14*, 700–711.

Capurro, M. I., Shi, W., and Filmus, J. (2012). J.Cell Sci., in press.

DeBaun, M. R., Ess, J., and Saunders, S. (2001). Simpson Golabi Behmel syndrome: progress toward understanding the molecular basis for overgrowth, malformation, and cancer predisposition. Mol Genet Metab *72*, 279–286.

De Cat, B., Muyldermans, S. Y., Coomans, C., et al. (2003). Processing by proprotein convertases is required for glypican-3 modulation of cell survival, Wnt signaling, and gastrulation movements. J Cell Biol *163*, 625–635.

Dejima, K., Kanai, M. I., Akiyama, T., Levings, D. C., and Nakato, H. (2011). Novel contact-dependent bone morphogenetic protein (BMP) signaling mediated by heparan sulfate proteoglycans. J Biol Chem *286*, 17103–17111.

Desbordes, S. C., and Sanson, B. (2003). The glypican Dally-like is required for hedgehog signalling in the embryonic epidermis of Drosophila. Development *130*, 6245–6255.

Di Tommaso, L., Destro, A., Seok, J. Y., et al. (2009). The application of markers (HSP70 GPC3 and GS) in liver biopsies is useful for detection of hepatocellular carcinoma. J Hepatol *50*, 746–754.

Eugster, C., Panakova, D., Mahmoud, A., and Eaton, S. (2007). Lipoprotein-heparan sulfate interactions in the Hh pathway. Dev Cell *13*, 57–71.

Feng, M., Kim, H., and Ho, M. (2011). Recombinant soluble glypican 3 protein inhibits the growth of hepatocellular carcinoma in vitro. Int J Cancer *128*, 2246–2247.

Filmus, J., Capurro, M., and Rast, J. (2008). Glypicans. Genome Biol *9*, 224.

Franch-Marro, X., Marchand, O., Piddini, E., Ricardo, S., Alexandre, C., and Vincent, J. P. (2005). Glypicans shunt the wingless signal between local signalling and further transport. Development *132*, 659–666.

Fujise, M., Takeo, S., Kamimura, K., et al. (2003). Dally regulates Dpp morphogen gradient formation in the Drosophila wing. Development *130*, 1515–1522.

Gallet, A., Staccini-Lavenant, L., and Therond, P. P. (2008). Cellular trafficking of the glypican Dally-like is required for full-strength hedgehog signaling and wingless transcytosis. Dev Cell *14*, 712–725.

Guo, Z., and Wang, Z. (2009). The glypican Dally is required in the niche for the maintenance of germline stem cells and short-range BMP signaling in the Drosophila ovary. Development *136*, 3627–3635.

Gutierrez, J., and Brandan, E. (2010). A novel mechanism of sequestering FGF-2 by glypican in lipid rafts, allowing skeletal muscle differentiation. Mol Cell Biol *30*, 1634–1649.

Hartwig, S., Hu, M. C., Cella, C., Piscione, T., Filmus, J., and Rosenblum, N. D. (2005). Glypican-3 modulates inhibitory Bmp2-Smad signaling to control renal development in vivo. Mech Dev *122*, 928–938.

Hayashi, Y., Kobayashi, S., and Nakato, H. (2009). Drosophila glypicans regulate the germline stem cell niche. J Cell Biol *187*, 473–480.

Hippo, Y., Watanabe, K., Watanabe, A., et al. (2004). Identification of soluble NH2-terminal fragment of glypican-3 as a serological marker for early-stage hepatocellular carcinoma. Cancer Res *64*, 2418–2423.

Ho, M., and Kim, H. (2010). Glypican-3: a new target for cancer immunotherapy. Eur J Cancer *47*, 333–338.

Hsu, H. C., Cheng, W., and Lai, P. L. (1997). Cloning and expression of a developmentally regulated transcript MXR7 in hepatocellular carcinoma: biological significance and temporospatial distribution. Cancer Res *57*, 5179–5184.

Ishiguro, T., Sugimoto, M., Kinoshita, Y., et al. (2008). Anti-glypican-3 antibody as a potential antitumor agent for human liver cancer. Cancer Res *68*, 9832–9838.

Jackson, S. M., Nakato, H., Sugiura, M., et al. (1997). Dally, a Drosophila glypican, controls celular responses to the TGF-beta-related morphogen Dpp. Development *124*, 4113–4120.

Jen, Y. H. L., Musacchio, M., and Lander, A. D. (2009). Glypican-1 controls brain size through regulation of fibroblast growth factor signaling in early neurogenesis. Neural Dev *4*, 33.

Johnson, K. G., Tenney, A. P., Ghose, A., et al. (2006). The HSPGs syndecan and Dallylike bind the receptor phophatase LAR and exert distinct effects on synaptic development. Neuron *49*, 517–531.

Kim, M. S., Saunders, A. M., Hamaoka, B., et al. (2011). Structure of the protein core of the glypican Dally-like and localization of a region important for hedgehog signaling. Porc. Natl. Acad. Sci. U.S.A. *108*, 13112–13117.

Kleeff, J., Ishiwata, T., Kumbasar, A., et al. (1998). The cell surface heparan sulfate proteoglycan glypican-1 regulates growth factor in pancreatic carcinoma cells and is overexpressed in human pancreatic cancer. J Clin Invest *102*, 1662–1673.

Komori, H., Nakatsura, T., Senju, S., et al. (2006). Identification of HLA-A2- or HLA-A24-restricted CTL epitopes possibly useful for glypican-3-specific immunotherapy of hepatocellular carcinoma. Clin Cancer Res *12*, 2689–2697.

Kreuger, J., Perez, L., Giraldez, A. J., and Cohen, S. M. (2004). Opposing activities of Dally-like glypican at high and low levels of Wingless morphogen activity. Dev Cell *7*, 503–512.

Li, F., Shi, W., Capurro, M., and Filmus, J. (2011). Glypican-5 stimulates rhabdomyosarcoma cell proliferation by activating Hedgehog signaling. J Cell Biol *192*, 691–704.

Libbrecht, L., Severi, T., Cassiman, D., et al. (2006). Glypican-3 expression distinguishes small hepatocellular carcinomas from cirrhosis, dysplastic nodules, and focal nodular hyperplasia-like nodules. Am J Surg Pathol *30*, 1405–1411.

Lin, X., and Perrimon, N. (1999). Dally cooperates with Drosophila Frizzled 2 to transduce Wingless signaling. Nature *400*, 281–284.

Lum, L., Yao, S., Mozer, B., et al. (2003). Identification of hedgehog pathway components by RNAi in Drosophila cultured cells. Science *299*, 2039–2045.

Luo, J. H., Ren, B., Keryanov, S., et al. (2006). Transcriptomic and genomic analysis of human hepatocellular carcinomas and hepatoblastomas. Hepatology *44*, 1012–1024.

Nakatsura, T., Yoshitake, Y., Senju, S., et al. (2003). Glypican-3, overexpressed specifically in human hepatocellular carcinoma, is a novel tumor marker. Biochem Biophys Res Commun *306*, 16–25.

Neri, G., Gurrieri, F., Zanni, G., and Lin, A. (1998). Clinical and molecular aspects of the Simpson-Golabi-Behmel syndrome. Am J Med Genet *79*, 279–283.

Ng, A., Wong, M., Viviano, B., et al. (2009). Loss of glypican-3 function causes growth factor-dependent defects in cardiac and coronary vascular development. Dev Biol *335*, 208–215.

Ohkawara, B., Yamamoto, T. S., Tada, M., and Ueno, N. (2003). Role of glypican 4 in the regulation of convergent extension movements during gastrulation in Xenopus laevis. Development *130*, 2129–2138.

Okamoto, K., Tokunaga, K., Doi, K., et al. (2011). Common variation in GPC5 is associated with acquired nephrotic syndrome. Nat Genet *43*, 459–463.

Paine-Saunders, S., Viviano, B. L., Zupicich, J., Skarnes, W. C., and Saunders, S. (2000). Glypican-3 controls cellular responses to Bmp4 in limb patterning and skeletal development. Dev Biol *225*, 179–187.

Patil, M. A., Chua, M. S., Pan, K. H., et al. (2005). An integrated data analysis approach to characterize genes highly expressed in hepatocellular carcinoma. Oncogene *24*, 3737–3747.

Pilia, G., Hughes-Benzie, R. M., MacKenzie, A., et al. (1996). Mutations in GPC3, a glypican gene, cause the Simpson-Golabi-Behmel overgrowth syndrome. Nat Genet *12*, 241–247.

Rawson, J. M., Dimitroff, B., Johnson, K. G., et al. (2005). The heparan sulfate proteoglycans Dally-like and Syndecan have distinct functions in axon guidance and visual-system assembly in Drosophila. Curr Biol *15*, 833–838.

Saunders, S., Paine-Saunders, S., and Lander, A. D. (1997). Expression of the cell surface proteoglycan glypican-5 is developmentally regulated in kidney, limb, and brain. Dev Biol *190*, 78–93.

Shao, M., Liu, Z. Z., Wang, C. D., et al. (2009). Down syndrome critical region protein 5 regulates membrane localization of Wnt receptors, Dishevelled stability and convergent eeextension in vertebrate embryos. Development *136*, 2121–2131.

Shiba, K., Hu, N., and Bronner-Fraser, M. (2010). Altering glypican-1 levels modulates canonical Wnt signaling during trigeminal placode development. Dev Biol *348*, 107–118.

Smart, A. D., Course, M. M., Rawson, J., Selleck, S. B., Van Vactor, D., and Johnson, K. G. (2011). Heparan sulfate proteoglycan specificity during axon pathway formation in Drosophila embryo. Dev Neurobiol *71*, 608–618.

Song, H. H., and Filmus, J. (2002). The role of glypicans in mammalian development. Biochim Biophys Acta *1573*, 241–246.

Song, H. H., Shi, W., Xiang, Y., and Filmus, J. (2005). The loss of glypican-3 induces alterations in Wnt signaling. J Biol Chem *280*, 2116–2125.

Su, G., Meyer, K., Nandini, C. D., Qiao, D., Salamat, S., and Friedl, A. (2006). Glypican-1 is frequently overexpressed in human gliomas and enhances FGF-2 signaling in glioma cells. Am J Pathol *168*, 2014–2026.

Svensson, G., Awad, W., Hakansson, M., Mani, K., and Logan, D. T. (2012). Crystal structure of N-glycosylated human glypican-1 core protein. J Biol Chem *287*, 14040–14051.

Taylor, D. R., Whitehouse, I. J., and Hooper, N. M. (2009). Glypican-1 mediates both prion protein lipid raft association and disease isoform formation. PLOS Pathog *5*, e1000666.

Topczewsky, J., Sepich, D. S., Myers, D. C., et al. (2001). The zebrafish glypican knypek controls cell polarity during gastrulation movements of convergent extension. Dev Cell *1*, 251–264.

Tsuda, M., Kamimura, K., Nakato, H., et al. (1999). The cell-surface proteoglycan Dally regulates Wingless signaling in Drosophila. Nature *400*, 276–280.

Vyas, N., Goswami, D., Manonmani, A., et al. (2008). Nanoscale organization of hedgehog is essential for long-range signaling. Cell *133*, 1214–1227.

Watanabe, K., Yamada, H., and Yamaguchi, Y. (1995). K-glypican: a novel GPI-linked heparan sulfate proteoglycan that is highly expressed in developing brain and kidney. J Cell Biol *130*, 1207–1218.

Williams, E. H., Pappano, W. N., Saunders, A. M., Kim, M. S., Leahy, D. J., and Beachy, P. A. (2010). Dally-like core protein and its mammalian homologues mediate stimulatory and inhibitory effects on Hedgehog signal response. Proc Natl Acad Sci U S A *107*, 5869–5874.

Williamson, D., Selfe, J., Gordon, T., et al. (2007). Role for amplification and expression of glypican-5 in rhabdomyosarcoma. Cancer Res *67*, 57–65.

Yan, D., and Lin, X. (2007). Drosophila glypican Dally-like acts in FGF-receiving cells to modulate FGF signaling during tracheal morphogenesis. Dev Biol *312* (1), 203–216.

Yan, D., Wu, Y., Feng, Y., Lin, S. C., and Lin, X. (2009). The core protein of glypican Dally-like determines its biphasic activity in wingless morphogen signaling. Dev Cell *17*, 470–481.

Yan, D., Wu, Y., Yang, Y., Belenkaya, T. Y., Tang, X., and Lin, X. (2010). The cell-surface proteins Dally-like and Ihog differentially regulate Hedgehog signaling strength and range during development. Development *137*, 2033–2044.

Zhang, F., Ronca, F., Linhardt, R. J., and Margolis, R. U. (2004). Structural determinants of heparan sulfate interactions with Slit proteins. Biochem Biophys Res Commun *317*, 352–357.

Zhu, A. X., Gold, P. J., El-Khoueiry, A. B., Abrams, T. A., Morikawa, H., and Philip, P. A. (2011). A phase 1 study of GC33, a recombinant humanized antibody against glypican-3, in patients with advanced hepatocellular carcinoma. J Clin Oncol *29* (Suppl., abstr. 4085).

Zittermann, S. I., Capurro, M. I., and Filmus, J. (2010). Soluble glypican 3 inhibits the growth of hepatocellular carcinoma in vitro and in vivo. Int J Cancer *126*, 1291–1301.

3.8 Serglycin proteoglycan: implications for thrombosis, inflammation, atherosclerosis, and metastasis

Barbara P. Schick

3.8.1 Introduction

Serglycin is the proteoglycan of secretory granules and vesicles in platelets, megakaryocytes (MKs), and all myeloid cells. Serglycin has also been identified in endothelial cells and other tissues. Numerous *in vitro* studies have suggested that the functions of serglycin include packaging proteins into the granules; protection of molecules (e.g. proteases or growth factors), with which they are secreted as complexes; and facilitation of delivery of these proteins to their targets after secretion. Heparan sulfate proteoglycans on cell surfaces and matrix have been implicated in the establishment of chemokine gradients thought to be involved in immune function, embryonic implantation and development, and pathologies such as inflammation, atherosclerosis, or metastasis.

We propose that serglycin/protein complexes, whether bound to cells or matrix or free in solution, are the first step in establishing the gradients because many of the proteins are presumably packaged into granules with and released as complexes with serglycin.

Deletion of serglycin leads to remarkable changes of platelet function *in vitro* and *in vivo*, as well as to effects on other hematopoietic cells *in vitro*. The implications of these findings for physiological events in which these cells play a significant role, such as thrombosis and normal blood coagulation, inflammation, atherosclerosis, and tumor metastasis, will be explored. This subject has been discussed previously in more detail (Schick, 2010).

3.8.2 Cloning and cell and tissue localization of serglycin

Rat L2 yolk sac tumor serglycin was the first proteoglycan gene to be cloned. The translated protein was 18 kDa and is the smallest known proteoglycan core protein. The name *serglycin* reflects the 42-amino-acid serine/glycine (Ser/Gly) repeat region. Each serine of a Ser/Gly repeat is a potential glycosaminoglycan (GAG) attachment site. This structure, which allows clustering of GAGs near the center of the protein, is unique to serglycin. The mouse and human cDNAs have only a 16-amino-acid Ser/Gly region. The human platelet serglycin core protein bears about 4 GAG chains, and rat serglycin has about 14 GAG chains.

Serglycin is the only proteoglycan of hematopoietic cell granules. Serglycin is in protease-containing granules in mast cells, eosinophils, and neutrophils and in platelet α granules. In these cells, serglycin and the granule contents are stored until the cell is activated. Serglycin is constitutively secreted from lymphocytes (Kolset and Gallagher,

1990) and many hematopoietic tumor cell lines (Schick and Senkowski-Richardson, 1992; Schick and Jacoby, 1995).

Serglycin is also expressed in nonhematopoietic cells: endothelial cells (Schick et al., 2001a), murine uterine mesometrial decidua (Ho et al., 2001), murine parietal endoderm (Ho et al., 2001), murine fetal liver but not yolk sac hematopoietic cells (Ho et al., 2001), murine embryonic stem cells (Schick et al., 2003), and chondrocytes (Zhang et al., 2010). The significance of serglycin in these cells is not known.

3.8.3 Cell-specific serglycin structure

Cell- and species-specific diversity of serglycin proteoglycan size (35 to >600 kDa), GAG chain length (5–60 kDa), GAG number, and GAG type are likely critical to the cell-specific interaction of serglycin with cytokines, chemokines, chymases, and other proteins and the resulting physiological effects. Normal rat and human platelet serglycin are about the same overall size, but rat platelet serglycin GAGs are about half as long as human serglycin GAGs (Schick et al., 1997).

Mast-cell serglycin has oversulfated chondroitin sulfate (CS), heparin (found only on mast-cell serglycin), or heparan sulfate depending on the tissue of residence. GAG chain type coordinates with cell-specific chymases (Sali et al., 1993). Mouse mastocytoma cell line serglycin is a CS/heparin hybrid (Lidholt et al., 1995). GAGs of human, rat, and mouse platelets are exclusively chondroitin-4-sulfate. Megakaryocytic tumor cells have hybrid CS/heparan sulfate serglycin (Schick and Jacoby, 1995). Uterine decidual and human umbilical vein endothelial cells (HUVEC) serglycins are CSs (Ho et al., 2001; Schick et al., 2001a). The serglycins and their GAG chains are smaller than in platelets.

3.8.4 Regulation of serglycin expression

We have described human serglycin gene regulation in detail (Schick et al., 2001b; Castronuevo et al., 2003). Serglycin mRNA expression and GAG chain length change during granule formation and cell maturation (Schick et al., 1988, 1989; Stellrecht et al., 1991), and in cell-specific response to various agents in culture (Grover et al., 1987; Schick and Senkowski-Richardson, 1992; Schick and Thornton, 1993; Schick and Jacoby, 1995a). The changes in serglycin expression may coincide with changes in granule proteins (Eklund et al., 1997).

3.8.5 Binding of cell-specific serglycin to biologically active proteins

The overall structure of the intact proteoglycan is essential for high-affinity binding. Intact CS serglycin binds to other proteins as strongly as heparin, but neither free core protein nor CS GAG chains isolated from serglycins have significant binding capability (Brennan et al., 1983; Schick and Jacoby, 1995a; Toyama-Sorimachi et al., 1995, 1997; Verrecchio et al., 2000). Proteins that bind to the CS form of serglycin include extracellular matrix proteins (fibronectin and collagen) (Brennan et al., 1983; Schick and Jacoby, 1995a, 1995b; Schick et al., 1997), growth factors/cytokines/chemokines (platelet factor-4 [PF4], macrophage inflammatory protein (MIP-1α), a bone morphogenetic protein [BMP]-like protein) (Kolset et al., 1996), the membrane protein/proteoglycan CD44 (Toyama-Sorimachi et al., 1995), and lysozyme (Kolset et al., 1996; Lemansky and Hasilik, 2001). Heparin serglycin was postulated to complex with specific chymases, different from those complexed to mast-cell CS serglycin, and to regulate the function of these enzymes after secretion (Sali et al., 1993). Several matrix metalloproteinases (MMPs) bind serglycin. Granule protein content and function in the *N*-deacetylase/*N*-sulfotransferase-2 (NDST2) knockout mast cells were abnormal (Forsberg et al., 1999; Humphries et al., 1999).

Interactions of platelets with monocytes and other leukocytes during coagulation and inflammation are modulated by chemokines and cytokines (e.g. PF4, regulated upon activation, normal T-cell expressed, and secreted (RANTES), and MIP-1α). PF4 (chemokine [C-X-C motif] ligand-3 [CXCL3]) activates neutrophils (Aziz et al., 1997), is synthesized only by MKs, and is associated with the development of atherosclerotic plaque (Nassar et al., 2003; Sachais et al., 2004). Understanding interactions of these proteins with serglycin may provide insight into blood-cell interactions in inflammatory disease, asthma, and human immunodeficiency virus (HIV) infection (Clemetson et al., 2000) or into the role of platelets in chemokine-related events in tumor metastasis and atherosclerosis.

3.8.6 Serglycin in hematopoietic cells

Platelets and megakaryocytes

Platelets are the first line of defense for sealing a wound following injury. They interact with the vessel wall, other blood cells, and coagulation proteins during hemostasis and thrombosis. Platelets are anucleate fragments of the cytoplasm of the bone marrow MK. Serglycin is stored in α granules. The α granules contain several hundred proteins; many induce signficant biological activity such as thrombus formation, inflammation, and atherosclerosis. Activated platelets release α-granule proteins that interact with other platelet components including the dense granules, which contain serotonin, adenosine diphosphate (ADP), adenosine triphosphate (ATP), Ca++, receptors on the highly intercalated surface membrane, cytoskeletal components, and several signaling pathways. The loss of serglycin affects these processes in unexpected ways described subsequently.

The only known natural model of abnormal serglycin is the Wistar Furth macrothrombocytopenic rats (platelets are abnormally large and severely reduced in number, the rats have a bleeding disorder, and there is loss of α-granule proteins) (Schick et al., 1997): serglycin has only 60% of normal GAG chain mass. The abnormal serglycin may contribute to the bleeding disorder by altering granule protein content.

Premature release of α-granule growth factors or chemokines from MK into the marrow in this rat and in human myeloproliferative disorders and myelofibrosis is thought to cause lethal marrow fibrosis. Normal MK and platelets store serglycin in α granules and release them only upon activation (Schick et al., 1988). In contrast, the human hematopoietic tumor cells constitutively secrete the proteoglycans. Lysozyme, which binds serglycin, is released constitutively from MK from patients with polycythemia vera but not from normal MK (Wickenhauser et al., 1999). Possibilities leading to matrix pathology in hematologic disorders are that serglycin/protein complexes may not form properly or may be directed to constitutive secretory vesicles instead of storage granules, or granule storage and/or release mechanisms may be defective.

Platelets have long been implicated in the development of atherosclerosis. PF4 has been found in human atherosclerotic plaques (De Meyer et al., 2002; Sachais et al., 2002), binds to LDL, and activates macrophages. Absence of serglycin might prevent deposition of these proteins into the blood vessels.

Mast cells

NDST2 deletion in mice showed the importance of the overall structure of heparin serglycin to the organization and function of mast-cell granules (Forsberg et al., 1999; Humphries et al., 1999), in agreement with *in vitro* data on chymases (Sali et al., 1993; Matsumoto et al., 1995). Mast-cell oversulfated CS serglycins bind different proteases from heparin serglycin. Uterine mast-cell protease-9 MCP9 complexes uniquely with CS serglycin (Hunt et al., 1997).

Leukocytes

CS serglycins aggregate specific proteins into neutrophil granules (Gullberg et al., 1997). Serglycin mRNA is synthesized at all stages and is presumed to be in all three types of granules (Cowland and Borregaard, 1999). However, we found serglycin only in isolated secondary and tertiary granules (Castronuevo et al., 2003). Only secondary granules are deficient in leukocytes of gray platelet syndrome (Drouin et al., 2001). Epstein-Barr virus (EBV) infection increases serglycin mRNA expression in T lymphocytes (Birkenbach et al., 1993). Serglycin may be involved in granzyme B–mediated apoptosis in natural killer (NK) cells (Raja et al., 2005).

Hematopoietic malignancies

Serglycin is a selective leukemic marker for immature myeloid cells (Niemann et al., 2007b), found in acute myelogenous leukemia (AML) but not Philadelphia chromosome-negative chronic myeloproliferative disorders. Serglycin was localized to immature blasts in marrow from patients with AML but not acute lymphocytic leukemia (ALL); plasma serglycin was 15-fold higher in AML than ALL patients. Serglycin is overexpressed in multiple myeloma (MM) (Theocharis et al., 2006) and is intracellular and on the cell surface. MM serglycin compromised bone mineralization by inhibiting the crystal growth rate of hydroxyapatite. The level of serglycin expression is associated with chemotherapy drug resistance in malignant hematopoietic cell lines (Beyer-Sehlmeyer et al., 1999). Serglycin may affect drug metabolism in malignant hematopoietic cells. Mast-cell heparin serglycin may serve as a carrier for doxorubicin and gentamicin delivery to macrophages and smooth muscle cells (Decorti et al., 2000).

3.8.7 Serglycin in nonhematopoietic cells

Endothelial and vascular smooth muscle cells

We identified serglycin mRNA, core protein, and proteoglycan in human umbilical vein endothelial cells (Schick et al., 2001a). Rat serosal mast-cell heparin serglycin inhibits proliferation of cultured rat aortic smooth muscle cells (Wang and Kovanen, 1999). Serglycin mRNA increases in HUVEC treated with tumor necrosis factor-alpha (TNF-α) and interleukin (IL)-1β. TNF-α-treated HUVECs secrete virtually all their serglycin (Kulseth et al., 1999; Schick et al., 2001a). Vascular smooth muscle cells (vSMCs) synthesize and secrete serglycin (Kulseth et al., 1999).

Reproduction and development

Serglycin may modulate communication between inner cell mass and trophectoderm cells (Ho et al., 2001), and between the decidua and placenta (Schick et al., 2003).

Chondrocytes

Several chondrocyte proteoglycans, including serglycin, bind matrix metalloproteinase-13 (MMP13). Serglycin mRNA expression by chondrocytes and localization of serglycin with MMP13 in cytoplasmic granules were demonstrated. Their interaction may play a role in cartilage degradation (Zhang et al., 2010).

Nasopharyngeal carcinoma

Serglycin regulates nasopharyngeal carcinoma (NPC) metastasis via autocrine and paracrine routes and serves as an independent prognostic indicator of metastasis-free survival and disease-free survival in NPC patients. Serglycin enhances cellular migration, cellular invasiveness, vimentin expression level, and *in vivo* spread of these cancer cells (Li et al., 2011).

3.8.8 The serglycin knockout mouse

Abrink et al (2004) created a serglycin knockout ($SG^{-/-}$) mouse. Cultured cells from this mouse have been used to study mRNA expression, protein synthesis, storage, degradation, and secretion of cell-specific proteins known to associate with serglycin.

Platelets and megakaryocytes

We described the striking effects of serglycin deletion on platelet function and MK structure (Woulfe et al., 2008).

Platelet and MK morphology are abnormal in $SG^{-/-}$ mice (Woulfe et al., 2008). Unusual scroll-like membranous inclusions in $SG^{-/-}$ MK and platelets also appear in platelets from patients with Medich syndrome, a gray platelet–like disorder, and in Wistar Furth rat platelets (White, 2004). Emperipolesis of neutrophils (taking up of the entire neutrophil) into the cytoplasm of MK was extensive. Emperipolesis occurs in MK from gray platelet syndrome and myelofibrosis, which are characterized by marrow fibrosis likely due to premature secretion of chemokines and growth factors from α granules forming in the MK (Schmitt et al., 2000).

Deletion of a number of α-granule or platelet-signaling proteins in mice has led to a reduced response to aggregating agents *in vitro* and defects in *in vivo* arterial clot formation or stability. Deletion of serglycin leads to deficient platelet aggregation and thrombus formation, which resembles that of many of these knockouts (Woulfe et al., 2008). $SG^{-/-}$ platelets showed a markedly defective aggregation response to agonists compared to wild-type (WT) platelets and a greater tendency to disaggregate. Surprisingly, release of dense granule contents was markedly reduced in platelets from $SG^{-/-}$ mice. We identified a defect in fibrinogen binding attributable to a defect in activation of integrin $\alpha IIb\beta 3$, the final critical step in aggregation.

Induced carotid artery thrombosis in mice is a measure of *in vivo* platelet function and the role of endothelial injury in thrombosis. The $SG^{-/-}$ mice had greatly reduced incidence of thrombus formation, comparable to knockouts of known coagulation-related proteins.

The levels of PF4, platelet-derived growth factor (PDGF), and thromboglobulin are dramatically reduced in platelets from $SG^{-/-}$ mice. However, PF4 mRNA is expressed equally in WT and $SG^{-/-}$ animals. Reduced α-granule storage capacity without serglycin, rather than reduced PF4 synthesis, is likely the cause of lower PF4 levels.

Clemetson has proposed that surface GAGs promote chemokine activation of platelets (Clemetson et al., 2000). PF4 likely binds to membrane proteoglycans, as do RANTES or MIP-1α. The absence of any of these proteins or other known activation-related proteins secondary to lack of serglycin might compromise the signaling, which requires binding of released α-granule proteins to the platelet surface.

Mast cells

Peritoneal mast cells were greatly reduced in number in $SG^{-/-}$ mice (Abrink et al., 2004). Surprisingly, the c-kit receptor was absent from the $SG^{-/-}$ cells. Mast-cell proteases (MCPs) and tryptases were absent from ear and peritoneal mast cells, but all were transcribed. Storage of MCPs in granules was impaired.

Another study (Henningsson et al., 2006) described defects in accumulation, storage, and secretion of proteases in $SG^{-/-}$ mouse bone-marrow-derived mast cells (mBMMC). MCPs enter the granules without depending on serglycin but must interact with serglycin in order to be retained.

Leukocytes

Zernichow et al. (2006) studied macrophages cultured from adherent cells derived from peritoneal and spleen cells. Lysozyme secretion after lipopolysaccharide (LPS) treatment was not affected by lack of serglycin, nor were there consistent effects on matrix metalloproteinase-9 (MMP9), MIP-1α, or IL-1α. The explanation might be that binding studies were performed with serglycin with much longer GAG chains than the very short macrophage serglycin. Unexpected increase and secretion of TNF-α was a secondary effect of the loss of serglycin. Serglycin was found to be necessary for storage of granzyme B but not granzyme A or perforin in cytotoxic T lymphocytes (Grujic et al., 2005). The contraction responses of CD8+ T cells from control and $SG^{-/-}$ mice to lymphocytic choriomeningitis virus infection suggests that serglycin is important for regulating the kinetics of antiviral CD8(+) T-cell responses (Grujic et al., 2008). Neutrophils from the $SG^{-/-}$ mice lack elastase but not myeloperoxidase, although elastase mRNA was present. The virulence of intraperitoneally injected gram-negative bacteria was increased in $SG^{-/-}$ mice compared to controls (Niemann et al., 2007a). Mice lacking serglycin spontaneously develop enlargement of multiple lymphoid organs, including the spleen, Peyer's patches (PP), and bronchus-associated lymphoid tissue. Serglycin deficiency causes multiple age-related effects on the lymphoid system (Wernersson et al., 2009).

In summary, platelets of $SG^{-/-}$ mice exhibit complex defects in aggregation and are missing several important α-granule proteins. Mast cells show defective development of electron-dense granules. One or more granule proteins specific to three types of mast cells, macrophages, CTLs, and neutrophils are absent or greatly reduced. $SG^{-/-}$ mice show abnormalities in response to *in vivo* challenges by thrombotic, infectious, or inflammatory stimuli. Other unanticipated effects were observed.

3.8.9 Challenges and future prospects

α-Granule protein content

A complete analysis of platelet α-granule proteins and their ability to bind to serglycin should be undertaken and compared to those of the $SG^{-/-}$ mouse.

Inflammation and atherosclerosis

Platelets are involved in inflammation (Wagner and Burger, 2003) and atherosclerosis (Gawaz et al., 2005) through complex interactions with endothelium and leukocytes. Endothelial activation results in platelet activation, release of proinflammatory chemokines, and leukocyte activation and extravasation. Different types of inflammatory challenges and infection in the $SG^{-/-}$ mouse can assess the effects of loss of serglycin and its conjugate proteins on different blood and blood vessel cells in inflammation and atherosclerosis.

Atherogenesis is a product of inflammation. Chemokines released from platelets and vessel wall attract leukocytes and monocytes to the inflamed vessel. Mediators include RANTES, MIP-1α, platelet-activating factor, transforming growth factor-beta (TGF-β), PDGF, serotonin, matrix-degrading proteases, and CD40 ligand. A major factor in the role of platelets in atherosclerosis is the deposition of PF4, in human atherosclerotic lesions. PF4 attracts monocytes and promotes their differentiation into macrophages and the retention of lipoproteins in the plaque. The close association of PF4 with serglycin, and its near absence from platelets and plasma of $SG^{-/-}$ mice, suggests that the $SG^{-/-}$ mice would be useful for experimental atherosclerosis.

Tumor growth and metastasis

Studies mentioned previously have shown a potential role for serglycin in adverse sequelae of multiple myeloma and involvement in leukemia. Many studies have suggested that platelets are involved in tumor metastasis, so we propose the possibility that platelet serglycin is involved in solid tumor metastasis. Serglycin may influence the balance of tumor growth-promoting chemokines, growth factors, and cytokines delivered to the environment of tumor cells. The release of the CXC chemokines PF4 and platelet basic protein may affect angiogenesis and therefore blood supply to the tumor, and release of MIP-1α may have other effects. The $SG^{-/-}$ mouse would be an excellent animal model for assessing the role of platelet serglycin and its associated proteins in tumor metastasis. The serglycin/chemokine interactions could conceivably be a target for antimetastasis drug development.

Tools for further investigation

A conditional serglycin knockout mouse would be invaluable in sorting out the cell-specific contributions of serglycin to these disorders.

3.8.10 Take-home message

Serglycin plays an important role in regulating platelet aggregation and thrombus formation. Absence of serglycin affects functions of all cells studied. It is proposed that serglycin is important in delivery of chemokines and other functionally important molecules to target cells and may interact via proteoglycans in the matrix or on the surface of the target cells to initiate or help regulate chemokine gradients. These phenomena may have significant impact on the processes of hemostasis/thrombosis, inflammation, atherosclerosis, and tumor metastasis.

References

Abrink, M., Grujic, M., and Pejler, G. (2004). Serglycin is essential for maturation of mast cell secretory granule. J Biol Chem *279*, 40897–40905.

Aziz, K. A., Cawley, J. C., Treweeke, A. T., and Zuzel, M. (1997). Sequential potentiation and inhibition of PMN reactivity by maximally stimulated platelets. J Leukoc Biol *61*, 322–328.

Beyer-Sehlmeyer, G., Hiddemann, W., Wormann, B., and Bertram, J. (1999). Suppressive subtractive hybridisation reveals differential expression of serglycin, sorcin, bone marrow proteoglycan and prostate-tumour-inducing gene I (PTI-1) in drug resistant and sensitive tumour cell lines of hematopoietic origin. Eur J Cancer *35*, 1735–1742.

Birkenbach, M., Josefsen, K., Yalamanchili, R., Lenoir, G., and Kieff, E. (1993). Epstein-Barr virus-induced genes: first lymphocyte-specific G protein- coupled peptide receptors. J Virol *67*, 2209–2220.

Brennan, M. J., Oldberg, A., Hayman, E. G., and Ruoslahti, E. (1983). Effect of a proteoglycan produced by rat tumor cells on their adhesion to fibronectin-collagen substrata. Cancer Res *43*, 4302–4307.

Castronuevo, P., Thornton, M. A., McCarthy, L. E., Klimas, J., and Schick, B. P. (2003). DNase I hypersensitivity patterns of the serglycin proteoglycan gene in resting and phorbol 12-myristate 13-acetate-stimulated human erythroleukemia (HEL), CHRF 288–11, and HL-60 cells compared with neutrophils and human umbilical vein endothelial cells. J Biol Chem *278*, 48704–48712.

Clemetson, K. J., Clemetson, J. M., Proudfoot, A. E., Power, C. A., Baggiolini, M., and Wells, T. N. (2000). Functional expression of CCR1, CCR3, CCR4, and CXCR4 chemokine receptors on human platelets. Blood *96*, 4046–4054.

Cowland, J. B., and Borregaard, N. (1999). The individual regulation of granule protein mRNA levels during neutrophil maturation explains the heterogeneity of neutrophil granules. J Leukoc Biol *66*, 989–995.

Decorti, G., Klugmann, F. B., Crivellato, E., et al. (2000). Biochemical and microscopic evidence for the internalization of drug-containing mast cell granules by macrophages and smooth muscle cells. Toxicol Appl Pharmacol *169*, 269–275.

De Meyer, G. R., De Cleen, D. M., Cooper, S., et al. (2002). Platelet phagocytosis and processing of beta-amyloid precursor protein as a mechanism of macrophage activation in atherosclerosis. Circ Res *90*, 1197–1204.

Drouin, A., Favier, R., Masse, J. M., et al. (2001). Newly recognized cellular abnormalities in the gray platelet syndrome. Blood *98*, 1382–1391.

Eklund, K. K., Humphries, D. E., Zia, Z., et al. (1997). Glucocorticoids inhibit the cytokine-induced proliferation of mast cells, the high affinity IgE receptor-mediated expression of TNF-alpha, and the IL-10 induced expression of chymases. J Immunol *158*, 4373–4380.

Forsberg, E., Pejler, G., Ringvall, M., et al. (1999). Abnormal mast cells in mice deficient in a heparin-synthesizing enzyme. Nature *400*, 773–776.

Gawaz, M., Langer, H., and May, A. E. (2005). Platelets in inflammation and atherogenesis. J Clin Invest *115*, 3378–3384.

Grover, A., Edwards, S. A., Bourdon, M., and Adamson, E. D. (1987). Proteoglycan-19, laminin and collagen type IV production is correlated with the levels of mRNA in F9 cell aggregates differentiating in the presence or absence of cyclic AMP. Differentation *36*, 138–144.

Grujic, M., Braga, T., Lukinius, A., et al. (2005). Serglycin-deficient cytotoxic T lymphocytes display defective secretory granule maturation and granzyme B storage. J Biol Chem *280* (39), 33411–33418.

Grujic, M., Christensen, J. P., Sorensen, M. R., Abrink, M., Pejler, G., and Thomsen, A. R. (2008). Delayed contraction of the CD8+ T cell response toward lymphocytic choriomeningitis virus infection in mice lacking serglycin. J Immunol *181*, 1043–1051.

Gullberg, U., Andersson, E., Garwicz, D., Lindmark, A., and Olsson, I. (1997). Biosynthesis, processing and sorting of neutrophil proteins: insight into neutrophil granule development. [Review]. Eur J Haematol *58*, 137–153.

Henningsson, F., Hergeth, S., Cortelius, R., Abrink, M., and Pejler, G. (2006). A role for serglycin proteoglycan in granular retention and processing of mast cell secretory granule components. FEBS J *273* (21), 4901–4912.

Ho, H.-C. K., McGrath, K. E., Brodbeck, K. C., Palis, J., and Schick, B. P. (2001). Serglycin proteoglycan synthesis in the murine uterine decidua and early embryo. Biol Reprod *64*, 1667–1676.

Humphries, D. E., Wong, G. W., Friend, D. S., et al. (1999). Heparin is essential for the storage of specific granule proteases in mast cells. Nature *400*, 769–772.

Hunt, J. E., Friend, D. S., Gurish, M. F., et al. (1997). Mouse mast cell protease 9, a novel member of the chromosome 14 family of serine proteases that is selectively expressed in uterine mast cells. J Biol Chem *272*, 29158–29155.

Kolset, S. O., and Gallagher, J. T. (1990). Proteoglycans in hematopoietic cells. Biochim Biophys Acta *1032*, 191–211.

Kolset, S. O., Mann, D. M., Uhlin-Hansen, L., Winberg, J. O., and Ruoslahti, E. (1996). Sergylcin-binding proteins in activated macrophages and platelets. J Leukoc Biol *59*, 545–554.

Kulseth, M. A., Kolset, S. O., and Ranheim, T. (1999). Stimulation of serglycin and CD44 mRNA expression in endothelial cells exposed to TNF-alpha and IL-1alpha. Biochim Biophys Acta *1428*, 225–232.

Lemansky, P., and Hasilik, A. (2001). Chondroitin sulfate is involved in lysosomal transport of lysozyme in U937 cells. J Cell Sci *114*, 345–352.

Li, X. J., Ong, C. K., Cao, Y., et al. (2011). Serglycin is a theranostic target in nasopharyngeal carcinoma that promotes metastasis. Cancer Res *71*, 3162–3172.

Lidholt, K., Eriksson, I., and Kjellen, L. (1995). Heparin proteoglycans synthesized by mouse mastocytoma contain chondroitin sulphate. Biochem J *311*, 233–238.

Matsumoto, R., Sali, A., Ghildyal, N., Karplus, M., and Stevens, R. L. (1995). Packaging of proteases and proteoglycans in the granules of mast cells and other hematopoietic cells. A cluster of histidines on mouse mast cell protease 7 regulates its binding to heparin serglycin proteoglycans. J BiolChem *270*, 19524–19531.

Nassar, T., Sachais, B. S., Akkawi, S., et al. (2003). Platelet factor 4 enhances the binding of oxidized low-density lipoprotein to vascular wall cells. J Biol Chem *278*, 6187–6193.

Niemann, C. U., Abrink, M., Pejler, G., et al. (2007a). Neutrophil elastase depends on serglycin proteoglycan for localization in granules. Blood *109* (10), 4478–4486.

Niemann, C. U., Kjeldsen, L., Ralfkiaer, E., Jensen, M. K., and Borregaard, N. (2007b). Serglycin proteoglycan in hematologic malignancies: a marker of acute myeloid leukemia. Leukemia *21*, 2406–2410.

Raja, S. M., Metkar, S. S., Honing, S., et al. (2005). A novel mechanism for protein delivery: granzyme B undergoes electrostatic exchange from serglycin to target cells. J Biol Chem *280*, 20752–20761.

Sachais, B. S., Higazi, A. A., Cines, D. B., Poncz, M., and Kowalska, M. A. (2004). Interactions of platelet factor 4 with the vessel wall. Semin Thromb Hemost *30*, 351–358.

Sachais, B. S., Kuo, A., Nassar, T., et al. (2002). Platelet factor 4 binds to low-density lipoprotein receptors and disrupts the endocytic machinery, resulting in retention of low-density lipoprotein on the cell surface. Blood *99*, 3613–3622.

Sali, A., Matsumoto, R., McNeil, H. P., Karplus, M., and Stevens, R. L. (1993). Three-dimensional models of four mouse mast cell chymases. J Biol Chem *268*, 9203–9034.

Schick, B. P. (2010). Serglycin proteoglycan deletion in mouse platelets: physiological effects and their implications for platelet contributions to thrombosis, inflammation, atherosclerosis, and metastasis. Prog Mol Biol Transl Sci *93*, 235–287.

Schick, B. P., Gradowski, J. F., and San Antonio, J. D. (2001a). Synthesis, secretion and subcellular localization of serglycin proteoglycan in human endothelial cells. Blood *97*, 449–458.

Schick, B. P., Ho, H. C., Brodbeck, K. C., Wrigley, C. W., and Klimas, J. (2003). Serglycin proteoglycan expression and synthesis in embryonic stem cells. Biochim Biophys Acta *1593*, 259–267.

Schick, B. P., and Jacoby, J. A. (1995). Serglycin and betaglycan proteoglycans are expressed in the megakaryocytic cell line CHRF 288–11 and normal human megakaryocytes. J Cell Physiol *165*, 96–106.

Schick, B. P., Pestina, T. I., San Antonio, J. D., Stenberg, P. E., and Jackson, C. W. (1997). Decreased serglycin proteoglycan size is associated with the platelet alpha granule storage defect in Wistar Furth hereditary macrothrombocytopenic rats. Serglycin binding affinity to type I collagen is unaltered. J Cell Physiol *172*, 87–93.

Schick, B. P., Petrushina, I., Brodbeck, K. C., and Castronuevo, P. (2001b). Promoter regulatory elements and DNase I-hypersensitive sites involved in serglycin proteoglycan gene expression in human erythroleukemia, CHRF 288–11, and HL-60 cells. J Biol Chem *276*, 24726–24735.

Schick, B. P., and Senkowski-Richardson, S. (1992). Proteoglycan synthesis in human erythroleukaemia (HEL) cells. Biochem J *282*, 651–658.

Schick, P. K., Schick, B. P., and Williams-Gartner, K. (1989). Characterization of guinea pig megakaryocyte subpopulations at different phases of maturation prepared with a Celsep separation system. Blood *73*, 1801–1808.

Schick, B. P., and Thornton, R. D. (1993). Expression of mRNA for serglycin core protein and other platelet alpha granule proteins is increased in human erythroleukemia cells by phorbol myristate acetate. Leukemia *7*, 1955–1959.

Schick, B. P., Walsh, C. W., and Jenkins-West, T. (1988). Sulfated proteoglycans and sulfated proteins in guinea pig megakaryocytes and platelets in vivo: relevance to megakaryocyte maturation and platelet activation. J Biol Chem *263*, 1052–1062.

Schmitt, A., Jouault, H., Guichard, J., Wendling, F., Drouin, A., and Cramer, E. M. (2000). Pathologic interaction between megakaryocytes and polymorphonuclear leukocytes in myelofibrosis. Blood *96*, 1342–1347.

Stellrecht, C. M., Mars, W. M., Miwa, H., Beran, M., and Saunders, G. F. (1991). Expression pattern of a hematopoietic proteoglycan core protein gene during human hematopoiesis. Differentiation *48*, 127–135.

Theocharis, A. D., Seidel, C., Borset, M., et al. (2006). Serglycin constitutively secreted by myeloma plasma cells is a potent inhibitor of bone mineralization in vitro. J Biol Chem *281* (46), 35116–35128.

Toyama-Sorimachi, N., Kitamura, F., Habuchi, H., Tobita, Y., Kimata, K., and Miyasaka, M. (1997). Widespread expression of chondroitin sulfate-type serglycins with CD44 binding ability in hematopoietic cells. J Biol Chem *272*, 26714–26719.

Toyama-Sorimachi, N., Sorimachi, H., Tobita, Y., et al. (1995). A novel ligand for CD44 is serglycin, a hematopoietic cell lineage-specific proteoglycan. Possible involvement in lymphoid cell adherence and activation. J Biol Chem *270*, 7437–7444.

Verrecchio, A., Germann, M. W., Schick, B. P., Kung, B., Twardowski, T., and San Antonio, J. D. (2000). Design of peptides with high affinities for heparin and endothelial cell proteoglycans. J Biol Chem *275*, 7701–7707.

Wagner, D. D., and Burger, P. C. (2003). Platelets in inflammation and thrombosis. Arterioscler Thromb Vasc Biol *23*, 2131–2137.

Wang, Y., and Kovanen, P. T. (1999). Heparin proteoglycans released from rat serosal mast cells inhibit proliferation of rat aortic smooth muscle cells in culture. Circ Res *84*, 74–83.

Wernersson, S., Braga, T., Sawesi, O., et al. (2009). Age-related enlargement of lymphoid tissue and altered leukocyte composition in serglycin-deficient mice. J Leukoc Biol *85*, 401–408.

White, J. G. (2004). Medich giant platelet disorder: a unique alpha granule deficiency I. Structural abnormalities. Platelets *15*, 345–353.

Wickenhauser, C., Thiele, J., Schmitz, B., et al. (1999). Polycythemia vera megakaryocytes store and release lysozyme to a higher extent than megakaryocytes in secondary polycythemia (polyglobuly). Leuk Res *23*, 299–306.

Woulfe, D. S., Lilliendahl, J. K., August, S., et al. (2008). Serglycin proteoglycan deletion induces defects in platelet aggregation and thrombus formation in mice. Blood *111*, 3458–3467.

Zernichow, L., Abrink, M., Hallgren, J., Grujic, M., Pejler, G., and Kolset, S. O. (2006). Serglycin is the major secreted proteoglycan in macrophages and has a role in the regulation of macrophage tumor necrosis factor-alpha secretion in response to lipopolysaccharide. J Biol Chem *281*, 26792–26801.

Zhang, L., Yang, M., Yang, D., Cavey, G., Davidson, P., and Gibson, G. (2010). Molecular interactions of MMP-13 C-terminal domain with chondrocyte proteins. Connect Tissue Res *51* (3), 230–239.

4 Matrix Proteinases: biological significance in health and disease

4.1 Introduction

Jan-Olof Winberg

Proteinases are hydrolases (EC.3.4) that cleave peptides and proteins either at their terminal ends (exopeptidases) or inside the peptide chain (endopeptidases). They have been found in all organisms and have fundamental roles in many biological processes. Dysregulation of one or several proteinases has been associated with a wide range of diseases (Cawston and Wilson, 2006; van Goor et al., 2009; Vassar et al., 2009; Artenstein and Opal, 2011; Gialeli et al., 2011; Hadler-Olsen et al., 2011; Murphy and Nagase, 2011). In addition, many of the venoms/toxins/poisons of animals, plants, fungi, and microorganisms contain proteinases (Yang et al., 2007; Fox and Serrano, 2009; Costa Jde et al., 2010; Harrison and Bonning, 2010; Shinoda and Miyoshi, 2011). These enzymes are localized either inside the cells in various organelles, inside membranes (intramembrane), or pericellularly within the extracellular matrix (ECM) or associated with the cell membrane.

Proteinases are classified into five classes based on the chemical group that participates in the hydrolysis, and these are aspartic, cysteine, threonine, serine, and metalloproteinases. The classes of proteinases can be further divided into clans and families based on their primary, secondary, and tertiary structures as described in the MEROPS database (http://www.merops.ac.uk) (Rawlings et al., 2010). There are assumed to be more than 66,000 proteinases based on protein sequences, and these have been classified into 50 clans and 184 families (Page and Di Cera, 2008; Artenstein and Opal, 2011).

The human degradome (i.e. the complete set of proteinases produced by human cells) contains at least 569 proteinases distributed between the five classes (Lopez-Otin and Matrisian, 2007; Overall and Blobel, 2007). The number of intracellular (including intramembrane) and extracellular proteinases from the different classes is shown in ▶Figure 4.1. Also shown are some of the known physiological functions in which proteinases are involved, as well as some pathological conditions that are associated with an increased expression of one or several proteinases. Rats and mice are often used as models to study the physiological and pathological roles of proteinases. The murine degradome contains a larger number of proteinases than the human degradome (i.e. at least 644 proteinases). Of these, 341 are extracellularly located, and 303 have an intracellular location of which 16 are intramembrane enzymes (Overall and Blobel, 2007).

One or several proteinases are elevated in various diseases, and some examples are listed in the figure. The role of these enzymes appears to be dual (i.e. the proteinase may either promote or prevent disease progression). Which of these two roles prevails depends on the time of the proteinase's expression and the substrate on which it acts. The substrate specificity has been characterized for a large number of the proteinases, and these characterizations have been mainly performed by *in vitro* studies. A protein that has been shown to be a substrate for a proteinase from in vitro studies may not necessarily be a substrate for the proteinase during *in vivo* conditions. Whether a given

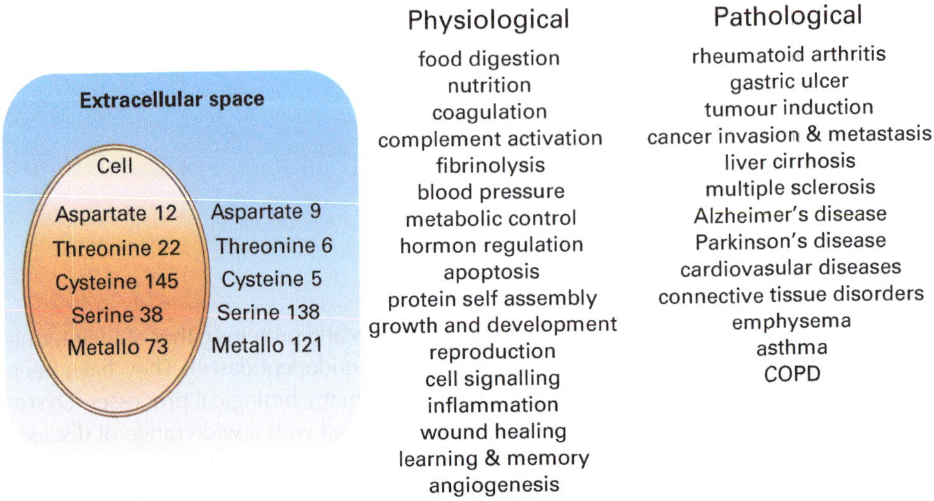

Figure 4.1 The human proteinase degradome and its function. In humans, there are at least 569 proteinases, and the number of extracellular- and intracellular-located enzymes is shown. It has been estimated that 16 of the 290 intracellular proteinases are located within membranes (Overall and Blobel, 2007). Also shown are some of the known physiological functions of proteinases, as well as some pathological conditions in which one or several proteinases are overexpressed. COPD, chronic obstructive pulmonary disease.

protein is a substrate for a given proteinase *in vivo* depends on several factors, such as a timely expression of the proteinase and its substrate at the same location. Furthermore, the interactions between the potential substrate and other molecules may either hide or generate new proteinase cleavage sites. Similarly, the interaction of a proteinase with other molecules, including ECM molecules, may alter both the activity and substrate specificity of the proteinase.

During the past decade, several new techniques have been developed to study *in vivo* proteolysis, and some of these will be presented in this part of the book (see Chapter 4.7 by Dufour and Overall). Another interesting discovery is that many intracellular proteins are potential substrates for extracellular proteinases such as matrix metalloproteinases (MMPs) (Butler and Overall, 2009; Cauwe and Opdenakker, 2010). One reason for this may be that during cell death in which intracellular proteins are released from the cells, the proteinases in the extracellular space act as cleaners and degrade the released proteins. However, proteinases that are normally secreted will under certain circumstances be found inside the cells and have a function there. One example is matrix metalloproteinase-2 (MMP2), which has been found inside cardiac muscle cells where it acts on specific intracellular proteins such as titin and plays a role in various cardiovascular diseases (Ali et al., 2010; Kandasamy et al., 2010). Another example is membrane-type 1 matrix metalloproteinase (MT1-MMP), which can be internalized and act on pericentrin. This promotes genomic instability and hence a proteolysis-driven oncogenesis (Strongin, 2006; Golubkov and Strongin, 2007). A third example of a mislocalized MMP is matrix metalloproteinase-26 (MMP26), which in the intracellular space can cleave estrogen receptor-beta (ER-β) and have an antitumorigenic effect (Strongin, 2006). This part of the book will not cover cases where extracellular proteinases act at so-called mislocated or unusual sites, but there are several recent reviews that describe this phenomenon (Strongin, 2006; Golubkov and Strongin, 2007; Butler and Overall, 2009; Cauwe and Opdenakker, 2010; Hadler-Olsen et al., 2011).

In this section of the book, we have focused on some of the most well-characterized extracellular-localized cysteine-, serine-, and metalloproteinases and their role in development, homeostasis, remodeling, inflammation, wound healing, and disease.

One chapter is devoted to cysteine proteinases (Chapter 4.2 by Johansen and Solberg), of which only approximately 3% of the enzymes are localized extracellularly (▶Figure 4.1). The various clans and families and their endogenous inhibitors are described. It is notable that an enzyme like legumain, which is mainly localized in the lysosomes and has a slightly acidic pH optimum (~5.5), has also been found secreted into the ECM in acidic tumor microenvironments. This cysteine proteinase is believed to have a regulatory role by cleaving/activating other proteinases such as procathepsins B and L and promatrix metalloproteinase-2 (pro-MMP2). Johansen and Solberg also describe known ECM substrates for cysteine proteinases in various diseases like cancer, atherosclerosis, and osteoporosis, as well as cysteine proteinases as pharmacological targets.

The chapter by Svineng et al. (Chapter 4.3) is solely devoted to the serine proteinases plasmin and their activators urokinase plasminogen activator (uPA) and tissue-type plasminogen activator (tPA) in addition to the endogenous inhibitors of these enzymes. This chapter also focuses on plasmin substrates and the role of the plasmin system in cancer, as well as the processing of plasminogen that generates the antiangiogenetic factor angiostatin.

The other four chapters are mainly devoted to metalloproteinases. The chapter by Fadnes et al. (Chapter 4.4) focuses on the complex formation of individual MMPs with other enzymes, ECM molecules as well as non-ECM molecules, and the biological role of these enzyme complexes.

The chapter by Nagase (Chapter 4.5) describes a more recently discovered group of metalloproteinases called ADAMTSs (a disintegrin and metalloproteinases with thrombospondin motifs). They are secreted multidomain metalloproteinases whose functions include N-terminal procollagen processing; cleavage of extracellular proteoglycans such as aggrecan, versican, and brevican; inhibition of angiogenesis; and blood-clotting homeostasis due to the proteinase that specifically cleaves a large multimeric von Willebrand factor. They also play a role in organogenesis, ovulation, fertility, and wound healing. Uncontrolled activities are associated with arthritis and thrombotic thrombocytopenic purpura. In this chapter, the structure and biological and pathological functions of ADAMTSs are also discussed.

The chapter by Toriseva and Kähäri (Chapter 4.6) has a focus on the role of some individual enzymes in the MMP, ADAM (a disintegrin and metalloproteinases), and ADAMTS families, as well as the plasminogen activator (PA)–plasmin system in different stages of wound healing.

The last chapter, by Dufour and Overall (Chapter 4.7), deals with metalloproteinases and describes the role of MMPs in cancer and inflammation, as well as MMPs as drug targets in clinical trials. In addition, a main focus of this article is on the proteinase web and new techniques that can be used to study the degradome.

References

Ali, M. A., Cho, W. J., Hudson, B., Kassiri, Z., Granzier, H., and Schulz, R. (2010). Titin is a target of matrix metalloproteinase-2: implications in myocardial ischemia/reperfusion injury. Circulation *122*, 2039–2047.

Artenstein, A. W., and Opal, S. M. (2011). Proprotein convertases in health and disease. N Engl J Med *365*, 2507–2518.

Butler, G. S., and Overall, C. M. (2009). Updated biological roles for matrix metalloproteinases and new "intracellular" substrates revealed by degradomics. Biochemistry *48*, 10830–10845.

Cauwe, B., and Opdenakker, G. (2010). Intracellular substrate cleavage: a novel dimension in the biochemistry, biology and pathology of matrix metalloproteinases. Crit Rev Biochem Mol Biol *45*, 351–423.

Cawston, T. E., and Wilson, A. J. (2006). Understanding the role of tissue degrading enzymes and their inhibitors in development and disease. Best Pract Res Clin Rheumatol *20*, 983–1002.

Costa Jde, O., Fonseca, K. C., Garrote-Filho, M. S., et al. (2010). Structural and functional comparison of proteolytic enzymes from plant latex and snake venoms. Biochimie *92*, 1760–1765.

Fox, J. W., and Serrano, S. M. (2009). Timeline of key events in snake venom metalloproteinase research. J Proteomics *72*, 200–209.

Gialeli, C., Theocharis, A. D., and Karamanos, N. K. (2011). Roles of matrix metalloproteinases in cancer progression and their pharmacological targeting. FEBS J *278*, 16–27.

Golubkov, V. S., and Strongin, A. Y. (2007). Proteolysis-driven oncogenesis. Cell Cycle *6*, 147–150.

Hadler-Olsen, E., Fadnes, B., Sylte, I., Uhlin-Hansen, L., and Winberg, J. O. (2011). Regulation of matrix metalloproteinase activity in health and disease. FEBS J *278*, 28–45.

Harrison, R. L., and Bonning, B. C. (2010). Proteases as insecticidal agents. Toxins (Basel) *2*, 935–953.

Kandasamy, A. D., Chow, A. K., Ali, M. A., and Schulz, R. (2010). Matrix metalloproteinase-2 and myocardial oxidative stress injury: beyond the matrix. Cardiovasc Res *85*, 413–423.

Lopez-Otin, C., and Matrisian, L. M. (2007). Emerging roles of proteases in tumour suppression. Nat Rev Cancer *7*, 800–808.

Murphy, G., and Nagase, H. (2011). Localizing matrix metalloproteinase activities in the pericellular environment. FEBS J *278*, 2–15.

Overall, C. M., and Blobel, C. P. (2007). In search of partners: linking extracellular proteases to substrates. Nat Rev Mol Cell Biol *8*, 245–257.

Page, M. J., and Di Cera, E. (2008). Serine peptidases: classification, structure and function. Cell Mol Life Sci *65*, 1220–1236.

Rawlings, N. D., Barrett, A. J., and Bateman, A. (2010). MEROPS: the peptidase database. Nucleic Acids Res *38*, D227–233.

Shinoda, S., and Miyoshi, S. (2011). Proteases produced by vibrios. Biocontrol Sci *16*, 1–11.

Strongin, A. Y. (2006). Mislocalization and unconventional functions of cellular MMPs in cancer. Cancer Metastasis Rev *25*, 87–98.

van Goor, H., Melenhorst, W. B., Turner, A. J., and Holgate, S. T. (2009). Adamalysins in biology and disease. J Pathol *219*, 277–286.

Vassar, R., Kovacs, D. M., Yan, R., and Wong, P. C. (2009). The beta-secretase enzyme BACE in health and Alzheimer's disease: regulation, cell biology, function, and therapeutic potential. J Neurosci *29*, 12787–12794.

Yang, J., Tian, B., Liang, L., and Zhang, K. Q. (2007). Extracellular enzymes and the pathogenesis of nematophagous fungi. Appl Microbiol Biotechnol *75*, 21–31.

4.2 Extracellular functions of cysteine proteases

Harald Thidemann Johansen and Rigmor Solberg

4.2.1 Introduction

Cysteine proteases are enzymes that use a cysteine residue as a nucleophile to cleave peptide bonds. The cleavage is hydrolytic by nature and irreversible. In addition to the cysteine, a histidine residue is also normally present in the catalytic site of this class of enzymes.

Cysteine proteases are numerous in all life-forms and have evolved since the appearance of life some 3 billion years ago. It is now possible to organize these proteins in families based on similarities in the primary structures. In addition, clusters of families with similarities in secondary and/or tertiary structures can be described. These clusters of families have been denoted "clans" and reflect a common evolutionary relationship that has been mostly lost in the sequences but retained in the protein structures (Barrett and Rawlings, 2001). A complete and updated overview of all proteases (as well as protease inhibitors) can be found in the MEROPS database (http://www.merops.ac.uk) (Rawlings et al., 2012), where information on single enzymes, families, clans, organisms, and nomenclature is easily accessible.

Proteins and peptides present in the extracellular matrix (ECM) are potential substrates for cysteine proteases. It is thus relevant to consider which cysteine proteases can be expressed by the mammalian genome, as well as their localization, activation, and substrate specificities. According to the MEROPS database, four clans of cysteine proteases are present in humans. The clan CA primarily comprises the families C1 (cathepsins) and C2 (calpains), but also a large number of deubiquitinating enzymes in the family C12. In clan CD, the families C13 (legumain), C14 (caspases), and C50 (separase) are included. There are notable differences in substrate specificities between clans CA and CD. Whereas cathepsins and calpains prefer a hydrophobic amino acid in position P2 (according to the nomeclature of Schechter and Berger, 1967), legumain and caspases prefer a specific amino acid in position P1 (asparagine and aspartate, respectively). To complete the picture of classification, clan CE is represented by family C48 (the small ubiquitin-like modifier [SUMO] proteases named sentrin-specific proteases [SENPs] that deconjugate SUMOylated proteins), and clan CF includes pyroglutamyl peptidase.

From a traditional point of view, a general picture emerges in which cysteine proteases primarily have intracellular localizations and functions. Cathepsins and legumain have been described as lysosomal enzymes, which generally require acidic environment for full activity. Caspases have a central role in cell death mechanisms (apoptosis), and numerous intracellular substrates have been described. Calpains have a cytosolic localization, and the enzymes involved in deubiquitination/deSUMOylation take part in highly regulated processes in the cytosol.

However, during the past decade or so, this restricted description of cysteine proteases has been modified, and it now seems clear that extracellular substrates are possible, real, and important in both physiology and pathophysiology. This raises the question of whether ECM and/or proteins/peptides attached to ECM could be modified or degraded by cysteine proteases in situations in which they are released to the extracellular space. When comparing the different families/clans of cysteine proteases mentioned previously, it varies to what extent they have been evaluated for possible extracellular functions. The reason could of course be that no such activity is present, but it could also be that such functions have not been addressed in a systematic way so far. Thus, recent reviews on calpain and caspases do not discuss the possibility that these enzymes could be released from cells to perform proteolytic activity extracellularly (Li and Yuan, 2008; Storr et al., 2011). However, it should be pointed out that an effect on ECM could be established by many indirect mechanisms. For example, it has been shown that inhibition of calpain reduces both expression and secretion of matrix metalloproteinase-2 (MMP-2) and matrix metalloproteinase-9 (MMP-9) in monocytic THP-1 cells (Popp et al., 2003). The one family of cysteine proteases that has been studied with respect to extracellular functions is the cysteine cathepsins. Therefore, most of the remainder of this chapter will concern this group of enzymes and their inhibitors.

4.2.2 Cysteine proteases and their inhibitors

Cathepsins

The family C1 of cysteine cathepsins comprises 11 members in humans, and in an extensive search for genes coding for new cysteine cathepsins, it was concluded that this is probably the final number (Rossi et al., 2004). Cathepsins B, C (also known as dipeptidyl peptidase I), F, H, and L are considered to be widely expressed in many tissues, while others such as cathepsins K and S have a more restricted and highly regulated expression. All these proteases are produced as preproenzymes and lose their signal peptides during transportation to the lysosomal system. Some proenzymes then undergo autoactivation stimulated by the lysosomal acidic pH and/or interactions with glycosaminoglycans, while others seem to be dependent on other proteases to acquire activity (Rozman et al., 1999; Vasiljeva et al., 2005; Pungercar et al., 2009). The proenzymes are inactive zymogens in which the propeptide blocks access of substrates to the active site. Accordingly, released propeptides have been shown to act as inhibitors of the active enzymes. The active forms of these enzymes have molecular weights of approximately 30 kDa and primarily lysosomal localizations.

Most cysteine cathepsins are endopeptidases, while some have exopeptidase activity (cathepsins B, C, H, and X). In fact, cathepsin B functions both as an exo- and endopeptidase (Barrett and Kirschke, 1981). The traditional view concerning the functions of these enzymes is that they participate in general degradation of cellular proteins (e.g. after cellular uptake of proteins; Barrett, 1992). However, distribution to cellular locations other than the lysosomes and to the extracellular environment has been described recently and will be discussed below. There seems to be a degree of redundancy between the cysteine cathepsins, meaning that the loss of activity of one particular enzyme, either by gene deletion or enzyme inhibition, does not necessarily result in changes in cellular functions or a characteristic phenotype. Still, some of the enzymes perform specialized functions and are considered interesting pharmacological targets. Loss-of-function mutations in the gene encoding cathepsin K result in pycnodysostosis (Gelb et al., 1996). But also loss of the more widely expressed cathepsins give rise to organ-specific symptoms, such as skin anomalies and hair loss in mice after inactivation of the gene encoding cathepsin L (Roth et al., 2000) and the clinical syndrome of Papillon-Lefevre due to cathepsin C-deficiency (Toomes et al., 1999). Finally, it should be mentioned that not all cathepsins are cysteine proteases. The term *cathepsin* is derived from the Greek word for digestion and has also been given to other proteolytic enzymes from other classes, such as serine (cathepsins A and G) and aspartic proteases (cathepsins D and E; see MEROPS for further information). In the following text, the term *cathepsin* will refer only to the cysteine class of these proteases.

Legumain

In 1997, a new cysteine protease was discovered in mammalian lysosomes (Chen et al., 1997). The enzyme was named legumain due to a similar enzyme already described in plants, but it is now also referred to as asparaginyl endopeptidase (AEP). The latter name indicates two characteristics of the enzyme: it is an endopeptidase, and it cleaves only carboxyterminally to the amino acid asparagine. This type of cleavage is very unusual for an endopeptidase and probably the reason for the late discovery of legumain. Legumain is unrelated to the cathepsins but still shares several of their characteristics. It has primarily a lysosomal localization, needs acidic pH for activation, undergoes autoactivation, and surprisingly, is also inhibited by cystatins (Alvarez-Fernandez et al., 1999). The closest relatives of legumain are the caspases, which also share the strict requirement for a specific amino acid in position P1 (Chen et al., 1998). The very selective and exclusive cleavage of asparagine bonds is at odds with a general digestion function in the lysosomes. Thus, legumain has been hypothesized to play a more upstream regulatory role by cleaving/activating other proteases. Such functions have been described in relation to processing of cathepsin B and L, and also the activation of proMMP-2 (Chen et al., 2001; Chan et al., 2009; Miller et al., 2011). Although legumain is primarily stored in the lysosomal compartments, the enzyme has also been reported to appear extracellularly in the acidic tumor microenvironment and associated with matrix, as well as colocalized to integrins on cell surfaces (Liu et al., 2003; Wu et al., 2006). Therefore, extracellular functions of legumain have been suggested.

4.2.3 Endogenous inhibitors of cysteine proteases

Natural inhibitors of cysteine proteases are members of the cystatin superfamily, and there are three major families or types of cystatins in mammals: type 1 (cystatins A and B, also called stefins), type 2 (cystatins C, D, E/M, F, G, S, SA, and SN), and type 3 (L- and H-kininogens). In addition, several proteins have been identified that contain cystatin-like domains but lack protease inhibitory activities (e.g. histidine-rich glycoprotein [HRG], cystatin-related protein [CRP], and fetuin A), as well as proteins with inhibitory potential toward cysteine proteases but without cystatin inhibitory motifs (e.g. cathelin). Also, cystatins are found in plants (phytocystatins), protozoa, and fungi, and probably also in bacteria (Kordis and Turk, 2009). All cystatins contain a common structure called a cystatin domain/motif or the "cystatin fold." The stefins (MEROPS subfamily I25A) are primarily located intracellularly in the cytosol but may also be secreted and appear in body fluids (Abrahamson et al., 2003). They are single-chain proteins of approximately 100 amino acids and with neither carbohydrate side chains nor disulfide bonds. The type 2 cystatins (MEROPS subfamily I25B) are mainly secreted and thus appear extracellularly in a variety of human body fluids; they will be further explored in the next section. These are synthesized as preproteins with a 20 to 26 amino acid export signal, whereas the single-chain mature proteins are of approximately 120 amino acids (13–14 kDa) and contain two conserved disulfide bridges in the C terminal. The kininogens (MEROPS subfamily I25B) are located intravascularly and are proteins of 60–120 kDa containing various domains, including three tandemly repeated type 2–like cystatin domains.

Type 2 cystatins

The first member of this superfamily of enzyme inhibitors (ovocystatin or chicken egg white cystatin) showed specific inhibition of papain and related cysteine proteases (Barrett 1986). All functional type 2 cystatins are cysteine protease inhibitors of the papain (C1) family, and some also inhibit the legumain (C13) family (Abrahamson et al., 2003), but their inhibition profiles are different (▶Table 4.1). Interestingly, some of the cystatins are able to inhibit C1- and C13-family proteases simultaneously because of different binding sites. Body fluids like spinal fluid and semen are especially rich in cystatins (Abrahamson et al., 2003).

The cystatins regulate normal body processes through regulation of the activity of cysteine proteases, which can cause diseases when overexpressed, unless their activity is firmly controlled. Thus, loss of cystatin function results in uncontrolled activity of cysteine proteases and may lead to a variety of disorders, including chronic inflammatory reactions (Henskens et al., 1996) and faulty differentiation in epidermis (Zeeuwen, 2004). In addition, it has been suggested that the change in the balance between cysteine proteases and cystatins could alter the susceptibility to cancer and promote atherosclerosis (Dickinson 2002; Liu et al., 2003; Shridhar et al., 2004; Zhang et al., 2004; Vigneswaran et al., 2006; Keppler, 2006; Werle et al., 2006).

The type 2 cystatins (cystatins C, D, E/M, F, G, S, SA, and SN) are encoded by genes mainly clustered on chromosome 20 (Dickinson et al., 1994). Cystatins E/M, F, and G have low sequence homology with the other type 2 members (<35% versus >50% sequence homology), but all type 2 cystatins have similar protein domain structures and form reversible, high-affinity complexes with their target cysteine proteases (Bobek and Levine, 1992). Although the secretory nature of type 2 cystatins does not make them obvious candidates to control the lysosomal active cysteine proteases, it has been shown that during tumorigenesis, these proteases can appear extracellularly on the surface of tumor cells and act together with matrix metalloproteases (MMPs) and serine and aspartic proteases to dissolve ECM proteins and thus are regulated by the cystatins in this microenvironment. Interestingly, cystatins occur in all body compartments and fluids in which MMPs are also present, and some cystatins are able to stabilize and protect the MMPs from autolytic degradation and thus prolong (and not inhibit) their activities.

Among the various type 2 cystatins, cystatin C (CST3) is most thoroughly studied in mammals and is a ubiquitously secreted protein found in all tissues and body fluids. Mature

Table 4.1 **Endogenous secreted cystatins and their inhibitory profiles (K_i) against papain and mammalian cysteine proteases.**

Inhibitor	*Enzymes inhibited*	K_i	*Reference*
Cystatin C (CST3)	Papain	0.000011 nM	Abrahamson (1994)
	Cathepsin L	<0.005 nM	Abrahamson (1994)
	Cathepsin S	0.008 nM	Abrahamson (1994)
	Legumain	0.20 nM	Alvarez-Fernandez et al. (1999)
	Cathepsin B		
	Cathepsin H	0.25 nM	Barrett et al. (1986)
	Cathepsin C (dipeptidyl	0.28 nM	Barrett et al. (1986)
	peptidase I)	3.5 nM	Barrett et al. (1986)
Cystatin D (CST5)	Papain	1.2 nM	Abrahamson (1994)
	Cathepsin S	0.24 nM	Abrahamson (1994)
	Cathepsin H	8.5 nM	Abrahamson (1994)
	Cathepsin L	25 nM	Abrahamson (1994)
	Cathepsin B	>1,000 nM	Abrahamson (1994)
Cystatin E/M (CST6)	Papain	0.39 nM	Ni et al. (1997)
	Legumain	0.0016 nM	Alvarez-Fernandez et al. (1999)
	Cathepsin V		
	Cathepsin L	0.47 nM	Cheng et al. (2006)
	Cathepsin B	1.78 nM	Cheng et al. (2006)
		32 nM	Ni et al. (1997)
Cystatin F (CST7)	Papain	1.1 nM	Ni et al. (1998)
	Cathepsin L	0.31 nM	Ni et al. (1998)
	Legumain	10 nM	Alvarez-Fernandez et al. (1999)
Cystatin S (CST4)	Papain	108 nM	Abrahamson (1994)
Cystatin SA (CST2)	Papain	0.32 nM	Abrahamson (1994)
Cystatin SN (CST1)	Papain	0.016 nM	Abrahamson (1994)
	Cathepsin B	19 nM	Abrahamson (1994)
	Cathepsin C (dipeptidyl peptidase I)	100 nM	Tseng et al. (2000)
Kininogen	Papain	0.015 nM	Barrett et al. (1986)
	Cathepsin L	0.017 nM	Barrett et al. (1986)
	Cathepsin H	1.2 nM	Barrett et al. (1986)
		0.72 nM	Abrahamson (1994)
	Calpain C	1 nM	Barrett et al. (1986)
	Cathepsin B	600 nM	Barrett et al. (1986)

human cystatin C is composed of 120 amino acids and is synthesized as a preprotein with a 26-residue signal peptide (Abrahamson et al., 1987). The concentration of cystatin C in normal human serum is approximately 77 nM (1.16 µg/mL), has low molecular weight (15 kDa), and is efficiently eliminated by glomerular filtration. Consequently, these and other features make cystatin C a suitable marker of the glomerular filtration rate (GFR).

Recently, the internalization of secreted cystatin C as a result of an uptake process was described (Ekström et al., 2008; Wallin et al., 2010), and this could be a feature of all secreted cystatins in controlling cellular protease activity. Cystatin C is considered to be a general extracellular cysteine protease inhibitor and a powerful inhibitor of cathepsins S and L (for K_i, see ▶Table 4.1). In addition to the pathophysiological conditions discussed subsequently, cystatin C has also been linked to neurogenerative diseases characterized by cerebrovascular amyloid deposition (cerebral amyloid angiopathy [CAA]) due to colocalization with amyloid-beta protein (Aβ) in sporadic CAA (Fujihara et al., 1989; Maruyama et al., 1990), although the role of this association is not understood. A mutation in cystatin C (L68Q) has also been linked to the dominantly inherited hereditary cystatin C amyloid angiopathy (HCCAA) in Iceland (reviewed in Olafsson and Grubb, 2000).

The other type 2 cystatins show a more restrictive expression. Cystatin D (CST5) is found in parotid glands, saliva, and tears (Freije et al., 1991,1993) and is believed to have a protective role against proteases present in the oral cavity. Also, cystatin D was recently shown to be a tumor suppressor candidate in colon cancer and to be inducible by vitamin D (Alvarez-Díaz et al., 2009). Cystatin E/M (CST6) is secreted in both a glycosylated (17 kDa) and an unglycosylated (14 kDa) form, the former due to an *N*-glycosylation site. In contrast to other type 2 cystatins, the *CST6* gene is located on chromosome 11 (Stenman et al., 1997). The mature protein contains 121 amino acids, and high expression is found in cutaneous epithelia (Zeeuwen et al., 2001), although it is also expressed in a variety of normal human cells and tissues (Ni et al., 1997; Sotiropolou et al., 1997). Cystatin E/M has the highest inhibitory potential versus legumain (K_i 0.0016 nM; ▶Table 4.1). Cystatin F (CST7; also known as leukocystatin or cystatin-like metastasis-associated protein [CMAP]) is found in serum and pleural fluid (Ni et al., 1998; Werle et al., 2003), but in contrast to the other type 2 cystatins, a large fraction is located intracellularly in lysosomal granules (Nathanson et al., 2002; Langerholc et al., 2005). The mature protein is composed of 126 amino acids (Abrahamson et al., 2003) being predominantly expressed in hematopoietic cells (Keppler, 2006) and has two *N*-glycosylation sites (Ni et al., 1998; Nathanson et al., 2002). Cystatin G is also known as cystatin-related epididymal and spermatogenic (CRES) protein, specifically expressed in mouse sex glands (Hsia and Cornwall, 2001). The function of cystatin G has not been fully explored, but it may function as an inhibitor of C1-cysteine proteases (Abrahamson et al., 2003). Cystatins S (CST4), SA (CST2), and SN (CST1) are called the "salivary cystatins" and appear in saliva, tears, urine, and seminal plasma (Abrahamson et al., 1986). Due to their expression profiles, these are thought to be glandular cystatins together with cystatin D and function as defense inhibitors (Wallin et al., 2010). Their amino acid sequence homology is high (90%), and they are believed to have evolved after gene duplication late in evolution. These cystatins are composed of 121 amino acids having export signal peptides but no glycosylation sites.

4.2.4 Cysteine proteases and their inhibitors in diseases

Cancer

Numerous studies have linked cathepsins to the development of cancer. The relevant literature was reviewed in 2006 (Mohamed and Sloane, 2006), and the role of cathepsin L was recently summarized (Lankelma et al., 2010). It is generally assumed that upregulation of cathepsins promotes the various aspects of cancer progression. Still, with the diverse functions of MMPs in mind, the possibility of an anticancer function of some cathepsins in specific cancer forms should not be overlooked. In fact, it was recently found that cathepsin L could function as an antitumor protease in keratinocytes (Dennemarker et al., 2010). Still, most studies conclude with a positive correlation between expression of cathepsins (especially B and L) and malignancy. The transcriptional regulation of cathepsins has not been described in detail so far. Transcript variants of cathepsins B and L have been studied, and the involvement of epigenetic mechanisms like methylation-dependent silencing of cathepsin L has been demonstrated in lymphoma cells (Jean et al., 2006).

When it comes to understanding how cathepsins could promote malignancy, several mechanisms of importance are not highlighted herein, including cell proliferation and apoptosis. In relation to direct effects on ECM, two aspects of tumor biology seem important: generation of new vessels to the tumor (angiogenesis) and direct solubilization of ECM necessary for invasive growth and metastasis.

A growing tumor needs to remodulate its microenvironment to expand in volume and to escape to the circulation for subsequent formation of distant metastases. Cathepsins are capable of degrading most proteins of the ECM (see ▶Table 4.2). Proteins can undergo endocytosis and be degraded in the lysosomal apparatus, but even more interesting is the documented release of cathepsins and legumain to the extracellular space

Table 4.2 Mammalian cysteine proteases and their known involvement in diseases and cleavage of extracellular matrix proteins.

Protease (gene)	*Known involvement in*	*Known extracellular matrix substrates (reference)*
Cathepsin B (*CTSB*)	Cancer, atherosclerosis, bone remodeling	Osteocalsin (Baumgrass et al., 1997)
Cathepsin F (*CTSF*)	Atherosclerosis	
Cathepsin H (*CTSH*)		Osteocalsin (Baumgrass et al., 1997)
Cathepsin K (*CTSK*)	Atherosclerosis, bone remodeling	Collagen type I (Kafienah et al., 1998) Collagen type II (Kafienah et al., 1998) Elastin (Yasuda et al., 2004) Osteonectin (Bossard et al., 1996)
Cathepsin L (*CTSL1*)	Cancer, atherosclerosis	Collagen type I (Nosaka et al., 1999) Osteocalsin (Baumgrass et al., 1997) Procollagen XVIII (Felbor et al., 2000)
Cathepsin S (*CTSS*)	Atherosclerosis, antigen presentation	Aggrecan (Hou et al., 2001) Elastin (Yasuda et al., 2004) Fibronectin (de Nooijer et al., 2009) Osteocalsin (Baumgrass et al., 1997)
Cathepsin V (*CTSL2*)		Elastin (Yasuda et al., 2004)
Legumain (*LGMN*)	Cancer, atherosclerosis	Fibronectin (Morita et al., 2007)

allowing a direct degradation of ECM components (Bailey et al., 2011). The latter is well documented for cathepsins B and L (Roshy et al., 2003; Collette et al., 2004) and was recently shown to take place in thyroid carcinoma cells (Tedelind et al., 2011). Secreted cathepsins are even present in extracellular fluids such as serum and cerebrospinal fluid (CSF) (Nagai et al., 2000; Sundelöf et al., 2010; Åkerfeldt and Larsson, 2011). As mentioned previously, cathepsins normally require an acidic environment for activation. In this regard, tumors have been described to be able to generate such an acidic microenvironment at the tumor border (Rozhin et al., 1994; Gatenby et al., 2006). Components of the ECM also seem to be of great importance for the activation of procathepsins, and glycosaminoglycans have been shown to facilitate the autoactivation of these proenzymes (Ishidoh and Kominami, 1995; Rozman et al., 1999; Caglic et al., 2007). Legumain has also been associated with cancer (Murthy et al., 2005; Gawenda et al., 2007; Briggs et al., 2010), and this cysteine endopeptidase has been suggested to play a role in the activation of proenzymes such as pro-MMP2 and processing of cathepsin B and L (Chen et al., 2001; Shirahama-Noda et al., 2003). Based on the unique cleavage specificity at asparagine, anticancer prodrugs exclusively activated by legumain have been developed (Liu et al., 2003; Stern et al., 2009; Bajjuri et al., 2011). The characterization of cathepsins and legumain as participants in invasive growth could establish these enzymes as pharmacological targets in cancer, although no drug has yet been marketed based on this mechanism of action.

To allow tumor growth beyond a few cubic millimeters, new blood vessels are required to supply nutrients. The process of angiogenesis is thus important for continued cancer growth and must be stimulated. This stimulation is thought to be established by paracrine factors from the tumor stimulating local endothelial cells. The formation of new vessels requires both the generation of new endothelial cells and opening of the ECM to allow the formation of vessels. Cathepsin B has been shown to degrade components of the basal membrane (Buck et al., 1992). Vascular endothelial growth factor (VEGF) is an important stimulator for generation of new endothelial cells, and several mechanisms have been described in which cathepsins such as cathepsin B could regulate the activity of this mediator (Im et al., 2005). Analyses of the secretome of early proangiogenic cells have documented secretion of a wide range of cathepsins and legumain (Urbich et al., 2011). A cathepsin L-inhibitor attenuated cathepsin L release and thus confirmed previous work showing that inhibition of this protease reduces angiogenesis (Urbich et al., 2005). When cathepsins degrade components of the ECM *in vitro,* both pro- and antiangiogenic factors can be generated or destroyed, which is a likely mechanism to explain the effects of cathepsins on angiogenesis. In a recent study, it was shown that knockdown of both cathepsin B and urokinase-type plasminogen receptor (uPAR) caused inhibition of angiogenesis in glioma cells (Malla et al., 2011). The explanation for this effect was a reduced expression of VEGF. Stimulation of uPAR activates the Janus kinase/signal transducer and activator of transcription (JAK/STAT) pathway-dependent expression of VEGF. The role of cathepsin B was hypothesized to be linked to the cleavage of prourokinase to active urokinase, the ligand for uPAR. Also, recently it was shown that basal fibroblastic growth factor (bFGF) induced the upregulation and release of cathepsin L from human skeletal muscle cells (Chung et al., 2011). The authors suggest that cathepsin L plays a direct and crucial role in endothelial cell migration and that the effect is mediated through activation of the c-Jun N-terminal kinase (JNK) pathway.

Expression of cystatin C in most premalignant and malignant cells does not appear to alter much. Nevertheless, some reports have shown elevated cystatin C levels in sera, pleural effusions, and ascitic fluids gathered from cancer patients (reviewed in Keppler, 2006). The clinical value of these high levels of cystatin C in cancer patients, however, has to be clarified. Increased levels of cathepsin B and decreased levels of cathepsin-B/cystatin C-complexes have been observed in lung cancer patients versus patients with noncancerous lung disease or healthy individuals (Zore et al., 2001). Interestingly, when cathepsins B and L are simultaneously present in serum, cathepsin L could displace cathepsin B from its complexes with cystatin C due to pseudo-irreversible binding (K_i 5 pM; Abrahamson 1994). Cystatin C has been shown to inhibit *in vitro* tumor-cell-mediated degradation and invasion of ECM (reviewed in Keppler, 2006). Also, cystatin C has been described as a transforming growth factor-beta (TGF-β) receptor antagonist and thus able to inhibit cellular signaling of TGF-β, and such potent anticancer properties have been observed in models of glioblastoma (Sokol and Schiemann, 2004). As TGF-β has both tumor-promoting and tumor-suppressing functions, cystatin C could thus also have dual effects in this respect. The tissue levels of both cystatin C and cystatin E/M were downregulated in non-small-cell lung cancer (NSCLC) tumors when compared to healthy lung tissue (Werle et al., 2006). However, these decreased levels of type 2 cystatins provided no prognostic information. In an attempt to explore the underlying molecular mechanism in prostate cancer cells *in vitro*, cystatin C was reported to inhibit invasion of cancer cells in cooperation with mitogen-activated protein kinase (MAPK)/extracellular signal-regulated kinase (ERK) and androgen receptor (AR) pathways (Wegiel et al., 2009).

Cystatin E/M has been reported to be strongly involved in cancer (reviewed in Keppler, 2006), and it was initially identified as a cysteine protease inhibitor downregulated in metastatic versus primary breast cancer cells (Ni et al., 1997; Sotiropoulou et al., 1997). Further studies have indicated that loss of cystatin E/M expression during the progression of breast cancer, glioma, and lung cancer results from epigenetic silencing due to hypermethylation of the *CST6* promoter (Ali et al., 2006; Kim et al., 2006; Rivenbark et al., 2006; Schagdarsurengin et al., 2007; Zhong et al., 2007). Restoration of cystatin E/M expression in breast cancer cells reduced growth, migration, and matrigel invasion *in vitro* (Shridhar et al., 2004), as well as tumor growth and metastatic burden *in vivo* (Zhang et al., 2004). In glioma, expression of cystatin E/M inhibited cell motility and invasion *in vitro* (Qiu et al., 2008). Overall, these studies have suggested a tumor suppressor function for this protease inhibitor.

In a highly tumorigenic and metastatic breast cancer cell line (MDA-MB-435S) stably transfected with a cystatin E/M expression vector, it was found that cystatin E/M overexpression reduced cell proliferation, migration, matrix invasion, and tumor-endothelial cell adhesion (Shridhar et al., 2004). Notably, cell migration and matrix invasion appeared to be dependent on cysteine proteases, as both recombinant cystatin E/M and *trans*-epoxysuccinyl-leucylamido(4-guanidino)butane (E64; a general cathepsin inhibitor) blocked such processes. In support of these determinations, the activity of legumain and cathepsins B and L, as well as cell proliferation and *in vitro* invasion were shown to be increased in metastatic oral cancer cell lines when cystatin E/M was silenced by siRNA (Vigneswaran et al., 2006).

Cystatin E/M is highly expressed in cutaneous epithelia (Zeeuwen et al., 2001), and reports have shown considerable reduction or loss of cystatin E/M expression in a number of skin cancer cell lines (Keppler, 2006). It has also been identified as a possible tumor suppressor in malignant melanoma by microarray analysis (Riker et al., 2008). In a recent study, we examined the secretion of cystatins C, D, E/M, and F, as well as the expression of legumain and cathepsins B and L, in various human melanoma cell lines (Briggs et al., 2010). Interestingly, we observed an inverse correlation between the secretion of cystatin C and cystatin E/M. Also, glycosylated cystatin E/M (17 kDa) was predominantly secreted compared to the unglycosylated form. Likewise, a similar inverse correlation between cystatin E/M and both legumain and cathepsin B was observed. Notably, the cell lines predominantly expressing the glycosylated form of cystatin E/M also expressed low levels of legumain. Also, overexpression of cystatin E/M in melanoma cells resulted in significant inhibition of intracellular legumain activity and decreased invasiveness into matrigel. Collectively, these data suggest that cystatin E/M contributes to inhibiting melanoma progression by suppressing legumain activity, but the interplay between the protease and its inhibitor is not fully elucidated. Because cystatin C has been shown to be internalized (Ekström et al., 2008; Wallin et al., 2010), it is tempting to speculate that a similar mechanism exists for other cystatins and that it is responsible for the previously mentioned observations.

Cystatin F has been reported to be highly expressed in a number of metastatic human cancer cell lines, and it has been associated with a higher rate of liver metastasis (reviewed in Keppler, 2006). Interestingly, suppression of this gene has shown reduction in metastasis of tumor cells in liver and spleen, and improved survival of mice bearing tumors. Thus, cystatin F acts in contrast to, and in a completely different manner from, cystatins C and E/M by stimulating rather than suppressing metastasis to the liver, and this could probably be linked to its predominantly intracellular rather than extracellular localization (Nathanson et al., 2002; Langerholc et al., 2005). In addition, a 5-year survival study confirmed the correlation of higher levels of cystatin F to a significantly worse survival rate of the patients (Keppler, 2006).

Cytoplasmic stefins A and B (type 1 cystatins) are regulators of initiation or propagation of the lysosomal cell death pathway, which will not be explored in this chapter. Also, secretion of low levels of stefins A and B are found in normal human serum and other biological fluids. In colorectal cancer patients, the serum stefin B level has been shown to correlate significantly with Dukes' stage, being highest in stage D (Kos et al., 2000). Also, increased levels of stefin B (and cystatin C) correlated with patient survival.

Atherosclerosis

Since the discovery that lysosomal cathepsins can be secreted in the active forms from macrophages (Reddy et al., 1995), the involvement of these proteases in atherosclerosis has been studied in great detail. A role of cysteine proteases in cardiovascular disease and atherosclerosis is based on several lines of evidence and has recently been reviewed (Lutgens et al., 2007; Bai et al., 2010). First, macrophages could perform phagocytosis of ECM components followed by lysosomal degradation. However, the mentioned release of active proteases from macrophages also makes it possible that these enzymes can degrade ECM in a direct manner. Most lysosomal proteases require an acidic environment to gain full activity. This could well be present in an inflamed arterial wall where atherosclerotic processes are in progress. It is well known that activated macrophages are equipped with pumps for protons and lactic acid (Leake, 1997), so the presence of acidic microenvironments is thus likely. The cellular origin of lysosomal cysteine proteases in an atherosclerotic lesion could be smooth muscle cells, endothelial cells, and/or immune cells such as invading macrophages (Liu et al., 2004; Lutgens et al., 2007). Quantitative immunohistochemical analysis of human carotid atherosclerotic lesions has shown that expression of lysosomal cathepsin L was associated mainly with cluster of differentiation (CD)68-positive macrophages (Li et al., 2009). The expressions of cathepsin S and legumain have also been correlated with the expression of CD68, suggesting that these proteases are largely expressed by macrophages in the atherosclerotic lesion (Papaspyridonos et al., 2006). Therefore, it seems reasonable to assume that most of the cathepsins and legumain in these areas originate from macrophages. Still, it has been shown that cathepsin S was present on the plasma membrane of smooth muscle cells and colocalized with αvβ3-integrin, which also serves as a receptor for MMPs (Cheng et al., 2006).

The next question is whether ECM proteins are substrates for the cysteine proteases. Collagen I constitutes 60%–70% of ECM in arteries, and cathepsin K has been shown to be a very efficient collagenase (Brömme and Okamoto, 1995; Garnero et al., 1998; Kafienah et al., 1998; Garnero et al., 2003). The collagenolytic activity of cathepsin K was documented by the generation of the C-terminal telopeptides of type I collagen (CTX1) specifically after cathepsin K cleavage in advanced atherosclerotic lesions. Also, a co-localization of cathepsin K and CTX1 was found. Both stimulated macrophages and foam cells were able to generate the CTX1-fragment from a collagen I–enriched matrigel. The inhibition of CTX1 formation by E64 demonstrated the involvement of cathepsins and pointed to extracellular localization of the protease activity since E64 is not cell permeable (Barascuk et al., 2010).

Elastin is another important component of ECM, and several cathepsins have elastase activity. It was shown that cysteine proteases could account for 60% of elastolytic activity, with approximately equal contributions from cathepsins K, S, and V (Yasuda et al., 2004). Also, *in vitro* data have suggested that both legumain and cathepsin S may mediate pericellular fibronectin breakdown (Morita et al., 2007; de Nooijer et al., 2009). In addition, because legumain can activate proMMP-2 and process cathepsin B and L (Chen et al., 2001; Shirahama-Noda et al., 2003), this protease could enhance the breakdown of ECM by other proteases.

If cathepsins really play an important role in atherosclerotic processes, we should expect that these enzymes are present in higher concentrations in atherosclerotic plaques than in healthy arteries. Also, the inhibition of these enzymes by deletion of genes in experimental animals, siRNA, or enzyme inhibitors would be expected to affect the progression of atherosclerosis. In 1998, the expression of cathepsins K and S was described in atherosclerotic lesions (Sukhova et al., 1998). Since then, numerous reports have shown that both cathepsins and legumain are overexpressed in atherosclerosis. Apolipoprotein E (ApoE) knockout mice fed a Western-type diet develop atherosclerosis, and this is a much favored animal model to study this disease. In 2002, cathepsin B was found to be expressed and upregulated in atherosclerotic lesions, documented by imaging, increased mRNA expression, and immunohistochemistry (Chen et al., 2002). Cathepsin F was also found to be localized to macrophage-rich coronary plaques and was shown to degrade apolipoprotein B-100 (Oorni et al., 2004). Immunohistochemical analysis of human carotid atherosclerotic lesions has also shown increased expression of cathepsin L and legumain (Li et al., 2009). Also, legumain has been shown to be upregulated in unstable, compared to stable, atherosclerotic plaques (Papaspyridonos et al., 2006; Mattock et al., 2010). Cathepsin K, L, or S knockout mice all show reduced atherosclerosis in mice also deficient in apoE or low-density lipoprotein (LDL) receptor (Reiser et al., 2010).

Serum cystatin C is a well-known marker of kidney function and has recently also been reviewed as a prognostic marker in acute coronary diseases due to its function in enhancement of atherosclerosis (Ferraro et al., 2011). Also recently, enhanced serum cystatin C level has been correlated with early stage coronary atherosclerotic plaques (Imai et al., 2011). Thus, cystatin C could be a potential therapeutic target in cardiovascular diseases. Kininogens are multidomain plasma proteins first described as precursors of the vasoactive peptide bradykinin. Numerous reports have also described cofactor functions and cell-binding properties of high-molecular-weight H-kininogen (reviewed in Lalmanach et al., 2010). The presence of three cystatin domains in kininogens was discovered in 1984, and two of these sites have been shown to be active; kininogens are described as potent inhibitors of cathepsins H, K, L, and S (Ohkubo et al., 1984; Muller-Esterl et al., 1985). Taking into account the high concentration of kininogens in plasma, which by far exceeds the concentration of cystatin C, it is tempting to speculate that kininogens could regulate the activity of cathepsins in the blood vessel wall and thus possibly affect the development of atherosclerosis. So far, no proof of such a function has been presented, and rats deficient in kininogens do not develop atherosclerotic lesions but have aneurysms in the abdominal aorta (Kaschina et al., 2005).

Vitronectin is an abundant adhesive glycoprotein in blood plasma and is found associated with different ECM sites, the vessel wall, and tumor cells (reviewed in Preissner and Reuning, 2011). Vitronectin acts as a potent matricellular factor, coordinating cell migration with pericellular proteolysis and growth factor signaling at sites of tissue remodeling or in tumors. Examples of vitronectin ligands are plasminogen activator inhibitor-1 (PAI1) and H-kininogen, which confer strong antiadhesive functions upon integrin- or urokinase-receptor-mediated cell interactions with vitronectin. Structure-function studies of such vitronectin-related ligands and receptors have stimulated the search for new antagonists in tumor angiogenesis, platelet aggregation, or atherosclerosis. Very recently, an H-kininogen fragment of 2,209 Da was identified as a marker for early gastric cancer by matrix-assisted laser desorption-ionization time-of-flight mass spectrometry (MALDI-TOF MS) peptidome profiling of sera from such patients and was also shown to be inversely correlated with the intact protein (Umemura et al., 2011). The protease responsible for the H-kininogen fragmentation is not known, but this indicates that the interplay between proteases and inhibitors is more complex and intricate than thought at first glance.

Bone and cartilage

The best example of a cysteine protease undergoing secretion and performing extracellular proteolysis is cathepsin K (Bossard et al., 1996). This cathepsin is highly expressed and secreted by osteoclasts and degrades the protein content of bone, primarily collagen I (Garnero et al., 1998). Although several factors take part in the regulation of cathepsin-K expression, the major stimulator has been identified as RANKL (receptor activator of nuclear factor kappa-B ligand). The activity of cathepsin K also requires an acidic environment, and the osteoclasts establish this by attaching to the bone surface, sealing off a closed compartment, and then secreting H^+ to establish a local pH of approximately 4.5 (recently reviewed by Brömme, 2011; Costa et al., 2011). The low pH serves a dual purpose, first solubilizing the mineral content of the bone and next securing the activity of cathepsin K. Bone resorption is necessary in securing a constant remodeling of bone. After local removal of old bone materials, osteoblasts enter the area to rebuild new bone and thus secure bone turnover. The most abundant protein in bone is collagen I, and cathepsin K is an excellent collagenase. Other proteases, such as various MMPs and enzymes belonging to the serine class of proteases, are also capable of degrading collagen I, but it now seems that cathepsin K is the most important enzyme in this respect. Also, collagen II present in cartilage and several other ECM proteins can be degraded by cathepsin K. Loss-of-function mutations in the *CTSK* gene encoding cathepsin K result in the disease pycnodysostosis characterized by numerous bone abnormalities (Gelb et al., 1996). In addition, investigations of the cathepsin K knockout mice have underlined the importance of this enzyme in bone homeostasis (Gowen et al., 1999). It is well known that a normally tuned balance between osteoclasts and osteoblasts is required for constant bone mass, and a gradual shift in favor of increased osteoclast activity results in bone loss, especially prominent in postmenopausal women, giving rise to osteoporosis. Both cathepsin B and cathepsin L have been found in the synovial fluid from patients with rheumatoid arthritis (Lemaire et al., 1997). Inflammatory cytokines induce an increased secretion of these enzymes, which then could participate in the degradation of bone and cartilage. This also applies to cathepsin K, which can be upregulated by interleukin-1α (IL-1α) and tumor necrosis factor-alpha (TNF-α) (Kudo et al., 2002).

4.2.5 Pharmacological targeting of cysteine proteases

Taking into account the possible detrimental consequences of cysteine protease activity in diseases like cancer, atherosclerosis, and osteoporosis, it is obvious that the exploration of cathepsins as pharmacological targets has received great attention (Vasiljeva et al., 2007). In general, inhibition of enzymes has been a successful strategy in pharmacology for many years, including drugs inhibiting proteolytic enzymes like the angiotensin-converting enzyme and the HIV protease. However, when it comes to the cysteine proteases, there is still inadequate knowledge of cellular functions, *in vivo* substrates, localization, and regulation of expression and activity to be able to predict what will happen if individual enzymes are targeted (Brix et al., 2008). Also, the cystatins could potentially be used as drugs to control cysteine protease activity, but they are not specific (▶Table 4.1). The epoxysuccinyl peptide derivatives (E64 and its synthetic analogues) have acquired most attention among the many cysteine protease inhibitors that have been isolated from microorganisms (Leung-Toung et al., 2006). Recently, several epoxysuccinyl-based inhibitors have been reported to have selective and potent inhibitory effects against papain-like cysteine proteases, both *in vivo* and *in vitro* (Sadaghiani et al., 2007). Probably the best strategy will be to develop inhibitors with high selectivity toward single cathepsins instead of global inhibitors that will affect many or all members in the family, like the reference compound E64. However, the effects of inhibiting a single cathepsin could be blunted by the phenomenon of redundancy, meaning that other enzymes may compensate for the lost activity.

The strategy for developing inhibitors of cysteine cathepsins has been based on targeting the active site of the enzymes. A peptide sequence that mimics known substrates of a particular enzyme can be coupled to a "warhead" group that reacts covalently with the cysteine in the active site. Such a strategy can be successful in effectively quenching the activity of an enzyme *in vitro*, but the prospect of irreversible inactivation of enzymes raises many concerns. It is expected that selectivity can be lost because given enough time during continuous drug treatment, a compound with good *in vitro* specificity against one protease could gradually also affect the activities of other members of the same protease family. Also, this field of drug development will not avoid encountering the many obstacles in the field of pharmacokinetics (absorption, distribution, metabolism, and excretion [ADME]), which is often overlooked by basic researchers.

As expected, the development of inhibitors of highly regulated cathepsins is in the forefront of drug development. Targeting ubiquitously expressed enzymes like cathepsins B and L is considered to be a more daring approach than targeting enzymes with a more restricted pattern of expression. Cathepsin S is mainly localized to lymphoid organs where it plays important roles in major histocompatibility complex class II (MHC II)–mediated antigen presentation by B lymphocytes, macrophages, and dendritic cells (Riese et al., 1996). Cathepsin S also degrades proteoglycans such as aggrecan and is secreted from macrophages and found in elevated levels in the synovial fluid from rheumatoid arthritis patients (Hou et al., 2001). The role of cathepsin S in MHC II antigen presentation is processing of the invariant chain. It is thus possible that inhibitors of cathepsin S could halt the development of autoimmune diseases, as shown for experimental autoimmune myasthenia gravis and Sjögren's syndrome (Saegusa et al., 2002; Yang et al., 2005). Since 2000, a large number of small molecule inhibitors of cathepsin S have been developed and patents applied for. Clinical trials were launched in 2006, but so far, no phase III trials for cathepsin-S inhibitors have been initiated (reviewed by Lee-Dutra et al., 2011). The company Johnson & Johnson has completed a phase II trial of their drug RWJ-445380 against rheumatoid arthritis, but the results have not been published so far.

With the prominent role of cathepsin K in bone remodeling in mind, inhibition of cathepsin K activity has been an obvious drug target in the prevention of osteoporosis (Drake et al., 1996; Costa et al., 2011). Initial compounds that were tested did not have sufficient specificity toward cathepsin K; they also inhibited cathepsins B, L, and S, probably accounting for the adverse skin reactions. The drug odanacatib from the company Merck is at present the most likely drug candidate for being the first cathepsin K-inhibitor to be marketed for prophylaxis of osteoporosis. Preliminary studies in humans have shown expected reduction in collagen I degradation products and a positive increase in bone mass density (Stoch et al., 2009; Bone et al., 2010). An ongoing trial over 5 years in more than 16,000 postmenopausal women will show whether these effects translate into reduction in bone fractures, which is the goal of osteoporosis prophylaxis (http://clinicaltrials.gov).

An alternative approach to drug development is to take advantage of the increased presence of specific enzymes at a location in cells or organs where pathophysiological processes take place. The activity of the enzyme can then be exploited to release active compounds, thus establishing both targeting and prodrug activation. The cleavage pattern of cysteine cathepsins is not very specific in the requirement for particular amino acids at the cleavage point, whereas legumain is more specific and will only perform cleavage carboxyterminally to asparagine. This characteristic of legumain has been used successfully to develop anticancer drugs in which doxorubicin, auristatin, and etoposide have been linked to a legumain-cleavable peptide as a prodrug (Liu et al., 2003; Stern et al., 2009; Bajjuri et al., 2011). In tumors, the prodrugs are cleaved and the active compounds released locally due to the increased legumain expression and activity at these sites. Thus, the cytotoxic drugs are targeted to the site of action and are less toxic, and lower doses could be used. Also, a legumain minigene vaccine has been developed and shown to suppress breast cancer growth and angiogenesis (Lewēn et al., 2007), but such vaccine strategy in cancer prevention needs to be clarified.

4.2.6 Take-home message

The interplay between cysteine proteases and their endogenous inhibitors is complex and not fully understood. Thus, the following points should be taken into consideration:

- Lysosomal cysteine proteases should not only be viewed as being lysosomal digestive enzymes, but their extracellular activities must be considered when studying mechanisms of disease.
- The established presence of active cysteine proteases outside the cell should result in the inclusion of this class of enzymes in the extracellular "protease web" where both activation of other proenzymes, processing of other proteases, or inactivation of inhibitors could take place.
- Since numerous interactions between ECM proteins and cysteine proteases have been described, further experimental studies should take into account the actual microenvironment where these enzymes perform their functions.
- The first breakthrough for cysteine cathepsins as pharmacological targets is expected to emerge from the development of specific inhibitors of cathepsin K used in osteoporosis prophylaxis. Clinical experience with such drugs will tell us more about *in vivo* functions of these enzymes.

References

Abrahamson, M. (1994). Cystatins. Methods Enzymol *244*, 685–700.

Abrahamson, M., Alvarez-Fernandez M., and Nathanson C. M. (2003). Cystatins. Biochem Soc Symp *70*, 179–199.

Abrahamson, M., Barrett, A. J., Salvesen, G., and Grubb, A. (1986). Isolation of six cysteine proteinase inhibitors from human urine. Their physicochemical and enzyme kinetic properties and concentrations in biological fluids. J Biol Chem *261*, 11282–11289.

Abrahamson, M., Grubb, A., Olafsson, I., and Lundwall, A. (1987). Molecular cloning and sequence analysis of cDNA coding for the precursor of the human cysteine proteinase inhibitor cystatin C. FEBS Lett *216*, 229–233.

Åkerfeldt, T., and Larsson, A. (2011). Inflammatory response is associated with increased cathepsin B and decreased cathepsin S concentrations in the circulation. Scand J Clin Lab Invest *71*, 203–207.

Ali, L. B., Kim, W. J., Kim, T. Y., et al. (2006). Epigenetic silencing of the tumor suppressor cystatin M occurs during breast cancer progression. Cancer Res *66*, 7899–7909.

Alvarez-Díaz, S., Valle, N., García, J. M., et al. (2009). Cystatin D is a candidate tumor suppressor gene induced by vitamin D in human colon cancer cells. J Clin Invest *119*, 2343–2358.

Alvarez-Fernandez, M., Barrett, A. J., Gerhartz, B., Dando, P. M., Ni, J., and Abrahamson, M. (1999). Inhibition of mammalian legumain by some cystatins is due to a novel second reactive site. J Biol Chem *274*, 19195–19203.

Bai, L., Lutgens, E., and Heeneman, S. (2010). Cathepsins in atherosclerosis. In Atherosclerosis: Molecular and Cellular Mechanisms, S. J. George and J. Johnson, eds. Weinheim, Germany: Wiley-VSH Verlag GmbH & Co., pp. 173–191.

Bailey, M. J., Lacey, D. C., de Kok, B. V. A., Veith, P. D., Reynolds, E. C., and Hamilton, J. A. (2011). Extracellular proteomes of M-CSF (CSF-1) and GM-CSF-dependent macrophages. Immunol Cell Biol *89*, 283–293.

Bajjuri, K. M., Liu, Y., and Sinha S. C. (2011). The legumain protease-activated auristatin prodrugs suppress tumor growth and metastasis without toxicity. Chem Med Chem *6*, 54–59.

Barascuk, N., Skjot-Arkil, H., Register, T. C., et al. (2010). Human macrophage foam cells degrade atherosclerotic plaques through cathepsin K mediated processes. BMC Cardiovasc Disord *10*, 19.

Barrett, A. J. (1986). The cystatins: a diverse superfamily of cysteine peptidase inhibitors. Biomed Biochim Acta *45*, 1363–1374.

Barrett, A. J. (1992). Cellular proteolysis: an overview. Ann N Y Acad Sci *674*, 1–15.

Barrett, A. J., and Kirschke, H. (1981). Cathepsin B, cathepsin H, and cathepsin L. Methods Enzymol *80*, 535–561.

Barrett, A. J., and Rawlings, N. D. (2001). Evolutionary lines of cysteine peptidases. Biol Chem *382*, 727–733.

Barrett, A. J., Rawlings, N. D., Davies, M. E., Machleidt, W., Salvesen, G., and Turk, V. (1986). Cysteine proteinase inhibitors of the cystatin superfamily. In Proteinase Inhibitors, 1st edn, A. J. Barrett and G. Salvesen, eds. Amsterdam: Elsevier, pp. 515–569.

Baumgrass, R., Williamson, M. K., and Price, P. A. (1997). Identification of peptide fragments generated by digestion of bovine and human osteocalcin with the lysosomal proteinases cathepsin B, D, L, H, and S. J Bone Miner Res *12*, 447–455.

Bobek, L. A., and Levine, M. J. (1992). Cystatins-inhibitors of cysteine proteinases. Crit Rev Oral Biol Med *3*, 307–332.

Bone, H. G., McClung, M. R., Roux, C., et al. (2010). Odanacatib, a cathepsin-K inhibitor for osteoporosis: a two-year study in postmenopausal women with low bone density. J Bone Miner Res *25*, 937–947.

Bossard, M. J., Tomaszek, T. A., Thompson, S. K., et al. (1996). Proteolytic activity of human osteoclast cathepsin K. Expression, purification, activation, and substrate identification. J Biol Chem *271*, 12517–12524.

Briggs, J. J., Haugen, M. H., Johansen, H. T., et al. (2010). Cystatin E/M suppresses legumain activity and invasion of melanoma. BMC Cancer *10*, 17.

Brix, K., Dunkhorst, A., Mayer, K., and Jordans, S. (2008). Cysteine cathepsins: cellular roadmap to different functions. Biochimie *90*, 194–207.

Brömme, D. (2011). Cysteine cathepsins and the skeleton. Clin Rev Bone Miner Metab *9*, 83–93.

Brömme, D., and Okamoto, K. (1995). Human cathepsin O2, a novel cysteine protease highly expressed in osteoclastomas and ovary molecular-cloning, sequencing and tissue distribution. Biol Chem Hoppe Seyler *376*, 379–384.

Buck, M. R., Karustis, D. G., Day, N. A., Honn, K. V., and Sloane, B. F. (1992). Degradation of extracellular-matrix proteins by human cathepsin B from normal and tumour tissues. Biochem J *282*, 273–278.

Caglic, D., Pungercar, J. R., Pejler, G., Turk, V., and Turk, B. (2007). Glycosaminoglycans facilitate procathepsin B activation through disruption of propeptide-mature enzyme interactions. J Biol Chem *282*, 33076–33085.

Chan, C.-B., Abe, M., Hashimoto, N., et al. (2009). Mice lacking asparaginyl endopeptidase develop disorders resembling hemophagocytic syndrome. Proc Natl Acad Sci U S A *106*, 468–473.

Chen, J. M., Dando, P. M., Rawlings, N. D., et al. (1997). Cloning, isolation, and characterization of mammalian legumain, an asparaginyl endopeptidase. J Biol Chem *272*, 8090–8098.

Chen, J. M., Fortunato, M., Stevens, R. A., and Barrett, A. J. (2001). Activation of progelatinase A by mammalian legumain, a recently discovered cystein proteinase. Biol Chem *382*, 777–783.

Chen, J. M., Rawlings, N. D., Stevens, R. A., and Barrett, A. J. (1998). Identification of the active site of legumain links it to caspases, clostripain and gingipains in a new clan of cysteine endopeptidases. FEBS Lett *441*, 361–365.

Chen, J. Q., Tung, C. H., Mahmood, U., et al. (2002). In vivo imaging of proteolytic activity in atherosclerosis. Circulation *105*, 2766–2771.

Cheng, X. W., Kuzuya, M., Nakamura, K., et al. (2006). Localization of cysteine protease, cathepsin S, to the surface of vascular smooth muscle cells by association with integrin αvβ3. Am J Pathol *168*, 685–694.

Chung, J. H., Im, E. K., Jin, T., et al. (2011). Cathepsin L derived from skeletal muscle cells transfected with bFGF promotes endothelial cell migration. Exp Mol Med *43*, 179–188.

Collette, J., Bocock, J. P., Ahn, K., et al. (2004). Biosynthesis and alternate targeting of the lysosomal cysteine protease cathepsin L. Int Rev Cytol *241*, 1–51.

Costa, A. G., Cusano, N. E., Silva, B. C., Cremers, S., and Bilezikian, J. P. (2011). Cathepsin K: its skeletal actions and role as a therapeutic target in osteoporosis. Nat Rev Rheumatol *7*, 447–456.

Dennemarker, J., Lohmuller, T., Mayerle, J., et al. (2010). Deficiency for the cysteine protease cathepsin L promotes tumor progression in mouse epidermis. Oncogene *29*, 1611–1621.

de Nooijer, R., Bot, I., von der Thusen, J. H., et al. (2009). Leukocyte cathepsin S is a potent regulator of both cell and matrix turnover in advanced atherosclerosis. Arterioscler Thromb Vasc Biol *29*, 188–194.

Dickinson, D. P. (2002). Cysteine peptidases of mammals: their biological roles and potential effects in the oral cavity and other tissues in health and disease. Crit Rev Oral Biol Med *13*, 238–275.

Dickinson, D. P., Zhao, Y., Thiesse, M., and Siciliano, M. J. (1994). Direct mapping of seven genes encoding human type 2 cystatins to a single site located at 20p11.2. Genomics *24*, 172–175.

Drake, F. H., Dodds, R. A., James, I. E., et al. (1996). Cathepsin K, but not cathepsins B, L, or S, is abundantly expressed in human osteoclasts. J Biol Chem *271*, 12511–12516.

Ekström, U., Wallin, H., Lorenzo, J., Holmqvist, B., Abrahamson, M., and Avilés, F. X. (2008). Internalization of cystatin C in human cell lines. FEBS J *275*, 4571–4582.

Felbor, U., Dreier, L., Bryant, R. A., Ploegh, H. L., Olsen, B. R., and Mothes, W. (2000). Secreted cathepsin L generates endostatin from collagen XVIII. EMBO J *19*, 1187–1194.

Ferraro, S., Marano, G., Biganzoli, E. M., Boracchi, P., and Bongo, A. S. (2011). Prognostic value of cystatin C in acute coronary syndromes: enhancer of atherosclerosis and promising therapeutic target. Clin Chem Lab Med *49*, 1397–1404.

Freije, J. P., Abrahamson, M., Olafsson, I., Velasco, G., Grubb, A., and López-Otín, C. (1991). Structure and expression of the gene encoding cystatin D, a novel human cysteine proteinase inhibitor. J Biol Chem *266*, 20538–20543.

Freije, J. P., Balbín, M., Abrahamson, M., et al. (1993). Human cystatin D. cDNA cloning, characterization of the Escherichia coli expressed inhibitor, and identification of the native protein in saliva. J Biol Chem *268*, 15737–15744.

Fujihara, S., Shimode, K., Nakamura, M., Kobayashi, S., and Tsunematsu, T. (1989). Cerebral amyloid angiopathy with the deposition of cystatin C (gamma-trace) and beta-protein. Prog Clin Biol Res *317*, 939–944.

Garnero, P., Borel, O., Byrjalsen, I., et al. (1998). The collagenolytic activity of cathepsin K is unique among mammalian proteinases. J Biol Chem *273*, 32347–32352.

Garnero, P., Ferreras, M., Karsdal, M. A., et al. (2003). The type I collagen fragments ICTP and CTX reveal distinct enzymatic pathways of bone collagen degradation. J Bone Miner Res *18*, 859–867.

Gatenby, R. A., Gawlinski, E. T., Gmitro, A. F., Kaylor, B., and Gillies, R. J. (2006). Acid-mediated tumor invasion: a multidisciplinary study. Cancer Res *66*, 5216–5223.

Gawenda, J., Traub, F., Lück, H. J., Kreipe, H., and von Wasielewski, R. (2007). Legumain expression as a prognostic factor in breast cancer patients. Breast Cancer Res Treat *102*, 1–6.

Gelb, B. D., Shi, G. P., Chapman, H. A., and Desnick, R. J. (1996). Pycnodysostosis, a lysosomal disease caused by cathepsin K deficiency. Science *273*, 1236–1239.

Gowen, M., Lazner, F., Dodds, R., et al. (1999). Cathepsin K knockout mice develop osteopetrosis due to a deficit in matrix degradation but not demineralization. J Bone Miner Res *14*, 1654–1663.

Henskens, Y. M., Veerman, E. C. I., and Amerongen, A. V. N. (1996). Cystatins in health and disease. Biol Chem Hoppe Seyler *377*, 71–86.

Hou, W. S., Li, Z., Gordon, R. E., et al. (2001). Cathepsin K is a critical protease in synovial fibroblast-mediated collagen degradation. Am J Pathol *159*, 2167–2177.

Hsia, N., and Cornwall, G. A. (2001). CCAAT/enhancer binding protein beta regulates expression of the cystatin-related epididymal spermatogenic (Cres) gene. Biol Reprod *65*, 1452–1461.

Im, E., Venkatakrishnan, A., and Kazlauskas, A. (2005). Cathepsin B regulates the intrinsic angiogenic threshold of endothelial cells. Mol Biol Cell *16*, 3488–3500.

Imai, A., Komatsu, S., Ohara, T., et al. (2011). Serum cystatin C is associated with early stage coronary atherosclerotic plaque morphology on multidetector computed tomography. Atherosclerosis *218* (2), 350–355.

Ishidoh, K., and Kominami, E. (1995). Procathepsin L degrades extracellular matrix proteins in the presence of glycosaminoglycans in vitro. Biochem Biophys Res Commun *217*, 624–631.

Jean, D., Rousselet, N., and Frade, R. (2006). Expression of cathepsin L in human tumor cells is under the control of distinct regulatory mechanisms. Oncogene *25*, 1474–1484.

Kafienah, W., Brömme, D., Buttle, D. J., Croucher, L. J. and Hollander, A. P. (1998). Human cathepsin K cleaves native type I and II collagens at the N-terminal end of the triple helix. Biochem J *331*, 727–732.

Kaschina, E., Stoll, M., Sommerfeld, M., Steckelings, U. M., Kreutz, R., and Unger, T. (2005). Genetic kininogen deficiency contributes to aortic aneurysm formation but not to atherosclerosis. Physiol Genomics *19*, 41–49.

Keppler, D. (2006). Towards novel anti-cancer strategies based on cystatin function. Cancer Lett *235*, 159–176.

Kim, T. Y., Zhong, S., Fields, C. R., Kim, J. H., and Robertson, K. D. (2006). Epigenomic profiling reveals novel and frequent targets of aberrant DNA methylation-mediated silencing in malignant glioma. Cancer Res *66*, 7490–7501.

Kordis, D., and Turk, V. (2009). Phylogenomic analysis of the cystatin superfamily in eukaryotes and prokaryotes. BMC Evol Biol *18*, 266.

Kos, J., Krasovec, M., Cimerman, N., Nielsen, H. J., Christensen, I. J., and Brünner, N. (2000). Cysteine proteinase inhibitors stefin A, stefin B, and cystatin C in sera from patients with colorectal cancer: relation to prognosis. Clin Cancer Res *6*, 505–511.

Kudo, O., Fujikawa, Y., Itonaga, I., Sabokbar, A., Torisu, T., and Athanasou, N. A. (2002). Proinflammatory cytokine (TNFα/IL-lα) induction of human osteoclast formation. J Pathol *198*, 220–227.

Lalmanach, G., Naudin, C., Lecaille, F., and Fritz, H. (2010). Kininogens: more than cysteine protease inhibitors and kinin precursors. Biochimie *92*, 1568–1579.

Langerholc, T., Zavasnik-Bergant, V., Turk, B., Turk, V., Abrahamson, M., and Kos, J. (2005). Inhibitory properties of cystatin F and its localization in U937 promonocyte cells. FEBS J *272*, 1535–1545.

Lankelma, J. M., Voorend, D. M., Barwari, T., et al. (2010). Cathepsin L, target in cancer treatment? Life Sci *86*, 225–233.

Leake, D. S. (1997). Does an acidic pH explain why low density lipoprotein is oxidised in atherosclerotic lesions? Atherosclerosis *129*, 149–157.

Lee-Dutra, A., Wiener, D. K., and Sun, S. (2011). Cathepsin S inhibitors: 2004–2010. Expert Opin Ther Pat *21*, 311–337.

Lemaire, R., Huet, G., Zerimech, F., et al. (1997). Selective induction of the secretion of cathepsins B and L by cytokines in synovial fibroblast-like cells. Br J Rheumatol *36*, 735–743.

Leung-Toung, R., Zhao, Y., Li, W., Tam, T. F., Karimian, K., and Spino, M. (2006). Thiol proteases: inhibitors and potential therapeutic targets. Curr Med Chem *13*, 547–581.

Lewēn, S., Hu, H. D., Cheng, T., et al. (2008). A legumain-based minigene vaccine targets the tumor stroma and suppresses breast cancer growth and angiogenesis. Cancer Immunol Immunother *57*, 507–515.

Li, J., and Yuan, J. (2008). Caspases in apoptosis and beyond. Oncogene *27*, 6194–6206.

Li, W., Lou, I., Jonasson, L., Forssell, C., and Yuan, X. M. (2009). Cathepsin L is significantly associated with apoptosis and plaque destabilization in human atherosclerosis. Atherosclerosis *202*, 92–102.

Liu, C., Sun, C., Huang, H., Janda, K., and Edgington, T. (2003). Overexpression of legumain in tumors is significant for invasion/metastasis and a candidate enzymatic target for prodrug therapy. Cancer Res *63*, 2957–2964.

Liu, J. A., Sukhova, G. K., Sun, J. S., Xu, W. H., Libby, P., and Shi, G. P. (2004). Lysosomal cysteine proteases in atherosclerosis. Arterioscler Thromb Vasc Biol *24*, 1359–1366.

Lutgens, S. P. M., Cleutjens, K., Daemen, M., and Heeneman, S. (2007). Cathepsin cysteine proteases in cardiovascular disease. FASEB J *21*, 3029–3041.

Malla, R. R., Gopinath, S., Gondi, C. S., et al. (2011). Cathepsin B and uPAR knockdown inhibits tumor-induced angiogenesis by modulating VEGF expression in glioma. Cancer Gene Ther *18*, 419–434.

Maruyama, K., Ikeda, S., Ishihara, T., Allsop, D., and Yanagisawa, N. (1990). Immunohistochemical characterization of cerebrovascular amyloid in 46 autopsied cases using antibodies to beta protein and cystatin C. Stroke *21*, 397–403.

Mattock, K. L., Gough, P. J., Humphries, J., et al. (2010). Legumain and cathepsin-L expression in human unstable carotid plaque. Atherosclerosis *208*, 83–89.

Miller, G., Matthews, S. P., Reinheckel, T., Fleming, S., and Watts, C. (2011). Asparagine endopeptidase is required for normal kidney physiology and homeostasis. FASEB J *25*, 1606–1617.

Mohamed, M. M., and Sloane, B. F. (2006). Cysteine cathepsins: multifunctional enzymes in cancer. Nat Rev Cancer *6*, 764–775.

Morita, Y., Araki, H., Sugimoto, T., et al. (2007). Legumain/asparaginyl endopeptidase controls extracellular matrix remodeling through the degradation of fibronectin in mouse renal proximal tubular cells. FEBS Lett *581*, 1417–1424.

Muller-Esterl, W., Fritz, H., and Machleidt, W. (1985). Human plasma kininogens are identical with α-cysteine proteinase inhibitors. Evidence from immunological, enzymological and sequence data. FEBS Lett *182*, 310–314.

Murthy, R. V., Arbman, G., Gao, J., Roodman, G. D., and Sun, X. F. (2005). Legumain expression in relation to clinicopathologic and biological variables in colorectal cancer. Clin Cancer Res *11*, 2293–2299.

Nagai, A., Murakawa, Y., Terashima, M., et al. (2000). Cystatin C and cathepsin B in CSF from patients with inflammatory neurologic diseases. Neurology *55*, 1828–1832.

Nathanson, C. M., Wassélius, J., Wallin, H., and Abrahamson, M. (2002). Regulated expression and intracellular localization of cystatin F in human U937 cells. Eur J Biochem *269*, 5502–5511.

Ni, J., Abrahamson, M., Zhang, M., et al. (1997). Cystatin E is a novel human cysteine proteinase inhibitor with structural resemblance to family 2 cystatins. J Biol Chem *272*, 10853–10858.

Ni, J., Fernandez, M. A., Danielsson, L., et al. (1998). Cystatin F is a glycosylated human low molecular weight cysteine proteinase inhibitor. J Biol Chem *273*, 24797–24804.

Nosaka, A. Y., Kanaori, K., Teno, N., et al. (1999). Conformational studies on the specific cleavage site of type I collagen (alpha-1) fragment (157–192) by cathepsins K and L by proton NMR spectroscopy. Bioorg Med Chem *7*, 375–379.

Ohkubo, I., Kurachi, K., and Takasawa, T. (1984). Isolation of a human cDNA for α2-thiol proteinase inhibitor and its identity with low molecular weight kininogen. Biochem *23*, 5691–5697.

Olafsson, I., and Grubb, A. (2000). Hereditary cystatin C amyloid angiopathy. Amyloid *7*, 70–79.

Oorni, K., Sneck, M., Brömme, D., et al. (2004). Cysteine protease cathepsin F is expressed in human atherosclerotic lesions, is secreted by cultured macrophages, and modifies low density lipoprotein particles in vitro. J Biol Chem *279*, 34776–34784.

Papaspyridonos, M., Smith, A., Burnand, K. G., et al. (2006). Novel candidate genes in unstable areas of human atherosclerotic plaques. Arterioscler Thromb Vasc Biol *26*, 1837–1844.

Popp, O., Heidinger, M., Ruiz-Heinrich, L., Ries, C., Jochum, M., and Gil-Parrado, S. (2003). The calpastatin-derived calpain inhibitor CP1B reduces mRNA expression of matrix metalloproteinase-2 and-9 and invasion by leukemic THP-1 cells. Biol Chem *384*, 951–958.

Preissner, K. T., and Reuning, U. (2011). Vitronectin in vascular context: facets of a multitalented matricellular protein. Semin Thromb Hemost *37*, 408–24.

Pungercar, J. R., Caglic, D., Sajid, M., et al. (2009). Autocatalytic processing of procathepsin B is triggered by proenzyme activity. FEBS J *276*, 660–668.

Qiu, J., Ai, L., Ramachandran, C., et al. (2008). Invasion suppressor cystatin E/M (CST6): high-level cell type-specific expression in normal brain and epigenetic silencing in gliomas. Lab Invest *88* (9), 910–925.

Rawlings, N.D., Barrett, A.J., and Bateman, A. (2012). MEROPS: the database of proteolytic enzymes, their substrates and inhibitors. Nucleic Acids Res *40*, D343–D350.

Reddy, V. Y., Zhang, Q. Y., and Weiss, S. J. (1995). Pericellular mobilization of the tissue-destructive cysteine proteinases, cathepsins B, L, and S, by human monocyte-derived macrophages. Proc Natl Acad Sci U S A *92*, 3849–3853.

Reiser, J., Adair, B., and Reinheckel, T. (2010). Specialized roles for cysteine cathepsins in health and disease. J Clin Invest *120*, 3421–3431.

Riese, R. J., Wolf, P. R., Brömme, D., et al. (1996). Essential role for cathepsin S in MHC class II-associated invariant chain processing and peptide loading. Immunity *4*, 357–366.

Riker, A. I., Enkemann, S. A., Fodstad, O., et al. (2008). The gene expression profiles of primary and metastatic melanoma yields a transition point of tumor progression and metastasis. BMC Med Genomics *1*, 13.

Rivenbark, A. G., Jones, W. D., and Coleman, W. B. (2006). DNA methylation-dependent silencing of CST6 in human breast cancer cell lines. Lab Invest *86* (12), 1233–1242.

Roshy, S., Sloane, B. F., and Moin, K. (2003). Pericellular cathepsin B and malignant progression. Cancer Metastasis Rev *22*, 271–286.

Rossi, A., Deveraux, Q., Turk, B., and Sali, A. (2004). Comprehensive search for cysteine cathepsins in the human genome. Biol Chem *385*, 363–372.

Roth, W., Deussing, J., Botchkarev, V. A., et al. (2000). Cathepsin L deficiency as molecular defect of furless: hyperproliferation of keratinocytes and pertubation of hair follicle cycling. FASEB J *14*, 2075–2086.

Rozhin, J., Sameni, M., Ziegler, G., and Sloane, B. F. (1994). Pericellular pH affects distribution and secretion of cathepsin B in malignant cells. Cancer Res *54*, 6517–6525.

Rozman, J., Stojan, J., Kuhelj, R., Turk, V., and Turk, B. (1999). Autocatalytic processing of recombinant human procathepsin B is a bimolecular process. FEBS Lett *459*, 358–362.

Sadaghiani, A. M., Verhelst, S. H. L., Gocheva, V., et al. (2007). Design, synthesis, and evaluation of in vivo potency and selectivity of epoxysuccinyl-based inhibitors of papain-family cysteine proteases. Chem Biol *14*, 499–511.

Saegusa, K., Ishimaru, N., Yanagi, K., et al. (2002). Cathepsin S inhibitor prevents autoantigen presentation and autoimmunity. J Clin Invest *110*, 361–369.

Schagdarsurengin, U., Pfeifer, G. P., and Dammann, R. (2007). Frequent epigenetic inactivation of cystatin M in breast carcinoma. Oncogene *26*, 3089–3094.

Schechter, I., and Berger, A. (1967). On the size of the active site in proteases. I. Papain. Biochem Biophys Res Commun *27*, 157–162.

Shirahama-Noda, K., Yamamoto, A., Sugihara, K., et al. (2003). Biosynthetic processing of cathepsins and lysosomal degradation are abolished in asparaginyl endopeptidase-deficient mice. J Biol Chem *278*, 33194–33199.

Shridhar, R., Zhang, J., Song, J., et al. (2004). Cystatin M suppresses the malignant phenotype of human MDA-MB-435S cells. Oncogene *23*, 2206–2215.

Sokol, J. P., and Schiemann, W. P. (2004). Cystatin C antagonizes transforming growth factor beta signaling in normal and cancer cells. Mol Cancer Res *2*, 183–195.

Sotiropoulou, G., Anisowicz, A., and Sager, R. (1997). Identification, cloning and characterization of cystatin M, a novel cysteine proteinase inhibitor, down-regulated in breast cancer. J Biol Chem *272*, 903–910.

Stenman, G., Aström, A. K., Röijer, E., Sotiropoulou, G., Zhang, M., and Sager, R. (1997). Assignment of a novel cysteine proteinase inhibitor (CST6) to 11q13 by fluorescence in situ hybridization. Cytogenet Cell Genet *76*, 45–46.

Stern, L., Perry, R., Ofek, P., Many, A., Shabat, D., and Satchi-Fainaro, R. (2009). A novel antitumor prodrug platform designed to be cleaved by the endoprotease legumain. Bioconjug Chem *20*, 500–510.

Stoch, S. A., Zajic, S., Stone, J., et al. (2009). Effect of the cathepsin K inhibitor odanacatib on bone resorption biomarkers in healthy postmenopausal women: two double-blind, randomized, placebo-controlled phase I studies. Clin Pharmacol Ther *86*, 175–182.

Storr, S. J., Carragher, N. O., Frame, M. C., Parr, T., and Martin, S. G. (2011). The calpain system and cancer. Nat Rev Cancer *11*, 364–374.

Sukhova, G. K., Shi, G. P., Simon, D. I., Chapman, H. A., and Libby, P. (1998). Expression of the elastolytic cathepsins S and K in human atheroma and regulation of their production in smooth muscle cells. J Clin Invest *102*, 576–583.

Sundelöf, J., Sundström, J., Hansson, O., et al. (2010). Higher cathepsin B levels in plasma in Alzheimer's disease compared to healthy controls. J Alzheimer's Dis *22*, 1223–1230.

Tedelind, S., Jordans, S., Resemann, H., et al. (2011). Cathepsin B trafficking in thyroid carcinoma cells. Thyroid Res *4* (Suppl. 1), S2

Toomes, C., James, J., Wood, A. J., et al. (1999). Loss-of-function mutations in the cathepsin C gene result in periodontal disease and palmoplantar keratosis. Nat Genet *23*, 421–424.

Tseng, C.C., Tseng, C.P., Levine, M.J., and Bobek, L.A. (2000). Differential effect toward inhibition of papain and cathepsin C by recombinant human salivary cystatin SN and its variants produced by a baculovirus system. Arch Biochem Biophys *380*, 133–140.

Umemura, H., Togawa, A., Sogawa, K., et al. (2011). Identification of a high molecular weight kininogen fragment as a marker for early gastric cancer by serum proteome analysis. J Gastroenterol *46*, 577–85.

Urbich, C., De Souza, A. I., Rossig, L., et al. (2011). Proteomic characterization of human early pro-angiogenic cells. J Mol Cell Cardiol *50*, 333–336.

Urbich, C., Heeschen, C., Aicher, A., et al. (2005). Cathepsin L is required for endothelial progenitor cell-induced neovascularization. Nat Med *11*, 206–213.

Vasiljeva, O., Dolinar, M., Pungercar, J. R., Turk, V., and Turk, B. (2005). Recombinant human procathepsin S is capable of autocatalytic processing at neutral pH in the presence of glycosaminoglycans. FEBS Lett *579*, 1285–1290.

Vasiljeva, O., Reinheckel, T., Peters, C., Turk, D., Turk, V., and Turk, B. (2007). Emerging roles of cysteine cathepsins in disease and their potential as drug targets. Curr Pharm Des *13*, 387–403.

Vigneswaran, N., Wu, J., Nagaraj, N., James, R., Zeeuwen, P., and Zacharias, W. (2006). Silencing of cystatin M in metastatic oral cancer cell line MDA-686Ln by siRNA increases cysteine proteinases and legumain activities, cell proliferation and in vitro invasion. Life Sci *78*, 898–907.

Wallin, H., Bjarnadottir, M., Vogel, L. K., Wassélius, J., Ekström, U., and Abrahamson, M. (2010). Cystatins – extra- and intracellular cysteine protease inhibitors: high-level secretion and uptake of cystatin C in human neuroblastoma cells. Biochimie *92*, 1625–1634.

Wegiel, B., Jiborn, T., Abrahamson, M., et al. (2009). Cystatin C is downregulated in prostate cancer and modulates invasion of prostate cancer cells via MAPK/Erk and androgen receptor pathways. PLoS One *4*, e7953.

Werle, B., Sauckel, K., Nathanson, C. M., et al. (2003). Cystatins C, E/M and F in human pleural fluids of patients with neoplastic and inflammatory lung disorders. Biol Chem *384*, 281–287.

Werle, B., Schanzenbächer, U., Lah, T. T., et al. (2006). Cystatins in non-small cell lung cancer: tissue levels, localization and relation to prognosis. Oncol Rep *16*, 647–655.

Wu, W., Luo, Y., Sun, C., et al. (2006). Targeting cell-impermeable prodrug activation to tumor microinvironment eradicates multiple-drug resistant neoplasms. Cancer Res *66*, 970–980.

Yang, H., Kala, M., Scott, B. G., Goluszko, E., Chapman, H. A., and Christadoss, P. (2005). Cathepsin S is required for murine autoimmune myasthenia gravis pathogenesis. J Immunol *174*, 1729–1737.

Yasuda, Y., Li, Z. Q., Greenbaum, D., Bogyo, M., Weber, E., and Brömme, D. (2004). Cathepsin V, a novel and potent elastolytic activity expressed in activated macrophages. J Biol Chem *279*, 36761–36770.

Zeeuwen, P. L. (2004). Epidermal differentiation: the role of proteases and their inhibitors. Eur J Cell Biol *83*, 761–773.

Zeeuwen, P. L., Van Vlijmen-Willems, I. M., Jansen, B. J., et al. (2001). Cystatin M/E expression is restricted to differentiated epidermal keratinocytes and sweat glands: a new skin-specific proteinase inhibitor that is a target for cross-linking by transglutaminase. J Invest Dermatol *116*, 693–701.

Zhang, J., Shridhar, R., Dai, Q., et al. (2004). Cystatin M: a novel candidate tumor suppressor gene for breast cancer. Cancer Res *64*, 6957–6964.

Zhong, S., Fields, C. R., Su, N., Pan, Y. X., and Robertson, K. D. (2007). Pharmacologic inhibition of epigenetic modifications, coupled with gene expression profiling, reveals novel targets of aberrant DNA methylation and histone deacetylation in lung cancer. Oncogene *26*, 2621–2634.

Zore, I., Krasovec, M., Cimerman, N., et al. (2001). Cathepsin B/cystatin C complex levels in sera from patients with lung and colorectal cancer. Biol Chem *382*, 805–810.

4.3 Plasmin and the plasminogen activator system in health and disease

Gunbjørg Svineng, Synnøve Magnussen, and Elin Hadler-Olsen

4.3.1 Introduction

Plasmin and the plasminogen activator system play important roles in normal physiology and disease. The interplay between all the components of the plasminogen activator system is highly regulated, and disturbance in any of these pathways may result in diseases such as thrombosis, bleeding disorders, neurological diseases, and cancer (Dano et al., 2005; Dass et al., 2008; Jacobsen et al., 2008). The plasminogen activator system works in concert with the other proteolytic networks, in particular the matrix metalloproteinases (MMPs) and the coagulation system. In addition, it interacts with and modulates other networks, such as the chemokines, cytokines, and integrins, all of which influence intracellular signaling (Smith and Marshall, 2010; Mason and Joyce, 2011; Schaller and Gerber, 2011). This review will cover the normal physiological functions of the main players of the plasmin system. In addition, the role of this system in diseases such as cancer will be discussed.

4.3.2 Plasmin

The serine protease plasmin is the main executor of the plasmin system and has several physiological roles: dissolving fibrin clots, in plasticity and tissue remodeling, and during embryogensis. In addition, plasmin has been found to be involved in a whole range of pathological conditions, such as thrombosis, cancer, and Alzheimer's disease. Plasminogen deficiency in humans is often the cause of ligneous conjunctivitis (Mehta and Shapiro, 2008), while mice deficient in plasminogen (Plg$^{-/-}$ mice) undergo normal embryogenesis and development, are fertile, and survive to adulthood. However, they suffer from severe thrombosis and impaired wound healing (Bugge et al., 1995).

Processing of plasminogen to plasmin

Human plasminogen (PLG) is 810 amino acids (aa) in size and is mainly expressed in the liver by hepatocytes and released into the circulation. The signal peptide of 19 aa is cleaved off before secretion, yielding a 791 aa inactive proenzyme Glu1-plasminogen of 92 kDa (▶Figure 4.2). Glu1-plasminogen consists of an N-terminal preactivation peptide (PAP) (aa 1–77), five kringle domains (aa 78–561), and a C-terminal catalytic domain (aa 562–791) with a serine protease triad constituted of His603, Asp646, and Ser741 (Petersen et al., 1990). Plasminogen exists in two main conformations,

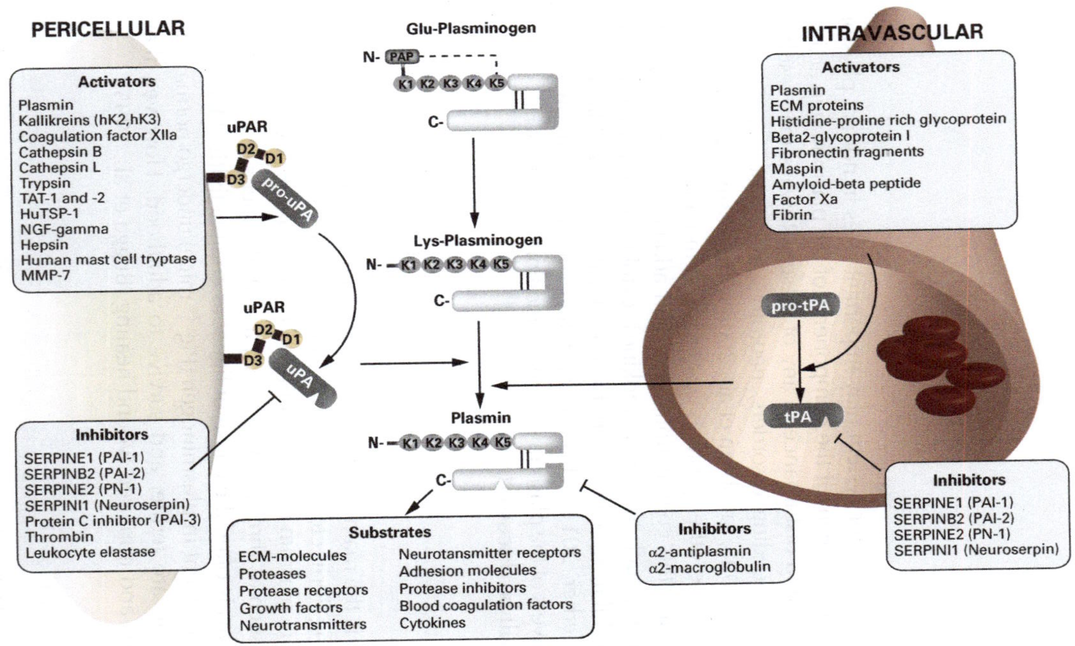

Figure 4.2 **Activation cascade of plasminogen to plasmin.** The secreted 92 kDa proenzyme Glu1-plasminogen consists of a preactivation peptide (PAP), five kringle (K1–K5) domains, and a C-terminal catalytic domain. Plasminogen exists in several conformations depending on the presence of intramolecular interactions between lysine residues in the PAP domain and lysine-binding sites in several of its kringle domains (here illustrated by a dashed line between PAP and K5). The activation of plasminogen works as a feedback loop, where active plasmin is involved in its own activation. Active plasmin can cleave the PAP domain, thus processing Glu1-plasminogen into the more readily activated Lys78-plasminogen. Both Glu1- and Lys78-plasminogen can be converted into the two-chain active plasmin enzyme by cleavage of the Arg561-Val562 peptide bond by the plasminogen activators uPA and tPA. Both uPA and tPA are themselves activated by plasmin, in addition to several other proteases. tPA is mainly involved in fibrinolysis, and uPA in pericellular proteolysis. Binding of tPA to fibrin increases its activity, whereas binding of pro-uPA to its glycosylphosphatidylinositol (GPI)–anchored receptor, urokinase plasminogen activator receptor (uPAR), increases its activation by several activators. Binding of uPA to uPAR is mediated through the amino terminal fragment (ATF) of uPA and all three domains of uPAR (D1–D3).

closed and open, depending on the presence of intramolecular interactions between the PAP domain and the kringle domains. Active plasmin can process Glu1-plasminogen to Lys78-plasminogen (84 kDa) by cleaving off the N-terminal PAP domain. This converts it from the closed conformation to the open conformation (An et al., 1998a, 1998b; Cockell et al., 1998; Ho Tin-Noe et al., 2005). Both Glu1-plasminogen and Lys78-plasminogen can be converted to active plasmin by the plasminogen activators urokinase plasminogen activator (uPA) or tissue-type plasminogen activator (tPA); however, Lys78-plasminogen is more readily converted than Glu1-plasminogen (Petersen et al., 1990; Ponting et al., 1992; Miles et al., 2003). Activation by uPA and tPA through cleavage of the peptide bond between Arg561 and Val562 induces a conformational change in plasmin but does not change the size of the protein since the two polypeptide chains are still held together by two disulfide bonds (Cys548-Cys666 and Cys558-Cys567) (Petersen et al., 1990).

Kringle domains and angiostatins

The five kringle domains in the N-terminal half of plasminogen and plasmin consist of two loops interconnected with three pairs of disulfide bonds (Petersen et al., 1990). Kringle domains can be found in many proteins, including uPA, tPA, apolipoproteins, prothrombin, factor XII, hepatocyte growth factor (HGF), and macrophage stimulating protein (MSP) (Cao and Xue, 2004). Kringle domains are typically 80 aa in size, and many have lysine-binding sites that can form bonds with lysine residues on a variety of proteins, thus mediating protein-protein interactions. Importantly, kringle domains of plasminogen and tPA have affinity for fibrin, leading to colocalization and enhanced plasmin activation and fibrin degradation (Cesarman-Maus and Hajjar, 2005; Mosesson, 2005). Isolated kringle domains may therefore be used to inhibit fibrinolysis, and recent findings demonstrate that only kringle domains with a lysine-binding activity inhibit fibrinolysis (Ahn et al., 2011).

The kringle domains of plasminogen or plasmin can be released from the catalytic domain by proteolytic processing by several proteases (▶Table 4.3). These kringle-containing fragments have been shown to have antiangiogenic properties and are called angiostatins (O'Reilly et al., 1994). Depending on the proteases involved, the angiostatins may contain kringle domain 1–3, 1–4, 1–4 plus 85% of 5 (named angiostatin4.5), or 1–5, exhibiting various degrees of antiangiogenic properties (Soff, 2000; Cao and Xue, 2004). Isolated kringle 5 has antiangiogenic properties that have been found to be more prominent than those of the other plasminogen/plasmin kringle domains (▶Table 4.3) (Cao et al., 1997; Cao and Xue, 2004). Nevertheless, the physiological relevance of many of the angiostatin isoforms, their site of generation, and their half-life *in vivo* are largely unresolved. Angiostatins have been purified from plasma and urine of both normal and tumor-bearing mice, from peritoneal cavities of mice following inflammation, and from cancer patients (O'Reilly et al., 1994; Falcone et al., 1998; Cao et al., 2000; Kassam et al., 2001). Also, many cell lines have the capacity to produce angiostatins, mainly through uPA- or tPA-activated plasmin activity (Westphal et al., 2000; Wang et al., 2004).

The angiostatins' antiangiogenic mechanisms include inhibition of endothelial cell migration and proliferation, as well as induction of apoptosis (Claesson-Welsh et al., 1998; Lucas et al., 1998). Kringle domains 2, 3, and 5 convey both antiproliferative and antimigratory effects, while kringle domain 4 predominantly inhibits endothelial

Table 4.3 Angiostain isoforms.

Enzymes	*Kringle domains*	*Fragments*	*Cells**	*References*
Plasmin	K1	Lys78-Thr181	HT1080, BCE, HeLa	(Kwon et al., 2001)
	K1–4	Lys78-Lys468	HT1080, BCE	(Kassam et al., 2001)
	K1–4	ND	Macrophages	(Falcone et al., 1998)
	K1–4 K1–4.5	ND Lys78-Arg530	PC3	(Gately et al., 1997; Cao et al., 1999; Wang et al., 2004)
MMP2	K1–4	Tyr80-	LLC-LM	(O'Reilly et al., 1999)
MMP3	K1–4 K5	Asn60-Pro447 Val448-Pro545	*In vitro*	(Lijnen et al., 1998a)
MMP7	K1–4	Lys78-Pro447	*In vitro*	(Patterson and Sang, 1997)
MMP9	K1–4	Lys78-Pro446	*In vitro*	(Patterson and Sang, 1997)
MMP12	K1–3 K1–4 K4	Lys78- Lys78- Val354-	Monocytes, macrophages	(Dong et al., 1997; Cornelius et al., 1998)
MMP19	K1–3 K1–4	ND[a]	*In vitro*	(Brauer et al., 2011)
Prostate-specific antigen (PSA)/ kallikrein-3	K1–4	Lys78-Glu439	*In vitro*	(Heidtmann et al., 1999)
Elastase	K1–3 K1–4 K4 K5	Tyr80-Val338 Val449 (Pro452)-Ala544	*In vitro*	(Takada et al., 1988; Cao et al., 1997; Lucas et al., 1998)
Cathepsin D	K1–4	Phe75-Leu451	PC3	(Morikawa et al., 2000)

[a]ND, not determined.
*BCE, bovine capillary endothelial cells; LLC-LM, Lewis lung carcinoma; HT1080, human fibrosarcoma cell line; HeLa, human cervical adenocarcinoma cell line; PC3, prostate carcinoma cell line-3.

cell migration (Cao et al., 1997; Ji et al., 1998a, 1998b). Several endothelial cell proteins have been identified as angiostatin-binding proteins, including adenosine triphosphate (ATP) synthase (Moser et al., 1999), angiomotin (Troyanovsky et al., 2001), and integrin αVβ3 (Tarui et al., 2001). Binding to ATP synthase or other cell surface proteins may lead to internalization of angiostatin, and it has been shown in both endothelial and tumor cells that internalized angiostatin localizes to the mitochondria where it binds both ATP synthase and malate dehydrogenase-2 (MDH2) causing apoptosis (Lee et al., 2009). In addition, angiostatin has been shown to bind tPA and thereby hinder tPA from mediating further plasmin generation and subsequent cell migration (Stack et al., 1999).

Cell surface plasminogen/plasmin-binding proteins

The proteolytic activity of the plasmin system is under strict control, and proper localization of the activity is crucial. Under physiological conditions, plasminogen is a poor substrate for its activators unless it is interacting with other proteins. Binding of soluble Glu1-plasminogen to fibrin induces a conformational change that enhances its activation to plasmin by tPA (Plow et al., 1995; Miles et al., 2003; Han et al., 2011). In addition, binding of Glu1-plasminogen to cell surface proteins via its kringle domains induces a conformational change, making it a better target for plasmin that converts it to Lys78-plasminogen (Han et al., 2011). Lys78-plasminogen is more readily activated to plasmin by uPA; thus plasminogen binding to cell surface proteins plays an important role in localizing plasmin activity to the pericellular region. Furthermore, cell-surface-localized plasmin is better protected from its inhibitors α2-macroglobulin and α2-antiplasmin than soluble plasmin (Hall et al., 1991). The lysine-binding sites within the plasminogen/plasmin kringle domains are commonly involved in the interactions with cell surface plasminogen-binding proteins (Hajjar et al., 1986). Hence Glu1-plasminogen, Lys78-plasminogen, active plasmin, and some isoforms of angiostatin can all bind to these cell surface proteins. In most cases, the interactions are of low affinity, but often with very high capacity. Cells that have a plasminogen-binding capacity are hematopoetic cells, endothelial cells, and several different types of cancer cells (Plow et al., 1995; Capello et al., 2011). The group of proteins reported to bind plasminogen can be divided into those that have defined functions as extracellular proteins (annexin A2 heterotetramer [AIIt], Plg-R_{KT}, and histidine-rich glycoprotein [HRG]) (Andronicos et al., 2010; Madureira et al., 2011; Poon et al., 2011) and those that have known intracellular functions but can translocate to the outer layer of the plasma membrane (α-enolase [ENOA], histone H2B, β-actin, cytokeratin-8) (Hembrough et al., 1996; Andronicos and Ranson, 2001; Das et al., 2007; Capello et al., 2011).

The protein complex AIIt is an important plasminogen-binding protein complex located on the surface of endothelial cells, macrophages, and cancer cells, and is composed of two molecules of annexin A2 (ANXA2) bound together by a dimer of S100A10 (Choi et al., 2003; Madureira et al., 2011). On endothelial cells, S100A10 binds kringle domains on both plasminogen and tPA, thus facilitating plasmin generation and fibrinolysis (Surette et al., 2011). In addition, S100A10 interacts with urokinase plasminogen activator receptor (uPAR), thus mediating colocalization of plasminogen with uPA (Choi et al., 2003; Madureira et al., 2011). Both macrophages and cancer cells express S100A10, and reduction of S100A10 levels in tumor-associated macrophages (TAMs) reduces primary tumor growth, while tail-vein-injected cancer cells with reduced

S100A10 expression show less propensity to establish metastasis in lung (Choi et al., 2003; Phipps et al., 2011).

Plg-R_{KT} is a recently identified transmembrane plasminogen-binding protein expressed on monocytes, macrophages, lymphocytes, and granulocytes where it is colocalized with uPAR. Plg-R_{KT} expression is increased when monocytes are differentiated into macrophages and is involved in macrophage recruitment to sites of peritonitis in a mouse model (Andronicos et al., 2010; Lighvani et al., 2011).

The plasma protein HRG binds plasminogen and tethers it to the cell surface at sites of tissue injury (Borza et al., 2004; Jones et al., 2004; Poon et al., 2011). HRG can also be cleaved by active plasmin, and this new conformation reduces its binding to cell surface heparan sulfate proteoglycans, simultaneously as its binding to plasminogen increases (Poon et al., 2009). Hence, plasmin activity can modulate the localization of HRG-bound plasminogen. Also, the role of HRG in fibrinolysis was demonstrated by the more rapid lysis of fibrin clots in $HRG^{-/-}$ mice than in wild-type mice (Tsuchida-Straeten et al., 2005). This suggests that HRG can act as a decoy receptor for plasminogen, sequestering plasminogen from its cell surface or fibrin-located activators.

ENOA was previously considered to have only an intracellular function in the synthesis of pyruvate (Capello et al., 2011). However, recent studies have shown that ENOA can locate to the cell surface of both cancer and hemapoetic cells upon stimulation, although the mechanism for its relocalization is unknown (Miles et al., 1991; Wygrecka et al., 2009; Capello et al., 2011). The C-terminal lysines in ENOA are predominantly responsible for the interaction with the kringle domains of plasminogen/plasmin. Interestingly, lipopolysaccharide (LPS) stimulation of monocytes induced relocalization of ENOA to the outer layer of the plasma membrane, which increased plasmin activation and migration of the monocytes into inflamed lung tissue (Wygrecka et al., 2009). Histone H2B is another intracellular protein that can relocate to the plasma membrane and function as a plasminogen-binding protein. In peritonitis, H2B histones were identified on monocytes recruited to the site of inflammation (Das et al., 2007). This suggests that monocytes can express different plasminogen-binding proteins depending on the cause and/or tissue of inflammation.

Globular β-actin has been found on the cell surface of endothelial cells and many cancer cells where it binds plasminogen (Hu et al., 1993; Dudani and Ganz, 1996; Andronicos and Ranson, 2001). The binding has been mapped to the kringle 5 domain (Wang et al., 2006), and hence both plasminogen and plasmin can bind β-actin. Interestingly, cleavage of plasminogen/plasmin, which generates angiostatin4.5, will abolish this binding and perhaps explains how angiostatin4.5 can be released from the cell surface into the circulation and thereby inhibit angiogenesis at a distant site (O'Reilly et al., 1994; Wang et al., 2006). Taken together, binding of plasminogen to cell surface proteins mediates tissue and cell specificity for trapping of plasminogen from the circulation and the interstitial fluid, and it is important for a specific localization of plasmin activity.

4.3.3 Plasminogen activators

The two main plasminogen activators, tPA and uPA, are both serine proteases and share some domain structures (Brunner et al., 1990). Although plasminogen is the main substrate for tPA and uPA, they are able to cleave and activate the plasminogen-related kringle proteins HGF/scatter factor (SF) and MSP, resulting in increased cell migration and proliferation (Mars et al., 1993; Vigna et al., 1994; Galimi et al., 2001). Both HGF/SF and MSP and their respective receptors, c-Met and recepteur d'origine nantais (RON), are often overexpressed in cancer (Benvenuti and Comoglio, 2007), suggesting a role for uPA and tPA in activation of HGF/SF and MSF mediated signaling in cancer. However, the ability of uPA to activate HGF/SF has recently been challenged – Owen et al. (2010) were unable to show a direct activation of HGF/SF by soluble active uPA. It still remains to be elucidated whether uPA bound to cell surface uPAR could cleave HGF/SF (Owen et al., 2010).

Tissue-type plasminogen activator

Human tPA (PLAT) is a glycoprotein (72 kDa, 530 aa) synthesized mainly by vascular endothelial cells, keratinocytes, melanocytes, and neurons (Myohanen and Vaheri, 2004; Cesarman-Maus and Hajjar, 2005; Gebbink, 2011). Certain stimuli, such as thrombin, bradykinin, and adrenaline, can increase the release of tPA. However, the half-life of tPA in plasma is only minutes (Cesarman-Maus and Hajjar, 2005). tPA is composed of an N-terminal finger domain (aa 4–50), a growth factor domain (GFD) (aa 50–87) with homology to the epidermal growth factor (EGF), two kringle domains (aa 87–262), and a C-terminal serine protease domain (aa 276–527) with the active site triad constituted of His322, Asp371, and Ser478 (Mignatti and Rifkin, 1993). tPA is secreted as a single-chain protein, which has some enzymatic activity, and cleavage of the Arg275-Ile276 bond by plasmin converts it into a fully active, two-chain molecule held together by a single disulfide bridge (Cys264-Cys395). Both single-chain tPA and active tPA can cleave plasminogen into active plasmin. tPA and plasminogen both have a strong affinity for fibrin, and their binding sites juxtapose each other, which results in enhanced plasminogen activation. Since tPA is poorly active in the absence of fibrin, it is generally accepted that tPA is mainly involved in the activation of plasmin for intravascular thrombolysis. Intravenous injection of tPA is used for treatment of acute ischemic stroke, myocardial infarction, and pulmonary embolism (Killer et al., 2010; Schaller and Gerber, 2011).

Urokinase plasminogen activator

Human uPA (PLAU) is a glycoprotein synthesized by endothelial cells, epithelial cells, leukocytes, monocytes, fibroblasts, and cancer cells (Vincenza Carriero et al., 2009; Hildenbrand et al., 2010). It consists of 411 aa and is secreted as an inactive 52 kDa proenzyme (pro-uPA or single-chain uPA). The amino terminal fragment (ATF)(130 aa) consists of the GFD (aa 1–46) and a kringle domain (aa 47–135). The carboxyl-terminal end of the protein comprises the serine protease domain (aa 159–411) with the active site triad constituted of His204, Asp255, and Ser356 (Spraggon et al., 1995). Binding of pro-uPA to its cell surface receptor uPAR occurs via the ATF domain. tPA lacks a similar ATF and is unable to bind uPAR (Mignatti and Rifkin, 1993). Both uPAR-bound pro-uPA and soluble pro-uPA can be activated by plasmin through cleavage of the peptide bond Lys158-Ile159. Several other proteases also activate pro-uPA (▶Table 4.4). Active uPA (high-molecular-weight [HMW] uPA, 52 kDa) is a two-chain protein linked by a single disulfide bond between Cys148 and Cys279. HMW uPA can catalyze the

Table 4.4 Activators and inactivators/inhibitors of uPA.

Activators	*Reference*
Plasmin	(Schaller and Gerber, 2011)
Plasma kallikrein	(Ichinose et al., 1986)
Coagulation factor XIIa	(Ichinose et al., 1986; Schaller and Gerber, 2011)
Tumor-associated trypsin(ogen) (TAT)-1 and -2	(Koivunen et al., 1989)
Trypsin	(Brunner et al., 1990)
T-cell-associated serine proteases, HuTSP1	(Brunner et al., 1990)
Cathepsin B	(Kobayashi et al., 1991)
Cathepsin L	(Goretzki et al., 1992)
Nerve growth factor-gamma (NGF-γ)	(Wolf et al., 1993)
Hepsin	(Moran et al., 2006)
Human mast cell tryptase	(Stack and Johnson, 1994)
PSA/human kallikrein-3 (hK3)	(Yoshida et al., 1995; disputed later by Frenette et al., 1997)
Kallikrein (human kallikrein-2 [hK2])	(Frenette et al., 1997)
Kallikrein-4	(Takayama et al., 2001)
MMP7 (pump-1)	(Marcotte et al., 1992)
Inactivators/inhibitors	
Plasminogen activator inhibitor (PAI)-1 (PAI1) (SERPINE1)	(Hekman and Loskutoff, 1985; Tang and Wei, 2008)
PAI2 (SERPINB2)	(Astedt et al., 1985; Kruithof et al., 1986)
Protase nexin-1 (PN1) (SERPINE2)	(Sprengers and Kluft, 1987; Naldini et al., 1992)
PAI3, protein-C inhibitor (PCI)	(Espana et al., 1993)
Thrombin	(Ichinose et al., 1986)
Leukocyte elastase	(Schmitt et al., 1989)

activation of plasminogen into active plasmin by cleaving the Arg560-Val561 polypeptide bond (Robbins et al., 1967; Andreasen et al., 1997), after which plasmin can create a feedback loop by activating pro-uPA. Plasmin can also cleave active HMW uPA C-terminal of the kringle domain, causing release of the proteolytically active 33 kDa low-molecular-weight form (LMW uPA), which is no longer able to bind uPAR (Schaller et al., 1982). The remaining ATF domain (17 kDa) can be found either in solution or bound to uPAR where it can function as a proteolytic inhibitor because it will block further binding and activation of pro-uPA. *In vivo* studies in mice have shown that ATF can function as an anticancer agent suppressing tumor growth, angiogenesis, and metastasis of breast and gastric cancer xenografts in nude mice (Hu et al., 2008).

The urokinase receptor

Cell surface localization of uPA is mediated by binding to its receptor, uPAR (PLAUR) (Blasi and Sidenius, 2010). uPAR is expressed at moderate levels in several tissues, while strong uPAR expression is often found in tissues that undergo remodeling, such as the placenta and during wound healing. In particular, endothelial cells, keratinocytes, monocytes and macrophages, granulocytes, myofibroblasts, and hematopoetic stem cells express uPAR (Solberg et al., 2001; Tjwa et al., 2009). In addition, uPAR expression is often upregulated in various cancers (Dano et al., 2005; Ulisse et al., 2009). Mature uPAR (also named CD87) is a cysteine-rich glycoprotein (55–60 kDa, 283 aa) bound to the plasma membrane via a glycosylphosphatidylinositol (GPI) anchor at Gly283 (Stoppelli et al., 1985; Vassalli et al., 1985). uPAR is composed of three homologous domains, D1 (aa 1–77), D2 (aa 93–177), and D3 (aa 193–272), connected with short linker regions. Pro-uPA binds uPAR through its ATF domain, and all three domains of uPAR are required for this interaction, which induces a significant conformational change in uPAR (Llinas et al., 2005, Barinka et al., 2006; Huai et al., 2006, 2008). Dissociation of uPA from the receptor is very slow, and uPA can remain active at the cell surface for several hours. uPA is therefore mainly involved in pericellular proteolysis and tissue-remodeling processes (Dano et al., 1985; Mignatti and Rifkin, 1993; Andreasen et al., 1997; Dass et al., 2008).

Soluble uPAR and uPAR fragments

uPAR can be shed from the cell surface by the action of phospholipase D (Wilhelm et al., 1999) or cleaved in the linker region between D1 and D2 by proteases such as active uPA on a neighboring uPAR, plasmin, matrix metalloproteinase-12 (MMP12), and human kallikrein 4 (hK4) (Koolwijk et al., 2001; Andolfo et al., 2002; Beaufort et al., 2006; Hoyer-Hansen and Lund, 2007; Tjwa et al., 2009). All soluble forms can be found in various body fluids, such as blood, cystic fluid, and urine. Interestingly, soluble D2-D3 is a chemoattractant for monocytes and basophils. Recently, detection of soluble uPAR or uPAR fragments has been used as diagnostic and prognostic markers for several cancers (Almasi et al., 2005, 2011; Rasch et al., 2008; Sorio et al., 2011). The rationale is that increased levels of soluble uPAR and uPAR fragments are indicative of high tumor-associated proteolytic activity. Indeed, removal of primary tumors has been shown to reduce the level of circulating uPAR. Likewise, soluble uPAR and uPAR fragments have also been suggested as markers for other diseases that induce inflammation, including chronic liver disease and infectious diseases such as HIV infection, malaria, and bacteraemia (Ostrowski et al., 2007; Huttunen et al., 2011; Yilmaz et al., 2011). Thus, soluble uPAR has been suggested as a biomarker for low-grade inflammation associated with cancer, cardiovascular diseases, and type 2 diabetes (Eugen-Olsen et al., 2010).

Binding of uPAR to vitronectin

Binding of uPA to uPAR increases the affinity of uPAR to the glycoprotein vitronectin (78 kDa, 459 aa) (Wei et al., 1994), and direct binding of uPAR to vitronectin has been shown to mediate cell adhesion (Madsen and Sidenius, 2008). Vitronectin in the blood is monomeric; however, when incorporated in the extracellular matrix (ECM), it is multimeric (Preissner and Seiffert, 1998). Multimeric vitronectin can bind the uPA inhibitor PAI1, thus providing a reservoir of PAI1 that can further inhibit plasminogen activation (Lindahl et al., 1989). In addition, several integrins bind to the recognition site Arginine-Glycine-Aspartic acid (RGD) (aa 45-47) of vitronectin (Hynes, 1992). Crystal structure analysis has revealed that uPA and vitronectin bind to two different parts of uPAR, thus allowing simultaneous binding of the two ligands (Huai et al., 2008). Vitronectin binds via its N-terminal somatomedin-B domain (SMB) (aa 1–44) to the D1 domain and to 2 aa in the D1-D2 linker region of uPAR. uPA-mediated proteolytic cleavage of uPAR releases D1 and abolishes the binding to vitronectin (Hoyer-Hansen et al., 1997).

Modulation of signal transduction by uPAR through interactions with transmembrane proteins

In addition to being a receptor for uPA, uPAR has been suggested to interact with several cell surface proteins such as integrins, epidermal growth factor receptor (EGFR), and uPAR-associated protein (uPARAP/Endo180) (Engelholm et al., 2009; Blasi and Sidenius, 2010; Smith and Marshall, 2010). Many laboratories have demonstrated colocalization of uPAR with integrins and subsequent activation of intracellular signaling cascades (reviewed in Blasi and Carmeliet, 2002; Madsen and Sidenius, 2008; Smith and Marshall, 2010). Whether this occurs through direct interactions or indirectly through mechanical alterations in focal adhesion sites induced by uPAR is still debated. Nevertheless, integrins $\alpha5\beta1$, $\alpha3\beta1$, $\alpha V\beta3$, $\alpha V\beta5$, and $\alpha M\beta2$ are all modulated by the presence of uPAR. Colocalization of uPA-uPAR and integrin $\alpha5\beta1$ enhances cell adhesion to fibronectin and increases cell proliferation and migration through focal adhesion kinase (FAK) and rat sarcoma (Ras)–extracellular signal-regulated kinase (ERK) activation (Tang et al., 1998; Aguirre-Ghiso et al., 2001). In some cells, an interplay between uPA-uPAR, integrin $\alpha5\beta1$, and the EGFR is necessary for the activation of the Ras-ERK pathway (Liu et al., 2002). Integrin $\alpha V\beta3$ binds the RGD site in vitronectin, and several studies have shown that simultaneous binding of uPAR and integrins to vitronectin leads to Ras-related C3 botulinum toxin substrate (Rac) activation via activation of FAK, Rous sarcoma oncogene (Src), and Crk-associated substrate (p130Cas) (Smith et al., 2008). Activated Rac stimulates actin polymerization and thereby increases cell motility and invasion (Kjoller and Hall, 2001; Wei et al., 2008).

4.3.4 Inhibitors of plasminogen activators

There are two main subtypes of plasminogen activator inhibitors, PAI1 (SERPINE1) and PAI2 (SERPINB2), both belonging to the serpin superfamily of serine protease inhibitors. Both PAI1 and PAI2 are able to form a one-to-one inhibitory complex with either uPA or tPA. However, PAI1 acts faster and is more widely expressed than PAI2 (Dass et al.,

2008; Lee et al., 2011a; Schaller and Gerber, 2011). In addition, neuroserpin (SERPINI1) and protease nexin-1 (PN1) (SERPINE2) are inhibitors of tPA and uPA (Miranda and Lomas, 2006). Neuroserpin is expressed by neurons, and PN1 by neurons, glial cells, fibroblasts, heart muscle cells, and epithelial cells.

Human PAI1 is a 379 aa protein of approximately 52 kDa, while PAI2 exists as either a secreted 60 kDa glycosylated protein or a 47 kDa nonglycosylated intracellular form (Czekay et al., 2011; Lee et al., 2011a). PAI1 exists as an active, latent, or cleaved form, where only the active form can bind and inhibit its target proteases (Schroeck et al., 2002). Most of the active PAI1 is bound to vitronectin, either to a low-affinity site in the monomeric form of vitronectin or to the high-affinity site at the N-terminal (SMB) domain of multimeric vitronectin (Madsen and Sidenius, 2008). In fact, binding of PAI1 to vitronectin has been shown to induce multimerization of vitronectin (Seiffert and Loskutoff, 1996). Free active PAI1 is converted into the latent form in 1–2 hours, while vitronectin-bound PAI1 stays active for much longer. Latent PAI1 is no longer able to bind vitronectin and is released. *In vitro*, the latent form can be reactivated; however, this has not been demonstrated *in vivo*. Interestingly, PAI1 bound to vitronectin has been shown to sterically interfere with integrin binding to the RGD site (Zhou et al., 2003). However, a ternary complex between vitronectin, PAI1, and integrin $\alpha V\beta 3$ can still occur as long as the actin cytoskeleton is intact (Stefansson et al., 2007). Thus, under certain conditions, PAI1 can modulate integrin-mediated cell adhesion to vitronectin.

Binding of active PAI1 to active uPAR-bound uPA induces a conformational change allowing the complex to bind to the low-density lipoprotein (LDL) receptor-related protein-1 (LRP1) (Herz and Strickland, 2001). This triggers a rapid internalization of the complex through interactions of the cytoplasmic tail of LRP1 with endocytic proteins. Both LRP1 and uPAR are recycled back to the plasma membrane, while uPA and PAI1 are degraded in the lysosomes (Cubellis et al., 1990; Nykjaer et al., 1992; Dass et al., 2008). Inhibition of this recycling of uPAR has been suggested as a mechanism to reduce pericellular proteolysis (Czekay et al., 2001).

PAI2 is mainly expressed by activated monocytes and macrophages, in addition to eosinophils, keratinocytes, microglia, and endothelial and epithelial cells (Croucher et al., 2008; Lee et al., 2011a). Its expression can be highly induced during pregnancy and by a large array of proinflammatory stimuli (Estelles et al., 1989). The high levels of PAI2 during pregnancy are suggested to decrease the overall fibrinolytic activity, thus protecting against premature placental separation and preeclampsia. The normal concentrations of PAI2 and tPA in gingival crecicular fluid are very high, and the function of PAI2 has been linked to regulation of proteolytic activity in relation to gingival inflammation (Kinnby, 2002). Although PAI2 has its main function as an inhibitor of tPA and uPA in the extracellular space and at the plasma membrane, it also has a role as an intracellular protein unrelated to its functions as a serpin (Lee et al., 2011a).

4.3.5 Plasmin substrates

According to the peptidase database MEROPS (http://merops.sanger.ac.uk/), the preferred cleavage site for plasmin is Lys-Xaa > Arg-Xaa, thus allowing for many different proteins to be cleaved by plasmin (▶Table 4.5) (Rawlings et al., 2010). The substrates can be grouped according to their functions: ECM and adhesion molecules, proteases

Table 4.5 Plasmin substrates.

Substrate	*Reference*
ECM and cell adhesion molecules	
Aggrecan	(Fosang et al., 1998)
Amyloid-beta peptide	(Tucker et al., 2000)
Fibrin	(Cesarman-Maus and Hajjar, 2005)
Fibrinogen	(Cesarman-Maus and Hajjar, 2005)
Fibulin-2	(Sasaki et al., 1996)
Fibronectin	(Gold et al., 1992; Cesarman-Maus and Hajjar, 2005)
Laminin	(Paulsson et al., 1988; Goldfinger et al., 1998; Ogura et al., 2008)
Neural cell adhesion molecule L1	(Nayeem et al., 1999)
Nidogen-1	(Mayer et al., 1993)
Osteocalcin	(Novak et al., 1997)
Osteopontin	(Christensen et al., 2010)
Syndecan-4	(Schmidt et al., 2005)
Thrombospondin (*in vitro*)	(Cesarman-Maus and Hajjar, 2005)
Vitronectin	(Chain et al., 1991)
Versican	(Ricciardelli et al., 2009)
Proteases, inhibitors, receptors	
Plasminogen	(Miles et al., 2003)
tPA	(Schaller and Gerber, 2011)
uPA	(Schaller et al., 1982; Stephens et al., 1989; Mason and Joyce, 2011)
uPAR	(Tjwa et al., 2009)
α2-Antiplasmin	(Holmes et al., 1987)
α2-Macroglobulin	(Sottrup-Jensen et al., 1981)
MMP1	(Murphy et al., 1999; Ramos-DeSimone et al., 1999; Saunders et al., 2005)
MMP2 (via MT1-MMP activation)	(Baramova et al., 1997; Mazzieri et al., 1997)
MMP3	(Ramos-DeSimone et al., 1999)
MMP9	(Goldberg et al., 1992; Baramova et al., 1997; Mazzieri et al., 1997; Liu et al., 2005)
MMP10	(Saunders et al., 2005)
MMP13	(Knauper et al., 1996)
MMP14 (MT1-MMP)	(Okumura et al., 1997)
Carboxypeptidase N	(Quagraine et al., 2005)
Carboxypeptidase U	(Schatteman et al., 2000)

(Continued)

Table 4.5 **Plasmin substrates. (*Continued*)**

Substrate	*Reference*
Ceruloplasmin	(Kingston et al., 1977)
Chromogranin-A	(Jiang et al., 2001)
Kallikrein-related peptidase 11 (failed to show complete cleavage)	(Luo et al., 2006)
Proteinase-activated receptors 1, 2, 4	(Loew et al., 2000; Steinberg, 2005)
Blood coagulation	
Coagulation factor X, light chain	(Pryzdial et al., 1999)
Coagulation factor Xa	(Pryzdial et al., 1999)
Coagulation factor XI	
Coagulation factors V, IX	(Cesarman-Maus and Hajjar, 2005)
Complement factor I	(Tsiftsoglou et al., 2005)
Other substrates	
Anti-Mullerian hormone	(Maggard et al., 1996)
β-2-Glycoprotein	(Ohkura et al., 1998)
HGF	(Shanmukhappa et al., 2009)
HRG	(Poon et al., 2009)
Immunoglobulin G (IgG) heavy chain	(Ryan et al., 2008)
Insulin-like growth factor–binding protein-3	(Lalou et al., 1997)
N-methyl-D-aspartate receptor (NMDA-R) subunit 2 (NR2A)	(Yuan et al., 2009)
Chemerin	(Yamaguchi et al., 2011)
Vascular endothelial growth factor (VEGF)-D and VEGF-C	(McColl et al., 2003)
LTBP1	(Taipale et al., 1992; Dallas et al., 1995)
TGF-β (*in vitro*)	(Lyons et al., 1990)

and their inhibitors and receptors, blood coagulation factors, growth factors, cytokines, and neurotransmitters and their receptors.

The major physiological substrate of plasmin is fibrin (Cesarman-Maus and Hajjar, 2005; Mosesson, 2005). The balance between coagulation and fibrinolysis secures adequate blood flow and is tightly regulated. Upon formation of a fibrin clot, both tPA and plasminogen are recruited to the site due to direct binding of their kringle domains to fibrin. Plasmin cleaves fibrin into soluble fragments, and plasmin is protected from its inhibitors (see Section 4.3.6 "Inhibitors of Plasmin") as long as it is bound to fibrin.

Plasmin can activate certain MMPs *in vitro* (Eeckhout and Vaes, 1977; Mazzieri et al., 1997; Lijnen, 2002), but whether plasmin is able to cleave and thereby activate MMPs *in vivo* is still debated (Lijnen, 2002; Ra and Parks, 2007). Plasmin cleaves matrix metalloproteinase-13 (MMP13) at several positions *in vitro*, but cellular model systems exclude a role for plasmin in MMP13 activation *in vivo* (Knauper et al., 1996). Plasmin-mediated activation of matrix metalloproteinase-2 (MMP2) and matrix metalloproteinase-9 (MMP9) was enhanced in the presence of cell membranes (Mazzieri et al., 1997), indicating that binding of the components to cell surface proteins is crucial for optimal activation. The pro-MMP2 activator matrix metalloproteinase-14 (MMP-14)/membrane-type 1-MMP (MT1-MMP) was later shown to be necessary for plasmin-mediated MMP2 activation (Baramova et al., 1997; Monea et al., 2002). However, controversies around this finding exist, and others have shown that plasmin does not activate pro-MMP2 or MMP14 (Lim et al., 1996), and that the level of active MMP2 is not reduced in $Plg^{-/-}$ mice (Lijnen et al., 1998b). Activation of MMP9 can be either plasmin dependent or plasmin independent. In $Plg^{-/-}$ mice, MMP9 was not activated in macrophages and smooth muscle cells, whereas fibroblast-derived MMP9 was active (Lijnen, 2002).

Plasmin can cleave and activate several growth factors. Transforming growth factor-beta (TGF-β) is a multifunctional growth factor that effects cell growth, immune cell function, and ECM formation, and it is either secreted as a small latent TGF-β or sequestered to the ECM through binding to the latent TGF-β-binding protein-1 (LTBP1) (Jenkins, 2008). Plasmin can cleave both small latent TGF-β and LTBP1 and thereby release active TGF-β (Taipale et al., 1992; Dallas et al., 2002).

During tissue repair, a concerted action of ECM degradation, cell proliferation, and cell migration is taking place. Using plasmin-deficient mice, it has been shown that plasmin-mediated activation of HGF/SF, and subsequent activation of its receptor c-Met, is important for hepatocyte migration and hence liver repair (Shanmukhappa et al., 2009).

The plasmin system plays an important role in brain development and function. Accumulation and deposition of amyloid-beta is one of the hallmarks of Alzheimer's disease, and lack of degradation by plasmin has been recognized as a plausible contributor to the disease (Jacobsen et al., 2008). Patients with Alzheimer's have lower levels of tPA actitvity in the brain, and this has been shown to be due to elevated levels of the tPA inhibitor neuroserpin (Fabbro et al., 2011). Thus, inhibition of tPA results in less active plasmin and reduced clearance of amyloid-beta plaques.

4.3.6 Inhibitors of plasmin

Only two physiological inhibitors of plasmin have been reported: the serpin α2-antiplasmin and the general protease inhibitor α2-macroglobulin (Schaller and Gerber, 2011).

α2-Antiplasmin

The human serpin α2-antiplasmin (SERPINF2), also called α2-plasmin inhibitor, is the main physiological inhibitor of plasmin and is a plasma glycoprotein of approximately 67 kDa (452 aa) synthesized by the liver (Coughlin, 2005; Schaller and Gerber, 2011). Unlike other serpins, α2-antiplasmin has both N-and C-terminal extensions. The C-terminal 55 aa extension, which extends beyond the inhibitory core (the serpin domain), is involved in the initial interaction with the lysine-binding site in the kringle domains of plasmin

(Frank et al., 2003), whereas the N-terminal extension is involved in fibrin-binding. α2-Antiplasmin has an important role during fibrinolysis as it is incorporated into the fibrin clot where it inhibits plasmin from digesting fibrin. Circulating antiplasmin-cleaving enzyme (APCE) is able to cleave off the N-terminal 12 aa, generating the two α2-antiplasmin isoforms found in human plasma: Met-α2-antiplasmin (464 aa) and Asn-α2-antiplasmin (452 aa). The normal plasma concentration is 1 μM, 30% Met-α2-antiplasmin and 70% Asn-α2-antiplasmin. Asn-α2-antiplasmin is cross-linked to fibrin by factor XIIIa 13 times faster than Met-α2-antiplasmin (Lee et al., 2011b); thus, APCE activity is modulating the level of plasmin inhibition. APCE is the soluble form of fibroblast activation protein (FAP), or seprase, which is a transmembrane serine protease expressed on pericytes and activated fibroblasts during wound healing, chronic inflammation, and fibrosis, and by tumor-associated fibroblasts in several carcinomas (O'Brien and O'Connor, 2008; Pure, 2009). Expression of FAP/seprase is highly regulated and restricted, and it is still unknown how APCE is generated from FAP/seprase. In the pericellular compartment, α2-antiplasmin can be cleaved and inactivated by MMP3 (Lijnen et al., 2001). MMP3 can also cleave and inhibit PAI1 and thus enhance pericellular plasmin activity by simultaneous inactivation of both α2-antiplasmin and PAI1.

α2-Macroglobulin

α2-Macroglobulin is a general protease inhibitor of the I39 inhibitor family and is mainly synthesized by the liver (Sottrup-Jensen, 1989; Schaller and Gerber, 2011). Mature human α2-macroglobulin is a large glycoprotein (1451 aa) and in plasma α2-macroglobulin is present as a tetramer of about 720 kDa. α2-Macroglobulin is a general inhibitor of all four types of proteases, and cleavage of a bait region by the target protease induces a conformational change of α2-macroglobulin that traps the protease in a central cavity (Sottrup-Jensen, 1989). The α2 macroglobulin with its trapped protease is then endocytosed by the LRP receptor and degraded (Kristensen et al., 1990; Herz and Strickland, 2001).

4.3.7 Plasmin system in cancer

Members of the plasminogen activator system are upregulated in several cancer types and may affect cancer progression through regulation of angiogenesis, cell proliferation, and apoptosis, as well as cancer invasion and metastasis (Dano et al., 2005; Hildenbrand et al., 2010; Mason and Joyce, 2011). Tumors in mice deficient in either uPA or plasminogen grow more slowly than in wild-type mice, and inhibitors of uPA reduce metastasis (Ossowski and Reich, 1983; Dano et al., 2005), suggesting that these proteins are cancer promoters. Depending on the biological setting, members of this proteolytic system can either promote or inhibit cancer progression.

In order to invade and metastasize, cancer cells need to remodel and move through the ECM. Plasmin is a broad spectrum protease that can degrade several ECM proteins, thereby promoting cancer cell migration. Furthermore, plasmin can activate several members of the MMP family, which are major contributors to ECM degradation, and a functional relationship between these families of proteases is thought to exist

(Dano et al., 1999; Ramos-DeSimone et al., 1999). Remodeling of the ECM can also release and activate growth factors and cytokines and thereby affect cell proliferation, differentiation, and apoptosis. Processing of the ECM and change in cell adhesion might also trigger intracellular signal transduction cascades and modify cellular behavior (Andreasen et al., 1997). In addition to the processing of plasminogen to plasmin, uPA can activate growth factors such as HGF/SF and MSP, which both may promote proliferation and migration of cancer cells (Mars et al., 1993; Vigna et al., 1994; Galimi et al., 2001).

Angiogenesis is fundamental for progression of most cancer types by providing the dividing cancer cells with oxygen and nutrients and to remove waste products. Furthermore, the blood vessels may function as highways for metastasizing cancer cells. The state of angiogenesis is determined by the balance between proangiogenic and antiangiogenic signals at a certain location (Folkman, 1990; Holmgren et al., 1995). Antiangiogenic molecules in the circulation may contribute to keeping micrometastasis dormant and thereby hinder cancer progression (Naumov et al., 2006). Depending on the substrate they act on, the plasminogen activator system may have contradicting effects on angiogenesis. Plasmin and other proteases can cleave plasminogen/plasmin at several positions, generating the antiangiogenic angiostatins. Interestingly, angiostatin was released into the circulation from an experimental lung tumor in mice, and removal of the primary tumor, and thereby the source of angiostatin, caused increased metastasis (O'Reilly et al., 1994). On the other hand, remodeling of the ECM by plasmin and other proteolytic enzymes may mobilize and activate potent proangiogenic factors such as basic fibroblast growth factor (bFGF) and transforming growth factor-beta 1 (TGF-β1), thereby promoting angiogenesis (Mignatti and Rifkin, 1993). The complex role of the plasminogen system in angiogenesis is further illustrated by PAI1, which can both enhance and inhibit angiogenesis, and hence tumor growth (McMahon et al., 2001). Accordingly, the protein was found to inhibit cancer progression in some studies (Soff et al., 1995; Bajou et al., 1998; Gutierrez et al., 2000; Stefansson et al., 2001), while it was associated with poor prognosis in breast and ovarian cancers, among others (Grondahl-Hansen et al., 1993; Foekens et al., 1994; Kuhn et al., 1994; Pedersen et al., 1994).

The different members of the plasminogen activator system may be produced by both cancer cells and tumor-associated stromal cells such as fibroblasts, endothelial cells, and infiltrating macrophages. In prostate and ductal breast cancers, uPA and uPAR are expressed by stromal cells, while in skin and oral squamous cell carcinomas, they are expressed by the cancer cells (Dano et al., 2005; Li and Cozzi, 2007; Shi and Stack, 2007; Dass et al., 2008; Ulisse et al., 2009).

Due to their involvement in cancer progression, members of the plasminogen activation system have been suggested as prognostic markers in cancer patients. High levels of soluble uPAR and/or uPA in plasma and other body fluids correlate with poor prognosis in patients with ovarian cancer (Sier et al., 1998; Wahlberg et al., 1998), breast cancer (Grondahl-Hansen et al., 1993, 1995), colorectal cancer (Stephens et al., 1999), and squamous cell lung cancer (Almasi et al., 2005). In addition, expression of uPAR and uPA in disseminated cancer cells in the bone marrow of patients with gastric and breast cancers, respectively, predicted an early relapse (Solomayer et al., 1997). In patients with oral squamous cell carcinomas, levels of both uPA and PAI1 have been found to be independent prognostic markers for recurrence-free survival (Hundsdorfer et al., 2005).

4.3.8 Take-home message

Plasminogen and plasmin have two major properties: the proteolytic properties of the plasmin serine protease domain and the antiangiogenic properties of the angiostatin domain. Plasminogen is mainly produced in the liver and released into the circulation. Thus, the trapping of plasminogen, and subsequent activation into an active protease (plasmin) or processing into angiostatin, is largely dependent on colocalization with the various proteases that have the capability to cleave plasminogen or plasmin. In both cases, colocalization through protein-protein interactions is a prerequisite for activation. tPA-mediated activation of plasmin in fibrin clots is facilitated by concomitant binding of both tPA and plasminogen to fibrin. Activation of plasmin in the pericellular environment is accomplished by the trapping of plasminogen by various plasma membrane proteins, such as Plg-R_{KT} and ENOA, and a subsequent colocalization with uPA bound to its plasma membrane receptor, uPAR. Active plasmin has many substrates, including plasminogen and plasmin itself. This autoproteolytic activity can release angiostatins and the proteolytic plasmin domain, or it can destroy the catalytic domain. Importantly, plasmin can, via a positive feedback loop, activate tPA and uPA and thereby escalate the proteolytic activity. Plasmin has a broad substrate specificity and can cleave a large range of ECM molecules, including fibrin, proteases such as MMPs, cytokines, and growth factors. Taken together, plasminogen and active plasmin have important functional roles in many processes, and dysregulation of their function may cause serious pathological conditions such as cardiovascular diseases and cancer.

References

Aguirre-Ghiso, J. A., Liu, D., Mignatti, A., Kovalski, K., and Ossowski, L. (2001). Urokinase receptor and fibronectin regulate the ERK(MAPK) to p38(MAPK) activity ratios that determine carcinoma cell proliferation or dormancy in vivo. Mol Biol Cell *12*, 863–879.

Ahn, J. H., Lee, H. J., Lee, E. K., et al. (2011). Antiangiogenic kringles derived from human plasminogen and apolipoprotein(a) inhibit fibrinolysis through a mechanism that requires a functional lysine-binding site. Biol Chem *392*, 347–356.

Almasi, C. E., Christensen, I. J., Hoyer-Hansen, G., et al. (2011). Urokinase receptor forms in serum from non-small cell lung cancer patients: relation to prognosis. Lung Cancer *74*, 510–515.

Almasi, C. E., Hoyer-Hansen, G., Christensen, I. J., Dano, K., and Pappot, H. (2005). Prognostic impact of liberated domain I of the urokinase plasminogen activator receptor in squamous cell lung cancer tissue. Lung Cancer *48*, 349–355.

An, S. S., Carreno, C., Marti, D. N., Schaller, J., Albericio, F., and Llinas, M. (1998a). Lysine-50 is a likely site for anchoring the plasminogen N-terminal peptide to lysine-binding kringles. Protein Sci *7*, 1960–1969.

An, S. S., Marti, D. N., Carreno, C., Albericio, F., Schaller, J., and Llinas, M. (1998b). Structural/functional properties of the Glu1-HSer57 N-terminal fragment of human plasminogen: conformational characterization and interaction with kringle domains. Protein Sci *7*, 1947–1959.

Andolfo, A., English, W. R., Resnati, M., Murphy, G., Blasi, F., and Sidenius, N. (2002). Metalloproteases cleave the urokinase-type plasminogen activator receptor in the D1-D2 linker region and expose epitopes not present in the intact soluble receptor. Thromb Haemost *88*, 298–306.

Andreasen, P. A., Kjoller, L., Christensen, L., and Duffy, M. J. (1997). The urokinase-type plasminogen activator system in cancer metastasis: a review. Int J Cancer *72*, 1–22.

Andronicos, N. M., Chen, E. I., Baik, N., et al. (2010). Proteomics-based discovery of a novel, structurally unique, and developmentally regulated plasminogen receptor, Plg-RKT, a major regulator of cell surface plasminogen activation. Blood *115*, 1319–1330.

Andronicos, N. M., and Ranson, M. (2001). The topology of plasminogen binding and activation on the surface of human breast cancer cells. Br J Cancer *85*, 909–916.

Astedt, B., Lecander, I., Brodin, T., Lundblad, A., and Low, K. (1985). Purification of a specific placental plasminogen activator inhibitor by monoclonal antibody and its complex formation with plasminogen activator. Thromb Haemost *53*, 122–125.

Bajou, K., Noel, A., Gerard, R. D., et al. (1998). Absence of host plasminogen activator inhibitor 1 prevents cancer invasion and vascularization. Nat Med *4*, 923–928.

Baramova, E. N., Bajou, K., Remacle, A., et al. (1997). Involvement of PA/plasmin system in the processing of pro-MMP-9 and in the second step of pro-MMP-2 activation. FEBS Lett *405*, 157–162.

Barinka, C., Parry, G., Callahan, J., et al. (2006). Structural basis of interaction between urokinase-type plasminogen activator and its receptor. J Mol Biol *363*, 482–495.

Beaufort, N., Debela, M., Creutzburg, S., et al. (2006). Interplay of human tissue kallikrein 4 (hK4) with the plasminogen activation system: hK4 regulates the structure and functions of the urokinase-type plasminogen activator receptor (uPAR). Biol Chem *387*, 217–222.

Benvenuti, S., and Comoglio, P. M. (2007). The MET receptor tyrosine kinase in invasion and metastasis. J Cell Physiol *213*, 316–325.

Blasi, F., and Carmeliet, P. (2002). uPAR: a versatile signalling orchestrator. Nat Rev Mol Cell Biol *3*, 932–943.

Blasi, F., and Sidenius, N. (2010). The urokinase receptor: focused cell surface proteolysis, cell adhesion and signaling. FEBS Lett *584*, 1923–1930.

Borza, D. B., Shipulina, N. V., and Morgan, W. T. (2004). Effects of histidine-proline-rich glycoprotein on plasminogen activation in solution and on surfaces. Blood Coagul Fibrinolysis *15*, 663–672.

Brauer, R., Beck, I. M., Roderfeld, M., Roeb, E., and Sedlacek, R. (2011). Matrix metalloproteinase-19 inhibits growth of endothelial cells by generating angiostatin-like fragments from plasminogen. BMC Biochem *12*, 38.

Brunner, G., Simon, M. M., and Kramer, M. D. (1990). Activation of pro-urokinase by the human T cell-associated serine proteinase HuTSP-1. FEBS Lett *260*, 141–144.

Bugge, T. H., Flick, M. J., Daugherty, C. C., and Degen, J. L. (1995). Plasminogen deficiency causes severe thrombosis but is compatible with development and reproduction. Genes Dev *9*, 794–807.

Cao, R., Wu, H. L., Veitonmaki, N., et al. (1999). Suppression of angiogenesis and tumor growth by the inhibitor K1–5 generated by plasmin-mediated proteolysis. Proc Natl Acad Sci U S A *96*, 5728–5733.

Cao, Y., Chen, A., An, S. S., Ji, R. W., Davidson, D., and Llinas, M. (1997). Kringle 5 of plasminogen is a novel inhibitor of endothelial cell growth. J Biol Chem *272*, 22924–22928.

Cao, Y., Veitonmaki, N., Keough, K., Cheng, H., Lee, L. S., and Zurakowski, D. (2000). Elevated levels of urine angiostatin and plasminogen/plasmin in cancer patients. Int J Mol Med *5*, 547–551.

Cao, Y., and Xue, L. (2004). Angiostatin. Semin Thromb Hemost *30*, 83–93.

Capello, M., Ferri-Borgogno, S., Cappello, P., and Novelli, F. (2011). α-Enolase: a promising therapeutic and diagnostic tumor target. FEBS J *278*, 1064–1074.

Cesarman-Maus, G., and Hajjar, K. A. (2005). Molecular mechanisms of fibrinolysis. Br J Haematol *129*, 307–321.

Chain, D., Kreizman, T., Shapira, H., and Shaltiel, S. (1991). Plasmin cleavage of vitronectin. Identification of the site and consequent attenuation in binding plasminogen activator inhibitor-1. FEBS Lett *285*, 251–256.

Choi, K. S., Fogg, D. K., Yoon, C. S., and Waisman, D. M. (2003). p11 regulates extracellular plasmin production and invasiveness of HT1080 fibrosarcoma cells. FASEB J *17*, 235–246.

Christensen, B., Schack, L., Klaning, E., and Sorensen, E. S. (2010). Osteopontin is cleaved at multiple sites close to its integrin-binding motifs in milk and is a novel substrate for plasmin and cathepsin D. J Biol Chem *285*, 7929–7937.

Claesson-Welsh, L., Welsh, M., Ito, N., et al. (1998). Angiostatin induces endothelial cell apoptosis and activation of focal adhesion kinase independently of the integrin-binding motif RGD. Proc Natl Acad Sci U S A *95*, 5579–5583.

Cockell, C. S., Marshall, J. M., Dawson, K. M., Cederholm-Williams, S. A., and Ponting, C. P. (1998). Evidence that the conformation of unliganded human plasminogen is maintained via an intramolecular interaction between the lysine-binding site of kringle 5 and the N-terminal peptide. Biochem J *333* (Pt 1), 99–105.

Cornelius, L. A., Nehring, L. C., Harding, E., et al. (1998). Matrix metalloproteinases generate angiostatin: effects on neovascularization. J Immunol *161*, 6845–6852.

Coughlin, P. B. (2005). Antiplasmin: the forgotten serpin? FEBS J *272*, 4852–4857.

Croucher, D. R., Saunders, D. N., Lobov, S., and Ranson, M. (2008). Revisiting the biological roles of PAI2 (SERPINB2) in cancer. Nat Rev Cancer *8*, 535–545.

Cubellis, M. V., Wun, T. C., and Blasi, F. (1990). Receptor-mediated internalization and degradation of urokinase is caused by its specific inhibitor PAI-1. EMBO J *9*, 1079–1085.

Czekay, R. P., Kuemmel, T. A., Orlando, R. A., and Farquhar, M. G. (2001). Direct binding of occupied urokinase receptor (uPAR) to LDL receptor-related protein is required for endocytosis of uPAR and regulation of cell surface urokinase activity. Mol Biol Cell *12*, 1467–1479.

Czekay, R. P., Wilkins-Port, C. E., Higgins, S. P., et al. (2011). PAI-1: an integrator of cell signaling and migration. Int J Cell Biol *2011*, 562481.

Dallas, S. L., Miyazono, K., Skerry, T. M., Mundy, G. R., and Bonewald, L. F. (1995). Dual role for the latent transforming growth factor-beta binding protein in storage of latent TGF-beta in the extracellular matrix and as a structural matrix protein. J Cell Biol *131*, 539–549.

Dallas, S. L., Rosser, J. L., Mundy, G. R., and Bonewald, L. F. (2002). Proteolysis of latent transforming growth factor-beta (TGF-beta)-binding protein-1 by osteoclasts. A cellular mechanism for release of TGF-beta from bone matrix. J Biol Chem *277*, 21352–21360.

Dano, K., Andreasen, P. A., Grondahl-Hansen, J., Kristensen, P., Nielsen, L. S., and Skriver, L. (1985). Plasminogen activators, tissue degradation, and cancer. Adv Cancer Res *44*, 139–266.

Dano, K., Behrendt, N., Hoyer-Hansen, G., et al. (2005). Plasminogen activation and cancer. Thromb Haemost *93*, 676–681.

Dano, K., Romer, J., Nielsen, B. S., et al. (1999). Cancer invasion and tissue remodeling–cooperation of protease systems and cell types. APMIS *107*, 120–127.

Das, R., Burke, T., and Plow, E. F. (2007). Histone H2B as a functionally important plasminogen receptor on macrophages. Blood *110*, 3763–3772.

Dass, K., Ahmad, A., Azmi, A. S., Sarkar, S. H., and Sarkar, F. H. (2008). Evolving role of uPA/uPAR system in human cancers. Cancer Treat Rev *34*, 122–136.

Dong, Z., Kumar, R., Yang, X., and Fidler, I. J. (1997). Macrophage-derived metalloelastase is responsible for the generation of angiostatin in Lewis lung carcinoma. Cell *88*, 801–810.

Dudani, A. K., and Ganz, P. R. (1996). Endothelial cell surface actin serves as a binding site for plasminogen, tissue plasminogen activator and lipoprotein(a). Br J Haematol *95*, 168–178.

Eeckhout, Y., and Vaes, G. (1977). Further studies on the activation of procollagenase, the latent precursor of bone collagenase. Effects of lysosomal cathepsin B, plasmin and kallikrein, and spontaneous activation. Biochem J *166*, 21–31.

Engelholm, L. H., Ingvarsen, S., Jurgensen, H. J., et al. (2009). The collagen receptor uPARAP/Endo180. Front Biosci *14*, 2103–2114.

Espana, F., Estelles, A., Fernandez, P. J., Gilabert, J., Sanchez-Cuenca, J., and Griffin, J. H. (1993). Evidence for the regulation of urokinase and tissue type plasminogen activators by the serpin, protein C inhibitor, in semen and blood plasma. Thromb Haemost *70*, 989–994.

Estelles, A., Gilabert, J., Aznar, J., Loskutoff, D. J., and Schleef, R. R. (1989). Changes in the plasma levels of type 1 and type 2 plasminogen activator inhibitors in normal pregnancy and in patients with severe preeclampsia. Blood *74*, 1332–1338.

Eugen-Olsen, J., Andersen, O., Linneberg, A., et al. (2010). Circulating soluble urokinase plasminogen activator receptor predicts cancer, cardiovascular disease, diabetes and mortality in the general population. J Int Med *268*, 296–308.

Fabbro, S., Schaller, K., and Seeds, N. W. (2011). Amyloid-beta levels are significantly reduced and spatial memory defects are rescued in a novel neuroserpin-deficient Alzheimer's disease transgenic mouse model. J Neurochem *118*, 928–938.

Falcone, D. J., Khan, K. M., Layne, T., and Fernandes, L. (1998). Macrophage formation of angiostatin during inflammation. A byproduct of the activation of plasminogen. J Biol Chem *273*, 31480–31485.

Foekens, J. A., Schmitt, M., van Putten, W. L., et al. (1994). Plasminogen activator inhibitor-1 and prognosis in primary breast cancer. J Clin Oncol *12*, 1648–1658.

Folkman, J. (1990). What is the evidence that tumors are angiogenesis dependent? J Natl Cancer Inst *82*, 4–6.

Fosang, A. J., Last, K., Fujii, Y., Seiki, M., and Okada, Y. (1998). Membrane-type 1 MMP (MMP-14) cleaves at three sites in the aggrecan interglobular domain. FEBS Lett *430*, 186–190.

Frank, P. S., Douglas, J. T., Locher, M., Llinas, M., and Schaller, J. (2003). Structural/functional characterization of the alpha 2-plasmin inhibitor C-terminal peptide. Biochemistry *42*, 1078–1085.

Frenette, G., Tremblay, R. R., Lazure, C., and Dube, J. Y. (1997). Prostatic kallikrein hK2, but not prostate-specific antigen (hK3), activates single-chain urokinase-type plasminogen activator. Int J Cancer *71*, 897–899.

Galimi, F., Cottone, E., Vigna, E., et al. (2001). Hepatocyte growth factor is a regulator of monocyte-macrophage function. J Immunol *166*, 1241–1247.

Gately, S., Twardowski, P., Stack, M. S., et al. (1997). The mechanism of cancer-mediated conversion of plasminogen to the angiogenesis inhibitor angiostatin. Proc Natl Acad Sci U S A *94*, 10868–10872.

Gebbink, M. F. (2011). Tissue-type plasminogen activator-mediated plasminogen activation and contact activation, implications in and beyond haemostasis. J Thromb Haemost *9* (Suppl. 1), 174–181.

Gold, L. I., Rostagno, A., Frangione, B., and Passalaris, T. (1992). Localization of the cleavage sites on fibronectin following digestion by urokinase. J Cell Biochem *50*, 441–452.

Goldberg, G. I., Strongin, A., Collier, I. E., Genrich, L. T., and Marmer, B. L. (1992). Interaction of 92-kDa type IV collagenase with the tissue inhibitor of metalloproteinases prevents dimerization, complex formation with interstitial collagenase, and activation of the proenzyme with stromelysin. J Biol Chem *267*, 4583–4591.

Goldfinger, L. E., Stack, M. S., and Jones, J. C. (1998). Processing of laminin-5 and its functional consequences: role of plasmin and tissue-type plasminogen activator. J Cell Biol *141*, 255–265.

Goretzki, L., Schmitt, M., Mann, K., et al. (1992). Effective activation of the proenzyme form of the urokinase-type plasminogen activator (pro-uPA) by the cysteine protease cathepsin L. FEBS Lett *297*, 112–118.

Grondahl-Hansen, J., Christensen, I. J., Rosenquist, C., et al. (1993). High levels of urokinase-type plasminogen activator and its inhibitor PAI-1 in cytosolic extracts of breast carcinomas are associated with poor prognosis. Cancer Res *53*, 2513–2521.

Grondahl-Hansen, J., Peters, H. A., van Putten, W. L., et al. (1995). Prognostic significance of the receptor for urokinase plasminogen activator in breast cancer. Clin Cancer Res *1*, 1079–1087.

Gutierrez, L. S., Schulman, A., Brito-Robinson, T., Noria, F., Ploplis, V. A., and Castellino, F. J. (2000). Tumor development is retarded in mice lacking the gene for urokinase-type plasminogen activator or its inhibitor, plasminogen activator inhibitor-1. Cancer Res *60*, 5839–5847.

Hajjar, K. A., Harpel, P. C., Jaffe, E. A., and Nachman, R. L. (1986). Binding of plasminogen to cultured human endothelial cells. J Biol Chem *261*, 11656–11662.

Hall, S. W., Humphries, J. E., and Gonias, S. L. (1991). Inhibition of cell surface receptor-bound plasmin by alpha 2-antiplasmin and alpha 2-macroglobulin. J Biol Chem *266*, 12329–12336.

Han, J., Baik, N., Kim, K. H., et al. (2011). Monoclonal antibodies detect receptor-induced binding sites in Glu-plasminogen. Blood *118*, 1653–1662.

Heidtmann, H. H., Nettelbeck, D. M., Mingels, A., Jager, R., Welker, H. G., and Kontermann, R. E. (1999). Generation of angiostatin-like fragments from plasminogen by prostate-specific antigen. Br J Cancer *81*, 1269–1273.

Hekman, C. M., and Loskutoff, D. J. (1985). Endothelial cells produce a latent inhibitor of plasminogen activators that can be activated by denaturants. J Biol Chem *260*, 11581–11587.

Hembrough, T. A., Kralovich, K. R., Li, L., and Gonias, S. L. (1996). Cytokeratin 8 released by breast carcinoma cells in vitro binds plasminogen and tissue-type plasminogen activator and promotes plasminogen activation. Biochem J *317* (Pt 3), 763–769.

Herz, J., and Strickland, D. K. (2001). LRP: a multifunctional scavenger and signaling receptor. J Clin Invest *108*, 779–784.

Hildenbrand, R., Allgayer, H., Marx, A., and Stroebel, P. (2010). Modulators of the urokinase-type plasminogen activation system for cancer. Expert opinion on investigational drugs *19*, 641–652.

Holmes, W. E., Nelles, L., Lijnen, H. R., and Collen, D. (1987). Primary structure of human alpha 2-antiplasmin, a serine protease inhibitor (serpin). J Biol Chem *262*, 1659–1664.

Holmgren, L., O'Reilly, M. S., and Folkman, J. (1995). Dormancy of micrometastases: balanced proliferation and apoptosis in the presence of angiogenesis suppression. Nat Med *1*, 149–153.

Ho-Tin-Noe, B., Rojas, G., Vranckx, R., Lijnen, H. R., and Angles-Cano, E. (2005). Functional hierarchy of plasminogen kringles 1 and 4 in fibrinolysis and plasmin-induced cell detachment and apoptosis. FEBS J *272*, 3387–3400.

Hoyer-Hansen, G., Behrendt, N., Ploug, M., Dano, K., and Preissner, K. T. (1997). The intact urokinase receptor is required for efficient vitronectin binding: receptor cleavage prevents ligand interaction. FEBS Lett *420*, 79–85.

Hoyer-Hansen, G., and Lund, I. K. (2007). Urokinase receptor variants in tissue and body fluids. Adv Clin Chem *44*, 65–102.

Hu, G. F., Strydom, D. J., Fett, J. W., Riordan, J. F., and Vallee, B. L. (1993). Actin is a binding protein for angiogenin. Proc Natl Acad Sci U S A *90*, 1217–1221.

Hu, X. W., Duan, H. F., Gao, L. H., et al. (2008). Inhibition of tumor growth and metastasis by ATF-Fc, an engineered antibody targeting urokinase receptor. Cancer Biol Ther *7*, 651–659.

Huai, Q., Mazar, A. P., Kuo, A., et al. (2006). Structure of human urokinase plasminogen activator in complex with its receptor. Science *311*, 656–659.

Huai, Q., Zhou, A., Lin, L., et al. (2008). Crystal structures of two human vitronectin, urokinase and urokinase receptor complexes. Nat Struct Mol Biol *15*, 422–423.

Hundsdorfer, B., Zeilhofer, H. F., Bock, K. P., et al. (2005). Tumour-associated urokinase-type plasminogen activator (uPA) and its inhibitor PAI-1 in normal and neoplastic tissues of patients with squamous cell cancer of the oral cavity – clinical relevance and prognostic value. J Craniomaxillofac Surg *33*, 191–196.

Huttunen, R., Syrjanen, J., Vuento, R., et al. (2011). Plasma level of soluble urokinase-type plasminogen activator receptor as a predictor of disease severity and case fatality in patients with bacteraemia: a prospective cohort study. J Int Med *270*, 32–40.

Hynes, R. O. (1992). Integrins: versatility, modulation, and signaling in cell adhesion. Cell *69*, 11–25.

Ichinose, A., Fujikawa, K., and Suyama, T. (1986). The activation of pro-urokinase by plasma kallikrein and its inactivation by thrombin. J Biol Chem *261*, 3486–3489.

Jacobsen, J. S., Comery, T. A., Martone, R. L., et al. (2008). Enhanced clearance of Abeta in brain by sustaining the plasmin proteolysis cascade. Proc Natl Acad Sci U S A *105*, 8754–8759.

Jenkins, G. (2008). The role of proteases in transforming growth factor-beta activation. Int J Biochem Cell Biol *40*, 1068–1078.

Ji, W. R., Barrientos, L. G., Llinas, M., et al. (1998a). Selective inhibition by kringle 5 of human plasminogen on endothelial cell migration, an important process in angiogenesis. Biochem Biophys Res Commun *247*, 414–419.

Ji, W. R., Castellino, F. J., Chang, Y., et al. (1998b). Characterization of kringle domains of angiostatin as antagonists of endothelial cell migration, an important process in angiogenesis. FASEB J *12*, 1731–1738.

Jiang, Q., Taupenot, L., Mahata, S. K., et al. (2001). Proteolytic cleavage of chromogranin A (CgA) by plasmin. Selective liberation of a specific bioactive CgA fragment that regulates catecholamine release. J Biol Chem *276*, 25022–25029.

Jones, A. L., Hulett, M. D., Altin, J. G., Hogg, P., and Parish, C. R. (2004). Plasminogen is tethered with high affinity to the cell surface by the plasma protein, histidine-rich glycoprotein. J Biol Chem *279*, 38267–38276.

Kassam, G., Kwon, M., Yoon, C. S., et al. (2001). Purification and characterization of A61. An angiostatin-like plasminogen fragment produced by plasmin autodigestion in the absence of sulfhydryl donors. J Biol Chem *276*, 8924–8933.

Killer, M., Ladurner, G., Kunz, A. B., and Kraus, J. (2010). Current endovascular treatment of acute stroke and future aspects. Drug Discov Today *15*, 640–647.

Kingston, I. B., Kingston, B. L., and Putnam, F. W. (1977). Chemical evidence that proteolytic cleavage causes the heterogeneity present in human ceruloplasmin preparations. Proc Natl Acad Sci U S A *74*, 5377–5381.

Kinnby, B. (2002). The plasminogen activating system in periodontal health and disease. Biol Chem *383*, 85–92.

Kjoller, L., and Hall, A. (2001). Rac mediates cytoskeletal rearrangements and increased cell motility induced by urokinase-type plasminogen activator receptor binding to vitronectin. J Cell Biol *152*, 1145–1157.

Knauper, V., Will, H., Lopez-Otin, C., et al. (1996). Cellular mechanisms for human procollagenase-3 (MMP-13) activation. Evidence that MT1-MMP (MMP-14) and gelatinase a (MMP-2) are able to generate active enzyme. J Biol Chem *271*, 17124–17131.

Kobayashi, H., Schmitt, M., Goretzki, L., et al. (1991). Cathepsin B efficiently activates the soluble and the tumor cell receptor-bound form of the proenzyme urokinase-type plasminogen activator (Pro-uPA). J Biol Chem *266*, 5147–5152.

Koivunen, E., Huhtala, M. L., and Stenman, U. H. (1989). Human ovarian tumor-associated trypsin. Its purification and characterization from mucinous cyst fluid and identification as an activator of pro-urokinase. J Biol Chem *264*, 14095–14099.

Koolwijk, P., Sidenius, N., Peters, E., et al. (2001). Proteolysis of the urokinase-type plasminogen activator receptor by metalloproteinase-12: implication for angiogenesis in fibrin matrices. Blood *97*, 3123–3131.

Kristensen, T., Moestrup, S. K., Gliemann, J., Bendtsen, L., Sand, O., and Sottrup-Jensen, L. (1990). Evidence that the newly cloned low-density-lipoprotein receptor related protein (LRP) is the alpha 2-macroglobulin receptor. FEBS Lett *276*, 151–155.

Kruithof, E. K., Vassalli, J. D., Schleuning, W. D., Mattaliano, R. J., and Bachmann, F. (1986). Purification and characterization of a plasminogen activator inhibitor from the histiocytic lymphoma cell line U-937. J Biol Chem *261*, 11207–11213.

Kuhn, W., Pache, L., Schmalfeldt, B., et al. (1994). Urokinase (uPA) and PAI-1 predict survival in advanced ovarian cancer patients (FIGO III) after radical surgery and platinum-based chemotherapy. Gynecol Oncol *55*, 401–409.

Kwon, M., Yoon, C. S., Fitzpatrick, S., et al. (2001). p22 is a novel plasminogen fragment with antiangiogenic activity. Biochemistry *40*, 13246–13253.

Lalou, C., Sawamura, S., Segovia, B., Ogawa, Y., and Binoux, M. (1997). Proteolytic fragments of insulin-like growth factor binding protein-3: N-terminal sequences and relationships between structure and biological activity. C R Acad Sci III *320*, 621–628.

Lee, J. A., Cochran, B. J., Lobov, S., and Ranson, M. (2011a). Forty years later and the role of plasminogen activator inhibitor type 2/SERPINB2 is still an enigma. Semin Thromb Hemost *37*, 395–407.

Lee, K. N., Jackson, K. W., Christiansen, V. J., Dolence, E. K., and McKee, P. A. (2011b). Enhancement of fibrinolysis by inhibiting enzymatic cleavage of precursor alpha2-antiplasmin. J Thromb Haemost *9*, 987–996.

Lee, T. Y., Muschal, S., Pravda, E. A., Folkman, J., Abdollahi, A., and Javaherian, K. (2009). Angiostatin regulates the expression of antiangiogenic and proapoptotic pathways via targeted inhibition of mitochondrial proteins. Blood *114*, 1987–1998.

Li, Y., and Cozzi, P. J. (2007). Targeting uPA/uPAR in prostate cancer. Cancer Treat Rev *33*, 521–527.

Lighvani, S., Baik, N., Diggs, J. E., Khaldoyanidi, S., Parmer, R. J., and Miles, L. A. (2011). Regulation of macrophage migration by a novel plasminogen receptor Plg-R KT. Blood *118* (20), 5622–5630.

Lijnen, H. R. (2002). Matrix metalloproteinases and cellular fibrinolytic activity. Biochemistry (Mosc) *67*, 92–98.

Lijnen, H. R., Ugwu, F., Bini, A., and Collen, D. (1998a). Generation of an angiostatin-like fragment from plasminogen by stromelysin-1 (MMP-3). Biochemistry *37*, 4699–4702.

Lijnen, H. R., Van Hoef, B., and Collen, D. (2001). Inactivation of the serpin alpha (2)-antiplasmin by stromelysin-1. Biochim Biophys Acta *1547*, 206–213.

Lijnen, H. R., Van Hoef, B., Lupu, F., Moons, L., Carmeliet, P., and Collen, D. (1998b). Function of the plasminogen/plasmin and matrix metalloproteinase systems after vascular injury in mice with targeted inactivation of fibrinolytic system genes. Arterioscler Thromb Vasc Biol *18*, 1035–1045.

Lim, Y. T., Sugiura, Y., Laug, W. E., Sun, B., Garcia, A., and DeClerck, Y. A. (1996). Independent regulation of matrix metalloproteinases and plasminogen activators in human fibrosarcoma cells. J Cell Physiol *167*, 333–340.

Lindahl, T. L., Sigurdardottir, O., and Wiman, B. (1989). Stability of plasminogen activator inhibitor 1 (PAI-1). Thromb Haemost *62*, 748–751.

Liu, D., Aguirre Ghiso, J., Estrada, Y., and Ossowski, L. (2002). EGFR is a transducer of the urokinase receptor initiated signal that is required for in vivo growth of a human carcinoma. Cancer Cell *1*, 445–457.

Liu, Z., Li, N., Diaz, L. A., Shipley, M., Senior, R. M., and Werb, Z. (2005). Synergy between a plasminogen cascade and MMP-9 in autoimmune disease. J Clin Invest *115*, 879–887.

Llinas, P., Le Du, M. H., Gardsvoll, H., et al. (2005). Crystal structure of the human urokinase plasminogen activator receptor bound to an antagonist peptide. EMBO J *24*, 1655–1663.
Loew, D., Perrault, C., Morales, M., et al. (2000). Proteolysis of the exodomain of recombinant protease-activated receptors: prediction of receptor activation or inactivation by MALDI mass spectrometry. Biochemistry *39*, 10812–10822.
Lucas, R., Holmgren, L., Garcia, I., et al. (1998). Multiple forms of angiostatin induce apoptosis in endothelial cells. Blood *92*, 4730–4741.
Luo, L. Y., Shan, S. J., Elliott, M. B., Soosaipillai, A., and Diamandis, E. P. (2006). Purification and characterization of human kallikrein 11, a candidate prostate and ovarian cancer biomarker, from seminal plasma. Clin Cancer Res *12*, 742–750.
Lyons, R. M., Gentry, L. E., Purchio, A. F., and Moses, H. L. (1990). Mechanism of activation of latent recombinant transforming growth factor beta 1 by plasmin. J Cell Biol *110*, 1361–1367.
Madsen, C. D., and Sidenius, N. (2008). The interaction between urokinase receptor and vitronectin in cell adhesion and signalling. Eur J Cell Biol *87*, 617–629.
Madureira, P. A., Surette, A. P., Phipps, K. D., Taboski, M. A., Miller, V. A., and Waisman, D. M. (2011). The role of the annexin A2 heterotetramer in vascular fibrinolysis. Blood *118*, 4789–4797.
Maggard, M. A., Catlin, E. A., Hudson, P. L., Donahoe, P. K., and MacLaughlin, D. T. (1996). Reduction of epidermal growth factor receptor phosphorylation by activated Mullerian inhibiting substance is vanadate-sensitive. Metabolism *45*, 190–195.
Marcotte, P. A., Kozan, I. M., Dorwin, S. A., and Ryan, J. M. (1992). The matrix metalloproteinase pump-1 catalyzes formation of low molecular weight (pro)urokinase in cultures of normal human kidney cells. J Biol Chem *267*, 13803–13806.
Mars, W. M., Zarnegar, R., and Michalopoulos, G. K. (1993). Activation of hepatocyte growth factor by the plasminogen activators uPA and tPA. Am J Pathol *143*, 949–958.
Mason, S. D., and Joyce, J. A. (2011). Proteolytic networks in cancer. Trends Cell Biol *21*, 228–237.
Mayer, U., Mann, K., Timpl, R., and Murphy, G. (1993). Sites of nidogen cleavage by proteases involved in tissue homeostasis and remodelling. Eur J Biochem *217*, 877–884.
Mazzieri, R., Masiero, L., Zanetta, L., et al. (1997). Control of type IV collagenase activity by components of the urokinase-plasmin system: a regulatory mechanism with cell-bound reactants. EMBO J *16*, 2319–2332.
McColl, B. K., Baldwin, M. E., Roufail, S., et al. (2003). Plasmin activates the lymphangiogenic growth factors VEGF-C and VEGF-D. J Exp Med *198*, 863–868.
McMahon, G. A., Petitclerc, E., Stefansson, S., et al. (2001). Plasminogen activator inhibitor-1 regulates tumor growth and angiogenesis. J Biol Chem *276*, 33964–33968.
Mehta, R., and Shapiro, A. D. (2008). Plasminogen deficiency. Haemophilia *14*, 1261–1268.
Mignatti, P., and Rifkin, D. B. (1993). Biology and biochemistry of proteinases in tumor invasion. Physiol Rev *73*, 161–195.
Miles, L. A., Castellino, F. J., and Gong, Y. (2003). Critical role for conversion of glu-plasminogen to Lys-plasminogen for optimal stimulation of plasminogen activation on cell surfaces. Trends Cardiovasc Med *13*, 21–30.
Miles, L. A., Dahlberg, C. M., Plescia, J., Felez, J., Kato, K., and Plow, E. F. (1991). Role of cell-surface lysines in plasminogen binding to cells: identification of alpha-enolase as a candidate plasminogen receptor. Biochemistry *30*, 1682–1691.
Miranda, E., and Lomas, D. A. (2006). Neuroserpin: a serpin to think about. Cell Mol Life Sci *63*, 709–722.
Monea, S., Lehti, K., Keski-Oja, J., and Mignatti, P. (2002). Plasmin activates pro-matrix metalloproteinase-2 with a membrane-type 1 matrix metalloproteinase-dependent mechanism. J Cell Physiol *192*, 160–170.
Moran, P., Li, W., Fan, B., Vij, R., Eigenbrot, C., and Kirchhofer, D. (2006). Pro-urokinase-type plasminogen activator is a substrate for hepsin. J Biol Chem *281*, 30439–30446.

Morikawa, W., Yamamoto, K., Ishikawa, S., et al. (2000). Angiostatin generation by cathepsin D secreted by human prostate carcinoma cells. J Biol Chem *275*, 38912–38920.

Moser, T. L., Stack, M. S., Asplin, I., et al. (1999). Angiostatin binds ATP synthase on the surface of human endothelial cells. Proc Natl Acad Sci U S A *96*, 2811–2816.

Mosesson, M. W. (2005). Fibrinogen and fibrin structure and functions. J Thromb Haemost *3*, 1894–1904.

Murphy, G., Stanton, H., Cowell, S., et al. (1999). Mechanisms for pro matrix metalloproteinase activation. APMIS *107*, 38–44.

Myohanen, H., and Vaheri, A. (2004). Regulation and interactions in the activation of cell-associated plasminogen. Cell Mol Life Sci *61*, 2840–2858.

Naldini, L., Tamagnone, L., Vigna, E., et al. (1992). Extracellular proteolytic cleavage by urokinase is required for activation of hepatocyte growth factor/scatter factor. EMBO J *11*, 4825–4833.

Naumov, G. N., Akslen, L. A., and Folkman, J. (2006). Role of angiogenesis in human tumor dormancy: animal models of the angiogenic switch. Cell Cycle *5*, 1779–1787.

Nayeem, N., Silletti, S., Yang, X., et al. (1999). A potential role for the plasmin(ogen) system in the posttranslational cleavage of the neural cell adhesion molecule L1. J Cell Sci *112* (Pt 24), 4739–4749.

Novak, J. F., Hayes, J. D., and Nishimoto, S. K. (1997). Plasmin-mediated proteolysis of osteocalcin. J Bone Miner Res *12*, 1035–1042.

Nykjaer, A., Petersen, C. M., Moller, B., et al. (1992). Purified alpha 2-macroglobulin receptor/LDL receptor-related protein binds urokinase.plasminogen activator inhibitor type-1 complex. Evidence that the alpha 2-macroglobulin receptor mediates cellular degradation of urokinase receptor-bound complexes. J Biol Chem *267*, 14543–14546.

O'Brien, P., and O'Connor, B. F. (2008). Seprase: an overview of an important matrix serine protease. Biochim Biophys Acta *1784*, 1130–1145.

Ogura, Y., Matsunaga, Y., Nishiyama, T., and Amano, S. (2008). Plasmin induces degradation and dysfunction of laminin 332 (laminin 5) and impaired assembly of basement membrane at the dermal-epidermal junction. Br J Dermatol *159*, 49–60.

Ohkura, N., Hagihara, Y., Yoshimura, T., Goto, Y., and Kato, H. (1998). Plasmin can reduce the function of human beta2 glycoprotein I by cleaving domain V into a nicked form. Blood *91*, 4173–4179.

Okumura, Y., Sato, H., Seiki, M., and Kido, H. (1997). Proteolytic activation of the precursor of membrane type 1 matrix metalloproteinase by human plasmin. A possible cell surface activator. FEBS Lett *402*, 181–184.

O'Reilly, M. S., Holmgren, L., Shing, Y., et al. (1994). Angiostatin: a novel angiogenesis inhibitor that mediates the suppression of metastases by a Lewis lung carcinoma. Cell *79*, 315–328.

O'Reilly, M. S., Wiederschain, D., Stetler-Stevenson, W. G., Folkman, J., and Moses, M. A. (1999). Regulation of angiostatin production by matrix metalloproteinase-2 in a model of concomitant resistance. J Biol Chem *274*, 29568–29571.

Ossowski, L., and Reich, E. (1983). Antibodies to plasminogen activator inhibit human tumor metastasis. Cell *35*, 611–619.

Ostrowski, S. R., Shulman, C. E., Peshu, N., et al. (2007). Elevated plasma urokinase receptor predicts low birth weight in maternal malaria. Parasite Immunol *29*, 37–46.

Owen, K. A., Qiu, D., Alves, J., et al. (2010). Pericellular activation of hepatocyte growth factor by the transmembrane serine proteases matriptase and hepsin, but not by the membrane-associated protease uPA. Biochem J *426*, 219–228.

Patterson, B. C., and Sang, Q. A. (1997). Angiostatin-converting enzyme activities of human matrilysin (MMP-7) and gelatinase B/type IV collagenase (MMP-9). J Biol Chem *272*, 28823–28825.

Paulsson, M., Saladin, K., and Landwehr, R. (1988). Binding of Ca2+ influences susceptibility of laminin to proteolytic digestion and interactions between domain-specific laminin fragments. Eur J Biochem *177*, 477–481.

Pedersen, H., Grondahl-Hansen, J., Francis, D., et al. (1994). Urokinase and plasminogen activator inhibitor type 1 in pulmonary adenocarcinoma. Cancer Res *54*, 120–123.

Petersen, T. E., Martzen, M. R., Ichinose, A., and Davie, E. W. (1990). Characterization of the gene for human plasminogen, a key proenzyme in the fibrinolytic system. J Biol Chem *265*, 6104–6111.

Phipps, K. D., Surette, A. P., O'Connell, P. A., and Waisman, D. M. (2011). Plasminogen receptor S100A10 is essential for the migration of tumor-promoting macrophages into tumor sites. Cancer Res *71*, 6676–6683.

Plow, E. F., Herren, T., Redlitz, A., Miles, L. A., and Hoover-Plow, J. L. (1995). The cell biology of the plasminogen system. FASEB J *9*, 939–945.

Ponting, C. P., Marshall, J. M., and Cederholm-Williams, S. A. (1992). Plasminogen: a structural review. Blood Coagul Fibrinolysis *3*, 605–614.

Poon, I. K., Olsson, A. K., Hulett, M. D., and Parish, C. R. (2009). Regulation of histidine-rich glycoprotein (HRG) function via plasmin-mediated proteolytic cleavage. Biochem J *424*, 27–37.

Poon, I. K., Patel, K. K., Davis, D. S., Parish, C. R., and Hulett, M. D. (2011). Histidine-rich glycoprotein: the Swiss army knife of mammalian plasma. Blood *117*, 2093–2101.

Preissner, K. T., and Seiffert, D. (1998). Role of vitronectin and its receptors in haemostasis and vascular remodeling. Thromb Res *89*, 1–21.

Pryzdial, E. L., Lavigne, N., Dupuis, N., and Kessler, G. E. (1999). Plasmin converts factor X from coagulation zymogen to fibrinolysis cofactor. J Biol Chem *274*, 8500–8505.

Pure, E. (2009). The road to integrative cancer therapies: emergence of a tumor-associated fibroblast protease as a potential therapeutic target in cancer. Expert Opin Ther Targets *13*, 967–973.

Quagraine, M. O., Tan, F., Tamei, H., Erdos, E. G., and Skidgel, R. A. (2005). Plasmin alters the activity and quaternary structure of human plasma carboxypeptidase N. Biochem J *388*, 81–91.

Ra, H. J., and Parks, W. C. (2007). Control of matrix metalloproteinase catalytic activity. Matrix Biol *26*, 587–596.

Ramos-DeSimone, N., Hahn-Dantona, E., Sipley, J., Nagase, H., French, D. L., and Quigley, J. P. (1999). Activation of matrix metalloproteinase-9 (MMP-9) via a converging plasmin/stromelysin-1 cascade enhances tumor cell invasion. J Biol Chem *274*, 13066–13076.

Rasch, M. G., Lund, I. K., Almasi, C. E., and Hoyer-Hansen, G. (2008). Intact and cleaved uPAR forms: diagnostic and prognostic value in cancer. Front Biosci *13*, 6752–6762.

Rawlings, N. D., Barrett, A. J., and Bateman, A. (2010). MEROPS: the peptidase database. Nucleic Acids Res *38*, D227–D233.

Ricciardelli, C., Sakko, A. J., Ween, M. P., Russell, D. L., and Horsfall, D. J. (2009). The biological role and regulation of versican levels in cancer. Cancer Metastasis Rev *28*, 233–245.

Robbins, K. C., Summaria, L., Hsieh, B., and Shah, R. J. (1967). The peptide chains of human plasmin. Mechanism of activation of human plasminogen to plasmin. J Biol Chem *242*, 2333–2342.

Ryan, M. H., Petrone, D., Nemeth, J. F., Barnathan, E., Bjorck, L., and Jordan, R. E. (2008). Proteolysis of purified IgGs by human and bacterial enzymes in vitro and the detection of specific proteolytic fragments of endogenous IgG in rheumatoid synovial fluid. Mol Immunol *45*, 1837–1846.

Sasaki, T., Mann, K., Murphy, G., Chu, M. L., and Timpl, R. (1996). Different susceptibilities of fibulin-1 and fibulin-2 to cleavage by matrix metalloproteinases and other tissue proteases. Eur J Biochem *240*, 427–434.

Saunders, W. B., Bayless, K. J., and Davis, G. E. (2005). MMP-1 activation by serine proteases and MMP-10 induces human capillary tubular network collapse and regression in 3D collagen matrices. J Cell Sci *118*, 2325–2340.

Schaller, J., and Gerber, S. S. (2011). The plasmin-antiplasmin system: structural and functional aspects. Cell Mol Life Sci *68*, 785–801.

Schaller, J., Nick, H., Rickli, E. E., Gillessen, D., Lergier, W., and Studer, R. O. (1982). Human low-molecular-weight urinary urokinase. Partial characterization and preliminary sequence data of the two polypeptide chains. Eur J Biochem *125*, 251–257.

Schatteman, K. A., Goossens, F. J., Scharpe, S. S., and Hendriks, D. F. (2000). Proteolytic activation of purified human procarboxypeptidase U. Clin Chim Acta *292*, 25–40.

Schmidt, A., Echtermeyer, F., Alozie, A., Brands, K., and Buddecke, E. (2005). Plasmin- and thrombin-accelerated shedding of syndecan-4 ectodomain generates cleavage sites at Lys(114)-Arg(115) and Lys(129)-Val(130) bonds. J Biol Chem *280*, 34441–34446.

Schmitt, M., Kanayama, N., Henschen, A., et al. (1989). Elastase released from human granulocytes stimulated with N-formyl-chemotactic peptide prevents activation of tumor cell prourokinase (pro-uPA). FEBS Lett *255*, 83–88.

Schroeck, F., Arroyo de Prada, N., Sperl, S., Schmitt, M., and Viktor, M. (2002). Interaction of plasminogen activator inhibitor type-1 (PAI-1) with vitronectin (Vn): mapping the binding sites on PAI-1 and Vn. Biol Chem *383*, 1143–1149.

Seiffert, D., and Loskutoff, D. J. (1996). Type 1 plasminogen activator inhibitor induces multimerization of plasma vitronectin. A suggested mechanism for the generation of the tissue form of vitronectin in vivo. J Biol Chem *271*, 29644–29651.

Shanmukhappa, K., Matte, U., Degen, J. L., and Bezerra, J. A. (2009). Plasmin-mediated proteolysis is required for hepatocyte growth factor activation during liver repair. J Biol Chem *284*, 12917–12923.

Shi, Z., and Stack, M. S. (2007). Urinary-type plasminogen activator (uPA) and its receptor (uPAR) in squamous cell carcinoma of the oral cavity. Biochem J *407*, 153–159.

Sier, C. F., Stephens, R., Bizik, J., et al. (1998). The level of urokinase-type plasminogen activator receptor is increased in serum of ovarian cancer patients. Cancer Res *58*, 1843–1849.

Smith, H. W., Marra, P., and Marshall, C. J. (2008). uPAR promotes formation of the p130Cas-Crk complex to activate Rac through DOCK180. J Cell Biol *182*, 777–790.

Smith, H. W., and Marshall, C. J. (2010). Regulation of cell signalling by uPAR. Nat Rev Mol Cell Biol *11*, 23–36.

Soff, G. A. (2000). Angiostatin and angiostatin-related proteins. Cancer Metastasis Rev *19*, 97–107.

Soff, G. A., Sanderowitz, J., Gately, S., et al. (1995). Expression of plasminogen activator inhibitor type 1 by human prostate carcinoma cells inhibits primary tumor growth, tumor-associated angiogenesis, and metastasis to lung and liver in an athymic mouse model. J Clin Invest *96*, 2593–2600.

Solberg, H., Ploug, M., Hoyer-Hansen, G., Nielsen, B. S., and Lund, L. R. (2001). The murine receptor for urokinase-type plasminogen activator is primarily expressed in tissues actively undergoing remodeling. J Histochem Cytochem *49*, 237–246.

Solomayer, E. F., Diel, I. J., Wallwiener, D., et al. (1997). Prognostic relevance of urokinase plasminogen activator detection in micrometastatic cells in the bone marrow of patients with primary breast cancer. Br J Cancer *76*, 812–818.

Sorio, C., Mafficini, A., Furlan, F., et al. (2011). Elevated urinary levels of urokinase-type plasminogen activator receptor (uPAR) in pancreatic ductal adenocarcinoma identify a clinically high-risk group. BMC Cancer *11*, 448.

Sottrup-Jensen, L. (1989). Alpha-macroglobulins: structure, shape, and mechanism of proteinase complex formation. J Biol Chem *264*, 11539–11542.

Sottrup-Jensen, L., Lonblad, P. B., Stepanik, T. M., Petersen, T. E., Magnusson, S., and Jornvall, H. (1981). Primary structure of the "bait" region for proteinases in alpha 2-macroglobulin. Nature of the complex. FEBS Lett *127*, 167–173.

Spraggon, G., Phillips, C., Nowak, U. K., et al. (1995). The crystal structure of the catalytic domain of human urokinase-type plasminogen activator. Structure *3*, 681–691.

Sprengers, E. D., and Kluft, C. (1987). Plasminogen activator inhibitors. Blood *69*, 381–387.

Stack, M. S., Gately, S., Bafetti, L. M., Enghild, J. J., and Soff, G. A. (1999). Angiostatin inhibits endothelial and melanoma cellular invasion by blocking matrix-enhanced plasminogen activation. Biochem J *340* (Pt 1), 77–84.

Stack, M. S., and Johnson, D. A. (1994). Human mast cell tryptase activates single-chain urinary-type plasminogen activator (pro-urokinase). J Biol Chem *269*, 9416–9419.

Stefansson, S., Petitclerc, E., Wong, M. K., McMahon, G. A., Brooks, P. C., and Lawrence, D. A. (2001). Inhibition of angiogenesis in vivo by plasminogen activator inhibitor-1. J Biol Chem *276*, 8135–8141.

Stefansson, S., Su, E. J., Ishigami, S., et al. (2007). The contributions of integrin affinity and integrin-cytoskeletal engagement in endothelial and smooth muscle cell adhesion to vitronectin. J Biol Chem *282*, 15679–15689.

Steinberg, S. F. (2005). The cardiovascular actions of protease-activated receptors. Mol Pharmacol *67*, 2–11.

Stephens, R. W., Nielsen, H. J., Christensen, I. J., et al. (1999). Plasma urokinase receptor levels in patients with colorectal cancer: relationship to prognosis. J Natl Cancer Inst *91*, 869–874.

Stephens, R. W., Pollanen, J., Tapiovaara, H., et al. (1989). Activation of pro-urokinase and plasminogen on human sarcoma cells: a proteolytic system with surface-bound reactants. J Cell Biol *108*, 1987–1995.

Stoppelli, M. P., Corti, A., Soffientini, A., Cassani, G., Blasi, F., and Assoian, R. K. (1985). Differentiation-enhanced binding of the amino-terminal fragment of human urokinase plasminogen activator to a specific receptor on U937 monocytes. Proc Natl Acad Sci U S A *82*, 4939–4943.

Surette, A. P., Madureira, P. A., Phipps, K. D., Miller, V. A., Svenningsson, P., and Waisman, D. M. (2011). Regulation of fibrinolysis by S100A10 in vivo. Blood *118*, 3172–3181.

Taipale, J., Koli, K., and Keski-Oja, J. (1992). Release of transforming growth factor-beta 1 from the pericellular matrix of cultured fibroblasts and fibrosarcoma cells by plasmin and thrombin. J Biol Chem *267*, 25378–25384.

Takada, A., Sugawara, Y., and Takada, Y. (1988). Degradation of Glu- and Lys-plasminogen by elastase in the presence or absence of tranexamic acid. Thromb Res *50*, 285–294.

Takayama, T. K., McMullen, B. A., Nelson, P. S., Matsumura, M., and Fujikawa, K. (2001). Characterization of hK4 (prostase), a prostate-specific serine protease: activation of the precursor of prostate specific antigen (pro-PSA) and single-chain urokinase-type plasminogen activator and degradation of prostatic acid phosphatase. Biochemistry *40*, 15341–15348.

Tang, C. H., and Wei, Y. (2008). The urokinase receptor and integrins in cancer progression. Cell Mol Life Sci *65*, 1916–1932.

Tang, H., Kerins, D. M., Hao, Q., Inagami, T., and Vaughan, D. E. (1998). The urokinase-type plasminogen activator receptor mediates tyrosine phosphorylation of focal adhesion proteins and activation of mitogen-activated protein kinase in cultured endothelial cells. J Biol Chem *273*, 18268–18272.

Tarui, T., Miles, L. A., and Takada, Y. (2001). Specific interaction of angiostatin with integrin alpha(v)beta(3) in endothelial cells. J Biol Chem *276*, 39562–39568.

Tjwa, M., Sidenius, N., Moura, R., et al. (2009). Membrane-anchored uPAR regulates the proliferation, marrow pool size, engraftment, and mobilization of mouse hematopoietic stem/progenitor cells. J Clin Invest *119*, 1008–1018.

Troyanovsky, B., Levchenko, T., Mansson, G., Matvijenko, O., and Holmgren, L. (2001). Angiomotin: an angiostatin binding protein that regulates endothelial cell migration and tube formation. J Cell Biol *152*, 1247–1254.

Tsiftsoglou, S. A., Willis, A. C., Li, P., et al. (2005). The catalytically active serine protease domain of human complement factor I. Biochemistry *44*, 6239–6249.

Tsuchida-Straeten, N., Ensslen, S., Schafer, C., et al. (2005). Enhanced blood coagulation and fibrinolysis in mice lacking histidine-rich glycoprotein (HRG). J Thromb Haemost *3*, 865–872.

Tucker, H. M., Kihiko, M., Caldwell, J. N., et al. (2000). The plasmin system is induced by and degrades amyloid-beta aggregates. J Neurosci *20*, 3937–3946.

Ulisse, S., Baldini, E., Sorrenti, S., and D'Armiento, M. (2009). The urokinase plasminogen activator system: a target for anti-cancer therapy. Curr Cancer Drug Targets *9*, 32–71.

Vassalli, J. D., Baccino, D., and Belin, D. (1985). A cellular binding site for the Mr 55,000 form of the human plasminogen activator, urokinase. J Cell Biol *100*, 86–92.

Vigna, E., Naldini, L., Tamagnone, L., et al. (1994). Hepatocyte growth factor and its receptor, the tyrosine kinase encoded by the c-MET proto-oncogene. Cell Mol Biol (Noisy-le-grand) *40*, 597–604.

Vincenza Carriero, M., Franco, P., Vocca, I., et al. (2009). Structure, function and antagonists of urokinase-type plasminogen activator. Front Biosci *14*, 3782–3794.

Wahlberg, K., Hoyer-Hansen, G., and Casslen, B. (1998). Soluble receptor for urokinase plasminogen activator in both full-length and a cleaved form is present in high concentration in cystic fluid from ovarian cancer. Cancer Res *58*, 3294–3298.

Wang, H., Doll, J. A., Jiang, K., et al. (2006). Differential binding of plasminogen, plasmin, and angiostatin4.5 to cell surface beta-actin: implications for cancer-mediated angiogenesis. Cancer Res *66*, 7211–7215.

Wang, H., Schultz, R., Hong, J., Cundiff, D. L., Jiang, K., and Soff, G. A. (2004). Cell surface-dependent generation of angiostatin4.5. Cancer Res *64*, 162–168.

Wei, C., Moller, C. C., Altintas, M. M., et al. (2008). Modification of kidney barrier function by the urokinase receptor. Nat Med *14*, 55–63.

Wei, Y., Waltz, D. A., Rao, N., Drummond, R. J., Rosenberg, S., and Chapman, H. A. (1994). Identification of the urokinase receptor as an adhesion receptor for vitronectin. J Biol Chem *269*, 32380–32388.

Westphal, J. R., Van't Hullenaar, R., Geurts-Moespot, A., et al. (2000). Angiostatin generation by human tumor cell lines: involvement of plasminogen activators. Int J Cancer *86*, 760–767.

Wilhelm, O. G., Wilhelm, S., Escott, G. M., et al. (1999). Cellular glycosylphosphatidylinositol-specific phospholipase D regulates urokinase receptor shedding and cell surface expression. J Cell Physiol *180*, 225–235.

Wolf, B. B., Vasudevan, J., Henkin, J., and Gonias, S. L. (1993). Nerve growth factor-gamma activates soluble and receptor-bound single chain urokinase-type plasminogen activator. J Biol Chem *268*, 16327–16331.

Wygrecka, M., Marsh, L. M., Morty, R. E., et al. (2009). Enolase-1 promotes plasminogen-mediated recruitment of monocytes to the acutely inflamed lung. Blood *113*, 5588–5598.

Yamaguchi, Y., Du, X. Y., Zhao, L., Morser, J., and Leung, L. L. (2011). Proteolytic cleavage of chemerin protein is necessary for activation to the active form, Chem157S, which functions as a signaling molecule in glioblastoma. J Biol Chem *286*, 39510–39519.

Yilmaz, G., Mentese, A., Kaya, S., Uzun, A., Karahan, S. C., and Koksal, I. (2011). The diagnostic and prognostic significance of soluble urokinase plasminogen activator receptor in Crimean-Congo hemorrhagic fever. J Clin Virol *50*, 209–211.

Yoshida, E., Ohmura, S., Sugiki, M., Maruyama, M., and Mihara, H. (1995). Prostate-specific antigen activates single-chain urokinase-type plasminogen activator. Int J Cancer *63*, 863–865.

Yuan, H., Vance, K. M., Junge, C. E., et al. (2009). The serine protease plasmin cleaves the amino-terminal domain of the NR2A subunit to relieve zinc inhibition of the N-methyl-D-aspartate receptors. J Biol Chem *284*, 12862–12873.

Zhou, A., Huntington, J. A., Pannu, N. S., Carrell, R. W., and Read, R. J. (2003). How vitronectin binds PAI-1 to modulate fibrinolysis and cell migration. Nat Struct Biol *10*, 541–544.

4.4 Matrix metalloproteinase complexes and their biological significance

Bodil Fadnes, Elin Hadler-Olsen, Ingebrigt Sylte, Lars Uhlin-Hansen and Jan-Olof Winberg

4.4.1 Introduction

In humans, 23 different matrix metalloproteinases (MMPs) have been described so far. These are central regulators of cell function under both normal and pathological conditions. Some of the biological activities mediated by MMP processing of extracellular matrix (ECM) and non-ECM proteins are cell migration, invasion, differentiation, proliferation, and apoptosis; pro- and anti-inflammatory reactions; angiogenesis; antiangiogenesis; release and activation of cytokines, chemokines, and growth factors; and platelet aggregation. MMPs are also known to be involved in numerous pathological processes that affect almost all organs. Their broad substrate specificity and various roles in development and homeostasis appear to be some of the reasons why MMPs also have complicated and dual roles in the development of various diseases (for reviews, see McCawley and Matrisian, 2001; Mott and Werb, 2004; Martin and Matrisian, 2007; Page-McCaw et al., 2007; Kandasamy et al., 2010; Gialeli et al., 2011; Hadler-Olsen et al., 2011; Hua et al., 2011; Murphy and Nagase, 2011).

The synthesis and activity of MMPs are regulated at several levels, including transcription, activation, inhibition, complex formation, and compartmentalization. At the transcriptional level, the synthesis of MMPs is regulated by the interaction with other cells, ECM molecules, growth factors, cytokines, and chemokines. All MMPs are synthesized as inactive proenzymes that need to be activated as described in the next section. Once activated, the activity of MMPs is also controlled by inhibitors in the ECM and the circulation. Among these are the four tissue inhibitors of MMPs (TIMPs), α2-macroglobulin, reversion-inducing cysteine-rich protein with Kazal motifs (RECK), tissue-factor pathway-inhibitor-2 (TFPI2), and degradation products of plasminogen and collagens IV, XV, and XVIII. Examples of such degradation products generated by MMPs and other proteinases are angiostatin, endostatin, tumstatin, arrestin, and canstatin (Rundhaug, 2005; Overall and Dean, 2006; Malla et al., 2008b; Brauer et al., 2011; Gialeli et al., 2011).

It has been shown that MMPs form complexes with other molecules, both macromolecules and smaller compounds. These complexes include MMP homodimers and heterodimers, as well as various larger heteromeric complexes. In this context, enzyme-substrate complexes and most enzyme-inhibitor complexes are not included. In the present review, we focus on some of the complexes formed by individual MMPs, the interactions involved in the complex formation, and whether these interactions hide or generate new exosites, as well as the biological implications of the complexes.

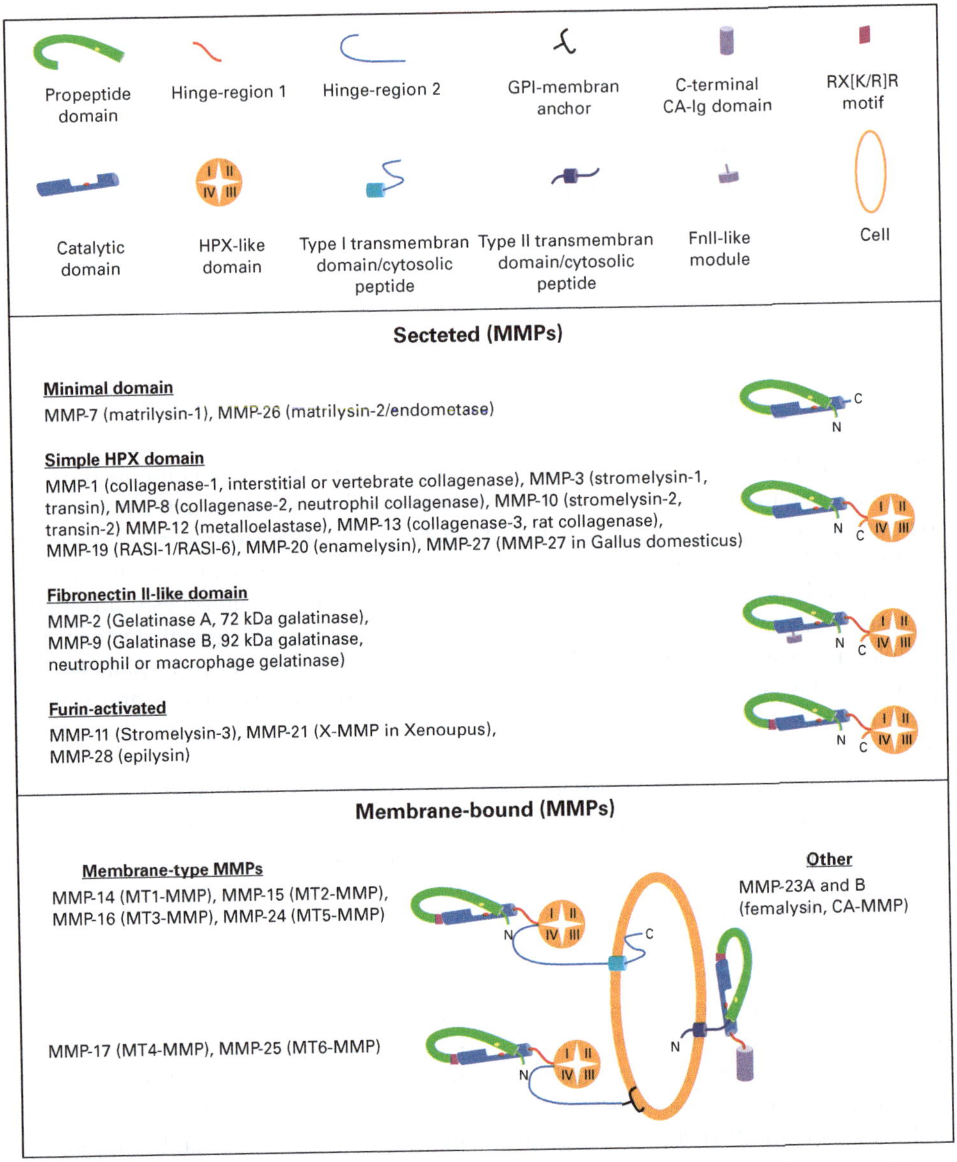

Figure 4.3 Domain structure of secreted and membrane-anchored MMPs. Shown along with the MMP number is the most commonly used name. Most MMPs contain a propeptide domain, a catalytic domain, a linker (hinge region), and an HPX-like domain. The hinge region varies in size between the different MMPs. In MMP9, this region contains a collagen V–like motif and is heavily *O*-glycosylated. The three furin-activated MMPs and all of the membrane-anchored MMPs have a basic RX[K/R]R motif at the C-terminal end of their prodomains. This motif can be cleaved inside the cells by furin-like proteinases. The two gelatinases (MMP2 and MMP9) contain three FNII-like repeats in their catalytic domain, N terminal to the catalytic zinc-binding site. Four of the six membrane-type MMPs (MT-MMPs) are anchored to the cell membranes through a type I transmembrane domain, and the other two through a GPI moiety. These membrane-anchoring regions are linked to the HPX domain

4.4.2 MMP structure and classification

MMPs belong to the metzincin superfamily of metalloproteinases and are zinc and calcium dependent. Historically, they were named and classified according to what was believed to be the main substrate they acted on, such as collagenases for the collagen-degrading enzymes, gelatinases for the gelatin-degrading enzymes, and so on (Lapière, 2005). Later, it was recognized that each of these enzymes was able to cleave and degrade several ECM proteins, and in the mid-1980s, these enzymes were named MMPs and numbered in order of their discovery (Okada et al., 1986; Nagase et al., 1992; Lapière, 2005). Today, we know that the MMPs together are able to process and degrade all ECM proteins, as well as a large number of other non-ECM proteins, such as growth factors, cytokines, chemokines, cell receptors, proteinases including other MMPs, proteinase inhibitors, and a cascade of intracellularly localized proteins (Butler and Overall, 2009; Rodriguez et al., 2010). This knowledge has given rise to discussions about whether the name MMP is adequate, but it is likely that the present nomenclature will remain as it is already well established.

The present MMP classification is based on the domains, modules, and motifs forming their structure (▶Figure 4.3). All MMPs contain an N-terminal signal peptide, which directs the enzymes to the secretory pathway, a prodomain with a conserved PRCGXPD sequence that confers the latency of the enzymes and a catalytic (CAT) domain containing a catalytic zinc ion. Furthermore, they also contain the characteristic of the metzincin superfamily – that is, the HEXXHXXGXXH zinc-binding motif and a conserved methionine (Met) located C-terminal to the zinc ligands that form a Met turn (Bode et al., 1993). The CAT domain in two of the MMPs (MMP2 and MMP9) also contains a module of three fibronectin II (FNII)–like inserts. All MMPs except the two matrilysins (MMP7 and MMP26) and MMP23 contain a C-terminal hemopexin (HPX)–like domain. The HPX domains consist of a four-blade β-propeller stabilized by calcium ions and a disulfide bridge between blades I and IV. The HPX domain is linked to the CAT domain through a hinge region, which varies in size between the MMPs (Sela-Passwell et al., 2010). In MMP9, this linker region contains a heavily glycosylated collagen V–like module. In six of the membrane-anchored MMP family members, the HPX region ends in either a type I transmembrane domain with a short intracellular sequence or a glycosylphosphatidylinosityl (GPI) moiety. MMP23 differs from the other MMPs by lacking the HPX domain, which is replaced by a C-terminal cystein array (CA) region and an immunoglobulin (Ig)–like domain, and instead of the N-terminal signal peptide, this enzyme contains an N-terminal type II transmembrane domain. The membrane-bound MMPs and three of the secreted MMPs (MMP11, MMP21, and MMP28) contain a short basic motif (RX[K/R]R) at the end of their prodomain. MMPs with this motif can be activated within the Golgi apparatus by furin, a serine proteinase that belongs to the convertase family.

Figure 4.3 (*Continued*)

through a second hinge region. The seventh membrane-anchored MMP, MMP23, has an N-terminal type II transmembrane domain. The two minimal domain MMPs and MMP23 lack the HPX domain, and in the latter enzyme, this domain is replaced by a C-terminal CA and an Ig-like domain.

Most of the secreted MMPs are synthesized in a latent proform and hence need to be activated after their release from the cells (Nagase, 1997; Hadler-Olsen et al., 2011). This activation involves proteolytic removal of the prodomain by proteinases from other families, active MMPs, or by autoactivation induced by binding to compounds such as reactive oxygen species (ROS), chaotropic agents, or organomercurials. Various studies indicate that the proenzymes may also be active and process substrates. This activation is referred to as allosteric activation, in which interactions with other molecules force the prodomain away from the active site without being cleaved off from the enzyme.

Processing of certain protein and peptide substrates by MMPs requires that the substrates interact with the active site and regions outside the active site. These non-catalytic sites, or exosites, can be motifs localized outside or within the catalytic domain. By the interaction with an exosite, a substrate is properly oriented for cleavage. In some cases, as for the cleavage of interstitial collagens by collagenases, binding to an exosite in the HPX domain is an absolute requirement for degradation (for review, see Overall, 2002; Overall et al., 2002; Sela-Passwell et al., 2010; Hadler-Olsen et al., 2011; Murphy and Nagase, 2011).

4.4.3 MMP complexes

Like other enzymes and structural proteins, MMPs form complexes with other macromolecules, such as proteins and glucosaminoglycan (GAG) chains of proteoglycans (PGs). Formation of MMP complexes may regulate the activity, substrate specificity, stability, and localization of the enzymes. Regulation of the activity can be achieved in various ways (e.g. by affecting the activation of the proenzyme or the access of MMP inhibitors to the active enzyme). Substrate specificity may be altered by blocking existing enzyme exosites or inducing new exosites into the formed complex. Binding to other macromolecules may further direct the enzyme to new locations, such as anchoring a secreted MMP to the cell membrane, or to new sites in the ECM and hence localize the enzyme close to a potential substrate. Membrane-type MMPs (MT-MMPs) are thought to be important mediators of pericellular proteolysis. Unlike most of the secreted MMPs, the MT-MMPs are activated intracellularly by the proprotein convertase furin, and thus are active as soon as they reach their destination at the cell surface. By being anchored to the cell membrane, these MMPs are subjected to unique regulatory mechanisms to control localization and the amount of proteolytic activity in the pericellular milieu.

The structure of TIMPs consists of two domains where the N-terminal domain binds to the active site of the MMPs and inhibits the protease activity (Murphy et al., 1991; Gomis-Ruth et al., 1997; Williamson et al., 1997; Fernandez-Catalan et al., 1998; Tuuttila et al., 1998). Furthermore, the C-terminal domain of TIMPs can form a complex with the HPX domain of some MMPs. Pro-matrix metalloproteinase 9 (pro-MMP9) forms a complex with TIMP1 and TIMP3; pro-MMP2, with TIMP2, TIMP3, and TIMP4; and pro-MMP13, with TIMP1 (Goldberg et al., 1989; Wilhelm et al., 1989; Brew and Nagase, 2010; Zhang et al., 2010). X-ray structure of the pro-MMP2/TIMP2 complex shows that the interactions forming the complex involve both hydrophobic clusters accompanied by polar and electrostatic contacts. The most C-terminal part of TIMP2 binds to

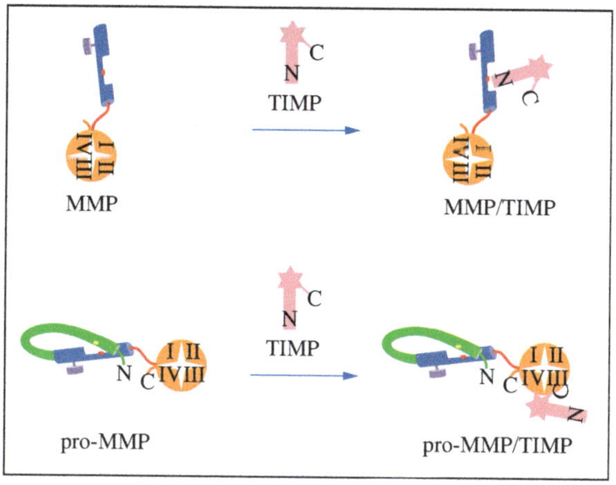

Figure 4.4 **Binding of TIMPs to MMPs.** All TIMPs bind to the active site of the MMPs by their N-terminal domain, where the side chain of the second amino acid binds to the enzyme's S1′ pocket. TIMPs can also bind to certain pro-MMPs, where the C-terminal part of the inhibitor binds to the enzyme's HPX region.

blade III of the HPX domain of MMP2, and residues around Met-149 of TIMP2 interact with blade IV of the HPX domain of MMP2 (Morgunova et al., 2002). ▶Figure 4.4 is a schematic presentation of TIMP interaction with pro- and active forms of MMPs.

Various MMPs also binds to PGs, and these interactions involve either the GAG chains or the core proteins of the PGs. The effects of the MMP/GAG-chain interaction will depend on the MMP and cell type involved, as described for the MT3-MMP/CSPG/pro-MMP2, MMP2/Syndecan-2 and MMP9/Heparan sulphate complexes and summarized in ▶Figure 4.5.

Mostly, formation of MMP complexes affects the ability of the MMP to act on specific substrates. However, it is also reported that the formation of MMP complexes induces biological activities independent of the proteolytic activity of the MMPs. There are also MMP complexes described in which the biological function of the complex so far is unknown. Here we will describe some of the characterized MMP complexes and what is known with respect to their function.

MMP complexes in which biological function involves proteolysis

MT1-MMP/MT1-MMP

The transmembrane-bound MMP, MT1-MMP, needs to form a homodimer (MT1-MMP_d) at the cell surface in order to activate pro-MMP2 or to cleave triple-helical collagen. It was recently shown by X-ray crystallography and expression of mutant enzymes that the dimer formation involves an interaction of the HPX blades II and III in molecule A with blades III and II in molecule B. Site-directed mutagenesis of amino acids involved in this interaction weakened the binding between the two HPX domains and resulted in reduced pro-MMP2 activation and collagen degradation but had no effect on dimer-independent functions such as gelatin film degradation and two-dimensional cell migration (Tochowicz et al., 2011).

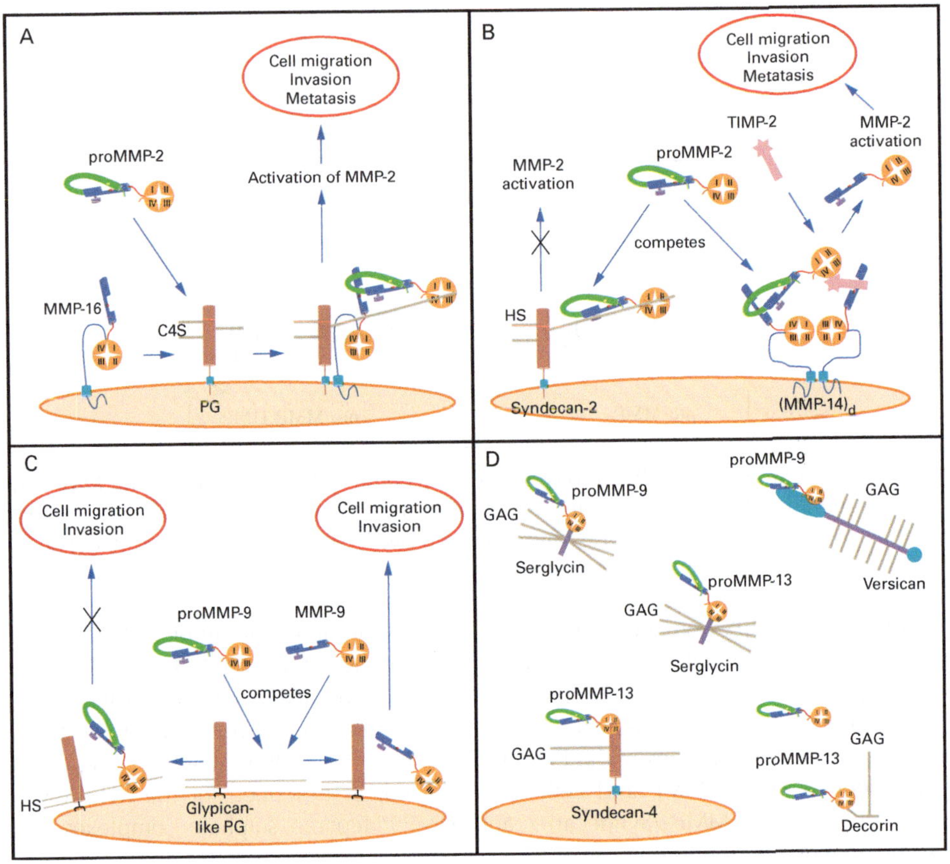

Figure 4.5 **Binding of MMPs to PGs.** The interaction between MMPs occurs either through the GAG chains or the core protein of the PG. (a) Binding of pro-MMP2 and MMP16 (MT3-MMP) to chondroitin 4-sulfate (C4S) chains at cell-surface-associated PGs in melanoma cells results in activation of MMP2 and increased cell migration, invasion, and metastasis. (b) Highly metastatic lung carcinoma cells produce MMP14 (MT1-MMP), TIMP2, and pro-MMP2. The latter MMP is activated at the cell membrane in a ternary complex, which also involves an MMP14 dimer and TIMP2. Low metastatic lung carcinoma cells also produced syndecan-2 at the cell membrane. Heparan sulfate (HS) chains at this PG competed with MMP14 for the binding of pro-MMP2, which prevented the activation of the latter enzyme and the following cell migration, invasion, and metastasis. (c) Highly metastatic colon carcinoma cells contained glypican-like PGs at their cell surface. Active MMP9 can bind to the HS chains of these PGs, which resulted in cell migration and invasion. The presence of pro-MMP9 competes with active MMP9 for binding to the HS chains. This competition inhibits cell migration and invasion. (d) Pro-MMP9 binds to the core protein of one or several chondroitin sulfate proteoglycans (CSPGs) secreted from phorbol 12-myristate 13-acetate (PMA)-stimulated leukemic monocytic cells (THP1) Among the PGs synthesized by these cells are serglycin and versican. In pro-MMP9, the interaction with the PG core proteins involves at least the HPX domain and the FNII-like module in the catalytic domain. The HPX domain of MMP13 binds to the core protein of syndecan-4, serglycin, and decorin.

MT1-MMP$_d$/TIMP2/pro-MMP2

One of the important roles of MT1-MMP is to function as a pro-MMP2 activator. This activation involves formation of several complexes. First, a "classical" enzyme-inhibitor complex is formed between the N-terminal domain of TIMP2 and the catalytic domain of a cell-membrane-anchored MT1-MMP. This complex then acts as a cell surface receptor where pro-MMP2 can bind the noninhibiting C-terminal domain of TIMP2 through its HPX domain. As the MT1-MMP in this ternary complex is inhibited by the TIMP2 molecule, the proteolytic activation of the MMP2 zymogen is dependent on an additional (TIMP2-free) MT1-MMP protein in close proximity to the ternary complex. It was found that activation of pro-MMP2 is promoted by formation of a tetramer involving the inactive ternary MT1-MMP complex and a TIMP2-free MT1-MMP enzyme. In this quaternary complex, the active site of the second MT1-MMP is available for proteolytic processing of the MMP2 prodomain (▶Figure 4.5b). The interaction between the two MT1-MMP molecules is thought to ensure a close proximity of the molecules in order to enhance the activation reaction. Experiments with human fibrosarcoma cells showed that prevention of the homodimerization of the MT-MMP reduced MMP2 activation and inhibited the invasive capacity of the cells (Itoh et al., 2001).

MT1-MMP/collagen I

Similarly to the ternary complex scenario described previously, an interaction between native type I fibrillar collagen and the HPX domain of MT1-MMP has been shown to promote cellular activation of pro-MMP2 in human gingival fibroblasts, possibly as an alternative way of facilitating physical clustering of MT1-MMP, thereby increasing the proximity of catalytically active MT1-MMPs to the trimeric activation complex (Tam et al., 2002).

MT-MMPs/claudins/pro-MMP2

Claudins are a family of small transmembrane proteins that are important components of tight junctions. Claudin-5 is an endothelial-specific member of this family and was found to be able to replace TIMP2 in mediating pro-MMP2 activation by MT1-MMP. Unlike TIMP2, claudin-5 could actually promote pro-MMP2 activation by all known MT-MMPs. Furthermore, claudin-1, -2, and -3 were also found to stimulate MT-MMP–mediated pro-MMP2 activation. A direct interaction of both pro-MMP2 and MT1-MMP with claudin-1 was demonstrated, and deletion analyses suggested that the interactions involved the ectodomain of claudin-1 and the catalytic domains of MT-MMPs and MMP2. The physiological significance of the MT-MMP/claudin interaction is not known but might be a way of achieving increased focal concentration of MT-MMPs and pro-MMP2 at the cell surface to enhance pro-MMP2 activation (Miyamori et al., 2001).

MT3-MMP/chondroitin sulfate proteoglycan /pro-MMP2

PGs with attached chondroitin sulfate (CS) chains are enhanced in certain chronically inflamed tissues such as arthritic joints and atherosclerotic plaques, as well as in many tumors. MT3-MMP–induced activation of pro-MMP2 required a pro-MMP2 with an intact HPX domain. Both MT3-MMP and pro-MMP2 bind to CS chains of chondroitin sulfate proteoglycans (CSPGs) on the surface of melanoma cells, which has been shown to enhance the activation of pro-MMP2 (see ▶Figure 4.5a). It was suggested that binding to the CS chains presented the gelatinase to its membrane-bound activator. Both the catalytic and the hinge regions of the MT3-MMP interacted with the CS chains, while pro-MMP2 interacted through the HPX domain. The sulfation pattern of the CS chains was important for the activation, as CS chains with the sulfate attached to the 4-position of the GAG chains (C4S), but not to the 6-position (C6S), enhanced the activation in the presence of suboptimal concentrations of MT3-MMP. Binding of pro-MMP2 to the CS chains without MT3-MMP did not result in activation. It was suggested that C4S at the surface of tumor cells producing MT3-MMP can enhance invasion and metastasis through recruitment and activation of pro-MMP2 (Ilda et al., 2007).

MT1-MMP/CD44

MT1-MMP is found in lamellipodia at the migrating front of invasive cancer cells. Here it is thought to be an important mediator of pericellular, precisely focused, proteolysis of the ECM to make way for the invading cells. This increased concentration of the enzyme to lammellipodia is found to promote homodimerization of the enzyme, which in turn facilitates pro-MMP2 activation. The localization of the enzyme to lamellipodia at the invasive front is thought to be regulated by the actin cytoskeleton through a complex between the HPX domain of MT1-MMP and the stem region of the CD44 cell receptor, where the cytoplasmic domain of CD44 connects the enzyme with the cytoskeleton. Furthermore, the MT1-MMP/CD44 complex was found to be essential for shedding of CD44, which is an important step in CD44-mediated migration (Mori et al., 2002; Suenaga et al., 2005). The ability to form a complex between the HPX domain and CD44 was conserved between the different members of the MT-MMP group, suggesting that CD44 plays an important role in assembling different MT-MMPs at the leading edge of migrating cells, thus forming an invasion machinery. In addition, MT2-, MT3-, and MT5-MMP could act as shedases of CD44, though less potent than MT1-MMP (Suenaga et al., 2005).

MT1-MMP/receptor for advanced glycation end products

Binding of advanced glycation end products (AGE) to its receptor (RAGE) results in signaling events that play a key role in diabetic vascular complications. RAGE is upregulated by protein glycation products, oxidant stress, and proinflammatory compounds. Furthermore, this receptor can signal via Rac1/NADPH oxidase, which stimulates ROS generation (▶Figure 4.6a) in vascular cells (Yoshida et al., 2006; Zhang et al., 2006). In both cultured smooth muscle cells (SMC) and diabetic rat aortas, it has been shown that RAGE can form a complex with MT1-MMP, which can affect the signaling through this receptor (Kamioka et al., 2011). Knockdown of MT1-MMP in SMC by siRNA supressed the AGE-triggered Rac1/p47^{phox} membrane translocation, NADPH oxidase activity, and ROS generation. TIMP2 inhibited AGE-triggered guanosine triphosphate (GTP) loading of Rac1, suggesting that the proteolytic activity of MT1-MMP is involved in AGE/RAGE-triggered signaling pathways (Kamioka et al., 2011).

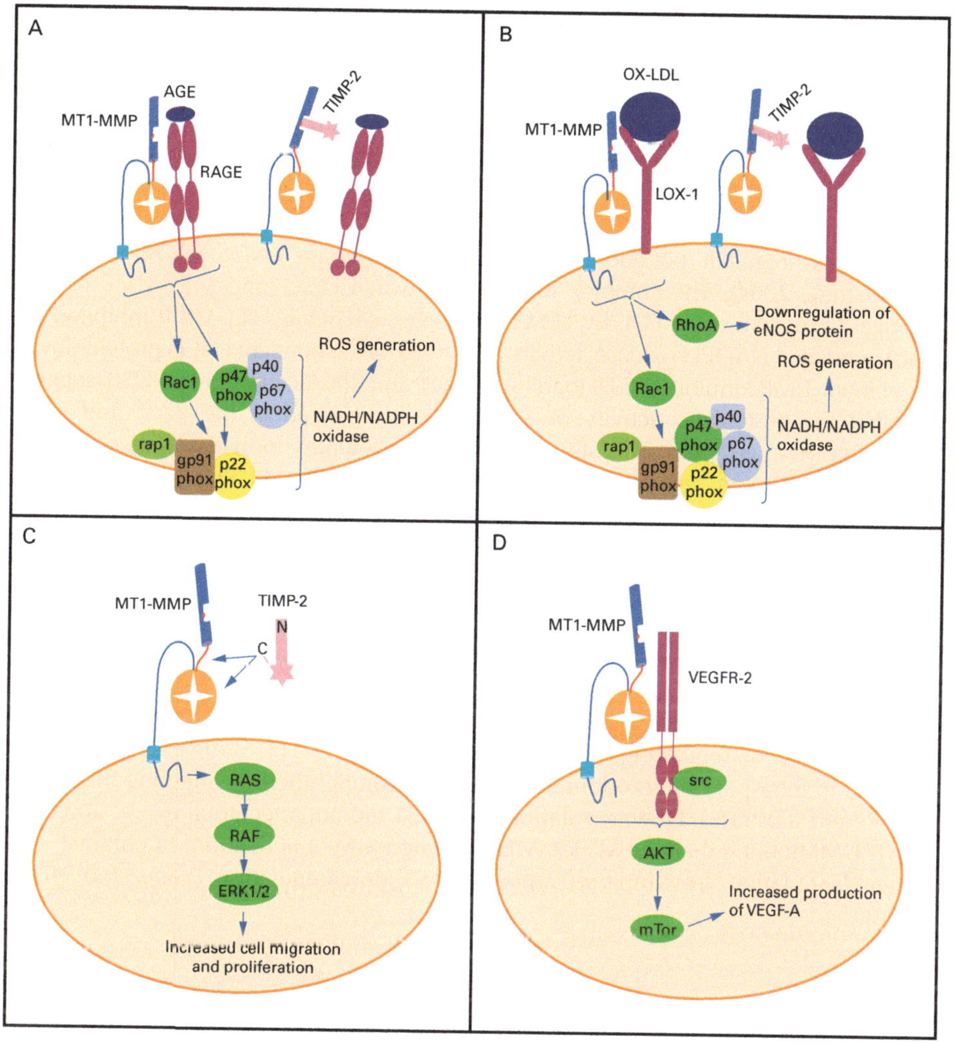

Figure 4.6 MT1-MMP complexes activate cell signaling pathways. (a) RAGE, which is a receptor for AGE, forms a complex with MT1-MMP in both cultured smooth muscle cells and diabetic rat aortas. The proteolytic activity of MT1-MMP plays a role in the Rac1/ $p47^{phox}$ membrane translocation, NADPH oxidase activation, and ROS generation in the RAGE/Rac1/NADPH oxidase-signaling pathways in AGE-exposed vasculature. (b) In endothelial cells, complex formation between lectin-like oxidized low-density lipoprotein receptor-1 (LOX1) and proteolytically active MT1-MMP was shown to regulate the oxidized low-density lipoprotein (ox-LDL)–mediated Rac1 and RhoA activation and their downstream signaling pathways. (c) The MT1-MMP/TIMP2 complex initiate ERK1/2 activation in breast carcinoma cells. Signaling is initiated by an interaction between the noninhibiting C terminal of TIMP2 and either the HPX or the hinge region of MT1-MMP. MT1-MMP/TIMP2 initiated ERK 1/2 signaling through Ras and Raf activation. ERK1/2 activation by TIMP2 binding to MT1-MMP required the cytoplasmic tail but not the catalytic activity of MT1-MMP. (d) In breast cancer cell lines, MT1-MMP was found to regulate the function of vascular endothelial

MT1-MMP/lectin-like oxidized low-density lipoprotein receptor-1

Lectin-like oxidized low-density lipoprotein receptor-1 (LOX1) is a major receptor for oxidized low-density lipoprotein (ox-LDL) in endothelial cells. Fluorescent immunostaining and immunopercipitation have shown that LOX1 and MT1-MMP form complexes on the cell surface of human aortic endothelial cells (Sugimoto et al., 2009). Ox-LDL impairs endothelial function by giving rise to ROS generation via Rac1-mediated NADPH oxidase activity. Ox-LDL also reduces the nitric oxide production through downregulation of endothelial nitric oxide synthase (eNOS) via RhoA (Stocker and Keaney, 2004). The complex formation between LOX1 and MT1-MMP did not change the uptake of ox-LDL by LOX1. However, silencing MT1-MMP inhibited the Rac1-mediated NADPH oxidase activity and the RhoA-dependent eNOS protein downregulation. TIMP2 inhibited GTP loading of Rac1 and RhoA caused by ox-LDL, suggesting that the proteolytic activity of MT1-MMP is involved in the Rac1 and RhoA activation (▶Figure 4.6b). This study shows that a complex formation between LOX1 and MT1-MMP can regulate the ox-LDL–mediated Rac1 and RhoA activation and their downstream signaling pathways in endothelial cells.

MT1-MMP/CD151/integrin α3β1

In primary endothelial cells, MT1-MMP is found in a ternary complex with CD151, a transmembrane protein of the tetraspanin family, and its associated integrin α3β1. The complex involves the HPX domain of MT1-MMP, and CD151 is thought to serve as a linker between MT1-MMP and integrin α3β1. This complex may represent a mechanism for regulating the enzymatic activity of MT1-MMP by directing the enzyme to tetraspanin microdomains in the membrane, thereby affecting the access to different substrates. Downregulation of CD151 did not affect total expression level of MT1-MMP but enhanced MT1-MMP–mediated MMP2 activation. In contrast, collagen degradation was impaired upon CD151 downregulation (Yanez-Mo et al., 2008).

MT1-MMP/CD63

MT1-MMP is found in a complex with another member of the tetraspanin family, CD63, a component of late endosomal and lysosomal membranes. This complex was also shown to involve the HPX domain of MT1-MMP and the N terminus of CD63. Ectopic expression of CD63 induced an accelerated degradation of MT-MMP, probably due to the lysosomal targeting domain of CD63. The result suggests that complex formation between CD63 and MT1-MMP is a way of regulating proteolytic activity at the cell surface (Takino et al., 2003).

Figure 4.6 ***(Continued)***
growth factor receptor-2 (VEGFR2) independently of its catalytic activity. The formation of an MT1-MMP/VEGFR2 complex associates with the proto-oncogene tyrosine-protein kinase Src and induces activation of the a serine/threonine protein kinases AKT and mTOR, leading to increased autocrine production of vascular endothelial growth factor (VEGF)-A.

MT6-MMP/MT6-MMP

A homodimer of MT6-MMP has been detected at the cell surface of neutrophiles, probably formed by disulfide bridges between Cys residues in the stem region of the enzymes. This dimerization did not seem to be essential for transport of the enzyme to the cell membrane, partitioning into lipid rafts or cleavage of the α1-proteinase inhibitor, but instead were found to have a protective effect against autolysis and metalloproteinase dependent degradation (Zhao et al., 2008).

MMP/α2(VI) collagen

Microfibrillar collagen VI (CVI) consists of three α chains, α1(VI), α2(VI), and α3(VI), and so does the pepsin-resistant triple-helical fragment (CVI/PR) (Baldock et al., 2003). Normally, CVI is only weakly expressed in human liver tissues, while it is highly upregulated during liver fibrosis where it is colocalized with fibrillar structures (Freise et al., 2009). Both CVI/PR and α2(VI) show nanomolar affinity for the three collagenases pro-MMP1, pro-MMP8, and pro-MMP13, as well as MMP3 and MMP9 (Freise et al., 2009). The strongest binding was to pro-MMP3. Pro-MMP2 also binds to the α2(VI) chain, but with a 10-fold weaker affinity than the other pro-MMPs. The affinity of the α2(VI) to the active form of the corresponding proteinases was weaker than to their proenzyme forms. For the two gelatinases, it appeared that the interaction to the α2 (VI) chain was through the FNII module in the enzyme's catalytic site. In contrast to MMP8, binding of active MMP1, MMP3, and MMP13 to either CVI/PR or the α2(VI) chain strongly reduced their capacity to process specific fluorogenic substrates. A similar reduction in activity occurred when just the catalytic domains of MMP3 and MMP13 were used. Furthermore, it was shown that binding of pro-MMP1, pro-MMP8, and pro-MMP13 to CVI/PR or the α2(VI) chain reduced the autolytic activation of these enzymes. In the case of MMP9, binding to either CVI/PR or the α2(VI) chain significantly reduced the autolytic activation of the proenzyme and the enzymatic activity of the active enzyme. In contrast to this, autoactivation of pro-MMP2 and enzymatic activity of MMP2 were significantly enhanced by the interaction with CVI/PR or the α2(VI) chain compared to the control. However, a high molar excess of the α2(VI) chain blocked both autolytic and *p*-aminophenylmercuric acetate (APMA)-catalyzed autoactivation and truncation of pro-MMP2 (Freise et al., 2009). *In vivo*, CVI is degraded by serine proteinases but is resistant to MMP degradation (Okada et al., 1990; Kielty et al., 1993). In contrast to this, it has been reported that MMP2 can slightly process the α3(VI) chain and MMP11 can process native CVI (Myint et al., 1996; Motrescu et al., 2008). In fibrosis, there is an accumulation of tissue proteins including CVI, which might be a result of an imbalance between ECM protein synthesis and degradation. The results from Friese et al. (2009) suggest that the increased amount of CVI can prevent activation and activity of matrix-degrading proteinases and hence block ECM degradation including CVI, which further stimulates the fibrotic process.

MMP9/Ku

During synthesis, pro-MMP9 becomes heavily *N*- and *O*-glycosylated prior to its secretion as a 92 kDa proenzyme. A nonsecreted 85 kDa form of pro-MMP9 that is devoid of complex carbohydrates has been detected associated to the cell surface (Mazzieri et al.,

1997; Toth et al., 1997; Monferran et al., 2004; Ries et al., 2007; Paupert et al., 2008). This MMP9 form has been shown to be associated with the cell membrane through its HPX domain, which binds to the Ku protein (Monferran et al., 2004; Paupert et al., 2008). The Ku protein is a heterodimer of two tightly associated proteins, Ku70 and Ku80. Ku plays a role in the repair of DNA double-strand breaks and is also expressed on the cell surface of a subset of primary human normal cells of haematopoietic origin, such as macrophages and activated monocytes (Monferran et al., 2004; Muller et al., 2005). A subfraction of acute myeloid leukemia (AML) blasts that displayed monocytic lineage markers (AML4/5) expressed the 85 kDa pro-MMP9/Ku complex at the cell surface. This expression was associated with cell invasion that could be inhibited by blocking antibodies directed against either Ku or MMP9 (Paupert et al., 2008). It was also shown that the MMP9/Ku complex is less sensitive to inhibition by TIMP1 compared to uncomplexed enzyme, while inhibition with GM6001 was not affected (Monferran et al., 2004).

MMP9/integrin α4β1/CD44v

B-cell chronic lymphocytic leukemia (B-CLL) is characterized by the accumulation of CD5+ B lymphocytes in the peripheral blood (Zent and Kay, 2004; Rai, 2007). B-CLL cells, as well as other B cells, express CD44H and integrin α4β1 at the cell surface, while only the B-CLL cells express the CD44v isoform. It was shown that CD44v and α4β1 integrin formed a complex at the B-CLL cell surface that binds both pro- and active MMP9 through the enzyme's HPX and hinge region (Redondo-Munoz et al., 2008). The CD44H isoform could not form this complex. Formation of this ternary complex with the active form of MMP9 induced cell migration through matrigel, while this migration was inhibited when this ternary complex involved pro-MMP9 or if MMP inhibitors were added.

MMP9/dentin matrix protein-1/integrins

Dentin matrix protein-1 (DMP1) belongs to the family of small integrin-binding ligands, *N*-linked glycoproteins (SIBLINGs). Binding of DMP1 to proMMP9 resulted in activation of the enzyme without chemical cleavage and removal of the enzyme's prodomain. This interaction also reduced the enzyme's sensitivity to TIMPs and synthetic inhibitors (Fedarko et al., 2004). It has also been shown that DMP1 could bridge MMP9 to the cell surface receptors CD44 and αVβ3 and αVβ5 integrins (Karadag et al., 2005). Formation of these ternary complexes on the cell surface of various cancer cell lines enhanced the *in vitro* invasiveness through a matrigel.

MMP9/heparan sulfate

Both active and pro-MMP9 bind to heparin and heparan sulfate (HS) (▶Figure 4.5c), where the active form showed a slightly stronger binding. A highly metastatic cell clone (LuM1) isolated from murine colon adenocarcinoma contained active MMP9 bound to HS chains of GPI-anchored glypican-like PGs on the cell surface. This MMP9/PG complex was concentrated at the leading edge of the LuM1 cells during *in vitro* migration on a fibronectin substrate. Localization of active MMP9 to the HS chains at the LuM1 cell surface mediated *in vitro* migration, as well as cell invasion through matrigel (Koyama et al., 2008).

MMP2/syndecan-2

A large difference in experimental and spontaneous metastatic potential was observed in various cell clones derived from Lewis lung carcinoma 3LL (Munesue et al., 2007). The metastatic potential correlated inversely with the expression of the heparan sulfate proteoglycan (HSPG), syndecan-2. Although all clones produced similar amounts of pro-MMP2 and the two components involved in the activation of the proenzyme, TIMP2 and MT1-MMP, only the highly metastatic clones contained active MMP2. Binding of pro-MMP2 to the HS chains of syndecan-2 prevented the activation of the enzyme by MT1-MMP, as well as the metastatic potential of the cancer cells (▶Figure 4.5b).

MMP2/bone sialoprotein/integrin αVβ3

Bone sialoprotein (BSP) is another SIBLING that was suggested to bridge MMP2 to the cell surface receptor αVβ3 integrin (Karadag and Fisher, 2006). The bridging of MMP2 to the integrin at the cell surface of bone marrow stromal cells (BMSCs) resulted in an enhanced migration of these cells through matrigel and increased amounts of denatured collagen I *in vitro*. These effects were caused by activation of MMP2. The binding of MMP2 to BSP, and the following activation of the proenzyme, is controversial as another group could not reproduce this result (Hwang et al., 2009).

MMP7/cholesterol sulfate

On the cell surface of the human colon cancer cell lines WiDr and Colo201, cholesterol sulfate has been demonstrated to be a major cell-surface-binding partner for active MMP7. The high-affinity interaction with cholesterol sulfate is mediated by amino acid residues Ile29, Arg33, Arg51, and Trp55 in the internal sequence and Arg171, Lys172, and Lys173 in the C terminus of MMP7. In the three-dimensional structure of MMP7, the residues essential for binding to cholesterol sulfate are located on the surface at the opposite side of the catalytic cleft. This suggests that MMP7 is bound to the cell with its active site directed away from the cell surface to the ECM (Higashi et al., 2008). The binding of MMP7 to cell surface cholesterol sulfate was found to be essential for the cell-membrane-associated proteolytic action of this protease (Yamamoto et al., 2006). It was shown that cell surface cholesterol sulfate promoted the proteolytic activities of MMP7 toward laminin-332 and fibronectin. Moreover, MMP7 tethered to the cancer cell surface via cholesterol sulfate degraded laminin-332 and fibronectin coated on a culture plate and abrogated their cell adhesion activities, thereby leading to cell detachment (Yamamoto et al., 2010). Furthermore, this complex induced loose cell aggregation that was followed by E-cadherin-mediated tight cell aggregation. When such tight cell aggregates produced *in vitro* were injected into the spleen of nude mice, they formed liver metastasis (Kioi et al., 2003). Since MMP7 has been detected specifically in tumor cells and not in stromal cells as seen with various other MMPs (Nagashima et al., 1997), formation of an MMP7/cholesterol sulfate complex in cancer cell lines may be one of the factors that contribute to cell migration and metastasis.

MMP3/ connective tissue growth factor promoter

Connective tissue growth factor (CTGF/CCN2) is a member of the CCN family of matricellular proteins, and CTGF/CCN2 promotes physiological chondrocytic proliferation and ECM formation. MMP3 has been detected in the nuclei of chondrocytic cells in culture and also in the nuclei of normal and osteoarthritic chondrocytes *in vivo*. Nuclear MMP3 was bound to a transcription enhancer sequence (TRENDIC) in the CTGF/CCN2 promoter and activated CTGF/CCN2 transcription. Both the full-length and the active form of MMP3 could activate the CTGF/CCN2 promoter. Furthermore, both the HPX and the CAT hinge domains alone activated the CTGF/CCN2 promoter, while the prodomain and the hinge region alone had no effect. To clarify whether the activation was dependent on the proteolytic activity of MMP3, catalytically dead mutants were constructed. The catalytically dead mutants decreased the activation ability of the promoter compared to the wild-type MMP3. These results indicate that both the CAT domain and the HPX domain can activate the promoter independently. MMP3 can bind to several nuclear proteins, including heterochromatin proteins, transcription coactivators/corepressors, RNA polymerase II, nucleosome/chromatin assembly protein, and others. One of these nuclear proteins, heterochromatin protein 1 γ (HP1γ), activated the CTGF/CCN2 promoter when it was bound to MMP3. Another nuclear protein that can interact with MMP3 is nuclear receptor corepressor 1 (NCoR1), which is a transcription repressor. It has been assumed that MMP3 might prevent transcription repression of the CTGF/CCN2 promoter by degrading NCoR1 (Eguchi et al., 2008).

MMP complexes with unknown biological function

MMP9/MMP9

Gelatinase B is known to form both heteromers as well as a homodimer (Goldberg et al., 1992; Triebel et al., 1992; Kjeldsen et al., 1993; Olson et al., 2000; Winberg et al., 2000). The homodimerization occurs already in the cell and is mediated by the C-terminal HPX domain of the enzyme (Olson et al., 2000). Recombinant MMP9 HPX domain (HPX9) also forms a reduction-sensitive dimer, and X-ray crystallography revealed that the dimer was held together by noncovalent and mainly hydrophobic interactions and a salt bridge between the C terminus of HPX9(A) and the side chain of R677 of HPX9(B) (Cha et al., 2002). The reduction sensitivity was due to an intramolecular disulfide bond between the conserved Cys516 and Cys704. This bridge connects blade I and blade IV in one HPX9 monomer and is critical for its structural integrity. Recombinant MMP9 that lacks the highly *O*-glycosylated hinge region occurs only as a monomer (Van den Steen et al., 2006). This suggests that the hinge region allows contact between the HPX domains of two full-length pro-MMP9 molecules due to the large size and flexibility of the hinge region, which prevents unfavorable steric interactions between the two molecules' pro- and CAT domains. Pro-MMP9 monomer and homodimer have somewhat different biochemical properties. The monomer is more rapidly activated by MMP3 and has a higher activity than the dimer (Olson et al., 2000). There are conflicting data concerning whether the pro-MMP9 homodimer is able to form a complex with TIMP1 (Goldberg et al., 1992; Olson et al., 2000; Dufour et al., 2010). So far, nothing is known about the biological significance of the MMP9 homodimer and to what extent this dimer affects enzyme activity and specificity. However, it appears that this dimer is important for the migration of simian SV40 (COS-1) kidney cells, but that this is independent of the enzyme's proteolytic function (see next paragraph).

MMP9/CSPG

Gelatinase B is known to form various types of heterodimers where the biological function of the complex has not yet been resolved (Goldberg et al., 1992; Triebel et al., 1992; Kjeldsen et al., 1993; Winberg et al., 2000). One such complex is the reduction-sensitive pro-MMP9/CSPG heteromer produced by the leukemic monocyte cell line THP1 (Winberg et al., 2000). Small amounts of this complex are produced by unstimulated cells, while stimulation by the tumor promoter PMA resulted in at least a 100-times increase in complex formation (Malla et al., 2011). This complex was shown to involve the core protein of one or several CSPGs produced by these cells (▶Figure 4.5d) and several parts of MMP9: the HPX domain, the FNII module, and probably the prodomain. The FNII module in the enzyme was hidden in the complex (Malla et al., 2008a). The FNII module in both the pro- and active form of the two gelatinases binds to various ECM molecules, such as gelatin and various collagen types. In addition, the FNII module acts as an exosite and is important for the ability of MMP2 and MMP9 to cleave various ECM molecules (Collier et al., 1992; Strongin et al., 1993; Murphy et al., 1994; Pourmotabbed, 1994; Allan et al., 1995; Steffensen et al., 1995; Shipley et al., 1996; O'Farrell and Pourmotabbed, 1998; Hornebeck et al., 2005; Xu et al., 2005). This suggests that formation of the MMP9/CSPG complex will alter the substrate specificity. It was also shown that a small fraction of the formed complex could bind to gelatin and collagen. This interaction involved either a new exosite generated by the complex formation or possibly a minor CSPG core protein variant with a gelatin- and collagen-binding motif. The gelatin-binding motif in the complex did not involve the FNII module of the enzyme and bound a different region in gelatin than free MMP9 (Malla et al., 2008a). The presence of calcium but not magnesium or sodium ions induced activation of the pro-MMP9 in the complex, but not of unbound pro-MMP9 monomer or homodimer (Winberg et al., 2003). This autoactivation appeared to be intramolecular and caused a stepwise processing of the enzyme in the complex. The prodomain was cleaved prior to a stepwise cleavage of both the HPX domain of the enzyme and parts of the CSPG core protein. Enzyme with only the prodomain removed was still bound to the CSPG. But successive processing of the HPX domain resulted in the release of activated and processed forms of the enzyme from the CSPG molecule. The biological relevance of these MMP9/CSPG complexes is so far unknown, but it has been speculated that a CSPG molecule can act as a linker between CD44 and MMP9 (Winberg et al., 2000) as CSPGs are known to bind CD44 (Toyama-Sorimachi et al., 1997).

MMP9/MMP1

Stimulation of the monocytic leukemia cell line U937 with a combination of the phorbolester TPA and lipopolysaccharide (LPS) resulted in a large production of pro-MMP9 and pro-MMP1, as well as TIMP1 and active forms of MMP1 (Goldberg et al., 1992). A fraction of the synthesized enzymes contained a heterodimer between pro-MMP9 and pro-MMP1, which involved the two enzymes' HPX domains. TIMP1 replaced pro-MMP1 when a preformed pro-MMP9/pro-MMP1 dimer was incubated with TIMP1, indicating that the sites of MMP1 and TIMP1 interaction with HPX in pro-MMP9 overlap. MMP3 alone could activate pro-MMP9 in the pro-MMP9/pro-MMP1 dimer, resulting in a complex able to degrade gelatin but not interstitial collagen. Plasmin could only partly activate the two enzymes in the complex, resulting in low gelatinase and collagenase activities. A combination of plasmin and MMP3 caused a rapid activation of both components in the complex and hence a high activity against both gelatin and interstitial fibrillar collagen. This suggested a mechanism for cooperative action of the two enzymes in the degradation of collagen fibrils to small peptides under physiological conditions. To what extent this heterodimeric enzyme complex is formed and has a role *in vivo* in physiological or pathological conditions is so far unknown.

MMP9/MMP8

Bone marrow cells from wild-type mice of a mixed C57BL/6J/129 background contained a heteromer between MMP9 and MMP8, which was not found in the corresponding cells from MMP9$^{-/-}$ (C57BL/6J background) and MMP8$^{-/-}$ (C57BL/6J/129 background) mice (Gutierrez-Fernandez et al., 2007). Although the study did not address which modules were engaged in the complex formation, it can be assumed that the two enzymes' HPX regions were involved. It has been suggested that the complex formation might be a way to regulate the activity of these enzymes or that both proteinases could act coordinately to process specific substrates, but the biological functions of the complex have not been studied.

MMP13/PG

Chondrocytes synthesize various PGs, including syndecan-4, decorin, and serglycin. The core protein of these three PGs was shown to interact with the HPX region of MMP13 (▶Figure 4.5d). Furthermore, this C-terminal region of MMP13 was also shown to form complexes with TIMP1, α2-macroglobulin, and xylosyltransferase-1 (Zhang et al., 2010). Xylosyltransferase-1 catalyses the transfer of xylose to specific serine residues in the core protein of PGs and initiates the GAG-chain formation. This occurs in the Golgi apparatus, and in spite of this location, more than 90% of the active xylosyltransferase-1 is found in the medium of cultured cells. The extracellular function of this enzyme is not known. Although the biological role and consequences of these MMP13 interactions were not investigated, Zhang and coworkers suggested that the interactions may have a role in the regulation of cartilage degradation.

MT1-MMP/gC1qR

gC1qR (p33/p32) is a multifunctional, ubiquitously expressed protein that may be found in mitochondria, in the cytosol, and at the cell membrane. The protein is thought to be a compartment-specific regulator of numerous cellular proteins. *In vitro*, the cytoplasmic tail of MT1-MMP was found to interact with gC1qR (Rozanov et al., 2002), but the relevance of this complex is unknown.

MT6-MMP/clustrin

In neutrophils, MT6-MMP has been found in complex with clustrin, a heterodimeric glycoprotein abundantly present in all tissue fluids in the body. The complex is thought to involve the HPX domain of MT6-MMP, and complex formations seem to inhibit the enzymatic activity of MT6-MMP. The specific functions of clustrin are not clear, but it can interact with a number of components in serum and at the cell surface. It can bind both bacteria and immunoglobulins and might play a role in inflammatory reactions. Clustrin-deficient mice show excessive tissue damage at inflammatory sites, which may be associated with increased/dysregulated MT6-MMP activity (Matsuda et al., 2003).

MMP complexes with biological function independent of proteolysis

MT1-MMP/TIMP2

Some functions of MT1-MMP seem to be independent of its proteolytic activity. In addition to acting as a cell surface receptor for pro-MMP2, the MT1-MMP/TIMP2 complex has been found to initiate cell signaling through activation of ERK1/2 in breast carcinoma cells (▶Figure 4.6c). This signaling was not associated with proteolytic activity and was mediated through the cytoplasmic tail of MT1-MMP. Signaling seemed to be initiated by an interaction between the noninhibiting C-terminal domain of TIMP2 and either the HPX or the hinge region of MT1-MMP. Furthermore, preformed TIMP2/pro-MMP2 complex could initiate ERK 1/2 signaling as effectively as TIMP2 alone. Thus, the interacting motifs in TIMP2 and MT1-MMP that initiate signaling are still an open question. MT1-MMP/TIMP2–initiated ERK1/2 signaling has been shown to upregulate cell migration and proliferation. *In vivo*, it was found that breast cancer cells modified to overexpress wild-type MT1-MMP or MT1-MMP devoid of proteolytic activity promoted tumor growth compared with breast cancer cells transfected with empty vector or MT1-MMP lacking the cytoplasmic tail. This indicates that the cytoplasmic tail of MT1-MMP is important for mediating the tumor-promoting effects of this enzyme, possibly through its involvement in signaling activity (D'Alessio et al., 2008).

MT1-MMP/CD44

MT1-MMP is reported to be involved in cell sorting during organ development. In a model of mammary gland morphogenesis, cells sorted according to their level of MT1-MMP, probably due to differences in cell motility. Cells expressing MT1-MMP were found in the leading edge of the engineered mammary ducts. This sorting was independent of the catalytic activity of the enzyme as well as of signaling through ERK or addition of exogenous growth factors, but dependent on signaling through the Rho–guanosine triphosphatase (GTPase) effector ROCK (Mori et al., 2009). Such signaling has previously been associated with an amoeboid mode of motility (Wolf et al., 2003). Deletion of the HPX domain of MT1-MMP impaired the sorting of the enzyme to the leading edge, and the signaling was found to be initiated by a complex between the receptor CD44 and the HPX domain of MT1-MMP (Mori et al., 2009).

MT1-MMP/vascular endothelial growth factor receptor-2/Src

In breast cancer cell lines, MT1-MMP was found to regulate the function of vascular endothelial growth factor receptor-2 (VEGFR2) independently of its catalytic activity by the formation of an MT1-MMP/VEGFR2 complex. This complex associates with Src and induces phosphorylation of AKT and mTor (▶Figure 4.6d), leading to increased autocrine production of vascular endothelial growth factor (VEGF)-A (Eisenach et al.,

2010). By using different MT1-MMP mutants in immunoprecipitation experiments, it was found that the interaction between MT1-MMP and VEGFR2 was dependent on the MT1-MMP HPX domain. MT1-MMP and Src also coimmunopercipitated; however, deletions in the intracellular domain of MT1-MMP did not alter the presence of the MT1-MMP/Src complex, indicating an indirect interaction between MT1-MMP and Src. VEGFR2 has previously been found to associate with Src (Chou et al., 2002; Olsson et al., 2006), suggesting that Scr is connected to VEGFR2 and not MT1-MMP in this complex.

MMP9/CD44

As with MT1-MMP, it has also been shown that MMP9 can exert biological functions that are independent of the enzyme's catalytic activity. Recently, it was shown that MMP9 can induce epithelial cell migration independent of its proteolytic activity (Dufour et al., 2008). The MMP9-induced cell migration was dependent on the COS1 cells' endogenous synthesis of the enzyme, but exogenous added MMP9 had no effect, suggesting an intracellular effect. The MMP-induced cell migration required formation of an MMP9 homodimer. Furthermore, it was also shown that binding of MMP9 to the cell receptor CD44 was necessary for cell migration (Dufour et al., 2010). However, homodimerization of MMP9 was not a prerequisite for binding to CD44. Two regions of the HPX domain in MMP9 were essential for the MMP9-induced cell migration. One was the outer strand four of blade IV, which is involved in the homodimer formation, and the other was the outer section (strand four) of blade I, which is involved in the binding to CD44. Use of peptides that mimicked either the outer strands of MMP9 HPX blades I or IV inhibited cell migration. Although not shown, it is tempting to assume that the interaction between an MMP9 homodimer and CD44 promotes receptor dimerization, which induces cell migration. Dufour et al. (2010) also showed that binding of MMP9 to CD44 triggered an epidermal growth factor receptor (EGFR) signaling pathway, which induced cell migration. Furthermore, these authors have recently shown that small compounds that bind specifically to the HPX region of MMP9 (but not other MMPs) and abrogate the MMP9 homodimer formation also retarded tumor growth and inhibited lung metastases in a tumor xenograft model (Dufour et al., 2011).

4.4.4 Take-home message

Both MT-MMPs and soluble MMPs form either homodimers or heterodimers. This complex formation is an important mode of regulation and localization of MMP activity, as well as activation of pro-MMPs. Cell motility and invasion is regulated by either recruiting soluble MMPs or concentrating MT-MMPs at specific areas of the cell membrane. Furthermore, complex formation between MMPs and other molecules is also a way to regulate substrate accessibility and specificity. Complex formation with other macromolecules can either generate new exosites or hide existing exosites. In the future, we can expect even more focus on research exploring new MMP-binding partners and the biochemical and biological/physiological consequences of such complexes in health and disease. In addition, we can also expect more efforts to reveal the nature of the epitopes involved in these interactions, as breaking or forming these specific interactions will be one of the approaches to generate new specific drugs against diseases in which specific MMP complexes are involved.

References

Allan, J. A., Docherty, A. J., Barker, P. J., Huskisson, N. S., Reynolds, J. J., and Murphy, G. (1995). Binding of gelatinases A and B to type-I collagen and other matrix components. Biochem J *309*, 299–306.

Baldock, C., Sherratt, M. J., Shuttleworth, C. A., and Kielty, C. M. (2003). The supramolecular organization of collagen VI microfibrils. J Mol Biol *330*, 297–307.

Bode, W., Gomis-Ruth, F. X., and Stockler, W. (1993). Astacins, serralysins, snake venom and matrix metalloproteinases exhibit identical zinc-binding environments (HEXXHXXGXXH and Met-turn) and topologies and should be grouped into a common family, the "metzincins." FEBS Lett *331*, 134–140.

Brauer, R., Beck, I. M., Roderfeld, M., Roeb, E., and Sedlacek, R. (2011). Matrix metalloproteinase-19 inhibits growth of endothelial cells by generating angiostatin-like fragments from plasminogen. BMC Biochem *12*, 38.

Brew, K., and Nagase, H. (2010). The tissue inhibitors of metalloproteinases (TIMPs): an ancient family with structural and functional diversity. Biochim Biophys Acta *1803*, 55–71.

Butler, G. S., and Overall, C. M. (2009). Updated biological roles for matrix metalloproteinases and new "intracellular" substrates revealed by degradomics. Biochemistry *48*, 10830–10845.

Cha, H., Kopetzki, E., Huber, R., Lanzendorfer, M., and Brandstetter, H. (2002). Structural basis of the adaptive molecular recognition by MMP9. J Mol Biol *320*, 1065–1079.

Chou, M. T., Wang, J., and Fujita, D. J. (2002). Src kinase becomes preferentially associated with the VEGFR, KDR/Flk-1, following VEGF stimulation of vascular endothelial cells. BMC Biochem *3*, 32.

Collier, I. E., Krasnov, P. A., Strongin, A. Y., Birkedal-Hansen, H., and Goldberg, G. I. (1992). Alanine scanning mutagenesis and functional analysis of the fibronectin-like collagen-binding domain from human 92-kDa type IV collagenase. J Biol Chem *267*, 6776–6781.

D'Alessio, S., Ferrari, G., Cinnante, K., et al. (2008). Tissue inhibitor of metalloproteinases-2 binding to membrane-type 1 matrix metalloproteinase induces MAPK activation and cell growth by a non-proteolytic mechanism. J Biol Chem *283*, 87–99.

Dufour, A., Sampson, N. S., Li, J., et al. (2011). Small-molecule anticancer compounds selectively target the hemopexin domain of matrix metalloproteinase-9. Cancer Res *71*, 4977–4988.

Dufour, A., Sampson, N. S., Zucker, S., and Cao, J. (2008). Role of the hemopexin domain of matrix metalloproteinases in cell migration. J Cell Physiol *217*, 643–651.

Dufour, A., Zucker, S., Sampson, N. S., Kuscu, C., and Cao, J. (2010). Role of matrix metalloproteinase-9 dimers in cell migration: design of inhibitory peptides. J Biol Chem *285*, 35944–35956.

Eguchi, T., Kubota, S., Kawata, K., et al. (2008). Novel transcription-factor-like function of human matrix metalloproteinase 3 regulating the CTGF/CCN2 gene. Mol Cell Biol *28*, 2391–2413.

Eisenach, P. A., Roghi, C., Fogarasi, M., Murphy, G., and English, W. R. (2010). MT1-MMP regulates VEGF-A expression through a complex with VEGFR-2 and Src. J Cell Sci *123*, 4182–4193.

Fedarko, N. S., Jain, A., Karadag, A., and Fisher, L. W. (2004). Three small integrin binding ligand N-linked glycoproteins (SIBLINGs) bind and activate specific matrix metalloproteinases. FASEB J *18*, 734–736.

Fernandez-Catalan, C., Bode, W., Huber, R., et al. (1998). Crystal structure of the complex formed by the membrane type 1-matrix metalloproteinase with the tissue inhibitor of metalloproteinases-2, the soluble progelatinase A receptor. EMBO J *17*, 5238–5248.

Freise, C., Erben, U., Muche, M., et al. (2009). The alpha 2 chain of collagen type VI sequesters latent proforms of matrix-metalloproteinases and modulates their activation and activity. Matrix Biol *28*, 480–489.

Gialeli, C., Theocharis, A. D., and Karamanos, N. K. (2011). Roles of matrix metalloproteinases in cancer progression and their pharmacological targeting. FEBS J *278*, 16–27.

Goldberg, G. I., Marmer, B. L., Grant, G. A., Eisen, A. Z., Wilhelm, S., and He, C. S. (1989). Human 72-kilodalton type IV collagenase forms a complex with a tissue inhibitor of metalloproteases designated TIMP-2. Proc Natl Acad Sci U S A *86*, 8207–8211.

Goldberg, G. I., Strongin, A., Collier, I. E., Genrich, L. T., and Marmer, B. L. (1992). Interaction of 92-kDa type IV collagenase with the tissue inhibitor of metalloproteinases prevents dimerization, complex formation with interstitial collagenase, and activation of the proenzyme with stromelysin. J Biol Chem *267*, 4583–4591.

Gomis-Ruth, F. X., Maskos, K., Betz, M., et al. (1997). Mechanism of inhibition of the human matrix metalloproteinase stromelysin-1 by TIMP-1. Nature *389*, 77–81.

Gutierrez-Fernandez, A., Inada, M., Balbin, M., et al. (2007). Increased inflammation delays wound healing in mice deficient in collagenase-2 (MMP-8). FASEB J *21*, 2580–2591.

Hadler-Olsen, E., Fadnes, B., Sylte, I., Uhlin-Hansen, L., and Winberg, J. O. (2011). Regulation of matrix metalloproteinase activity in health and disease. FEBS J *278*, 28–45.

Higashi, S., Oeda, M., Yamamoto, K., and Miyazaki, K. (2008). Identification of amino acid residues of matrix metalloproteinase-7 essential for binding to cholesterol sulfate. J Biol Chem *283*, 35735–35744.

Hornebeck, W., Bellon, G., and Emonard, H. (2005). Fibronectin type II (FnII)-like modules regulate gelatinase A activity. Pathol Biol (Paris) *53*, 405–410.

Hua, H., Li, M., Luo, T., Yin, Y., and Jiang, Y. (2011). Matrix metalloproteinases in tumorigenesis: an evolving paradigm. Cell Mol Life Sci *68*, 3853–3868.

Hwang, Q., Cheifetz, S., Overall, C. M., McCulloch, C. A., and Sodek, J. (2009). Bone sialoprotein does not interact with pro-gelatinase A (MMP-2) or mediate MMP-2 activation. BMC Cancer *9*, 121.

Ilda, J., Wilhelmson, K. L., Ng, J., et al. (2007). Cell surface chondroitin sulfate glycosaminoglycan in melanoma: role in the activation of pro-MMP-2 (pro-gelatinase A). Biochem J *403*, 553–563.

Itoh, Y., Takamura, A., Ito, N., et al. (2001). Homophilic complex formation of MT1-MMP facilitates proMMP-2 activation on the cell surface and promotes tumor cell invasion. EMBO J *20*, 4782–4793.

Kamioka, M., Ishibashi, T., Ohkawara, H., et al. (2011). Involvement of membrane type 1-matrix metalloproteinase (MT1-MMP) in RAGE activation signaling pathways. J Cell Physiol *226*, 1554–1563.

Kandasamy, A. D., Chow, A. K., Ali, M. A., and Schulz, R. (2010). Matrix metalloproteinase-2 and myocardial oxidative stress injury: beyond the matrix. Cardiovasc Res *85*, 413–423.

Karadag, A., Fedarko, N. S., and Fisher, L. W. (2005). Dentin matrix protein 1 enhances invasion potential of colon cancer cells by bridging matrix metalloproteinase-9 to integrins and CD44. Cancer Res *65*, 11545–11552.

Karadag, A., and Fisher, L. W. (2006). Bone sialoprotein enhances migration of bone marrow stromal cells through matrices by bridging MMP-2 to alpha(v)beta3-integrin. J Bone Miner Res *21*, 1627–1636.

Kielty, C. M., Lees, M., Shuttleworth, C. A., and Woolley, D. (1993). Catabolism of intact type VI collagen microfibrils: susceptibility to degradation by serine proteinases. Biochem Biophys Res Commun *191*, 1230–1236.

Kioi, M., Yamamoto, K., Higashi, S., Koshikawa, N., Fujita, K., and Miyazaki, K. (2003). Matrilysin (MMP-7) induces homotypic adhesion of human colon cancer cells and enhances their metastatic potential in nude mouse model. Oncogene *22*, 8662–8670.

Kjeldsen, L., Johnsen, A. H., Sengelov, H., and Borregaard, N. (1993). Isolation and primary structure of NGAL, a novel protein associated with human neutrophil gelatinase. J Biol Chem *268*, 10425–10432.

Koyama, Y., Naruo, H., Yoshitomi, Y., et al. (2008). Matrix metalloproteinase-9 associated with heparan sulphate chains of GPI-anchored cell surface proteoglycans mediates motility of murine colon adenocarcinoma cells. J Biochem *143*, 581–592.

Lapière, Ch. M. (2005). Tadpole collagenase, the single parent of such a large family. Biochimie *87*, 243–247.

Malla, N., Berg, E., Moens, U., Uhlin-Hansen, L., and Winberg, J. O. (2011). Biosynthesis of promatrix metalloproteinase-9/chondroitin sulphate proteoglycan heteromer involves a Rottlerin-sensitive pathway. PLoS One *6*, e20616.

Malla, N., Berg, E., Uhlin-Hansen, L., and Winberg, J. O. (2008a). Interaction of pro-matrix metalloproteinase-9/proteoglycan heteromer with gelatin and collagen. J Biol Chem *283*, 13652–13665.

Malla, N., Sjoli, S., Winberg, J. O., Hadler-Olsen, E., and Uhlin-Hansen, L. (2008b). Biological and pathobiological functions of gelatinase dimers and complexes. Connect Tissue Res *49*, 180–184.

Martin, M. D., and Matrisian, L. M. (2007). The other side of MMPs: protective roles in tumor progression. Cancer Metastasis Rev *26*, 717–724.

Matsuda, A., Itoh, Y., Koshikawa, N., Akizawa, T., Yana, I., and Seiki, M. (2003). Clusterin, an abundant serum factor, is a possible negative regulator of MT6-MMP/MMP-25 produced by neutrophils. J Biol Chem *278*, 36350–36357.

Mazzieri, R., Masiero, L., Zanetta, L., et al. (1997). Control of type IV collagenase activity by components of the urokinase-plasmin system: a regulatory mechanism with cell-bound reactants. EMBO J *16*, 2319–2332.

McCawley, L. J., and Matrisian, L. M. (2001). Matrix metalloproteinases: they're not just for matrix anymore! Curr Opin Cell Biol *13*, 534–540.

Miyamori, H., Takino, T., Kobayashi, Y., et al. (2001). Claudin promotes activation of pro-matrix metalloproteinase-2 mediated by membrane-type matrix metalloproteinases. J Biol Chem *276*, 28204–28211.

Monferran, S., Paupert, J., Dauvillier, S., Salles, B., and Muller, C. (2004). The membrane form of the DNA repair protein Ku interacts at the cell surface with metalloproteinase 9. EMBO J *23*, 3758–3768.

Morgunova, E., Tuuttila, A., Bergmann, U., and Tryggvason, K. (2002). Structural insight into the complex formation of latent matrix metalloproteinase 2 with tissue inhibitor of metalloproteinase 2. Proc Natl Acad Sci U S A *99*, 7414–7419.

Mori, H., Gjorevski, N., Inman, J. L., Bissell, M. J., and Nelson, C. M. (2009). Self-organization of engineered epithelial tubules by differential cellular motility. Proc Natl Acad Sci U S A *106*, 14890–14895.

Mori, H., Tomari, T., Koshikawa, N., et al. (2002). CD44 directs membrane-type 1 matrix metalloproteinase to lamellipodia by associating with its hemopexin-like domain. EMBO J *21*, 3949–3959.

Motrescu, E. R., Blaise, S., Etique, N., et al. (2008). Matrix metalloproteinase-11/stromelysin-3 exhibits collagenolytic function against collagen VI under normal and malignant conditions. Oncogene *27*, 6347–6355.

Mott, J. D., and Werb, Z. (2004). Regulation of matrix biology by matrix metalloproteinases. Curr Opin Cell Biol *16*, 558–564.

Muller, C., Paupert, J., Monferran, S., and Salles, B. (2005). The double life of the Ku protein: facing the DNA breaks and the extracellular environment. Cell Cycle *4*, 438–441.
Munesue, S., Yoshitomi, Y., Kusano, Y., et al. (2007). A novel function of syndecan-2, suppression of matrix metalloproteinase-2 activation, which causes suppression of metastasis. J Biol Chem *282*, 28164–28174.
Murphy, G., Houbrechts, A., Cockett, M. I., Williamson, R. A., O'Shea, M., and Docherty, A. J. (1991). The N-terminal domain of tissue inhibitor of metalloproteinases retains metalloproteinase inhibitory activity. Biochemistry *30*, 8097–8102.
Murphy, G., and Nagase, H. (2011). Localizing matrix metalloproteinase activities in the pericellular environment. FEBS J *278*, 2–15.
Murphy, G., Nguyen, Q., Cockett, M. I., et al. (1994). Assessment of the role of the fibronectin-like domain of gelatinase A by analysis of a deletion mutant. J Biol Chem *269*, 6632–6636.
Myint, E., Brown, D. J., Ljubimov, A. V., Kyaw, M., and Kenney, M. C. (1996). Cleavage of human corneal type VI collagen alpha 3 chain by matrix metalloproteinase-2. Cornea *15*, 490–496.
Nagase, H. (1997). Activation mechanisms of matrix metalloproteinases. Biol Chem *378*, 151–160.
Nagase, H., Barrett, A. J., and Woessner, J. F., Jr. (1992). Nomenclature and glossary of the matrix metalloproteinases. Matrix Suppl *1*, 421–424.
Nagashima, Y., Hasegawa, S., Koshikawa, N., et al. (1997). Expression of matrilysin in vascular endothelial cells adjacent to matrilysin-producing tumors. Int J Cancer *72*, 441–445.
O'Farrell, T. J., and Pourmotabbed, T. (1998). The fibronectin-like domain is required for the type V and XI collagenolytic activity of gelatinase B. Arch Biochem Biophys *354*, 24–30.
Okada, Y., Nagase, H., and Harris, E. D., Jr. (1986). A metalloproteinase from human rheumatoid synovial fibroblasts that digests connective tissue matrix components. Purification and characterization. J Biol Chem *261*, 14245–14255.
Okada, Y., Naka, K., Minamoto, T., et al. (1990). localization of type VI collagen in the lining cell layer of normal and rheumatoid synovium. Lab Invest *63*, 647–656.
Olson, M. W., Bernardo, M. M., Pietila, M., et al. (2000). Characterization of the monomeric and dimeric forms of latent and active matrix metalloproteinase-9. Differential rates for activation by stromelysin 1. J Biol Chem *275*, 2661–2668.
Olsson, A. K., Dimberg, A., Kreuger, J., and Claesson-Welsh, L. (2006). VEGF receptor signalling – in control of vascular function. Nat Rev Mol Cell Biol *7*, 359–371.
Overall, C. M. (2002). Molecular determinants of metalloproteinase substrate specificity: matrix metalloproteinase substrate binding domains, modules, and exosites. Mol Biotechnol *22*, 51–86.
Overall, C. M., and Dean, R. A. (2006). Degradomics: systems biology of the protease web. Pleiotropic roles of MMPs in cancer. Cancer Metastasis Rev *25*, 69–75.
Overall, C. M., McQuibban, G. A., and Clark-Lewis, I. (2002). Discovery of chemokine substrates for matrix metalloproteinases by exosite scanning: a new tool for degradomics. Biol Chem *383*, 1059–1066.
Page-McCaw, A., Ewald, A. J., and Werb, Z. (2007). Matrix metalloproteinases and the regulation of tissue remodelling. Nat Rev Mol Cell Biol *8*, 221–233.
Paupert, J., Mansat-De Mas, V., Demur, C., Salles, B., and Muller, C. (2008). Cell-surface MMP-9 regulates the invasive capacity of leukemia blast cells with monocytic features. Cell Cycle *7*, 1047–1053.
Pourmotabbed, T. (1994). Relation between substrate specificity and domain structure of 92-kDa type IV collagenase. Ann N Y Acad Sci *732*, 372–374.
Rai, K. R. (2007). Pathophysiologic mechanisms of chronic lymphocytic leukemia and their application to therapy. Exp Hematol *35*, 134–136.

Redondo-Munoz, J., Ugarte-Berzal, E., Garcia-Marco, J. A., et al. (2008). Alpha4beta1 integrin and 190-kDa CD44v constitute a cell surface docking complex for gelatinase B/MMP-9 in chronic leukemic but not in normal B cells. Blood *112*, 169–178.

Ries, C., Pitsch, T., Mentele, R., et al. (2007). Identification of a novel 82 kDa proMMP-9 species associated with the surface of leukaemic cells: (auto-)catalytic activation and resistance to inhibition by TIMP-1. Biochem J *405*, 547–558.

Rodriguez, D., Morrison, C. J., and Overall, C. M. (2010). Matrix metalloproteinases: what do they not do? New substrates and biological roles identified by murine models and proteomics. Biochim Biophys Acta *1803*, 39–54.

Rozanov, D. V., Ghebrehiwet, B., Ratnikov, B., Monosov, E. Z., Deryugina, E. I., and Strongin, A. Y. (2002). The cytoplasmic tail peptide sequence of membrane type-1 matrix metalloproteinase (MT1-MMP) directly binds to gC1qR, a compartment-specific chaperone-like regulatory protein. FEBS Lett *527*, 51–57.

Rundhaug, J. E. (2005). Matrix metalloproteinases and angiogenesis. J Cell Mol Med *9*, 267–285.

Sela-Passwell, N., Rosenblum, G., Shoham, T., and Sagi, I. (2010). Structural and functional bases for allosteric control of MMP activities: can it pave the path for selective inhibition? Biochim Biophys Acta *1803*, 29–38.

Shipley, J. M., Doyle, G. A., Fliszar, C. J., et al. (1996). The structural basis for the elastolytic activity of the 92-kDa and 72-kDa gelatinases. Role of the fibronectin type II-like repeats. J Biol Chem *271*, 4335–4341.

Steffensen, B., Wallon, U. M., and Overall, C. M. (1995). Extracellular matrix binding properties of recombinant fibronectin type II-like modules of human 72-kDa gelatinase/type IV collagenase. High affinity binding to native type I collagen but not native type IV collagen. J Biol Chem *270*, 11555–11566.

Stocker, R., and Keaney, J. F., Jr. (2004). Role of oxidative modifications in atherosclerosis. Physiol Rev *84*, 1381–1478.

Strongin, A. Y., Collier, I. E., Krasnov, P. A., Genrich, L. T., Marmer, B. L., and Goldberg, G. I. (1993). Human 92 kDa type IV collagenase: functional analysis of fibronectin and carboxyl-end domains. Kidney Int *43*, 158–162.

Suenaga, N., Mori, H., Itoh, Y., and Seiki, M. (2005). CD44 binding through the hemopexin-like domain is critical for its shedding by membrane-type 1 matrix metalloproteinase. Oncogene *24*, 859–868.

Sugimoto, K., Ishibashi, T., Sawamura, T., et al. (2009). LOX-1-MT1-MMP axis is crucial for RhoA and Rac1 activation induced by oxidized low-density lipoprotein in endothelial cells. Cardiovasc Res *84*, 127–136.

Takino, T., Miyamori, H., Kawaguchi, N., Uekita, T., Seiki, M., and Sato, H. (2003). Tetraspanin CD63 promotes targeting and lysosomal proteolysis of membrane-type 1 matrix metalloproteinase. Biochem Biophys Res Commun *304*, 160–166.

Tam, E. M., Wu, Y. I., Butler, G. S., Stack, M. S., and Overall, C. M. (2002). Collagen binding properties of the membrane type-1 matrix metalloproteinase (MT1-MMP) hemopexin C domain. The ectodomain of the 44-kDa autocatalytic product of MT1-MMP inhibits cell invasion by disrupting native type I collagen cleavage. J Biol Chem *277*, 39005–39014.

Tochowicz, A., Goettig, P., Evans, R., et al. (2011). The dimer interface of the membrane type 1 matrix metalloproteinase hemopexin domain: crystal structure and biological functions. J Biol Chem *286*, 7587–7600.

Toth, M., Gervasi, D. C., and Fridman, R. (1997). Phorbol ester-induced cell surface association of matrix metalloproteinase-9 in human MCF10A breast epithelial cells. Cancer Res *57*, 3159–3167.

Toyama-Sorimachi, N., Kitamura, F., Habuchi, H., Tobita, Y., Kimata, K., and Miyasaka, M. (1997). Widespread expression of chondroitin sulfate-type serglycins with CD44 binding ability in hematopoietic cells. J Biol Chem *272*, 26714–26719.

Triebel, S., Blaser, J., Reinke, H., and Tschesche, H. (1992). A 25 kDa alpha 2-microglobulin-related protein is a component of the 125 kDa form of human gelatinase. FEBS Lett *314*, 386–388.

Tuuttila, A., Morgunova, E., Bergmann, U., et al. (1998). Three-dimensional structure of human tissue inhibitor of metalloproteinases-2 at 2.1 A resolution. J Mol Biol *284*, 1133–1140.

Van den Steen, P. E., Van Aelst, I., Hvidberg, V., et al. (2006). The hemopexin and O-glycosylated domains tune gelatinase B/MMP-9 bioavailability via inhibition and binding to cargo receptors. J Biol Chem *281*, 18626–18637.

Wilhelm, S. M., Collier, I. E., Marmer, B. L., Eisen, A. Z., Grant, G. A., and Goldberg, G. I. (1989). SV40-transformed human lung fibroblasts secrete a 92-kDa type IV collagenase which is identical to that secreted by normal human macrophages. J Biol Chem *264*, 17213–17221.

Williamson, R. A., Carr, M. D., Frenkiel, T. A., Feeney, J., and Freedman, R. B. (1997). Mapping the binding site for matrix metalloproteinase on the N-terminal domain of the tissue inhibitor of metalloproteinases-2 by NMR chemical shift perturbation. Biochemistry *36*, 13882–13889.

Winberg, J. O., Berg, E., Kolset, S. O., and Uhlin-Hansen, L. (2003). Calcium-induced activation and truncation of promatrix metalloproteinase-9 linked to the core protein of chondroitin sulfate proteoglycans. Eur J Biochem *270*, 3996–4007.

Winberg, J. O., Kolset, S. O., Berg, E., and Uhlin-Hansen, L. (2000). Macrophages secrete matrix metalloproteinase 9 covalently linked to the core protein of chondroitin sulphate proteoglycans. J Mol Biol *304*, 669–680.

Wolf, K., Mazo, I., Leung, H., et al. (2003). Compensation mechanism in tumor cell migration: mesenchymal-amoeboid transition after blocking of pericellular proteolysis. J Cell Biol *160*, 267–277.

Xu, X., Chen, Z., Wang, Y., Yamada, Y., and Steffensen, B. (2005). Functional basis for the overlap in ligand interactions and substrate specificities of matrix metalloproteinases-9 and -2. Biochem J *392*, 127–134.

Yamamoto, K., Higashi, S., Kioi, M., Tsunezumi, J., Honke, K., and Miyazaki, K. (2006). Binding of active matrilysin to cell surface cholesterol sulfate is essential for its membrane-associated proteolytic action and induction of homotypic cell adhesion. J Biol Chem *281*, 9170–9180.

Yamamoto, K., Miyazaki, K., and Higashi, S. (2010). Cholesterol sulfate alters substrate preference of matrix metalloproteinase-7 and promotes degradations of pericellular laminin-332 and fibronectin. J Biol Chem *285*, 28862–28873.

Yanez-Mo, M., Barreiro, O., Gonzalo, P., et al. (2008). MT1-MMP collagenolytic activity is regulated through association with tetraspanin CD151 in primary endothelial cells. Blood *112*, 3217–3226.

Yoshida, T., Yamagishi, S., Nakamura, K., et al. (2006). Pigment epithelium-derived factor (PEDF) inhibits advanced glycation end product (AGE)-induced C-reactive protein expression in hepatoma cells by suppressing Rac-1 activation. FEBS Lett *580*, 2788–2796.

Zent, C. S., and Kay, N. E. (2004). Advances in the understanding of biology and prognosis in chronic lymphocytic leukemia. Curr Oncol Rep 6, 348–354.

Zhang, L., Yang, M., Yang, D., Cavey, G., Davidson, P., and Gibson, G. (2010). Molecular interactions of MMP-13 C-terminal domain with chondrocyte proteins. Connect Tissue Res *51*, 230–239.

Zhang, M., Kho, A. L., Anilkumar, N., et al. (2006). Glycated proteins stimulate reactive oxygen species production in cardiac myocytes: involvement of Nox2 (gp91phox)-containing NADPH oxidase. Circulation *113*, 1235–1243.

Zhao, H., Sohail, A., Sun, Q., et al. (2008). Identification and role of the homodimerization interface of the glycosylphosphatidylinositol-anchored membrane type 6 matrix metalloproteinase (MMP25). J Biol Chem *283*, 35023–35032.

4.5 The ADAMTS family of metalloproteinases

Hideaki Nagase

4.5.1 Introduction

The turnover of extracellular matrix (ECM) molecules is essential in embryogenesis, morphogenesis, tissue remodeling, angiogenesis, and wound healing (Brinckerhoff and Matrisian, 2002; Page-McCaw et al., 2007). While a number of proteolytic enzymes including serine, cysteine, and metalloproteinases are involved in these processes, matrix metalloproteinases (MMPs) have been considered to play key roles in ECM degradation, as exemplified by the discovery of collagenase in tadpoles undergoing metamorphosis by Gross and Lapiere (1962). This stimulated researchers to search for collagenases in mammalian tissues as they play important roles not only in biological processes but also in diseases such as skin ulcers, arthritis, cancer, and periodontal disease (Woolley and Evanson, 1980). Human collagenase was first purified from the culture medium of rheumatoid arthritis synovium (Woolley et al., 1975). The enzyme has been designated matrix metalloproteinase-1 (collagenase 1), or MMP1, as the first member of the MMP family, which now comprises a large number of members including those in animals and plants (humans have 23 MMPs) (Visse and Nagase, 2003). All MMPs are normally secreted or plasma membrane anchored, and they act primarily on the cell surface or in the extracellular space (Murphy and Nagase, 2011).

The determination of three-dimensional (3D) structures of human collagenases (MMP1 and MMP8) (Bode et al., 1994; Lovejoy et al., 1994) has revealed that they have similar polypeptide folds to those of the previously solved structures of the crayfish metalloproteinase astacin, the snake venom metalloproteinases adamalysin (Gomis-Rüth et al., 1993) and atrolysin (Zhang et al., 1994), and the bacterial metalloproteinase serralysin (Baumann, 1994). They all have the common zinc-binding motif HEXXHXXGXXH and the unique methionine turn (Met-turn), although amino acid sequence homology among these families of metalloproteinases indicates that they are distantly related. These structurally related zinc metalloproteinases are collectively called "metzincins" (Bode et al., 1993; Gomis-Rüth, 2003). The mammalian metalloproteinases homologous to snake venom adamalysins are called ADAMs (a disintegrin and metalloproteinases), and they are plasma-membrane-anchored multidomain proteins. In humans, there are 21 genes encoding ADAMs, but only 13 of them are proteolytically functional as the other 8 members possess an imperfect catalytic zinc-binding motif (Edwards et al., 2008). The functional ADAM metalloproteinases are involved in ectodomain shedding of membrane-anchored cytokines, growth factors, receptors, and adhesion molecules. One of the well-characterized enzymes is ADAM17, also called tumor necrosis factor α-converting enzyme (TACE), which processes pro-tumor necrosis factor (pro-TNF-α) and releases tumor necrosis factor-α (TNF-α) from the cell surface (Black et al., 1997; Moss et al., 1997). ADAM17, however, can process a large number of other cell surface molecules including epidermal growth factor (EGF) receptor ligands (e.g. heparin-binding EGF, transforming growth factor-α [TGF-α], epiregulin, etc.) and other cell-surface-anchored molecules (Blobel, 2005). Other proteolytically active ADAMs can also function as sheddases (Edwards et al., 2008). These metalloproteinases, therefore, change cell-cell and cell-matrix interactions, as well as cell signaling in paracrine, autocrine, and juxacrine modes of action (Edwards et al., 2008).

The ADAMTSs (a disintegrin and metalloproteinase with thrombospondin motifs) are metalloproteinases with a catalytic metalloproteinase (M) domain closely related to those of ADAMs (Apte, 2004). Like ADAMs, they harbor the so-called "disintegrin (Dis)" domain, which is followed by a thrombospondin type I (TS) domain, a cysteine-rich (CysR) domain, a spacer (Sp) domain, and an additional 0–14 TS domains at the C terminus (Apte, 2004, 2009; Porter et al., 2005). Some ADAMTSs have additional unique domains at, or toward, the C terminus (▶Figure 4.7). Unlike ADAMs, however, ADAMTSs are secreted from the cell, and all have the complete zinc-binding motif HEXXHXX[G/N/S]XXH and the Met-turn in the M domain. Therefore, all ADAMTSs

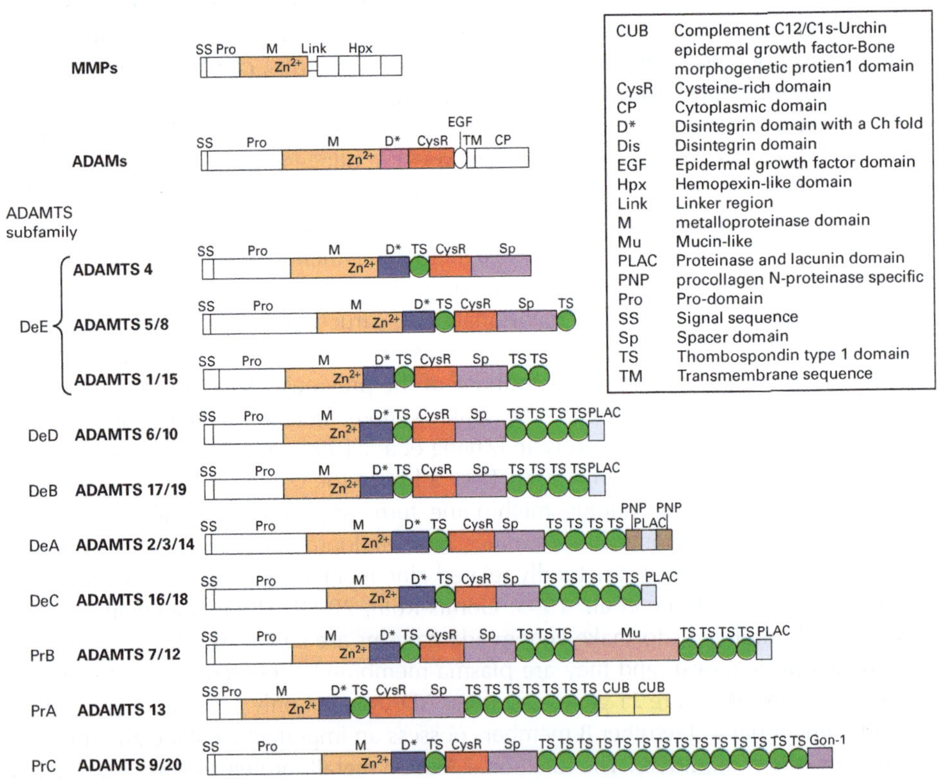

Figure 4.7 Domain arrangement of ADAMTSs. ADAMTS metalloproteinases are compared with a general domain arrangement of MMPs and ADAMs. ADAMTSs are subgrouped by evolutionarily related clades.

are considered to be functional proteinases, but biological substrates have not been well characterized for many of the members. The well-characterized enzymes include so-called "aggrecanases," ADAMTSs that process the N-propeptide of procollagens I, II, III, and/or V, and ADAMTS13 that selectively cleaves the ultralarge von Willebrand factor (VWF). More recent genetic approaches have characterized biological and pathological functions of ADAMTSs. This review discusses the structure and function of ADAMTSs and recent progress made toward our understanding of their biological and pathological functions.

4.5.2 The ADAMTS family

The human genome encodes 19 ADAMTSs. Substrates cleaved by human ADAMTSs are listed in the ▶Table 4.6. The first ADAMTS, ADAMTS1, was discovered by Kuno et al. (1997) by cDNA cloning from mouse colon adenocarcinoma cells. The deduced amino acid sequence revealed a signal peptide, a prodomain, an M domain, a Dis domain, a TS domain, a CysR domain, an Sp domain, and two additional TS domains. The M and Dis domains are related to those of ADAMs, but TS domains are unique, which distinguished it from ADAMs. Thus, the name ADAMTS was coined. Ciona intestinalis (sea squirt) contains 6 ADAMTS genes (Huxley-Jones et al., 2005), and the phylogenetic grouping based on sequence conservation of the catalytic M domain of representative protostomes, deuterostomes, and human orthologs independently supports grouping the genes that share similar noncatalytic domain arrangements in the C-terminal region (▶Figure 4.7). Those genes cluster with two or more vertebrate ADAMTS genes, and based on their origination in protostomes (Pr) or deuterostomes (De), they are defined into subfamilies: PrA–C and DeA–E (Huxley-Jones et al., 2005). For example, ADAMTS1 and ADAMTS15 have two C-terminal TS domains. ADAMTS5 and ADAMTS8 have one C-terminal TS domain, and ADAMTS4 has none (▶Figure 4.7). All these enzymes have an ability to cleave aggrecan core protein and they form a large clade DeE of the ADAMTS family. ADAMTS1, 4, and 5 are inhibited by tissue inhibitor of metalloproteinases-3 (TIMP-3) (Brew and Nagase, 2010).

The second enzyme cloned was ADAMTS2, which processes the N-propeptide of procollagens I, II, and III (Colige et al., 1997). The enzyme is also called procollagen N-proteinase. It has four TS domains after the spacer domain and a unique domain specific to procollagen N-proteinase and a proteinase and lacunin (PLAC) domain at the C terminus. ADAMTS3 (Fernandes et al., 2001) and ADAMTS14 (Bolz et al., 2001) have the same domain arrangement, and they also exhibited procollagen N-proteinase activity, forming the DeA clade. ADAMTS2 is weakly inhibited by TIMP-3, but their affinity is enhanced in the presence of heparin (Wang et al., 2006).

ADAMTS6 and 10 and ADAMTS17 and 19 have four TS domains and a PLAC domain after the spacer domain, but these two pairs are grouped into separate clades (DeD and DeB, respectively) because of the sequence similarity of their M domains (Huxley-Jones et al., 2005).

ADAMTS16 and 18 have five TS domains and a PLAC domain (clade DeC). Although the metalloproteinase domains of these enzymes were reported to degrade aggrecan core protein at the Glu373-Ala374 bond (Zeng et al., 2006), a signature cleavage of aggrecanases, their activities on aggrecan are very weak. It is not known whether they

Table 4.6 ADAMTS metallproteinases.

	Potential functions	*Expression*	*Substrates*	*Chromosome location*
ADAMTS1	Organ morphogenesis (kidney, adrenal glands, urinary tract, female reproductive organs), ovulation, keratinocyte differentiation, antiangiogenesis	Urinary epithelium, ovary, muscle, kidney, aorta, many other tissues	Aggrecan, versican, α2-macroglobulin, thrombospondin-1 and -2, nidogen, tissue factor pathway inhibitor-2 (Kunitz-type serine protease inhibitor), shed syndecan-4	21q21
ADAMTS2	Procollagen N-propeptide processing	Skin, uterus, arteries, fibroblasts	N-propeptides of procollagens I, II, III, and V	5q35
ADAMTS3	Procollagen N-propeptide processing	Cartilage, bone and musculotendinous tissue during development	N-propeptide of procollagens I and II	4q21
ADAMTS4	Tissue breakdown	Synovium, cartilage, central nervous system (CNS)	Aggrecan, versican, brevican, decorin biglycan, fibromodulin, COMP, α2-macroglobulin, S-carboxymethylated transferrin	1q23
ADAMTS5	Tissue breakdown, dermal wound healing	Placenta, uterus, cervix, cartilage	Aggrecan, versican, brevican, decorin biglycan, fibromodulin, S-carboxymethylated transferrin	21q21
ADAMTS6		Placenta		5q12
ADAMTS7		Synovium, cartilage	COMP	15q24
ADAMTS8		(Not expressed in cartilage)	Aggrecan	11q25

ADAMTS9	Closure of the mouse palate in cooperation with ADAMTS20, antiangiogenic	Chondrocytes, fibroblasts, embryonic tissues, many tissues, microvascular endothelial cells, multiple myeloma cells	Aggrecan, versican	3p14
ADAMTS10	Microfibril biogenesis (mutation causes Weill-Marchesani syndrome)	Kidney, liver, heart, placenta	α_2-Macroglobulin, fibrillin-1	19p13
ADAMTS12		Cartilage, synovium	COMP	5q35
ADAMTS13	Processing of ultralarge von Willebrand factor under shear stress	Liver, prostate, brain	von Willebrand factor	9q34
ADAMTS14	Procollagen N-propeptide processing	Prostate, brain, liver, fetal lung	Propeptide of procollagens I and III	10q21
ADAMTS15		Cartilage, fetal kidney, fetal lung	Aggrecan	11q25
ADAMTS16		Cartilage, brain, ovary, fetal lung, fetal kidney	Aggrecan	5p15
ADAMTS17		Ovary, fetal lung	Aggrecan	15q24
ADAMTS18	Platelet fragmentation and dissolution	Brain, submaxillary gland, fetal lung, fetal kidney, endothelial cell		16q23
ADAMTS19		Osteosarcoma, fetal lung		5c31
ADAMTS20	Neural-crest-derived melanoblast migration through dermal mesenchymal, or survival and proliferation within the ectoderm; closure of the mouse palate in cooperation with ADAMTS9	Subectodermal mesenchyme during development	Versican	12q12

have better activity on aggrecan and other ECM substrates when expressed as full-length proteins.

ADAMTS7 and 12 (clade PrB) have seven TS domains, a mucin-like domain that was inserted between the fourth and fifth TS domains, and a PLAC domain after the spacer domain (Somerville et al., 2004). Both enzymes were reported to cleave cartilage oligomeric matrix protein (COMP) at the same site (Liu et al., 2006a, 2006b). This reaction was inhibited by granulin-epithelin precursor though its direct binding to COMP and four C-terminal TS motifs of ADAMTS7 and ADAMTS12 (Guo et al., 2010)

ADAMTS13 is unique in both structure and enzymatic activity (clade PrA). It is the only ADAMTS that has two C-terminal CUB (complement C1r/C1s-urchin epidermal growth factor-bone morphogenetic protein-1) domains along with seven TS domains after the Sp domain. Multimeric VWF is the only substrate, and ADAMTS13 cleaves it under shear stress.

ADAMTS9 and 20 possess 14 TS domains and a Gon1 domain after the Sp domain (Somerville et al., 2003) (clade PrC). They are orthologs of GON1, an ADAMTS proteinase in Caenorhabditis elegans, which is essential for migration of distal tip cells during gonadal morphogenesis (Blelloch and Kimble, 1999). Human ADAMTS9 was characterized to cleave versican core protein at the Glu441-Ala442 bond and the Glu1771-Ala1772 bond of aggrecan (Somerville et al., 2003). Mouse ADAMTS20 cleaves versican and participates in melanoblast survival (Silver et al., 2008).

4.5.3 Three-dimensional structures of ADAMTSs

The first 3D structure reported in this family of proteins was the M domain with the Dis domain of ADAMTS1 in 2007 by Gerhardt et al. (2007). The structures of the M-Dis domains of ADAMTS4 and ADAMTS5 (Mosyak et al., 2008) and the M domain of ADAMTS5 (Shieh et al., 2008) were also reported. Full-length ADAMTS protein structures have not been solved yet, but Akiyama et al. (2009) reported the crystal structure of the Dis-TS-CysR-Sp domains of ADAMTS13. The composite structure of the M-Dis domain of ADAMTS5 (Mosyak et al., 2008) and the TS-CysR-Sp of ADAMTS13 (Akiyama et al., 2009) shown in ▶Figure 4.8 illustrates a possible topological arrangement of the domains in this family of enzymes.

The M domain of ADAMTS1 consists of a twisted central β sheet of five strands, five α helices, and connecting loops. The overall polypeptide fold of the M domain is a typical metzincin metalloproteinase fold (Gerhardt et al., 2007), similar to those of MMPs and ADAMs. The catalytic zinc ion interacts with three His residues in the HEXXHXXGXXH motif. Unlike MMPs, there is no structural zinc in the M domain of ADAMTS1. However, there are four disulfide bridges, and the fourth bridge, Cys379-Cys462, connects the loop region between strands β4 and β5 with a 22-residue connector loop after helix α5, which wraps around the back of the metalloproteinase domain and connects with the Dis domain. This arrangement places the Dis domain on the prime side of the active site (the sites that interact with substrate residues located at the C-terminal side of the scissile bond). An interesting feature noted for the M domains of ADAMTS4 and ADAMTS5 (Mosyak et al., 2008) is that, in the presence of an active-site-directed inhibitor, they form a unique S2′ loop defined by a short disulfide-containing loop with the motif CGXXXCDTL (322–330 in ADAMTS4) near the S1′ pocket of the active site. This

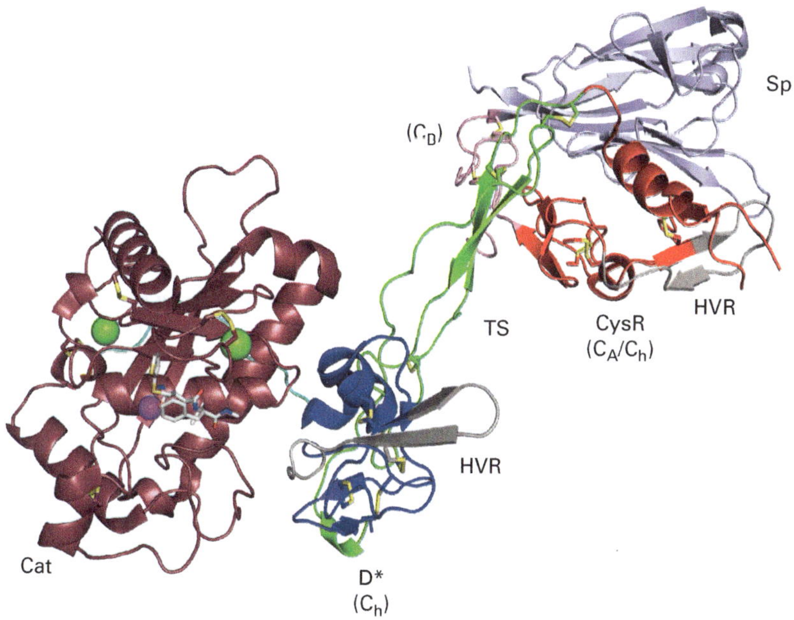

Figure 4.8 A 3D structural model of ADAMTS. The structure was modeled based on the M (raspberry) and D* (blue and gray) domains of ADAMTS5 (PDBID: 2RJQ) and the D*, TS (green), CysR (red and pink), and Sp (light blue) domains of ADAMTS13 (PDBID: 3GHN). Catalytic zinc ion is in a red sphere, and calcium ions are in a green sphere. Disulfide bonds are shown in yellow. HVR, hypervariable region (gray).

loop adopts a compact β-turn structure, and it is stabilized by both the disulfide bond and a calcium ion. Without the inhibitor, the S2´ loop is in a markedly different conformation by losing a calcium ion. These changes bring Asp328 and Thr329 closer to the catalytic zinc, so that the carboxylate group of Asp328 chelates the zinc ion and Thr329 fills the space at the mouth of the S1´ pocket. The S2´ loop, therefore, behaves like an autoinhibitor of the metalloproteinases. This interaction is reminiscent of the "cysteine switch" found in the prodomain of MMPs that interacts with the catalytic zinc ion and maintains their inactive zymogen form (Van Wart and Birkedal-Hansen, 1990; Becker et al., 1995). Noncatalytic domains of ADAMTSs play important roles in substrate recognition and catalysis. It may therefore be speculated that the binding of a natural substrate through the noncatalytic ancillary domains of the enzyme may induce such changes of the S2´-loop to open and activate the enzyme. This may also explain why only certain substrates that interact in the noncatalytic domain are substrates of this group of enzymes.

The 3D structure of the Dis domain of ADAMTS1 (Gerhardt et al., 2007) did not show structural homology to the Dis domain of snake venom metalloproteinase vascular apoptosis-inducing protein-1 (VAP1) (Takeda et al., 2006) and that of ADAM10 (Janes et al., 2005). It was similar to the structure of the "hand" (C_h) segment of the CysR domain of VAP1. The CysR domain of VAP1 and P-III snake venom metalloproteinases is structurally subdivided into "wrist" (C_w) and C_h subdomains, where the C_w subdomain is tightly associated with the Dis loop, which contains a potential integrin-interaction motif, making the Dis loop unavailable to protein-protein interactions due to steric hindrance (Takeda et al., 2006, 2012). The C_h subdomain has a core of α/β-fold structure consisting of two antiparallel β strands packed against two of the three α helices and five disulfide bonds (Takeda et al., 2006). The C_h fold of VAP1 is unique, but it is similar to the fold of the Dis domain and an N-terminal part of the CysR domain called "C_A" subdomain of ADAMTS13, even though their amino acid sequence identity is low (Akiyama et al., 2009). Part of the Dis domains of ADAMTS4 and ADAMTS5 also forms a C_h fold (Mosyak et al., 2008), and they are superimposed with that of ADAMTS1. The Dis domain of ADAMTSs is designated as "D*" (Takeda, 2009; Takeda et al., 2012) to distinguish its structure from the canonical Dis domains found in snake venom metalloproteinases and ADAMs, and this distinction is adopted in this review (see ▶Figure 4.8). The D* domains in ADAMTSs are located in close proximity to the active site and form part of the S3′ pocket; therefore, they participate in substrate interaction. This may be the reason why the M domain alone has little proteolytic activity (Kashiwagi et al., 2004; Gendron et al., 2007).

The TS domain has an antiparallel three-stranded fold with three disulfide bonds, (Akiyama et al., 2009), a very similar structure of that of the prototypical thrombospondin type I repeat, TSR2 in thrombospondin-1 (Tan et al., 2002).

The structure of the CysR domain has revealed two subdomains, the N-terminal C_A subdomain and the C-terminal C_B subdomain (Akiyama et al., 2009). The structure of the C_A domain is similar to that of D* (or C_h) of ADAMTS13, although they have only 17% identity in their amino acid sequences. The C_B subdomain does not have an apparent secondary structure, but it has a series of turns stabilized by a pair of disulfide bonds at its N and C termini and forms a rod shape. The C_A subdomain contains an RGD (498–500) sequence. The side chain of Arg498 is buried and unavailable for protein-protein interaction, but Asp500 is exposed to the solvent.

The Sp domain is a long cysteineless segment, and its primary structure shows no apparent homology to known structural motifs, but it folds into a single globular domain with 10 β strands in a jelly-roll topology, forming two antiparallel β sheets that lie almost parallel to each other (Akiyama et al., 2009). Many key residues of the Sp domains are conserved among ADAMTS proteins, particularly those that maintain the structure, and it is predicted that ADAMTSs share similar Sp domain architecture as that in ADAMTS13. In addition, the N and C termini of the Sp domain are in close proximity, and therefore, the TS2 domain after the Sp domain should be close to the C_A/Sp domain junction. The general topological arrangements of D*-TS-CysR-Sp in ADAMTSs are considered to be similar to that of ADAMTS13 (shown in ▶Figure 4.8) as the residues located at the domain-domain interface are conserved (Akiyama et al., 2009).

4.5.4 Procollagen N-proteinases (ADAMTS2, 3, and 14)

Fibril-forming interstitial collagens I, II, III, and V are synthesized as procollagens with the N-propetides and C-propeptides flanking the 300 nm long triple-helical collagens. To form collagen fibrils and fibrillar bundles that provide the tissue with mechanical resistance and tissue architecture, both propeptides of the procollagen must be removed: firstly, the C-propeptide removal by bone morphogenetic protein-1 or related mammalian Tolloid-like metalloproteinases belonging to the astacin family, and then the N-propeptides by procollagen N-proteinases (i.e. ADAMTS2, ADAMTS3, and ADAMTS14). Copolymerization of collagen I retaining the N-propeptide with fully processed collagen I results in altered fibril organization *in vitro* (Hulmes et al., 1989) and *in vivo* (Nusgens et al., 1992).

Mutations of ADAMTS2 cause dermatosparaxis in cattle and Ehlers-Danlos syndrome VIIC in humans, an autosomal recessive disorder with skin fragility and joint laxity (Colige et al., 1999). ADAMTS2-null mice are essentially normal, but they develop skin fragility in 1–2 months after birth, and male mice are sterile, although the females have normal fertility (Li et al., 2001). More recently, it was reported that ADAMTS2 has antitumor and antiangiogenic activity (Dubail et al., 2010).

4.5.5 Aggrecanases

The treatment of bovine articular cartilage in culture with a proinflammatory cytokine interleukin-1 (IL-1) increases the breakdown of aggrecan and releases the proteolytic fragments into the medium; one of the cleavage sites was identified at the Glu373-Ala374 bond located in the interglobular domain (IGD) between the G1 and G2 globular domains of the core protein (Sandy et al., 1991) (▶Figure 4.9). At about the same time, two other groups also found the same site and other sites located in the chondroitin sulfate attachment site 2 (CS-2 region): KFFE1666-GLGS, TAQE1771-AGEG, and VSQE1871 -LGQR (Ilic et al., 1992; Loulakis et al., 1992). These fragments are found in synovial fluid and serum of age-matched animals (Ilic et al., 1992), suggesting that the enzymes responsible for these cleavages are important in both physiological and pathological catabolism of aggrecan. Indeed, aggrecan fragments resulting from the cleavage of the Glu373-Ala374 bond accumulate in synovial fluids of patients with osteoarthritis (OA) and inflammatory joint disease (Sandy et al., 1992; Lohmander et al., 1993). This enzymic activity, referred to as "aggrecanase," is distinct form the action of MMPs, which cleave the different sites including IPEN341-FFGV and TSED441-LVVQ site of the IGD (▶Figure 4.9) (Nagase and Kashiwagi, 2003). The first two aggrecanases were cloned by the group in DuPont in 1999 and identified as ADAMTS4 for aggrecanase-1 (Tortorella et al., 1999) and ADAMTS5 for aggrecanase-2 (Abbaszade et al., 1999). Aggrecanase-2 was originally assigned as ADAMTS11 (Abbaszade et al., 1999), but it was found to be identical to ADAMTS5 (Hurskainen et al., 1999); therefore, ADAMTS11 is no longer used.

ADAMTS5 is an approximately 30-fold more potent aggrecanase than ADAMTS4 (Fushimi et al., 2008). Both enzymes cleave aggrecan core protein at several sites (Tortorella et al., 2002). ADAMTS5 also digests bovine aggrecan at a unique RPAE2047-ARLE site located close to the C-terminal CH2 region, adjacent to the G3 domain (Durigova et al., 2008). Both aggrecanases can cleave versican, brevican, decorin, fibromodulin, and S-carboxymethlyated transferrin (▶Table 4.6).

ADAMTS1 (Rodriguez-Manzaneque et al., 2002), ADAMTS8 (Collins-Racie et al., 2004), ADAMTS9 (Demircan et al., 2005), ADAMTS15 (Collins-Racie et al., 2004), and ADAMTS16 and ADAMTS18 (Zeng et al., 2006) were also reported to cleave aggrecan core protein at the sites identified for the first two aggrecanases, but their activities on aggrecan core protein are much weaker than those of ADAMTS4 and ADAMTS5. It should be noted, however, that many of those studies were carried out

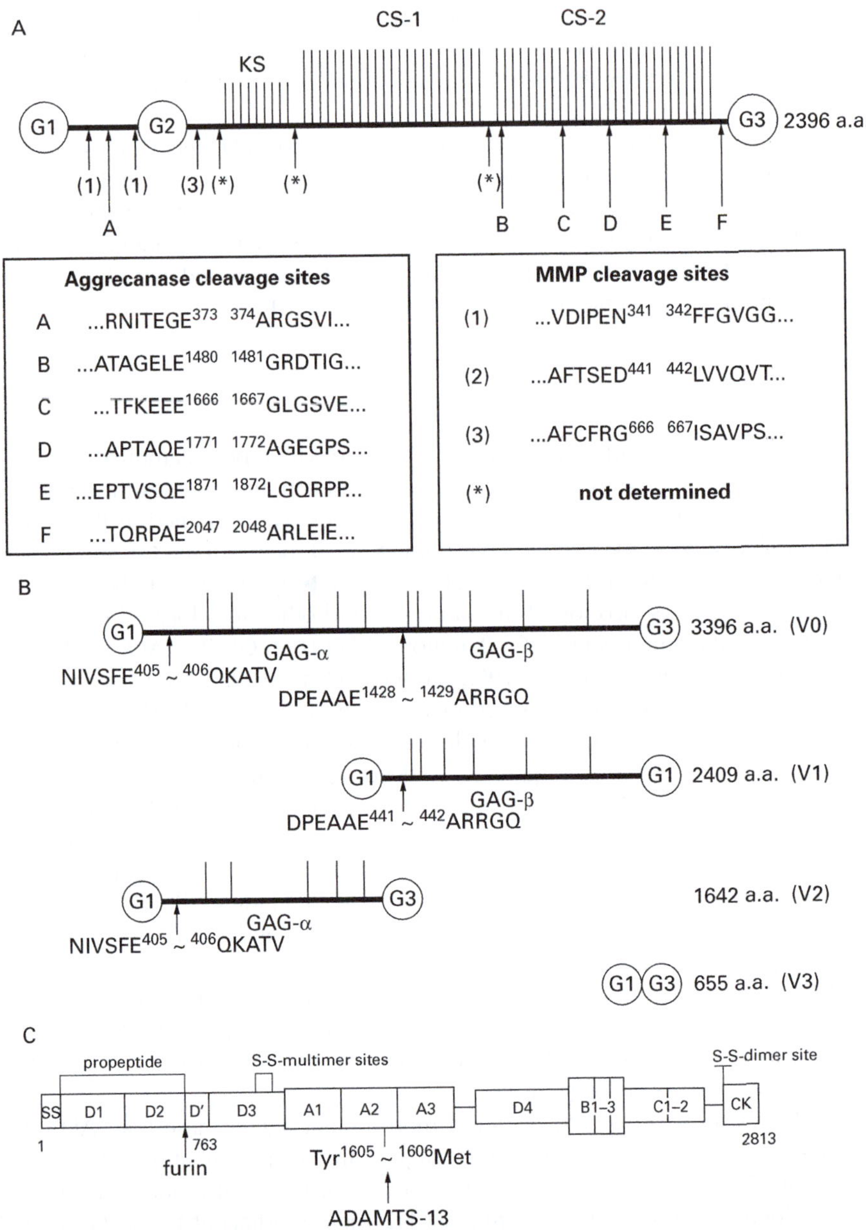

Figure 4.9 **ADAMTS cleavage sites in aggrecan, versican, and von Willebrand factor.** (A) Aggrecan and aggrecanase (ADAMTSs) and MMP cleavage sites in aggrecan. (B) Versican variants and ADAMTS cleavage sites. (C) Von Willebrand factor and the ADAMTS13 cleavage site.

with the enzymes lacking the C-terminal ancillary domains, which play key roles in recognition and cleavage of aggrecan. Therefore, it is not known whether full-length enzymes of these ADAMTSs synthesized in the tissue have better proteolytic activities against aggrecan or other proteoglycans such as versican and brevican.

TIMP-3 is a potent inhibitor of aggrecanases, and it blocks aggrecan breakdown of articular cartilage treated with inflammatory cytokines in culture (Gendron et al., 2003).

The role of noncatalytic domains of ADAMTS4 and ADAMTS5 in aggrecanolysis

Both ADAMTS4 and ADAMTS5 are synthesized as pre-proenzyme, and therefore they are destined to be secreted from the cell. Prodomains of these enzymes may assist folding of the proteinase as shown for other ADAMTSs (Longpré and Leduc, 2004; Koo et al., 2007), and they are most likely processed by a proprotein convertase intracellularly (Longpré and Leduc, 2004; Wang et al., 2004), although it has been suggested that proADAMTS5 may be activated on the cell surface (Longpré et al., 2009), as in the case of proADAMTS9 (Koo et al., 2006). The mature full-length enzymes cleave aggrecan core protein effectively, but the enzymes consisting of the M-D* domains alone showed little aggrecanase activity (Kashiwagi et al., 2004; Gendron et al., 2007), suggesting that noncatalytic ancillary domains regulate their aggrecanolytic activities.

Systematic domain deletion studies of ADAMTS4 by Kashiwagi et al. (2004) showed that the full-length ADAMTS4 has the greatest aggrecanase activity using an aggrecan-incorporated bead assay, and the deletion of the Sp domain reduces only about 20% of the activity. Further deletion of the CysR decreased the activity by approximately 60%. The subsequent deletion of the TS domain showed only 3% of the full activity. Furthermore, Kashiwagi et al. (2004) and Gao et al. (2004) reported that the full-length ADAMTS4 showed only a very weak activity to cleave the Glu373-Ala374 bond in the IGD of aggrecan, but it readily cleaved chondroitin-sulfate-rich CS-2 (Kashiwagi et al., 2004). Upon removal of the Sp domain by proteolysis (Gao et al., 2004) or mutagenesis (Kashiwagi et al., 2004), ADAMTS4 gained the ability to cleave the Glu373-Ala374 bond, and activity toward S-carboxymethylated transferrin, fibromodulin, and decorin (Kashiwagi et al., 2004). Gao et al. (2004) found that membrane-type 4 MMP (MT4-MMP/MMP17) on the cell surface processes the C-terminal Sp domain and activates ADAMTS4 to cleave the Glu373-Ala374 bond. Those data were interpreted as showing that the Sp domain inhibits ADAMTS4 activity toward specific cleavage sites.

However, Fushimi et al. (2008) found that the suppressed cleavage of the Glu373-Ala374 bond by full-length ADAMTS4 was due to the presence of heparin that was used to prevent the enzyme from binding to ECM. Recombinant ADAMTS4 expressed in chondrosarcoma cells tightly binds to the ECM, and heparin was used to compete for the enzyme-matrix interaction. When heparin was chromatographically removed from the full-length ADAMTS4 preparation, it showed about 20-times higher activity on the Glu373-Ala374 bond than the Sp-domain-deleted form. It was found that heparin preferentially blocks the activity of full-length ADAMTS4 toward the Glu373-Ala374 bond, but it is less effective on its activity for the chondroitin-sulfate-rich region. Those observations suggest that when full-length ADAMTS-4 is bound to heparan sulfate proteoglycan on the cell surface or the ECM in the tissue, the activity toward the IGD site is likely to be largely suppressed. When the Sp domain is removed, the enzyme can be released from the cell surface or ECM, and it gains activity toward decorin, biglycan, and fibromodulin, and broader proteolytic activity as shown with S-carboxymethylated transferrin. However, the truncation of the Sp domain from ADAMTS4 results in approximately 95% loss of aggrecanase activity on both the Glu373-Ala374 bond and the chondroitin-sulfate-rich region (Fushimi et al., 2008). Thus, this shifts the preference of its substrates, as well as the site of its action in the tissue.

ADAMTS5 also binds to the cell surface and the ECM, and it is not released into the medium when it is overexpressed in chondrosarcoma cells. The key ECM-binding site of ADAMTS5 is located in the CysR domain (Gendron et al., 2007). Domain deletion studies have indicated that the C-terminal TS domain has little effect on aggrecanase activity. However, the deletion of the Sp domain reduces the activity for the chondroitin sulfate region by approximately 100-fold, and the Glu373-Ala374 bond cleavage by only about 2-fold. Further deletion of the CysR domain reduced the activity on the Glu373-Ala374 bond by approximately 50-fold (Gendron et al., 2007).

Pentosan polysulfate is an exosite inhibitor of ADAMTS4 and ADAMTS5

Pentosan polysulfate (PPS) is a chemically sulfated xylosan from beechwood with molecular masses of 4,000–6,000 Da. It has been shown to be an effective antiarthritic agent in animal models (Ghosh, 1999). PPS inhibits ADAMTS4 and ADAMTS5 activity on native aggrecan, but not nonaggrecan substrate (Troeberg et al., 2008). This inhibition was due to the binding of PPS to the Sp domain of ADAMTS4 and the Sp and CysR domains of ADAMTS5, but not to the binding of the catalytic domain of the enzymes. PPS is therefore an exosite inhibitor of aggrecanases. It was also noted that in the presence of PPS, the affinity between aggrecanases and TIMP-3 increases by more than 100-fold (Troeberg et al., 2008). It has been shown that a single chain of PPS binds to both ADAMTS5 and TIMP-3 by forming a trimolecular complex, and the primary PPS-binding site in ADAMTS5 is located in the Sp domain. This effect of PPS depends on the size, and a saccharide length of 11 is sufficient for the trimolecular complex formation, which is driven electrostatically (Troeberg et al., 2012).

The treatment of the cartilage with IL-1 induces aggrecanases activity, which results in aggrecan degradation. Addition of PPS in this system reduces aggrecan breakdown. This was found to be associated with an increased amount of TIMP-3 in the cartilage without changing its mRNA levels. Troeberg et al. (2008) showed that PPS blocks the low-density lipoprotein receptor-related protein (LRP)–mediated endocytosis of TIMP-3 by binding to TIMP-3. The PPS-bound TIMP3 is far more effective to inhibit ADAMTS4 and ADAMTS5. PPS is ineffective to block aggrecan degradation of the cartilage from TIMP-3-nulll mice, indicating that chondroprotective activity of PPS is dependent on the endogenously produced TIMP-3.

ADAMTSs in arthritis

The role of ADAMTSs in arthritis is their participation in catabolism of aggrecan and possibly other ECM molecules in cartilage. The treatment of human cartilage with IL-1 or TNF-α increases aggrecanase activity, but it has no effect on mRNA levels for ADAMTS1, 4, and 5 (Flannery et al., 1999). mRNA levels for ADAMTS4 and ADAMTS5 in OA cartilage are not significantly elevated compared to that in normal cartilage (Kevorkian et al., 2004), but ADAMTS5 mRNA levels are higher than ADAMTS4 mRNA levels (Bau et al., 2002). Treatment of human chondrocytes in culture with IL-1 increased ADAMTS4 mRNA levels but did not alter the levels of ADAMTS5 mRNA (Bau et al., 2002). While some studies confirmed the unaltered ADAMTS5 mRNA level in human chondrocytes upon IL-1 treatment, other studies reported increased ADAMTS5 mRNA levels (see Fosang et al., 2008 for review). The inconsistency among these reports may be due to the age of the tissue, culture conditions of isolated chondrocytes, the time of the transcriptional activity, and stability of the mRNA. Nonetheless, an increase in the aggrecan fragments generated by aggrecanases was found in rheumatoid arthritis (RA) and OA cartilage (Lark et al., 1997) and in synovial fluids (Sandy et al., 1992; Lohmander et al., 1993; Struglics et al., 2006), suggesting that ADAMTSs are important enzymes in aggrecanolysis. ADAMTS4-null mice and ADAMTS5-null mice show no obvious abnormality, but when challenged either via surgically induced joint instability (Glasson et al., 2004, 2005) or antigen-induced arthritis (Stanton et al., 2005), the degradation of aggrecan in the cartilage of ADAMTS5-null mice was protected, but that of ADAMTS4-null mice was not. This finding indicates that ADAMTS5 is a major aggrecan-degrading enzyme in cartilage, at least in mice. Whether ADAMTS5 is a key aggrecanase in the development of human OA will only be determined by further investigation.

While the importance of aggrecanases in aggrecanolysis is well recognized, aggrecan can be degraded by MMPs as well as ADAMTSs (▶Figure 4.9). Using neoepitope antibodies that detect either MMP-cleaved fragments or ADAMTS-cleaved fragments, Lark et al. (1997) showed that both RA and OA cartilages contain aggrecan fragments generated by MMPs (G1-NITEGE373) and ADAMTSs (G1-VDIPEN360). Struglics et al. (2006) confirm these observations in OA cartilage and suggest that MMP-mediated aggrecanolysis is mostly pericellular, while ADAMTSs are both pericellular and in the interterritorial regions. Based on the fact that MT1-MMP–null mice and MMP9-null mice result in destruction of articular cartilage, impairment of endochondral-ossification and fracture repairs, Sandy (2006) suggests that some MMPs may be important in cartilage matrix homeostasis. Kevorkian et al. (2004) showed that ADAMTS16 is elevated in late human OA cartilage, but its function is not known.

4.5.6 Inhibition of angiogenesis by ADAMTSs

Thrombospondin-1 and -2 (TSP1 and TSP2, respectively) are trimeric extracellular molecules and have antiangiogenic activity *in vivo* (Good et al., 1990; Tolsma et al., 1993) and *in vitro* (Bagavandoss and Wilks, 1990; Iruela-Arispe et al., 1991). The region responsible for this activity is located to the 385–522 amino acid residues consisting of three type I TS repeats, of which repeats 2 and 3 in this region inhibit endothelial cell chemotaxis and proliferation (Vogel et al., 1993). Screening a human cDNA database of expressed sequence tags (ESTs) for sequences homologous to the second type I repeat of TSP1, Vazquez et al. (1999) obtained two homologous full-length cDNA clones. The deduced amino acid sequences revealed the presence of metalloproteinase and a disintegrin (D*) domain and two to three repeats of type I TS repeats in the C-terminal to the disintegrin domain. The proteins were called "metalloprotease and thrombospondin domains," referred to as METH1 and METH2. METH1 is a human ortholog of mouse ADAMTS1 (Kuno et al., 1997), and METH2 is assigned as ADAMTS8. Both proteins suppressed fibroblast growth factor-2 (FGF2)–induced vascularization in the cornea pocket assay and vascular endothelial growth factor (VEGF)–induced angiogenesis in the chorioallantric membrane assay. Their antiangiogenic activities are similar to, or slightly better than, endostatin or TSP1 (Vazquez et al., 1999).

ADAMTS1 binds to VEG165 and modulates the bioavailability of VEGF, and the responsible domains for this interaction are mapped in the TS-CysR-Spacer-two TS domains (Luque et al., 2003). While this is one of the mechanisms, the catalytic activity of ADAMTS1 also plays a role in antiangiogenic activity by cleaving trimeric TSP1 and TSP2 and releasing antiangiogenic fragments from the matrix (Lee et al., 2006). The cleavage sites are the Glu311-Leu312 bond of TSP1 and the Glu306-Leu307 bond of TSP2. This releases the C-terminal 110 kDa fragment from TSPs. Neoepitope antibodies that reacted to the cleaved TSP1 indicated that TSP1 was specifically cleaved at the leading edge of the wound epithelium and hair follicles in the wild-type mice, while ADAMTS1-null mice exhibited reduced staining (Lee et al., 2006). Human umbilical vein endothelial cells cultured with ADAMTS1 generated TSP1 fragments and showed reduced endothelial cell proliferation, but TSP1-null cells exhibited less inhibition in the presence of ADAMTS1 (Lee et al., 2006).

ADAMTS2 also expresses antitumor and antiangiogenic activity (Dubail et al., 2010). It inhibits proliferation of microvascular and umbilical vein endothelial cells at similar concentrations as TSP2, ADAMTS1, and ADAMTS8. It is not effective on proliferation of human fibroblasts or smooth muscle cells. It does not bind to a growth factor nor interfere with a growth factor pathway. The relative repression exerted by ADAMTS2 is similar in the presence or absence of VEGF or FGF2. In 3D culture models, ADAMTS2 reduces branching morphogenesis of capillary-like structures formed by endothelial cells. Antitumoral activity was also observed catalytically inactive enzyme. ADAMTS2 binds to the surface of endothelial cells, and the regions responsible for antiangiogenic activity are located at the short N-terminal extremity of the metalloproteinase domain and/or the 132-amino-acid-long spacer domain. ADAMTS2 binds to nucleolin on the cell surface. An antibody targeting the acidic domain of nucleolin causes apoptosis of endothelial cells and exhibits antiangiogenic *in vitro* and *in vivo* (Fogal et al., 2009). Thus, nucleolin is considered to be a potential receptor mediating the antiangiogenic properties of ADAMTS2.

ADAMTS12 also exhibits antitumorigenic and antiangiogenic properties by inhibiting the Ras-dependent extracellular signal-regulated kinase (ERK) signaling pathway (Llamazares et al., 2007). These effects are considered to be mediated by the C-terminal TS domains of the protein.

ADAMTS9 is expressed in capillary endothelial cells in embryonic and adult tissues. Heterotopic B.16-F10 melanomas in ADAMTS9$^{+/-}$ mice exhibited a greater vascularization and induction than wild-type littermates, suggesting its antiangiogenic activity (Koo et al., 2010). Unlike ADAMTS1, it does not cleave TSP1, nor bind to VEGFA165. The overexpression of ADAMTS9 is associated with esophageal squamous cell carcinoma and nasopharyngeal carcinoma, but it was shown to have antiangiogenic and tumor-suppressive functions *in vitro* and in a nude mice model (Lo et al., 2010).

4.5.7 Von Willebrand Factor–cleaving proteinase: ADAMTS13

VWF is essential in platelet adhesion to damaged blood vessels by forming a bridge between platelet surface glycoproteins and damaged subendothelium. Mature VWF generated after furin processing contains 2,050 amino acids with an apparent molecular mass of 250 kDa, and it is released from endothelial cells as "ultralarge" VWF (UL-VWF) multimers through an alternate head-to-head and tail-to-tail disulfide-bonded arrangement with a molecular mass of approximately 30,000 kDa (Sadler, 1998). In healthy individuals, UL-VWF is processed in the circulation to a series of multimers ranging from 500 to 15,000 kDa (Ruggeri and Zimmerman, 1981). Under sheer stress conditions, VWF becomes more susceptible to proteolysis, resulting in generation of 176 kDa and 140 kDa fragments, and the cleavage was postulated to be due to a specific proteinase. The enzyme was purified (Fujikawa et al., 2001; Gerritsen et al., 2001; Soejima et al., 2001) and cloned (Soejima et al., 2001; Zheng et al., 2001) and designated as ADAMTS13. The enzyme cleaves the Tyr1605-Met1606 bond within the A2 domain of VWF (▶Figure 4.9C). The control of the size by proteolysis, as well as distribution, is important for normal hemostasis, and larger multimers are hemostatically more active than smaller ones (Sadler, 1998). On the other hand, UL-VWF has been found in plasma from patients with thrombotic thrombocytopenic purpura (TTP) (Moake et al., 1982).

TTP is characterized by thrombocytopenia, microangiopathic hemolytic anemia, renal failure, neurologic dysfunction, and fever. It is caused by deficiencies of ADAMTS13 activity resulting in formation of blood clots in microcapillaries. ADAMTS13 is produced primarily in the liver and is released into the circulation. The deficiency of its activity is caused either by genetic mutation of the gene or by autoantibodies that inhibit the ADAMT13 activity. More than 70 causative mutations for congenital TTP in the ADAMTS13 gene (Levy et al., 2001; Lotta et al., 2010) have been reported, and they are found throughout the enzyme molecule. Some autoantibodies were characterized to react to the CysR and the Sp domains but still inhibit ADAMTS13 activity on VWF (Soejima et al., 2003; Luken et al., 2006), indicating that noncatalytic domains are exosites and key to recognizing the specific substrate. Domain deletion mutagenesis studies have indicated that the carboxy-terminal eight TS domains and two CUB domains of ADAMTS13 are dispensable for the VWF-cleaving activity, but the Sp domain plays a critical role in binding of the enzyme to the substrate (Soejima et al., 2003; Zheng et al., 2003; Majerus et al., 2005).

The ADAMTS13 cleavage site, Tyr1605-Met1606 bond in VWF is buried within the core of the globular A2 domain under static conditions (Zhang et al., 2009a). When exposed to fluid shear stress or denaturants, the A2 domain unfolds and adopts a partially extended conformation, which then interacts with multiple exosites of ADAMTS13 and place the Tyr1605-Met1606 bond to the active site. Based on mutagenesis studies, Gao et al. proposed that the Sp domain contains an exosite that interacts with helix α 6 of the unfolded A2 domain (Gao et al., 2006) and three other segments of VWF interact with MD*, TS1, and CysR domains of ADAMTS13 (Gao et al., 2008). Structural and mutagenesis studies of Akiyama et al. (2009) support this hypothesis. There are three exosites that extended linearly throughout the D*-TS1-CysR-Sp domains, and they bind collaboratively to multiple discontinuous region of unfolded VWF.

4.5.8 ADAMTS18 and dissolution of platelet aggregates

Li et al. (2009) found that the C-terminal 385-amino-acid fragment of ADAMTS18 induces platelet fragmentation and dissolution of platelet aggregates by generating reactive oxygen species through activation of NADPH oxidase and 2-lipoxygenase. Patients with early onset of human immunodeficiency virus type 1 (HIV1) or hepatitis C virus-related thrombocytopenia develop a unique platelet glycoprotein IIIa (GPIIIa) antibody against the GPIIIa49–66 epitope, which causes complement-independent plate fragmentation and death (Nardi et al., 2001; 2004). Searching for physiological ligands of GPIIIa, Li et al. (2009) screened a phage peptide library with the GPIIIa49–66 peptide and found peptides homologous to the C-terminal residues (1,148–1,152) of ADAMTS18. ADAMTS18 is constitutively expressed in endothelial cells, and thrombin enhances the production of ADAMTS18 and cleaves it to produce the C-terminal 45 kDa fragment (Li et al., 2009). This C-terminal fragment protects against $FeCl_3$-induced carotid artery thrombus formation and cerebral infarction in a postischaemic stroke model (Li et al., 2009).

4.5.9 Atherosclerosis

Atherosclerosis is characterized by formation of a plaque composed of cholesterol, lipids, inflammatory cells, debris resulting from cellular apoptosis, and proliferation of migratory smooth muscle cells, and it is associated with ECM remodeling. While MMPs have been characterized to play key roles in the progression of atherosclerosis and plaque rapture (Newby, 2005), recent studies have shown that ADAMTS4, 5, and 8 are localized in macrophages and that ADAMTS1 is upregulated in endothelial cells in the intima of atherosclerotic lesions and is particularly high in proliferating/migrating primary aortic vascular smooth muscle cells in culture (Jönsson-Rylander et al., 2005). Mice overexpressing ADAMTS1 crossed with apoliprotein E (ApoE)-deficient mice show an increased thickening of arterial intima (Jönsson-Rylander et al., 2005). ADAMTS1 is highly increased in human umbilical vein and cardiac microvascular endothelial cells under shear stress (Bongrazio et al., 2000), suggesting its role in flow-dependent vascular matrix remodeling. It is proposed that ADAMTS1 may promote atherosclerosis by cleaving versican and stimulate vascular smooth muscle cell migration and proliferation. The expression of ADAMTS4 colocalized with macrophages of the lesions is increased by TNF-α or interferon-γ (Jönsson-Rylander et al., 2005). In a baboon vascular graft model, high blood flow causes cell death and loss of ECM, which is accompanied by an increase of ADAMTS-cleaved versican products along with an increase in ADAMTS4 mRNA (Kenagy et al., 2009).

ADAMTS7 has been also implicated in atherosclerosis (Wang et al., 2009). It was found to accumulate in the neointima wall of the carotid artery wall after balloon injury and to facilitate the vascular smooth muscle cell migration and intima thickening by cleaving COMP.

4.5.10 ADAMTSs and morphogenesis

A number of ADAMTSs play key roles in organ morphogenesis. There are four ADAMTSs in *Caenorhabditis elegans*: GON1, MIG17, ADT1, and ADT2. GON1 (Blelloch and Kimble, 1999) and MIG17 (Nishiwaki et al., 2000) are involved in distal tip cell migration during gonad morphogenesis in larval development, possibly by cleaving the basement membrane ECM or processing extracellular cues required for cell migration (Ihara and Nishiwaki, 2008). ADT1 is essential for the ray morphogenesis of male copulatory organs where a rapid EMC remodeling is required (Kuno et al., 2002). ADT2 synthesized in glia-like cells of sensory neurons and in the vulva is necessary for normal cuticle collagen fibril organization, and it regulates the body size of *C. elegans* by maintaining cuticle structures and modulating the transforming growth factor-β (TGF-β) signaling pathway (Fernando et al., 2011). The latter was demonstrated by reduction in the TGF-β-responsive transcriptional reporter activity by *adt-2* inactivation (Fernando et al., 2011). Aberrant ECM assembly is likely to influence the cell signaling pathway as shown in *Drosophila* in which type IV collagens act as regulators of bone morphogenetic protein (BMP) signaling (Wang et al., 2008).

In vertebrates, many morphogenetic events involve the turnover of verscian, and a number of ADAMTSs have been reported to play a key role in this process. Versican is one of the members of hyaluronan-binding chondroitin sulfate proteoglycans, called lecticans which include aggrecan, brevican, and neurocan. It is found in many tissues where it serves as a hydrated matrix component. Alternative splicing of versican pre-mRNA results in transcripts that encode four variants (V0, V1, V2, and V3), and ADAMTS proteinases cleave at the Glu405-Gln406and the Glu1428-Ala1429bonds of α and β of chondroitin sulfate attachment regions in V0, the Glu441-Ala442 bond in V1, and the Glu405-Gln406 bond in V2 (▶Figure 4.9B). A neoepitope antibody that recognizes the cleaved C-terminal fragment sequence DPEAAE is used to detect the ADAMTS activity on versican.

ADAMTS1 is considered to be important in limb joint development. Versican is highly expressed in developing mouse joint interzones during limb morphogenesis, and the DPEAAE neoepitope is detected in the joint interzones along with ADAMTS1 colocalization (Capehart, 2010).

During bone development in the rat mandible and hind limb, expression of versican and ADAMTS1, 4, and 5 mRNA is detected in woven bone matrix, which is less ordered and weaker than the more mature lamellar bone (Nakamura et al., 2005). While versican protein is abundant in the woven bone matrix, it decreases in the lamellar bone, suggesting the degradation of versican by ADAMTSs during bone development.

The versican-ADAMTS axes are also important in heart development, and myocardial trabeculation is a key morphogenic event. It begins at embryonic day E9.0 in mice, when clusters of myocardial cells in the ventricles protrude into the ECM (cardiac jelly) consisting of hyaluronan and versican, which separate the myocardium from the endocardium. The nascent trabeculae continue to expand, forming long, thin projections, and ventricular endocardial cells invaginate between the myocardial projections. By E14.5, the cardiac ECM jelly is degraded, and trabeculae collapse and thicken the compact layer of myocardium. Abnormal trabeculation results in cardiomyopathies (Jenni et al., 1999). The regression of the ECM of the cardiac jelly requires a timely induction of ADAMTS1, whose expression is repressed by the Brg1-associated-factor (BAF) complex, a chromatin remodeling complex, in endocardium (Stankunas et al., 2008). How the repressive activity of the BAF complex is removed is not known, but versican degradation was detected by a 70 kDa versican fragment with DPEAAE neoepitope, which coincided with endocardial cells undergoing epithelial-mesenchymal transformation (EMT) and colocalization of ADAMTS1 and its cofactor fibrillin-1 (Kern et al., 2006).

Versican degradation is also important in cardiac valve development. In this process, the versican-rich matrix of endocardial cushions needs to be replaced by a stratified ECM including collagens and elastin. ADAMTS5 was found expressed predominantly in valve endocardium during cardiac development, and *Adamts5*$^{-/-}$ mice exhibited enlarged valves due to a reduced versican cleavage (Dupuis et al., 2011). *In vivo* reduction of versican in *Adamts5*$^{-/-}$ mice, which was achieved by versican heterozygocity, substantially reduced the valve anomalies. The lack of versican cleavage during the fetal valve development is considered to be a potential cause of myxomatous valve disease (Gupta et al., 2009).

Palatogenesis, closure of the secondary palate, segregates the oral and nasal cavities, which permits compartmentalization of digestive and respiratory processes. This developmental process involves interaction between pharyngeal ectoderm (palatal epithelium) and mesenchyme derived from craniofacial neural crest cells (Dudas et al., 2007). The study of Enomoto et al. (2010) demonstrated that ADAMTS9 expressed in craniofacial mesoderm and microvascular endothelial cells and ADAMTS20 expressed in mesenchyme cooperatively participate in this process by cleaving versican. It is proposed that the two ADAMTSs, belonging to the same subfamily with GON1 in *C. elegans*, are not required solely for versican clearance, but that they may generate versican fragments that influence palate mesenchymal proliferation required for palate closure.

McCulloch et al. (2009) used genetic approaches to investigate the role of ADAMTSs in regression of the interdigital web during mouse limb morphogenesis and found that ADAMTS5, ADAMTS9, and ADAMTS20 cooperatively participate in this process by digesting versican. Expression of *Adamts5, Adamts9, Adamts20,* and *Versican* mRNAs is temporally and spatially overlapped with the ADAMTS-cleaved versican fragments in the regressing interdigital web that fibulin-1, which binds to the C-terminal G3 domain of versican, is also colocalized and facilitates versican degradation by ADAMTS5. The recombinant versican fragment similar to that generated by ADAMTS proteolysis induces apoptosis in interdigital webs of *Adamts5*$^{-/-}$/*Adamts20* mutation (bt/bt) mice. Those data highlight that action of these ADAMTSs not only clears versican but also generates bioactive versican fragments.

4.5.11 Wound healing

A skin excision healing model showed that ADAMTS5-null mice exhibit impaired wound healing (Velasco et al., 2011). This is caused by the injury-induced aggrecan synthesis and the accumulation of aggrecan in the pericellular matrix around fibroblast progenitor cells in the skin due to ADAMTS5 deficiency. Alteration of transforming growth factor-β 1 (TGF-β1) signaling under such conditions has been investigated, and it is proposed that the removal of pericellular aggrecan by ADAMTS5 promotes the binding of the TGF-β1/TGF-β receptor II (TGF-βRII) complex to activin receptor-like kinase 5 (ALK5), which activates Smad2/3, leading to a fibroblastic/contractile phenotype. In the absence of ADAMTS5, aggrecan is retained and accumulated on CD44-bound hyaluronan. This creates the ECM environment in which the TGF-β1/TGF-βRII complex associates with ALK1, leading to Smad1/5/8 phosphorylation. This promotes further commitment of the progenitor cells to an epithelioid/aggregating phenotype. Aggrecan degradation is required for effective postwound dermal restoration, but the effect of ADAMTS5 on versican turnover is not central (Velasco et al., 2011).

In contrast, Hattori et al. (2011) reported that pericellular versican regulates the fibroblast-myofibroblast transition in which ADAMTS5 plays a role in controlling versican proteolysis. Dermal fibroblasts from *Adamts5*$^{-/-}$ mice exhibited reduced versican proteolysis, increased pericellular matrix, altered cell shaped, enhanced α-smooth muscle actin, and increased contractility within three-dimensional collagen gels. This myofibroblast-like phenotype was associated with TGF-β signaling. Fibroblasts from versican haploinsufficient mice had reduced contractility relative to wild-type fibroblasts, and addition of exogenous ADAMTS5 to *Adamts5*$^{-/-}$ fibroblasts restored normal fibroblast contractility.

ADAMTS1 has been implicated in dermal wound healing. Krampert et al. (2005) reported that ADAMTS1 was upregulated in the skin excision healing model and it was more prominent in healing-impaired genetically diabetic mice. They showed that, in early wound, the source of ADAMTS1 was macrophages and at a later stage keratinocytes and fibroblasts were the source. The role of ADAMTS1 in keratinocyte differentiation was suggested since *Adamts1*-null mice exhibited abnormal dermis with hyperthickening and parakeratosis, while ADAMTS1 expression was distributed in normal and wounded epidermis in the wild-type mice. It is postulated that ADAMTS1 cleaves proteoglycans, which release growth factors, but this is altered in *Adamts1*-null mice. ADAMTS1 also exhibited proteolytic-activity-dependent promigratory effects on fibroblasts and endothelial cells, but it was at a low concentration (1–2 nM). At a higher concentration (4–10 nM), this activity was blocked as ADAMTS1 binds to FGF2 and VEGF. Such antiangiogenic effects may be relevant to impaired wound healing in diabetic mice, which overexpress ADAMTS1.

4.5.12 Ovulation

The ablation of the *Adamts1* gene in mice has indicated that ADAMTS1 is essential for normal growth, structure, and function of the kidney, adrenal gland, and female reproductive organs (Shindo et al., 2000). About 45% of newborn null mice die most likely due to kidney and/or heart malfunction; surviving female mice are subfertile, whereas male reproduction is normal (Mittaz et al., 2004). Ovulation in *Adamts1*-null females was impaired because the mature oocytes remained trapped in ovarian follicles.

Ovulation is a complex luteinizing hormone-induced process, and it requires the proper formation of an extracellular hyaluronan-rich matrix along with hyaluronan-binding proteins and versican proteoglycans by the cumulus oocyte complex (Richards, 2005). Adult ovaries do not express ADAMTS1;however, it is induced after administration of chorionic gonadotropin, and this expression is dependent on normal synthesis of progesterone as *Adamts1* transcripts were barely detected after treatment in mice lacking progesterone receptor (Robker et al., 2000). The DPEAAE neoepitope of versican was detected in ovulated cumulus oocyte complex (Russell et al., 2003), indicating that ADAMTS1 plays a key role in this process by cleaving versican.

4.5.13 Future prospects

When the first ADAMTS, ADAMTS1, was discovered in 1997, the substrate for this enzyme was not identified. Now it is clear that the major substrate of ADAMTS1 is versican and that this cleavage is associated with many biological events, such as heart morphogenesis, wound healing, and ovulation; however, its substrate specificity is restricted. Other ADAMTSs also have highly selective substrate specificities – for example, ADAMTS2, 3, and 14 for procollagens; ADAMTS13 for multimeric VWF; and ADAMTS4, 5, 9, and 20 for lecticans. Nonetheless, these specialized activities are involved in many important biological and pathological processes. Highly specialized activities of ADAMTSs have arisen from each enzyme's unique composition of ancillary domains. The metalloproteinase domains alone have little ability to cleave peptide bonds. This is in contrast to the MMPs whose metalloproteinase domains are proteolytically active, although the ancillary domains play important roles in enhancing substrate specificity in some MMPs (Nagase et al., 2006). The crystal structures of the metalloproteinase domains of three ADAMTSs and those of ADAMTS13 noncatalytic ancillary domains have provided important insights into our understanding of the structure and function of ADAMTSs, but we still need to wait for further investigation into how various metalloproteinase domains and their ancillary domains cooperate to cleave selected substrates such as aggrecan and versican and N-propeptides of procollagens. Such information will be vital to design highly selective inhibitors for a specific ADAMTS. However, one of the challenges is to express a sufficient amount of recombinant ADAMTSs, as many of them are difficult to express in a large quantity, particularly in full-length form.

Genetic approaches have been fruitful to elucidate the function of a number of ADAMTSs, and they have shown that several ADAMTSs play an important role in organ morphogenesis and wound healing. The primary substrates in these cases are versican and aggrecan, which are expressed in many tissues during development. It is clear that ADAMTSs can alter the pericellular environment by cleaving those ECM components, and in some cases, the proteolytic fragments generated express new biological effects. This is an important area that needs further investigation as the combination of different ADAMTSs and their substrates and fragments and receptors may work together intricately. Such investigations will be necessary to understand the functions of these metalloproteinases under biological and pathological situations. How does ADAMTS activity relate to other types of tissue proteinases under such conditions? Timing of proteolysis will be key as only a short window of proteolytic activity may be crucial, particularly in organ morphogenesis during development.

4.5.14 Take-home message

The ADAMTS metalloproteinases are multidomain proteins with highly selective protein substrate specificities, which are found in pericellular and extracellular matrix. This derived from the cooperation between the metalloproteinase domain with the active site and a number of noncatalytic ancillary domains. The enzymes play important roles in organ morphogenesis during development, would healing, and ovulation, and they participate in the progression of diseases such as arthritis and atherosclerosis. Some ADAMTSs have antiangiogenic and antitumorigenic activities.

Acknowledgments

The author thanks Rob Visse, Ngee Han Lim, and Alan Lyons for preparation of figures, and Gill Murphy for critical reading of the manuscript. This work was in part supported by the NIH/NIAMS grant AR40994 and the Arthritis Research UK core grant to the Kennedy Institute of Rheumatology.

References

Abbaszade, I., Liu, R. Q., Yang, F., et al. (1999). Cloning and characterization of ADAMTS11, an aggrecanase from the ADAMTS family. J Biol Chem *274*, 23443–23450.

Akiyama, M., Takeda, S., Kokame, K., Takagi, J., and Miyata, T. (2009). Crystal structures of the noncatalytic domains of ADAMTS13 reveal multiple discontinuous exosites for von Willebrand factor. Proc Natl Acad Sci U S A *106*, 19274–19279.

Apte, S. S. (2004). A disintegrin-like and metalloprotease (reprolysin type) with thrombospondin type 1 motifs: the ADAMTS family. Int J Biochem Cell Biol *36*, 981–985.

Apte, S. S. (2009). A disintegrin-like and metalloprotease (reprolysin-type) with thrombospondin type 1 motif (ADAMTS) superfamily: functions and mechanisms. J Biol Chem *284*, 31493–31497.

Bagavandoss, P., and Wilks, J. W. (1990). Specific inhibition of endothelial cell proliferation by thrombospondin. Biochem Biophys Res Commun *170*, 867–872.

Bau, B., Gebhard, P. M., Haag, J., Knorr, T., Bartnik, E., and Aigner, T. (2002). Relative messenger RNA expression profiling of collagenases and aggrecanases in human articular chondrocytes in vivo and in vitro. Arthritis Rheum *46*, 2648–2657.

Baumann, U. (1994). Crystal structure of the 50 kDa metallo protease from Serratia marcescens. J Mol Biol *242*, 244–251.

Becker, J. W., Marcy, A. I., Rokosz, L. L., et al. (1995). Stromelysin-1: three-dimensional structure of the inhibited catalytic domain and of the C-truncated proenzyme. Protein Sci *4*, 1966–1976.

Black, R. A., Rauch, C. T., Kozlosky, C. J., et al. (1997). A metalloproteinase disintegrin that releases tumour-necrosis factor-alpha from cells. Nature *385*, 729–733.

Blelloch, R., and Kimble, J. (1999). Control of organ shape by a secreted metalloprotease in the nematode Caenorhabditis elegans. Nature *399*, 586–590.

Blobel, C. P. (2005). ADAMs: key components in EGFR signalling and development. Nat Rev Mol Cell Biol *6*, 32–43.

Bode, W., Gomis-Rüth, F. X., and Stöckler, W. (1993). Astacins, serralysins, snake venom and matrix metalloproteinases exhibit identical zinc-binding environments (HEXXHXXGXXH and Met-turn) and topologies and should be grouped into a common family, the "metzincins." FEBS Lett *331*, 134–140.

Bode, W., Reinemer, P., Huber, R., Kleine, T., Schnierer, S., and Tschesche, H. (1994). The X-ray crystal structure of the catalytic domain of human neutrophil collagenase inhibited by a substrate analogue reveals the essentials for catalysis and specificity. EMBO J *13*, 1263–1269.

Bolz, H., Ramirez, A., von Brederlow, B., and Kubisch, C. (2001). Characterization of ADAMTS14, a novel member of the ADAMTS metalloproteinase family. Biochim Biophys Acta *1522*, 221–225.

Bongrazio, M., Baumann, C., Zakrzewicz, A., Pries, A. R., and Gaehtgens, P. (2000). Evidence for modulation of genes involved in vascular adaptation by prolonged exposure of endothelial cells to shear stress. Cardiovasc Res *47*, 384–393.

Brew, K., and Nagase, H. (2010). The tissue inhibitors of metalloproteinases (TIMPs): an ancient family with structural and functional diversity. Biochim Biophys Acta *1803*, 55–71.

Brinckerhoff, C. E., and Matrisian, L. M. (2002). Matrix metalloproteinases: a tail of a frog that became a prince. Nat Rev Mol Cell Biol *3*, 207–214.

Capehart, A. A. (2010). Proteolytic cleavage of versican during limb joint development. Anat Rec *293*, 208–214.

Colige, A., Li, S. W., Sieron, A. L., Nusgens, B. V., Prockop, D. J., and Lapiere, C. M. (1997). cDNA cloning and expression of bovine procollagen I N-proteinase: a new member of the superfamily of zinc-metalloproteinases with binding sites for cells and other matrix components. Proc Natl Acad Sci U S A *94*, 2374–2379.

Colige, A., Sieron, A. L., Li, S. W., et al. (1999). Human Ehlers-Danlos syndrome type VII C and bovine dermatosparaxis are caused by mutations in the procollagen I N-proteinase gene. Am J Hum Genet *65*, 308–317.

Collins-Racie, L. A., Flannery, C. R., Zeng, W., et al. (2004). ADAMTS-8 exhibits aggrecanase activity and is expressed in human articular cartilage. Matrix Biol *23*, 219–230.

Demircan, K., Hirohata, S., Nishida, K., et al. (2005). ADAMTS-9 is synergistically induced by interleukin-1β and tumor necrosis factor α in OUMS-27 chondrosarcoma cells and in human chondrocytes. Arthritis Rheum *52*, 1451–1460.

Dubail, J., Kesteloot, F., Deroanne, C., et al. (2010). ADAMTS-2 functions as anti-angiogenic and anti-tumoral molecule independently of its catalytic activity. Cell Mol Life Sci *67*, 4213–4232.

Dudas, M., Li, W. Y., Kim, J., Yang, A., and Kaartinen, V. (2007). Palatal fusion – where do the midline cells go? A review on cleft palate, a major human birth defect. Acta Histochem *109*, 1–14.

Dupuis, L. E., McCulloch, D. R., McGarity, J. D., et al. (2011). Altered versican cleavage in ADAMTS5 deficient mice; a novel etiology of myxomatous valve disease. Dev Biol *357*, 152–164.

Durigova, M., Soucy, P., Fushimi, K., Nagase, H., Mort, J. S., and Roughley, P. J. (2008). Characterization of an ADAMTS-5-mediated cleavage site in aggrecan in OSM-stimulated bovine cartilage. Osteoarthritis Cartilage *16*, 1245–1252.

Edwards, D. R., Handsley, M. M., and Pennington, C. J. (2008). The ADAM metalloproteinases. Mol Aspects Med *29*, 258–289.

Enomoto, H., Nelson, C. M., Somerville, R. P., et al. (2010). Cooperation of two ADAMTS metalloproteases in closure of the mouse palate identifies a requirement for versican proteolysis in regulating palatal mesenchyme proliferation. Development *137*, 4029–4038.

Fernandes, R. J., Hirohata, S., Engle, J. M., et al. (2001). Procollagen II amino propeptide processing by ADAMTS-3. Insights on dermatosparaxis. *J* Biol Chem *276*, 31502–31509.

Fernando, T., Flibotte, S., Xiong, S., et al. (2011). C. elegans ADAMTS ADT-2 regulates body size by modulating TGFbeta signaling and cuticle collagen organization. Dev Biol *352*, 92–103.

Flannery, C. R., Little, C. B., Hughes, C. E., and Caterson, B. (1999). Expression of ADAMTS homologues in articular cartilage. Biochem Biophys Res Commun *260*, 318–322.

Fogal, V., Sugahara, K. N., Ruoslahti, E., and Christian, S. (2009). Cell surface nucleolin antagonist causes endothelial cell apoptosis and normalization of tumor vasculature. Angiogenesis *12*, 91–100.

Fosang, A. J., Rogerson, F. M., East, C. J., and Stanton, H. (2008). ADAMTS-5: the story so far. Eur Cell Mater *15*, 11–26.

Fujikawa, K., Suzuki, H., McMullen, B., and Chung, D. (2001). Purification of human von Willebrand factor-cleaving protease and its identification as a new member of the metalloproteinase family. Blood *98*, 1662–1666.

Fushimi, K., Troeberg, L., Nakamura, H., Lim, N. H., and Nagase, H. (2008). Functional differences of the catalytic and non-catalytic domains in human ADAMTS-4 and ADAMTS-5 in aggrecanolytic activity. J Biol Chem *283*, 6706–6716.

Gao, G., Plaas, A., Thompson, V. P., Jin, S., Zuo, F., and Sandy, J. D. (2004). ADAMTS4 (aggrecanase-1) activation on the cell surface involves C-terminal cleavage by glycosylphosphatidyl inositol-anchored membrane type 4-matrix metalloproteinase and binding of the activated proteinase to chondroitin sulfate and heparan sulfate on syndecan-1. J Biol Chem *279*, 10042–10051.

Gao, W., Anderson, P. J., Majerus, E. M., Tuley, E. A., and Sadler, J. E. (2006). Exosite interactions contribute to tension-induced cleavage of von Willebrand factor by the antithrombotic ADAMTS13 metalloprotease. Proc Natl Acad Sci U S A *103*, 19099–19104.

Gao, W., Anderson, P. J., and Sadler, J. E. (2008). Extensive contacts between ADAMTS13 exosites and von Willebrand factor domain A2 contribute to substrate specificity. Blood *112*, 1713–1719.

Gendron, C., Kashiwagi, M., Hughes, C., Caterson, B., and Nagase, H. (2003). TIMP-3 inhibits aggrecanase-mediated glycosaminoglycan release from cartilage explants stimulated by catabolic factors. FEBS Lett *555*, 431–436.

Gendron, C., Kashiwagi, M., Lim, N. H., et al. (2007). Proteolytic activities of human ADAMTS-5: comparative studies with ADAMTS-4. J Biol Chem *282*, 18294–18306.

Gerhardt, S., Hassall, G., Hawtin, P., et al. (2007). Crystal structures of human ADAMTS-1 reveal a conserved catalytic domain and a disintegrin-like domain with a fold homologous to cysteine-rich domains. J Mol Biol *373*, 891–902.

Gerritsen, H. E., Robles, R., Lammle, B., and Furlan, M. (2001). Partial amino acid sequence of purified von Willebrand factor-cleaving protease. Blood *98*, 1654–1661.

Ghosh, P. (1999). The pathobiology of osteoarthritis and the rationale for the use of pentosan polysulfate for its treatment. Semin Arthritis Rheum *28*, 211–267.

Glasson, S. S., Askew, R., Sheppard, B., et al. (2004). Characterization of and osteoarthritis susceptibility in ADAMTS-4-knockout mice. Arthritis Rheum *50*, 2547–2558.

Glasson, S. S., Askew, R., Sheppard, B., et al. (2005). Deletion of active ADAMTS5 prevents cartilage degradation in a murine model of osteoarthritis. Nature *434*, 644–648.

Gomis-Rüth, F. X. (2003). Structural aspects of the metzincin clan of metalloendopeptidases. Mol Biotechnol *24*, 157–202.

Gomis-Rüth, F. X., Kress, L. F., and Bode, W. (1993). First structure of a snake venom metalloproteinase: a prototype for matrix metalloproteinases/collagenases. EMBO J *12*, 4151–4157.

Good, D. J., Polverini, P. J., Rastinejad, F., et al. (1990). A tumor suppressor-dependent inhibitor of angiogenesis is immunologically and functionally indistinguishable from a fragment of thrombospondin. Proc Natl Acad Sci U S A *87*, 6624–6628.

Gross, J., and Lapiere, C. M. (1962). Collagenolytic activity in amphibian tissues: a tissue culture assay. Proc Natl Acad Sci U S A *48*, 1014–1022.

Guo, F., Lai, Y., Tian, Q., Lin, E. A., Kong, L., and Liu, C. (2010). Granulin-epithelin precursor binds directly to ADAMTS-7 and ADAMTS-12 and inhibits their degradation of cartilage oligomeric matrix protein. Arthritis Rheum *62*, 2023–2036.

Gupta, V., Barzilla, J. E., Mendez, J. S., et al. (2009). Abundance and location of proteoglycans and hyaluronan within normal and myxomatous mitral valves. Cardiovasc Pathol *18*, 191–197.

Hattori, N., Carrino, D. A., Lauer, M. E., et al. (2011). Pericellular versican regulates the fibroblast-myofibroblast transition: a role for ADAMTS5 protease-mediated proteolysis. J Biol Chem *286*, 34298–34310.

Hulmes, D. J., Kadler, K. E., Mould, A. P., et al. (1989). Pleomorphism in type I collagen fibrils produced by persistence of the procollagen N-propeptide. J Mol Biol *210*, 337–345.

Hurskainen, T. L., Hirohata, S., Seldin, M. F., and Apte, S. S. (1999). ADAM-TS5, ADAM-TS6, and ADAM-TS7, novel members of a new family of zinc metalloproteases. General features and genomic distribution of the ADAM-TS family. J Biol Chem *274*, 25555–25563.

Huxley-Jones, J., Apte, S. S., Robertson, D. L., and Boot-Handford, R. P. (2005). The characterisation of six ADAMTS proteases in the basal chordate Ciona intestinalis provides new insights into the vertebrate ADAMTS family. Int J Biochem Cell Biol *37*, 1838–1845.

Ihara, S., and Nishiwaki, K. (2008). Stage-specific activation of MIG-17/ADAMTS controls cell migration in Caenorhabditis elegans. FEBS J *275*, 4296–4305.

Ilic, M. Z., Handley, C. J., Robinson, H. C., and Mok, M. T. (1992). Mechanism of catabolism of aggrecan by articular cartilage. Arch Biochem Biophys *294*, 115–122.

Iruela-Arispe, M. L., Bornstein, P., and Sage, H. (1991). Thrombospondin exerts an antiangiogenic effect on cord formation by endothelial cells in vitro. Proc Natl Acad Sci U S A *88*, 5026–5030.

Janes, P. W., Saha, N., Barton, W. A., et al. (2005). Adam meets Eph: an ADAM substrate recognition module acts as a molecular switch for ephrin cleavage in trans. Cell *123*, 291–304.

Jenni, R., Rojas, J., and Oechslin, E. (1999). Isolated noncompaction of the myocardium. N Engl J Med *340*, 966–967.

Jönsson-Rylander, A. C., Nilsson, T., Fritsche-Danielson, R., et al. (2005). Role of ADAMTS-1 in atherosclerosis: remodeling of carotid artery, immunohistochemistry, and proteolysis of versican. Arterioscler Thromb Vasc Biol *25*, 180–185.

Kashiwagi, M., Enghild, J. J., Gendron, C., et al. (2004). Altered proteolytic activities of ADAMTS-4 expressed by C-terminal processing. J Biol Chem *279*, 10109–10119.

Kenagy, R. D., Min, S. K., Clowes, A. W., and Sandy, J. D. (2009). Cell death-associated ADAMTS4 and versican degradation in vascular tissue. J Histochem Cytochem *57*, 889–897.

Kern, C. B., Twal, W. O., Mjaatvedt, C. H., et al. (2006). Proteolytic cleavage of versican during cardiac cushion morphogenesis. Dev Dyn *235*, 2238–2247.

Kevorkian, L., Young, D. A., Darrah, C., et al. (2004). Expression profiling of metalloproteinases and their inhibitors in cartilage. Arthritis Rheum *50*, 131–141.

Koo, B. H., Coe, D. M., Dixon, L. J., et al. (2010). ADAMTS9 is a cell-autonomously acting, anti-angiogenic metalloprotease expressed by microvascular endothelial cells. Am J Pathol *176*, 1494–1504.

Koo, B. H., Longpré, J. M., Somerville, R. P., Alexander, J. P., Leduc, R., and Apte, S. S. (2006). Cell-surface processing of pro-ADAMTS9 by furin. J Biol Chem *281*, 12485–12494.

Koo, B. H., Longpré, J. M., Somerville, R. P., Alexander, J. P., Leduc, R., and Apte, S. S. (2007). Regulation of ADAMTS9 secretion and enzymatic activity by its propeptide. J Biol Chem *282*, 16146–16154.

Krampert, M., Kuenzle, S., Thai, S. N., Lee, N., Iruela-Arispe, M. L., and Werner, S. (2005). ADAMTS1 proteinase is up-regulated in wounded skin and regulates migration of fibroblasts and endothelial cells. J Biol Chem *280*, 23844–23852.

Kuno, K., Baba, C., Asaka, A., Matsushima, C., Matsushima, K., and Hosono, R. (2002). The Caenorhabditis elegans ADAMTS family gene adt-1 is necessary for morphogenesis of the male copulatory organs. J Biol Chem *277*, 12228–12236.

Kuno, K., Kanada, N., Nakashima, E., Fujiki, F., Ichimura, F., and Matsushima, K. (1997). Molecular cloning of a gene encoding a new type of metalloproteinase-disintegrin family protein with thrombospondin motifs as an inflammation associated gene. J Biol Chem *272*, 556–562.

Lark, M. W., Bayne, E. K., Flanagan, J., et al. (1997). Aggrecan degradation in human cartilage – evidence for both matrix metalloproteinase and aggrecanase activity in normal, osteoarthritic, and rheumatoid joints. J Clin Invest *100*, 93–106.

Lee, N. V., Sato, M., Annis, D. S., et al. (2006). ADAMTS1 mediates the release of antiangiogenic polypeptides from TSP1 and 2. EMBO J *25*, 5270–5283.

Levy, G. G., Nichols, W. C., Lian, E. C., et al. (2001). Mutations in a member of the ADAMTS gene family cause thrombotic thrombocytopenic purpura. Nature *413*, 488–494.

Li, S. W., Arita, M., Fertala, A., et al. (2001). Transgenic mice with inactive alleles for procollagen N-proteinase (ADAMTS-2) develop fragile skin and male sterility. Biochem J *355*, 271–278.

Li, Z., Nard, M. A., Li, Y. S., et al. (2009). C-terminal ADAMTS-18 fragment induces oxidative platelet fragmentation, dissolves platelet aggregates, and protects against carotid artery occusion and cerebral stroke. Blood *113*, 6051–6060.

Liu, C. J., Kong, W., Ilalov, K., et al. (2006a). ADAMTS-7: a metalloproteinase that directly binds to and degrades cartilage oligomeric matrix protein. FASEB J *20*, 988–990.

Liu, C. J., Kong, W., Xu, K., et al. (2006b). ADAMTS-12 associates with and degrades cartilage oligomeric matrix protein. J Biol Chem *281*, 15800–15808.

Llamazares, M., Obaya, A. J., Moncada-Pazos, A., et al. (2007). The ADAMTS12 metalloproteinase exhibits anti-tumorigenic properties through modulation of the Ras-dependent ERK signalling pathway. J Cell Sci *120*, 3544–3552.

Lo, P. H., Lung, H. L., Cheung, A. K., et al. (2010). Extracellular protease ADAMTS9 suppresses esophageal and nasopharyngeal carcinoma tumor formation by inhibiting angiogenesis. Cancer Res *70*, 5567–5576.

Lohmander, L. S., Neame, P. J., and Sandy, J. D. (1993). The structure of aggrecan fragments in human synovial fluid. Evidence that aggrecanase mediates cartilage degradation in inflammatory joint disease, joint injury, and osteoarthritis. Arthritis Rheum *36*, 1214–1222.

Longpré, J. M., and Leduc, R. (2004). Identification of prodomain determinants involved in ADAMTS-1 biosynthesis. J Biol Chem *279*, 33237–33245.

Longpré, J. M., McCulloch, D. R., Koo, B. H., Alexander, J. P., Apte, S. S., and Leduc, R. (2009). Characterization of proADAMTS5 processing by proprotein convertases. Int J Biochem Cell Biol *41*, 1116–1126.

Lotta, L. A., Garagiola, I., Palla, R., Cairo, A., and Peyvandi, F. (2010). ADAMTS13 mutations and polymorphisms in congenital thrombotic thrombocytopenic purpura. Hum Mutat *31*, 11–19.

Loulakis, P., Shrikhande, A., Davis, G., and Maniglia, C. A. (1992). N-terminal sequence of proteoglycan fragments isolated from medium of interleukin-1-treated articular-cartilage cultures. Putative site(s) of enzymic cleavage. Biochem J *284*, 589–593.

Lovejoy, B., Cleasby, A., Hassell, A. M., et al. (1994). Structure of the catalytic domain of fibroblast collagenase complexed with an inhibitor. Science *263*, 375–377.

Luken, B. M., Turenhout, E. A., Kaijen, P. H., et al. (2006). Amino acid regions 572–579 and 657–666 of the spacer domain of ADAMTS13 provide a common antigenic core required for binding of antibodies in patients with acquired TTP. Thromb Haemost *96*, 295–301.

Luque, A., Carpizo, D. R., and Iruela-Arispe, M. L. (2003). ADAMTS1/METH1 inhibits endothelial cell proliferation by direct binding and sequestration of VEGF165. J Biol Chem *278*, 23656–23665.

Majerus, E. M., Anderson, P. J., and Sadler, J. E. (2005). Binding of ADAMTS13 to von Willebrand factor. J Biol Chem *280*, 21773–21778.

McCulloch, D. R., Le Goff, C., Bhatt, S., Dixon, L. J., Sandy, J. D., and Apte, S. S. (2009). Adamts5, the gene encoding a proteoglycan-degrading metalloprotease, is expressed by specific cell lineages during mouse embryonic development and in adult tissues. Gene Expr Patterns *9*, 314–323.

Mittaz, L., Russell, D. L., Wilson, T., et al. (2004). Adamts-1 is essential for the development and function of the urogenital system. Biol Reprod *70*, 1096–1105.

Moake, J. L., Rudy, C. K., Troll, J. H., et al. (1982). Unusually large plasma factor VIII: von Willebrand factor multimers in chronic relapsing thrombotic thrombocytopenic purpura. N Engl J Med *307*, 1432–1435.

Moss, M. L., Jin, S. L. C., Milla, M. E., et al. (1997). Cloning of a disintegrin metalloproteinase that processes precursor tumour-necrosis factor-alpha. Nature *385*, 733–736.

Mosyak, L., Georgiadis, K., Shane, T., et al. (2008). Crystal structures of the two major aggrecan degrading enzymes, ADAMTS4 and ADAMTS5. Protein Sci *17*, 16–21.

Murphy, G., and Nagase, H. (2011). Localizing matrix metalloproteinase activities in the pericellular environment. FEBS J *278*, 2–15.

Nagase, H., and Kashiwagi, M. (2003). Aggrecanases and cartilage matrix degradation. Arthritis Res Ther *5*, 94–103.

Nagase, H., Visse, R., and Murphy, G. (2006). Structure and function of matrix metalloproteinases and TIMPs. Cardiovasc Res *69*, 562–573.

Nakamura, M., Sone, S., Takahashi, I., Mizoguchi, I., Echigo, S., and Sasano, Y. (2005). Expression of versican and ADAMTS1, 4, and 5 during bone development in the rat mandible and hind limb. J Histochem Cytochem *53*, 1553–1562.

Nardi, M., Feinmark, S. J., Hu, L., Li, Z., and Karpatkin, S. (2004). Complement-independent Ab-induced peroxide lysis of platelets requires 12-lipoxygenase and a platelet NADPH oxidase pathway. J Clin Invest *113*, 973–980.

Nardi, M., Tomlinson, S., Greco, M. A., and Karpatkin, S. (2001). Complement-independent, peroxide-induced antibody lysis of platelets in HIV-1-related immune thrombocytopenia. Cell *106*, 551–561.

Newby, A. C. (2005). Dual role of matrix metalloproteinases (matrixins) in intimal thickening and atherosclerotic plaque rupture. Physiol Rev *85*, 1–31.

Nishiwaki, K., Hisamoto, N., and Matsumoto, K. (2000). A metalloprotease disintegrin that controls cell migration in Caenorhabditis elegans. Science *288*, 2205–2208.

Nusgens, B. V., Verellen-Dumoulin, C., Hermanns-Le, T., et al. (1992). Evidence for a relationship between Ehlers-Danlos type VII C in humans and bovine dermatosparaxis. Nat Genet *1*, 214–217.

Page-McCaw, A., Ewald, A. J., and Werb, Z. (2007). Matrix metalloproteinases and the regulation of tissue remodelling. Nat Rev Mol Cell Biol *8*, 221–233.

Porter, S., Clark, I. M., Kevorkian, L., and Edwards, D. R. (2005). The ADAMTS metalloproteinases. Biochem J *386*, 15–27.

Richards, J. S. (2005). Ovulation: new factors that prepare the oocyte for fertilization. Mol Cell Endocrinol *234*, 75–79.

Robker, R. L., Russell, D. L., Espey, L. L., Lydon, J. P., O'Malley, B. W., and Richards, J. S. (2000). Progesterone-regulated genes in the ovulation process: ADAMTS-1 and cathepsin L proteases. Proc Natl Acad Sci U S A *97*, 4689–4694.

Rodriguez-Manzaneque, J. C., Westling, J., Thai, S. N., et al. (2002). ADAMTS1 cleaves aggrecan at multiple sites and is differentially inhibited by metalloproteinase inhibitors. Biochem Biophys Res Commun *293*, 501–508.

Ruggeri, Z. M., and Zimmerman, T. S. (1981). The complex multimeric composition of factor VIII/von Willebrand factor. Blood *57*, 1140–1143.

Russell, D. L., Doyle, K. M., Ochsner, S. A., Sandy, J. D., and Richards, J. S. (2003). Processing and localization of ADAMTS-1 and proteolytic cleavage of versican during cumulus matrix expansion and ovulation. J Biol Chem *278*, 42330–42339.

Sadler, J. E. (1998). Biochemistry and genetics of von Willebrand factor. Annu Rev Biochem *67*, 395–424.

Sandy, J. D. (2006). A contentious issue finds some clarity: on the independent and complementary roles of aggrecanase activity and MMP activity in human joint aggrecanolysis. Osteoarthritis Cartilage *14*, 95–100.

Sandy, J. D., Flannery, C. R., Neame, P. J., and Lohmander, L. S. (1992). The structure of aggrecan fragments in human synovial fluid. Evidence for the involvement in osteoarthritis of a novel proteinase which cleaves the Glu 373-Ala 374 bond of the interglobular domain. J Clin Invest *89*, 1512–1516.

Sandy, J. D., Neame, P. J., Boynton, R. E., and Flannery, C. R. (1991). Catabolism of aggrecan in cartilage explants. Identification of a major cleavage site within the interglobular domain. J Biol Chem *266*, 8683–8685.

Shieh, H. S., Mathis, K. J., Williams, J. M., et al. (2008). High resolution crystal structure of the catalytic domain of ADAMTS-5 (aggrecanase-2). J Biol Chem *283*, 1501–1507.

Shindo, T., Kurihara, H., Kuno, K., et al. (2000). ADAMTS-1: a metalloproteinase-disintegrin essential for normal growth, fertility, and organ morphology and function. J Clin Invest *105*, 1345–1352.

Silver, D. L., Hou, L., Somerville, R., Young, M. E., Apte, S. S., and Pavan, W. J. (2008). The secreted metalloprotease ADAMTS20 is required for melanoblast survival. PLoS Genet *4*, e1000003.

Soejima, K., Matsumoto, M., Kokame, K., et al. (2003). ADAMTS-13 cysteine-rich/spacer domains are functionally essential for von Willebrand factor cleavage. Blood *102*, 3232–3237.

Soejima, K., Mimura, N., Hirashima, M., et al. (2001). A novel human metalloprotease synthesized in the liver and secreted into the blood: possibly, the von Willebrand factor-cleaving protease? J Biochem *130*, 475–480.

Somerville, R. P., Longprē, J. M., Apel, E. D., et al. (2004). ADAMTS7B, the full-length product of the ADAMTS7 gene, is a chondroitin sulfate proteoglycan containing a mucin domain. J Biol Chem *279*, 35159–35175.

Somerville, R. P., Longpre, J. M., Jungers, K. A., et al. (2003). Characterization of ADAMTS-9 and ADAMTS-20 as a distinct ADAMTS subfamily related to Caenorhabditis elegans GON-1. J Biol Chem *278*, 9503–9513.

Stankunas, K., Hang, C. T., Tsun, Z. Y., et al. (2008). Endocardial Brg1 represses ADAMTS1 to maintain the microenvironment for myocardial morphogenesis. Dev Cell *14*, 298–311.

Stanton, H., Rogerson, F. M., East, C. J., et al. (2005). ADAMTS5 is the major aggrecanase in mouse cartilage in vivo and in vitro. Nature *434*, 648–652.

Struglics, A., Larsson, S., Pratta, M. A., Kumar, S., Lark, M. W., and Lohmander, L. S. (2006). Human osteoarthritis synovial fluid and joint cartilage contain both aggrecanase- and matrix metalloproteinase-generated aggrecan fragments. Osteoarthritis Cartilage *14*, 101–113.

Takeda, S. (2009). Three-dimensional domain architecture of the ADAM family proteinases. Semin Cell Dev Biol *20*, 146–152.

Takeda, S., Igarashi, T., Mori, H., and Araki, S. (2006). Crystal structures of VAP1 reveal ADAMs' MDC domain architecture and its unique C-shaped scaffold. EMBO J *25*, 2388–2396.

Takeda, S., Takeya, H., and Iwanaga, . (2012). Snake venom metalloproteinases: structure, function and relevance to the mammalian ADAM/ADAMTS family proteins. Biochim Biophys Acta *1824*, 164–176.

Tan, K., Duquette, M., Liu, J. H., et al. (2002). Crystal structure of the TSP-1 type 1 repeats: a novel layered fold and its biological implication. J Cell Biol *159*, 373–382.

Tolsma, S. S., Volpert, O. V., Good, D. J., Frazier, W. A., Polverini, P. J., and Bouck, N. (1993). Peptides derived from two separate domains of the matrix protein thrombospondin-1 have anti-angiogenic activity. J Cell Biol *122*, 497–511.

Tortorella, M. D., Burn, T. C., Pratta, M. A., et al. (1999). Purification and cloning of aggrecanase-1: a member of the ADAMTS family of proteins. Science *284*, 1664–1666.

Tortorella, M. D., Liu, R. Q., Burn, T., Newton, R. C., and Arner, E. (2002). Characterization of human aggrecanase 2 (ADAM-TS5): substrate specificity studies and comparison with aggrecanase 1 (ADAM-TS4). Matrix Biol *21*, 499–511.

Troeberg, L., Fushimi, K., Khokha, R., Emonard, H., Ghosh, P., and Nagase, H. (2008). Calcium pentosan polysulfate is a multifaceted exosite inhibitor of aggrecanases. FASEB J *22*, 3515–3524.

Troeberg, L., Mulloy, B., Ghosh, P., Lee, M. H., Murphy, G., and Nagase, H. (2012). Pentosan polysulfate increases affinity between ADAMTS-5 and TIMP-3 through formation of an electrostatically driven trimolecular complex. Biochem J *443*, 307–315.

Van Wart, H. E., and Birkedal-Hansen, H. (1990). The cysteine switch: a principle of regulation of metalloproteinase activity with potential applicability to the entire matrix metalloproteinase gene family. Proc Natl Acad Sci U S A *87*, 5578–5582.

Vazquez, F., Hastings, G., Ortega, M. A., et al. (1999). METH-1, a human ortholog of ADAMTS-1, and METH-2 are members of a new family of proteins with angio-inhibitory activity. J Biol Chem *274*, 23349–23357.

Velasco, J., Li, J., DiPietro, L., Stepp, M. A., Sandy, J. D., and Plaas, A. (2011). Adamts5 deletion blocks murine dermal repair through CD44-mediated aggrecan accumulation and modulation of transforming growth factor beta1 (TGFbeta1) signaling. J Biol Chem *286*, 26016–26027.

Visse, R., and Nagase, H. (2003). Matrix metalloproteinases and tissue inhibitors of metalloproteinases: structure, function, and biochemistry. Circ Res *92*, 827–839.

Vogel, T., Guo, N. H., Krutzsch, H. C., et al. (1993). Modulation of endothelial cell proliferation, adhesion, and motility by recombinant heparin-binding domain and synthetic peptides from the type I repeats of thrombospondin. J Cell Biochem *53*, 74–84.

Wang, L., Zheng, J., Bai, X., et al. (2009). ADAMTS-7 mediates vascular smooth muscle cell migration and neointima formation in balloon-injured rat arteries. Circ Res *104*, 688–698.

Wang, P., Tortorella, M., England, K., et al. (2004). Proprotein convertase furin interacts with and cleaves pro-ADAMTS4 (Aggrecanase-1) in the trans-Golgi network. J Biol Chem *279*, 15434–15440.

Wang, W. M., Ge, G., Lim, N. H., Nagase, H., and Greenspan, D. S. (2006). TIMP-3 inhibits the procollagen N-proteinase ADAMTS-2. Biochem J *398*, 515–519.

Wang, X., Harris, R. E., Bayston, L. J., and Ashe, H. L. (2008). Type IV collagens regulate BMP signalling in Drosophila. Nature *455*, 72–77.

Woolley, D. E., and Evanson, J. M. (1980). Collagenase in Normal and Pathological Connective Tissue. Chichester: John Wiley.

Woolley, D. E., Glanville, R. W., Crossley, M. J., and Evanson, J. M. (1975). Purification of rheumatoid synovial collagenase and its action on soluble and insoluble collagen. Eur J Biochem *54*, 611–622.

Zeng, W., Corcoran, C., Collins-Racie, L. A., LaVallie, E. R., Morris, E. A., and Flannery, C. R. (2006). Glycosaminoglycan-binding properties and aggrecanase activities of truncated ADAMTSs: comparative analyses with ADAMTS-5, -9, -16 and -18. Biochim Biophys Acta *1760*, 517–524.

Zhang, D., Botos, I., Gomis-Rüth, F. X., et al. (1994). Structural interaction of natural and synthetic inhibitors with the venom metalloproteinase, atrolysin C (form d). Proc Natl Acad Sci U S A *91*, 8447–8451.

Zhang, Q., Zhou, Y. F., Zhang, C. Z., Zhang, X., Lu, C., and Springer, T. A. (2009a). Structural specializations of A2, a force-sensing domain in the ultralarge vascular protein von Willebrand factor. Proc Natl Acad Sci U S A *106*, 9226–9231.

Zheng, X., Chung, D., Takayama, T. K., Majerus, E. M., Sadler, J. E., and Fujikawa, K. (2001). Structure of von Willebrand factor-cleaving protease (ADAMTS13), a metalloprotease involved in thrombotic thrombocytopenic purpura. J Biol Chem *276*, 41059–41063.

Zheng, X., Nishio, K., Majerus, E. M., and Sadler, J. E. (2003). Cleavage of von Willebrand factor requires the spacer domain of the metalloprotease ADAMTS13. J Biol Chem *278*, 30136–30141.

4.6 Proteinases in wound healing

Mervi Toriseva and Veli-Matti Kähäri

4.6.1 Introduction

Wound healing is a fundamental process evolved in multicellular organisms that reconstitutes structure and function of tissue after injury. Wound repair involves several cell types orchestrated by growth factors and cytokines, as well as by interactions with extracellular matrix (ECM) molecules (Gurtner et al., 2008; Shaw and Martin, 2009).

Proteolysis is an essential feature in wound healing (Toriseva and Kähäri, 2009). In addition to degrading ECM barriers in the way of migrating cells, proteinases coordinate cellular functions by regulating the availability and activity of various bioactive molecules in wound tissue. Moreover, proteinases are involved in regulating cell motility and proliferation by modifying cell-cell and cell-ECM contacts. In general, strictly controlled proteolysis is important for normal wound healing, and alterations in proteinase activity are associated with aberrant wound closure and scar formation (Sternlicht and Werb, 2001; Toriseva and Kähäri, 2009).

The principal proteinases in healing wounds include the serine proteinases plasmin and plasminogen activators (PAs), and a variety of metalloproteinases. The roles of plasmin and matrix metalloproteinases (MMPs) in wound repair have been extensively examined, but increasing data are also emerging about the roles of a disintegrin and metalloproteinases (ADAM) and a disintegrin and metalloproteinase with thrombospondin motifs (ADAMTS) proteinases in wound healing (Moali and Hulmes, 2009; Toriseva and Kähäri, 2009). Moreover, several other proteinases, such as the serine proteinase neutrophil elastase and metalloproteinases Tolloids, meprins, and pappalysins may also participate in wound healing (Moali and Hulmes, 2009). In this chapter, we discuss specifically the roles of plasmin, MMP, ADAM, and ADAMTS proteinases in cutaneous wound healing. We focus on normally healing adult human skin wounds and also present insight into aberrant cutaneous wound repair. Finally, we discuss the future prospects for therapy of aberrant wound healing by regulating proteolysis.

4.6.2 Overview of cutaneous wound repair

In general, wound healing can be divided into three major phases: (1) hemostasis and inflammation, (2) reepithelialization and granulation tissue formation, and (3) tissue remodeling (Clark, 1995). These phases are histologically and functionally distinct but overlap temporally. In general, the time required for wound closure always depends on the wound size (▶Figure 4.10). Transcriptional profiling of *in vivo* wounds in mice and humans has resulted in identification of hundreds of differentially regulated

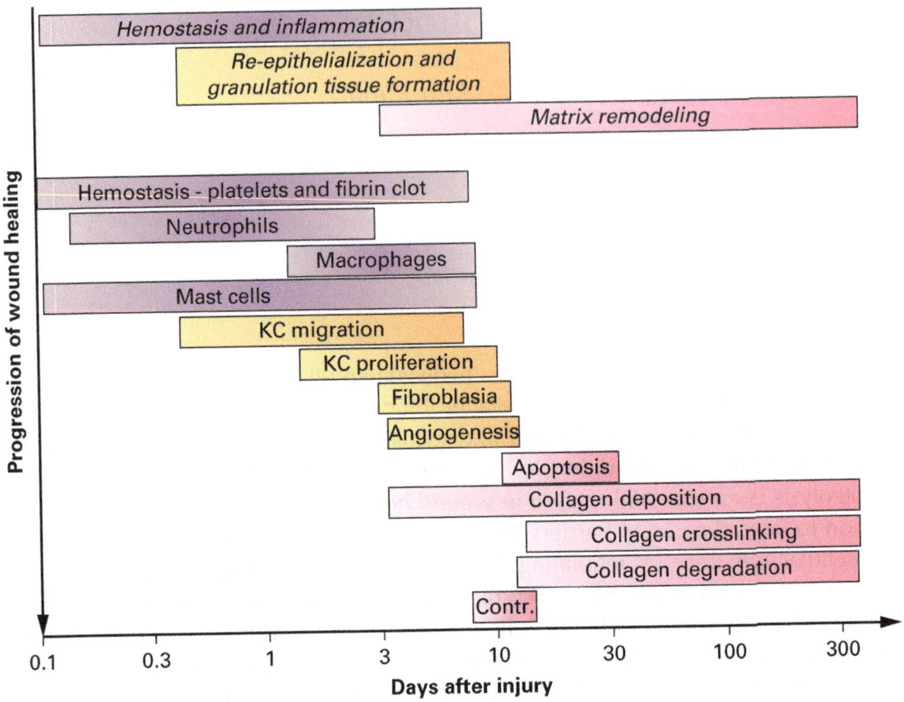

Figure 4.10 The phases and functional processes in cutaneous wound repair. Wound healing consists of three phases: (1) hemostasis and inflammation, (2) reepithelialization and granulation tissue formation, and (3) matrix remodeling. These phases are temporally overlapping, but functionally and histologically distinct. The rate of wound closure in general is dependent on the wound size. An incisional skin wound is normally closed in 2 weeks, but tissue remodeling and wound maturation continue for several months. The end result of cutaneous wound repair is typically a flat, collagenous, relatively acellular scar. KC, keratinocyte; Contr., contraction. Modified from Toriseva and Kähäri (2009).

genes at different stages of wound healing. These include genes associated in the inflammatory response, pathogen recognition, proteolysis, ECM composition, and a variety of cellular regulatory processes (Shaw and Martin, 2009).

4.6.3 Hemostasis and inflammation

Immediately following injury of skin extending into vascularized dermis, blood extravasates to the wound from disrupted blood vessels, and numerous circulating vasoactive factors are released into tissue. The inflammatory phase is initiated by vasoconstriction, and adhesion, aggregation, and degranulation of platelets. Activation of the coagulation cascade leads to the cleavage of fibrinogen by thrombin and formation of the fibrin network, which provides a scaffold for platelet adhesion and serves as a physical plug to stop bleeding, and provides chemotactic stimuli for inflammatory cells, epidermal keratinocytes, fibroblasts, and endothelial cells. As the wound healing progresses, provisional matrix composed of fibrin, plasma fibronectin, and vitronectin is gradually degraded by proteinases, especially plasmin, and finally disengaged from the wound site as a scab (Shaw and Martin, 2009).

The innate immune system is activated at the wound site in a few hours after injury. The first cells migrating to the wound are neutrophils recruited in part by the resident mast cells (Weller et al., 2006). The main function of the neutrophils is phagocytosis of infectious agents and devitalized tissue. Upon neutrophil degranulation, reactive oxygen species, proteinases, and factors further amplifying inflammation are released into the wound bed (Eming et al., 2009). Macrophages are essential for normal wound healing (Leibovich and Ross, 1975). Monocytes migrate to the wound site within 2 days of injury and become activated macrophages. Macrophages function as antigen-presenting cells and phagocytes and contribute to the regulation of wound healing by secreting numerous growth factors, such as transforming growth factor-beta (TGF-β), transforming growth factor-alpha (TGF-α), basic fibroblast growth factor (bFGF; FGF2), platelet-derived growth factor (PDGF), and vascular endothelial growth factor (VEGF) (Eming et al., 2009). In normally healing skin wounds, inflammation is resolved during the first 2 weeks due to the reduction of proinflammatory molecules and increase in anti-inflammatory factors promoting apoptosis and phagocytosis of leukocytes (Eming et al., 2009).

4.6.4 Reepithelialization

Epidermal reepithelialization and dermal granulation tissue formation are initiated during the inflammatory phase of wound healing. As early as a few hours after injury, epidermal keratinocytes at the wound edge and in the remnants of the skin appendages (e.g. hair follicles), detach from the underlying basement membrane and from neighboring cells and start to migrate into the wound, typically underneath the scab. Within 2 days, wound keratinocytes distant from the wound edge obtain hyperproliferative phenotype, fill the gap in the epithelium, and restore skin integrity (Shaw and Martin, 2009). After epidermal closure, the basement membrane is reestablished, the cellular contacts are reformed, and keratinocytes differentiate to reconstitute the multilayered epidermis of skin.

4.6.5 Granulation tissue formation

Formation of granulation tissue begins in a few days of injury by fibroplasia implying fibroblast proliferation and invasion to wound provisional matrix. Wound granulation tissue contains numerous fibroblasts and leukocytes embedded in wound ECM. Blood vessels sprouting into the wound give granulation tissue its characteristic porous appearance. Fibroplasia is promoted by platelet- and inflammatory-cell-derived growth factors such as PDGF and TGF-β, and ligation with ECM molecules (Werner and Grose, 2003; Shaw and Martin, 2009). Fibroblasts deposit most of the wound granulation tissue ECM initially consisting mainly of fibronectin and hyaluronan, which promote cell migration. At later stage, proteoglycans and type III and I collagens are deposited and become the major components of wound ECM (Clark, 1995).

In addition, granulation tissue contains matricellular proteins, such as syndecans, thrombospondins, secreted protein acidic and rich in cysteine (SPARC), tenascins, and vitronectin, which modulate adhesion and receptor function and consequently regulate cellular behavior (Midwood et al., 2004).

Angiogenesis is required to ensure supply of oxygen and nutrients to the wound tissue. In response to tissue destruction and hypoxia, endothelial cells of existing blood vessels start to proliferate and migrate into wound provisional matrix stimulated by blood clot components and soluble factors, such as bFGF, VEGF, and TGF-β (Werner and Grose, 2003; Hynes, 2007; Gurtner et al., 2008). Proteolytic activity of plasmin and MMPs is essential for wound angiogenesis (Pepper, 2001).

4.6.6 Tissue remodeling and wound maturation

The remodeling of ECM in granulation tissue is indispensable for restoration of structure and function of dermis. Granulation tissue fibroblasts deposit collagen, which remains in large quantities in the mature dermis. The remodeling of collagenous granulation tissue continues for months and involves traction force exerted by fibroblasts and collagenase-mediated collagen cleavage. This results in proper orientation and assembly of collagen fibrils in a manner that provides tensile strength for skin (Clark, 1995). TGF-β and mechanical tension generated by the open wound induce fibroblast differentiation to myofibroblasts during the second week of healing (Hinz, 2007). Myofibroblasts are able to contract wound tissue and pull wound edges together, thus contributing to wound closure. Wound maturation is also characterized by reduction in fibroblast number, elimination of myofibroblasts, and disintegration of the majority of the blood vessels via apoptosis (Desmouliere et al., 1995; Hinz, 2007). Usually, the end result of wound healing and maturation is a flat and relatively acellular collagenous scar.

4.6.7 Growth factors and cytokines regulating cutaneous wound healing

Tissue recovery during wound healing requires prompt and highly coordinated function of distinct cell types. Signals regulating cellular activities are induced by cell-cell and cell-ECM contacts, changes in cellular mechanical stress, and different growth factors and cytokines. Many of these factors are secreted in soluble form, but some are proteolytically released and activated from the cell surface or ECM (Werner and Grose, 2003). The effects of the best-characterized growth factors and cytokines involved in wound healing and their main cellular sources are listed in ▶Table 4.7. As growth factors and cytokines regulate the expression of proteinases by various cell types involved in wound healing, active proteinases regulate the bioavailability of a variety of soluble factors by shedding, releasing them from ECM, and activating them (Kessenbrock et al., 2010).

Table 4.7 The major growth factors and cytokines affecting wound repair.

Growth factor/cytokine	*Major source*	*Regulated wound-healing events*	*References*
PDGF	Platelets, macrophages, KC	Fibroblasia, leukocyte recruitment, ECM production, contraction, angiogenesis	(Werner and Grose, 2003; Uutela et al., 2004)
TGF-β	Platelets, macrophages, KC, fibroblasts	Inhibition of KC proliferation, fibroblasia, ECM production, contraction, leukocyte recruitment	(Werner et al., 2007)
FGF1	KC	Pleiotropic mitogen, angiogenesis	(Werner and Grose, 2003)
bFGF	KC, fibroblasts, macrophages, endothelial cells	Pleiotropic mitogen, angiogenesis, ECM production	(Werner and Grose, 2003)
FGF7 (KGF)	Fibroblasts	KC proliferation and differentiation	(Werner, 1998; Andreadis et al., 2001)
EGF	Platelets, leukocytes	Epithelialization, fibroblasia, ECM production and degradation	(Laato et al., 1987; Brown et al., 1989; Werner and Grose, 2003; Mimura et al., 2006)
TGF-α	Macrophages, granulocytes, KC	Epithelialization	(Werner and Grose, 2003; Li et al., 2006)
HB-EGF	KC	KC migration	(Shirakata et al., 2005)
HGF/SF	Fibroblasts, KC	Epithelialization, leukocyte recruitment, angiogenesis	(Bevan et al., 2004; Chmielowiec et al., 2007)
VEGF	Platelets, leukocytes, SMCs, KC, fibroblasts	Angiogenesis, leukocyte recruitment, expression of growth factors	(Bao et al., 2009)
CTGF	Platelets, fibroblasts	Fibroblasia, ECM production, angiogenesis	(Werner and Grose, 2003)

(*Continued*)

Table 4.7 The major growth factors and cytokines affecting wound repair. (*Continued*)

Growth factor/cytokine	*Major source*	*Regulated wound-healing events*	*References*
IGF1	Fibroblasts, KC, plasma	Epithelialization, ECM production	(Edmondson et al., 2003)
TNF-α	Leukocytes, KC	Expression of growth factors, leukocyte recruitment, ECM degradation	(Werner et al., 2007)
IL-1	Leukocytes, KC	Expression of growth factors, leukocyte recruitment, ECM degradation, regulation of contraction	(Werner et al., 2007)

Modified from Toriseva and Kähäri (2009). CTGF, connective tissue growth factor; ECM, extracellular matrix; EGF, epidermal growth factor; FGF, fibroblast growth factor; HB-EGF, heparin-binding EGF-like growth factor; HGF/SF, hepatocyte growth factor/scatter factor; IGF, insulin-like growth factor; IL, interleukin; KC, keratinocyte; KGF, keratinocyte growth factor; PDGF, platelet-derived growth factor; SMC, smooth muscle cell; TGF, transforming growth factor; TNF, tumor necrosis factor; VEGF, vascular endothelial growth factor.

4.6.8 Proteolysis in cutaneous wound healing

The four major groups of proteinases involved in proteolytic processes during wound healing are *PA-plasmin-system* proteins, *MMPs, ADAMs*, and *ADAMTSs*. They participate in the clearance of the ECM for migrating cells, generate bioactive peptides with different properties, and regulate cell-cell and cell-ECM contacts. The changes in strictly regulated proteolysis are associated with scarring and aberrant wound repair. Here, we discuss in detail the current view of the roles of PA-plasmin system, MMPs, ADAMs, and ADAMTSs in cutaneous wound repair. While the roles and expression patterns of PA-plasmin-system proteinases and MMPs have been extensively studied in skin wounds, detailed expression analysis of ADAM and ADAMTS proteinases is still incomplete, and the data from skin-wound-healing studies using ADAM or ADAMTS transgenic animals are also limited. The wound-healing phenotypes with genetically modified mouse models targeting proteinases discussed here are listed in ▶Table 4.8.

4.6.9 PA-plasmin system

The serine endopeptidase plasmin is responsible for fibrin homeostasis in the body, and it is also capable of cleaving certain ECM components (Goldfinger et al., 1998; Bonnefoy and Legrand, 2000). Plasmin is a potent activator of several latent MMPs, including MMP1, 8, 13, 9, 3, and 7, and membrane-type 1-MMP (MT1-MMP) (Vihinen and Kähäri, 2002), and it may also enhance ECM turnover by promoting the production

Table 4.8 Proteinase gene targeting – mouse skin wound phenotypes.

GENE	*Modification*	*Wound phenotype*	*Ref.*
hMMP1	Overexpression in KCs	Delayed reepithelialization	(Di Colandrea et al., 1998)
MMP8	Knockout	Delayed reepithelialization, delayed onset and persistent inflammation	(Gutierrez-Fernandez et al., 2007)
MMP13	Knockout	(1) Unaltered (small excisional) (2) Delayed reepithelialization, reduced vascularization, reduced wound contraction (large excisional)	(1) (Hartenstein et al., 2006) (2) (Hattori et al., 2009)
MMP2	Knockout	Unaltered	(Frøssing et al., 2010)
MMP9	Knockout	(1) Delayed reepithelialization (2) Delayed reepithelialization, reduced clearance of fibrin clots	(1) (Hattori et al., 2009) (2) (Kyriakides et al., 2009)
MMP9/13	Double knockout	Delayed reepithelialization, reduced vascularization, reduced wound contraction	(Hattori et al., 2009)
MMP3	Knockout	Impaired wound contraction	(Bullard et al., 1999)
MMP10	Overexpression in KCs	Unaltered closure, scattered epithelial sheet	(Krampert et al., 2004)
MT1-MMP	Knockout	Unaltered, impaired epithelialization ex vivo	(Mirastschijski et al., 2004b)
ADAM9	Knockout	Enhanced reepithelialization	(Mauch et al., 2010)
ADAMTS1	Knockout	Delayed reepithelialization, increased angiogenesis	(Krampert et al., 2005)
ADAMTS5	Knockout	Impaired closure, accumulation of aggrecan, defected response to TGF-β by fibroblasts	(Velasco et al., 2011)
PLG	Knockout	Severely impaired closure	(Rømer et al., 1996)
PLG/ MMP13	Double knockout	Increased effect compared to PLG-deficiency	(Juncker-Jensen and Lund, 2011)

Modified from Toriseva and Kähäri (2009). ADAM, a disintegrin and metalloproteinase; ADAMTS, a disintegrin and metalloproteinase with thrombospondin motifs; KC, keratinocyte; MMP, matrix metalloproteinase; hMMP, human MMP; MT-MMP, membrane type–MMP; PLG, plasminogen; TGF, transforming growth factor.

of MMPs (Menshikov et al., 2002; Zhang et al., 2007). Plasma-derived plasminogen (PLG) is activated to plasmin mainly by urokinase plasminogen activator (uPA) in normal skin wounds. The activity of plasmin is regulated by serine proteinase inhibitors (serpins) such as α2-antiplasmin, α2-macroglobulin, and PA inhibitors (PAI) (Cesarman-Maus and Hajjar, 2005). The expression of the key components of the PLG activation system, that is uPA and its receptor (uPAR), as well as its inhibitor PAI1, is induced upon initiation of reepithelialization in the migrating epithelial sheet in murine cutaneous wound (Rømer et al., 1991, 1994, 1996). In acute human wounds, uPA expression coincides with the expression of MMP1 (see "Collagenases" in Section 4.6.10) (Schäfer et al., 1994; Vaalamo et al., 1996). In addition, uPA is detected in macrophages and fibroblasts, and uPAR in macrophages *in vivo* in human wound granulation tissue (Schäfer et al., 1994).

The pivotal role of plasmin in cutaneous wound healing was demonstrated using PLG-null mice, which displayed severely impaired wound closure (Rømer et al., 1996). In these animals, disability of migrating keratinocytes to dissect the fibrin-containing provisional matrix results in dramatically affected wound reepithelialization, and additional deletion of the fibrinogen gene in PLG-deficient mice largely rescues the wound phenotype (Bugge et al., 1996). Although wound repair is delayed, skin wounds of PLG-deficient mice eventually close in 60 days, but if these mice are simultaneously treated with the broad-spectrum metalloproteinase inhibitor galardin, wound healing is virtually blocked (Lund et al., 1999), demonstrating the parallel roles of plasmin and metalloproteinases in reepithelialization. Recently, specific MMPs, such as MMP13 and MMP2, were found to promote proteolysis in mouse skin wounds together with plasmin (Frøssing et al., 2010; Juncker-Jensen and Lund, 2011). Interestingly, wound closure in mice double deficient of uPA and tissue-type plasminogen activator (tPA), another plasminogen activator primarily functioning in the vascular system, is not as severely impaired as in PLG-deficient mice, and treatment with galardin further delays but does not prevent wound healing in these mice (Lund et al., 2006). This appears to be due to the effect of the serine proteinase plasma kallikrein, which can directly activate PLG, thus resulting in plasmin activity in the wounds of uPA/tPA-null mice (Lund et al., 2006).

Wound granulation tissue formation is unaltered in PLG-deficient mice probably due to adequate enzymatic redundancy (Rømer et al., 1996). However, induction of uPA and PAI1 expression in endothelial cells and especially by accompanying stromal cells during angiogenesis has been documented, suggesting a role for plasmin in granulation tissue formation as well (Rømer et al., 1991; Schäfer et al., 1994; Vaalamo et al., 1996; Pepper, 2001). *In vitro*, the expression of uPA, uPAR, and PAI1 is upregulated by the potent angiogenic factors VEGF and bFGF (Pepper, 2001). *In vivo*, plasmin degrades fibrin clots, which may interfere with angiogenesis, and it can also release FGF2 and VEGF bound to ECM associated with blood vessels, generating an autocrine loop for the PA-plasmin system during angiogenesis (Park et al., 1993; Whitelock et al., 1996). In addition, PAI1 appears to be important in tumor angiogenesis, as well as in physiological angiogenesis. PAI1 may protect ECM from excessive proteolysis and in this way maintain the scaffold for endothelial cell migration. Alternatively, it may regulate adhesion of endothelial cells via the interaction network of PAI1, uPAR, integrins, and vitronectin (Pepper, 2001). Finally, angiostatin, which consists of the first four kringles of PLG, is generated by proteolytic cleavage of PLG by, for example, certain MMPs and inhibits angiogenesis by regulating endothelial cell migration and proliferation and by promoting apoptosis (O'Reilly et al., 1999).

4.6.10 Matrix metalloproteinases

MMPs comprise a group of structurally related metalloendopeptidases, which are produced and secreted as inactive zymogens referred as pro-MMPs. Collectively, MMPs cleave practically all types of ECM molecules and also various soluble and membrane-bound substrates. These comprise other proteinases, proteinase inhibitors, growth factors, and cytokines (McCawley and Matrisian, 2001; Kessenbrock et al., 2010). The levels of MMPs are low in intact tissues, and their expression and activity are elevated in various dynamic physiological situations such as fetal tissue development and postnatal tissue repair (Page-McCaw et al., 2007; Toriseva and Kähäri, 2009).

In vivo, the activity of MMPs is regulated by proteolytic activation of the zymogen and by specific inhibitors, tissue inhibitors of metalloproteinases (TIMPs) (see Section 4.6.13). In addition to TIMPs, most MMPs are also inhibited by unspecific proteinase inhibitors, such as α2-macroglobulin (Murphy and Nagase, 2008). Factors abundantly present in cutaneous wounds that regulate MMP expression include TGF-β, PDGF, tumor necrosis factor-α (TNF-α), interleukin-1β (IL-1β), FGF2, epidermal growth factor (EGF), and keratinocyte growth factor (KGF; FGF7). For example, the expression of MMP1, which is induced by contact with collagen in migrating keratinocytes *in vivo*, is downregulated *in vitro* by FGF2 and KGF in keratinocytes (Pilcher et al., 1997b). Moreover, the expression of MMP1 by human fibroblasts is enhanced by, for example, PDGF, while TGF-β downregulates MMP1 expression (Bauer et al., 1985; Yuan and Varga, 2001).

The expression level of MMPs in intact skin is low. Only MMP7 and MMP19 are constitutively produced in sweat and sebaceous glands (Saarialho-Kere et al., 1995; Sadowski et al., 2003b). In addition, MMP19 is expressed by basal keratinocytes and in hair follicles, as well as in endothelial and smooth muscle cells of the veins and arteries (Sadowski et al., 2003b). Moreover, low expression levels of MMP2 and MT1-MMP has been detected in normal dermis in animal models (Ågren, 1994; Madlener et al., 1998). Due to the chemical and physical changes in the cellular environment after cutaneous injury, the expression of multiple MMPs is induced. MMPs expressed in normally healing skin wounds include collagenases MMP1 and MMP8 (Inoue et al., 1995; Nwomeh et al., 1999), gelatinases MMP2 and MMP9 (Mirastschijski et al., 2002a), stromelysins MMP3 and MMP10 (Vaalamo et al., 1996), metalloelastase MMP12 (reported in mice) (Madlener et al., 1998), MT1-MMP (Mirastschijski et al., 2002a), MMP19 (Hieta et al., 2003), MMP26 (Ahokas et al., 2005), and MMP28 (Lohi et al., 2001). The expression and cellular source of MMPs and other major proteinases in an acute cutaneous wound are illustrated in ▶Figure 4.11.

Collagenases

In cutaneous wounds, in which the basement membrane is disrupted, temporarily and spatially restricted expression of MMP1 (collagenase-1) is induced in migrating keratinocytes that come in direct contact with the dermal compartment of skin and lack the contact with intact basement membrane. The expression peaks 24 h after wounding and subsides by completion of reepithelialization (Saarialho-Kere et al., 1993; Inoue et al., 1995). It has been shown that contact with native type I collagen induces MMP1 expression in migrating keratinocytes *in vitro*, whereas contact with basement

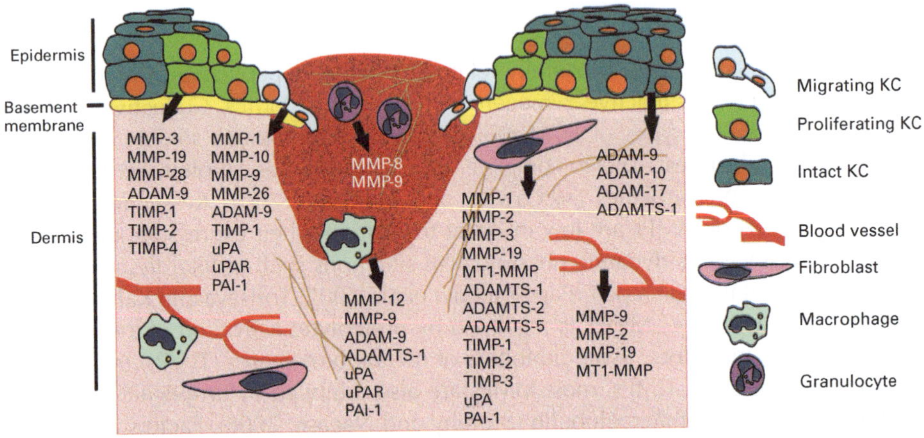

Figure 4.11 The cellular source of metalloproteinases and their inhibitors, and the main components of PA-plasmin system in a normal cutaneous wound. In wound repair, several cell types efficiently function together resulting in tissue recovery and epithelial closure. Strictly regulated proteolysis is essential for normal wound closure, and a variety of proteinases are secreted by cells involved in processes such as inflammation (granulocytes and macrophages), reepithelialization (keratinocytes), granulation tissue formation (fibroblasts and endothelial cells), and tissue remodeling and wound maturation (fibroblasts). Please note that in murine skin MMP1 is functionally substituted by MMP13. ADAM, a disintegrin and metalloproteinase; ADAMTS, a disintegrin and metalloproteinase with thrombospondin motifs; KC, keratinocyte; MMP, matrix metalloproteinase; MT-MMP, membrane type–MMP; PAI, PA inhibitor; TIMP, tissue inhibitor of metalloproteinases; uPA, urokinase plasminogen activator; uPAR, uPA receptor.

membrane proteins does not (Sudbeck et al., 1997). This is important in the initiation of reepithelialization *in vivo* when keratinocytes have detached from the basement membrane and come into contact with type I collagen at the time when the provisional matrix has not yet been formed. The contact of keratinocytes to type I collagen is mediated via collagen receptor integrin α2β1 resulting in MMP1 upregulation, which appears to be indispensable for keratinocyte migration on native type I collagen (Pilcher et al., 1997a). It has been demonstrated that pro-MMP1 and activated MMP1 can bind to cell surface integrin α2β1 in keratinocytes, suggesting specifically directed collagenolysis by MMP1 at the pericellular sites where its activity is required (Dumin et al., 2001; Stricker et al., 2001). MMP1 cleaves type I collagen, generating fragments, which at body temperature denature to gelatin, a less adhesive ligand for α2β1 integrin compared to native collagen and therefore a more suitable substrate for migration. Thus, in humans, α2β1-MMP1 complex is suggested to function as a motor stimulating migration of keratinocytes on type I collagen during reepithelialization.

Following reepithelialization and establishment of new basement membrane, the expression of epidermal MMP1 ceases as a result of keratinocyte adhesion to basement membrane proteins including laminin-111 (Inoue et al., 1995; Sudbeck et al., 1997). The importance of the strict regulation of MMP1 activity during wound healing has been demonstrated by overexpressing human MMP1 in mouse epidermis, which resulted in markedly delayed wound closure and hyperproliferative epidermis (Di Colandrea et al., 1998). On the other hand, in mice with collagenase-resistant mutation in type I collagen, the closure of an incisional skin wound was also severely impaired due to delayed wound contraction and reepithelialization (Beare et al., 2003). Thus, controlled epidermal and dermal collagenolysis is required for proper wound healing.

MMP1 is also expressed by fibroblasts in granulation tissue (Inoue et al., 1995; Vaalamo et al., 1997), where it is believed to participate in remodeling of collagenous ECM (Pins et al., 2000). MMP13 (collagenase-3) displays a similar expression pattern in mouse skin as MMP1 in human skin (Madlener et al., 1998), and in humans, MMP13 is expressed by fibroblasts in fetal cutaneous wounds (Ravanti et al., 2001). MMP1 and MMP13 are important in mediating collagen reorganization (Toriseva et al., 2007) and may also regulate survival of fibroblasts during dermal wound healing by affecting fibroblast-mediated matrix contraction and matrix rigidity, and by revealing cryptic binding sites from native collagen for integrins such as αV integrin, which promotes cell survival in collagen lattice (Bao and Strömblad, 2004; Toriseva et al., 2007).

Using an MMP13-deficient mouse model, MMP13 has been implicated in several aspects of wound healing. Hattori et al. (2009) reported delayed reepithelialization, reduced density of granulation tissue blood vessels, and impaired formation of myofibroblasts associated with reduced wound contraction in the skin wounds of MMP13-null mice (Hattori et al., 2009). Our recent observations verify the effects of MMP13 activity on myofibroblast formation and suggest wide regulatory role for MMP13 in blood vessel morphology, MMP expression, and general granulation tissue growth (Toriseva et al, unpublished). Interestingly, another study reported completely comparable wound healing in MMP13-null mice with the wild-type mice (Hartenstein et al., 2006). These two studies were performed using mouse strains with different genetic backgrounds, which may explain the difference. Moreover, as wounds in the latter study were relatively small, they were likely to heal with minimal granulation tissue formation (Hartenstein et al., 2006). Moreover, lack of wound-healing phenotype in this study can be explained by enzymatic redundancy (e.g. of MMP8) (Hartenstein et al., 2006). Specific overlap of MMP13 activity with plasmin was also recently examined in a mouse skin wound (Juncker-Jensen and Lund, 2011). In this study, additional deletion of the MMP13 gene further impaired severely retarded healing in PLG-deficient mice, although not as efficiently as addition of broad-spectrum metalloproteinase inhibitor (Lund et al., 1999; Juncker-Jensen and Lund, 2011). These observations suggest functional overlap of plasmin with MMP13 but still emphasize the role of other metalloproteinases in wound healing.

Moreover, MMP2 (gelatinase-A) and MT1-MMP, which both are also capable of cleaving fibrillar collagens, are present in granulation tissue of human and murine skin wounds (Madlener et al., 1998; Mirastschijski et al., 2002b). MT1-MMP is considered one of the collagenases in fibroblasts (Lee et al., 2006a; Sabeh et al., 2009), and it is likely to mediate fibroblast migration (Sabeh et al., 2004). Moreover, its role in TIMP2-mediated activation of pro-MMP2 is well characterized (Strongin et al., 1995), and MMP2 plays a role in ECM remodeling and angiogenesis (see the following section "Gelatinases").

MMP8 (collagenase-2) is mainly expressed by wound neutrophils, in which it is stored in cellular granules and secreted on neutrophil activation (Hasty et al., 1986). In human excisional skin wounds, MMP8 was shown to be the most abundant collagenase (Nwomeh et al., 1999). Its relevance in wound healing has been elucidated by studies with MMP13-deficient mice suggesting the enzymatic compensation of MMP13 by MMP8 (Hartenstein et al., 2006). Subsequently, it was also reported that wound healing was significantly delayed in MMP8-null mice due to impaired reepithelialization associated with a lag in neutrophil infiltration and persistent inflammation (Gutierrez-Fernandez et al., 2007). Thus, as a regulator of inflammation, MMP8 is able to modulate other processes such as keratinocyte migration and proliferation, and it may also participate in ECM remodeling with other MMPs present in wound tissue.

Stromelysins

Stromelysins MMP3 and MMP10 are also both expressed by epidermal keratinocytes during wound repair in human and mouse wounds. MMP3 is expressed by the basal proliferating keratinocytes behind the migrating edge, while MMP10 is exclusively produced at the tip of the migrating keratinocyte sheet (Vaalamo et al., 1996; Madlener et al., 1998; Rechardt et al., 2000). Additionally, MMP3 is produced by wound fibroblasts (Vaalamo et al., 1996; Madlener et al., 1998). MMP3-deficient mice display delayed wound closure associated with impaired wound contraction, providing evidence for the role of MMP3 in assembly and organization of wound fibroblasts to facilitate wound contraction (Bullard et al., 1999). Despite epidermal expression of MMP3 during wound closure, reepithelialization *per se* was not affected in these mice (Bullard et al., 1999). However, MMP3 can activate several pro-MMPs and increase activity and bioavailability of cytokines and growth factors (e.g. heparin-binding EGF-like growth factor [HB-EGF] and TGF-β) (Imai et al., 1997; Visse and Nagase, 2003). MMP3 is also able to digest several ECM substrates, including basement membrane proteins, and it has been shown to modulate intercellular contacts stimulating epithelial cell invasion (Noe et al., 2001; Visse and Nagase, 2003). Thus, MMP3 may promote detachment of epidermal keratinocytes, as a prerequisite for their proliferation and migration.

Epidermal MMP10 is induced 3 days postwounding in humans, and at least *in vitro*, its expression is upregulated by EGF, TGF-β1, and TNF-α (Rechardt et al., 2000). MMP10 is proposed to regulate keratinocyte organization and migration during reepithelialization, since the overexpression of constitutively active MMP10 in basal keratinocytes of transgenic mice severely scatters the migrating epithelial sheet. This was thought to be due to the overprocessing of laminin-332 by MMP10, resulting in alterations in cell-ECM adhesion and keratinocyte migration. However, wound healing in these mice is virtually unaffected and characterized by normal closure rate and normally established basement membrane (Krampert et al., 2004).

Gelatinases

MMP2 and MMP9 (gelatinase-A and -B, respectively) exhibit a distinct expression pattern during wound repair in skin. MMP9 is detected at the migrating epithelial front, whereas MMP2 is exclusively expressed in the dermal compartment of the skin by fibroblasts, especially adjacent to regenerating epidermis, and by endothelial cells (Madlener et al., 1998; Mirastschijski et al., 2002a). MMP9 is also present in inflammatory cells including T cells, macrophages, and neutrophils (Leppert et al., 1995; Okada et al., 1997; Inkinen et al., 2000). For MMP2, the relatively lengthy and consistent upregulation in porcine skin wound suggests that MMP2 plays a role in prolonged ECM remodeling (Ågren, 1994; Frøssing et al., 2010). Indeed, in addition to direct processing of dermal ECM proteins, several studies have implicated MMP2 in the activation of TGF-β, a potent growth factor regulating matrix deposition and remodeling, by several mechanisms (Imai et al., 1997; Yu and Stamenkovic, 2000; Dallas et al., 2002). Nevertheless, mice deficient for MMP2 display only slight and insignificant delay in wound closure and otherwise normal architecture of reconstituted skin (Frøssing et al., 2010), reflecting enzymatic redundancy by other proteinases.

Wound-healing studies with MMP9-deficient mice have provided contradictory results. In an earlier study, MMP9 was reported to play an inhibitory role in epidermal wound healing, and MMP9-deficient mice showed accelerated reepithelialization of cornea and skin probably due to the enhanced proliferation of keratinocytes (Mohan et al., 2002). This could be explained by, for example, the ability of MMP9 to activate TGF-β, which inhibits keratinocyte proliferation (Yu and Stamenkovic, 2000; Dallas et al., 2002). In more recent studies, MMP9 deficiency was, however, reported to significantly delay reepithelialization, and this was suggested to result from impaired migration of keratinocytes (Hattori et al., 2009; Kyriakides et al., 2009). Interestingly, degradation of fibrin-containing provisional matrix was reported defective in wounds of MMP9-deficient mice (Mohan et al., 2002; Kyriakides et al., 2009). Accordingly, MMP9 has been shown to potently digest fibrin (Lelongt et al., 2001). Thus, MMP9 may function by, for example, activating TGF-β, which regulates keratinocyte proliferation and at the same time stimulates ECM deposition and remodeling. In addition, MMP9 appears to act in parallel with plasmin for resorption of provisional matrix from the wound bed to allow keratinocyte migration.

Both MMP2 and MMP9, as well as MT1-MMP and MMP19, are expressed in endothelial cells (Oikarinen et al., 1993; Mirastschijski et al., 2002a; Hieta et al., 2003; Aplin et al., 2009). MMP2 and MMP9 have been shown to play pivotal roles in both physiological and tumorigenic angiogenesis (Itoh et al., 1998; Vu et al., 1998; Bergers et al., 2000; Kato et al., 2001). Gelatinases digest components of the vascular basement membrane, which is obligatory for blood vessel sprouting. Importantly, gelatinases also contribute to angiogenesis by releasing angiogenic cytokines and growth factors, such as TNF-α (Gearing et al., 1994) and VEGF (Bergers et al., 2000; Dean et al., 2007). Interestingly, gelatinases, as well as several other MMPs, not only promote angiogenesis but also may inhibit blood vessel formation by generating antiangiogenic peptides from other proteins. MMP3, 7, 9, 13, and 20 have been shown to generate endostatin from type XVIII collagen (Heljasvaara et al., 2005); and MMP2, 3, 7, 9, 12, and 19, angiostatin from plasminogen *in vitro* (O'Reilly et al., 1994; Cornelius et al., 1998; Brauer et al., 2011). In addition, MMP2 may regulate FGF-mediated mitogenic and angiogenic signals by shedding cellular FGF receptor-1 while still preserving the receptor capable of binding FGF (Levi et al., 1996). It is conceivable that potent angiogenic stimulation in physiological situations, such as wound repair, is strictly controlled, and in this respect, MMPs play an important role in generating antiangiogenic peptides.

MT1-MMP, matrilysin-2, and the other MMPs

MT1-MMP appears to be pivotal for angiogenesis (Chun et al., 2004). The important role of MT1-MMP in wound angiogenesis may relate to its fibrinolytic and collagenolytic activity needed for vessel invasion through fibrin barriers and collagenous obstruction in the tissue stroma (Hiraoka et al., 1998; Chun et al., 2004; Genis et al., 2007). Again, its role in activation of pro-MMP2 potentiates its proteolytic effect (Strongin et al., 1995). *In vitro*, MT1-MMP colocalizes with β1 and αVβ3 integrins to the intercellular contacts, suggesting a regulatory role for endothelial cell adhesion and migration (Galvez et al., 2002). Although MT1-MMP–null mice display severe abnormalities in bone development and defective angiogenesis in cartilage and cornea (Zhou et al., 2000), cutaneous wound healing is unaffected in 3-day-old animals (Mirastschijski et al., 2004b).

During epidermal wound repair in humans, MMP19 has been detected in proliferating epithelium, microvascular endothelial cells, fibroblasts, and macrophages. In keratinocytes, MMP19 has been implicated in regulation of migration and proliferation via releasing insulin-like growth factor (IGF) by IGF-binding protein-3 cleavage and via laminin-332 cleavage (Sadowski et al., 2003a, 2005). Regulation of MMP19 expression by POU transcription factors Tst-1 (Oct-6) and Skn-1a (Oct-11) suggests that its expression is linked to cellular differentiation (Beck et al., 2007). Increasing evidence suggests that MMP19 plays an inhibitory role in angiogenesis by destabilizing ECM necessary for capillary morphogenesis, as well as by generating antiangiogenic peptides (Jost et al., 2006; Brauer et al., 2011).

MMP12 or metalloelastase is produced by perivascular macrophages in acute murine excisional wounds with the maximum expression noted at the time of completion of the reepithelialization (Madlener et al., 1998). In humans, abundant expression of MMP12 has been detected in various cutaneous granulomas, but the authors could not find MMP12 in majority of the examined acute or chronic skin wounds despite the presence of macrophages (Vaalamo et al., 1999a). MMP12 exerts potent antibacterial activity (Houghton et al., 2009), and its ability to generate angiostatin and degrade fibrinogen suggests a role as potential regulator of angiogenesis (Cornelius et al., 1998; Hiller et al., 2000).

MMP26 and MMP28 (matrilysin-2/endometase and epilysin, respectively) are detected in epidermis during wound reepithelialization. MMP26 was reported in epithelial tip bordering the wound gap (Ahokas et al., 2005) and it is suggested to coordinate cell-cell adhesion involved in keratinocyte migration (Pirilä et al., 2007). MMP26 expression is regulated by Wnt signaling (Marchenko et al., 2004), and altered migration of mucosal keratinocytes treated with MMP26 antibody (Pirilä et al., 2007) suggest that MMP26 may be involved in keratinocyte proliferation and migration.

MMP28 is expressed by suprabasal keratinocytes distal to the epithelial tip in the region occupying virtually intact basement membrane in human skin excisional wounds and in suction blisters (Saarialho-Kere et al., 2002). Thus, the spatial expression pattern of MMP28 and its ability to modify intercellular adhesion suggest a regulatory role for initiation or keratinocyte migration during reepithelialization.

4.6.11 ADAM proteinases

ADAMs are a family of transmembrane proteinases possessing adhesive and proteinase activities in their ectodomains and putative signaling activities in the cytoplasmic domain (Edwards et al., 2008). A typical function of ADAMs is shedding of cytokines and growth factors. ADAM17 (i.e., TNF-α-converting enzyme [TACE]) was first shown to release and activate TNF-α, but currently, more than 30 substrates for TACE have been identified (Huovila et al., 2005; Edwards et al., 2008). One important set of substrates for ADAM10 or ADAM17 is the ligands of EGF receptor (EGFR, ErbB), namely, TGF α, HB-EGF, amphiregulin, epiregulin, EGF, betacellulin, and epigen, all produced as membrane-associated inactive molecules (Sahin et al., 2004; Sahin and Blobel, 2007). ADAM9, 10, 12, 15, and 17 cleave ECM components, including fibronectin, laminin, collagen types IV and XVII, and gelatin (Millichip et al., 1998; Schwettmann and Tschesche, 2001; Martin et al., 2002; Edwards et al., 2008), and in this way, may regulate release of growth factors from ECM, as well as cell migration.

ADAM9, 10, and 17 are expressed by keratinocytes in intact human epidermis, and ADAM9 also by fibroblasts *in vitro* (Franzke et al., 2002; Zigrino et al., 2007). In addition, epidermal expression of ADAM9 is enhanced upon injury in mouse skin (Mauch et al., 2010). Thus, the sheddase activity of these epidermal ADAMs may play an important role in keratinocyte biology. As mentioned previously, ADAM9, 10, and 17 are potent sheddases/activators of EGF receptor ligands (Izumi et al., 1998; Sahin et al., 2004; Peduto et al., 2005). EGF receptor ligands are key regulators of keratinocyte proliferation and migration, as well as of granulation tissue formation during wound repair (Werner and Grose, 2003). Shedding of HB-EGF ectodomain is especially important for keratinocyte migration in skin wound repair (Tokumaru et al., 2000). In addition, ADAM10 plays a significant role in regulating keratinocyte adhesion, migration, and proliferation *in vitro* by shedding cell surface E-cadherin (Maretzky et al., 2005). The ability of ADAM10 and ADAM17 to shed, for example, CD44 may also regulate keratinocyte migration and proliferation during wound healing (Nagano et al., 2004). On the other hand, shedding of hemidesmosome-associated collagen XVII from keratinocytes by ADAM9, 10, and 17 is implicated in reduced motility of the cells due to the inhibitory effect of the shed collagen XVII ectodomain on migration (Franzke et al., 2002). This is likely to explain accelerated closure of skin wounds in ADAM9-deficient mice due to enhanced migration of keratinocytes associated with decreased collagen XVII processing (Mauch et al., 2010). ADAM9 may also have an opposite effect on keratinocyte migration through binding to β1 integrin on adjacent cells and subsequent induction of MMP9 expression (Zigrino et al., 2007). These observations provide evidence that ADAMs modulate keratinocyte migration during wound repair.

ADAM10 and 17 may also play a role in wound repair via regulation of inflammation. They have been shown to shed endothelial cell transmembrane chemokines and promote detachment of leukocytes bound to endothelium, thereby regulating leukocyte trafficking (Garton et al., 2001; Hundhausen et al., 2007).

4.6.12 ADAMTS proteinases

ADAMTS proteinases are structurally related to ADAMs. ADAMTS proteinases are secreted proteins that cleave ECM proteins as their main substrates. ADAMTS proteinases are widely expressed in tissues, and their involvement in several biological processes, such as regulation in collagen fibrillogenesis, ECM proteolysis, blood clotting, inflammation, angiogenesis, and cell migration, suggests a role for this proteinase family in wound healing (Porter et al., 2005; Apte, 2009). For instance, ADAMTS2, which is highly expressed in skin, cleaves the aminoterminal propeptide of procollagens I, II, and III, which is essential for collagen fibril formation. ADAMTS2 was also shown to cleave the propeptide of fibrillar type V procollagen, which, in turn, is also needed in type I collagen fibrillogenesis (Colige et al., 2005).

Among ADAMTS proteinases, only ADAMTS1 and ADAMTS5 have been implicated in cutaneous wound healing. ADAMTS1 is constitutively expressed in mouse epidermis, and mice deficient in the *Adamts1* gene show abnormal epidermal differentiation (Krampert et al., 2005). During wound healing in mouse skin, the expression of ADAMTS1 is enhanced in the basal keratinocytes after closure of the wound, and the expression pattern appears to follow the differentiation level of keratinocytes (Krampert et al., 2005). The regulation of ADAMTS1 expression is induced by TGF-β1 and TGF-β3 in cultured keratinocyte cell line (Krampert et al., 2005). In wound granulation tissue, ADAMTS1 is upregulated initially in macrophages, and by day 5, the expression is noted in fibroblasts, in which it is suggested to regulate cell migration (Krampert et al., 2005). ADAMTS1 has also been implicated in the inhibition of angiogenesis, as demonstrated by enhanced angiogenesis associated with delayed wound healing in ADAMTS1-null mice (Iruela-Arispe et al., 2003; Lee et al., 2006b). Inhibition of blood vessel formation attributed to ADAMTS1 involves proteolytic release of angiostatic polypeptides from thrombospondin-1 and -2, and sequestration of angiogenic VEGF and bFGF (Luque et al., 2003; Krampert et al., 2005; Lee et al., 2006b).

Recently, ADAMTS5 was shown to play an important role in mouse wound healing. ADAMTS5 is constituently expressed by skin fibroblasts (Hattori et al., 2011), and the expression is elevated during wound repair (Velasco et al., 2011). *In vitro*, fibroblasts from ADAMTS5-deficient mice expressed elevated levels of myofibroblast-specific α-smooth muscle actin (α-SMA) and enhanced ability to contract collagen matrix (Hattori et al., 2011). However, *in vivo* during wound healing in ADAMTS5-deficient mice, wound contraction was restrained (Velasco et al., 2011). In these mice, fibroblast aggregates surrounded by aggrecan, versican, and hyaluronan developed in the granulation tissue, resulting in impaired deposition of collagen and severely defective wound closure. The observations were proposed to be due to altered TGF-β signaling (Velasco et al., 2011). The differences between the two studies employing ADAMTS5-deficient mice and derived fibroblasts need to be explored further to gain an in-depth understanding of the role of ADAMTS5 in dermal wound healing.

4.6.13 TIMPs and chemical targeting of metalloproteinases

There are four known mammalian TIMPs, namely, TIMP1, 2, 3, and 4, which bind and inhibit MMPs with some differences in affinities (Murphy and Nagase, 2008). In addition, TIMP3 inhibits the activity of ADAM10, 12, 17, 28, and 33, and ADAM10 is inhibited by TIMP1 (Baker et al., 2002; Edwards et al., 2008). ECM-associated TIMP3 inhibits the activity of ADAMTS4 and 5, as well as ADAMTS2 and ADAMTS1, the latter of which is also inhibited by TIMP2 (Kashiwagi et al., 2001; Rodriguez-Manzaneque et al., 2002; Wang et al., 2006). Thus, especially TIMP3 appears to be an important regulator of ADAM and ADAMTS activity in tissues.

The expression of TIMP1, 2, and 3 has been detected in wound epithelium and dermal fibroblasts and macrophage-like cells in humans (Stricklin et al., 1993; Vaalamo et al., 1996, 1999b). In fibroblasts, TIMP1 was detected especially around blood vessels, suggesting a role in vessel stability or angiogenesis (Stricklin et al., 1993; Vaalamo et al., 1999b). Occasional vascular vessel structures of acute wounds were also found to be positive for TIMP1 and TIMP3. TIMP4 expression was not detected in acute human wounds (Vaalamo et al., 1999b). It has become obvious that the function of TIMPs is not limited to metalloproteinase inhibition but that they also play other roles in human physiology and pathology (e.g. by regulating proliferation and survival of variety of normal and tumor cells); this should also be recognized when clarifying regulatory mechanisms of wound healing (Lambert et al., 2004).

Finally, although the single metalloproteinase gene knockout studies have not typically revealed tremendous wound-healing defects, various studies have demonstrated the importance of collective action of metalloproteinases in processes involved in wound repair using broad-spectrum chemical inhibitors of metalloproteinases. For instance, treatment of fibroblasts cultured in three-dimensional collagen matrix with marimastat (BB-2516) or galardin (GM6001, ilomastat) impairs fibroblast-mediated collagen contraction (Scott et al., 1998; Martin-Martin et al., 2011), and batimastat (BB-94) totally blocked TGF-β-stimulated migration of keratinocytes *in vitro* (Mäkelä et al., 1999). Furthermore, systemic administration of galardin (GM6001, ilomastat) to murine skin wounds attenuates wound contraction and interferes with migration of keratinocytes (Lund et al., 1999; Mirastschijski et al., 2004a). It must be kept in mind that although these small-molecule inhibitors are sometimes referred to as MMP inhibitors, the specificity of these inhibitors is not restricted to MMPs; for example, batimastat and galardin can also inhibit ADAMs and meprin metalloproteinases, capable of cleaving basement membrane proteins, among of some other proteinases (Solorzano et al., 1997; Kruse et al., 2004).

4.6.14 Proteolysis in aberrant cutaneous wound healing

Normally, a wound is initially closed in a few days, and in adult human skin, the end result is a flat, collagenous, relatively acellular scar, which, however, will never quite reach the tensile strength of uninjured skin. Under certain circumstances, wound closure can be delayed, resulting in chronic ulceration. On the other hand, defective regulation of wound healing may also lead to excessive scar formation, which in some locations may severely interfere with normal function of skin.

Chronic skin wounds

The majority of chronic wounds are located in lower extremities and typically involve ischemia associated with insufficiency of venous or arterial circulation, diabetic vasculapathy, or pressure. Also, increasing age adds susceptibility for development of chronic ulcers due to impaired wound repair in general (Eming et al., 2007). Chronic wounds are typically characterized by pathological inflammation, fibroblast senescence, and uncontrolled proteolysis (Eming et al., 2007; Menke et al., 2007). Excessive proteolysis may lead to destruction of wound ECM, disturbance of cell migration, or degradation of growth factors and their receptors.

A number of studies have shown, that in a poorly healing chronic ulcer, the activity of MMPs is upregulated while the expression of TIMPs is decreased compared to acute wounds. The levels of MMP1, 2, 8, and 9 are elevated in wound tissues or wound fluid of chronic ulcers of nondiabetic and diabetic patients (Wysocki et al., 1993; Yager et al., 1996; Nwomeh et al., 1999; Lobmann et al., 2002; Muller et al., 2008; Rayment et al., 2008;). While immunoreactivity of MMP26 starts to decline in acute skin wounds 1 day after injury, in chronic wounds MMP26 is consistently detected in stroma beneath the basement membrane at the ulcer margin, where it may participate in the activation of MMP9 and release of IGF from insulin-like growth factor-binding protein 1 (IGFBP1) (Pirilä et al., 2007). A distinct feature of chronic ulcers is the expression of MMP13 by wound fibroblasts embedded in collagenous stroma (Vaalamo et al., 1997). This may provide a survival mechanism for the fibroblasts and contribute to the remodeling of the ECM of a chronic ulcer (Toriseva et al., 2007).

Partially, the increased amount of MMPs in chronic wounds can be explained by dramatic infiltration of inflammatory cells secreting MMPs such as MMP8 and MMP9 (Nwomeh et al., 1999; Mirastschijski et al., 2002a) and a variety of growth factors, which in turn are likely to regulate MMP expression by cells. Moreover, compounds derived from *Staphylococcus aureus*, a common infectious microbe in chronic leg ulcers, may upregulate the expression of multiple MMPs, namely, MMP1, 2, 3, 7, 10, 11, and 13, as well as TIMP1 and TIMP2 by normal dermal fibroblasts (Kanangat et al., 2006). Nonhealing ulcers also possess less TIMP1 and TIMP2, as compared to normally healing wounds, which abrogates the inhibition of MMP activity, but also, in the case of TIMP2, may alter activation of MMP2 (Bullen et al., 1995; Nwomeh et al., 1999; Vaalamo et al., 1999b). Moreover, while acute wound fibroblasts express only TIMP1, 2, and 3, fibroblasts from chronic ulcers also express TIMP4 (Vaalamo et al., 1999b). Despite general upregulation of MMP activity in chronic wounds, stromal fibroblasts from chronic leg ulcers have been reported to express lower levels of active MMP2 and pro-MMP1, and more TIMP1 and 2 compared to fibroblasts from an acute wound, when cultured inside three-dimensional collagen gel (Cook et al., 2000). Thus, it must be noted that the total MMP levels detected in wound fluids reflect the sum of MMPs expressed by different cell types.

In contrast to human chronic wounds, genetically diabetic mice, a well-established model for impaired wound repair, show lower levels of MMP2 and MMP9 in wound tissue extracts, and increased expression of ADAMTS1 mRNA compared to control animals during early wound healing (up to day 7 and day 1, respectively) (Wall et al., 2002; Krampert et al., 2005).

Keratinocytes at the edge of a chronic ulcer express uPA comparable with keratinocytes at the leading edge of acute wound epidermis (Vaalamo et al., 1996). In contrast, abundant tPA mRNA and protein were found in the basal and suprabasal keratinocytes at the margin of chronic venous leg ulcers, while tPA in acute wound epidermis was weakly expressed (Weckroth et al., 2004). In addition, the higher expression level of tPA in the fibroblast- and macrophage-like cells of acute wound granulation tissue compared to chronic ulcer stroma was also reported. Thus, both stromal and epidermal compartments differ in acute and chronic wounds in their fibrinolytic capacity with respect to PLG activation, which in turn may affect activation of MMPs.

Fibrotic cutaneous wounds

Fibrotic wounds are characterized by excessive formation of collagenous scar tissue. Excessive scarring may occur, for example, in healing of burn wounds or skin grafts. As a result, skin loses elasticity, and in extremities, this may lead to contractures and functional disability. Hypertrophic scars and keloids are two types of local fibrotic conditions that may develop in skin after defective wound maturation. Keloids are fibrous, reddish, and firm nodules that grow beyond the borders of the original wound. Hypertrophic scars are elevated scars that are smaller than keloids, and they are typically restricted to the boundaries of the initial injury. They may, however, develop severe contractures affecting tissue functionality. Keloids are less frequents than hypertrophic scars and are characterized by genetic predisposition (Bran et al., 2009). Large area, depth, delayed closure, and tension attributable to motion or loss of tissue are common risks for scar formation in a skin wound. However, the molecular events behind the shift from normal healing to excessive scarring are still incompletely understood.

Hypertrophic scars and keloids share common properties, such as existence only in humans and accumulation of collagen and fibroblasts. Interestingly, in keloids, but not in hypertrophic scars, fibroblasts proliferate actively, while myofibroblasts colonize hypertrophic scars (Ehrlich et al., 1994; Bran et al., 2009). Profibrotic growth factor TGF-β is commonly proposed to play a role in these conditions due to upregulation of TGF-β in hypertrophic scars and keloid fibroblasts (Peltonen et al., 1991; Ghahary et al., 1993, Lee et al., 1999; Wang et al., 2000), as well as elevated levels of TGF-β receptors in keloid fibroblasts (Chin et al., 2001). This is likely to contribute to enhanced collagen synthesis detected in keloid and hypertrophic scar fibroblasts (Ghahary et al., 1993; Fujiwara et al., 2005) and may stimulate differentiation of fibroblasts to myofibroblast in hypertrophic scars.

Moreover, hypertrophic scar fibroblasts show a reduced level of MMP1 mRNA expression (Ghahary et al., 1996), while upregulation of MMP1, MMP2, and TIMP1 proteins has been reported in keloid fibroblasts (Fujiwara et al., 2005) compared to normal dermal fibroblasts. A study comparing gene expression profiles of keloid fibroblasts and normal scar fibroblasts identified downregulation of both MMP1 and MMP3 in keloid fibroblasts (Smith et al., 2008). Tissue extracts from both lesions show markedly increased levels of MMP2 activity compared to normal skin samples, and low levels of MMP9 activity (Neely et al., 1999). Moreover, the expression level of MMP13 is elevated in keloid tissues, and certain treatments inducing keloid regression are reported to further upregulate its expression (Kuo et al., 2005). Thus, the expression of collagenolytic enzymes MMP13, MMP1, and MMP2 by keloid fibroblasts may reflect their attempt to cleave the excess collagen in tissue. In hypertrophic scar fibroblasts, the downregulation of MMP1 is suggested to be due to response to IGF1 (Ghahary et al., 1996). It is also possible that cell surface collagen receptor integrins on keloid fibroblasts are defective in collagen ligation, resulting in abnormal regulation of collagenase and collagen synthesis, which in keloid fibroblasts may counteract negative regulation of MMP1 expression by TGF-β (Heino, 2000). On the other hand, certain MMPs, ADAMs, and ADAMTSs may contribute to fibrosis (e.g. by activation of TGF-β1) (Ribeiro et al., 1999; Annes et al., 2003). Furthermore, a marked increase of PAI1 with a concomitant decrease of uPA expression levels is observed in keloid fibroblasts.

This reduces the ability of keloid fibroblasts to degrade fibrin but may also reduce activation of other proteinases such as MMPs (Tuan et al., 2003).

Scarless wound healing

Numerous observations have revealed that unlike postnatal human skin wounds, fetal skin wounds before 24 weeks of gestation, as well as adult oral mucosal wounds, heal rapidly and without scarring or with minimal scarring. There are a number of external and internal factors that differ between adult skin and fetal skin and adult oral mucosal wounds. The latter two heal in a moist environment rich in various soluble factors. Fetal skin wounds are also located in a sterile environment and practically lack the inflammatory response during epidermal wound healing, while oral wounds face the oral microbiome (Häkkinen et al., 2000; Buchanan et al., 2009). Fetal skin fibroblasts also differ from adult skin fibroblasts by exhibiting more rapid migration and more dynamic production of ECM during wound healing. As a consequence, fetal and postnatal wounds differ with respect to composition of the ECM (Buchanan et al., 2009).

The histological landmark for scarring is dermal accumulation of thick and tense collagen bundles; thus, a potent element in scarless repair should be efficient ECM remodeling. Many studies have shown that fetal skin fibroblasts display increased ability to remodel collagen matrix *in vitro* in comparison to adult skin fibroblasts (Irwin et al., 1998; Sandulache et al., 2007). Moreover, the high expression ratio of TGF-β3 to fibrogenic isoforms TGF-β1 and -β2 in fetal skin at early gestation time is associated with scarless repair (Nath et al., 1994; Chen et al., 2005). Accordingly, exogenously added TGF-β3 and extinguishing both TGF-β1 and -β2 with function-blocking antibodies reduce scarring in a rat model (Shah et al., 1995).

Several studies indicate upregulation of MMPs and TIMPs in intact fetal skin as the function of gestation time and from scarless healing to healing with scar formation. For instance, in human fetal skin, the levels of MMP2, 9, and 14, as well as TIMP1 and TIMP2, increase during gestation time (Chen et al., 2007). In rat fetal skin, the levels of interstitial collagenase; MMP2, 3, 9, and 14; and TIMP2 were shown to increase in a similar manner. The level of TIMP3 appeared to decrease during gestation time (Peled et al., 2002; Dang et al., 2003). However, in early gestation fetal rat skin, injury induces higher expression of interstitial collagenase, MMP2, MMP9, and MMP14 and lower expression of TIMPs than in late gestation rat skin. Thus, a markedly higher MMP:TIMP ratio is detected in scarless wounds than in scarring wounds (Dang et al., 2003). In wounded human fetal skin (gestational age 16–20 weeks) grafted onto SCID mice, fibroblasts express MMP13 (Ravanti et al., 2001), which also can proteolytically activate TGF-β3 (Deng et al., 2000). MMP13 is also expressed by gingival wound fibroblasts, while it is absent in acute skin wounds in adults (Ravanti et al., 1999; Vaalamo et al., 1997). Both gingival and fetal skin fibroblasts also express MMP13 in response to TGF-β in culture (Ravanti et al., 1999, 2001). Finally, when compared to adult skin fibroblasts, oral fibroblasts inside three-dimensional collagen exhibit markedly elevated activation of MMP2, while the levels of TIMP1 and 2 are significantly lower (Stephens et al., 2001). In fetal skin, the level of uPA is decreased over time during development, while the expression of its inhibitor PAI1 is increased (Huang et al., 2002). Therefore, it is likely that different proteolytic expression profiles may contribute to scarless wound repair in fetal skin and oral mucosa.

4.6.15 Targeting proteolysis – applications for wound-healing therapy

Proteolysis plays an important role in all phases of physiological wound repair. However, alterations in proteolysis may severely impair wound closure. Even after extensive research, it is not known why increased expression levels or activity of certain MMPs (e.g. MMP2 and MMP9) are detected in chronic wounds. Moreover, in some cases, as for MMP13 expression by chronic wound fibroblasts, it remains unclear whether MMP production is contributing to poor healing or whether it is actually induced because of the altered stimuli in the wound striving for tissue repair. However, a general view concerning the chronic wounds appears to be that controlled inhibition of excessive proteolytic activity would stimulate healing.

Persistent and magnified inflammation is one of the main characteristics of chronic cutaneous ulcers. One explanation for high MMP levels in chronic ulcers is the increased infiltration of leukocytes, which secrete high amounts of neutrophil elastase, MMP8, and gelatinases (Nwomeh et al., 1999; Mirastschijski et al., 2002a; Pirilä et al., 2007; Rayment et al., 2008). Increased activity could be due to the higher number of zymogen activators present in a chronic wound. Also, bacteria-derived molecules can enhance expression of several MMPs in fibroblasts (Kanangat et al., 2006). Thus, reduction of inflammation and fulminant bacterial infection is a pivotal objective during therapeutic intervention in chronic wounds.

Clinical trials for reducing overwhelming proteolytic activity in chronic venous leg ulcers and diabetic ulcers using proteinase absorbing matrices have shown somewhat promising results, especially combined with autologous growth factors (Veves et al., 2002; Vin et al., 2002; Kakagia et al., 2007). Furthermore, an approach to neutralize the detrimental effect of high proteinase levels in chronic wounds has been taken in a pilot study by topical administration of doxycyline, an antibiotic with MMP- and TACE-inhibitor properties, on diabetic foot ulcers (Chin et al., 2003). It is also suggested that glycosaminoglycan-containing compounds, such as sulodexide, which is a fibrinolytic compound successfully used for venous ulcer management (Scondotto et al., 1999; Coccheri et al., 2002), could mediate its function partially by inhibiting secretion and activation of MMPs (Ra and Parks, 2007; Mannello and Raffetto, 2011).

Some of the alterations in MMP expression of chronic wounds, such as the expression of MMP13 by fibroblasts (Vaalamo et al., 1997), the prominent expression of MMP26 by stromal cells (Pirilä et al., 2007), and the expression of tPA in epidermis (Weckroth et al., 2004), are not explained by the changes in cell number but rather by changed levels of regulatory stimuli in the wound bed or altered response for growth factors. Understanding the regulatory mechanisms of specific proteinase expression could give us possibilities to modify the pool of proteinases and provide a window for therapeutics to chronic ulcer treatment (Toriseva and Kähäri, 2009). Finally, targeted inhibition of specific proteinases would be of great importance, as MMP and other proteinase activity in general is clearly beneficial for wound closure. In this respect, it is critical to understand the roles of individual proteinases in normal and aberrant wound healing and further define the mechanisms of the diseases.

Excessive scarring may be a major cosmetic problem, but in severe cases it, can also interfere with mobility of extremities and normal function of skin. There has been a remarkable progression in understanding the mechanisms behind scarring. The role of the TGF-β family in fibrosis of skin and other organs is well established (Verrecchia and Mauviel, 2007). In scars, such as keloids and hypertrophic scars, inhibition of excessive collagen deposition is of great interest, as well as reduction of persistent contraction in the latter. In addition, enhancing collagen degradation and remodeling is a tempting approach for reducing scar formation. The concept of using collagenases in treatment of fibrotic conditions has recently been taken into clinical practice by using *Clostridium histolyticum* collagenase injections to reduce tissue fibrosis in Dupuytren's contracture, a fibroproliferative disorder causing flexion contracture of joints in palm and fingers resulting in severe disability of hand function (Hurst et al., 2009).

Elucidation of the mechanisms of scarless healing may give us clues about how to reduce scarring. Recently, scarless skin wound healing of athymic nude mice deficient of forkhead box protein 1 (FOXN1) was associated with increased levels of MMP13 and MMP9 (Gawronska-Kozak, 2011). Moreover, MMP-13 is expressed by fibroblasts in scarless human wounds (Ravanti et al., 1999, 2001). Thus, existence of MMP13 in a variety of wounds showing healing with minimal scarring suggests that MMP13 may play a role in scarless healing, which may involve effective collagen remodeling and activation of antiscarring growth factor TGF-β3 (Deng et al., 2000; Toriseva et al., 2007). Indeed, there is evidence that MMP13 could mediate the resolution of tissue fibrosis (Fallowfield et al., 2007; Endo et al., 2011). In an appropriate model of skin fibrosis, it would be tempting to utilize gene transfer for inducing gene expression of a specific proteinase in fibrotic tissue to explore the potential of gene therapy in tissue fibrosis (Kangasniemi et al., 2009). Finally, modern molecular techniques such as siRNAs, cell-type-specific expression vectors, or biomaterials as a vehicle for certain substances provide tools for more systematic studies concerning the roles of MMPs, plasmin and PAs, and the newcomers in the field, ADAMs and ADAMTSs, that are needed for generating novel therapeutics for aberrant wound healing.

4.6.16 Take-home message

While physiological wound healing in skin efficiently rescues normal tissue function, chronic ulcers and fibrotic scars impair the quality of life of millions of people worldwide. An aberrant proteolytic profile is connected to both of these abnormal wounds, and controlling it could pose potential therapeutics for impaired wound repair. Regardless of general interest for decreasing excessive MMP activity in chronic wounds, there is still little data available on therapeutic feasibility of MMP inhibitors for these wounds. Moreover, data on controlled expression of certain proteinases in fibrotic wounds would be very useful. At present, a variety of more or less specific vehicles are available for delivery of inhibitors or gene vectors *in vivo*. As understanding of the specific roles of individual MMPs and other proteinases in wound healing is increasing, the possibilities for using targeted inhibition of specific proteinases are becoming available. Finally, although they are associated with several wound-healing-related events, little is known about the roles of ADAM and ADAMTS proteinases, specifically in wounds. Elucidation of their roles and expression pattern in normal and aberrant wounds could reveal novel mechanisms of the diseases and provide more possibilities for developing personalized therapy for patients with chronic ulcers or excessive scarring of cutaneous wounds.

References

Ågren, M. S. (1994). Gelatinase activity during wound healing. Br J Dermatol *131*, 634–640.

Ahokas, K., Skoog, T., Suomela, S., et al. (2005). Matrilysin-2 (matrix metalloproteinase-26) is upregulated in keratinocytes during wound repair and early skin carcinogenesis. J Invest Dermatol *124*, 849–856.

Andreadis, S. T., Hamoen, K. E., Yarmush, M. L., and Morgan, J. R. (2001). Keratinocyte growth factor induces hyperproliferation and delays differentiation in a skin equivalent model system. FASEB J *15*, 898–906.

Annes, J. P., Munger, J. S., and Rifkin, D. B. (2003). Making sense of latent TGFβ activation. J Cell Sci *116*, 217–224.

Aplin, A. C., Zhu, W. H., Fogel, E., and Nicosia, R. F. (2009). Vascular regression and survival are differentially regulated by MT1-MMP and TIMPs in the aortic ring model of angiogenesis. Am J Physiol Cell Physiol *297*, C471–C480.

Apte, S. S. (2009). A disintegrin-like and metalloprotease (reprolysin-type) with thrombospondin type 1 motif (ADAMTS) superfamily: functions and mechanisms. J Biol Chem *284*, 31493–31497.

Baker, A. H., Edwards, D. R., and Murphy, G. (2002). Metalloproteinase inhibitors: biological actions and therapeutic opportunities. J Cell Sci *115*, 3719–3727.

Bao, P., Kodra, A., Tomic-Canic, M., Golinko, M. S., Ehrlich, H. P., and Brem, H. (2009). The role of vascular endothelial growth factor in wound healing. J Surg Res *153*, 347–358.

Bao, W., and Strömblad, S. (2004). Integrin αv-mediated inactivation of p53 controls a MEK1-dependent melanoma cell survival pathway in three-dimensional collagen. J Cell Biol *167*, 745–756.

Bauer, E. A., Cooper, T. W., Huang, J. S., Altman, J., and Deuel, T. F. (1985). Stimulation of in vitro human skin collagenase expression by platelet-derived growth factor. Proc Natl Acad Sci U S A *82*, 4132–4136.

Beare, A. H., O'Kane, S., Krane, S. M., and Ferguson, M. W. (2003). Severely impaired wound healing in the collagenase-resistant mouse. J Invest Dermatol *120*, 153–163.

Beck, I. M., Muller, M., Mentlein, R., et al. (2007). Matrix metalloproteinase-19 expression in keratinocytes is repressed by transcription factors Tst-1 nd Skn-1a: implications for keratinocyte differentiation. J Invest Dermatol *127*, 1107–1114.

Bergers, G., Brekken, R., McMahon, G., et al. (2000). Matrix metalloproteinase-9 triggers the angiogenic switch during carcinogenesis. Nat Cell Biol *2*, 737–744.

Bevan, D., Gherardi, E., Fan, T. P., Edwards, D., and Warn, R. (2004). Diverse and potent activities of HGF/SF in skin wound repair. J Pathol *203*, 831–838.

Bonnefoy, A., and Legrand, C. (2000). Proteolysis of subendothelial adhesive glycoproteins (fibronectin, thrombospondin, and von Willebrand factor) by plasmin, leukocyte cathepsin G, and elastase. Thromb Res *98*, 323–332.

Bran, G. M., Goessler, U. R., Hormann, K., Riedel, F., and Sadick, H. (2009). Keloids: current concepts of pathogenesis (review). Int J Mol Med *24*, 283–293.

Brauer, R., Beck, I. M., Roderfeld, M., Roeb, E., and Sedlacek, R. (2011). Matrix metalloproteinase-19 inhibits growth of endothelial cells by generating angiostatin-like fragments from plasminogen. BMC Biochem *12*, 38.

Brown, G. L., Nanney, L. B., Griffen, J., et al. (1989). Enhancement of wound healing by topical treatment with epidermal growth factor. N Engl J Med *321*, 76–79.

Buchanan, E. P., Longaker, M. T., and Lorenz, H. P. (2009). Fetal skin wound healing. Adv Clin Chem *48*, 137–161.

Bugge, T. H., Kombrinck, K. W., Flick, M. J., Daugherty, C. C., Danton, M. J., and Degen, J. L. (1996). Loss of fibrinogen rescues mice from the pleiotropic effects of plasminogen deficiency. Cell *87*, 709–719.

Bullard, K. M., Lund, L., Mudgett, J. S., et al. (1999). Impaired wound contraction in stromelysin-1-deficient mice. Ann Surg *230*, 260–265.

Bullen, E. C., Longaker, M. T., Updike, D. L., et al. (1995). Tissue inhibitor of metalloproteinases-1 is decreased and activated gelatinases are increased in chronic wounds. J Invest Dermatol *104*, 236–240.

Cesarman-Maus, G., and Hajjar, K. A. (2005). Molecular mechanisms of fibrinolysis. Br J Haematol *129*, 307–321.

Chen, W., Fu, X., Ge, S., Sun, T., and Sheng, Z. (2007). Differential expression of matrix metalloproteinases and tissue-derived inhibitors of metalloproteinase in fetal and adult skins. Int J Biochem Cell Biol *39*, 997–1005.

Chen, W., Fu, X., Ge, S., et al. (2005). Ontogeny of expression of transforming growth factor-β and its receptors and their possible relationship with scarless healing in human fetal skin. Wound Repair Regen *13*, 68–75.

Chin, G. A., Thigpin, T. G., Perrin, K. J., Moldawer, L. L., and Schultz, G. S. (2003). Treatment of chronic ulcers in diabetic patients with a topical metalloproteinase inhibitor, doxycycline. Wounds *15*, 315–323.

Chin, G. S., Liu, W., Peled, Z., et al. (2001). Differential expression of transforming growth factor-beta receptors I and II and activation of Smad 3 in keloid fibroblasts. Plast Reconstr Surg *108*, 423–429.

Chmielowiec, J., Borowiak, M., Morkel, M., et al. (2007). c-Met is essential for wound healing in the skin. J Cell Biol *177*, 151–162.

Chun, T. H., Sabeh, F., Ota, I., et al. (2004). MT1-MMP-dependent neovessel formation within the confines of the three-dimensional extracellular matrix. J Cell Biol *167*, 757–767.

Clark, R. A. F. (1995). Wound repair: overview and general considerations. In The Molecular and Cellular Biology of Wound Repair, 2nd edn. New York: Plenum Press, pp. 513–560.

Coccheri, S., Scondotto, G., Agnelli, G., Aloisi, D., Palazzini, E., and Zamboni, V. (2002). Randomised, double blind, multicentre, placebo controlled study of sulodexide in the treatment of venous leg ulcers. Thromb Haemost *87*, 947–952.

Colige, A., Ruggiero, F., Vandenberghe, I., et al. (2005). Domains and maturation processes that regulate the activity of ADAMTS-2, a metalloproteinase cleaving the aminopropeptide of fibrillar procollagens types I-III and V. J Biol Chem *280*, 34397–34408.

Cook, H., Davies, K. J., Harding, K. G., and Thomas, D. W. (2000). Defective extracellular matrix reorganization by chronic wound fibroblasts is associated with alterations in TIMP-1, TIMP-2, and MMP-2 activity. J Invest Dermatol *115*, 225–233.

Cornelius, L. A., Nehring, L. C., Harding, E., et al. (1998). Matrix metalloproteinases generate angiostatin: effects on neovascularization. J Immunol *161*, 6845–6852.

Dallas, S. L., Rosser, J. L., Mundy, G. R., and Bonewald, L. F. (2002). Proteolysis of latent transforming growth factor-β (TGF-β)-binding protein-1 by osteoclasts. A cellular mechanism for release of TGF-beta from bone matrix. J Biol Chem *277*, 21352–21360.

Dang, C. M., Beanes, S. R., Lee, H., Zhang, X., Soo, C., and Ting, K. (2003). Scarless fetal wounds are associated with an increased matrix metalloproteinase-to-tissue-derived inhibitor of metalloproteinase ratio. Plast Reconstr Surg *111*, 2273–2285.

Dean, R. A., Butler, G. S., Hamma-Kourbali, Y., et al. (2007). Identification of candidate angiogenic inhibitors processed by MMP-2 in cell based proteomic screens: disruption of VEGF/HARP (pleiotrophin) and VEGF/CTGF angiogenic inhibitory complexes by MMP-2 proteolysis. Mol Cell Biol *27*, 8454–8465.

Deng, S. J., Bickett, D. M., Mitchell, J. L., et al. (2000). Substrate specificity of human collagenase 3 assessed using a phage-displayed peptide library. J Biol Chem *275*, 31422–31427.

Desmouliere, A., Redard, M., Darby, I., and Gabbiani, G. (1995). Apoptosis mediates the decrease in cellularity during the transition between granulation tissue and scar. Am J Pathol *146*, 56–66.

Di Colandrea, T., Wang, L., Wille, J., D'Armiento, J., and Chada, K. K. (1998). Epidermal expression of collagenase delays wound-healing in transgenic mice. J Invest Dermatol *111*, 1029–1033.

Dumin, J. A., Dickeson, S. K., Stricker, T. P., et al. (2001). Pro-collagenase-1 (matrix metalloproteinase-1) binds the α(2)β(1) integrin upon release from keratinocytes migrating on type I collagen. J Biol Chem *276*, 29368–29374.

Edmondson, S. R., Thumiger, S. P., Werther, G. A., and Wraight, C. J. (2003). Epidermal homeostasis: the role of the growth hormone and insulin-like growth factor systems. Endocr Rev *24*, 737–764.

Edwards, D. R., Handsley, M. M., and Pennington, C. J. (2008). The ADAM metalloproteinases. Mol Aspects Med *29*, 258–589.

Ehrlich, H. P., Desmouliere, A., Diegelmann, R. F., et al. (1994). Morphological and immunochemical differences between keloid and hypertrophic scar. Am J Pathol *145*, 105–113.

Eming, S. A., Hammerschmidt, M., Krieg, T., and Roers, A. (2009). Interrelation of immunity and tissue repair or regeneration. Semin Cell Dev Biol *20*, 517–527.

Eming, S. A., Krieg, T., and Davidson, J. M. (2007). Inflammation in wound repair: molecular and cellular mechanisms. J Invest Dermatol *127*, 514–525.

Endo, H., Niioka, M., Sugioka, Y., et al. (2011). Matrix metalloproteinase-13 promotes recovery from experimental liver cirrhosis in rats. Pathobiology *78*, 239–252.

Fallowfield, J. A., Mizuno, M., Kendall, T. J., et al. (2007). Scar-associated macrophages are a major source of hepatic matrix metalloproteinase-13 and facilitate the resolution of murine hepatic fibrosis. J Immunol *178*, 5288–5295.

Franzke, C. W., Tasanen, K., Schacke, H., et al. (2002). Transmembrane collagen XVII, an epithelial adhesion protein, is shed from the cell surface by ADAMs. EMBO J *21*, 5026–5035.

Frøssing, S., Rønø, B., Hald, A., Rømer, J., and Lund, L. R. (2010). Skin wound healing in MMP2-deficient and MMP2/plasminogen double-deficient mice. Exp Dermatol *19*, e234–240.

Fujiwara, M., Muragaki, Y., and Ooshima, A. (2005). Keloid-derived fibroblasts show increased secretion of factors involved in collagen turnover and depend on matrix metalloproteinase for migration. Br J Dermatol *153*, 295–300.

Galvez, B. G., Matias-Roman, S., Yanez-Mo, M., Sanchez-Madrid, F., and Arroyo, A. G. (2002). ECM regulates MT1-MMP localization with β1 or αvβ3 integrins at distinct cell compartments modulating its internalization and activity on human endothelial cells. J Cell Biol *159*, 509–521.

Garton, K. J., Gough, P. J., Blobel, C. P., et al. (2001). Tumor necrosis factor-α-converting enzyme (ADAM17) mediates the cleavage and shedding of fractalkine (CX3CL1). J Biol Chem *276*, 37993–38001.

Gawronska-Kozak, B. (2011). Scarless skin wound healing in FOXN1 deficient (nude) mice is associated with distinctive matrix metalloproteinase expression. Matrix Biol. *30* (4), 290–300.

Gearing, A. J., Beckett, P., Christodoulou, M., et al. (1994). Processing of tumour necrosis factor-α precursor by metalloproteinases. Nature *370*, 555–557.

Genis, L., Gonzalo, P., Tutor, A. S., et al.(2007). Functional interplay between endothelial nitric oxide synthase and membrane type 1 matrix metalloproteinase in migrating endothelial cells. Blood *110*, 2916–2923.

Ghahary, A., Shen, Y. J., Nedelec, B., Wang, R., Scott, P. G., and Tredget, E. E. (1996). Collagenase production is lower in post-burn hypertrophic scar fibroblasts than in normal fibroblasts and is reduced by insulin-like growth factor-1. J Invest Dermatol *106*, 476–481.

Ghahary, A., Shen, Y. J., Scott, P. G., Gong, Y., and Tredget, E. E. (1993). Enhanced expression of mRNA for transforming growth factor-β, type I and type III procollagen in human post-burn hypertrophic scar tissues. J Lab Clin Med *122*, 465–473.

Goldfinger, L. E., Stack, M. S., and Jones, J. C. (1998). Processing of laminin-5 and its functional consequences: role of plasmin and tissue-type plasminogen activator. J Cell Biol *141*, 255–265.

Gurtner, G. C., Werner, S., Barrandon, Y., and Longaker, M. T. (2008). Wound repair and regeneration. Nature *453*, 314–321.

Gutierrez-Fernandez, A., Inada, M., Balbin, M., et al. (2007). Increased inflammation delays wound healing in mice deficient in collagenase-2 (MMP-8). FASEB J *21*, 2580–2591.

Häkkinen, L., Uitto, V. J., and Larjava, H. (2000). Cell biology of gingival wound healing. Periodontol 2000 *24*, 127–152.

Hartenstein, B., Dittrich, B. T., Stickens, D., et al. (2006). Epidermal development and wound healing in matrix metalloproteinase 13-deficient mice. J Invest Dermatol *126*, 486–496.

Hasty, K. A., Hibbs, M. S., Kang, A. H., and Mainardi, C. L. (1986). Secreted forms of human neutrophil collagenase. J Biol Chem *261*, 5645–5650.

Hattori, N., Carrino, D. A., Lauer, M. E., et al. (2011). Pericellular versican regulates the fibroblast-myofibroblast transition. A role for ADAMTS5 protease-mediated proteolysis. J Biol Chem *286*, 34298–34310.

Hattori, N., Mochizuki, S., Kishi, K., et al. (2009). MMP-13 plays a role in keratinocyte migration, angiogenesis, and contraction in mouse skin wound healing. Am J Pathol *175*, 533–546.

Heino, J. (2000). The collagen receptor integrins have distinct ligand recognition and signaling functions. Matrix Biol *19*, 319–323.

Heljasvaara, R., Nyberg, P., Luostarinen, J., et al. (2005). Generation of biologically active endostatin fragments from human collagen XVIII by distinct matrix metalloproteases. Exp Cell Res *307*, 292–304.

Hieta, N., Impola, U., Lopez-Otin, C., Saarialho-Kere, U., and Kähäri, V.-M. (2003). Matrix metalloproteinase-19 expression in dermal wounds and by fibroblasts in culture. J Invest Dermatol *121*, 997–1004.

Hiller, O., Lichte, A., Oberpichler, A., Kocourek, A., and Tschesche, H. (2000). Matrix metalloproteinases collagenase-2, macrophage elastase, collagenase-3, and membrane type 1-matrix metalloproteinase impair clotting by degradation of fibrinogen and factor XII. J Biol Chem *275*, 33008–33013.

Hinz, B. (2007). Formation and function of the myofibroblast during tissue repair. J Invest Dermatol *127*, 526–537.

Hiraoka, N., Allen, E., Apel, I. J., Gyetko, M. R., and Weiss, S. J. (1998). Matrix metalloproteinases regulate neovascularization by acting as pericellular fibrinolysins. Cell *95*, 365–377.

Houghton, A. M., Hartzell, W. O., Robbins, C. S., Gomis-Ruth, F. X., and Shapiro, S. D. (2009). Macrophage elastase kills bacteria within murine macrophages. Nature *460*, 637–641.

Huang, E. Y., Wu, H., Island, E. R., et al. (2002). Differential expression of urokinase-type plasminogen activator and plasminogen activator inhibitor-1 in early and late gestational mouse skin and skin wounds. Wound Repair Regen *10*, 387–96.

Hundhausen, C., Schulte, A., Schulz, B., et al. (2007). Regulated shedding of transmembrane chemokines by the disintegrin and metalloproteinase 10 facilitates detachment of adherent leukocytes. J Immunol *178*, 8064–8072.

Huovila, A. P., Turner, A. J., Pelto-Huikko, M., Kärkkäinen, I., and Ortiz, R. M. (2005). Shedding light on ADAM metalloproteinases. Trends Biochem Sci *30*, 413–422.

Hurst, L. C., Badalamente, M. A., Hentz, V. R., et al. (2009). Injectable collagenase clostridium histolyticum for Dupuytren's contracture. N Engl J Med *361*, 968–979.

Hynes, R. O. (2007). Cell-matrix adhesion in vascular development. J Thromb Haemost *5* (Suppl. 1), 32–40.

Imai, K., Hiramatsu, A., Fukushima, D., Pierschbacher, M. D., and Okada, Y. (1997). Degradation of decorin by matrix metalloproteinases: identification of the cleavage sites, kinetic analyses and transforming growth factor-β1 release. Biochem J *322* (Pt 3), 809–814.

Inkinen, K., Turakainen, H., Wolff, H., Ravanti, L., Kähäri, V.-M., and Ahonen, J. (2000). Expression and activity of matrix metalloproteinase-2 and -9 in experimental granulation tissue. APMIS *108*, 318–328.

Inoue, M., Kratz, G., Haegerstrand, A., and Stahle-Backdahl, M. (1995). Collagenase expression is rapidly induced in wound-edge keratinocytes after acute injury in human skin, persists during healing, and stops at re-epithelialization. J Invest Dermatol *104*, 479–483.

Iruela-Arispe, M. L., Carpizo, D., and Luque, A. (2003). ADAMTS1: a matrix metalloprotease with angioinhibitory properties. Ann N Y Acad Sci *995*, 183–190.

Irwin, C. R., Myrillas, T., Smyth, M., Doogan, J., Rice, C., and Schor, S. L. (1998). Regulation of fibroblast-induced collagen gel contraction by interleukin-1beta. J Oral Pathol Med *27*, 255–259.

Itoh, T., Tanioka, M., Yoshida, H., Yoshioka, T., Nishimoto, H., and Itohara, S. (1998). Reduced angiogenesis and tumor progression in gelatinase A-deficient mice. Cancer Res *58*, 1048–1051.

Izumi, Y., Hirata, M., Hasuwa, H., et al. (1998). A metalloprotease-disintegrin, MDC9/meltrin-gamma/ADAM9 and PKCdelta are involved in TPA-induced ectodomain shedding of membrane-anchored heparin-binding EGF-like growth factor. EMBO J *17*, 7260–7272.

Jost, M., Folgueras, A. R., Frerart, F., et al. (2006). Earlier onset of tumoral angiogenesis in matrix metalloproteinase-19-deficient mice. Cancer Res *66*, 5234–5241.

Juncker-Jensen, A., and Lund, L. R. (2011). Phenotypic overlap between MMP-13 and the plasminogen activation system during wound healing in mice. PLoS One *6*, e16954.

Kakagia, D. D., Kazakos, K. J., Xarchas, K. C., et al. (2007). Synergistic action of protease-modulating matrix and autologous growth factors in healing of diabetic foot ulcers. A prospective randomized trial. J Diabetes Complications *21*, 387–391.

Kanangat, S., Postlethwaite, A., Hasty, K., et al. (2006). Induction of multiple matrix metalloproteinases in human dermal and synovial fibroblasts by Staphylococcus aureus: implications in the pathogenesis of septic arthritis and other soft tissue infections. Arthritis Res Ther *8*, R176.

Kangasniemi, L., Koskinen, M., Jokinen, M., et al. (2009). Extended release of adenovirus from silica implants in vitro and in vivo. Gene Ther *16*, 103–110.

Kashiwagi, M., Tortorella, M., Nagase, H., and Brew, K. (2001). TIMP-3 is a potent inhibitor of aggrecanase 1 (ADAM-TS4) and aggrecanase 2 (ADAM-TS5). J Biol Chem *276*, 12501–12504.

Kato, T., Kure, T., Chang, J. H., et al. (2001). Diminished corneal angiogenesis in gelatinase A-deficient mice. FEBS Lett *508*, 187–190.

Kessenbrock, K., Plaks, V., and Werb, Z. (2010). Matrix metalloproteinases: regulators of the tumor microenvironment. Cell *141*, 52–67.

Krampert, M., Bloch, W., Sasaki, T., et al. (2004). Activities of the matrix metalloproteinase stromelysin-2 (MMP-10) in matrix degradation and keratinocyte organization in wounded skin. Mol Biol Cell *15*, 5242–5254.

Krampert, M., Kuenzle, S., Thai, S. N., Lee, N., Iruela-Arispe, M. L., and Werner, S. (2005). ADAMTS1 proteinase is up-regulated in wounded skin and regulates migration of fibroblasts and endothelial cells. J Biol Chem *280*, 23844–23852.

Kruse, M. N., Becker, C., Lottaz, D., et al. (2004). Human meprin α and β homo-oligomers: cleavage of basement membrane proteins and sensitivity to metalloprotease inhibitors. Biochem J *378*, 383–389.

Kuo, Y. R., Wu, W. S., Jeng, S. F., et al. (2005). Suppressed TGF-β1 expression is correlated with up-regulation of matrix metalloproteinase-13 in keloid regression after flashlamp pulsed-dye laser treatment. Lasers Surg Med *36*, 38–42.

Kyriakides, T. R., Wulsin, D., Skokos, E. A., et al. (2009). Mice that lack matrix metalloproteinase-9 display delayed wound healing associated with delayed reepithelization and disordered collagen fibrillogenesis. Matrix Biol *28*, 65–73.

Laato, M., Kähäri, V.-M., Niinikoski, J., and Vuorio, E. (1987). Epidermal growth factor increases collagen production in granulation tissue by stimulation of fibroblast proliferation and not by activation of procollagen genes. Biochem J *247*, 385–388.

Lambert, E., Dasse, E., Haye, B., and Petitfrere, E. (2004). TIMPs as multifacial proteins. Crit Rev Oncol Hematol *49*, 187–198.

Lee, H., Overall, C. M., McCulloch, C. A., and Sodek, J. (2006a). A critical role for the membrane-type 1 matrix metalloproteinase in collagen phagocytosis. Mol Biol Cell *17*, 4812–4826.

Lee, N. V., Sato, M., Annis, D. S., et al. (2006b). ADAMTS1 mediates the release of antiangiogenic polypeptides from TSP1 and 2. EMBO J *25*, 5270–5283.

Lee, T. Y., Chin, G. S., Kim, W. J., Chau, D., Gittes, G. K., and Longaker, M. T. (1999). Expression of transforming growth factor β 1, 2, and 3 proteins in keloids. Ann Plast Surg *43*, 179–184.

Leibovich, S. J., and Ross, R. (1975). The role of the macrophage in wound repair. A study with hydrocortisone and antimacrophage serum. Am J Pathol *78*, 71–100.

Lelongt, B., Bengatta, S., Delauche, M., Lund, L. R., Werb, Z., and Ronco, P. M. (2001). Matrix metalloproteinase 9 protects mice from anti-glomerular basement membrane nephritis through its fibrinolytic activity. J Exp Med *193*, 793–802.

Leppert, D., Waubant, E., Galardy, R., Bunnett, N. W., and Hauser, S. L. (1995). T cell gelatinases mediate basement membrane transmigration in vitro. J Immunol *154*, 4379–4389.

Levi, E., Fridman, R., Miao, H. Q., Ma, Y. S., Yayon, A., and Vlodavsky, I. (1996). Matrix metalloproteinase 2 releases active soluble ectodomain of fibroblast growth factor receptor 1. Proc Natl Acad Sci U S A *93*, 7069–7074.

Li, Y., Fan, J., Chen, M., Li, W., and Woodley, D. T. (2006). Transforming growth factor-α: a major human serum factor that promotes human keratinocyte migration. J Invest Dermatol *126*, 2096–2105.

Lobmann, R., Ambrosch, A., Schultz, G., Waldmann, K., Schiweck, S., and Lehnert, H. (2002). Expression of matrix-metalloproteinases and their inhibitors in the wounds of diabetic and non-diabetic patients. Diabetologia *45*, 1011–1016.

Lohi, J., Wilson, C. L., Roby, J. D., and Parks, W. C. (2001). Epilysin, a novel human matrix metalloproteinase (MMP-28) expressed in testis and keratinocytes and in response to injury. J Biol Chem *276*, 10134–10144.

Lund, L. R., Green, K. A., Stoop, A. A., et al. (2006). Plasminogen activation independent of uPA and tPA maintains wound healing in gene-deficient mice. EMBO J *25*, 2686–2697.

Lund, L. R., Rømer, J., Bugge, T. H., et al. (1999). Functional overlap between two classes of matrix-degrading proteases in wound healing. EMBO J *18*, 4645–4656.

Luque, A., Carpizo, D. R., and Iruela-Arispe, M. L. (2003). ADAMTS1/METH1 inhibits endothelial cell proliferation by direct binding and sequestration of VEGF165. J Biol Chem *278*, 23656–23665.

Madlener, M., Parks, W. C., and Werner, S. (1998). Matrix metalloproteinases (MMPs) and their physiological inhibitors (TIMPs) are differentially expressed during excisional skin wound repair. Exp Cell Res *242*, 201–210.

Mäkelä, M., Larjava, H., Pirilä, E., et al. (1999). Matrix metalloproteinase 2 (gelatinase A) is related to migration of keratinocytes. Exp Cell Res *251*, 67–78.

Mannello, F., and Raffetto, J. D. (2011). Matrix metalloproteinase activity and glycosaminoglycans in chronic venous disease: the linkage among cell biology, pathology and translational research. Am J Transl Res *3*, 149–158.

Marchenko, N. D., Marchenko, G. N., Weinreb, R. N., et al. (2004). Beta-catenin regulates the gene of MMP-26, a novel metalloproteinase expressed both in carcinomas and normal epithelial cells. Int J Biochem Cell Biol *36*, 942–956.

Maretzky, T., Reiss, K., Ludwig, A., et al. (2005). ADAM10 mediates E-cadherin shedding and regulates epithelial cell-cell adhesion, migration, and beta-catenin translocation. Proc Natl Acad Sci U S A *102*, 9182–9187.

Martin, J., Eynstone, L. V., Davies, M., Williams, J. D., and Steadman, R. (2002). The role of ADAM 15 in glomerular mesangial cell migration. J Biol Chem *277*, 33683–33689.

Martin-Martin, B., Tovell, V., Dahlmann-Noor, A. H., Khaw, P. T., and Bailly, M. (2011). The effect of MMP inhibitor GM6001 on early fibroblast-mediated collagen matrix contraction is correlated to a decrease in cell protrusive activity. Eur J Cell Biol *90*, 26–36.

Mauch, C., Zamek, J., Abety, A. N., Grimberg, G., Fox, J. W., and Zigrino, P. (2010). Accelerated wound repair in ADAM-9 knockout animals. J Invest Dermatol *130*, 2120–2130.

McCawley, L. J., and Matrisian, L. M. (2001). Matrix metalloproteinases: they're not just for matrix anymore! Curr Opin Cell Biol *13*, 534–450.

Menke, N. B., Ward, K. R., Witten, T. M., Bonchev, D. G., and Diegelmann, R. F. (2007). Impaired wound healing. Clin Dermatol *25*, 19–25.

Menshikov, M., Elizarova, E., Plakida, K., et al. (2002). Urokinase upregulates matrix metalloproteinase-9 expression in THP-1 monocytes via gene transcription and protein synthesis. Biochem J *367*, 833–839.

Midwood, K. S., Williams, L. V., and Schwarzbauer, J. E. (2004). Tissue repair and the dynamics of the extracellular matrix. Int J Biochem Cell Biol *36*, 1031–1037.

Millichip, M. I., Dallas, D. J., Wu, E., Dale, S., and McKie, N. (1998). The metallo-disintegrin ADAM10 (MADM) from bovine kidney has type IV collagenase activity in vitro. Biochem Biophys Res Commun *245*, 594–598.

Mimura, Y., Ihn, H., Jinnin, M., Asano, Y., Yamane, K., and Tamaki, K. (2006). Epidermal growth factor affects the synthesis and degradation of type I collagen in cultured human dermal fibroblasts. Matrix Biol *25*, 202–212.

Mirastschijski, U., Haaksma, C. J., Tomasek, J. J., and Ågren, M. S. (2004a). Matrix metalloproteinase inhibitor GM 6001 attenuates keratinocyte migration, contraction and myofibroblast formation in skin wounds. Exp Cell Res *299*, 465–475.

Mirastschijski, U., Impola, U., Jahkola, T., Karlsmark, T., Ågren, M. S., and Saarialho-Kere, U. (2002a). Ectopic localization of matrix metalloproteinase-9 in chronic cutaneous wounds. Hum Pathol *33*, 355–364.

Mirastschijski, U., Impola, U., Karsdal, M. A., Saarialho-Kere, U., and Ågren, M. S. (2002b). Matrix metalloproteinase inhibitor BB-3103 unlike the serine proteinase inhibitor aprotinin abrogates epidermal healing of human skin wounds ex vivo. J Invest Dermatol *118*, 55–64.

Mirastschijski, U., Zhou, Z., Rollman, O., Tryggvason, K., and Ågren, M. S. (2004b). Wound healing in membrane-type-1 matrix metalloproteinase-deficient mice. J Invest Dermatol *123*, 600–602.

Moali, C., and Hulmes, D. J. (2009). Extracellular and cell surface proteases in wound healing: new players are still emerging. Eur J Dermatol *19*, 552–564.

Mohan, R., Chintala, S. K., Jung, J. C., et al. (2002). Matrix metalloproteinase gelatinase B (MMP-9) coordinates and effects epithelial regeneration. J Biol Chem *277*, 2065–2072.

Muller, M., Trocme, C., Lardy, B., Morel, F., Halimi, S., and Benhamou, P. Y. (2008). Matrix metalloproteinases and diabetic foot ulcers: the ratio of MMP-1 to TIMP-1 is a predictor of wound healing. Diabet Med *25*, 419–426.

Murphy, G., and Nagase, H. (2008). Progress in matrix metalloproteinase research. Mol Aspects Med *29*, 290–308.

Nagano, O., Murakami, D., Hartmann, D., et al. (2004). Cell-matrix interaction via CD44 is independently regulated by different metalloproteinases activated in response to extracellular Ca(2+) influx and PKC activation. J Cell Biol *165*, 893–902.

Nath, R. K., LaRegina, M., Markham, H., Ksander, G. A., and Weeks, P. M. (1994). The expression of transforming growth factor type β in fetal and adult rabbit skin wounds. J Pediatr Surg *29*, 416–421.

Neely, A. N., Clendening, C. E., Gardner, J., Greenhalgh, D. G., and Warden, G. D. (1999). Gelatinase activity in keloids and hypertrophic scars. Wound Repair Regen *7*, 166–171.

Noe, V., Fingleton, B., Jacobs, K., et al. (2001). Release of an invasion promoter E-cadherin fragment by matrilysin and stromelysin-1. J Cell Sci *114*, 111–118.

Nwomeh, B. C., Liang, H. X., Cohen, I. K., and Yager, D. R. (1999). MMP-8 is the predominant collagenase in healing wounds and nonhealing ulcers. J Surg Res *81*, 189–195.

Oikarinen, A., Kylmäniemi, M., Autio-Harmainen, H., Autio, P., and Salo, T. (1993). Demonstration of 72-kDa and 92-kDa forms of type IV collagenase in human skin: variable expression in various blistering diseases, induction during re-epithelialization, and decrease by topical glucocorticoids. J Invest Dermatol *101*, 205–210.

Okada, S., Kita, H., George, T. J., Gleich, G. J., and Leiferman, K. M. (1997). Migration of eosinophils through basement membrane components in vitro: role of matrix metalloproteinase-9. Am J Respir Cell Mol Biol *17*, 519–528.

O'Reilly, M. S., Holmgren, L., Shing, Y., et al. (1994). Angiostatin: a novel angiogenesis inhibitor that mediates the suppression of metastases by a Lewis lung carcinoma. Cell *79*, 315–328.

O'Reilly, M. S., Wiederschain, D., Stetler-Stevenson, W. G., Folkman, J., and Moses, M. A. (1999). Regulation of angiostatin production by matrix metalloproteinase-2 in a model of concomitant resistance. J Biol Chem *274*, 29568–29571.

Page-McCaw, A., Ewald, A. J., and Werb, Z. (2007). Matrix metalloproteinases and the regulation of tissue remodelling. Nat Rev Mol Cell Biol *8*, 221–233.

Park, J. E., Keller, G. A., and Ferrara, N. (1993). The vascular endothelial growth factor (VEGF) isoforms: differential deposition into the subepithelial extracellular matrix and bioactivity of extracellular matrix-bound VEGF. Mol Biol Cell *4*, 1317–1326.

Peduto, L., Reuter, V. E., Shaffer, D. R., Scher, H. I., and Blobel, C. P. (2005). Critical function for ADAM9 in mouse prostate cancer. Cancer Res *65*, 9312–9319.

Peled, Z. M., Phelps, E. D., Updike, D. L., et al. (2002). Matrix metalloproteinases and the ontogeny of scarless repair: the other side of the wound healing balance. Plast Reconstr Surg *110*, 801–811.

Peltonen, J., Hsiao, L. L., Jaakkola, S., et al. (1991). Activation of collagen gene expression in keloids: co-localization of type I and VI collagen and transforming growth factor-β 1 mRNA. J Invest Dermatol *97*, 240–248.

Pepper, M. S. (2001). Role of the matrix metalloproteinase and plasminogen activator-plasmin systems in angiogenesis. Arterioscler Thromb Vasc Biol *21*, 1104–1117.

Pilcher, B. K., Dumin, J. A., Sudbeck, B. D., Krane, S. M., Welgus, H. G., and Parks, W. C. (1997a). The activity of collagenase-1 is required for keratinocyte migration on a type I collagen matrix. J Cell Biol *137*, 1445–1457.

Pilcher, B. K., Gaither-Ganim, J., Parks, W. C., and Welgus, H. G. (1997b). Cell type-specific inhibition of keratinocyte collagenase-1 expression by basic fibroblast growth factor and keratinocyte growth factor. A common receptor pathway. J Biol Chem *272*, 18147–18154.

Pins, G. D., Collins-Pavao, M. E., Van De Water, L., Yarmush, M. L., and Morgan, J. R. (2000). Plasmin triggers rapid contraction and degradation of fibroblast-populated collagen lattices. J Invest Dermatol *114*, 647–653.

Pirilä, E., Korpi, J. T., Korkiamäki, T., et al. (2007). Collagenase-2 (MMP-8) and matrilysin-2 (MMP-26) expression in human wounds of different etiologies. Wound Repair Regen *15*, 47–57.

Porter, S., Clark, I. M., Kevorkian, L., and Edwards, D. R. (2005). The ADAMTS metalloproteinases. Biochem J *386*, 15–27.

Ra, H. J., and Parks, W. C. (2007). Control of matrix metalloproteinase catalytic activity. Matrix Biol *26*, 587–596.

Ravanti, L., Häkkinen, L., Larjava, H., et al. (1999). Transforming growth factor-β induces collagenase-3 expression by human gingival fibroblasts via p38 mitogen-activated protein kinase. J Biol Chem *274*, 37292–37300.

Ravanti, L., Toriseva, M., Penttinen, R., et al. (2001). Expression of human collagenase-3 (MMP-13) by fetal skin fibroblasts is induced by transforming growth factor β via p38 mitogen-activated protein kinase. FASEB J *15*, 1098–1100.

Rayment, E. A., Upton, Z., and Shooter, G. K. (2008). Increased matrix metalloproteinase-9 (MMP-9) activity observed in chronic wound fluid is related to the clinical severity of the ulcer. Br J Dermatol *158*, 951–961.

Rechardt, O., Elomaa, O., Vaalamo, M., et al. (2000). Stromelysin-2 is upregulated during normal wound repair and is induced by cytokines. J Invest Dermatol *115*, 778–787.

Ribeiro, S. M., Poczatek, M., Schultz-Cherry, S., Villain, M., and Murphy-Ullrich, J. E. (1999). The activation sequence of thrombospondin-1 interacts with the latency-associated peptide to regulate activation of latent transforming growth factor-β. J Biol Chem *274*, 13586–13593.

Rodriguez-Manzaneque, J. C., Westling, J., Thai, S. N., et al. (2002). ADAMTS1 cleaves aggrecan at multiple sites and is differentially inhibited by metalloproteinase inhibitors. Biochem Biophys Res Commun *293*, 501–508.

Rømer, J., Bugge, T. H., Pyke, C., et al. (1996). Impaired wound healing in mice with a disrupted plasminogen gene. Nat Med *2*, 287–292.

Rømer, J., Lund, L. R., Eriksen, J., Pyke, C., Kristensen, P., and Dano, K. (1994). The receptor for urokinase-type plasminogen activator is expressed by keratinocytes at the leading edge during re-epithelialization of mouse skin wounds. J Invest Dermatol *102*, 519–522.

Rømer, J., Lund, L. R., Eriksen, J., et al. (1991). Differential expression of urokinase-type plasminogen activator and its type-1 inhibitor during healing of mouse skin wounds. J Invest Dermatol *97*, 803–811.

Saarialho-Kere, U., Kerkela, E., Jahkola, T., Suomela, S., Keski-Oja, J., and Lohi, J. (2002). Epilysin (MMP-28) expression is associated with cell proliferation during epithelial repair. J Invest Dermatol *119*, 14–21.

Saarialho-Kere, U. K., Crouch, E. C., and Parks, W. C. (1995). Matrix metalloproteinase matrilysin is constitutively expressed in adult human exocrine epithelium. J Invest Dermatol *105*, 190–196.

Saarialho-Kere, U. K., Kovacs, S. O., Pentland, A. P., Olerud, J. E., Welgus, H. G., and Parks, W. C. (1993). Cell-matrix interactions modulate interstitial collagenase expression by human keratinocytes actively involved in wound healing. J Clin Invest *92*, 2858–2866.

Sabeh, F., Li, X. Y., Saunders, T. L., Rowe, R. G., and Weiss, S. J. (2009). Secreted versus membrane-anchored collagenases: relative roles in fibroblast-dependent collagenolysis and invasion. J Biol Chem *284*, 23001–23011.

Sabeh, F., Ota, I., Holmbeck, K., et al. (2004). Tumor cell traffic through the extracellular matrix is controlled by the membrane-anchored collagenase MT1-MMP. J Cell Biol *167*, 769–781.

Sadowski, T., Dietrich, S., Koschinsky, F., et al. (2005). Matrix metalloproteinase 19 processes the laminin 5 gamma 2 chain and induces epithelial cell migration. Cell Mol Life Sci *62*, 870–880.

Sadowski, T., Dietrich, S., Koschinsky, F., and Sedlacek, R. (2003a). Matrix metalloproteinase 19 regulates insulin-like growth factor-mediated proliferation, migration, and adhesion in human keratinocytes through proteolysis of insulin-like growth factor binding protein-3. Mol Biol Cell *14*, 4569–4580.

Sadowski, T., Dietrich, S., Muller, M., et al. (2003b). Matrix metalloproteinase-19 expression in normal and diseased skin: dysregulation by epidermal proliferation. J Invest Dermatol *121*, 989–996.

Sahin, U., and Blobel, C. P. (2007). Ectodomain shedding of the EGF-receptor ligand epigen is mediated by ADAM17. FEBS Lett *581*, 41–44.

Sahin, U., Weskamp, G., Kelly, K., et al. (2004). Distinct roles for ADAM10 and ADAM17 in ectodomain shedding of six EGFR ligands. J Cell Biol *164*, 769–779.

Sandulache, V. C., Parekh, A., Dohar, J. E., and Hebda, P. A. (2007). Fetal dermal fibroblasts retain a hyperactive migratory and contractile phenotype under 2-and 3-dimensional constraints compared to normal adult fibroblasts. Tissue Eng *13*, 2791–2801.

Schäfer, B. M., Maier, K., Eickhoff, U., Todd, R. F., and Kramer, M. D. (1994). Plasminogen activation in healing human wounds. Am J Pathol *144*, 1269–1280.

Schwettmann, L., and Tschesche, H. (2001). Cloning and expression in Pichia pastoris of metalloprotease domain of ADAM 9 catalytically active against fibronectin. Protein Expr Purif *21*, 65–70.

Scondotto, G., Aloisi, D., Ferrari, P., and Martini, L. (1999). Treatment of venous leg ulcers with sulodexide. Angiology *50*, 883–889.

Scott, K. A., Wood, E. J., and Karran, E. H. (1998). A matrix metalloproteinase inhibitor which prevents fibroblast-mediated collagen lattice contraction. FEBS Lett *441*, 137–140.

Shah, M., Foreman, D. M., and Ferguson, M. W. (1995). Neutralisation of TGF-β 1 and TGF-β 2 or exogenous addition of TGF-β 3 to cutaneous rat wounds reduces scarring. J Cell Sci *108* (Pt 3), 985–1002.

Shaw, T. J., and Martin, P. (2009). Wound repair at a glance. J Cell Sci *122*, 3209–3213.

Shirakata, Y., Kimura, R., Nanba, D., et al. (2005). Heparin-binding EGF-like growth factor accelerates keratinocyte migration and skin wound healing. J Cell Sci *118*, 2363–2370.

Smith, J. C., Boone, B. E., Opalenik, S. R., Williams, S. M., and Russell, S. B. (2008). Gene profiling of keloid fibroblasts shows altered expression in multiple fibrosis-associated pathways. J Invest Dermatol *128*, 1298–1310.

Solorzano, C. C., Ksontini, R., Pruitt, J. H., et al. (1997). A matrix metalloproteinase inhibitor prevents processing of tumor necrosis factor α (TNF α) and abrogates endotoxin-induced lethality. Shock *7*, 427–431.

Stephens, P., Davies, K. J., Occleston, N., et al. (2001). Skin and oral fibroblasts exhibit phenotypic differences in extracellular matrix reorganization and matrix metalloproteinase activity. Br J Dermatol *144*, 229–237.

Sternlicht, M. D., and Werb, Z. (2001). How matrix metalloproteinases regulate cell behavior. Annu Rev Cell Dev Biol *17*, 463–516.

Stricker, T. P., Dumin, J. A., Dickeson, S. K., et al. (2001). Structural analysis of the alpha(2) integrin I domain/procollagenase-1 (matrix metalloproteinase-1) interaction. J Biol Chem *276*, 29375–29381.

Stricklin, G. P., Li, L., Jancic, V., Wenczak, B. A., and Nanney, L. B. (1993). Localization of mRNAs representing collagenase and TIMP in sections of healing human burn wounds. Am J Pathol *143*, 1657–1666.

Strongin, A. Y., Collier, I., Bannikov, G., Marmer, B. L., Grant, G. A., and Goldberg, G. I. (1995). Mechanism of cell surface activation of 72-kDa type IV collagenase. Isolation of the activated form of the membrane metalloprotease. J Biol Chem *270*, 5331–5338.

Sudbeck, B. D., Pilcher, B. K., Welgus, H. G., and Parks, W. C. (1997). Induction and repression of collagenase-1 by keratinocytes is controlled by distinct components of different extracellular matrix compartments. J Biol Chem *272*, 22103–22110.

Tokumaru, S., Higashiyama, S., Endo, T., et al. (2000). Ectodomain shedding of epidermal growth factor receptor ligands is required for keratinocyte migration in cutaneous wound healing. J Cell Biol *151*, 209–220.

Toriseva, M., and Kähäri, V.-M. (2009). Proteinases in cutaneous wound healing. Cell Mol Life Sci *66*, 203–224.

Toriseva, M. J., Ala-aho, R., Karvinen, J., et al. (2007). Collagenase-3 (MMP-13) enhances remodeling of three-dimensional collagen and promotes survival of human skin fibroblasts. J Invest Dermatol *127*, 49–59.

Tuan, T. L., Wu, H., Huang, E. Y., et al. (2003). Increased plasminogen activator inhibitor-1 in keloid fibroblasts may account for their elevated collagen accumulation in fibrin gel cultures. Am J Pathol *162*, 1579–1589.

Uutela, M., Wirzenius, M., Paavonen, K., et al. (2004). PDGF-D induces macrophage recruitment, increased interstitial pressure, and blood vessel maturation during angiogenesis. Blood *104*, 3198–3204.

Vaalamo, M., Kariniemi, A. L., Shapiro, S. D., and Saarialho-Kere, U. (1999a). Enhanced expression of human metalloelastase (MMP-12) in cutaneous granulomas and macrophage migration. J Invest Dermatol *112*, 499–505.

Vaalamo, M., Leivo, T., and Saarialho-Kere, U. (1999b). Differential expression of tissue inhibitors of metalloproteinases (TIMP-1, -2, -3, and -4) in normal and aberrant wound healing. Hum Pathol *30*, 795–802.

Vaalamo, M., Mattila, L., Johansson, N., et al. (1997). Distinct populations of stromal cells express collagenase-3 (MMP-13) and collagenase-1(MMP-1) in chronic ulcers but not in normally healing wounds. J Invest Dermatol *109*, 96–101.

Vaalamo, M., Weckroth, M., Puolakkainen, P., et al. (1996). Patterns of matrix metalloproteinase and TIMP-1 expression in chronic and normally healing human cutaneous wounds. Br J Dermatol *135*, 52–59.

Velasco, J., Li, J., DiPietro, L., Stepp, M. A., Sandy, J. D., and Plaas, A. (2011). Adamts5 deletion blocks murine dermal repair through CD44-mediated aggrecan accumulation and modulation of transforming growth factor β1 (TGFβ1) signaling. J Biol Chem *286*, 26016–26027.

Verrecchia, F., and Mauviel, A. (2007). Transforming growth factor-β and fibrosis. World J Gastroenterol *13*, 3056–3062.

Veves, A., Sheehan, P., and Pham, H. T. (2002). A randomized, controlled trial of Promogran (a collagen/oxidized regenerated cellulose dressing) vs standard treatment in the management of diabetic foot ulcers. Arch Surg *137*, 822–827.

Vihinen, P., and Kähäri, V.-M. (2002). Matrix metalloproteinases in cancer: prognostic markers and therapeutic targets. Int J Cancer *99*, 157–166.

Vin, F., Teot, L., and Meaume, S. (2002). The healing properties of Promogran in venous leg ulcers. J Wound Care *11*, 335–341.

Visse, R., and Nagase, H. (2003). Matrix metalloproteinases and tissue inhibitors of metalloproteinases: structure, function, and biochemistry. Circ Res *92*, 827–839.

Vu, T. H., Shipley, J. M., Bergers, G., et al. (1998). MMP-9/gelatinase B is a key regulator of growth plate angiogenesis and apoptosis of hypertrophic chondrocytes. Cell *93*, 411–422.

Wall, S. J., Bevan, D., Thomas, D. W., Harding, K. G., Edwards, D. R., and Murphy, G. (2002). Differential expression of matrix metalloproteinases during impaired wound healing of the diabetes mouse. J Invest Dermatol *119*, 91–98.

Wang, R., Ghahary, A., Shen, Q., Scott, P. G., Roy, K., and Tredget, E. E. (2000). Hypertrophic scar tissues and fibroblasts produce more transforming growth factor-beta1 mRNA and protein than normal skin and cells. Wound Repair Regen *8*, 128–137.

Wang, W. M., Ge, G., Lim, N. H., Nagase, H., and Greenspan, D. S. (2006). TIMP-3 inhibits the procollagen N-proteinase ADAMTS-2. Biochem J *398*, 515–519.

Weckroth, M., Vaheri, A., Virolainen, S., Saarialho-Kere, U., Jahkola, T., and Siren, V. (2004). Epithelial tissue-type plasminogen activator expression, unlike that of urokinase, its receptor, and plasminogen activator inhibitor-1, is increased in chronic venous ulcers. Br J Dermatol *151*, 1189–1196.

Weller, K., Foitzik, K., Paus, R., Syska, W., and Maurer, M. (2006). Mast cells are required for normal healing of skin wounds in mice. FASEB J *20*, 2366–2368.

Werner, S. (1998). Keratinocyte growth factor: a unique player in epithelial repair processes. Cytokine Growth Factor Rev *9*, 153–165.

Werner, S., and Grose, R. (2003). Regulation of wound healing by growth factors and cytokines. Physiol Rev *83*, 835–870.

Werner, S., Krieg, T., and Smola, H. (2007). Keratinocyte-fibroblast interactions in wound healing. J Invest Dermatol *127*, 998–1008.

Whitelock, J. M., Murdoch, A. D., Iozzo, R. V., and Underwood, P. A. (1996). The degradation of human endothelial cell-derived perlecan and release of bound basic fibroblast growth factor by stromelysin, collagenase, plasmin, and heparanases. J Biol Chem *271*, 10079–10086.

Wysocki, A. B., Staiano-Coico, L., and Grinnell, F. (1993). Wound fluid from chronic leg ulcers contains elevated levels of metalloproteinases MMP-2 and MMP-9. J Invest Dermatol *101*, 64–68.

Yager, D. R., Zhang, L. Y., Liang, H. X., Diegelmann, R. F., and Cohen, I. K. (1996). Wound fluids from human pressure ulcers contain elevated matrix metalloproteinase levels and activity compared to surgical wound fluids. J Invest Dermatol *107*, 743–748.

Yu, Q., and Stamenkovic, I. (2000). Cell surface-localized matrix metalloproteinase-9 proteolytically activates TGF-beta and promotes tumor invasion and angiogenesis. Genes Dev *14*, 163–176.

Yuan, W., and Varga, J. (2001). Transforming growth factor-β repression of matrix metalloproteinase-1 in dermal fibroblasts involves Smad3. J Biol Chem *276*, 38502–38510.

Zhang, Y., Zhou, Z. H., Bugge, T. H., and Wahl, L. M. (2007). Urokinase-type plasminogen activator stimulation of monocyte matrix metalloproteinase-1 production is mediated by plasmin-dependent signaling through annexin A2 and inhibited by inactive plasmin. J Immunol *179*, 3297–3304.

Zhou, Z., Apte, S. S., Soininen, R., et al. (2000). Impaired endochondral ossification and angiogenesis in mice deficient in membrane-type matrix metalloproteinase I. Proc Natl Acad Sci U S A *97*, 4052–4057.

Zigrino, P., Steiger, J., Fox, J. W., et al. (2007). Role of ADAM-9 disintegrin-cysteine rich domains in human keratinocyte migration. J Biol Chem *282*, 30785–30793.

4.7 Rock, paper, and molecular scissors: regulating the game of extracellular matrix homeostasis, remodeling, and inflammation

Antoine Dufour and Christopher M. Overall

4.7.1 Proteases

Proteolysis is the cleavage and sometimes degradation of proteins. Accomplished by hydrolases termed *proteases*, proteolysis occurs in virtually all biological processes, including DNA replication, transcription, translation, cell proliferation, angiogenesis, neurogenesis, cell migration and invasion, embryogenesis, and apoptosis (López-Otín and Bond, 2008). The biology of proteolysis was first reported more than a century ago in the first issue of the *Journal of Biological Chemistry* in 1905 by P. A. Levene. It was later demonstrated that proteases hydrolyze peptide bonds that link the amino acids of the polypeptide chain that constitutes a protein. Since the beginning of the twentieth century, our view of proteases has greatly expanded. Proteases can be either exopeptidases, cleaving the terminal peptide bond of a substrate, or endopeptidases, cleaving an internal peptide bond. Mammalian endopeptidases have been divided into five groups based on the different mechanisms of catalysis: metallo, cysteine, serine, threonine, and aspartic. The different classes are further grouped into families, based on amino acid sequence homology, and into clans, based on three-dimensional structural similarities (López-Otín and Bond, 2008). MEROPS (http://merops.sanger.asc.uk), the definitive database of proteases and inhibitors, provides information for many proteases and homologs in a variety of species. Another useful tool is the Degradome Database in which protease pseudogenes and retrovirus-derived sequences have been excluded. In this database, the number of human proteases and homologs totals 569, and they are classified into 68 families (López-Otín and Matrisian, 2007): the metalloprotease superfamily contains 194 members, while there are 176 serine proteases, 150 cysteine proteases, 28 threonine proteases, and 21 aspartic proteases. Within the metalloproteases, the metzincin family includes astacins, adamalysins (a disintegrin and metalloproteinases [ADAMs] and a disintegrin and metalloproteinase with thrombospondin motifs [ADAMTSs]), serralysins, snapalysins, leishmanolysins, and matrixins (also known as matrix metalloproteinases, or MMPs). Recently, a new knowledge base has been released called TopFIND that provides all substrates and N and C termini of all proteins of five species to form a very powerful integrated portal for the global analysis of substrates, proteases, and inhibitors (Lange and Overall, 2011).

4.7.2 Matrix metalloproteinases

In this book chapter, we will focus on one family of the metalloproteases: the MMPs. Notably, the same concepts and biochemical approaches are also applicable to studying other proteases (e.g. cathepsins, ADAMs). In humans, the MMP family consists of 24 members that require Zn^{2+} in their active site for proteolytic activity (Sternlicht and Werb, 2001; Kessenbrock et al., 2010). Crystal structures of various domains of MMP1, 2, 3, 7, 8, 9, 11, 12, 13, and 14 have been solved (Visse and Nagase, 2003). Typically, MMPs consist of five domains: (1) an N-terminal signal peptide (or "predomain"), which directs MMPs, like most secreted proteins, to the secretory pathway; (2) an ~80–90 residue prodomain that confers latency to the enzyme that is secreted as an inactive zymogen (pro-MMP); (3) a zinc-containing catalytic domain of ~165 residues in length; (4) a 15–65 residue flexible linker that connects the catalytic and the hemopexin domains; and (5) an ~200 residue hemopexin domain with a three-dimensional disklike β-propeller structure, which mediates interactions with substrates and confers specificity to the enzymes (▶Figure 4.12) (Overall and López-Otín, 2002). Each propeller blade is constituted of four antiparallel β strands and one α helix linked by two conserved cysteine residues. The hemopexin domain of MMP9 and MMP14/membrane type-1 matrix metalloproteinase (MT1-MMP) forms homodimers and/or heterodimers, which are required for cell migration, invasion, adhesion, and cell signaling (Cha et al., 2002; Cao et al., 2004; Dufour et al., 2010, 2011; Zarrabi et al., 2011).

Besides these five archetypal domains of secreted MMPs, there are six membrane-anchored members (MT-MMPs) that contain an additional hydrophobic transmembrane sequence and a short cytoplasmic domain at the C terminus or are glycosylphosphatidylinositol (GPI) anchored (▶Figure 4.12).

MMPs contain two highly conserved motifs, the "cysteine switch" and a zinc-binding motif. The "cysteine switch" motif PRCGXPD (where X is a variable amino acid) is located in the prodomain. The sulfhydryl group of the cysteine residue is critical in coordinating the zinc atom at the active site, blocking access of the essential nucleophilic H_2O molecule and thus conferring latency to the proenzyme (Vanwart and Birkedal-hansen, 1990; Nagase and Woessner, 1999). The crystal structures of MMP2, 3, and 9 show that the prodomain forms three α helices and connecting loops (Visse and Nagase, 2003). The current model suggests that the first loop between the helix 1 and helix 2 serves as a protease-sensitive "bait region" that upon cleavage destabilizes the prodomain leading to autolytic cleavage and enzyme activation, whereas the region after helix 3 lies in the catalytic domain substrate-binding pocket keeping the enzyme inactive (Visse and Nagase, 2003). Cleavage of the prodomain enables access of the catalytic water molecule and substrate molecules to the active site cleft. Within the catalytic domain, the extended zinc-binding motif, HEBXHXBGXHS (where B is a bulky hydrophobic amino acid), provides three zinc-binding histidines and a glutamate that polarizes the zinc-bound H_2O molecule to provide the nucleophile that cleaves peptide bonds (Birkedal-Hansen et al., 1993). Notably, a single mutation of any of the histidines or the glutamate ablates catalytic activity rendering the enzyme proteolytically inactive (Windsor et al., 1994).

There are several other nonproteolytic ways of activating MMPs that have been widely utilized *in vitro*: (1) chemical agents such as thiol-modifying agents ($HgCl_2$, 4-aminophenylmercuric acetate [APMA] and *N*-ethylmaleimide), (2) oxidized glutathione,

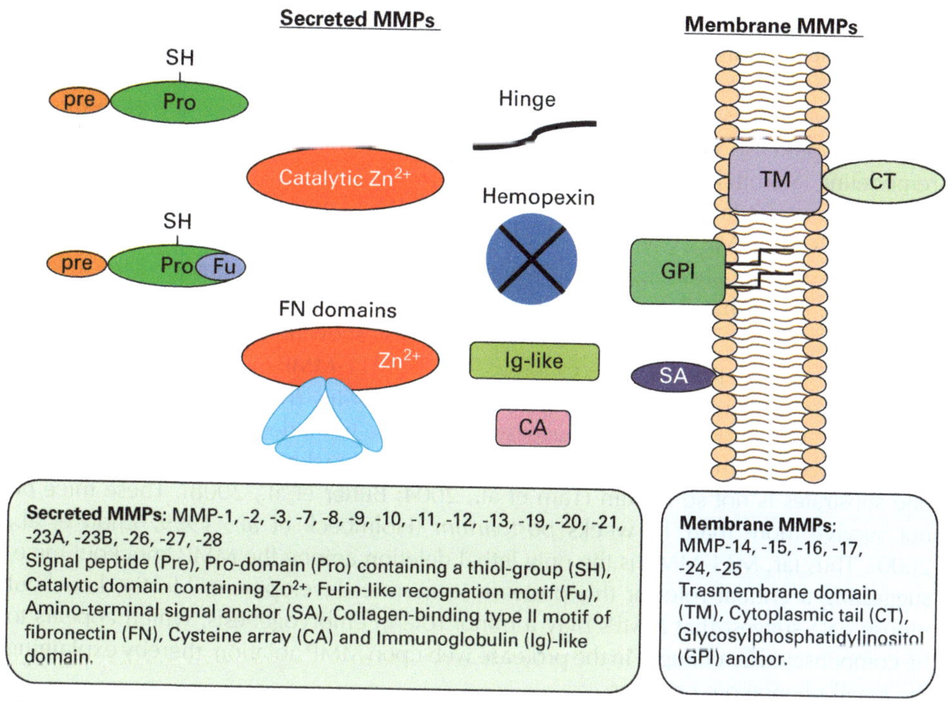

Figure 4.12 **Human MMPs.** Schematic representation of various domains of the 24 human matrix metalloproteinases that can be divided into four groups: archetypal MMPs, matrilysins, gelatinases, and convertase-activated MMPs. The archetypal MMPs (MMP1, 3, 8, 10, 12, 13, 19, 20, and 27) contain a signal peptide, a propeptide, a catalytic domain (Zn^{2+}) and a hemopexin domain. Matrilysins (MMP7 and 26) lack a hemopexin domain. Gelatinases (MMP2 and 9) contain fibronectin (FN) type II modules within their catalytic domains. Convertase-activated MMPs (MMP11, 14, 15, 16, 17, 21, 23A, 23B, 24, 25, and 28) have a domain that is targeted by furin-like proteases (convertase cleavage site). There are 6 membrane MMPs (MMP14, 15, 16, 17, 24, and 25) containing a glycosylphosphatidylinositol (GPI) anchor or transmembrane (TM) segments. MMP23A and MMP23B have a unique cysteine array (CA) and immunoglobin (Ig)–like domains.

(3) sodium dodecyl sulfate (SDS), (4) chaotropic agents, (5) reactive oxygen species (ROS), (6) low pH, and (7) heat treatment (Visse and Nagase, 2003). However, the mechanism for *in vivo* activation of secreted MMPs remains widely unknown, an exception being MMP2, which has been well characterized. Activation of secreted pro-MMP2 is mediated by a cell-surface complex consisting of a homodimer of MT1-MMP with a single molecule of tissue inhibitor of metalloproteinases-2 (TIMP2). TIMP2 binds to the catalytic domain of one of the two MT1-MMP molecules of the homodimer and to the hemopexin domain of pro-MMP2 (Overall et al., 2000), thereby facilitating cleavage and activation of pro-MMP2 by the second MT1-MMP molecule of the homodimer (Sato et al., 1996; Butler et al., 1998). Pro-MMP2 can also be activated by other membrane MMPs: MT2-MMP (independent of TIMP2) (Morrison et al., 2001), MT3-MMP (Takino et al., 1995), MT5-MMP (Pei, 1999), and MT6-MMP (Velasco et al., 2000), but interestingly, not by MT4-MMP (English et al., 2000). Proteases of other

classes can also activate MMPs – for example, the serine protease furin can mediate intracellular proteolytic activation of MMPs containing a furin recognition motif (KX [R/K]R): MMP11 (Pei and Weiss, 1995), MT1-MMP (Sato et al., 1996), MMP23 (Pei et al., 2000), and MMP28 (Lohi et al., 2001). Thus, an interdependence of MMP and the serine protease family is implied in extracellular matrix (ECM) degradation and remodeling (Sternlicht and Werb, 2001).

To better understand the specific role of each MMP, knockout mice have been studied extensively. MMP3 was the first knockout to be generated; against expectations, these mice did not show notable phenotypic embryological or reproductive defects, despite the belief that these processes required extensive matrix remodeling (Mudgett et al., 1998). MMP2, 7, 11 and 12 knockout mice also show limited phenotypic changes compared to wild-type mice. Interestingly, MMP14/MT1-MMP–null mice exhibit several dysfunctional phenotypes, including osteopenia, arthritis, dwarfism, craniofacial dysmorphism, and fibrotic synovitis, suggested to be due to inadequate collagen turnover; however, with the myriad of substrates of MT1-MMP now known, this single linkage to one substrates is not so certain (Tam et al., 2004; Butler et al., 2008). These mice do not survive more than 13 weeks *postpartum* (Holmbeck et al., 1999; Zhou et al., 2000). Thus far, MT1-MMP is the only lethal deletion among the MMP knockout mice, suggesting a critical role for this protease in murine development. Mostly, knockout mouse data suggest that MMPs play a minor role in embryogenesis, which appears to be compensated by changes in the protease web upon MMP ablation, thereby explaining the small phenotypic changes.

4.7.3 Natural inhibitors of MMPs

Since activated MMPs can cleave a large array of substrates and facilitate remodeling of the extracellular environment, it is essential that they be tightly regulated. The tissue inhibitors of metalloproteinases (TIMP1, 2, 3, and 4) comprise a family of protease inhibitors that inhibit the activity of all MMPs in a 1:1 stoichiometric fashion (Sternlicht and Werb, 2001). One exception is the inability of TIMP1 to inhibit the MT-MMPs (Remacle et al., 2011). TIMPs are composed of two domains: an N-terminal domain of ~125 amino acids and a C-terminal domain of ~65 residues stabilized by three disulfide bonds (Williamson et al., 1990). The TIMPs are relatively small proteins of ~21–34 kDa and are highly glycosylated. Nuclear magnetic resonance (NMR) (Williamson et al., 1997) and X-ray crystallography (GomisRuth et al., 1997) have shown that, not unexpectedly, both TIMP1 and TIMP2 interact with the catalytic domain of MMPs. Multiple studies have revealed that the TIMP N-terminal domain tightly coordinates in an extended conformation with the active site cleft, including the catalytic Zn^{2+}, and thus is responsible for the inhibitory activity. In addition, the C-terminal hemopexin domains of MMP2 selectively bind TIMP2 and TIMP4 (Bigg et al., 1997), whereas TIMP1 binds the hemopexin domain of MMP9 (Goldberg et al., 1989; Morgunova et al., 2002; Dufour et al., 2010). The total concentration of TIMPs in tissue and extracellular fluids generally exceeds the concentration of MMPs, thereby limiting proteolytic activity to focal pericellular sites. TIMPs have additional functions: TIMP1 was initially named EPA for its erythroid-potentiating activity, and TIMP1, 2, and 3 have mitogenic activity that functions independent of the protease inhibitory activity (Gomez et al., 1997).

Other endogenous inhibitors of MMPs include β-amyloid precursor protein, α2-macroglobulin, tissue factor pathway inhibitor-2, endostatin, the reversion-inducing cysteine-rich protein with Kazal motifs (RECK), the noncollagenous NC1 domain of type IV collagen, and procollagen C-terminal proteinase enhancer (Netzer et al., 1998; Takahashi et al., 1998; Kim et al., 2000; Sternlicht and Werb, 2001; Liepinsh et al., 2003; Overall and Kleifeld, 2006a). α2-Macroglobulin, an abundant plasma protein of ~725 kDa molecular mass, represents the major inhibitor of MMPs in body fluids, whereas the TIMPs act more locally in tissues.

4.7.4 MMPs in cancer

For more than 40 years, the role of MMPs to degrade the ECM and promote cancer invasion and metastasis has been widely investigated in cells and animal models. Studies have shown that (1) invasion and migration of cancer cells is modulated by transfection with expression vectors encoding the cDNA of MMPs; (2) tumor progression correlates with enhanced secretion of MMPs by both tumor cells and stromal cells; and (3) tumor growth and metastasis *in vivo* can be reduced using TIMPs, synthetic MMP inhibitors (MMPIs) including peptidomimetic compounds, antisense oligonucleotides, small molecules, and neutralizing antibodies (Overall and Kleifeld, 2006b). However, recently a number of MMPs have been recognized to have beneficial roles in host defense in cancer, with MMP gene ablation leading to increased tumorigenesis (Overall and Kleifeld, 2006b), which was first demonstrated for MMP8 (Balbin et al., 2003). In fact, it is likely that more MMPs are drug antitargets in cancer than targets due to the practicalities and risk of drug treatment of known or candidate MMP antitargets.

In human tumors, MMPs are produced by cancer cells as well as surrounding stromal cells and immune cells (Sternlicht and Werb, 2001; Kessenbrock et al., 2010). Several MMPs (MMP1, 2, 3, 7, 9, 13, and 14) actively contribute to cancer progression at various stages, including tumor growth, invasion, metastasis, and angiogenesis (Sternlicht and Werb, 2001; Deryugina and Quigley, 2006). Most human cancers are epithelial in origin, and the loss of the epithelial cells' organizational integrity through phenotypic changes is an initiator of cancer invasion and metastasis. The switch of an epithelial cell into a mesenchymal-like cell, termed *epithelial-to-mesenchymal transition* (EMT) requires several alterations in morphology, migration, adhesion, and cellular architecture. EMT is marked by a loss or decrease of E-cadherin, cytokeratins, and the acquisition of mesenchymal proteins such as vimentin, fibronectin, and N-cadherin. The overexpression of several MMPs (MMP1, 2, 3, 7, 9, 13, 14, and 28) is associated with EMT (Orlichenko and Radisky, 2008); MMPs cleave E-cadherin, leading to loss of epithelial integrity and gain of a more aggressive invasion program by cancer cells and initiating a metastatic cascade (Lochter et al., 1997).

Lung, liver, and bone are prone to the development of micrometastases followed by gross metastatic tumors. Metastasis depends on the type of primary cancer cell but also requires the formation of a receptive environment, a metastatic niche, which is a specifically suited host of distant tumor cells. MMP9 is critical for the formation of a metastatic niche by either liberating vascular endothelial growth factor (VEGF) and promoting angiogenesis (Bergers et al., 2000), or by releasing Kit-ligand, thereby recruiting stem cells from the bone marrow to the site of metastasis (Kaplan et al., 2005). Certain primary tumor cells can release soluble factors that recruit a specific population of nonmalignant hematopoietic cells to distant organs (Kaplan et al., 2006). The secretion of growth factors and chemokines and proteolytic ECM turnover create a receptive microenvironment or "premetastatic niche" for cancer cells (Hiratsuka et al., 2006).

Angiogenesis, the formation of new blood vessels from existing ones, is crucial for both development and tumor progression (Kraling et al., 1999). Angiogenesis is characterized by matrix degradation and remodeling, deposition of new basement membrane, and the induction of endothelial cell quiescence. MMPs regulate angiogenesis by (1) promoting vessel formation by releasing angiogenic factors from matrix or inhibitory binding proteins; (2) generating antiangiogenic breakdown products, and (3) remodeling the ECM to enable endothelial cell migration through tissues. $CD45^+$-myeloid cells increase VEGF bioavailability through MMP9 activity by inducing angiogenesis and regulate tumor cell invasiveness (Du et al., 2008), as demonstrated in two transgenic mouse models of tumor progression (RIP1-Tag2 insulinoma model; Bergers et al., 2000) and K14-HPV16 skin cancer model (Coussens et al., 2000). Although the substrate cleaved by MMP9 that liberates VEGF was not identified, the cleavage of connective tissue growth factor (CTGF) or pleiotrophin (also known as heparin affinity regulatory peptide [HARP]) when bound as inhibitory complexes to VEGF by MMP2 (Dean et al., 2007) and MMP3 (Hashimoto et al., 2002) suggests that MMP9 might also function in this capacity to mobilize VEGF. Additionally, MMP3, 7, and 16 can affect the vascular patterning of tumors *in vivo* by cleaving VEGF molecules (Lee et al., 2005). MT1-MMP can degrade the fibrin matrix surrounding newly formed vessels, therefore promoting angiogenesis by enhancing cell migration, invasion, and capillary-tube formation (Galvez et al., 2001; Devy et al., 2009). To the contrary, several MMPs (MMP2, 3, 7, 9, and 12) can negatively regulate angiogenesis by generating antiangiogenic peptides like angiostatin (a cleaved product of plasminogen) (O'Reilly et al., 1999), tumstatin and endostatin (cleaved products of collagens IV and XVIII, respectively) (Maeshima et al., 2000). Therefore, deregulation of MMPs in angiogenesis has been linked to several cardiovascular diseases, such as atherosclerosis, myocardial infarction, and development and rupture of aneurysms (Baker et al., 2002). Collectively, MMPs seem to play a "yin-yang" role in angiogenesis; rather than exclusively stimulating or inhibiting new blood vessel development, they fine-tune angiogenesis.

From extensive proteomic analyses, less than 20% of all known MMP substrates are ECM members (Morrison et al., 2011). Hence, the functions of MMPs extend well beyond matrix degradation, and many aspects of these newer physiological roles remain unknown and are presently under investigation. Cleavage of a variety of (non-ECM) substrates such as chemokines and growth factors, their binding proteins, and proteins mediating cell-cell or cell-ECM adhesion and ectodomain shedding affect signal transduction pathways that regulate homeostasis in the extracellular space. In disease, some of these signaling functions lead to cell motility, invasion and metastasis, EMT, skeletal homeostasis, angiogenic regulation, inflammatory response, regulation of apoptosis, initiation of neoplastic progression, and formation of the metastatic niche (Kessenbrock et al., 2010). For example, MMP2 and MT1-MMP can cleave the $\gamma 2$ chain of laminin-5 causing a release of domain III, in turn, enhancing mammary epithelial cell migration (Gilles et al., 2001; Koshikawa et al., 2004); the shedding of E-cadherin from the cell surface by MMP3 or MMP7 enhances lung epithelial cell migration (Noe et al., 2001; McGuire et al., 2003). Thus, MMPs are not just matrix degraders but important signaling proteases for the homeostasis of the extracellular environment (Morrison et al., 2009). Since MMPs have numerous biological roles, they are associated with a large variety of human diseases other than cancer. For example, the first human genetic disease to be linked with MMPs was Sorby's fundus dystrophy, caused by a mutation in the TIMP3 gene (Weber et al., 1994).

4.7.5 MMPs in Inflammation

MMPs precisely cleave most if not all chemokines, which in humans form a family of 54 chemotactic proteins involved in inflammatory and immune cell recruitment and trafficking that act through G-protein-coupled receptors (Starr et al., 2012). Proteolysis of chemokine termini results in critical functional changes regulating immune cell recruitment and inflammation *in vivo* (Mortier et al., 2011). For instance, in neutrophils, MMP9 processes interleukin-8 (IL-8 or chemokine (C-X-C motif) ligand-8 [CXCL8]), to increase its activity 10-fold (Opdenakker et al., 2001; Tester et al., 2007), whereas processing of CXCL7, CXCL4, and CXCL1 by MMP9 (Opdenakker et al., 2001) and processing of all 7 CXCL chemokines that are chemoattractants for polymorphonuclear neutrophils (PMNs) by MMP12 (Dean et al., 2008) potently inactivates these chemokines. CXCL12 (stromal-cell-derived factor-1, or SDF1) is cleaved by several MMPs (MMP1, 2, 3, 9, 13, and 14), which abrogates binding to the C-X-C chemokine receptor type 4 (CXCR4) receptor and thus destroys chemoattractant activity (McQuibban et al., 2001); in so doing, the cleaved chemokine switches receptors to CXCR3 and thus becomes neurotoxic (Zhang et al., 2003; Zhu et al., 2009). Recently, all 14 chemokine (C-C motif) ligand (CCL) chemokines that are chemoattractant for monocytes/macrophages were screened for cleavage by a number of MMPs with 149 cleavage sites identified by mass spectrometry (Starr et al., 2012). Notably, several of these chemokines displayed altered glycosaminoglycan (GAG) binding properties, and 2, the long N-terminal CCL17 and CCL23, were activated by MMP cleavage to 10-fold more potent forms than the classic monocyte chemoattractant protein-3 (MCP3)/CCL7. Other CCL chemokines are converted from agonists to potent antagonists that shut down macrophage recruitment *in vitro* and *in vivo* (McQuibban et al., 2000, 2002; Starr et al., 2012).

Rheumatoid arthritis is an autoimmune disorder that is characterized by the inflammation of joints and degradation of joint cartilage. The collagenases (MMP1, 8, and 13) have been tightly associated with this pathology by cleaving collagen II, a major structural constituent of cartilage, and the remaining fragments are proposed to be degraded by MMP9 (Van den Steen et al., 2002). In a spontaneous model of arthritis, MMP8 deficiency led to an exaggerated accumulation of neutrophil infiltrates at the sites of arthritic lesions and a decrease of neutrophil apoptosis leading to a worsening of the disease; notably, using neoepitope antibodies that detect MMP8-cleaved aggrecan and type II collagen, no detectable amounts were found in both MMP8 wild-type and knockout mice (Cox et al., 2010). Since MMPs also affect the migration of leukocytes to the site of inflammation by regulating chemokine cleavage, it is still ambiguous whether MMPIs can be useful in the treatment of rheumatoid arthritis. For example,

BAY12–9566 inhibited neutrophil infiltration and reduced swelling of the rats' paws in an adjuvant-induced arthritis (AA) model (Hamada et al., 2000), but recent results investigating the effects of MMP8 knockout mice showed a worse prognosis in a slow progressive model of the disease (Cox et al., 2010). Such antitarget results dampen enthusiasm for the use of MMPIs in inflammatory disease considering their pleiotropic effects and the demonstrated protective effects of MMP8 (Cox et al., 2010) and MMP12 (Bellac et al., unpublished).

4.7.6 MMP inhibitors and clinical trials

Due to the critical importance of MMPs in several human diseases, MMPs became a prime target for drug development in the 1990s. Many approaches were taken ranging from pseudopeptides that can mimic MMP substrates to nonpeptidic compounds that can coordinate with the catalytic zinc ion (Overall and Kleifeld, 2006a). The TIMPs, being natural inhibitors of MMPs, have also been considered for treatment of diseases because of their picomolar affinities. However, the lack of selectivity and multiple biological activities of TIMPs have tended to exclude this strategy from the clinic. The first line of compounds primarily focused on a hydroxymate moiety as the zinc-binding group, which led to the development of compounds with diverse functional groups. British Biotech first took Batimastat, a broad-spectrum hydroxamic-acid peptide inhibitor based on the collagen cleavage site, into a human clinical trial (Wang et al., 1994). Early administration of Batimastat in the pancreatic islet cell carcinogenesis model significantly inhibited angiogenesis, whereas no effect was observed when Batimastat was administered in the late stage of the model (Bergers et al., 1999). In order to improve bioavailability, Batimastat was replaced by the orally available compound Marimastat. However, treatment with Marimastat lead to musculoskeletal problems in some patients, including stiffening of the joints, pain in the hands, skin discoloration, and inflammation linked to tendonitis, and so several phase III cancer clinical trials were deemed to have failed (Pavlaki and Zucker, 2003) despite matching conventional therapy for pancreatic cancer (Overall and Kleifeld, 2006b).

Groups interested in cancer applications designed MMPIs with selectivity toward gelatinases MMP2 and 9, whereas others targeting arthritis designed inhibitors of collagenases (MMP1, 8, and 13) (Whittaker et al., 1999). Another hydroxamic-acid derivative, Prinomastat (AG-3340), was designed to be selective for gelatinases but displayed similar side effects to Marimastat and was not effective in use. Bayer designed a butanoic-acid analogue (BAY12–9566) that did not cause any musculoskeletal side effects, but it was clinically ineffective.

After the failure of MMPIs to treat cancer in clinical trials, three caveats should be addressed when designing future inhibitors before entering the clinic: (1) focus on preventative therapies in early stage cancers, (2) develop more specific drugs due to antitarget activities of some MMPs in disease, and (3) overcome unanticipated long-term drug intolerance. In addition, several MMPs can display antitumor (MMP3, 8, 9, and 12) and proangiogenic (MMP9) activity depending on the stage of tumor progression and on the cell types present (Overall and Kleifeld, 2006b). One example of this methodology for novel inhibitors was demonstrated by a highly specific antibody to MT1-MMP (DX-2400) that reacts only with the activated form of the enzyme and not the latent form, developed by Dyax Corporation. This antibody inhibits tumor invasion, metastasis, and angiogenesis in animal models, like many of the previous MMPI drugs, and is destined for clinical trials (Devy et al., 2009).

Tetracyclines inhibit antimicrobial activity by interfering with inflammatory cell migration and chemotaxis to sites of inflammation (Sorsa et al., 1998). These small molecules also inhibit MMP expression and activity, and thus several tetracycline derivatives have entered clinical trials as MMPIs (Cianfrocca et al., 2002). Tetracycline compounds are currently FDA-approved for the treatment of periodontitis, which occurs when bacterial induction of gingival tissue release of inflammatory cytokines mediates connective-tissue destruction by MMPs, leading to loss of teeth. Periostat (doxycycline hydrate) is the only MMPI approved in the United States for the treatment of a human disease (periodontitis) by reducing the activity of host-derived collagenases (Wynn, 1999). MMPs have been examined for several therapeutic targets in diseases other than cancer, such as corneal ulcers, multiple sclerosis, glomerulonephritis, bacterial meningitis, uveorentinitus, emphysema, aortic aneurysm, and atherosclerosis (Galardy et al., 1994; Whittaker et al., 1999). Given the previous problems with MMPIs and the risk of antitarget activities of MMPs that are now known, short-term clinical indications seem the most advisable for use of new MMPIs in the near term.

4.7.7 The protease web

To design a successful drug against a protease, a clear understanding of its roles and substrates must first be characterized. Complications arise as proteases act in the context of complex cascades, circuits, networks, and pathways, thereby creating a multifaceted universe of protein-protein interactions termed the *protease web* (Overall and Dean, 2006; Butler et al., 2008; Kruger, 2009). This web of protein interactions is not static but dynamic and is further modulated in disease states. The net activity of a protease, like the web itself, relies on the regulated activities of many proteases and inhibitors. Interpretation of data when analyzing the *in vitro* roles of a single protease can sometimes lead to biochemical artifacts. By viewing proteolysis as a multidimensional system, as in genetic knockouts and animal models, it is apparent that protease overexpression affects a myriad of processes that can lead to unexpected effects. Collectively, a "proteolytic signature" can be identified in an *in vivo* environment that reveals the global activity of a protease within the system. For example, 87 protease and inhibitor mRNA, proteins, and activity level changes were found in an inflammatory model of *Mmp2*$^{-/-}$ mouse as compared to wild type (A. D. Keller et al., unpublished). Thus, viewing the protease web in the context of a disease can lead to a better understanding of cellular processes and successful targeting of the different proteolytic signatures within the network for effective therapeutics.

4.7.8 Degradomics

The emergence of genomics and proteomics in the twenty-first century has revolutionized the fields of medicine and biological sciences. Degradomics is a specific subfield whereby the application of genomics and proteomics is used to identify proteases and protease-substrate repertoires (also known as degradomes) on an organism-wide scale. High content, unbiased screens are essential to unravel novel *in vivo* roles of proteases and to gain knowledge of proteolytic processes, to determine how the protease web is modulated, and to identify novel drug targets and antitargets for treatment of diseases (López-Otín and Overall, 2002). There are several tools available to examine the modulation of the protease web at both the gene and protein levels in tissues and on a patient-by-patient basis, and these will be covered in the remainder of this chapter.

4.7.9 The CLIP-CHIP, a dedicated and focused microarray for every protease and inhibitor

Microarrays are designed to analyze mRNA expression levels of thousands of genes simultaneously. Such large-scale data accumulation can often be challenging to analyze; therefore, focusing on a specific class or family of genes can facilitate analysis of these genes of interest. The availability of complete genomes (i.e. human, murine) has enabled the identification of the degradome, the whole set of proteases and their natural inhibitors in an organism (Puente et al., 2003). Currently, there are three methods to analyze this: the Affymetrix-based Hu/Mu Protln Chip (Schwartz et al., 2007), the microfuidic taqman low-density qRT-PCR (Swingler et al., 2009), and the customized CLIP-CHIP (Kappelhoff and Overall, 2007). The CLIP-CHIP is a dedicated oligonucleotide-based microarray containing unique 70-mer oligonucleotides representing all human and murine proteases, nonproteolytic homologs, and inhibitor transcripts. The CLIP-CHIP has been successfully used to investigate the role of MMP8 in arthritis. By comparing *Mmp8* $^{-/-}$ mice on a Fas ligand-defective Murphy Roths Large (MRL)/*lpr* background to wild-type MRL mice, expression of the apoptosis initiator caspase-11 was absent, and this was demonstrated to lead to delayed apoptosis and increased neutrophil accumulation at sites of inflammation (Cox et al., 2010). Recently the CLIP-CHIP was used to identify changes in the protease and inhibitor expression profile of cocultured breast cancer and osteoblast cells (Morrison et al., 2011).

4.7.10 Classic biochemical approaches

In the classic game of rock/paper/scissors, the rules are simple: scissors cut paper, paper envelops rock, and rock blunts scissors. Our view of proteases can be translated to a simple game of rock/paper/scissors. As shown in ▶Figure 4.13 (green arrows), in this "biochemical" version of the game, proteases are molecular scissors capable of cutting (enzymatic cleavages) ECM proteins, depicted as paper, affect the physiology of the cells (rock); and the cells (rock) are producing the proteases (molecular scissors). Although, biochemical techniques present a view of the protease web that is mainly one-dimensional, they can still reveal useful information to better understand the protease web and are described in Sections 4.7.11, 4.7.12, 4.7.13, and 4.7.14. In ▶Figure 4.13 (blue arrows), we describe how a simple game of rock/paper/scissors can become more complex, multidimensional, and thus resemble *in vivo* proteolysis.

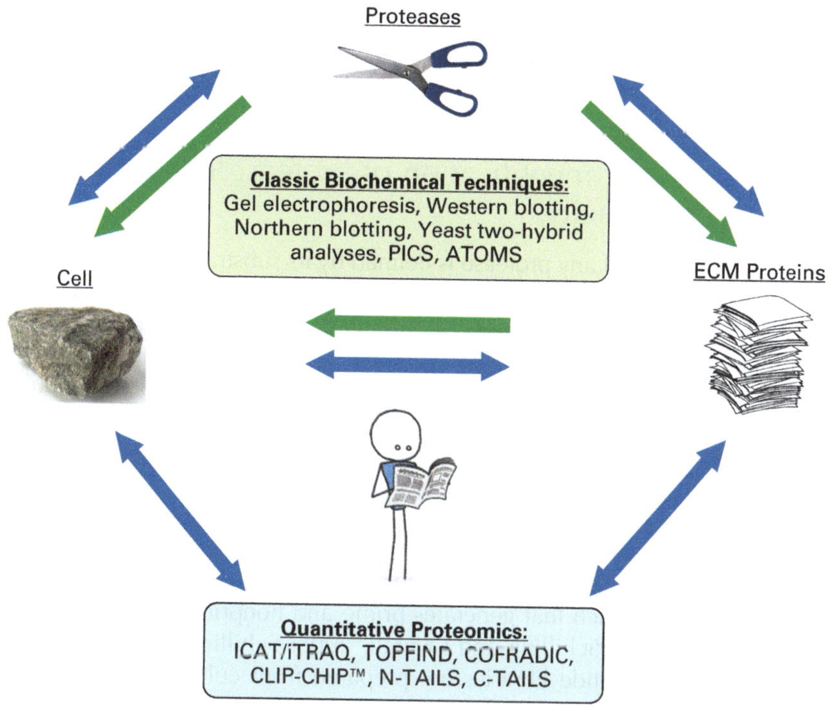

Figure 4.13 **Biochemical game of rock/paper/scissors.** Schematic representation of proteases (scissors), ECM proteins (paper), and cells (rock): the green arrows illustrate the game using classic biochemistry techniques, and the blue arrows represent the additional multidirectional options when using quantitative proteomics.

4.7.11 Sodium dodecyl sulfate polyacrylamide gel electrophoresis, zymography, mass spectrometry, and high-performance liquid chromatography

Several biochemical techniques and tools are routinely utilized for characterization of a protease in a given sample (i.e. recombinant proteins, cell lysates, conditioned media, tumor tissue, blood, etc.), and proteases and proteolytic activity can be detected using sodium dodecyl sulfate polyacrylamide gel electrophoresis (SDS-PAGE), Western blotting, gelatin or casein zymography, fluorogenic peptide assays, or mass spectrometry (Overall and Blobel, 2007). These techniques are useful for the identification of specific proteases, determining cleavage sites, calculating enzyme kinetics (e.g. k_{cat}/K_M) and substrate validation. However, there are several limitations to these techniques when analyzing complex samples. *In vitro* studies can identify candidate *in vivo* substrates, but an *in vitro* cleavage does not always correlate to an *in vivo* product – "just because it can, does not mean it does" (Overall and Blobel, 2007) – and discrepancies often exist between the two. Thus, newly discovered substrates must be carefully validated in cell culture and three-dimensional cell culture assays, followed by robust animal models and human samples. Before validation in more complex systems, basic information is needed to facilitate later analyses. To this end, there are several techniques that provide valuable information about protease cleavage and binding sites.

4.7.12 Proteomic identification of protease cleavage site specificity

The biological function of any protease is defined by its substrates. Therefore, by investigating the cleavage specificities of a single protease, important information can be achieved for the development of novel inhibitors. Multiple techniques, such as phage display, random peptide library or alanine-scanning mutagenesis of a cleavage site, have been designed to elucidate the consensus cleavage sites of a particular protease (Overall and Blobel, 2007). A caveat of peptide-library approaches is that they usually probe only nonprime residues. For example, although phage display can provide useful information on a stretch of sequence that can be cleaved, cleavage sites are not directly identified. Follow-up techniques are needed for this. One option to circumvent such limitations is the use of a technique termed *proteomic identification of protease cleavage site specificity* (PICS) technique. PICS is a recently introduced substrate-profiling approach that generates prime and nonprime side (from P6′ to P6) cleavage site specificity (Schilling and Overall, 2008; Schilling et al., 2011). Diverse biologically relevant peptide libraries are prepared from cultured cells. Trypsin and *Staphylococcus aureus* protease V8 (GluC) are used to prepare tryptic and GluC peptides from proteomes collected from, for example, cell lysates, conditioned media, or *Escherichia coli*. These biologically relevant peptide libraries are diverse and, critically, are database searchable. The protease of interest is incubated with the library that has been N-terminally blocked first. After protease digestion, cleaved peptides display neo–N termini with a free primary amine. These are chemically reactive unlike the N-terminally blocked peptides in the library. This property is used to biotinylate the cleaved peptides, which are then isolated on streptavidin beads. Processed peptides are eluted, and the sequence identified using mass spectrometry. The nonprime side of the cleavage sites is filled in using bioinformatics. The cleaved peptide sequences can be organized into heat maps and sequence logos. Positional occurrences can be corrected for natural amino acid abundance. PICS offers a large array of substrate profiling (P6′ to P6), whereas most methods are limited to only four nonprime side residues. Thus, PICS is very useful for the design of novel protease-activity-based probes and inhibitors.

4.7.13 Yeast two-hybrid analyses: exosite scanning and inactive-catalytic-domain capture

MMPs exosites are critical for initial substrate binding and recognition before proteolysis (Overall, 2002), as well as for the cleavage mechanism in the case of MMP1, 8, 13, and 14 in which a hemopexin domain deletion mutant fails to cleave native triple-helix collagen (Nagase and Woessner, 1999). Utilizing this knowledge on the essential role of the hemopexin domain for collagen cleavage, a yeast two-hybrid screen was performed utilizing the hemopexin domain of MMP to scan for new substrates

(McQuibban et al., 2000) in an approach that has come to be known as "exosite scanning" (Overall et al., 2002). MMP2 was shown to bind the chemokine MCP3 through interaction of its hemopexin domain that potentiated its cleavage (McQuibban et al., 2000; Overall et al., 2002). For proteases that have no ancillary substrate-binding exosite domains, an alternative approach was devised whereby the yeast two-hybrid screen was performed utilizing the inactive protease mutant as bait. Termed *inactive-catalytic-domain capture* (ICDC), a novel substrate of MT1-MMP was discovered: Wnt1-inducible signaling-pathway protein-2 (WISP2) (Overall et al., 2002). However, ICDC is not necessarily feasible for all proteases: high K_M values of certain substrates and differences in the catalytic mechanisms of protease families can limit the discovery of substrates.

4.7.14 Amino-terminal-oriented mass spectrometry of substrates

The identification of cleavage sites in protease substrates can also be performed by SDS-PAGE, Edman degradation, and sequencing. Edman sequencing is an efficient method that labels the amino-terminal residue before cleavage from the peptide of interest without disruption of the peptide bonds (Edman and Begg, 1967). However, limitations of this technique are that proteolytic fragments have to be analyzed separately and approximately 100 pmoles (minimum) protein is required. Thus, the analysis of complex large modular proteins can be quite challenging. One way to avoid the limitations of Edman sequencing is amino-terminal-oriented mass spectrometry of substrates (ATOMS), which is broadly applicable in identifying N termini of high-molecular-weight proteins. The work flow of ATOMS is as follows: (1) the substrate of interest is incubated with the protease (or buffer alone as a control sample); (2) proteins are denatured, cysteines are alkylated, and amine groups (N termini and lysine side chains) are isotopically labeled; (3) the two differentially labeled samples (protease-treated vs control) are mixed and digested with trypsin; (4) peptides are identified and quantified by mass spectrometry, where labeled singletons in the protease-treated sample represent peptides generated by the protease of interest (Doucet and Overall, 2011). MMP2 and 8 have been demonstrated to cleave fibronectins and laminins, implicating these two proteases in inflammation, tissue healing, and cancer (Overall and Kleifeld, 2006b; Cox et al., 2008). MEROPS, the protease databases, lists a total of 10 cleavage sites for these MMPs, whereas ATOMS identified 34 cleavage sites (Doucet and Overall, 2011). For the cleavage of fibronectin alone by MMP2 and 8, ATOMS detected 11 cleavage sites and 6 by N-terminal sequencing; for neutrophil elastase cleavage of laminin-1 and fibronectin-1, the results were impressive with 55 cleavages sites identified (Doucet and Overall, 2011). Hence, ATOMS offers an efficient complementation to Edman sequencing for the discovery of novel biological peptide cleavages.

4.7.15 Quantitative N- and C-terminal proteomics for substrate discovery

Our view of proteolysis has greatly extended over the past decade. ▶Figure 4.13 (green arrows) illustrates a one-dimensional perspective of the game rock/paper/scissors in context of the protease web, and with the emergence of recent proteomics techniques, a more multifaceted set of rules can be integrated into ▶Figure 4.13 (blue arrows). By using global approaches to view the complexity of the protease web, novel roles and interactions can be unraveled, and a better understanding will be achieved.

Biological samples are highly complex, containing vast arrays of proteins and peptides at a wide range of concentrations. In recent years, gel-free proteomics procedures have emerged, and these techniques facilitate the study of proteolysis and α-*N*-acetylation. In a proteomics work flow, proteins need to be reduced to smaller peptides for MS analysis, usually by trypsin cleavage in the sample preparation work flow, but often the terminal peptides of interest tend to be vastly outnumbered by the high number of internal tryptic peptides and so are mostly undetectable. Liquid chromatography is useful for fractionation of complex samples to reduce the number of peptides that need to be analyzed at the same time, but abundant peptides from abundant proteins still tend to mask low abundance ones in the mass spectrometer. There are different ways to circumvent these limitations, such as enriching the protein terminal peptides prior to MS analysis, therefore simplifying the proteome (Gevaert et al., 2007). Another way is to use differential isotope tagging of samples (e.g. stable isotope labeling by/with amino acids in cell culture [SILAC], isotope-coded affinity tag [ICAT], isotopic dimethylation, isobaric tags for relative and absolute quantitation [iTRAQ]) to facilitate relative quantification of protein amounts that can be useful for *in vivo* samples, which, importantly, allows the comparison of multiple samples (e.g. +/– protease) (Bantscheff et al., 2007; Boersema et al., 2009). Such an approach has been very successfully used for MMP substrate discovery using ICAT (Tam et al., 2004; Dean et al., 2007; Butler et al., 2008) and iTRAQ (Dean and Overall, 2007; Morrison et al., 2011) in which two or up to four samples are labeled and compared as to whether they differ in the presence of the protease or not.

While highly successful for substrate discovery, with more than 150 new MMP substrates discovered by these approaches (Morrison et al., 2009), the actual cleavage site is not necessarily identified without subsequent follow-up experiments. Terminomics is a specific subfield of degradomics that enriches for the cleaved neo-N- and neo-C-terminal peptides following protease cleavage. There are two types of enrichment: positive and negative selection. An example of positive selection is the enzyme-mediated labeling of protein N termini: enzymatic biotinylation by using subtiligase, thus resulting in biotinylation of an N-terminal α-amine (Mahrus et al., 2008; Agard et al., 2010). Peptides are released for analysis using the highly specific tobacco etch virus (TEV) protease. Negative selection can be achieved using terminal amine isotopic labeling of substrates (TAILS) or acetylation of N termini and combined fractional diagonal chromatography (COFRADIC) and will be discussed in this chapter (McDonald and Beynon, 2006; Staes et al., 2008; Kleifeld et al., 2010, 2011).

4.7.16 N-terminal combined fractional diagonal chromatography

Gevaert et al. (2003) developed an N-terminomics method for the study of proteolysis and α-*N*-acetylation by selection of terminal peptides based on diagonal electrophoresis and chromatography (Staes et al., 2011). After reduction and alkylation of the sample, free primary amines are blocked by an *N*-acylation reaction before proteins are digested with trypsin. Using low pH, the peptides are prefractioned by reversed-phase (RP) high-performance liquid chromatography (HPLC). To increase the hydrophobicity of the peptides, the fractions are subjected to 2,4,6-trinitrobenzene-sulfonic acid (TNBS) to modify the α-amines of the internal peptides. The fractions are then subjected to sorting by stong cation exchange (SCX)-HPLC, which allows for separation of N-terminal and internal peptides. Quantitation of the differentially cleaved peptides between +/– protease samples can be achieved by using trideutero-acetylation, and metabolic labeling by SILAC or heavy isotopes (^{13}C, ^{15}N, or ^{18}O) can be introduced at different steps in the work flow (Staes et al., 2011). N-terminal COFRADIC using ^{18}O isotope peptide and SILAC labeling has been utilized to identify substrates in different cell extracts (Impens et al., 2010). While being a successful method, COFRADIC suffers from different limitations: the MS step can be labor intensive, and there are risks of high carryover of internal tryptic peptides.

4.7.17 N-terminal amine isotopic labeling of substrates

TAILS is a quantitative proteomics approach for labeling and isolating N-terminal peptides for annotation of the N terminome (all protein N termini) and for the global analysis of protease cleavage products in complex *in vitro* and *in vivo* samples (i.e. cell culture, blood samples, and tumors). A key feature of the approach is that the mature original N termini of proteins are also identified, from which the loss of substrates can be inferred and statistical models generated iteratively for each experiment to determine the isotope ratio cutoffs needed to reliably identify cleaved neo–N termini and hence protease substrates. The primary amines of the N termini (NH_2) and lysines of each sample to be compared are differently labeled using ICAT, iTRAQ, or isotopically labeled formaldehyde. The labeled proteomes are combined at a 1:1 ratio and digested with trypsin. Newly generated unblocked internal and C-terminal tryptic peptides react with and are removed using a water-soluble polymer to selectively enrich (by negative selection) the original blocked N-terminal peptides from each sample (Kainthan et al., 2006). These will be the original N termini, modified N termini (e.g. acetylated, and uniquely, N-terminal cyclized glutamine, glutamate, and cysteine at the P1′ position) and neo–N termini in samples with the protease of interest. These peptides can be easily identified by mass spectrometry/mass spectrometry (MS/MS) (Kleifeld et al., 2010).

One clear advantage of using N-terminal amine isotopic labeling of substrates (N-TAILS) is the increase in efficiency of the analysis of a complex proteome. Analysis of a GluC-exposed murine fibroblastic secretome before TAILS identified 1,501 peptides, 73 of which resulted from GluC cleavage, whereas 531 GluC-cleaved peptides were identified after TAILS. Thus, by using TAILS, the number of identified neo–N termini was increased from 4.8% to 62% (Kleifeld et al., 2010). A total of 113 substrates of MMP2 were successfully identified using TAILS to analyze a secretome from fibroblasts expressing endogenous MMP2 compared to a *Mmp2*$^{-/-}$ fibroblasts secretome (Kleifeld et al., 2010). Using a similar approach, 7 novel substrates of MMP9 were discovered and validated: cystatin C, galectin-1, insulin-like growth factor-binding protein 4 (IGFBP4), thrombospondin-2, Dickkopf-related protein 3 (Dkk-3), pyruvate kinase isozyme, and peptidyl-prolyl *cis-trans* isomerase A (Prudova et al., 2010). The metalloproteinase meprin has also been profiled by TAILS with the identification of 5 substrates (Becker-Pauly et al., 2011). The biological implications of the processing of these protein substrates is currently being investigated for the involvement of MMP9 in angiogenesis and carcinogenesis.

TAILS can also be applied to systems exposed to native protease compared with catalytically inactive mutant. Comparing the secretomes of transfected MCF7 breast cancer cells with either MMP11 or the inactive mutant MMP11/E216A, galectin-1 was among the several newly identified MMP11 substrates (Kleifeld et al., 2010). The loss of function of galectin-1 by MMP11 cleavage may contribute to a modulation of tumor aggressiveness, T-cell apoptosis, and the immune response (He and Baum, 2004). TAILS has also been performed on bronchoalveolar lavage fluid from lipopolysaccharide-induced lung inflammation and recently of whole arthritic joints in murine models of inflammation (Bellac et al., unpublished). A large number of proteins have been identified spanning a range greater than six orders of magnitude, thus clearly illustrating the potential of TAILS over previous methods to solve the dynamic range problem for complex samples.

4.7.18 C terminomics and C-terminal amine-based isotope labeling of substrates

The lower chemical reactivity of carboxyl groups, as compared with amine groups, has rendered the study of C-terminal posttranslational modifications difficult for decades by any strategy. C terminomics aims at proteome-wide C-termini enrichment: Current methods for simple samples are performed by capturing tryptic or Lys-C internal peptides using anhydrotrypsin column, a covalent linkage to diisothiocyanate-coupled beads, diagonal electrophoresis, or carboxypeptidase ladder sequencing, but these methods are not suitable for complex proteomes (Impens et al., 2010). To this end, C-terminal amine-based isotope labeling of substrates (C-TAILS) was developed to study important C-terminal proteolysis events, such as chemokine processing and fibrinolysis, in complex samples (Reznik and Fricker, 2001; Cox et al., 2008). C-TAILS selectively enriches for original and protease-generated C termini in control versus a protease-exposed sample: Amines and carboxyl groups are chemically protected to block the original C termini in the sample. Following trypsin digestion, a second round of amine protection is carried out. Then, trypsin-generated C-terminally unblocked peptides are coupled to a polymer (pollyallylamine) via the free C terminus and removed by ultracentrifugation. Selectively enriched original C termini are analyzed by LC-MS/MS. As with N-TAILS, stable isotope labeling means that C-TAILS offers a reliable quantification of neo-C-terminal sequence and protease substrates. Following C-TAILS analysis of GluC-treated versus untreated native *E. Coli* cell lysate, more than 90% of the identified peptides had C-terminal residues corresponding to a GluC signature cleavage (aspartic acid and glutamic acid) (Schilling et al., 2010). C-TAILS can be used as a complement to N-TAILS to extend the substrate coverage of a proteome.

4.7.19 Perspectives and Take-home message

To understand the biological roles of proteases in pathobiology, it is essential to globally integrate all the players within the protease web (proteases, inhibitors, cofactors, receptors, matrix components, etc.) and investigate all potential cleavage sites. It is often the case that multiple proteases can cleave the same substrates giving rise to redundancy within a system. But more important than the individual proteases themselves is the network of interactions they can form and how they trigger diseases. Utilizing "-omics" experimental approaches, particular diseases can be examined at multiple levels – genomics, transcriptomics, proteomics, and degradomics – by many platforms, some of which have been discussed here, including the CLIP-CHIP, Affymetrix-based Hu/Mu ProtIn Chip, N-TAILS, C-TAILS, biotinylation, and COFRADIC, to globally assess the state of a particular disease (▶Figure 4.14). Up to eight samples (8-plex) can now be analyzed in parallel in a single TAILS experiment as studied in arthritic and skin inflammation animal models (A. D. Keller A. D. et al., unpublished; Bellac et al., unpublished). Human blood and tissues are now being analyzed using N-TAILS. These kinds of proteomics experiments generate huge amounts of data, which often renders the analysis difficult. Current bioinformatics technologies can simplify data interpretation, for example, TopFIND (http://clipserve.clip.ubc.ca/topfind/) facilitates the interpretation of novel functions of proteins by providing information on translated protein N and C termini, their amino acid modifications (acetylation, pyroglutamate [pGlu] formation, etc.), and their cleavage sites (Lange and Overall, 2011). TopFIND integrates experimental studies (e.g. N-TAILS, C-TAILS, COFRADIC, etc.) with information from databases (i.e. MEROPS, UniProtKB) from five different organisms (*Homo sapiens, Mus musculus, E. coli, Saccharomyces cerevisiae*, and *Arabidopsis*), with more being added. Bioinformatics provides novel ways of interpreting data but, most importantly, facilitates intricate analysis of experimental data from complex samples. With the battery of highly sophisticated new proteomics techniques in place,

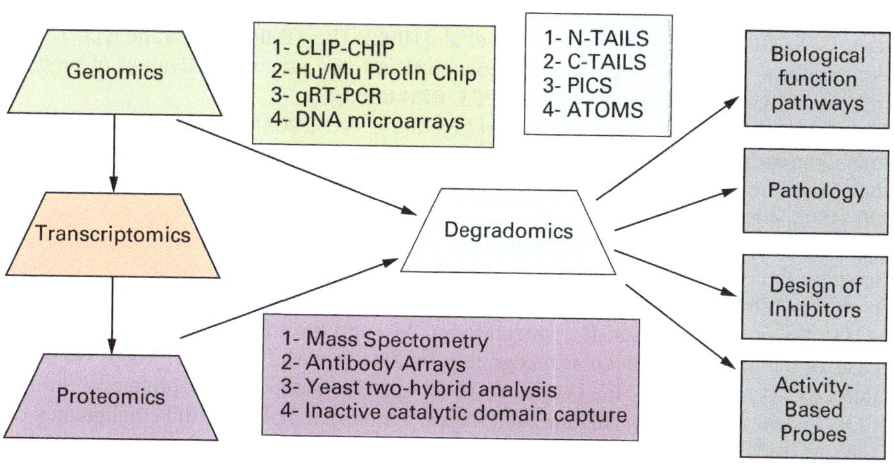

Figure 4.14 Linking genomics, transcriptomics, proteomics, and degradomics: the big picture. Diagram displaying the different outcomes when combining various genomics, proteomics, and degradomics techniques.

the frontier now moves to improving bioinformatics analyses. Interpretations of the changes of the protease web in a pathological state highlights new and essential pathways and interactions used to maintain cell homeostasis. Such analyses of the normal are needed along with the diseased to unravel novel directions for effective drug therapy of protease diseases.

References

Agard, N. J., Maltby, D., and Wells, J. A. (2010). Inflammatory stimuli regulate caspase substrate profiles. Mol Cell Proteomics *9*, 880–893.

Baker, A. H., Edwards, D. R., and Murphy, G. (2002). Metalloproteinase inhibitors: biological actions and therapeutic opportunities. J Cell Sci *115*, 3719–3727.

Balbin, M., Fueyo, A., Tester, A. M., et al. (2003). Loss of collagenase-2 confers increased skin tumor susceptibility to male mice. Nat Genet *35*, 252–257.

Bantscheff, M., Schirle, M., Sweetman, G., Rick, J., and Kuster, B. (2007). Quantitative mass spectrometry in proteomics: a critical review. Anal Bioanal Chem *389*, 1017–1031.

Becker-Pauly, C., Barre, O., Schilling, O., et al. (2011). Proteomic analyses reveal an acidic prime side specificity for the astacin metalloprotease family reflected by physiological substrates. Mol Cell Proteomics *10* (9), M111.009233.

Bergers, G., Brekken, R., McMahon, G., et al. (2000). Matrix metalloproteinase-9 triggers the angiogenic switch during carcinogenesis. Nat Cell Biol *2*, 737–744.

Bergers, G., Javaherian, K., Lo, K. M., Folkman, J., and Hanahan, D. (1999). Effects of angiogenesis inhibitors on multistage carcinogenesis in mice. Science *284*, 808–812.

Bigg, H. F., Shi, Y. E., Liu, Y. E., Steffensen, B., and Overall, C. M. (1997). Specific, high affinity binding of tissue inhibitor of metalloproteinases-4 (TIMP-4) to the COOH-terminal hemopexin-like domain of human gelatinase A. TIMP-4 binds progelatinase A and the COOH-terminal domain in a similar manner to TIMP-2. J Biol Chem *272*, 15496–15500.

Birkedal-Hansen, H., Moore, W. G., Bodden, M. K., et al. (1993). Matrix metalloproteinases: a review. Crit Rev Oral Biol Med *4*, 197–250.

Boersema, P. J., Raijmakers, R., Lemeer, S., Mohammed, S., and Heck, A. J. (2009). Multiplex peptide stable isotope dimethyl labeling for quantitative proteomics. Nat Protoc *4*, 484–494.

Butler, G. S., Butler, M. J., Atkinson, S. J., et al. (1998). The TIMP2 membrane type 1 metalloproteinase "receptor" regulates the concentration and efficient activation of progelatinase A – a kinetic study. J Biol Chem *273*, 871–880.

Butler, G. S., Dean, R. A., Tam, E. M., and Overall, C. M. (2008). Pharmacoproteomics of a metalloproteinase hydroxamate inhibitor in breast cancer cells: dynamics of membrane type 1 matrix metalloproteinase-mediated membrane protein shedding. Mol Cell Biol *28*, 4896–4914.

Cao, J., Kozarekar, P., Pavlaki, M., Chiarelli, C., Bahou, W. F., and Zucker, S. (2004). Distinct roles for the catalytic and hemopexin domains of membrane type 1-matrix metalloproteinase in substrate degradation and cell migration. J Biol Chem *279*, 14129–14139.

Cha, H., Kopetzki, E., Huber, R., Lanzendorfer, M., and Brandstetter, H. (2002). Structural basis of the adaptive molecular recognition by MMP9. J Mol Biol *320*, 1065–1079.

Cianfrocca, M., Cooley, T. P., Lee, J. Y., et al. (2002). Matrix metalloproteinase inhibitor COL-3 in the treatment of AIDS-related Kaposi's sarcoma: a phase I AIDS malignancy consortium study. J Clin Oncol *20*, 153–159.

Coussens, L. M., Tinkle, C. L., Hanahan, D., and Werb, Z. (2000). MMP-9 supplied by bone marrow-derived cells contributes to skin carcinogenesis. Cell *103*, 481–490.

Cox, J. H., Dean, R. A., Roberts, C. R., and Overall, C. M. (2008). Matrix metalloproteinase processing of CXCL11/I-TAC results in loss of chemoattractant activity and altered glycosaminoglycan binding. J Biol Chem *283*, 19389–19399.

Cox, J. H., Starr, A. E., Kappelhoff, R., Yan, R., Roberts, C. R., and Overall, C. M. (2010). Matrix metalloproteinase 8 deficiency in mice exacerbates inflammatory arthritis through delayed neutrophil apoptosis and reduced caspase 11 expression. Arthritis Rheum *62*, 3645–3655.

Dean, R. A., Butler, G. S., Hamma-Kourbali, Y., et al. (2007). Identification of candidate angiogenic inhibitors processed by matrix metalloproteinase 2 (MMP-2) in cell-based proteomic screens: disruption of vascular endothelial growth factor (VEGF)/heparin affin regulatory peptide (pleiotrophin) and VEGF/Connective tissue growth factor angiogenic inhibitory complexes by MMP-2 proteolysis. Mol Cell Biol *27*, 8454–8465.

Dean, R. A., Cox, J. H., Bellac, C. L., Doucet, A., Starr, A. E., and Overall, C. M. (2008). Macrophage-specific metalloelastase (MMP-12) truncates and inactivates ELR+ CXC chemokines and generates CCL2, -7, -8, and -13 antagonists: potential role of the macrophage in terminating polymorphonuclear leukocyte influx. Blood *112*, 3455–3464.

Dean, R. A., and Overall, C. M. (2007). Proteomics discovery of metalloproteinase substrates in the cellular context by iTRAQ labeling reveals a diverse MMP-2 substrate degradome. Mol Cell Proteomics *6*, 611–623.

Deryugina, E. I., and Quigley, J. P. (2006). Matrix metalloproteinases and tumor metastasis. Cancer Metastasis Rev *25*, 9–34.

Devy, L., Huang, L., Naa, L., et al. (2009). Selective inhibition of matrix metalloproteinase-14 blocks tumor growth, invasion, and angiogenesis. Cancer Res *69*, 1517–1526.

Doucet, A., and Overall, C. M. (2011). Broad coverage identification of multiple proteolytic cleavage site sequences in complex high molecular weight proteins using quantitative proteomics as a complement to Edman sequencing. Mol Cell Proteomics *10*, M110 003533.

Du, R., Lu, K. V., Petritsch, C., et al. (2008). HIF1alpha induces the recruitment of bone marrow-derived vascular modulatory cells to regulate tumor angiogenesis and invasion. Cancer Cell *13*, 206–220.

Dufour, A., Sampson, N. S., Li, J., et al. (2011). Small-molecule anticancer compounds selectively target the hemopexin domain of matrix metalloproteinase-9. Cancer Res *71*, 4977–4988.

Dufour, A., Zucker, S., Sampson, N. S., Kuscu, C., and Cao, J. A. (2010). Role of matrix metalloproteinase-9 dimers in cell migration design of inhibitory peptides. J Biol Chem *285*, 35944–35956.

Edman, P., and Begg, G. (1967). A protein sequenator. Eur J Biochem *1*, 80–91.

English, W. R., Puente, X. S., Freije, J. M., et al. (2000). Membrane type 4 matrix metalloproteinase (MMP17) has tumor necrosis factor-alpha convertase activity but does not activate pro-MMP2. J Biol Chem *275*, 14046–14055.

Galardy, R. E., Cassabonne, M. E., Giese, C., et al. (1994). Low molecular weight inhibitors in corneal ulceration. Ann N Y Acad Sci *732*, 315–323.

Galvez, B. G., Matias-Roman, S., Albar, J. P., Sanchez-Madrid, F., and Arroyo, A. G. (2001). Membrane type 1-matrix metalloproteinase is activated during migration of human endothelial cells and modulates endothelial motility and matrix remodeling. J Biol Chem *276*, 37491–37500.

Gevaert, K., Goethals, M., Martens, L., et al. (2003). Exploring proteomes and analyzing protein processing by mass spectrometric identification of sorted N-terminal peptides. Nat Biotechnol *21*, 566–569.

Gevaert, K., Van Damme, P., Ghesquiere, B., et al. (2007). A la carte proteomics with an emphasis on gel-free techniques. Proteomics *7*, 2698–2718.

Gilles, C., Polette, M., Coraux, C., et al. (2001). Contribution of MT1-MMP and of human laminin-5 gamma2 chain degradation to mammary epithelial cell migration. J Cell Sci *114*, 2967–2976.

Goldberg, G. I., Marmer, B. L., Grant, G. A., Eisen, A. Z., Wilhelm, S., and He, C. S. (1989). Human 72-kilodalton type-IV collagenase forms a complex with a tissue inhibitor of metalloproteases designated TIMP-2. Proc Natl Acad Sci U S A *86*, 8207–8211.

Gomez, D. E., Alonso, D. F., Yoshiji, H., and Thorgeirsson, U. P. (1997). Tissue inhibitors of metalloproteinases: structure, regulation and biological functions. Eur J Cell Biol *74*, 111–122.

GomisRuth, F. X., Maskos, K., Betz, M., et al. (1997). Mechanism of inhibition of the human matrix metalloproteinase stromelysin-1 by TIMP-1. Nature *389*, 77–81.

Hamada, T., Arima, N., Shindo, M., Sugama, K., and Sasaguri, Y. (2000). Suppression of adjuvant arthritis of rats by a novel matrix metalloproteinase-inhibitor. Br J Pharmacol *131*, 1513–1520.

Hashimoto, G., Inoki, I., Fujii, Y., Aoki, T., Ikeda, E., and Okada, Y. (2002). Matrix metalloproteinases cleave connective tissue growth factor and reactivate angiogenic activity of vascular endothelial growth factor 165. J Biol Chem *277*, 36288–36295.

He, J., and Baum, L. G. (2004). Presentation of galectin-1 by extracellular matrix triggers T cell death. J Biol Chem *279*, 4705–4712.

Hiratsuka, S., Watanabe, A., Aburatani, H., and Maru, Y. (2006). Tumour-mediated upregulation of chemoattractants and recruitment of myeloid cells predetermines lung metastasis. Nat Cell Biol *8*, 1369–1375.

Holmbeck, K., Bianco, P., Caterina, J., et al. (1999). MT1-MMP-deficient mice develop dwarfism, osteopenia, arthritis, and connective tissue disease due to inadequate collagen turnover. Cell *99*, 81–92.

Impens, F., Colaert, N., Helsens, K., et al. (2010). MS-driven protease substrate degradomics. Proteomics *10*, 1284–1296.

Kainthan, R. K., Muliawan, E. B., Hatzikiriakos, S. G., and Brooks, D. E. (2006). Synthesis, characterization, and viscoelastic properties of high molecular weight hyperbranched polyglycerols. Macromolecules *39*, 7708–7717.

Kaplan, R. N., Rafii, S., and Lyden, D. (2006). Preparing the "soil": the premetastatic niche. Cancer Res *66*, 11089–11093.

Kaplan, R. N., Riba, R. D., Zacharoulis, S., et al. (2005). VEGFR1-positive haematopoietic bone marrow progenitors initiate the pre-metastatic niche. Nature *438*, 820–827.

Kappelhoff, R., and Overall, C. (2007). The CLIP-CHIP oligonucleotide microarray: dedicated array for analysis of all protease, nonproteolytic homolog, and inhibitor gene transcripts in human and mouse. Curr Protoc Protein Sci *Chapter 21*, Unit 21 19.

Kessenbrock, K., Plaks, V., and Werb, Z. (2010). Matrix metalloproteinases: regulators of the tumor microenvironment. Cell *141*, 52–67.

Kim, Y. M., Jang, J. W., Lee, O. H., et al. (2000). Endostatin inhibits endothelial and tumor cellular invasion by blocking the activation and catalytic activity of matrix metalloproteinase. Cancer Res *60*, 5410–5413.

Kleifeld, O., Doucet, A., auf dem Keller, U., et al. (2010). Isotopic labeling of terminal amines in complex samples identifies protein N-termini and protease cleavage products. Nat Biotechnol *28*, 281–288.

Kleifeld, O., Doucet, A., Prudova, A., et al. (2011). Identifying and quantifying proteolytic events and the natural N terminome by terminal amine isotopic labeling of substrates. Nat Protoc *6*, 1578–1611.

Koshikawa, N., Schenk, S., Moeckel, G., et al. (2004). Proteolytic processing of laminin-5 by MT1-MMP in tissues and its effects on epithelial cell morphology. FASEB J *18*, 364–366.

Kraling, B. M., Wiederschain, D. G., Boehm, T., Rehn, M., Mulliken, J. B., and Moses, M. A. (1999). The role of matrix metalloproteinase activity in the maturation of human capillary endothelial cells in vitro. J Cell Sci *112* (Pt 10), 1599–1609.

Kruger, A. (2009). Functional genetic mouse models: promising tools for investigation of the proteolytic internet. Biol Chem *390*, 91–97.

Lange, P. F., and Overall, C. M. (2011). TopFIND, a knowledgebase linking protein termini with function. Nat Methods *8*, 703–704.

Lee, S., Jilani, S. M., Nikolova, G. V., Carpizo, D., and Iruela-Arispe, M. L. (2005). Processing of VEGF-A by matrix metalloproteinases regulates bioavailability and vascular patterning in tumors. J Cell Biol *169*, 681–691.

Levene, P. A. (1905). The cleavage products of proteoses. J Biol Chem *1*, 45–58.

Liepinsh, E., Banyai, L., Pintacuda, G., Trexler, M., Patthy, L., and Otting, G. (2003). NMR structure of the netrin-like domain (NTR) of human type I procollagen C-proteinase enhancer defines structural consensus of NTR domains and assesses potential proteinase inhibitory activity and ligand binding. J Biol Chem *278*, 25982–25989.

Lochter, A., Galosy, S., Muschler, J., Freedman, N., Werb, Z., and Bissell, M. J. (1997). Matrix metalloproteinase stromelysin-1 triggers a cascade of molecular alterations that leads to stable epithelial-to-mesenchymal conversion and a premalignant phenotype in mammary epithelial cells. J Cell Biol *139*, 1861–1872.

Lohi, J., Wilson, C. L., Roby, J. D., and Parks, W. C. (2001). Epilysin, a novel human matrix metalloproteinase (MMP-28) expressed in testis and keratinocytes and in response to injury. J Biol Chem *276*, 10134–10144.

López-Otín, C., and Bond, J. S. (2008). Proteases: multifunctional enzymes in life and disease. J Biol Chem *283*, 30433–30437.

López-Otín, C., and Matrisian, L. M. (2007). Tumour microenvironment – opinion: Emerging roles of proteases in tumour suppression. Nat Rev Cancer *7*, 800–808.

López-Otín, C., and Overall, C. M. (2002). Protease degradomics: a new challenge for proteomics. Nat Rev Mol Cell Biol *3*, 509–519.

Maeshima, Y., Colorado, P. C., and Kalluri, R. (2000). Two RGD-independent alpha vbeta 3 integrin binding sites on tumstatin regulate distinct anti-tumor properties. J Biol Chem *275*, 23745–23750.

Mahrus, S., Trinidad, J. C., Barkan, D. T., Sali, A., Burlingame, A. L., and Wells, J. A. (2008). Global sequencing of proteolytic cleavage sites in apoptosis by specific labeling of protein N termini. Cell *134*, 866–876.

McDonald, L., and Beynon, R. J. (2006). Positional proteomics: preparation of amino-terminal peptides as a strategy for proteome simplification and characterization. Nat Protoc *1*, 1790–1798.

McGuire, J. K., Li, Q., and Parks, W. C. (2003). Matrilysin (matrix metalloproteinase-7) mediates E-cadherin ectodomain shedding in injured lung epithelium. Am J Pathol *162*, 1831–1843.

McQuibban, G. A., Butler, G. S., Gong, J. H., et al. (2001). Matrix metalloproteinase activity inactivates the CXC chemokine stromal cell-derived factor-1. J Biol Chem *276*, 43503–43508.

McQuibban, G. A., Gong, J. H., Tam, E. M., McCulloch, C. A., Clark-Lewis, I., and Overall, C. M. (2000). Inflammation dampened by gelatinase A cleavage of monocyte chemoattractant protein-3. Science *289*, 1202–1206.

McQuibban, G. A., Gong, J. H., Wong, J. P., Wallace, J. L., Clark-Lewis, I., and Overall, C. M. (2002). Matrix metalloproteinase processing of monocyte chemoattractant proteins generates CC chemokine receptor antagonists with anti-inflammatory properties in vivo. Blood *100*, 1160–1167.

Morgunova, E., Tuuttila, A., Bergmann, U., and Tryggvason, K. (2002). Structural insight into the complex formation of latent matrix metalloproteinase 2 with tissue inhibitor of metalloproteinase 2. Proc Natl Acad Sci U S A *99*, 7414–7419.

Morrison, C., Mancini, S., Cipollone, J., Kappelhoff, R., Roskelley, C., and Overall, C. (2011). Microarray and proteomic analysis of breast cancer cell and osteoblast co-cultures: role of osteoblast matrix metalloproteinase (MMP)-13 in bone metastasis. J Biol Chem *286*, 34271–34285.

Morrison, C. J., Butler, G. S., Bigg, H. F., Roberts, C. R., Soloway, P. D., and Overall, C. M. (2001). Cellular activation of MMP-2 (gelatinase A) by MT2-MMP occurs via a TIMP-2-independent pathway. J Biol Chem *276*, 47402–47410.

Morrison, C. J., Butler, G. S., Rodriguez, D., and Overall, C. M. (2009). Matrix metalloproteinase proteomics: substrates, targets, and therapy. Curr Opin Cell Biol *21*, 645–653.

Mortier, A., Gouwy, M., Van Damme, J., and Proost, P. (2011). Effect of posttranslational processing on the in vitro and in vivo activity of chemokines. Exp Cell Res *317*, 642–654.

Mudgett, J. S., Hutchinson, N. I., Chartrain, N. A., et al. (1998). Susceptibility of stromelysin 1-deficient mice to collagen-induced arthritis and cartilage destruction. Arthritis Rheum *41*, 110–121.

Nagase, H., and Woessner, J. F., Jr. (1999). Matrix metalloproteinases. J Biol Chem *274*, 21491–21494.

Netzer, K. O., Suzuki, K., Itoh, Y., Hudson, B. G., and Khalifah, R. G. (1998). Comparative analysis of the noncollagenous NC1 domain of type IV collagen: identification of structural features important for assembly, function, and pathogenesis. Protein Sci *7*, 1340–1351.

Noe, V., Fingleton, B., Jacobs, K., et al. (2001). Release of an invasion promoter E-cadherin fragment by matrilysin and stromelysin-1. J Cell Sci *114*, 111–118.

Opdenakker, G., Van den Steen, P. E., and Van Damme, J. (2001). Gelatinase B: a tuner and amplifier of immune functions. Trends Immunol *22*, 571–579.

O'Reilly, M. S., Wiederschain, D., Stetler-Stevenson, W. G., Folkman, J., and Moses, M. A. (1999). Regulation of angiostatin production by matrix metalloproteinase-2 in a model of concomitant resistance. J Biol Chem *274*, 29568–29571.

Orlichenko, L. S., and Radisky, D. C. (2008). Matrix metalloproteinases stimulate epithelial-mesenchymal transition during tumor development. Clin Exp Metastasis *25*, 593–600.

Overall, C. M. (2002). Molecular determinants of metalloproteinase substrate specificity: matrix metalloproteinase substrate binding domains, modules, and exosites. Mol Biotechnol *22*, 51–86.

Overall, C. M., and Blobel, C. P. (2007). In search of partners: linking extracellular proteases to substrates. Nat Rev Mol Cell Biol *8*, 245–257.

Overall, C. M., and Dean, R. A. (2006). Degradomics: systems biology of the protease web. Pleiotropic roles of MMPs in cancer. Cancer Metastasis Rev *25*, 69–75.

Overall, C. M., and Kleifeld, O. (2006a). Towards third generation matrix metalloproteinase inhibitors for cancer therapy. Br J Cancer *94*, 941–946.

Overall, C. M., and Kleifeld, O. (2006b). Tumour microenvironment – opinion: validating matrix metalloproteinases as drug targets and anti-targets for cancer therapy. Nat Rev Cancer *6*, 227–239.

Overall, C. M., and López-Otín, C. (2002). Strategies for MMP inhibition in cancer: innovations for the post-trial era. Nat Rev Cancer *2*, 657–672.

Overall, C. M., McQuibban, G. A., and Clark-Lewis, I. (2002). Discovery of chemokine substrates for matrix metalloproteinases by exosite scanning: a new tool for degradomics. Biol Chem *383*, 1059–1066.

Overall, C. M., Tam, E., McQuibban, G. A., et al. (2000). Domain interactions in the gelatinase A.TIMP-2.MT1-MMP activation complex. The ectodomain of the 44-kDa form of membrane type-1 matrix metalloproteinase does not modulate gelatinase A activation. J Biol Chem *275*, 39497–39506.

Pavlaki, M., and Zucker, S. (2003). Matrix metalloproteinase inhibitors (MMPIs): the beginning of phase I or the termination of phase III clinical trials. Cancer Metastasis Rev *22*, 177–203.

Pei, D. Q. (1999). Identification and characterization of the fifth membrane-type matrix metalloproteinase MT5-MMP. J Biol Chem *274*, 8925–8932.

Pei, D. Q., Kang, T. B., and Qi, H. X. (2000). Cysteine array matrix metalloproteinase (CA-MMP)/MMP-23 is a type II transmembrane matrix metalloproteinase regulated by a single cleavage for both secretion and activation. J Biol Chem *275*, 33988–33997.

Pei, D. Q., and Weiss, S. J. (1995). Furin-dependent intracellular activation of the human stromelysin-3 zymogen. Nature *375*, 244–247.

Prudova, A., auf dem Keller, U., Butler, G. S., and Overall, C. M. (2010). Multiplex N-terminome analysis of MMP-2 and MMP-9 substrate degradomes by iTRAQ-TAILS quantitative proteomics. Mol Cell Proteomics *9*, 894–911.

Puente, X. S., Sanchez, L. M., Overall, C. M., and López-Otín, C. (2003). Human and mouse proteases: a comparative genomic approach. Nat Rev Genet *4*, 544–558.

Remacle, A. G., Shiryaev, S. A., Radichev, I. A., Rozanov, D. V., Stec, B., and Strongin, A. Y. (2011). Dynamic interdomain interactions contribute to the inhibition of matrix metalloproteinases (MMPs) by tissue inhibitors of metalloproteinases (TIMPs). J Biol Chem *286* (23), 21002–21012.

Reznik, S. E., and Fricker, L. D. (2001). Carboxypeptidases from A to Z: implications in embryonic development and Wnt binding. Cell Mol Life Sci *58*, 1790–1804.

Sato, H., Kinoshita, T., Takino, T., Nakayama, K., and Seiki, M. (1996). Activation of a recombinant membrane type 1-matrix metalloproteinase (MT1-MMP) by furin and its interaction with tissue inhibitor of metalloproteinases (TIMP)-2. FEBS Lett *393*, 101–104.

Schilling, O., Barre, O., Huesgen, P. F., and Overall, C. M. (2010). Proteome-wide analysis of protein carboxy termini: C terminomics. Nat Methods *7*, 508–511.

Schilling, O., Huesgen, P. F., Barre, O., Auf dem Keller, U., and Overall, C. M. (2011). Characterization of the prime and non-prime active site specificities of proteases by proteome-derived peptide libraries and tandem mass spectrometry. Nat Protoc *6*, 111–120.

Schilling, O., and Overall, C. M. (2008). Proteome-derived, database-searchable peptide libraries for identifying protease cleavage sites. Nat Biotechnol *26*, 685–694.

Schwartz, D. R., Moin, K., Yao, B., et al. (2007). Hu/Mu protin oligonucleotide microarray: dual-species array for profiling protease and protease inhibitor gene expression in tumors and their microenvironment. Mol Cancer Res *5*, 443–454.

Sorsa, T., Ramamurthy, N. S., Vernillo, A. T., et al. (1998). Functional sites of chemically modified tetracyclines: inhibition of the oxidative activation of human neutrophil and chicken osteoclast pro-matrix metalloproteinases. J Rheumatol *25*, 975–982.

Staes, A., Impens, F., Van Damme, P., et al. (2011). Selecting protein N-terminal peptides by combined fractional diagonal chromatography. Nat Protoc *6*, 1130–1141.

Staes, A., Van Damme, P., Helsens, K., Demol, H., Vandekerckhove, J., and Gevaert, K. (2008). Improved recovery of proteome-informative, protein N-terminal peptides by combined fractional diagonal chromatography (COFRADIC). Proteomics *8*, 1362–1370.

Starr, A. E., Dufour, A., Maier, J., and Overall, C. M. (2012). Biochemical analysis of matrix metalloproteinase activation of chemokines CCL15 and CCL23 and increased glycosaminoglycan binding of CCL16. J Biol Chem *287* (8), 5848–5860.

Sternlicht, M. D., and Werb, Z. (2001). How matrix metalloproteinases regulate cell behavior. Annu Rev Cell Dev Biol *17*, 463–516.

Swingler, T. E., Waters, J. G., Davidson, R. K., et al. (2009). Degradome expression profiling in human articular cartilage. Arthritis Res Ther *11*, R96.

Takahashi, C., Sheng, Z., Horan, T. P., et al. (1998). Regulation of matrix metalloproteinase-9 and inhibition of tumor invasion by the membrane-anchored glycoprotein RECK. Proc Natl Acad Sci U S A *95*, 13221–13226.

Takino, T., Sato, H., Shinagawa, A., and Seiki, M. (1995). Identification of the second membrane-type matrix metalloproteinase (MT-MMP-2) gene from a human placenta

cDNA library. MT-MMPs form a unique membrane-type subclass in the MMP family. J Biol Chem *270*, 23013–23020.
Tam, E. M., Morrison, C. J., Wu, Y. I., Stack, M. S., and Overall, C. M. (2004). Membrane protease proteomics: isotope-coded affinity tag MS identification of undescribed MT1-matrix metalloproteinase substrates. Proc Natl Acad Sci U S A *101*, 6917–6922.
Tester, A. M., Cox, J. H., Connor, A. R., et al. (2007). LPS responsiveness and neutrophil chemotaxis in vivo require PMN MMP-8 activity. PLoS One *2*, e312.
Van den Steen, P. E., Proost, P., Grillet, B., et al. (2002). Cleavage of denatured natural collagen type II by neutrophil gelatinase B reveals enzyme specificity, post-translational modifications in the substrate, and the formation of remnant epitopes in rheumatoid arthritis. FASEB J *16*, 379–389.
Vanwart, H. E., and Birkedalhansen, H. (1990). The cysteine switch – a principle of regulation of metalloproteinase activity with potential applicability to the entire matrix metalloproteinase gene family. Proc Natl Acad Sci U S A *87*, 5578–5582.
Velasco, G., Cal, S., Merlos-Suarez, A., et al. (2000). Human MT6-matrix metalloproteinase: identification, progelatinase A activation, and expression in brain tumors. Cancer Res *60*, 877–882.
Visse, R., and Nagase, H. (2003). Matrix metalloproteinases and tissue inhibitors of metalloproteinases – structure, function, and biochemistry. Circ Res *92*, 827–839.
Wang, X., Fu, X., Brown, P. D., Crimmin, M. J., and Hoffman, R. M. (1994). Matrix metalloproteinase inhibitor BB-94 (batimastat) inhibits human colon tumor growth and spread in a patient-like orthotopic model in nude mice. Cancer Res *54*, 4726–4728.
Weber, B. H., Vogt, G., Pruett, R. C., Stohr, H., and Felbor, U. (1994). Mutations in the tissue inhibitor of metalloproteinases-3 (TIMP3) in patients with Sorsby's fundus dystrophy. Nat Genet *8*, 352–356.
Whittaker, M., Floyd, C. D., Brown, P., and Gearing, A. J. (1999). Design and therapeutic application of matrix metalloproteinase inhibitors. Chem Rev *99*, 2735–2776.
Williamson, R. A., Carr, M. D., Frenkiel, T. A., Feeney, J., and Freedman, R. B. (1997). Mapping the binding site for matrix metalloproteinase on the N-terminal domain of the tissue inhibitor of metalloproteinases-2 by NMR chemical shift perturbation. Biochemistry *36*, 13882–13889.
Williamson, R. A., Marston, F. A. O., Angal, S., et al. (1990). Disulfide bond assignment in human tissue inhibitor of metalloproteinases (TIMP). Biochem J *268*, 267–274.
Windsor, L. J., Bodden, M. K., Birkedal-Hansen, B., Engler, J. A., and Birkedal-Hansen, H. (1994). Mutational analysis of residues in and around the active site of human fibroblast-type collagenase. J Biol Chem *269*, 26201–26207.
Wynn, R. L. (1999). Latest FDA approvals for dentistry. Gen Dent *47*, 19–22.
Zarrabi, K., Dufour, A., Li, J., et al. (2011). Inhibition of matrix metalloproteinase 14 (MMP-14)-mediated cancer cell migration. J Biol Chem *286*, 33167–33177.
Zhang, K., McQuibban, G. A., Silva, C., et al. (2003). HIV-induced metalloproteinase processing of the chemokine stromal cell derived factor-1 causes neurodegeneration. Nat Neurosci *6*, 1064–1071.
Zhou, Z. J., Apte, S. S., Soininen, R., et al. (2000). Impaired endochondral ossification and angiogenesis in mice deficient in membrane-type matrix metalloproteinase I. Proc Natl Acad Sci U S A *97*, 4052–4057.
Zhu, Y., Vergote, D., Pardo, C., et al. (2009). CXCR3 activation by lentivirus infection suppresses neuronal autophagy: neuroprotective effects of antiretroviral therapy. FASEB J *23*, 2928–2941.

5 ECM cell surface receptors

5.1 Introduction

Donald Gullberg

Traditionally, cell adhesion receptors have been regarded as mediators of the linkage of cells to the surrounding or adjacent ECM. In this role, they mediate cell attachment, cell spreading, matrix assembly, and remodeling. In the early days of ECM receptor research, the view of ECM receptors mainly being needed for connective tissue assembly and structural integrity reinforced the picture of the ECM as a relatively biologically inert filling/support substance (Gullberg and Ekblom, 1995).

With the finding that ECM receptors can generate signals (Hynes, 1992), the picture of ECM receptors as simple linker molecules has changed dramatically. We now know that the activity of cell adhesion receptors can be regulated (Askari et al., 2009), and once activated, these receptors are an integrated part of the cell-sensing mechanism of the outside extracellular world (Schwartz and Ginsberg, 2002; Yamada and Even-Ram, 2002). An emerging theme is that they take part in mechanosensing and mechanotransduction in diverse processes such as stem cell differentiation, fibrosis, and tumor growth (Engler et al., 2006; Hinz, 2009; Levental et al., 2009). Cell adhesion receptor signaling occurs in complex ways (receptors acting as coreceptors, signal amplifiers, and signal attenuators) and involves the regulation of matrix turnover (ECM synthesis and matrix metalloproteinase [MMP] activity), cell proliferation, cell differentiation, and apoptosis.

Combined with the traditional roles of matrix receptors, this places adhesion receptor function among a large variety of process during tissue morphogenesis, homeostasis, and pathology (Pure and Assoian, 2009; Choi et al., 2011; Leitinger, 2011; Weber et al., 2011). Without doubt, cell adhesion receptors have entered the center stage of cell biology.

Despite the fact that the genomes of several model organisms were sequenced some time ago, the number of matrix receptors and the identification of their respective ligands do not seem to have reached equilibrium yet. Two recent examples include the discovery of nonintegrin collagen receptors with specialized functions in brain and in bone (Barrow et al., 2011; Luo et al., 2011).

The amount of data that is published on matrix receptors is vast, and we have therefore chosen to select a few specific topics confined to integrins, syndecans, discoidin domain receptors (DDRs), and CD44, with the following objectives:

1. Illustrate the diversity of biological systems in which different types of matrix receptors are attracting interest: Chapter 5.2, "Integrin Function in Heart Fibrosis: Mechanical Strain, Transforming Growth Factor-Beta 1 Activation, and Collagen Glycation," by B. Hinz and C. A. McCulloch; Chapter 5.3, "Cancer-Associated Fibroblast Integrins as Therapeutic Targets in the Tumor Micronevironment," by N. Lu, C. Brakebusch, and D. Gullberg; and Chapter 5.6, "CD44: A Sensor of Tissue Damage Critical for Restoring Homeostasis," by E. Puré.

2. Discuss new roles (Chapter 5.4, "Discoidin Domain Receptors: Nonintegrin Collagen Receptors on the Move," by A. S. Rocca and M. P. Bendeck).
3. Highlight new emerging areas where these receptors will be interesting in the future.

We refer to recent reviews on integrins (Barczyk et al., 2010; Campbell and Humphries, 2011; Weber et al., 2011; Wickstrom et al., 2011), collagen receptors (Leitinger, 2011), syndecans (Couchman, 2010; Choi et al., 2011), and CD44 (Pure and Assoian, 2009; Zoller, 2011) for a general background on these receptor classes.

The increase in the number of published articles pertaining to these four receptor groups is illustrated in ▶Figure 5.1. As shown in this figure, some receptor groups have the potential to generate much more data in the years to come, whereas other fields already produce a steady level of new research reports.

I look forward to a new edition of the book in a few years' time, when, hopefully, our knowledge about these receptors classes will have increased, if not exponentially, at least considerably.

With the help of new technology and a new generation of enthusiastic scientists, we should reach the next level of understanding – a better appreciation of how these receptors classes interact, cooperate, synergize, and antagonize each other to generate the harmonized equilibrium that under normal physiological conditions often exists in the ECM.

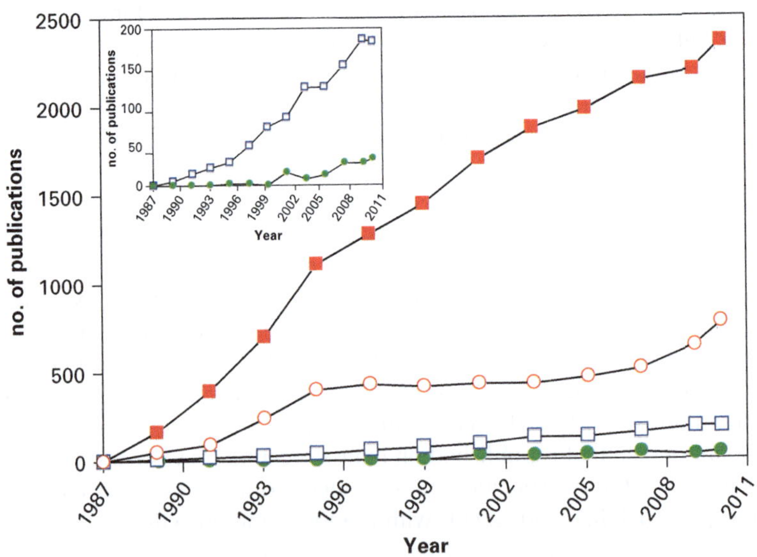

Figure 5.1 Graphic illustration of the number of articles published per year on integrins, syndecans, discoidin domain receptors (DDRs) 1 and 2, and CD44. Source PubMed. Red squares: integrin publications; white circles: syndecan publications; white squares: CD44 publications; and green circles: DDR publications.

References

Askari, J. A., Buckley, P. A., Mould, A. P., and Humphries, M. J. (2009). Linking integrin conformation to function. J Cell Sci *122*, 165–170.

Barczyk, M., Carracedo, S., and Gullberg, D. (2010). Integrins. Cell Tissue Res *339*, 269–280.

Barrow, A. D., Raynal, N., Andersen, T. L., et al. (2011). OSCAR is a collagen receptor that costimulates osteoclastogenesis in DAP12-deficient humans and mice. J Clin Invest *121*, 3505–3516.

Campbell, I. D., and Humphries, M. J. (2011). Integrin structure, activation, and interactions. Cold Spring Harb Perspect Biol *3*, a004994.

Choi, Y., Chung, H., Jung, H., Couchman, J. R., and Oh, E. S. (2011). Syndecans as cell surface receptors: unique structure equates with functional diversity. Matrix Biol *30*, 93–99.

Couchman, J. R. (2010). Transmembrane signaling proteoglycans. Annu Rev Cell Dev Biol *26*, 89–114.

Engler, A. J., Sen, S., Sweeney, H. L., and Discher, D. E. (2006). Matrix elasticity directs stem cell lineage specification. Cell *126*, 677–689.

Gullberg, D., and Ekblom, P. (1995). Extracellular matrix and its receptors during development. Int J Dev Biol *39*, 845–854.

Hinz, B. (2009). Tissue stiffness, latent TGF-beta1 activation, and mechanical signal transduction: implications for the pathogenesis and treatment of fibrosis. Curr Rheumatol Rep *11*, 120–126.

Hynes, R. O. (1992). Integrins: versatility, modulation, and signaling in cell adhesion. Cell *69*, 11–25.

Leitinger, B. (2011). Transmembrane collagen receptors. Annu Rev Cell Dev Biol *27*, 265–290.

Levental, K. R., Yu, H., Kass, L., et al. (2009). Matrix crosslinking forces tumor progression by enhancing integrin signaling. Cell *139*, 891–906.

Luo, R., Jeong, S. J., Jin, Z., Strokes, N., Li, S., and Piao, X. (2011). G protein-coupled receptor 56 and collagen III, a receptor-ligand pair, regulates cortical development and lamination. Proc Natl Acad Sci U S A *108*, 12925–12930.

Pure, E., and Assoian, R. K. (2009). Rheostatic signaling by CD44 and hyaluronan. Cell Signal *21*, 651–655.

Schwartz, M. A., and Ginsberg, M. H. (2002). Networks and crosstalk: integrin signalling spreads. Nat Cell Biol *4*, E65–E68.

Weber, G. F., Bjerke, M. A., and DeSimone, D. W. (2011). Integrins and cadherins join forces to form adhesive networks. J Cell Sci *124*, 1183–1193.

Wickstrom, S. A., Radovanac, K., and Fässler, R. (2011). Genetic analyses of integrin signaling. Cold Spring Harb Perspect Biol *3*, a005116.

Yamada, K. M., and Even-Ram, S. (2002). Integrin regulation of growth factor receptors. Nat Cell Biol *4*, E75–E76.

Zoller, M. (2011). CD44: can a cancer-initiating cell profit from an abundantly expressed molecule? Nat Rev Cancer *11*, 254–267.

5.2 Integrin function in heart fibrosis: mechanical strain, transforming growth factor-beta 1 activation, and collagen glycation

Boris Hinz and Christopher A. McCulloch

5.2.1 Introduction

Fibroblasts are the principal cell type in the synthesis and remodeling of the extracellular matrix (ECM) and play a central role in pathological events such as fibrosis, which has important consequences for kidney, liver, corneal, and cardiac function (Brown et al., 2005). The replacement of the normal organ stroma with disorganized ECM proteins impacts normal organ structure in a wide variety of tissues in which the organization of the ECM is critical for the function of specialized cells or processes (e.g. contraction, filtration, gas exchange, and vision). The formation of profibrotic cells in many types of organs and tissues is currently considered a central event in the generation of fibrosis, but it is important to note that in normal wound healing, profibrotic cells such as myofibroblasts also play an important role in acute tissue repair, such as skin wound closure. Myofibroblasts, which can express abundant collagen, α-smooth muscle actin (α-SMA), and the extradomain A (ED-A) fibronectin splice variant, transiently form in response to transforming growth factor-beta 1 (TGF-β1) and cell-generated mechanical tension (Desmouliere et al., 1993; Arora et al., 1999). These contractile cells help to close open wounds, and, 4–7 days after wounding, they usually disappear by apoptosis (Gabbiani, 2003). However, in many pathological situations, including fibrosis of lung, liver, kidney, heart, and periodontium, myofibroblasts persist and continue to synthesize and remodel the ECM, which results in the formation of scar tissue and disorganized collagen (Tomasek et al., 2002). While some of the cytokines and factors that drive fibrosis are known (e.g. TGF-β1, endothelin-1, and connective tissue growth factor [CTGF]/CCN2) (Shi-Wen et al., 2007), it is not known whether these cytokines are activated to regulate the conversion of fibroblasts in all types of fibrotic lesions. Further, the role of integrin-dependent activation of these profibrotic cytokines is only now beginning to emerge. One organ that has generated a great deal of interest in the context of fibrosis is cardiac muscle, in part because of the high prevalence of cardiac disease in human populations and the high morbidity and mortality that is associated with these diseases. In this chapter, we explore cardiac fibrosis in response to pressure overload and to the alterations that occur in integrin-dependent signaling as examples of how fibrosis is generated.

5.2.2 Cardiac fibrosis – the players

The cellular players in cardiac fibrosis – fibroblasts and myofibroblasts

While cardiac myocytes comprise the largest volume fraction of the adult heart, they represent <25% of the cell number (Grove et al., 1969). By far, the most abundant nonmyocytic cell in the myocardium is the fibroblast (Eghbali, 1992) (30%–50% of the cell number), which, along with endothelial and smooth muscle cells, pericytes, neurons, and blood-borne cells, makes up the remaining 75%. The fibroblast is the principal cell involved in the synthesis and remodeling of the cardiac ECM. Therefore, the fibroblast and its more profibrotic offspring, the myofibroblast (Tomasek et al., 2002; Hinz et al., 2004, 2007;), are likely to play central roles in the fibrotic response in a variety of important diseases that affect the heart, including heart failure, postmyocardial infarction scarring, and diabetic cardiomyopathy (DCM) (Kelly et al., 2007). When we consider the central role of the fibroblast in fibrotic diseases of the heart, it is notable that the cardiac fibroblast plays pivotal roles in regulating cardiac function that are crucial for the development of fibrosis in heart failure but also in the pathological alterations that accompany DCM.

Notably, in heart failure or after myocardial infarction, the cell populations of the ventricular wall undergo marked changes in which there is a substantial conversion of resident fibroblasts to myofibroblasts (Leslie et al., 1991). Other potential precursors of myofibroblasts in the heart are endothelial cells, which can undergo endothelial-to-mesenchymal transition (Zeisberg et al., 2007); bone-marrow-derived circulating fibrocytes; hematopoietic cells; and mesenchymal stem cells (Haudek et al., 2006; Mollmann et al., 2006; Frangogiannis, 2008; van Amerongen et al., 2008). With the exception of the valve leaflet, myofibroblasts are not found in the normal heart but have been identified in fibrotic myocardium (Leslie et al., 1991; Sun and Weber, 2000) and are locally increased at sites of myocardial infarction (Campbell et al., 1995; Campbell and Katwa, 1997; Suurmeijer et al., 2003).

ECM stiffness contributes to heart fibrosis and dysfunction

Myofibroblasts, in contrast to fibroblasts or other precursors, produce excessive, poorly organized collagen ECM (Kuwahara et al., 2002). This collagenous scar tissue has important implications for reduced heart function. One hallmark of the fibrotic scar is its high stiffness compared to the compliant texture of the healthy heart. Tissue stiffness can be measured, and the stiffness expressed as Young's modulus (in pascals, Pa), which is defined as the force per unit area (stress) that is required to strain a material. High stress levels are required to deform materials with high Young's modulus. The Young's modulus of normal heart muscle is ~10 kPa as measured by atomic force microscopy (AFM) (Berry et al., 2006; Engler et al., 2008; Wells and Discher, 2008). Fibrotic tissue is up to 10 times stiffer (20–100 kPa), which is similar to the stiffness of collagen-dense tendon or osteoid that precedes formation of bone (Goffin et al., 2006; Engler et al., 2008; Wells and Discher, 2008). What are the consequences of myofibroblast-generated fibrotic scars being stiffer than heart muscle? Scar tissue impedes cardiac compliance and thus ventricular function. This contributes to diastolic and systolic dysfunction and to left ventricular hypertrophy (Kitamura et al., 2001; Brown et al., 2005; Ouzounian et al., 2008).

When prolonged hypertension contributes to pathological hypertrophy, scar stiffness and hypertensive stress amplify each other synergistically (Wakatsuki et al., 2004). Scar stiffness also promotes arrhythmias and heart failure by directly influencing cardiomyocyte function. This notion is supported by recent findings that embryonic cardiomyocytes beat periodically when cultured on soft substrates that replicate the stiffness of normal cardiac tissues but not on stiffer substrates that model fibrosis (Engler et al., 2008). Scar stiffness has even more profound influence on the progression of cardiac fibrosis by promoting the conversion of healthy precursor cells into fibrotic myofibroblasts (Hinz et al., 2007). As the generation of myofibroblasts is a critical element in the development of disorganized collagen matrices in the myocardium in cardiac overload (Leslie et al., 1991) and after infarction (Tomasek et al., 2002), the role of integrins and the generation of profibrotic signals is an excellent example of the importance of fibrosis in organ dysfunction.

Mechanical signals received via integrins promote myofibroblast differentiation

Mechanical stress is one of the most potent factors controlling myofibroblast fate and development (Arora et al., 1999; Hinz, 2009) (▶Figure 5.2). In the heart, the mechanoresponsiveness of cardiac fibroblasts and myofibroblasts is an important precondition for the onset and progression of fibrosis (Sun and Weber, 2000; Squires et al., 2005; Hinz et al., 2007; Porter and Turner, 2009). Stress arises when the heart is excessively stretched, which may occur in pressure overload and is associated with a hypertrophic response that leads to cardiac remodeling. Reversible remodeling, which occurs as a result of exercise, augments cardiomyocyte and cardiac fibroblast populations. Pathological irreversible remodeling, however, which is associated with chronic hypertension, is associated with increases of the fibroblastic population, induces myofibroblast differentiation, and often leads to fibrosis (Leslie et al., 1991; Sun and Weber, 2000; Wakatsuki et al., 2004). After a myocardial infarct, loss of the mechanoprotective architecture of the ECM is associated with diminished resistance to load and maintenance of the tensile strength of the ventricular wall. As a result, cardiac fibroblasts develop tension and, when modeled *in vitro*, form de novo contractile microfilament bundles (stress fibers) (Sussman et al., 2002; Baudino et al., 2006). In the precisely controlled mechanical feedback of cell remodeling and the resulting increased stiffening of the ECM, myofibroblasts become more contractile by expression of α-SMA in stress fibers (Hinz et al., 2001a), which is the most widely used molecular marker of the myofibroblast (Tomasek et al., 2002) (▶Figure 5.2).

Many previous studies have demonstrated that stress transmitted from ECM to integrin governs myofibroblast differentiation. Application of mechanical force to ECM protein-coated magnetite beads that are attached to cultured cardiac fibroblasts induces α-SMA expression (Wang et al., 2003). Similarly, application of cyclic stretching forces to cultured aortic valve fibroblasts enhances α-SMA expression (Merryman et al., 2007). Expression of α-SMA is augmented in myofibroblasts cultured on stiff two-dimensional culture substrates but suppressed on soft polymer substrates (Goffin et al., 2006; Solon et al., 2007). Myofibroblasts express α-SMA in attached (stressed) three-dimensional collagen gel culture but not in free-floating (relaxed) gels (Grinnell, 1994; Arora et al., 1999; Hinz, 2006). In healing rat skin wounds, expression of α-SMA is accelerated by splinting the edges of full-thickness rat skin wounds with plastic frames, which

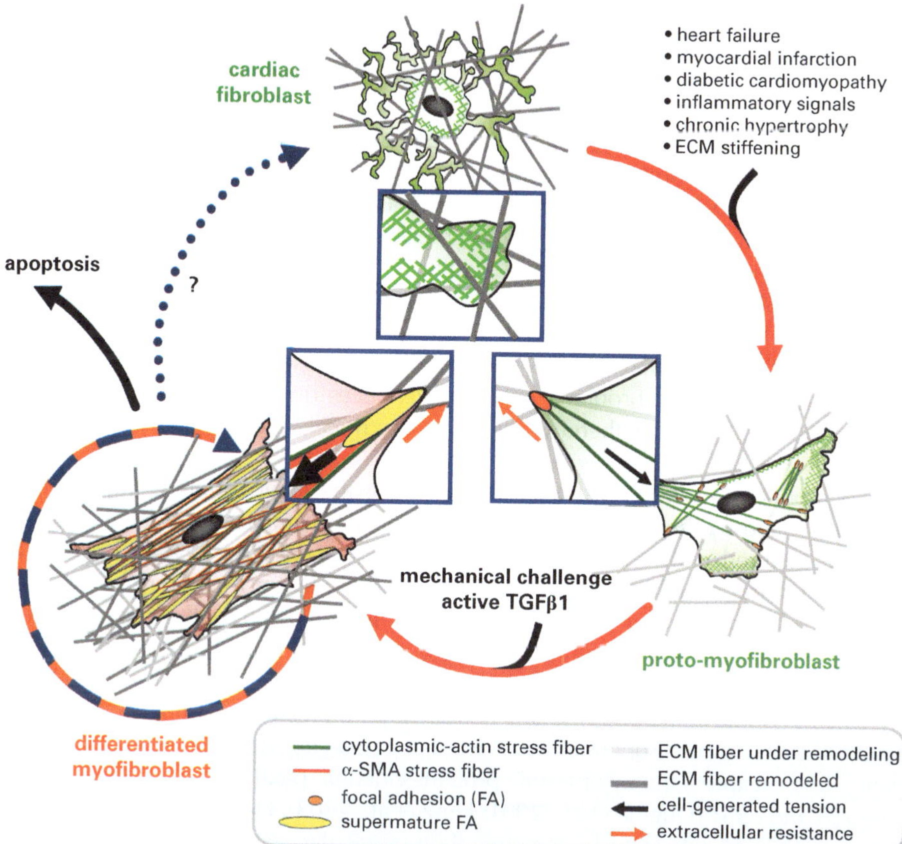

Figure 5.2 The mechanical feedback loop in myofibroblast development. Fibroblast-to-myofibroblast differentiation is controlled by a careful balance of intracellular (contractility) and extracellular (e.g. ECM stiffness) stresses. Fibroblasts in the intact heart are stress-shielded by an organized collagen matrix; they do not develop contractile features and cell-ECM adhesions. In response to injury, chronic overload, or ECM stiffening, fibroblasts become activated to spread and migrate. Fibroblast remodeling activity leads to gradual increase in global matrix stiffness that counteracts cell traction forces. The resulting formation of small focal adhesionsfocal adhesions (FAs) and stress fibers that contain only cytoplasmic actins characterize the protomyofibroblast. Continuing ECM fiber alignment creates larger surfaces for adhesion formation; larger adhesions permit development of stronger stress fibers and generation of higher contractile forces. In the presence of active TGF-β1, protomyofibroblasts begin to express α-SMA that is gradually incorporated into the preexisting stress fibers. The force generated by α-SMA-containing stress fiber is significantly higher compared to cytoplasmic actin stress fibers, leading to further FA supermaturation and ECM contraction, thereby establishing a mechanical feed-forward loop. Myofibroblasts may exit this cycle when the original structure of the ECM is reconstituted and again takes over the mechanical load. Failure of stress-released myofibroblasts to undergo apoptosis results in their persisting activity and ultimately leads to fibrosis (modified after Hinz, 2007).

increases tissue tension; the release of tension has the opposite effect (Hinz et al., 2001b). A similar result is found after stretching human burn scar tissue in situ (Junker et al., 2008). Comparable controlled experiments with animal and human cardiac tissues have not yet been performed.

Although the relationship between increased stress following pressure overload and the formation of myofibroblasts in the heart has been described (Leslie et al., 1991), the mechanisms of mechanotransduction are unclear. Myofibroblast mechanoregulation is thought to occur on different levels. First, mechanical load determines the intracellular stress fiber localization of α-SMA (Goffin et al., 2006). Second, stress directly modulates α-SMA promoter activity and protein expression (Wang et al., 2003; Zhao et al., 2007). Third, stress modulates the bioactivity of TGF-β1, the major cytokine inducing myofibroblast differentiation. We will discuss subsequently the recent findings demonstrating that myofibroblasts exert tensile forces on ECM-stored latent TGF-β1 via specific integrins and thereby activate this critical, profibrotic cytokine (Wipff et al., 2007).

Myofibroblasts sense ECM stiffness by assessing the level of intracellular tension, which is limited by the size of their cell-ECM contacts (similar to the focal adhesions [FAs]) of cultured cells). Stiff ECM induces large "supermature" FAs (Goffin et al., 2006) (▶Figure 5.2). Inhibition of supermature FA formation by culture of myofibroblasts on substrates with a compliance ≤11 kPa is associated with disappearance of α-SMA from stress fibers (Goffin et al., 2006). Restricting myofibroblast adhesions to small FAs (≤6 μm) using microcontact printing also results in the loss of α-SMA from stress fibers (Goffin et al., 2006). α-SMA is reincorporated into stress fibers by stretching small FA islets (6 μm) to supermature FA islets (8 μm) on extendable membranes. Supermature FAs coexpress the cytoplasmic proteins vinculin, paxillin, and tensin and integrins αvβ3 and α5β1 (Dugina et al., 2001; Hinz et al., 2003). Hence, interfering with the formation of ECM adhesions by targeting their molecular components is one potential strategy to interrupt the mechanical feedback-loop of myofibroblast differentiation.

The onset of the vicious cycle of ECM stiffening and myofibroblast formation

Tissue stiffness increases as a consequence of the ECM-remodeling activities of fibroblasts and myofibroblasts (Tomasek et al., 2002; Grinnell and Petroll, 2010). Hence, actively contracting cells generate the conditions that make them more contractile in a detrimental feed-forward loop. The reoccurring chicken-and-egg question is whether ECM stiffening precedes myofibroblast formation or whether myofibroblast formation leads to ECM stiffening. The answer is simple: *both* (▶Figure 5.2). In an animal model of liver fibrosis, tissue stiffening occurring during inflammation has been shown to precede activation of fibroblastic cells and accumulation of collagen; it was suggested that such early mechanical changes would be sufficient to trigger the contracture cascade (Georges et al., 2007). Collagen cross-linking by lysyl oxidases appears to be one factor responsible for early structural changes and tissue stiffening in these conditions (Georges et al., 2007). Fibrosis-specific and stable collagen cross-links are further formed by lysyl hydroxylase and pro-collagen-lysine, 2-oxoglutarate 5-dioxygenase (PLOD2), which are responsible for pyridinoline and aldehyde-derived collagen cross-links (van der Slot et al., 2003). Further, inhibition of the lysyl oxidase–like 2 (LOXL2) enzyme has been shown to reduce formation and progression of fibrosis in a variety of organ systems

(Barry-Hamilton et al., 2010). It remains to be shown whether this effect is due to reduced ECM stiffness. Another factor leading to stiffening of the collagen ECM in the heart is glycation, a posttranslational modification strongly associated with diabetes.

5.2.3 ECM posttranslational modifications in fibrosis: type 1 and type 2 diabetes

Diabetic cardiomyopathy as a disease model of posttranslational modification of the ECM

Alterations of the ECM that are caused by posttranslational modifications are associated with a wide range of disorders and diseases; some of these alterations may contribute to the development of fibrosis. As a model for study of posttranslational modification of the ECM, we consider here DCM. This condition is frequently defined as the sum of alterations induced by diabetes mellitus in cardiac structure and function in the absence of ischemic heart disease, hypertension, or other comorbidities (Maya and Villarreal, 2010). In patients with diabetes, there is often damage to the microcirculation, as well as vascular complications and impaired renal function (Pessina, 2007). As the prevalence of diabetes mellitus escalates (King et al., 1998) and progressively impacts younger individuals, the incidence of DCM is expected to rise dramatically. Indeed, DCM is often referred to as the "frequent, forgotten and often fatal" complication of diabetes (Bell, 2003). Currently available data from Canada indicate that diabetes confers a substantial increased risk of congestive heart failure, partly because of DCM, and particularly in the <65-year-old age group (15- to 25-fold) (Chow et al., 2005). Furthermore, the presence of diabetes confers a greatly increased risk of death.

Left ventricular hypertrophy and perivascular and interstitial fibrosis are fundamental, histopathological features in hearts of subjects with DCM, regardless of the presence or absence of ischemic heart disease (Frustaci et al., 2000; van Heerebeek et al., 2006, 2008; Srivastava et al., 2008). Detailed studies of chamber compliance in diabetic patients with heart failure but normal renal function demonstrate increased myocyte stiffness and interstitial fibrosis as potential mediators of cardiac stiffness (Asbun and Villarreal, 2006). Reduced chamber compliance leads to diastolic dysfunction, for which diabetics are particularly prone (Poirier et al., 2000). Indeed, the ability to noninvasively assess cardiac function by echocardiography has enabled the realization that ~30%–50% of hospitalizations for heart failure occur in patients with preserved left ventricular systolic function but with diastolic dysfunction of the left ventricle ("heart failure with preserved ejection fraction" [Bhatia et al., 2006; Owan et al., 2006]). Recent studies documenting the syndrome of heart failure with preserved ejection fraction demonstrate mortality similar to subjects with systolic heart failure (Bhatia et al., 2006; Owan et al., 2006). As diastolic dysfunction is an important cause of symptomatic heart failure (Burlew and Weber, 2002), it is notable that in DCM there is adverse accumulation and structural remodeling of the cardiac fibrillar collagen ECM (Asbun and Villarreal, 2006). Despite these findings, there are no evidence-based therapies to target diabetes-induced cardiac fibrosis, and treatment remains empirical (Howlett et al., 2009). Urgent therapies and an improved understanding of the pathophysiological processes are therefore required to address this issue.

Animal models of DCM

Several animal models have been used to study DCM (Boudina et al., 2007; Hsueh et al., 2007; Bugger and Abel, 2009). The cardiac phenotypes in rodent models of type 1 and type 2 diabetes show significant overlap of increased fatty acid utilization, decreased glucose utilization, impaired Ca^{2+} handling, compromised mitochondrial energetics, and increased connective tissue content in the heart (Bugger and Abel, 2009). Differences between the type 1 and 2 diabetes models include variations of mitochondrial reactive oxygen species production (Boudina et al., 2007) and fatty-acid-induced mitochondrial uncoupling (Bugger et al., 2008).

In brief, rodent models of type 1 diabetes (e.g. streptozotocin [STZ]–treated rats or mice, or genetic nonobese diabetic mice) exhibit systolic and diastolic dysfunction (Ban and Twigg, 2008) and several other pathognomonic features of human DCM, including left ventricular hypertrophy and impaired inotropic response and tolerance to ischemia (Boudina and Abel, 2007). After single injections of STZ (50 mg/kg i.p.), typically within days, blood glucose rises, islet β-cells in the pancreas are depleted, and, after 10–16 weeks, there is marked formation of interstitial fibrosis and advanced glycation end products in the heart (Chang et al., 2006; Van den Bergh et al., 2008; El-Seweidy et al., 2010; Widyantoro et al., 2010; Xue et al., 2010; Diao et al., 2011; Huynh et al., 2011; Khong et al., 2011). Since rodents are resistant to atherosclerosis, the rat or mouse STZ model is useful for studying the development of DCM (Ban and Twigg, 2008) and for studies of fibrosis in the myocardium, the effect of advanced glycation end products on cardiac function (Kato et al., 2008), and the breakdown of advanced glycation end products by aminoguanidine (Chang et al., 2006). Notably, side effects of STZ toxicity must be carefully evaluated to avoid spurious conclusions from these models (Bugger and Abel, 2009), but the ability to rapidly induce cardiac fibrosis in a wide variety of genetic backgrounds is useful.

Several genetically defined mice effectively model human type 2 diabetes and DCM in terms of cardiac size, function, efficiency, mitochondrial energetics, lipid storage, fatty acid oxidation, glucose oxidation, Ca^{2+} handling, oxidative stress, diastolic dysfunction, and interstitial fibrosis (Bugger and Abel, 2009). For example, the *db/db* mouse is a leptin receptor–defective mouse (Stolen et al., 2009) on the C57BL/Ks background; *ob/ob* mice on the C57BL/6 background develop diabetes as a consequence of recessive mutations in the *ob* gene (Zhang et al., 1994). By 15 weeks, *ob/ob* mice are severely obese, develop type 2 diabetes, and exhibit diastolic dysfunction and interstitial fibrosis (Zaman et al., 2004; Van den Bergh et al., 2008; Zibadi et al., 2011).

Impact of diabetes on the cardiac ECM

The cardiac interstitium is composed of nonmyocytic cells and a structural protein network that plays a dominant role in governing the architecture and mechanical behavior of the myocardium (Brilla et al., 1995). The cardiac ECM is composed predominantly of collagen fibers and a variety of other ECM proteins, including fibronectin, laminins, and tenascin-C. Cardiac muscle contains about sixfold more collagen than skeletal muscle. Accordingly, differences of resting tension relationships in cardiac and skeletal muscle may result largely from alterations in the structure of the connective tissue ECM (Covell, 1990). In health, the fibrillar elements of the myocardial wall form a stress-tolerant network that facilitates the distribution of forces generated in the heart and provide for appropriate alignment of cardiac myocytes to optimize function (Carver et al., 1991). The long-term performance of cardiac muscle is regulated through a complex but poorly understood group of feedback mechanisms in which mechanical loading controls the organization of myofibrils, the size of muscle fibers, the expression of muscle-specific genes, and the synthesis and secretion of a wide variety of ECM products and trophic factors (Yamazaki et al., 1998). These mechanisms are tightly regulated during growth or adaptive responses but are dysregulated in diabetes (Dyntar et al., 2006), in part because of inappropriate remodeling of the ECM by matrix metalloproteinases (Van Linthout et al., 2008), which affect the cardiac ECM architecture.

Diabetes is associated with overproduction of collagen (Singh et al., 2008), which contributes to diastolic dysfunction (Aragno et al., 2008). Contemporaneous with the increased collagen synthesis is a reduction of collagen degradation, possibly caused by reduced collagenolytic activity (Dyntar et al., 2006). The net increase of collagen in the interstitium and the adventitia is an important determinant of pathological hypertrophy because it may account for abnormal myocardial stiffness (Wilke et al., 1996). Over the long term, this adaptive response can contribute to impairment of cardiac function and heart failure (Keating and Sanguinetti, 1996). In the context of the fibrous tissue response in DCM, these alterations are of primary clinical interest in the pathogenesis of the disease (Tsujino et al., 2006). Alterations in collagen cross-linking attributable to diabetic metabolites have been shown to have a marked negative impact on cardiac function (Herrmann et al., 2003). Consistent with these observations, increased serum levels of advanced glycation end products are associated with greatly reduced left ventricular diastolic function in patients with type 1 diabetes (Berg et al., 1999). Diabetes, which is known to be an important risk factor for cardiac hypertrophy (Adeghate, 2004), increases the stiffness of the ECM, contributes to reduced cardiac compliance (Zieman and Kass, 2004), and alters electromechanical connectivity in the myocardium (Casis and Echevarria, 2004).

What we do not know about fibrosis in DCM

Currently, we do not understand the relative importance of diabetic metabolite-induced collagen cross-linking versus the synthesis of disorganized collagen by cardiac fibroblasts and profibrotic myofibroblasts in contributing to diastolic dysfunction. Consequently, an improved understanding of how cardiac fibroblasts and myofibroblasts mediate inappropriate remodeling of myocardial collagen could lead to improved prevention and identification of treatments that interfere with fibrosis. Such an approach has been suggested in animal models using drugs that break advanced glycation end-product-induced cross-links, thereby enhancing cardiac function (Asif et al., 2000).

The effect of glycated collagen on cardiac fibroblasts and myofibroblasts

Notably, in cardiac fibroblasts, pathological levels of advanced glycation end products in the myocardium induce the following: (1) activation of the extracellular signal-regulated kinase (ERK) mitogen-activated protein kinase pathway to promote the production of type I and type III collagens (Tang et al., 2007); (2) increased proliferation of fibroblasts (Tokudome et al., 2004); (3) TGF-β1-dependent fibrotic changes (Martin et al., 2005); and (4) angiotensin-II-dependent increases of ECM production (Singh et al., 2008). The enhanced angiotensin-II response in turn stimulates fibroblast proliferation (Sadoshima and Izumo, 1997), enhances the autocrine production of TGF-β1 in adult cardiac fibroblasts (Lee et al., 1995; Campbell and Katwa, 1997), and strongly increases integrin-mediated contraction of the ECM (Burgess et al., 1994).

In vitro models have been used to assess the impact of glycated collagen on cell adhesion and on the activation of the α-SMA gene (Yuen et al., 2010). Collagen glycation can be readily induced by the diabetes metabolite methylglyoxal (MGO) *in vitro*. Cardiac fibroblasts respond to glycated ECM collagens by increased proliferation (Tokudome et al., 2004), elevated remodeling of collagen (Martin et al., 2005; Tang et al., 2007), and switch to the myofibroblast phenotype (Yuen et al., 2010). Cells cultured on MGO-treated collagen show increased activation of the α-SMA promoter (Yuen et al., 2010), increased expression of α-SMA protein, but decreased adhesion compared to cells on control collagen, as has been shown previously (Chong et al., 2007). These data indicate that collagen glycation may enhance the formation of myofibroblasts through its effect on integrin-mediated cell adhesion, a potentially important process for maintaining the structure and function of the cardiac interstitium (▶Figure 5.3).

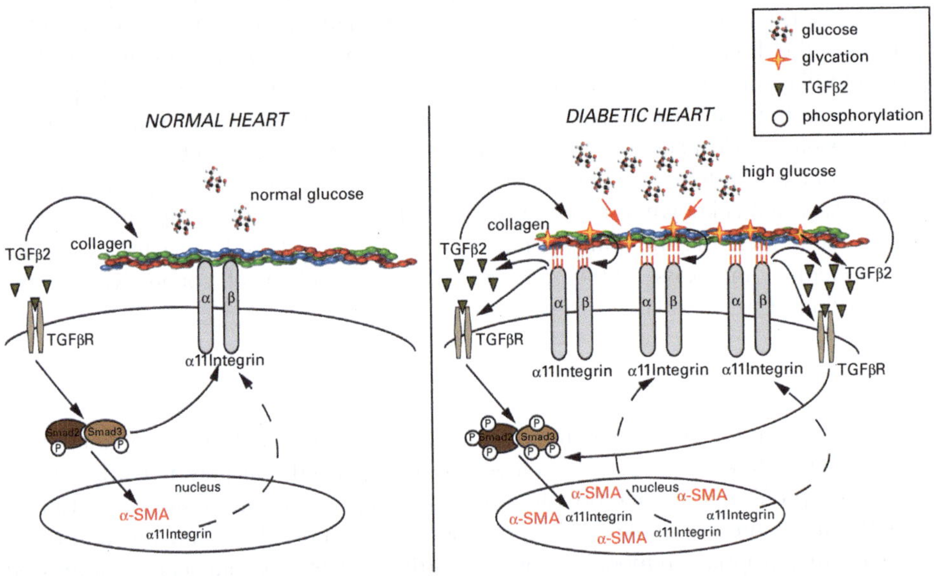

Figure 5.3 A diagram demonstrating the signaling and adhesion pathways that are involved in fibroblast adhesion to normal collagen (left panel; normal heart) and to glycated collagen (right panel; diabetic heart). Glycated collagen promotes adhesion through the α 11 β 1 integrin, which is mediated by activation of transforming growth factor-beta 2 (TGF-β2) and transforming growth factor-beta (TGF-β) receptors and the further enhancement of α11β1-integrin expression. Adhesion through the α11β1 integrin strongly increases expression of α-SMA as a result of TGF-β2 signaling through SMADs and promotion of the myofibroblast phenotype (the SMAD proteins are homologs of both the drosophila protein, mothers against decapentaplegic [MAD], and the Caenorhabditis elegans protein SMA). Notably, the α11β1 integrin may act as a compensatory adhesion system to adjust for the reduced expression of fibroblasts to the α2β1 integrin, which is normally expressed by cardiac fibroblasts. This diagram was created by Ilana Talior.

5.2.4 Interaction of integrins with glycated collagen

Collagen-binding integrins

In addition to the well-recognized role of the ECM as a structural framework, the ECM is also involved in bidirectional and dynamic processes such as cell adhesion, cell migration, and the modulation of cytokine signaling. Collagen-fibril-binding integrins are centrally important for mediating cell-ECM interactive processes (Carracedo et al., 2010) because they provide adhesion of cardiac fibroblasts (Ross, 2002) to the most abundant protein of mammals, type I collagen (Perez-Tamayo, 1978), which also is the principal protein of the cardiac interstitium (Adeghate, 2004). A large number of studies have shown that collagen-binding integrins participate in many different cellular processes, such as cell adhesion, cell migration, control of collagen synthesis, matrix metalloproteinase synthesis, and remodeling of collagen matrices, and that they influence cell proliferation, cell differentiation, angiogenesis, platelet adhesion/aggregation, and epithelial tubulogenesis (Bouvard et al., 2001; Tiger et al., 2001; Gullberg and Lundgren-Akerlund, 2002; Zhang et al., 2003). A critical role of integrins in cardiac tissues is their ability to serve as mechanotransducers during normal development and in response to physiological and pathophysiological signals (Simpson et al., 1999), including diabetes (Westermann et al., 2007).

A specific role for $\alpha 11\beta 1$ integrin in diabetes-associated cardiac fibrosis?

The type I collagen–binding $\alpha 11$ integrin emerges as an important player in diabetes-stimulated differentiation of cardiac fibroblasts into myofibroblasts (Talior-Volodarsky, I., Gullberg, D., and McCulloch, C. A., unpublished). The $\alpha 11$ integrin is a member of the collagen-binding integrin subfamily, which includes the $\alpha 1\beta 1$, $\alpha 2\beta 1$, and $\alpha 10\beta 1$ integrins (Popova et al., 2007; Barczyk et al., 2009). We have used the STZ rat diabetic model and *in vitro* glycation of collagen with MGO (Yuen et al., 2010) to examine how cell adhesions to collagen may be involved in the conversion of fibroblasts to myofibroblasts in the diabetic heart. Our data showed that the $\alpha 11$ integrin is upregulated in the rat diabetic heart and that the $\alpha 11$ integrin is centrally involved in the signaling processes that lead to myofibroblast development (Talior-Volodarsky et al., unpublished). These results are consistent with earlier data indicating that the $\alpha 11$ integrin is involved in type I collagen–mediated myofibroblast differentiation in human corneal fibroblasts (Carracedo et al., 2010). The $\alpha 11$-integrin chain was initially detected in differentiating human fetal muscle cells (Gullberg et al., 1995), but subsequent analyses of protein and mRNA expression in human embryos revealed that $\alpha 11$-integrin expression is localized to mesenchymal, nonmuscle cells in sites of highly organized interstitial collagen networks. Consistent with the expression of the $\alpha 11$ integrin in areas rich in interstitial collagens, the $\alpha 11$ integrin binds more efficiently to type I collagen than to type IV collagen (Tiger et al., 2001). While the $\alpha 11$ integrin is also expressed in odontoblasts and various fibroblast subtypes (Popova et al., 2004, 2007), there is very low expression of $\alpha 11$ integrin in fibroblasts in control hearts, while experimental diabetes and plating human cardiac fibroblasts on glycated collagen increases $\alpha 11$-integrin expression (Talior-Volodarsky et al., unpublished).

Recent data show that the α11 integrin controls myofibroblast differentiation in human corneal fibroblasts and may play important roles in collagen reorganization mediated by myofibroblasts (Carracedo et al., 2010), but currently nothing is known about the role of the α11 integrin in DCM or the differentiation of cardiac fibroblasts into profibrotic, myofibroblastic cells. We have used STZ-induced diabetes in rats as an animal model and human cardiac fibroblasts plated on MGO collagen as an *in vitro* model to examine myofibroblast differentiation in response to glycated collagen (Talior-Volodarsky et al., unpublished). The data show that glycated collagen upregulates α-SMA and α11-integrin expression in parallel. Further, cellular adhesions mediated by α11 integrins and interactions between the α11-integrin and Smad-dependent TGF-β2 pathways promote myofibroblast differentiation, which may be important for cardiac fibrosis in the development of DCM. Collectively, these data indicate that the α11 integrin and TGF-β2 mediate myofibroblast differentiation in cardiac fibroblasts that adhere to glycated collagen in the interstitium of the diabetic heart.

Since glycation of collagen inhibits the strength of fibroblast adhesion (Howard et al., 1996; Chong et al., 2007; Yuen et al., 2010), there may be increased expression of a compensatory adhesion system in cardiac fibroblasts that maintains cell attachment and mediates myofibroblast differentiation (▶Figure 5.4). Increased formation of cadherin-mediated adherens junctions may be one possible compensation mechanism (Hinz et al., 2004; Solon et al., 2007; Pittet et al., 2008). Accordingly, in cardiac fibroblasts plated on glycated collagen, microarray analysis has shown specific upregulation of α11-integrin expression (but no change of the α2 and α10 integrins) (Talior-Volodarsky et al., unpublished). This is an important finding as the α11 integrin is a critical adhesion receptor for myofibroblast differentiation (Carracedo et al., 2010) and may be a pathologically expressed adhesion receptor in DCM.

The functional importance of the α11 integrin in the conversion of fibroblasts to myofibroblasts has been assessed by siRNA knockdown of α11-integrin expression in human cardiac fibroblasts. Compared to controls, knockdown of the α11 integrin blocked MGO-collagen-induced α-SMA expression (Talior-Volodarsky et al., unpublished). As TGF-β1 can increase expression of some integrin subunits in cardiac tissues (Thibault et al., 2001), TGF-β1 may regulate α11-integrin expression in diabetic hearts. Notably, TGF-β2 (but not TGF-β1 or transforming growth factor-beta 3 [TGF-β3]) (Talior-Volodarsky et al., unpublished) dose-dependently upregulates α11 integrin in human cardiac fibroblasts. Knockdown of the α11 integrin in human cardiac fibroblasts blocks MGO-collagen-induced expression of TGF-β2, supporting the importance of the α11 integrin in TGF-β2 expression by cardiac fibroblasts (Talior-Volodarsky et al., unpublished). Further, knockdown of the TGF-β receptor 1 and TGF-β receptor 2 by siRNA blocks upregulation of the α11 integrin and α-SMA in human cardiac fibroblasts plated on MGO collagen, and this signaling system involved SMAD3 phosphorylation (Talior-Volodarsky et al., unpublished). Collectively, these data indicate that at least for the specific disease situation of DCM, the interaction of fibroblasts with glycated ECMs has an important impact on the integrin signaling that mediates fibrosis, specifically by the conversion of fibroblasts to profibrotic myofibroblasts.

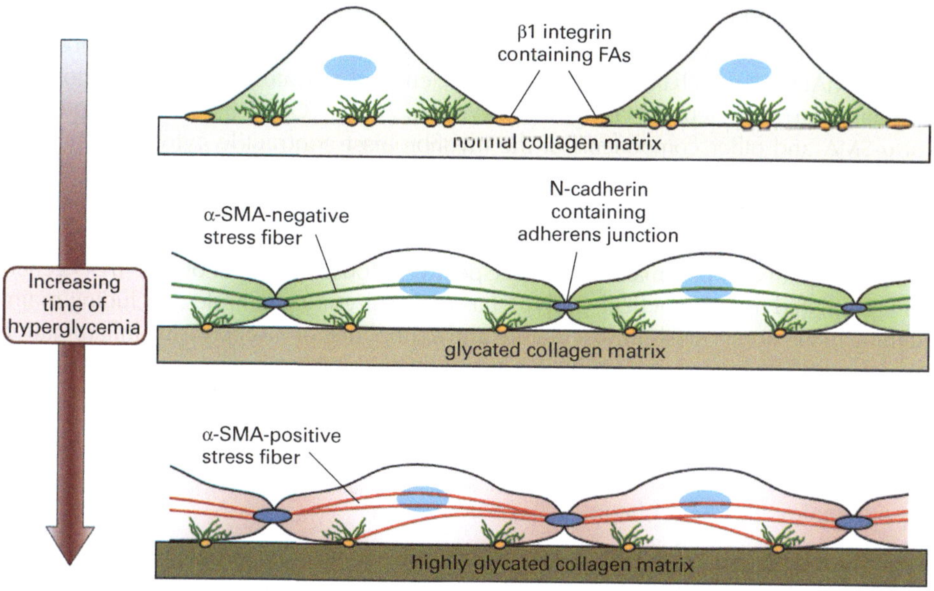

Figure 5.4 Drawing depicting glycated collagen model to induce myofibroblast differentiation. Cells are grown on collagen or collagen that has been glycated with MGO, a highly reactive glucose metabolite that is prominently expressed in diabetics. Cell adhesions through the α2β1 integrin are weakened on attachment to glycated collagen, but intercellular adhesions are enhanced (Yuen et al., 2010), and there is increased adhesion to glycated collagen because of the enhanced expression of the α11β1 integrin (Talior-Volodarsky et al., unpublished).

5.2.5 TGF-β and integrins – a close relationship

The TGF-β isoforms, TGF-β1, TGF-β2, and TGF-β3 control a wide variety of cellular processes, including cell proliferation, differentiation, apoptosis, and specification of developmental fate during embryogenesis (Shi and Massague, 2003). Moreover, TGF-β regulates transcriptional control of integrin expression, as well as expression of integrin-associated proteins in several cell types and tissues (Margadant and Sonnenberg, 2010). In particular, TGF-β1 plays a fundamental role throughout the development of cardiac fibrosis. Tissue expression levels of TGF-β1 correlate with the degree of cardiac fibrosis. TGF-β1 mRNA is increased in patients with idiopathic hypertrophic cardiomyopathy and in animal models of cardiac fibrosis (Thompson et al., 1988; Casscells et al., 1990; Wunsch et al., 1991; Villarreal and Dillmann, 1992; Takahashi et al., 1994; Li et al., 1998; Hao et al., 1999).

TGF-β1 is primarily upregulated during the transition from stable hypertrophy to irreversible remodeling and heart failure in animals and humans (Boluyt et al., 1994; Hein et al., 2003). Overexpression of TGF-β1 in transgenic mice causes cardiac hypertrophy and interstitial fibrosis (Nakajima et al., 2000; Rosenkranz et al., 2002). The predominant detrimental role of TGF-β1 in cardiac fibrosis is the induction of fibroblast-to-myofibroblast transition (Petrov et al., 2002; Rosenkranz, 2004; Bujak and

Frangogiannis, 2007; Leask, 2007). Active TGF-β1 assembles a complex of TGF-β1-receptor types I and II in the myofibroblast membrane. The kinase activity of the complex phosphorylates SMAD 2/3, which translocates to the nucleus and activates transcription factors that regulate fibrogenic genes (Massague and Wotton, 2000). TGF-β1 induces the expression of α-SMA and other components of the myofibroblast contractile cytoskeleton (Malmstrom et al., 2004; Zhang et al., 2005), as well as collagen production (Breen, 2000).

Blocking TGF-β1 function with neutralizing antibodies prevents development of fibrosis in a rat model of a pressure-overloaded heart (Kuwahara et al., 2002). However, from a clinical and pharmacological perspective, global inhibition of TGF-β1 is generally estimated as not practical. TGF-β1 has many physiological roles, including maintaining tissue homeostasis in vasculature and epithelium, and suppressing inflammation (Leask, 2007; Frangogiannis, 2008; Yang and Moses, 2008; Goumans et al., 2009; Taylor, 2009). The challenge is to therapeutically inhibit the potentially harmful actions of TGF-β1 without impeding its beneficial effects.

The complex biology of latent TGF-β1 activation

Myofibroblasts and other cells secrete latent TGF-β1 in a large latent complex (LLC) together with the latency-associated peptide (LAP) and the latent TGF-β1-binding protein-1 (LTBP-1) (▶Figure 5.5). LTBP-1 is a member of the fibrillin family of ECM proteins that binds ECM proteins like fibronectin, fibrillin-1, and vitronectin. This provides a reservoir of latent TGF-β1 in the ECM (Annes et al., 2004; Jenkins, 2008; Wipff and Hinz, 2008). Both TGF-β1 and LTBP-1 are upregulated during cardiac fibrosis (Waltenberger et al., 1993a, 1993b). Activation of latent TGF-β1 by its dissociation from LAP is promoted by various mechanisms, which differ according to cell type and physiological context (Annes et al., 2003; Jenkins, 2008). Transmembrane integrins have been shown to bind to and activate latent TGF-β1. Both LAP and LTBP-1 contain arginine-glycine-aspartate (RGD) motifs for integrin docking (Derynck et al., 1986; Munger et al., 1998; Lu et al., 2002). Binding comprises all αv integrins (αvβ1, αvβ3, αvβ5, αvβ6, and αvβ8) and integrins α5β1, α8β1, and αIIbβ3 (Ludbrook et al., 2003; Sheppard, 2005). Integrins αvβ5, αvβ6, αvβ8, and αvβ3 and a β1 integrin were shown to activate latent TGF-β1 using two different modes of action (Sheppard, 2005; Wipff and Hinz, 2008). The first depends on proteases that seem to be guided to the LLC by association mainly with integrin αvβ8 (Jenkins, 2008; Nishimura, 2009).

The second mechanism is independent of proteolysis and involves transmission of cell traction forces to the LLC (Sheppard, 2005; Jenkins, 2008; Wipff and Hinz, 2008). We and others have previously shown that cultured myofibroblasts can activate latent TGF-β1 from a sufficiently stiffened ECM by integrin-mediated contraction (Wipff et al., 2007; Zhou et al., 2010). Integrin binding to LAP is essential for contraction-mediated TGF-β1 activation and is inhibited by RGD- (integrin epitope) and LAP-competitive peptides (Wipff et al., 2007). TGF-β1-activation efficiency increases with increasing ECM stiffness. Whereas compliant substrates appear to absorb the large deformations generated by myofibroblast contraction and thus protect the latent complex against conformational changes, the complex is stretchable on stiff substrates (Wipff et al., 2007). The currently available data suggest that myofibroblast traction on the LLC induces a conformational change in LAP that liberates active TGF-β1. This purely mechanically driven mechanism of profibrotic growth factor activation is supported by the recently revealed structure of latent TGF-β1 (Shi et al., 2011) and establishes a direct link between the mechanical and chemical factors regulating myofibroblast differentiation and ultimately causing fibrosis (Wipff and Hinz, 2008; Hynes, 2009; Ramirez and Rifkin, 2009). Hence, activation of TGF-β1 via integrin-mediated myofibroblast contraction is a potential checkpoint in the progression of fibrosis, restricting autocrine generation of myofibroblasts to a sufficiently remodeled, stiff ECM (▶Figure 5.5).

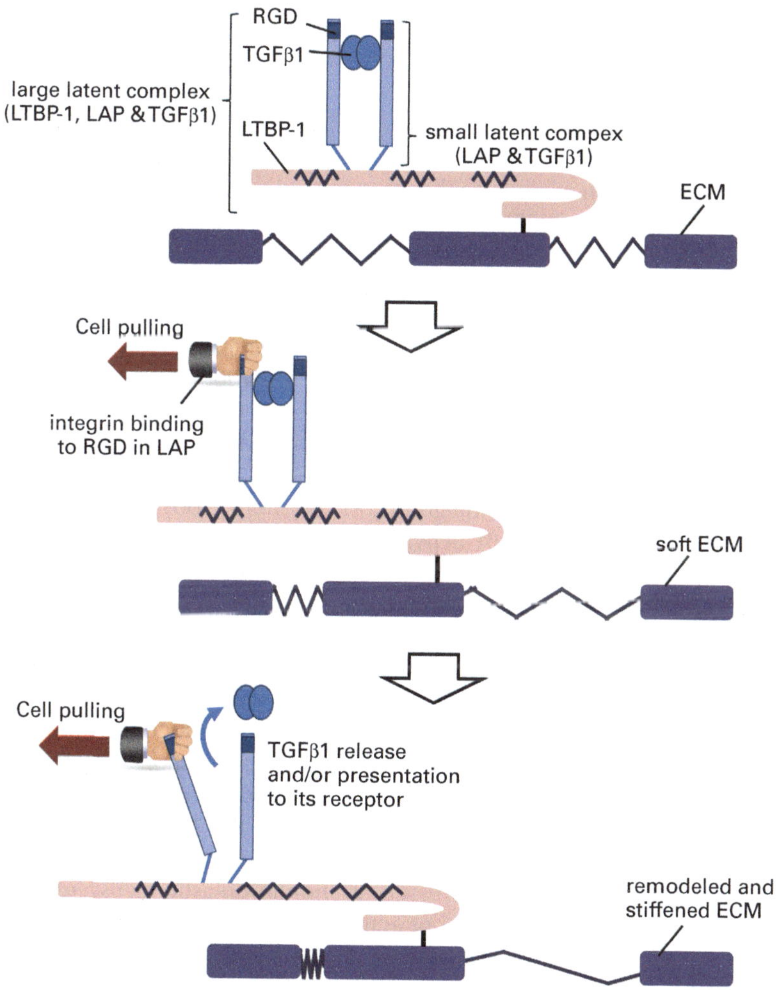

Figure 5.5 A simplified view on latent TGF-β1 activation by cell traction. Integrin binding to a specific RGD site in the LAP portion of the LLC transmits intracellular force to the latent TGF-β1 complex. In the context of a soft ECM (unloaded springs), cell pulling will simply drag the latent complex. When bound to a remodeled stiff ECM (loaded springs), integrin-mediated force exertion can trigger a conformation change in the LAP and make TGF-β1 available for its receptor (modified after Hinz, 2010).

Inhibition of integrin-mediated activation of TGF-β1 reduces fibrosis

Inhibition of TGF-β1-activating integrins has promising therapeutic potential (Aluwihare and Munger, 2008). Knocking out integrin subunits that activate TGF-β1, including β6 (Huang et al., 1996), αv (Bader et al., 1998), and β8 (Zhu et al., 2002), and mutation of the integrin-binding site in LAP (Yang et al., 2007), recapitulates the TGF-β1 knockout mouse (Shull et al., 1992). Knockout mice have defective vasculogenesis, are devoid of Langerhans cells, and spontaneously develop lung and skin inflammation. Importantly, the lungs of β6-integrin knockout mice are protected from bleomycin-induced fibrosis (Munger et al., 1999). Controlled delivery of -integrin-blocking antibodies reduces lung fibrosis without negative side effects (Horan et al., 2008). Integrin αvβ6 is the best studied in the context of TGF-β1 activation and fibrosis (Munger et al., 1999; Annes et al., 2004; Jenkins et al., 2006; Yang et al., 2007). However, only epithelial cells express this integrin. Epithelial cells are important in the onset of fibrosis in epithelialized tissues like the lung. Progression to end-stage pulmonary fibrosis depends on αvβ6-integrin-negative myofibroblasts (Phan, 2002; Thannickal et al., 2004). In the heart, which exhibits neither pronounced epithelium nor cells that express αvβ6 integrin (Hinz et al., 2007), myofibroblasts may indeed be the major population of TGF-β1-producing and activating cells.

In lung myofibroblast cultures, contraction-mediated TGF-β1 activation is reduced significantly by inhibition of αvβ5 integrin and moderately by blocking the function of αvβ3 and β1 integrin. The αvβ5 integrin is expressed in cardiac fibroblasts and is upregulated by TGF-β1 (Graf et al., 2000; Kawano et al., 2000); however, little data are available concerning its role in cardiac fibrosis. The phenotype of the αvβ5 knockout-mouse is mild, but these animals have yet to be challenged in fibrosis models where TGF-β1 activation relies on myofibroblasts. Another integrin prominently expressed in cardiac fibroblasts is the αvβ3 integrin (Ross and Borg, 2001; Babbitt et al., 2002; Ross, 2002), but its role in TGF-β1 activation or cardiac myofibroblast differentiation remains elusive (Graf et al., 2000; Thibault et al., 2001). The β3-integrin knockout mouse spontaneously develops cardiac inflammation, which becomes severe in conditions of pressure-overload-induced cardiac hypertrophy (Ren et al., 2007). The fibroblasts in these β3-integrin knockout mice do not express the αvβ3 integrin, because the β3 integrin only pairs with the αv subunit in these cells (Ballestrem et al., 2001; Hinz et al., 2003). Another possible candidate for TGF-β1 activation in the fibrotic heart is the α8β1 integrin, which is upregulated in conjunction with myofibroblast differentiation in cardiac (Bouzeghrane et al., 2004), pulmonary, and hepatic fibrosis (Levine et al., 2000). In cultured cardiac fibroblasts, the α8β1 integrin is induced by angiotensin-II (Thibault et al., 2001), but the function of this integrin in heart fibrosis is unclear. When overexpressed in nonfibroblastic cells, α8β1 integrin strongly interacts with LAP but does not activate TGF-β1 (Lu et al., 2002). Currently, it is unknown whether this function is gained when expressed in highly contractile myofibroblasts.

5.2.6 Conclusions

Cardiac fibrosis is one of the most common causes of heart failure. Fibrosis arises from pathological, previous conditions that involve mechanical stress and that are characterized by stiff and collagenous scar tissue. The stiff fibrotic scar promotes diastolic and systolic dysfunction, supports arrhythmia, and contributes to progressive loss of functional heart muscle. All of these detrimental conditions are contributed in part by the myofibroblast, a cell type that is associated with the excessive secretion of collagen and by the generation of contractile forces. The increasing tissue stress produced by the myofibroblast feeds back on this mechanoresponsive cell and leads to continued accumulation of scar tissue. One potential therapeutic strategy is to interrupt this vicious circle by interfering with the integrins that serve as mechanoreceptors and mechanotransducers in myofibroblast activation.

An improved understanding of how the ECM interacts with cell adhesion receptors could impact a wide range of high-prevalence diseases. Here we have focused on DCM. The relationship of cardiac fibrosis with diabetes-induced alterations in the cardiac interstitium is critical for our understanding of diastolic dysfunction in DCM. The α11-integrin-dependent activation of profibrotic cells is evidently important for diabetes-associated fibrosis but may also suggest novel approaches for clinical management of the fibrotic response in DCM. For example, the use of aminoguanidine and other inhibitors of collagen cross-linkers to prevent fibrosis could be optimized if we had a better understanding of fibrotic cell responses to diabetes, which could be estimated from α11-integrin levels in the heart. Further, the potential development of a soluble, specific, small molecule inhibitor of α11-integrin function, which hopefully would not impact α2-integrin function, could provide a potential therapeutic approach for preventing the pathological initiation of fibrosis in DCM.

There are potential antifibrotic targets that involve mechanical activation of TGF-β1. As noted previously, myofibroblasts appear to use specific integrins to liberate active TGF-β1 by pulling on the LLC. It will be important to identify the specific integrins in cardiac fibroblasts that mediate mechanical activation of TGF-β1 and to obtain insights into their activation mechanism. Pharmacological interference with TGF-β1-activating integrins could specifically counteract the harmful activity of active TGF-β1 but without inhibiting its beneficial effects on other cell types. The inhibition of specific integrins has already entered preclinical and clinical trials as therapy for other diseases, such as tumor treatment (Tucker, 2002; Weinreb et al., 2004). Hence, most of the antagonistic peptides and antibodies that have been developed for cancer therapies would be readily available for the treatment of cardiac fibrosis.

5.2.7 Take-home message

Various integrins are expressed and upregulated on fibroblasts and myofibroblasts during development of cardiac fibrosis. (1) Specific integrins are responsible for the transmission of the high-contractile myofibroblast forces that lead to tissue stiffening. (2) Specific integrins promote the development of the myofibroblast phenotype by sensing the stiff environment and by providing adhesion to glycated collagen, a product of long-standing diabetes. (3) Another set of integrins contributes to expression and activation of the profibrotic growth factors TGF-β1 and TGF-β2. These key roles make them attractive therapeutic targets to counteract fibrosis.

Acknowledgments

The research leading to this work has been supported by the Heart and Stroke Foundation of Ontario, grants #NA-7086 (B. H.) and NA-6736 (C. A. M.); the Canadian Institutes of Health Research, grants #210820 and #219974 (B. H.); the Canada Research Chairs Program (C. A. M.); and the Swiss National Science Foundation, grant #3200–067254 (B. H.).

References

Adeghate, E. (2004). Molecular and cellular basis of the aetiology and management of diabetic cardiomyopathy: a short review. Mol Cell Biochem *261*, 187–191.

Aluwihare, P., and Munger, J. S. (2008). What the lung has taught us about latent TGF-beta activation. Am J Respir Cell Mol Biol *39*, 499–502.

Annes, J. P., Chen, Y., Munger, J. S., and Rifkin, D. B. (2004). Integrin {alpha}V{beta}6-mediated activation of latent TGF-{beta} requires the latent TGF-{beta} binding protein-1. J Cell Biol *165*, 723–734.

Annes, J. P., Munger, J. S., and Rifkin, D. B. (2003). Making sense of latent TGFbeta activation. J Cell Sci *116*, 217–224.

Aragno, M., Mastrocola, R., Alloatti, G., et al. (2008). Oxidative stress triggers cardiac fibrosis in the heart of diabetic rats. Endocrinology *149*, 380–388.

Arora, P. D., Narani, N., and McCulloch, C. A. (1999). The compliance of collagen gels regulates transforming growth factor-beta induction of alpha-smooth muscle actin in fibroblasts. Am J Pathol *154*, 871–882.

Asbun, J., and Villarreal, F. J. (2006). The pathogenesis of myocardial fibrosis in the setting of diabetic cardiomyopathy. J Am Coll Cardiol *47*, 693–700.

Asif, M., Egan, J., Vasan, S., et al. (2000). An advanced glycation endproduct cross-link breaker can reverse age-related increases in myocardial stiffness. Proc Natl Acad Sci U S A *97*, 2809–2813.

Babbitt, C. J., Shai, S. Y., Harpf, A. E., Pham, C. G., and Ross, R. S. (2002). Modulation of integrins and integrin signaling molecules in the pressure-loaded murine ventricle. Histochem Cell Biol *118*, 431–439.

Bader, B. L., Rayburn, H., Crowley, D., and Hynes, R. O. (1998). Extensive vasculogenesis, angiogenesis, and organogenesis precede lethality in mice lacking all alpha v integrins. Cell *95*, 507–519.

Ballestrem, C., Hinz, B., Imhof, B. A., and Wehrle-Haller, B. (2001). Marching at the front and dragging behind: differential alphaVbeta3-integrin turnover regulates focal adhesion behavior. J Cell Biol *155*, 1319–1332.

Ban, C. R., and Twigg, S. M. (2008). Fibrosis in diabetes complications: pathogenic mechanisms and circulating and urinary markers. Vasc Health Risk Manag *4*, 575–596.

Barczyk, M., Carracedo, S., and Gullberg, D. (2009). Integrins. Cell Tissue Res *339*, 269–280.

Barry-Hamilton, V., Spangler, R., Marshall, D., et al. (2010). Allosteric inhibition of lysyl oxidase-like-2 impedes the development of a pathologic microenvironment. Nat Med *16*, 1009–1017.

Baudino, T. A., Carver, W., Giles, W., and Borg, T. K. (2006). Cardiac fibroblasts: friend or foe? Am J Physiol Heart Circ Physiol *291*, H1015–H1026.

Bell, D. S. (2003). Heart failure: the frequent, forgotten, and often fatal complication of diabetes. Diabetes Care *26*, 2433–2441.

Berg, T. J., Snorgaard, O., Faber, J., et al. (1999). Serum levels of advanced glycation end products are associated with left ventricular diastolic function in patients with type 1 diabetes. Diabetes Care *22*, 1186–1190.

Berry, M. F., Engler, A. J., Woo, Y. J.,et al. (2006). Mesenchymal stem cell injection after myocardial infarction improves myocardial compliance. Am J Physiol Heart Circ Physiol *290*, H2196–2203.

Bhatia, R. S., Tu, J. V., Lee, D. S., et al. (2006). Outcome of heart failure with preserved ejection fraction in a population-based study. N Engl J Med *355*, 260–269.

Boluyt, M. O., O'Neill, L., Meredith, A. L., et al. (1994). Alterations in cardiac gene expression during the transition from stable hypertrophy to heart failure. Marked upregulation of genes encoding extracellular matrix components. Circ Res *75*, 23–32.

Boudina, S., and Abel, E. D. (2007). Diabetic cardiomyopathy revisited. Circulation *115*, 3213–3223.

Boudina, S., Sena, S., Theobald, H., et al. (2007). Mitochondrial energetics in the heart in obesity-related diabetes: direct evidence for increased uncoupled respiration and activation of uncoupling proteins. Diabetes *56*, 2457–2466.

Bouvard, D., Brakebusch, C., Gustafsson, E., et al. (2001). Functional consequences of integrin gene mutations in mice. Circ Res *89*, 211–223.

Bouzeghrane, F., Mercure, C., Reudelhuber, T. L., and Thibault, G. (2004). Alpha8beta1 integrin is upregulated in myofibroblasts of fibrotic and scarring myocardium. J Mol Cell Cardiol *36*, 343–353.

Breen, E. C. (2000). Mechanical strain increases type I collagen expression in pulmonary fibroblasts in vitro. J Appl Physiol *88*, 203–209.

Brilla, C. G., Maisch, B., Rupp, H., Funck, R., Zhou, G., and Weber, K. T. (1995). Pharmacological modulation of cardiac fibroblast function. Herz *20*, 127–134.

Brown, R. D., Ambler, S. K., Mitchell, M. D., and Long, C. S. (2005). The cardiac fibroblast: therapeutic target in myocardial remodeling and failure. Annu Rev Pharmacol Toxicol *45*, 657–687.

Bugger, H., and Abel, E. D. (2009). Rodent models of diabetic cardiomyopathy. Dis Model Mech *2*, 454–466.

Bugger, H., Boudina, S., Hu, X. X., et al. (2008). Type 1 diabetic akita mouse hearts are insulin sensitive but manifest structurally abnormal mitochondria that remain coupled despite increased uncoupling protein 3. Diabetes *57*, 2924–2932.

Bujak, M., and Frangogiannis, N. G. (2007). The role of TGF-beta signaling in myocardial infarction and cardiac remodeling. Cardiovasc Res *74*, 184–195.

Burgess, M. L., Carver, W. E., Terracio, L., Wilson, S. P., Wilson, M. A., and Borg, T. K. (1994). Integrin-mediated collagen gel contraction by cardiac fibroblasts. Effects of angiotensin II. Circ Res *74*, 291–298.

Burlew, B. S., and Weber, K. T. (2002). Cardiac fibrosis as a cause of diastolic dysfunction. Herz *27*, 92–98.

Campbell, S. E., Janicki, J. S., and Weber, K. T. (1995). Temporal differences in fibroblast proliferation and phenotype expression in response to chronic administration of angiotensin II or aldosterone. J Mol Cell Cardiol *27*, 1545–1560.

Campbell, S. E., and Katwa, L. C. (1997). Angiotensin II stimulated expression of transforming growth factor-beta1 in cardiac fibroblasts and myofibroblasts. J Mol Cell Cardiol *29*, 1947–1958.

Carracedo, S., Lu, N., Popova, S. N., Jonsson, R., Eckes, B., and Gullberg, D. (2010). The fibroblast integrin alpha11beta1 is induced in a mechanosensitive manner involving activin A and regulates myofibroblast differentiation. J Biol Chem *285*, 10434–10445.

Carver, W., Nagpal, M. L., Nachtigal, M., Borg, T. K., and Terracio, L. (1991). Collagen expression in mechanically stimulated cardiac fibroblasts. Circ Res *69*, 116–122.

Casis, O., and Echevarria, E. (2004). Diabetic cardiomyopathy: electromechanical cellular alterations. Curr Vasc Pharmacol *2*, 237–248.

Casscells, W., Bazoberry, F., Speir, E., et al. (1990). Transforming growth factor-beta 1 in normal heart and in myocardial infarction. Ann N Y Acad Sci *593*, 148–160.

Chang, K. C., Hsu, K. L., Tseng, C. D., Lin, Y. D., Cho, Y. L., and Tseng, Y. Z. (2006). Aminoguanidine prevents arterial stiffening and cardiac hypertrophy in streptozotocin-induced diabetes in rats. Br J Pharmacol *147*, 944–950.

Chong, S. A., Lee, W., Arora, P. D., et al. (2007). Methylglyoxal inhibits the binding step of collagen phagocytosis. J Biol Chem *282*, 8510–8520.

Chow, C. M., Donovan, L., Manuel, D., Johansen, H., and Tu, J. V. (2005). Regional variation in self-reported heart disease prevalence in Canada. Can J Cardiol *21*, 1265–1271.

Covell, J. W. (1990). Factors influencing diastolic function. Possible role of the extracellular matrix. Circulation *81*, III155– III158.

Derynck, R., Jarrett, J. A., Chen, E. Y., and Goeddel, D. V. (1986). The murine transforming growth factor-beta precursor. J Biol Chem *261*, 4377–4379.

Desmouliere, A., Geinoz, A., Gabbiani, F., and Gabbiani, G. (1993). Transforming growth factor-beta 1 induces alpha-smooth muscle actin expression in granulation tissue myofibroblasts and in quiescent and growing cultured fibroblasts. J Cell Biol *122*, 103–111.

Diao, X., Shen, E., Wang, X., and Hu, B. (2011). Differentially expressed microRNAs and their target genes in the hearts of streptozotocin-induced diabetic mice. Mol Med Rep *4*, 633–640.

Dugina, V., Fontao, L., Chaponnier, C., Vasiliev, J., and Gabbiani, G. (2001). Focal adhesion features during myofibroblastic differentiation are controlled by intracellular and extracellular factors. J Cell Sci *114*, 3285–3296.

Dyntar, D., Sergeev, P., Klisic, J., Ambuhl, P., Schaub, M. C., and Donath, M. Y. (2006). High glucose alters cardiomyocyte contacts and inhibits myofibrillar formation. J Clin Endocrinol Metab *91*, 1961–1967.

Eghbali, M. (1992). Cardiac fibroblasts: function, regulation of gene expression, and phenotypic modulation. Basic Res Cardiol *87* (Suppl 2), 183–189.

El-Seweidy, M. M., Sadik, N. A., and Shaker, O. G. (2010). Role of sulfurous mineral water and sodium hydrosulfide as potent inhibitors of fibrosis in the heart of diabetic rats. Arch Biochem Biophys *506*, 48–57.

Engler, A. J., Carag-Krieger, C., Johnson, C. P., et al. (2008). Embryonic cardiomyocytes beat best on a matrix with heart-like elasticity: scar-like rigidity inhibits beating. J Cell Sci *121*, 3794–3802.

Frangogiannis, N. G. (2008). The immune system and cardiac repair. Pharmacol Res *58*, 88–111.

Frustaci, A., Kajstura, J., Chimenti, C., et al. (2000). Myocardial cell death in human diabetes. Circ Res *87*, 1123–1132.

Gabbiani, G. (2003). The myofibroblast in wound healing and fibrocontractive diseases. J Pathol *200*, 500–503.

Georges, P. C., Hui, J. J., Gombos, Z., et al. (2007). Increased stiffness of the rat liver precedes matrix deposition: implications for fibrosis. Am J Physiol Gastrointest Liver Physiol *293*, G1147–G1154.

Goffin, J. M., Pittet, P., Csucs, G., Lussi, J. W., Meister, J. J., and Hinz, B. (2006). Focal adhesion size controls tension-dependent recruitment of alpha-smooth muscle actin to stress fibers. J Cell Biol *172*, 259–268.

Goumans, M. J., Liu, Z., and ten Dijke, P. (2009). TGF-beta signaling in vascular biology and dysfunction. Cell Res *19*, 116–127.

Graf, K., Neuss, M., Stawowy, P., Hsueh, W. A., Fleck, E., and Law, R. E. (2000). Angiotensin II and alpha(v)beta(3) integrin expression in rat neonatal cardiac fibroblasts. Hypertension *35*, 978–984.

Grinnell, F. (1994). Fibroblasts, myofibroblasts, and wound contraction. J Cell Biol *124*, 401–404.

Grinnell, F., and Petroll, W. M. (2010). Cell motility and mechanics in three-dimensional collagen matrices. Annu Rev Cell Dev Biol *26*, 335–361.

Grove, D., Zak, R., Nair, K. G., and Aschenbrenner ,V. (1969). Biochemical correlates of cardiac hypertrophy. IV. Observations on the cellular organization of growth during myocardial hypertrophy in the rat. Circ Res *25*, 473–485.

Gullberg, D., Velling, T., Sjoberg, G., and Sejersen, T. (1995). Up-regulation of a novel integrin alpha-chain (alpha mt) on human fetal myotubes. Dev Dyn *204*, 57–65.

Gullberg, D. E., and Lundgren-Akerlund, E. (2002). Collagen-binding I domain integrins–what do they do? Prog Histochem Cytochem *37*, 3–54.

Hao, J., Ju, H., Zhao, S., Junaid, A., Scammell-La Fleur, T., and Dixon, I. M. (1999). Elevation of expression of Smads 2, 3, and 4, decorin and TGF-beta in the chronic phase of myocardial infarct scar healing. J Mol Cell Cardiol *31*, 667–678.

Haudek, S. B., Xia, Y., Huebener, P., et al. (2006). Bone marrow-derived fibroblast precursors mediate ischemic cardiomyopathy in mice. Proc Natl Acad Sci U S A *103*, 18284–18289.

Hein, S., Arnon, E., Kostin, S., et al. (2003). Progression from compensated hypertrophy to failure in the pressure-overloaded human heart: structural deterioration and compensatory mechanisms. Circulation *107*, 984–991.

Herrmann, K. L., McCulloch, A. D., and Omens, J. H. (2003). Glycated collagen cross-linking alters cardiac mechanics in volume-overload hypertrophy. Am J Physiol Heart Circ Physiol *284*, H1277–H1284.

Hinz, B. (2006). Masters and servants of the force: the role of matrix adhesions in myofibroblast force perception and transmission. Eur J Cell Biol *85*, 175–181.

Hinz, B. (2007). Formation and function of the myofibroblast during tissue repair. J Invest Dermatol *127*, 526–537.

Hinz, B. (2009). Tissue stiffness, latent TGF-beta1 activation, and mechanical signal transduction: implications for the pathogenesis and treatment of fibrosis. Curr Rheumatol Rep *11*, 120–126.

Hinz, B. (2010). The myofibroblast: paradigm for a mechanically active cell. J Biomech *43*, 146–155.

Hinz, B., Celetta, G., Tomasek, J. J., Gabbiani, G., and Chaponnier, C. (2001a). Alpha-smooth muscle actin expression upregulates fibroblast contractile activity. Mol Biol Cell *12*, 2730–2741.

Hinz, B., Dugina, V., Ballestrem, C., Wehrle-Haller, B., and Chaponnier, C. (2003). Alpha-smooth muscle actin Is crucial for focal adhesion maturation in myofibroblasts. Mol Biol Cell *14*, 2508–2519.

Hinz, B., Mastrangelo, D., Iselin, C. E., Chaponnier, C., and Gabbiani, G. (2001b). Mechanical tension controls granulation tissue contractile activity and myofibroblast differentiation. Am J Pathol *159*, 1009–1020.

Hinz, B., Phan, S. H., Thannickal, V. J., Galli, A., Bochaton-Piallat, M. L., and Gabbiani, G. (2007). The myofibroblast: one function, multiple origins. Am J Pathol *170*, 1807–1816.

Hinz, B., Pittet, P., Smith-Clerc, J., Chaponnier, C., and Meister, J. J. (2004). Myofibroblast development is characterized by specific cell-cell adherens junctions. Mol Biol Cell *15*, 4310–4320.

Horan, G. S., Wood, S., Ona, V., et al. (2008). Partial inhibition of integrin alpha(v)beta6 prevents pulmonary fibrosis without exacerbating inflammation. Am J Respir Crit Care Med *177*, 56–65.

Howard, E. W., Benton, R., Ahern-Moore, J., and Tomasek, J. J. (1996). Cellular contraction of collagen lattices is inhibited by nonenzymatic glycation. Exp Cell Res *228*, 132–137.

Howlett, J. G., McKelvie, R. S., Arnold, J. M., et al. (2009). Canadian Cardiovascular Society Consensus Conference guidelines on heart failure, update 2009: diagnosis and management of right-sided heart failure, myocarditis, device therapy and recent important clinical trials. Can J Cardiol *25*, 85–105.

Hsueh, W., Abel, E. D., Breslow, J. L., et al. (2007). Recipes for creating animal models of diabetic cardiovascular disease. Circ Res *100*, 1415–1427.

Huang, X. Z., Wu, J. F., Cass, D., et al. (1996). Inactivation of the integrin beta 6 subunit gene reveals a role of epithelial integrins in regulating inflammation in the lung and skin. J Cell Biol *133*, 921–928.

Huynh, K., McMullen, J. R., Julius, T. L., et al. (2011). Cardiac-specific IGF-1 receptor transgenic expression protects against cardiac fibrosis and diastolic dysfunction in a mouse model of diabetic cardiomyopathy. Diabetes *59*, 1512–1520.

Hynes, R. O. (2009). The extracellular matrix: not just pretty fibrils. Science *326*, 1216–1219.

Jenkins, G. (2008). The role of proteases in transforming growth factor-beta activation. Int J Biochem Cell Biol *40*, 1068–1078.

Jenkins, R. G., Su, X., Su, G., et al. (2006). Ligation of protease-activated receptor 1 enhances alpha(v)beta6 integrin-dependent TGF-beta activation and promotes acute lung injury. J Clin Invest *116*, 1606–1614.

Junker, J. P., Kratz, C., Tollback, A., and Kratz, G. (2008). Mechanical tension stimulates the transdifferentiation of fibroblasts into myofibroblasts in human burn scars. Burns *34* (7), 942–946.

Kato, T., Yamashita, T., Sekiguchi, A., et al. (2008). AGEs-RAGE system mediates atrial structural remodeling in the diabetic rat. J Cardiovasc Electrophysiol *19*, 415–420.

Kawano, H., Cody, R. J., Graf, K., et al. (2000). Angiotensin II enhances integrin and alpha-actinin expression in adult rat cardiac fibroblasts. Hypertension *35*, 273–279.

Keating, M. T., and Sanguinetti, M. C. (1996). Molecular genetic insights into cardiovascular disease. Science *272*, 681–685.

Kelly, D. J., Zhang, Y., Connelly, K., et al. (2007). Tranilast attenuates diastolic dysfunction and structural injury in experimental diabetic cardiomyopathy. Am J Physiol Heart Circ Physiol *293*, H2860–H2869.

Khong, F. L., Zhang, Y., Edgley, A. J., et al. (2011). 3′,4′-Dihydroxyflavonol antioxidant attenuates diastolic dysfunction and cardiac remodeling in streptozotocin-induced diabetic m(Ren2)27 rats. PLoS One *6*, e22777.

King, H., Aubert, R. E., and Herman, W. H. (1998). Global burden of diabetes, 1995–2025: prevalence, numerical estimates, and projections. Diabetes Care *21*, 1414–1431.

Kitamura, M., Shimizu, M., Ino, H., et al. (2001). Collagen remodeling and cardiac dysfunction in patients with hypertrophic cardiomyopathy: the significance of type III and VI collagens. Clin Cardiol *24*, 325–329.

Kuwahara, F., Kai, H., Tokuda, K., et al. (2002). Transforming growth factor-beta function blocking prevents myocardial fibrosis and diastolic dysfunction in pressure-overloaded rats. Circulation *106*, 130–135.

Leask, A. (2007). TGFbeta, cardiac fibroblasts, and the fibrotic response. Cardiovasc Res *74*, 207–212.

Lee, A. A., Dillmann, W. H., McCulloch, A. D., and Villarreal, F. J. (1995). Angiotensin II stimulates the autocrine production of transforming growth factor-beta 1 in adult rat cardiac fibroblasts. J Mol Cell Cardiol *27*, 2347–2357.

Leslie, K. O., Taatjes, D. J., Schwarz, J., vonTurkovich, M., and Low, R. B. (1991). Cardiac myofibroblasts express alpha smooth muscle actin during right ventricular pressure overload in the rabbit. Am J Pathol *139*, 207–216.

Levine, D., Rockey, D. C., Milner, T. A., Breuss, J. M., Fallon, J. T., and Schnapp, L. M. (2000). Expression of the integrin alpha8beta1 during pulmonary and hepatic fibrosis. Am J Pathol *156*, 1927–1935.

Li, G., Li, R. K., Mickle, D. A., et al. (1998). Elevated insulin-like growth factor-I and transforming growth factor-beta 1 and their receptors in patients with idiopathic hypertrophic obstructive cardiomyopathy. A possible mechanism. Circulation *98*, II144–II149; discussion II149–II150.

Lu, M., Munger, J. S., Steadele, M., Busald, C., Tellier, M., and Schnapp, L. M. (2002). Integrin alpha8beta1 mediates adhesion to LAP-TGFbeta1. J Cell Sci *115*, 4641–4648.

Ludbrook, S. B., Barry, S. T., Delves, C. J., and Horgan, C. M. (2003). The integrin alphavbeta3 is a receptor for the latency-associated peptides of transforming growth factors beta1 and beta3. Biochem J *369*, 311–318.

Malmstrom, J., Lindberg, H., Lindberg, C., et al. (2004). Transforming growth factor-beta 1 specifically induce proteins involved in the myofibroblast contractile apparatus. Mol Cell Proteomics *3*, 466–477.

Margadant, C., and Sonnenberg, A. (2010). Integrin-TGF-beta crosstalk in fibrosis, cancer and wound healing. EMBO Rep *11*, 97–105.

Martin, J., Kelly, D. J., Mifsud, S. A., et al. (2005). Tranilast attenuates cardiac matrix deposition in experimental diabetes: role of transforming growth factor-beta. Cardiovasc Res *65*, 694–701.

Massague, J., and Wotton, D. (2000). Transcriptional control by the TGF-beta/Smad signaling system. EMBO J *19*, 1745–1754.

Maya, L., and Villarreal, F. J. (2010). Diagnostic approaches for diabetic cardiomyopathy and myocardial fibrosis. J Mol Cell Cardiol *48*, 524–529.

Merryman, W. D., Lukoff, H. D., Long, R. A., Engelmayr, G. C., Jr, Hopkins, R. A., and Sacks, M. S. (2007). Synergistic effects of cyclic tension and transforming growth factor-beta1 on the aortic valve myofibroblast. Cardiovasc Pathol *16*, 268–276.

Mollmann, H., Nef, H. M., Kostin, S., et al. (2006). Bone marrow-derived cells contribute to infarct remodelling. Cardiovasc Res *71*, 661–671.

Munger, J. S., Harpel, J. G., Giancotti, F. G., and Rifkin, D. B. (1998). Interactions between growth factors and integrins: latent forms of transforming growth factor-beta are ligands for the integrin alphavbeta1. Mol Biol Cell *9*, 2627–2638.

Munger, J. S., Huang, X., Kawakatsu, H., et al. (1999). The integrin alpha v beta 6 binds and activates latent TGF beta 1: a mechanism for regulating pulmonary inflammation and fibrosis. Cell *96*, 319–328.

Nakajima, H., Nakajima, H. O., Salcher, O., et al. (2000). Atrial but not ventricular fibrosis in mice expressing a mutant transforming growth factor-beta(1) transgene in the heart. Circ Res *86*, 571–579.

Nishimura, S. L. (2009). Integrin-mediated transforming growth factor-beta activation, a potential therapeutic target in fibrogenic disorders. Am J Pathol *175*, 1362–1370.

Ouzounian, M., Lee, D. S., and Liu, P. P. (2008). Diastolic heart failure: mechanisms and controversies. Nat Clin Pract Cardiovasc Med *5*, 375–386.

Owan, T. E., Hodge, D. O., Herges, R. M., Jacobsen, S. J., Roger, V. L., and Redfield, M. M. (2006). Trends in prevalence and outcome of heart failure with preserved ejection fraction. N Engl J Med *355*, 251–259.

Perez-Tamayo, R. (1978). Pathology of collagen degradation. A review. Am J Pathol *92*, 508–566.

Pessina, A. C. (2007). Target organs of individuals with diabetes caught between arterial stiffness and damage to the microcirculation. J Hypertens Suppl *25*, S13–S18.

Petrov, V. V., Fagard, R. H., and Lijnen, P. J. (2002). Stimulation of collagen production by transforming growth factor-beta1 during differentiation of cardiac fibroblasts to myofibroblasts. Hypertension *39*, 258–263.

Phan, S. H. (2002). The myofibroblast in pulmonary fibrosis. Chest *122*, 286S–289S.

Pittet, P., Lee, K., Kulik, A. J., Meister, J. J., and Hinz, B. (2008). Fibrogenic fibroblasts increase intercellular adhesion strength by reinforcing individual OB-cadherin bonds. J Cell Sci *121*, 877–886.

Poirier, P., Garneau, C., Bogaty, P., et al. (2000). Impact of left ventricular diastolic dysfunction on maximal treadmill performance in normotensive subjects with well-controlled type 2 diabetes mellitus. Am J Cardiol *85*, 473–477.

Popova, S. N., Lundgren-Akerlund, E., Wiig, H., and Gullberg, D. (2007). Physiology and pathology of collagen receptors. Acta Physiol (Oxf) *190*, 179–187.

Popova, S. N., Rodriguez-Sanchez, B., Liden, A., Betsholtz, C., Van Den Bos, T., and Gullberg, D. (2004). The mesenchymal alpha11beta1 integrin attenuates PDGF-BB-stimulated chemotaxis of embryonic fibroblasts on collagens. Dev Biol *270*, 427–442.

Porter, K. E., and Turner, N. A. (2009). Cardiac fibroblasts: at the heart of myocardial remodeling. Pharmacol Ther *123* (2), 255–278.

Ramirez, F., and Rifkin, D. B. (2009). Extracellular microfibrils: contextual platforms for TGFbeta and BMP signaling. Curr Opin Cell Biol *21*, 616–622.

Ren, J., Avery, J., Zhao, H., Schneider, J. G., Ross, F. P., and Muslin, A. J. (2007). Beta3 integrin deficiency promotes cardiac hypertrophy and inflammation. J Mol Cell Cardiol *42*, 367–377.

Rosenkranz, S. (2004). TGF-beta1 and angiotensin networking in cardiac remodeling. Cardiovasc Res *63*, 423–432.

Rosenkranz, S., Flesch, M., Amann, K., et al. (2002). Alterations of beta-adrenergic signaling and cardiac hypertrophy in transgenic mice overexpressing TGF-beta(1). Am J Physiol Heart Circ Physiol *283*, H1253–H1262.

Ross, R. S. (2002). The extracellular connections: the role of integrins in myocardial remodeling. J Card Fail *8*, S326–S331.

Ross, R. S., and Borg, T. K. (2001). Integrins and the myocardium. Circ Res *88*, 1112–1119.

Sadoshima, J., and Izumo, S. (1997). The cellular and molecular response of cardiac myocytes to mechanical stress. Annu Rev Physiol *59*, 551–571.

Sheppard, D. (2005). Integrin-mediated activation of latent transforming growth factor beta. Cancer Metastasis Rev *24*, 395–402.

Shi, M., Zhu, J., Wang, R., et al. (2011). Latent TGF-beta structure and activation. Nature *474*, 343–349.

Shi, Y., and Massague, J. (2003). Mechanisms of TGF-beta signaling from cell membrane to the nucleus. Cell *113*, 685–700.

Shi-Wen, X., Renzoni, E. A., Kennedy, L., et al. (2007). Endogenous endothelin-1 signaling contributes to type I collagen and CCN2 overexpression in fibrotic fibroblasts. Matrix Biol *26*, 625–632.

Shull, M. M., Ormsby, I., Kier, A. B., et al. (1992). Targeted disruption of the mouse transforming growth factor-beta 1 gene results in multifocal inflammatory disease. Nature *359*, 693–699.

Simpson, D. G., Majeski, M., Borg, T. K., and Terracio, L. (1999). Regulation of cardiac myocyte protein turnover and myofibrillar structure in vitro by specific directions of stretch. Circ Res *85*, e59–69.

Singh, V. P., Baker, K. M., and Kumar, R. (2008). Activation of the intracellular renin-angiotensin system in cardiac fibroblasts by high glucose: role in extracellular matrix production. Am J Physiol Heart Circ Physiol *294*, H1675–H1684.

Solon, J., Levental, I., Sengupta, K., Georges, P. C., and Janmey, P. A. (2007). Fibroblast adaptation and stiffness matching to soft elastic substrates. Biophys J *93*, 4453–4461.

Squires, C. E., Escobar, G. P., Payne, J. F., et al. (2005). Altered fibroblast function following myocardial infarction. J Mol Cell Cardiol *39*, 699–707.

Srivastava, P. M., Calafiore, P., Macisaac, R. J., et al. (2008). Prevalence and predictors of cardiac hypertrophy and dysfunction in patients with type 2 diabetes. Clin Sci (Lond) *114*, 313–320.

Stolen, T. O., Hoydal, M. A., Kemi, O. J., et al. (2009). Interval training normalizes cardiomyocyte function, diastolic Ca2+ control, and SR Ca2+ release synchronicity in a mouse model of diabetic cardiomyopathy. Circ Res *105*, 527–536.

Sun, Y., and Weber, K. T. (2000). Infarct scar: a dynamic tissue. Cardiovasc Res *46*, 250–256.

Sussman, M. A., McCulloch, A., and Borg, T. K. (2002). Dance band on the Titanic: biomechanical signaling in cardiac hypertrophy. Circ Res *91*, 888–898.

Suurmeijer, A. J., Clement, S., Francesconi, A., et al. (2003). Alpha-actin isoform distribution in normal and failing human heart: a morphological, morphometric, and biochemical study. J Pathol *199*, 387–397.

Takahashi, N., Calderone, A., Izzo, N. J., Jr, Maki, T. M., Marsh, J. D., and Colucci, W. S. (1994). Hypertrophic stimuli induce transforming growth factor-beta 1 expression in rat ventricular myocytes. J Clin Invest *94*, 1470–1476.

Tang, M., Zhang, W., Lin, H., Jiang, H., Dai, H., and Zhang, Y. (2007). High glucose promotes the production of collagen types I and III by cardiac fibroblasts through a pathway dependent on extracellular-signal-regulated kinase 1/2. Mol Cell Biochem *301*, 109–114.

Taylor, A. W. (2009). Review of the activation of TGF-{beta} in immunity. J Leukoc Biol *85*, 29–33.

Thannickal, V. J., Toews, G. B., White, E. S., Lynch, J. P., III, and Martinez, F. J. (2004). Mechanisms of pulmonary fibrosis. Annu Rev Med *55*, 395–417.

Thibault, G., Lacombe, M. J., Schnapp, L. M., Lacasse, A., Bouzeghrane, F., and Lapalme, G. (2001). Upregulation of alpha(8)beta(1)-integrin in cardiac fibroblast by angiotensin II and transforming growth factor-beta1. Am J Physiol Cell Physiol *281*, C1457–C1467.

Thompson, N. L., Bazoberry, F., Speir, E. H., et al. (1988). Transforming growth factor beta-1 in acute myocardial infarction in rats. Growth Factors *1*, 91–99.

Tiger, C. F., Fougerousse, F., Grundstrom, G., Velling, T., and Gullberg, D. (2001). Alpha11-beta1 integrin is a receptor for interstitial collagens involved in cell migration and collagen reorganization on mesenchymal nonmuscle cells. Dev Biol *237*, 116–129.

Tokudome, T., Horio, T., Yoshihara, F., et al. (2004). Direct effects of high glucose and insulin on protein synthesis in cultured cardiac myocytes and DNA and collagen synthesis in cardiac fibroblasts. Metabolism *53*, 710–715.

Tomasek, J. J., Gabbiani, G., Hinz, B., Chaponnier, C., and Brown, R. A. (2002). Myofibroblasts and mechano-regulation of connective tissue remodelling. Nat Rev Mol Cell Biol *3*, 349–363.

Tsujino, T., Kawasaki, D., and Masuyama, T. (2006). Left ventricular diastolic dysfunction in diabetic patients: pathophysiology and therapeutic implications. Am J Cardiovasc Drugs *6*, 219–230.

Tucker, G. C. (2002). Inhibitors of integrins. Curr Opin Pharmacol *2*, 394–402.

van Amerongen, M. J., Bou-Gharios, G., Popa, E., et al. (2008). Bone marrow-derived myofibroblasts contribute functionally to scar formation after myocardial infarction. J Pathol *214*, 377–386.

Van den Bergh, A., Vanderper, A., Vangheluwe, P., et al. (2008). Dyslipidaemia in type II diabetic mice does not aggravate contractile impairment but increases ventricular stiffness. Cardiovasc Res *77*, 371–379.

van der Slot, A. J., Zuurmond, A. M., Bardoel, A. F., et al. (2003). Identification of PLOD2 as telopeptide lysyl hydroxylase, an important enzyme in fibrosis. J Biol Chem *278*, 40967–40972.

van Heerebeek, L., Borbely, A., Niessen, H. W., et al. (2006). Myocardial structure and function differ in systolic and diastolic heart failure. Circulation *113*, 1966–1973.

van Heerebeek, L., Hamdani, N., Handoko, M. L., et al. (2008). Diastolic stiffness of the failing diabetic heart: importance of fibrosis, advanced glycation end products, and myocyte resting tension. Circulation *117*, 43–51.

Van Linthout, S., Seeland, U., Riad, A., et al. (2008). Reduced MMP-2 activity contributes to cardiac fibrosis in experimental diabetic cardiomyopathy. Basic Res Cardiol *103*, 319–327.

Villarreal, F. J., and Dillmann, W. H. (1992). Cardiac hypertrophy-induced changes in mRNA levels for TGF-beta 1, fibronectin, and collagen. Am J Physiol *262*, H1861–H1866.

Wakatsuki, T., Schlessinger, J., and Elson, E. L. (2004). The biochemical response of the heart to hypertension and exercise. Trends Biochem Sci *29*, 609–617.

Waltenberger, J., Lundin, L., Oberg, K., et al. (1993a). Involvement of transforming growth factor-beta in the formation of fibrotic lesions in carcinoid heart disease. Am J Pathol *142*, 71–78.

Waltenberger, J., Wanders, A., Fellstrom, B., Miyazono, K., Heldin, C. H., and Funa, K. (1993b). Induction of transforming growth factor-beta during cardiac allograft rejection. J Immunol *151*, 1147–1157.

Wang, J., Chen, H., Seth, A., and McCulloch, C. A. (2003). Mechanical force regulation of myofibroblast differentiation in cardiac fibroblasts. Am J Physiol Heart Circ Physiol *285* (5), H1871–H1881

Weinreb, P. H., Simon, K. J., Rayhorn, P., et al. (2004). Function-blocking integrin alphavbeta6 monoclonal antibodies: distinct ligand-mimetic and nonligand-mimetic classes. J Biol Chem *279*, 17875–17887.

Wells, R. G., and Discher, D. E. (2008). Matrix elasticity, cytoskeletal tension, and TGF-β: the insoluble and soluble meet. Sci Signal *1*, pe13.

Westermann, D., Van Linthout, S., Dhayat, S., et al. (2007). Tumor necrosis factor-alpha antagonism protects from myocardial inflammation and fibrosis in experimental diabetic cardiomyopathy. Basic Res Cardiol *102*, 500–507.

Widyantoro, B., Emoto, N., Nakayama, K., et al. (2010). Endothelial cell-derived endothelin-1 promotes cardiac fibrosis in diabetic hearts through stimulation of endothelial-to-mesenchymal transition. Circulation *121*, 2407–2418.

Wilke, A., Funck, R., Rupp, H., and Brilla, C. G. (1996). Effect of the renin-angiotensin-aldosterone system on the cardiac interstitium in heart failure. Basic Res Cardiol *91* (Suppl 2), 79–84.

Wipff, P. J., and Hinz, B. (2008). Integrins and the activation of latent transforming growth factor beta1 – an intimate relationship. Eur J Cell Biol *87*, 601–615.

Wipff, P. J., Rifkin, D. B., Meister, J. J., and Hinz, B. (2007). Myofibroblast contraction activates latent TGF-beta1 from the extracellular matrix. J Cell Biol *179*, 1311–1323.

Wunsch, M., Sharma, H. S., Markert, T., et al. (1991). In situ localization of transforming growth factor beta 1 in porcine heart: enhanced expression after chronic coronary artery constriction. J Mol Cell Cardiol *23*, 1051–1062.

Xue, W., Cai, L., Tan, Y., et al. (2010). Cardiac-specific overexpression of HIF-1{alpha} prevents deterioration of glycolytic pathway and cardiac remodeling in streptozotocin-induced diabetic mice. Am J Pathol *177*, 97–105.

Yamazaki, T., Komuro, I., and Yazaki, Y. (1998). Signalling pathways for cardiac hypertrophy. Cell Signal *10*, 693–698.

Yang, L., and Moses, H. L. (2008). Transforming growth factor beta: tumor suppressor or promoter? Are host immune cells the answer? Cancer Res *68*, 9107–9111.

Yang, Z., Mu, Z., Dabovic, B., et al. (2007). Absence of integrin-mediated TGF{beta}1 activation in vivo recapitulates the phenotype of TGF{beta}1-null mice. J Cell Biol *176*, 787–793.

Yuen, A., Laschinger, C., Talior, I., et al. (2010). Methylglyoxal-modified collagen promotes myofibroblast differentiation. Matrix Biol *29*, 537–548.

Zaman, A. K., Fujii, S., Goto, D., et al. (2004). Salutary effects of attenuation of angiotensin II on coronary perivascular fibrosis associated with insulin resistance and obesity. J Mol Cell Cardiol *37*, 525–535.

Zeisberg, E. M., Tarnavski, O., Zeisberg, M., et al. (2007). Endothelial-to-mesenchymal transition contributes to cardiac fibrosis. Nat Med *13*, 952–961.

Zhang, A., Liu, X., Cogan, J. G., et al. (2005). YB-1 coordinates vascular smooth muscle alpha-actin gene activation by transforming growth factor beta1 and thrombin during differentiation of human pulmonary myofibroblasts. Mol Biol Cell *16*, 4931–4940.

Zhang, W. M., Kapyla, J., Puranen, J. S., et al. (2003). Alpha 11beta 1 integrin recognizes the GFOGER sequence in interstitial collagens. J Biol Chem *278*, 7270–7277.

Zhang, Y., Proenca, R., Maffei, M., Barone, M., Leopold, L., and Friedman, J. M. (1994). Positional cloning of the mouse obese gene and its human homologue. Nature *372*, 425–432.

Zhao, X. H., Laschinger, C., Arora, P., Szaszi, K., Kapus, A., and McCulloch, C. A. (2007). Force activates smooth muscle alpha-actin promoter activity through the Rho signaling pathway. J Cell Sci *120*, 1801–1809.

Zhou, Y., Hagood, J. S., Lu, B., Merryman, W. D., and Murphy-Ullrich, J. E. (2010). Thy-1-integrin alphavbeta5 interactions inhibit lung fibroblast contraction-induced latent TGF-beta1 activation and myofibroblast differentiation. J Biol Chem *285* (29), 22382–22393.

Zhu, J., Motejlek, K., Wang, D., Zang, K., Schmidt, A., and Reichardt, L. F. (2002). beta8 integrins are required for vascular morphogenesis in mouse embryos. Development *129*, 2891–2903.

Zibadi, S., Cordova, F., Slack, E. H., Watson, R. R., and Larson, D. F. (2011). Leptin's regulation of obesity-induced cardiac extracellular matrix remodeling. Cardiovasc Toxicol *11* (4), 325–333.

Zieman, S., and Kass, D. (2004). Advanced glycation end product cross-linking: pathophysiologic role and therapeutic target in cardiovascular disease. Congest Heart Fail *10*, 144–149; quiz 150–151.

5.3 Cancer-associated fibroblast integrins as therapeutic targets in the tumor microenvironment

Ning Lu, Cord Brakebusch and Donald Gullberg

5.3.1 Introduction

A solid cancer consists of mutated tumor cells, extracellular matrix (ECM), and various nontransformed cells such as fibroblasts, endothelial cells, and immune cells. Collectively, these nontransformed cells and the ECM are referred to as *tumor stroma*, and in recent years, the paramount importance of the tumor stroma for tumor growth and metastasis has become increasingly clear. A major cell type within the stroma is the *cancer-associated fibroblast* (CAF) (Kalluri and Zeisberg, 2006; Franco et al., 2009; Haviv et al., 2009). This term in fact describes a group of fibroblastic cells of different origin, some of which share characteristics with myofibroblasts in granulation tissue during wound healing. CAFs are the major producers of ECM in the tumor, but they are also involved in the dynamic reorganization and remodeling of the ECM, which affects proliferation, survival, and migration of the tumor cells by binding to ECM receptors on their cell surface. In addition, the ECM serves as a reservoir of growth factors and cytokines that take part in bidirectional communication that occurs between the stroma and the tumor cells (Östman and Augsten, 2009; Pietras and Östman, 2010). Furthermore, recent data demonstrate that the stiffness of the ECM is an important parameter regulating tumor progression by modulating signal transduction of mechanosensitive ECM receptors (Levental et al., 2009; Samuel et al., 2011). Integrins are heterodimeric cell surface receptors (Barczyk et al., 2010) that mediate cell adhesion and mechanoreception, and that contribute in a complex manner to growth factor signaling and proteolytic remodeling of the ECM.

In this chapter, we will discuss how integrins on CAFs are involved in the generation and maintenance of CAFs and how they contribute to different cancer-promoting CAF functions. We will give an overview of recent findings in the field and point out areas of research that remain to be explored.

5.3.2 CAF Biology

CAFs are heterogeneous, which can be understood in terms of tissue specificity and cellular origin. Fibroblasts in different tissues are different, and their origins vary (Parsonage et al., 2005; Rinn et al., 2006; Sorrell et al., 2007). CAFs in different tissues are thought to be derived from tissue resident fibroblasts, adult tissue stem cells, bone marrow–derived mesenchymal stem cells and fibrocytes that enter the tumor via the blood stream, and epithelial and endothelial cells by epithelial-mesenchymal transition (EMT) and endothelial-mesenchymal transition (EnMT) (Haviv et al., 2009; Kalluri and Weinberg, 2009; Östman and Augsten, 2009; Xouri and Christian, 2010). However, before discussing integrin function in CAFs, it is important to define CAFs more carefully.

A fundamental problem with respect to the origin of CAFs in the tumor stroma relates to the definition of a fibroblast (Schmitt-Graff et al., 1994). What is a fibroblast, and what markers are there to unequivocally distinguish fibroblasts from other cell types? The identification of mesenchymal adult stem cells has contributed to the confusion surrounding what a fibroblast is and how it is best defined but has also highlighted the need to separate these cell types. In vertebrates, fibroblasts are derived from different types of mesoderm or ectomesenchyme (Weston et al., 2004). Unlike tissue stem cells, fibroblasts are terminally differentiated cells. Although forced expression of a combination of transcription factors can induce fibroblasts to become pluripotent stem cells, suggesting a certain plasticity (Woltjen et al., 2009; Yu et al., 2009), it is very unclear whether such reprogramming occurs under physiological conditions *in vivo*. Thus, there is a need to find markers/signatures that define the fibroblast lineage.

In contrast to the tumor cells, which are clonally derived from a single, mutated cell, the CAFs in a given tumor are polyclonal and from different origins as described previously. Several earlier publications reported mutations and genetic instability in the tumor stroma, but later, more careful analyses cast doubt on some of these data and suggested they were technical artifacts (Campbell et al., 2009). The tumor-promoting functions of CAFs are therefore not derived by genetic mutations, but rather by "education" of cancer cells mediated by microenvironmental factors including soluble factors. Some of these factors have recently been identified, and these will be discussed later. Epigenetic changes in CAFs seem to be important for the maintenance of CAF properties during cell divisions because CAFs retain tumor-promoting functions when cultured *in vitro* in the absence of cancer cells (Gonda et al., 2010; McGee et al., 2011).

Many CAFs display an activated, myofibroblastic phenotype similar to fibroblasts during wound healing and fibrotic conditions, illustrating the description of cancer as a "wound that does not heal" (Dvorak, 1986). Markers such as α-smooth muscle actin (α-SMA) or the spliced-in extra type III domain A of fibronectin (FN ED-A) are generally used to define myofibroblasts, yet none of them is unique to this cell population (▶Table 5.1).

Identification of myofibroblastoid CAFs by single markers such as α-SMA is useful for *in vitro* differentiation studies of homogenous cell populations, but great care should be taken when characterizing CAFs in complex cell mixtures or in tissues by single markers.

5.3.3 Integrins on CAFs

In vitro culture of fibroblasts changes the integrin repertoire. Integrins identified under in vitro conditions are, therefore, not necessarily the ones that are expressed and active in CAFs *in vivo*. Accumulated data from different types of fibroblasts have

Table 5.1 Integrin expression on various populations of differentiated fibroblasts.

Integrins	*Ligands*	*Myofibroblasts*	*CAFs*	*References*
α1β1	collagen IV	+, dermal fibroblasts (in vitro)	++, colon	(Ng et al., 2005; Rodriguez et al., 2008)
α2β1	collagen I	+, dermal fibroblasts (in vitro)	+, colon	(Arora et al., 1999; Rodriguez et al., 2008)
α11β1	collagen I	+++, corneal fibroblasts, dermal fibroblasts (in vitro)	++, lung	(Zhu et al., 2007; Carracedo et al., 2010; Honda et al., 2010)
α3β1	laminins		++, head and neck, vulva	(Gaggioli et al., 2007)
α5β1	fibronectin		++, head and neck, vulva, colon	(Gaggioli et al., 2007; Rodriguez et al., 2008)
α4β7	fibronectin ED-A, VCAM-1, MadCAM-1	+, lung	?	(Kohan et al., 2010)
α8β1	fibronectin, vitronectin, nephronectin	++, heart, lung, liver	?	(Bouzeghrane et al., 2004; Levine et al., 2000)
α9β1	tenascin-C, emilin, vascular endothelial growth factor (VEGF)–C, VEGF-D	+, skin	?	(Nakayama et al., 2010; Danussi et al., 2011)
αvβ1	vitronectin, fibronectin	?	?	
αvβ3	vitronectin, fibronectin, transforming growth factor-beta (TGF-β)/latency-associated protein (LAP)	+, dermal fibroblasts (in vitro)	?	(Honda et al., 2010)
αvβ5	vitronectin, fibronectin, TGF-β/LAP	++, dermal fibroblasts (in vitro)	++, lung, colon	(Shinde et al., 2008)
αvβ8	vitronectin, fibronectin, TGF-β/LAP	++, dermal fibroblasts (in vitro)	?	(Payne and Jones, 2011)

VCAM-1 = Vascular Cell Adhesion Molecule-1; MadCAM-1 = Mucosal addressin Cell Adhesion Molecule-1.

shown that integrins on fibroblasts include the collagen-binding integrins α1β1, α2β1, and α11β1; the arginine-glycine-aspartic acid (RGD)-binding integrins α4β1, α4β7 (on lung fibroblasts), α5β1, α9β1, αvβ3, αvβ5, and αvβ8; and the laminin-binding integrin α3β1. In general, substantially more work has been performed on myofibroblast integrins in relation to wound healing and fibrotic conditions in comparison to what has been published on CAF integrins.

In ▶Table 5.2, we have listed some available data on integrins on myofibroblasts and CAFs.

The α1 (Ng et al., 2005) and α2 (Arora et al., 1999) integrin chains have been described on dermal myofibroblasts, but data on their expression in CAFs are limited. In one histochemical study of colon adenocarcinomas, α-SMA–positive, PDGFβR–negative stromal cells expressed high levels of α1β1 and α5β1 and some α2β1 (Rodriguez et al., 2008). In vitro differentiation of human fetal lung fibroblasts to myofibroblasts using transforming growth factor-beta (TGF-β) have identified increased levels of the β1-associated integrin chains αv and α11, as well as the αvβ3 integrin (Honda et al., 2010). An elegant study of lung fibroblasts identified α4β7 as the integrin receptor that mediates FN ED-A–dependent myofibroblast differentiation (Kohan et al., 2010). Corneal fibroblasts upregulate α11 during myofibroblast differentiation, and knockdown of α11 on the same cells attenuates myofibroblast differentiation in vitro (Carracedo et al., 2010). *In vivo* α11 is expressed in skin wounds, but the exact cell type expressing α11 in wounds has not been described (Zweers et al., 2007). In one study, non-small-cell lung cancers expressed α11 in activated stroma (Zhu et al., 2007). It will be important to determine whether this α11 expression applies to CAFs in other tumors, or whether this is a lung-specific CAF expression of α11.

The α8 integrin subunit is upregulated on some myofibroblasts *in vivo* (Bouzeghrane et al., 2004), but its role in myofibroblast-mediated fibrosis and CAF activation is unclear (Hartner et al., 2009).

The α9 integrin appears to play a role during skin granulation tissue formation via a mechanism involving fibroblast migration rather than having a direct role in myofibroblast differentiation (Nakayama et al., 2010). The α9β1 binds a cryptic site in FN ED-A, but no data exist to suggest that this binding occurs/is important during myofibroblast differentiation (Shinde et al., 2008).

In an experimental three-dimensional system in vitro, human CAFs from ovary carcinomas were found to express αv integrins and a limited set of β1 integrins (Gaggioli et al., 2007). In the same experimental system, only the α3β1 and α5β1 integrins were found to mediate CAF migration through a complex matrix composed of collagen I and laminin-111–enriched matrix.

It should be kept in mind that fibroblasts cultured in vitro become activated and thus at some stage assume the activated myofibroblastic phenotype. In this way, integrins are not different from many other stromal markers, whose expression varies in a manner that is tissue, cell, and culture-condition specific. The knowledge about integrins in different types of CAFs is thus limited, and much remains to be done in this area. Most likely, the emerging picture will reveal tumor- and tissue-specific integrin repertoires on different types of CAFs. When discussing the roles of CAF integrins during different steps of tumorigenesis, we will divide these into three different, arbitrary phases: establishment (phase 1), growth (phase 2), and metastasis (phase 3) (▶Figure 5.6).

Table 5.2 **Survey of selected biomarkers expressed on cancer-associated fibroblasts.**

Biomarkers	*Category*	*Fibroblasts*	*CAFs*	*Activated CAFs*	*Fibroblast-specific marker in vivo?*	*Selected reference(s)*
FN ED-A	ECM	+	+	+++	No, carcinoma cells	(Pujuguet et al., 1996; Matsumoto et al., 1999)
periostin	ECM	+	+	++	No, some types of carcinoma cells	(Maschler et al., 2005)
tenascin-C	ECM	+	+	++	No, carcinoma cells and vascular cells	(Kohan et al., 2010)
cadherin-11	cell surface	+	+	+++	No, smooth muscle cells	(Monahan et al., 2007)
N-cadherin	cell surface	++	++	+	No, myoepithelial cells, muscle cells, some epithelial cells	(Tsuchiya et al., 2006)
integrin α11	cell surface	+	+	+++	Yes	(Popova et al., 2004, 2007)
fibroblast activation protein (FAP)	cell surface	+	+	++	No, pericytes	(Pure, 2009)
platelet-derived growth factor α receptor (PDGFαR)	cell surface	+	+	+	No, smooth muscle cells, carcinoma cells	(Chaudhry et al., 1992)
platelet-derived growth factor β receptor (PDGFβR)	cell surface	+	++	++	No, pericytes, smooth muscle cells	(Pietras et al., 2008)
thymus cell antigen-1 (Thy-1)	cell surface	+	+	+	No, microvascular endothelium	(Rege and Hagood, 2006)

desmin	cytoskeletal	-	-	+	No, muscle cells	(Van Muijen et al., 1987)
vimentin	cytoskeletal	+	+	+	No, lymphocytes, monocytes/macrophages	(Hornbeck et al., 1993; Nieminen et al., 2006)
α-SMA	cytoskeletal	-	-	+++	No, smooth muscle cells, pericytes	(Verbeek et al., 1994)
endosialin	intracellular	+	+	++	No, endothelial cells, pericytes	(Christian et al., 2008)
fibroblast specific protein-1 (FSP-1)/ S100 calcium binding protein A4 (S100A4)	intracellular	+	+	+	No, immune cells	(Cabezon et al., 2007)

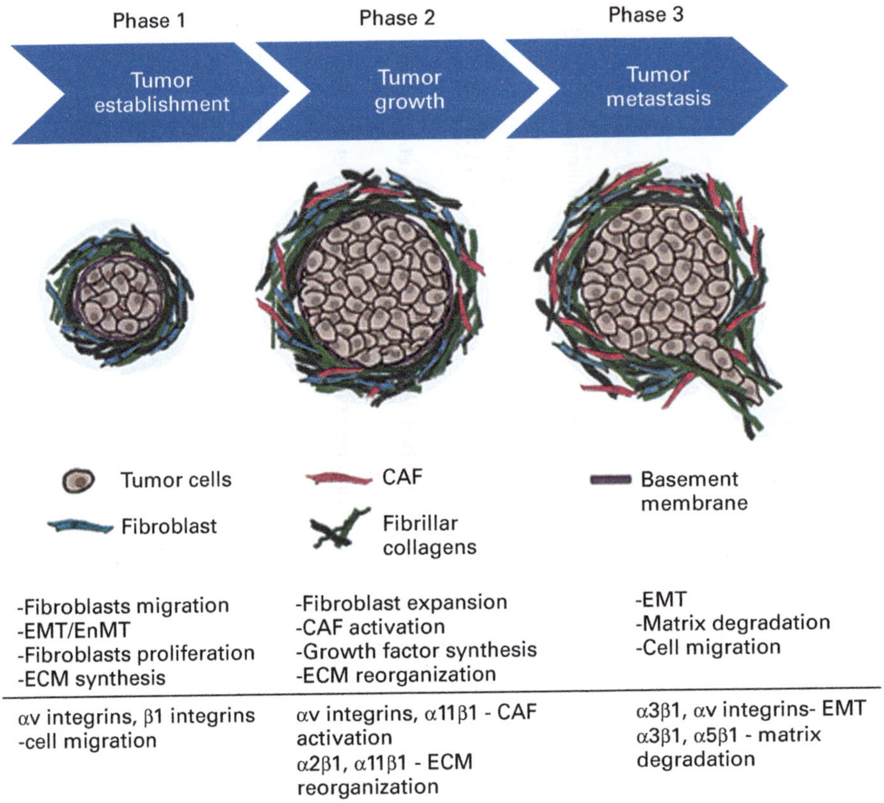

Figure 5.6 Schematic illustration of different steps of tumorigenesis where integrins have been shown, or are suggested to, play a role.

5.3.4 Integrin function on CAF precursors

During the initial phase of tumor growth, the tumor establishes a tumor stroma composed of a variety of cell types and a variable amount of ECM. As discussed previously, the source of the CAFs varies depending on the organ, and their properties depend on the type of tumor with which they are interacting. In ▶Figure 5.6, we have attempted to list some of the major processes that are affected by CAFs during different stages of tumor growth and spread. In the context of cancer, most of the data on the role of integrins in regulating cell proliferation and cell migration have been collected from various tumor cells (Desgrosellier and Cheresh, 2010). In comparison, again, relatively little is known about CAF integrins in these processes.

During the establishment phase, cells of different origins are recruited to the growing tumor, and these heterogeneous CAF precursors become the CAFs of the tumor. Independent of the source of cells, endogenous tissue cells or recruited from the circulation, cell migration is a crucial step during the establishment phase. In some cases, where EMT and EnMT are involved (see discussion on EMT in Section 5.3.7), integrins are predicted to be important during this step. Already at this step, bidirectional paracrine signaling affecting tumor growth is likely to be important.

Furthermore, another important aspect of these events, especially during the establishment phase, concerns the dynamic nature of cell-matrix interactions and integrin expression levels and functions are likely to change when the CAF niche goes from an immature stroma into a more mature stroma.

Recent data suggest that integrin turnover might be a central regulatory process in cell migration (Ivaska and Heino, 2011) and that it most likely plays an important role in CAFs.

Elegant studies have been performed demonstrating that integrins synergize with growth factor (GF) receptors to generate signals of importance for cell proliferation (Ivaska and Heino, 2011; Kim et al., 2011). Corresponding studies need to be performed on CAFs.

Other cell types in the tumor miroenvironment include endothelial cells and pericytes, both involved in forming the tumor vasculature (Robinson and Hodivala-Dilke, 2011). Emerging studies suggest that neuron/glia type 2 antigen (NG2)-positive pericytes are important cells in the production of ECM and are actually the cellular source of myofibroblast-like cells in certain tissues during fibrosis (Kida and Duffield, 2011). Just as tumor cells and CAFs communicate to influence each other, so do forming blood vessels and CAFs, and further studies of the pericytes and their role in the tumor stroma are needed.

5.3.5 Integrin function in CAF differentiation

Once the tumor has reached a certain size and the tumor stroma is well established, the major function of the stroma seems to be to nurture and support the growth of the tumor. An essential step for the growth of the tumor is that the CAFs differentiate into myofibroblastic-like cells. Once differentiated, they more actively affect cell proliferation, matrix reorganization, and tumor angiogenesis.

For differentiation of a variety of progenitors into myofibroblasts, two processes are of major importance: TGF-β signaling and cell contraction (Hinz, 2007). Integrins are important for both of these functions. As mentioned before, the source of the fibroblasts most likely has bearing on the integrin repertoire of activated fibroblasts. In mouse skin wounds, as much as 30% of the myofibroblast-like cells in skin wounds are estimated to be contributed by activated NG2-positive pericytes (Liu et al., 2010a).

A variety of fibroblasts express the collagen-binding integrins α1β1 (Gardner et al., 1999), α2β1 (Eckes et al., 2006), and α11β1 (Tiger et al., 2001). As mentioned earlier, both α1β1 and α2β1 were shown in independent studies to be expressed on myofibroblasts (Arora et al., 1999; Ng et al., 2005; Rodriguez et al., 2008). Expression of the collagen-binding α11β1 integrin is strongly increased during stress-induced myofibroblast differentiation, mediated by activin A and Smad3 (Carracedo et al., 2010). The α11β1 expression further promotes myofibroblast differentiation and is thus acting as a positive feedback loop. It is interesting to note that α11 is expressed in mechanoresponsive cells, with a special function in the mechanoresponsive periodontal ligament fibroblasts (Popova et al., 2007). It is plausible that α11 has evolved to meet the demands of a collagen receptor expressed in an environment subjected to mechanical strain. The existing data thus suggest that α11β1 in embryonic fibroblasts acts in a paracrine manner to regulate activin-A secretion, bypassing the need for activation seen with the TGF-β/latency-associated protein (LAP) complex. It remains to be determined whether similar α11-dependent autocrine loops are active in CAFs.

More recently, fibroblast-specific deletion of the β1-integrin gene in mice, which leads to a loss of all collagen-binding integrin receptors, was found to interfere with myofibroblast differentiation during skin wound healing (Liu et al., 2010b). This was described to be due to a decreased activation of latent TGF-β. The exact mechanism for this defect was not elucidated.

The importance of β1-integrin function for TGF-β signaling was underlined by the phenotype of mice lacking the β1-integrin-associated protein integrin linked kinase (ILK) in fibroblasts (Honda et al., 2009). In these mice, myofibroblast differentiation was severely impaired, corresponding to a defective TGF-β signaling. Blumbach et al. (2010) observed an impaired maintenance of active TGF-β in the extracellular space but normal responsiveness to TGF-β, while Vi et al. (2011) detected a reduced Smad2 phosphorylation of ILK-null fibroblasts after treatment with TGF-β. Furthermore, in the same studies, loss of ILK function in fibroblasts resulted in increased ras homolog gene family, member A (RhoA)/rho-associated protein kinase (ROCK) activation and decreased activation of ras-related C3 botulinum toxin substrate-1 (Rac1). This was not expected, because the RhoA/ROCK pathway is part of the Smad2-independent pathways that are activated downstream of TGF-β receptors, and cell contraction, which is dependent on RhoA/ROCK, was shown to increase the amount of active TGF-β. Here, however, activation of RhoA/ROCK is also required for the classical TGF-β-induced Smad2 phosphorylation. In ILK-null fibroblasts, inhibition of ROCK by Y27632 partially restored the amounts of active TGF-β in the supernatant, indicating a complex role of RhoA/ROCK in TGF-β signaling (Blumbach et al., 2010).

The αv-containing integrins αvβ1, αvβ3, and, in particular, αvβ5 in a variety of fibroblasts were reported to activate latent TGF-β by binding to the RGD site of the LAP, which is bound to TGF-β and the latent TGF-β-binding protein (LTBP), forming the large latent complex (LLC). LTBP anchors the LLC firmly to the stretch-resistant ECM. Cell contraction then "cracks" the ECM-bound TGF-β/LAP complex mechanically, releasing active TGF-β in a protease-independent manner (Wipff et al., 2007).

More recently, αvβ8 on lung fibroblasts has been suggested to take part in lung fibrosis, via a mechanism involving TGF-β activation by myofibroblasts (Kitamura et al., 2011). It is tempting to speculate that αvβ8 might have similar functions on subsets of CAFs.

Stretching of fibroblasts results in increased actin assembly and cell contraction counteracting the external force. Integrin-dependent activation of RhoA is crucial for this response. While the RhoA effector ROCK is required for cell contraction and thus for mechanical-force-induced actin assembly (Zhao et al., 2007), the RhoA effector mammalian diaphanous-related formin protein (mDia) plays an important role in the increased actin polymerization triggered by the mechanosensing pathway (Chan et al., 2010). The formation of actin filaments leads furthermore to the release of the transcription cofactor megakaryocytic acute leukemia (MAL) protein (myocardin-related transcription factor A [MRTF-A]), which is released from the polymerizing actin monomers, translocates to the nucleus, and induces the transcription of α-SMA (Morita et al., 2007). In this context, recent data suggest that several mechanical-strain-induced genes are regulated through the binding of certain transcription factor complexes, including serum response factor (SRF) and its cofactor MRTF-A, to so-called CArG boxes (Asparuhova et al., 2011).

Recent elegant studies have shown that the activated phenotype in CAFs is maintained in vitro due to TGF-β/activin A receptor type II-like kinase (ALK5) and stromal cell derived factor1 (SDF1)/chemokine (C-X-C motif) receptor 4 (CXCR4) loops (Kojima et al., 2010). TGF-β activates the SDF-1/CXCR4 pathway. The potential role of integrins in these processes in this particular context is unexplored, but most likely it involves TGF-β-activating integrins. As mentioned before, it is also likely that epigenetic mechanisms are active to maintain the activated phenotype.

5.3.6 CAF integrins and tumor cell proliferation

A number of autocrine and paracrine loops exist in the tumor microenvironment, influencing tumor growth directly or indirectly. These have been reviewed (Shimoda et al., 2010; Bremnes et al., 2011; Mishra et al., 2011) elsewhere and are not the focus of this chapter.

Integrin function during these processes in CAFs might be to synergize with GF-signaling pathways, but also to activate TGF-β. We have mentioned that the α11β1 is involved in paracrine signaling (Zhu et al., 2007), but more examples of this modus operandi are likely to exist. In addition to the two central loops maintaining the activated state of CAFs, other paracrine mechanisms affecting the tumor cells or tumor angiogenesis are active. An example of this is that the bidirectional platelet-derived growth factor-CC (PDGF-CC) secreted by the tumor has been shown to recruit PDGFαR-positive fibroblasts and to affect synthesis of fibroblast growth factor-2 (FGF2) and fibroblast growth factor-7 (FGF7) by CAFs, in turn affecting tumor angiogenesis (Pietras et al., 2008).

Several studies suggest that the mechanical stiffness of the matrix is a major determinant of tumor growth (Levental et al., 2009; Samuel et al., 2011). One suggested role of the CAFs in this context is to reorganize the matrix and facilitate increased cross-linking of the fibrillar collagen matrix In this process, integrin-dependent matrix metalloproteinase (MMP) activation is likely to be involved, as well as integrin-mediated contractile force. The close association between external and internal force in the context of tumor stroma has recently been highlighted in a study of squamous carcinomas (Hooper et al., 2010).

A second role of matrix stiffness might be to regulate TGF-β activation on CAFs. It is very likely that the CAFs take a central part in the TGF-β activation in the CAF-rich stroma. As already mentioned, possible candidates for this activation are found within the αv subfamily of integrins and within β1-integrin family (Hinz, 2009, 2010; Wipff and Hinz, 2009; Hinz and Gabbiani, 2010).

5.3.7 Integrin function in CAF-promoted invasion and metastasis

Epithelial-mesenchymal transition

CAFs change the ECM of the tumor stroma not only by producing large amounts of ECM molecules, causing in some cancers a servere desmoplasia, but also by affecting the activity of ECM modifying proteases, which is regulated by integrin expressed on CAFs

(Stefanidakis and Koivunen, 2006; Gonzalo et al., 2010). In addition, ECM-binding integrins are crucial to mediate CAF contraction to increased stiffness of the tumor, which is emerging as an important regulator of tumor cell invasion as described subsequently.

Before, however, we would like to emphasize the importance of CAF-dependent ECM remodeling by having a closer look at the ability of the cancer cells to interact with the ECM, which changes dramatically during EMT. These two processes, EMT and ECM remodeling, are in fact crucial for the selection of invasive cancer cells, which are able to migrate through the stroma and eventually form a metastasis at a distant location.

During metastasis, tumor cells at the edges have been suggested to undergo EMT, whereby they loose E-cadherin expression and start expressing markers of mesenchymal cells (Kalluri and Weinberg, 2009).

TGF-β is an efficient inducer of EMT in many tumor types. Since TGF-β by itself regulates integrin synthesis (Heino et al., 1989), it is important to establish a casual relationship between changes in integrin levels and the actual effect on the EMT process.

In agreement with integrins being important during EMT, a number of integrin-related signaling molecules are activated during EMT, including ILK (Huang et al., 2007), phosphatidylinositol 3-kinase (PI3K) (Bakin et al., 2000), and sarcoma tyrosine kinase (Src) (Nagaharu et al., 2011).

In general, epithelial cells express the laminin receptor α6β4, which during EMT becomes downregulated and nonfunctional as cells loose contact with the basement membrane (Maschler et al., 2005). Another laminin receptor on epithelial cells, integrin α3β1, has been reported to play a prerequisite role to facilitate the EMT-induced transcriptional changes by forming a tripartite complex with TGF-β receptor in lung fibrosis (Kim et al., 2009a, 2009b). In lung alveolar cells, a direct interaction of α3β1 with transforming growth factor-beta 1 receptor I (TGF-βRI) was reported to interfere with E-cadherin function and promoting EMT. In lung alveolar cells, however, Smad2 activation was not altered by the absence of α3 integrin. A number of nonintegrin receptors, including neural cell adhesion molecule (N-CAM) (Frame and Inman, 2008), are central in the complex cross-talk between integrins and cadherins during EMT.

During EMT, as cells assume a fibroblastic phenotype, mesenchymal integrins like α2 (Shintani et al., 2008) and α5 (Maschler et al., 2005) become upregulated. Blocking both of these integrins has been shown to attenuate the EMT process. In two recent studies, the mRNA for integrin α11 was also upregulated during EMT events (Ke et al., 2008; Fernando et al., 2010), but the role of this collagen-binding integrin during EMT is still unknown.

For the TGF-β-induced EMT, integrins play a role during the TGF-β-activation step. The role of the αv integrins αvβ3, αvβ5, and αvβ6 (Bates et al., 2005; Galliher and Schiemann, 2006; Ramos et al., 2009; Bianchi et al., 2010) in EMT probably relates to the TGF-β-activating function of these integrins as discussed previously. The role of the surrounding CAF integrins (present on CAFs in the activated stroma) in this context (i.e. activating TGF-β and presenting it to the tumor cells at the tumor front to affect EMT) is largely an unexplored area that warrants further studies.

CAF-assisted matrix degradation

Elegant in vitro studies have suggested that CAF integrins mediate tunneling of the matrix, paving the way for invading tumor cells (Gaggioli et al., 2007; Hooper et al.,

2010). A tissue-, tumor-, and stage-specific induction of different MMPs accompanies this process. In the future, it will be important to determine whether collective and single-cell migration involves similar mechanisms of dissemination and to what degree an active CAF phenotype is needed for the proteolytic activity.

CAF-assisted metastasis during establishment of secondary tumors

Recent data suggest that primary tumor cells bring along CAFs to the new metastatic site (Duda et al., 2010). In terms of the soil-and-seed terminology, the tumor cells bring their soil along. This might serve to condition the new environment for the future expansion of the tumor cells.

In the future, it will be important to characterize integrin profiles and integrin-specific signaling on tumor- and tissue-specific CAFs, as well as their relative contribution to various downstream events during the three different phases of tumorigenesis described in ▶Figure 5.6.

CAFs as therapeutic targets

Why CAFs?

Because CAFs are important for tumor growth and spread, they represent an alternative target in tumor therapy. Fibroblasts are genetically more stable than the tumor cells, so from this respect as well they might be a good target. Moreover, because angiogenesis and immune surveillance depend on CAFs, this is yet another reason to target CAFs. Finally, chemoresistance has been shown to depend on CAFs, so from this point of view, it would also be desirable to target the CAFs.

We have discussed a wide spectrum of roles for CAFs in the stroma: regulating CAF activation, regulating paracrine signaling, regulating matrix stiffness, contributing to EMT, and regulating directed matrix degradation.

Targeting integrins

Because integrins in turn are important for the recruitment, differentiation, and even for several of the tumor- and invasion-promoting functions of CAFs, integrins are potential drug targets to interfere with the tumor-growth-promoting and metastasis-promoting function of CAFs. Integrin function can be targeted as follows:

1. By preventing binding of ligands.
2. By reducing integrin expression.
3. By targeting integrin effectors.
4. Finally, integrins on CAFs can be used to target and eliminate the CAFs altogether.

Depending on the type and stage of the cancer, different integrin-dependent processes in CAFs are blocked. Antibodies as well as low-molecular-weight inhibitors have been generated against different integrin heterodimers, and some of them have already been tested in clinical cancer studies.

Cilengitide is a cyclic pentapeptide that inhibits αvβ3 and αvβ5 (Tucker, 2003). It was designed to block tumor angiogenesis, but in most cancers tested, no beneficial effect was found. To the contrary, low concentration of cilengitide rather increased angiogenesis in mouse melanoma and Lewis lung carcinoma (Reynolds et al., 2009). Glioblastomas, however, seem to respond to cilengitide, and currently, several phase I–III studies are ongoing to test the efficacy of cilengitide in combination with other therapies (Tabatabai et al., 2011).

It would be interesting to assess whether formation of α-SMA and CAFs was altered in desmoplastic cancers treated with Cilengitide.

An α5β1-integrin inhibitor was generated by Jerini; in animal experiments, it was shown to inhibit neovascularization (Zahn et al., 2009), but no successful or ongoing studies of this drug with respect to tumor formation have been published.

One problem with the use of integrin inhibitors in clinical studies is that they act not only on CAFs but also on tumor cells and stroma cells other than CAFs. This makes it difficult to evaluate the effect of integrin inhibition on CAF function in human patients. This would also speak in favor of integrins expressed exclusively on CAFs, such as α11.

Genetically modified mice appear to be an excellent option to circumvent this problem in mouse models. Cancer studies in mice with fibroblast- or even CAF-restricted inactivation of integrin genes would elucidate the specific contribution of integrins to the tumor-promoting function of CAFs. Knockout induction after tumor development will furthermore address whether integrin inhibition on CAFs is a treatment option in cancer.

Another problem to address is the redundancy of different integrin heterodimers with respect to ligand binding and downstream signaling. Combined inhibition of different integrins might therefore be much more effective than single-target drugs. More preclinical research therefore appears mandatory to evaluate the potential of integrin inhibitors in anti-CAF therapy.

5.3.8 Summary

In summary, we have tried to highlight some processes in which integrins are likely to have central functions on CAFs. In the years to come, we see it as essential to identify the integrins involved and to characterize the mechanisms whereby they operate in these different CAF contexts. Much work remains before a therapy approach targeting CAFs will be available in the clinic, but we believe that an increased focus on understanding cell adhesive mechanisms on this cell type will help in this endeavor.

Acknowledgments

The authors' work is supported by grants to D. G. from the Research Council of Norway (183258/O10; 197066, F20) and the Norwegian Cancer Society (107485-PR-2007–0103/0710326b), and to C. B. from Novo Nordisk Foundation. N. L. is supported by a postdoctoral scholarship from the Research Council of Norway.

References

Arora, P. D., Narani, N., and McCulloch, C. A. (1999). The compliance of collagen gels regulates transforming growth factor-beta induction of alpha-smooth muscle actin in fibroblasts. Am J Pathol *154*, 871–882.

Asparuhova, M. B., Ferralli, J., Chiquet, M., and Chiquet-Ehrismann, R. (2011). The transcriptional regulator megakaryoblastic leukemia-1 mediates serum response factor-independent activation of tenascin-C transcription by mechanical stress. FASEB J *25* (10), 3477–3488.

Bakin, A. V., Tomlinson, A. K., Bhowmick, N. A., Moses, H. L., and Arteaga, C. L. (2000). Phosphatidylinositol 3-kinase function is required for transforming growth factor beta-mediated epithelial to mesenchymal transition and cell migration. J Biol Chem *275*, 36803–36810.

Barczyk, M., Carracedo, S., and Gullberg, D. (2010). Integrins. Cell Tissue Res *339*, 269–280.

Bates, R. C., Bellovin, D. I., Brown, C., et al. (2005). Transcriptional activation of integrin beta6 during the epithelial-mesenchymal transition defines a novel prognostic indicator of aggressive colon carcinoma. J Clin Invest *115*, 339–347.

Bianchi, A., Gervasi, M. E., and Bakin, A. (2010). Role of beta5-integrin in epithelial-mesenchymal transition in response to TGF-beta. Cell Cycle *9*, 1647–1659.

Blumbach, K., Zweers, M. C., Brunner, G., et al. (2010). Defective granulation tissue formation in mice with specific ablation of integrin-linked kinase in fibroblasts – role of TGFbeta1 levels and RhoA activity. J Cell Sci *123*, 3872–3883.

Bouzeghrane, F., Mercure, C., Reudelhuber, T. L., and Thibault, G. (2004). Alpha8beta1 integrin is upregulated in myofibroblasts of fibrotic and scarring myocardium. J Mol Cell Cardiol *36*, 343–353.

Bremnes, R.M., Donnem, T., Al-Saad, S., et al. (2011). The role of tumor stroma in cancer progression and prognosis: Emphasis on carcinoma-associated fibroblasts and non-small cell lung cancer. J Thorac Oncol *6*, 209–217.

Cabezon, T., Celis, J. E., Skibshoj, I., et al. (2007). Expression of S100A4 by a variety of cell types present in the tumor microenvironment of human breast cancer. Int J Cancer *121*, 1433–1444.

Campbell, I., Polyak, K., and Haviv, I. (2009). Clonal mutations in the cancer-associated fibroblasts: the case against genetic coevolution. Cancer Res *69*, 6765–6768; discussion 6769.

Carracedo, S., Lu, N., Popova, S. N., Jonsson, R., Eckes, B., and Gullberg, D. (2010). The fibroblast integrin {alpha}11{beta}1 is induced in a mechanosensitive manner involving activin A and regulates myofibroblast differentiation. J Biol Chem *285*, 10434–10445.

Chan, M.W., Hinz, B., and McCulloch, C.A. (2010). Mechanical induction of gene expression in connective tissue cells. Methods Cell Biol *98*, 178–205.

Chaudhry, A., Papanicolaou, V., Oberg, K., Heldin, C. H., and Funa, K. (1992). Expression of platelet-derived growth factor and its receptors in neuroendocrine tumors of the digestive system. Cancer Res *52*, 1006–1012.

Christian, S., Winkler, R., Helfrich, I., et al. (2008). Endosialin (tem1) is a marker of tumor-associated myofibroblasts and tumor vessel-associated mural cells. Am J Pathol *172*, 486–494.

Danussi, C., Petrucco, A., Wassermann, B., et al. (2011). EMILIN1-{alpha}4/{alpha}9 integrin interaction inhibits dermal fibroblast and keratinocyte proliferation. J Cell Biol *195*, 131–145.

Desgrosellier, J. S., and Cheresh, D. A. (2010). Integrins in cancer: biological implications and therapeutic opportunities. Nat Rev Cancer *10*, 9–22.

Duda, D. G., Duyverman, A. M., Kohno, M., et al. (2010). Malignant cells facilitate lung metastasis by bringing their own soil. Proc Natl Acad Sci U S A *107*, 21677–21682.

Dvorak, H. F. (1986). Tumors: wounds that do not heal. Similarities between tumor stroma generation and wound healing. N Engl J Med *315*, 1650–1659.

Eckes, B., Zweers, M. C., Zhang, Z. G., et al. (2006). Mechanical tension and integrin alpha 2 beta 1 regulate fibroblast functions. J Investig Dermatol Symp Proc *11*, 66–72.

Fernando, R. I., Litzinger, M., Trono, P., Hamilton, D. H., Schlom, J., and Palena, C. (2010). The T-box transcription factor Brachyury promotes epithelial-mesenchymal transition in human tumor cells. J Clin Invest *120*, 533–544.

Frame, M. C., and Inman, G. J. (2008). NCAM is at the heart of reciprocal regulation of E-cadherin- nd integrin-mediated adhesions via signaling modulation. Dev Cell *15*, 494–496.

Franco, O. E., Shaw, A. K., Strand, D. W., and Hayward, S. W. (2009). Cancer associated fibroblasts in cancer pathogenesis. Semin Cell Dev Biol *21*, 33–39.

Gaggioli, C., Hooper, S., Hidalgo-Carcedo, C., et al. (2007). Fibroblast-led collective invasion of carcinoma cells with differing roles for RhoGTPases in leading and following cells. Nat Cell Biol *9*, 1392–1400.

Galliher, A. J., and Schiemann, W. P. (2006). Beta3 integrin and Src facilitate transforming growth factor-beta mediated induction of epithelial-mesenchymal transition in mammary epithelial cells. Breast Cancer Res *8*, R42.

Gardner, H., Broberg, A., Pozzi, A., Laato, M., and Heino, J. (1999). Absence of integrin a1b1 in the mouse causes loss of feedback regulation of collagen synthesis in normal and wounded dermis. J Cell Sci *112*, 263–272.

Gonda, T. A., Varro, A., Wang, T. C., and Tycko, B. (2010). Molecular biology of cancer-associated fibroblasts: can these cells be targeted in anti-cancer therapy? Semin Cell Dev Biol *21*, 2–10.

Gonzalo, P., Moreno, V., Galvez, B.G., and Arroyo, A.G. (2010). MT1-MMP and integrins: Hand-to-hand in cell communication. Biofactors *36*, 248–254.

Hartner, A., Cordasic, N., Rascher, W., and Hilgers, K. F. (2009). Deletion of the alpha8 integrin gene does not protect mice from myocardial fibrosis in DOCA hypertension. Am J Hypertens *22*, 92–99.

Haviv, I., Polyak, K., Qiu, W., Hu, M., and Campbell, I. (2009). Origin of carcinoma associated fibroblasts. Cell Cycle *8*, 589–595.

Heino, J., Ignotz, R. A., Hemler, M. E., Crouse, C., and Massague, J. (1989). Regulation of cell adhesion receptors by transforming growth factor-beta. Concomitant regulation of integrins that share a common beta 1 subunit. J Biol Chem *264*, 380–388.

Hinz, B. (2007). Formation and function of the myofibroblast during tissue repair. J Invest Dermatol *127*, 526–537.

Hinz, B. (2009). Tissue stiffness, latent TGF-beta1 activation, and mechanical signal transduction: implications for the pathogenesis and treatment of fibrosis. Curr Rheumatol Rep *11*, 120–126.

Hinz, B. (2010). The myofibroblast: paradigm for a mechanically active cell. J Biomech *43*, 146–155.

Hinz, B., and Gabbiani, G. (2010). Fibrosis: recent advances in myofibroblast biology and new therapeutic perspectives. F1000 Biol Rep *2*, 78.

Honda, E., Yoshida, K., and Munakata, H. (2010). Transforming growth factor-beta upregulates the expression of integrin and related proteins in MRC-5 human myofibroblasts. Tohoku J Exp Med *220*, 319–327.

Honda, S., Shirotani-Ikejima, H., Tadokoro, S., et al. (2009). Integrin-linked kinase associated with integrin activation. Blood *113*, 5304–5313.

Hooper, S., Gaggioli, C., and Sahai, E. (2010). A chemical biology screen reveals a role for Rab21-mediated control of actomyosin contractility in fibroblast-driven cancer invasion. Br J Cancer *102*, 392–402.

Hornbeck, P. V., Garrels, J. I., Capetanaki, Y., and Heimer, S. (1993). Vimentin expression is differentially regulated by IL-2 and IL-4 in murine T cells. J Immunol *151*, 4013–4021.

Huang, H. C., Hu, C. H., Tang, M. C., Wang, W. S., Chen, P. M., and Su, Y. (2007). Thymosin beta4 triggers an epithelial-mesenchymal transition in colorectal carcinoma by upregulating integrin-linked kinase. Oncogene *26*, 2781–2790.

Ivaska, J., and Heino, J. (2011). Cooperation between integrins and growth factor receptors in signaling and endocytosis. Annu Rev Cell Dev Biol *27*, 291–320.

Kalluri, R., and Weinberg, R. A. (2009). The basics of epithelial-mesenchymal transition. J Clin Invest *119*, 1420–1428.

Kalluri, R., and Zeisberg, M. (2006). Fibroblasts in cancer. Nat Rev Cancer *6*, 392–401.

Ke, X., Qu, Y., Goldfinger, N., et al. (2008). Epithelial to mesenchymal transition of a primary prostate cell line with switches of cell adhesion modules but without malignant transformation. PLoS One *3*, e3368.

Kida, Y., and Duffield, J. S. (2011). Pivotal role of pericytes in kidney fibrosis. Clin Exp Pharmacol Physiol *38*, 417–423.

Kim, C., Ye, F., and Ginsberg, M. H. (2011). Regulation of integrin activation. Annu Rev Cell Dev Biol *27*, 321–345.

Kim, K. K., Wei, Y., Szekeres, C., et al. (2009a). Epithelial cell alpha3beta1 integrin links beta-catenin and Smad signaling to promote myofibroblast formation and pulmonary fibrosis. J Clin Invest *119*, 213–224.

Kim, Y., Kugler, M. C., Wei, Y., et al. (2009b). Integrin alpha3beta1-dependent beta-catenin phosphorylation links epithelial Smad signaling to cell contacts. J Cell Biol *184*, 309–322.

Kitamura, H., Cambier, S., Somanath, S., et al. (2011). Mouse and human lung fibroblasts regulate dendritic cell trafficking, airway inflammation, and fibrosis through integrin alphavbeta8-mediated activation of TGF-beta. J Clin Invest *121*, 2863–2875.

Kohan, M., Muro, A. F., White, E. S., and Berkman, N. (2010). EDA-containing cellular fibronectin induces fibroblast differentiation through binding to alpha4beta7 integrin receptor and MAPK/Erk 1/2-dependent signaling. FASEB J *24*, 4503–4512.

Kojima, Y., Acar, A., Eaton, E. N., et al. (2010). Autocrine TGF-beta and stromal cell-derived factor-1 (SDF-1) signaling drives the evolution of tumor-promoting mammary stromal myofibroblasts. Proc Natl Acad Sci U S A *107*, 20009–20014.

Levental, K. R., Yu, H., Kass, L., et al. (2009). Matrix crosslinking forces tumor progression by enhancing integrin signaling. Cell *139*, 891–906.

Levine, D., Rockey, D.C., Milner, T.A., Breuss, J.M., Fallon, J.T., and Schnapp, L.M. (2000). Expression of the integrin alpha8beta1 during pulmonary and hepatic fibrosis. Am J Pathol *156*, 1927–1935.

Liu, S., Taghavi, R., and Leask, A. (2010a). Connective tissue growth factor is induced in bleomycin-induced skin scleroderma. J Cell Commun Signal *4*, 25–30.

Liu, S., Xu, S. W., Blumbach, K., et al. (2010b). Expression of integrin beta1 by fibroblasts is required for tissue repair in vivo. J Cell Sci *123*, 3674–3682.

Maschler, S., Wirl, G., Spring, H., et al. (2005). Tumor cell invasiveness correlates with changes in integrin expression and localization. Oncogene *24*, 2032–2041.

Matsumoto, E., Yoshida, T., Kawarada, Y., and Sakakura, T. (1999). Expression of fibronectin isoforms in human breast tissue: production of extra domain A+/extra domain B+ by cancer cells and extra domain A+ by stromal cells. Jpn J Cancer Res *90*, 320–325.

McGee, K. M., Vartiainen, M. K., Khaw, P. T., Treisman, R., and Bailly, M. (2011). Nuclear transport of the serum response factor coactivator MRTF-A is downregulated at tensional homeostasis. EMBO Rep *12*, 963–970.

Mishra, P., Banerjee, D., and Ben-Baruch, A. (2011). Chemokines at the crossroads of tumor-fibroblast interactions that promote malignancy. J Leukoc Biol *89*, 31–39.

Monahan, T. S., Andersen, N. D., Panossian, H., et al. (2007). A novel function for cadherin 11/osteoblast-cadherin in vascular smooth muscle cells: modulation of cell migration and proliferation. J Vasc Surg *45*, 581–589.

Morita, T., Mayanagi, T., and Sobue, K. (2007). Dual roles of myocardin-related transcription factors in epithelial mesenchymal transition via slug induction and actin remodeling. J Cell Biol *179*, 1027–1042.

Nagaharu, K., Zhang, X., Yoshida, T., et al. (2011). Tenascin C induces epithelial-mesenchymal transition-like change accompanied by SRC activation and focal adhesion kinase phosphorylation in human breast cancer cells. Am J Pathol *178*, 754–763.

Nakayama, Y., Kon, S., Kurotaki, D., Morimoto, J., Matsui, Y., and Uede, T. (2010). Blockade of interaction of alpha9 integrin with its ligands hinders the formation of granulation in cutaneous wound healing. Lab Invest *90*, 881–894.

Ng, C. P., Hinz, B., and Swartz, M. A. (2005). Interstitial fluid flow induces myofibroblast differentiation and collagen alignment in vitro. J Cell Sci *118*, 4731–4739.

Nieminen, M., Henttinen, T., Merinen, M., Marttila-Ichihara, F., Eriksson, J. E., and Jalkanen, S. (2006). Vimentin function in lymphocyte adhesion and transcellular migration. Nat Cell Biol *8*, 156–162.

Östman, A., and Augsten, M. (2009). Cancer-associated fibroblasts and tumor growth – bystanders turning into key players. Curr Opin Genet Dev *19*, 67–73.

Parsonage, G., Filer, A. D., Haworth, O., et al. (2005). A stromal address code defined by fibroblasts. Trends Immunol *26*, 150–156.

Payne, S. J., and Jones, L. (2011). Influence of the tumor microenvironment on angiogenesis. Future Oncol *7*, 395–408.

Pietras, K., and Östman, A. (2010). Hallmarks of cancer: interactions with the tumor stroma. Exp Cell Res *316*, 1324–1331.

Pietras, K., Pahler, J., Bergers, G., and Hanahan, D. (2008). Functions of paracrine PDGF signaling in the proangiogenic tumor stroma revealed by pharmacological targeting. PLoS Med. *5*, e19.

Popova, S. N., Barczyk, M., Tiger, C. F., et al. (2007). Alpha11 beta1 integrin-dependent regulation of periodontal ligament function in the erupting mouse incisor. Mol Cell Biol *27*, 4306–4316.

Popova, S. N., Rodriguez-Sanchez, B., Liden, A., Betsholtz, C., Van Den Bos, T., and Gullberg, D. (2004). The mesenchymal alpha11beta1 integrin attenuates PDGF-BB-stimulated chemotaxis of embryonic fibroblasts on collagens. Dev Biol *270*, 427–442.

Pujuguet, P., Hammann, A., Moutet, M., Samuel, J. L., Martin, F., and Martin, M. (1996). Expression of fibronectin ED-A+ and ED-B+ isoforms by human and experimental colorectal cancer. Contribution of cancer cells and tumor-associated myofibroblasts. Am J Pathol *148*, 579–592.

Pure, E. (2009). The road to integrative cancer therapies: emergence of a tumor-associated fibroblast protease as a potential therapeutic target in cancer. Expert Opin Ther Targets *13*, 967–973.

Ramos, D. M., Dang, D., and Sadler, S. (2009). The role of the integrin alpha v beta6 in regulating the epithelial to mesenchymal transition in oral cancer. Anticancer Res *29*, 125–130.

Rege, T. A., and Hagood, J. S. (2006). Thy-1, a versatile modulator of signaling affecting cellular adhesion, proliferation, survival, and cytokine/growth factor responses. Biochim Biophys Acta *1763*, 991–999.

Reynolds, A. R., Hart, I. R., Watson, A. R., et al. (2009). Stimulation of tumor growth and angiogenesis by low concentrations of RGD-mimetic integrin inhibitors. Nat Med *15*, 392–400.

Rinn, J. L., Bondre, C., Gladstone, H. B., Brown, P. O., and Chang, H. Y. (2006). Anatomic demarcation by positional variation in fibroblast gene expression programs. PLoS Genet *2*, e119.

Robinson, S. D., and Hodivala-Dilke, K. M. (2011). The role of beta3-integrins in tumor angiogenesis: context is everything. Curr Opin Cell Biol *23*, 630–637.

Rodriguez, A., Karen, J., Gardner, H., Gerdin, B., Rubin, K., and Sundberg, C. (2008). Integrin alpha1beta1 is involved in the differentiation into myofibroblasts in adult reactive tissues in vivo. J Cell Mol Med *13*, 3449–3462.

Samuel, M. S., Lopez, J. I., McGhee, E. J., et al. (2011). Actomyosin-mediated cellular tension drives increased tissue stiffness and beta-catenin activation to induce epidermal hyperplasia and tumor growth. Cancer Cell *19*, 776–791.

Schmitt-Graff, A., Desmouliere, A., and Gabbiani, G. (1994). Heterogeneity of myofibroblast phenotypic features: an example of fibroblastic cell plasticity. Virchows Arch *425*, 3–24.

Shimoda, M., Mellody, K.T., and Orimo, A. (2010). Carcinoma-associated fibroblasts are a rate-limiting determinant for tumour progression. Semin Cell Dev Biol *21*, 19–25.

Shinde, A. V., Bystroff, C., Wang, C., et al. (2008). Identification of the peptide sequences within the EIIIA (EDA) segment of fibronectin that mediate integrin alpha9beta1-dependent cellular activities. J Biol Chem *283*, 2858–2870.

Shintani, Y., Fukumoto, Y., Chaika, N., Svoboda, R., Wheelock, M. J., and Johnson, K. R. (2008). Collagen I-mediated up-regulation of N-cadherin requires cooperative signals from integrins and discoidin domain receptor 1. J Cell Biol *180*, 1277–1289.

Sorrell, J. M., Baber, M. A., and Caplan, A. I. (2007). Clonal characterization of fibroblasts in the superficial layer of the adult human dermis. Cell Tissue Res *327*, 499–510.

Stefanidakis, M., and Koivunen, E. (2006). Cell-surface association between matrix metalloproteinases and integrins: Role of the complexes in leukocyte migration and cancer progression. Blood *108*, 1441–1450.

Tabatabai, G., Tonn, J. C., Stupp, R., and Weller, M. (2011). The role of integrins in glioma biology and anti-glioma therapies. Curr Pharm Des *17*, 2402–2410.

Tiger, C. F., Fougerousse, F., Grundstrom, G., Velling, T., and Gullberg, D. (2001). Alpha11-beta1 integrin is a receptor for interstitial collagens involved in cell migration and collagen reorganization on mesenchymal nonmuscle cells. Dev Biol *237*, 116–129.

Tsuchiya, B., Sato, Y., Kameya, T., Okayasu, I., and Mukai, K. (2006). Differential expression of N-cadherin and E-cadherin in normal human tissues. Arch Histol Cytol *69*, 135–145.

Tucker, G. C. (2003). Alpha v integrin inhibitors and cancer therapy. Curr Opin Investig Drugs *4*, 722–731.

Van Muijen, G. N., Ruiter, D. J., and Warnaar, S. O. (1987). Coexpression of intermediate filament polypeptides in human fetal and adult tissues. Lab Invest *57*, 359–369.

Verbeek, M. M., Otte-Holler, I., Wesseling, P., Ruiter, D. J., and de Waal, R. M. (1994). Induction of alpha-smooth muscle actin expression in cultured human brain pericytes by transforming growth factor-beta 1. Am J Pathol *144*, 372–382.

Vi, L., de Lasa, C., DiGuglielmo, G.M., and Dagnino, L. (2011). Integrin-linked kinase is required for TGF-beta1 induction of dermal myofibroblast differentiation. J Invest Dermatol *131*, 586–593.

Weston, J. A., Yoshida, H., Robinson, V., Nishikawa, S., and Fraser, S. T. (2004). Neural crest and the origin of ectomesenchyme: neural fold heterogeneity suggests an alternative hypothesis. Dev Dyn *229*, 118–130.

Wipff, P. J., and Hinz, B. (2009). Myofibroblasts work best under stress. J Bodyw Mov Ther. *13*, 121–127.

Wipff, P. J., Rifkin, D. B., Meister, J. J., and Hinz, B. (2007). Myofibroblast contraction activates latent TGF-beta1 from the extracellular matrix. J Cell Biol *179*, 1311–1323.

Woltjen, K., Michael, I. P., Mohseni, P., et al. (2009). piggyBac transposition reprograms fibroblasts to induced pluripotent stem cells. Nature *458*, 766–770.

Xouri, G., and Christian, S. (2010). Origin and function of tumor stroma fibroblasts. Semin Cell Dev Biol *21*, 40–46.

Yu, J., Hu, K., Smuga-Otto, K., et al. (2009). Human induced pluripotent stem cells free of vector and transgene sequences. Science *324*, 797–801.

Zahn, G., Vossmeyer, D., Stragies, R., et al. (2009). Preclinical evaluation of the novel small-molecule integrin alpha5beta1 inhibitor JSM6427 in monkey and rabbit models of choroidal neovascularization. Arch Ophthalmol *127*, 1329–1335.

Zhao, X.H., Laschinger, C., Arora, P., et al. (2007). Force activates smooth muscle alpha-actin promoter activity through the Rho signaling pathway. J Cell Sci *120,* 1801–1809.

Zhu, C. Q., Popova, S. N., Brown, E. R., et al. (2007). Integrin alpha 11 regulates IGF2 expression in fibroblasts to enhance tumorigenicity of human non-small-cell lung cancer cells. Proc Natl Acad Sci U S A *104*, 11754–11759.

Zweers, M. C., Davidson, J. M., Pozzi, A., et al. (2007). Integrin a2b1 is required for regulation of murine wound angiogenesis but is dispensable for reepithelialization. J Invest Dermatol *127*, 467–478.

5.4 Discoidin domain receptors: non-integrin collagen receptors on the move

Antonio S. Rocca and Michelle P. Bendeck

5.4.1 Introduction

The organization of complex structure in specialized tissues depends on the intricate relationships between the underlying scaffolding patterns provided by matrix molecules and the overlying cell layer. Collagen exists in a variety of forms that provide both structural integrity and functional diversity within tissues capable of withstanding varying mechanical and tensile loads. Cells overlying the collagen scaffold can respond to perturbations in force with alterations in cellular phenotype and differentiation. This mechanotransduction is mediated by several collagen receptors, including a recently discovered family of collagen receptors named the discoidin domain receptors (DDRs). These collagen receptors are a subfamily of the receptor tyrosine kinases and demonstrate unique interactions with collagen. The functions of the DDRs are only beginning to be elucidated. Emerging roles in diverse physiological contexts ranging from atherosclerosis, skeletal formation, fibrotic diseases, and cancer cell migration will be highlighted in this chapter.

5.4.2 Collagen and collagen receptors

Structure and function of collagens

Collagens are the most abundant proteins in mammalian organisms. Long known as important structural components of the extracellular matrix, they provide tensile strength to diverse tissues such as long bones and cardiac valves. In addition to the mechanical properties of collagen molecules, it has become increasingly apparent that collagens provide signals to cells that can alter cell behavior.

Collagen family

The collagen family contains up to 28 different collagen molecules that are composed of three polypeptide chains termed α chains. The α chains are encoded by 46 different genes and are packaged in either homo- or heterotrimeric helical bundles to form the procollagen triple-helix molecule. These undergo self-assembly and are packaged to form collagen microfibrils. Continued cross-linking leads to the formation of large collagen fibers. Of the 28 collagens, some are fibril forming, such as collagen I, II, III, V, and XI. Other collagens, such as IV, VI, and VIII, are capable of forming networks found in basement membranes. Collagens also exist in related forms that include fibril-associated collagen with interrupted triple helices (FACIT) and membrane-associated collagen with interrupted triple helices (MACIT) (Shoulders and Raines, 2009). The collagen triple helix is formed from repeating amino acid units of glycine-any amino acid-any amino acid (G-X-X) within each α strand. The major function of collagens is to provide structural support to multicellular, complex tissues. For example, within the human aortic valve, collagen is the major mediator of the constantly changing tensile force imposed on the valve apparatus by alterations in blood pressure (Schoen, 2008). Several human diseases result from the improper assembly of collagen molecules, including osteogenesis imperfecta, Ehlers-Danlos syndrome, and arterial aneurysms (Myllyharju and Kivirikko, 2004). In blood vessels, fibrillar collagens are extremely stable proteins with low turnover rates; however, in disease conditions, the normally quiescent structural substrate is modified in adaptive fashion. For example, when vessels are subjected to high stress as occurs in hypertension, collagen synthesis is triggered (Lemarié et al., 2010). However, in atherosclerosis, matrix-degrading enzymes remodel the collagen matrix, which can contribute to plaque rupture (Adiguzel et al., 2009).

Collagen receptors

Four types of receptor that interact directly with collagen have been described. Integrins, leukocyte-associated immunoglobulin-like receptor (LAIR), and glycoprotein VI are limited to vertebrates, whereas DDRs are found throughout evolution and are well conserved, indicating their importance. The integrins and DDRs are structurally distinct but can have interrelated effects on various cellular functions.

Integrins

Integrin Adhesion receptors mediating cell-matrix interactions are important in the development of complex structures. This family of transmembrane receptors binds to extracellular matrix through the interaction of two heterodimeric subunits. There are 18 α subunits and 8 β subunits present in mammals that can combine to form 24 different integrins, each with different tissue and matrix-binding specificity. Integrins can signal information from the extracellular matrix to the cell interior by coordinating with other intracellular linkage molecules such as focal adhesion kinase and Src. This process is known as outside-in signaling. In addition, integrin activation can occur in an inside-out fashion, whereby intracellular signals are transmitted to the outside of the cell by influencing integrin affinity for the extracellular matrix. The structure and function of integrin binding and activation is comprehensively reviewed elsewhere (Hynes, 2002; Kim and Ginsberg, 2010); the focus of the following discussion will be on integrin interactions with collagen.

Integrin binding to collagens is mediated by four integrins, $\alpha1\beta1$, $\alpha2\beta1$, $\alpha10\beta1$, and $\alpha11\beta1$. Fibrillar collagens are bound preferentially by the $\alpha2\beta1$ and the $\alpha11\beta1$ integrins, whereas $\alpha1\beta1$ and $\alpha10\beta1$ bind nonfibrillar collagens such as collagen IV. The determination of the binding sequences for integrins in collagen has been aided by the study of synthetic triple-helical proteins (Farndale et al., 2008). These involve the coordination of a metal ion–dependent adhesion site (MIDAS) along with critical GFOGER sequences of collagens. Binding exerts complex conformational changes in the integrin that ultimately lead to signaling across the cell membrane. However, the exact mechanisms of how integrins differentiate between fibrillar and nonfibrillar collagens are yet to be fully elucidated (Farndale et al., 2008; Orgel et al., 2010). Signaling from $\alpha1\beta1$ and $\alpha2\beta1$ integrins has been investigated in many diseases, including inflammatory, wound-healing, and fibrotic disorders (Zutter and Edelson, 2007; Evans et al., 2009; Patsenker and Stickel, 2011). By contrast, the involvement of the $\alpha10\beta1$ and $\alpha11\beta1$ integrins in disease progression is poorly understood at present.

Leukocyte-associated immunoglobulin-like receptor-1

LAIR-1 interacts with collagen via a single extracellular immunoglobulin-like domain. The cytoplasmic domain contains two immunoreceptor tyrosine-based inhibitory motifs. As part of the inhibitory immunoglobulin superfamily, LAIR-1 acts to regulate immune function through interaction with collagens I to III, IV, and certain transmembrane collagens (Lebbink et al., 2006, 2009). The related LAIR-2 molecule shares high homology to the extracellular domain of LAIR-1; however, LAIR-2 protein does not contain the transmembrane or cytoplasmic domains. LAIR-2 is capable of binding collagens, thus regulating the interaction between collagen and LAIR-1. In fact, LAIR-2 is a potent inhibitor of collagen-induced platelet aggregation and is upregulated by activation of T cells (Lebbink et al., 2008; Lenting et al., 2010). Interestingly, LAIR-2 and LAIR-1 both recognize GPO repeats in the collagen triple helix in a similar fashion to glycoprotein VI; LAIR-1 can suppress glycoprotein VI signaling as a result (Tomlinson et al., 2007). LAIR-2 can also suppress glycoprotein VI signaling, as well as platelet interactions with von Willebrand factor, which utilizes a separate collagen-binding motif. It is not known whether LAIR-2 can suppress collagen-induced DDR activation.

Glycoprotein VI

The glycoprotein VI molecule is present on platelets and consists of an ectodomain that contains two immunoglobulin-like domains. This receptor requires a stabilizing interaction with the Fc (fragment crystallizable region of antibody structure) receptor γ chain for full function (Moroi and Jung, 2004). Platelets are activated when they are exposed to fibrillar collagen types I or III. This is a major cause of thrombotic disease and commonly occurs in acute myocardial infarction. As such, the glycoprotein VI molecule is increasingly implicated in disorders of platelet activation (Kato et al., 2003; Ruggeri and Mendolicchio, 2007). The exact molecular-binding sequence in collagen is yet to be fully determined but may involve GPO sequences that may be altered by conformational changes.

Discoidin domain receptors

The collagen-binding proteins discussed so far have all emerged in vertebrates; however, collagen interaction was necessary in order to develop advanced tissue structure and gain biomechanical information from the environment well before the vertebrate distinction. The DDRs are receptor tyrosine kinases, which have been highly conserved throughout evolution. They are distinguished from the other receptor tyrosine kinases by an extracellular discoidin domain and a longer juxtamembrane region encompassing approximately 170 amino acids for discoidin domain receptor-1 (DDR1). The discoidin domain was first identified in the slime mold *Dictyostelium discoideum* in which it functions as a galactose-binding lectin (Vogel et al., 2006). Discoidin domains are also found in several other proteins, including the coagulation factors V and VIII, as well as von Willebrand factor (Baumgartner et al., 1998). The DDRs bind to collagens independently of the integrins. These novel collagen receptors are now implicated in the control of cell proliferation, migration, and matrix remodeling (Vogel et al., 2006). The role of the DDR family in these processes and possible implications for human disease will be the focus of the remainder of the chapter.

5.4.3 Discoidin domain receptor subfamily of receptor tyrosine kinases

Structure of the discoidin domain receptors

Two forms of the DDRs are expressed, the products of two distinct genes (Vogel, 1999). Alternative splicing of the juxtamembrane and cytoplasmic regions generates five different isoforms of DDR1 (Vogel et al., 2006) (▶Table 5.3). DDR1c is the longest isoform, totaling 919 amino acids. DDR1a lacks 37 amino acids of the juxtamembrane region, whereas 6 amino acids from the kinase domain are spliced out in DDR1b. DDR1d and DDR1e are variants that lack the entire kinase region or parts of the juxtamembrane region containing the adenosine triphosphate (ATP)–binding site, respectively (Alves et al., 2001). The expression pattern and functional significance of these isoforms have not been fully determined; however, the DDR1b isoform is predominantly expressed during embryogenesis. The various isoforms may trigger different downstream signaling pathways; for example, only the DDR1b isoform contains the phosphotyrosine-residue which interacts with the ShcA adapter protein (Vogel et al., 1997). In human monocytic THP1 cells overexpressing either DDR1a or DDR1b, a phenotypic difference was observed upon collagen stimulation. DDR1a-expressing cells formed pseudopods associated with increased migration, whereas DDR1b-expressing cells had reduced migration rates, suggesting that the isoforms of DDR possess diverse roles dependent on the cell type (Kamohara et al., 2001). To date, only one isoform of discoidin domain receptor-2 (DDR2) has been described.

Table 5.3 Isoforms of DDR1.

DDR1 isoform	*Characteristic*
DDR1a	Lacks 37 amino acids of the juxtamembrane region containing the ShcA-binding region; interacts with fibroblast growth factor receptor substrate-2 and leads to migration of leukocytes
DDR1b	Lacks 6 amino acids of the kinase domain; interacts with ShcA
DDR1c	919 amino acids, full-length receptor
DDR1d	Lacks entire kinase domain
DDR1e	Lacks portion of ATP-binding site of juxtamembrane region

The DDR1 protein can be modified by proteolytic cleavage that results in shedding of the extracellular domain. This cleavage gives rise to a soluble 54 kDa extracellular domain and a membrane-anchored 62 kDa subunit. It appears that this process may be mediated by a sheddase enzyme from the membrane-type matrix metalloproteinase (MT-MMP) or a disintegrin and metalloproteinase with thrombospondin motifs (ADAM) families (Vogel, 2002). It is not clear whether cleavage results in alterations in receptor signaling.

Localization

The DDRs are widely expressed. DDR1 generally is found in epithelial cell types, whereas DDR2 is largely confined to mesenchymal cells. The receptors are involved in normal fetal development and provide important regulatory and signaling functions in adults. In addition, the DDRs are expressed in several cancers, including breast, brain, and lung. The full extent of DDR localization is summarized in a review by Vogel (Vogel et al., 2006).

DDR1

DDR1 expression in epithelium appears to be important for aligning cells with underlying collagen scaffolding to impart specialized tissue architecture. This interaction is mediated by DDR1 binding of collagen IV, an integral component of basement membranes. In human bronchial epithelium, DDR1 is localized to the basolateral aspects of the cell in association with collagen IV-containing basement membranes (Sakamoto et al., 2001). DDR1 is also important for normal branching patterns of the mammary gland and plays a crucial role in the successful implantation of blastocysts (Vogel et al., 2001). Within the complex structure of the renal parenchyma, DDR1 is localized to the podocytes, forming slit diaphragms of the glomerular basement membrane (Girgert et al., 2010). DDR1 is also found within the cells of the cortical collecting tubules (Lee et al., 2004). DDR1 demonstrates a critical role in neurite outgrowth within the oligodendrocytes of the developing nervous system. In the adult human cerebral cortex, DDR1 is associated with myelin and is present within oligodendrocytes (Roig et al., 2010). The localization of DDR1 in the various epithelial contacts with basement membrane has led to novel efforts in understanding the pathophysiology of a diverse set of disorders.

DDR2

DDR2 is localized predominantly in mesenchymal tissues such as skeletal muscle and fibroblasts of the heart and skin. Within the developing heart, cardiac fibroblasts can be characterized by the expression of DDR2 (Goldsmith et al., 2004). Furthermore, DDR2 expression within the embryonic mouse heart revealed localization within the epicardium on embryonic day 11, which then progressed to expression within the myocardium and septal walls by embryonic day 16. This is consistent with a migratory role of cardiac fibroblasts in the establishment of a tensile network of extracellular matrix within the developing heart (Morales et al., 2005). The function of DDR2 in the regulation of cardiac fibroblasts in the adult heart in disease states represents an interesting avenue of future investigation. Recent studies have demonstrated that DDR2 is a key regulator of bone growth through the control of chondrocyte maturation and ossification by regulating osteoblast differentiation (Zhang et al., 2011).

Collagen-binding specificity

DDR1

Several collagens serve as ligands for to the DDR receptors. Both of the DDR receptors require triple-helical conformation of the collagen to induce binding and activation. Denatured collagens do not activate DDR. Both DDR1 and DDR2 can bind various fibrillar collagens; however, the nonfibrillar collagen type IV is bound by DDR1 but not by DDR2. The specific binding site for DDRs within collagens was recently determined utilizing recombinant libraries of triple-helical peptides derived from collagen II and collagen III. Both DDR1 and DDR2 bind to a GVMGFO sequence in collagens I–III. The specificity of DDR1 adhesion to collagen IV is determined by five amino acids that are adjacent to the GVMGFO-sensing region of the discoidin domain (Xu et al., 2011). Additionally, it appears that the DDR receptors dimerize prior to binding to collagen (▶Figure 5.7). This is unlike other receptor tyrosine kinases, which dimerize upon ligand binding (Leitinger, 2003; Mihai et al., 2009).

DDR2

Although the discoidin domains of DDR1 and DDR2 are highly conserved (59% sequence identity), the two receptors bind to collagens with different specificities. In

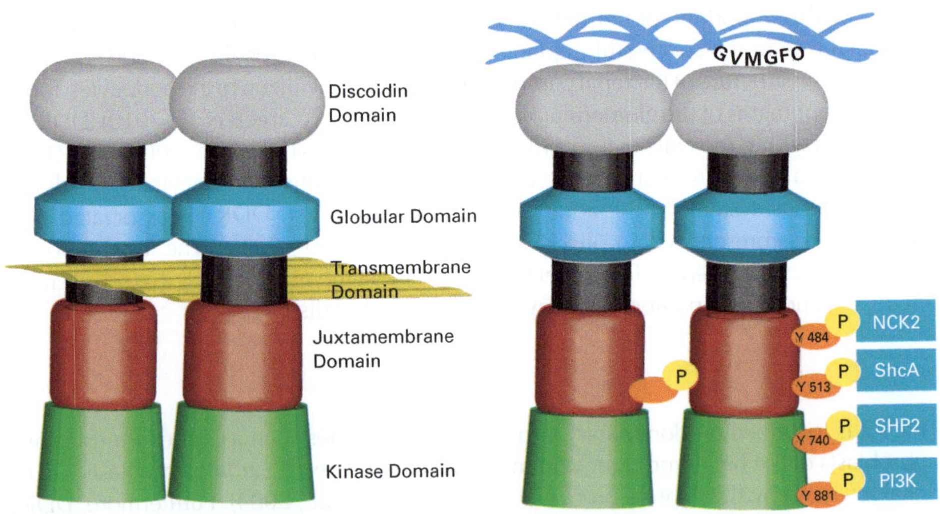

Figure 5.7 Structure and function of the discoidin domain receptors. The DDRs are composed of an N-terminal discoidin domain (160 amino acids) followed by a globular domain. The intracellular domains include a long juxtamembrane domain (174 amino acids), which contains the ATP-binding site, followed by the kinase domain. The DDRs constitutively form dimers in the absence of collagen interaction (A). The DDRs bind to short GVMGFO sequences within triple-helical forms of collagen (blue helix). This induces autophosphorylation of the dimer leading to phosphorylation of several tyrosine residues within the juxtamembrane and kinase domains. These residues are then able to interact with various phosphotyrosine-interacting molecules leading to downstream signaling.

contrast to DDR1, DDR2 does not bind to collagen IV, and DDR2 binds collagens II and X with higher affinity than DDR1. The chondrocyte-specific collagen II contains three high-affinity binding sites for DDR2, which are distinct from the integrin and glycoprotein VI-binding sites. The DDR2-binding site overlaps with the von Willebrand factor–binding site on collagen III; however, several subtle amino acid differences have become evident (Konitsiotis et al., 2008). Nevertheless, the overlap of binding sites within collagens raises interesting possibilities of interactions between von Willebrand factor and DDR2 for collagen adhesion.

Signaling

Collagen stimulation of the DDR leads to receptor autophosphorylation; however, this process is slower but more sustained when compared to the other members of the receptor tyrosine kinase family (Vogel et al., 1997). DDR1 receptor stimulation leads to the activation of a variety of intracellular signaling cascades, which are dependent on the cell type under study. Phosphorylation of critical tyrosine residues in the DDR cytoplasmic domain leads to the creation of docking sites for adaptor molecules, which include ShcA, a noncatalytic region of tyrosine kinase adaptor protein 2 (Nck2), and other molecules containing phosphotyrosine-binding domains.

DDR1

Of the 15 tyrosine residues of the cytoplasmic domain of DDR1, three major and five minor phosphorylation sites were identified (Koo et al., 2006) (▶Figure 5.7). The DDR1b/c isoforms contain the consensus sequence for the adaptor molecule ShcA. In human breast cancer cells, collagen was capable of inducing DDR1b receptor activation and ShcA interaction (Vogel et al., 1997). Within human embryonic kidney 293 (HEK293) cells that were transfected to express DDR1, Shp-2, a phosphatase, was demonstrated to interact with tyrosine 740 on DDR1. In addition, Nck2 was also important in DDR1 signaling within human breast cancer cells (Koo et al., 2006). In vascular smooth muscle cells stimulated with type I collagen, extracellular-signal-regulated kinases (ERK1/2) was activated downstream of DDR1. Src coimmunoprecipitated with DDR1, and Src inhibitors blocked DDR1 phosphorylation and cell migration (Lu et al., 2011).

DDR2

Within hepatic stellate cells, phosphorylation of DDR2 requires the presence of Src for full autophosphorylation (Ikeda et al., 2002). In contrast to DDR1, the interaction of ShcA with DDR2 is mediated via the Src homology 2 (SH2) domain of ShcA rather than the phosphotyrosine-binding domain (Yang et al., 2005).

Cross talk with integrin collagen receptors

As both the integrin and the DDR families of collagen receptors are capable of sensing extracellular matrix and relaying this information to the cell affecting cellular phenotype, it is interesting to speculate on whether these two collagen receptors engage in cross talk of signaling pathways. One such example has been elucidated in a model of pancreatic cancer. As collagen exposure can influence the epithelial-to-mesenchymal transition in some cancerous states, the roles of collagen receptors are important considerations in the search to prevent this switch to metastatic potential. In pancreatic cancer cells, it was determined that collagen I can signal simultaneously via the β1-integrin and DDR1 receptors. The downstream effectors of DDR1 activation included the activation of proline-rich tyrosine kinase 2 (Pyk2) via interaction with the 130 kD Crk-associated substrate (p130CAS) scaffold protein. The signals from both the β1-integrin-induced focal adhesion kinase activation and the DDR1-induced Pyk2 activation were necessary for the upregulation of N-cadherin and the epithelial-to-mesenchymal transition (Shintani et al., 2008).

Depending on cellular context, inhibitory cross talk between integrins and DDRs has also been described. In Madin-Darby canine kidney (MDCK) cells, collagen binding to the α2β1 integrin stimulated cell spreading by activating the guanosine triphosphatase (GTPase) Cdc42, a known regulator of the actin cytoskeleton important in cell migration and spreading. Conversely, collagen-dependent DDR1 activation downregulated Cdc42 and inhibited cell spreading (Yeh et al., 2009). Similarly, collagen-induced DDR1 activation was associated with inhibition of α2β1-integrin-mediated phosphorylation of signal transducers and activators of transcription (Stat 1/3) in MDCK cells leading to decreased cell migration. This effect of DDR1 required the phosphatase activity of Shp-2 (Wang et al., 2006).

Inhibitors of DDR

Several novel inhibitors of the breakpoint cluster region–Abelson kinase oncoprotein (BCR-ABL) have been found to act on other tyrosine kinases, including members of the receptor tyrosine kinase family. Imatinib is used in the treatment of chronic myeloid leukemia to suppress the tyrosine kinase activity of the BCR-ABL protein. Imatinib has in fact been demonstrated to be a much more potent inhibitor of DDR1 and DDR2 compared to BCR-ABL. Nilotinib is another inhibitor of BCR-ABL formulated to treat Imatinib-resistant forms of chronic myeloid leukemia. Nilotinib at concentrations less than 1 μM resulted in significant inhibition of DDR1 to a greater extent than DDR2 as demonstrated by biochemical assays, as well as cellular assays in HEK293 cells using receptor autophosphorylation as the major output measure (Manley et al., 2010; Blay and von Mehren, 2011). A related inhibitor of the BCR-ABL oncoprotein, dasatinib, along with imatinib was found to have potent effects in the inhibition of collagen-induced secretion of monocyte chemoattractant protein from THP1 monocytic cells overexpressing DDR1b (Day et al., 2008). Interestingly, the effectiveness of imatinib in preventing the progression of fibrotic lung disease was demonstrated in an animal model of bleomycin-induced pulmonary fibrosis (Daniels et al., 2004). However, the role of DDR signaling in this model was not examined.

5.4.4 Functions of DDRs

The roles of the DDRs have been elucidated by a variety of techniques, including genetic ablation in knockout mice, RNA silencing, and human genetic-screening analysis. The DDR family has many diverse roles in modulating cellular proliferation, adhesion, migration and regulation of the extracellular matrix. (▶Table 5.4)

Atherosclerosis

The atherosclerotic lesion is a complex environment involving multiple cell types, such as macrophages and vascular smooth muscle cells in close contact with dynamic extracellular matrix products. Collagens play pivotal roles in the stability of the fibrous cap within these lesions and are implicated in regulating the cellular responses of elements within the plaque (Franco et al., 2002; Adiguzel et al., 2009). DDR1 is a positive regulator of the development of atherogenesis, as deletion of DDR1 in a LDLR-deficient mouse model of atherosclerosis led to reduced plaque size and macrophage accumulation concomitant with an increase in the matrix content demonstrated by enhanced fibrillar collagen and elastin accumulation (Franco et al., 2008). Together with the observed increase in matrix accumulation in the DDR1-deficient animals, a reduction of matrix metalloproteinase (MMP) activity was observed in the atherosclerotic lesions of

Table 5.4 Roles of the DDRs in various physiological systems.

System	*DDR1*	*DDR2*
Pulmonary	Regulates bronchial matrix repair Implicated in the development of pulmonary fibrosis	
Cardiac		Marker of cardiac fibroblast Fibroblast migration in the developing heart
Vascular	Contributes to increases in plaque size in a low density lipoprotein receptor ($LDLR^{-/-}$) model of atherosclerosis	
Renal	Contributes to renal inflammation and fibrosis in obstructive uropathy Promotes homeostasis of the slit diaphragm of the glomerular basement membrane	
Reproductive	Promotes blastocyst implantation into uterine wall Critical in mammary gland branching and milk production	
Immune	Regulates T-cell migration in tissues, mediating inflammation and fibrosis Critical for macrophage migration	
Skeletal	Knockout results in dwarfism	Knockout results in dwarfism Human mutation leads to spondylo-meta-epiphyseal dysplasia with short limbs and abnormal calcifications (SMED-SL)

these mice. In order to further investigate the role of DDR1 in this context, elegant bone marrow transplantation experiments were conducted to achieve specific DDR1 deletion either within bone-marrow-derived cells or host cells within the vessel wall. These demonstrated reduced infiltration of DDR1-depleted macrophages into the neointima and resulted in decreased lesion size (Franco et al., 2009). In addition, the vascular smooth muscle cells of the vessel wall require DDR1 in order to regulate extracellular matrix turnover within these plaques. The deletion of DDR1 in host vessel wall cells led to increases in the number of vessel-wall-derived cells migrating into the plaques. This resulted in enhanced matrix accumulation as evidenced by increased collagen production (Franco et al., 2010). This is supported by DDR1 overexpression experiments, which resulted in decreased collagen production in smooth muscle cells (Ferri et al., 2004). In addition to plaque composition, DDR1 is also a key regulator of intimal calcification, a complication of the atherosclerotic process. Mice lacking DDR1 have reduced rates of intimal calcification (Ahmad et al., 2009). Overall, DDR1 is a critical regulator of extracellular matrix turnover and macrophage infiltration within atherosclerotic plaques, and this is an interesting avenue for further investigations that focus on therapeutic considerations.

Cellular migration

The role of DDRs in cell migration is complex and dependent on cell type, interactions with integrins, and the specific composition of the extracellular matrix. In an experimental model of mechanical injury to the vessel wall, DDR1 was upregulated following balloon injury. This was associated with increased rates of cell proliferation, migration, and collagen expression, suggesting that DDR1 is an important regulator of these processes. This was confirmed by experiments showing that DDR1-deficient smooth muscle cells demonstrated reduced attachment to and migration toward collagen (Hou et al., 2001). In addition, the matrix-remodeling enzymes, matrix metalloproteinase-2 (MMP2) and matrix metalloproteinase-9 (MMP9), were downregulated as a result of the loss of DDR1 in vascular smooth muscle cells. The kinase domain of DDR1 was a critical element necessary for the restoration of migration and elaboration of MMP2 and MMP9 within this model; however, it was not required for collagen binding (Hou et al., 2002). DDR1 has also been implicated in the upregulation of MMP2 and MMP9 activity in breast cancer cells stimulated with type IV collagen (Castro-Sanchez et al., 2011). Migratory capability within cells is a function of the balance between cellular contractile and protrusive forces. In this context, DDR1 has been demonstrated to interact with nonmuscle myosin IIA heavy chain via the kinase domain of DDR1. This interaction lead to inhibition of cell spreading but stimulated cell migration, indicating that complex signaling pathways relaying information from the extracellular context are important regulators of cell migration (Huang et al., 2009).

Collagen-independent functions

Novel roles for the DDRs have been defined in the absence of collagen binding. There is growing evidence that DDR may function as a scaffolding molecule by coordinating the subcellular localization of various regulators. The first example involves the coordination of collective cell migration and invasion of cancer cells through the regulation of cell polarity. DDR1 is involved in coordinating cohesive collective cell migration by interacting with the cell polarity regulators protease activated receptors 3 and 6 (Par3 and Par6), which resulted in decreased actomyosin contractility at cell-cell junctions (Hidalgo-Carcedo et al., 2011). In this context, DDR1 was localized to cell-cell junctions, implying a role in the localization of Par3 and Par6 to cell-cell junctions.

Scaffolding and interaction with cadherin

E-cadherin, an important regulator of cell-cell adhesions and epithelial polarity, was recently found to form complexes with both DDR1 and DDR2 via the interaction of the extracellular domains. The finding that surface expression of E-cadherin was reduced in MDCK cells depleted of DDR1 indicated that the DDR functioned as a membrane-stabilizing agent. This occurs in the absence of collagen stimulation, indicating that DDRs have roles in addition to their classical collagen receptor function (Eswaramoorthy et al., 2010; Yeh et al., 2011). Interestingly, a reciprocal relationship appears to exist between E-cadherin and DDR-mediated effects as E-cadherin homophilic adhesions suppressed DDR1 activation in MDCK cells. Furthermore, a ligand-independent interaction was observed between E-cadherin and DDR1 that regulated the localization of DDR1 to the lateral aspects of the cell, which was abolished upon E-cadherin knockdown (Wang et al., 2009).

Knockout animals

The generation of DDR1 and DDR2 knockout mice has proved to be a major advance in understanding DDR functions in development and disease. Mice lacking either of the DDRs exhibit dwarfism. Within DDR2 knockout mice, this is the result of shortened long bones resulting from reduced chondrocyte proliferation (Labrador et al., 2001). DDR1 knockout mice are viable but have an impaired ability for blastocyst implantation leading to infertility. In addition DDR1 is a critical regulator of mammary gland branching and milk production (Vogel et al., 2001). DDR1 is an important regulator of extracellular matrix architecture, and this role is observed in DDR1-deficient mice as evidenced by abnormal development of glomerular basement membranes leading to the expansion of subepithelial matrix and the loss of slit diaphragms resulting in proteinuria (Gross et al., 2004). Thus, mice deficient in the DDRs have an immense potential in the development of animal models of a variety of fibrotic and inflammatory disorders.

Inflammation and fibrosis

DDRs contribute to many diseases characterized by inflammation and fibrosis. DDR1 knockout mice are less susceptible to fibrosis following bleomycin-induced lung injury, which is thought to be the result of impairments in both inflammatory and fibroblast cellular responses (Avivi-Green et al., 2006). Accordingly, DDR1 is required for T-lymphocyte migration through type I collagen gels (Hachehouche et al., 2010). Furthermore, a recent study describes an autoregulatory pathway for the DDRs in lung fibroblasts, whereby stimulation of DDR2 with collagen upregulates DDR1 expression (Ruiz and Jarai, 2011). Similarly, inflammation and fibrosis in the kidney is reduced in

DDR1 knockout mice subject to ureteral obstruction (Guerrot et al., 2011), renal hypertension (Flamant et al., 2006), and hereditary type IV collagen disease (Gross et al., 2010). DDR2, expressed on the hepatic stellate cell, has been implicated in liver fibrosis secondary to MMP production, which mediates cell migration, invasion, and matrix turnover (Olaso et al., 2001).

Role in human disease

Missense mutations of the human *Ddr2* have been identified as the cause of a rare human genetic disorder affecting skeletal growth, the autosomal recessive chondrodysplasia, spondylo-meta-epiphyseal dysplasia with short limbs and abnormal calcifications (SMED-SL). (Bargal et al., 2009; Ali et al., 2010). In addition, overexpression of DDR2 has been associated with human arthritic conditions, including osteoarthritis (Xu et al., 2010) and rheumatoid arthritis (Wang et al., 2002; Su et al., 2009). The overexpression of DDR in many human cancers, such as ovarian cancer and non-small-cell lung carcinoma (Yang et al., 2010), is a common finding, and thus DDR1 has been touted as a marker of disease association.

5.4.5 Conclusions

The discoidin domain receptors, DDR1 and DDR2, are a subfamily of the receptor tyrosine kinase family with unique ability to sense the extracellular matrix, specifically collagens, and integrate this information in order to impact a diverse array of cellular phenotypes. In addition, evidence is accumulating that demonstrates that these novel collagen receptors can also have important scaffolding roles in the absence of collagen binding. Given the near ubiquitous role of collagens in many human disorders ranging from pancreatic cancer to the inflammation of vascular obstructive disorders, the DDRs offer an interesting avenue for further investigations, which may have an impact on therapeutic opportunities for a wide variety of human afflictions.

5.4.6 Take-home message

Collagens, as constituents of the extracellular matrix, can profoundly influence a wide variety of cellular behaviors in processes ranging from atherosclerosis to tumor biology. The DDRs are unique collagen-binding receptor tyrosine kinases that have been found to the transfer of information from the matrix to cells in several pathological states. Further understanding of the signaling mechanisms may lead to important advances in several conditions.

References

Adiguzel, E., Ahmad, P. J., Franco, C., and Bendeck, M. P. (2009). Collagens in the progression and complications of atherosclerosis. Vasc Med *14*, 73–89.

Ahmad, P. J., Trcka, D., Xue, S., et al. (2009). Discoidin domain receptor-1 deficiency attenuates atherosclerotic calcification and smooth muscle cell-mediated mineralization. Am J Pathol *175*, 2686–2696.

Ali, B. R., Xu, H., Akawi, N. A., et al. (2010). Trafficking defects and loss of ligand binding are the underlying causes of all reported DDR2 missense mutations found in SMED-SL patients. Hum Mol Genet *19*, 2239–2250.

Alves, F., Saupe, S., Ledwon, M., Schaub, F., Hiddemann, W., and Vogel, W. F. (2001). Identification of two novel, kinase-deficient variants of discoidin domain receptor 1: differential expression in human colon cancer cell lines. FASEB J *15*, 1321–1323.

Avivi-Green, C., Singal, M., and Vogel, W. F. (2006). Discoidin domain receptor 1-deficient mice are resistant to bleomycin-induced lung fibrosis. Am J Respir Crit Care Med *174*, 420–427.

Bargal, R., Cormier-Daire, V., and Ben-Neriah, Z., et al. (2009). Mutations in DDR2 gene cause SMED with short limbs and abnormal calcifications. Am J Hum Genet *84*, 80–84.

Baumgartner, S., Hofmann, K., Chiquet-Ehrismann, R., and Bucher, P. (1998). The discoidin domain family revisited: new members from prokaryotes and a homology-based fold prediction. Protein Sci *7*, 1626–1631.

Blay, J. Y., and von Mehren, M. (2011). Nilotinib: a novel, selective tyrosine kinase inhibitor. Semin Oncol *38* (Suppl 1), S3–S9.

Castro-Sanchez, L., Soto-Guzman, A., Guaderrama-Diaz, M., Cortes-Reynosa, P., and Salazar, E. (2011). Role of DDR1 in the gelatinases secretion induced by native type IV collagen in MDA-MB-231 breast cancer cells. Clin Exp Metastasis *28*, 463–477.

Daniels, C. E., Wilkes, M. C., Edens, M., et al. (2004). Imatinib mesylate inhibits the profibrogenic activity of TGF-beta and prevents bleomycin-mediated lung fibrosis. J Clin Invest *114*, 1308–1316.

Day, E., Waters, B., Spiegel, K., et al. (2008). Inhibition of collagen-induced discoidin domain receptor 1 and 2 activation by imatinib, nilotinib and dasatinib. Eur J Pharmacol *599*, 44–53.

Eswaramoorthy, R., Wang, C. K., Chen, W. C., et al. (2010). DDR1 regulates the stabilization of cell surface E-cadherin and E-cadherin-mediated cell aggregation. J Cell Physiol *224*, 387–397.

Evans, R., Patzak, I., Svensson, L., et al. (2009). Integrins in immunity. J Cell Sci *122*, 215–225.

Farndale, R. W., Lisman, T., Bihan, D., et al. (2008). Cell-collagen interactions: the use of peptide Toolkits to investigate collagen-receptor interactions. Biochem Soc Trans *36*, 241–250.

Ferri, N., Carragher, N. O., and Raines, E. W. (2004). Role of discoidin domain receptors 1 and 2 in human smooth muscle cell-mediated collagen remodeling: potential implications in atherosclerosis and lymphangioleiomyomatosis. Am J Pathol *164*, 1575–1585.

Flamant, M., Placier, S., Rodenas, A., et al. (2006). Discoidin domain receptor 1 null mice are protected against hypertension-induced renal disease. J Am Soc Nephrol *17*, 3374–3381.

Franco, C., Ahmad, P. J., Hou, G., Wong, E., and Bendeck, M. P. (2010). Increased cell and matrix accumulation during atherogenesis in mice with vessel wall-specific deletion of discoidin domain receptor 1. Circ Res *106*, 1775–1783.

Franco, C., Britto, K., Wong, E., et al. (2009). Discoidin domain receptor 1 on bone marrow-derived cells promotes macrophage accumulation during atherogenesis. Circ Res *105*, 1141–1148.

Franco, C., Hou, G., Ahmad, P. J., et al. (2008). Discoidin domain receptor 1 (ddr1) deletion decreases atherosclerosis by accelerating matrix accumulation and reducing inflammation in low-density lipoprotein receptor-deficient mice. Circ Res *102*, 1202–1211.

Franco, C. D., Hou, G., and Bendeck, M. P. (2002). Collagens, integrins, and the discoidin domain receptors in arterial occlusive disease. Trends Cardiovasc Med *12*, 143–148.

Girgert, R., Martin, M., Kruegel, J., et al. (2010). Integrin alpha2-deficient mice provide insights into specific functions of collagen receptors in the kidney. Fibrogenesis Tissue Repair *3*, 19.

Goldsmith, E. C., Hoffman, A., Morales, M. O., et al. (2004). Organization of fibroblasts in the heart. Dev Dyn *230*, 787–794.

Gross, O., Beirowski, B., Harvey, S. J., et al. (2004). DDR1-deficient mice show localized subepithelial GBM thickening with focal loss of slit diaphragms and proteinuria. Kidney Int *66*, 102–111.

Gross, O., Girgert, R., Beirowski, B., et al. (2010). Loss of collagen-receptor DDR1 delays renal fibrosis in hereditary type IV collagen disease. Matrix Biol *29*, 346–356.

Guerrot, D., Kerroch, M., Placier, S., et al. (2011). Discoidin domain receptor 1 is a major mediator of inflammation and fibrosis in obstructive nephropathy. Am J Pathol *179*, 83–91.

Hachehouche, L. N., Chetoui, N., and Aoudjit, F. (2010). Implication of discoidin domain receptor 1 in T cell migration in three-dimensional collagen. Mol Immunol *47*, 1866–1869.

Hidalgo-Carcedo, C., Hooper, S., Chaudhry, S. I., et al. (2011). Collective cell migration requires suppression of actomyosin at cell-cell contacts mediated by DDR1 and the cell polarity regulators Par3 and Par6. Nat Cell Biol *13*, 49–58.

Hou, G., Vogel, W., and Bendeck, M. P. (2001). The discoidin domain receptor tyrosine kinase DDR1 in arterial wound repair. J Clin Invest *107*, 727–735.

Hou, G., Vogel, W. F., and Bendeck, M. P. (2002). Tyrosine kinase activity of discoidin domain receptor 1 is necessary for smooth muscle cell migration and matrix metalloproteinase expression. Circ Res *90*, 1147–1149.

Huang, Y., Arora, P., McCulloch, C. A., and Vogel, W. F. (2009). The collagen receptor DDR1 regulates cell spreading and motility by associating with myosin IIA. J Cell Sci *122*, 1637–1646.

Hynes, R. O. (2002). Integrins: bidirectional, allosteric signaling machines. Cell *110*, 673–687.

Ikeda, K., Wang, L. H., Torres, R., et al. (2002). Discoidin domain receptor 2 interacts with Src and Shc following its activation by type I collagen. J Biol Chem *277*, 19206–19212.

Kamohara, H., Yamashiro, S., Galligan, C., and Yoshimura, T. (2001). Discoidin domain receptor 1 isoform-a (DDR1alpha) promotes migration of leukocytes in three-dimensional collagen lattices. FASEB J *15* (14), 2724–2726.

Kato, K., Kanaji, T., Russell, S., et al. (2003). The contribution of glycoprotein VI to stable platelet adhesion and thrombus formation illustrated by targeted gene deletion. Blood *102*, 1701–1707.

Kim, C., Ye, F., and Ginsberg, M. H. (2010). Regulation of integrin activation. Annu Rev Cell Dev Biol *27*, 321–345.

Konitsiotis, A. D., Raynal, N., Bihan, D., Hohenester, E., Farndale, R. W., and Leitinger, B. (2008). Characterization of high affinity binding motifs for the discoidin domain receptor DDR2 in collagen. J Biol Chem *283*, 6861–6868.

Koo, D. H., McFadden, C., Huang, Y., Abdulhussein, R., Friese-Hamim, M., and Vogel, W. F. (2006). Pinpointing phosphotyrosine-dependent interactions downstream of the collagen receptor DDR1. FEBS Lett *580*, 15–22.

Labrador, J. P., Azcoitia, V., Tuckermann, J., et al. (2001). The collagen receptor DDR2 regulates proliferation and its elimination leads to dwarfism. EMBO Rep *2*, 446–452.

Lebbink, R. J., de Ruiter, T., Adelmeijer, J., et al. (2006). Collagens are functional, high affinity ligands for the inhibitory immune receptor LAIR-1. J Exp Med *203*, 1419–1425.

Lebbink,R. J., Raynal, N., de Ruiter, T., Bihan, D. G., Farndale, R. W., and Meyaard, L. (2009). Identification of multiple potent binding sites for human leukocyte associated Ig-like receptor LAIR on collagens II and III. Matrix Biol *28*, 202–210.

Lebbink, R. J., van den Berg, M. C., de Ruiter, T., et al. (2008). The soluble leukocyte-associated Ig-like receptor (LAIR)-2 antagonizes the collagen/LAIR-1 inhibitory immune interaction. J Immunol *180*, 1662–1669.

Lee, R., Eidman, K. E., Kren, S. M., Hostetter, T. H., and Segal, Y. (2004). Localization of discoidin domain receptors in rat kidney. Nephron Exp Nephrol *97*, e62–e70.

Leitinger, B. (2003). Molecular analysis of collagen binding by the human discoidin domain receptors, DDR1 and DDR2. J Biol Chem *278*, 16761–16769.

Leitinger, B. (2010). Transmembrane collagen receptors. Annu Rev Cell Dev Biol *27*, 265–290.

Lemarié, C. A., Tharaux, P. L., and Lehoux, S. (2010). Extracellular matrix alterations in hypertensive vascular remodeling. J Mol Cell Cardiol *48*, 433–439.

Lenting, P. J., Westerlaken, G. H., Denis, C. V., Akkerman, J. W., and Meyaard, L. (2010). Efficient inhibition of collagen-induced platelet activation and adhesion by LAIR-2, a soluble Ig-like receptor family member. PLoS One *5*, e12174.

Lu, K. K., Trcka, D., and Bendeck, M. P. (2011). Collagen stimulates discoidin domain receptor 1-mediated migration of smooth muscle cells through Src. Cardiovasc Pathol *20*, 71–76.

Manley, P. W., Drueckes, P., Fendrich, G., et al. (2010). Extended kinase profile and properties of the protein kinase inhibitor nilotinib. Biochim Biophys Acta *1804*, 445–453.

Mihai, C., Chotani, M., Elton, T. S., and Agarwal, G. (2009). Mapping of DDR1 distribution and oligomerization on the cell surface by FRET microscopy. J Mol Biol *385*, 432–445.

Morales, M. O., Price, R. L., and Goldsmith, E. C. (2005). Expression of discoidin domain receptor 2 (DDR2) in the developing heart. Microsc Microanal *11*, 260–267.

Moroi, M., and Jung, S. M. (2004). Platelet glycoprotein VI: its structure and function. Thromb Res *114*, 221–233.

Myllyharju, J., and Kivirikko, K. I. (2004). Collagens, modifying enzymes and their mutations in humans, flies and worms. Trends Genet *20*, 33–43.

Olaso, E., Ikeda, K., Eng, F. J., et al. (2001). DDR2 receptor promotes MMP-2-mediated proliferation and invasion by hepatic stellate cells. J Clin Invest *108*, 1369–1378.

Orgel, J. P., San Antonio, J. D., and Antipova, O. (2010). Molecular and structural mapping of collagen fibril interactions. Connect Tissue Res *52*, 2–17.

Patsenker, E., and Stickel, F. (2011). Role of integrins in fibrosing liver diseases. Am J Physiol Gastrointest Liver Physiol *301* (3), G425–G434.

Roig, B., Franco-Pons, N., Martorell, L., Tomαs, J., Vogel, W. F., and Vilella, E. (2010). Expression of the tyrosine kinase discoidin domain receptor 1 (DDR1) in human central nervous system myelin. Brain Res *1336*, 22–29.

Ruggeri, Z. M., and Mendolicchio, G. L. (2007). Adhesion mechanisms in platelet function. Circ Res *100*, 1673–1685.

Ruiz, P. A., and Jarai, G. (2011). Collagen I induces discoidin domain receptor (DDR) 1 expression through DDR2 and a JAK2-ERK1/2-mediated mechanism in primary human lung fibroblasts. J Biol Chem *286*, 12912–12923.

Sakamoto, O., Suga, M., Suda, T., and Ando, M. (2001). Expression of discoidin domain receptor 1 tyrosine kinase on the human bronchial epithelium. Eur Respir J *17*, 969–974.

Schoen, F. J. (2008). Evolving concepts of cardiac valve dynamics: the continuum of development, functional structure, pathobiology, and tissue engineering. Circulation *118*, 1864–1880.

Shintani, Y., Fukumoto, Y., Chaika, N., Svoboda, R., Wheelock, M. J., and Johnson, K. R. (2008). Collagen 1-mediated up-regulation of N-cadherin requires cooperative signals from integrins and discoidin domain receptor 1. J Cell Biol *180*, 1277–1289.

Shoulders, M. D., and Raines, R. T. (2009). Collagen structure and stability. Annu Rev Biochem *78*, 929–958.

Su, J., Yu, J., Ren, T., et al. (2009). Discoidin domain receptor 2 is associated with the increased expression of matrix metalloproteinase-13 in synovial fibroblasts of rheumatoid arthritis. Mol Cell Biochem *330*, 141–152.

Tomlinson, M. G., Calaminus, S. D., Berlanga, O., et al. (2007). Collagen promotes sustained glycoprotein VI signaling in platelets and cell lines. J Thromb Haemost *5*, 2274–2283.

Vogel, W., Gish, G. D., Alves, F., and Pawson, T. (1997). The discoidin domain receptor tyrosine kinases are activated by collagen. Mol Cell *1*, 13–23.

Vogel, W. F. (1999). Discoidin domain receptors: structural relations and functional implications. FASEB J *13*, 77–82.

Vogel, W. F. (2002). Ligand-induced shedding of discoidin domain receptor 1. FEBS Lett *514*, 175–180.

Vogel, W. F., Abdulhussein, R., and Ford, C. E. (2006). Sensing extracellular matrix: an update on discoidin domain receptor function. Cell Signal *18*, 1108–1116.

Vogel, W. F., Aszodi, A., Alves, F., and Pawson, T. (2001). Discoidin domain receptor 1 tyrosine kinase has an essential role in mammary gland development. Mol Cell Biol *21*, 2906–2917.

Wang, C. Z., Su, H. W., Hsu, Y. C., Shen, M. R., and Tang, M. J. (2006). A discoidin domain receptor 1/SHP-2 signaling complex inhibits {alpha}2beta1-integrin-mediated signal transducers and activators of transcription 1/3 activation and cell migration. Mol Biol Cell *17*, 2839–2852.

Wang, C. Z., Yeh, Y. C., and Tang, M. J. (2009). DDR1/E-cadherin complex regulates the activation of DDR1 and cell spreading. Am J Physiol Cell Physiol *297*, C419–C429.

Wang, J., Lu, H., Liu, X., et al. (2002). Functional analysis of discoidin domain receptor 2 in synovial fibroblasts in rheumatoid arthritis. J Autoimmun *19*, 161–168.

Xu, H., Raynal, N., Stathopoulos, S., Myllyharju, J., Farndale, R. W., and Leitinger, B. (2011). Collagen binding specificity of the discoidin domain receptors: binding sites on collagens II and III and molecular determinants for collagen IV recognition by DDR1. Matrix Biol *30*, 16–26.

Xu, L., Servais, J., Polur, I., et al. (2010). Attenuation of osteoarthritis progression by reduction of discoidin domain receptor 2 in mice. Arthritis Rheum *62*, 2736–2744.

Yang, K., Kim, J. H., Kim, H. J., Park, I. S., Kim, I. Y., and Yang, B. S. (2005). Tyrosine 740 phosphorylation of discoidin domain receptor 2 by Src stimulates intramolecular autophosphorylation and Shc signaling complex formation. J Biol Chem *280*, 39058–39066.

Yang, S. H., Baek, H. A., Lee, H. J., et al. (2010). Discoidin domain receptor 1 is associated with poor prognosis of non-small cell lung carcinomas. Oncol Rep *24*, 311–319.

Yeh, Y. C., Wang, C. Z., and Tang, M. J. (2009). Discoidin domain receptor 1 activation suppresses α2β1 integrin-dependent cell spreading through inhibition of Cdc42 activity. J Cell Physiol *218*, 146–156.

Yeh, Y. C., Wu, C. C., Wang, Y. K., and Tang, M. J. (2011). DDR1 triggers epithelial cell differentiation by promoting cell adhesion through stabilization of E-cadherin. Mol Biol Cell *22*, 940–953.

Zhang, Y., Su, J., Yu, J., et al. (2011). An essential role of discoidin domain receptor 2 (DDR2) in osteoblast differentiation and chondrocyte maturation via modulation of Runx2 activation. J Bone Miner Res *26*, 604–617.

Zutter, M. M., and Edelson, B. T. (2007). The α2β1 integrin: a novel collectin/C1q receptor. Immunobiology *212*, 343–353.

5.5 Syndecans as receptors for pericellular molecules

Xiaojie Xian, James R. Whiteford, and John R. Couchman

5.5.1 Introduction

Syndecans are type I transmembrane cell surface proteoglycans expressed ubiquitously in multicellular animals. Invertebrates possess only one syndecan gene, while higher vertebrates possess four (Couchman, 2003). The extracellular core proteins of syndecans are covalently substituted with glycosaminoglycan (GAG) chains, most commonly heparan sulfate (HS) but additionally with chondroitin sulfate (CS) or dermatan sulfate chains in some cases. Syndecans have crucial roles in cell adhesion, migration, wound healing, and angiogenesis, mainly through interactions with extracellular ligands (Couchman, 2003; Alexopoulou et al., 2007). They may also have important roles in pathogenesis of important diseases, such as cancer and vascular disease (Alexopoulou et al., 2007; Manon-Jensen et al., 2010).

The syndecan core proteins

Syndecans are type I membrane proteins, comprising an extracellular domain, single transmembrane domain, and cytoplasmic domain. Ligands may bind either or both the core protein and GAG chains, though interactions of the latter are more widespread. The cytoplasmic domains (described elsewhere in the volume) are small, have no intrinsic enzymatic activity, but undertake a number of protein-protein and protein-phospholipid interactions. A key common denominator is interaction with actin-associated proteins, providing a linkage to, and regulation of, the cytoskeleton. Further interactions with post-synaptic density protein 95/discs large/zonula occludens-1 (PSD-95/Dlg1/zo-1) domain proteins (Couchman, 2010) provide possibilities for endocytosis and recycling.

The binding properties of HS chains

The length of HS chains and the extent of their sulfation are tissue and cell-type specific. Tissue-specific differences include the fine structure of HS chains in terms of size and frequency of highly sulfated regions (Esko and Selleck, 2002). Moreover, the number and size of the HS chains can also vary. These differences result in alterations in ligand-binding properties and thus diversity in syndecan function (Kreuger et al., 2006). Syndecan HS chains interact with a wide range of molecules, including growth factors, extracellular matrix (ECM) macromolecules, chemokines, morphogens, lipid-regulating enzymes, cell-cell adhesion molecules, and blood-coagulation factors

(▶Table 5.5). Syndecans can serve as matrix and cell surface receptors, coreceptors for growth factor signaling, internalization receptors, and soluble paracrine effectors (▶Figure 5.8). In this chapter, the interactions of syndecans with extracellular ligands, including ECM components and growth factors, are considered, along with the consequences of these interactions. The structure of syndecans and their GAG chains, and their signaling properties are discussed earlier in this volume.

Table 5.5 Glycosaminoglycan-binding proteins (incomplete list). List derived from previous reviews (Lander, 1994; Esko and Selleck, 2002; Götte, 2003; Cool and Nurcombe, 2006; Taylor and Gallo, 2006).

Growth factors	Fibroblast growth factors (FGFs; except 15, 19, 21, and 23) and FGF receptors Vascular endothelial growth factors Platelet-derived growth factors Wnts Hedgehogs Hepatocyte growth factor Transforming growth factor-beta (TGF-β; TGF-β1 and TGF-β2, bone morphogenetic protein-2 and -4 [BMP2 and BMP4]) Glial cell line–derived neurotrophic factor (GDNF) Granulocyte-macrophage colony-stimulating factor (GM-CSF) Heparin-binding EGF-like growth factor (HB-EGF) Insulin-like growth factor (IGF) Growth factor–binding protein (IGF BP-3 and -5, TGF-β BP)
Extracellular matrix components	Laminins, fibronectins, collagens, vitronectin, tenascins, thrombospondins Fibrillin, von Willebrand factor
Chemokines and cytokines	Interleukins (IL-2, -3, -4, -5, -7, -8, -10, and -12) Tumor necrosis factor-alpha (TNF-α) Interferon-gamma (IFN-γ) IFN-γ-inducible protein-10 (IP-10) Monocyte chemoattractant protein-1 and -4 (MCP1 and MCP4) GM-CSF Chemokine (C-C motif) ligand-5 (CCL5) Migration inhibitory factor–related protein-14 (MRP14)
Cell-cell adhesion molecules	Neural cell adhesion molecule (NCAM), L1-cell adhesion molecule (L1CAM), platelet endothelail cell adhesion molecule-1 (PECAM1), selectins (L, E, and P), myelin-associated glycoprotein, macrophage antigen complex-1/αMβ2 integrin (MAC1)
Enzymes and inhibitors	Thrombin, tissue plasminogen activator, heparin cofactor II, lipoprotein lipase, hepatic lipase, neutrophil elastase, urokinase plasminogen activator, clotting factors IX and XI, antithrombin III, protease nexin I, extracellular superoxide dismutase, plasminogen activator inhibitor, protein C inhibitor
Others	Angiostatin, endostatin, platelet factor-4 and -5, microbes and their products (*Pseudomonas serugimosa, Staphylococcus aureus, Bordetella pertussis, Neisseria gonorrhoeae, Mycobacterium tuberculosis*), herpes simplex virus gD

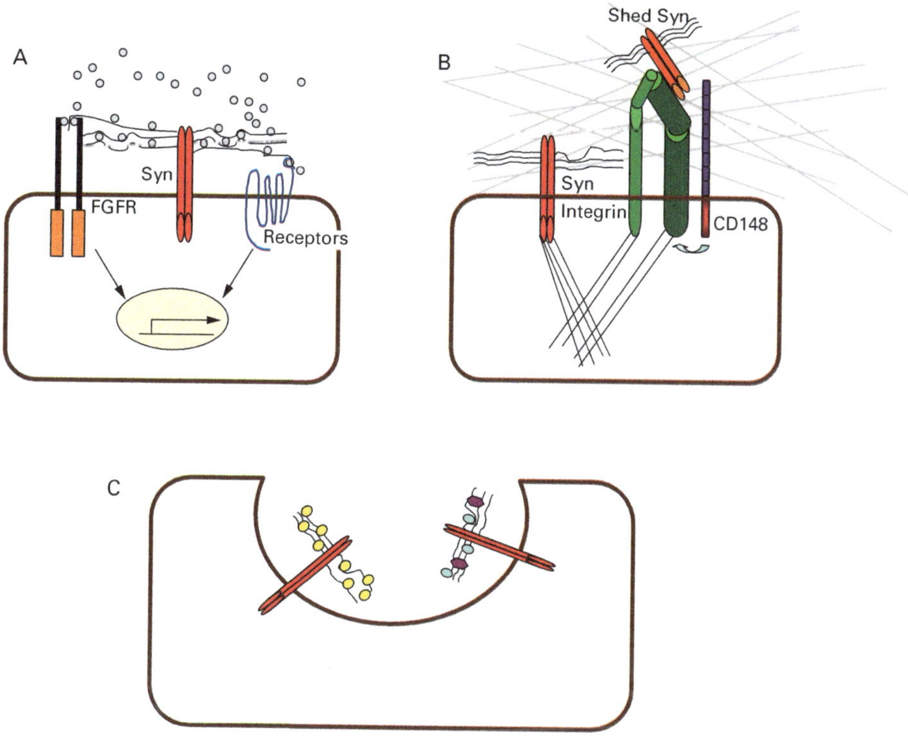

Figure 5.8 **Syndecans as receptors and coreceptors.** Syndecans interact with a variety of pericellular proteins and function in multiple ways. In many cases, the HS chains are required, but there are contributions of core proteins through extracellular interactions and/or signaling from intracellular domains. (A) Syndecans associate with growth factors either as low-affinity binding partners or by forming complexes with high-affinity receptors (e.g. fibroblast growth factor receptor [FGFR]) mediating several signaling pathways. (B) Syndecan-ECM interactions induce intracellular signals, which synchronize with integrin signaling, regulating cytoskeleton rearrangement, cell adhesion, and migration. In some cases, shed syndecan can associate with integrins, either directly or indirectly. (C) Syndecans serve as clearance receptors mediating internalization of lipoproteins and a number of pathogens, through interaction with their HS chains.

5.5.2 Syndecans as cell surface ECM receptors

All organs contain ECM, with a spectrum of collagen, glycoprotein, and proteoglycan expression that is tissue specific. Collagen I is by far the most abundant ECM molecule in the mammalian body, but the ECM contains numerous glycoproteins, proteoglycans, and the GAG hyaluronan. An important principle is, however, that in addition to physical function, ECM macromolecules can interact with, and signal through, specific cell surface receptors. Many ECM glycoproteins have heparin-binding domains that facilitate interaction with syndecans. Moreover, most collagens also share this property. A recent review discusses details of the heparin-binding sites within ECM molecules (Xian et al., 2010).

Fibronectin

Fibronectin is a modular protein consisting of type I, II, and III domains. It is an abundant glycoprotein of serum and is a constituent of matrices formed during plasma clotting, together with fibrin. Contained within the tenth type III repeat of fibronectin is an arginine-glycine-aspartate (RGD) motif that is a ligand for several integrins. Several mRNA splice variants can result in fibronectin protein variants that may include an additional α4β1-integrin ligand–binding site. Fibronectin has three heparin-binding sites, a low-affinity N-terminal site (HepI), a high-affinity site contained within type III repeats 13 and 14 (HepII), and a more cryptic domain contained within type III repeats 4 and 5 (Xian et al., 2010). Considerable attention has focused on the interaction between the HS chains of syndecan-4 and the HepII domain of fibronectin. Experiments with fibronectin fragments reveal that the RGD containing fibronectin type III (FNIII) repeat can support fibroblast attachment and spreading; however, it is only the addition of fragments containing the HepII domain of fibronectin that stimulate focal adhesion formation. This requires a coordinated signaling pathway with input from both α5β1 integrin and syndecan-4 (Bass et al., 2008; Okina et al., 2009). The HS chains of syndecan-4 mediate this interaction, in which *N*-sulfation is particularly important (Mahalingam et al., 2007). Syndecans bear multiple HS attachment sites, and for syndecan-4, at least two of its three HS chains are required to stimulate focal adhesion formation and concomitant actin cytoskeleton organization (Gopal et al., 2010).

Laminin

Laminins are heterotrimeric proteins composed of three distinct subunits (α, β, and γ) forming cross- or rod-shaped molecules. They are predominantly localized to basement membranes. Syndecan HS chains interact with laminin α chains. The heparin-binding sites are located at the laminin G-like (LG4-5) modules of the C-terminal large globular domain (Yamashita et al., 2004). Syndecans-1, -2, and -4 all interact with the LG4 module of the α1 and α3 chains, whereas syndecan-3 has not been examined (Xian et al., 2010). The 150 kDa γ2 chain of laminin-332 (α3β3γ2) can be processed to a 105 kDa form, and this is mediated by proteinases like membrane-type matrix metalloproteinase-1 MT-MMP1 (matrix metalloproteinase-14 [MMP14]) (Sugawara et al., 2008). Syndecan-1, but not syndecan-2, has been identified as a membrane receptor of the processed γ2 chain in EJ-1 cells (Ogawa et al., 2007). In keratinocytes, α3β1 integrin works closely with syndecan-1 for binding the precursor of laminin-332. Their collaboration induces the formation of fascin-containing protrusions, which is dependent on activation of guanosine triphosphatases (GTPases) Rac and Cdc42 (Bachy et al., 2008). Syndecan-2 plays a pivotal role in laminin assembly because expression of a syndecan-2 mutant with a truncated cytoplasmic domain in Chinese hamster ovary (CHO) cells fails to assemble laminin into fibrillar matrix (Klass et al., 2000).

Fibrillin

As with fibronectin, interactions between cells and the RGD (integrin binding) and heparin-binding sites of fibrillin-1 can influence cell adhesion responses. Fibrillins are a family of large glycoproteins that assemble into higher-order extended structures, microfibrils, which can serve as a template for elastin deposition and formation of elastin fibers (Kielty et al., 2005). The aortic wall of fibrillin-1–null mice fails to develop properly, leading to dissecting aneurysms and neonatal death. In humans, genetic mutations in fibrillin-1 lead to Marfan syndrome (Robinson et al., 2006).

In terms of molecular organization, the fibrillins share common characteristics in that they are composed of between 46 and 47 epidermal growth factor (EGF)–like repeats, of which 42/43 are of the calcium-binding type (cb-EGF). Interspersed among them are seven modules containing a characteristic eight-cysteine motif similar to those of latent TGF-β-binding proteins (LTBPs). Fibrillin-1 contains an RGD motif in the fourth transforming growth factor β binding protein-like (TB) module, recognized by both αVβ3 and α5β1 integrins, though α5β1 may be the principle fibrillin receptor (Bax et al., 2003). HS interactions with fibrillin-1 have been well documented; there are four heparin-binding sites in fibrillin-1. Similar to fibronectin, engagement of HS chains (most likely from syndecan-4) by the heparin-binding domain in the TB5 module enhances focal adhesion formation in conjunction with engagement of the RGD site in the nearby TB4 domain by α5β1 integrin (Bax et al., 2007). Additionally, the heparin-binding regions toward the N terminus of fibrillin can also support cell adhesion, although it is unclear which proteoglycan is involved (Cain et al., 2008). Heparin can compete with interactions between the fibrillin N-terminal heparin-binding domains and microfibrillar-associated protein-1, as well as C-terminal domain interactions with elastin. This suggests that the N- and C-terminal heparin-binding domains may regulate fibrillin assembly into supramolecular microfibril assemblies (Cain et al., 2005).

Collagen

Collagens are the most abundant ECM protein, comprising at least 29 different types in several different classes. Collagen fibrils confer integrity and tensile strength to tissues, while numerous other ECM molecules such as proteoglycans, other collagens, and glycoproteins such as laminin interact with collagens (Sweeney et al., 2008). In addition to forming supramolecular fibrillar structures, they interact with cells through integrins such as α1β1, α2β1, α10β1, and α11β1. A feature of many collagens is that they contain heparin and HS-binding domains (Sweeney et al., 2008). In keeping with other syndecan/ECM interactions, syndecan-1 appears to be a coreceptor with integrins, in this case α2β1 (Vuoriluoto et al., 2008). This collaboration enhances the transcription of matrix metalloproteinase-1 in response to collagen binding (Vuoriluoto et al., 2008) with implications for tissue remodeling.

Matricellular proteins tenascin-C and thrombospondin

Tenascin-C affects processes such as cell adhesion, migration, proliferation, and survival, and its effects on cell spreading and attachment are particularly interesting because it can promote cell spreading in certain cell types and be inhibitory to others (Chiquet-Ehrismann, 2004). Mature tenascin-C consists of six N-terminally linked polypeptide chains, which contain multiple EGF-like repeats, FNIII repeats, and a terminal globular domain similar to that of the β and γ chains of fibrinogen. The six chains are linked by an N-terminal tenascin assembly domain (Chiquet-Ehrismann, 2004; Orend and Chiquet-Ehrismann, 2006). Heparin-binding sites in tenascin-C have been identified in both the FNIII repeats and the C-terminal globular domain (Salmivirta et al., 1991). Tenascin-C often localizes to fibronectin fibrils both *in vitro* and *in vivo* (Chiquet-Ehrismann and Chiquet, 2003; Midwood and Orend, 2009). Several lines of evidence support a role for tenascin-C in regulating the interactions between syndecan-4 and fibronectin. Tenascin-C can bind both HS and the heparan-binding FNIII repeat 13 (FNIII 13) of fibronectin. These interactions are inhibitory to fibronectin-fibrinogen matrix contraction by fibroblasts and cell anchorage to fibronectin, resulting in cessation of cell cycle progression (Orend et al., 2003). Tenascin-C may therefore modulate syndecan-4 signaling, acting both as a competitor to fibronectin for the HS chains of syndecan-4 or for its interaction with the HepII domain of fibronectin (FNIII 13) (Midwood et al., 2004). Intriguingly, tenascin-C–null and syndecan-4–null mice both exhibit defective wound healing, suggestive of a mutual involvement. Moreover, other glycoproteins expressed during wound healing, such as fibulin-1, also interact with FNIII 13 of fibronectin, influencing syndecan-4 interactions with fibronectin (Williams and Schwarzbauer, 2009).

Thrombospondin-1 is a homotrimeric protein implicated in cancer cell adhesion, migration, and invasion; angiogenesis; and the activation of latent TGF-β (Isenberg et al., 2008; Kazerounian et al., 2008). Toward the N-terminus are two heparin-binding domains that compete, as in tenascin-C, with the HepII domain of fibronectin for engagement of syndecan-4 HS chains. Clustering of syndecan-4 with monoclonal antibodies blocks the protubulogenic effects of endothelial cells in response to the N-terminal (1–174) residues of thrombospondin-1 (Nunes et al., 2008). It is hypothesized that the N-terminal proangiogenic activity of thrombospondin-1 is linked to its capacity to interfere with syndecan-4 functions in the course of cell adhesion (Nunes et al., 2008). Cellular adhesion to thrombospondin-1 results in the formation of fascin spikes with a requirement for syndecan-1 and its cytoplasmic domain (Chakravarti et al., 2005). However, the signaling pathway has not been resolved.

Syndecans as ECM components

The HS-bearing extracellular core proteins of syndecans can be shed from the cell surface by the action of matrix metalloproteinases (MMPs), and this occurs principally at sites of inflammation and tissue damage. This property is reviewed in depth elsewhere in this volume. It has been speculated that syndecan shedding is a means by which cells release not only syndecan HS chains but also the numerous growth factors and chemokines associated with them. However, it is emerging that cell adhesion regulatory domains also reside in the extracellular core proteins of syndecans, which can trigger integrin-mediated responses. Adhesion regulatory domains have been identified in the syndecan-1, -2, and -4 ectodomains. Syndecan-2 and -4 ectodomains can promote adhesion of some mesenchymal and leukocytic cells in the absence of other ECM components. This process occurs independently of HS but requires β1 integrin and in fibroblasts can stimulate focal adhesion formation (McFall and Rapraeger, 1998; Whiteford et al., 2007). The syndecan-2 and -4 extracellular core proteins are not β1-integrin ligands, and the involvement of another cell surface receptor is likely (Whiteford et al., 2007). Cell adhesion to the syndecan-2 ectodomain requires the protein tyrosine phosphatase receptor CD148. The interaction between CD148 and the syndecan-2 ectodomain is direct and promotes β1-integrin-mediated cell adhesion through a pathway involving the CD148 phosphatase module, Src kinase, and isoforms of phosphatidylinositol 3-kinase (PI3K) (▶Figure 5.8) (Whiteford et al., 2011). The core protein of syndecan-1 can strongly influence αVβ3- and αVβ5-integrin-driven cell adhesion and migration via a direct interaction between the two molecules (Beauvais et al., 2004, 2009). The active part of the syndecan core protein corresponds to amino acids 88–121 of the mature protein, and the corresponding peptide (termed *synstatin*) can inhibit processes such as tumor growth and development through inhibition of angiogenesis. Recent work also suggests a requirement for syndecan-1 core protein with the insulin-like growth factor-1 receptor in inside-out integrin activation (Beauvais and Rapraeger, 2010). This concept of receptor cross talk leading to alterations in integrin-mediated responses deserves further attention and may reveal targets of therapeutic value.

5.5.3 Syndecans as receptors mediating endocytosis

Vitronectin is an abundant glycoprotein found both in serum and the ECM. One-third of the mature form of vitronectin is composed of carbohydrate, and it exists either in monomeric form or multimeric form. Contained within the N-terminal portion of the molecule is an RGD site, and this confers cell adhesive properties mediated largely by αV integrins (Charo et al., 1990). A heparin-binding domain has been identified toward the C terminus of vitronectin, which is known to interact with syndecan-1, -2, and -4 and can support cell adhesion. This interaction facilitates receptor-mediated endocytosis and the clearance of vitronectin (Wilkins-Port et al., 2003).

Syndecans serving as clearance receptors is recapitulated in the murine liver. Genetic evidence shows that syndecan-1 is the major HS proteoglycan (HSPG) required for uptake of triglyceride-rich lipoproteins (Stanford et al., 2009). Syndecan-1 is a major hepatocyte HSPG, and its HS chains are essential for this role in endocytic clearance of plasma lipoproteins (▶Figure 5.8). This ability of syndecans to mediate endocytosis has, on the other hand, been conscripted by a variety of pathogens to gain entry into tissues (▶Figure 5.8). Syndecans may mediate entry of *Neisseria gonorrhoeae, Listeria monocytogenes,* and *Orientia tsutsugamushi* (scrub typhus) and possibly *Helicobacter pylori. Pseudomonas aeruginosa* pathogenesis and human immunodeficiency and other virus internalization may also require HS and syndecans (see review by Bartlett and Park, 2010).

5.5.4 Syndecans as receptors for growth factors and chemokines

Numerous growth factor signaling pathways control aspects of embryonic development, homeostasis, and various pathologies. Most growth factors are secreted proteins, which interact with their corresponding high-affinity cell surface receptor kinases with subsequent downstream signaling. During development, some secreted growth factors, like Hedgehog (Hh), Wnt, and bone morphogenetic proteins (BMPs), form concentration gradients to specify different cell fates. Many growth factors, interleukins, and morphogens have heparin/HS-binding properties (▶Table 5.5), and it has been known for some time that HSPGs regulate both morphogen gradient formation and the downstream signaling pathways (▶Figure 5.8).

Fibroblast growth factors

Fibroblast growth factors (FGFs) and their receptor tyrosine kinases (FGFRs) comprise a complex signaling system, which regulates a wide range of cellular functions and behavior, with key roles in development and homeostasis. FGF binding to its receptor (FGFR) leads to receptor dimerization, which triggers (trans-) autophosphorylation of the FGFR intracellular domain. The interaction between FGFs and HS is well documented. Most of the secreted FGFs have high affinity for HS/heparin (Asada et al., 2009). However, both FGF and FGFR can bind HS through specific sites, forming ternary complexes. The formation of this complex significantly lowers the concentration of FGF needed for signaling and receptor dimerization. Crystal structures reveal two models of FGF-FGFR complexed with fully sulfated heparin decasaccharides (Pellegrini et al., 2000; Schlessinger et al., 2000; Harmer et al., 2004; Robinson et al., 2005). The fibroblast growth factor-1 (FGF1):fibroblast growth factor receptor-2 (FGFR2):HS model yielded an asymmetric stoichiometry of 2:2:1 (Schlessinger et al., 2000), whereas the fibroblast growth factor-2 (FGF2):fibroblast growth factor receptor-1 (FGFR1):HS complex showed a 2:2:2 symmetrical stoichiometry (Pellegrini et al., 2000; Schlessinger et al., 2000). However, the methods were not identical in these two cases. Size exclusion chromatography of complexes in solution suggests that the 2:2:1 stoichiometry is favored with highly sulfated oligosaccharides of ≥8 sugar units. In contrast, selectively desulfated heparin 12-mers bound FGF1 monomerically and formed 1:1:1 of FGF1:FGFR2c:heparin ternary complexes (Harmer et al., 2004; Robinson et al., 2005). Goodger et al. (2008) suggested that both structures are equally stable in solution, although 2:1 FGF:heparin complexes were highly favored over 1:1 complexes (Goodger et al., 2008). A recent study identified nonreducing-end *N*-sulfated domains as the minimal unit of HS for FGF2 binding, which favored the symmetrical architecture of the complex (Naimy et al., 2011).

HS chains from many different proteoglycans can bind FGF. However, at least in endothelial cells, syndecan-4 signaling is a key component of FGF signaling. Experiments with syndecan core protein chimeras and syndecan-4 deletion mutants demonstrate that the syndecan-4 cytoplasmic domain can be required for FGF2 signaling. Expression of syndecan-4 with mutations in the V region (phosphoinsotide binding) or C2 (PDZ binding) region indicate that both are required for FGF signaling (Horowitz et al., 2002; Volk et al., 1999). The absence of the PDZ-binding C2 region of syndecan-4 core protein prevents interaction with Gα-interacting protein [GAIP]-interacting protein C terminus (GIPC)/synectin. In endothelial cells, this is accompanied by a failure to activate Rac1 and defective migration (Tkachenko et al., 2006). A study of transgenic mice overexpressing syndecan-4 in cardiac endothelial cells, in which FGF2 induced nitric oxide release (Zhang et al., 2003), also suggested a key role for this interaction with GIPC/synectin (Dedkov et al., 2007). In *Xenopus*, the importance of the syndecan-4/protein kinase C-alpha (PKC-α) interaction for FGF signaling in neural induction was observed, applicable to both PKC-α and extracellular signal-regulated kinase (ERK)-dependent pathways (Kuriyama and Mayor, 2009).

Syndecans are also involved in FGF-regulated processes during embryonic development. Syndecan-4, but not syndecan-1, responds to FGF1, FGF2, and fibroblast growth factor-7 (FGF7) during mouse vibrissae growth, with the result that syndecan-4–null mice possess shorter and smaller vibrissae (Iwabuchi and Goetinck, 2006). Other syndecans have also been implicated in FGF signaling. Overexpression of syndecan-1 in myoblasts results in a dramatic increase in sensitivity toward FGF2-dependent inhibition of myogenin expression (Larrain et al., 1998). Proteomic studies with multiple myeloma cells involving the activation of fibroblast growth factor receptor-3 (FGFR3) with FGF1 identified syndecan-1 as having a role (St-Germain et al., 2009). Reexpression of syndecan-1 in murine mammary carcinoma cells decreased fibroblast growth factor 8b (FGF8b)-induced ERK phosphorylation, therefore limiting cellular responses to FGF receptor activation (Viklund et al., 2006). In activated human macrophages, syndecan-2, as the only syndecan expressed at the cell surface, bound FGF2 and transactivated FGFR1 on neighboring cells (Clasper et al., 1999). Syndecan-3 has also been implicated in FGF signaling during chondrocyte proliferation (Kirsch et al., 2002) and in FGF-mediated limb bud formation in chicken embryos (Dealy et al., 1997). With the exception of syndecan-4, however, little is known about how syndecans convert FGF signals into cellular responses. This is an area in which further studies are needed.

Wnt and Hedgehog signaling

Wnt morphogens play a pivotal role in development and disease through interactions with their Frizzled (Fz) receptors. Wnts can transduce their signals through either a "canonical" (β-catenin-dependent pathway) or "noncanonical" pathways (planar cell polarity [PCP] pathway or Wnt/Ca^{2+} pathway) (Wodarz and Nusse, 1998; Veeman et al., 2003). Hedgehog (Hh) is a secreted signaling molecule, which forms concentration gradients from sites of their secretion and regulates tissue patterning during development (Beachy et al., 2010). Hh signaling is tightly regulated in a spatial and temporal manner but is frequently upregulated in diseases such as pancreatic adenocarcinoma and basal cell carcinoma of the skin. There are numerous studies in model organisms in which either HS synthesis is ablated or the sulfation patterns of HS chains are radically altered. This has shown a key role for HS in Wnt and Hh signaling and, in particular, in the establishment of morphogen gradients (Hacker et al., 2005).

A major HSPG coreceptor for Wnt and Hh signaling is glypican, but the involvement of syndecan core proteins in Wnt signaling has been demonstrated by syndecan-4 depletion in *Xenopus* embryos. Here, syndecan-4 interacts with the Wnt receptor Frizzled7 and its downstream signal transducer Dishevelled. This leads to activation of the noncanonical Wnt signaling pathway (Munoz et al., 2006). In further studies, it was demonstrated that syndecan-4 plays a key role in PCP signaling, which underpins directional migration of neural crest cells through Rho family small GTPases (Matthews et al., 2008).

Transforming growth factor-beta signaling

The TGF-β superfamily comprises a large number of different polypeptides, including the TGF-β polypeptides themselves, as well as BMPs (e.g. bone morphogenetic protein-2 [BMP2] and bone morphogenetic protein-4 [BMP4]), and many other growth and differentiation factors. There are three mammalian TGF-β isoforms that signal through the TGF-β receptors I, II, and III (TGF-βRI, TGF-βRII, and TGF-βRIII, respectively).

Upon binding to TGF-β, TGF-βRII transphosphorylates TGF-βRI and TGF-βRIII (betaglycan) and, in turn, activates downstream targets (Massague et al., 1992). Syndecan-2 is reported to regulate TGF-β signaling accompanied by a direct interaction between the syndecan-2 extracellular core protein and TGF-β (Chen et al., 2004b). Transfection of renal papillary fibroblasts with syndecan-2 promotes TGF-β-mediated matrix deposition. Surface levels of TGF-βRI and TGF-βRII are also increased. However, a cytoplasmically truncated form of syndecan-2 does not support such effects, suggesting a role for signaling.

The involvement of HS in BMP signaling has long been suspected because BMPs were originally isolated using heparin affinity techniques (Cool and Nurcombe, 2006). Biochemical studies in *Xenopus* suggest that a basic core of only three amino acids in the N-terminal region of BMP4 is responsible for binding to HSPGs, with retention at the cell surface (Ohkawara et al., 2002). The *Drosophila* decapentaplegic (*Dpp*) gene, which is the functional ortholog of vertebrate BMP2 and BMP4, confirms the participation of HS in Dpp signaling. Although the major regulator for Dpp signaling is glypican, more recent studies in *Xenopus* reveal a role for syndecan in BMP signaling. The effect of syndecan-1 on BMP signaling is biphasic (i.e. BMP is inactive at high syndecan-1 concentrations and is activated at low syndecan-1 concentrations). More interestingly, the effect of syndecan-1 on dorsoventral patterning and BMP signaling is specific because knockdown of syndecan-4 has no effect (Olivares et al., 2009). Overexpression of syndecan-3 also impairs the ability of BMP2 to promote cartilage differentiation in the chicken limb bud. BMP2-mediated Smad phosphorylation is disrupted by syndecan-3, suggesting impaired interactions between BMP2 and its high-affinity receptors (Fisher et al., 2006).

Vascular endothelial growth factor and platelet-derived growth factor

Vascular endothelial growth factor (VEGF) and platelet-derived growth factor (PDGF) belong to the same family of growth factors, and the VEGFs are key regulators of vascular and lymphatic development and function (Grunewald et al., 2010). PDGF is a potent mitogen for mesenchymal cells such as fibroblasts and smooth muscle cells. During vascular development, VEGFs and PDGFs regulate the migration and proliferation of endothelial cells and their supporting mural cells (i.e. pericytes and vascular smooth muscle cells [vSMCs]) (Andrae et al., 2008). Among all VEGF-A isoforms, the interaction of VEGF-A165 with HS is the most studied, for which both *N*-sulfation and 6-*O*-sulfation of the HS are crucial (Robinson et al., 2006). Interaction between HS and tyrosine receptor kinase of VEGF (VEGFR) has been shown, but it is not yet clear if HS, VEGF, and VEGFR can form a complex (Grunewald et al., 2010). Genetic analysis in zebra fish indicates that the VEGF-A165 interacts with syndecan-2, since ectopic expression of the proteoglycan potentiates VEGF-A165–induced neovascularization (Chen et al., 2004a).

Both PDGF-A and PDGF-B possess a positively charged C-terminal motif (referred to as retention motif). Several studies demonstrate the interaction between the retention motif of PDGF and HS chains (Zafiropoulos et al., 2008). This interaction is mediated via *N*-sulfated saccharide domains containing both 2-*O*- and 6-*O*-sulfate groups (Feyzi et al., 1997). Syndecan-1–null mice develop large neointimal lesions after carotid injury, while syndecan-1–null vSMCs exhibit increased proliferation and migration in response to PDGF-BB compared to wild-type cells. This is probably related to the increased expression level of PDGF-B in syndecan-1–null cells (Fukai et al., 2009). Syndecan-1 may therefore act as a suppressor of PDGF signaling during wound healing. Another study with rat embryonic fibroblasts (REFs) shows that syndecan-4 can mediate PDGF-induced reactive oxygen species (ROS) generation and mitogen-activated protein (MAP) kinase activation (Kim et al., 2008).

Syndecans in inflammation

Acute inflammation involves two major aspects: vascular changes (increased blood flow and vascular permeability) and cellular events. Cellular events occur in complex multistep stages mediated by a variety of cytokines and chemokines. An essential feature of these is rapid recruitment of leukocytes from the bloodstream to the site of inflammation. HS has been demonstrated to participate in almost all steps of leukocyte recruitment and activation, and evidence for roles of syndecans, as opposed to other HSPGs such as glypicans, is accumulating (Götte, 2003).

Syndecan-1–deficient mice exhibit increased inflammatory responses and enhanced leukocyte–endothelial cell interactions, indicating that syndecan-1 is a negative regulator of leukocyte–endothelial cell adhesion (Götte et al., 2002, 2005). The impact of syndecans in maintaining chemokine gradients may be mediated by syndecan shedding. Matrix metalloproteinase-7 (MMP7)–mediated shedding of syndecan-1 ectodomain generates a transepithelial gradient of keratinocyte-derived chemokine (KC), an CXC chemokine. Shed syndecan-1/KC complexes from the mucosal surface direct and confine the neutrophil influx to sites of injury (Li et al., 2002). Further studies have shown that syndecan-1 shedding was activated in systemic inflammation and that shed syndecan-1 functions in removal of tissue-bound CXC chemokines and resolution of accumulated neutrophils (Hayashida et al., 2009). During inflammation, the expression level of syndecans can be regulated by chemokines. For instance, syndecan-2 expression is induced during differentiation of monocytes into macrophages (Clasper et al., 1999), whereas high levels of syndecan-1 are selectively expressed when B cells differentiate into plasma cells (Götte, 2003).

5.5.5 Perspective – specificity of syndecans and their signaling responses

The list of HS-interacting growth factors, cytokines, chemokines, and morphogens continues to grow. However, much has to be learned regarding which proteoglycans are involved and when. Redundancy seems to be a key issue in mammals, in which deletion of genes encoding vital steps in HS synthesis often leads to more severe phenotypes than those of syndecan core proteins. In invertebrates and lower vertebrates, developmental defects are apparent where syndecan expression is suppressed or mutant forms are expressed. In mammals, however, it is unclear where other cell surface HSPGs such as glypicans, neuropilin-1, or CD44 can replace a syndecan.

Techniques that permit visualization of syndecan-ligand interaction *in vivo* have not been developed, and similarly, it is not clear under what circumstances syndecans signal independently from, or linked to, that of high-affinity coreceptors with which they associate. Syndecan signaling is considered elsewhere in this volume but is known to occur in response to interactions with ECM molecules (Couchman, 2010). In the examples given previously, for example, the cytoplasmic domain of syndecan-1 is required for fascin spike formation in response to thrombospondin-1, while angiogenesis defects in zebra fish embryos (as a result of syndecan-2 depletion) can be restored only by forms of the syndecans bearing a cytoplasmic domain. It is also well known that adhesion responses to fibronectin are in part mediated by the cytoplasmic domain of syndecan-4. A common denominator seems to be linkage between all syndecan cytoplasmic domains and the actin cytoskeleton. At least for some FGF-syndecan interactions in vascular tissue, signaling through syndecan-4 can follow (Dedkov et al., 2007; Murakami et al., 2008; Kuriyama and Mayor, 2009). HS is involved in many inflammatory and pathological processes; however, when syndecans have specific roles, and whether their cytoplasmic domains are essential, remains to be elucidated.

As described elsewhere in this volume, syndecans can be cleaved in membrane proximal domains by a variety of cell surface proteinases. This can liberate an almost intact ectodomain bearing its GAG contingent, which may compete with transmembrane proteoglycans or have entirely separate functions. Since inflammation, vascular disease, and particularly tumor microenvironments may harbor elevated proteinase levels, this aspect of syndecan biology is likely to be important. In summary, syndecans can function as coreceptors but also have a capacity to signal independently in response to ligands. Syndecans appear to have specific signaling that is syndecan-type specific, but also generic linkage to the actin-associated cytoskeleton. It seems likely that further generic functions common to all syndecans await description.

5.5.6 Take-home message

Syndecans interact with multiple pericellular molecules, including ECM components, numerous growth factors, and related proteins, mainly through their extracellular HS chains. While collaborating with other cell surface receptors, syndecans influence various signaling pathways involved in cell adhesion, migration, and proliferation, and they may also impact development and disease pathogenesis.

References

Alexopoulou, A. N., Multhaupt, H. A., and Couchman, J. R. (2007). Syndecans in wound healing, inflammation and vascular biology. Int J Biochem Cell Biol *39*, 505–528.

Andrae, J., Gallini, R., and Betsholtz, C. (2008). Role of platelet-derived growth factors in physiology and medicine. Genes Dev *22*, 1276–1312.

Asada, M., Shinomiya, M., Suzuki, M., et al. (2009). Glycosaminoglycan affinity of the complete fibroblast growth factor family. Biochim Biophys Acta *1790*, 40–48.

Bachy, S., Letourneur, F., and Rousselle, P. (2008). Syndecan-1 interaction with the LG4/5 domain in laminin-332 is essential for keratinocyte migration. J Cell Physiol *214*, 238–249.

Bartlett, A. H., and Park, P. W. (2010). Proteoglycans in host-pathogen interactions: molecular mechanisms and therapeutic implications. Expert Rev Mol Med *12*, e5.

Bass, M. D., Morgan, M. R., Roach, K. A., Settleman, J., Goryachev, A. B., and Humphries, M. J. (2008). p190RhoGAP is the convergence point of adhesion signals from alpha 5 beta 1 integrin and syndecan-4. J Cell Biol *181*, 1013–1026.

Bax, D. V., Bernard, S. E., Lomas, A., et al. (2003). Cell adhesion to fibrillin-1 molecules and microfibrils is mediated by alpha 5 beta 1 and alpha v beta 3 integrins. J Biol Chem *278*, 34605–34616.

Bax, D. V., Mahalingam, Y., Cain, S., et al. (2007). Cell adhesion to fibrillin-1: identification of an Arg-Gly-Asp-dependent synergy region and a heparin-binding site that regulates focal adhesion formation. J Cell Sci *120*, 1383–1392.

Beachy, P. A., Hymowitz, S. G., Lazarus, R. A., Leahy, D. J., and Siebold, C. (2010). Interactions between Hedgehog proteins and their binding partners come into view. Genes Dev *24*, 2001–2012.

Beauvais, D. M., Burbach, B. J., and Rapraeger, A. C. (2004). The syndecan-1 ectodomain regulates alphavbeta3 integrin activity in human mammary carcinoma cells. J Cell Biol *167*, 171–181.

Beauvais, D. M., Ell, B. J., McWhorter, A. R., and Rapraeger, A. C. (2009). Syndecan-1 regulates alphavbeta3 and alphavbeta5 integrin activation during angiogenesis and is blocked by synstatin, a novel peptide inhibitor. J Exp Med *206*, 691–705.

Beauvais, D. M., and Rapraeger, A. C. (2010). Syndecan-1 couples the insulin-like growth factor-1 receptor to inside-out integrin activation. J Cell Sci *123*, 3796–3807.

Cain, S. A., Baldock, C., Gallagher, J., et al. (2005). Fibrillin-1 interactions with heparin. Implications for microfibril and elastic fiber assembly. J Biol Chem *280*, 30526–30537.

Cain, S. A., Baldwin, A. K., Mahalingam, Y., et al. (2008). Heparan sulfate regulates fibrillin-1 N- and C-terminal interactions. J Biol Chem *283*, 27017–27027.

Chakravarti, R., Sapountzi, V., and Adams, J. C. (2005). Functional role of syndecan-1 cytoplasmic V region in lamellipodial spreading, actin bundling, and cell migration. Mol Biol Cell *16*, 3678–3691.

Charo, I. F., Nannizzi, L., Smith, J. W., and Cheresh, D. A. (1990). The vitronectin receptor alpha v beta 3 binds fibronectin and acts in concert with alpha 5 beta 1 in promoting cellular attachment and spreading on fibronectin. J Cell Biol *111*, 2795–2800.

Chen, E., Hermanson, S., and Ekker, S. C. (2004a). Syndecan-2 is essential for angiogenic sprouting during zebrafish development. Blood *103*, 1710–1719.

Chen, L., Klass, C., and Woods, A. (2004b). Syndecan-2 regulates transforming growth factor-beta signaling. J Biol Chem *279*, 15715–15718.

Chiquet-Ehrismann, R. (2004). Tenascins. Int J Biochem Cell Biol *36*, 986–990.

Chiquet-Ehrismann, R., and Chiquet, M. (2003). Tenascins: regulation and putative functions during pathological stress. J Pathol *200*, 488–499.

Clasper, S., Vekemans, S., Fiore, M., et al. (1999). Inducible expression of the cell surface heparan sulfate proteoglycan syndecan-2 (fibroglycan) on human activated macrophages can regulate fibroblast growth factor action. J Biol Chem *274*, 24113–24123.

Cool, S. M., and Nurcombe, V. (2006). Heparan sulfate regulation of progenitor cell fate. J Cell Biochem *99*, 1040–1051.

Couchman, J. R. (2003). Syndecans: proteoglycan regulators of cell-surface microdomains? Nat Rev Mol Cell Biol *4*, 926–937.

Couchman, J. R. (2010). Transmembrane signaling proteoglycans. Annu Rev Cell Dev Biol *26*, 89–114.

Dealy, C. N., Seghatoleslami, M. R., Ferrari, D., and Kosher, R. A. (1997). FGF-stimulated outgrowth and proliferation of limb mesoderm is dependent on syndecan-3. Dev Biol *184*, 343–350.

Dedkov, E. I., Thomas, M. T., Sonka, M., et al. (2007). Synectin/syndecan-4 regulate coronary arteriolar growth during development. Dev Dyn *236*, 2004–2010.

Esko, J. D., and Selleck, S. B. (2002). Order out of chaos: assembly of ligand binding sites in heparan sulfate. Annu Rev Biochem *71*, 435–471.

Feyzi, E., Lustig, F., Fager, G., Spillmann, D., Lindahl, U., and Salmivirta, M. (1997). Characterization of heparin and heparan sulfate domains binding to the long splice variant of platelet-derived growth factor A chain. J Biol Chem *272*, 5518–5524.

Fisher, M. C., Li, Y., Seghatoleslami, M. R., Dealy, C. N., and Kosher, R. A. (2006). Heparan sulfate proteoglycans including syndecan-3 modulate BMP activity during limb cartilage differentiation. Matrix Biol *25*, 27–39.

Fukai, N., Kenagy, R. D., Chen, L., Gao, L., Daum, G., and Clowes, A. W. (2009). Syndecan-1: an inhibitor of arterial smooth muscle cell growth and intimal hyperplasia. Arterioscler Thromb Vasc Biol *29*, 1356–1362.

Goodger, S. J., Robinson, C. J., Murphy, K. J., et al. (2008). Evidence that heparin saccharides promote FGF2 mitogenesis through two distinct mechanisms. J Biol Chem *283*, 13001–13008.

Gopal, S., Bober, A., Whiteford, J. R., Multhaupt, H. A., Yoneda, A., and Couchman, J. R. (2010). Heparan sulfate chain valency controls syndecan-4 function in cell adhesion. J Biol Chem *285*, 14247–14258.

Götte, M. (2003). Syndecans in inflammation. FASEB J *17*, 575–591.

Götte, M., Bernfield, M., and Joussen, A. M. (2005). Increased leukocyte-endothelial interactions in syndecan-1-deficient mice involve heparan sulfate-dependent and -independent steps. Curr Eye Res *30*, 417–422.

Götte, M., Joussen, A. M., Klein, C., et al. (2002). Role of syndecan-1 in leukocyte-endothelial interactions in the ocular vasculature. Invest Ophthalmol Vis Sci *43*, 1135–1141.

Grunewald, F. S., Prota, A. E., Giese, A., and Ballmer-Hofer, K. (2010). Structure-function analysis of VEGF receptor activation and the role of coreceptors in angiogenic signaling. Biochim Biophys Acta *1804*, 567–580.

Hacker, U., Nybakken, K., and Perrimon, N. (2005). Heparan sulphate proteoglycans: the sweet side of development. Nat Rev Mol Cell Biol *6*, 530–541.

Harmer, N. J., Ilag, L. L., Mulloy, B., Pellegrini, L., Robinson, C. V., and Blundell, T. L. (2004). Towards a resolution of the stoichiometry of the fibroblast growth factor (FGF)-FGF receptor-heparin complex. J Mol Biol *339*, 821–834.

Hayashida, K., Parks, W. C., and Park, P. W. (2009). Syndecan-1 shedding facilitates the resolution of neutrophilic inflammation by removing sequestered CXC chemokines. Blood *114*, 3033–3043.

Horowitz, A., Tkachenko, E., and Simons, M. (2002). Fibroblast growth factor-specific modulation of cellular response by syndecan-4. J Cell Biol *157*, 715–725.

Isenberg, J. S., Frazier, W. A., and Roberts, D. D. (2008). Thrombospondin-1: a physiological regulator of nitric oxide signaling. Cell Mol Life Sci *65*, 728–742.

Iwabuchi, T., and Goetinck, P. F. (2006). Syndecan-4 dependent FGF stimulation of mouse vibrissae growth. Mech Dev *123*, 831–841.

Kazerounian, S., Yee, K. O., and Lawler, J. (2008). Thrombospondins in cancer. Cell Mol Life Sci *65*, 700–712.

Kielty, C. M., Sherratt, M. J., Marson, A., and Baldock, C. (2005). Fibrillin microfibrils. Adv Protein Chem *70*, 405–436.

Kim, J., Lee, J. H., Park, H. S., et al. (2008). Syndecan-4 regulates platelet-derived growth factor-mediated MAP kinase activation by altering intracellular reactive oxygen species. FEBS Lett *582*, 2725–2730.

Kirsch, T., Koyama, E., Liu, M., Golub, E. E., and Pacifici, M. (2002). Syndecan-3 is a selective regulator of chondrocyte proliferation. J Biol Chem *277*, 42171–42177.

Klass, C. M., Couchman, J. R., and Woods, A. (2000). Control of extracellular matrix assembly by syndecan-2 proteoglycan. J Cell Sci *113*, 493–506.

Kreuger, J., Spillmann, D., Li, J. P., and Lindahl, U. (2006). Interactions between heparan sulfate and proteins: the concept of specificity. J Cell Biol *174*, 323–327.

Kuriyama, S., and Mayor, R. (2009). A role for syndecan-4 in neural induction involving ERK- and PKC-dependent pathways. Development *136*, 575–584.

Lander, A. D. (1994). Targeting the glycosaminoglycan-binding sites on proteins. Chem Biol *1*, 73–78.

Larrain, J., Carey, D. J., and Brandan, E. (1998). Syndecan 1 expression inhibits myoblast differentiation through a basic fibroblast growth factor-dependent mechanism. J Biol Chem *273*, 32288–32296.

Li, Q., Park, P. W., Wilson, C. L., and Parks, W. C. (2002). Matrilysin shedding of syndecan-1 regulates chemokine mobilization and transepithelial efflux of neutrophils in acute lung injury. Cell *111*, 635–646.

Mahalingam, Y., Gallagher, J. T., and Couchman, J. R. (2007). Cellular adhesion responses to the heparin-binding (HepII) domain of fibronectin require heparan sulfate with specific properties. J Biol Chem *282*, 3221–3230.

Manon-Jensen, T., Itoh, Y., and Couchman, J. R. (2010). Proteoglycans in health and disease: the multiple roles of syndecan shedding. FEBS J *277*, 3876–3889.

Massague, J., Andres, J., Attisano, L., et al. (1992). TGF-beta receptors. Mol Reprod Dev *32*, 99–104.

Matthews, H. K., Marchant, L., Carmona-Fontaine, C., et al. (2008). Directional migration of neural crest cells in vivo is regulated by syndecan-4/Rac1 and non-canonical Wnt signaling/RhoA. Development *135*, 1771–1780.

McFall, A. J., and Rapraeger, A. C. (1998). Characterization of the high affinity cell-binding domain in the cell surface proteoglycan syndecan-4. J Biol Chem *273*, 28270–28276.

Midwood, K. S., and Orend, G. (2009). The role of tenascin-C in tissue injury and tumorigenesis. J Cell Commun Signal *3*, 287–310.

Midwood, K. S., Valenick, L. V., Hsia, H. C., and Schwarzbauer, J. E. (2004). Coregulation of fibronectin signaling and matrix contraction by tenascin-C and syndecan-4. Mol Biol Cell *15*, 5670–5677.

Munoz, R., Moreno, M., Oliva, C., Orbenes, C., and Larrain, J. (2006). Syndecan-4 regulates non-canonical Wnt signalling and is essential for convergent and extension movements in Xenopus embryos. Nat Cell Biol *8*, 492–500.

Murakami, M., Elfenbein A., and Simons M. (2008). Non-canonical fibroblast growth factor signalling in angiogenesis. Cardiovasc Res *78*, 223–231.

Naimy, H., Buczek-Thomas, J. A., Nugent, M. A., Leymarie, N., and Zaia, J. (2011). Highly sulfated nonreducing end-derived heparan sulfate domains bind fibroblast growth factor-2 with high affinity and are enriched in biologically active fractions. J Biol Chem *286*, 19311–19319.

Nunes, S. S., Outeiro-Bernstein, M. A., Juliano, L., et al. (2008). Syndecan-4 contributes to endothelial tubulogenesis through interactions with two motifs inside the pro-angiogenic N-terminal domain of thrombospondin-1. J Cell Physiol *214*, 828–837.

Ogawa, T., Tsubota, Y., Hashimoto, J., Kariya, Y., and Miyazaki, K. (2007). The short arm of laminin gamma2 chain of laminin-5 (laminin-332) binds syndecan-1 and regulates cellular adhesion and migration by suppressing phosphorylation of integrin beta4 chain. Mol Biol Cell *18*, 1621–1633.

Ohkawara, B., Iemura, S., ten Dijke, P., and Ueno, N. (2002). Action range of BMP is defined by its N-terminal basic amino acid core. Curr Biol *12*, 205–209.

Okina, E., Manon-Jensen, T., Whiteford, J. R., and Couchman, J. R. (2009). Syndecan proteoglycan contributions to cytoskeletal organization and contractility. Scand J Med Sci Sports *19*, 479–489.

Olivares, G. H., Carrasco, H., Aroca, F., Carvallo, L., Segovia, F., and Larrain, J. (2009). Syndecan-1 regulates BMP signaling and dorso-ventral patterning of the ectoderm during early Xenopus development. Dev Biol *329*, 338–349.

Orend, G., and Chiquet-Ehrismann, R. (2006). Tenascin-C induced signaling in cancer. Cancer Lett *244*, 143–163.

Orend, G., Huang, W., Olayioye, M. A., Hynes, N. E., and Chiquet-Ehrismann, R. (2003). Tenascin-C blocks cell-cycle progression of anchorage-dependent fibroblasts on fibronectin through inhibition of syndecan-4. Oncogene *22*, 3917–3926.

Pellegrini, L., Burke, D. F., von Delft, F., Mulloy, B., and Blundell, T. L. (2000). Crystal structure of fibroblast growth factor receptor ectodomain bound to ligand and heparin. Nature *407*, 1029–1034.

Robinson, C. J., Harmer, N. J., Goodger, S. J., Blundell, T. L., and Gallagher, J. T. (2005). Cooperative dimerization of fibroblast growth factor 1 (FGF1) upon a single heparin saccharide may drive the formation of 2:2:1 FGF1.FGFR2c.heparin ternary complexes. J Biol Chem *280*, 42274–42282.

Robinson, P. N., Arteaga-Solis, E., Baldock, C., et al. (2006). The molecular genetics of Marfan syndrome and related disorders. J Med Genet *43*, 769–787.

Salmivirta, M., Elenius, K., Vainio, S., et al. (1991). Syndecan from embryonic tooth mesenchyme binds tenascin. J Biol Chem *266*, 7733–7739.

Schlessinger, J., Plotnikov, A. N., Ibrahimi, O. A., et al. (2000). Crystal structure of a ternary FGF-FGFR-heparin complex reveals a dual role for heparin in FGFR binding and dimerization. Mol Cell *6*, 743–750.

St-Germain, J. R., Taylor, P., Tong, J., et al. (2009). Multiple myeloma phosphotyrosine proteomic profile associated with FGFR3 expression, ligand activation, and drug inhibition. Proc Natl Acad Sci U S A *106*, 20127–20132.

Stanford, K. I., Bishop, J. R., Foley, E. M., et al. (2009). Syndecan-1 is the primary heparan sulfate proteoglycan mediating hepatic clearance of triglyceride-rich lipoproteins in mice. J Clin Invest *119*, 3236–3245.

Sugawara, K., Tsuruta, D., Ishii, M., Jones, J. C., and Kobayashi, H. (2008). Laminin-332 and -511 in skin. Exp Dermatol *17*, 473–480.

Sweeney, S. M., Orgel, J. P., Fertala, A., et al. (2008). Candidate cell and matrix interaction domains on the collagen fibril, the predominant protein of vertebrates. J Biol Chem *283*, 21187–21197.

Taylor, K. R., and Gallo, R. L. (2006). Glycosaminoglycans and their proteoglycans: host-associated molecular patterns for initiation and modulation of inflammation. FASEB J *20*, 9–22.

Tkachenko, E., Elfenbein, A., Tirziu, D., and Simons, M. (2006). Syndecan-4 clustering induces cell migration in a PDZ-dependent manner. Circ Res *98*, 1398–1404.

Veeman, M. T., Axelrod, J. D., and Moon, R. T. (2003). A second canon. Functions and mechanisms of beta-catenin-independent Wnt signaling. Dev Cell *5*, 367–377.

Viklund, L., Vorontsova, N., Henttinen, T., and Salmivirta, M. (2006). Syndecan-1 regulates FGF8b responses in S115 mammary carcinoma cells. Growth Factors *24*, 151–157.

Volk, R., Schwartz, J. J., Li, J., Rosenberg, R. D., and Simons, M. (1999). The role of syndecan cytoplasmic domain in basic fibroblast growth factor-dependent signal transduction. J Biol Chem *274*, 24417–24424.

Vuoriluoto, K., Jokinen, J., Kallio, K., Salmivirta, M., Heino, J., and Ivaska, J. (2008). Syndecan-1 supports integrin alpha2beta1-mediated adhesion to collagen. Exp Cell Res *314*, 3369–3381.

Whiteford, J. R., Behrends, V., Kirby, H., Kusche-Gullberg, M., Muramatsu, T., and Couchman, J. R. (2007). Syndecans promote integrin-mediated adhesion of mesenchymal cells in two distinct pathways. Exp Cell Res *313*, 3902–3913.

Whiteford, J. R., Xian, X., Chaussade, C., Vanhaesebroeck, B., Nourshargh, S., and Couchman, J. R. (2011). Syndecan-2 is a novel ligand for the protein tyrosine phosphatase receptor CD148. Mol Biol Cell *22*, 3609–3624.

Wilkins-Port, C. E., Sanderson, R. D., Tominna-Sebald, E., and McKeown-Longo, P. J. (2003). Vitronectin's basic domain is a syndecan ligand which functions in trans to regulate vitronectin turnover. Cell Commun Adhes *10*, 85–103.

Williams, S. A., and Schwarzbauer, J. E. (2009). A shared mechanism of adhesion modulation for tenascin-C and fibulin-1. Mol Biol Cell *20*, 1141 1149.

Wodarz, A., and Nusse, R. (1998). Mechanisms of Wnt signaling in development. Annu Rev Cell Dev Biol *14*, 59–88.

Xian, X., Gopal, S., and Couchman, J. R. (2010). Syndecans as receptors and organizers of the extracellular matrix. Cell Tissue Res *339*, 31–46.

Yamashita, H., Goto, A., Kadowaki, T., and Kitagawa, Y. (2004). Mammalian and Drosophila cells adhere to the laminin alpha4 LG4 domain through syndecans, but not glypicans. Biochem J *382*, 933–943.

Zafiropoulos, A., Fthenou, E., Chatzinikolaou, G., and Tzanakakis, G. N. (2008). Glycosaminoglycans and PDGF signaling in mesenchymal cells. Connect Tissue Res *49*, 153–156.

Zhang, Y., Li, J., Partovian, C., Sellke, F. W., and Simons, M. (2003). Syndecan-4 modulates basic fibroblast growth factor 2 signaling in vivo. Am J Physiol Heart Circ Physiol *284*, H2078–H2082 .

5.6 CD44: a Sensor of tissue damage critical for restoring homeostasis

Ellen Puré

5.6.1 Introduction

Cells and tissues must adapt to a constantly changing microenvironment through homeostatic mechanisms to maintain the integrity of the organism. Perturbations of tissue integrity due to microbial infection, sterile inflammation, physical injury, or tissue dysfunction lead to the activation of host defense mechanisms involving immune, inflammatory, and fibrotic responses that are designed to eliminate the infectious agent and intrinsic instigators (typically excess or modified self components) that accumulate due to injury, organ dysfunction, or neoplastic transformation. Ultimately, these reparative responses evolve to restore homeostasis through the elimination of the instigating stimuli and through engagement of intrinsic regulatory mechanisms that limit the extent and duration of the host defense response. Failure to restore the system to homeostasis can result in organ damage and disease.

Parenchymal cells and resident leukocytes act as sentinels that detect perturbations in the tissue microenvironment. These sentinel cells are activated by microbial products and/or the accumulation of modified forms or excess amounts of endogenous components through receptor-mediated signal transduction and mechanotransduction. Consequently, resident cells initiate host defense mechanisms and the recruitment of circulating and bone marrow–derived inflammatory and immune cells. Microbial products engage so-called pattern recognition receptors (PRRs) such as Toll-like receptors (TLRs), but numerous endogenous ligands of TLRs have also been identified, including oxidized lipids, stress response proteins, and modified matrix components. A number of other cell surface receptors, including CD44, are also now recognized as sensors that convey the status of the microenvironment, including changes in the extracellular matrix (ECM).

CD44 is a cell surface receptor for hyaluronan (HA), a major glycosaminoglycan (GAG) of peri- and ECMs that is expressed by both parenchymal cells and leukocytes. In this chapter, we discuss the complex but elegant way by which the structure and function of CD44 is regulated in a cell context and activation-dependent fashion and how this, together with differential CD44-mediated signaling by the distinct forms of HA present under normal conditions versus those which accumulate at sites of injury, inflammation, and fibrosis and in tumors, defines the role of CD44 in homeostasis, the response to tissue damage, resolution of the inflammatory response, and restoration of homeostasis.

5.6.2 CD44 structure and processing

CD44 is a type I transmembrane glycoprotein. The extracellular domain (ECD) of CD44 contains an HA-binding motif (BX_7B) in the N terminus, a "link" domain that shares homology with cartilage link protein, and six cysteine residues that form intramolecular disulfide bonds (Kohda et al., 1996). The standard ~80 kDa form of CD44 is the isoform found in the vast majority of normal tissues, but the plasma membrane proximal stem structure can be extended by inclusion of up to 10 alternatively spliced exons in multiple combinations, generating variants referred to, based on the inclusion of specific exons, as CD44v1–v10, which can be as large as 200 kDa (Screaton et al., 1992; Hirano et al., 1993; Naor et al., 1997; Ponta et al., 2003). CD44 undergoes cell-context-specific post-translational modifications that modulate CD44 function. Modifications of CD44, including attachment of heparan sulfate GAG, sulfation, phosphorylation, and glycosylation, are cell-context dependent and vary between cell types, stages of differentiation, and states of activation.

An interesting aspect of CD44 is its susceptibility to proteolytic processing (Cichy and Puré, 2003). Sequential cleavage of CD44 by membrane-type matrix metalloproteinase-1 (MT-MMP1) and presinilin-dependent gamma secretase leads to the release of three fragments of the receptor (Okamoto et al., 1999, 2001; Lammich et al., 2002; Cichy and Puré, 2003): a soluble form of the ECD that retains ligand-binding capacity and can be incorporated as an integral component of the ECM (Cichy et al., 2002), a so-called CD44β peptide of largely unknown function (Lammich et al., 2002), and the intracellular domain (ICD) that has been suggested to play a role in signal transduction and may be translocated to the nucleus where it has been suggested to play a role in transcription (Okamoto et al., 2001; Nagano and Saya, 2004). CD44 intracellular domain (CD44ICD) has also been suggested to enhance the activity of the transcriptional coactivator CREB binding protein (CBP)/protein 300 (p300) to upregulate transcription of CD44 itself (Okamoto et al., 2001; Nagano and Saya, 2004). Further studies are required to identify bona fide gene targets of CD44ICD.

5.6.3 Hyaluronan and other ligands of CD44

The principle ligand of CD44 is the matrix GAG HA, and the best understood functions of CD44 depend on HA binding. Thus, CD44 functions in matrix remodeling through its essential role in HA turnover. It mediates attachment and migration on ECM substratum via binding to HA, HA acts as a ligand bridge to mediate CD44-dependent cell-cell interactions, and HA binding induces CD44-mediated signal transduction that can regulate cell survival, growth, and function. Importantly, although CD44 is broadly expressed on virtually all cells of neuroectodermal origin, its function as a receptor for HA is cell-context-dependent with its affinity for HA determined by cell type, stage of differentiation, and state of cellular activation. CD44 also exercises HA-independent functions that are likely differentially regulated from its HA receptor function. HA is secreted as an anionic unbranched GAG composed of repeating disaccharide units of *D*-glucuronic acid and *N*-acetylglucosamine. Its repetitive structure means that a single molecule of HA can engage multiple receptors on individual cells. By virtue of its multivalency, HA binding mediates the oligomerization of its receptors as required for CD44-mediated signal transduction (Lesley et al., 1992, 2000; Sleeman et al., 1996).

The multivalency of HA also underlies its capacity to simultaneously engage receptors on multiple cells and thus act as a ligand bridge to mediate cell-cell interactions.

HA is synthesized at the plasma membrane by members of the HA synthase family (HAS1–3). Once extruded, HA can remain associated with HAS or be secreted and re-associated with the plasma membrane by binding to cell surface HA receptors such as CD44, thereby contributing to the structure of the pericellular coat. Secreted CD44 can also be incorporated into the ECM (Cichy et al., 2002). HA is found as large, high-molecular-weight (HMW) HA polymers of several to tens of millions of daltons (Da) under physiological conditions. However, HA undergoes major remodeling during wound healing, at sites of inflammation and in tumor-associated matrices, leading to the accumulation of low-molecular-weight (LMW) forms of on average 250,000 Da or less. LMW forms can be synthesized de novo or generated enzymatically (by hyaluronidases) or nonenzymatically by oxidative hydrolysis in the face of reactive oxygen/nitrogen species. Furthermore, there is a high rate of turnover of HA even under normal conditions. Thus, the HA content of tissue microenvironments is subject to dynamic changes, including in the ratio of HMW to LMW HA. As the HMW and LMW species of HA can mediate opposing effects on cells (as discussed throughout this chapter), at least in part by inducing differential signal transduction (Kothapalli et al., 2008; Puré and Assoian, 2009), it appears likely that the ratio of HMW to LMW HA is an important factor in determining cellular responses to the HA content of the microenvironment under normal conditions versus injured tissues.

Although all alternatively spliced variants of CD44 have the potential to bind HA, the alternatively spliced variants exhibit affinity for additional ligands such as collagen, fibronectin, and osteopontin (Jalkanen and Jalkanen, 1992; Lesley et al., 1993; Henke et al., 1996; Weber et al., 1996). Specific isoforms of CD44 also serve as docking sites for matrix metalloproteinases (MMPs) (Yu and Stamenkovic, 2000), promoting the activation of matrix metalloproteinase-7 (MMP7) and matrix metalloproteinase-9 (MMP9) and thereby promoting their function in matrix remodeling. Variant isoforms of CD44 also have the capacity to interact with other cell surface proteins, including growth factor receptors such as c-met and ErbB2 (van der Voort et al., 1999; Gumireddy et al., 2007). With few exceptions, such as the expression of the largest isoform of CD44 that comprises variants 1 through 10 by normal epithelial cells, expression of variant isoforms is associated with cellular activation and pathological conditions (Naor et al., 1997).

5.6.4 CD44-mediated signaling

Intracellularly, the cytoplasmic tail of CD44 interacts with ankyrin and ezrin, radixin, and moesin (ERM) proteins that provide a link between the actin cytoskeleton and CD44, as well as Merlin (the product of the *NF2* neurofibromatosis gene) to regulate cell proliferation, motility and survival (Stamenkovic and Yu, 2010). The cytoplasmic domain can also undergo posttranslational modification (phosphorylation/dephosphorylation) (Camp et al., 1991; Puré et al., 1995) that may mediate inside-out signaling to regulate CD44 receptor function, can affect the subcellular localization of CD44, and in turn, can affect cell migration. Indeed, CD44 has been shown in several cell types to be required for maintenance of cell polarity and directional migration (Isacke, 1994;

Acharya et al., 2008; Mrass et al., 2008; Martin-Villar et al., 2010). In at least some cases, although the ECD may play a regulatory role, its ICD is sufficient, and the ERM-binding region required, to mediate CD44-dependent directional migration, at least in T lymphocytes (Mrass et al., 2008). In addition to its link to the cytoskeleton, important for cell morphology, adhesion, and migration, the cytoplasmic domain of CD44 also mediates signal transduction. CD44 cytoplasmic domain can associate with Rho guanosine triphosphatases (GTPases) (Rac, Rho, and Cdc42) and modulate downstream signaling events such as MEK/ERK signaling, phosphatidylinositol 3-kinase (PI3K)/AKT activation, activation of Src kinases, Ca^{2+} mobilization, transcriptional up-regulation, and cytoskeletal rearrangement (Taher et al., 1996; Ilangumaran et al., 1999; Ishiwatari-Hayasaka et al., 1999; Bourguignon et al., 2000, 2003; Oliferenko et al., 2000; Cheng et al., 2006; Hamilton et al., 2007; Kothapalli et al., 2008; Vigetti et al., 2008). Most recently, CD44 has been implicated in tumor growth and resistance to cytotoxic therapies in a model of glioblastoma through the regulation of the hippo signaling pathway (Xu et al., 2010).

5.6.5 CD44 function in mesenchymal stromal cells

Differential HA-mediated signaling of CD44 on mesenchymal stromal cells in normal and damaged tissue

Mesenchymal-derived stromal cells such as fibroblasts and vascular and airway smooth muscle cells were at one time thought to function mainly in tissue structure by providing the cellular scaffold and ECM comprising the connective tissue. It is now evident, however, that stromal cells contribute to the multiple dynamic mechanisms engaged to maintain homeostasis, mediate stress responses, and restore homeostasis following perturbation of tissues. Expression of PRRs and their coupling to signaling pathways that mediate inflammatory gene expression provide evidence of the sentinel function of these cells. In addition to providing the machinery required to respond to microbial invasion, the affinity of PRRs for endogenous ligands that accumulate in damaged tissues and the expression of receptors such as CD44, receptor for advanced glycation endproducts (RAGE), and integrins that also act as sensors of tissue perturbations confer stromal cells with the capacity to act as first responders to changes in their microenvironment. Multiple mechanisms have evolved to regulate the function of receptors that serve as sensors of the microenvironment. Interestingly, although CD44 is constitutively expressed in the vast majority of cells of neuroectodermal origin, different mechanisms have evolved to regulate the function of CD44 in different cell types.

Like most cells, stromal cells, including fibroblasts and vascular and airway smooth muscle cells, express CD44 constitutively. However unlike leukocytes, in which the high-affinity state of CD44 for HA is induced only following cellular activation, there is evidence that in adherent stromal cells, CD44 exhibits constitutive affinity for HA. The constitutive function of CD44 raises the interesting questions of why CD44 is constitutively active, and this being the case, how CD44 can act as a sensor of tissue damage in the context of stromal cells. As to the first question, the principle ligand of CD44, HA, is a prevalent component of matrices under normal conditions, and large amounts of HA are turned over daily to maintain homeostasis. CD44 mediates HA turnover (Culty et al., 1992; Underhill et al., 1993), and thus homeostasis requires that CD44 in stromal cells be maintained in its high-affinity state, although the mechanisms involved are not fully understood. Although CD44 can actively mediate HA uptake under normal conditions, HA binding to CD44 under normal conditions does not lead to activation of stromal cells. This can be explained by the fact that HMW HA predominates under normal conditions, and it has been shown that HMW HA induces anti-inflammatory signals and antimitogenic signals in stromal cells (Kothapalli et al., 2007, 2008; Puré and Assoian, 2009). Specifically, HMW HA inhibits growth factor–induced activation of cyclin D1 and S phase entry of stromal cells by inhibiting activation of Rac (▶Figure 5.9) (Kothapalli et al., 2007).

Under pathological conditions involving inflammation and in tumors, the levels of CD44 expression on stromal cells are increased and LMW HA is generated and accumulates in the microenvironment. In contrast to the antimitogenic properties of HMW HA, LMW HA induces CD44-dependent promitogenic signals in stromal cells (▶Figure 5.9B) through ERK-dependent activation of cyclin D1 leading to S phase entry (Kothapalli et al., 2008; Puré and Assoian, 2009). Thus, it has been proposed that CD44 acts as a rheostat through differential signaling to cyclin D1 as it senses changes in the ratio of HMW HA to LMW HA that result from perturbation of tissue homeostasis (Puré and Assoian, 2009). Concomitant increases in inflammation and growth factor production following tissue injury and in tumors may also regulate interactions of CD44 with the cytoskeleton and thereby regulate CD44-mediated signaling in stromal cells. CD44 can interact with ERM proteins and also differentially interacts with Merlin depending on its state of phosphorylation (Morrison et al., 2001). Thus, it is possible that under homeostatic conditions, Merlin is hypophosphorylated and binds to CD44 to displace ERM proteins from the receptor, while perturbation leading to inflammation and increased growth factor production is associated with hyperphosphorylation of Merlin and its dissociation from CD44 and association of ERM proteins with CD44 in its place. Thus, the transition between the ERM- and Merlin-associated states of CD44 may also contribute to the differential signaling observed in stromal cells under homeostatic and activated conditions. Furthermore, it will be of interest to determine whether LMW HA and HMW HA differentially regulate phosphorylation of Merlin.

Taken together, current evidence suggests that CD44 is indeed constitutively active on stromal cells as required for continuous HA turnover. In addition, binding of HMW HA under normal conditions actively mediates antimitogenic signals to maintain stromal cells in a quiescent state, while in contrast, CD44 mediates promitogenic signaling in stromal cells when LMW HA accumulates in the microenvironment as the result of infection, inflammation, injury, tissue dysfunction, and in the context of neoplasia.

The role of CD44 in fibrosis – activation of transforming growth factor-beta, fibroblast migration, and HA turnover

Excess or chronic inflammation leads to organ damage and disease. To avoid these consequences and restore normal tissue function and homeostasis, inflammation must be regulated and resolved by elimination of extrinsic and/or intrinsic inflammatory stimuli and activation of mechanisms that negatively regulate the inflammatory response. Activation of transforming growth factor-beta (TGF-β) is central to induction of fibrotic repair mechanisms and also regulates inflammation through multiple mechanisms, including the induction of T regulatory cells that inhibit the function of immune effector

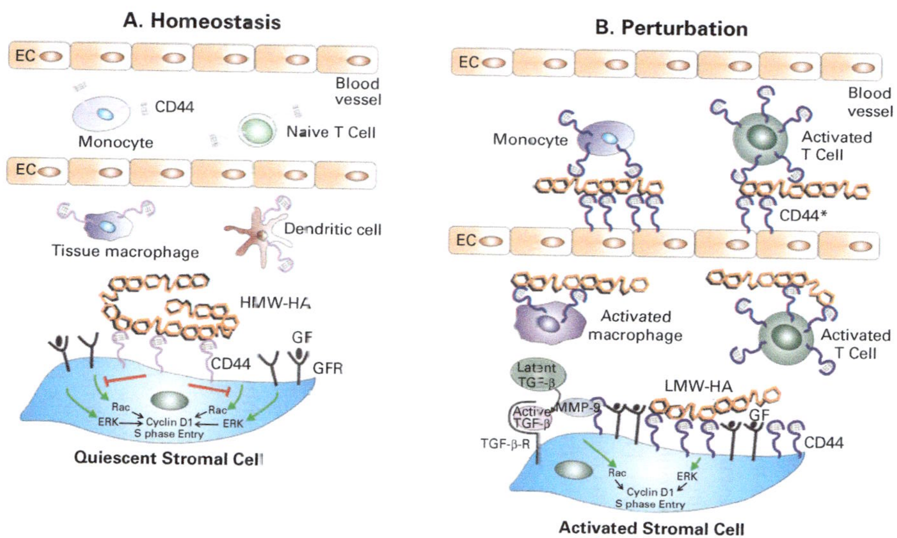

Figure 5.9 Status and function of CD44 in homeostasis and following tissue injury. (A) Under normal conditions, where the high-molecular-weight (HMW) form of HA predominates, CD44 functions on stromal cells to mediate HA turnover and transduce antimitogenic signals by inhibiting the induction of cyclin D1 and S phase entry through inhibition of Rac, to maintain homeostasis and stromal cell quiescence. Resident leukocytes express CD44 possibly in its high-affinity state, but even so, the preponderance of HMW HA in normal tissue would lead to CD44-mediated anti-inflammatory signaling rather than activation of the leukocytes resident in normal tissue. The expression of little, if any, CD44 and low levels of HA associated with endothelial cell (EC) under normal conditions, together with low levels of expression of CD44 in its low-affinity state on circulating naïve lymphocytes and other unactivated leukocytes, prevents their recruitment to normal tissues. (B) The response to injury leads to the accumulation of the promitogenic and proinflammatory low-molecular-weight forms of HA that stimulate stromal cell proliferation by inducing cyclin D1 and S phase entry via activation of ERK. Variant forms of CD44 on activated stromal cells provide a docking site for MMPs and promote their activity leading to the activation of latent transforming growth factor-beta (TGF-β) (green) to the active form of TGF-β (designated TGF-β* in purple) in proximity of the cell surface where it subsequently binds to TGF-β receptor (TGF-βR) and induces signaling. At sites of inflammation, HA production is upregulated on vascular EC where it binds to CD44 and is available to bind activated leukocytes expressing CD44 in its high-affinity state due to antigen receptor and/or cytokine–induced post-translational modification of the receptor. In concert with ligation of other adhesion receptors and their ligands between EC and leukocytes, the leukocytes undergo transendothelial migration where CD44 can mediate their interstitial migration and activation.

cells (Li and Flavell, 2008). Thus, activation of TGF-β is necessary for restoration of homeostasis following tissue damage. Activation of TGF-β requires processing of latent TGF-β associated with matrix, in part through the action of metalloproteinases leading to the liberation of active TGF-β and its binding to TGF-β receptors. Recent studies suggest that activation of TGF-β can be localized to the cell surface by docking to cell surface molecules including integrins (Munger et al., 1996; Keski-Oja et al., 2004; Wipff and Hinz, 2008; Margadant and Sonnenberg, 2010). More recent studies suggest that CD44 also plays a role in the local activation of TGF-β (▶Figure 5.9B) (Yu and Stamenkovic, 1999, 2000; Acharya et al., 2008).

CD44-deficient mice were found to exhibit an exacerbated response to bleomycin-induced lung injury (Teder et al., 2002). CD44 appears to contribute to the response to lung injury through multiple mechanisms, among them, through enhancing activation of TGF-β (Teder et al., 2002). It was then shown that fibroblasts derived from CD44-deficient mice had a decreased capacity to mediate MMP-dependent TGF-β activation and that the forced expression of CD44 rescued this phenotype (Acharya et al., 2008). In addition, CD44 was shown to be required for sustaining the polarity and directional migration of fibroblasts, which may also contribute to the altered fibrotic response in CD44-deficient mice (Acharya et al., 2008). Finally, an excess accumulation of LMW HA was observed in the lungs of bleomycin-challenged CD44-deficient mice compared to wild-type controls, consistent with the role of CD44 in HA turnover (Teder et al., 2002). Clearance of LMW HA and shifting the balance in favor of HMW HA may be required for transition from inflammation to fibrosis and, ultimately, restoration of homeostasis. In addition to revealing multiple potential roles for stromal cell CD44 in the response to lung injury, inflammatory cell CD44 was shown to be critical as discussed in section 5.6.7.

In an analogous fashion, CD44 appears to play a dual role in cardiovascular disease through its differential signaling in vascular smooth muscle cells by HMW and LMW HA, likely depending on the degree of inflammation in the local milieu, and its multiple functions on smooth muscle cells and infiltrating leukocytes. Atherosclerosis is associated with vascular inflammation and the accumulation of LMW HA that has promitogenic activity on vascular smooth muscle cells and regulates vascular smooth muscle cell gene expression and stimulates smooth muscle cell migration at least in part through CD44 (Cuff et al., 2001; Puré, 2006; Zhao et al., 2007, 2008). LMW HA also delivers proinflammatory and proatheorgenic signals through CD44-dependent leukocyte recruitment and activation such that deficiency in CD44 on either vascular cells or bone marrow–derived leukocytes partially protected apolipoprotein E (apoE)-deficient mice from atherosclerosis. Interestingly, deficiency of CD44 in both compartments, which conferred the greatest protection against atherosclerosis, was associated with a marked increase in fibrosis in residual lesions (Zhao et al., 2008). Consistent with this finding, global deletion of CD44 also exacerbated the fibrotic response to wire-induced vascular injury leading to increased stenosis. This model, which is associated with minimal inflammation, and therefore presumably only a modest reduction in the ratio of HMW to LMW HA in the injured vessel, suggested that deletion of CD44 released vascular smooth muscle cells from CD44-mediated antimitogenic signals (Kothapalli et al., 2007).

5.6.6 CD44 function in leukocytes

Regulation of CD44 affinity for HA and its role in leukocyte recruitment, interstitial migration, and activation

CD44 is expressed on all leukocytes. Strikingly, however, in the absence of activation, leukocytes isolated from peripheral blood and secondary lymphoid organs such as spleen and lymph nodes do not exhibit affinity for HA. This suggests that in contrast to its role in stromal cells, CD44 does not play a role in homeostasis of inflammatory or immune cells. Tissue leukocytes, such as alveolar macrophages, may prove to be an exception to this rule and express CD44 in a high-affinity state (Culty et al., 1994; Hodge-Dufour et al., 1997), but the status of CD44 on resident leukocytes in normal tissue needs to be examined more broadly going forward. Activation of leukocytes by, for example, cytokines or antigen receptor engagement or adhesion (DeGrendele et al., 1997b; Hodge-Dufour et al., 1997; Levesque and Haynes, 1997), however, results in increased expression of CD44 and transition of the receptor to its high-affinity state for HA, which facilitates their recruitment to sites of inflammation where CD44 is also induced on activated endothelial cells (ECs) (▶Figure 5.9) (DeGrendele et al., 1996, 1997a; Mohamadzadeh et al., 1998; Cuff et al., 2001). Inflammation also up regulates HA production by local ECs. The HA produced decorates the surface of activated ECs by binding to CD44 (DeGrendele et al., 1996; Mohamadzadeh et al., 1998; Siegelman et al., 1999) (▶Figure 5.9B). By virtue of its multivalency, HA presented by CD44 on the luminal surface of vascular ECs can also bind CD44 on circulating activated leukocytes and serve as a ligand bridge to facilitate localized binding of activated leukocytes to ECs followed by transendothelial migration (▶Figure 5.9B).

CD44 remains in its high-affinity state on leukocytes once they have traversed the endothelium where it can regulate leukocyte function, survival, and interstitial migration (Clark et al., 1996; Puré and Cuff, 2001; Gee et al., 2004; Mrass et al., 2008). Even in conditions where CD44 is not involved in leukocyte recruitment, it can still regulate the function of leukocytes once they have been recruited. For example, recruitment of tumor antigen-specific CD44-deficient T cells into transplanted tumors was comparable to that of CD44-expressing antigen-specific T cells, but the interstitial migration of the CD44-deficient cells was defective and translated into reduced antitumor immunity (Mrass et al., 2008). Similar to results described previously for fibroblasts, CD44 was found to be required to maintain T-cell polarity, which is required for directional interstitial migration. The role of CD44 in T-cell migration within tumors was dependent on the ICD of CD44 and its interaction with ERM proteins but did not require the ECD of CD44. Thus, CD44's role in interstitial migration in this setting does not require interaction with HA, but it is not know whether HA binding to the ECD can regulate CD44-dependent interstitial T-cell migration. On the other hand, lymphocyte trafficking to lymph nodes is also CD44 independent, but migration within lymph nodes was found to be mediated by CD44-HA interactions (Clark et al., 1996).

Finally, binding of HMW HA and LMW HA to CD44 on activated leukocytes mediates anti-inflammatory and proinflammatory signaling, respectively. HMW HA inhibits leukocyte phagocytosis and inflammatory gene expression, while LMW HA stimulates phagocytosis and induced inflammatory gene expression (Lesley et al., 1997; Puré and Cuff, 2001; Gee et al., 2004; Liang et al., 2007; Kawana et al., 2008; Zhao et al., 2008).

In summary, as depicted in ▶Figure 5.9, under normal conditions where HMW HA predominates and leukocytes are in their unactivated states, CD44 does not appear to be engaged by ligand (▶Figure 5.9A). In the face of inflammation, the HA content shifts to a preponderance of the LMW-HA form, and leukocyte CD44 transitions to its high-affinity state; together, they promote inflammatory and immune responses (▶Figure 5.9B). As inflammation resolves and the excess of LMW HA associated with inflammation is cleared, even activated leukocytes still expressing CD44 in the high-affinity state would transition to mediating anti-inflammatory signals as the ratio of HMW to LMW HA increases toward levels found in normal tissue. Finally, with receptor turnover and the removal of activated leukocytes as inflammation resolves, there would be a steady decline in leukocytes capable of binding HA.

5.6.7 Role of CD44 in the resolution of inflammation

Although CD44 can facilitate leukocyte recruitment to sites of tissue damage and activate proinflammatory pathways, CD44 can also play a critical role in resolving inflammation. Specifically, CD44 was shown to be required for the resolution of lung inflammation in mice treated with bleomycin (Teder et al., 2002). The excessive inflammation observed in bleomycin-treated CD44-deficient mice compared to wild-type mice was attributed to three causes: (1) An excess and protracted accumulation of LMW HA would be expected to contribute to the failure to resolve inflammation by maintaining a low ratio of HMW to LMW forms of HA and thus providing a positive feedback inflammatory loop. Clearance of HA has also been suggested to be required for evolution of the response to injury to a reparative fibrotic phase. (2) The excessive inflammation was also attributed to a defect in clearance of apoptotic neutrophils consistent with the role of CD44 in phagocytosis of apoptotic cells (Hart et al., 1997). Like LMW HA, accumulation of apoptotic cells in damaged tissues exacerbates inflammation. (3) Finally, decreased local activation of TGF-β, which regulates inflammation and induces the transition to a reparative fibrotic response, as discussed previously, was thought to be a cause. Experiments using bone marrow chimeric mice demonstrated that CD44 on bone marrow–derived cells plays a critical role in the resolution of inflammation (Teder et al., 2002), but the potential for stromal cell CD44 in resolving inflammation was not directly tested.

5.6.8 CD44 in disease and as a potential therapeutic target

CD44 has been implicated in a wide range of diseases including cancer, cardiovascular disease, and organ-specific fibrosis. The role of CD44 in each of these settings is complex with potential to exacerbate or attenuate disease depending on the cellular context. In this regard, extending the studies of deletion of CD44 in mouse models from global knockouts to tissue-specific knockouts in the future should be revealing. The most consistent data are in inflammation where CD44 initially promotes the inflammatory processes mediated by leukocytes, stromal cells, and ECs and has therefore been considered a potential therapeutic target. Although the function of CD44 is indeed best understood in the context of inflammation and in the context of pathological changes in the tissue environment, it will be important to proceed cautiously in developing CD44-targeted therapies for inflammatory disease in view of the recent evidence that CD44 also appears to play a role in maintaining homeostasis. Nonetheless, it may still be a rational approach if rather than taking the conventional approach of targeting activated CD44 ligand binding, instead the mechanisms involved in the transition of leukocyte CD44 to the high-affinity state could be selectively targeted without affecting the homeostatic function of CD44 on stromal cells in uninvolved organs. Another important consideration is the recent evidence that in addition to initially promoting inflammation, CD44 may also be required for the resolution of inflammation and restoration of homeostasis. Again, cell-type selective targeting of CD44 or targeting CD44 in patients at specific stages of inflammatory disease progression may circumvent this issue. Targeting specific variant isoforms of CD44 may also circumvent the potential risks of targeting CD44 by disrupting isoform-specific disease-related functions

Whether therapeutic targeting of CD44 is a rational approach to the treatment of fibrosis is far less clear. In part, this stems from the fact that in most of the animal models of disease in which CD44 has been studied to date, a significant inflammatory phase precedes progression to fibrosis. As the most convincing approach has been to compare disease progression in these mouse models in wild-type and global CD44 knockout animals, it is not yet possible to distinguish the impact of CD44 on fibrosis independent of its impact on inflammation. Cell-type and inducible CD44 knockouts or models of fibrosis in which there is an insignificant inflammatory component will be required to definitively address the role of CD44 and its potential as a therapeutic target in fibrosis.

Perhaps most complex is the role of CD44 and its potential as a therapeutic target in cancer. Although the preponderance of data suggests that high levels of expression of CD44 are associated with increased aggressiveness and poorer prognosis of some tumor types, this is not the case in others. Specific variant isoforms of CD44 have also been implicated in various tumors types. As in inflammation, the potentially opposing pro- and antitumorigenic roles of CD44 on the multiple cell types involved in tumorigenesis, including the neoplastic cells, stromal cells, inflammatory cells, and immune cells, and the multiple functions of CD44 even within cell types, need to be delineated, and these may vary in different tumor types. Interestingly, in spite of the evidence that CD44 may promote tumor progression and metastasis, a number of recent studies have established that in fact downregulation of CD44 is required for metastasis (Asao et al., 2000; Lopez et al., 2005; Huang et al., 2008; Liu et al., 2011). Perhaps, as in inflammation, CD44 may have opposing roles not only in different tumor types and different cell types involved in tumorigenesis but also at different stages of cancer progression. Based on current data, at least in some tumor types, CD44 might prove to promote certain stages of tumorigenesis, including invasiveness, but then suppress later stages of metastasis. Thus, although currently there are many challenges, further understanding of its role in distinct tumor types (and subtypes) at different stages of tumor progression could still provide the way forward for the development of CD44-targeted therapies.

5.6.9 Concluding remarks

As necessitated by the broad expression of CD44 and the ubiquitous distribution of its principle ligand in matrices and in the circulation, elegant mechanisms involving alternative splicing, posttranslational modifications, and varied protein-protein interactions have evolved to regulate the specificity and affinity of CD44, depending on cell type, stage of differentiation, and state of activation. Combined with the differential CD44-dependent signaling induced by distinct forms of HA that predominate under normal conditions versus at sites of inflammation and in tumors, the constitutive expression of CD44 in its high-affinity state on parenchymal stromal cells and resident leukocytes confers CD44 with the capacity to act as a rheostatic sensor of perturbations in the microenvironment. Together with the requirement for cellular activation to induce transition of CD44 from an inactive to its high-affinity state for HA on circulating leukocytes and to upregulate the expression of CD44 and HA by ECs, these mechanisms endow CD44 with the capacity to serve distinct functions during hoemostasis and under stress conditions including infection, injury, sterile inflammation, and cancer. Further delineating the molecular basis of the regulated function of CD44 and dissecting its potentially opposing functions on various cell types and at different stages of the response to injury and disease progression should provide opportunities for the development of CD44-targeted therapeutics.

5.6.10 Take-home message

Molecular pathways involved in pathogenic processes that are not essential to cellular function in normal tissues are theoretically the ideal targets for therapeutic intervention in disease states. In spite of its broad expression, the function of CD44 is highly regulated. The altered affinity of CD44 for HA on activated versus resting or naive leukocytes; the expression of variant isoforms of CD44, expressed mainly under pathological conditions that confer affinity for additional ligands; and differential CD44-mediated signal transduction depending on the status of the tissue milieu suggest that targeting the activation or function of CD44 in a cell-type-selective or isoform-specific manner at appropriate stages of disease progression may be a feasible approach to therapy in some disease settings.

Acknowledgments

The author's research pertaining to this subject is supported by grants from the NIH (HL093283) and the Pennsylvania Department of Health to the Wistar Institute.

References

Acharya, P. S., Majumdar, S., Jacob, M., et al. (2008). Fibroblast migration is mediated by CD44-dependent TGF{beta} activation. J Cell Sci *121*, 1393–1402.

Asao, T., Nakamura, J., Shitara, Y., et al. (2000). Loss of standard type of CD44 expression in invaded area as a good indicator of lymph-node metastasis in colorectal carcinoma. Dis Colon Rectum *43*, 1250–1255.

Bourguignon, L. Y., Singleton, P. A., Zhu, H., and Diedrich, F. (2003). Hyaluronan-mediated CD44 interaction with RhoGEF and Rho kinase promotes Grb2-associated binder-1 phosphorylation and phosphatidylinositol 3-kinase signaling leading to cytokine (macrophage-colony stimulating factor) production and breast tumor progression. J Biol Chem *278*, 29420–29434.

Bourguignon, L. Y., Zhu, H., Shao, L., and Chen, Y. W. (2000). CD44 interaction with Tiam1 promotes Rac1 signaling and hyaluronic acid-mediated breast tumor cell migration. J Biol Chem *275*, 1829–1838.

Camp, R. L., Kraus, T. A., and Puré, E. (1991). Variations in the cytoskeletal interaction and posttranslational modification of the CD44 homing receptor in macrophages. J Cell Biol *115*, 1283–1292.

Cheng, C., Yaffe, M. B., and Sharp, P. A. (2006). A positive feedback loop couples Ras activation and CD44 alternative splicing. Genes Dev *20*, 1715–1720.

Cichy, J., Bals, R., Potempa, J., Mani, A., and Puré, E. (2002). Proteinase-mediated release of epithelial cell-associated CD44: extracellular CD44 complexes with components of cellular matrices. J Biol Chem *277*, 44440–44447.

Cichy, J., and Puré, E. (2003). The liberation of CD44. J Cell Biol *161*, 839–843.

Clark, R. A., Alon, R., and Springer, T. A. (1996). CD44 and hyaluronan-dependent rolling interactions of lymphocytes on tonsillar stroma. J Cell Biol *134*, 1075–1087.

Cuff, C. A., Kothapalli, D., Azonobi, I., et al. (2001). The adhesion receptor CD44 promotes atherosclerosis by mediating inflammatory cell recruitment and vascular cell activation. J Clin Invest *108*, 1031–1040.

Culty, M., Nguyen, H. A., and Underhill, C. B. (1992). The hyaluronan receptor (CD44) participates in the uptake and degradation of hyaluronan. J Cell Biol *116*, 1055–1062.

Culty, M., O'Mara, T. E., Underhill, C. B., Yeager, H., Jr, and Swartz, R.P. (1994). Hyaluronan receptor (CD44) expression and function in human peripheral blood monocytes and alveolar macrophages. J Leukoc Biol *56*, 605–611.

DeGrendele, H. C., Estess, P., Picker, L. J., and Siegelman, M. H. (1996). CD44 and its ligand hyaluronate mediate rolling under physiologic flow: a novel lymphocyte-endothelial cell primary adhesion pathway. J Exp Med *183*, 1119–1130.

DeGrendele, H. C., Estess, P., and Siegelman, M. H. (1997a). Requirement for CD44 in activated T cell extravasation into an inflammatory site. Science *278*, 672–675.

DeGrendele, H. C., Kosfiszer, M., Estess, P., and Siegelman, M. H. (1997b). CD44 activation and associated primary adhesion is inducible via T cell receptor stimulation. J Immunol *159*, 2546–2553.

Gee, K., Kryworuchko, M., and Kumar, A. (2004). Recent advances in the regulation of CD44 expression and its role in inflammation and autoimmune diseases. Arch Immunol Ther Exp (Warsz) *52*, 13–26.

Gumireddy, K., Sun, F., Klein-Szanto, A. J., et al. (2007). In vivo selection for metastasis promoting genes in the mouse. Proc Natl Acad Sci U S A *104*, 6696–6701.

Hamilton, S. R., Fard, S. F., Paiwand, F. F., et al. (2007). The hyaluronan receptors CD44 and Rhamm (CD168) form complexes with ERK1,2 that sustain high basal motility in breast cancer cells. J Biol Chem *282* (22), 16667–16680.

Hart, S. P., Dougherty, G. J., Haslett, C., and Dransfield, I. (1997). CD44 regulates phagocytosis of apoptotic neutrophil granulocytes, but not apoptotic lymphocytes, by human macrophages. J Immunol *159*, 919–925.

Henke, C. A., Roongta, U., Mickelson, D. J., Knutson, J. R., and McCarthy, J. B. (1996). CD44-related chondroitin sulfate proteoglycan, a cell surface receptor implicated with tumor cell invasion, mediates endothelial cell migration on fibrinogen and invasion into a fibrin matrix. J Clin Invest *97*, 2541–2552.

Hirano, H., Screaton, G. R., Bell, M. V., Jackson, D. G., Bell, J. I., and Hodes, R. J. (1993). CD44 isoform expression mediated by alternative splicing: tissue-specific regulation in mice. Int Immunol *6*, 49–59.

Hodge-Dufour, J., Noble, P. W., Horton, M. R., et al. (1997). Induction of IL-12 and chemokines by hyaluronan requires adhesion-dependent priming of resident but not elicited macrophages. J Immunol *159*, 2492–2500.

Huang, Q., Gumireddy, K., Schrier, M., et al. (2008). The microRNAs miR-373 and miR-520c promote tumour invasion and metastasis. Nat Cell Biol *10*, 202–210.

Ilangumaran, S., Borisch, B., and Hoessli, D. C. (1999). Signal transduction via CD44: role of plasma membrane microdomains. Leuk. Lymphoma *35*, 455–469.
Isacke, C. M. (1994). The role of cytoplasmic domain in regulating CD44 function. J Cell Sci *107*, 2353–2359.
Ishiwatari-Hayasaka, H., Fujimoto, T., Osawa, T., Hirama, T., Toyama-Sorimachi, N., and Miyasaka, M. (1999). Requirements for signal delivery through CD44: analysis using CD44-Fas chimeric proteins. J Immunol *163*, 1258–1264.
Jalkanen, S., and Jalkanen, M. (1992). Lymphocyte CD44 binds the COOH-terminal heparin-binding domain of fibronectin. J Cell Biol *116*, 817–825.
Kawana, H., Karaki, H., Higashi, M., et al. (2008). CD44 suppresses TLR-mediated inflammation. J Immunol *180*, 4235–4245.
Keski-Oja, J., Koli, K., and von Melchner, H. (2004). TGF-beta activation by traction? Trends Cell Biol *14*, 657–659.
Kohda, D., Morton, C. J., Parkar, A. A., et al. (1996). Solution structure of the link module: a hyaluronan-binding domain involved in extracellular matrix stability and cell migration. Cell *86*, 767–775.
Kothapalli, D., Flowers, J., Xu, T., Puré, E., and Assoian, R. K. (2008). Differential activation of ERK and Rac mediates the proliferative and anti-proliferative effects of hyaluronan and CD44. J Biol Chem *283*, 31823–31829.
Kothapalli, D., Zhao, L., Hawthorne, E. A., et al. (2007). Hyaluronan and CD44 antagonize mitogen-dependent cyclin D1 expression in mesenchymal cells. J Cell Biol *176*, 535–544.
Lammich, S., Okochi, M., Takeda, M., et al. (2002). Presenilin-dependent intramembrane proteolysis of CD44 leads to the liberation of its intracellular domain and the secretion of an Abeta-like peptide. J Biol Chem *277*, 44754–44759.
Lesley, J., English, N., Charles, C., and Hyman, R. (2000). The role of the CD44 cytoplasmic and transmembrane domains in constitutive and inducible hyaluronan binding. Eur J Immunol *30*, 245–53.
Lesley, J., He, Q., Miyake, K., Hamann, A., Hyman, R., and Kincade, P. W. (1992). Requirements for hyaluronic acid binding by CD44: a role for the cytoplasmic domain and activation by antibody. J Exp Med *175*, 257–266.
Lesley, J., Hyman, R., English, N., Catterall, J. B., and Turner, G. A. (1997). CD44 in inflammation and metastasis. Glycoconj J *14*, 611–622.
Lesley, J., Hyman, R., and Kincade, P. W. (1993). CD44 and its interaction with extracellular matrix. Adv Immunol *54*, 271–335.
Levesque, M. C., and Haynes, B. F., (1997). Cytokine induction of the ability of human monocyte CD44 to bind hyaluronan is mediated primarily by TNF-alpha and is inhibited by IL-4 and IL-13. J Immunol *159*, 6184–6194.
Li, M. O., and Flavell, R. A. (2008). Contextual regulation of inflammation: a duet by transforming growth factor-beta and interleukin-10. Immunity *28*, 468–476.
Liang, J., Jiang, D., Griffith, J., et al. (2007). CD44 is a negative regulator of acute pulmonary inflammation and lipopolysaccharide-TLR signaling in mouse macrophages. J Immunol *178*, 2469–2475.
Liu, C., Kelnar, K., Liu, B., et al. (2011). The microRNA miR-34a inhibits prostate cancer stem cells and metastasis by directly repressing CD44. Nat Med *17*, 211–215.
Lopez, J. I., Camenisch, T. D., Stevens, M. V., Sands, B. J., McDonald, J., and Schroeder, J. A. (2005). CD44 attenuates metastatic invasion during breast cancer progression. Cancer Res *65*, 6755–6763.
Margadant, C., and Sonnenberg, A. (2010). Integrin-TGF-beta crosstalk in fibrosis, cancer and wound healing. EMBO Rep *11*, 97–105.
Martin-Villar, E., Fernandez-Munoz, B., Parsons, M., et al. (2010). Podoplanin associates with CD44 to promote directional cell migration. Mol Biol Cell *21*, 4387–4399.

Mohamadzadeh, M., DeGrendele, H., Arizpe, H., Estess, P., and Siegelman, M. (1998). Proinflammatory stimuli regulate endothelial hyaluronan expression and CD44/HA-dependent primary adhesion. J Clin Invest *101*, 97–108.

Morrison, H., Sherman, L. S., Legg, J., et al. (2001). The NF2 tumor suppressor gene product, merlin, mediates contact inhibition of growth through interactions with CD44. Genes Dev *15*, 968–980.

Mrass, P., Kinjyo, I., Ng, L. G., Reiner, S. L., Puré, E., and Weninger, W. (2008). CD44 mediates successful interstitial navigation by killer T cells and enables efficient antitumor immunity. Immunity *29*, 971–985.

Munger, J. S., Huang, X., Kawakatsu, H., et al. (1996). The integrin alpha v beta 6 binds and activates latent TGF beta 1: a mechanism for regulating pulmonary inflammation and fibrosis. Cell *96*, 319–328.

Nagano, O., and Saya, H. (2004). Mechanism and biological significance of CD44 cleavage. Cancer Sci *95*, 935.

Naor, D., Sionov, R. V., and Ish-Shalom, D. (1997). CD44: structure, function, and association with the malignant process. Adv Cancer Res *71*, 241–319.

Okamoto, I., Kawano, Y., Murakami, D., et al. (2001). Proteolytic release of CD44 intracellular domain and its role in the CD44 signaling pathway. J Cell Biol *255*, 755–762.

Okamoto, I., Kawano, Y., Tsuiki, H., et al. (1999). CD44 cleavage induced by a membrane-associated metalloprotease plays a critical role in tumor cell migration. Oncogene *18*, 1435–1446.

Oliferenko, S., Kaverina, I., Small, J. V., and Huber, L. A. (2000). Hyaluronic acid (HA) binding to CD44 activates Rac1 and induces lamellipodia outgrowth. J Cell Biol *148*, 1159–1164.

Ponta, H., Sherman, L., and Herrlich, P. A. (2003). CD44: from adhesion molecules to signalling regulators. Nat Rev Mol Cell Biol *4*, 33–45.

Puré, E. (2006). Role of CD44 in atherogenesis and its potential role as a therapeutic target. In Advances in Translational Medicine: Lipids and Atherosclerosis, C. J. Packard and D. J. Rader, eds., 191–206. London: Taylor and Francis Medical Publishing.

Puré, E., and Assoian, R. K. (2009). Rheostatic signaling by CD44 and hyaluronan. Cell Signal *21*, 651–655.

Puré, E., Camp, R. L., Peritt, D., Panettieri, R. A., Jr, Lazaar, A. L., and Nayak, S. (1995). Defective phosphorylation and hyaluronate binding of CD44 with point mutations in the cytoplasmic domain. J Exp Med *181*, 55–62.

Puré, E., and Cuff, C. A. (2001). A crucial role for CD44 in inflammation. Trends Mol Med *7*, 213–221.

Screaton, G. R., Bell, M. V., Jackson, F. B., Gerth, U., and Bell, J. I. (1992). Genomic structure of DNA encoding the lymphocyte homing receptor CD44 reveals at least 12 alternatively spliced exons. Proc Natl Acad Sci U S A *89*, 12160–12164.

Siegelman, M. H., DeGrendele, H. C., and Estess, P. (1999). Activation and interaction of CD44 and hyaluronan in immunological systems. J Leukoc Biol *66*, 315–321.

Sleeman, J., Rudy, W., Hofmann, M., Moll, J., Herrlich, P., and Ponta, H. (1996). Regulated clustering of variant CD44 proteins increases their hyaluronate binding capacity. J Cell Biol *135*, 1139–1150.

Stamenkovic, I., and Yu, Q. (2010). Merlin, a "Magic" linker between the extracellular cues and intracellular signaling pathways that regulate cell motility, proliferation, and survival. Curr Protein Pept Sci *11*, 471–484.

Taher, T. E., Smit, L., Griffioen, A. W., Schilder-Tol, E. J., Borst, J., and Pals, S. T. (1996). Signaling through CD44 is mediated by tyrosine kinases. Association with p56lck in T lymphocytes. J Biol Chem *271*, 2863–2867.

Teder, P., Vandivier, R. W., Jiang, D., et al. (2002). Resolution of lung inflammation by CD44. Science *296*, 155–158.

Underhill, C. B., Nguyen, H. A., Shizari, M., and Culty, M. (1993). CD44 positive macrophages take up hyaluronan during lung development. Dev Biol *155*, 324–336.

van der Voort, R., Taher, T. E., Wielenga, V. J. M., et al. (1999). Heparan sulfate-modified CD44 promotes hepatocyte growth factor/scatter factor-induced signal transduction through the receptor tyrosine kinase c-Met. J Biol Chem *274*, 6499–6506.

Vigetti, D., Viola, M., Karousou, E., et al. (2008). Hyaluronan-CD44-ERK1/2 regulate human aortic smooth muscle cell motility during aging. J Biol Chem *283*, 4448–4458.

Weber, G. F., Ashkar, S., Glimcher, M. J., and Cantor, H. (1996). Receptor-ligand interaction between CD44 and osteopontin (Eta-1). Science *271*, 509–512.

Wipff, P. J., and Hinz, B. (2008). Integrins and the activation of latent transforming growth factor [beta]1 – an intimate relationship. Eur J Cell Biol *87*, 601–615.

Xu, Y., Stamenkovic, I., and Yu, Q. (2010). CD44 attenuates activation of the hippo signaling pathway and is a prime therapeutic target for glioblastoma. Cancer Res *70*, 2455–2464.

Yu, Q., and Stamenkovic, I. (1999). Localization of matrix metalloproteinase 9 to the cell surface provides a mechanism for CD44-mediated tumor invasion. Genes Dev *13*, 35–48.

Yu, Q., and Stamenkovic, I. (2000). Cell surface-localized matrix metalloproteinase-9 proteolytically activates TGF-beta and promotes tumor invasion and angiogenesis. Genes Dev *14*, 163–176.

Zhao, L., Hall, A., Levenkova, N., et al. (2007). CD44 regulates vascular gene expression in a proatherogenic environment. Arterioscler Thromb Vasc Biol *27*, 886–892.

Zhao, L., Lee, E., Zukas, A. M., et al. (2008). CD44 expressed on both bone marrow-derived and non-bone marrow-derived cells promotes atherogenesis in apoE-deficient mice. Arterioscler Thromb Vasc Biol *28*, 1283–1289.

6 Collagens: insights into the folding, assembly and functions

6.1 Introduction

Ruggero Tenni

The collagen field is surely coming of age in many respects, thanks to the effort of scientists and collaborations around the world, and technological advances in the past two decades.

The advances concern, without pretense of completeness, the number of α chains coded in the human genome and the number of collagen types, and the increasing number of pathologies due to defects in genes of collagen α chains or in proteins (e.g. chaperones) or enzymes involved in collagen biosynthesis, maturation, and degradation. These aspects do not cover the whole field; broadly speaking, they deal with collagen itself, its molecular structure (the triple helix), and the supramolecular assemblies of the different collagen types.

Collagens have a further relevant property: they possess a high degree of connectivity, being able to interact with "almost countless" ligands, namely, proteins (e.g. secreted protein acidic and rich in cysteine [SPARC]), cell receptors (e.g. integrins and discoidin domain receptors), matrix proteoglycans (PGs) (e.g. small leucine-rich PGs [SLRPs]), and so on.

Many interactions have been determined *in vitro*. For instance, the presence of decorin or its core, biglycan, and some glycosaminoglycans, in an *in vitro* fibril formation assay, may affect fibril diameter and layout (Raspanti et al., 2008). What is difficult, however, is to determine when and how collagen assembles and functions when it is cosynthesized with ligand(s), and the influence of such interactions on collagen maturation. Cells themselves may affect fibril formation (e.g. tendon cells can handle early aggregates of type I collagen in fibripositors) (Canty et al., 2004).

Collagen structural properties and supramolecular assemblies can thus be much better understood only with a vision that integrates them with the effect of collagen ligands and vice versa.

Recent discoveries on pathologies caused by mutations in collagen genes or in the genes for proteins involved in collagen metabolism are seminal contributions to the understanding of basic collagen biology. For example, the discovery that prolyl-3-hydroxylase-1 (P3H1) acts as a complex with cyclophilin B (CYPB) and cartilage-associated protein (CRTAP) elucidated the molecular basis for a significant fraction of osteogenesis imperfecta mutations (Vranka et al., 2004; Morello et al., 2006; Marini et al., 2007a; see Makareeva et al., 2011 for a recent review on collagen chaperones) and could, in the future, help clarify the role of select posttranslational modifications of the triple helix in collagen fibril assembly and function.

This part of the book contains six chapters, dealing with several broad facets of collagen biology.

Chapter 6.2 on collagen trimerization by Sergei P. Boudko and Hans Peter Bächinger focuses on collagens IX, XV, XVIII, and XIX. Collagen trimerization is an essential step for triple-helix formation. In several instances, cells synthesize more than one collagen type. Specific chain selection is therefore mandatory; in addition, a high trimerization potential of the trimerization domain(s) at very low concentrations (down to picomolar) can be necessary. Furthermore, the mechanism of chain selection and trimerization of nonfibrillar collagens differs from that of fibrillar collagens that have special C-propeptides (http://www.ncbi.nlm.nih.gov/pubmed/19631658).

Chapter 6.3 by Anton V. Persikov and Barbara Brodsky looks at studies of how collagen mutations affect triple-helix structure, using model peptides, computational analyses, and a recombinant bacterial collagen system. The model peptide studies follow a long series of experiments of Brodsky and colleagues using a host-guest approach to study the propensity of individual tripeptides to form a triple helix (Persikov et al., 2005) and to incorporate mutations that lead to collagen diseases (Baum and Brodsky, 2002). Model triple-helical peptides are amenable to physical chemical measurements, such as thermal stability and kinetic parameters, and can be correlated with high-resolution crystallographic structures and nuclear magnetic resonance (NMR) dynamics. Important insights from the high-resolution structures included the extensive hydration network and resolution of the hydrogen-bonding pattern in the triple helix. Reviews of these studies are described in Brodsky and Persikov (2005) and Brodsky et al. (2008). More recently, mutations have been studied within a recombinant system based on a bacterial collagen-like protein, to complement the peptide studies.

The topic of Chapter 6.4 by Catherine Moali and David J. S. Hulmes deals with roles and regulation of bone morphogenetic protein-1 (BMP1)/Tolloid-like proteinases. These enzymes play major roles in collagen/matrix assembly, both in normal physiology and in pathological conditions, through processing of precursor forms of collagens and other extracellular matrix (ECM) components, thereby orchestrating fibril formation, cross-linking, and other aspects of matrix assembly. Another aspect is activation of different growth factors involved in tissue development and repair. Unlike the matrix metalloproteinases (MMPs) whose activities are regulated mainly by endogenous inhibitors (tissue inhibitors of metalloproteinases, or TIMPs), BMP1/Tolloid-like proteinases are regulated mainly by endogenous activators, many of which are substrate specific. This group has made important contributions in understanding the mechanism of action of one such activator, called procollagen C-proteinase enhancer (Kronenberg et al., 2009). It has also been found recently that closely related proteinases, the meprins, contribute to procollagen processing (Kronenberg et al., 2010).

Chapter 6.5 by Mario Raspanti discusses the state of the art and open questions in the field of supramolecular assemblies of fibrillar collagens and fibrillogenesis. Raspanti's experimental approach is mainly at the ultrastructural level, leveraging the potential of high-resolution microscopy (transmission electron microscopy [TEM], scanning electron microscopy [SEM], atomic force microscopy [AFM], and freeze-etching) to reveal novel views of collagen-based superstructures.

Two further chapters focus on the state of the art and insights of collagen interactions. James D. San Antonio's past work has dealt with creating an interactome, or "road map," of the type I collagen fibril (Di Lullo et al., 2002; Sweeney et al., 2008). The distribution of many ligand-binding sites on the fibril were discussed in these papers, as well as how they may functionally relate to osteogenesis imperfecta mutations (Marini et al., 2007b). In Chapter 6.6, San Antonio and collaborators discuss insights from the interactome of collagen I and its fibrils, proposing that the fibril may be organized into cell and matrix interaction domains and attempting to visualize how the most crucial sites of collagen are organized on a three-dimensional X-ray diffraction model of the microfibril (i.e. on the "living fibril"). Further discussion focuses on their recently published collagen IV interactome, which also suggests a functional domain organization, and a speculative model is put forth for the arrangement of type IV collagen in basement membranes. The remainder of the chapter summarizes the potential for creating interactomes of collagens II, III, and X, among others.

The localization of binding sites in collagen molecules and supramolecular aggregates has always been an important but difficult task because the collagen triple helix is not a completely rigid entity and because the loss of some weak bonds within the trimer is (more than) compensated for by weak bonds with the ligand. Many methods have been exploited, but conflicting results have been found in some cases (e.g. the collagen binding sites for the decorin core protein).

In Chapter 6.7, Takako Sasaki reports and discusses some ligands whose sites of interaction with the collagen molecule have been ascertained. The detailed knowledge of this issue is indeed relevant. In the past, Sasaki worked on SPARC (BM-40) and its interaction with collagen and was able to localize one binding site in collagen I by using a multidisciplinary approach (Giudici et al., 2008), which was confirmed and extended by cocrystallization of SPARC with a collagen model triple-helical peptide (Hohenester et al., 2008).

I would like to briefly comment on an aspect, raised by some papers, that does not directly involve collagen but can play, directly or indirectly, an important role in understanding bone pathophysiology (Wise et al., 2007; Duer et al., 2008, 2009; Reid et al., 2011).

These papers show that some mineral phases, in particular hydroxyapatite, are able to interact with some glycosaminoglycans. Let's take in to account that the biglycan-null mice show mainly a bone phenotype (osteoporosis) and that the decorin and biglycan double-null mice have a worse phenotype with an earlier onset (Young et al., 2002). Since both of these SLRPs are able to interact with collagen I through their protein core, it is possible to postulate that one or both SLRPs connect bone collagen to the mineral phase, acting as a bridge between the collagen fibrils and the hydroxyapatite crystals. As a consequence, we can also postulate that select mutations in type I collagen may reduce the affinity of the mutated collagen for the SLRP core protein, uncoupling or weakening collagen-mineral crystal associations. This can be involved in the mechanism of bone pathology; for instance, it could underlie or contribute to the phenotype of many patients affected by osteogenesis imperfecta.

In conclusion, the previous example, and other issues as well, tell us that further work is necessary and worth performing. Interesting and meaningful results, probably regarding basic issues and also the pathophysiology of collagens, are foreseeable in the future. Researchers currently working in the ECM field and on the collagen issue put to use their skills, passion, and curiosity, as will future scientists. Therefore, as the take-home message, future research in this field looks promising, despite budgetary drops for science in some countries because of worldwide financial troubles in the past few years.

Acknowledgments

I would finally like to warmly and heartily thank all colleagues that contributed to this, the collagen part of the book, for their continuous attention, input, and help. I also thank the editor, Nikos Karamanos, for his advice and careful attention to this project. Last but not least, I appreciated very much the work done with my valued colleagues of the Department of Biochemistry "A. Castellani," University of Pavia, in particular Antonio Rossi, Antonella Forlino, and M. Enrica Tira.

References

Baum, J., and Brodsky, B. (2002). Folding of the collagen triple-helix and its naturally occurring mutants. In Mechanisms of Protein Folding: Frontiers in Molecular Biology, R. H. Pain, ed. Oxford: Oxford Univ. Press, pp. 330–351.

Brodsky, B., and Persikov, A. V. (2005). Molecular structure of the collagen triple helix. Adv Protein Chem *70*, 301–339.

Brodsky, B., Thiagarajan, G., Madhan, B., and Kar, K. (2008). Triple-helical peptides: an approach to collagen conformation, stability, and self-association. Biopolymers *89*, 345–353.

Canty, E. G., Lu, Y., Meadows, R. S., Shaw, M. K., Holmes, D. F., and Kadler, K. E. (2004). Coalignment of plasma membrane channels and protrusions (fibripositors) specifies the parallelism of tendon. J Cell Biol *165*, 553–563.

Di Lullo, G. A., Sweeney, S. M., Körkkö, J., Ala-Kokko, L., and San Antonio, J. D. (2002) Mapping the ligand-binding sites and disease-associated mutations on the most abundant protein in human, type I collagen. J Biol Chem *277*, 4223–4231.

Duer, M. J., Friscić, T., Murray, R. C., Reid, D. G., and Wise, E. R. (2009). The mineral phase of calcified cartilage: its molecular structure and interface with the organic matrix. Biophys J *96*, 3372–3378.

Duer, M. J., Friscić, T., Proudfoot, D., et al. (2008). Mineral surface in calcified plaque is like that of bone: further evidence for regulated mineralization. Arterioscler Thromb Vasc Biol *28*, 2030–2034.

Giudici, C., Raynal, N., Wiedemann, H., et al. (2008). Mapping of SPARC/BM-40/osteonectin-binding sites on fibrillar collagens. J Biol Chem *283*, 19551–19560.

Hohenester, E., Sasaki, T., Giudici, C., Farndale, R. W., and Bächinger, H. P. (2008). Structural basis of sequence-specific collagen recognition by SPARC. Proc Natl Acad Sci U S A *105*, 18273–18277.

Kronenberg, D., Bruns, B. C., Moali, C., et al. (2010). Processing of procollagen III by meprins: new players in extracellular matrix assembly? J Invest Dermatol *130*, 2727–2735.

Kronenberg, D., Vadon-Le Goff, S., Bourhis, J. M., et al. (2009). Strong cooperativity and loose geometry between CUB domains are the basis for procollagen C-proteinase enhancer activity. J Biol Chem *284*, 33437–33446.

Makareeva, E., Aviles, N. A., and Leikin, S. (2011). Chaperoning osteogenesis: new protein-folding disease paradigms. Trends Cell Biol *21*, 168–176.

Marini, J. C., Cabral, W. A., Barnes, A. M., and Chang, W. (2007a). Components of the collagen prolyl 3-hydroxylation complex are crucial for normal bone development. Cell Cycle *15*, 1675–1681.

Marini, J. C., Forlino, A., Cabral, W. A., et al. (2007b). Consortium for osteogenesis imperfecta mutations in the helical domain of type I collagen: regions rich in lethal mutations align with collagen binding sites for integrins and proteoglycans. Hum Mutat *28*, 209–221.

Morello, R., Bertin, T. K., Chen, Y., et al. (2006). CRTAP is required for prolyl 3-hydroxylation and mutations cause recessive osteogenesis imperfecta. Cell *127*, 291–304.

Persikov, A. V., Ramshaw, J. A., and Brodsky, B. (2005). Prediction of collagen stability from amino acid sequence. J Biol Chem *280*, 19343–19349.

Raspanti, M., Viola, M., Forlino, A., Tenni, R., Gruppi, C., and Tira, M. E. (2008). Glycosaminoglycans show a specific periodic interaction with type I collagen fibrils. J Struct Biol *164*, 134–139.

Reid, D. G., Jackson, G. J., Duer, M. J., and Rodgers, A. L. (2011). Apatite in kidney stones is a molecular composite with glycosaminoglycans and proteins: evidence from nuclear magnetic resonance spectroscopy, and relevance to Randall's plaque, pathogenesis and prophylaxis. J Urol *185*, 725–730.

Sweeney, S. M., Orgel, J. P., Fertala, A., et al. (2008). Candidate cell and matrix interaction domains on the collagen fibril, the predominant protein of vertebrates. J Biol Chem *283*, 21187–21197.

Vranka, J. A., Sakai, L. Y., and Bächinger, H. P. (2004). Prolyl 3-hydroxylase 1, enzyme characterization and identification of a novel family of enzymes. J Biol Chem *279*, 23615–23621.

Wise, E. R., Maltsev, S., Davies, M. E., et al. (2007). The organic-mineral interface in bone is predominantly polysaccharide. Chem Mater *19*, 5055–5057.

Young, M. F., Bi, Y., Ameye, L., and Chen, X. D. (2002). Biglycan knockout mice: new models for musculoskeletal diseases. Glycoconj J *19*, 257–262.

6.2 Trimerization domains in collagens: chain selection, folding initiation, and triple-helix stabilization

Sergei P. Boudko and Hans Peter Bächinger

6.2.1 Introduction

Folding is the most challenging problem in protein science. The efficient formation of specific interactions is essential for structure formation of proteins. For monomeric proteins, these are exclusively intrachain interactions, whereas for oligomeric proteins interchain interactions also become crucial. Collagens are trimeric molecules that combine three chains of either the same or different types (homo- or heterotrimeric collagens, respectively). Collagens are usually highly elongated, fibrous molecules, in which the major part represents a triple helix. The triple helix is a tightly packed structure in which three chains with -G-X-Y- repeating sequence form an extensive network of interchain interactions. Short regions of -(G-X-Y)$_N$- sequences have a tendency to trimerize with each other among different chains in a random order, thus forming huge aggregates, known as gelatin. Fortunately, this process is relatively slow, and more importantly, it becomes rather inefficient at physiological temperature. Then how can three chains find each other and form a single triple-helical rod with correctly aligned sequences? And how can newly synthesized chains distinguish among different collagen types and compositions?

Two fundamentally different views of how proteins fold are now being debated. Do proteins fold through multiple unpredictable routes directed only by the energetically downhill nature of the folding landscape, or do they fold through specific intermediates in a defined pathway that systematically puts predetermined pieces of the target native protein into place (Englander et al., 2007)? In 1997, Panchenko et al. (1997) introduced independently folding units, so-called foldons, that formed in the early phase of the folding process and could be observed as stable intermediates under appropriate experimental conditions. It now becomes more acceptable that cooperative native-like foldon units and the sequential stabilization process together generate predetermined stepwise pathways in protein folding (Englander et al., 2007).

Chain selection, alignment, and oligomerization in collagen are the earliest events that precede the folding of the major triple-helical portion of the molecule. It becomes more evident that for each collagen type there is a specific region that fulfills these functions. This region is usually found outside of the triple-helical portion and is termed *trimerization domain*. In other words, the trimerization domain is an early stage collagen foldon.

6.2.2 Chain selection and trimerization

There are about 30 different types of collagen summarized in ▶Table 6.1. For most of them, the trimerization domains were identified at the C-terminal end of the last (counted from the N terminus) collagenous domain (COL). The trimerization domains can constitute either the entire C-terminal noncollagenous domain (NC) or only the small N-terminal portion of it, as found for collagens XV and XVIII. Numbering of COL and NC domains starts from the C terminus, which somewhat reflects the initiation of folding at NC1 and the triple-helix propagation toward the N terminus.

The length of trimerization domains varies dramatically from very short, ~35 residues, in type IX collagen to ~350 residues in fibrillar collagens (types I, II, III, V/XI, XXIV, and XXVII). It is worth mentioning though that for fibrillar collagens the real core of the trimerization domain can be significantly smaller due to the trilobe structural organization as derived from the small-angle X-ray scattering (SAXS) experiments (Bernocco et al., 2001). For several collagen types (IV, VIII, X, XV, and XVIII) the crystal structures of the trimerization domains are now available. They have different folds and type of the trimer association (▶Figure 6.1).

Now, we will briefly review data on the trimerization domains of fibrillar, basement membrane, and network-forming collagens as these are the most studied ones in terms of chain association and structural organization of the trimerization domains. Afterward, we will discuss very recent data on trimerization domains in multiplexins and fibril-associated collagens with interrupted triple helices (FACITs) that open great new possibilities for studying collagen molecules. Data on trimerization domains in other collagen types can be found in the references given in ▶Table 6.1.

Fibrillar, basement membrane, and network-forming collagens

Fibril-forming collagens share a long uninterrupted collagenous domain flanked by N- and C-terminal NC propeptides. The chains of fibrillar collagens assemble into certain homo- and heterotrimeric protomers. Mutational studies strongly suggest that the NC1 domain (short C-telopeptide with C-propeptide) provides correct alignment and registration for the nucleation of the triple helix (Bateman et al., 1989; Chessler et al., 1993; Pace et al., 2001). The C-propeptide is further cleaved by C-proteinase to form the mature collagen that is incorporated into fibrils (Miyahara et al., 1982). The best available model of the NC1 domain obtained from the SAXS experiments represents an elongated molecule with three large lobes and one minor lobe, where each of the major lobes corresponds to one of the three component polypeptide chains that come together in a junction region to connect to the rest of the procollagen molecule (Bernocco et al., 2001).

Assembly of basement membrane collagen IV was studied by electron microscopy. The folding was always initiated at the C terminus by forming a characteristic globular NC1 domain and followed by triple-helix propagation toward the N terminus (Dölz et al., 1988; Davis et al., 1989). It is thus highly possible that the NC1 domain provides chain selection and nucleates the folding of the entire molecule. The NC1 domain also plays a critical role in tail-to-tail association of two collagen molecules and formation of a hexamer with two triple helices protruding in an antiparallel manner. An additional linkage of four molecules at their N termini leads to formation of a two-dimensional network (Risteli et al., 1980).

Table 6.1 Trimerization domains in collagens.

Class	*Collagen types*	*Trimerization domain location (size in residues)*	*Trimerization domain atomic structure (Protein Data Bank identification number [PDB ID])*	*References for trimerization domain*
Fibrillar collagens	I, II, III, V/XI, XXIV, XXVII	NC1 (~350)	N/A[1]	(Bateman et al., 1989; Chessler et al., 1993; Bernocco et al., 2001; Pace et al., 2001)
Basement membrane collagens	IV	NC1 (~230)	1LI1, 1M3D, 1T60,1T61	(Dölz et al., 1988; Davis et al., 1989; Sundaramoorthy et al., 2002; Than et al., 2002)
Beaded filament-forming collagens	VI	N-terminal VWFA (von Willebrand factor A) domain of NC1 (~190)	N/A	(Ball et al., 2001; Lamande et al., 2002)
Anchoring fibril-forming collagens	VII	NC2?[2]	N/A	(Chen et al., 2001; Bogin et al., 2002; Kvansakul et al., 2003)
Network-forming collagens	VIII, X	NC1 (~165)	1O91, 1GR3	(Zhang and Chen, 1999; Bogin et al., 2002; Kvansakul et al., 2003)
Fibril-associated collagens with interrupted triple helices (FACIT) and FACIT-like collagens	IX, XII, XIV, XVI, XIX, XX, XXI, XXII	NC2 (35–50)	N/A	(Boudko et al., 2008, 2010)
Multiplexins	XV, XVIII	N-terminal fragment of NC1 (~55)	3N3F, 3HSH, 3HON	(Sasaki et al., 1998; Boudko et al., 2009; Wirz et al., 2011)
Transmembrane collagens	XIII, XVII, XXIII, XXV	Multiple?	N/A	(Snellman et al., 2000, 2007; Areida et al., 2001; Latvanlehto et al., 2003)
Others	XXVI, XXVIII	Unknown	N/A	N/A

[1]N/A – not available.
[2]Numbered from the N terminus. NC2 domain of collagen VII is the very C-terminal noncollagenous domain.

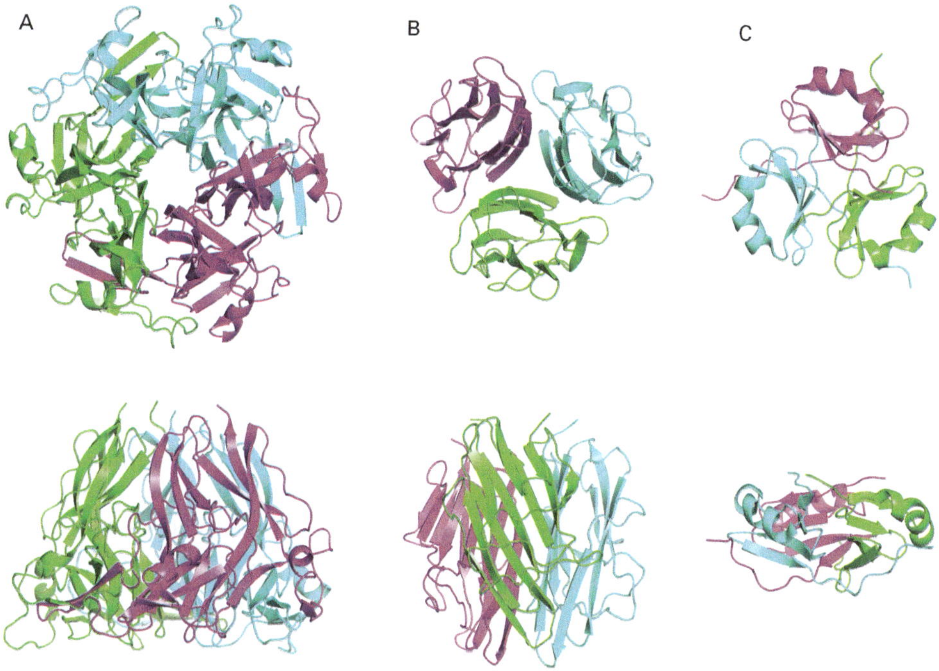

Figure 6.1 Crystal structure of collagen trimerization domains. Ribbon diagrams of representative trimerization domains with reported crystal structures. The upper panel represents N-terminal top views, and the lower panel represents side views. (A) Collagen IV (PDB ID: 1LI1, chains A,B,C), (B) collagen VIII (PDB ID: 1O91), and (C) collagen XVIII (PDB ID: 3HSH, chains A,B,C) trimerization domains. The figure was generated using the PyMol program (http://www.pymol.org/).

Network-forming collagens VIII and X share high sequence similarity within their NC1 domains and also with the trimeric globular domain (gC1q) of the complement protein C1q (Shapiro and Scherer, 1998) and C1q-like domains in other proteins. The NC1 domain of collagen X was shown to be sufficient to direct trimerization without a triple helix (Zhang and Chen, 1999). As in collagen IV, the NC1 domains of collagens VIII and X are retained in their suprastructures. A possible mechanism for polygonal lattice formation in type VIII and X collagens was suggested in which the NC1 trimer initiates the supramolecular assembly (Bogin et al., 2002; Kvansakul et al., 2003).

Whereas the structural organization of the trimerization domain in fibrillar collagens still remains elusive (Malone et al., 2005), the crystal structures of the trimerization domains for basement membrane (type IV) and network-forming collagens (types VIII and X) are already available (▶Table 6.1, ▶Figure 6.1).

Multiplexins

Collagens XV and XVIII are the only members of the multiplexin family (multiple triple-helix domains with interruptions) (Abe et al., 1993; Oh et al., 1994; Rehn and Pihlajaniemi, 1994; Rehn et al., 1994), identified as chondroitin sulfate and heparan sulfate proteoglycans, respectively (Halfter et al., 1998; Li et al., 2000). Both collagens are homotrimers composed of a single α chain that contains a central highly interrupted collagenous domain flanked by N- and C-terminal noncollagenous domains. Type XVIII collagen localizes in epidermal and vascular basement membrane, similar to other heparan sulfate proteoglycans found in basement membranes (Marneros and Olsen, 2005). Type XV is not an integral basement membrane component; rather, it localizes to areas peripheral to basement membranes and associates directly with collagen fibers/fibrils in a manner consistent with other chondroitin sulfate proteoglycans (Amenta et al., 2005).

An angiogenesis inhibitor produced by hemangioendothelioma and named endostatin was identified as a 20 kDa C-terminal fragment of collagen XVIII (O'Reilly et al., 1997). Endostatin was also identified as circulating fragments in serum (Standker et al., 1997). The NC1 domains of collagens XV and XVIII are organized into three different subdomains: an N-terminal region is followed by a protease-labile segment and the endostatin domain (Sasaki et al., 1998, 2000).

Recombinantly expressed NC1 domain of collagen XVIII was found to undergo a rapid proteolysis and generation of a number of segments, most of which were related to monomeric endostatin, but a small N-terminal fragment appeared to be trimeric (Sasaki et al., 1998). The similar trimeric fragment was also observed in recombinant preparations of the collagen XV NC1 domain (Sasaki et al., 2000). A more systematic approach of limited proteolysis of bacterially produced N-terminal fragments of the collagen XVIII NC1 domain helped to identify a minimal stable trimeric domain (Boudko et al., 2009). The sequence of this domain (only 54 residues per chain) is sufficient for self-assembly of the trimeric domain and thus determines the trimerization domain of collagen XVIII. Based on a sequence alignment between collagens XVIII and XV, the similar region of collagen XV was also demonstrated to be the trimerization domain (Wirz et al., 2011).

Collagen XVIII and XV trimerization domains can be recombinantly expressed at very high levels in soluble form (Boudko et al., 2009; Wirz et al., 2011). The final yields of purified samples are several tens of milligrams out of 1 liter of medium. The thermal stability is quite remarkable, the trimerization domain of collagen XVIII shows no signs of dissociation and unfolding up to 70°C. Guanidine-hydrochloride-induced unfolding/refolding equilibrium experiments demonstrate complete reversibility. The most striking feature is the trimerization potential, which was found to be ~60 pM or ~120 pM for XVIII or XV, respectively. In other words, the trimerization domains retain their trimeric structures at very dilute solutions down to the picomolar range. The trimerization potential of the well-known T4 phage foldon domain can be estimated only at ~14 nM (Güthe et al., 2004). Remarkably, the refolding rate of the collagen XVIII trimerization domain is in a subsecond interval (Boudko et al., 2009). Production and characterization of different collagenous sequences fused to the collagen XVIII trimerization domain is now in progress and very encouraging (S. P. Boudkos, unpublished data). All these advantages make the trimerization domains of multiplexins very attractive in terms of efficient recombinant production of homotrimeric triple-helical peptides, in which folding and stability will be assured by the presence of the natural collagen-specific trimerization domain. The trimerization domains can also be used in other collagen-unrelated fields. A trivalent chimeric antibody was produced and successfully used by fusing a single-chain antibody to the collagen XVIII trimerization domain (Sanchez-Arevalo Lobo et al., 2006).

The crystal structures of both collagen XV and XVIII trimerization domains are now available (Boudko et al., 2009; Wirz et al., 2011). They share a unique trimerization fold (▶Figure 6.1C). Each monomeric subunit has a single β sheet formed by mixed parallel and antiparallel β strands, a single α helix with a short 3_{10}-helix, and an extended tail going around the neighboring unit. The trimer-forming interface has a hydrophobic core, which is not connected to the hydrophobic core of the monomeric unit. There are also a number of other intra- and interchain interactions observed, which include hydrogen and ionic bonds. Sequences of collagens XV and XVIII share only ~30% of identical residues within their trimerization domains. Strikingly, a pattern (exact residue positions without gaps or inserts, but not identities!) of hydrophobic residues for both monomeric and trimeric cores is the same. Even more strikingly, the hydrophobic pattern is evolutionarily preserved in all species where multiplexin analogues were reported, from Echinodermata to vertebrates (Boudko et al., 2009).

FACIT collagens

The class of extracellular matrix molecules *FACIT* originally included types IX, XII, and XIV (Shaw and Olsen, 1991). With discovery of new collagen genes, this class was extended with types XVI, XIX, XX, XXI, and XXII. Since there is no information available on their association with collagen fibrils, these newer types are usually referred to as FACIT-like collagens. The NC1 domains of FACIT and FACIT-like collagens are rather short (only ~20 residues in type XIX) and have a conserved pattern of a GXCXXXXC sequence, which follows the last triple-helical domain (COL1). The cysteine residues form disulfide bridges and cross-link all three chains (Boulegue et al., 2008). The effectiveness of interchain disulfide bond formation depends on the presence of a preceding triple-helical part, as was shown for collagen XIV and IX sequences (Lesage et al., 1996; Mechling et al., 1996). A direct correlation between the stability of the preceding triple helix and the correctness of disulfide bridging in the NC1 domain of collagen XIX was established (Boudko et al., 2008). The NC1 domain of a FACIT collagen cannot provide chain trimerization, and its folding depends entirely on the triple-helix formation in COL1.

Based on a sequence analysis, it was suggested that the NC2 domain in FACIT can form an α-helical coiled coil structure and provide oligomerization (McAlinden et al., 2003). It was the NC2 domain of homotrimeric collagen XIX that was demonstrated experimentally to be the trimerization domain (Boudko et al., 2008). The NC2 domain sequence (~50 residues) forms a trimer without any additional triple-helical part. It has a high α-helical content as measured by circular dichroism spectroscopy, which also supports the suggested coiled coil organization. Nevertheless, the predicted coiled coil heptad repeat pattern is unusual and consists of two partially overlapping assignments. It is also possible that it forms a novel fold that can accommodate the triple-helix stagger.

The collagen XIX NC2 domain is the independent folding unit that also has a remarkable stability (Boudko et al., 2008). It withstands heating up to 90°C without dissociation and requires up to 5 M of denaturant guanidine hydrochloride to destroy its trimeric structure. Thermal unfolding/refolding transitions in the presence of moderate concentrations of guanidine hydrochloride are fully reversible. In addition to providing the folding nucleus by bringing three chains of collagen in the right register, this domain has a tremendous potential to stabilize the trimeric structure of the molecule. The NC2 domain is a unique collagen trimerization domain due to its special position between two collagenous domains, COL2 and COL1. It provides bidirectional folding of the triple helix and also stabilizes it on both ends. The stabilizing effect was experimentally observed upon recombinant fusion of a collagen-like $(GlyProPro)_{10}$ sequence either N or C terminally to the NC2 domain (see Section 6.2.3). The NC2 domain provides chain association and triple-helix folding in one direction toward the N terminus as most of the other known trimerization domains and also in the opposite direction, which also assures the NC1 domain folding. Interchain disulfide bridging within the NC1 domain then ultimately mimics the stabilizing role of C-terminal trimerization domains in other collagen types.

Among FACIT and FACIT-like collagens, type IX forms a heterotrimeric molecule, which consists of three distinct chains. A key question was raised: does the NC2 domain provide chain selection and heterotrimerization in collagen IX? To address this question, two complementary systems were designed (Boudko et al., 2010). System 1 contained all three sequences of the NC2 domain without additional triple helix. In system 2, the sequences were extended with $(GPP)_5$GPCCGGV at the C termini. The triple-helical extension cannot provide trimerization by itself, because of insufficient length; it can form triple helix only in the presence of a fused exogenous trimerization domain. The GPCCGGV tail (a collagen III cystine knot) served as a covalent linker of all three chains in triple-helical conformation (Boudko and Engel, 2004). The NC2 domain sequences have single cysteine residues in the α1 and α3 chains, but none in the α2. All constructs for both systems were expressed separately in soluble form, mixed, and reoxidized. System 1 efficiently (>80%) produced a trimeric complex as analyzed by analytical ultracentrifugation. The mass spectroscopy identified two species, disulfide-bonded α1-α3 chains and α2 chain. The analytical high-performance liquid chromatography (HPLC) under denaturing conditions provided separation of these two species in an equimolar ratio. System 2 (with the same efficiency) produced a single species when analyzed by mass spectroscopy, which consisted of disulfide-bonded α1-α2-α3 chains with triple-helical extensions. The NC2 domain of collagen IX provides effective chain selection and heterotrimerization by itself.

Circular dichroism spectroscopy also demonstrated the predominant α-helical nature of the collagen IX NC2 domain (Boudko et al., 2010), as was the case for type XIX. The thermal stability of type IX was found to be lower. Fully reversible thermal transitions were already observed without adding denaturants. The midpoint transition temperature (51°C–66°C) was concentration dependent (2–19 μM), which indicated coupling between dissociation and unfolding. The dissociation constant of the NC2 domain in oxidized form can be estimated in the 1–30 nM range for the 25°C–37°C temperature interval (Boudko et al., 2010), comparable to a good antibody/antigen complex.

The formation of collagen molecules with all three different chains was only reported for three other types (i.e. IV, V, and VI). While the trimerization domain in collagen VI still remains uncharacterized, recombinant production of the NC1 domains (responsible for heterotrimerization) of collagens IV and V is rather challenging. Moreover, the NC1 domain of collagen IV is involved in the hexamer formation, and that of collagen V is prone to cleavage. The NC2 heterotrimerization domain of collagen IX appears to be an attractive choice for recombinant production of heterotrimeric stagger-specific triple-helical peptides.

6.2.3 Trimerization domains and triple-helix folding and stabilization

Two systems were successfully used to study the triple-helix folding in naturally trimerized chains of collagen IV (Dölz et al., 1988; Davis et al., 1989) and collagen III (Bächinger et al., 1980). The latter contains a cystine knot at the C-terminal end of the triple helix, which bridges the three chains after initial trimerization by NC1 (Bruckner et al., 1978). Without NC1 or the cystine knot, collagen triple-helix formation is slow, and mismatched triple helices are formed (Engel and Prockop, 1991); however, with proper linkage kinetics proceeds in a fast and correct way. The cystine knot covalently links all three chains and thus mimics the trimerization domain. In both systems, folding of the triple helix starts in a zipper-like manner from the C terminus, the site where three chains are associated. The rate of triple-helix formation is limited by the *cis-trans* isomerization of peptide bonds in which proline or hydroxyproline residues are involved (Bächinger et al., 1978). These bonds are very frequent in collagens and show a high fraction of *cis* state in randomly coiled chains. Only the *trans* state can be incorporated into the triple helix. Propagation of the triple helix after nucleation by the NC or other nucleation domains is independent of collagen concentration, and for long collagens, the time course of the reaction is of zero order during the initial phase (Bächinger et al., 1980).

Collagen chains that lack the trimerization domain show large concentration dependence on the rate of triple-helix formation, indicating the need of forming a nucleus from several chains. A detailed study was performed using non-cross-linked collagen model peptide chains $(ProProGly)_{10}$ and $(ProHypGly)_{10}$ (Boudko et al., 2002). It was demonstrated that at very low peptide concentrations the reaction order was 3 but decreased at higher concentrations. To investigate the role of trimerization domains in triple-helix folding, an artificial trimerization domain was introduced by fusion of $(GlyProPro)_{10}$ with the trimeric foldon domain of T4 phage or the cross-linking cystine knot of collagen III. These domains were recombinantly attached to the C terminus of $(GlyProPro)_{10}$ to link the three chains (Boudko et al., 2002). It is worth mentioning here that most of natural collagens have significantly longer triple helices, ~1,000 residues for fibrillar collagens. A system with much shorter trimerized peptide $(GlyProPro)_{10}$ allowed the study of early kinetics steps in triple-helix formation. A first-order reaction (concentration independent!) was observed for the folding of the triple helix in $(GlyProPro)_{10}$ foldon with a half-time of 8.3 minutes as compared to a very long half-time of 15 hours at 60 mM chain concentration of peptide $(ProProGly)_{10}$. The trimerization domains play multiple roles in the triple-helix folding; in addition to chain selection and registration, they dramatically accelerate the triple-helix nucleation step that otherwise would be a critically slow concentration-dependent rate-limiting step in collagen folding. Triple-helix nucleation by trimerization domains appears to be effective in both N- and C-terminal directions as revealed using fusions of $(GlyProPro)_{10}$ on either ends of the foldon domain or the cystine knot (Frank et al., 2003).

Trimerization domains also have a dramatic effect on the thermal stability of adjacent triple helices, especially on short sequences. $(ProProGly)_{10}$ peptides form collagen triple helices of only moderate stability (25°C–30°C). When fused to the foldon domain, however, triple-helix formation of $(GlyProPro)_{10}$ becomes concentration independent, and the midpoint temperature of the triple-helix unfolding is significantly increased (65°C)

(Frank et al., 2001). Similar stabilization of $(GlyProPro)_{10}$ peptide was observed in fusions with the NC2 trimerization domain of collagen XIX (Boudko et al., 2008). The midpoint thermal transition temperature values were 68°C and 55°C for $(GPP)_{10}$NC2 and NC2 $(GPP)_{10}$, respectively. The difference between N- or C-terminally fused $(GlyProPro)_{10}$ peptide could either reflect different structural organizations at the N or C terminus of the NC2 domain or represent a polarity in the triple-helix stabilization. Interestingly, the same tendency was also monitored in a system with N- or C-terminal fusion of $(GlyProPro)_{10}$ with either the foldon domain or the cystine knot, 70°C and 53°C or 82°C and 67°C, respectively (Frank et al., 2003). The stabilizing function of the trimerization domain is due to an entropic effect or can be explained by the close vicinity of its termini, which induce a high local concentration of collagen chains. For $(GlyProPro)_{10}$ foldon, the concentration for the C termini of the collagen-like peptide was estimated in the range of 1 M (Frank et al., 2001).

The effect of stabilization of the triple helix depends on the nature of a trimerization domain. Even for natural collagen trimerization domains, the midpoint transition temperature was higher by 20°C for the same triple-helical sequence recombinantly fused to the FACIT rather than multiplexin trimerization domain (S. P. Boudko, unpublished data).

6.2.4 Pathologies associated with trimerization domains

Heritable disorders

Mutations in the trimerization domain of collagen are expected to have multiple consequences. Mutations that affect trimerization can lead to complete loss of folded collagen or at least a delay in its folding. Other mutations can affect formation of collagen networks or prevent efficient cleavage of the C-propeptide. Most of reported mutations are found in fibrillar collagens, which lead to osteogenesis imperfecta, Ehlers-Danlos syndrome, Stickler syndrome, and spontaneous cervical artery dissection (▶Table 6.2). Mutations in the collagen X NC1 domain lead to Schmid metaphyseal chondrodysplasia, whereas mutations in the collagen IV NC1 domain cause different forms of Alport syndrome (▶Table 6.2).

Other disorders

Goodpasture's disease is an autoimmune disorder characterized by rapidly progressive glomerulonephritis and pulmonary hemorrhage, with linear deposits of antibodies along the glomerular basement membrane (Bergs, 2005). In patients with Goodpasture's disease, autoantibodies to the distinct epitopes on the α3NC1 and α5NC1 monomers of collagen IV were identified (Pedchenko et al., 2010). Surprisingly, the autoantibodies did not bind to the native cross-linked α3α4α5NC1 hexamer. It was thus suggested that the development of Goodpasture's disease may be considered an autoimmune "conformeropathy" that involves perturbation of the quaternary structure of the α3α4α5NC1 hexamer of collagen IV, inducing a pathogenic conformational change in the α3NC1 and α5NC1 subunits, which in turn elicits an autoimmune response.

Alport syndrome is a heritable disorder that is associated with mutations in COL4A5 gene (▶Table 6.2). Following renal transplantation, an average of 6% of male Alport

Table 6.2 Mutations in collagen trimerization domains.

Gene	*Disease*	*Collagen-folding manifestations*	*References*
COL1A1, COL1A2	Osteogenesis imperfecta	Lethality, delayed folding, overmodification, retarded disulfide bond formation, reduced collagen production	(Chessler et al., 1993; Lamande et al., 1995; Pace et al., 2001, 2008)
COL2A1	Stickler syndrome	N/A	(Ahmad et al., 1995)
COL3A1	Vascular Ehlers-Danlos syndrome	Endoplasmic reticulum (ER) retention	(Schwarze et al., 2001; Pope et al., 2010; Pickup and Pollanen, 2011)
COL4A5	Alport syndrome	Formation of α3α4α5 heterotrimer abolished, no secretion	(Nakazato et al., 1993; Hertz et al., 1995; Barker et al., 1996; Wilson et al., 2007; Kobayashi et al., 2008)
COL5A1, COL5A2	Vascular Ehlers-Danlos syndrome	Reduced collagen production	(De Paepe et al., 1997; Grond-Ginsbach et al., 2002; Mitchell et al., 2009)
COL10A1	Schmid metaphyseal chondrodysplasia	Aberrant disulfide bond formation, ER retention, unfolded protein response	(Bateman et al., 2005; Wilson et al., 2005)

syndrome patients develop antiglomerular basement membrane nephritis, which was primarily associated with the production of alloantibodies against α5NC1 (Dehan et al., 1996). Later, all three chains of the α3α4α5NC1 complex were identified as alloantigens for Alport posttransplant antiglomerular basement membrane antibodies (Kalluri et al., 2000). Interestingly, alloantibodies against α5NC1 subunit were shown to interact with the intact hexamer more strongly than with the dissociated subunit (Pedchenko et al., 2010).

Very recently mutations in COL1A1 and COL1A2 were identified within the C-propeptide (NC1) cleavage site that lead to procollagen incorporation into matrix and subsequently cause high bone mass osteogenesis imperfecta (Lindahl et al., 2011).

6.2.5 Future prospects and challenges

Since trimerization domains initiate folding of the entire collagen molecule, they appear to be handy tools to design miniature recombinant variants with only certain regions of interest – containing binding sites for chaperones, connective tissue components or cell receptors, protease sensitive regions, interesting structural elements (e.g.

interruptions of the triple helix), sites for posttranslational modifications or cross-linking, and so forth. Those miniature versions are on high demand for two major reasons: first, they allow studying an isolated specific region disregarding other competing or nonspecific areas; and, second, for certain collagen types, it is the only way to produce a reasonable amount of material. Collagens are often multidomain proteins with multiple interacting sites for other connective tissue components that make them practically "insoluble" material under physiological conditions. This is the major obstacle on the way to extract or recombinantly produce pure and soluble material for biochemical and biophysical studies. The miniature technique will allow a divide-and-conquer approach to studying collagen molecules at a detailed molecular level.

The use of trimerization domains should help to clarify a long-standing question on the triple-helical stagger in heterotrimeric types of collagen. It is still a wide open question whether the order of the leading, middle, and trailing chains is specific, preferred, or random. If they are ordered, what determines their positions? Is it the trimerization domain or a helical part? Does the trimerization domain only bring three specific chains into close proximity and then let them find the way to nucleate the triple helix, or does it preposition them for a certain stagger? Structural and biophysical work on the interface between the trimerization domain and the beginning of the triple-helical part is required to resolve this issue.

For certain types of collagen, the nature of the trimerization domain remains unknown or unclear. Very little is known on the folding kinetics of the trimerization domains and their stability. The role of general and collagen-specific chaperones is also opaque.

6.2.6 Take-home message

Collagen trimerization domains are independently folding units that select and bring type-specific chains in the right register in order to initiate the triple-helix nucleation and propagation. After completing this function, the trimerization domains are either removed by specific proteases in some collagen types or remain incorporated into the final structure, thus stabilizing the triple helix and performing other type-specific functions, such as network formation.

References

Abe, N., Muragaki, Y., Yoshioka, H., Inoue, H., and Ninomiya, Y. (1993). Identification of a novel collagen chain represented by extensive interruptions in the triple-helical region. Biochem Biophys Res Commun *196*, 576–582.

Ahmad, N. N., Dimascio, J., Knowlton, R. G., and Tasman, W. S. (1995). Stickler syndrome. A mutation in the nonhelical 3′ end of type II procollagen gene. Arch Ophthalmol *113*, 1454–1457.

Amenta, P. S., Scivoletti, N. A., Newman, M. D., Sciancalepore, J. P., Li, D., and Myers, J. C. (2005). Proteoglycan-collagen XV in human tissues is seen linking banded collagen fibers subjacent to the basement membrane. J Histochem Cytochem *53*, 165–176.

Areida, S. K., Reinhardt, D. P., Müller, P. K., et al. (2001). Properties of the collagen type XVII ectodomain. Evidence for N- to C-terminal triple helix folding. J Biol Chem *276*, 1594–1601.

Bächinger, H. P., Bruckner, P., Timpl, R., and Engel, J. (1978). The role of cis-trans isomerization of peptide bonds in the coil ↔ triple helix conversion of collagen. Eur J Biochem *90*, 605–613.

Bächinger, H. P., Bruckner, P., Timpl, R., Prockop, D. J., and Engel, J. (1980). Folding mechanism of the triple helix in type-III collagen and type-III pN-collagen. Role of disulfide bridges and peptide bond isomerization. Eur J Biochem *106*, 619–632.

Ball, S. G., Baldock, C., Kielty, C. M., and Shuttleworth, C. A. (2001). The role of the C1 and C2 a-domains in type VI collagen assembly. J Biol Chem *276*, 7422–7430.

Barker, D. F., Pruchno, C. J., Jiang, X., et al. (1996). A mutation causing Alport syndrome with tardive hearing loss is common in the western United States. Am J Hum Genet *58*, 1157–1165.

Bateman, J. F., Lamande, S. R., Dahl, H. H., Chan, D., Mascara, T., and Cole, W. G. (1989). A frameshift mutation results in a truncated nonfunctional carboxyl-terminal pro alpha 1(I) propeptide of type I collagen in osteogenesis imperfecta. J Biol Chem *264*, 10960–10964.

Bateman, J. F., Wilson, R., Freddi, S., Lamande, S. R., and Savarirayan, R. (2005). Mutations of COL10A1 in Schmid metaphyseal chondrodysplasia. Hum Mutat *25*, 525–534.

Bergs, L. (2005). Goodpasture syndrome. Crit Care Nurse *25*, 50–54, 56, 57–58.

Bernocco, S., Finet, S., Ebel, C., et al. (2001). Biophysical characterization of the C-propeptide trimer from human procollagen III reveals a tri-lobed structure. J Biol Chem *276*, 48930–48936.

Bogin, O., Kvansakul, M., Rom, E., Singer, J., Yayon, A., and Hohenester, E. (2002). Insight into Schmid metaphyseal chondrodysplasia from the crystal structure of the collagen X NC1 domain trimer. Structure *10*, 165–173.

Boudko, S., Frank, S., Kammerer, R. A., et al. (2002). Nucleation and propagation of the collagen triple helix in single-chain and trimerized peptides: transition from third to first order kinetics. J Mol Biol *317*, 459–470.

Boudko, S. P., and Engel, J. (2004). Structure formation in the C terminus of type III collagen guides disulfide cross-linking. J Mol Biol *335*, 1289–1297.

Boudko, S. P., Engel, J., and Bächinger, H. P. (2008). Trimerization and triple helix stabilization of the collagen XIX NC2 domain. J Biol Chem *283*, 34345–34351.

Boudko, S. P., Sasaki, T., Engel, J., et al. (2009). Crystal structure of human collagen XVIII trimerization domain: a novel collagen trimerization fold. J Mol Biol *392*, 787–802.

Boudko, S. P., Zientek, K. D., Vance, J., Hacker, J. L., Engel, J., and Bächinger, H. P. (2010). The NC2 domain of collagen IX provides chain selection and heterotrimerization. J Biol Chem *285*, 23721–23731.

Boulegue, C., Musiol, H. J., Götz, M. G., Renner, C., and Moroder, L. (2008). Natural and artificial cystine knots for assembly of homo- and heterotrimeric collagen models. Antioxid Redox Signal *10*, 113–125.

Bruckner, P., Bächinger, H. P., Timpl, R., and Engel, J. (1978). Three conformationally distinct domains in the amino-terminal segment of type III procollagen and its rapid triple helix ↔ coil transition. Eur J Biochem *90*, 595–603.

Chen, M., Keene, D. R., Costa, F. K., Tahk, S. H., and Woodley, D. T. (2001). The carboxyl terminus of type VII collagen mediates antiparallel dimer formation and constitutes a new antigenic epitope for epidermolysis Bullosa acquisita autoantibodies. J Biol Chem *276*, 21649–21655.

Chessler, S. D., Wallis, G. A., and Byers, P. H. (1993). Mutations in the carboxyl-terminal propeptide of the pro alpha 1(I) chain of type I collagen result in defective chain association and produce lethal osteogenesis imperfecta. J Biol Chem *268*, 18218–18225.

Davis, J. M., Boswell, B. A., and Bächinger, H. P. (1989). Thermal stability and folding of type IV procollagen and effect of peptidyl-prolyl cis-trans-isomerase on the folding of the triple helix. J Biol Chem *264*, 8956–8962.

Dehan, P., Van den Heuvel, L. P., Smeets, H. J., Tryggvason, K., and Foidart, J. M. (1996). Identification of post-transplant anti-alpha 5 (IV) collagen alloantibodies in X-linked Alport syndrome. Nephrol Dial Transplant *11*, 1983–1988.
De Paepe, A., Nuytinck, L., Hausser, I., Anton-Lamprecht, I., and Naeyaert, J. M. (1997). Mutations in the COL5A1 gene are causal in the Ehlers-Danlos syndromes I and II. Am J Hum Genet *60*, 547–554.
Dölz, R., Engel, J., and Kühn, K. (1988). Folding of collagen IV. Eur J Biochem *178*, 357–366.
Engel, J., and Prockop, D. J. (1991). The zipper-like folding of collagen triple helices and the effects of mutations that disrupt the zipper. Annu Rev Biophys Biophys Chem *20*, 137–152.
Englander, S. W., Mayne, L., and Krishna, M. M. (2007). Protein folding and misfolding: mechanism and principles. Q Rev Biophys *40*, 287–326.
Frank, S., Boudko, S., Mizuno, K., Schulthess, T., Engel, J., and Bächinger, H. P. (2003). Collagen triple helix formation can be nucleated at either end. J Biol Chem *278*, 7747–7750.
Frank, S., Kammerer, R. A., Mechling, D., et al. (2001). Stabilization of short collagen-like triple helices by protein engineering. J Mol Biol *308*, 1081–1089.
Grond-Ginsbach, C., Wigger, F., Morcher, M., et al. (2002). Sequence analysis of the COL5A2 gene in patients with spontaneous cervical artery dissections. Neurology *58*, 1103–1105.
Güthe, S., Kapinos, L., Möglich, A., Meier, S., Grzesiek, S., and Kiefhaber, T. (2004). Very fast folding and association of a trimerization domain from bacteriophage T4 fibritin. J Mol Biol *337*, 905–915.
Halfter, W., Dong, S., Schurer, B., and Cole, G. J. (1998). Collagen XVIII is a basement membrane heparan sulfate proteoglycan. J Biol Chem *273*, 25404–25412.
Hertz, J. M., Heiskari, N., Zhou, J., Jensen, U. B., and Tryggvason, K. (1995). A nonsense mutation in the COL4A5 collagen gene in a family with X-linked juvenile Alport syndrome. Kidney Int *47*, 327–332.
Kalluri, R., Torre, A., Shield, C. F., III, et al. (2000). Identification of alpha3, alpha4, and alpha5 chains of type IV collagen as alloantigens for Alport posttransplant anti-glomerular basement membrane antibodies. Transplantation *69*, 679–683.
Kobayashi, T., Kakihara, T., and Uchiyama, M. (2008). Mutational analysis of type IV collagen alpha5 chain, with respect to heterotrimer formation. Biochem Biophys Res Commun *366*, 60–65.
Kvansakul, M., Bogin, O., Hohenester, E., and Yayon, A. (2003). Crystal structure of the collagen alpha1(VIII) NC1 trimer. Matrix Biol *22*, 145–152.
Lamande, S. R., Chessler, S. D., Golub, S. B., et al. (1995). Endoplasmic reticulum-mediated quality control of type I collagen production by cells from osteogenesis imperfecta patients with mutations in the pro alpha 1 (I) chain carboxyl-terminal propeptide which impair subunit assembly. J Biol Chem *270*, 8642–8649.
Lamande, S. R., Mörgelin, M., Selan, C., Jobsis, G. J., Baas, F., and Bateman, J. F. (2002). Kinked collagen VI tetramers and reduced microfibril formation as a result of Bethlem myopathy and introduced triple helical glycine mutations. J Biol Chem *277*, 1949–1956.
Latvanlehto, A., Snellman, A., Tu, H., and Pihlajaniemi, T. (2003). Type XIII collagen and some other transmembrane collagens contain two separate coiled-coil motifs, which may function as independent oligomerization domains. J Biol Chem *278*, 37590–37599.
Lesage, A., Penin, F., Geourjon, C., Marion, D., and van der Rest, M. (1996). Trimeric assembly and three-dimensional structure model of the FACIT collagen COL1-NC1 junction from CD and NMR analysis. Biochemistry *35*, 9647–9660.
Li, D., Clark, C. C., and Myers, J. C. (2000). Basement membrane zone type XV collagen is a disulfide-bonded chondroitin sulfate proteoglycan in human tissues and cultured cells. J Biol Chem *275*, 22339–22347.
Lindahl, K., Barnes, A. M., Fratzl-Zelman, N., et al. (2011). COL1 C-propeptide cleavage site mutations cause high bone mass osteogenesis imperfecta. Hum Mutat *32*, 598–609.

Malone, J. P., Alvares, K., and Veis, A. (2005). Structure and assembly of the heterotrimeric and homotrimeric C-propeptides of type I collagen: significance of the alpha2(I) chain. Biochemistry *44*, 15269–15279.

Marneros, A. G., and Olsen, B. R. (2005). Physiological role of collagen XVIII and endostatin. FASEB J *19*, 716–728.

McAlinden, A., Smith, T. A., Sandell, L. J., Ficheux, D., Parry, D. A., and Hulmes, D. J. (2003). Alpha-helical coiled-coil oligomerization domains are almost ubiquitous in the collagen superfamily. J Biol Chem *278*, 42200–42207.

Mechling, D. E., Gambee, J. E., Morris, N. P., et al. (1996). Type IX collagen NC1 domain peptides can trimerize in vitro without forming a triple helix. J Biol Chem *271*, 13781–13785.

Mitchell, A. L., Schwarze, U., Jennings, J. F., and Byers, P. H. (2009). Molecular mechanisms of classical Ehlers-Danlos syndrome (EDS). Hum Mutat *30*, 995–1002.

Miyahara, M., Njieha, F. K., and Prockop, D. J. (1982). Formation of collagen fibrils in vitro by cleavage of procollagen with procollagen proteinases. J Biol Chem *257*, 8442–8448.

Nakazato, H., Hattori, S., Matsuura, T., Koitabashi, Y., Endo, F., and Matsuda, I. (1993). Identification of a single base insertion in the COL4A5 gene in Alport syndrome. Kidney Int *44*, 1091–1096.

Oh, S. P., Warman, M. L., Seldin, M. F., et al. (1994). Cloning of cDNA and genomic DNA encoding human type XVIII collagen and localization of the alpha 1(XVIII) collagen gene to mouse chromosome 10 and human chromosome 21. Genomics *19*, 494–499.

O'Reilly, M. S., Boehm, T., Shing, Y., et al. (1997). Endostatin: an endogenous inhibitor of angiogenesis and tumor growth. Cell *88*, 277–285.

Pace, J. M., Kuslich, C. D., Willing, M. C., and Byers, P. H. (2001). Disruption of one intra-chain disulphide bond in the carboxyl-terminal propeptide of the proalpha1(I) chain of type I procollagen permits slow assembly and secretion of overmodified, but stable procollagen trimers and results in mild osteogenesis imperfecta. J Med Genet *38*, 443–449.

Pace, J. M., Wiese, M., Drenguis, A. S., et al. (2008). Defective C-propeptides of the proalpha2(I) chain of type I procollagen impede molecular assembly and result in osteogenesis imperfecta. J Biol Chem *283*, 16061–16067.

Panchenko, A. R., Luthey-Schulten, Z., Cole, R., and Wolynes, P. G. (1997). The foldon universe: a survey of structural similarity and self-recognition of independently folding units. J Mol Biol *272*, 95–105.

Pedchenko, V., Bondar, O., Fogo, A. B., et al. (2010). Molecular architecture of the Goodpasture autoantigen in anti-GBM nephritis. N Engl J Med *363*, 343–354.

Pickup, M. J., and Pollanen, M. S. (2011). Traumatic subarachnoid hemorrhage and the COL3A1 gene: emergence of a potential causal link. Forensic Sci Med Pathol *7*, 192–197.

Pope, F. M., Holden, S., Ferguson, D. J. P., et al. (2010). Abstracts of the British Human Genetics Conference 2010. University of Warwick, 6–8 September 2010. J Med Genet *47* (Suppl. 1), 20.

Rehn, M., Hintikka, E., and Pihlajaniemi, T. (1994). Primary structure of the alpha 1 chain of mouse type XVIII collagen, partial structure of the corresponding gene, and comparison of the alpha 1(XVIII) chain with its homologue, the alpha 1(XV) collagen chain. J Biol Chem *269*, 13929–13935.

Rehn, M., and Pihlajaniemi, T. (1994). Alpha 1(XVIII), a collagen chain with frequent interruptions in the collagenous sequence, a distinct tissue distribution, and homology with type XV collagen. Proc Natl Acad Sci U S A *91*, 4234–4238.

Risteli, J., Bächinger, H. P., Engel, J., Furthmayr, H., and Timpl, R. (1980). 7-S collagen: characterization of an unusual basement membrane structure. Eur J Biochem *108*, 239–250.

Sanchez-Arevalo Lobo, V. J., Cuesta, A. M., Sanz, L., et al. (2006). Enhanced antiangiogenic therapy with antibody-collagen XVIII NC1 domain fusion proteins engineered to exploit matrix remodeling events. Int J Cancer *119*, 455–462.

Sasaki, T., Fukai, N., Mann, K., Göhring, W., Olsen, B. R., and Timpl, R. (1998). Structure, function and tissue forms of the C-terminal globular domain of collagen XVIII containing the angiogenesis inhibitor endostatin. EMBO J *17*, 4249–4256.

Sasaki, T., Larsson, H., Tisi, D., Claesson-Welsh, L., Hohenester, E., and Timpl, R. (2000). Endostatins derived from collagens XV and XVIII differ in structural and binding properties, tissue distribution and anti-angiogenic activity. J Mol Biol *301*, 1179–1190.

Schwarze, U., Schievink, W. I., Petty, E., et al. (2001). Haploinsufficiency for one COL3A1 allele of type III procollagen results in a phenotype similar to the vascular form of Ehlers-Danlos syndrome, Ehlers-Danlos syndrome type IV. Am J Hum Genet *69*, 989–1001.

Shapiro, L., and Scherer, P. E. (1998). The crystal structure of a complement-1q family protein suggests an evolutionary link to tumor necrosis factor. Curr Biol *8*, 335–338.

Shaw, L. M., and Olsen, B. R. (1991). FACIT collagens: diverse molecular bridges in extracellular matrices. Trends Biochem Sci *16*, 191–194.

Snellman, A., Tu, H., Vaisanen, T., Kvist, A. P., Huhtala, P., and Pihlajaniemi, T. (2000). A short sequence in the N-terminal region is required for the trimerization of type XIII collagen and is conserved in other collagenous transmembrane proteins. EMBO J *19*, 5051–5059.

Snellman, A., Tuomisto, A., Koski, A., Latvanlehto, A., and Pihlajaniemi, T. (2007). The role of disulfide bonds and alpha-helical coiled-coils in the biosynthesis of type XIII collagen and other collagenous transmembrane proteins. J Biol Chem *282*, 14898–14905.

Standker, L., Schrader, M., Kanse, S. M., Jurgens, M., Forssmann, W. G., and Preissner, K. T. (1997). Isolation and characterization of the circulating form of human endostatin. FEBS Lett *420*, 129–133.

Sundaramoorthy, M., Meiyappan, M., Todd, P., and Hudson, B. G. (2002). Crystal structure of NC1 domains. Structural basis for type IV collagen assembly in basement membranes. J Biol Chem *277*, 31142–31153.

Than, M. E., Henrich, S., Huber, R., et al. (2002). The 1.9-A crystal structure of the noncollagenous (NC1) domain of human placenta collagen IV shows stabilization via a novel type of covalent Met-Lys cross-link. Proc Natl Acad Sci U S A *99*, 6607–6612.

Wilson, J. C., Yoon, H. S., Walker, R. J., and Eccles, M. R. (2007). A novel Cys1638Tyr NC1 domain substitution in alpha5(IV) collagen causes Alport syndrome with late onset renal failure without hearing loss or eye abnormalities. Nephrol Dial Transplant *22*, 1338–1346.

Wilson, R., Freddi, S., Chan, D., Cheah, K. S., and Bateman, J. F. (2005). Misfolding of collagen X chains harboring Schmid metaphyseal chondrodysplasia mutations results in aberrant disulfide bond formation, intracellular retention, and activation of the unfolded protein response. J Biol Chem *280*, 15544–15552.

Wirz, J. A., Boudko, S. P., Lerch, T. F., Chapman, M. S., and Bächinger, H. P. (2011). Crystal structure of the human collagen XV trimerization domain: a potent trimerizing unit common to multiplexin collagens. Matrix Biol *30*, 9–15.

Zhang, Y., and Chen, Q. (1999). The noncollagenous domain 1 of type X collagen. A novel motif for trimer and higher order multimer formation without a triple helix. J Biol Chem *274*, 22409–22413.

6.3 Structural basis of collagen missense mutations

Anton V. Persikov and Barbara Brodsky

6.3.1 Introduction: collagens and disease

The first cases of mutations in collagen leading to a hereditable disease were reported in the 1970s and 1980s, where mutations in type I collagen were found to lead to the bone disorder osteogenesis imperfecta (OI) (Byers and Cole, 2002). Since then mutations in many different collagen genes have been associated with a range of clinical pathologies (Myllyharju and Kivirikko, 2004). Disease-causing mutations have been identified in all of the fibrillar collagen genes and in many nonfibrillar collagens as well. The most frequent types of mutations found are missense mutations and splicing errors, but frameshift mutations, deletions, and all other possibilities are observed as well. The presence of a triple-helix domain is a defining feature of all collagens, with its unique $(\text{Gly-Xaa-Yaa})_n$ repeating sequence. A common class of collagen mutations involves a single base substitution that leads to the replacement of one Gly in the $(\text{Gly-Xaa-Yaa})_n$ repeating sequence by a larger residue. This chapter will focus on Gly missense mutations in fibrillar type I collagen chains that lead to OI.

Although more than 1,000 Gly missense mutations in various collagens have been reported to result in disease, the sequence of events that lead from a single amino acid change to a pathological phenotype is not clear. Different approaches have been used to clarify the consequences of collagen mutations (▶Figure 6.2). These include studying collagens expressed by OI fibroblasts and OI cell cultures (Byers and Cole, 2002), characterizing OI mouse models (Forlino et al., 2007), studies on model peptides containing Gly missense mutations (Baum and Brodsky, 1999), molecular modeling and dynamics simulations of triple helices with Gly mutations (Mooney and Klein, 2002), and computational analysis of mutation data to obtain biological correlations (Persikov et al., 2004a; Marini et al., 2007). These studies have identified alterations in structure and stability, defects in molecular folding, secretion problems, abnormal fibrils, and potential abnormal interactions with proteoglycans or other matrix molecules. In this chapter, peptide and computational studies will be considered, together with a recent approach of incorporating mutations within a recombinant bacterial collagen system.

6.3.2 Peptide models of collagen mutations

Full-length type I collagen has not proved an easy sample for modern biophysical investigations. Synthetic peptides, however, which adopt the collagen triple-helix structure,

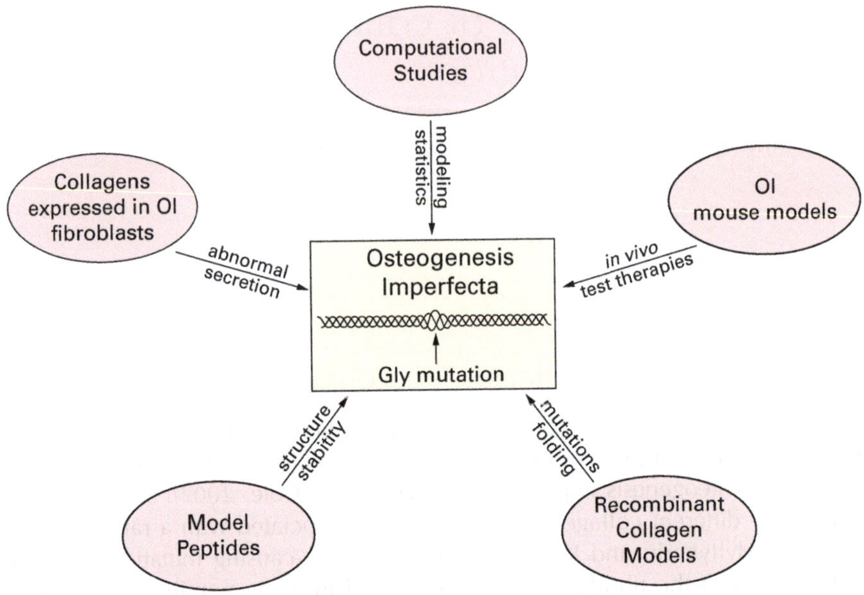

Figure 6.2 An overview of different approaches used to understand how Gly missense mutations in type I collagen lead to osteogenesis imperfecta.

are amenable to many techniques and have served as good models. Even relatively short peptides (~24–30 residues long) with Gly as every third residue and a high imino acid content can adopt a triple-helical conformation similar to that of collagen at neutral pH, as well as at acid pH values (Suto and Noda, 1974; Brodsky et al., 2008). While a crystal structure for purified native collagen has not been obtained yet, collagen peptides can be crystallized and are amenable to X-ray structure determination. The structure of the first collagen peptide was reported by Okuyama in 1972 and as of this date, more than 20 high-resolution triple-helical peptide structures have been determined by X-ray crystallography (Kramer et al., 2001; Berisio et al., 2002; Brodsky and Persikov, 2005; Boudko et al., 2008; Okuyama, 2008). In addition, multidimensional nuclear magnetic resonance (NMR) studies on peptides with selectively labeled amino acids has provided residue specific information on dynamics, folding and conformation (Baum and Brodsky, 1999; Buevich et al., 2000).

The ease of obtaining synthetic peptides of well-defined sequence has made it possible to study the effect of systematic variations in Xaa and Yaa positions on thermal stability. This has allowed determination of individual position-specific amino acid propensities (Persikov et al., 2000) and evaluation of amino acid pairwise interactions for the collagen triple helix (Persikov et al., 2002, 2005b), which were used to create an algorithm to predict triple-helix stability for any (Gly-Xaa-Yaa)$_n$ sequence (Collagen Stability Calculator is available at http://compbio.cs.princeton.edu/csc/) (Persikov et al., 2005a). Detailed investigations have shown that some triple-helical peptides are capable of reversible folding/unfolding, allowing rigorous thermodynamic analysis (Persikov et al., 2004b). Peptides were also used to analyze the importance of posttranslation proline hydroxylation, as well as the role of various proline variants on collagen conformation and stability (Persikov et al., 2004b; Doi et al., 2005; Mizuno et al., 2008; Shoulders et al., 2010). In addition, the Toolkit peptide sets from Richard Farndale's laboratory, which represent overlapping sequences from type II and III collagens, showed stability variations (Slatter et al., 2011) and have been key in determining the collagen-binding sites for receptors (integrins, discoidin domain receptor [DDR], and leukocyte-associated Ig-like receptor [LAIR]), as well as other extracellular matrix molecules such as secreted protein acidic and rich in cysteine (SPARC) (Farndale et al., 2008). More details are found in recent reviews on collagen peptides (Brodsky and Persikov, 2005; Koide, 2007; Brodsky et al., 2008; Fallas et al., 2010).

Triple-helical peptides allow the definition of the structural consequences of replacing one Gly in the repeating (Gly-Xaa-Yaa)$_n$ sequence by a larger residue and can serve as models for Gly missense mutations in collagens that cause OI, Ehlers-Danlos syndrome (EDS-IV), and other collagen diseases. The structural perturbation resulting from a Gly substitution has been defined for several peptides and the physical chemical consequences of having such mutations in different sequence environments has been examined. A high-resolution (1.9Å) crystal structure was obtained for a collagen triple helix with a Gly→Ala mutation: (Pro-Hyp-Gly)$_4$-Pro-Hyp-Ala-(Pro-Hyp-Gly)$_5$ (PDB ID 1CAG) (Bella et al., 1994). This structure confirmed the basic features of the collagen superhelix and showed that a Gly→Ala substitution interrupted the regular pattern of direct interchain Gly NH...CO (Xaa) hydrogen bonds. At the mutation site, the direct hydrogen bonds were replaced by water mediated hydrogen bonds involving the Ala NH, and led to local alterations in the φ,ψ angles. These changes resulted in a local untwisting of the triple helix, which put the superhelix at the N-terminal end of the chain out of register with the C-terminal end (Bella et al., 1994). The conformational disruptions observed could contribute to quality control problems in the endoplasmic reticulum and may interfere with binding at the molecular or fibrillar level. The loss of registration could disrupt the overlapped packing of type I collagen molecules to form D periodic fibrils. The structural and dynamic consequences of a Gly→Ser mutation were later addressed by NMR studies (Li et al., 2009). This study indicated that while the three Ser residues are still packed in the center of the triple helix, the distortion of having the side chains present leads to disruption of hydrogen bonding, with the greater perturbation seen C terminal to the Ser mutation site when compared with the effects N terminal to the site (▶Figure 6.3). This is consistent with recent suggestions relating clinical severity with an asymmetric effect of residues C terminal versus N terminal to a mutation site (Bodian et al., 2008).

A systematic study on the effect of the identity of the residue replacing Gly was carried out using a set of host-guest peptides. The Gly residue (Zaa position) in the middle of the peptide (Gly-Pro-Hyp)$_3$-Zaa-Pro-Hyp-(Gly-Pro-Hyp)$_4$ was systematically replaced by all 8 possible amino acids resulting from a single nucleotide substitution in the Gly codon (Beck et al., 2000). Any Gly→Zaa replacement resulted in a dramatic destabilization, but the decrease in thermal stability depended on which residue type replaced the Gly. The destabilizing effect of Zaa = Ala or Ser was the least, while other residues resulted in greater destabilization of the peptide, requiring addition of the molecular chaperone trimethylamine N-oxide (TMAO) for the peptides to form a collagen helix. The order of disruption of different Gly replacements in these peptides was determined as Ala ≤ Ser < Cys < Arg < Val < Glu ≤ Asp. The rank of destabilization for amino acids replacing Gly showed a good correlation with the severity of OI cases

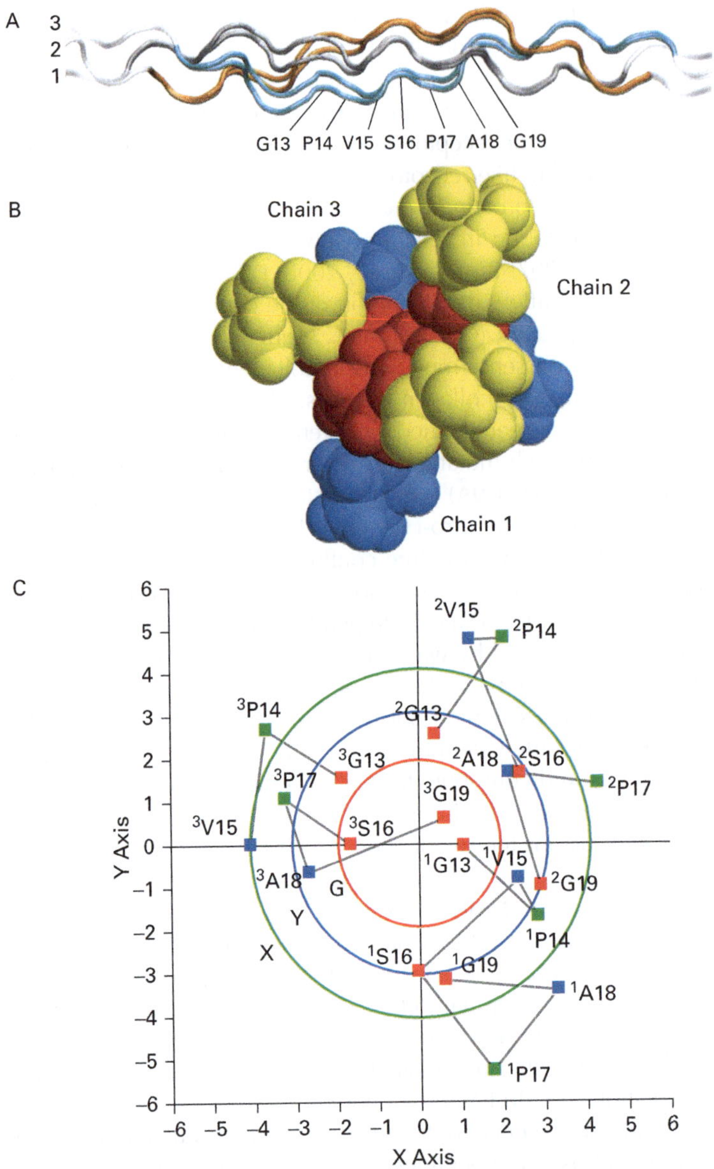

Figure 6.3 Structure of a peptide containing a Gly to Ser mutation (peptide T1–898) obtained by modeling to fit NMR data. (A) A ribbon model; (B) space-filling model; and (C) plot of the projected cross-section positions (taken from Li et al., 2009).

caused by corresponding mutations in the α1 chain of type I collagen. For example, while G382S mutation leads to the mild type of OI, the Gly→Arg mutation at the same position leads to the lethal type II form of OI (Marini et al., 2007). In contrast to the peptide results, studies on collagens secreted by OI fibroblasts in the Leikin laboratory (Makareeva et al., 2008) did not find any relation between the identity of the residue replacing Gly and the drop in global thermal stability of the OI collagen, which may reflect the much greater triple-helix length of type I collagen compared with peptides. But the local disruptions that are magnified in studies on short peptides may have consequences for folding and/or binding, which could be important in the etiology of OI and other collagen diseases.

The same missense mutation may lead to different clinical phenotypes when found at different sites in the collagen molecule; for example, Gly→Ser at position 964 in the α1 (I) chain leads to a lethal phenotype, while Gly→Ser at residue 832 leads to a mild phenotype (▶Table 6.3). Makareeva et al. (2008) have suggested the importance of the location with respect to folding domains, and it had been earlier proposed that the sequence environment of the Gly missense mutation could influence the level of structural perturbation in collagens (Bächinger et al., 1993). The impact of local environment and the identity of the residue replacing Gly has been investigated by changing the sequence surrounding the mutation in synthetic peptides (Bhate et al., 2002; Hyde et al., 2006). Recently, it was reported that the peptide destabilization due to a Gly→Ala or Gly→Ser change was greater in more rigid $(\text{Pro-Hyp-Gly})_{10}$ peptides compared to the same mutations introduced within a less stable environment, which better mimics natural type I collagen sequences (Bryan et al., 2010). In both peptide environments, a Gly→Arg replacement prevented the formation of a fully folded triple helix (▶Figure 6.4).

Table 6.3 Effect of Gly missense mutations on the stability changes of the synthetic peptide models for mutations observed in human genetic disorders. ΔT_m values represent the difference between the melting temperature of the wild-type and mutant peptides; $\Delta\Delta H$ shows the decrease of unfolding enthalpy upon mutation; the plus in GP column indicates the presence of Pro next to the Gly being mutated.

Peptide	*Internal sequence*	*GP*	*ΔT_m, °C*	*$\Delta\Delta H$, kJ/mol*	*Phenotype observed*
A1-G382S	GPOGPA**S**QDGRO		12.8	167	OI type IV
A1-G448S	GPAGPA**S**ERGEQ		14.1	n/a	OI type IV
A1-G661S	GAKGDA**S**POGPA	+	23.5	256	OI type III
A2-G661S	GPKGEN**S**VVGPT		19.0	67	Osteoporosis
A1-G832S	GLAGPO**S**ESGRE		13.9	n/a	OI type IV
A1-G862S	GPKGDRGET**S**PA	+	26.2	288	OI type II/III
A1-G964S	GPAGPR**S**POGSA	+	24.4	277	OI type II
A1-G382R	GPOGPA**R**QDGRO		22.3	n/a	OI type II
A1-G382S	GPOGPA**S**QDGRO		12.8	167	OI type IV
T3-G16S	GPOGPO**S**POGTS	+	22.4	103	EDS-IV
T3-G19S	GPOGPO**S**TSGPO		18.0	59	EDS-IV
T3-G70R	GESGROR**R**OGER		32.7	229	EDS-IV
T3-G277R	GAKGEO**R**PRGER	+	44.6	302	EDS-IV
T3-G415S	GVMGFO**S**PKGND	+	25.2	163	EDS-IV
T3-G415R	GVMGFO**R**PKGND	+	34.1	321	EDS-IV
T3-G619R	GEGGAO**R**LOGIA		28.9	324	Non-EDS

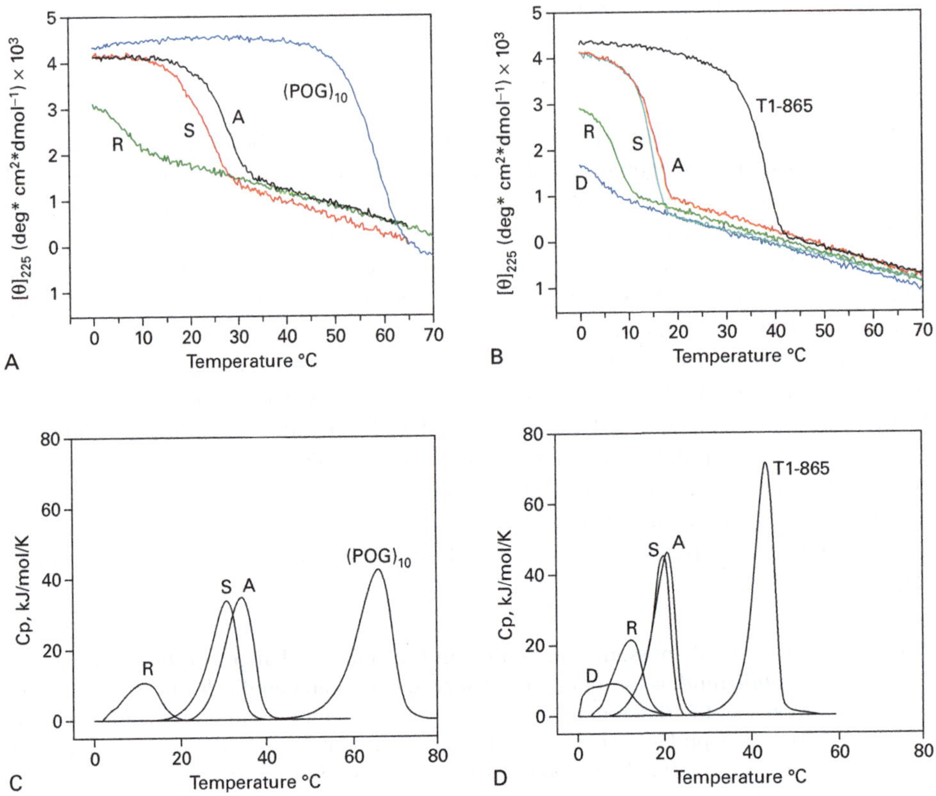

Figure 6.4 The effects of Gly substitutions in different peptide environments. (A) Within a (Pro-Hyp-Gly)$_{10}$ contest ((POG)$_{10}$ in short), the circular dichroism (CD) thermal folding profiles are shown for the control peptide; a Gly to Ala substitution (A); a Gly to Ser substitution (S); and a Gly to Arg substitution (R). (B) Within the context of a peptide containing a sequence from α1(I) chain, T1–865, the CD thermal folding profiles are shown for the control peptide (black); a Gly to Ala substitution (A); a Gly to Ser substitution (S); a Gly to Arg substitution (R); and a Gly to Asp substitution (D). (C) Differential scanning calorimetry (DSC) transitions of (POG)$_{10}$ with Gly substitutions denoted by the same notations used in (A). (D) DSC transitions of T1–865 control and with Gly substitutions denoted by the same notations used in (B) (based on Bryan et al., 2010).

A systematic study was recently performed studying the effect of local sequence environment on the triple-helix destabilization caused by mild (Gly→Ser) and severe (Gly→Arg) mutations. In this study natural type I or type III collagen regions including four tripeptide units surrounding an OI or EDS-IV mutation site were introduced in the middle of a longer synthetic peptide using the following design:

Ac-GPOGPOGPOGPO*GXYGXY**G**XYGXY*GPOGPOGPOGPO-GY-[NH$_2$]

where the middle 12 residues (Gly-X-Y)$_4$ were taken from the natural collagen sequence for one peptide and where one Gly (bold) was replaced by Ser or Arg in a second homologous peptide. The underlined flanking Gly-Pro-Hyp sequences were constant to ensure complete folding of normal and mutant peptides into triple helices even when a severe mutation, such as Gly→Arg is introduced.

The decrease in thermal stability (ΔT_m) as a consequence of a Gly substitution was measured by circular dichroism spectroscopy and differential scanning calorimetry for each set of peptide pairs modeling 16 OI- or EDS-causing mutations and their native matching peptide (▶Table 6.3) (Persikov, unpublished data).

The results on these peptide models confirmed that Gly→Arg mutations (average ΔT_m = 35.1 ± 6.7°C) are more destabilizing for the triple helix when compared to Gly→Ser (average ΔT_m = 21.9 ± 3.6°C), regardless of the amino acid environment. Consistent with previous observations (Beck et al., 2000; Bryan et al., 2010), more destabilizing substitutions cause more severe phenotypes. For example, a Gly382Arg mutation observed in lethal OI type II leads to greater peptide destabilization compared with Gly382Ser, which causes milder type IV OI. In general, a given type of missense mutation (e.g. Gly→Ser) leads to more severe phenotypes when it leads to greater peptide destabilization. As a rule of thumb, if ΔT_m < 20°C, it results in milder OI phenotypes, while destabilization values greater than 20°C lead to severe type II/III cases of OI. It was further determined that the same mutation is more destabilizing when the original Gly site of the mutation is followed by a Pro residue (▶Table 6.3). When a Gly→Ser mutation is followed by Pro residue (GP sequence), the average destabilization is ΔT_m = 24.3 ± 1.5°C, while the average for all other sequences is ΔT_m = 15.8 ± 2.9°C.

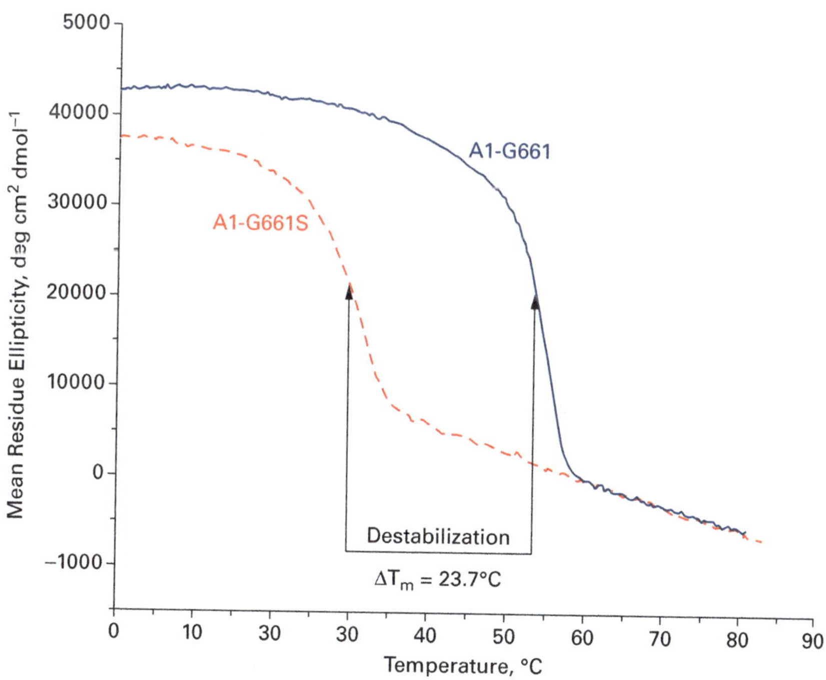

Figure 6.5 Thermal destabilization of OI peptides containing Gly→Ser mutations. Thermal unfolding is monitored by CD at 225 nm. Destabilization is measured as the difference between meting temperatures for wild-type (blue solid line) and mutant (red dashed line) peptides (Persikov, unpublished data).

The presence of Pro next to a Gly mutation may place more restrictions on the φ and ψ angles, resulting in more severe distortion of interchain hydrogen bonding. This result is confirmed and discussed further in the context of computational analyses in Section 6.3.3 of this chapter.

A mutation in a peptide creates self-associating homotrimeric molecules containing the mutation in all three chains, while OI and many other collagen diseases are dominant disorders that result in molecules with only one or two chains containing the mutation. Recently, a novel design was utilized to model heterotrimeric synthetic peptides by utilizing an electrostatic recognition motif (Gauba and Hartgerink, 2008; Fallas et al., 2010). This design can force the interaction of any three peptides, including AAA (all same), AAB (two same and one different), or ABC (all different) triple helices. An accurate register of the synthetic heterotrimers was also confirmed by NMR studies (Fallas et al., 2009). The set of component peptides was designed in such a way that Gly to Ser mutations are present in zero, one, two, or all three chains of the triple helix (Brodsky and Baum, 2008; Gauba and Hartgerink, 2008). The trimer-to-monomer thermal transition showed that a single chain Gly→Ser mutation substantially reduces the triple-helix stability, while subsequent mutations in the second and third chains lead to less drastic destabilization. This approach offers promise in better modeling the heterotrimeric nature of collagen molecules with mutations.

In summary, synthetic peptides currently remain as useful models for studying collagen missense mutations: they can fold and unfold reversibly into a collagen triple-helical structure; they are suitable for thermodynamic analysis; and the structural perturbation caused by the missense mutations has been determined by both X-ray crystallography and solution NMR. By allowing high resolution structural data, such model peptides can lead to a better understanding of the structural changes underlying collagen genetic disorders.

6.3.3 Computational analysis

While substantial progress has been made in elucidating molecular mechanisms leading to pathological phenotypes in collagen genetic disorders, many questions have not been answered experimentally. For example, why does the same missense mutation at different sites results in a wide phenotypic spectrum from normal to prenatally lethal? Modern computational methods offer a complementary approach to analyze the relationships between pathological alterations in collagen molecules and phenotypes observed in growing clinical databases. The computational approaches for studying collagen genetic disorders can be subdivided in several groups: (1) molecular modeling and molecular dynamics, showing how pathological mutations affect the triple-helical structure and dynamics; (2) statistical analysis of clinical databases to locate correlations between expected and observed mutation spectra; and (3) development of methods for predicting phenotypic changes resulting from a given mutation, based on clinical databases.

Molecular modeling and dynamics utilize theoretical and computational methods to simulate the behavior of molecules. New force fields developed for the simulation of proteins and exponentially increasing power of computer systems since the turn of the century can be applied to the current high resolution structures of collagen, to better model the consequences of mutations in collagen molecules. Molecular mechanics calculations using the ZMM program with Monte Carlo minimization showed that hydrogen bonding network disruption and hydration changes depend strongly on the identity of the amino acid replacement for Gly in clinically observed EDS-IV-associated type III collagen mutations (Milchevsky et al., 2001). The Klein group, doing molecular dynamics simulations of type I collagen OI mutations utilizing the Assisted Model Building with Energy Refinement (AMBER) force field, showed the degree of perturbation as measured by root-mean-square deviation (RMSD) depended on the residue replacing Gly and found lethal mutations had fewer backbone hydrogen bonds than nonlethal mutations (Mooney and Klein, 2002). A positive correlation was observed between the severity of the OI and amino acid residues near the site of the mutation, including weaker interchain hydrogen bonding in the proline-rich regions of the collagen (Radmer and Klein, 2004). Systematic stochastic boundary molecular dynamics simulations were used to estimate the energetic changes of having one or two residues replacing Gly, as observed in OI (Lee et al., 2011a, 2011b). In general, the destabilization scale for residues replacing Gly correlated well with the experimental disruption scale observed in synthetic peptides (Beck et al., 2000) (see Section 6.3.2 of this chapter). Moreover, the calculated energetic changes indicated that the introduction of a mutation into the first chain is more destabilizing than introduction of a second mutation (Lee et al., 2011a, 2011b), consistent with the experimental measurements on heterotrimeric peptides containing mutations (Gauba and Hartgerink, 2008).

A number of studies have employed statistical analysis of clinical databases to try to establish correlations that may clarify the molecular features leading to disease (Byers and Cole, 2002; Marini et al., 2007). One systematic analysis was performed to analyze the expected spectrum of Gly missense mutations compared to those observed for the following collagen genetic disorders and their corresponding genes: OI (COL1A1 and COL1A2), EDS-IV (COL3A1), Alport syndrome (COL4A5), and dystrophic epidermolysis bullosa (COL7A1) (Persikov et al., 2004a). The question was raised whether all Gly missense mutations occurring in these genes result in pathological phenotypes that come to clinical attention. Expected spontaneous mutation spectra of different Gly replacements were computed on the basis of known single base-pair mutation rates, including the effect of nearest neighbors, and these were compared with observed missense mutations known to cause disease. The spectrum of amino acids replacing Gly was significantly different from that expected for the COL1A1 and COL3A1 collagen genes (Persikov et al., 2004a). For α1(III) collagen chains, the most destabilizing mutations were identified more often than expected in cases of EDS-IV. For α1(I), the most destabilizing residues, Val, Glu, and Asp, and the least destabilizing residue, Ala, were underrepresented in OI cases. The underrepresentation of Ala is consistent with a phenotype so mild that it is not clinically detected, while the underrepresentation of the most destabilizing residues may represent nonviable situations. It is suggested that only intermediate outcomes between these extremes are clinically observed for OI (▶Figure 6.6). These results and interpretation support the hypothesis that the destabilization of the triple helix is the main determinant of the phenotypes observed in OI and EDS-IV.

A number of studies have investigated the frequency of residues in the environment surrounding mutations, to see if there are unusual residues that may play a role in the lethal versus nonlethal phenotype (Radmer and Klein, 2004). A recent study on type III collagen sequence investigating the eight X and eight Y positions around the mutation

Order of destabilization→

←Clinical Severity		Gly→	Ala	Ser	Cys	Arg	Val	Glu	Asp
		Normal	[17]	[0]	[1]	[1]	[1]	[0]	[0]
	Clinically detectable	Non-lethal	6	30	14	15	2	1	1
		Lethal	3	29	5	13	10	1	8
		Not viable	[0]	[0]	[0]	[8]	[10]	[9]	[36]
		Expected Total	26	59	20	37	23	11	45

Figure 6.6 Illustration of the hypothetical distribution of Gly substitutions in COL1A1 into four clinical outcomes, together with the order of their corresponding destabilization of the triple helix (taken from Persikov et al., 2004a).

site showed that almost all the amino acid residues distributed randomly (Persikov, unpublished data). Surprisingly, Pro residue in X position next to the substituted Gly was significantly different from that expected on the basis of random occurrences. Computational analysis showed that when Gly mutation is followed by Pro residue (GP sequence), it more often leads to EDS-IV phenotypes (Persikov, unpublished data).

A large number of experimental studies, together with computational analyses, suggest that factors that play a role in determining the phenotypic outcome include the following: (1) the identity of the residue replacing Gly, (2) the position of the mutation along the triple helix with respect to folding domains and with respect to the N and C terminus, (3) the local environment of the mutation site, (4) any alteration of the functionally important regions (such as integrin-binding sites). This information should provide a basis for the development of an algorithm to predict the clinical outcome of any Gly missense mutation observed in collagens. Decision trees were developed to make such a predictions (Bodian et al., 2008). A computational model that predicts the outcome of substitutions for Gly within the COL1A1 gene predicted the lethality in OI cases with ~90% accuracy.

6.3.4 Collagen mutations in a recombinant bacterial system

A recombinant bacterial collagen system has been utilized recently to investigate the effects of a Gly missense mutation on the collagen triple helix. Using expression of a bacterial collagen-like protein in *Escherichia coli*, it is easy to vary the residue replacing Gly, the location of the mutation, and the environment of the mutation, so that the influence of such factors can be investigated. First, the bacterial collagen system will be described, followed by recent results on mutations in this context.

Recombinant bacterial collagen system

The extracellular matrix collagen is a defining feature of multicellular animals, and the collagen triple-helix motif was considered to be absent in bacteria and viruses, as well as plants. However, genome sequencing revealed that more than 100 hypothetical proteins in different bacteria contain long collagen-like domains with Gly as every third residue and a high Pro content (Rasmussen et al., 2003). Studies on recombinant bacterial collagen-like proteins indicate they adopt a typical triple-helical structure, with a characteristic collagen CD spectrum and high thermal stabilities (T_m = ~34°C–37°C) (Mohs et al., 2007; Xu et al., 2002, 2010). The functions of bacterial collagen domains are not known, although there is evidence that some are cell surface proteins that may operate as collagen mimics to bind to α2β1 integrins and enhance infectivity (Caswell et al., 2008). Other collagen-like proteins may be involved in immunological masking and structural protection. The gram positive bacterium *Streptococcus pyogenes* contains two collagen-like proteins (Xu et al., 2002), defined as *Streptococcus* collagen-like proteins 1 and 2 (Scl1 and Scl2), which have been studied extensively in the Lukomski laboratory, and a construct containing a portion of the Scl2 protein has been used for structural and mutation studies.

Proteins based on the *S. pyogenes* Scl2 protein contain an N-terminal globular domain, denoted as V, adjacent to a collagen domain $(\text{Gly-Xaa-Yaa})_{79}$, denoted as CL. The V-CL protein, and a longer version, V-CL-CL have been efficiently expressed and purified from a high yield cold shock vector system (Mohs et al., 2007; Yoshizumi et al., 2009; Yu et al., 2010, 2011). The N-terminal V domain has been shown to have substantial α-helix content and to act as a trimerization domain that is essential for the refolding of the CL triple helix. This domain organization suggests that the folding of V-CL constructs may serve as a good model for the folding of human fibril-forming collagens that also contain a globular domain at one end that mediates trimerization. Scl2 and other bacterial collagen-like proteins do not contain hydroxyproline (Hyp), since bacteria in general do not appear to have the enzyme prolyl hydroxylase needed for this posttranslational modification (Myllyharju, 2009). Hyp residues make significant contributions to the thermal stability of all animal collagens, but the highly charged Scl2 protein achieves a T_m ~ 36°C–37°C stability in the absence of Hyp. Studies on V-CL indicate electrostatic interactions play an important stabilizing role in the Scl2 triple helix, as they appear to do in a number of bacterial collagen proteins (Mohs et al., 2007; Xu et al., 2010).

It is problematic to express recombinant animal collagens in bacteria because they require posttranslational formation of Hyp for triple helix stability. The ability to achieve a stable triple helix without Hyp allows high yields and easy modification of the V-CL based constructs in an *E. coli* expression system. This bacterial system allows the characterization of a pure well defined homotrimeric triple helix containing Gly missense mutations.

Introduction of Gly missense mutations within V-CL collagen domain

Single base substitutions leading to a Gly missense mutation were introduced into the CL domain of the V-CL protein using site-directed mutagenesis (Cheng et al., 2011). Initially, a Gly→Ser mutation was introduced at position 199 in the protein, changing the sequence Gly-Pro-Met-<u>Gly</u>-Pro-Ala to Gly-Pro-Met-<u>Ser</u>-Pro-Ala. Residue 199 is located within the main triple-helical domain, with 34 tripeptide units N terminal to the site, and 43 tripeptide units C terminal to it (▶Figure 6.7). The protein, denoted as V-CL (G199S), was expressed and compared to the control V-CL protein. The CD spectra of V-CL and V-CL(G199S) both showed a characteristic collagen maximum at 220nm and a minimum near 198nm. The magnitudes of these extrema were lower than found for a pure triple helix, such as the isolated CL domain, because of the opposing contributions of the α helix containing V domain (Yoshizumi et al., 2009; Yu et al., 2010). Trypsin digestion of V-CL and V-CL(G199S) cleaved off the V domain, leaving intact CL domains in both cases (▶Figure 6.8). The features of the CD spectrum and the resistance of the CL domain to trypsin both indicate that the introduction of a Gly→Ser mutation within the center of CL did not lead to any significant disruption of the triple helix.

The thermal stability of the proteins was monitored by CD and DSC and sharp thermal transitions were observed in both cases. A small but highly reproducible and significant decrease in the thermal stability was observed as a result of the Gly to Ser replacement in both the V-CL protein and the isolated CL domain. The CD showed a decrease of 1.5°C–2°C, while DSC, which has a faster heating rate, indicated a decrease of 2°C–3°C. All proteins expressed in the Scl2 V-CL system are homotrimers, compared to the heterotrimeric composition of human type I collagens and the heterotrimers present when there is a dominant mutation in one allele. The similarity of the ΔT_m values for this homotrimeric recombinant bacterial collagen molecule with Gly mutations in all 3 chains to the those reported by Makareeva et al. (2008) for OI type I heterotrimers containing one or two chains with mutations ($\Delta T_m = 1.5$°C–5°C) is surprising. These results suggest the number of chains within a molecule containing a missense mutation may not play a major role in the magnitude of the stability decrease, consistent with the observations on heterotrimeric peptides (Gauba and Hartgerink, 2008).

It has been known since early studies that OI collagens show increased hydroxylation and glycosylation of Lys, consistent with a pause or delay in folding at the mutation

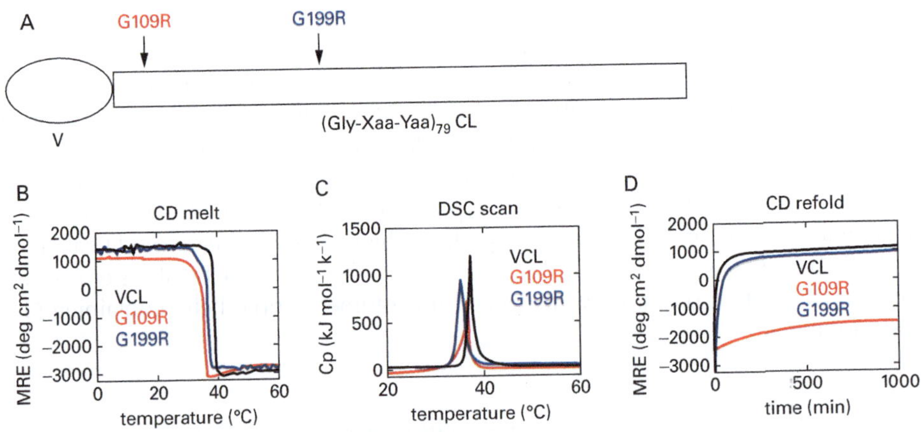

Figure 6.7 (A) Schematic drawing of the location of Gly to Arg substitutions at residue 199, near the center of the bacterial collagen domain, and at residue 109, which is very close to the N terminus of the triple helix and the V trimerization domain. (B) CD thermal transitions, (C) DSC scans, and (D) refolding profiles are shown for the control V-CL (black), V-CL (G199R) (blue), and V-CL(G109R) (red) (based on Cheng et at., 2011).

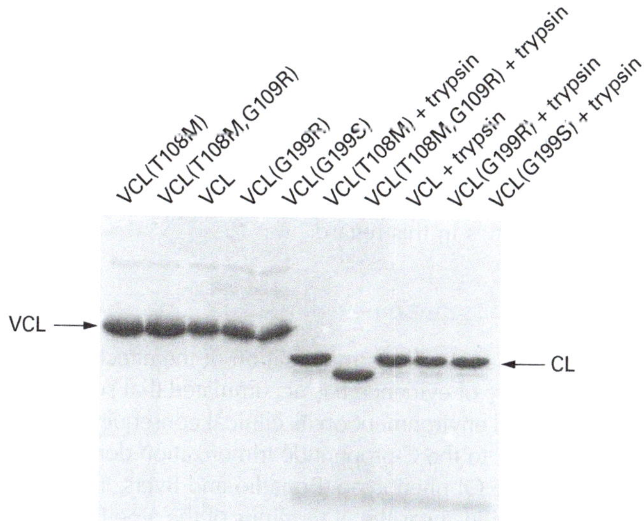

Figure 6.8 Sodium dodecyl sulfate polyacrylamide gel electrophoresis (SDS-PAGE) gels showing V-CL constructs (left lanes) and the products of trypsin digestion (right lanes). Trypsin digestion removes the V domain, resulting in a CL band. The CL bands are intact for all constructs except for V-CL(G109R), which is further digested, indicating that the Gly to Arg replacement near the triple-helix N terminus results in local unfolding (taken from Cheng et al., 2011).

site (Bonadio and Byers, 1985). A folding disruption is also supported by studies on cell cultures (Raghunath et al., 1994), and folding problems have been implicated in endoplasmic reticulum (ER) stress responses and ER retention of mutant molecules in OI mouse models (Forlino et al., 2007). A major limitation in studying folding has been the heterogeneous nature of OI collagens, which are a mixture of molecular species containing 0, 1, or 2 chains with mutations. The availability of significant quantities of well-defined homotrimeric molecules with a Gly to Ser mutation allows direct observation of the effect of mutations on folding. The refolding of the V-CL protein showed a significant delay due to the presence of the Gly to Ser mutation, going from a $t_{1/2}$ value of 10 minutes in V-CL to 50 minutes for V-CL (G199S) (▶Figure 6.7). These results provide quantitation of the folding delay and are consistent with an important role for such misfolding in the etiology of OI.

Effect of the identity of the residue replacing Gly

Analysis of the large number of OI mutations suggests a correlation between the identity of the residue replacing Gly and the severity of the OI case (Persikov et al., 2004a; Marini et al., 2007). To consider this question, a second construct was made replacing the Gly at position 199 by Arg, a larger residue found more often in lethal OI cases than Ser. This protein construct V-CL(G199R) showed the same results seen for V-CL(G199S) within experimental error, including the same CD spectrum, resistance to trypsin, thermal stability and folding delay. The replacement of a Gly by an Arg did not introduce a new trypsin cleavage site, again supporting a tightly folded, protease resistant triple helix for the molecule with a missense mutation in all three chains.

The lack of effect of a Ser versus Arg mutation contrasts with peptide results, where the T_m value was highly dependent on the identity of the residue replacing Gly. It appears that the smaller peptides are far more destabilized by and more sensitive to a Gly missense mutation, magnifying any effects. The bacterial collagen results agree with the studies on OI collagens purified from fibroblasts, which found no consistent effect of the residue replacing Gly on the T_m values of full-length collagens (Makareeva et al., 2008). It appears that homotrimer bacterial collagen models are long enough to mimic the collagen properties in this regard.

Effect of location of Gly missense mutation

The bacterial collagen constructs allow investigation of the effect of location of the Gly missense mutation. A variety of evidence has accumulated that supports the importance of the mutation position and environment on its clinical consequences. Early studies suggested that mutations closer to the C-propeptide trimerization domain in type I collagen are related to a more severe OI phenotype (Bonadio and Byers, 1985), and more recent studies indicated that mutations near the N terminus of the type I triple helix have a unique clinical phenotype and distinct structural effects (Makareeva et al., 2006). Site directed mutagenesis was used to introduce a Gly to Arg mutation at residue 109, a site close to the N terminus of the bacterial triple helix, and this protein was compared with that containing a V-CL(G199R) mutation. To keep the immediate amino acid triplets surrounding residue 109 (originally GPT**G**PA) the same as that for residue 199 (GPM**G**PA), the Thr108 N terminal to Gly109 was changed to Met, to yield the final construct.

The effect of a Gly to Arg substitution near the N terminus was very different from the Gly to Arg mutation nearer the middle of the triple helix. The CD spectra were very similar, indicating that the Gly substitution did not result in any major perturbation of the overall conformation. While the V-CL(G199S) and V-CL(G199R) both had collagenous domains completely resistant to trypsin, the V-CL(G109R) protein had a trypsin susceptible site within the (Gly-Xaa-Yaa)$_n$ sequence at Arg114, five residues C terminal to the Gly to Arg substitution site (▶Figure 6.8). The replacement of a Gly to Arg so close to the N terminus of the triple helix appears to have resulted in significant loosening or unwinding of the structure. The highly local nature of this disruption is supported by the normal CD spectra and by the similar small drop in stability ~2°C (see ▶Table 6.4). A most significant difference was observed for the refolding (▶Figure 6.7). The Gly to

Table 6.4 Mean residue ellipticities (MRE), stabilities, and the half-times of folding for bacterial collagen constructs with Gly substitutions.

Protein	*MRE_{220} (deg. $cm^2 \cdot dmol^{-1}$)*	*T_m (°C) by CD*	*T_m (°C) by DSC*	*$t_{1/2}$ (min)*
V-CL	1,340	38.5	37.3	10
V-CL(G199R)	1,320	36.5	34.4	55
V-CL(G199S)	1,275	37	34.5	50
V-CL(G109R)	1,250	36.7	36.6	>1,000

Arg change close to the N terminus dramatically decreased the rate and extent of refolding. The N-terminal Gly-Xaa-Yaa sequence is enriched in Pro compared with the overall sequence, and the extreme disruption of folding suggests that the Gly109 residue is part of a triple helix nucleation sequence. The marked difference in V-CL (G109R) versus V-CL(G199R) suggests that location may play an important role in determining the effect of a mutation on folding, a factor that could impact responses in the endoplasmic reticulum, secretion, and clinical phenotype. In particular, interference with triple-helix nucleation could be an important factor for mutations near the trimerization domain.

6.3.5 Summary and take-home message

A large variety of approaches have been applied to understanding the molecular etiology of how Gly missense mutations within the collagen triple helix lead to OI. Peptide studies and computational approaches have allowed the definition of alterations in molecular features, dynamics and local stabilities. Recent studies using a bacterial system quantitated folding delays for mutations at different sites. The study of additional recombinant bacterial constructs can be used to further explore the effects of the identity of the residue replacing Gly, the immediate sequence environment of the Gly substitution sites, and its location within or near the ends of the triple-helix domain. Better heterotrimer models are needed for both peptides and the recombinant bacterial system to elucidate the effects of different molecular species, and to clarify whether molecules containing one mutation and those with two mutations are both equally responsible for further pathology. Also more information is needed about the response in the endoplasmic reticulum to the altered molecular features and delayed folding of the collagen triple helix.

Peptides, recombinant bacterial collagens, and computational approaches have all proved important in defining structural consequences of a Gly missense mutation within a triple helix and provide a starting point for elucidating the chain of events leading to fragile bones and osteogenesis imperfecta.

Acknowledgments

We are grateful to Geetha Thiagarajan for her assistance on the bacterial collagen section (Section 6.3.4). This work was supported by NIH grant GM60048 to B. B.

References

Bächinger, H. P., Morris, N. P., and Davis, J. M. (1993). Thermal stability and folding of the collagen triple helix and the effects of mutations in osteogenesis imperfecta on the triple helix of type I collagen. Am J Med Genet *45*, 152–162.

Baum, J., and Brodsky, B. (1999). Folding of peptide models of collagen and misfolding in disease. Curr Opin Struct Biol *9*, 122–128.

Beck, K., Chan, V. C., Shenoy, N., Kirkpatrick, A., Ramshaw, J. A., and Brodsky, B. (2000). Destabilization of osteogenesis imperfecta collagen-like model peptides correlates with the identity of the residue replacing glycine. Proc Natl Acad Sci U S A *97*, 4273–4278.

Bella, J., Eaton, M., Brodsky, B., and Berman, H. M. (1994). Crystal and molecular structure of a collagen-like peptide at 1.9 A resolution. Science *266*, 75–81.

Berisio, R., Vitagliano, L., Mazzarella, L., and Zagari, A. (2002). Crystal structure of the collagen triple helix model [(Pro-Pro-Gly)(10)](3). Protein Sci *11*, 262–270.

Bhate, M., Wang, X., Baum, J., and Brodsky, B. (2002). Folding and conformational consequences of glycine to alanine replacements at different positions in a collagen model peptide. Biochemistry *41*, 6539–6547.

Bodian, D. L., Madhan, B., Brodsky, B., and Klein, T. E. (2008). Predicting the clinical lethality of osteogenesis imperfecta from collagen glycine mutations. Biochemistry *47*, 5424–5432.

Bonadio, J., and Byers, P. H. (1985). Subtle structural alterations in the chains of type I procollagen produce osteogenesis imperfecta type II. Nature *316*, 363–366.

Boudko, S. P., Engel, J., Okuyama, K., Mizuno, K., Bächinger, H. P., and Schumacher, M. A. (2008). Crystal structure of human type III collagen Gly991-Gly1032 cystine knot-containing peptide shows both 7/2 and 10/3 triple helical symmetries. J Biol Chem *283*, 32580–32589.

Brodsky, B., and Baum, J. (2008). Structural biology: modelling collagen diseases. Nature *453*, 998–999.

Brodsky, B., and Persikov, A. V. (2005). Molecular structure of the collagen triple helix. Adv Protein Chem *70*, 301–339.

Brodsky, B., Thiagarajan, G., Madhan, B., and Kar, K. (2008). Triple-helical peptides: an approach to collagen conformation, stability, and self-association. Biopolymers *89*, 345–353.

Bryan, M. A., Cheng, H., and Brodsky, B. (2010). Sequence environment of mutation affects stability and folding in collagen model peptides of osteogenesis imperfecta. Biopolymers *96*, 4–13.

Buevich, A. V., Dai, Q. H., Liu, X., Brodsky, B., and Baum, J. (2000). Site-specific NMR monitoring of cis-trans isomerization in the folding of the proline-rich collagen triple helix. Biochemistry *39*, 4299–4308.

Byers, P. H., and Cole, W. G. (2002). Osteogenesis imperfecta. In Connective Tissue and Its Heritable Disorders: Molecular, Genetic, and Medical Aspects, P. M. Royce and B. U. Steinmann, eds. New York: Wiley-Liss, pp. xvii, 1201.

Caswell, C. C., Barczyk, M., Keene, D. R., Lukomska, E., Gullberg, D. E., and Lukomski, S. (2008). Identification of the first prokaryotic collagen sequence motif that mediates binding to human collagen receptors, integrins alpha2beta1 and alpha11beta1. J Biol Chem *283*, 36168–36175.

Cheng, H., Rashid, S., Yu, Z., Yoshizumi, A., Hwang, E., and Brodsky, B. (2011). Location of glycine mutations within a bacterial collagen protein affects degree of disruption of triple-helix folding and conformation. J Biol Chem *286*, 2041–2046.

Doi, M., Nishi, Y., Uchiyama, S., et al. (2005). Collagen-like triple helix formation of synthetic (Pro-Pro-Gly)10 analogues: (4(S)-hydroxyprolyl-4(R)-hydroxyprolyl-Gly)10, (4(R)-hydroxyprolyl-4(R)-hydroxyprolyl-Gly)10 and (4(S)-fluoroprolyl-4(R)-fluoroprolyl-Gly)10. J Pept Sci *11*, 609–616.

Fallas, J. A., Gauba, V., and Hartgerink, J. D. (2009). Solution structure of an ABC collagen heterotrimer reveals a single-register helix stabilized by electrostatic interactions. J Biol Chem *284*, 26851–26859.

Fallas, J. A., O'Leary, L. E., and Hartgerink, J. D. (2010). Synthetic collagen mimics: self-assembly of homotrimers, heterotrimers and higher order structures. Chem Soc Rev *39*, 3510–3527.

Farndale, R. W., Lisman, T., Bihan, D., et al. (2008). Cell-collagen interactions: the use of peptide Toolkits to investigate collagen-receptor interactions. Biochem Soc Trans *36*, 241–250.

Forlino, A., Kuznetsova, N. V., Marini, J. C., and Leikin, S. (2007). Selective retention and degradation of molecules with a single mutant alpha1(I) chain in the Brtl IV mouse model of OI. Matrix Biol *26*, 604–614.

Gauba, V., and Hartgerink, J. D. (2008). Synthetic collagen heterotrimers: structural mimics of wild-type and mutant collagen type I. J Am Chem Soc *130*, 7509–7515.

Hyde, T. J., Bryan, M. A., Brodsky, B., and Baum, J. (2006). Sequence dependence of renucleation after a Gly mutation in model collagen peptides. J Biol Chem *281*, 36937–36943.

Koide, T. (2007). Designed triple-helical peptides as tools for collagen biochemistry and matrix engineering. Philos Trans R Soc Lond B Biol Sci *362*, 1281–1291.

Kramer, R. Z., Bella, J., Brodsky, B., and Berman, H. M. (2001). The crystal and molecular structure of a collagen-like peptide with a biologically relevant sequence. J Mol Biol *311*, 131–147.

Lee, K. H., Kuczera, K., and Banaszak Holl, M. M. (2011a). The severity of osteogenesis imperfecta: a comparison to the relative free energy differences of collagen model peptides. Biopolymers *95*, 182–193.

Lee, K. H. K., Kuczera, K., and Holl, M. M. (2011b). Effect of osteogenesis imperfecta mutations on free energy of collagen model peptides: a molecular dynamics simulation. Biophys Chem *156*, 146–152.

Li, Y., Brodsky, B., and Baum, J. (2009). NMR conformational and dynamic consequences of a Gly to Ser substitution in an osteogenesis imperfecta collagen model peptide. J Biol Chem *284*, 20660–20667.

Makareeva, E., Cabral, W. A., Marini, J. C., and Leikin, S. (2006). Molecular mechanism of alpha 1(I)-osteogenesis imperfecta/Ehlers-Danlos syndrome: unfolding of an N-anchor domain at the N-terminal end of the type I collagen triple helix. J Biol Chem *281*, 6463–6470.

Makareeva, E., Mertz, E. L., Kuznetsova, N. V., et al. (2008). Structural heterogeneity of type I collagen triple helix and its role in osteogenesis imperfecta. J Biol Chem *283*, 4787–4798.

Marini, J. C., Forlino, A., Cabral, W. A., et al. (2007). Consortium for osteogenesis imperfecta mutations in the helical domain of type I collagen: regions rich in lethal mutations align with collagen binding sites for integrins and proteoglycans. Hum Mutat *28*, 209–221.

Milchevsky, J. V., Ramensky, V. E., Esipova, N. G., Tumanyan, V. G., and Zorov, B. S. (2001). Molecular modelling of disease-causing single-nucleotide polymorphisms in collagen. SAR QSAR Environ Res *12*, 383–399.

Mizuno, K., Peyton, D. H., Hayashi, T., Engel, J., and Bächinger, H. P. (2008). Effect of the -Gly-3(S)-hydroxyprolyl-4(R)-hydroxyprolyl- tripeptide unit on the stability of collagen model peptides. FEBS J *275*, 5830–5840.

Mohs, A., Silva, T., Yoshida, T., et al. (2007). Mechanism of stabilization of a bacterial collagen triple helix in the absence of hydroxyproline. J Biol Chem *282*, 29757–29765.

Mooney, S. D., and Klein, T. E. (2002). Structural models of osteogenesis imperfecta-associated variants in the COL1A1 gene. Mol Cell Proteomics *1*, 868–875.

Myllyharju, J. (2009). Recombinant collagen trimers from insect cells and yeast. Methods Mol Biol *522*, 51–62.

Myllyharju, J., and Kivirikko, K. I. (2004). Collagens, modifying enzymes and their mutations in humans, flies and worms. Trends Genet *20*, 33–43.

Okuyama, K. (2008). Revisiting the molecular structure of collagen. Connect Tissue Res *49*, 299–310.

Persikov, A. V., Pillitteri, R. J., Amin, P., Schwarze, U., Byers, P. H., and Brodsky, B. (2004a). Stability related bias in residues replacing glycines within the collagen triple helix (Gly-Xaa-Yaa) in inherited connective tissue disorders. Hum Mutat *24*, 330–337.

Persikov, A. V., Ramshaw, J. A., and Brodsky, B. (2005a). Prediction of collagen stability from amino acid sequence. J Biol Chem *280*, 19343–19349.

Persikov, A. V., Ramshaw, J. A., Kirkpatrick, A., and Brodsky, B. (2000). Amino acid propensities for the collagen triple-helix. Biochemistry *39*, 14960–14967.

Persikov, A. V., Ramshaw, J. A., Kirkpatrick, A., and Brodsky, B. (2002). Peptide investigations of pairwise interactions in the collagen triple-helix. J Mol Biol *316*, 385–394.

Persikov, A. V., Ramshaw, J. A., Kirkpatrick, A., and Brodsky, B. (2005b). Electrostatic interactions involving lysine make major contributions to collagen triple-helix stability. Biochemistry *44*, 1414–1422.

Persikov, A. V., Xu, Y., and Brodsky, B. (2004b). Equilibrium thermal transitions of collagen model peptides. Protein Sci *13*, 893–902.

Radmer, R. J., and Klein, T. E. (2004). Severity of osteogenesis imperfecta and structure of a collagen-like peptide modeling a lethal mutation site. Biochemistry *43*, 5314–5323.

Raghunath, M., Bruckner, P., and Steinmann, B. (1994). Delayed triple helix formation of mutant collagen from patients with osteogenesis imperfecta. J Mol Biol *236*, 940–949.

Rasmussen, M., Jacobsson, M., and Bjorck, L. (2003). Genome-based identification and analysis of collagen-related structural motifs in bacterial and viral proteins. J Biol Chem *278*, 32313–32316.

Shoulders, M. D., Satyshur, K. A., Forest, K. T., and Raines, R. T. (2010). Stereoelectronic and steric effects in side chains preorganize a protein main chain. Proc Natl Acad Sci U S A *107*, 559–564.

Slatter, D. A., Bihan, D. G., and Farndale, R. W. (2011). The effect of purity upon the triple-helical stability of collagenous peptides. Biomaterials *32*, 6621–6632.

Suto, K., and Noda, H. (1974). Conformational change of the triple-helical structure. II. Conformation of (Pro-Pro-Gly)n and (Pro-Pro-Gly)n (Ala-Pro-Gly)m(Pro-Pro-Pro-Gly)n in an aqueous solution. Biopolymers *13*, 2391–2404.

Xu, C., Yu, Z., Inouye, M., Brodsky, B., and Mirochnitchenko, O. (2010). Expanding the family of collagen proteins: recombinant bacterial collagens of varying composition form triple-helices of similar stability. Biomacromolecules *11*, 348–356.

Xu, Y., Keene, D. R., Bujnicki, J. M., Hook, M., and Lukomski, S. (2002). Streptococcal Scl1 and Scl2 proteins form collagen-like triple helices. J Biol Chem *277*, 27312–27318.

Yoshizumi, A., Yu, Z., Silva, T., et al. (2009). Self-association of streptococcus pyogenes collagen-like constructs into higher order structures. Protein Sci *18*, 1241–1251.

Yu, Z., Brodsky, B., and Inouye, M. (2011). Dissecting a bacterial collagen domain from Streptococcus pyogenes: sequence and length-dependent variations in triple helix stability and folding. J Biol Chem *286*, 18960–18968.

Yu, Z., Mirochnitchenko, O., Xu, C., Yoshizumi, A., Brodsky, B., and Inouye, M. (2010). Noncollagenous region of the streptococcal collagen-like protein is a trimerization domain that supports refolding of adjacent homologous and heterologous collagenous domains. Protein Sci *19*, 775–785.

6.4 Roles and regulation of BMP1/Tolloid-like proteinases: collagen/matrix assembly, growth factor activation, and beyond

Catherine Moali and David J. S. Hulmes

6.4.1 Introduction

It is now 40 years since the discovery of procollagen (Bellamy and Bornstein, 1971; Church et al., 1971; Layman et al., 1971), the soluble form of collagen synthesized by cells that is converted to the mature form by proteolytic removal of the N- and C-propeptides, thereby triggering the process of collagen fibrillogenesis. This chapter focuses on an important family of enzymes involved in procollagen processing, the bone morphogenetic protein-1 (BMP1)/Tolloid-like proteinases. First identified as procollagen C-proteinases, the proteinases that remove the C-propeptides from the major fibrillar procollagens (types I, II, and III), the substrate repertoire of BMP1/Tolloid-like proteinases has since expanded to include several proteins involved in different aspects of extracellular matrix (ECM) assembly, as well as, but not limited to, an increasing number of growth factors and growth factor antagonists. These enzymes therefore play key roles in tissue morphogenesis and tissue repair. In addition, the activities of BMP1/Tolloid-like proteinases are regulated by other extracellular proteins, most of which have the property of enhancing BMP1/Tolloid-like proteinase activity in a substrate-specific manner. This provides a mechanism for fine tuning proteolytic activity according to specific tissue requirements. Here we review current knowledge of these proteinases, their structures, substrates, and mechanisms of regulation with particular attention to collagen/ECM assembly.

6.4.2 BMP1/Tolloid-like proteinases

The founder member of the BMP1/Tolloid-like proteinase family was discovered in the 1980s in extracts of demineralized bone (Wozney et al., 1988) along with other proteins capable of inducing ectopic bone formation in soft tissues. The proteinase was therefore named BMP1, a name that has caused some confusion since this protein is the only proteinase among the BMPs and the only one not to be a member of the transforming growth factor-beta (TGF-β) family of growth factors. Shortly after this discovery, a genetic screening carried out in *Drosophila* led to the demonstration that the protein Tolloid, a product of the *tld* gene involved in the formation of the dorsoventral axis during early embryonic development, had a strong sequence homology with BMP1

(Shimell et al., 1991). Since then, numerous orthologs have been characterized in different species such as chicken, *Xenopus*, zebrafish, and sea urchin. In mammals, there are four main BMP1/Tolloid-like proteinases: the short form BMP1 and the long forms mammalian tolloid (mTLD), mammalian tolloid-like 1 (mTLL1), and mammalian tolloid-like 2 (mTLL2) (▶Figure 6.9).

While BMP1 and mTLD are products of alternative splicing of the same gene (*Bmp1*) (Takahara et al., 1994b), mTLL1 and mTLL2 are encoded by distinct genes (Takahara et al., 1996; Scott et al., 1999). These enzymes belong to the astacin family within the metzincin clan of zinc metalloproteinases, which, in mammals, also comprises the matrix metalloproteinases (MMPs), the ADAM(TS)s, and the pappalysins (Stöcker et al., 1995; Sterchi et al., 2008). Other members of the astacin family in humans include meprins α and β, and ovastacin.

In addition to the astacin-like catalytic domain, BMP1/Tolloid-like proteinases include a propeptide, which maintains latency prior to cleavage by the subtilisin-like proprotein convertases (SPCs, such as furin) in the Golgi, as well as several complement-Uegf-BMP1 (CUB) domains (three or five depending on the isoform) and one or two epidermal growth factor (EGF) domains (▶Figure 6.9). After secretion, therefore, these proteinases are mainly in processed and active form, unlike the MMPs, for example, which can be secreted in latent form and for which processing plays a key role in controlling proteolytic activity.

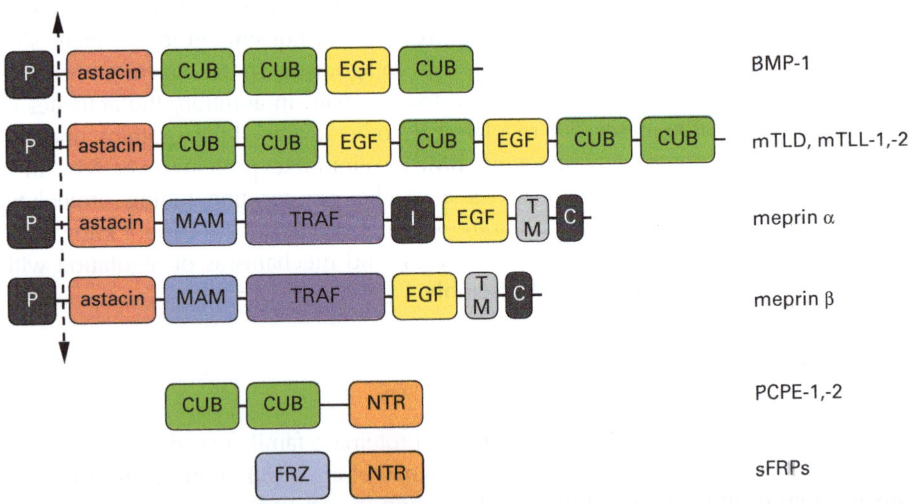

Figure 6.9 **Domain structures of BMP1/Tolloid-like proteinases and meprins, as well as procollagen C-proteinase enhancers (PCPEs) and secreted frizzled-related proteins (sFRPs).** P = propeptide; CUB = complement Uegf BMP1; EGF = epidermal growth factor; MAM = meprin, A5 protein, receptor protein tyrosine phosphatase μ; TRAF = tumor necrosis factor receptor-associated factor; I = inserted sequence; TM = transmembrane domain; C = cytoplasmic domain; FRZ = frizzled; NTR = netrin-like. Propeptides are removed (arrow) by subtilisin-like proprotein convertases (for BMP1/Tolloid-like proteinases) or trypsin-like enzymes (for meprins) to obtain fully active proteases.

In 1996, two groups (Kessler et al., 1996; Li et al., 1996) showed independently that BMP1 and mTLD are different forms of the protein previously known as procollagen C-proteinase (PCP; Hojima et al., 1985; Prockop et al., 1998), which removes the C-propeptides from procollagens. These proteinases trigger fibrillogenesis of fibrillar collagens by lowering the solubility of procollagen precursors, along with ADAMTS2, 3, and 14 which cleave the N-propeptides (Canty and Kadler, 2005; Hulmes, 2008; Apte, 2009). While failure to cleave the N-propeptides, as occurs in Ehlers-Danlos syndrome type VII, has less effect on collagen solubility than the lack of C-propeptide cleavage, the shape of the fibrils that do form are highly abnormal (Hulmes et al., 1989), leading to extreme tissue fragility and severe joint laxity (Colige et al., 1999).

Mouse studies show expression of both BMP1 and mTLD in most adult tissues: heart, lung, liver, skeletal muscle, kidney, pancreas, placenta, and so forth. Interestingly, mTLD is expressed in brain while BMP1 is not (Takahara et al., 1994b), this being the sole notable reported difference in expression between these two isoforms. BMP1 and mTLD are also strongly expressed early during embryogenesis, much more so than mTLL1 and mTLL2 for which the expression pattern is relatively limited. Early expression of mTLL1, for example, up to 10.5 days post conception (dpc), is limited to the cardiovascular system, consistent with the defects observed in *Tll1*$^{-/-}$ mice that lead to death at ~13.5 dpc (Ge and Greenspan, 2006b). While expression of mTLL1 after 10.5 dpc becomes more generalized, that of mTLL2 remains limited to the central nervous system and skeletal muscle (Scott et al., 1999), suggesting that its roles could be very different from those of the other isoforms.

Bmp1$^{-/-}$ knockout mice die around birth, essentially due to malformations at the junction with the umbilical cord that could be due to the presence of abnormal collagen fibrils and/or the consequences of perturbations of ventralization (Suzuki et al., 1996). Surprisingly, even the double knockout *Bmp1*$^{-/-}$/*Tll1*$^{-/-}$ shows no major defects in the dorsoventral axis as found in *Drosophila* and *Xenopus* (Marques et al., 1997; Piccolo et al., 1997). On the other hand, these embryos show cardiac malformations identical to those of *Tll1*$^{-/-}$ embryos and even though low levels of residual PCP activity could be found, the collagen fibrils that are deposited have a "barbed-wire" appearance, which is likely to correspond to the presence of non-cleaved C-propeptides (Pappano et al., 2003; Steiglitz et al., 2006). This result is consistent with the observation that procollagen can form fibrils at high concentration (Mould et al., 1990).

In 2008, the crystal structures of the catalytic domains of BMP1 and mTLL1 produced in bacteria were published (MacSweeney et al., 2008). These proteins are characterized by considerable flexibility in the active site region, which requires a ligand (in this case dimethyl sulfoxide) to become sufficiently ordered for structure determination. The structure of the BMP1 catalytic domain shows a disulfide bonding pattern different to that of astacin with, in particular, a disulfide bridge involving vicinal cysteine residues just above the active site. This is extremely rare and suggests relatively high reactivity of the residues involved. Nevertheless, these proteins having been produced in the form of inclusion bodies and then refolded *in vitro* without extensive functional validation, the reality of this disulfide bond is debatable and should be validated using proteins produced in a eukaryotic system.

Using recombinant human proteinases expressed in 293 EBNA cells, the group of C. Baldock has recently obtained low resolution structural data for the full-length proteinases BMP1, mTLD, and mTLL1 (Berry et al., 2009, 2010). Interestingly, BMP1 seems to behave differently to the long isoforms mTLD and mTLL1; while BMP1 remains monomeric, mTLD and mTLL1 form calcium-dependent dimers wherein the mode of association could limit access to the active site and hence explain why these isoforms are less active (on procollagen and chordin, for example) than BMP1 (Berry et al., 2009). This result can be compared to the only other proteinases to share the CUB-EGF-CUB motif, the complement proteinases C1r and C1s, and the mannan-binding lectin-associated serine proteases (MASPs), which have been shown, by X-ray crystallography, to assemble into dimers with an antiparallel arrangement (Teillet et al., 2008). By analogy, it was proposed that the most likely form of association for mTLD or mTLL1 would be that in which the two monomers are superimposed in an antiparallel fashion (Berry et al., 2009, 2010).

6.4.3 Substrates

Currently, 24 substrates of BMP1/Tolloid-like substrates are listed in the literature (▶Table 6.5), considering the heterotrimeric collagens as single substrates and excluding the broad-spectrum inhibitor α2-macroglobulin, which acts as a suicide substrate (Zhang et al., 2006). With the exception of apolipoprotein A1 and gliomedin, which play unique roles in lipid metabolism and the peripheral nervous system, these substrates can be classified into seven partially overlapping categories: fibrillar procollagens, small leucine-rich proteoglycans, cross-linking enzymes, basement membrane components, mineralization factors, growth factors and antagonists, and (anti)angiogenic factors. Concerning features of the cleavage site, analysis of the corresponding 34 unique sequences (▶Table 6.5) shows a strong specificity of the BMP1/Tolloid-like proteinases for an aspartate residue in position P1′ (▶Figure 6.10), as well as a preference (when normalized to natural amino acid abundance) for methionine in P3, glutamine in P2, and proline in P3′.

Though reliable quantitative data are lacking for physiological substrates, the catalytic efficiency can differ significantly between isoforms. For example, mTLL2 seems to have barely detectable PCP or chordinase activity (Petropoulou et al., 2005) but can cleave pro–lysyl oxidase (pro-LOX) (Uzel et al., 2001), osteoglycin (Ge et al., 2004), laminin-332 (Veitch et al., 2002), procollagen VII (Rattenholl et al., 2002), myostatin (Wolfman et al., 2003), and dentin matrix protein-1 (DMP1) (Steiglitz et al., 2004). Similarly, mTLD and mTLL1 seem generally to be less active than BMP1, the most active proteinase on all known substrates.

Extracellular matrix assembly

Control of collagen fibrillogenesis

There are 28 types of collagen in humans, formed from 43 separate α chains, the most abundant being the major fibrillar procollagens (types I, II, and III), which have the ability to self-assemble into fibrils (▶Figure 6.11A) and play major roles in the organization of the ECM. Intracellular steps in procollagen biosynthesis are complex, including specific posttranslational modifications (hydroxylation of proline and lysine, *O*-and *N*-glycosylation) as well as controlled formation of homo- and hetero-trimers, often in the same cell, involving specific chaperones such as Hsp47 (Myllyharju and Kivirikko,

Table 6.5 **Substrates of BMP1/Tolloid-like proteinases described in the literature and sequences cleaved.** Sequence numbering is for the full-length human protein.

Human protein	*Accession number (UniProtKB)*	*Sequence cleaved*	*Reference*
Fibrillar procollagens			
Procollagen I (α1)	P02452	YYRA ^{1219}DDAN	(Dickson et al., 1981)
Procollagen I (α2)	P08123	FYRA ^{1120}DQPR	(Dickson et al., 1981)
Procollagen II (α1)	P02458	YMRA ^{1242}DQAA	(Sandell et al., 1984)
Procollagen III (α1)	P02461	PYYG ^{1222}DEPM	(Yamada et al., 1983)
Procollagen V (α1)	P20908	QLLD ^{1595}DGNG	(Kessler et al., 2001)
		TPQS ^{0255}QDPN	(Imamura et al., 1998)
Procollagen V (α2)	P05997	EFTE ^{1254}DQAA	(Unsold et al., 2002)
Procollagen V (α3)	P25940	SFQQ ^{0464}AQAQ	(Gopalakrishnan et al., 2004)
Procollagen XI (α1)	Q149N0	GMQA ^{1448}DADD	Putative
		AAQA ^{0254}QEPQ	(Pappano et al., 2003)
Procollagen XI (α2)	P13942	LMQE ^{1525}DEAI	Putative
		RPQN ^{0239}QQPH	Putative
Small leucine-rich proteoglycans			
Probiglycan	P21810	FMMN ^{0038}DEEA	(Scott et al., 2000)
Osteoglycin (mimecan)	P20774	QLQK ^{0076}DEAI	(Ge et al., 2004)
Decorin	P07585	FMLE ^{0031}DEAS	(von Marschall and Fisher, 2010a)
Cross-linking enzymes			
Lysyl oxidase (LOX)	P28300	RMVG ^{0169}DDPY	(Panchenko et al., 1996)
Lysyl oxidase–like (LOXL)	Q08397	VAVG ^{0135}DSTG VRSS ^{0338}DTPP	(Borel et al., 2001)

(Continued)

Table 6.5 **Substrates of BMP1/Tolloid-like proteinases described in the literature and sequences cleaved.** Sequence numbering is for the full-length human protein. (*Continued*)

Human protein	*Accession number (UniProtKB)*	*Sequence cleaved*	*Reference*
Basement membrane components			
Procollagen VII (α1)	Q02388	SYAA ^{2822}DTAG	(Rattenholl et al., 2002)
Laminin-332 (γ2)	Q13753	CYSG ^{0435}DENP	(Amano et al., 2000)
Laminin-332 (α3)	Q13753	not determined	(Amano et al., 2000; Veitch et al., 2002)
Mineralization factors			
Dentin matrix protein-1 (DMP1)	Q13316	EKQS ^{0218}DDPE	(Steiglitz et al., 2004)
Dentin sialophosphoprotein (DSPP)	Q9NZW4	SMQG ^{0463}DDPN	(von Marschall and Fisher, 2010b; Tsuchiya et al., 2011)
Growth factors and antagonists			
Chordin	Q9H2X0	RSYS ^{0154}DRGE PMQA ^{0867}DGPR	(Scott et al., 1999)
Myostatin (GDF8)	O14793	DVQR ^{0099}DDSS	(Wolfman et al., 2003)
GDF11 (BMP11)	O95390	DFQG ^{0122}DALQ	(Ge et al., 2005)
Latent TGF-β-binding protein-1 (LTBP1)	Q14766	IPSL ^{0812}DQEK YFIQ ^{1611}DRFL	(Ge and Greenspan, 2006a)
Insulin-like growth factor binding protein-3	P17936	ESQS ^{0195}TDTQ	(Kim et al., 2011)
(Anti)angiogenic factors			
Perlecan*	P98160	SGGN ^{4196}DAPG	(Gonzalez et al., 2004)
Prolactin	P01236	LQMA ^{0188}DEES	(Ge et al., 2007)
Growth hormone (somatotropin)	P01241	NSHN ^{0179}DDAL	(Ge et al., 2007)
Others			
Apolipoprotein A1	P02647	FWQQ ^{0025}DEPP	(Chau et al., 2007)
Gliomedin	Q6ZMI3	AIPN ^{0281}DDTL	(Maertens et al., 2007)

*Perlecan is also a basement membrane protein.

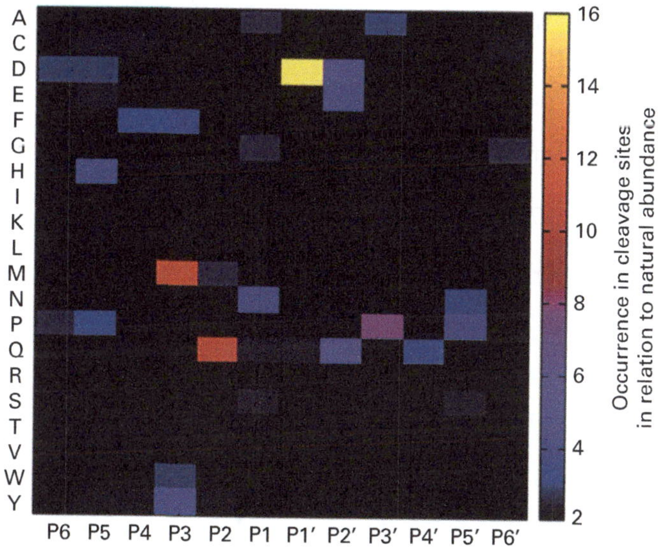

Figure 6.10 Specificity of BMP1/Tolloid-like proteinases. Heat map generated using gnuplot (http://www.gnuplot.info) showing the specificity of BMP1, based on the 34 cleavages listed in Table 6.5, between P6 and P6′ (nomenclature of Schechter and Berger, 1967). Relative occurrence is normalized to amino-acid natural abundance.

2004). Outside the cell, the situation is no less complex, fibril assembly being controlled by numerous binding partners (integrins, fibronectin, small leucine-rich proteoglycans, matricellular proteins, etc.), heterotypic collagen interactions as well as procollagen processing and covalent cross-linking (Hulmes, 2008; Kadler et al., 2008; Bradshaw, 2009) (see also Chapter 6.6 by J. D. San Antonio and coworkers). There is evidence that in some cases procollagen processing can begin in membrane bound compartments during intracellular transit from the Golgi to the cell surface (Canty and Kadler, 2005). This is sometimes accompanied by the appearance of fibrils in protruding cell surface invaginations, called fibripositors, though this seems to depend on cell type (Humphries et al., 2008). In general, however, most of the processing seems to occur in non-membrane-bound compartments, presumably at the cell surface or in the ECM.

In the case of the quantitatively minor fibrillar procollagens (types V and XI), the role of BMP1/Tolloid-like proteinases is more complex (▶Figure 6.11A). Here, in addition to the C-propeptides, these enzymes are also involved in cleavage of the N-propeptides, which are important for collagen fibril diameter regulation (Linsenmayer et al., 1993; Wenstrup et al., 2011). Collagen V forms heterotypic fibrils with collagen I (skin, tendons, ligaments, cornea, etc.), while collagen XI is expressed mainly in cartilage where it associates with collagen II. While the most common form of the collagen V heterotrimer is $\alpha1(V)_2\alpha2(V)$, both the homotrimer $\alpha1(V)_3$ and the heterotrimer $\alpha1(V)\alpha2(V)\alpha3(V)$ have also been reported. BMP1/Tolloid-like proteinases cleave the N-propeptides of the $\alpha1(V)$ and $\alpha3(V)$ chains (after the TSPN domain or between the short and long triple-helical regions, respectively; Imamura et al., 1998; Gopalakrishnan et al., 2004). In both cases, the cleavage site lacks the usual aspartate at P1′, but contains a glutamine

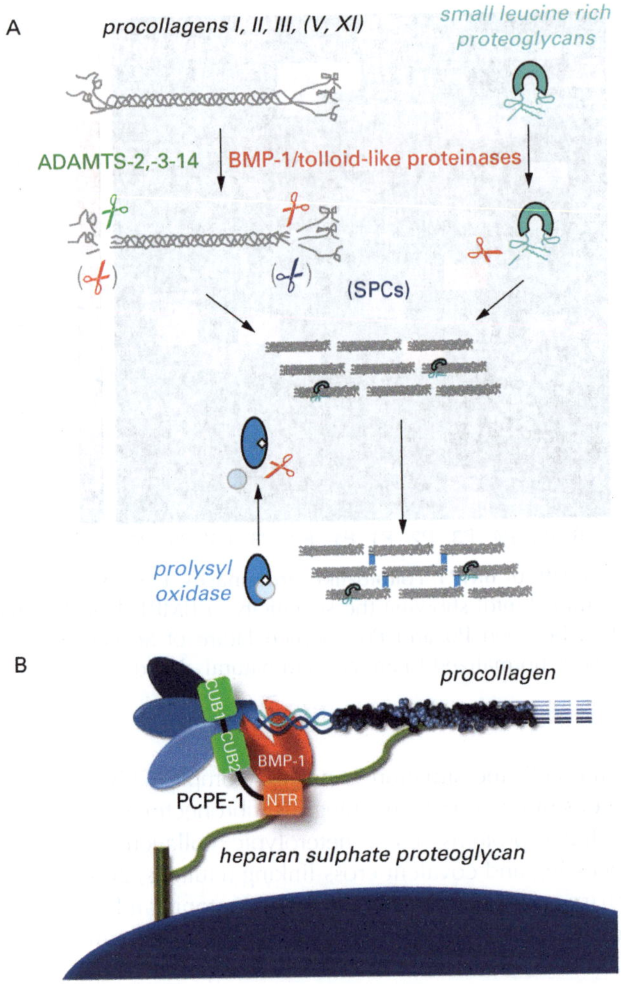

Figure 6.11 **Proteolytic control of collagen fibril assembly.** (A) BMP1/Tolloid proteinases trigger collagen fibril formation by cleaving the C-propeptides from the major fibrillar procollagens (types I, II, and III), while the N-propeptides are removed by ADAMTS2, 3, and 14. In the case of the minor fibrillar procollagens V and XI, BMP1/Tolloid proteinases also cleave the N-propeptides, while subtilisin-like proprotein convertases (SPCs) play major roles in C-terminal processing (see text). BMP1/Tolloid proteinases also remove N-terminal peptides from the small leucine-rich proteoglycans biglycan, decorin, and osteoglycin, which themselves control fibril formation, and they activate precursor forms of lysyl oxidase and lysyl oxidase–like, triggering covalent cross-linking. After Moali and Hulmes (2009). (B) Model of the putative procollagen/BMP1/procollagen C-proteinase enhancer-1 (PCPE1)/heparan sulfate proteoglycan complex involved in superstimulation of BMP1 activity. After (Bekhouche et al., 2010).

at P2, as in the procollagen α1(XI) chain (Pappano et al., 2003) and also insulin-like growth factor binding protein 3 (Kim et al., 2011). In the case of the C-propeptides, the α1(V) and α3(V) chains are cleaved by SPCs (Imamura et al., 1998; Gopalakrishnan et al., 2004), albeit that there is a BMP1/Tolloid-like proteinase cleavage site just after the SPC site in the α1(V) chain (Kessler et al., 2001), while the α2(V) C-propeptide is cleaved by BMP1/Tolloid-like proteinases only (Unsold et al., 2002).

The diameter, spacing, and kinetics of fibril formation are controlled by members of small leucine-rich proteoglycan family, as demonstrated in mice where the corresponding genes have been inactivated (Danielson et al., 1997; Corsi et al., 2002). These proteoglycans interact directly with collagen fibrils thereby affecting the transparency of the cornea, bone mass, or the tensile strength of tendons and skin. By cleaving the N-terminal regions of biglycan, decorin, or osteoglycin (▶Figure 6.11A; Scott et al., 2000; Ge et al., 2004; von Marschall and Fisher, 2010a), BMP1/Tolloid-like proteinases may change the ability of these molecules to interact with their partners. For example, the rate of formation of collagen fibrils appears to be affected differently by the full-length and mature forms of osteoglycin (Ge et al., 2004).

Other proteins matured by BMP1/Tolloid-like proteinases and playing a role in establishing the three-dimensional network of collagen are lysyl oxidase (LOX; ▶Figure 6.11A; Cronshaw et al., 1995; Panchenko et al., 1996) and lysyl oxidase–like (LOXL; Borel et al., 2001). These copper-dependent amine oxidases modify the amino groups in the side chains of specific lysine residues found in the C-telopeptides of the α1 chains of collagens I, II, and III. This oxidative deamination has the effect of initiating the formation of covalent cross-links between these modified lysines and other lysine or hydroxylysine residues present in neighboring collagen molecules.

Basement membrane assembly

Basement membranes (BMs) are widely distributed extracellular matrices that coat the basal aspect of epithelial and endothelial cells and surround muscle, fat, and Schwann cells, ensuring cohesion to the underlying mesenchyme and often with a filtering or protective barrier function. The composition of the BM (and associated structures) is very specific (collagens IV, VII, XVII, XVIII, laminins, nidogens, perlecan, etc.) and allows the formation of multi-molecular complexes (hemidesmosomes, anchoring filaments, and anchoring fibrils) interacting to ensure the cohesion of the whole. Among these components, BMP1/Tolloid-like proteinases cleave procollagen VII, laminin-332, and perlecan (see further on in this chapter).

Collagen VII associates into antiparallel dimers, stabilized by disulfide bonds, after cleavage of its C-terminal extremity by BMP1/Tolloid-like proteinases (Rattenholl et al., 2002); this allows its assembly into anchoring fibrils, which can interact through their NC1 domains with components of the BM. The anchoring fibrils are intermingled with the collagen fibers of the stroma to ensure the cohesion of the stroma with the BM.

On the epithelial side, laminin-332 (previously laminin-5) interacts with receptors expressed by epithelial cells (integrins and heparan sulfate proteoglycans) as well as other BM components, ensuring adherence of epithelial cells to the underlying tissue. Laminin-332 can be cleaved at several sites by proteinases, especially BMP1/Tolloid-like proteinases ,which control its biological activity by cleaving its α3 and γ2 chains. Fully mature laminin-332 is inserted into the basement membrane to stabilize basement membrane/epithelium interactions while retention of the LG4 and LG5 domains on the α3 chain seems to promote migration of epithelial cells (Bachy et al., 2008). However, while cleavage of the γ2 chain seems to be controlled exclusively by BMP1/Tolloid-like proteinases (at least for the human form), cleavage of the α3 chain can also be carried out by plasmin and proteinases of the MMP family such as MMP2 and MMP14 (Amano et al., 2000; Veitch et al., 2002). Note that the cleavage site on the α3 chain has not been determined.

Mineralization

The last role relating to synthesis of the ECM so far described for BMP1/Tolloid-like proteinases concerns cleavage of proteins of the small integrin-binding ligand *N*-linked glycoprotein (SIBLING) family found primarily in bone and dentin. Dentin matrix protein-1 (DMP-1) and dentin sialophosphoprotein (DSPP) are the major constituents of the matrix of dentin after collagen, but neither of these two proteins is found in intact form in tissues (Ge and Greenspan, 2006b). The corresponding fragments start for the most part with aspartate residues; these proteins have been tested and confirmed to be substrates for BMP1/Tolloid-like proteinases or other proteinases of the astacin family such as meprins for DSPP (Steiglitz et al., 2004; von Marschall and Fisher, 2010b; Tsuchiya et al., 2011). These proteins may play a role in the initiation of mineralization of osteoid and predentin with the presence of acidic domains that can bind calcium.

Activation of growth factors

Activation of BMP2 and 4 by cleavage of an antagonist

Along with the processing of fibrillar procollagens, one of the first functions associated with BMP1/Tolloid-like proteinases was the potentiation of BMP signaling during formation of the dorsoventral axis, organogenesis, and development of the nervous system (Ge and Greenspan, 2006b). Chordin (or its analog Sog in *Drosophila*) associates with BMP growth factors (Dpp) to form a latent complex that diffuses from the ventral BMP synthesis site toward the dorsal part of the embryo, thus creating a gradient of inhibition. BMP1/Tolloid-like proteinases also present on the ventral side allow the release of BMPs (Dpp) by cleavage of their antagonist (Marques et al., 1997; Piccolo et al., 1997). Very recently, a similar mechanism of activation by BMP1/Tolloid-like proteinases of another growth factor, insulin-like growth factor-1 (IGF1), important for development, growth, and metabolism, has been described. This involves cleavage of IGF-like binding protein-3, which interestingly also interacts with BMP2 and 4 (Kim et al., 2011).

Activation of GDF8 and 11 by cleavage of their propeptides

All growth factors of the TGF-β superfamily are synthesized in precursor form where the prodomain is cleaved by SPC type proteinases during secretion. However, for some family members of this family including TGF-β itself, this prodomain may remain associated non-covalently with the active part of the protein to form a latent complex. This is also the case for GDF8 (known more as myostatin, an inhibitor of muscle growth) and GDF11 (also called BMP11, which plays various roles in development) where the propeptides can be cleaved at an internal site by BMP1/Tolloid-like proteinases (Wolfman et al., 2003; Ge et al., 2005). This second proteolytic event leads to the release of the mature, active growth factor.

Activation of TGF-β by cleavage of a sequestrating protein in the matrix

Activation of the three isoforms of TGF-β present in humans (β1, β2, and β3) is of special interest for matrix pathobiology since several disease states such as fibrotic conditions, which are characterized by excessive/disorganized deposition of collagen fibers, are known to be highly dependent on the level of TGF-β. TGF-β activity is regulated in an extremely complex manner by a multitude of mechanisms (Jenkins, 2008; Wipff and Hinz, 2008; Worthington et al., 2011), the case most studied being that of transforming growth factor-beta 1 (TGF-β1). In addition to being secreted in latent form, like GDF8 and GDF11, in association with its propeptide called latency associated peptide (LAP), TGF-β is usually secreted as a large latent complex (LLC) by covalent association of LAP with proteins called latent TGF-β-binding proteins, or LTBPs (LTBP1–4). The LLC is directed to the ECM where it is sequestered by interaction of fibrillin with LTBP (stabilized by transglutaminase forming covalent cross-links between LTBP and fibrillin). Dissociation of active TGF-β from its LAP can occur by proteolytic and non-proteolytic mechanisms (interaction with thrombospondin-1 and integrins), and also by mechanical effects, and is the key step in the activation of TGF-β. In the case of proteolytic activation, cleavage within the LAP by MMPs and plasmin-type proteases, leading to liberation of active TGF-β, requires preliminary release of the LLC from the matrix, a process carried out by BMP1/Tolloid-like proteinases (Ge and Greenspan, 2006a). Interestingly, TGF-β induces the expression of BMP1/Tolloid-like proteinases as well as many matrix proteins in fibroblasts and keratinocytes (Lee et al., 1997), which could potentially lead to amplification of TGF-β signaling and synthesis of ECM.

Other roles

BMP1/Tolloid-like proteinases are also involved in the control of angiogenesis. For example, a major role of the basement membrane component perlecan, a large heparan sulfate proteoglycan, concerns the control of the activity of certain growth factors likely to interact with glycosaminoglycans and involved in angiogenesis (fibroblast growth factor [FGF] and vascular endothelial growth factor [VEGF]). While the effect of the full-length molecule seems to be pro-angiogenic, proteolytic fragments released from the C-terminal extremity (endorepellin or its LG3 domain) have anti-angiogenic activity. While BMP1/Tolloid-like proteinases are responsible for the release of the C-terminal LG3 domain, the proteases involved in release of the endorepellin fragment itself are unknown (Gonzalez et al., 2004).

Similarly, prolactin and growth hormone are substrates for BMP1, as well as other Tolloid-like proteinases, and among their multiple functions (lactation, reproduction, growth, metabolism, etc.), these also show pro-angiogenic activity. After cleavage, a fragment of 17 kDa with anti-angiogenic activity is produced (Nguyen et al., 2006; Ge et al., 2007). Thus, it seems to be a general property of BMP-1/Tolloid-like proteinases to produce anti-angiogenic fragments from pro-angiogenic parent molecules.

Two other, somewhat atypical substrates have been described in the literature: gliomedin and apolipoprotein A1. Gliomedin is a transmembrane protein belonging to the collagen superfamily with, in its ectodomain, a collagen-like region followed by an olfactomedin domain. This ectodomain can be released by a process of SPC shedding, while the olfactomedin domain may also be released through an additional proteolytic step controlled by BMP1/Tolloid-like proteinases (Maertens et al., 2007). The olfactomedin domain thus liberated can form insoluble aggregates that act to stabilize mature nodes of Ranvier in the peripheral nervous system.

Finally, apolipoprotein A1 is the major protein of high-density lipoprotein (HDL) in serum, involved in the regulation of cholesterol efflux from cells. This protein must be matured by removal of 6 amino acids at its N-terminal extremity in order to interact with a cellular carrier (adenosine triphosphate [ATP]–binding cassette transport A1), change conformation and acquire the ability to bind phospholipids. This maturation appears to be carried out by BMP1 (Chau et al., 2007), which gives the family of BMP1/Tolloid-like proteinases an unexpected role in lipid metabolism.

6.4.4 Endogenous regulators of activity

Virtually all extracellular proteases are associated with one or more regulatory molecules, usually inhibitors that normally prevent the tissue destruction seen in certain pathologies such as osteoarthritis or chronic wounds. While most of these regulatory molecules are specific for a particular protease or a family of proteases, there are also broad-spectrum inhibitors able to inhibit many proteases, such as α2-macroglobulin, which can irreversibly inactivate both serine proteases and metalloproteinases, including BMP1 (Zhang et al., 2006). Surprisingly, however, no really specific endogenous inhibitor of human BMP1/Tolloid-like proteinases has so far been reported. Mechanisms of inhibition have been described in *Xenopus* and zebrafish but their transposition to humans remains highly controversial. In contrast, BMP1/Tolloid-like proteinases function with an exceptionally large panel of enhancers, described further on.

Inhibitors

In 2006, Lee et al. reported that the sizzled protein, which belongs to the family of secreted frizzled-related proteins (sFRPs) but is present only in *Xenopus*, zebrafish, and chicken, efficiently and competitively inhibits the BMP1/Tolloid-like proteinases of *Xenopus* (Lee et al., 2006). Extinction of sizzled expression led to ventralization of embryos in a manner similar to that seen after extinction of chordin. Inhibition seems to involve the catalytic domain of BMP1 and the frizzled domain of sizzled.

Since then, several groups have looked to see whether a similar BMP1/Tolloid-like proteinase control mechanism exists in mammals, but with widely varying results. Kobayashi et al. (2009) reported that sFRP2, a human homologue of sizzled, far from inhibiting the activity of BMP1/Tolloid-like proteinase, activates the cleavage of procollagen I by BMP1 and that *sfrp*$^{-/-}$ mice develop more limited fibrosis after myocardial infarction than normal mice (an effect assumed to be correlated with the reduction of collagen deposition). In contrast, He et al. (2010) have recently reported that sFRP2 enhances cleavage of procollagen I by BMP1 at low concentrations but inhibits cleavage at higher concentrations, resulting in their hands in a decrease in cardiac fibrosis when sFRP2 was injected in several places in the heart! Finally, for von Marschall and Fisher (2010b), sFRP2 has no effect on the cleavage of procollagen I by BMP1.

Another interesting controversy concerns the potential role of BMP growth factors in controlling BMP1/Tolloid-like proteinases. First, Jasuja et al. (2007) reported that BMP2 and 4 could interact with the prodomain of BMP1 but not with the mature form of the proteinase. In contrast, Lee et al. (2009) showed that BMP4 interacted with the mature region of human and *Xenopus* BMP1. This interaction led in both cases to the inhibition of BMP4 signaling but the effect on BMP1/Tolloid-like proteinase activity was less clear. For the moment, the debate continues but this indicates the considerable complexity in the possible feedback pathways during the formation of the dorsoventral axis as well as other processes controlled by BMPs.

Enhancers

Putting to one side the current controversy about the sFRPs, five types of activators of BMP1/Tolloid-like proteinases have been described to date (▶Figure 6.12). All of these, with the exception of fibronectin, appear to be capable of enhancing BMP1/Tolloid-like proteinase activity in a substrate-specific manner, targeting either collagen assembly/cross-linking (procollagen C-proteinase enhancers [PCPEs], periostin) or BMP growth factor activation (twisted gastrulation, olfactomedin-like protein-3).

Fibronectin

In the conditioned medium of *fibronectin*$^{-/-}$ mouse embryo fibroblasts (MEFs), Fogelgren et al. (2005) observed a significant reduction in the conversion of pro-LOX to mature LOX, as well as tight binding of LOX to the cell form of fibronectin (cFN).

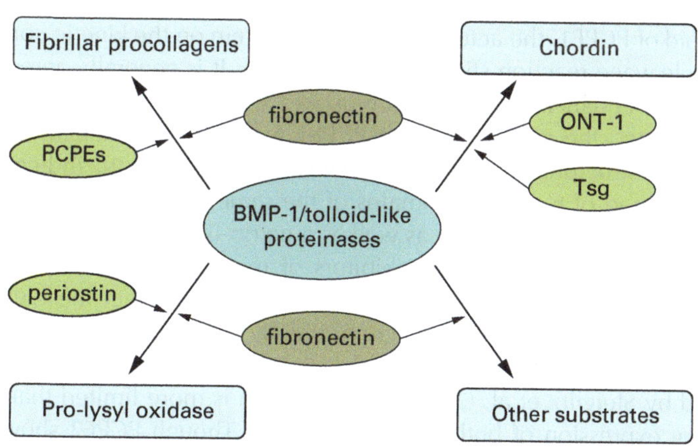

Figure 6.12 **Proteins that enhance the activity of BMP1/Tolloid-like proteinases.** With the exception of fibronectin, which can enhance the cleavage of several substrates, all the other proteins seem to be substrate specific. The protein sFRP2 is not included here as its precise effects on BMP1/Tolloid-like proteinase activity are unclear.

Subsequently, Huang et al. (2009) reported that fibronectin, in either its cFN or plasma (pFN) form, binds also to BMP1, with moderate affinity, and that the cFN form of fibronectin enhances the cleavage of chordin, biglycan and procollagen I by BMP1, both *in vitro* and in cell cultures. Fibronectin also stimulates BMP1 processing of IGF-like binding protein-3 in cell culture (Kim et al., 2011).

Periostin

A lack of cross-linking of collagen fibrils has been observed in several *periostin*$^{-/-}$ mouse tissues, particularly in heart (Norris et al., 2007), tendons and periosteum. Furthermore, overexpression of periostin in calvaria osteoblasts (COB cells) leads to a very significant increase in LOX maturation (Maruhashi et al., 2010). Periostin seems to be able to interact with the catalytic domain of BMP1 via its Fas1 domains but its direct effect on pro-LOX processing by BMP1 was not studied *in vitro*. On the other hand, it seems that periostin can help concentrate BMP1 on the network formed by fibronectin and thus promote its interaction with pro-LOX (Maruhashi et al., 2010), which is also able to interact with fibronectin (Fogelgren et al., 2005).

Procollagen C-proteinase enhancers

The protein PCPE1 was first described in 1986 as an enhancer of BMP1/Tolloid-like proteinases (called PCP at the time; Adar et al., 1986; Kessler and Adar, 1989). This 55 kDa glycoprotein or its proteolytic fragment (CUB1-CUB2; ▶Figure 6.9) was copurified with BMP1 from mouse 3T6 fibroblast culture medium and when separated from the proteinase led to a decrease in the activity of BMP1 on procollagen I. In this initial study, the kinetics of the cleavage reaction of procollagen I were measured, and showed an increase in K_m and k_{cat} in the presence of PCPE1. Subsequently, it became clear that these data were questionable due to the difficulty in obtaining preparations of proteinase completely free of PCPE1. Despite the evidence accumulated since then on PCPE and the development of systems for the expression of BMP1/Tolloid-like proteinases devoid of PCPE1, the actual effect of this protein on the kinetic constants of the procollagen cleavage reaction still remains unknown. It is generally agreed, however, that the extent of enhancement can reach up to 15- to 20-fold.

A few years later, the gene coding for PCPE1 was cloned (Takahara et al., 1994a), and the domain structure of PCPE1 was revealed (▶Figure 6.9), indicating that the protein is composed of two CUB domains of the same type as those found in BMP1/Tolloid-like proteinases themselves, as well as a netrin-like (NTR) domain homologous to the N-terminal domain of tissue inhibitors of metalloproteinases (TIMPs) (Banyai and Patthy, 1999). It was also shown that PCPE1 did not affect the cleavage of the N-propeptides of fibrillar procollagens controlled by ADAMTS2 (Hulmes et al., 1997).

The PCPE2 isoform was first identified about 10 years ago (Xu et al., 2000) and was characterized by Steiglitz et al. (2002). Its expression is more limited than PCPE1, but there is strong expression of both isoforms in heart. Though PCPE2 shows only 43% identity with PCPE1, it also enhances processing of fibrillar procollagen (types I and II; Steiglitz et al., 2002; Zhu et al., 2009) but differs from PCPE1 by its relatively high affinity for heparin. These results, in cell culture conditions, show preferential association of PCPE2 with the cell layer, while PCPE1 is found mainly in the medium.

The enhancing activity observed for both isoforms of PCPE on fibrillar procollagen processing suggested early on that these proteins could also play a role in fibrosis. Examples of their involvement in tissue repair and fibrosis are indeed more numerous than for BMP1/Tolloid-like proteinases, for which two studies show potential roles in cardiac and renal fibrosis (Kobayashi et al., 2009; Grgurevic et al., 2011). Thus, a first study showed an overexpression of PCPE1 in a model of liver fibrosis induced by CCl_4 (Ogata et al., 1997); a second, in a model of infarction (Kessler-Icekson et al., 2006); and a third, in a model of plaque formation following arterial injury (Kanaki et al., 2000). When PCPE1 expression is genetically extinguished, *Pcolce*$^{-/-}$ mice are viable and fertile but have defects in long bones (geometry, mechanical properties, etc.) as well as tendon collagen fibrils with abnormal appearance (Steiglitz et al., 2006).

To investigate the specificity of PCPE1, we compared its effect on BMP1 processing of several substrates, including procollagen III, the procollagen $\alpha1(V)_3$ homotrimer (N-terminal processing), procollagen VII, pro-LOX, laminin-332, osteoglycin, and chordin (Moali et al., 2005). We found C-terminal fibrillar procollagen processing to be the only BMP1 activity to be enhanced by PCPE1, showing for the first time that this enhancer is substrate-specific. C-terminal processing of the other major fibrillar procollagens (types I and II) is also enhanced by PCPEs, as is release of the C-propeptides from the procollagen $\alpha2(V)$ chain (which unlike the $\alpha1[V]$ and $\alpha3[V]$ chains is not cleaved by SPCs), since this chain was relatively poorly cleaved in fibroblast cultures from *Pcolce*$^{-/-}$ mice (Steiglitz et al., 2006). This specificity of action of PCPE1 has been confirmed by several other studies (Petropoulou et al., 2005; von Marschall and Fisher, 2010b), with the exception of a report by (Symoens et al., 2010), who observed enhancement of N-terminal processing of procollagen V. While the case of procollagen V requires further study, these latest data do not call into question the observed specificity of PCPEs for fibrillar procollagens.

Concerning the molecular mechanism of PCPE1, early observations showed that the CUB1-CUB2 region was sufficient for BMP1 enhancing activity and that it bound to the procollagen C-propeptide region (Adar et al., 1986; Kessler and Adar, 1989). To investigate precisely which domains are involved in the enhancer-substrate interaction, we carried out systematic studies on the roles of domains in both procollagen III and PCPE1. Concerning PCPE1, the contiguous CUB1-CUB2 region was found to be the minimally active form, with the CUB1 and CUB2 domains binding to the procollagen molecule in a cooperative manner, the linker between them acting as a flexible tether (Kronenberg et al., 2009). In addition, a surface-exposed phenylalanine residue in CUB1, found only in PCPEs and not in any other CUB-domain-containing proteins, is essential for activity, as are residues in the region of a putative calcium-binding site (Blanc et al., 2007). With regard to the NTR domain, contrary to earlier suggestions in the literature (Mott et al., 2000), we found no evidence for TIMP-like inhibitory activity after testing various metzincin metalloproteinases (including the MMP, ADAMTS, and astacin families). Instead, however, this domain may play a role *in vivo*, through a stable interaction with heparin and heparan sulfate glycosaminoglycan chains, in specific pericellular microenvironments, giving rise to an even more efficient enhancement of BMP1 (▶Figure 6.11B; Bekhouche et al., 2010; Weiss et al., 2010).

With regard to regions in the procollagen molecule involved in PCPE enhancing activity, we showed earlier that removal of the triple-helical region has no effect on enhancement (Moali et al., 2005). This left open the question of the possible involvement of the

C-telopeptide region (the nonhelical region remaining after C-propeptide cleavage). More recently, using surface plasmon resonance, ELISA and coimmunoprecipitation approaches, along with new constructs, we have shown unequivocally that PCPE1 binds to the C-propeptide only (Vadon-Le Goff et al., 2011). Detailed structural characterization of the PCPE1/C-propeptide complex, currently in progress, should give further insights into the mechanism of PCPE action.

Twisted gastrulation and olfactomedin noelin tiarin factor-1

The enhancer most studied along with the PCPEs is twisted gastrulation (Tsg). This protein is ventrally expressed in *Xenopus* embryos like BMP2 and BMP4 (Oelgeschlager et al., 2000). It interacts simultaneously with BMP growth factors and with their antagonist chordin, thereby favoring the cleavage of chordin by BMP1/Tolloid-like proteinases and thus promoting the activation of BMPs (Scott et al., 2001). A protein with similar activity is olfactomedin noelin tiarin factor-1 (ONT1) or olfactomedin-like protein-3, this time expressed on the dorsal side of *Xenopus* embryos, which also interacts with both BMP1 and chordin leading to modest enhancement of chordin cleavage (Inomata et al., 2008).

6.4.5 Meprins and matrix assembly

Meprins α and β constitute with ovastacin the other members of the family of human astacins. The meprins differ from other members of the family by the presence of a transmembrane domain (▶Figure 6.9), albeit that the β subunit can be released in soluble form under the action of proteases such as ADAM17 (Hahn et al., 2003). The quaternary structure of meprins varies depending on which subunits are expressed, occurring as homodimers, heterodimers, or very large oligomers (up to 6000 kDa; Sterchi et al., 2008). The meprins are activated by "trypsin-like" proteases (trypsin, kallikreins, etc.).

The principal substrates currently described for meprins are bioactive peptides (gastrin, substance P, interleukin-1β, etc.) but more and more matrix proteins have been found to be substrates of meprins (collagen IV, laminin-111, fibronectin, nidogen-1, etc.). Despite the fact that both meprins also favor acidic residues in the P1′ position (Becker-Pauly et al., 2011), the substrate specificity of the two meprin subunits differs more between isoforms and is broader than for the BMP1/Tolloid-like proteinases especially for meprin alpha which also cleaves substrates with neutral or aromatic residues in P1′. The cleavages are thus generally less specific than those made by the BMP1/Tolloid-like proteinases and in some substrates such as collagen IV, this can lead to degradation of the protein. Meprins are abundant in epithelial cells of the kidney, intestine, and skin. They are also expressed by leukocytes and many types of cancer cells (Sterchi et al., 2008).

We have recently shown that meprins, which have in common with BMP1/Tolloid-like proteinases an astacin-like catalytic domain, can also cleave the C-propeptide of procollagen III at an identical site (Kronenberg et al., 2010). These results challenge the established dogma that the BMP1/Tolloid-like proteinases are the only procollagen C-proteinases. Interestingly, PCPE1 inhibits the cleavage of procollagen III by meprins, probably because its interaction with the procollagen sterically hinders access of meprins to the cleavage site. Although the physiological consequences of these results have not yet been fully analyzed, they could be particularly relevant in the context of pathological tissue repair, as for example in keloids (Kronenberg et al., 2010).

6.4.6 Conclusions and take-home message

In conclusion, the BMP1/Tolloid-like proteinases have emerged as major players orchestrating matrix assembly and growth factor activation, two key events central to tissue morphogenesis and repair. It is probably no accident that these two functions have evolved in such harmony. In addition, there is increasing evidence that these proteinases are also involved in the control of angiogenesis and new functions are constantly coming to light. It is likely that the full list of substrates is far from complete.

Unlike other proteinases involved in tissue homeostasis and remodeling, BMP1/Tolloid-like proteinases rely almost exclusively on endogenous enhancers, rather than inhibitors, for regulation of activity. In addition, it seems that most of these enhancers are substrate specific. These peculiarities of the BMP1/Tolloid-like proteinases suggest that these enzymes present minor risks for tissues when normally expressed, and that some of the proteolytic events they control must be accelerated rather than slowed down. Substrate-specific enhancement of activity is an excellent way of focusing activity to particular physiological situations, explaining, for example, how high rates of collagen deposition can be achieved during development or wound healing

BMP1/Tolloid-like proteinases are of particular interest as therapeutic targets (Turtle and Ho, 2004; Fish et al., 2007), for example, in the control of muscle growth, by inhibition of myostatin cleavage, or in the prevention of fibrosis, by targeting collagen maturation and TGF-β activation. However, the ongoing discovery of novel substrates with unrelated functions may complicate the use of BMP1/Tolloid-like proteinase inhibitors. Local/controlled administration of drugs or targeting of specific proteolytic events through substrate-specific enhancers may provide ways to circumvent potential problems.

References

Adar, R., Kessler, E., and Goldberg, B. (1986). Evidence for a protein that enhances the activity of type I procollagen C-proteinase. Coll Relat Res *6*, 267–277.

Amano, S., Scott, I. C., Takahara, K., et al. (2000). Bone morphogenetic protein 1 is an extracellular processing enzyme of the laminin 5 gamma 2 chain. J Biol Chem *275*, 22728–22735.

Apte, S. S. (2009). A disintegrin-like and metalloprotease (reprolysin-type) with thrombospondin type 1 motif (ADAMTS) superfamily: functions and mechanisms. J Biol Chem *284*, 31493–31497.

Bachy, S., Letourneur, F., and Rousselle, P. (2008). Syndecan-1 interaction with the LG4/5 domain in laminin-332 is essential for keratinocyte migration. J Cell Physiol *214*, 238–249.

Banyai, L., and Patthy, L. (1999). The NTR module: domains of netrins, secreted frizzled related proteins, and type I procollagen C-proteinase enhancer protein are homologous with tissue inhibitors of metalloproteases. Protein Sci *8*, 1636–1642.

Becker-Pauly, C., Barré, O., Schilling, O., et al. (2011). Proteomic analyses reveal an acidic prime side specificity for the astacin metalloprotease family reflected by physiological substrates. Mol Cell Proteomics *10* (9), M111.009233.

Bekhouche, M., Kronenberg, D., Vadon-Le, G. S., et al. (2010). Role of the netrin-like domain of procollagen C-proteinase enhancer-1 in the control of metalloproteinase activity. J Biol Chem *285*, 15950–15959.

Bellamy, G., and Bornstein, P. (1971). Evidence for procollagen, a biosynthetic precursors of collagen. Proc Natl Acad Sci U S A *68*, 1138–1142.

Berry, R., Jowitt, T. A., Ferrand, J., et al. (2009). Role of dimerization and substrate exclusion in the regulation of bone morphogenetic protein-1 and mammalian tolloid. Proc Natl Acad Sci U S A *106*, 8561–8566.

Berry, R., Jowitt, T. A., Garrigue-Antar, L., Kadler, K. E., and Baldock, C. (2010). Structural and functional evidence for a substrate exclusion mechanism in mammalian tolloid like-1 (TLL-1) proteinase. FEBS Lett *584*, 657–661.

Blanc, G., Font, B., Eichenberger, D., et al. (2007). Insights into how CUB domains can exert specific functions while sharing a common fold: conserved and specific features of the CUB1 domain contribute to the molecular basis of procollagen C-proteinase enhancer-1 activity. J Biol Chem *282*, 16924–16933.

Borel, A., Eichenberger, D., Farjanel, J., et al. (2001). Lysyl oxidase-like protein from bovine aorta: isolation and maturation to an active form by bone morphogenetic protein-1. J Biol Chem *276*, 48944–48949.

Bradshaw, A. D. (2009). The role of SPARC in extracellular matrix assembly. J Cell Commun Signal *3*, 239–246.

Canty, E. G., and Kadler, K. E. (2005). Procollagen trafficking, processing and fibrillogenesis. J Cell Sci *118*, 1341–1353.

Chau, P., Fielding, P. E., and Fielding, C. J. (2007). Bone morphogenetic protein-1 (BMP-1) cleaves human proapolipoprotein A1 and regulates its activation for lipid binding. Biochemistry *46*, 8445–8450.

Church, R. L., Pfeiffer, S. E., and Tanzer, M. L. (1971). Collagen biosynthesis: synthesis and secretion of a high molecular weight collagen precursor (procollagen). Proc Natl Acad Sci U S A *68*, 2638–2642.

Colige, A., Sieron, A. L., Li, S. W., et al. (1999). Human Ehlers-Danlos syndrome type VII C and bovine dermatosparaxis are caused by mutations in the procollagen I N-proteinase gene. Am J Hum Genet *65*, 308–317.

Corsi, A., Xu, T., Chen, X. D., et al. (2002). Phenotypic effects of biglycan deficiency are linked to collagen fibril abnormalities, are synergized by decorin deficiency, and mimic Ehlers-Danlos-like changes in bone and other connective tissues. J Bone Miner Res *17*, 1180–1189.

Cronshaw, A. D., Fothergill-Gilmore, L. A., and Hulmes, D. J. S. (1995). The proteolytic processing site of the precursor of lysyl oxidase. Biochem J *306*, 279–284.

Danielson, K. G., Baribault, H., Holmes, D. F., Graham, H., Kadler, K. E., and Iozzo, R. V. (1997). Targeted disruption of decorin leads to abnormal collagen fibril morphology and skin fragility. J Cell Biol *136*, 729–743.

Dickson, L. A., Ninomiya, Y., Bernard, M. P., et al. (1981). The exon/intron structure of the 3′-region of the pro alpha 2(I) collagen gene. J Biol Chem *256*, 8407–8415.

Fish, P. V., Allan, G. A., Bailey, S., et al. (2007). Potent and selective nonpeptidic inhibitors of procollagen C-proteinase. J Med Chem *50*, 3442–3456.

Fogelgren, B., Polgar, N., Molnarne, S. K., et al. (2005). Cellular fibronectin binds to lysyl oxidase with high affinity and is critical for its proteolytic activation. J Biol Chem *280*, 24690–24697.

Ge, G., Fernandez, C. A., Moses, M. A., and Greenspan, D. S. (2007). Bone morphogenetic protein 1 processes prolactin to a 17-kDa antiangiogenic factor. Proc Natl Acad Sci U S A *104*, 10010–10015.

Ge, G., and Greenspan, D. S. (2006a). BMP1 controls TGFbeta1 activation via cleavage of latent TGFbeta-binding protein. J. Cell Biol. *175*, 111–120.

Ge, G., and Greenspan, D. S. (2006b). Developmental roles of the BMP1/TLD metalloproteinases. Birth Defects Res C Embryo Today *78*, 47–68.

Ge, G., Hopkins, D. R., Ho, W. B., and Greenspan, D. S. (2005). GDF11 forms a bone morphogenetic protein 1-activated latent complex that can modulate nerve growth factor-induced differentiation of PC12 cells. Mol Cell Biol *25*, 5846–5858.

Ge, G., Seo, N. S., Liang, X., Hopkins, D. R., Hook, M., and Greenspan, D. S. (2004). Bone morphogenetic protein-1/tolloid-related metalloproteinases process osteoglycin and enhance its ability to regulate collagen fibrillogenesis. J Biol Chem *279*, 41626–41633.

Gonzalez, E. M., Reed, C. C., Bix, G., et al. (2004). BMP-1/Tolloid-like metalloproteases process endorepellin, the angiostatic C-terminal fragment of perlecan. J Biol Chem *280*, 7080–7087.

Gopalakrishnan, B., Wang, W. M., and Greenspan, D. S. (2004). Biosynthetic processing of the pro-alpha1(V)pro-alpha2(V)pro-alpha3(V) procollagen heterotrimer. J Biol Chem *279*, 30904–30912.

Grgurevic, L., Macek, B., Healy, D. R., et al. (2011). Circulating bone morphogenetic protein 1–3 isoform increases renal fibrosis. J Am Soc Nephrol *22*, 681–692.

Hahn, D., Pischitzis, A., Roesmann, S., et al. (2003). Phorbol 12-myristate 13-acetate-induced ectodomain shedding and phosphorylation of the human meprinbeta metalloprotease. J Biol Chem *278*, 42829–42839.

He, W., Zhang, L., Ni, A., et al. (2010). Exogenously administered secreted frizzled related protein 2 (Sfrp2) reduces fibrosis and improves cardiac function in a rat model of myocardial infarction. Proc Natl Acad Sci U S A *107*, 21110–21115.

Hojima, Y., van der, R. M., and Prockop, D. J. (1985). Type I procollagen carboxyl-terminal proteinase from chick embryo tendons. Purification and characterization. J Biol Chem *260*, 15996–16003.

Huang, G., Zhang, Y., Kim, B., et al. (2009). Fibronectin binds and enhances the activity of bone morphogenetic protein 1. J Biol Chem *284*, 25879–25888.

Hulmes, D. J. S., Kadler, K. E., Mould, A. P., et al. (1989). Pleomorphism in type I collagen fibrils produced by persistence of the procollagen N-propeptide. J Mol Biol *210*, 337–345.

Hulmes, D. J. S., Mould, A. P., and Kessler, E. (1997). The CUB domains of procollagen C-proteinase enhancer control collagen assembly solely by their effect on procollagen C-proteinase/bone morphogenetic protein-1. Matrix Biol *16*, 41–45.

Hulmes, D. J. S. (2008). Collagen diversity, synthesis and assembly. In Collagen – Structure and Mechanics, P. Fratzl, ed. New York: Springer, pp. 15–47.

Humphries, S. M., Lu, Y., Canty, E. G., and Kadler, K. E. (2008). Active negative control of collagen fibrillogenesis in vivo. Intracellular cleavage of the type I procollagen propeptides in tendon fibroblasts without intracellular fibrils. J Biol Chem *283*, 12129–12135.

Imamura, Y., Steiglitz, B. M., and Greenspan, D. S. (1998). Bone morphogenetic protein-1 processes the NH2-terminal propeptide, and a furin-like proprotein convertase processes the COOH-terminal propeptide of pro-alpha1(V) collagen. J Biol Chem *273*, 27511–27517.

Inomata, H., Haraguchi, T., and Sasai, Y. (2008). Robust stability of the embryonic axial pattern requires a secreted scaffold for chordin degradation. Cell *134*, 854–865.

Jasuja, R., Ge, G., Voss, N. G., et al. (2007). Bone morphogenetic protein 1 prodomain specifically binds and regulates signaling by bone morphogenetic proteins 2 and 4. J Biol Chem *282*, 9053–9062.

Jenkins, G. (2008). The role of proteases in transforming growth factor-beta activation. Int J Biochem Cell Biol *40*, 1068–1078.

Kadler, K. E., Hill, A., and Canty-Laird, E. G. (2008). Collagen fibrillogenesis: fibronectin, integrins, and minor collagens as organizers and nucleators. Curr Opin Cell Biol *20*, 495–501.

Kanaki, T., Morisaki, N., Bujo, H., Takahashi, K., Ishii, I., and Saito, Y. (2000). The regulatory expression of procollagen COOH-terminal proteinase enhancer in the proliferation of vascular smooth muscle cells. Biochem Biophys Res Commun *270*, 1049–1054.

Kessler, E., and Adar, R. (1989). Type I procollagen C-proteinase from mouse fibroblasts. Purification and demonstration of a 55-kDa enhancer glycoprotein. Eur J Biochem *186*, 115–121.

Kessler, E., Fichard, A., Chanut-Delalande, H., Brusel, M., and Ruggiero, F. (2001). Bone morphogenetic protein-1 (BMP-1) mediates C-terminal processing of procollagen V homotrimer. J Biol Chem *276*, 27051–27057.

Kessler, E., Takahara, K., Biniaminov, L., Brusel, M., and Greenspan, D. S. (1996). Bone morphogenetic protein-1: the type I procollagen C-proteinase. Science *271*, 360–362.

Kessler-Icekson, G., Schlesinger, H., Freimann, S., and Kessler, E. (2006). Expression of procollagen C-proteinase enhancer-1 in the remodeling rat heart is stimulated by aldosterone. Int J Biochem Cell Biol *38*, 358–365.

Kim, B., Huang, G., Ho, W. B., and Greenspan, D. S. (2011). Bone morphogenetic protein-1 processes insulin-like growth factor-binding protein 3. J Biol Chem *286*, 29014–29025.

Kobayashi, K., Luo, M., Zhang, Y., et al. (2009). Secreted Frizzled-related protein 2 is a procollagen C proteinase enhancer with a role in fibrosis associated with myocardial infarction. Nat Cell Biol *11*, 46–55.

Kronenberg, D., Bruns, B. C., Moali, C., et al. (2010). Processing of procollagen III by meprins: new players in extracellular matrix assembly? J Invest Dermatol *130*, 2727–2735.

Kronenberg, D., Vadon-Le, G. S., Bourhis, J. M., et al. (2009). Strong cooperativity and loose geometry between CUB domains are the basis for procollagen C-proteinase enhancer activity. J Biol Chem *284*, 33437–33446.

Layman, D. L., McGoodwin, E. B., and Martin, G. R. (1971). The nature of the collagen synthesized by cultured human fibroblasts. Proc Natl Acad Sci U S A *68*, 454–458.

Lee, H. X., Ambrosio, A. L., Reversade, B., and De Robertis, E. M. (2006). Embryonic dorsal-ventral signaling: secreted frizzled-related proteins as inhibitors of tolloid proteinases. Cell *124*, 147–159.

Lee, H. X., Mendes, F. A., Plouhinec, J. L., and De Robertis, E. M. (2009). Enzymatic regulation of pattern: BMP4 binds CUB domains of Tolloids and inhibits proteinase activity. Genes Dev *23*, 2551–2562.

Lee, S., Solow-Cordero, D. E., Kessler, E., Takahara, K., and Greenspan, D. S. (1997). Transforming growth factor-beta regulation of bone morphogenetic protein-1/procollagen C-proteinase and related proteins in fibrogenic cells and keratinocytes. J Biol Chem *272*, 19059–19066.

Li, S. W., Sieron, A. L., Fertala, A., Hojima, Y., Arnold, W. V., and Prockop, D. J. (1996). The C-proteinase that processes procollagens to fibrillar collagens is identical to the protein previously identified as bone morphogenic protein-1. Proc Natl Acad Sci U S A *93*, 5127–5130.

Linsenmayer, T. F., Gibney, E., Igoe, F., et al. (1993). Type V collagen: molecular structure and fibrillar organization of the chicken alpha 1(V) NH2-terminal domain, a putative regulator of corneal fibrillogenesis. J Cell Biol *121*, 1181–1189.

MacSweeney, A., Gil-Parrado, S., Vinzenz, D., et al. (2008). Structural basis for the substrate specificity of bone morphogenetic protein 1/tolloid-like metalloproteases. J Mol Biol *384*, 228–239.

Maertens, B., Hopkins, D., Franzke, C. W., et al. (2007). Cleavage and oligomerization of gliomedin, a transmembrane collagen required for node of Ranvier formation. J Biol Chem *282*, 10647–10659.

Marques, G., Musacchio, M., Shimell, M. J., Wunnenberg-Stapleton, K., Cho, K. W. Y., and O'Connor, M. B. (1997). Production of a DPP activity gradient in the early drosophila embryo through the opposing actions of the SOG and TLD proteins. Cell *91*, 417–426.

Maruhashi, T., Kii, I., Saito, M., and Kudo, A. (2010). Interaction between periostin and BMP-1 promotes proteolytic activation of lysyl oxidase. J Biol Chem *285*, 13294–13303.

Moali, C., Font, B., Ruggiero, F., et al. (2005). Substrate-specific modulation of a multisubstrate proteinase. C-terminal processing of fibrillar procollagens is the only BMP-1-dependent activity to be enhanced by PCPE-1. J Biol Chem *280*, 24188–24194.

Moali, C., and Hulmes, D. J. S. (2009). Extracellular and cell surface proteases in wound healing: new players are still emerging. Eur J Dermatol *19*, 552–564.

Mott, J. D., Thomas, C. L., Rosenbach, M. T., Takahara, K., Greenspan, D. S., and Banda, M. J. (2000). Post-translational proteolytic processing of procollagen C-terminal proteinase enhancer releases a metalloproteinase inhibitor. J Biol Chem *275*, 1384–1390.

Mould, A. P., Hulmes, D. J. S., Holmes, D. F., Cummings, C., Sear, C. H., and Chapman, J. A. (1990). D-periodic assemblies of type I procollagen. J Mol Biol *211*, 581–594.

Myllyharju, J., and Kivirikko, K. I. (2004). Collagens, modifying enzymes and their mutations in humans, flies and worms. Trends Genet *20*, 33–43.

Nguyen, N. Q., Tabruyn, S. P., Lins, L., et al. (2006). Prolactin/growth hormone-derived antiangiogenic peptides highlight a potential role of tilted peptides in angiogenesis. Proc Natl Acad Sci U S A *103*, 14319–14324.

Norris, R. A., Damon, B., Mironov, V., et al. (2007). Periostin regulates collagen fibrillogenesis and the biomechanical properties of connective tissues. J Cell Biochem *101*, 695–711.

Oelgeschlager, M., Larrain, J., Geissert, D., and De Robertis, E. M. (2000). The evolutionarily conserved BMP-binding protein Twisted gastrulation promotes BMP signalling. Nature *405*, 757–763.

Ogata, I., Auster, A. S., Matsui, A., et al. (1997). Up-regulation of type I procollagen C-proteinase enhancer protein messenger RNA in rats with CCl4-induced liver fibrosis. Hepatology *26*, 611–617.

Panchenko, M. V., Stetler-Stevenson, W. G., Trubetskoy, O. V., Gacheru, S. N., and Kagan, H. M. (1996). Metalloproteinase activity secreted by fibrogenic cells in the processing of prolysyl oxidase. Potential role of procollagen C- proteinase. J Biol Chem *271*, 7113–7119.

Pappano, W. N., Steiglitz, B. M., Scott, I. C., Keene, D. R., and Greenspan, D. S. (2003). Use of Bmp1/Tll1 doubly homozygous null mice and proteomics to identify and validate in vivo substrates of bone morphogenetic protein 1/tolloid-like metalloproteinases. Mol Cell Biol *23*, 4428–4438.

Petropoulou, V., Garrigue-Antar, L., and Kadler, K. E. (2005). Identification of the minimal domain structure of bone morphogenetic protein-1 (BMP-1) for chordinase activity: chordinase activity is not enhanced by procollagen C-proteinase enhancer-1 (PCPE-1). J Biol Chem *280*, 22616–22623.

Piccolo, S., Agius, E., Lu, B., Goodman, S., Dale, L., and De Robertis, E. M. (1997). Cleavage of Chordin by Xolloid metalloprotease suggests a role for proteolytic processing in the regulation of Spemann organizer activity. Cell *91*, 407–416.

Prockop, D. J., Sieron, A. L., and Li, S. W. (1998). Procollagen N-proteinase and procollagen C-proteinase. Two unusual metalloproteinases that are essential for procollagen processing probably have important roles in development and cell signaling. Matrix Biol *16*, 399–408.

Rattenholl, A., Pappano, W. N., Koch, M., et al. (2002). Proteinases of the bone morphogenetic protein-1 family convert procollagen VII to mature anchoring fibril collagen. J Biol Chem *277*, 26372–26378.

Sandell, L. J., Prentice, H. L., Kravis, D., and Upholt, W. B. (1984). Structure and sequence of the chicken type II procollagen gene. Characterization of the region encoding the carboxyl-terminal telopeptide and propeptide. J Biol Chem *259*, 7826–7834.

Schechter, I., and Berger, A. (1967). On the size of the active site in proteases. I. Papain. Biochem Biophys Res Commun *27*, 157–162.

Scott, I. C., Blitz, I. L., Pappano, W. N., et al. (1999). Mammalian BMP-1/Tolloid-related metalloproteinases, including novel family member mammalian Tolloid-like 2, have differential enzymatic activities and distributions of expression relevant to patterning and skeletogenesis. Dev Biol *213*, 283–300.

Scott, I. C., Blitz, I. L., Pappano, W. N., Maas, S. A., Cho, K. W., and Greenspan, D. S. (2001). Homologues of Twisted gastrulation are extracellular cofactors in antagonism of BMP signalling. Nature *410*, 475–478.

Scott, I. C., Imamura, Y., Pappano, W. N., et al. (2000). Bone morphogenetic protein-1 processes probiglycan. J Biol Chem *275*, 30504–30511.

Shimell, M. J., Ferguson, E. L., Childs, S. R., and O'Connor, M. B. (1991). The Drosophila dorsal-ventral patterning gene tolloid is related to human bone morphogenetic protein 1. Cell *67*, 469–481.

Steiglitz, B. M., Ayala, M., Narayanan, K., George, A., and Greenspan, D. S. (2004). Bone morphogenetic protein-1/Tolloid-like proteinases process dentin matrix protein-1. J Biol Chem *279*, 980–986.

Steiglitz, B. M., Keene, D. R., and Greenspan, D. S. (2002). PCOLCE2 encodes a functional procollagen C-proteinase enhancer (PCPE2) that is a collagen-binding protein differing in distribution of expression and post-translational modification from the previously described PCPE1. J Biol Chem *277*, 49820–49830.

Steiglitz, B. M., Kreider, J. M., Frankenburg, E. P., et al. (2006). Procollagen C proteinase enhancer 1 genes are important determinants of the mechanical properties and geometry of bone and the ultrastructure of connective tissues. Mol Cell Biol *26*, 238–249.

Sterchi, E. E., Stöcker, W., and Bond, J. S. (2008). Meprins, membrane-bound and secreted astacin metalloproteinases. Mol Aspects Med *29*, 309–328.

Stöcker, W., Grams, F., Baumann, U., et al. (1995). The metzincins–topological and sequential relations between the astacins, adamalysins, serralysins, and matrixins (collagenases) define a superfamily of zinc-peptidases. Protein Sci *4*, 823–840.

Suzuki, N., Labosky, P. A., Furuta, Y., et al. (1996). Failure of ventral body wall closure in mouse embryos lacking a procollagen C-proteinase encoded by Bmp1, a mammalian gene related to Drosophila tolloid. Development *122*, 3587–3595.

Symoens, S., Renard, M., Bonod-Bidaud, C., et al. (2010). Identification of binding partners interacting with the alpha1-N-propeptide of type V collagen. Biochem J *433*, 371–381.

Takahara, K., Brevard, R., Hoffman, G. G., Suzuki, N., and Greenspan, D. S. (1996). Characterization of a novel gene product (mammalian tolloid-like) with high sequence similarity to mammalian tolloid/bone morphogenetic protein-1. Genomics *34*, 157–165.

Takahara, K., Kessler, E., Biniaminov, L., et al. (1994a). Type I procollagen COOH-terminal proteinase enhancer protein: identification, primary structure, and chromosomal localization of the cognate human gene (PCOLCE). J Biol Chem *269*, 26280–26285.

Takahara, K., Lyons, G. E., and Greenspan, D. S. (1994b). Bone morphogenetic protein-1 and a mammalian tolloid homologue (mTld) are encoded by alternatively spliced transcripts which are differentially expressed in some tissues. J Biol Chem *269*, 32572–32578.

Teillet, F., Gaboriaud, C., Lacroix, M., Martin, L., Arlaud, G. J., and Thielens, N. M. (2008). Crystal structure of the CUB1-EGF-CUB2 domain of human MASP-1/3 and identification of its interaction sites with mannan-binding lectin and ficolins. J Biol Chem *283*, 25715–25724.

Tsuchiya, S., Simmer, J. P., Hu, J. C., Richardson, A. S., Yamakoshi, F., and Yamakoshi, Y. (2011). Astacin proteases cleave dentin sialophosphoprotein (Dspp) to generate dentin phosphoprotein (Dpp). J Bone Miner Res *26*, 220–228.

Turtle, E. D., and Ho, W. B. (2004). Inhibition of procollagen C-proteinase: fibrosis and beyond. Expert Opin Ther Pat *14*, 1185–1197.

Unsold, C., Pappano, W. N., Imamura, Y., Steiglitz, B. M., and Greenspan, D. S. (2002). Biosynthetic processing of the pro-alpha 1(V)2pro-alpha 2(V) collagen heterotrimer by bone morphogenetic protein-1 and furin-like proprotein convertases. J Biol Chem *277*, 5596–5602.

Uzel, M. I., Scott, I. C., Babakhanlou-Chase, H., et al. (2001). Multiple bone morphogenetic protein 1-related mammalian metalloproteinases process pro-lysyl oxidase at the correct physiological site and control lysyl oxidase activation in mouse embryo fibroblast cultures. J Biol Chem *276*, 22537–22543.

Vadon-Le Goff, S., Kronenberg, D., Bourhis, J. M., et al. (2011). Procollagen C-proteinase enhancer stimulates procollagen processing by binding to the C-propeptide region only. J Biol Chem *286* (45), 38932–38938.

Veitch, D. P., Nokelainen, P., McGowan, K. A., et al. (2002). Mammalian Tolloid metalloproteinase, and not matrix metalloprotease 2 nor membrane type 1 metalloprotease, processes laminin-5 in keratinocytes and skin. J Biol Chem *278*, 15661–15668.

von Marschall, Z., and Fisher, L. W. (2010a). Decorin is processed by three isoforms of bone morphogenetic protein-1 (BMP1). Biochem Biophys Res Commun *391*, 1374–1378.

von Marschall, Z., and Fisher, L. W. (2010b). Dentin sialophosphoprotein (DSPP) is cleaved into its two natural dentin matrix products by three isoforms of bone morphogenetic protein-1 (BMP1). Matrix Biol *29*, 295–303.

Weiss, T., Ricard-Blum, S., Moschcovich, L., Wineman, E., Mesilaty, S., and Kessler, E. (2010). Binding of procollagen C-proteinase enhancer-1 (PCPE-1) to heparin/heparan sulfate: properties and role in PCPE-1 interaction with cells. J Biol Chem *285*, 33867–33874.

Wenstrup, R. J., Smith, S. M., Florer, J. B., et al. (2011). Regulation of collagen fibril nucleation and initial fibril assembly involves coordinate interactions with collagens V and XI in developing tendon. J Biol Chem *286*, 20455–20465.

Wipff, P. J., and Hinz, B. (2008). Integrins and the activation of latent transforming growth factor beta1 – an intimate relationship. Eur J Cell Biol *87*, 601–615.

Wolfman, N. M., McPherron, A. C., Pappano, W. N., et al. (2003). Activation of latent myostatin by the BMP-1/tolloid family of metalloproteinases. Proc Natl Acad Sci U S A *100*, 15842–15846.

Worthington, J. J., Klementowicz, J. E., and Travis, M. A. (2011). TGFbeta: a sleeping giant awoken by integrins. Trends Biochem Sci. *36*, 47–54.

Wozney, J. M., Rosen, V., Celeste, A. J., et al. (1988). Novel regulators of bone formation: molecular clones and activities. Science *242*, 1528–1534.

Xu, H., Acott, T. S., and Wirtz, M. K. (2000). Identification and expression of a novel type I procollagen C- proteinase enhancer protein gene from the glaucoma candidate region on 3q21-q24. Genomics 66, 264–273.

Yamada, Y., Kühn, K., and de Crombrugghe, B. (1983). A conserved nucleotide sequence, coding for a segment of the C-propeptide, is found at the same location in different collagen genes. Nucleic Acids Res *11*, 2733–2744.

Zhang, Y., Ge, G., and Greenspan, D. S. (2006). Inhibition of bone morphogenetic protein 1 by native and altered forms of alpha2-macroglobulin. J Biol Chem *281*, 39096–39104.

Zhu, J., Gardner, J., Pullinger, C. R., Kane, J. P., Thompson, J. F., and Francone, O. L. (2009). Regulation of apoAI processing by procollagen C-proteinase enhancer-2 and bone morphogenetic protein-1. J Lipid Res *50*, 1330–1339.

6.5 Supramolecular assembly of type I collagen

Mario Raspanti

6.5.1 Introduction

A collagen primer and a bit of nomenclature

Of the large (and ever-growing) collagen family, the members able to form fibrils, or *fibrillar collagens*, are just five: the types I, II, III, V, and XI. Other types, defined fibril-associated collagens with interrupted triple helices (FACITs), can be found on the fibril surface but are unable to form fibrils by themselves. Others still, defined membrane-associated collagens with interrupted triple helices (MACITs) or multiple triple-helix domains and interruptions (MULTIPLEXINs), have other roles in basal membranes, microfibrils and the like (see Shoulders and Raines, 2009).

The archetypal collagen, the type I, is a triple-helical heterotrimer of two α1(I) and an α2(I) chains. Each α chain exceeds 1,000 amino acids and is coiled into an helix with an average interresidue spacing of 108°, or 0.286 nm, and terminated by two nonhelical domains named *telopeptides*. The three α chains are then supercoiled with a 1-residue staggering into a flexible rod, approximately 300 nm long and 1.5 nm thick. Other nonhelical domains, the *propeptides*, are cleaved from the mature molecule prior to its aggregation into fibrils.

All collagen fibrils share the same essential structure, being formed by triple-helix collagen molecules laid parallel side-by-side with the same N→C orientation and progressively staggered by 234 residues, or 67 nm. This is the familiar *Hodge and Petruska* scheme, usually but erroneously called *quarter-stagger*. In fact, the 234-residue stagger divides the molecule into five molecular segments, one of which is shorter than the others and leaves an empty space, or *gap*. The elementary 67 nm cell (the *D-period*) can be therefore divided in two parts with a mass ratio of 5:4, named respectively *overlap zone* and *gap zone*, separated by two *step zones*. Depending on the experimental conditions, the fibrils can reveal more prominently either the cross-banding due to the succession of the gap- and overlap zones, or the lateral packing of the fibril subunits, or both. The collagen *fibrils* then aggregate, with the intermediation of small proteoglycans and other noncollagenous molecules into collagen *fibers*, which are the functional unit usually found in tissues.

Beyond this common basic architecture, the variety of fibrillar structures that type I collagen can form is surprisingly high. This is what we will see in the following pages.

6.5.2 The multimodal fibrils: tendon, bone, and ligaments

Structure

Because of its regular structure and simple composition, the tendon always was the tissue of choice for the study of the extracellular matrix. As a result, the overwhelming majority of published studies on collagen structure, fibrillogenesis and even mineralization were carried out on tendon, and the tendon fibril became the archetypal collagen fibril.

The collagen fibrils of tendon are extremely variable in size and they show an evident multimodal distribution of diameters, which range from almost nil to more than 300 nm (▶Figure 6.13). Their length has only been estimated, but most studies agree on a millimeter range (Craig et al., 1989). Similar fibrils are present not only in tendons and ligaments, but also in some other dense connective tissues such as the sclera (which is a sort of spherical tendon) and – with smaller diameters – in bone and in dentin.

Treatment with chaotropic agents loosens the fibril and reveals a tight packing of subfibrils slowly spiraling around the fibril axis at an very shallow angle. The reported thickness of these subfibrils is widely variable, ranging from 4 nm to more than 20 nm in different studies (see Ottani et al., 2001), while the winding angle is difficult to measure precisely and is often reported as "under 5 degrees."

Staining with electron-opaque ionic solutions such as uranyl acetate and lead citrate (*positive staining*) causes the appearance of a distinctive array of intraperiod bands, directly related to the intraperiod distribution of charged residues and named with the lowercase letters *a* to *e* (in the C→N direction). Staining with neutral solutions of heavy metals, typically buffered phosphotungstic acid or uranyl nitrate (*negative staining*), reveals a clearer distinction of gap- and overlap zones together with three intraperiod dense bands, named X_1, X_2, and X_3. The latter two are located on the gap-overlap

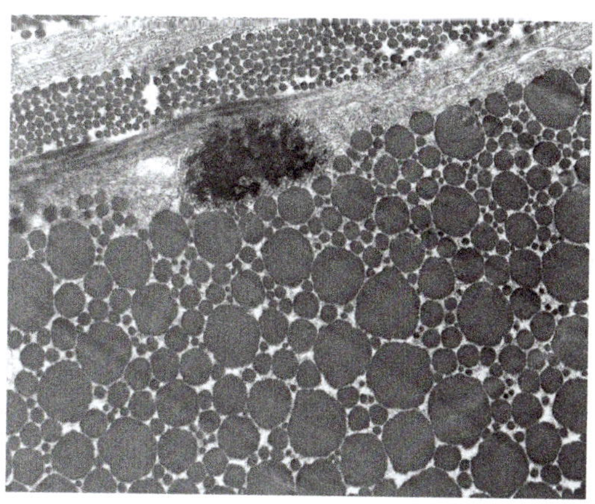

Figure 6.13 TEM micrograph of a cross-sectioned tendon. Most of the image shows the large, heterogeneous fibrils distinctive of this tissue. The tendon sheath, which appears near the top of the image, is made of a completely different fibrillar population. The black irregular structure beneath the tendon sheath is a small elastic fiber. Stained with tannic acid, uranyl acetate, and lead citrate. Original magnification 12,000x.

transition (the *step zones*) and correspond to the nonhelical N-terminal and C-terminal regions respectively, while the X_1 band is located within the gap zone and does not have a clear causation. Freeze-etching repeats the same banding as the negative staining, and both techniques lend ample support to the 4:5 gap:overlap ratio of the Hodge and Petruska scheme.

It must be noted that the Hodge and Petruska scheme is inherently bidimensional. Huge efforts (as always, all carried out on tendon) have been devoted to extend this basic scheme to the third dimension, and they yielded an infinite variety of structural models (see Veis et al., 1979; Squire and Freundlich, 1980). The two most successful concepts were the Smith microfibril, which essentially is a Hodge and Petruska five-stranded cell rolled into a cylinder, and a crystalline quasi-hexagonal packing of the molecules with no intermediate structures. All the morphological techniques supported the microfibrillar structure, while the crystalline packing was supported by most diffraction studies. The two models, supported by different techniques, were of course irreconcilable. Attempts to merge these incompatible views into a compromising "compressed microfibril" were initially met with skepticism, but recent very high resolution diffraction studies indicate the inner architecture of tendon fibrils as composed of five-stranded microfibrils compressed into a quasi-hexagonal packing and interconnected by stable cross-links (Orgel et al., 2006).

If a single objection can be moved to this elegant model, is that it takes for granted a radial symmetry of the fibril itself. This hypothesis is consistent with most, but not all of the experimental data; for instance, it cannot justify some observations on mineralizing tissues. Abstract models are always at risk of specifying a degree of precision exceeding biological significance. The real world is always imperfect, as any microscopist knows.

Fibrillogenesis

The processes that aggregate threadlike collagen molecules into almost-visible fibrils have of course been the subject of extensive research, both *in vitro* and *in vivo* (usually again on embryonic tendon: Birk and Trelstad, 1986; Birk et al., 1995; Hulmes, 2002; Starborg et al., 2008).

In vivo the formation of tendon begins with the appearance of slender fibrils quite different from the final ones (Banos et al., 2008; Starborg et al., 2008). Some fibrils are bipolar, with two N-terminal ends and a central short region of polarity inversion; branching fibrils were also demonstrated by three-dimensional (3D) reconstruction (Graham et al., 2000; Starborg et al., 2009). Neither bipolar nor branching fibrils, however, are present in mature tendon, so they must be superseded by other mechanisms of fibril formation.

During the fibrillogenesis process the cells also take care of the ordered layout of tendon fibrils, which is critical for the tensile requirement the tissue has to withstand. Early fibrils are individually spun in deep invaginations of the fibroblast cytoplasm, or fibripositors, and then deposited in parallel bundles in extracellular compartments, which form either between winglike protrusions of the cell body or in channels between adjacent cells (Birk and Trelstad, 1986; Starborg et al., 2008).

Fibrillogenesis *in vitro*, being this condition cell-free by definition, results instead in a spatially isotropic, weblike matrix (Raspanti et al., 2007). Again there are differences due to the experimental conditions: warming before neutralization leads to the

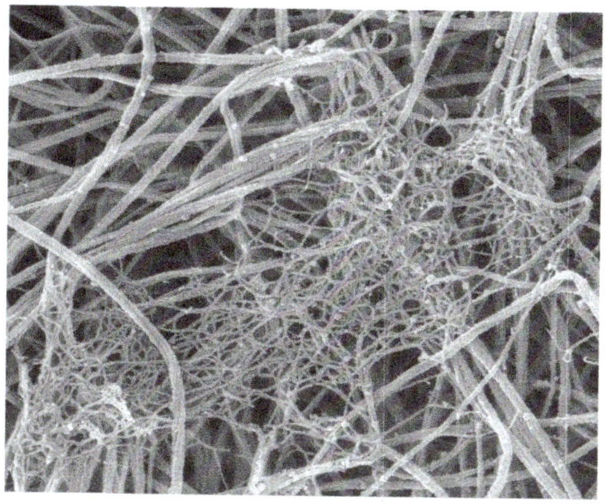

Figure 6.14 SEM micrograph of type I collagen fibrils reconstituted *in vitro*. Several banded fibrils are spun from the same network of slender, uniform protofibrils. The horizontal field of view is about 6 μm.

formation of short, banded fibrils with a typical diameter of 25 nm, similar to the early fibrils observed in embryonic connective tissues. In contrast, neutralization prior to warming leads to the formation of unbanded fine filaments, which then coalesce into mature fibrils (▶Figure 6.14).

Also unclear are the events that terminate the fibril accretion and determine the fibril final size. A number of molecules, including uncleaved propeptides, fibril-bound small leucine-rich proteoglycans (SLRPs), FACITs, and other factors have been credited with some effect (Capaldi and Chapman, 1982; Danielson et al., 1997; Svensson et al., 1999; Neame et al., 2000; Reed and Iozzo, 2003). Again, it is likely that all of them can have some effect, more or less important in different conditions, and it is difficult to isolate their contribution. As we will see later on, it is perfectly possible that the fibrils size is defined, at least in part, simply by the kinetics of the fibrillogenesis process.

Superfibrillar organization

In most tissues collagen fibrils are present in thick bundles, which act as the real functional units. There is ample evidence that small, fibril-bound proteoglycans act as potent organizers of the fibrils themselves, interconnecting them and keeping them in register quite effectively (Pins et al., 1997; Matheson et al., 2005; Zhang et al., 2006; Raspanti et al., 2007; Iwasaki et al., 2008). All SLRPs have specific binding sites on the fibril surface, for which they have an high affinity; recent data, however, showed that even isolated glycosaminoglycans are able to recognize specific binding sites and to stick to the collagen fibril surface in a specific, D-periodic pattern (Raspanti et al., 2008).

Since tendons and ligaments in most animals are measured in centimeters while a single collagen fibril may be shorter by an order of magnitude or more, the interfibrillar substance (which was once called *amorphous substance* but is now known to be an ordered lattice of fibril-bound glycoconjugates) must necessarily have a role in the interfibrillar mechanical coupling, both in lateral and in longitudinal direction (Sasaki and Odajima, 1996; Raspanti et al., 2002; Redaelli et al., 2003).

A tensile role of proteoglycans at first can seem implausible because of the jelly consistence of hydrated glycosaminoglycans. But we must take into account the sheer number of interacting proteoglycans, each strongly bound to a specific site within each D-period. Since there are approximately 15,000 D-periods per millimeter, and in each D-period several proteoglycan chains radiate from the fibril, the cumulative force they can exert along a fibril may easily exceed the ultimate tensile strength of the fibril itself. At this point, the functional behavior of short fibrils becomes undistinguishable from that of continuous fibrils.

Although not all the authors agree (Provenzano and Vanderby, 2006; Svensson et al., 2011), this hypothesis is consistent with diffraction studies on tendons, revealing that only part of the strain exerted on the tissue is withstood by the fibrils (Fratzl et al., 1997), the rest being evidently borne by interfibrillar fraction; with functional studies showing that the tensile strength of a tissue is greatly reduced by proteoglycan removal (Osborne et al., 1998); and with morphological observations showing the shearing of interfibrillar proteoglycan bridges under tension (Liao and Vesely, 2007). This means, by the way, that part of the tension exerted on the tissue is initially spent in viscoelastic realignment prior to being passed to the fibrils, and this has an obvious protective significance toward the fibrils, which come to bear the tensile load more gradually.

But this is by no way the only such mechanism. Another obvious way is the realignment of the fibrils themselves: if the fibrils (or the fibers) at rest are not yet aligned with the tensile stress, the tissue is initially compliant and becomes less and less extensible as soon as more and more subunits are aligned and put under tension. In other words the Young's modulus of the tissue increases steadily until all the fibers are under load. The *efficiency of reinforcement* is formally defined as $\eta = \Sigma a_n \cos^4(\theta)$, where a_n is the fraction of fibers lying along any given direction and θ is the angle they form with the tension axis. In tendons the collagen fibers are plaited in such a way that each fiber at rest follows a zigzag course, and straightens under tension. To this end, a special structure is present in tendons: the crimp.

The presence of crimps has been recognized long time ago, but recent observations by scanning electron microscopy and scanning probe microscopy, which make possible a high magnification three-dimensional visualization of the specimen, have better elucidated their structure (Raspanti et al., 2005; Franchi et al., 2007).

Tendon fibrils are relatively thick and compact and are scarcely flexible. They behave, in fact, more like solid rods than like ropes and, at least for short distances, tend to run stiff and straight (▶Figure 6.15). These straight segments are joined by sharp bends, often combined with some amount of axial twisting and acting as flexible joints. The crimp region does not show any cross-banding in electron micrographs: this suggests that in the short and sharp crimp region, which spans just a few D-periods, the collagen microfibrils cannot maintain the regular Hodge and Petruska quarter-stagger. Even when these fibril are straightened under tension a permanent local deformation persists, still betraying the original crimp location (Raspanti et al., 2005).

Figure 6.15 **Three-dimensional rendering of a fluid-tapping mode-atomic force microscopy dataset of a fully hydrated tendon, imaged in immersion.** The crimps are here unusually close to each other; the collagen fibrils clearly appear as a sequence of short, straight segment. The image spans 3 x 3 μm.

6.5.3 The unimodal fibrils: cornea, sheaths, and blood vessels

Structure

As we have seen, most of the research on the structural biology of collagen was carried out on tendon. For an unfortunate circumstance this tissue represents a sort of exception, so that the knowledge gained on tendon fibrils may not apply, as is often incorrectly taken for granted, to all and any connective tissue.

In most tissues the collagen fibrils appear indeed very different. To begin with, these fibrils are much more flexible than those of tendon. They usually run in slender, flexuous bundles and never show fibrillar crimps. Their diameter is always small and, although variable from tissue to tissue, is locally extremely uniform (see the upper portion of ▶Figure 6.13), so much to suggest that an entirely different, and much more stringent, mechanism of growth control must be at work. When imaged after negative staining or freeze-etching they show an additional fourth band, located between the X_1 and X_3 and named Y_1. In addition to collagen type I, these fibrils contain also important proportions of collagen types III and V; both types, but in particular type V, have elicited in recent years a great interest and have been credited with important functional roles in the formation of these fibrils (Birk et al., 1988; Birk and Mayne, 1997), either as initiators (Wenstrup et al., 2004), regulators (Birk, 2001) or terminators of the fibrillogenesis (Birk et al., 1990; Chanut-Delalande, 2001).

More important, however, is that the subunits of these fibrils are curved to follow an evident spiral course around the fibril, with an angle close to 17 degrees to the fibril axis (as always, the precise figure varies with the technique used to measure it). As a consequence of this winding the D-period is shortened to approximately 64 nm (64 nm $\approx$ 67 nm × $cos[17°]$), and the collagen molecules form a distinctive cross-link, involving an histidine residue that would not be available otherwise.

Although this helical winding that sets most tissues apart from tendon was confirmed long ago by electron microscopy (Marchini et al., 1986; Raspanti et al., 1989), X-ray diffraction (Folkhard et al., 1987), cross-link analysis (Mechanic et al., 1987), and atomic force microscopy (Yamamoto et al., 2000), it received rather scarce attention in front of all the research devoted to tendon. Of the tissues having this structure, only the cornea was the object of some serious investigations because of its higher clinical relevance.

An outstanding application of electron tomography (Holmes et al., 2001) eventually revealed that corneal fibrils are made of coaxial layers of microfibrils, which in all layers maintain a constant angle of approximately 17° to the fibril axis. These findings were later confirmed by X-ray diffraction (Cameron et al., 2002).

A constant angle implies inevitably a variable pitch (the axially projected length of the helix that each subfibril describes around the fibril axis), this latter being $p = 2\pi r \times tan(\pi/2-\alpha)$, where α is the angle with respect to the fibril axis, and r the distance from the fibril axis. All other factors being constant, the pitch is therefore linearly variable with the radius.

This structure has several direct consequences. The spiral course of their microfibrils, together with their small diameter, make these fibrils highly flexible, and in their tissues of origin they are often gathered in thin, flexuous bundles following a wavy course. The variation of the pitch with the radius means that lateral relationship among adjoining microfibrils is not maintained along the fibril, so that a network of intermolecular cross-links can extend to the whole fibril. As a consequence these fibrils can withstand even extreme and/or repeated flexion without splaying or splitting axially as tendon fibrils do.

Fibrillogenesis

Another important difference with tendon fibrils that so far has gone unnoticed involves the fibrillogenesis process. As we have seen in the previous pages, there is no doubt that tendon fibrils grow by lateral apposition of smaller proto-fibrils, or even small fibrils, until the process is terminated by fibril-bound SLRPs or FACITs. When two immature tendon fibrils touch, being all their subfibrils more or less straight and parallel, they can simply flow into each other. But when two helical fibrils touch they cannot possibly merge, because their helical structure implies that in order to do so they would have to unwind and rewind completely, an event made highly implausible by the very entropic forces that rule the fibrillogenesis process. Moreover, the microfibrils that come in contact run in opposite directions (Raspanti, 2010). (▶Figure 6.16)

All this means that, while tendon fibrils can grow both by apposition of single collagen microfibrils and by coalescence with other (sub)fibrils, these unimodal fibrils can only grow by apposition of single microfibrils, the coalescence being banned by their inner structure.

Unexpectedly, this can also explain the remarkable uniformity of their diameters. Quite simply, these fibrils precipitate from a supersaturated solution of collagen molecules where each growing fibril competes with its neighbors for the available subunits until all these have been depleted, at which point the accretion stops. The extracellular environment being the same for all fibrils growing in a given location, they all end up of a similar size. Somewhat similar happens in nature to raindrops or snowflakes. By contrast, in fibrils with straight microfibrils the lateral fusion occurs as a stochastic, inherently unpredictable process, which ends up in a population of widely different aggregates (▶Figure 6.17).

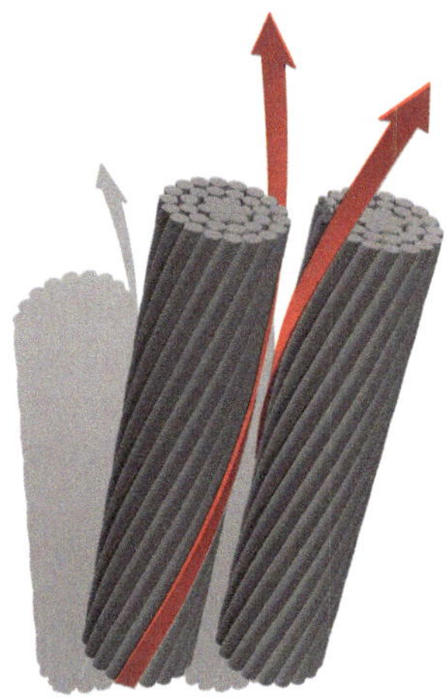

Figure 6.16 **A 3D model of collagen fibrils with helically wound subfibrils.** Each fibril is made of coaxial layers of microfibrils winding in a right-handed helix. As a consequence, where two fibrils touch their microfibrils run in opposite directions. Any lateral fusion is clearly impossible.

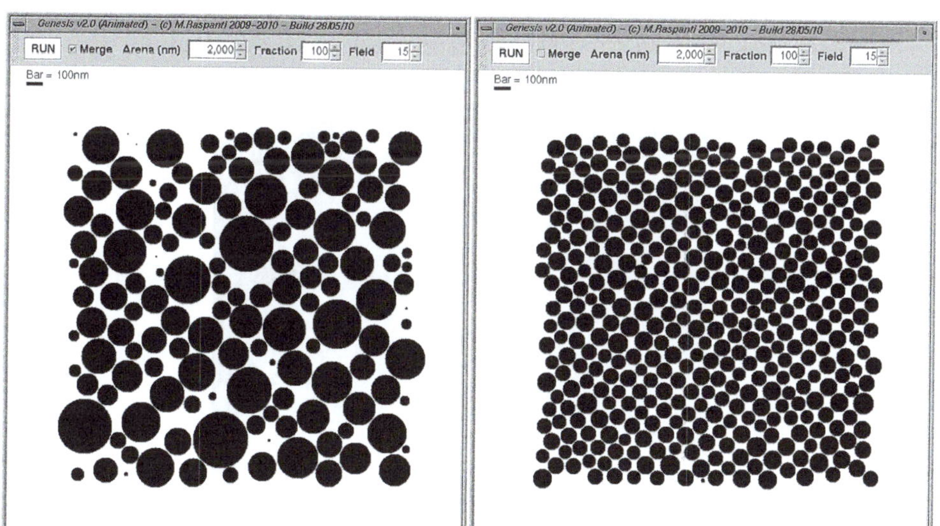

Figure 6.17 **The panels show a computer simulation of the fibrillogenesis process and represent a cross-section across a bundle of parallel fibrils.** All other factors being equal, the outcome is completely different depending from the lateral fusion being allowed (left) or disallowed (right). Picture taken from Raspanti M. (2010). J Biomed Sci Eng *3*, 1169–1174.

Surprising as it can be, the striking uniformity of the unimodal fibrils may not require different and purposeful mechanisms of growth control; their aspect could be simply dictated by their inner architecture and by the laws of kinetics. *Entia non sunt multiplicanda praeter necessitatem.* Some mechanism that stops the fibril accretion by preventing lateral fusion is still indispensable to limit the lateral growth of tendon-like fibrils, but it is not necessary for fibrils with helical subfibrils. Not surprisingly, decorin$^{-/-}$ organisms have abnormal fibrils in tendon but *not* in cornea (Danielson et al., 1997).

The coaxial structure of these fibrils has one last implication: the dependence of the microfibrils pitch from their intrafibrillar position implies that the microfibrils closer to the fibril axis, where both *r* and *p* tend to zero, should warp up to an implausible extent. This may look as a critical shortcoming of this model.

It must be noted, however, that this architecture lends itself quite naturally to an epitaxial layout where successive layers of microfibrils wrap around an axial core of different nature. Tubular fibrils with an electron-lucent core are also observable in invertebrates and lower vertebrates (see Figure 5/C in Ottani et al., 2002), but they were never investigated in depth.

There is, however, some evidence that, at least in some tissues, ever type I–based mammalian fibrils have a distinct axial domain. A single electron-dense central spot has been demonstrated at the center of the unimodal fibrils of nerve sheath and tendon sheath by means of the Seligman reaction, which uses OsO_4 vapors to highlight sugars (▶Figure 6.18). In our observations this domain was never found in tendon fibrils, and observations of other authors were never confirmed. The absence of conventional counterstaining procedures in our research gives quite a clean specimen and rules out any possible confusion with other structures. On the other hand, it implies that the micrographs obtained have a very low contrast and are of a very poor quality.

Further research are under way to identify the highly glycosylated core of these unimodal fibrils. The most important objective, however, is to discover the factors, so far

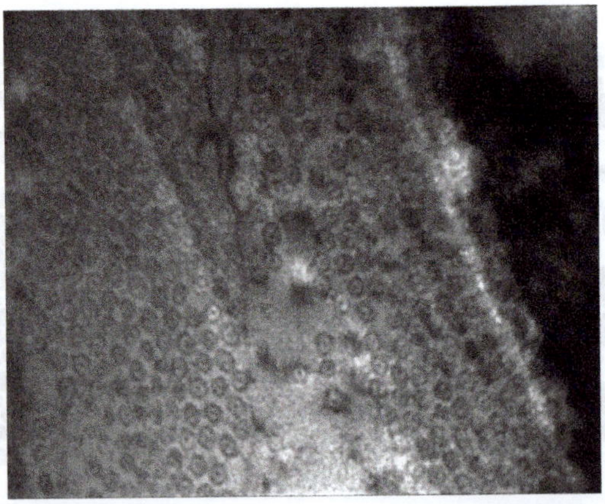

Figure 6.18 Transmission electron microscopy (TEM) micrograph of cross-sectioned endoneurial sheath of the sciatic nerve of rat. The specimen was left unstained except for the Thiéry-Seligman reaction, which reveals the location of polysaccharides. Each fibril shows a dark ring in periphery, presumably corresponding to fibril-bound proteoglycans, and a single dark spot on the fibril axis. Original magnification 56,000x.

unknown, that during fibrillogenesis determine the onset of one or the other of these two subfibrillar architectures. This appears to be a daunting task, because the assembly of collagen molecules into fibrils is basically an entropy-driven process that can be influenced by countless parameters (concentration, pH, temperature, ionic strength, etc.), while several noncollagenous molecules can act as organizers or nucleators (Douglas et al., 2006; Kvist et al., 2006; Kadler et al., 2008; Kalamajski and Oldberg, 2010) without remaining in the final product. Moreover, all these factors seem to act differently *in vivo* and *in vitro*, because fibrillogenesis *in vitro* takes place happily even in complete absence of some molecular factors that *in vivo* are apparently indispensable (Bornstein et al., 2000; Rühland et al., 2007).

As a result, in spite of all the efforts devoted, many molecular details of the process remain incompletely understood. They are the holy grail of our ongoing research.

6.5.4 Take-home message

If we must distil a conclusion from these observations, then a few points emerge quite clearly:

- The fibrillogenesis process is susceptible of being influenced by countless factors and parameters, both *in vivo* and *in vitro*. As a consequence, normal, functional collagen fibrils exist in a wide – and usually tissue-specific – range of sizes and shapes.
- Collagen I is known to form mixed fibrils with collagens III and V. Although their mutual interactions within the fibril are still unclear, their presence can be reasonably expected to have an effect on the fibril structure, morphology, and functional behavior.
- Most of the collagen-related research has been carried out on a single tissue. This is not a problem in itself. The inclination to extrapolate to other tissues the knowledge gained, however, definitely is. The inability of "helical" fibrils (i.e. fibrils with helical subunits) to undergo lateral fusion, and the consequent need that we reassess the fibrillogenesis process for these fibrils, is an adequate case in point. There is no shortage of other examples.

References

Banos, C. C., Thomas, A. H., and Kuo, C. K. (2008). Collagen fibrillogenesis in tendon development: current models and regulation of fibril assembly. Birth Defects Res C Embryo Today *84*, 228–244.

Birk, D. E. (2001). Type V collagen: heterotypic type I/V collagen interactions in the regulation of fibril assembly. Micron *32*, 223–237.

Birk, D. E., Fitch, J. M., Babiarz, J. P., Doane, K. J., and Linsenmayer, T. F. (1990). Collagen fibrillogenesis in vitro: interaction of types I and V collagen regulates fibril diameter. J Cell Sci *95*, 649–657.

Birk, D. E., Fitch, J. M., Babiarz, J. P., and Linsenmayer, T. F. (1988). Collagen type I and type V are present in the same fibril in the avian corneal stroma. J Cell Biol *106*, 999–1008.

Birk, D. E., and Mayne, R. (1997) Localization of collagen types I, III and V during tendon development. Changes in collagen types I and III are correlated with changes in fibril diameter. Eur J Cell Biol *72*, 352–361.

Birk, D. E., Nurminskaya, M. V., and Zycband, E. (1995). Collagen fibrillogenesis in situ: fibril segments undergo post-depositional modifications resulting in linear and lateral growth during matrix development. Dev Dyn *202*, 229–243.

Birk, D. E., and Trelstad, R. L. (1986). Extracellular compartments in tendon morphogenesis: collagen fibril, bundle, and macroaggregate formation. J Cell Biol *103*, 231–240.

Bornstein, P., Kyriakides, T. R., Yang, Z., Armstrong, L. C., and Birk, D. E. (2000). Thrombospondin 2 modulates collagen fibrillogenesis and angiogenesis. J Invest Dermatol *5*, 61–66.

Cameron, G. J., Alberts, I. L., Laing, J. H., and Wess, T. J. (2002). Structure of type I and type III heterotypic collagen fibrils: an X-ray diffraction study. J Struct Biol *137*, 15–22.

Capaldi, M. J., and Chapman, J. A. (1982). The C-terminal extrahelical peptide of type I collagen and its role in fibrillogenesis in vitro. Biopolymers *21*, 2291–2313.

Chanut-Delalande, H., Fichard, A., Bernocco, S., Garrone, R., Hulmes, D. J. S., and Ruggiero, F. (2001). Control of heterotypic fibril formation by collagen V is determined by chain stoichiometry. J Biol Chem *276*, 24352–24359.

Craig, A. S., Birtles, M. J., Conway, J. F., and Parry, D. A. D. (1989). An estimate of the mean length of collagen fibrils in rat tail tendon as a function of age. Connect Tissue Res *19*, 51–62.

Danielson, K. G., Baribault, H., Holmes, D. F., Graham, H., Kadler, K. E., and Iozzo, R. V. (1997). Targeted disruption of decorin leads to abnormal collagen fibril morphology and skin fragility. J Cell Biol *136*, 729–743.

Douglas, T., Heinemann, S., Bierbaum, S., Scharnweber, D., and Worch, H. (2006). Fibrillogenesis of collagen types I, II and III with small leucine-rich proteoglycans decorin and biglycan. Biomacromolecules *7*, 2388–2393.

Folkhard, W., Christmann, D., Geercken, W., et al. (1987). Twisted fibrils are a structural principle in the assembly of interstitial collagens, chordae tendineae included. Z Naturforsch C *42*, 1303–1306.

Franchi, M., Fini, M., Quaranta, M., et al. (2007). Crimp morphology in relaxed and stretched rat Achilles tendon. J Anat *210*, 1–7.

Fratzl, P., Misof, K., Zizak, I., Rapp, G., Amenitsch, H., and Bernstorff, S. (1997). Fibrillar structure and mechanical properties of collagen. J Struct Biol *122*, 119–122.

Graham, H. K., Holmes, D. F., Watson, R. B., and Kadler, K. E. (2000). Identification of collagen fibril fusion during vertebrate tendon morphogenesis. The process relies on unipolar fibrils and is regulated by collagen-proteoglycan interaction. J Mol Biol *295*, 891–902.

Holmes, D. F., Gilpin, C. J., Baldock, C., Ziese, U., Koster, A. J., and Kadler, K. E. (2001). Corneal collagen fibril structure in three dimensions: structural insights into fibril assembly, mechanical properties, and tissue organization. Proc Natl Acad Sci U S A *98*, 7307–7312.

Hulmes, D. J. S. (2002). Building collagen molecules, fibrils, and suprafibrillar structures. J Struct Biol *137*, 2–10.

Iwasaki I., Hosaka Y., Iwasaki T., et al. (2008). The modulation of collagen fibril assembly and its structure by decorin: an electron microscopic study. Arch Histol Cytol *71*, 37–44.

Kadler, K. E., Hill, A., and Canty-Laird, E. G. (2008). Collagen fibrillogenesis: fibronectin, integrins, and minor collagens as organizers and nucleators. Curr Opin Cell Biol *20*, 495–501.

Kalamajski, S., and Oldberg, Å. (2010). The role of small leucine-rich proteoglycans in collagen fibrillogenesis. Matrix Biol *29*, 248–253.

Kvist, A. J., Johnson, A. E., Mörgelin, M., et al. (2006). Chondroitin sulfate perlecan enhances collagen fibril formation. Implications for perlecan chondrodysplasias. J Biol Chem *281*, 33127–33139.

Liao, J., and Vesely, I. (2007). Skewness angle of interfibrillar proteoglycans increases with applied load on mitral valve chordae tendineae. J Biomech *40*, 390–398.

Marchini, M., Morocutti, M., Ruggeri, A., Koch, M. H. J., Bigi, A., and Roveri, N. (1986). Differences in the fibril structure of corneal and tendon collagen. An electron microscopy and X-ray diffraction investigation. Connect Tissue Res *15*, 269–281.

Matheson, S., Larjava, H., and Häkkinen, L. (2005). Distinctive localization and function for lumican, fibromodulin and decorin to regulate collagen fibril organization in periodontal tissues. J Periodontal Res *40*, 312–324.

Mechanic, G. L., Katz, E. P., Henmi, M., Noyes, C., and Yamauchi, M. (1987). Locus of a histidine-based, stable trifunctional, helix to helix collagen cross-link: stereospecific structure of type I skin fibrils. Biochemistry *26*, 3500–3509.

Neame, P. J., Kay, C. J., McQuillan, D. J., Beales, M. P., and Hassell, J. R. (2000). Independent modulation of collagen fibrillogenesis by decorin and lumican. Cell Mol Life Sci *57*, 859–863.

Orgel, J. P. R. O., Irving, T. C., Miller, A., and Wess, T. J. (2006). Microfibrillar structure of type I collagen in situ. Proc Natl Acad Sci U S A *103*, 9001–9005.

Osborne, C. S., Barbenel, J. C., Smith, D., Savakis, M., and Grant, M. H. (1998). Investigation into the tensile properties of collagen/chondritin-6-sulphate gels: the effect of crosslinking agents and diamines. Med Biol Eng Comput *36*, 129–134.

Ottani, V., Martini, D., Franchi, M., Ruggeri, A., and Raspanti, M. (2002). Hierarchical structures in fibrillar collagens. Micron *33*, 587–596.

Ottani, V., Raspanti, M., and Ruggeri, A. (2001) Collagen structure and functional implications. Micron *32*, 251–260.

Pins, G. D., Christiansen, D. L., Patel, R., and Silver, F. H. (1997). Self-assembly of collagen fibers. Influence of fibrillar alignment and decorin on mechanical properties. Biophys J *73*, 2164–2172.

Provenzano, P. P., and Vanderby, R., Jr. (2006). Collagen fibril morphology and organization: implications for force transmission in ligament and tendon. Matrix Biol *25*, 71–84.

Raspanti, M. (2010). Different architectures of collagen fibrils enforce different fibrillogenesis mechanisms. J Biomed Sci Eng *3*, 1169–1174.

Raspanti, M., Congiu, T., and Guizzardi, S. (2002). Structural aspects of the extracellular matrix of the tendon: an atomic force and scanning electron microscopy study. Arch Histol Cytol *65*, 37–43.

Raspanti, M., Manelli, A., Franchi, M., and Ruggeri, A. (2005). The 3D structure of crimps in the rat Achilles tendon. Matrix Biol *24*, 503–507.

Raspanti, M., Ottani, V., and Ruggeri, A. (1989). Different architectures of the collagen fibril: morphological aspects and functional implications. Int J Biol Macromol *11*, 367–371.

Raspanti, M., Viola, M., Forlino, A., Tenni, R., Gruppi, C., and Tira, M. E. (2008). Glycosaminoglycans show a specific periodic interaction with type I collagen fibrils. J Struct Biol *164*, 134–139.

Raspanti, M., Viola, M., Sonaggere, M., Tira, M. E., and Tenni, R. (2007). Collagen fibril structure is affected by collagen concentration and decorin. Biomacromolecules *8*, 2087–2091.

Redaelli, A., Vesentini, S., Soncini, M., Vena, P., Mantero, S., and Montevecchi, F. M. (2003). Possible role of decorin glycosaminoglycans in fibril to fibril force transfer in relative mature tendons. A computational study from molecular to microstructural level. J Biomech *36*, 1555–1569.

Reed, C. C., and Iozzo, R. V. (2003). The role of decorin in collagen fibrillogenesis and skin homeostasis. Glycoconj J *19*, 249–255.

Rühland, C., Schönherr, E., Robenek, H., et al. (2007). The glycosaminoglycan chain of decorin plays an important role in collagen fibril formation at the early stages of fibrillogenesis. FEBS J *274*, 4246–4255.

Sasaki, N., and Odajima, S. (1996). Elongation mechanism of collagen fibrils and force-strain relations of tendon at each level of structural hierarchy. J Biomech *29*, 1131–1136.

Shoulders, M. D., and Raines, R. T. (2009). Collagen structure and stability. Annu Rev Biochem *78*, 929–958.

Squire, J. M., and Freundlich, A. (1980). Direct observation of a transverse periodicity in collagen fibrils. Nature *288*, 410–413.

Starborg, T., Lu, Y., Huffman, A., Holmes, D. F., and Kadler, K. E. (2009). Electron microscope 3D reconstruction of branched collagen fibrils in vivo. Scand J Med Sci Sports *19*, 547–552.

Starborg, T., Lu, Y., Kadler, K. E., and Holmes, D. F. (2008). Electron microscopy of collagen fibril structure in vitro and in vivo including three-dimensional reconstruction. Methods Cell Biol *88*, 319–345.

Svensson, L., Aszodi, A., Reinholt, F. P., Fassler, R., Heinegård, D., and Oldberg, A. (1999). Fibromodulin-null mice have abnormal collagen fibrils, tissue organization, and altered lumican deposition in tendon. J Biol Chem *274*, 9636–9647.

Svensson, R. B., Hassenkam, T., Hansen, P., Kjaer, M., and Magnusson, S. P. (2011). Tensile force transmission in human patellar tendon fascicles is not mediated by glycosaminoglycans. Connect Tissue Res *52*, 415–421.

Veis, A., Miller, A., Leibovich, S. J., and Traub, W. (1979). The minimum structure demonstrating native axial periodicity. Biochim Biophys Acta *576*, 88–98.

Wenstrup, R. J., Florer, J. B., Brunskill, E. W., Bell, S. M., Chervoneva, I., and Birk, D. E. (2004). Type V collagen controls the initiation of collagen fibril assembly. J Biol Chem *279*, 53331–53337.

Yamamoto, S., Hashizume, H., Hitomi, J., et al. (2000). The subfibrillar arrangement of corneal and scleral collagen fibrils as revealed by scanning electron and atomic force microscopy. Arch Histol Cytol *63*, 127–135.

Zhang, G., Ezura, E., Chervoneva, I., et al. (2006). Decorin regulates assembly of collagen fibrils and acquisition of biomechanical properties during tendon development. J Cell. Biochem *98*, 1436–1449.

6.6 Collagen interactomes: mapping functional domains and mutations on fibrillar and network-forming collagens

James D. San Antonio, J. Des Parkin, Judy Savige, Joseph P. R. O. Orgel, and Olena Jacenko

6.6.1 Collagen interactomes

Collagens are among the most ubiquitous and complex of the extracellular matrix (ECM) molecules of vertebrates and invertebrates (Piez and Reddi, 1984; Jacenko et al., 1991; Kadler et al., 2007; Marini et al., 2007 for review of the following). Yet, how they accomplish their crucial roles in maintaining tissue structure and function remains incompletely understood. This is because the three-dimensional (3D) structures of the proteins, and their interactions with other ECM macromolecules have not been fully elucidated. The primary protein sequences of more than twenty five of the collagens are known, hundreds of interacting partners have been identified, and more than a thousand mutations associated with a wide spectrum of human diseases have been mapped to these proteins. Therefore, there is great interest in defining the structure-function relationship for the collagens.

To be classified as a collagen, a protein must meet three criteria (Jacenko et al., 1991). First, it must be destined for deposition in the extracellular space and contribute to the structural integrity of the ECM. Second, it must form supramolecular aggregates (e.g. fibrils, filaments, and networks) either alone, or together with other matrix components. Third, it must contain at least one triple-helical domain. For most collagens, the majority of their sequence exists as a triple helix, which makes them unique among proteins. These domains are rigid, ropelike cylindrical structures, which, depending on the collagen type, are sometimes interspersed between small flexible non-triple-helical regions, or larger, sometimes globular noncollagenous domains. The triple-helical domains are composed of contiguous glycine-X-Y tripeptide repeats with the obligate glycine as the only residue with a side chain small enough to be accommodated within the coiled coil of the triple helix.

The unique molecular structure of collagens allows the construction of maps wherein triple-helical domains may be represented as two-dimensional (2D) linear arrays of three polypeptide chains. Annotating such sequences with positions of functional domains such as posttranslational modifications, sites mediating cell or ligand binding, proteolysis, or amino acids associated with mutations results in the creation of collagen interactomes. Over the past decade interactomes were constructed for type I collagen and three type IV collagen isoforms (Di Lullo et al., 2002; Sweeney et al., 2008; Parkin et al., 2011). The intent of these projects was to archive protein data, but unexpectedly, they provided deeper insights into the structure-function relationships for these collagens. These projects, and assessments of creating interactomes for cartilage collagens are reviewed.

Simple maps of proteins, showing their dimensions, shapes, domain features, and positions of mutations are commonplace in the scientific literature. Among the first map of type I collagen included the primary protein sequence and presented a mechanism of charge-based monomer alignment consistent with the "quarter stagger" fibril structure (Chapman, 1974). The next generation of maps included our study wherein the sites of heparin/heparan sulfate proteoglycan (HSPG) binding to collagen monomers and fibrils were physically mapped on the fibril D-period, and their spatial relationships relative to several cell interaction sequences were discussed (San Antonio et al., 1994). Yet another map showed the positions of chemically reactive amino acids and GPP motifs on the D-period (Heidemann, 1988). (G-X-Y; single letter amino acid designations are used throughout) In the 1990s, with reports of more than 100 cell and ligand-binding sites and substitution mutations mapping to type I collagen, we began constructing a type I collagen interactome. Note that although crystal protein structures are becoming commonplace, native collagens cannot be crystallized and therefore X-ray diffraction studies have yet to produce molecular models of the collagens at ultra-high resolution. Thus, 2D interactomes are a useful complement to ongoing structural analyses of collagens.

6.6.2 Type I collagen interactome

Protein structure

Type I collagen is the most abundant protein in vertebrates, making up much of the fibrous matrix scaffold of connective tissues like skin, tendons, ligaments, and bones (see Piez and Reddi, 1984, for review of the following). Type I collagen is synthesized in the endoplasmic reticulum as $\alpha1$ and $\alpha2$ procollagen chains, each encoded by separate genes and translated into proteins slightly longer than 1,000 amino acid residues. Cotranslational modifications include hydroxylations of proline and lysines, and glycosylation of certain lysines. Nucleation domains on the C-terminal propeptides promote the polymerization of two $\alpha1$ and one $\alpha2$ chain into the procollagen triple-helical monomer (▶Figure 6.19A and B). In the extracellular space, N- and C-proteinases remove the globular termini of procollagen, and every about 67 nm along the fiber axis, five tropocollagen molecules, or monomers (M), assemble in a quarter-staggered fashion to form part of the supramolecular helix, the microfibril (▶Figure 6.19C and D). Each microfibril, the subunit of the fibril (▶Figure 6.19D), and its close neighbors are covalently joined by N- and C-terminal intermolecular cross-links to form fibrils (▶Figure 6.19E). The basic repeating structure of the fibril is the D-period, which is about 67 nm long and composed of an overlap and gap zone. Each D-period contains the complete monomer sequence derived from overlapping consecutive elements of five monomers. The type I collagen interactome is an expanded, 2D view of the fibril D-period (▶Figures 6.19C and 6.20).

Identifying relationships between sites

Functionally interrelated sites on the interactome were identified as follows: (1) sequences to which more than one ligand is proposed to bind, or which are near neighbors

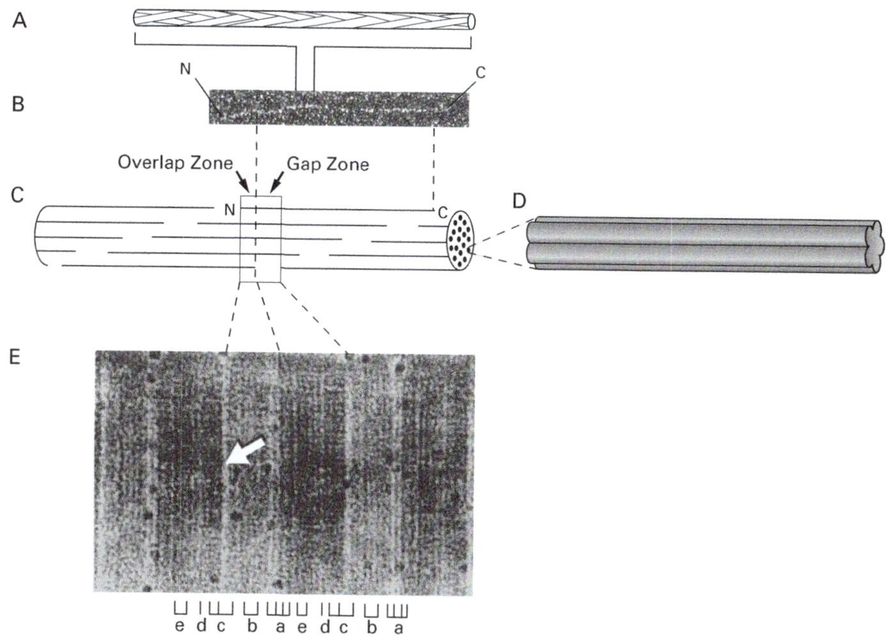

Figure 6.19 Type I collagen assembly and structure. Portion of single type I collagen triple helix (monomer) is depicted (A). Procollagen monomers are secreted then cleaved by proteinases (dashed vertical lines, [B]) and the tropocollagen monomers assemble into collagen fibrils (C) where one D-period repeat (expanded two-dimensional view of 67 nm segment of microfibril, box) contains the complete collagen sequence from elements of the five monomers, including an overlap and gap zone. The subunit structure of the fibril is considered a 5-mer microfibril (D). Collagen fibrils appear as periodic banded structures by electron microscopy; arrow, left border of overlap zone; particles are heparin-albumin gold conjugates used to map the heparin-binding site (E). The collagen map in Figure 6.20 represents an expanded view of one D-period. (This and Figures 6.20 and 6.21B were originally published in Sweeney et al. [2008]. J Biol Chem 283, 21187–21197).

on the same monomer; (2) sequences that fall within a broad protein region shown to bind a particular ligand; and (3) binding sites on neighboring monomers that might impact upon each other (Di Lullo et al., 2002). In the native fibril the monomers are ≤1.5 nm apart (Trus and Piez, 1980). Therefore, interactions between ligands on adjacent monomers may be possible if their binding sites align vertically within the D-period, and if they may simultaneously bind one triple helix and reach another ligand or ligand-binding site on an adjacent monomer.

Distribution of cell- and ligand-binding sites, and mutations on type I collagen

Several ligand-binding hot spots are apparent (▶Figure 6.20), the most notable being "major ligand-binding region-2" (MLBR2) including amino acid residues 680–830 (Di Lullo et al., 2002). This zone is also rich in lethal mutations, implying its crucial role in fibril function (Marini et al., 2007). Moreover, most of the major ligands are

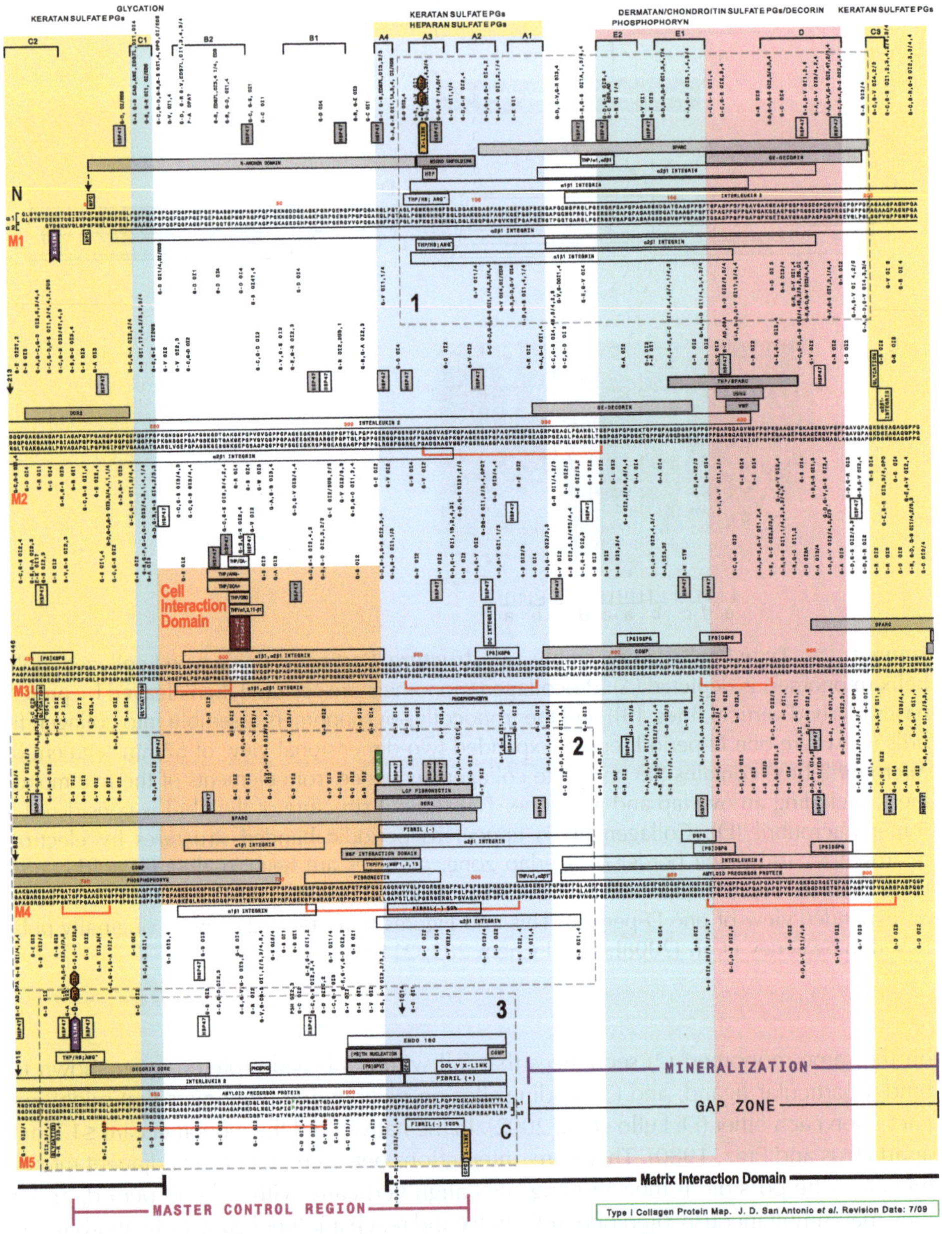

Figure 6.20 Type I collagen interactome. Human collagen primary sequences were obtained from GenBank, accession numbers – α1(I), NP000079; α2(I), NP000080 – and aligned as described (Chapman, 1974). Cross-fibril ligand-binding regions are indicated by colored overlays. Ligand-binding sites are indicated by rectangular boxes adjacent to relevant collagen sequences. Gray boxes denote ligand binding to the monomer. Nonshaded boxes denote ligand binding to one α chain. Major ligand-binding regions are designated 1, 2, and 3. Disease-associated mutations are indicated next to affected residues. See Sweeney et al. (2008) for abbreviations of mapped sites and associated literature citations. Figure 6.21A shows translation of select interactome elements onto a model of the native protein.

proposed to interact with multiple sites on several monomers; thus, intermonomer multivalency in binding may be a hallmark of ligand-collagen interactions.

Human mutation patterns

The most prevalent mutations on the map are associated with osteogenesis imperfecta (OI). OI is typically divided into four clinical types: type I (mild), II (lethal), III (severe), and IV (moderately severe) (Marini et al., 2007). The consequences of these mutations are generally thought to arise from their disruption of the folding or stability of the triple helix. The "gradient" model suggests that, because the helix folds from the C to the N terminus, more C-terminal mutations affect collagen modification and assembly more profoundly, resulting in a more severe phenotype, which is generally seen (Byers et al., 1991). "Regional models" suggest the distribution of some mutations may correlate with their impact upon various functional landmarks on the protein (Marini et al., 1993; Scott and Tenni, 1997). Thus, on the alpha 2(I) chain clusters of lethal OI mutations are interspersed with nonlethals, with the former corresponding with proteoglycan (PG)–fibril interaction zones (Marini et al., 2007). The interactome also supports a regional model: (1) consecutive runs of glycines associated exclusively with lethal, nonlethal, or no mutations exist throughout the protein; and (2) non-OI, or "atypical," mutations do not exhibit a gradient of phenotype severity according to their N- to C-terminal position along the protein, but rather, cluster to distinct zones on the monomer and fibril.

GFPGER: centerpiece of type I collagen

The interactome highlights the importance of the $GFPGER_{502-507}$ integrin-binding site to cell-collagen interactions (▶Figure 6.20) (subscripted numbers denote sequence position of amino acid residues in the native protein). This region binds $\alpha1\beta1/\alpha2\beta1$ integrins and supports angiogenesis (Sweeney et al., 2003), cell adhesion (Knight et al., 1998, 2000), activation (Sweeney et al., 1995), and osteoblast differentiation (Reyes and Garcia, 2003). Notably, GFPGER is located in the middle of a ligand-free zone, consistent with its dedicated function – no other ligands colocalize with it on the monomer, nor does it fall under the influence of major fibril-binding ligands such as PGs. The integrin-binding site is also a near neighbor of the predominant glycation site proposed to occur in diabetes and aging.

Interactome suggests a domain organization of the human collagen fibril

Master control region

Most of the functionally crucial sites of the fibril are concentrated in a relatively narrow zone defined by the region including sequences on M 1–5, falling on the overlap zone between the c2 and a3 bands, which we named the master control region (MCR) (Orgel et al., 2011a) (▶Figure 6.21A).

Fibril domains

The most critical functional sites of collagen including $GFPGER_{502-507}$, the matrix metalloproteinase-1 (MMP1) cleavage site, and other prominent elements of MLBR2 localize to a small region of M3 and 4 of the fibril that we propose regulates dynamic

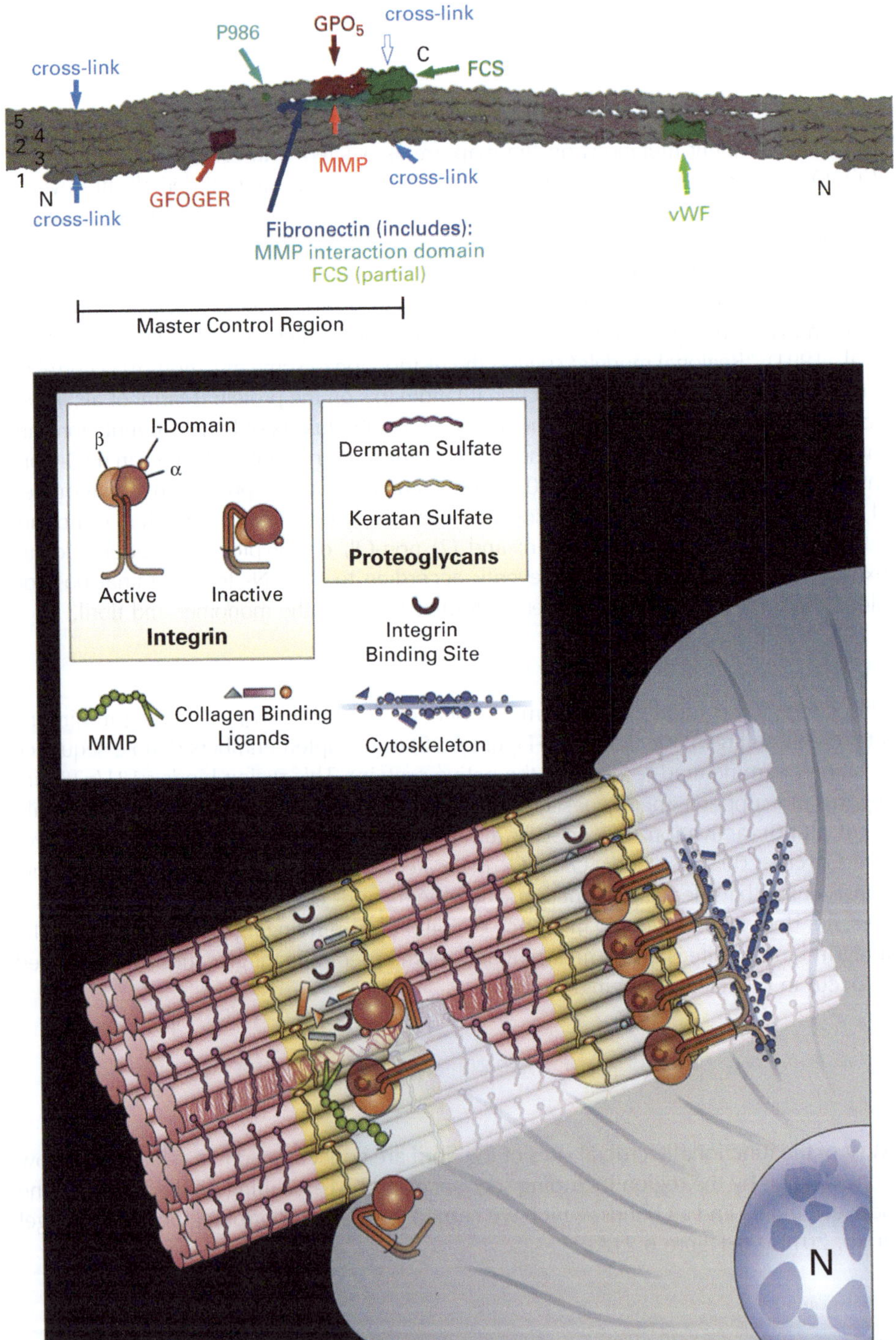

Figure 6.21 (A) Mapping functional sites on the X-ray diffraction structure of the type I collagen microfibril in situ. (B) Domain model of fibril function (see text). Originally published in Orgel et al. (2011a). Connective Tissue Res. *52*, 18–24.

aspects of collagen biology, thereby comprising the "cell interaction domain" (▶Figures 6.20 and 6.21B) (Sweeney et al., 2008). The remainder of the fibril contains sequences mediating intermolecular cross-links, PG binding, and mineralization and is proposed to assume structural duties, thereby comprising the "matrix interaction domain" (▶Figures 6.20 and 6.21B) (Sweeney et al., 2008). We speculate that similar domains are universal among the fibrillar and fibril-associated collagens.

Translating the interactome to the living collagen fibril

We mapped crucial functional sites of type I collagen on the X-ray diffraction model of the native fibril, including $GFPGER_{502-507}$; the N- and C-terminal intermolecular cross-links; the MMP1 cleavage site; P_{986}, for which inappropriate hydroxylation is associated with a significant fraction of OI mutations; and the C-terminal GPP(6) sequence with dual roles in platelet glycoprotein VI (GPVI) receptor binding and triple-helix nucleation (▶Figure 6.21A) (Orgel et al., 2011b). The fibrils' most exposed aspect includes the intermolecular cross-links and GPP(6). Thus, the exterior of the native fibril may present a structurally tough, hemostatic face. Yet under the surface exists a constellation of sites including $GFPGER_{502-507}$, the MMP1 cleavage site, and other hemostasis domains that may become available in collagen assembly and remodeling.

The interactome suggests how cells may interact with collagen (▶Figure 6.21B) (Sweeney et al., 2008). Thus, the overlap zone of the fibril contains $\alpha 2\beta 1$-integrin-binding sequences within each D-period as "landing strips" for cells. Based on the dimensions of integrins and collagen fibrils (Piez and Reddi, 1984; Emsley et al., 2004), the fibril is an optimal substrate for integrin receptor clustering, activation, and signaling. The proximity of the integrin-binding site to the MMP1 cleavage site and predominant ligand-binding region (only 3–11 nm apart) suggests that collagen assembly, function, and degradation may be achieved in an integrated fashion. The remainder of the fibril is densely decorated by PGs with anionic glycosaminoglycan (GAG) chains constrained via binding to the electropositive "bands" of the fibril surface.

Dinosaur peptides help validate collagen model

An unexpected source of support for our interactome was provided through analysis of a population of collagen peptides isolated from exceptionally well preserved *Tyrannosaurus rex* and *Brachylophosaurus canadensis* dinosaur fossil bones (San Antonio et al., 2011). That such protein fragments should survive through the estimated 60–80 million years remains controversial. Nonetheless, we mapped the positions of the dinosaur peptides on our models of the human and rat proteins. Interestingly, all of the surviving peptides mapped to the core, or more sheltered region of the microfibril. The peptides also had a distinct chemical nature, being less acidic than expected for an average population of collagen peptides. Moreover, at least three peptides from each dinosaur mapped to identical or overlapping positions on the collagen model. Four of the eleven peptides included functionally crucial sequences – those mediating integrin binding and MMP1 cleavage. Finally, the majority of the peptides aligned within several common regions of the fibril model, suggesting their preservation *en bloc*. These results support the authenticity of the peptides, propose physicochemical mechanisms underlying their selective preservation, and help validate the structure-function relationship suggested by the type I collagen interactome.

6.6.3 Type IV collagen Interactome

Basement membranes are 50–350 nm thick sheets of specialized ECM found throughout the body. In most tissues (blood vessels, bowel, testis, skin, alveolus, breast, thyroid, and liver) they separate epithelial or endothelial cells from the underlying stroma. However, they also separate apposing cell layers (glomerulus and retina) and surround cardiac, smooth and skeletal muscle cells, adipocytes, and Schwann and perineural cells. Basement membranes provide structural support to tissues, but are also highly biologically active, and interact with many cells and ligands in a complex array.

The composition and function of a basement membrane depends on the tissue in which it is found and the developmental stage. The major basement membrane-related structural proteins include collagen IV, laminin, nidogen, PGs (perlecan and agrin), and other collagens. In addition to providing structural support and tissue compartmentalization, basement membranes affect cell growth, adhesion, migration and differentiation, and are important in embryogenesis, tissue maintenance, regeneration, and repair. Basement membranes also have tissue-specific functions such as angiogenesis, hemostasis, protein filtration, and the support of muscle and nerve function. They are commonly affected in disease: by glycation in diabetes and ageing, and by infection, tumor spread, inherited disease, and autoimmune processes.

Type IV collagen

Collagen IV is the major component of basement membranes and forms a "chicken wire"–like network (Timpl, 1989). It comprises six isoforms, α1(IV)–α6(IV), encoded by COL4A1–COL4A6 (genes encoding the collagen IV protein chains). Collagen IV occurs as three distinct heterotrimers (α1, α1, α2; α3, α4, α5; and α5, α5, α6) in separate networks in different tissues (reviewed Hudson et al., 2003). In adults, the α1, α1, α2 network is present in all membranes, but especially the vasculature; the α3, α4, α5 network is found in membranes of the glomerulus, cochlea, and retina, and the α5, α5, α6 network is present in the skin.

Protein structure

Each collagen IV α chain consists of an intermediate collagenous sequence of about 1,400 residues of Gly-X-Y residues (see Khoshnoodi et al., 2008, for review of the following). There are noncollagenous (NC) domains at the N and C termini, as well as numerous short interruptions in the intermediate collagenous sequence. The type IV collagen chains associate intracellularly into triple helices. Assembly begins at the C terminus by disulfide bridge formation and progresses toward the N terminus. Triple helices are secreted to form a supramolecular network through dimerization at the C terminus and tetramerization at the N terminus. Basement membranes are intricately organized, with the collagen IV and laminin networks connected by nidogen as well as PGs (Pihlajaniemi, 1996; Timpl and Brown, 1996).

Identifying relationships between sites

The basement membrane proteins are well-characterized biochemically, but their structural features, binding partners, and missense mutations have not previously been integrated into a form that facilitates the prediction of functional domains, interactions, and genotype-phenotype correlations. We thus constructed linear protein maps of the three collagen IV heterotrimers where the three chains were aligned according to their carboxy terminal sequences, cysteines, and intervening noncollagenous sequences (Parkin et al., 2011). We then searched the literature and web-based databases for major structural domains and ligands, and their corresponding binding sites and motifs. Structural landmarks included normal splice variants, kinks, cross-links, and intrachain disulfide bonds, phosphoserines, and sites for glycosylation and hydroxylation, chaperone-binding, and protein cleavage. We also looked for integrin-binding sites for endothelial and epithelial, and other cells. Binding to ECM molecules was often approximated from measurements made on rotary shadowing assuming the average distance between amino acids in the triple helix to be 0.238 nm. Ligand interactions were predicted based on their physical proximity, and the size of the ligand including side chains. Ligand-binding sites involved in the same biological processes were often colocated and designated "functional" domains. The type IV collagen interactomes were published and are not shown here. Interactions with cells and other matrix proteins occurred throughout the triple helical and NC1 domains of at least the α1, α1, α2 heterotrimer (▶Figure 6.22A).

Interactome suggests organization of type IV collagen network in basement membranes

Examination of binding sites on the map enabled us to create a preliminary model of type IV collagen orientation within the membrane (▶Figure 6.22B). Thus, the typical basement membrane varies from 50 to 400 nm wide, and collagen IV is 200 nm long and has a kink 40 nm from the amino terminus. The α1(IV) chain has an endothelial-binding site at the amino terminus and an epithelial site at the carboxy terminus. This suggested the amino terminus lay flat against the endothelial surface but the "postkink" region extended to the epithelial surface with "corrugations" from flexibility conferred by the noncollagenous interruptions (Parkin et al., 2011).

The collagen IV maps highlight the networks' different biological activities, and suggest functional and disease-associated domains. The α1, α1, α2 heterotrimer has a functional domain for hemostasis, and the NC1 domains are particularly important in cell binding, angiogenesis, angiogenesis inhibition, and inhibition of cell growth.

Collagen IV is also critical in microbial infections (via lectin-binding sites), autoimmune disease (Goodpasture syndrome, and postallograft in Alport syndrome), tumor invasion (angiogenesis, integrin-binding, and canstatin/tumstatin sites), and advanced glycation end-product modification in diabetes and ageing (Paul and Bailey, 1996; Reigle et al., 2008). Sites for bacterial adhesion were also distributed throughout the chains (Vercellotti et al., 1985; Flugel et al., 1994; Alonso et al., 2001; Kajimura et al., 2004). The NC1 domains of the α3, α4, and α5 chains were all involved in auto- or alloimmune disease. Several sites at or near the NC1 domains of different chains were involved in tumor adhesion but the NC1 domains or fragments comprising integrin-binding sites

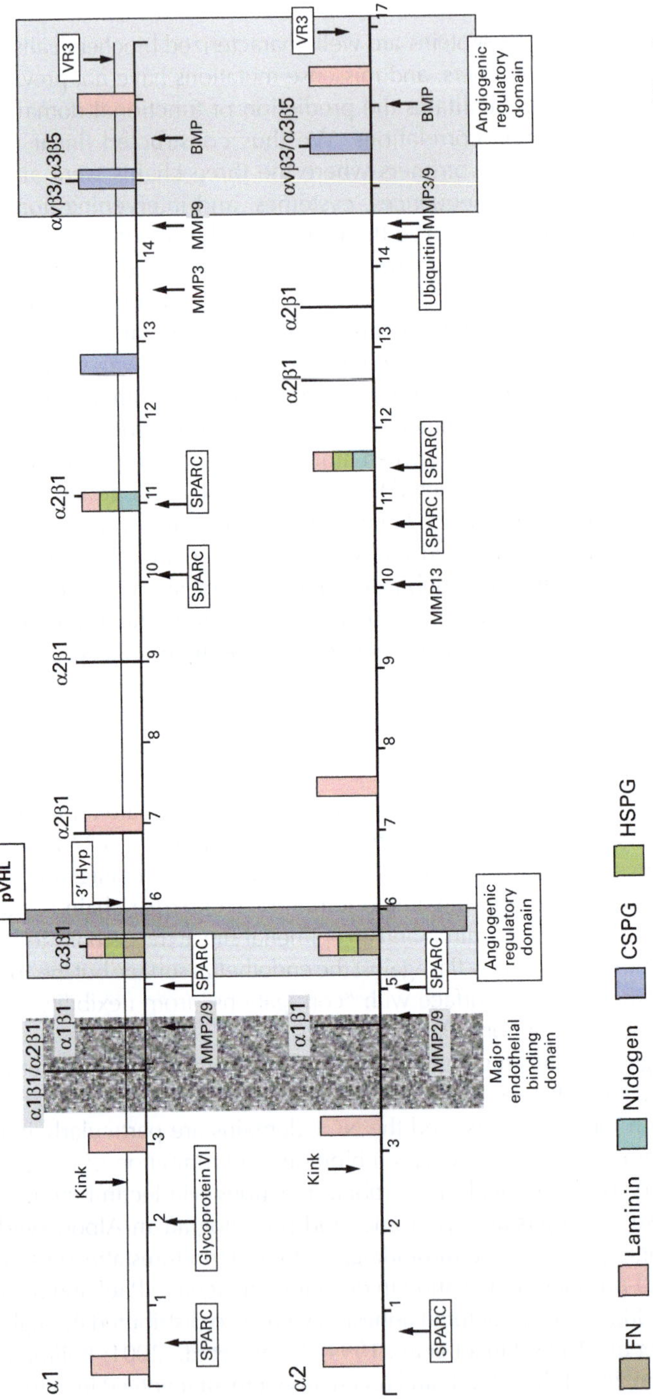
α1
α2
SPARC
Glycoprotein VI
Kink
α1β1/α2β1
α1β1
MMP2/9
SPARC
α3β1
pVHL
3′ Hyp
α2β1
α2β1
SPARC
SPARC
α2β1
MMP3
MMP9
αvβ3/α3β5
BMP
VR3
MMP13
SPARC
SPARC
α2β1
α2β1
Ubiquitin
MMP3/9
αvβ3/α3β5
BMP
VR3
Major endothelial binding domain
Angiogenic regulatory domain
Angiogenic regulatory domain
FN
Laminin
Nidogen
CSPG
HSPG

(Continued)

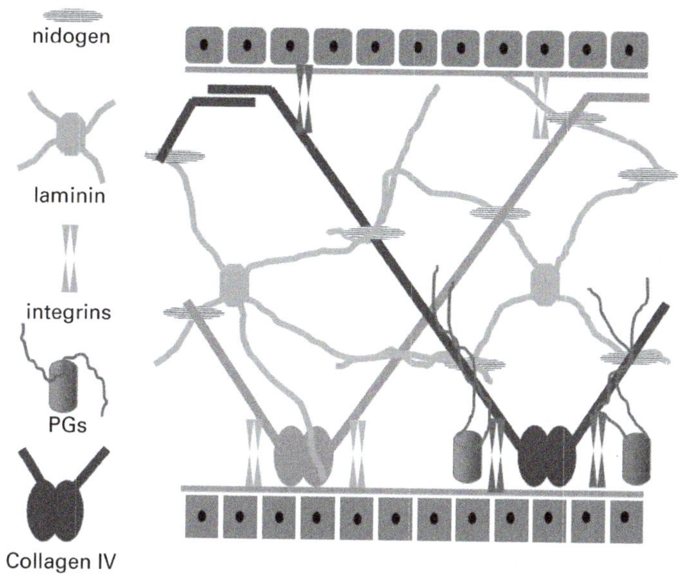

Figure 6.22 (A) Analysis of the type α1, α1, α2 type IV collagen interactome (Parkin et al., 2011) suggests domains for endothelial cell interactions, angiogenesis, and binding to other structural macromolecules of basement membranes. See Parkin et al. (2011) for abbreviations of mapped sites and associated literature citations. (B) The type IV collagen interactome suggests collagen orientation in basement membranes (see text).

inhibited tumor growth. One such fragment is currently in clinical trials for the treatment of human renal cell carcinoma (Eikesdal et al., 2008). Glycation occurred throughout the heterotrimers but was concentrated in the 7S and NC1 domains (Raabe et al., 1996).

Human mutation patterns

All published missense mutations were also indicated on the interactomes (Parkin et al., 2011). These are the commonest collagen IV mutation, and since they change only a single residue, their clinical consequences may in some cases correlate with the nature of the binding site affected. For example, the maps demonstrated that missense mutations affecting the von Hippel Lindau protein–binding site in the α1 chain resulted in a phenotype hereditary angiopathy with nephropathy and muscle cramps (HANAC) (Plaisier et al., 2007), a condition that shares some characteristics (renal cysts and tortuous blood vessels) with von Hippel Lindau syndrome. Very few missense mutations affected the α2 or α6 chains, possibly because they resulted in disease too rare or too mild to come to medical attention, or too severe to be viable. In X-linked Alport syndrome, large rearrangements, and deletions resulting in a frameshift, as well as nonsense mutations, all produce early onset renal failure (Jais et al., 2000; Gross et al., 2002). Renal failure can be delayed with the use of angiotensin converting enzyme inhibitors even before the onset of proteinurea (Gross et al., 2003). Missense mutations near the carboxy terminus, and those substituted with a larger or more highly charged residue also result in early onset renal failure (Persikov et al., 2004). However, the clinical phenotype of ≅ 50% of the missense mutations in X-linked Alport syndrome cannot currently be predicted, but may possibly manifest their phenotypic effects by interfering with a major structural or ligand-binding site. The maps also suggested that the distribution of missense mutations in the α5 chain (that cause X-linked Alport syndrome) was nonrandom with variants segregating in exons 25 and 26. Furthermore the two mutations that affected an integrin-binding site both resulted in early onset renal failure, although this observation has still to be confirmed in a larger series. This approach is likely to yield answers also in the autosomal forms of Alport syndrome. Heterozygous mutations in the COL4A3 or COL4A4 genes result in autosomal dominant Alport syndrome with renal failure or Thin basement membrane nephropathy with hematuria and normal renal function (Savige et al., 2003). It is unclear why only some mutations result in renal failure, but it may be due to mutation-specific effects on the functions of major structural domains or ligand-binding sites of the type IV collagen network.

Comparison of collagen IV and I interactomes

Both collagens have a GFPGER cell-binding motif, as well as binding sites for matrix molecules, secreted protein acidic and rich in cysteine (SPARC), bone morphogenetic protein (BMP), decorin core, heat shock protein 47 (HSP47), platelet GPVI receptor, and various MMPs. Some motifs bind at similar locations on both molecules, and multivalent ligand binding is common for both. Collagen I has critical ligand-binding sites (GFPGER, MMP1–2, and intermolecular cross-links) at regular intervals and "in register," which allows integrin receptor clustering and signaling, and the presentation of a single surface to the environment. Many fewer ligands are known for collagen IV, but at least the integrin- and SPARC-binding sites on the α1 chain also appeared to be in register. There were also notable differences – collagen IV forms a network through its amino and carboxy domains and has cell- and matrix-binding sites throughout the triple helix and NC1. Collagen I forms fibrils after the removal of the N- and C-terminal domains so the major cell-binding site is located midway between the amino and carboxy cross-links, and ligand-binding domains are within the triple helix. Collagen I has no known αv- or α3-integrin-binding sites, and ligand specificity reflects its role as a structural protein in bone and connective tissue.

6.6.4 Type III collagen interactome

Collagen type III is an abundant fibrillar collagen important in embryogenesis (Rong et al., 2008), hemostasis (Savage et al., 1998; Ruggeri, 2001), and wound healing (Lehto et al., 1985; Oliveira et al., 2009), and has a critical structural role in blood vessels and distensible organs, such as the large bowel and uterus. Mutations in the collagen III gene generally result in Ehlers-Danlos syndrome type IV (or "vascular" type) characterized by extensive bruising, and sometimes organ or vascular rupture (Pepin et al., 2000; Germain, 2007). Construction of an interactome of the collagen III α1 homotrimer is underway, based on the D-period repeat of collagen I, since embryonic collagen III is often replaced by collagen I after birth, and collagens III and I commonly coexist (Di Lullo et al., 2002; Sweeney et al., 2008). Preliminary data reveal a similar domain structure to collagen I and possible clues regarding its association with type I collagen in the heterotypic fibril (data not shown).

6.6.5 Type II collagen

Collagen II is the major structural component in cartilage. It is the predominant component of fibrils in hyaline and articular cartilage as well as in the eye vitreous, the inner ear, and the nucleus pulposus of vertebrae. The protein is a homotrimer of three α1(II) chains, and its biosynthesis and structure are similar to that of fibrillar collagens I, III, V, and XI (Jacenko et al., 1991). Collagen II mutations result in a wide spectrum of disease manifestations, ranging from lethal chondrodysplasias to short stature conditions, to degenerative joint diseases. For many type II collagenopathies, there is significant variability in disease onset and severity, making it difficult to predict the phenotype of the majority of the mutations (Kannu et al., 2010). Moreover, more than 100 collagen II mutations have been described to cause autosomal dominant diseases including the following: achondrogenesis II, hypochondrogenesis, spondyloepiphyseal dysplasia congenital, spondyloepimetaphyseal dysplasia (SEMD) Strudwick type, Kniest dysplasia, Stickler syndrome, multiple epiphyseal dysplasia, arthritis, osteonecrosis of the femoral head, and spondyloarthropathies (Jacenko et al., 1991; Kannu et al., 2010). Given its homology to other fibrillar collagens, we predict that a type II collagen fibril interactome would likely show a similar domain structure to that of collagens I and III (Sweeney et al., 2008). Moreover, as for collagen I, localization of the numerous collagen II mutations on the linear and D-period protein map may provide insights into the genotype-phenotype correlation for mutations on this protein. However, relatively few ligand-binding sites have been identified on type II collagen, so functional mapping of the protein may best be achieved using triple-helical peptide (THP) libraries of type II collagen sequences to localize binding sites for well-known collagen ligands, for example (Farndale et al., 2008; Fields, 2010).

6.6.6 Type X collagen

The "short-chain" collagen X is the product of a condensed gene of three exons (Chan and Jacenko, 1998). It is a homotrimer of α1(X) chains that contain a short central triple-helical domain flanked by globular amino- (NC2) and carboxyl- (NC1) terminal domains. Collagen X trimerization proceeds from the NC1 domain, and its globular domains are retained upon secretion into the ECM. This protein is proposed to form networks around hypertrophic chondrocytes, and is their major biosynthetic product. More than 40 collagen X mutations within the NC1, and 2 mutations within the signal peptide cleavage site, have been associated with metaphyseal chondrodysplasia type Schmid, an autosomal dominant skeletal dysplasia characterized by disproportionate short stature (Ho et al., 2007). Interestingly, while no human mutations have been identified in the central triple helix, dominant interference mutations within this domain in transgenic mice have led to skeleto-hematopoietic defects (Jacenko et al., 2002). Further studies with collagen X murine models have proposed that an endochondral ossification-derived matrix containing collagen X may represent a transient hematopoietic niche (Sweeney et al., 2010). As for type II collagen, few functional domains have been mapped on type X collagen, which also could be remedied using the THP library approach. Of interest is whether this lattice-forming collagen displayed domain functions and mutation distributions reminiscent of those revealed by the type IV collagen interactome.

Finally, other collagens such as types V and VI provide sufficient mutation and/or ligand-binding data for interactomes, but space limitations in this review preclude discussion of these intriguing projects.

6.6.7 Future perspectives

Internet databases were recently created that include much of the ligand-binding and mutation data present on the type I collagen interactome (Bodian and Klein, 2009), and a comprehensive "Matrix DB" site allows the user to visualize interactomes for many ECM molecules, which are represented as 2D nodal interaction networks (Chautard et al., 2009). We aim to extend these concepts to create a database that includes 3D high-resolution models of collagen molecules, upon which the user will superimpose 3D models of many of the predominant ligands and map positions of mutations. Currently, type II and III collagen 2D interactomes are under construction and may be correlated with 3D protein structures, once solved. Moreover, we intend to correlate the type IV 2D map with type IV collagen molecular models derived from X-ray diffraction analysis of intact basement membranes, although the complexity of these matrix superstructures may present unique challenges.

6.6.8 Take-home message

Interactomes of type I collagen and several isoforms of type IV collagen were constructed including all reported ligand-binding sites, functional landmarks, and locations of mutations associated with human diseases. The interactomes indicate these proteins may be organized into distinct functional domains, suggest potential mechanistic relationships between protein landmarks and mutations, and corroborate accepted views of the structure-function relationships for these proteins. The promise of interactome mapping of other fibrillar and network-forming collagens is also addressed herein.

References

Alonso, R., Llopis, I., Flores, C., Murgui, A., and Timoneda, J. (2001). Different adhesins for type IV collagen on Candida albicans: identification of a lectin-like adhesin recognizing the 7S(IV) domain. Microbiology *147*, 1971–1981.

Bodian, D. L., and Klein, T. E. (2009). COLdb, a database linking genetic data to molecular function in fibrillar collagens. Hum Mutat *30*, 946–951.

Byers, P. H., Wallis, G. A., and Willing, M. C. (1991). Osteogenesis imperfecta: translation of mutation to phenotype. J Med Genet *28*, 433–442.

Chan, D., and Jacenko, O. (1998). Phenotypic and biochemical consequences of collagen X mutations in mice and humans. Matrix Biol *17*, 169–184.

Chapman, J. A. (1974). The staining pattern of collagen fibrils. I. An analysis of electron micrographs. Connect Tissue Res *2*, 137–150.

Chautard, E., Ballut, L., Thierry-Mieg, N., and Ricard-Blum, S. (2009). MatrixDB, a database focused on extracellular protein-protein and protein-carbohydrate interactions. Bioinformatics *25*, 690–691.

Di Lullo, G. A., Sweeney, S. M., Korkko, J., Ala-Kokko, L., and San Antonio, J. D. (2002). Mapping the ligand-binding sites and disease-associated mutations on the most abundant protein in the human, type I collagen. J Biol Chem *277*, 4223–4231.

Eikesdal, H. P., Sugimoto, H., Birrane, G., et al. (2008). Identification of amino acids essential for the antiangiogenic activity of tumstatin and its use in combination antitumor activity. Proc Natl Acad Sci U S A *105*, 15040–15045.

Emsley, J., Knight, C. G., Farndale, R. W., and Barnes, M. J. (2004). Structure of the integrin alpha2beta1-binding collagen peptide. J Mol Biol *335*, 1019–1028.

Farndale, R. W., Lisman, T., Bihan, D., et al. (2008). Cell-collagen interactions: the use of peptide Toolkits to investigate collagen-receptor interactions. Biochem Soc Trans *36*, 241–250.

Fields, G. B. (2010). Synthesis and biological applications of collagen-model triple-helical peptides. Org Biomol Chem *8*, 1237–1258.

Flugel, A., Schulze-Koops, H., Heesemann, J., et al. (1994). Interaction of enteropathogenic Yersinia enterocolitica with complex basement membranes and the extracellular matrix proteins collagen type IV, laminin-1 and -2, and nidogen/entactin. J Biol Chem *269*, 29732–29738.

Germain, D. P. (2007). Ehlers-Danlos syndrome type IV. Orphanet J Rare Dis *2*, 32.

Gross, O., Beirowski, B., Koepke, M. L., et al. (2003). Preemptive ramipril therapy delays renal failure and reduces renal fibrosis in COL4A3-knockout mice with Alport syndrome. Kidney Int *63*, 438–446.

Gross, O., Netzer, K. O., Lambrecht, R., Seibold, S., and Weber, M. (2002). Meta-analysis of genotype-phenotype correlation in X-linked Alport syndrome: impact on clinical counselling. Nephrol Dial Transplant *17*, 1218–1227.

Heidemann, E. (1988). The chemistry of tanning. In Collagen. Vol. III, Biotechnology, M. Nimni, ed. Boca Raton, FL: CRC Press, pp. 39–61.

Ho, M. S., Tsang, K. Y., Lo, R. L., et al. (2007). COL10A1 nonsense and frame-shift mutations have a gain-of-function effect on the growth plate in human and mouse metaphyseal chondrodysplasia type Schmid. Hum Mol Genet *16*, 1201–1215.

Hudson, B. G., Tryggvason, K., Sundaramoorthy, M., and Neilson, E. G. (2003). Alport's syndrome, Goodpasture's syndrome, and type IV collagen. N Engl J Med *348*, 2543–2556.

Jacenko, O., Olsen, B. R., and LuValle, P. (1991). Organization and regulation of collagen genes. Crit Rev Eukaryot Gene Expr *1*, 327–353.

Jacenko, O., Roberts, D. W., Campbell, M. R., McManus, P. M., Gress, C. J., and Tao, Z. (2002). Linking hematopoiesis to endochondral skeletogenesis through analysis of mice transgenic for collagen X. Am J Pathol *160*, 2019–2034.

Jais, J. P., Knebelmann, B., Giatras, I., et al. (2000). X-linked Alport syndrome: natural history in 195 families and genotype- phenotype correlations in males. J Am Soc Nephrol *11*, 649–657.

Kadler, K. E., Baldock, C., Bella, J., and Boot-Handford, R. P. (2007). Collagens at a glance. J Cell Sci *120*, 1955–1958.

Kajimura, D., Takahashi, S., Yoshikawa, K., et al. (2004). Non-helical type IV collagen polypeptides in human placenta. Biochem Biophys Res Commun *314*, 11–16.

Kannu, P., Bateman, J. F., Randle, S., et al. (2010). Premature arthritis is a distinct type II collagen phenotype. Arthritis Rheum 62, 1421–1430.

Khoshnoodi, J., Pedchenko, V., and Hudson, B. G. (2008). Mammalian collagen IV. Microsc Res Tech *71*, 357–370.

Knight, C. G., Morton, L. F., Onley, D. J., et al. (1998). Identification in collagen type I of an integrin alpha2 beta1-binding site containing an essential GER sequence. J Biol Chem *273*, 33287–33294.

Knight, C. G., Morton, L. F., Peachey, A. R., Tuckwell, D. S., Farndale, R. W., and Barnes, M. J. (2000). The collagen-binding A-domains of integrins alpha(1)beta(1) and alpha(2) beta(1) recognize the same specific amino acid sequence, GFOGER, in native (triple-helical) collagens. J Biol Chem *275*, 35–40.

Lehto, M., Sims, T. J., and Bailey, A .J. (1985). Skeletal muscle injury–molecular changes in the collagen during healing. Res Exp Med (Berl) *185*, 95–106.

Marini, J. C., Forlino, A., Cabral, W. A., et al. (2007). Consortium for osteogenesis imperfecta mutations in the helical domain of type I collagen: regions rich in lethal mutations align with collagen binding sites for integrins and proteoglycans. Hum Mutat *28*, 209–221.

Marini, J. C., Lewis, M. B., Wang, Q., Chen, K. J., and Orrison, B. M. (1993). Serine for glycine substitutions in type I collagen in two cases of type IV osteogenesis imperfecta (OI). Additional evidence for a regional model of OI pathophysiology. J Biol Chem *268*, 2667–2673.

Oliveira, G. V., Hawkins, H. K., Chinkes, D., et al. (2009). Hypertrophic versus non hypertrophic scars compared by immunohistochemistry and laser confocal microscopy: type I and III collagens. Int Wound J *6*, 445–452.

Orgel, J. P., Antipova, O., Sagi, I., et al. (2011a). Collagen fibril surface displays a constellation of sites capable of promoting fibril assembly, stability, and hemostasis. Connect Tissue Res *52*, 18–24.

Orgel, J. P., San Antonio, J. D., and Antipova, O. (2011b). Molecular and structural mapping of collagen fibril interactions. Connect Tissue Res *52*, 2–17.

Parkin, J. D., San Antonio, J. D., Pedchenko, V., Hudson, B., Jensen, S. T., and Savige, J. (2011). Mapping structural landmarks, ligand binding sites, and missense mutations to the collagen IV heterotrimers predicts major functional domains, novel interactions, and variation in phenotypes in inherited diseases affecting basement membranes. Hum Mutat *32*, 127–143.

Paul, R. G., and Bailey, A. J. (1996). Glycation of collagen: the basis of its central role in the late complications of ageing and diabetes. Int J Biochem Cell Biol *28*, 1297–1310.

Pepin, M., Schwarze, U., Superti-Furga, A., and Byers, P. H. (2000). Clinical and genetic features of Ehlers-Danlos syndrome type IV, the vascular type. N Engl J Med *342*, 673–680.

Persikov, A. V., Pillitteri, R. J., Amin, P., Schwarze, U., Byers, P. H., and Brodsky, B. (2004). Stability related bias in residues replacing glycines within the collagen triple helix (Gly-Xaa-Yaa) in inherited connective tissue disorders. Hum Mutat *24*,330–337.

Piez, K. A., and Reddi, A. H. (1984). Extracellular Matrix Biochemistry. New York: Elsevier.

Pihlajaniemi, T. (1996). Molecular properties of the glomerular basement membrane. Contrib Nephrol *117*, 46–79.

Plaisier, E., Gribouval, O., Alamowitch, S., et al. (2007). COL4A1 mutations and hereditary angiopathy, nephropathy, aneurysms, and muscle cramps. N Engl J Med *357*, 2687–2695.

Raabe, H. M., Molsen, H., Mlinaric, S. M., et al. (1996). Biochemical alterations in collagen IV induced by in vitro glycation. Biochem J *319*, 699–704.

Reigle, K. L., Di Lullo, G., Turner, K. R., et al. (2008). Non-enzymatic glycation of type I collagen diminishes collagen-proteoglycan binding and weakens cell adhesion. J Cell Biochem *104*, 1684–1698.

Reyes, C. D., and Garcia, A. J. (2003). Engineering integrin-specific surfaces with a triple-helical collagen-mimetic peptide. J Biomed Mater Res A *65*, 511–523.

Rong, Y. H., Zhang, G. A., Wang, C., and Ning, F. G. (2008). [Quantification of type I and III collagen content in normal human skin in different age groups]. Zhonghua Shao Shang Za Zhi *24*, 51–53.

Ruggeri, Z. M. (2001). Structure of von Willebrand factor and its function in platelet adhesion and thrombus formation. Best Pract Res Clin Haematol *14*, 257–279.

San Antonio, J. D., Lander, A. D., Karnovsky, M. J., and Slayter, H. S. (1994). Mapping the heparin-binding sites on type I collagen monomers and fibrils. J Cell Biol *125*, 1179–1188.

San Antonio, J. D., Schweitzer, M. H., Jensen, S. T., Kalluri, R., Buckley, M., and Orgel J. P. (2011). Dinosaur peptides suggest mechanisms of protein survival. PLoS One *6*, e20381.

Savage, B., Almus-Jacobs, F., and Ruggeri, Z. M. (1998). Specific synergy of multiple substrate-receptor interactions in platelet thrombus formation under flow. Cell *94*, 657–666.

Savige, J., Rana, K., Tonna, S., Buzza, M., Dagher, H., and Wang, Y. Y. (2003). Thin basement membrane nephropathy. Kidney Int *64*, 1169–1178.

Scott, J. E., and Tenni, R. (1997). Osteogenesis imperfecta mutations may probe vital functional domains (e.g. proteoglycan binding sites) of type 1 collagen fibrils. Cell Biochem Funct *15*, 283–286.

Sweeney, E., Roberts, D., Corbo, T., and Jacenko, O. (2010). Congenic mice confirm that collagen X is required for proper hematopoietic development. PLoS One *5*, e9518.

Sweeney, S. M., DiLullo, G., Slater, S. J., et al. (2003). Angiogenesis in collagen I requires alpha2beta1 ligation of a GFP*GER sequence and possibly p38 MAPK activation and focal adhesion disassembly. J Biol Chem *278*, 30516–30524.

Sweeney, S. M., Orgel, J. P., Fertala, A., et al. (2008). Candidate cell and matrix interaction domains on the collagen fibril, the predominant protein of vertebrates. J Biol Chem *283*, 21187–21197.

Sweeney, T. M., Opperman, L. A., Persing, J. A., and Ogle, R. C. (1995). Repair of critical size rat calvarial defects using extracellular matrix protein gels. J Neurosurg *83*, 710–715.

Timpl, R. (1989). Structure and biological activity of basement membrane proteins. Eur J Biochem *180*, 487–502.

Timpl, R., and Brown, J. C. (1996). Supramolecular assembly of basement membranes. Bioessays *18*, 123–132.

Trus, B. L., and Piez, K. A. (1980). Compressed microfibril models of the native collagen fibril. Nature *286*, 300–301.

Vercellotti, G. M., McCarthy, J. B., Lindholm, P., Peterson, P. K., Jacob, H. S., and Furcht, L. T. (1985). Extracellular matrix proteins (fibronectin, laminin, and type IV collagen) bind and aggregate bacteria. Am J Pathol *120*, 13–21.

6.7 Collagen-binding proteins

Takako Sasaki

6.7.1 Introduction

Various molecules have been shown to bind collagens and these interactions have crucial roles in a variety of biological processes including collagen biosynthesis, fibrillogenesis, matrix assembly, organogenesis, hemostasis, and immunity. Furthermore, pathogenic bacteria express binding proteins for several extracellular matrix molecules, and blood proteins and these bacterial proteins/receptors have important roles in the initial steps of host invasion. The classification of proteins as collagen-binding proteins is largely operational and defined by the experimental approach and stringency of *in vitro* binding conditions, or by colocalization experiments in tissues using immunohistological tools at the light or electron microscopic level. Partially conflicting reports are often due to differences in purity and homogeneity of binding proteins, as well as the structural heterogeneity and aggregate status of collagens.

Newly synthesized procollagens receive multistep posttranslational modifications including hydroxylation of proline and lysine residues, glycosylation of hydroxylysine, disulfide bond formation, and the *cis-trans* isomerization of peptidyl-proline bonds. Procollagen molecules need chaperones for proper folding, trafficking, and processing. These chaperones include not only the general endoplasmic reticulum (ER) chaperones, calnexin, calreticulin, protein disulfide isomerase, glucose-regulated protein, 94kDa (GRP94), and Bip, but also collagen prolyl-4-hydroxylase, heat-shock protein-47 (HSP47), FK506-binding protein 10 (FKBP10, previously known as FKBP65; FK506, also known as tacrolimus, is an immunosuppressive drug), and the collagen 3-hydroxylation complex composed of prolyl 3-hydroxylase 1 (P3H1), cartilage-associated protein (CRTAP), and peptidyl-prolyl *cis-trans* isomerase B (PPIB, also known as cyclophilin B) (Makareeva et al., 2011). Recessive osteogenesis imperfecta is caused by deficiency of HSP47, FKBP10, P3H1, CRTAP, and PPIB (Forlino et al., 2011). Many of these modifying enzymes and chaperones are specific to collagens and transiently interact with the triple-helix-forming region of procollagen molecules. Also in the extracellular space, numerous protein interactions regulate collagen assembly into fibrils, filaments and networks or other supramolecular structures. Whereas fibrillar collagens I, II, and III can form fibrils spontaneously *in vitro*, the fibrillogenesis *in vivo* is more complex and requires many molecules including fibronectin, integrins, collagen V/XI, small leucine-rich proteoglycans (SLRPs) and secreted protein acidic and rich in cysteine (SPARC). The N-propeptide of collagen V is shown to be essential for the assembly of type I collagen fibrils, and mice lacking collagen V die at approximately embryonic day 10 displaying a lack of collagen fibrils (Wenstrup et al., 2004). The *Col5a1* $^{+/-}$ mice are viable; however, 50% reduction in fibril numbers and collagen content is observed, together with an abnormal structure of collagen fibers in the dermis. Analogously to the situation with collagens V and I, the N-propeptdide of collagen XI is important for collagen II fibrillogenesis. SLRPs are important for the assembly of proper and specialized collagen matrices and distinct SLRPs regulate collagen assembly temporally and in a tissue-specific manner (Kalamajski and Oldberg, 2010). The collagen-binding sites on some of SLRPs have been identified (Kalamajski and Oldberg, 2007, 2009; Kalamajski et al., 2007, 2009), but most of SLRP-binding sites on collagen are not known. How SLRPs control collagen fibrillogenesis is largely unknown, but there is evidence that different SLRPs control fibril diameter and intermolecular cross-linking.

One of the earliest known cases of specific protein interactions with collagens occurs in hemostasis. Blood vessel wall injury triggers platelet activation and platelet plug formation, followed by the activation of the coagulation cascade and the formation of thrombi. Crucial for the initiation of this process are specific interactions of von Willebrand factor (VWF), glycoprotein VI (GPVI), and α2β1 integrin with intima collagens exposed in the injured vessel (Nieswandt and Watson, 2003). Hemostasis is vital to limit blood loss, but it may also cause plug formation in diseased vessels. As arterial thrombosis is one of major clinical problems in developed countries, it is important to understand the detailed mechanism of how these proteins recognize collagens.

Nearly 50 molecules have been found to interact with collagen I, and their binding sites have been elucidated by rotary shadowing electron microscopy and/or using cyanogen bromide (CNBr) peptides derived from α1(I) and α2(I) chains (Di Lullo et al., 2002; Sweeney et al., 2008). Although collagen-binding proteins do not necessarily recognize only collagenous peptide sequences, most protein interactions with collagens are dependent on the triple-helical collagen structure. Recent progress was made by using synthetic triple-helical peptides based on the host-guest strategy (Shah et al., 1996; Farndale et al., 2008). The guest sequence from primary collagen sequence are placed between (GPP)n or (GPO)n as hosts (G is Gly, P is proline, and O is hydroxyproline) in order to form and stabilize triple-helical structures.

This chapter will describe the current knowledge about major collagen-binding proteins for which specific binding epitopes have been identified. For a few of these proteins, X-ray crystal structures are available of complexes with collagen peptides and those interactions with collagen at the atomic level will be described. Given the limited scope of this chapter, it should be selective.

6.7.2 Heat-shock protein-47

HSP47 is an ER resident molecular chaperone and essential for procollagen biosynthesis in mammals. It is a 47 kDa glycoprotein that belongs to the serpin (serine protease inhibitor) family but lacks protease inhibitory function (Ishida and Nagata, 2011). The same protein was found as collagen-binding protein-2 (CBP2) and colligin-2. Its gene symbol for human is *SERPINH1* and HSP47 is also called serpin H1. Gene deletion of HSP47 results in early embryonic lethality (Nagai et al., 2000). The collagen I secreted by HSP47-null fibroblasts is susceptible to proteolysis indicating that the triple helix is not properly formed. Recently missense mutation in *SERPINH1* was identified to result in severe autosomal recessive osteogenesis imperfecta (OI) (Christiansen et al., 2010; Forlino et al., 2011). How HSP47 functions as a chaperone has not been clarified yet. HSP47 is thought to monitor and stabilize procollagen triple helices by inhibiting

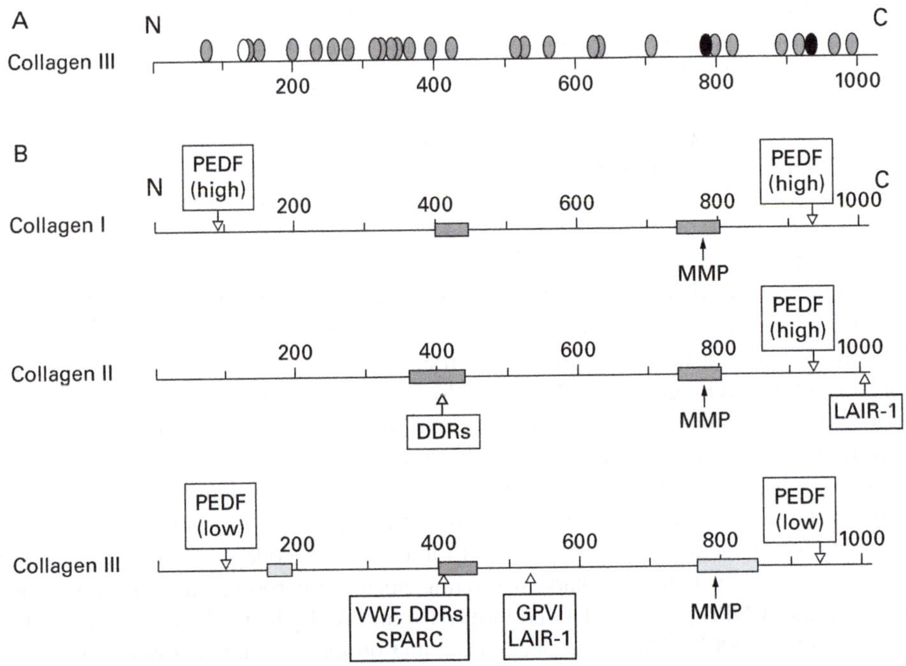

Figure 6.23 Schematic representation of potential HSP47-binding sites in collagen III (A) and of the binding sites identified using triple-helical collagen peptides (B). Only the triple-helical domains of collagens are shown, and the amino acid residues are numbered from the first glycine of the triple-helical domain. (A) Ovals with black, gray, and white indicate high, moderate, and low binding affinities, respectively. (B) The binding sites for pigment epithelium-derived factor (PEDF) with high and low affinities are shown. The solid arrows indicate the matrix metalloproteinase (MMP) cleavage site. The bars show the binding regions of SPARC found by rotary shadowing electron microscopy, and the bars with light gray indicate SPARC was found with lower frequency.

thermal denaturation of partly folded procollagens. *In vitro* study suggested that HSP47 prevents procollagen aggregation in the ER.

Studies using synthetic collagenous peptides revealed that HSP47 only recognizes correctly folded triple helices; the minimal binding motif is Yaa-Gly-Pro-Arg-Gly, with the Arg residue being essential for binding. The residue at the y position determines the affinity; the highest affinity is observed when Yaa is Thr (Koide et al., 2006). The Thr-Gly-Pro-Arg-Gly motif occurs several times in various collagens and collagen-related proteins, but whether these sites are actually binding HSP47 is not known. Possible binding sites in human collagen III are shown in ▶Figure 6.23.

6.7.3 Pigment epithelium-derived factor

Pigment epithelium-derived factor (PEDF), also referred as to caspin (collagen-associated serpin) and EPC1 (early population doubling level cDNA1), is a 50 kDa secreted collagen-binding glycoprotein and belongs to the noninhibitory serpin family

(Gettins et al., 2002). This protein is expressed in various tissues including eye, cartilage, bone, liver, kidney, and brain and also found in blood. PEDF exhibits antiangiogenic and neurotrophic activities. The crystal structure of PEDF revealed a striking asymmetric charge distribution; a high density of basic residues concentrated on one surface and acidic residues on the opposite surface of the molecule. The site-directed mutagenesis study determined that the acidic amino acid cluster is responsible for the collagen binding, whereas the basic amino acid cluster is involved in the heparin/heparin sulfate-binding (Yasui et al., 2003). The interaction of PEDF with collagen was shown to be important for its antiangiogenic activity (Hosomichi et al., 2005). Interestingly, three truncating mutations in *SERPINF1*, which encodes PEDF, were identified in the patients with severe recessive OI (Becker et al., 2011). The analyses with cultured dermal fibroblasts from those patients showed no evidence for impaired posttranslational modification and secretion of type I collagen. PEDF-deficient mice displayed increased microvasculature in several organs and epithelial cell hyperplasia in prostate and pancreas; however, no bone phenotype was reported (Doll et al., 2003).

The binding to collagen is conformation dependent; one of the binding sites is located in the α1(I) chain at residues 929-IKGHRGFSGL-938, which is also a binding site for heparin/heparin sulfate proteoglycan (HSPG). Screening with alanine substitutions identified KGXRGFXGL as the PEDF-binding motif and also KGHRG(F/Y) for heparin/HSPG. In type I collagen, there is one more binding site located at α1(I) chain residues 86 MKGHRGFSGLDG-97 (Sekiya et al., 2011). Corresponding regions of α1(II) and α1(III) chains also bind to PEDF and heparin/HSPG with different affinities. The Lys residues at α1(I) 87 and 930 and similar sites in α1(II) and α1(III) chains are known to be chemical cross-linking sites suggesting potential modulation of those bindings by the modification of these Lys residues.

6.7.4 Fibronectin

Fibronectin (FN) is a widely expressed large glycoprotein (molecular weight [Mr] 450 kDa) and plays important roles in cell adhesion, growth, differentiation and migration (Hynes, 1990; Pankov and Yamada, 2002). The multidomain protein contains fibrin-, collagen-, heparin-, and cell-binding domains and exists in 20 different splice variants; in serum it is a disulfide-bonded homodimer, while it assembles into fibrils on the cell surface and in the extracellular matrix. Each monomer is composed of three types of repeating units called FN type I, type II, and type III repeats. FN binds with a higher affinity to denatured collagen (gelatin) than to native collagen. The gelatin-binding domain (GBD) is located near the N terminus and consists of the sixth FNI repeat, the first and second FNII repeats, and the seventh to ninth FNI repeats (^{6}FNI$^{1-2}$FNII$^{7-9}$FNI). Both subfragments of the GBD (6FNI$^{1-2}$FNII^{7}FNI and $^{8-9}$FNI) bind the peptide 778-GQRGVVGLOGQRGERGFOGLOG-799 of the α1(I) collagen chain, which is located just C terminally to the MMP1 cleavage site of collagen I between G775 and I776. The core $^{8-9}$FNI-binding site, GLOGQRGER, was deduced from a crystal structure of $^{8-9}$FNI in complex with α1(I) G778-G799 peptide. Further putative fibronectin-binding sites were found at residues 82-GLPGHKGHR-90 in the α1(I) chain and at 82-GLPGFKGIR-90 in the α2(I) chain (Erat et al., 2009). How do ^{6}FNI$^{1-2}$FNII^{7}FNI and $^{8-9}$FNI cooperate in collagen binding? Erat et al. (2010) proposed a model in which ^{6}FNI$^{1-2}$FNII^{7}FNI binds adjacent to the MMP1 cleavage site and shifts the local fluctuations of the collagen triple helix toward the unwound state, so that $^{8-9}$FNI can bind tightly to the single-stranded collagen peptide. This model suggests a biological role for the FN-collagen interaction in collagen proteolysis and tissue remodeling.

6.7.5 Von Willebrand factor

VWF is a large glycoprotein found in blood plasma, platelet α granules, and subendothelial connective tissue. It consists of four types of repeated domains, A, B, C, and D and is assembled from 250 kDa monomers into disulfide-linked multimers of Mr > 20 MDa. VWF has two essential roles in hemostasis: VWF binds blood clotting factor VIII, thereby protecting it from degradation, and VWF mediates the adhesion of platelets to the subendothelium. Platelet adhesion to the injured blood vessel is initiated by the interaction of the platelet receptor glycoprotein complex, GP1b-X-V, with collagen-bound VWF. This transient interaction slows down the platelets and facilitates firm adhesion through collagen receptors, glycoprotein VI, and integrin α2β1. The binding site in VWF for collagens I and III is located in the VWF A3 domain, and the GP1bα component of the GP1b-X-V complex binds to the VWF A1 domain. Using a set of synthetic triple-helical peptides, the "Collagen III Toolkit," the minimal VWF-binding sequence was identified as α1(III) 405-RGQOGVMGF-413 (Lisman et al., 2006). Recently, aegyptin, a 30 kDa mosquito salivary gland protein was reported to bind the same peptide with a high affinity (Calvo et al., 2010). This sequence also overlaps with the binding sequences for discoidin domain receptors (DDRs) and SPARC (see Sections 6.7.8 and 6.7.9).

6.7.6 Glycoprotein VI

GPVI consists of two C2-type immunoglobulin-like (IgC2) domains, a mucin-like stalk, a transmembrane domain and a short intracellular domain that associates with the γ-chain of the Fc receptor. GPVI is expressed on megakaryocytes and platelets. Following the initial interaction of the collagen-bound VWF with the platelet receptor glycoprotein complex GP1b-X-V, GPVI binds to collagen, thereby leading to inside-out activation of the platelet integrins α2β1 and αIIbβ3 and release of soluble mediators from the platelet granule, which cause platelet aggregation and thrombus formation. The clustering of GPVI on the platelet surface by collagens is thought to be required for the activation process.

GPVI binds to the collagen-related peptide (CRP) containing $(GPO)_{10}$ but not $(GPP)_{10}$ repeats. One GPO triplet can support the adhesion of platelets, but platelet aggregation and protein tyrosine phosphorylation are only induced by cross-linked peptides, which have more than two GPO triplets. However, higher numbers of GPO triplet are more efficient (Smethurst et al., 2007). GPVI binding to the triple-helical peptides of the Collagen III Toolkit was tested using platelets and the recombinant ectodomain of GPVI. It was found that peptide III-30 was the best substrate (Jarvis et al., 2008). The guest sequence of III-30 is GAOGLRGGAGPOGPEGGKGAAGPOGPO (α1(III) chain residues 523–540). Replacement of the GPO sequences in the fourth and last triplets of III-30 by

GPP abolished platelet binding and aggregation, whereas a change in the penultimate triplet had no effect.

Based on the three-dimensional structure of rat tail tendon collagen I obtained by fiber diffraction experiments (Orgel et al., 2006), the binding sites of VWF, GPVI, and integrin α2β1 were mapped in the intact collagen fiber (Herr and Farndale, 2009).

6.7.7 Leukocyte-associated immunoglobulin-like receptor-1

The genes for leukocyte-associated immunoglobulin-like receptor-1 (LAIR1) and GPVI are located in leukocyte-associated immunoglobulin-like receptor complex on human chromosome 19. These cell-surface receptors are structurally related having extracellular IgC2 domain(s). LAIR1 is an immune inhibitory receptor and expressed on almost all immune cells as a type I transmembrane glycoprotein containing a single extracellular IgC2 domain and two immunoreceptor tyrosine-based inhibitory motifs (ITIMs) in the cytoplasmic tail. LAIR1 binds to various collagens including transmembrane collagens through its extracellular IgC2 domain (Lebbink et al., 2006). Binding to collagen induces cross-linking of LAIR1, which transmits an inhibitory signal. The inhibitory receptors are thought to be important in the prevention of autoimmunity.

Like GPVI, LAIR1 was found to bind $(GPO)_{10}$ and not to $(GPP)_{10}$, and K562 cells expressing LAIR1 bind to five peptides and eight peptides of the Collagen II and Collagen III Toolkits, respectively. Among those peptides, peptides III-30 (see Section 6.7.6) and II-56 (GPRGRSGETGPAGPOGNOGPOGPOGPO, located at the C terminus of the triple-helical domain of collagen II) were more efficient ligands than CRP in inhibiting an immune response (Lebbink et al., 2009). As described in the previous section, peptide III-30 was found to be the best ligand also for GPVI. Although the peptides found to bind to GPVI and LAIR1 are overlapping, the binding requirements are not identical. Substitutions in III-30, which had an effect on the binding and activation of GPVI, did not show a significant reduction in LAIR1 binding and inhibitory activity. However, deletion of the C terminal two GPO triplets from peptide III-30 almost abolished adhesion and inhibition by LAIR1. Furthermore, the absence of the N-terminal GAOGLR sequence reduced GPVI binding to 30%, whereas the LAIR1-induced inhibition was less affected. Thus, although homologous domains of LAIR1 and PGVI bind to the same peptide, the collagen residues involved in the interaction are different. Moreover, the collagen-binding site of LAIR1, identified by NMR experiments and site-directed mutagenesis, is located in a different region from the proposed collagen-binding site in GPVI (Horii et al., 2006; Brondijk et al., 2010).

6.7.8 Discoidin domain receptors (DDR)

DDR1 and DDR2 are receptor tyrosine kinases that are activated by their interaction with collagens. Both DDRs consist of an extracellular discoidin homology (DS) domain, a second globular domain, a transmembrane domain, a large juxtamembrane domain, and a C-terminal tyrosine kinase domain. DDRs are involved in organogenesis, remodeling of extracellular matrices, cancer and atherosclerosis (Vogel et al., 2006; Leitinger and Hohenester, 2007). The binding site for collagens is located in the DS domain and binding requires a native triple-helical collagen conformation. The two DDRs have different preferences for collagen types; DDR1 binds to collagen I and IV, and DDR2 binds to collagen I, II, and X, but not to collagen IV. Within the Collagen II Toolkit, DDR1 bound strongly to only one peptide, II-22, in contrast to DDR2, which showed binding to several peptides including II-22. Similarly, within the Collagen III Tookit DDR1 bound only to III-23, whereas DDR2 bound to several other sites (Konitsiotis et al, 2008; Xu et al., 2011). The common binding site of both DDRs was studied further by truncation and alanine substitution analysis, revealing that the minimum binding motif is GVMGFO. Triple-helical peptides containing this motif act as receptor agonists and induce autophosphorylation of DDR2 expressed in cells. The structural basis of the DDR-collagen interaction was elucidated by a crystal structure of the DDR2 DS domain in complex with a GVMGFO-containing peptide (Carafoli et al, 2009). This structure showed that the Met and Phe side chains of the collagen peptide are bound in an amphiphilic pocket delineated by the critical Trp52 residue and a salt bridge between Arg105 and Glu113. The latter residue also interacts with the hydroxyl group of the hydroxyproline of the GVMGFO motif. How collagen binding results in DDR activation is not known. Collagen binding is accompanied by only minor structural changes in the DS domain, and it is likely that changes within the DDR dimer, or aggregation of DDR dimers, are involved in signal transduction. The crystal structure also permitted a mutational analysis of the different collagen specificities of DDR1 and DDR2. Substitution of five amino acids in DDR2 close to the collagen-binding site by their DDR1 counterparts was sufficient to impart collagen-IV-binding activity (Xu et al., 2011).

6.7.9 Secreted protein acidic and rich in cysteine

SPARC, which is also referred to as osteonectin or BM-40, is a 33 kDa extracellular calcium-binding glycoprotein expressed in various tissues. Originally the protein was identified as a bone-specific protein that binds to both hydroxyapatite and collagen, and accordingly named osteonectin (Termine et al., 1981). Independently, the same protein was found to be highly expressed in mouse embryo parietal endoderm and termed *SPARC* (Mason et al., 1986). A third study identified the protein as BM-40 in Engelbreth-Holm-Swarm sarcoma, which produces a basement membrane–like extracellular matrix (Mann et al., 1987). SPARC exhibits antiadhesive and antiproliferative properties, and influences the synthesis of extracellular matrix proteins. Several extracellular ligands have been identified including some collagen types and cytokines (Brekken and Sage, 2001). Interestingly, SPARC-null mice develop early onset cataracts, indicating its important role in lens transparency. Other phenotypes found were osteopenia, an alteration of collagen fibrils in the dermis and accelerated dermal wound healing, suggesting that SPARC is involved in collagen fibrillogenesis. Further analyses of SPARC-null mice revealed an important role in adipogenesis and adipose tissue hyperplasia. Moreover, SPARC-null mice exhibit increased cardiac rupture and dysfunction after acute cardiac infarction indicating that SPARC is essential for maintaining the integrity of the cardiac extracellular matrix. Very recently Rivera and Brekken (2011) reported that SPARC promotes pericyte migration by blocking endoglin association with αV integrin, thereby diminishing αV-integrin-mediated transforming growth factor-beta (TGF-β) activation.

SPARC consists of three domains, a flexible acidic N-terminal domain I, a follistatin-like (FS) domain and a C-terminal extracellular calcium-binding (EC) domain, which contains a pair of EF-hands (Hohenester et al., 1996, 1997). SPARC binds to the fibril-forming collagens, I, II, III and V and basement membrane collagen IV through its EC domain and this binding is dependent on calcium and the triple-helical conformation of collagen. A single cleavage in helix C of the EC domain by unknown proteases or by several MMPs resulted in a 10-fold increase in the affinity for collagen mainly due to an increase in the association rate constant. Inspection of the crystal structure of the EC domain (Hohenester et al., 1996) suggested that the cleavage in helix C might expose the underlying helix A of the EC domain, thereby enhancing collagen binding. Residues in helix A were speculated to be involved in collagen binding, and this hypothesis was confirmed experimentally by site-directed mutagenesis, which showed that human SPARC residues Arg149 and Asn156 in helix A, and Leu242, Met245 and Glu246 in a loop region connecting the two EF hands are crucial for collagen binding (Sasaki et al., 1998).

The peptide bonds cleaved either endogenously (Leu197-Leu198) or by MMPs (Glu196-Leu197) are located next to each other (Sasaki et al., 1997). Neoepitope antibodies were obtained by immunizing rabbits with synthetic peptides containing either Leu197 or Leu198 at the N terminus. In mouse tissues, these neoepitope-specific antibodies detected a variable degree of cleavage mainly at Leu198 site and only a weak reaction for Leu197 epitope (Sasaki et al., 1999). These observations indicate that these neoepitopes are generated *in vivo* and that there could well be a tissue-specific modulation of the SPARC-collagen interaction.

The SPARC-binding sites on collagens were analyzed using rotary shadowing electron microscopy, as well as binding studies with CNBr peptides and the Collagen III Toolkit. These studies identified a major binding site at approximately 180 nm from the C terminus of collagens I, II, and III, corresponding to residues 397-GPOGPSGPRG QOGVMGFOGPKGNDGAO-423 in human collagen III (Giudlci et al., 2008). A crystal structure of SPARC complexed to a peptide containing this sequence showed that the interaction is centered on the GVMGFO motif, with the Met, Phe and Hyp residues making important contacts with the EC domain of SPARC (Hohenester et al, 2008). The collagen-binding site revealed by the structure is in excellent agreement with the previous mutagenesis study of Sasaki et al. (1998). Arg149 and Glu243 form a salt bridge that is involved in recognition of the hydroxyproline and the hydrophobic residues Leu242 and Met245 are involved in the recognition of the Met and Phe residues of the collagen peptide. Unexpectedly, the collagen-binding region undergoes substantial structural changes upon collagen binding, such that the major specificity pocket is obstructed in free SPARC. The SPARC-binding site in collagens I–III overlaps with the major DDR-binding site (see above) and the structural mechanisms of GVMGFO recognition are remarkably similar in the two proteins (Carafoli et al, 2009). It is possible that SPARC binding to this interaction hotspot in collagens is related to its regulation of collagen fibrillogenesis.

Several SPARC-related proteins containing both FS and EC domains exist and these include TSC36/Flik/FRP, SC1/hevin/QR1, testicans, and SMOCs. Like SPARC, hevin/SC1 has been shown to regulate collagen fibrillogenesis (Sullivan et al., 2006).

6.7.10 Take-home message

The binding sites described in this chapter are summarized in ▶Figure 6.23, except those of fibronectin, for which the binding analysis was mainly performed using single-stranded collagen peptides. As described, synthetic triple-helical collagen peptides are powerful tools for identification of binding sequences and for structural studies. The proteins localized in extracellular matrix interact with collagens formed fibrils and complex supramolecular structures and some proteins may interact with collagens only specific stage(s) of fibrillogenesis. Therefore, future analyses should be extended to deal with dynamic and higher order architecture of collagens.

References

Becker, J., Semler, O., Gilissen, C., et al. (2011). Exome sequencing identifies truncating mutations in human *SERPNF1* in autosomal-recessive osteogenesis imperfecta. Am J Hum Genet *88*, 362–371.

Brekken, R. A., and Sage, E. H. (2001). SPARC, a matricellular protein: at the crossroads of cell-matrix communication. Matrix Biol *19*, 816–827.

Brondijk, T. H. C., de Ruiter, T., Ballering, J., et al. (2010). Crystal structure and collagen-binding site of immune inhibitory receptor LAIR-1: unexpected implications for collagen binding by platelet receptor GPVI. Blood *115*, 1364–1373.

Calvo E., Tokumasu, F., Mizurini, D. M., et al. (2010). Aegyptin displays high-affinity for the von Willebrand factor binding site (RGQOGVMGF) in collagen and inhibits carotid thrombus formation in vivo. FEBS J *277*, 413–427.

Carafoli, F., Bihan, D., Stathopoulos, S., et al. (2009). Crystallographic insight into collagen recognition by discoidin domain receptor 2. Structure *17*, 1573–1581.

Christiansen, H. E., Schwarze, U., Pyott, S. M., et al. (2010). Homozygosity for a missense mutation in SERPINH1, which encodes the collagen chaperone protein HSP47, results in severe recessive osteogenesis imperfecta. Am J Hum Genet *86*, 389–398.

Di Lullo, G. A., Sweeney, S. M., Körkkö, J., Ala-Kokko, L., and San Antonio, J. D. (2002). Mapping the ligand-binding sites and disease-associated mutations on the most abundant protein in human, type I collagen. J Biol Chem *277*, 4223–4231.

Doll J. A., Stellmach V. M., Bouck N. P., et al. (2003). Pigment epithelium-derived factor regulates the vasculature and mass of the prostate and pancreas. Nat Med *9*, 774–780.

Erat, M. C., Schwarz-Linek, U., Pickford, A. R., Farndale, E. W., Campbell, I. D., and Vakonakis, I. (2010). Implications for collagen binding from the crystallographic structure of fibronectin 6FnI$^{1-2}$FnII7FnI. J Biol Chem *285*, 33754–33770.

Erat, M. C., Slatter, D. A., Lowe, E. D., et al. (2009). Identification and structural analysis of type I collagen sites in complex with fibronectin fragments. Proc Natl Acad Sci U S A *106*, 4195–4200.

Farndale, R. W., Lisman, T., Bihan, D., et al. (2008). Cell-collagen interactions: the use of peptide Toolkits to investigate collagen-receptor interactions. Biochem Soc Trans *36*, 241–250.

Forlino, A., Cabral, W. A., Barnes, A. M., and Marini, J. C. (2011). New perspectives on osteogenesis imperfecta. Nat Rev Endocrinol *7*, 540–547.

Gettins, P. G. W., Simonovic, M., and Volz, K. (2002). Pigment epithelium-derived factor (PEDF), a serpin with potent anti-angiogenic and neurite outgrowth-promoting properties. Biol Chem *383*, 1677–1682.

Giudici, C., Raynal, N., Wiedemann, H., et al. (2008). Mapping of SPARC/BM-40/osteonectin-binding sites on fibrillar collagens. J Biol Chem *283*, 19551–19560.

Herr, A. B., and Farndale, R. W. (2009). Structural insights into interactions between platelet receptors and fibrillar collagen. J Biol Chem *284*, 19781–19785.

Hohenester, E., Maurer, P., Hohenadl C., Timpl, R., Jansonius, J. N., and Engel, J. (1996). Structure of a novel extracellular Ca^{2+}-binding module in BM-40. Nat Struct Biol *3*, 67–73.

Hohenester, E., Maurer, P., and Timpl, R. (1997). Crystal structure of a pair of follistatin-like and EF-hand calcium-binding domains in BM-40. EMBO J *16*, 3778–3786.

Hohenester, E., Sasaki, T., Giudici, C., Farndale, R. W., and Bächinger, H. P. (2008). Structural basis of sequence-specific collagen recognition by SPARC. Proc Natl Acad Sci U S A *105*, 18273–18277.

Horii, K., Kahn, M. L., and Herr, A. B. (2006). Structural basis for platelet collagen responses by the immune-type receptor glycoprotein VI. Blood *108*, 936–942.

Hosomichi, J., Yasui, N., Koide, T., Soma, K., and Morita, I. (2005). Involvement of the collagen I-binding motif in the anti-angiogenic activity of pigment epithelium-derived factor. Biochem Biophys Res Commun *335*, 756–761.

Hynes, R. O. (1990). Fibronectin. New York: Springer-Verlag.

Ishida, Y., and Nagata, K. (2011). Hsp47 as a collagen-specific molecular chaperone. Methods Enzymol *499*, 167–182.

Jarvis, G. E., Raynal, N., Langford, J. P., et al. (2008). Identification of a major GpVI-binding locus in human type III collagen. Blood *118*, 4986–4996.

Kalamajski, S., Aspberg, A., and Oldberg, Å. (2007). The decorin sequence SYIRIADTNIT binds collagen type I. J Biol Chem *282*, 16062–16067.

Kalamajski, S., Aspberg, A., Lindblom, K., Heinegård, D., and Oldberg, Å. (2009). Asporin competes with decorin for collagen binding, binds calcium and promotes osteoblast collagen mineralization. Biochem J *423*, 53–59.

Kalamajski, S., and Oldberg, Å. (2007). Fibromodulin binds collagen type I via Glu-353 and Lys-355 in leucine-rich repeat 11. J Biol Chem *282*, 26740–26745.

Kalamajski, S., and Oldberg, Å. (2009). Homologous sequence in lumican and fibromodulin leucine-rich repeat 5–7 competes for collagen binding. J Biol Chem *284*, 534–539.

Kalamajski, S., and Oldberg, Å. (2010). The role of small leucine-rich proteoglycan in collagen fibrillogenesis. Matrix Biol *29*, 248–253.

Koide, I., Nishikawa, Y., Asada, S., et al. (2006). Specific recognition of the collagen triple helix by chaperone HSP47. II The HSP47-binding structural motif in collagens and related proteins. J Biol Chem *281*, 11177–11185.

Konitsiotis, A. D., Raynal, N., Bihan, D., Hohenester, E., Farndale, R. W., and Leitinger, B. (2008). Characterization of high affinity binding motifs for the discoidin domain receptor DDR2 in collagen. J Biol Chem *283*, 6861–6868.

Lebbink, R. J., de Ruiter, T., Adelmeijer, J., et al. (2006). Collagens are functional, high affinity ligands for the inhibitory immune receptor LAIR-1. J Exp Med *203*, 1419–1425.

Lebbink, R. J., Raynal, N., de Ruiter, T., Bihan, D. G., Farndale, R. W., and Meyaard, L. (2009). Identification of multiple potent binding sites for human leukocyte associated Ig-like receptor LAIR on collagen II and III. Matrix Biol *28*, 202–210.

Leitinger, B., and Hohenester, H. (2007). Mammalian collagen receptors. Matrix Biol *26*, 146–155.

Lisman, T., Raynal, N., Groeneveld, D., et al. (2006). A single high-affinity binding site for von Willebrand factor in collagen III, identified using synthetic triple-helical peptides. Blood *108*, 3753–3756.

Makareeva, E., Aviles, N. A., and Leikin, S. (2011). Chaperoning osteogenesis: new protein-folding disease paradigms. Trends Cell Biol *21*, 168–176.

Mann, K., Deutzmann, R., Paulsson, M., and Timpl, R. (1987). Solubilization of protein BM-40 from a basement membrane tumor with chelating agents and evidence for its identity with osteonectin and SPARC. FEBS Lett *218*, 167–172.

Mason, I. J., Taylor, A., Williams, J. G., Sage, H., and Hogan, B. L. M. (1986). Evidence from molecular cloning that SPARC, a major product of mouse embryo parietal endoderm, is related to an endothelial cell "culture shock" glycoprotein of Mr 43,000. EMBO J *5*, 1465–1472.

Nagai, N., Hosokawa, M., Itohara, S., et al. (2000). Embryonic lethality of molecular chaperone Hsp47 knockout mice is associated with defects in collagen biosynthesis. J Cell Biol *150*, 1499–1506.

Nieswandt, B., and Watson, S. (2003). Platelet-collagen interaction: is GPVI the central receptor? Blood *102*, 449–461.

Orgel, J. P., Irving, T. C., Miller, A., and Wess, T. J. (2006). Microfibrillar structure of type I collagen in situ. Proc Natl Acad Sci U S A *103*, 9001–9005.

Pankov, R., and Yamada, K. M. (2002). Fibronectin at a glance. J Cell Sci *115*, 3861–3863.

Rivera L. B., and Brekken, R. A. (2011). SPARC promotes pericyte recruitment via inhibition of endoglin-dependent TGF-β1 activity. J Cell Biol *193*, 1305–1319.

Sasaki, T., Göhring, W., Mann, K., et al. (1997). Limited cleavage of extracellular matrix protein BM-40 by matrix metalloproteinases increases its affinity for collagens. J Biol Chem *272*, 9237–9243.

Sasaki, T., Hohenester, E., Göhring, W., and Timpl, R. (1998). Crystal structure and mapping by site-directed mutagenesis of the collagen-binding epitope of an activated form of BM-40/SPARC/osteonectin. EMBO J *17*, 1625–1634.

Sasaki, T., Miosge, N., and Timpl, R. (1999). Immunochemical and tissue analysis of protease generated neoepitopes of BM-40 (osteonectin, SPARC) which are correlated to a higher affinity binding to collagens. Matrix Biol *18*, 499–508.

Sekiya, A., Okano-Kosugi, H., Yamazaki, C. M., and Koide, T. (2011). Pigment epithelium-derived factor (PEDF) shares binding sites in collagen with heparin/heparin sulphate proteoglycans. J Biol Chem *286*, 26364–26374.

Shah, N. K., Ramshaw, J. A., Kirkpatrick, A., Shah, C., and Brodsky, B. (1996). A host-guest set of triple-helical peptides: stability of Gly-X-Y triplets containing common nonpolar residues. Biochemistry *35*, 10262–10268.

Smethurst, P. A., Onley, D. J., Jarvis, G. E., et al. (2007). Structural basis for the platelet-collagen interaction. The smallest motif within collagen that recognizes and activates platelet glycoprotein VI contains two glycine-proline-hydroxyproline triplets. J Biol Chem *282*, 1296–1304.

Sullivan, M. M., Barker, T. H., Funk, S. E., et al. (2006). Matricellular hevin regulates decorin production and collagen assembly. J Biol Chem *281*, 27621–27632.

Sweeney, S. M., Orgel, J. P., Fertala, A., et al. (2008). Candidate cell and matrix interaction domains on the collagen fibril, the predominant protein of vertebrates. J Biol Chem *283*, 21187–21197.

Termine, J. D., Kleinman, H. K., Whitson, S. W., Conn, K. M., McGarvey, M. L., and Martin, G. R. (1981). Osteonectin, a bone-specific protein linking mineral to collagen. Cell *26*, 99–105.

Vogel, W. F., Abdulhussein, R., and Ford, D. E. (2006). Sensing extracellular matrix: an update on discoidin domain receptor function. Cell Signal *18*, 1108–1116.

Wenstrup, R. J., Florer, J. B., Brunskill, E. W., Bell, S. M., Chervoneva, I., and Birk, D. E. (2004). Type V collagen controls the initiation of collagen fibril assembly. J Biol Chem *279*, 53331–53337.

Xu, H., Raynal, N., Stathopoulos, S., Myllyharju, J., Farndale, R. W., and Leitinger, B. (2011). Collagen binding specificity of the discoidin domain receptors: binding sites on collagen II and III and molecular determinants for collagen IV recognition by DDR1. Matrix Biol *30*, 16–26.

Yasui, N., Mori, T., Morito, D., et al. (2003). Dual-site recognition of different extracellular matrix components by anti-angiogenic/neurotrophic serpin, PEDF. Biochemistry *42*, 3160–3167.

7 Emerging aspects in extracellular matrix pathobiology

7.1 Introduction

Achilleas Theocharis

Extracellular matrix (ECM) is composed of a large variety of macromolecules that are synthesized by all cells present in a certain tissue. Matrix molecules are capable of interacting with each other assembled into an organized three-dimensional multimolecular meshwork, adapted to the functional requirements of the particular tissue (Frantz et al., 2010).

It is well recognized that ECM is a dynamic structure with multiple functional properties that regulates cell homing and behavior through its binding to numerous cell surface receptors and, consequently, transmission of the various chemical and mechanical signals to cells. Cell-matrix binding is mediated via specific cell surface receptors such as integrins, cell surface proteoglycans (PGs) (syndecans and glypicans), and hyaluronan (HA) receptor cluster of differentiation 44 (CD44). The different binding affinities between cell surface receptors and the various matrix effectors greatly affect gene expression and diverse cellular responses such as proliferation, adhesion, migration, polarization, differentiation, survival, and apoptosis (Frantz et al., 2010). For example, the specific location of microbes such as staphylococci to skeletal structures involves several adhesion molecules, which recognize adhesive matrix molecules including collagen and bone sialoprotein (BSP).

Matrix remodeling occurs in physiological and pathological conditions, such as malignant development and metastasis, and infections. During such processes, new molecules are deposited in the ECM, whereas simultaneously matrix molecules are degraded by the coordinated action of enzymes released within matrix. In the case of cancer, ECM remodeling mimics the initial steps of tissue repair, which end in wound healing. However, this process is deregulated and continues as a wound that fails to heal (Dvorak, 1986). The composition and architecture of ECM dramatically change as a consequence of altered biosynthesis of ECM molecules by cancer and stromal cells. Deposition of collagens, PGs, HA, and other important matrix molecules results in the formation of a modified matrix with altered functional and mechanical properties. Reorganization of tumor stroma may facilitate or prevent cancer progression, creating a permissive or restrictive microenvironment, respectively. For example, the density and stiffness of collagen-enriched tumor stroma contribute to tumor initiation, invasion, and metastasis in solid tumors (Levental et al., 2009; Gao et al., 2010).

PG expression is markedly modified during ECM remodeling in malignancies. Tumor microenvironment is characterized by altered composition and structure of PGs. Furthermore, altered expression of PGs on tumor and stromal cells affects cancer cell signaling, growth and survival, cell adhesion, migration, and angiogenesis (▶Figure 7.1). For example, increased expression and secretion of serglycin by tumor cells and accumulation of matrix PGs such as versican in the tumor stroma promote cancer cell

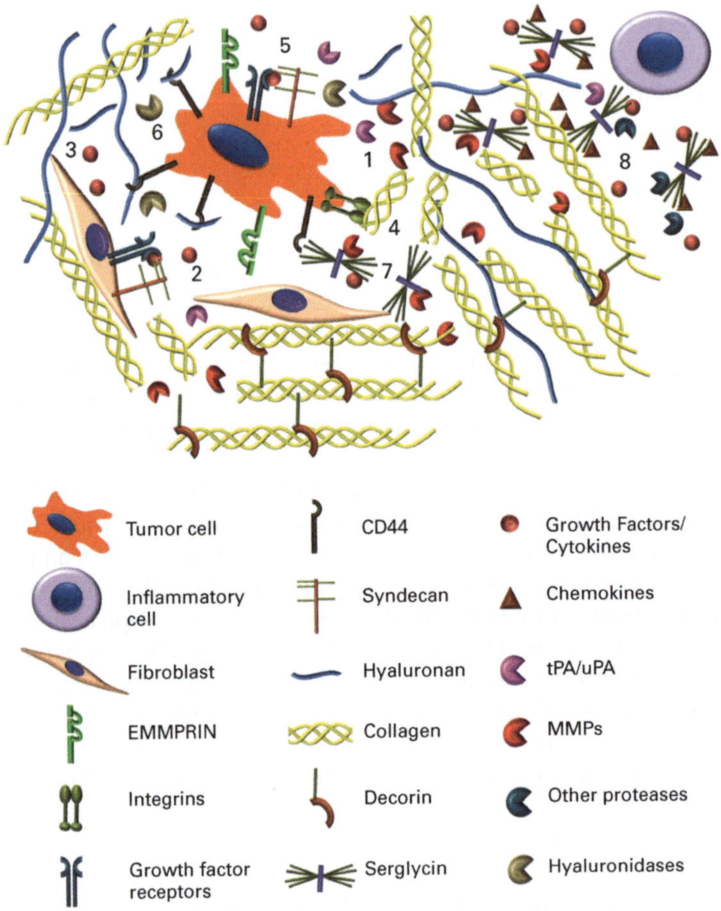

Figure 7.1 Functions of extracellular matrix in the tumor microenvironment. Tumor cells overexpress proteolytic enzymes, which break down ECM and in this way promote cell migration and spread (1). They also express extracellular matrix metalloproteinase inducer (EMMPRIN) and release growth factors and cytokines, which activate stromal fibroblasts to secrete proteases to accelerate ECM degradation (2). Activated fibroblasts synthesize and secrete in the tumor microenvironment growth factors and cytokines promoting tumor cell growth and matrix molecules such as collagen, PGs, and HA leading to ECM reorganization (3). Tumor cells interact with matrix molecules through cell surface receptors such as integrins, CD44, and syndecans regulating tumor cell functions. For example, integrins mediate the adhesion to matrix molecules such as collagen and promote tumor cell migration (4). Syndecans participate in various signaling events since they interact with growth factors and present them to their receptors in order to signal (5). CD44 interacts with HA of various sizes, which is accumulated in the tumor microenvironment and degraded by hyaluronidases secreted by tumor cells (6). Binding of different HA fragments with CD44 results in activation of discrete signaling pathways that modulate tumor cell behavior. Tumor cells also express serglycin, which is associated with increased cell motility and invasion (7). Serglycin is also secreted by inflammatory cells present in the tumor microenvironment. Serglycin is important for storage and secretion of various proteases implicated in matrix degradation, as well as chemokines and growth factors that may create a chemotactic gradient promoting tumor cell spread (8).

proliferation, migration, and survival, whereas deposition of decorin suppresses tumor growth by interfering with various signaling pathways (Theocharis et al., 2010; Iozzo and Sanderson, 2011; Kolset and Pejler, 2011). HA is a key glycosaminoglycan, which is often accumulated in the tumor stroma promoting tumorigenesis (Misra et al., 2011). HA content and size are regulated by three HA-synthesizing enzymes and HA-degrading enzymes such as hyaluronidases (Hyal1, 2, and 3 and PH-20), which degrade HA, thereby producing fragments with diverse biological functions (▶Figure 7.1).

Continuous degradation of ECM involves several proteolytic enzymes, among them matrix metalloproteinases (MMPs), tissue-type plasminogen activator (tPA) and urokinase plasminogen activator (uPA). Cancer cells secrete soluble factors acting either in an autocrine manner upregulating the expression of such enzymes by themselves or triggering their biosynthesis by stromal cells (Gialeli et al., 2011) (▶Figure 7.1). Another mechanism involves the transmembrane glycoprotein extracellular matrix metalloproteinase inducer (EMMPRIN), which is enriched in tumor cell membrane and can interact with stromal cells either directly or through its shedding, promoting the biosynthesis of MMPs, uPA, and angiogenic mediators (Huet et al., 2008; Bougatef et al., 2009) (▶Figure 7.1).

The action of proteolytic enzymes in the tumor microenvironment generates space for the cells to migrate, producing cryptic peptides with novel biological activities. In such a way, the ECM-stored growth factors could be released and stimulate malignant and endothelial cells, thereby promoting tumor growth, angiogenesis, and spread of cancer cells (Kessenbrock et al., 2010). Cell surface PGs, such as syndecans, act as coreceptors for growth factors working in concert with growth factor receptors to properly transmit intracellular signals (▶Figure 7.1). Syndecans promote proangiogenic signaling by binding fibroblast growth factor-2 (FGF2) and vascular endothelial growth factor (VEGF) and presenting them to their high-affinity receptors, as well as by protecting them from inactivation. Syndecans may also regulate adhesion and migration of tumor cells through binding to various ECM components (Couchman, 2010; Choi et al., 2011).

A substantial amount of data has been published at the level of the pathophysiological role of ECM. The chapters in this section have been selected to present some recent developments on the physiological roles of ECM molecules in tissue organization and function and their involvement in the development and progression of pathological situations, such as cancer and infections. This part begins with Chapter 7.2, entitled "Extracellular Matrix in Breast Cancer: Permissive and Restrictive Influences Emanating from the Stroma" by Pucci-Minafra. This chapter presents the extracellular context and the physiological roles of ECM in the mammary gland, as well as the modifications present in the tumor stroma during cancer progression. The contribution of collagen types I and IV and decorin, which are upregulated in the tumor stroma and differentially affect cancer biology, is also discussed.

Chapter 7.3 by Huet et al., entitled "EMMPRIN/CD147: Potential Functions in Tumor Microenvironment and Therapeutic Target for Human Cancer" introduces the reader to the structure and biological roles of EMMPRIN in human pathology with special focus in cancer. EMMPRIN has a key regulatory role in cancer biology. The induction of proteolytic enzymes, angiogenic factors, and differentiation of stromal fibroblasts by EMMPRIN highlights the importance of this molecule in disease targeting and treatment.

Chapter 7.4 by Triantaphyllidou et al., entitled "Implication of Hyaluronidases in Cancer Growth, Metastasis, Diagnosis, and Treatment" covers the expression and involvement of hyaluronidases in cancer pathobiology. The dual role of hyaluronidases in the progression or suppression of cancer and their prognostic and diagnostic significance is shown. Furthermore, the regulation of cellular functions by hyaluronidases and their targeting in disease treatment is also illuminated.

Chapter 7.5 by Zong et al., entitled "Structure-Function Relationship of Syndecan-1, with Focus on Nuclear Translocation and Tumor Cell Behavior" presents the structural domains of syndecan-1 that regulate the cellular distribution and specific biological functions of this PG. It also discusses the importance of nuclear translocation of syndecan-1, its possible roles in this cell compartment, and the involvement in tumorigenesis.

The novel roles of serglycin in malignancies are presented in Chapter 7.6 by Korpetinou et al., entitled "Serglycin: A Novel Player in the Terrain of Neoplasia". It discusses the biological role of serglycin expressed by tumor cells in the establishment of an aggressive and drug-resistant malignant phenotype, as well as the ability of serglycin to modulate the immune responses in the tumor microenvironment. Furthermore, it highlights the role of serglycin to regulate the inflammatory process and its importance in tumor progression.

The multiple interactions between cancer cells and ECM that create a three-dimensional matrix during cancer progression *in vivo* and the importance of using three-dimensional cell culture systems to study the role of ECM is highlighted in Chapter 7.7 by Maffei and Zaman. In this chapter, entitled "Quantifying Cell-ECM Pathobiology in 3D", the research progression and the emerging methods used to quantitatively analyze how the ECM affects cancer pathobiology are illuminated.

In Chapter 7.8 by Rydén, entitled "Diabetic Foot Infections", the involvement of BSP in skeletal infections caused by *Staphylococcus aureus* is presented. It is demonstrated that *S. aureus* strains causing osteitis specifically bind to BSP through BSP-binding protein compared to those causing soft tissue infection only. The possible laboratory parameters used for diagnosis of osteitis in diabetic foot are presented. The utility of measuring antibodies in serum against BSP-binding protein of *S. aureus*, which are increased in patients with diabetic foot osteitis compared to diabetes patients with soft tissue infection only, are discussed as a helpful tool in diagnosing diabetic foot osteitis.

References

Bougatef, F., Quemener, C., Kellouche, S., et al. (2009). EMMPRIN promotes angiogenesis through hypoxia-inducible factor-2alpha-mediated regulation of soluble VEGF isoforms and their receptor VEGFR-2. Blood *114*, 5547–5556.

Choi, Y., Chung, H., Jung, H., Couchman, J. R., and Oh, E. S. (2011). Syndecans as cell surface receptors: unique structure equates with functional diversity. Matrix Biol *30*, 93–99.

Couchman, J. R. (2010). Transmembrane signaling proteoglycans. Annu Rev Cell Dev Biol *26*, 89–114.

Dvorak, H. F. (1986). Tumors: wounds that do not heal. Similarities between tumor stroma generation and wound healing. N Engl J Med *315*, 1650–1659.

Frantz, C., Stewart, K. M., and Weaver, V. M. (2010). The extracellular matrix at a glance. J Cell Sci *123*, 4195–4200.

Gao, Y., Xiao, Q., Ma, H., et al. (2010). LKB1 inhibits lung cancer progression through lysyl oxidase and extracellular matrix remodeling. Proc Natl Acad Sci *107*, 18892–18897.

Gialeli, C., Theocharis, A. D., and Karamanos, N. K. (2011). Roles of matrix metalloproteinases in cancer progression and their pharmacological targeting. FEBS J *278*, 16–27.

Huet, E., Gabison, E. E., Mourah, S., and Menashi, S. (2008). Role of emmprin/CD147 in tissue remodeling. Connect Tissue Res *49*, 175–179.

Iozzo, R. V., and Sanderson, R. D. (2011). Proteoglycans in cancer biology, tumour microenvironment and angiogenesis. J Cell Mol Med *15*, 1013–1031.

Kessenbrock, K., Plaks, V., and Werb, Z. (2010). Matrix metalloproteinases: regulators of the tumor microenvironment. Cell *141*, 52–67.

Kolset, S. O., and Pejler, G. (2011). Serglycin: a structural and functional chameleon with wide impact on immune cells. J Immunol *187*, 4927–4933.

Levental, K. R., Yu, H., Kass, L., et al. (2009). Matrix crosslinking forces tumor progression by enhancing integrin signaling. Cell *139*, 891–906.

Misra, S., Heldin, P., Hascall, V. C., et al. (2011). Hyaluronan-CD44 interactions as potential targets for cancer therapy. FEBS J *278*, 1429–1443.

Theocharis, A. D., Skandalis, S. S., Tzanakakis, G. N., and Karamanos, N. K. (2010). Proteoglycans in health and disease: novel roles for proteoglycans in malignancy and their pharmacological targeting. FEBS J *277*, 3904–3923.

7.2 Extracellular matrix in breast cancer: permissive and restrictive influences emanating from the stroma

Ida Pucci-Minafra

7.2.1 Introduction

The mammary gland is the only organ that undergoes cyclical morphogenesis (growth and differentiation, cell death and involution) in the adult life. The extracellular matrix (ECM), with its assortment of factors and remodeling enzymes, plays a fundamental role in these processes. Since the pionieristic research of the 1980s and 1990s, the mammary gland has represented a powerful model to study cellular responses to matrix interaction in normal development, differentiation and cancer (Barcellos-Hoff et al., 1989; Berdichevsky et al., 1994; Lochter and Bissell, 1995; Rønnov-Jessen et al., 1996).

Correct stromal-epithelial interactions are coordinated temporally and spatially by various signaling pathways throughout the female reproductive life. Perturbations in this equilibrium are directly or indirectly responsible for the onset of breast cancer and its progression. The past decades of breast cancer research have focused attention on the epithelial component of the gland, searching for markers within the genes and proteins with roles in the cellular compartments. In more recent years the instructive role of the ECM for cells and tissues placed on it, or embedded within it, is being more and more recognized. The challenge is now to discern the relevant factors and design among the intricate network of signals emanating from different components of the extracellular matrix.

7.2.2 The extracellular context in the mammary gland

An important aspect, not always properly underlined, is the existence of at least two distinct ECM compartments, having different roles in normal development and cancer progression: the *basal lamina* (BL), also known as the *basement membrane*, and the *interstitial stroma*. The latter is further subdivided into two functional territories: the *interlobular stroma*, running between lobules, and the *intralobular stroma* within lobules (▶Figure 7.2).

The basal lamina

The BL represents the first and most relevant boundary underlying the mammary parenchymal cells. It represents the first frontier that separates the epithelium from the

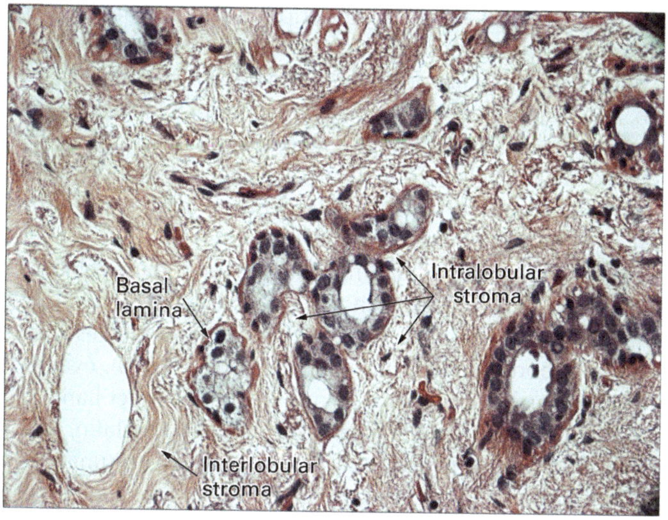

Figure 7.2 **Histological section of a nontumoral mammary gland, illustrating the presence of the basal lamina around the terminal end buds and the territories of intra- and interlobular stroma (Hematoxylin-eosin, 50x objective).**

underneath connective tissue. The macromolecular organization of BL includes laminin and type IV collagen, as major constituents, together with heparan sulfate proteoglycans, fibronectin, nidogen/entactin, netrin-1 and other minor components, which will not be discussed in this context.

One of the critical factors responsible for tissue homeostasis is the correct adhesion of the epithelial sheet on its underlying basal lamina (Alcaraz et al., 2008). On the other hand, BL matrix disassembly typically occurs during mammary morphogenesis and in the phase of regression following pregnancy and lactation. As discussed in other chapters of this book, the key element responsible for matrix remodeling is the machinery of matrix degrading enzymes, which operates under precise regulatory controls, performed by a balance of enzyme activators, modulators and inhibitors. Perturbations of this system, correlated to disruption, or even to local degradation of the BL and aberrant cell-laminin interactions, are known to occur from tumor appearance to the metastatic progression (Patarroyo et al., 2002).

In vitro studies have documented that when nonmalignant cells from epithelial tissues are cultured as 2D monolayers, they rapidly lose tissue-specific functions (Bissell, 1981; Walpita and Hay, 2002). On the contrary, mammary epithelial cells cultured on a laminin-rich extracellular matrix respond to a number of signals operating *"in vivo"* and regarding gene expression, tissue architecture and function. Moreover, nonmalignant mammary epithelial cells cultured in, or on, three-dimensional (3D) matrix, after an initial proliferative phase, undergo a process of acinar morphogenesis but only in the presence of laminin (reviewed by Muschler and Streuli, 2010). Laminin-1 (laminin-111), together with lactogenic hormones and nidogen, is also required for beta-casein expression in mammary epithelial cells (Streuli et al., 1991; Pujuguet et al., 2000). More recently it has been shown (Spencer et al., 2011) that cell quiescence induced by the addition of laminin-111 rapidly decreases nuclear β-actin, destabilizes the binding of RNA polymerase II and III to transcription sites, finally leading to a dramatic drop in DNA synthesis and transcription; forced overexpression of nuclear actin levels reverses these effects, and cells are prevented from becoming quiescent even in the presence of laminin. The results emphasize the role of laminin-111 as the regulator of epithelial cell quiescence through the action of nuclear β-actin as a key mediator. Loss of BL architecture due to high levels of oncogenic raf-induced matrix metalloproteinase-9 (MMP9) during carcinogenesis restores proliferation and suppresses polarity and phenotypic cell expression (Beliveau et al., 2010).

Type IV collagen, along with laminin, plays an important role in cell adhesion, growth, migration, and differentiation. In this context, the degradation of type IV collagen occurring during the invasive stage of cancer, destabilizes tissue architecture and cell polarity, as reported for laminin (Beliveau et al., 2010), even if the molecular networks from and towards cells are still unknown. On the other hand, proteolytic degradation of type IV collagen by MMP9 is known to release tumstatin, a 28 kDa fragment that displays both antiangiogenic and proapoptotic activity, so functioning not only as an endogenous inhibitor of angiogenesis, but also as a tumor suppressor. Tumstatin is a member of the group of bioactive ECM fragments (matricryptins) released, or unmasked, by cryptic sites cleaved from collagens, proteoglycans and glycosaminoglycans. Most matricryptins play key roles in controling various physiopathological processes including angiogenesis, tissue remodeling, wound healing, inflammation, tumor growth, and metastasis (Ricard-Blum and Ballut, 2011). These findings also show that MMP9 may perform opposing roles at different stages of tumor progression; for instance, it can promote the angiogenic switch that favors the initial burst of tumor growth, but it can also generate endogenous inhibitors of angiogenesis, such as tumstatin, which restrains further growth of the tumor (Hamano et al., 2003).

The interstitial stroma

As already mentioned, the interstitial stroma of the adult mammary gland is composed of two distinct territories: the interlobular and the intralobular connective tissue. The first, running between lobules, consists mainly of a dense collagenous matrix within which there are varying amounts of adipose tissue and scattered fibroblasts. On the contrary, the intralobular stroma that surrounds the terminal end buds (TEB) consists of a loose connective tissue including a row of hormone-sensitive fibroblasts surrounding the lobular epithelial components. These fibroblasts are thought to take an active part in the inductive interactions between epithelium, basement membrane and stroma occurring during morphogenesis and differentiation.

During tumor invasion, the ECM architecture is dramatically changed. The first step is the local degradation of the basal lamina and the underlying intralobular stroma. In this phase, several cascade processes start to be activated, involving from one side the progressive transition of the epithelial phenotype (stationary and polarized) into a mesenchymal motile phenotype. On the other side, the stroma undergoes severe changes, including local degradation and new matrix deposition (Pucci-Minafra et al., 1986). Clinical correlations indicate that in many cases the host stroma reaction, rather than refraining tumor growth, induces cancer cell dissemination, host cells recruitment and tumor progression (Provenzano et al., 2008).

This tumor stroma remodeling is similar to the initial steps of wound repair, with the great difference that wound healing is a temporary and regulated response to injury, while in tumor invasion the cell-stroma interaction becomes degenerated; therefore, tumors have also been related to wounds that fail to heal (Dvorak, 1986).

7.2.3 The physical role of connective tissue stroma

Studies using inert matrix substrates have suggested that the matrix stiffness may have profound effects on cell fate and behavior (Discher et al., 2009), as for instance on stem cell lineage specification, cell migration, proliferation, and survival (Wells, 2008).

It is believed that the stiffening of the stroma tumor is induced by ECM deposition and remodeling, partly due to the activity of recruited fibroblasts or by the cancer cells themselves. This reaction, also called desmoplastic response, is known from classical histopathology. For instance, a prolonged release of growth factors (GFs) and chemokines potentiates inflammation that in turn modifies the repertoire of infiltrating T lymphocytes (Tan and Coussens, 2007), the recruitment of stromal fibroblasts and induces their transdifferentiation into myofibroblasts, thus exacerbating and promoting tissue desmoplasia (Desmouliere et al., 2004; De Wever et al., 2008). It has also been reported that newly deposited collagen and elastic fibers are remodeled and cross-linked by lysyl oxidase and transglutaminase, thus generating larger, more rigid fibrils that may be less sensitive to the action of matrix proteases and further stiffen the tissue ECM (Lucero and Kagan, 2006; Levental et al., 2009).

Clinically, the mammographic analysis is a consolidated procedure as a first approach to breast cancer diagnosis. Both glandular parenchyma and surrounding connective tissue display variably attenuated (absorbed or scattered) radiological images. The "denser" the tissue, the more the X-rays are attenuated, and the structures emerge light: an appearance that is referred to as "mammographic density." These variations in radiological appearance are reported to be associated with breast cancer risk (Boyd et al., 1998). A prototype of a bilateral mammography of a normal breast is shown in ▶Figure 7.3.

The biological basis of mammographic density

Studies seeking to understand the biological basis of mammographic density have focused, from time to time, on associations with epithelial changes, stromal alterations, proteoglycans increment and also with significantly increased fibrillar collagen deposition (Alowami et al., 2003; Skandalis et al., 2011). In addition, it has been reported that levels of total collagen density promotes mammary tumor initiation and progression (Provenzano et al., 2008).

Such mechanical signals arising from increased density or rigidity of the microenvironment are believed to play a role in the transformed phenotype of breast epithelial cells (Paszek et al., 2005).

Collagen changes: type I trimer collagen, a permissive substrate

Concerning the nature of collagen changes during carcinogenesis, several reports have produced evidence for a shift in collagen synthesis towards a fetal homotrimeric isoform

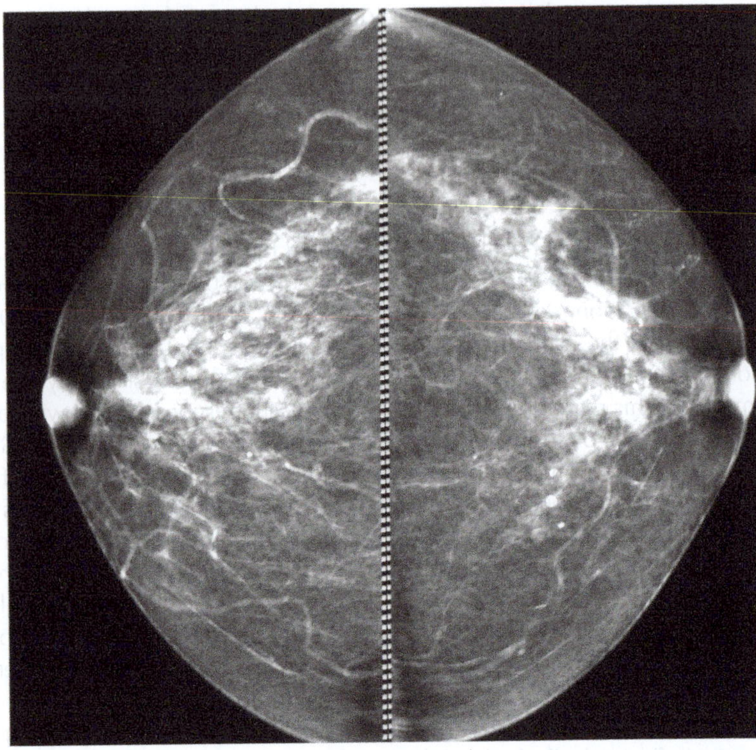

Figure 7.3 A prototype of a bilateral mammography of a normal breast, where the lighter areas are given by the parenchyma and the underlying texture of moderate density is the supporting stroma.

of type I collagen, composed of three α1(I) chains and also named oncofetal laminin binding (OF/LB) for its ability to bind to laminin (Pucci-Minafra et al., 1993).

The existence of a homotrimer collagen of type I (α1[I]3) was initially identified in some embryonic and fetal tissues (Jimenez et al., 1977; Crouch and Bornstein, 1978; Little and Church, 1978), and subsequently in carcinoma tissues (Minafra et al., 1984; Pucci-Minafra et al., 1985, 1993, 1998) and cultured cancer cells (DeClerck et al., 1987; Minafra et al., 1988; Rupard et al., 1988; Ishikoh et al., 1994).

Furthermore, it was assumed and subsequently proved *in vitro* (Schillaci et al., 1989; Luparello et al., 1991) and in mouse models (Pucci-Minafra et al., 1995) that tumoral OF/LB collagen could play a permissive role in the invasive growth, by enhancing cell proliferation and facilitating cell migration and invasion, in opposition to the canonical type I, which apparently exerts a restraining effect on the neoplastic cell growth and migration (Conklin et al., 2011; Sasaki et al., 2011).

Regarding the molecular basis for the aberrant homotrimer synthesis in place of the normal heterotrimer form, some Authors have demonstrated that the absence of α2(I) chains, observed also in some tumor tissues (Pucci-Minafra et al., 1998) may be due to gene silencing mechanisms, suppressing collagen pro-α2(I) gene transcription (Xu et al., 2006). Recently, the importance of this collagen in cancer has been further acknowledged following the experimental contribution of Nagase group (Imperial

College, London, UK), who furnished evidence that the homotrimer present in carcinomas is resistant to the action of human matrix metalloproteinase MMP1, MMP2, MMP8, or MMP13 at temperatures from 20°C to 35°C, and that this collagen isoform exerts selective support to cell invasion (Han et al., 2010; Makareeva et al., 2010). According to the Authors, a slower cleavage of homotrimer molecules may be explained by their resistance to the local triple helix unwinding at the cleavage site, since the unwinding is required for the collagen cleavage by all collagenases. These results have fully confirmed previous findings of our group, and add strength to the hypothesis that this oncofetal collagen isoform is a powerful tumor-promoting agent.

Two examples of stromal molecules with anticancer effects: type V collagen and decorin

Type V collagen, a "minor" component of the total collagen in normal breast stroma, was found to be overdeposited in the ECM of breast carcinoma (Barsky et al., 1982; Luparello et al., 1988) and in other form of tumors (Marian and Danner, 1987; Iwahashi and Muragaki, 2011).

Generally, in normal connective tissues, type V collagen copolymerizes with type I and type III collagen forming heterotypic fibers (Pucci-Minafra and Luparello, 1991; Spiess and Zorn, 2007). *In vitro* experiments, using neoplastic cells exposed to the presence of type V collagen, alone or in combination with type I collagen, demonstrated that type V induces a prominent delay of the proliferation rate, an inhibition of cell motility and invasion (Pucci-Minafra and Luparello, 1991) as well as modifications of the proteomic profile (Fontana et al., 2004; Pucci-Minafra et al., 2008a). More recent results provided evidence that type V collagen could trigger apoptosis via caspase-9 in breast cancer tissues (Parra et al., 2010).

Another minority member of the connective tissue stroma is the decorin, a prototype of small leucine-rich proteoglycans (SLRPs) (Schaefer and Schaefer, 2010), which is proposed to perform a number of important regulative roles in normal and pathological tissues, by the interaction with both collagen and cells of different origin. It can be hypothesized that an increase of available decorin at the tumor boundary, due either to an incremented synthesis by fibroblasts, or to a release from the collagen frame, due to proteolytic activities, may play roles on neoplastic cell behavior. Indeed, in the last decades decorin has been extensively taken into consideration also for possible clinical applications as a putative regulator of tumor growth and metastatic spreading through binding to and activation of several growth factor receptors (epidermal growth factor receptor [EGFR], insulin-like growth factor-I receptor [IGF-IR], low-density lipoprotein receptor-related protein-1 [LRP1], and transforming growth factor-beta [TGF-β]) (Iozzo and Karamanos, 2010; Iozzo and Schaefer, 2010).

To investigate the possible direct effects of decorin on breast cancer cells, we transfected the 8701-BC breast cancer line derived from a ductal infiltrating carcinoma of the breast (Minafra et al., 1989). Major decorin responses included a net decrease of the proliferation rate along with the reduction of the exaggerated membrane ruffling and vesiculation, typical of the parental neoplastic cells (Palazzolo et al., 2012). Concurrently, the ectopic decorin appeared to induce a more stationary cell behavior by the increment of cell-cell adhesiveness and, finally, to elicit changes in the proteomic profile (Pucci-Minafra et al., 2008c). Taken together, these results provide new important elements in support of the potential antioncogenic role exerted independently by type V collagen and decorin, and raise an additional interest for their possible role in clinical applications.

7.2.4 The proteomic lesson

Collagen influences

Due to the great complexity of the signaling network flowing between tumor cells and their microenvironment, several laboratories, including ours, have applied the proteomic strategy, utilizing appropriate cellular models, to study global cellular responses to individual signals emanating from the microenvironment. As already mentioned, one goal of our laboratory has been the investigation of the influences exerted by collagens and other microenvironment factors on the neoplastic cell behavior. We extended previous investigations on the effects individually exerted on cell proteome by different collagens, namely, type I, type IV, type V, and the homotrimer of type I, used as culture substrates (Fontana et al., 2004; Pucci-Minafra et al., 2008a).

The proteomic approach was aimed at the possible identification of proteins that, individually or as clusters, could be correlated with the phenotypic responses induced by the substrates, acting as final targets of the modulatory effects exerted by different extracellular stimuli. As a model system we have used the well-characterized 8701-BC breast cancer cell line cultured in the presence of the previously mentioned collagen substrates. The extrapolation of trustable proteomic profiles was carried out by triplicate assays for each experiment. Spot identity was assessed by mass spectrometry and/or Western blot assays; double normalization (technical and biological) for quantitative and comparative analyses was also performed (Pucci-Minafra et al., 2008b). Ambiguity was discharged and a threshold of 1.5-fold was established as cutoff for spot intensity differences.

The most interesting observation was that cells plated and grown respectively on different collagen substrates responded by modulating a large percentage of their proteomic profile, while only a minor component remained unvaried. Indeed, a collective evaluation of the capability of responses to the substrate solicitations indicates that the quantity of proteins modulated by the different substrates accounts for about 70% of the total proteomic collection. This number includes proteins individually modulated and proteins that are modulated by two or three or even by all four substrates. This resumptive view of the results is shown by the Venn diagram in ▶Figure 7.4, which shows subproteomic profiling consisting of protein clusters exclusively modulated by each substrate, as well as proteins that appear responsive to more than one substrate.

The proteins that appeared preferentially modulated by type I trimer (even if some of them not exclusively) fall into three main clusters, selected according to DAVID Bioinformatics Resources (Huang et al., 2007): metabolic enzymes (enolase, phosphoglycerate kinase, phosphoglycerate mutase, pyruvate kinase, and aconitase), antiapoptotic proteins (annexin A1, annexin A4, heat-shock 70 kDa, nucleophosmin, translationally controlled tumor protein, cofilin, heat-shock 27 kDa, nucleophosmin, and superoxide dismutase), and proteins of cytoskeleton and vesicular transport (thymosin, tropomyosin, vimentin, vinculin, cofilin, and aldolase A).

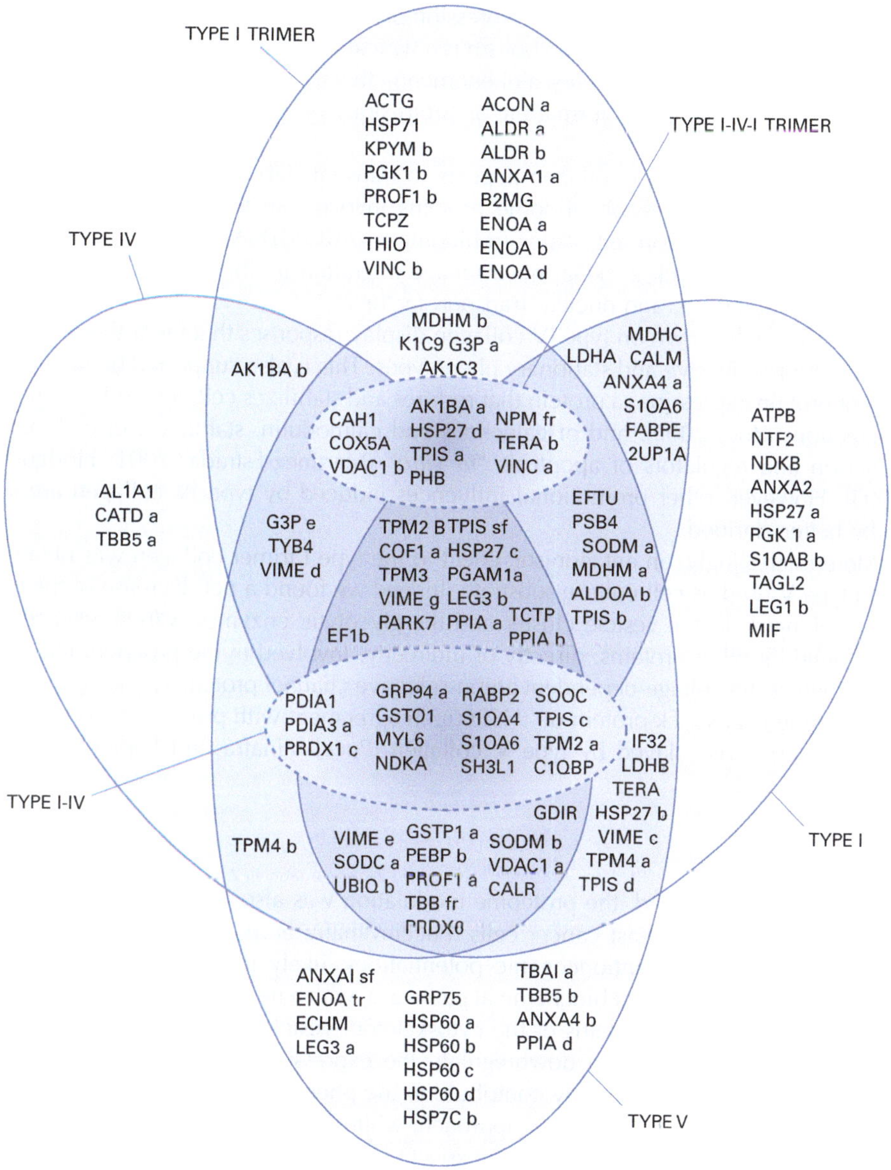

Figure 7.4 A Venn diagram showing the relationships between the collection of sets of proteins modulated by different collagen substrates used for the cell growth. Proteins responding to a single substrate are in the nonintersecting areas, while proteins sensible to more than one substrate are indicated in the proper intersections. The elliptical areas surrounded by dotted lines correspond respectively to the intersection of indicated substrates.

All clusters converge towards the expression of a high malignant phenotype. Indeed, the increment of carbohydrate catabolism is a typical attitude of highly malignant cells, as well as the shedding of vesicles, a phenomenon that is correlated with the membrane perturbation following the formation of protrusions or spikes from migratory cells (Palazzolo et al., 2012).

On the contrary, type I collagen appears to transmit stabilizing signals, essentially by the enhanced expression of some proteins involved in the cellular homeostasis and cell death regulation (e.g. calcium-binding S10AB (S100A11) protein, galectin-1, annexin A2, heat-shock 27 kDa, macrophage migration inhibitory factor, nucleoside diphosphate kinase B, and nuclear transport factor-2).

Similarly, cells grown on type IV collagen display responses that favor the establishment of more cohesive and stationary phenotypes. This is also supported by the induction of profilin expression, a protein that restores and stabilizes cell-cell and cell-matrix interactions (Zou, 2009), and of galectin-1 and calmodulin, stabilizers of cell-matrix adhesion and regulators of apoptosis *"in vitro"* (Martinez-Estrada, 2001; Horiguchi, 2003). However, other promotional influences induced by type IV collagen are still to be better clarified.

More interestingly, an antagonistic effect to the type I trimer collagen was obtained from type V used as cell culture substrate. Indeed we found a net decrease of the proteins belonging to the vesicle cluster and the glycolytic enzymes. While an increase was found for other proteins, directly or indirectly, involved in the proapoptotic pathway, such as the voltage-dependent anion-selective channel protein-1, and several isoforms of the heat shock protein 60 (HSP60), in agreement with previous reports on the antioncogenic role played by type V collagen (Pucci-Minafra and Luparello, 1991; Parra et al., 2010).

Decorin effects

As previously mentioned, the proteome modulation was also investigated in decorin-transfected 8701-BC breast cancer cells (Pucci-Minafra et al., 2008c). The results revealed new important antioncogenic potentialities, likely to be exerted by decorin through a variety of distinct biochemical pathways, with a net decrease of the glycolytic enzymes and of many proteins of the cytoskeleton, vesicle, and cell motility classes. Moreover, decorin seems to downregulate the expression of the oncogenes c-Myc and c-ErbB2, which in turn may contribute to the phenotype reprogramming.

At the same time, the proteomic approach, while disclosing new putative pathways for the decorin action, presented unexpected responses concerning a number of proteins, which deserve future investigations.

Paracrine influences

Finally, the proteomic approach was also applied to achieve a better understanding of the potential effects exerted by fibroblasts on cancer cells. At the beginning of a tumor development, fibroblasts recruited into the injury site, named cancer-associated fibroblast (CAF), start to aid the tissue recovery. However, in the absence of negative feedback signals, activated fibroblasts can actually hyperproliferate and overproduce bioactive factors (Erez et al., 2010). When applying the coculture method to primary human fibroblasts and 8701-BC breast cancer cells, it was observed that the neoplastic cells exposed to fibroblast media exhibited significantly increased growth rates and migratory and invasive abilities, compared with unstimulated cells (Cancemi et al., 2010).

Comparative proteomic analyses of control and cocultured cells revealed significant upregulation of several glycolytic enzymes (glyceraldehyde-3-phosphate dehydrogenase, triosephosphate isomerase, and phosphoglycerate kinase-1), which indicates a considerable rising of the so-called Warbourg effect, typical of very aggressive cells. The second group of proteins upregulated by fibroblast factors belong to the category of cytoskeleton and cell motion (Rho GDP-dissociation inhibitor, tropomyosin-2, tropomyosin-4, alpha tubulin, and vimentin). This suggests also a dynamic remodeling of the cytoskeleton of cocultured cells, which in turn may be responsible for the observed enhancement of migratory and invasive activities, according to what has also been reported in other systems (Heylen et al., 1998). Additionally, the fibroblast-conditioned medium increased neoplastic cell proliferation and invasion with a concurrent upregulation of the c-Myc oncogene.

Collectively these results suggest that fibroblast stimulation may enhance the malignant potential of breast cancer cells *in vitro*.

7.2.5 Concluding remarks

The short survey presented here, strongly reaffirms the importance of the microenvironment in cancer progression. The precognitive "soil and seeds" hypothesis of Paget, first described in 1889, today more than before, appears realistic and bright!

Remarkable is the existence of antagonistic effects generated by different components of the tumor microenvironment. One type being "permissive" for cell proliferation, migration, and metabolic adaptation, and the second being "restrictive" to limit cell proliferation, promote apoptosis, and favor cell adhesion and a more stationary state of neoplastic cells (▶Figure 7.5).

Among collagens, type I trimer collagen appears the most important substrate to play a permissive role in the invasive growth, by enhancing cell proliferation and facilitating cell migration and invasion. On the contrary, type V collagen apparently exerts a reversing effect on tumor progression by limiting neoplastic cell growth and migration and by inducing apoptotic pathways, while type I and type IV collagens behave as stabilizing substrates. All these effects have been observed *in vitro*. It is expected that the *in vivo* condition is further complicated by the interplay of many other factors present locally in the primary tumor. In first place are the host cells, among which fibroblasts that are known to occur numerous in the tumor site. From the *in vitro* stimulation, the effects of fibroblasts appear to favor neoplastic cells survival and invasion, while another component of the tumor matrix, the decorin, exerts an opposite effect by limiting cell proliferation and cell migration. How many other components do we expect to take an active part in cancer progression or regression? We do not have any clear answers at present.

So far much has been learned, but still more has to be achieved from collective efforts made by scientists utilizing in vitro models that mimic tumor microenvironment, but also ex vivo samples and animal models.

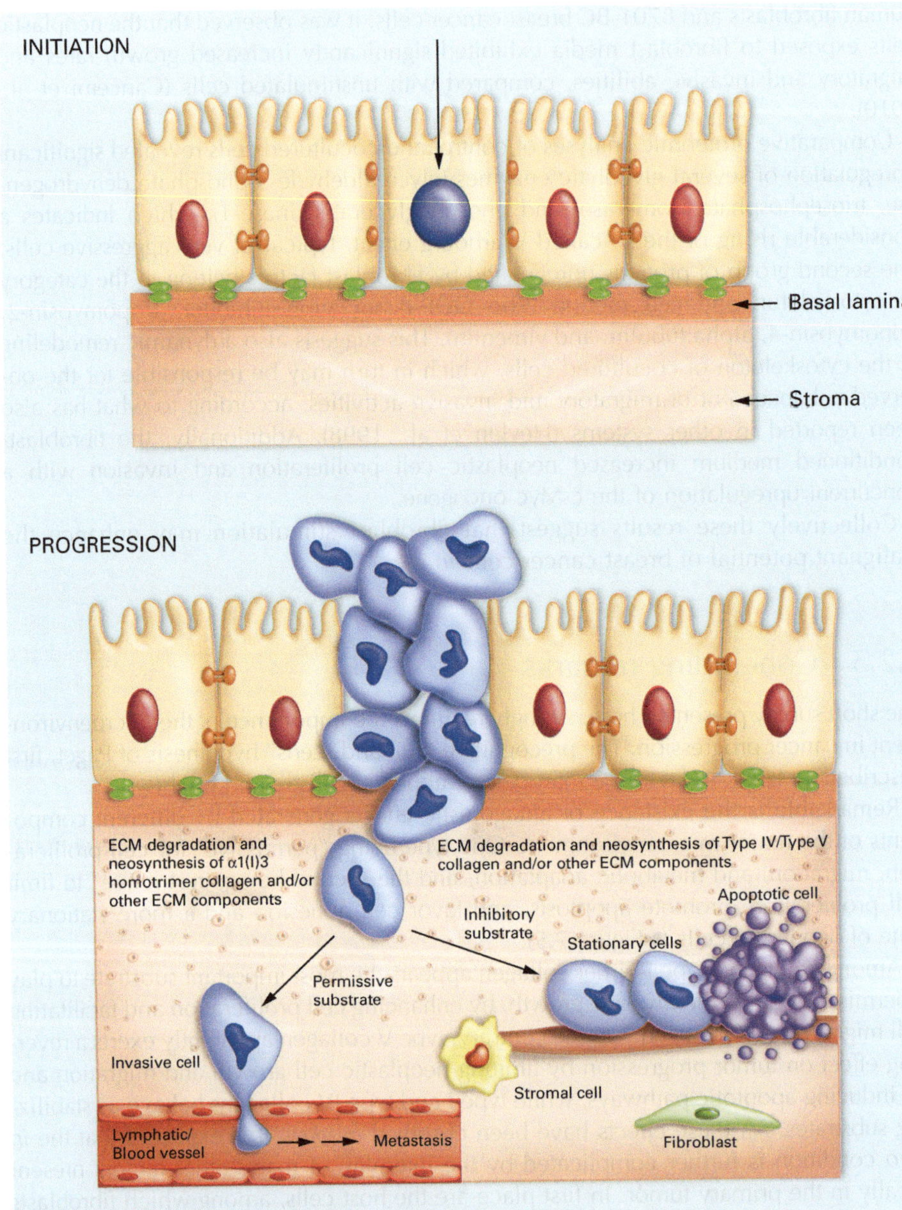

Figure 7.5 Schematic illustration of the carcinogesis focusing on the transition from the initiation (top) to open progression of the cancer. Here two possible routes are described: a permissive one, which may favor neoplastic cell dissemination and metastasis formation, and an inhibitory one, in which cells are inhibited from migration and become more stationary and polarized or, alternatively, are induced to enter in apoptosis.

7.2.6 Challenges and future prospects

The bulk of the information available to scientists today derives from experimental approaches addressed individually to solve a small piece of the larger puzzle. We know that too many pieces are still missing, but paradoxically, information on individual genes and proteins is too extensive for unequivocal interpretations. As we have observed in the past few years, there is a great redundancy of data and too many interlacing functions and pathways for each gene/protein. Therefore, we are facing a sort of paradoxical ambiguity in our approach to the modern research. On one hand, we need to know more about our genes and proteins, including how many there are and how they work in health and disease, as the human proteome is very far from conclusive. On the other hand, we are facing an almost unmanageable amount of data, and in order to understand and utilize it, we need powerful informatic supports. Above all, we need innovative intellectual landscapes.

7.2.7 Take-home message

An interesting message is that neoplastic cells, even at advanced stages of malignancy, are still able to respond to a number of extracellular signals. Efforts in the next years will make it possible to evaluate the therapeutic application and efficacy of a growing number of molecules emerging from the microenvironment, as well as their activity at different stages of cancer progression of certain tumors. A second message in this context is that the power in fighting against cancer strongly depends on the quality of the alliance between clinical medicine and molecular cell biology.

Acknowledgments

I thank all the members of my laboratory, past and present, who have actively contributed to the research reported in this paper. The support of the Oncology Department La Maddalena is gratefully acknowledged, with particular reference to the President Prof. Guido Filosto and Drs Elena Roz and Debora Castrogiovanni for providing clinical images.

I would like to express my sincere gratitude for the intelligent and generous support offered for the preparation of this chapter to the following people: Dr Patrizia Cancemi, for the critical reading of the manuscript and useful suggestions; Dr Gianluca Di Cara, for protein clustering; Dr Nadia N. Albanese, for the skilful preparation of diagrams; Dr Maria Rita Marabeti, for the careful editing of the text and for bibliography preparation; and above all, Prof. Salvatore Minafra who shares work and life with me.

References

Alcaraz, J., Xu, R., Mori, H., et al. (2008). Laminin and biomimetic extracellular elasticity enhance functional differentiation in mammary epithelia. EMBO J *27*, 2829–2838.

Alowami, S., Troup, S., Al-Haddad, S., Kirkpatrick, I., and Watson, P. H. (2003). Mammographic density is related to stroma and stromal proteoglycan expression. Breast Cancer Res *5*, R129–R135.

Barcellos-Hoff, M. H., Aggeler, J., Ram, T. G., and Bissell., M. J. (1989). Functional differentiation and alveolar morphogenesis of primary mammary cultures on reconstituted basement membrane. Development *105*, 223–235.

Barsky, S. H., Rao, C. N., Grotendorst, G. R., and Liotta, L. A. (1982). Increased content of type V collagen in desmoplasia of human breast carcinoma. Am J Pathol *108*, 276–283.

Beliveau, A., Mott, J. D., Lo, A., et al. (2010). Raf-induced MMP9 disrupts tissue architecture of human breast cells in three-dimensional culture and is necessary for tumor growth in vivo. Genes Dev *24*, 2800–2811.

Berdichevsky, F., Alford, D., D'Souza, B., and Taylor-Papadimitriou, J. (1994). Branching morphogenesis of human mammary epithelial cells in collagen gels. J Cell Sci *107*, 3557–3568.

Bissell, M. J. (1981). The differentiated state of normal and malignant cells or how to define a "normal" cell in culture. Int Rev Cytol *70*, 27–100.

Boyd, N. F., Lockwood, G. A., Byng, J. W., Tritchler, D. L., and Yaffe, M. J. (1998). Mammographic densities and breast cancer risk. Cancer Epidemiol Biomarkers Prev *7*, 1133–1144.

Cancemi, P., Albanese, N. N., Di Cara, G., et al. (2010). Multiple changes induced by fibroblasts on breast cancer cells. Connect Tissue Res *51*, 88–104.

Conklin, M. W., Eickhoff, J. C., Riching, K. M., et al. (2011). Aligned collagen is a prognostic signature for survival in human the breast carcinoma. Am J Pathol *178*, 1221–1232.

Crouch, E., and Bornstein, P. (1978). Collagen synthesis by human amniotic fluid cells in culture: characterization of a procollagen with three identical proalpha1(I) chains. Biochemistry *17*, 5499–5509.

DeClerck, Y. A., Bomann, E. T., Spengler, B. A., and Biedler, J. L. (1987). Differential collagen biosynthesis by human neuroblastoma cell variants. Cancer Res *47*, 6505–6510.

Desmoulière, A., Guyot, C., and Gabbiani, G. (2004). The stroma reaction myofibroblast: a key player in the control of tumor cell behavior. Int J Dev Biol *48*, 509–517.

De Wever, O., Pauwels, P., De Craene, B., et al. (2008). Molecular and pathological signatures of epithelial–mesenchymal transitions at the cancer invasion front. Histochem Cell Biol *130*, 481–494.

Discher, D. E., Mooney, D. J., and Zandstra, P. W. (2009). Growth factors, matrices, and forces combine and control stem cells. Science *324*, 1673–1677.

Dvorak, H. F. (1986) Tumors: wounds that do not heal. Similarities between tumor stroma generation and wound healing. N Engl J Med *315*, 1650–1659.

Erez, N., Truitt, M., Olson, P., Arron, S. T., and Hanahan, D. (2010). Cancer-associated fibroblasts are activated in incipient neoplasia to orchestrate tumor-promoting inflammation in an NF-κB-dependent manner. Cancer Cell *17*, 135–147.

Fontana, S., Pucci-Minafra, I., Becchi, M., Freyria, A. M., and Minafra, S. (2004). Effect of collagen substrates on proteomic modulation of breast cancer cells. Proteomics *4*, 849–860.

Hamano, Y., Zeisberg, M., Sugimoto, H., et al. (2003). Physiological levels of tumstatin, a fragment of collagen IV alpha3 chain, are generated by MMP-9 proteolysis and suppress angiogenesis via alphaV beta3 integrin. Cancer Cell *3*, 589–601.

Han, S., Makareeva, E., Kuznetsova, N. V., et al. (2010). Molecular mechanism of type I collagen homotrimer resistance to mammalian collagenases. J Biol Chem *285*, 22276–22281.

Heylen, N., Baurain, R., Remacle, C., and Trouet, A. (1998). Effect of MRC-5 fibroblast conditioned medium on breast cancer cell motility and invasion in vitro. Clin Exp Metastasis *16*, 193–203.

Horiguchi, N., Arimoto, K., Mizutani, A., Endo-Ichikawa, Y., Nakada, H., and Taketani, S. (2003). Galectin-1 induces cell adhesion to the extracellular matrix and apoptosis of non-adherent human colon cancer Colo201 cells. J Biochem *134*, 869–874.

Huang, da W., Sherman, B. T., Tan, Q., et al. (2007). DAVID Bioinformatics Resources: expanded annotation database and novel algorithms to better extract biology from large gene lists. Nucleic Acids Res *35* (Web Server issue), W169–W175.

Iozzo, R. V., and Karamanos, N. (2010). Proteoglycans in health and disease: emerging concepts and future directions. FEBS J *277*, 3863.

Iozzo, R. V., and Schaefer, L. (2010). Proteoglycans in health and disease: novel regulatory signaling mechanisms evoked by the small leucine-rich proteoglycans. FEBS J *277*, 3864–3875.

Ishikoh, A. U., Hayashi, A., Tokimitsu, I., Tajima, S., and Nishikawa, T. (1994). Coordinate modulation of melanogenesis and type I collagen secretion by type I collagen substratum during reversible conversion between melanotic and amelanotic cells in mouse B16 melanoma. J Biochem *116*, 610–614.

Iwahashi, M., and Muragaki, Y. (2011). Increased type I and V collagen expression in uterine leiomyomas during the menstrual cycle. Fertil Steril *95*, 2137–2139.

Jimenez, S. A., Bashey, R. I., Benditt, M., and Yankowski, R. (1977). Identification of collagen alpha1(I) trimer in embryonic chick tendons and calvaria. Biochem Biophys Res Commun *78*, 1354–1361.

Levental, K. R., Yu, H., Kass, L., et al. (2009). Matrix crosslinking forces tumor progression by enhancing integrin signaling. Cell *139*, 891–906.

Little, C. D., and Church, R. L. (1978). A new type of collagen produced by mouse cells derived from early embryonic sources. Arch Biochem Biophys *190*, 632–639.

Lochter, A., and Bissell, M. J. (1995). Involvement of extracellular matrix constituents in breast cancer. Semin Cancer Biol *6*, 165–173.

Lucero, H. A., and Kagan, H. M. (2006). Lysyl oxidase: an oxidative enzyme and effector of cell function. Cell Mol Life Sci *63*, 2304–2316.

Luparello, C., Rizzo, C. P., Schillaci, R., and Pucci-Minafra, I. (1988). Fractionation of type V collagen from carcinomatous and dysplasic breast in the presence of alkaline potassium chloride. Anal Biochem *169*, 26–32.

Luparello, C., Sheterline, P., Pucci-Minafra, I., and Minafra, S. (1991). A comparison of spreading and motility behaviour of 8701-BC breast carcinoma cells on type I, I-trimer and type V collagen substrata. Evidence for a permissive effect of type I-trimer collagen on cell locomotion. J Cell Sci *100*, 179–85.

Makareeva, E., Han, S., Vera, J. C., et al. (2010). Carcinomas contain a matrix metalloproteinase–resistant isoform of type I collagen exerting selective support to invasion. Cancer Res *70*, 4366–4374.

Marian, B., and Danner, M. W. (1987). Skin tumor promotion is associated with increased type V collagen content in the dermis. Carcinogenesis *8*, 151–154.

Martinez-Estrada, O. M., Villa, A., Breviario, F., Orsenigo, F., Dejana, E., and Bazzoni, G. (2001). Association of junctional adhesion molecule with calcium/calmodulin-dependent serine protein kinase (CASK/LIN-2) in human epithelial caco-2 cells. J Biol Chem *276*, 9291–9296.

Minafra, S., Luparello, C., Rallo, F., and Pucci-Minafra, I. (1988). Collagen biosynthesis by a breast carcinoma cell strain and biopsy fragments of the primary tumour. Cell Biol Int Rep *12*, 895–905.

Minafra, S., Morello, V., Glorioso, F., et al. (1989). A new cell line (8701-BC) from primary ductal infiltrating carcinoma of human breast. Br J Cancer *60*, 185–192.

Minafra, S., Pucci-Minafra, I., Tomasino, R. M., and Sciarrino, S. (1984). Collagen composition in the ductal infiltrating carcinoma of the human breast. Cell Biol Intern Rep *8*, 79.

Muschler, J., and Streuli, C. H. (2010) Cell-matrix interactions in mammary gland development and breast cancer. Cold Spring Harb Perspect Biol 2, a003202.

Paget, S. (1889). The distribution of secondary growths in cancer of the breast. Lancet *1*, 571–573.

Palazzolo, G., Albanese, N.N., Di Cara, G., Gygax, D., Maria Letizia Vittorelli, M.L., and Pucci-Minafra, I. (2012). Proteomic Analysis of Exosome-like Vesicles Derived from Breast Cancer. Cells Anticancer Research *32*, 847–860

Parra, E. R., Bielecki, L. C., Ribeiro, J. M., Andrade Balsalobre, F., Teodoro, W. R., and Capelozzi, V. L. (2010). A significant direct association was found between type V and III collagen fibers and epithelial apoptosis. Clinics (Sao Paulo) *65*, 425–432.

Paszek, M. J., Zahir, N., Johnson, K. R., et al. (2005). Tensional homeostasis and the malignant phenotype. Cancer Cell *8*, 241–254.

Patarroyo, M., Tryggvason, K., and Virtanen, I. (2002). Laminin isoforms in tumor invasion, angiogenesis and metastasis. Semin Cancer Biol *12*, 197–207.

Provenzano, P. P., Inman, D. R., Eliceiri, K. W., et al. (2008). Collagen density promotes mammary tumor initiation and progression. BMC Med *6*, 11.

Pucci-Minafra, I., Albanese, N. N., Di Cara, G., Minafra, L., Marabeti, M. R., and Cancemi, P. (2008a). Breast cancer cells exhibit selective modulation induced by different collagen substrates. Connect Tissue Res *49*, 252–256.

Pucci-Minafra, I., Andriolo, M., Basiricò, L., et al. (1998). Absence of regular alpha2(I) collagen chains in colon carcinoma biopsy fragments. Carcinogenesis *19*, 575–84.

Pucci-Minafra, I., Cancemi, P. Albanese, N. N., et al. (2008b). New protein clustering of breast cancer tissue proteomics using actin content as a cellularity indicator. J Proteome Res *7*, 1412–1418.

Pucci-Minafra, I., Cancemi, P., Di Cara, G., et al. (2008c). Decorin transfection induces proteomic and phenotypic modulation in breast cancer cells 8701-BC. Connect Tissue Res *49*, 30–41.

Pucci-Minafra, I., and Luparello, C. (1991). Type V/type I collagen interactions in vitro and growth-inhibitory effect of hybrid substrates on 8701-BC carcinoma cells. J Submicrosc Cytol Pathol *23*, 67–74.

Pucci-Minafra, I., Luparello, C., Andriolo, M., Basiricò, L., Aquino, A., and Minafra, S. (1993). A new form of tumor and fetal collagen that binds laminin. Biochemistry *32*, 7421–7427.

Pucci-Minafra, I., Luparello, C., Aquino, A., et al. (1995). Of/Lb collagen promotes chemoinvasion of breast-cancer cells and directs epithelial-cell migration into granulation-tissue of experimental dermal wounds. Int J Oncol *6*, 1015–1020.

Pucci-Minafra, I., Luparello, C., Sciarrino, S., Tomasino, R. M., and Minafra, S. (1985). Quantitative determination of collagen types present in the ductal infiltrating carcinoma of human mammary gland. Cell Biol Int Rep *9*, 291–296.

Pucci-Minafra, I., Minafra, S., Tomasino, R. M., Sciarrino, S., and Tinervia, R. (1986). Collagen changes in the ductal infiltrating (scirrhous) carcinoma of the human breast. A possible role played by type I trimer collagen on the invasive growth. J Submicrosc Cytol *18*, 795–805.

Pujuguet, P., Simian, M., Liaw, J., Timpl, R., Werb, Z., and Bissell, M. J. (2000). Nidogen-1 regulates laminin-1-dependent mammary-specific gene expression. J Cell Sci *113*, 849–858.

Ricard-Blum, S., and Ballut, L. (2011). Matricryptins derived from collagens and proteoglycans. Front Biosci *16*, 674–697.

Rønnov-Jessen, L., Petersen, O. W., and Bissell, M. J. (1996). Cellular changes involved in conversion of normal to malignant breast: importance of the stromal reaction. Physiol Rev *76*, 69–125.

Rupard, J. H., Dimari, S. J., Damjanov, I., and Haralson, M. A. (1988). Synthesis of type I homotrimer collagen molecules by cultured human lung adenocarcinoma cells. Am J Pathol *133*, 316–326.

Sasaki, J., Fujisaki, H., Adachi, E., Irie, S., and Hattori, S. (2011). Delay of cell cycle progression and induction of death of cancer cells on type I collagen fibrils. Connect Tissue Res *52*, 167–77.

Schaefer, L., and Schaefer, R. M. (2010). Proteoglycans: from structural compounds to signaling molecules. Cell Tissue Res *339*, 237–246.

Schillaci, R., Luparello, C., and Minafra, S. (1989). Type I and I-trimer collagens as substrates for breast carcinoma cells in culture. Effect on growth rate, morphological appearance and actin organization. Eur J Cell Biol *48*, 135–141.

Skandalis, S. S., Labropoulou, V. T., Ravazoula, P., et al. (2011). Versican but not decorin accumulation is related to malignancy in mammographically detected high density and malignant-appearing microcalcifications in non-palpable breast carcinomas. BMC Cancer *11*, 314.

Spencer, V. A., Costes, S., Inman, J. L., et al. (2011). Depletion of nuclear actin is a key mediator of quiescence in epithelial cells. J Cell Sci *124*, 123–132.

Spiess, K., and Zorn, T. M. (2007). Collagen types I, III, and V constitute the thick collagen fibrils of the mouse decidua. Microsc Res Tech *70*, 18–25.

Streuli, C. H., Bailey, N., and Bissell, M. J. (1991). Control of mammary epithelial differentiation: basement membrane induces tissue-specific gene expression in the absence of cell–cell interaction and morphological polarity. J Cell Biol *115*, 1383–1395.

Tan, T. T., and Coussens, L. M. (2007). Humoral immunity, inflammation and cancer. Curr Opin Immunol *19*, 209–216.

Walpita, D., and Hay, E. (2002). Studying actin-dependent processes in tissue culture. Nat Rev Mol Cell Biol *3*, 137–141.

Wells, R. G. (2008). The role of matrix stiffness in regulating cell behavior. Hepatology *47*, 1394–1400.

Xu, Y., Sengupta, P. K., Seto, E., and Smith, B. D. (2006). Regulatory factor for X-box family proteins differentially interact with histone deacetylases to repress collagen alpha2(I) gene (COL1A2) expression. J Biol Chem *281*, 9260–9270.

Zou, L., Hazan, R., and Roy, P. (2009). Profilin-1 overexpression restores adherens junctions in MDA-MB-231 breast cancer cells in R-cadherin-dependent manner. Cell Motil Cytoskeleton *66*, 1048–1056.

7.3 EMMPRIN/CD147: potential functions in tumor microenvironment and therapeutic target for human cancer

Eric Huet, Eric E Gabison, Samia Mourah, and Suzanne Menashi

7.3.1 Introduction

The host microenvironment undergoes extensive change during the evolution and progression of cancer. It has recently become the focus of intense research with the understanding that the alterations that occur in the stroma around the tumor can provide important prognostic information, independent of the tumor cell phenotype, and generate new therapeutic targets. The alterations that occur in the tumor microenvironment are thought to benefit the cancer cell and to promote its progression. They are triggered by the cross talk between the tumor cells and the host stromal cells and include the recruitment and activation of neighboring fibroblasts, enhanced angiogenesis, and infiltration of inflammatory cells, thus profoundly modifying the nature of the stroma, often referred to as "reactive stroma."

The importance of tumor-stroma interactions was first recognized back in the 1990s with the realization that the stromal cells at the vicinity of tumor cells were the origin of the overexpressed matrix metalloproteinases (MMPs) often observed in cancer tissues. Stroma-derived MMPs represent important mediators of tumor stroma cooperation as they can facilitate tumor cell invasion and spread by breaking down extracellular matrix (ECM). Extracellular matrix metalloproteinase inducer (EMMPRIN) was then identified (Kataoka et al., 1993; Biswas et al., 1995) as the tumor cell factor responsible for this induction of MMPs in neighboring stromal cells through direct tumor-stromal interactions. EMMPRIN was also shown to increase expression of the plasminogen activation system, including urokinase and its receptor, hence further increasing its proteolytic potential in the stroma (Quemener et al., 2007). Other malignant properties of EMMPRIN associated with tumor cell metabolism, survival and anchorage-independent growth have been since described (Marieb et al., 2004; Gallagher et al., 2007). However, accumulating evidence increasingly suggests that EMMPRIN also has a prominent role in the alteration of tumor microenvironment. In addition to increasing protease production in the stroma, EMMPRIN was shown to promote the differentiation of fibroblasts to myofibroblasts, numerous in the tumor stroma and to increase angiogenesis, an important feature in tumor microenvironment necessary for tumor growth. These functions are described in more details in the following sections.

General features of EMMPRIN

EMMPRIN, also commonly referred to as cluster of differentiation 147 (CD147) or basigin, is a widely expressed membrane glycoprotein that belongs to the immunoglobulin (Ig) superfamily (Biswas et al., 1995). It is composed of two C2-like immunoglobulin extracellular domains, a transmembrane domain and a short cytoplasmic domain (Miyauchi et al., 1991). The first Ig domain is required for counterreceptor activity, involved in MMP induction. The extracellular region contains three conserved *N*-glycosylation sites that are variably glycosylated (Muramatsu and Miyauchi, 2003). Glycosylation was shown to determine its MMP stimulating activity, as the purified deglycosylated EMMPRIN not only failed to induce MMP activity but also antagonized the activity of native molecule (Li et al., 2001; Sun and Hemler, 2001). A stretch of 29 amino acids (aa) in the transmembrane region is completely conserved among human, mouse and chicken, indicating the importance of this region in the function of the molecule. The presence of the charged amino acid glutamic acid and leucine zipper-like sequences within the hydrophobic sequence of the transmembrane domain suggests that intramembrane interactions are likely to occur with other transmembrane proteins (Muramatsu and Miyauchi, 2003). Using tagged expression vectors and cross-linking experiments, EMMPRIN molecules were also shown to associate with each other on the plasma membrane, forming homo-oligomers in a *cis*-dependent manner, with the potential to increase the overall avidity of EMMPRIN on the cell surface (Fadool and Linser, 1996; Yoshida et al., 2000).

This ability of EMMPRIN to associate with different partners within the membrane may in fact determine its function in different biological situations. Indeed, it is clear

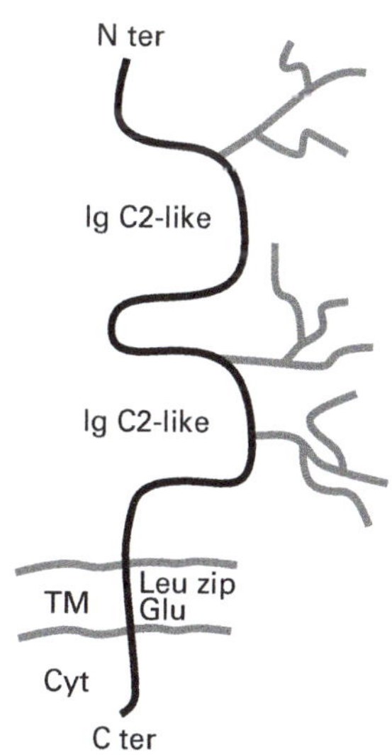

Figure 7.6 Scheme of EMMPRIN structure. EMMPRIN contains an extracellular domain composed of two Ig C2-like domains with three glycosylation sites, a short single transmembrane domain (TM) with glutamic acid residue and Leu zip domain, and a cytoplasmic domain (Cyt). The first Ig domain is required for counterreceptor activity, involved in MMP induction.

from numerous studies of various tissues and organs that EMMPRIN is multifunctional, and beyond MMPs induction, it was shown to be involved in immune cell interactions and activation (Chiampanichayakul et al., 2006), cell-cell interaction in the nervous system (Fadool and Linser, 1993), and cellular metabolism (Kirk et al., 2000). These functions may be best illustrated by the different developmental defects observed in the EMMPRIN knockout mouse. The null mutant is small and usually unable to undergo implantation. The embryos that implant survive past birth, but the offspring are sterile, demonstrating deficiencies in spermatogenesis and fertilization. In addition, these null mice are blind due to defective retinal development and have faulty sensory and memory functions (Muramatsu and Miyauchi, 2003). Experimental challenges to the adult EMMPRIN knockout mouse, which are still missing, may eventually clarify EMMPRIN's functions that are more relevant to diseases such as cancer.

EMMPRIN in pathology

As EMMPRIN was first identified in tumor cells where it is particularly enriched, its role in cancer progression through MMPs induction was more particularly emphasized (Yan et al., 2005). Immunohistochemical analysis detected EMMPRIN mainly at the periphery of invasive tumor clusters, corresponding to the leading edge of tumor invasion and compatible with the concept that EMMPRIN plays a role in tumor-stroma interaction (Caudroy et al., 1999). The assumption that EMMPRIN may be implicated in tumorigenesis has been strengthened by the demonstration of high levels of EMMPRIN compared with their normal counterparts in numerous malignant tumors including bladder, skin, lung, and breast carcinoma, and lymphoma (Polette et al., 1997; Bordador et al., 2000; Thorns et al., 2002). EMMPRIN levels were also shown to correlate with tumor progression (Sameshima et al., 2000a; Zucker et al., 2001; Kanekura et al., 2002) and in some cases was also associated with poor prognosis (Kanekura et al., 2002; Davidson et al., 2003b; Rosenthal et al., 2003; Ishibashi et al., 2004). Furthermore, EMMPRIN was found to be the most frequently upregulated mRNA and protein in micrometastatic cells isolated from the bone marrow of cancer patients supporting a key role for EMMPRIN in the processes of tumorigenesis and metastasis (Reimers et al., 2004).

Several mechanisms by which EMMPRIN promotes tumor progression have been suggested involving its different effects on both the EMMPRIN overexpressing tumor cells and on the surrounding stroma. The role of EMMPRIN in the induction of MMPs was confirmed by studies on a wide range of tumors showing that EMMPRIN expression level correlated with the degree of MMP expression by stromal fibroblasts (Davidson et al., 2003a). However, experimental studies have demonstrated that in addition to increasing invasion through proteinase induction, EMMPRIN induces several other malignant properties. These include the stimulation of angiogenesis by the upregulation vascular endothelial growth factor (VEGF) expression (Tang et al., 2005) as well as the stimulation of cell survival signaling, including AKT, extracellular signal-regulated kinase (ERK), and focal adhesion kinase (FAK), through the increased production of the pericellular polysaccharide hyaluronan (Toole and Slomiany, 2008). These effects of EMMPRIN would be further accentuated by its ability to induce its own expression by positive feedback regulatory mechanism (Tang et al., 2004).

In addition to cancer, EMMPRIN has been implicated in many other pathological processes. Its upregulation has been identified in tissues such as lung injury (Foda et al., 2001), rheumatoid arthritis (Konttinen et al., 2000), chronic liver disease (Shackel et al., 2002), heart failure (Spinale et al., 2000), corneal ulceration (Gabison et al., 2005), and atherosclerosis (Major et al., 2002). The induction of certain MMPs (e.g. MMP1, MMP2, and MMP3) as well as of urokinase plasminogen activator (uPA) has been shown to be central in the tissue destruction associated with these diseases.

7.3.2 Protease-inducing activity of EMMPRIN: role in tumor cell invasion

EMMPRIN as an inducer of MMPs

The increased MMP activity in cancerous tissues was primarily thought to be due to tumor cell production of the enzymes that aided the invasion and metastatic spread of the primary tumor. It was later realized, however, that many of the MMPs present in the tumor tissue are particularly observed at the tumor-stroma interface and are in fact expressed by the stromal cells rather than by the tumor cells themselves. A search for an MMP-inducing factor in tumor cell led to the identification of EMMPRIN as the tumor cell membrane glycoprotein responsible for stimulating MMP synthesis in neighboring fibroblasts (Kataoka et al., 1993). It was shown to stimulate the production of several matrix metalloproteinases, MMP1 (interstitial collagenase), MMP2 (gelatinase A), MMP3 (stromelysin 1), and MMP9 (gelatinase B), but has no effect on their physiological inhibitors, tissue inhibitor of metalloproteinases-1 (TIMP1) and tissue inhibitor of metalloproteinases-2 (TIMP2), hence increasing the proteolytic potential of the cell (Toole, 2003). The stimulation of the membrane type MMPs (MMP14 and MMP15) leading to an increased activation of MMP2 was also reported (Sameshima et al., 2000b). However, the role of EMMPRIN in inducing MMP production is not limited to fibroblasts. EMMPRIN had also paracrine effect on MMP production by endothelial cells (Caudroy et al., 2002; Menashi et al., 2003), suggesting the potential implication of tumor-produced EMMPRIN in angiogenesis. Beyond its well-known paracrine effect, EMMPRIN was shown to induce MMPs within the same populations of cells, in both tumoral and nontumoral cells, suggesting a more widespread role for EMMPRIN as an inducer of MMPs.

The increased synthesis of MMPs in fibroblasts by EMMPRIN occurs at the transcription level and is at least in part mediated by a mitogen-activated protein kinase (MAPK) p38 signaling pathway as MMP1 stimulation was blocked by the p38 inhibitor SB203580 (Lim et al., 1998). In another study, however, the induction of MMP2 was shown to be mediated by activation of phospholipase A2 and 5-lipoxygenase (Taylor et al., 2002). Whether EMMPRIN utilizes distinct signaling pathways in the regulation of different MMPs remains to be determined. It is possible that several distinct pathways can be stimulated by EMMPRIN depending on the cell system.

EMMPRIN as an inducer of urokinase plasminogen activator

Among the other proteases shown to contribute to matrix degradation and invasion, the uPA system, which includes the protease urokinase, its specific receptor (uPA receptor

[uPAR]), and its primary inhibitor (plasminogen activator inhibitor-1 [PAI1]) has been shown to be a key actor (Dano et al., 2005). uPA catalyzes the conversion of the zymogen plasminogen to the active broad-spectrum plasmin, which degrades a number of extracellular matrix proteins and also activates other proteases including some MMPs. Immunolocalization of this proteolytic system in human breast cancer tissues by in situ hybridization and immunohistochemistry showed that like the MMPs, the mRNA and protein of uPA and uPAR were also expressed by the stromal cells surrounding the tumors (Frandsen et al., 2001). The demonstration that EMMPRIN is able to upregulate both uPA and uPAR (Quemener et al., 2007) suggests that EMMPRIN may be responsible for the localization of this protease system in the tumor-stroma invasive front. It also implies that EMMPRIN has a larger proteolytic function than previously thought, beyond MMP inducing activity, and that the induction of the uPA system represents an additional degradation pathway enhancing its tumor invasion potential. Furthermore, the inoculation of EMMPRIN overexpressing breast tumor cells to nude mice produced bigger tumors containing higher levels of not only MMPs but also uPA, supporting the role of EMMPRIN as a major regulator of both systems. Whether the signaling events downstream of EMMPRIN are common for MMP and uPA stimulation is yet to be established. It is of interest that MAPK p38, which has been implicated in EMMPRIN mediated induction of MMP1 production, also regulates uPA expression as well as the stability of uPA and uPAR mRNA (Montero and Nagamine, 1999) and it may therefore represent a common pathway for both protease systems in response to EMMPRIN.

7.3.3 Role of EMMPRIN in myofibroblast differentiation

Myofibroblasts are differentiated host fibroblasts that have been traditionally studied in the context of tissue repair, as they are known to regulate matrix remodeling during normal and pathological wound healing (Gabbiani, 2003). However, differentiated myofibroblasts are also found in the stroma of different epithelial tumors and are thought to play a central role in tumor associated tissue remodeling (De Wever et al., 2008). Several studies suggested that these cells facilitate tumorigenesis and progression of carcinomas of the prostate, breast, and keratinocytes (Camps et al., 1990; Olumi et al., 1999; Yamashita et al., 2012). Their numbers in the tumor stroma has been shown to increase with disease progression suggesting the clinical value of myofibroblasts as a potentially important marker with respect to diagnosis, treatment, and prognosis of cancer (Tsujino et al., 2007; Yamashita et al., 2012).

The differentiation of fibroblast to myofibroblast can be identified by the expression of alpha-smooth muscle actin (α-SMA). This isoform of actin is typical of vascular smooth muscle cells but also characterizes the myofibroblasts both *in vitro* and *in vivo* and is accountable for the contractile activities of the myofibroblasts (Jester et al., 1999; Jester and Ho-Chang, 2003). Among the cytokines, transforming growth factor-beta (TGF-β) is generally considered to be the major actor in the differentiation of fibroblasts into myofibroblasts (Desmouliere et al., 2005).

EMMPRIN was recently shown to actively participate in the process of myofibroblasts differentiation (Huet et al., 2008). This was demonstrated by different experimental approaches. Increasing EMMPRIN expression by cDNA transfection or by treatment with exogenously added recombinant EMMPRIN resulted in an upregulation of α-SMA expression. EMMPRIN also increased the contractile properties of the treated fibroblasts as demonstrated by the immunohistochemical appearance of stress fibers and by the accelerated contraction of fibroblasts embedded collagen lattices. Furthermore, EMMPRIN and α-SMA colocalized to the same cells in the stroma of pathological tissues, thus supporting a role for EMMPRIN also in the differentiation of myofibroblasts *in vivo*. This effect of EMMPRIN on myofibroblast differentiation was supported by the observed inhibitory effect of EMMPRIN-siRNA on TGF-β induced α-SMA expression and collagen gel contraction (Huet et al., 2008), further suggesting that EMMPRIN mediates some of TGF-β effects in the differentiation process.

The mechanism by which EMMPRIN promotes myofibroblast differentiation remains to be elucidated. However, the possibility that EMMPRIN's effect may involve an MMP-dependent activation of latent TGF-β, leading in turn to an active TGFβ-mediated α-SMA induction was ruled out, as the presence of MMPs inhibitors had no effect, further implying that this function of EMMPRIN in myofibroblast differentiation is unrelated to its MMP inducing activity.

7.3.4 Role of EMMPRIN in angiogenesis

Angiogenesis, which is an important feature of tumoral stroma, is essential for the primary and metastatic growth of tumors, and has become a target for anticancer therapy. The ability of tumors to recruit endothelial cells and stimulate their proliferation, migration, or survival is thought to be central in tumor-induced angiogenesis. Much attention has been focused on the VEGF) family of growth factors and the receptor tyrosine kinases (vascular endothelial growth factor receptor-1 [VEGFR1] and vascular endothelial growth factor receptor [VEGFR2]), which appear to be the key mediators of tumor angiogenesis. Indeed, upregulation of VEGF has been observed in many human tumors, and VEGF expression is closely correlated with tumor progression and less favorable prognosis (Carmeliet, 2005).

The potential role of EMMPRIN in tumor angiogenesis was initially evoked by the observations of greatly enhanced angiogenesis in EMMPRIN overexpressing tumors (Menashi et al., 2003). EMMPRIN may regulate angiogenesis through the regulation of proteases needed to degrade the basement membrane of the original vessel and remodel the ECM around neovasculature sites, and both MMPs and uPA have been largely implicated in this process. More recently, however, EMMPRIN was also shown to mediate angiogenesis in a protease independent manner by directly regulating the VEGF/VEGFR system. EMMPRIN can increase VEGF transcriptional expression in tumor cells via the phosphatidylinositol 3-kinase/AKT signaling pathway (Tang et al., 2005). Tumor derived VEGF can then activate the VEGF receptors on endothelial cells in a paracrine manner. EMMPRIN can also upregulate the expression of VEGF as well as its receptor VEGFR2 directly in endothelial cells and as a consequence increases both migration and tube formation (Bougatef et al., 2009). The effect of EMMPRIN on VEGF production in tumor cells (Tang et al., 2005) and in endothelial cells (Bougatef et al., 2009) would be expected to increase the pool of VEGF in the tumoral tissue leading to increase in angiogenesis. Indeed, EMMPRIN overexpression in breast tumor cells stimulated tumor angiogenesis and growth *in vivo*; both were significantly inhibited by antisense suppression of EMMPRIN (Tang et al., 2005). Knockdown of EMMPRIN in B16 melanoma model caused reduced VEGF production *in vivo* accompanied by reduced blood vessel formation (Voigt et al., 2009). Furthermore, a correlation between EMMPRIN and microvessel density was found in salivary gland tumors from human patients (Huang et al., 2010). These data indicate that EMMPRIN plays an important role in cancer angiogenesis.

7.3.5 Shedding of EMMPRIN

The effect of tumoral EMMPRIN on endothelial cells raises an important question of proximity considering that EMMPRIN is an adhesion glycoprotein localized on tumor cell surface and therefore thought to act through direct cell-cell interaction. However, the fact that EMMPRIN was also shown to be released from the cell surface as a soluble form may suggest that it can also exert its effect at a distance. This would be in

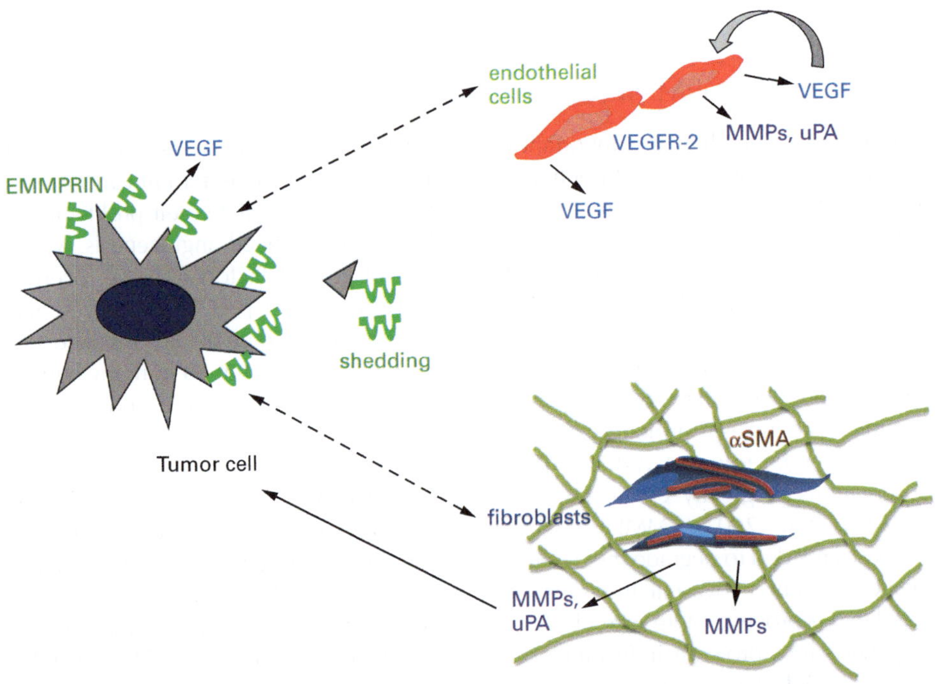

Figure 7.7 EMMPRIN functions in tumor microenvironment. EMMPRIN is enriched on tumor cell surface and can interact with stromal cells either directly or through its shedding. Active form of EMMPRIN release may occur after proteolytic cleavage or by shedding of microvesicles. Tumoral EMMPRIN can activate resident fibroblasts promoting their differentiation into myofibroblasts as evidenced by the induction of α-SMA. These activated fibroblasts then greatly increase secretion of MMPs and uPA, proteases necessary for tumor cells migration and spreading as they can break down extracellular matrix. EMMPRIN can also promote angiogenesis through the production of VEGF by both the tumor cells themselves and endothelial cells. Furthermore, EMMPRIN increases in a paracrine manner the expression of endothelial cells VEGFR2 favoring their proliferation, migration, and new tube formation.

agreement with the histopathological findings that stromal cells that are not in direct contact with EMMPRIN-expressing cells can still upregulate MMPs in or around the tumor (Kanekura et al., 2002; Nabeshima et al., 2006).

Two mechanisms for the generation of soluble form of EMMPRIN have been demonstrated in culture systems. Membrane microvesicle shedding, which is an active process in many tumor cells, was shown to account for a small proportion (approximately 2%) of the total full-length cellular EMMPRIN but was greatly accelerated by phorbol 12-myristate 13-acetate (PMA), the tumor promoter known to stimulate pathways involving the protein kinase C (PKC), Ca^{2+}, and ERK1/2 (Sidhu et al., 2004). Metalloproteinase-dependent EMMPRIN proteolytic shedding was also described, releasing a soluble form of EMMPRIN that lacked the intracellular carboxyl terminus (Tang et al., 2004). It may be that in addition to the low levels of full-length EMMPRIN released by tumor cell through vesicular shedding, EMMPRIN cleavage can also occur in situations where MMP activity is sufficiently increased, such as during tumor-stroma interactions. Significantly, circulating tumor-derived microvesicles isolated from the plasma of gastric cancer patients were shown to be enriched with EMMPRIN (Baran et al., 2010). Whatever the mechanism of EMMPRIN release, the fact that both the soluble full-length or the cleaved extracellular domain are functionally active suggests that soluble EMMPRIN may well be of profound biological importance as it would be able to exert its effect on fibroblasts or endothelial cells at distant sites to promote myofibroblasts differentiation or angiogenesis.

7.3.6 EMMPRIN as a therapeutic target for human cancer

Approaches targeting the tumor stroma attract increasing attention as anticancer therapy. To date, inhibitors to molecules contributing to angiogenesis (e.g. VEGF and its receptors) have been tested in preclinical or clinical studies with considerable success (Backer et al., 2009). Proteins involved in remodeling of the extracellular matrix (e.g. MMPs, uPA/PAR system) have also been silenced or functionally inhibited. However, the inhibition of one factor or even a group of molecules is often not sufficient to significantly influence tumor growth and progression. Inhibiting EMMPRIN represents a potential alternative strategy providing the possibility to target simultaneously both the tumor cells and the tumor microenvironment.

As EMMPRIN is overexpressed in most cancers, preclinical studies so far employed either siRNA or monoclonal antibody (mAb)–based approaches to inhibit its cell surface expression, in association or not with conventional antitumoral therapy, and have generally shown reduced tumor growth (Weidle et al., 2010). A chimeric EMMPRIN antibody, CNTO 3899, was evaluated for the treatment of head and neck squamous cell carcinoma (HNSCC) (Dean et al., 2009) and was shown to inhibit tumor growth in nude mice. The mechanism of action of this mAb was shown to involve the inhibition of cytokines, MMPs, and VEGF. Licatrin, a ^{131}I-labeled $F(ab')_2$ fragment of EMMPRIN mAb was more particularly evaluated for hepatocellular carcinoma (HCC) and was shown to reduce growth and metastasis as well as the expression of stromal factors such as MMPs, VEGF, and fibroblast surface protein (Xu et al., 2007b). In a clinical study, combining Licatrin and radiotherapy was shown very effective for HCC with more than 80% of patients showing a progression-free disease after completing two cycles in a phase II trial (Chen et al., 2006). In another trial, Licartin was used for prevention of recurrence of hepatoma after liver transplantation. The recurrence rate was reduced by 30%, and survival rate increased by 21% (Xu et al., 2007a). This anti-EMMPRIN antibody was well tolerated with only limited toxicity observed, warranting future trial in different types of cancer. The benefit of such anti-EMMPRIN strategies is yet to be demonstrated and confirmed in further studies in different types of cancer.

7.3.7 Take-home message

The tumor microenvironment undergoes extensive changes during the evolution and progression of cancer in a way that benefits the cancer cells. In addition to promoting the malignant properties of the tumor cells, EMMPRIN may have a prominent role in the alteration of the tumor microenvironment. This makes it a particularly attractive as a target to pursue in the development of anticancer treatments.

References

Backer, M. V., Hamby, C. V., and Backer, J. M. (2009). Inhibition of vascular endothelial growth factor receptor signaling in angiogenic tumor vasculature. Adv Genet *67*, 1–27.

Baran, J., Baj-Krzyworzeka, M., Weglarczyk, K., et al. (2010). Circulating tumour-derived microvesicles in plasma of gastric cancer patients. Cancer Immunol Immunother *59*, 841–850.

Biswas, C., Zhang, Y., DeCastro, R., et al. (1995). The human tumor cell-derived collagenase stimulatory factor (renamed EMMPRIN) is a member of the immunoglobulin superfamily. Cancer Res *55*, 434–439.

Bordador, L. C., Li, X., Toole, B., et al. (2000). Expression of emmprin by oral squamous cell carcinoma. Int J Cancer *85*, 347–352.

Bougatef, F., Quemener, C., Kellouche, S., et al. (2009). EMMPRIN promotes angiogenesis through hypoxia-inducible factor-2alpha-mediated regulation of soluble VEGF isoforms and their receptor VEGFR-2. Blood *114*, 5547–5556.

Camps, J. L., Chang, S. M., Hsu, T. C., et al. (1990). Fibroblast-mediated acceleration of human epithelial tumor growth in vivo. Proc Natl Acad Sci U S A *87*, 75–79.

Carmeliet, P. (2005). VEGF as a key mediator of angiogenesis in cancer. Oncology *69* (Suppl. 3), 4–10.

Caudroy, S., Polette, M., Nawrocki-Raby, B., et al. (2002). EMMPRIN-mediated MMP regulation in tumor and endothelial cells. Clin Exp Metastasis *19*, 697–702.

Caudroy, S., Polette, M., Tournier, J. M., et al. (1999). Expression of the extracellular matrix metalloproteinase inducer (EMMPRIN) and the matrix metalloproteinase-2 in bronchopulmonary and breast lesions. J Histochem Cytochem *47*, 1575–1580.

Chen, Z. N., Mi, L., Xu, J., et al. (2006). Targeting radioimmunotherapy of hepatocellular carcinoma with iodine (131I) metuximab injection: clinical phase I/II trials. Int J Radiat Oncol Biol Phys *65*, 435–444.

Chiampanichayakul, S., Peng-in, P., Khunkaewla, P., Stockinger, H., and Kasinrerk, W. (2006). CD147 contains different bioactive epitopes involving the regulation of cell adhesion and lymphocyte activation. Immunobiology *211*, 167–178.

Dano, K., Behrendt, N., Hoyer-Hansen, G., et al. (2005). Plasminogen activation and cancer. Thromb Haemost *93*, 676–681.

Davidson, B., Givant-Horwitz, V., Lazarovici, P., et al. (2003a). Matrix metalloproteinases (MMP), EMMPRIN (extracellular matrix metalloproteinase inducer) and mitogen-activated protein kinases (MAPK): co-expression in metastatic serous ovarian carcinoma. Clin Exp Metastasis *20*, 621–631.

Davidson, B., Goldberg, I., Berner, A., Kristensen, G. B., and Reich, R. (2003b). EMMPRIN (extracellular matrix metalloproteinase inducer) is a novel marker of poor outcome in serous ovarian carcinoma. Clin Exp Metastasis *20*, 161–169.

Dean, N. R., Newman, J. R., Helman, E. E., et al. (2009). Anti-EMMPRIN monoclonal antibody as a novel agent for therapy of head and neck cancer. Clin Cancer Res *15*, 4058–4065.

Desmouliere, A., Chaponnier, C., and Gabbiani, G. (2005). Tissue repair, contraction, and the myofibroblast. Wound Repair Regen *13*, 7–12.

De Wever, O., Demetter, P., Mareel, M., and Bracke, M. (2008). Stromal myofibroblasts are drivers of invasive cancer growth. Int J Cancer *123*, 2229–2238.

Fadool, J. M., and Linser, P. J. (1993). 5A11 antigen is a cell recognition molecule which is involved in neuronal-glial interactions in avian neural retina. Dev Dyn *196*, 252–262.

Fadool, J. M., and Linser, P. J. (1996). Evidence for the formation of multimeric forms of the 5A11/HT7 antigen. Biochem Biophys Res Commun *229*, 280–286.

Foda, H. D., Rollo, E. E., Drews, M., et al. (2001). Ventilator-induced lung injury upregulates and activates gelatinases and EMMPRIN: attenuation by the synthetic matrix metalloproteinase inhibitor, Prinomastat (AG3340). Am J Respir Cell Mol Biol *25*, 717–724.

Frandsen, T. L., Holst-Hansen, C., Nielsen, B. S., et al. (2001). Direct evidence of the importance of stromal urokinase plasminogen activator (uPA) in the growth of an experimental human breast cancer using a combined uPA gene-disrupted and immunodeficient xenograft model. Cancer Res *61*, 532–537.

Gabbiani, G. (2003). The myofibroblast in wound healing and fibrocontractive diseases. J Pathol *200*, 500–503.

Gabison, E. E., Mourah, S., Steinfels, E., et al. (2005). Differential expression of extracellular matrix metalloproteinase inducer (CD147) in normal and ulcerated corneas: role in epithelio-stromal interactions and matrix metalloproteinase induction. Am J Pathol *166*, 209–219.

Gallagher, S. M., Castorino, J. J., Wang, D., and Philp, N. J. (2007). Monocarboxylate transporter 4 regulates maturation and trafficking of CD147 to the plasma membrane in the metastatic breast cancer cell line MDA-MB-231. Cancer Res *67*, 4182–4189.

Huang, Z. Q., Chen, W. L., Li, H. G., Li, J. S., Xu, Z. Y., and Lin, Z. Y. (2010). Extracellular matrix metalloproteinase inducer expression in salivary gland tumors: a correlation with microvessel density. J Craniofac Surg *21*, 1855–1860.

Huet, E., Vallee, B., Szul, D., et al. (2008). Extracellular matrix metalloproteinase inducer/CD147 promotes myofibroblast differentiation by inducing alpha-smooth muscle actin expression and collagen gel contraction: implications in tissue remodeling. FASEB J *22*, 1144–1154.

Ishibashi, Y., Matsumoto, T., Niwa, M., et al. (2004). CD147 and matrix metalloproteinase-2 protein expression as significant prognostic factors in esophageal squamous cell carcinoma. Cancer *101*, 1994–2000.

Jester, J. V., and Ho-Chang, J. (2003). Modulation of cultured corneal keratocyte phenotype by growth factors/cytokines control in vitro contractility and extracellular matrix contraction. Exp Eye Res *77*, 581–592.

Jester, J. V., Petroll, W. M., and Cavanagh, H. D. (1999). Corneal stromal wound healing in refractive surgery: the role of myofibroblasts. Prog Retin Eye Res *18*, 311–356.

Kanekura, T., Chen, X., and Kanzaki, T. (2002). Basigin (CD147) is expressed on melanoma cells and induces tumor cell invasion by stimulating production of matrix metalloproteinases by fibroblasts. Int J Cancer *99*, 520–528.

Kataoka, H., DeCastro, R., Zucker, S., and Biswas, C. (1993). Tumor cell-derived collagenase-stimulatory factor increases expression of interstitial collagenase, stromelysin, and 72-kDa gelatinase. Cancer Res *53*, 3154–3158.

Kirk, P., Wilson, M. C., Heddle, C., Brown, M. H., Barclay, A. N., and Halestrap, A. P. (2000). CD147 is tightly associated with lactate transporters MCT1 and MCT4 and facilitates their cell surface expression. EMBO J *19*, 3896–3904.

Konttinen, Y. T., Li, T. F., Mandelin, J., et al. (2000). Increased expression of extracellular matrix metalloproteinase inducer in rheumatoid synovium. Arthritis Rheum *43*, 275–280.

Li, R., Huang, L., Guo, H., and Toole, B. P. (2001). Basigin (murine EMMPRIN) stimulates matrix metalloproteinase production by fibroblasts. J Cell Physiol *186*, 371–379.

Lim, M., Martinez, T., Jablons, D., et al. (1998). Tumor-derived EMMPRIN (extracellular matrix metalloproteinase inducer) stimulates collagenase transcription through MAPK p38. FEBS Lett *441*, 88–92.

Major, T. C., Liang, L., Lu, X., Rosebury, W., and Bocan, T. M. (2002). Extracellular matrix metalloproteinase inducer (EMMPRIN) is induced upon monocyte differentiation and is expressed in human atheroma. Arterioscler Thromb Vasc Biol *22*, 1200–1207.

Marieb, E. A., Zoltan-Jones, A., Li, R., et al. (2004). Emmprin promotes anchorage-independent growth in human mammary carcinoma cells by stimulating hyaluronan production. Cancer Res *64*, 1229–1232.

Menashi, S., Serova, M., Ma, L., Vignot, S., Mourah, S., and Calvo, F. (2003). Regulation of extracellular matrix metalloproteinase inducer and matrix metalloproteinase expression by amphiregulin in transformed human breast epithelial cells. Cancer Res *63*, 7575–7580.

Miyauchi, T., Masuzawa, Y., and Muramatsu, T. (1991). The basigin group of the immunoglobulin superfamily: complete conservation of a segment in and around transmembrane domains of human and mouse basigin and chicken HT7 antigen. J Biochem (Tokyo) *110*, 770–774.

Montero, L., and Nagamine, Y. (1999). Regulation by p38 mitogen-activated protein kinase of adenylate- and uridylate-rich element-mediated urokinase-type plasminogen activator (uPA) messenger RNA stability and uPA-dependent in vitro cell invasion. Cancer Res *59*, 5286–5293.

Muramatsu, T., and Miyauchi, T. (2003). Basigin (CD147): a multifunctional transmembrane protein involved in reproduction, neural function, inflammation and tumor invasion. Histol Histopathol *18*, 981–987.

Nabeshima, K., Iwasaki, H., Koga, K., Hojo, H., Suzumiya, J., and Kikuchi, M. (2006). Emmprin (basigin/CD147): matrix metalloproteinase modulator and multifunctional cell recognition molecule that plays a critical role in cancer progression. Pathol Int *56*, 359–367.

Olumi, A. F., Grossfeld, G. D., Hayward, S. W., Carroll, P. R., Tlsty, T. D., and Cunha, G. R. (1999). Carcinoma-associated fibroblasts direct tumor progression of initiated human prostatic epithelium. Cancer Res *59*, 5002–5011.

Polette, M., Gilles, C., Marchand, V., et al. (1997). Tumor collagenase stimulatory factor (TCSF) expression and localization in human lung and breast cancers. J Histochem Cytochem *45*, 703–709.

Quemener, C., Gabison, E. E., Naimi, B., et al. (2007). Extracellular matrix metalloproteinase inducer up-regulates the urokinase-type plasminogen activator system promoting tumor cell invasion. Cancer Res *67*, 9–15.

Reimers, N., Zafrakas, K., Assmann, V., et al. (2004). Expression of extracellular matrix metalloproteases inducer on micrometastatic and primary mammary carcinoma cells. Clin Cancer Res *10*, 3422–3428.

Rosenthal, E. L., Shreenivas, S., Peters, G. E., Grizzle, W. E., Desmond, R., and Gladson, C. L. (2003). Expression of extracellular matrix metalloprotease inducer in laryngeal squamous cell carcinoma. Laryngoscope *113*, 1406–1410.

Sameshima, T., Nabeshima, K., Toole, B. P., et al. (2000a). Expression of emmprin (CD147), a cell surface inducer of matrix metalloproteinases, in normal human brain and gliomas. Int J Cancer *88*, 21–27.

Sameshima, T., Nabeshima, K., Toole, B. P., et al. (2000b). Glioma cell extracellular matrix metalloproteinase inducer (EMMPRIN) (CD147) stimulates production of membrane-type matrix metalloproteinases and activated gelatinase A in co-cultures with brain-derived fibroblasts. Cancer Lett *157*, 177–184.

Shackel, N. A., McGuinness, P. H., Abbott, C. A., Gorrell, M. D., and McCaughan, G. W. (2002). Insights into the pathobiology of hepatitis C virus-associated cirrhosis: analysis of intrahepatic differential gene expression. Am J Pathol *160*, 641–654.

Sidhu, S. S., Mengistab, A. T., Tauscher, A. N., LaVail, J., and Basbaum, C. (2004). The microvesicle as a vehicle for EMMPRIN in tumor-stromal interactions. Oncogene *23*, 956–963.

Spinale, F. G., Coker, M. L., Heung, L. J., et al. (2000). A matrix metalloproteinase induction/activation system exists in the human left ventricular myocardium and is upregulated in heart failure. Circulation *102*, 1944–1949.

Sun, J., and Hemler, M. E. (2001). Regulation of MMP-1 and MMP-2 production through CD147/extracellular matrix metalloproteinase inducer interactions. Cancer Res *61*, 2276–2281.

Tang, Y., Kesavan, P., Nakada, M. T., and Yan, L. (2004). Tumor-stroma interaction: positive feedback regulation of extracellular matrix metalloproteinase inducer (EMMPRIN) expression and matrix metalloproteinase-dependent generation of soluble EMMPRIN. Mol Cancer Res *2*, 73–80.

Tang, Y., Nakada, M. T., Kesavan, P., et al.. (2005). Extracellular matrix metalloproteinase inducer stimulates tumor angiogenesis by elevating vascular endothelial cell growth factor and matrix metalloproteinases. Cancer Res *65*, 3193–3199.

Taylor, P. M., Woodfield, R. J., Hodgkin, M. N., et al. (2002). Breast cancer cell-derived EMMPRIN stimulates fibroblast MMP2 release through a phospholipase A(2) and 5-lipoxygenase catalyzed pathway. Oncogene *21*, 5765–5772.

Thorns, C., Feller, A. C., and Merz, H. (2002). EMMPRIN (CD 174) is expressed in Hodgkin's lymphoma and anaplastic large cell lymphoma. An immunohistochemical study of 60 cases. Anticancer Res *22*, 1983–1986.

Toole, B. P. (2003). Emmprin (CD147), a cell surface regulator of matrix metalloproteinase production and function. Curr Top Dev Biol *54*, 371–389.

Toole, B. P., and Slomiany, M. G. (2008). Hyaluronan, CD44 and emmprin: partners in cancer cell chemoresistance. Drug Resist Updat *11*, 110–121.

Tsujino, T., Seshimo, I., Yamamoto, H., et al. (2007). Stromal myofibroblasts predict disease recurrence for colorectal cancer. Clin Cancer Res *13*, 2082–2090.

Voigt, H., Vetter-Kauczok, C. S., Schrama, D., Hofmann, U. B., Becker, J. C., and Houben, R. (2009). CD147 impacts angiogenesis and metastasis formation. Cancer Invest *27*, 329–333.

Weidle, U. H., Scheuer, W., Eggle, D., Klostermann, S., and Stockinger, H. (2010). Cancer-related issues of CD147. Cancer Genomics Proteomics *7*, 157–169.

Xu, J., Shen, Z. Y., Chen, X. G., et al. (2007a). A randomized controlled trial of Licartin for preventing hepatoma recurrence after liver transplantation. Hepatology *45*, 269–276.

Xu, J., Xu, H. Y., Zhang, Q., et al. (2007b). HAb18G/CD147 functions in invasion and metastasis of hepatocellular carcinoma. Mol Cancer Res *5*, 605–614.

Yamashita, M., Ogawa, T., Zhang, X., et al. (2012). Role of stromal myofibroblasts in invasive breast cancer: stromal expression of alpha-smooth muscle actin correlates with worse clinical outcome. Breast Cancer *19* (2), 170–176.

Yan, L., Zucker, S., and Toole, B. P. (2005). Roles of the multifunctional glycoprotein, emmprin (basigin; CD147), in tumour progression. Thromb Haemost *93*, 199–204.

Yoshida, S., Shibata, M., Yamamoto, S., et al. (2000). Homo-oligomer formation by basigin, an immunoglobulin superfamily member, via its N-terminal immunoglobulin domain. Eur J Biochem *267*, 4372–4380.

Zucker, S., Hymowitz, M., Rollo, E. E., et al. (2001). Tumorigenic potential of extracellular matrix metalloproteinase inducer. Am J Pathol *158*, 1921–1928.

7.4 Implication of hyaluronidases in cancer growth, metastasis, diagnosis, and treatment

Irene-Eva Triantaphyllidou, Serafoula Filou, Helen Bouga, Constantine Kolliopoulos, and Demitrios H. Vynios

7.4.1 Introduction

Hyaluronidases (Hyals) are present in several toxins and venoms, helping their spread in the body. Testicular Hyal, present in mammals in the sperm acrosome, is necessary for the fertilization of the ovum. Although bacterial, invertebrate and testicular Hyals have been extensively studied, a connection between Hyals and cancer was clearly established, and the functional significance of Hyals in cancer was demonstrated in recent years (Jacobson et al., 2002; Simpson and Lokeshwar, 2008).

Hyals are a class of enzymes that predominantly degrade hyaluronan (HA), although they can also degrade chondroitin sulfate and chondroitin, albeit at a slower rate. Hyals are endoglycosidases, as they degrade the β-*N*-acetyl-D-glucosaminidic linkages in the HA polymer. Six Hyal genes are present in the human genome and these occur in two linked triads. *HYAL1*, *2*, and *3* genes are clustered in the chromosome 3p21.3 locus (▶Figure 7.8A), whereas *HYAL4*, *HYALP1*, and *PH-20* (encodes testicular Hyal) reside in the chromosome 7q31.3 locus. It seems that the six mammalian Hyal genes have arisen through gene duplication events, since they share a significant amino acid identity (Stern and Jedrzejas, 2006). Based on their pH activity profiles, Hyals are divided into two categories. HYAL1, 2, and 3 are acidic Hyals as they are active at acidic pH and PH-20 is a neutral active Hyal as it has an activity profile in a broader pH range 3.0–9.0.

The well-characterized mammalian Hyals are HYAL1, 2 and PH-20. The earliest known is PH-20, which is necessary for ovum fertilization. It also exists in other normal tissues and in malignancy as well. PH-20 and HYAL2 are glycosylphosphatidylinositol (GPI)-linked proteins. HYAL2 degrades HA into 20 kDa oligosaccharides (50 disaccharide units), suggesting a significant role of this isoform in many biological events. HYAL1 is the serum Hyal and is expressed in several somatic tissues, with the highest expression in liver followed by kidney, spleen and heart. It is also expressed in lung and placenta at very low levels. In humans, two transcripts at 2.4 and 3.0 kb are detected, that result from alternative splicing of the 5′ untranslated region of *HYAL1* gene (▶Figure 7.8B). HYAL1 has also been purified from human urine, where it is expressed as two molecular forms. Although HYAL1 has high specific activity for degrading HA, its concentration in human serum is low (60 ng/mL) (Stern and Jedrzejas, 2006).

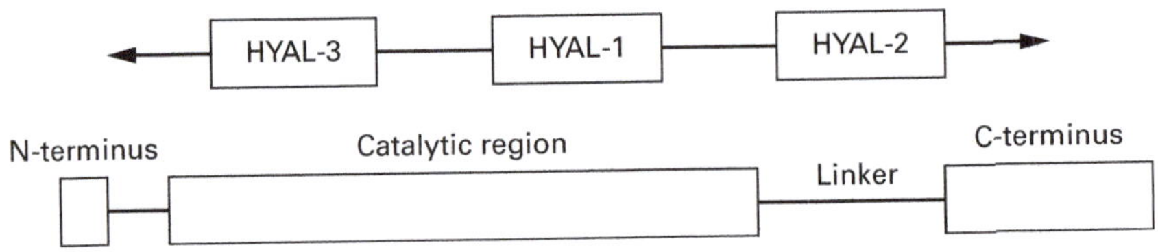

Figure 7.8 (A) Organization of *HYAL1*, *2*, and *3* genes in chromosome 3. (B) Organization of HYAL1 domains.

7.4.2 Hyaluronidases in cancer

HA has an extraordinarily high turnover rate in vertebrates. Moreover, HA deposition and turnover is even more abundant and rapid in malignant tissues. The proportion of low molecular weight fragments of HA is greater in tumors and tumor patients than in the normal (Kumar et al., 1989; Papadas et al. 2002). While ROS-directed degradation of HA is considered to be the major pathway to generate HA fragments in inflammatory and autoimmune diseases, Hyal-mediated degradation is often involved in tumor progression (Stern et al., 2006).

Measurement of Hyal activity became possible because of an enzyme linked immunosorbent assay (ELISA)-like assay (Stern and Stern, 1992) used to determine the Hyal levels in prostate and bladder carcinoma tissues, cells, and in the urine of bladder cancer patients (Simpson and Lokeshwar, 2008). Notably, the observation that the Hyal levels are elevated in prostate cancer tissues as compared to normal prostate and benign prostatic hyperplasia tissues linked Hyal levels to tumor progression for first time. In addition, Hyal levels correlate with prostate cancer grade and metastatic prostate cancer lesions have even higher Hyal levels than the high-grade primary tumor. Hyal levels are also elevated in high-grade bladder tumor tissues and in the urine of patients with high-grade bladder cancer. However, in low-grade bladder tumor tissues and urine the levels are comparable to those found in normal bladder tissues and urine. These data established a link between Hyal and the tumor invasive/metastatic phenotype. In addition to genitourinary tumors, Hyal levels are elevated in head and neck squamous cell carcinoma (HNSCC), colon cancer, breast tumors, metastatic tumors, and glioma cells.

The major tumor-derived Hyal expressed in prostate and bladder cancer is HYAL1, a ~55–60 kDa protein consisting of 435 amino acids. This Hyal is expressed by tumor cells and correlates with their invasive/metastatic potential (Simpson and Lokeshwar, 2008). No HYAL1 expression is observed in the tumor-associated stroma, although HYAL1 expression appears to correlate and perhaps induce HA production by the tumor associated stroma. Patients with HNSCC present elevated Hyal levels in their saliva, and HYAL1 is the major Hyal that is expressed in these tumor tissues (Franzmann et al., 2003). In addition to HYAL1, reverse transcriptase polymerase chain reaction (RT-PCR) analysis has revealed PH-20 expression in HNSCC, especially laryngeal carcinoma (Victor et al., 1999; Godin et al., 2000; Christopoulos et al., 2006), and colon cancer (Bouga et al., 2010). Hyal levels are also elevated in breast tumors. Moreover, PH-20, HYAL2, and HYAL3 are expressed in breast cancer. Similarly to prostate and bladder carcinomas, Hyal levels in metastatic breast tumors are higher than those expressed in primary tumors (Bertrand et al., 1997). Moreover, HYAL1 overexpression correlates with the malignant behavior of human breast cancer (Tan et al., 2011). Estrogen receptor-negative, ER(–), breast cancer cells found to secrete significantly more Hyal than ER(+) cells (Wang et al., 2009). Similarly, Hyal levels were higher in brain metastatic lesions of carcinomas other than primary glioblastomas (Delpech et al., 2002). Given that functionally inactive splice variants of HYAL1 and HYAL3 are previously reported, the expression of *HYAL* genes at the transcript level does not necessarily translate into Hyal activity.

Contrary to the previous discussion, it has been shown that the chromosome locus 3p21.3, where *HYAL1*, *-2*, and *-3* genes are clustered, is deleted at the DNA level in lung and some breast carcinomas at a higher frequency; however, the tumor suppressor gene in this region is Ras association domain-containing protein 1 (*RASSF1*). HYAL1 is also inactivated in many tobacco-related HNSCC by aberrant splicing of the mRNA. In addition, HYAL2 expression correlates with lymphoma diagnosis, but the expression actually decreases in high-grade lymphomas, when compared to low-grade lymphomas.

Involvement of hyaluronidases in tumor progression

Extensive digestion of HA by Hyal generates tetrasaccharides, whereas limited digestion generates variously sized HA fragments, some of which are angiogenic (3–25 disaccharide units). HA oligomers of 10–15 disaccharide units stimulate endothelial cell proliferation, adhesion and capillary formation (Lokeshwar and Selzer, 2000; Takahashi et al., 2005). Such angiogenic HA fragments are found in several cancers and in body fluids of cancer patients, suggesting that the HA-Hyal system is active in high-grade invasive tumors.

HYAL1 is involved in tumor growth, muscle infiltration by cancer cells, and tumor angiogenesis (Simpson and Lokeshwar, 2008). Blocking of HYAL1 expression in bladder and prostate cancer cells decreases tumor cell proliferation by ~4-fold, due to cell cycle arrest in the G2-M phase and decreases invasive activity. In xenografts, inhibition of HYAL1 expression decreases tumor growth by 9- to 17-fold. Although HYAL1 expressing tumors infiltrated muscle and blood vessels, tumors lacking HYAL1 expression resembles benign neoplasm and had 4- to 9-fold less microvessel density and smaller capillaries. Notably, the contribution of HYAL1 expression to muscle invasion has been observed in bladder cancer patients. These patients exhibit poor prognosis, as 60% of the patients with muscle invasive bladder cancer will have metastasis within 2 years and two-thirds will die within 5 years (Vaidya et al., 2001).

HYAL2 is also involved in tumor formation. It could function either as oncogene or a tumor suppressor gene product. At the one hand, overexpression of HYAL2 accelerates the formation of murine astrocytoma cells (Novak et al., 1999) and, on the other hand, HYAL2 accelerates apoptosis (Chang et al., 2010).

In addition to HYAL1 and 2, PH-20 is expressed in a series of malignant tumors and also in cell lines derived from melanomas, glioblastomas and colon carcinomas. It is therefore plausible to suggest that Hyals may well represent a potential target for novel therapies. This is in accordance with the suggestions that Hyals on tumor cells may provide a target for antineoplastic drugs (Ghatak et al., 2002).

Hyaluronidases as tumor suppressors

While HYAL1 levels expressed in tumor tissues and cells promote tumor growth, invasion, and angiogenesis, high Hyal levels, exceeding 100 milliunits/10^6 cells (not naturally expressed by tumor cells), significantly reduce tumor incidence and growth due to induction of apoptosis (Simpson and Lokeshwar, 2008). Therefore, the function of Hyal as a tumor promoter or a suppressor is concentration-dependent, and further *in vitro* and *in vivo* studies will improve our understanding on such activities of Hyals.

Hyaluronidases as diagnostic and prognostic indicators

The diagnostic potential of Hyal, either alone or together with HA, has been extensively explored in bladder cancer. In a side-by-side comparison, the HA-Hyal urinary test was superior to a variety of bladder tumor markers approved by the U.S. Food and Drug Administration (FDA) (Simpson and Lokeshwar, 2008). It has also been shown a correlation between increased tumor associated HYAL1 and HA in tumor tissues and a positive HA-Hyal test. This suggests that tumor-associated HYAL1 and HA are released into the urine when it comes in contact with a tumor in the bladder. In addition, measurement of HYAL1 mRNA levels in exfoliated cells found in urine are elevated in patients with invasive and poorly differentiated carcinoma (Aboughalia, 2006) and also appears to be a marker for bladder cancer with >90% accuracy (Eissa et al., 2005). These studies show that Hyal is a highly accurate marker for detecting high-grade bladder cancer, especially when combined with HA.

The prognostic potential of HYAL1 has been explored in prostate cancer. HYAL1 staining in radical prostatectomy specimens appears to be an independent predictor of biochemical recurrence (increase of serum prostate specific antigen). Furthermore, HYAL1 staining when combined with HA staining has 87% accuracy in predicting disease progression (Ekici et al., 2004). These results show that consistent with the function of HYAL1 in tumor growth, infiltration, and angiogenesis, it is most likely a prognostic indicator for disease progression.

Regarding bladder and prostate cancer, it should be noted that the presence of selected isoforms' variants, such as HYAL1-v3 and HYAL3-v1 and 2, could be predictors of good prognosis, since they seem to act as negative regulators of the malignancy (Simpson and Lokeshwar, 2008; de Sá et al., 2009).

Hyals expression has also been studied in other carcinomas. HYAL1 might be a useful marker for HNSCC since its levels in salivary are elevated in HNSCC patients (Franzmann et al., 2003). PH-20 mRNA levels are also elevated in primary and lymph node metastatic lesions of laryngeal carcinoma when compared to normal laryngeal tissues (Victor et al., 1999; Godin et al., 2000; Christopoulos et al., 2006). In colon cancer, there is a stage-related increase of HYAL1 activity, together with its overexpression in cancerous samples compared to normal tissue. HYAL1 mRNA levels are increased especially at late stages of cancer, whereas PH-20 mRNA levels are elevated at early stages; considerable PH-20 mRNA levels are observed in benign tumors (Bouga et al., 2010). In breast cancer, a correlation between HYAL1 overexpression and the malignant behavior was reported (Tan et al., 2011). In contrast to the previous observations, increased HYAL2 expression inversely correlates with invasion in B-cell lymphomas and may serve as a prognostic indicator (Bertrand et al., 2005).

7.4.3 Regulation of hyaluronidase activity

Cellular Hyal activity is regulated through different mechanisms. One of these involves loss of the chromosome 3p21.3 locus, which occurs at a higher frequency in some epithelial tumors (see Section 7.4.2). Another mechanism is alternative mRNA splicing. A common internal splicing event occurs in the 5′ untranslated region present in exon 1. HYAL1 protein levels and Hyal activity in tumor cells correlate with a HYAL1 transcript in which this 5′ untranslated region is spliced. In addition, HYAL1 protein is not detected in tumor cells expressing a HYAL1 transcript that retains the 5′ untranslated region, indicating that this transcript is not translated (Frost et al., 2000; Junker et al., 2003). It is unclear how and why the 5′-untranslated region in the HYAL1 mRNA prevents translation. Moreover, five and three alternatively spliced variants of HYAL1 and HYAL3 transcripts, respectively, exist, which are generated by alternative splicing occurring in the coding regions of HYAL1 and HYAL3 transcripts and encode truncated proteins that lack Hyal activity. Most of these alternatively spliced variants have been identified in normal and bladder tumor tissues and bladder and prostate cancer cells (Simpson and Lokeshwar, 2008) and in normal and colorectal tumor tissues (Bouga and Vynios, unpublished data), possessing different expression between normal and tumor tissues. HYAL1-v1 expression reduces Hyal activity because of a complex formation between HYAL1 and HYAL1-v1. HYAL1-v1 expression induces apoptosis in bladder cancer cells and reduces tumor growth, infiltration and angiogenesis (Lokeshwar et al., 2006). This suggests that a critical balance between the levels of HYAL1 and its variants may regulate HYAL1 function in cancer.

A third mechanism involves the presence of Hyals inhibitors. They were first detected over 70 years ago in blood and are present in tissue extracts. These activities are magnesium dependent, are acute phase substances synthesized by the liver, some of which are members of the Kunitz type of inter-alpha-inhibitor family (Mio and Stern, 2002). Hyals inhibitors have been detected in normal tissues, but they were absent in cancer, In the case of laryngeal cancer (Christopoulos et al., 2006). An ever-present inhibitor of HYAL2 activity on cell surfaces would have to be invoked in normal, healthy tissues, to preserve ECM integrity. A combination of cluster of differentiation 44 (CD44), HYAL2, and the Na^+/H^+ exchanger-1 (NHE1) are complexed within lipid rafts on cell surfaces for binding high-molecular-weight (HMW) HA (Bourguignon et al., 2004). Alternatively, a block of any one of these components could function as a potential inhibitor, for preserving HA integrity. An entirely different class of inhibitors of Hyals is found in the circulation of cancer patients, activities that are independent of magnesium. They have not been further characterized, but are clearly important for cancer progression.

Regulation of hyaluronidase gene expression

Mechanisms regarding the regulation of Hyal gene expression in normal and cancer cells are not yet well known. The minimal promoter region of HYAL2 has been identified, but the regulation of HYAL2 expression is unknown (Chow and Knudson, 2005). The basal promoter elements are in the region between nucleotides +959 and +1,158 (within intron-1) and the region between nucleotides +224 and +958 contain negative elements that may control the basal expression level of HYAL2. HYAL2 promoter contains a GATA-binding region and lacks a TATA-binding site.

HYAL1 expression is regulated by promoter methylation. HYAL1 promoter contains a TACAAA sequence and nucleotides –73 to –50 contain overlapping binding consensus sites for specificity protein 1 (SP1), early growth response protein 1 (Egr-1), and activating enhancer binding protein 2 (AP-2). The dinucleotides C^{-71}pG and C^{-59}pG and a nuclear factor-kappa B (NF-κB)–binding site at position –15 play also important role. Both C^{-71} and C^{-59} are methylated and SP1 binds to the promoter in non-HYAL1-expressing cells, whereas, in HYAL1 expressing cells, C^{-71} and C^{-59} are unmethylated and Egr-1/AP-2 binds to the promoter (Lokeshwar et al., 2008).

7.4.4 Hyaluronidases and cell cycle progression

Blocking HYAL1 expression in bladder and prostate cancer cells induces cell cycle arrest in the G2-M phase, due to the down regulation of the positive regulators of G2-M transition (Simpson and Lokeshwar, 2008). In HSC3 oral carcinoma cells, HYAL1 expression caused a 145% increase in the S-phase fraction, with a concomitant decrease in the G0-G1 phase (Lin and Stern, 2001). It is unknown the mechanism by which HYAL1 induces cell cycle transition and upregulates the levels of positive regulators of G2-M transition; however, PH-20 has been shown to induce phosphorylation of c-jun N-terminal kinases (JNKs)-1 and -2 and p44/42 extracellular signal-regulated kinase (ERK) in murine fibroblasts cells L929 (Chang et al., 2010). ERK is required for G2-M and G1-S transitions (Liu et al., 2004). It has also been shown that cell surface interaction between HA oligosaccharides and hyaluronan-mediated motility receptor (RHAMM) stimulates phosphorylation of p42/p44 ERK (activated p42/44ERK) and focal adhesion kinase in human endothelial cells (Lokeshwar and Selzer, 2000). RHAMM coimmunoprecipitates with sarcoma tyrosine kinase (Src) and ERK and contains recognition sequences for these kinases, suggesting a direct interaction (Hall et al., 1996; Zhang et al., 1998). Activated focal adhesion kinase (FAK) also activates ERK through growth factor receptor-bound protein 2 (Grb2) and SH2 homology 2 domain containing (Shc) and phosphatidylinositol 3-kinase (PI3K) through a direct interaction (McLean et al., 2005; Mitra et al., 2005). In addition to ERK activity, transient activation of JNKs is required for G2-M transition. For example, activated JNK may phosphorylate cdc25c and modulate its activity (Mingo-Sion et al., 2004). Furthermore, activated JNKs phosphorylate c-jun, which then increases cdc2 expression (Goss et al., 2003). However, at the present time it is unknown whether HYAL-mediated regulation of the cell cycle involves JNK and/or ERK pathways.

Hyaluronidases and apoptosis

The super high expression of HYAL1 induces apoptosis in prostate cancer cells. The apoptosis induction by HYAL1 involves mitochondrial depolarization and induction of a proapoptotic protein, WW domain-containing murine oxidoreductase (WOX1). Transient transfection of the murine fibroblast line L929, by HYAL1 or HYAL2 cDNA or ectopic addition of bovine PH-20 enhances tumor necrosis factor (TNF)–induced cytotoxicity, which is mediated by increased WOX1 expression and prolonged NF-κB activation. WOX1 induces apoptosis in a p53 independent manner involving WOX1 activation, its translocation to mitochondria and down regulation of antiapoptotic proteins bcl2 and bclxL (Chang et al., 2010). Phosphorylation of WOX1 at Tyr33 is mediated by Src tyrosine kinases and thereafter WOX1 interacts with p53, JNK1, murine double minute 2 (MDM2), Zfra and HYAL2. JNK1 is also associated with the mitochondria mediated apoptotic pathway, as it phosphorylates bcl-2 and bclxL, and suppresses their antiapoptotic activity (Deng et al., 2001; Basu and Haldar, 2003). Transforming growth factor-beta 1 (TGF-β1) reverses the effects of Hyals related with their apoptotic properties, and in addition, in many types of cells, but not breast cancer cells, TGF-β1 utilizes cell surface HYAL2 as a cognate receptor for signaling with WOX1 and SMAD4 to control gene transcription, growth and death (Chang et al., 2010).

The expression of HYAL1-v1 in bladder cancer cells, expressing wild-type HYAL1, induces G2-M arrest and apoptosis. HYAL1 and HYAL1-v1 form a noncovalent complex, which is enzymatically inactive. The HYAL1-v1-induced apoptosis involves the extrinsic pathway, since HYAL1-v1 expression induces activation of caspases-8, -9, and -3, TNF receptor superfamily member 6 (Fas) and Fas-associated death domain (FADD) upregulation, and BH3 interacting-domain death agonist (BID) activation. Moreover, inhibition of Fas expression by Fas siRNA inhibits HYAL1-v1-induced apoptosis (Ehrlich, 2002). These reports suggest that HYAL1 and its variants are capable of inducing apoptotic pathways, the understanding of which has only recently begun (see ▶Figure 7.9).

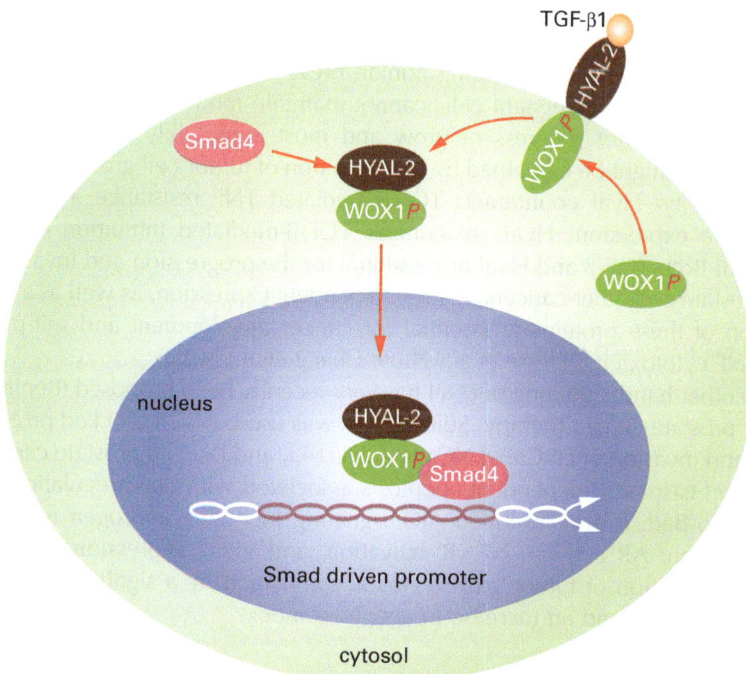

Figure 7.9 The TGF-β1/HYAL2/WOX1/Smad signal pathway. Three signaling paths are proposed: (1) TGF-β1 binds HYAL2 in cell membrane, which subsequently recruits cytosolic phosphorylated WOX1 the membrane. The complex translocates to the nucleus for enhancing the Smad promoter activation. Overly activated Smad promoter induces cell death. (2) The TGF-β1/HYAL2 complex is internalized via endocytic vesicles, followed by releasing of HYAL2 to the cytoplasm for interacting with the cytosolic phosphorylated WOX1, and the complex relocates to the nuclei. (3) Smad4 is recruited to the HYAL2/WOX1p complex, followed by relocating to the nuclei. Ectopic WOX1 and Smad4 dramatically increase the Smad promoter activation.

7.4.5 Anticancer properties of hyaluronidases

It is well documented that HA oligomers of defined sizes injected into cancer sites markedly inhibit tumor growth. This may be achieved by competing with high molecular weight chains for HA receptors. In addition, HA oligomers *in vitro* inhibit anchorage-independent growth of several tumor cell types. They induce apoptosis and stimulate caspase-3 activity through the PI3K/AKT cell survival pathway (Ghatak et al., 2002).

Based on these observations, the anticancer effects of Hyals were examined in experimental model systems. Human cancers grown in severe combined immunodeficiency (SCID) mice regress dramatically following administration of purified PH-20 (Shuster et al., 2002). In addition, Hyal administration delays the appearance of carcinogen-induced tumors, prevents lymph node invasion in a murine model for T-cell lymphoma, reverses multidrug resistance, alters cell cycle kinetics of chemoresistant carcinomas, and blocks TNF-mediated cancer cell death (Chang et al., 2010). Moreover, overexpression of HYAL1 suppresses tumorigenicity in a model for colon carcinoma (Jacobson et al., 2002).

It has also been proposed that Hyal eliminates the cancer cells' ability to move, by converting the long cancer form of CD44 to the short normal immobile form. It is under investigation how Hyal regulates this alternative splicing of CD44. This may open a new area to develop a new generation of anticancer drugs for preventing tumor associated metastasis.

It is also well known that the enzyme causes established breast cancers to shrink. Inoculation of Hyal in mice suffering human breast cancer shrinks the tumor to half its size in 4 days. The malignant cells cannot maintain motility and thus invasiveness and the tumor cannot progress or grow and must completely apparently undergo regression. This might be explained by the inhibition of tumor cell growth by increased TNF cytotoxicity. Hyal counteracts TGF-β-mediated TNF resistance and suppresses TGF-β1 gene expression. Hyal antagonizes TGF-β-mediated inhibition of epithelial cell growth. Both TGF-β and Hyal are essential for the progression and invasiveness of breast, prostate and other cancers. A stage dependent expression, as well as a balanced production of these proteins is essential for cancer development and self-protection against TNF cytotoxicity (Etesse et al., 2009; Chang et al., 2010).

On the other hand, targeting of Hyal has very recently been proposed (Benitez et al., 2011) for prostate cancer therapy. Sulfated HA was used, which blocked proliferation, motility, and invasion of LNCaP, LNCaP-AI, DU145, and LAPC-4 prostate cancer cells, and induced caspase-8-dependent apoptosis associated with downregulation of Bcl-2 and phospho-Bad. It also inhibited AKT signaling including androgen receptor (AR) phosphorylation, AR activity, NF-κB activation, and VEGF expression. In an animal model the inhibition of tumor growth was accompanied by a significant decrease in tumor angiogenesis and an increase in apoptosis index.

Hyaluronidases and cancer therapeutics

Before the observations of the intrinsic antitumor activity of Hyal, PH-20 has long been used in anticancer regimens. Tumors previously resistant to chemotherapy become sensitive when Hyal is added (Baumgartner et al., 1998; Klocker et al., 1998). The enzyme may decrease intratumoral pressure, permitting drugs to penetrate the malignancy.

Tumor cells growing in three-dimensional multicellular masses acquire resistance to chemotherapeutic drugs (Green et al., 2004). Treatment by Hyal of multicellular spheroids of EMT6 abolished their resistance to 4-hydroperoxycyclophosphamide (4-HC) (Croix et al., 1996; Kerbel et al., 1996). In accordance with the findings that Hyal is necessary for cell cycle progression (Lin and Stern, 2001; Simpson and Lokeshwar, 2008), Hyal treatment increases recruitment of disaggregated cells into the cycling pool, and thus renders them more sensitive to a cell-cycle-dependent drug (Croix et al., 1996, 1998; Kerbel et al., 1996). The enzyme, in addition, enhances the anticancer effects of adriamycin *in vitro* (Beckenlehner et al., 1992).

In clinical studies, Hyal, used in extremely high concentrations (1×10^5 to 2×10^5 IU), has been applied to enhance the efficacy of vinblastin in the treatment of malignant melanoma and Kaposi's sarcoma, boron neutron therapy of glioma, intravesical mitomycin treatment for bladder cancer, and chemotherapy involving cisplatin and vindesine in the treatment of HNSCC (Simpson and Lokeshwar, 2008). It is unlikely that at these concentrations the infused Hyal will act as a tumor promoter.

Regulation of HYAL1 promoter activity by methylation raises an interesting question about cancer therapeutics involving DNA demethylating agents. However, if DNA hypomethylation turns on the genes, such as *HYAL1* (and also heparanase, uPA, and MMP2) that function in tumor growth and metastasis, then DNA hypomethylation-inducing therapies may only have short-term efficacy, as they could speed up the progression of surviving cancer cells (Ehrlich, 2002).

The efficacy of Hyal in cancer treatment has also been investigated in experimental animal models, together with pharmacokinetics and toxicity studies. In these studies bovine PH-20 was used with very promising conclusions, since it enhanced doxorubicin efficacy when applied in the vicinity of a murine mammary carcinoma (Beckenlehner et al., 1992). It delayed the growth of a transplanted melanoma when applied subcutaneously (Spruss et al., 1995) and increased accumulation of melphalan in tumors after intraperitoneal application (Muckenschnabel et al., 1996). These findings indicate that Hyal is able to cross the blood vessel barrier and reach the tumor, even when applied at some distance away from it. This also indicates an extremely short time of decay of Hyal in plasma, in the range of few minutes, due to a very efficient distribution to all tissues and especially to a larger extent to tumor tissue.

The mechanisms responsible for the synergism of Hyals with cytostatics *in vivo* are not clear. Breakdown of the physical barrier shielding tumor cells against drugs is the simplest explanation. Solid tumors are partially protected against cytotoxic drugs by a limited physical access due to irregular vascularization and blood supply and dense stromal areas that impede the diffusion of chemotherapeutics, in addition to mechanisms of resistance at the cellular level. Transient disintegration of the intercellular structures, such as the degradation of HA by Hyals, would be expected to facilitate the transport of drugs to tumor cells and thereby increase their therapeutic efficacy. Degradation of HA by Hyals yields a heterogeneous mixture of HA fragments of different sizes, which possess different biological functions. These fragments may be responsible for effects of Hyals distant from a local injection site, since the enzyme may become rapidly inactivated in the peripheral circulation. It has been demonstrated that transient normalization of tumor vascularization creates an opportunity for chemotherapeutic intervention (Jain, 2005). At the tumor cell level, it has been shown an adhesion-dependent multicellular tissue chemoresistance, which is sensitive to Hyal (Croix et al., 1996).

7.4.6 Further medical applications of hyaluronidases

Hyals thereby increase in tissues the membrane permeability, reduce the viscosity and render the tissues more readily permeable to injected fluids (spreading effect). Their effects enable this enzyme to be used therapeutically to increase the speed of absorption and to diminish discomfort due to subcutaneous or intramuscular injection of fluid, to promote resorption of excess fluids and extravasated blood in the tissues, and to increase the effectiveness of local anesthesia. Hyals are widely used not only in oncology, but in many fields (i.e. in orthopedics, surgery, ophthalmology, internal medicine, dermatology, gynecology, etc.). Moreover, the PH-20 protein plays a major role in mammalian fertilization. The male and female guinea pig immunized with PH-20 showed 100% contraception and the effect was long lasting and reversible (Primakoff et al., 1988), indicating that PH-20 can be effectively used as a contraceptive vaccine.

Three formulations of Hyal are commercially available. Amphadase is bovine PH-20; Vitrase is sheep PH-20 and contains lactose; and Hylenex is human recombinant Hyal and contains human albumin. They can be injected subcutaneously or intramuscularly but not intravenously because the enzyme is rapidly inactivated in the blood.

There are three FDA-approved indications for the use of Hyal. It can function as an adjunct to increase the absorption and dispersion of other injected drugs. Most commonly, Hyal has been used in ophthalmologic surgery, since it facilitates the diffusion of anesthetics via degradation of HA and it decreases the duration of their action because it increases their absorption. Hyal is also approved for hypodermoclysis, which is subcutaneous infusion of fluids, primarily used for mildly to moderately dehydrated older adult patients, especially those in nursing homes. Common infusion sites include the chest, abdomen, thighs, and upper arms. The third FDA-approved use for Hyal is as an adjunct for subcutaneous urography for improving resorption of radiopaque agents, especially in infants and young children when intravenous administration cannot be successfully accomplished. Furthermore, Hyal has gained popularity as an off-label medication among physicians to manage the adverse effects of HA soft tissue fillers, because of its safety, efficacy, and ease of use. Additional less well-known unlabeled uses of Hyal in dermatology include for sclerodermoid lesions, lymphedema, and keloids (Lee et al., 2010). Preliminary skin testing for hypersensitivity before use of any of these Hyals is recommended.

Hyal is also applied to help of electrotransfer of plasmid into skeletal muscle, a procedure greatly enhancing the levels of transfected muscle fibers without increasing muscle damage associated with this process (McMahon et al., 2001) and a combination product for enhanced gene delivery is proposed (Braun, 2001). In addition, combination of Hyal to caudal steroid and hypertonic saline injection in patients with failed back surgery syndrome results in improved long-term pain relief (Al-Maksoud et al., 2010). Furthermore, Hyal may be used in the treatment of inflammation associated with increased local synthesis of HA, since it reduces inflammatory cell infiltration and a drug formulation has been proposed (Johnsson et al., 2003). The investigation of Hyal inhibitors that may be used in pharmaceutical and cosmetic products or for treating specific disorders is another field of interest together with the use of HA oligosaccharides in drugs for similar or even more applications (Asari et al., 2009).

7.4.7 Challenges and future prospects

From the results of the studies during the past 25 years, it is quite clear that Hyals, in combination with chemotherapy may be helpful, at least in specific tumors. Nevertheless, a reevaluation of the clinical trials with human recombinant Hyals to avoid any side effects of bovine PH-20 (immunogenicity, anaphylaxia or allergy have been reported in limited cases) will be very helpful for both cancer treatment and understanding the mechanisms involved in the improvement of the efficacy of specific anticancer drugs. In addition, stabilized formulations of Hyals (a pegylated [polyethylene glycol cross-linked] version of a Hyal is now in clinical trials) will provide more evidence on their effects in cancer chemotherapy.

7.4.8 Take-home message

Hyal belongs to endoglycosidases and functions in tumor growth, infiltration, and angiogenesis. At concentrations that are present in tumor tissues, Hyal acts as a tumor promoter. However, by increasing artificially these concentrations, Hyal functions as a tumor suppressor, and therefore it can be used in cancer treatment. Hyals either alone or together with HA are potentially accurate diagnostic and prognostic indicators for cancer detection and tumor metastasis. Hyals are also used in various medications and as drugs additives to help their spreading within tissues or organs. Natural or synthetic inhibitors of Hyals are used as contraceptives.

References

Aboughalia, A. H. (2006). Elevation of hyaluronidase-1 and soluble intercellular adhesion molecule-1 helps select bladder cancer patients at risk of invasion. Arch Med Res *37*, 109–116.

Al-Maksoud, Y. A. A., EL-deen, A. S., and Al-deeb, A. E. (2010). The role of adding hyaluronidase to fluoroscopically guided caudal steroid and hypertonic saline injection in patients with failed back surgery syndrome: a prospective, double-blinded, randomized study. Pain Pract *10*, 548–553.

Asari, A., Kurihara, H., Shibata, T., et al. (2009). Hyaluronic acid oligosaccharide fractions and drugs containing the same. U.S. Patent 7,507,723.

Basu, A., and Haldar, S. (2003). Identification of a novel Bcl-xL phosphorylation site regulating the sensitivity of taxol- or 2-methoxyestradiol-induced apoptosis. FEBS Lett *538*, 41–47.

Baumgartner, G., Gomar-Höss, C., Sakr, L., Ulsperger, E., and Wogritsch, C. (1998). The impact of extracellular matrix on the chemoresistance of solid tumors – experimental and clinical results of hyaluronidase as additive to cytostatic chemotherapy. Cancer Lett *131*, 85–99.

Beckenlehner, K., Bannke, S., Spruss, T., Bernhardt, G., Schönenberg, H., and Schiess, W. (1992). Hyaluronidase enhances the activity of adriamycin in breast cancer models in vitro and in vivo. J Cancer Res Clin Oncol *118*, 591–596.

Benitez, A., Yates, T. J., Lopez, L. E., Cerwinka, W. H., Bakkar, A., and Lokeshwar, V. B. (2011). Targeting hyaluronidase for cancer therapy: antitumor activity of sulfated hyaluronic acid in prostate cancer cells. Cancer Res *71*, 4085–4095.

Bertrand, P., Courel, M. N., Maingonnat, C., Jardin, F., Tilly, H., and Bastard, C. (2005). Expression of HYAL-2 mRNA, hyaluronan and hyaluronidase in B-cell non-Hodgkin lymphoma: relationship with tumor aggressiveness. Int J Cancer *113*, 207–212.

Bertrand, P., Girard, N., Duval, C., et al. (1997). Increased hyaluronidase levels in breast tumor metastases. Int J Cancer *73*, 327–346.

Bouga, H., Tsouros, I., Bounias, D., et al. (2010). Involvement of hyaluronidases in colorectal cancer. BMC Cancer *10*, 499.

Bourguignon, L. Y., Singleton, P. A., Diedrich, F., Stern, R., and Gilad, E. (2004). CD44 interaction with Na^+-H^+ exchanger (NHE1) creates acidic microenvironments leading to hyaluronidase-2 and cathepsin B activation and breast tumor cell invasion. J Biol Chem *279*, 26991–27007.

Braun, S. (2001). Combination product for enhanced gene delivery comprising a hyaluronidase. U.S. Patent 6,258,791.

Chang, J. Y., He, R. Y., Lin, H. P., et al. (2010). Signaling from membrane receptors to tumor suppressor WW domain-containing oxidoreductase. Exp Biol Med *235*, 796–804.

Chow, G., and Knudson, W. (2005). Characterization of promoter elements of the human HYAL-2 gene. J Biol Chem *280*, 26904–26912.

Christopoulos, T. A., Papageorgakopoulou, N., Theocharis, D. A., Mastronikolis, N. S., Papadas, T. A., and Vynios, D. H. (2006). Hyaluronidase and CD44 hyaluronan receptor expression in squamous cell laryngeal carcinoma. Biochim Biophys Acta *1760*, 1039–1045.

Croix, B. S., Man, S., and Kerbel, R. S. (1998). Reversal of intrinsic and acquired forms of drug resistance by hyaluronidase treatment of solid tumors. Cancer Lett *131*, 35–44.

Croix, B. S., Rak, J. W., Kapitain, S., Sheehan, C., Graham, C. H., and Kerbel, R. S. (1996). Reversal by hyaluronidase of adhesion dependent multicellular drug resistance in mammary carcinoma cells. J Natl Cancer Inst *88*, 1285–1296.

Delpech, B., Laquerriere, A., Maingonnat, C., Bertrand, P., and Freger, P. (2002). Hyaluronidase is more elevated in human brain metastases than in primary brain tumours. Anticancer Res *22*, 2423–2427.

Deng, X., Xiao, L., Lang, W., Gao, F., Ruvolo, P., and May, W. S., Jr. (2001). Novel role for JNK as a stress-activated Bcl2 kinase. J Biol Chem *276*, 23681–23688.

de Sá, V. K., Canavez, F. C., Silva, I. A., Srougi, M., and Leite, K. R. M. (2009). Isoforms of hyaluronidases can be a predictor of a prostate cancer of good prognosis. Urol Oncol *27*, 377–381.

Ehrlich, M. (2002). DNA methylation in cancer, too much, but also too little. Oncogene *21*, 5400–5413.

Eissa, S., Kassim, S. K., Labib, R. A., et al. (2005). Detection of bladder carcinoma by combined testing of urine for hyaluronidase and cytokeratin 20 RNAs. Cancer *103*, 1356–1362.

Ekici, S., Cerwinka, W. H., Duncan, R., et al. (2004). Comparison of the prognostic potential of hyaluronic acid, hyaluronidase (HYAL-1), CD44v6 and microvessel density for prostate cancer. Int J Cancer *112*, 121–129.

Etesse, B., Beaudroit, L., Deleuze, M., Nouvellon, E., and Ripart, J. (2009). Hyaluronidase: here we go again. Ann Fr Anesth Reanim *28*, 658–665.

Franzmann, E. J., Schroeder, G. L., Goodwin, W. J., Weed, D. T., Fisher, P., and Lokeshwar, V. B. (2003). Expression of tumor markers hyaluronic acid and hyaluronidase (HYAL1) in head and neck tumors. Int J Cancer *106*, 438–445.

Frost, G. I., Mohapatra, G., Wong, T. M., Csóka, A. B., Gray, J. W., and Stern, R. (2000). HYAL-1 LUCA-1, a candidate tumor suppressor gene on chromosome 3p21.3, is inactivated in head and neck squamous cell carcinomas by aberrant splicing of pre-mRNA. Oncogene *19*, 870–877.

Ghatak, S., Misra, S., and Toole, B. P. (2002). Hyaluronan oligosaccharides inhibit anchorage-independent growth of tumor cells by suppressing the phosphoinositide 3-kinase/Akt cell survival pathway. J Biol Chem *277*, 38013–38020.

Godin, D. A., Fitzpatrick, P. C., Scandurro, A. B., et al. (2000). PH-20: a novel tumor marker for laryngeal cancer. Arch Otolaryngol Head Neck Surg *126*, 402–404.

Goss, V. L., Cross, J. V., Ma, K., Qian, Y., Mola, P. W., and Templeton, D. J. (2003). SAPK/JNK regulates cdc2/cyclin B kinase through phosphorylation and inhibition of cdc25c. Cell Signal *15*, 709–718.

Green, S. K., Francia, G., Isidoro, C., and Kerbel, R. S. (2004). Antiadhesive antibodies targeting E-cadherin sensitize multicellular tumor spheroids to chemotherapy in vitro. Mol Cancer Ther *3*, 149–159.

Hall, C. L., Lange, L. A., Prober, D. A., Zhang, S., and Turley, E. A. (1996). pp60(c-src) is required for cell locomotion regulated by the hyaluronan receptor RHAMM. Oncogene *13*, 2213–2224.

Jacobson, A., Rahmanian, M., Rubin, K., and Heldin, P. (2002). Expression of hyaluronan synthase 2 or hyaluronidase 1 differentially affect the growth rate of transplantable colon carcinoma cell tumors. Int J Cancer *102*, 212–219.

Jain, R. K. (2005). Normalization of tumor vasculature: an emerging concept in antiangiogenic therapy. Science *307*, 58–62.

Johnsson, C., Tufveson, G., Hallgren, R., and Gerdin, B. (2003). Use of hyaluronidase in the manufacturing of a drug for the treatment of inflammation. U.S. Patent 6,572,855.

Junker, N., Latini, S., Petersen, L. N., and Kristjansen, P. E. (2003). Expression and regulation patterns of hyaluronidases in small cell lung cancer and glioma lines. Oncol Rep *10*, 609–616.

Kerbel, R. S., Croix, B. S., Florenes, V. A., and Rak, J. (1996). Induction and reversal of cell adhesion-dependent multicellular drug resistance in solid breast tumors. Hum Cell *9*, 257–264.

Klocker, J., Sabitzer, H., Raunik, W., Wieser, S., and Schumer, J. (1998). Hyaluronidase as additive to induction chemotherapy in advanced squamous cell carcinoma of the head and neck. Cancer Lett *131*, 113–115.

Kumar, S., West, D. C., Ponting, J. M., and Gattamaneni, H. R. (1989). Sera of children with renal tumours contain low-molecular-mass hyaluronic acid. Int J Cancer *44*, 445–448.

Lee, A., Grummer, S. E., Kriegel, D., and Marmur, E. (2010). Hyaluronidase. Dermatol Surg *36*, 1071–1077.

Lin, G., and Stern, R. (2001). Plasma hyaluronidase (HYAL-1) promotes tumor cell cycling. Cancer Lett *163*, 95–101.

Liu, X., Yan, S., Zhou, T., Terada, Y., and Erikson, R. L. (2004). The MAP kinase pathway is required for entry into mitosis and cell survival. Oncogene *23*, 763–776.

Lokeshwar, V. B., Estrella, V., Lopez, L., et al. (2006). HYAL1-v1, an alternatively spliced variant of HYAL1 hyaluronidase: a negative regulator of bladder cancer. Cancer Res *66*, 11219–11227.

Lokeshwar, V. B., Gomez, P., Kramer, M., et al. (2008). Epigenetic regulation of HYAL-1 hyaluronidase expression, identification of HYAL-1 promoter. J Biol Chem *283*, 29215–29227.

Lokeshwar, V. B., and Selzer, M. G. (2000). Differences in hyaluronic acid-mediated functions and signaling in arterial, microvessel, and vein-derived human endothelial cells. J Biol Chem *275*, 27641–27649.

McLean, G. W., Carragher, N. O., Avizienyte, E., Evans, J., Brunton, V. G., and Frame, M. C. (2005). The role of focal-adhesion kinase in cancer – a new therapeutic opportunity. Nat Rev Cancer *5*, 505–515.

McMahon, J. M., Signori, E., Wells, K. E., Fazio, V. M., and Wells, D. J. (2001). Optimisation of electrotransfer of plasmid into skeletal muscle by pretreatment with hyaluronidase – increased expression with reduced muscle damage. Gene Ther *8*, 1264–1270.

Mingo-Sion, A. M., Marietta, P. M., Koller, E., Wolf, D. M., and Van Den Berg, C. L. (2004). Inhibition of JNK reduces G2/M transit independent of p53, leading to endoreduplication, decreased proliferation, and apoptosis in breast cancer cells. Oncogene *23*, 596–604.

Mio, K., and Stern, R. (2002). Inhibitors of the hyaluronidases. Matrix Biol *21*, 31–37.

Mitra, S. K., Hanson, D. A., and Schlaepfer, D. D. (2005). Focal adhesion kinase, in command and control of cell motility. Nat Rev Mol Cell Biol *6*, 56–68.

Muckenschnabel, I., Bernhardt, G., Spruss, T., and Buschauer, A. (1996). Hyaluronidase pretreatment produces selective melphalan enrichment in malignant melanoma implanted in nude mice. Cancer Chemother Pharmacol *38*, 88–94.

Novak, U., Stylli, S. S., Kaye, A. H., and Lepperdinger, G. (1999). Hyaluronidase-2 overexpression accelerates intracerebral but not subcutaneous tumor formation of murine astrocytoma cells. Cancer Res *59*, 6246–6250.

Papadas, T. A., Stylianou, M., Mastronikolis, N. S., et al. (2002). Alterations in the content and composition of glycosaminoglycans in human laryngeal carcinoma. Acta Otolaryngol *122*, 330–337.

Primakoff, P., Lathrop, W., Woolman, L., Cowan, A., and Myles, D. (1988). Fully effective contraception in male and female guinea pigs immunized with the sperm protein PH-20. Nature *335*, 543–546.

Shuster, S., Frost, G. I., Csóka, A. B., Formby, B., and Stern, R. (2002). Hyaluronidase reduces human breast cancer xenografts in SCID mice. Int J Cancer *102*, 192–197.

Simpson, M. A., and Lokeshwar, V. B. (2008). Hyaluronan and hyaluronidase in genitourinary tumors. Front Biosci *13*, 5664–5680.

Spruss, T., Bernhardt, G., Schönenberger, H., and Schiess, W. (1995). Hyaluronidase significantly enhances the efficacy of regional vinblastine chemotherapy of malignant melanoma. J Cancer Res Clin Oncol *121*, 193–202.

Stern, M., and Stern, R. (1992). An ELISA-like assay for hyaluronidase and hyaluronidase inhibitors. Matrix *12*, 397–403.

Stern, R., Asari, A. A., and Sugahara, K. N. (2006). Hyaluronan fragments: an information rich system. Eur J Cell Biol *85*, 699–715.

Stern, R., and Jedrzejas, M. J. (2006). Hyaluronidases: their genomics, structures, and mechanisms of action. Chem Rev *106*, 818–839.

Takahashi, Y., Li, L., Kamiryo, M., et al. (2005). Hyaluronan fragments induce endothelial cell differentiation in a CD44- and CXCL1/GRO1-dependent manner. J Biol Chem *280*, 24195–24204.

Tan, J.-X., Wang, X.-Y., Li, H.-Y., et al. (2011). HYAL1 overexpression is correlated with the malignant behavior of human breast cancer. Int J Cancer *128*, 1303–1315.

Vaidya, A., Soloway, M. S., Hawke, C., Tiguert, R., and Civantos, F. (2001). De novo muscle invasive bladder cancer: is there a change in trend? J Urol *165*, 47–50.

Victor, R., Chauzy, C., and Girard, N. (1999). Human breast-cancer metastasis formation in a nude-mouse model, studies of hyaluronidase, hyaluronan and hyaluronan-binding sites in metastatic cells. Int J Cancer *82*, 77–83.

Wang, X.-Y., Tan, J.-X., Vasse, M., Delpech, B., and Ren, G.-S. (2009). Comparison of hyaluronidase expression, invasiveness and tubule formation promotion in ER (–) and ER (+) breast cancer cell lines in vitro. Chin Med J *122*, 1300–1304.

Zhang, S., Chang, M. C., Zylka, D., Turley, S., Harrison, R., Turley, E. A. (1998). The hyaluronan receptor RHAMM regulates extracellular-regulated kinase. J Biol Chem *273*, 11342–11348.

7.5 Structure-function relationship of syndecan-1, with focus on nuclear translocation and tumor cell behavior

Fang Zong, Ilona Kovalszky, Anders Hjerpe, and Katalin Dobra

7.5.1 Syndecans

Syndecans are a major family of transmembrane proteoglycans (PGs), composed of a core protein to which one or more glycosaminoglycan (GAG) chains of variable structure are covalently attached. These complex molecules are generally present on the cell surface, linking the extracellular matrix (ECM) to the cytoskeleton (Saunders et al., 1989; Carey, 1997), though there are also intracellular (Brockstedt et al., 2002) and shed, circulating forms (Seidel et al., 2000; Perrimon and Bernfield 2001; Joensuu et al., 2002; Vassilakopoulos et al., 2005). There are four members of syndecan family, namely, syndecan-1 (Saunders et al., 1989), syndecan-2 (Marynen et al., 1989), syndecan-3 (Carey et al., 1992; Gould et al., 1992), and syndecan-4 (David et al., 1992; Kojima et al., 1992). Syndecan-1 and -3 on one hand, and syndecan-2 and -4 on the other, belong to two subfamilies, based on the similarities in their core protein sequences (Bernfield et al., 1999).

Most cells express one or multiple syndecan family members. Syndecan-1 is the major syndecan on epithelial cells, syndecan-2 is present mainly on cells of mesenchymal origin, syndecan-3 is primarily in neuronal tissue and cartilage, whereas syndecan-4 is ubiquitously expressed. However, each syndecan is expressed in highly regulated cell-, tissue-, and development-specific patterns (Kim et al., 1994; Xian et al., 2010).

The characteristic structure of the syndecans endows them with multiple cellular functions. Their interactions with a number of cellular and extracellular ligands are covered elsewhere in this book. They are key regulators in the interplay between cells and their microenvironment, as "tuners of transmembrane signaling" (Zimmermann and David, 1999). Through these interactions they become involved in a number of basic functions of the cell.

7.5.2 Structural organization

The four syndecans share a general structural organization. Their core proteins have a short C-terminal cytoplasmic domain, a single-pass transmembrane domain and a large N-terminal extracellular domain (Bernfield et al., 1999; Xian et al., 2010).

The cytoplasmic domains contain two highly conserved regions, denoted as C1 and C2. They are conserved both within a specific syndecan across all species, as well as between all four syndecan members (Carey et al., 1994). A variable region (V), between C1 and C2, is unique for each family member but shows sequence conservation across species (Xian et al., 2010). The single-pass transmembrane domain is highly conserved among the syndecans. The ectodomains are variable in length and sequence between syndecan members but contain conserved motifs for GAG attachment, cell interaction, proteolytic cleavage, and oligomerization, described elsewhere in this book.

Syndecan GAGs exhibit great structural diversity including the type, number, length, and fine structure (Carey, 1997). Heparan sulfate (HS) is the principal GAG present in all four syndecans. In addition, syndecan-1 (Rapraeger et al., 1985), syndecan-4 (Shworak et al., 1994), and possibly syndecan-3 (Bernfield et al., 1999) are also substituted by chondroitin sulfate (CS). In syndecan-1 there are three consecutive motifs for HS attachment (Bourdon and Ruoslahti, 1989; Kokenyesi and Bernfield, 1994; Esko and Zhang, 1996), while attachment sites for CS are found near the cell membrane. In contrast, the N terminus of syndecan-4 is capable of bearing either HS or CS, without obvious preference (Shworak et al., 1994). Furthermore, GAG modifications generate complex sulfation patterns (Lindahl et al., 1998; Habuchi, 2000; Silbert and Sugumaran, 2002). Consequently, sulfation is critically important for GAG binding; the highly sulfated and IdoA-rich domains are the main regions for the recognition of growth factors and other proteins by HS (Lambaerts et al., 2009).

7.5.3 Functional domains and cellular interactions

The specific structure of syndecans determines their many varied functions. Different parts of the syndecan molecule interact with a number of ligands via their GAG chains and/or core proteins. The syndecans are in this way involved in a wide variety of functions, including cell proliferation, adhesion and migration. Regulation of syndecan expression will therefore affect all these cellular functions.

The attached GAGs provide binding sites for different extracellular ligands. The binding of proteins to GAGs is to a large extent electrostatic, due to the high content of negatively charged groups interacting with positively charged amino acid side chains, although other types of interaction such as hydrogen bonding also contribute (Kjellen and Lindahl, 1991). HS binds a large number of ligands including ECM components and cell-surface adhesion molecules (Woods et al., 2000; Yoneda and Couchman, 2003), chemokines (Gotte, 2003), growth factors, and growth factor receptors (Carey, 1997; Schlessinger et al., 2000; Xian et al., 2010). For example, it is well documented that HS interacts with both fibroblast growth factor-2 (FGF2) and fibroblast growth factor receptor-1 (FGFR1), thereby forming a ternary complex to exert efficient cell growth signaling (Schlessinger et al., 2000).

The role of CS in syndecans is less clear but it can modify the binding to proteins (Yamagata et al., 1989). It has been suggested that CS chains can cooperate with HS chains in ligating growth factors (Deepa et al., 2004). Adhesion of cells to the ECM protein laminin also involves binding to both HS and CS chains of syndecan-1 (Okamoto et al., 2003).

Discrete domains of the core proteins can themselves bind or associate with other components (▶Figure 7.10). The ectodomains of syndecan-2 and -4 promote the adhesion of mesenchymal cells (Whiteford et al., 2007). The region of the syndecan-1

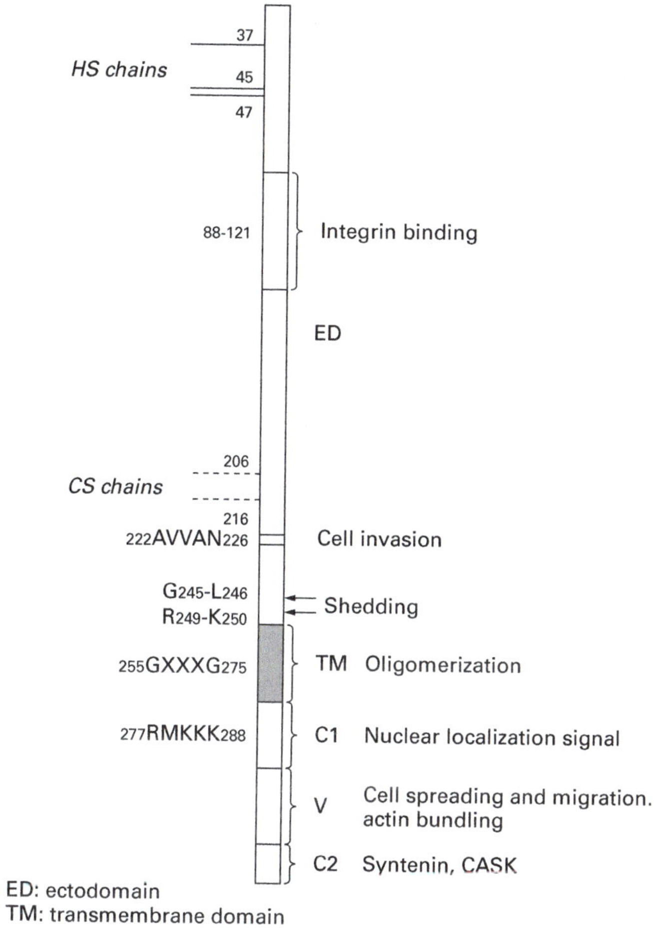

Figure 7.10 Schematic representation of human syndecan-1, indicating the positions of specific peptide sequences or domains that have known ligand interactions or functions.

ectodomain, close to the cell membrane, inhibits tumor cell invasion (Langford et al., 2005). Another site has recently been shown to bind and activate integrins independently of the GAG chains (Beauvais et al., 2009).

Together with the juxtamembrane tetrapeptide xRxE, the transmembrane domain is important for syndecan self-oligomerization. This oligomerization will enlarge the interaction surfaces of involved PGs (Klemm et al., 1998), a mechanism by which syndecans associate with the actin cytoskeleton (Carey et al., 1994; Carey, 1997; Bernfield et al., 1999). The oligomerization is also essential for activating the cytoplasmic domain for downstream signaling (Woods and Couchman, 2001; Alexopoulou et al., 2007).

The cytoplasmic domain of syndecan-1 interacts with the small guanosine triphosphatase (GTPase) Rab5, and this interaction regulates syndecan-1 ectodomain shedding (Hayashida et al., 2008). The C1 and C2 regions bind cytoskeletal and PSD-95/discs large/ZO-1 (PDZ)-domain proteins, respectively, thus influencing the dynamics of the actin cytoskeleton and membrane trafficking. These interactions control syndecan recycling through endosomal compartments, promote internalization of accompanying protein cargo, and regulate cell adhesion and various signaling systems (Woods and Couchman, 2001; Zimmermann et al., 2005; Alexopoulou et al., 2007). The conserved juxtamembrane peptide RMKKK has been demonstrated to be a nuclear localization signal (NLS) of importance for the tubulin dependent nuclear translocation of syndecan-1 (Zong et al., 2009). The V region of syndecan-1 has been ascribed essential roles in lamellopodial spreading, actin bundling, and cell migration (Chakravarti et al., 2005).

The intact ectodomains of each mammalian syndecan are constitutively shed from the cell surface by endogenous proteolytic cleavage (Kim et al., 1994; Hooper et al., 1997) as part of normal cell surface PG turnover (Yanagishita and Hascall, 1992). Syndecan shedding is often accelerated in response to pathophysiological cues and the shed fragment can be recovered in the blood. Elevated levels of soluble syndecan-1 ectodomain have thus been demonstrated in sera from patients with lung cancer (Joensuu et al., 2002), multiple myeloma (Seidel et al., 2000), and Hodgkin's lymphoma (Vassilakopoulos et al., 2005). Shedding can be accelerated by a variety of physiological stimuli, such as growth factors, chemokines, bacteria, and cellular stress (Lambaerts et al., 2009). Recent studies showed that heparanase enhances syndecan-1 shedding by stimulating the expression of protease. This suggests a novel mechanism for the stimulation of tumor growth and metastasis (Yang et al., 2007; Purushothaman et al., 2008).

The release of syndecan ectodomains from the cell surface by shedding has functional consequences. Membrane-bound and soluble forms of syndecan-1 have different roles in breast cancer cells. Overexpression of membrane-bound syndecan-1 stimulates proliferation, but inhibits invasiveness, of adenocarcinoma cells, whereas overexpression of a constitutively shed syndecan-1 decreases the proliferation, but promotes invasion, of the same cells (Nikolova et al., 2009). Soluble syndecan-1 promotes growth of myeloma tumors *in vivo*. This can be demonstrated by transfection of myeloma cells with a syndecan-1 construct lacking its transmembrane and cytoplasmic domains, which thus structurally and functionally mimics the shed syndecan-1 ectodomain *in vivo*. When injected into the marrow of human bones implanted in severe combined immunodeficient mice, the cells expressing this truncated form of syndecan-1 grow much faster than the cells transfected with full-length syndecan-1 (Yang et al., 2002).

7.5.4 Cellular distribution and nuclear translocation

Syndecans can be found in different cell compartments. Traditionally, syndecans have been thought to exert their functions at the cell surface due to their membrane localizations (Rapraeger et al., 1986; Shimazu et al., 1996; Bernfield et al., 1999). In myeloma cells, syndecan-1 is found on the uropod, a discrete membrane domain located at the trailing edge of the migrating cell (Borset et al., 2000). However, they have also been demonstrated in the cytoplasm and in the nucleus in several malignant and benign cell types. In malignant mesothelioma (▶Figure 7.11) and carcinoma cell lines syndecan-1 shows abundant nuclear and nucleolar immunoreactivity (Brockstedt et al., 2002).

Similar nuclear localization has been shown also with syndecan-2 in the injured cerebral cortex (Leadbeater et al., 2006) and in chondrosarcoma (Schrage et al.,

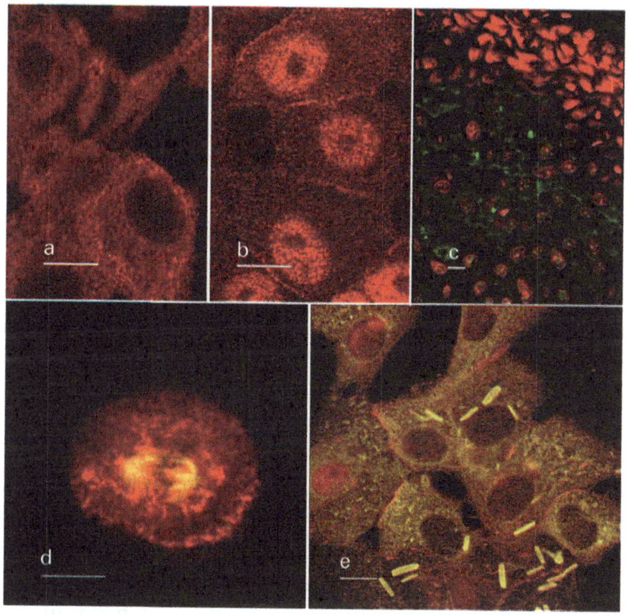

Figure 7.11 **Syndecan-1 (red) in cultured mesothelioma cells (a, b).** The reactivity to the Syndecan-1 epitope (CD138) first appears in the cytoplasm (a) and 2 days later (b) in the nucleus and in the cell membrane. Nuclear syndecan-1 reactivity (green) is commonly seen in the less differentiated cells of a squamous carcinoma of the uterine cervix containing papilloma virus (c). (Red = propidium iodide). Syndecan binds to tubulin throughout the mitosis (d). When the tubulin structure is deranged and precipitated into paracrystalline granules by vinblastine treatment (e), then syndecan-1 coprecipitates with tubulin. Syndecan = red, tubulin = green, and colocalization in overlay = yellow; bar = 10 μm.

2009). Glypican and biglycan have been shown in the nuclei of neurons and glioma cells (Liang et al., 1997). There are now an increasing number of publications about nuclear heparan sulfate PGs (HSPGs) in various cell types (Liang et al., 1997; Richardson et al., 2001; Leadbeater et al., 2006; Schrage et al., 2009). The mechanisms behind the nuclear localization of PGs are still incompletely understood and await further studies. Evidence point to possible functions of the nuclear PGs in differentiation and in cell proliferation, but the precise role(s) remains to be further clarified.

As visualized in cell culture, syndecan-1 is initially present in the cytoplasm and localizes to the cell membrane later, usually at cell-cell contact sites. Syndecan-1 also shows prominent nuclear and nucleolar immunoreactivity, and this nuclear translocation occurs in a time- and tubulin-dependent manner (▶Figure 7.11). In the nucleus the ectodomain epitopes strictly colocalizes with both endodomain and HS, indicating that the entire molecule is present. The nuclear localization of syndecan-1, and its colocalization with tubulin in the mitotic spindle, has also been shown in various carcinomas, neuroblastoma, and benign mesothelial and endothelial cells (Brockstedt et al., 2002).

Syndecan-1 and FGF2 share a tubulin-mediated transport route and colocalize with heparanase in the nucleus, whereas FGFR1 seems to be enriched in the perinuclear area (Zong et al., 2009), suggesting a different transport route. Observations of nuclear

PGs, and the implication of their functional importance, have triggered investigations on the underlying mechanisms of their nuclear translocation.

Structural requirement for nuclear translocation and functional implications

The nuclear import of large proteins depends on the presence of a NLS corresponding to a short peptide. The typical peptide NLS consists of one or more short sequences of positively charged lysines and/or arginines. The first NLS discovered was *PKKKRK* (Kalderon et al., 1984), and the sequence *K-K/R-X-K/R* is proposed as a consensus monopartite NLS (Dingwall et al., 1988). A bipartite NLS consists of two clusters of basic amino acids, separated by a spacer of variable length; the prototype is *KR [PAATKKAGQA]KKKK* (Robbins et al., 1991).

The minimal sequence required for the nuclear translocation of syndecan-1 is the RMKKK motif within the C1 region of core protein, serving as the NLS (Zong et al., 2009). Deletion of this sequence from the syndecan core prevents its nuclear translocation (Zong et al., 2011). This conserved, juxtamembrane RMKKK sequence in the cytoplasmic domains of human syndecan-1, -3, and -4 (but conservatively altered to RMRKK in syndecan-2) thus seems to act as an NLS for the syndecans. Functional NLSs have also been identified in the core proteins of glypican and biglycan (Liang et al., 1997). Interestingly, overexpression or addition of heparanase decreases the level of syndecan-1 in the nucleus of myeloma cells in a concentration-dependent manner (Chen and Sanderson, 2009), suggesting that also the HS chains may be important for the nuclear translocation or degradation of syndecan-1.

The role of nuclear PGs and GAGs has been linked to the control of cell division (Fedarko et al., 1989), the shuttling of FGF2 into the nucleus (Hsia et al., 2003), and inhibition of DNA topoisomerase-I activity (Kovalszky et al., 1998). The latter inhibitory effect suggests that the presence of HS in the nucleus may inhibit gene transcription (Kovalszky et al., 1998). Intriguingly, nuclear syndecan-1 seems to influence cell proliferation; TGF-β2 delays the nuclear translocation of syndecan-1 concomitantly with an antiproliferative effect on malignant mesothelioma cells (Dobra et al., 2003). The presence of PGs/GAGs in the nucleus raises interesting questions: How do PGs get to the nuclear destination, and what are their molecular partners along the route? What are their functions in the nucleus?

7.5.5 Nuclear interactions

The presence of heparan sulfate in the cell nucleus and its potential regulatory role in cell proliferation were reported as early as 25 years ago. It is, however, only recently that this unusual localization of GAG chains alone, or as part of PG molecules, became generally accepted. As described previously, evidence points to possible functions of the nuclear PGs in cell differentiation and proliferation, but other possible effects must also be considered.

The physical characteristics of HSs (e.g. their negative charge) also make these molecules suitable for actions inside the cell nucleus. HS can compete with DNA for its binding to proteins such as transcription factors and DNA-binding enzymes. It has been demonstrated *in vitro* that heparin not only binds to but also inhibits topoisomerase-I activity. The efficacy depends on the source of HS, HS isolated from normal liver being much more effective than that isolated from hepatocellular cancer. This indicates that the fine structure of the sugar chain determines the outcome of interaction. In harmony with this, it has recently been shown that the effect on topoisomerase-1 is modulated by nuclear heparanase. Activation of epidermal growth factor receptor (EGFR) initiates nuclear translocation of heparanase and degradation of HS, resulting in increased activity of topoisomerase-I. This regulatory cascade represents a new mechanism for EGFR induced cell proliferation (Zhang et al., 2010).

At present, we cannot exactly determine if the active nuclear HS is in the form of a PG or a free GAG. Experiments with biotinylated HS or heparin show that the GAGs are easily taken up by hepatoma cells (Dudas et al., 2000). Loss of topoisomerase-I activity in nuclear extract of Hep3 B cells treated by heparin tends to corroborate that this effect depends on the GAG chain. In a gel shift assay, heparin is capable of displacing topoisomerase-I from a DNA fragment that contains eight putative topoisomerase-I-binding sites, proving that HS not only competes for topoisomerase-I but is also able to dissociate the DNA topoisomerase-I complex.

The picture is, however, even more complex. Not only topoisomerase-I but also topoisomerase-II can be influenced by heparin. Furthermore, inside the nucleus, topoisomerase-I and FGF2 compete for the binding to HS in dose-dependent manner. As HS hampers the interaction between topoisomerase-I and DNA, it can also interfere with camptotechin, a topoisomerase-I inhibitory drug, whose mechanism of action requires physical contact between DNA and the enzyme. In this way, heparin hampers the camptotechin-induced fragmentation of linearized plasmid DNA (unpublished data).

The functional role of GAGs inside the nucleus is, however, not restricted to topoisomerases. There are several candidate proteins described to be regulated by heparin or HS. Thus, various transcription factors, casein kinase II, midkines, and histones have been reported as target molecules (Dudas et al., 2000). Inhibition of kinase activity may lead to decreased phosphorylation of nuclear proteins, such as Topoisomerase II or p53. Jun-Fos-mediated transcription was repressed by heparin in a reporter gene assay. This was the first demonstration of the direct effect of heparin on the protein kinase C (PKC)/mitogen-activated protein kinase (MAPK)/AP1 signal transduction pathway, and it was later shown that heparin induces posttranslational modification of Jun-B (Au et al., 1994).

The effect of HS and heparin is mainly attributed to their negative charge, but the fact that the efficacy of HS from normal liver is significantly higher than that from hepatocellular cancer sheds light on the importance of their fine structure. Disaccharide analysis of these HS species could, however, not uncover major alterations, indicating that more refined differences in their molecular assembly and sequence are responsible for their various efficacies (Tatrai et al., 2010).

Transcription factors are another group of nonhistone proteins that are capable of interacting with HS. Thus, heparin and liver HS inhibit the binding of consensus oligo-DNA to AP1, erythroblastosis oncogene homologe-1 (Ets1), specificity protein 1 (Sp1), transcription factor II D (TFIID), and nuclear factor-kappa B (NF-kB), and liver HS exerts inhibitory effect on an AP1 reporter gene assay. Here again malignant hepatoma-derived HS exerts no, or different, potential indicating binding specificity depending on the HS fine structure. It is, however, noteworthy, that the PG profile is considerably modified when the hepatocyte becomes malignant (Tatrai et al., 2006). It therefore remains to be shown if the observed differences between HS moieties of benign and malignant hepatocytes depends on their different core protein origins or an altered pattern of GAG modification reactions.

A critical factor regulating the level of both HS and syndecan-1 in the nucleus is heparanase, an endoglycosidase that cleaves HS and causes a dramatic reduction of the amount of nuclear syndecan-1 (Chen and Sanderson, 2009). Overexpression of heparanase is in many cancer types associated with poor survival (Vlodavsky et al., 1988, 2006, 2011; Ilan et al., 2006). In addition to its localization in the cytoplasm and cell membrane, heparanase is also identified in the nuclei of normal epithelial and tumor cells (Ohkawa et al., 2004; Schubert et al., 2004; Zong et al., 2009), and the nuclear heparanase seems to be related to cell differentiation (Nobuhisa et al., 2005). The decrease or loss of nuclear syndecan-1 in heparanase overexpressing myeloma cells is also accompanied with a simultaneous increase in histone acetyltransferase (HAT) activity. This will in turn enhance the transcription of a number of proteins such as matrix metalloproteinase-9 (MMP9), vascular endothelial growth factor (VEGF), hepatocyte growth factor (HGF), and receptor activator of nuclear factor-kappa B ligand (RANKL), which drive an aggressive tumor phenotype (Purushothaman et al., 2011). Conversely, restoration of the syndecan-1 level diminishes this HAT reactivity, pointing to an inhibitory role of nuclear syndecan-1.

7.5.6 Syndecan-1 expression in normal tissues

In the adult individual, syndecan-1 is expressed predominantly in epithelial tissues, especially in normal squamous and transitional epithelia, where it localizes to the basolateral surfaces of simple epithelial cells, and cell-cell contacts of stratified epithelial cells. It is generally faintly expressed or absent in most mesenchymal cells, although it is expressed on Leydig cells and some plasma cells (Hayashi et al., 1987). However, during embryonic development, syndecan-1 is transiently expressed also in the condensing mesenchymal cells in some tissues, especially during epithelial-mesenchymal transition (Carey, 1997). In addition, a small amount of uniformly distributed syndecan-1 has been seen in cultured fibroblasts (Bernfield et al., 1992). It is also expressed in pre–B cells in the bone marrow, but lost immediately before the mature B cells are released into the circulation (Sanderson et al., 1989).

Syndecan-1 expression is induced in response to wound healing; its expression is increased in the proliferating keratinocytes at the wound margins (Gallo et al., 1996) and in the endothelial cells within the granulation tissue (Elenius et al., 1991). This expression can be modified in cultured cells by treatment with various growth factors and cytokines, and the responses are cell-type dependent. For example, syndecan-1 expression is induced in 3T3 fibroblasts by FGF2 (Elenius et al., 1992), and in vascular smooth muscle cells by platelet-derived growth factor (PDGF) (Cizmeci-Smith et al., 1993). In contrast, syndecan-1 expression is decreased in endothelial cells by tumor necrosis factor-alpha (TNF-α) (Kainulainen et al., 1996), and in mesothelioma cells by epidermal growth factor (EGF), insulin-like growth factor (IGF), and transforming growth factor-beta 2 (TGF-β2) (Dobra et al., 2003). On the other hand, PDGF and TGF-β increase its expression in human periodontal fibroblasts and osteoblasts (Worapamorn et al., 2002). Interestingly, the effect of syndecan-1 is quite different in embryonal skin. Overexpression of the molecule results in overproduction of embryonal keratinocytes, whereas it hinders the proliferation of keratinocytes in the wound healing model (Ojeh et al., 2008).

As a matter of fact not much is known about the regulation of human syndecan-1 expression. The promoter region of syndecan-1 contains peroxisome proliferator-activated receptor γ (PPARγ) and downregulator of transcription 1 (DR1) responsive elements in breast and prostate cancer cells, which has been associated with the protective role of ω-3 polyunsaturated fatty acid against breast cancer (Edwards et al., 2008; Sun et al., 2008). During wound healing, the FGF responsive element, driven by AP1, is involved in migration, but not in the proliferation of keratinocytes (Jaakkola et al., 1998). Transfection with another transcription factor – Wilms tumor suppressor gene 1 (WT1) – activates syndecan-1 expression (Cook et al., 1996). Such WT1-induced activation can be linked to the epithelial differentiation, and the opposite decreased expression may relate to malignant transformation of kidney cells. Overexpression of truncated syndecan-1 downregulates the expression of ETS1 transcription factor, whereas silencing of this transcription factor results in the decreased expression of syndecan-1, indicating a negative feedback between the two molecules. These scattered data demonstrate that the actual effect of syndecan-1 is strongly influenced by the developmental state, localization, and additional circumstances; however, we are far from fully understanding its role.

HS can serve as receptors for viruses, including oncogenic ones. A cell line deficient in HSPGs could not be infected by Human Papilloma Virus (HPV)-like particles, and the infectivity was restored by syndecan-1 transfection (Shafti-Keramat et al., 2003; Letian and Tianyu, 2010). This efficacy of syndecan-1 was eightfold higher than that of glypican or syndecan-4. The syndecan-1 molecule is also capable of potentiating the penetration of human immunodeficiency virus (HIV) into T lymphocytes (Bobardt et al., 2003). Immunochemically, syndecan-1 colocalizes with the L1 protein of HPV16 virus in the nuclei of cervical cancer cells, indicating that the interaction is important not only for internalization but also for the nuclear translocation of the virus.

7.5.7 Syndecan-1 in cancers

One of the largest enigmas related to syndecan-1 is its significance in tumors. As growth factors and cytokines can elicit opposing effects depending on the cell type, it is conceivable that after malignant transformation syndecans act differently in various cell types. The tumor microenvironment, the changes in cell surface receptors, signal transduction proteins, and nuclear regulatory elements will altogether determine the final effects of syndecan-1.

Syndecan-1 expression is altered in pathological states, especially after neoplastic transformation (Bernfield et al., 1992), and it is differentially expressed in various human cancers; for review, see Fears and Woods (2006). This PG is highly expressed in many different forms of cancer. The presence of syndecan-1 in myeloma cells (Ridley et al., 1993) is used as a standard diagnostic marker for this tumor, and a high level of shed syndecan-1 in the serum is an independent predictor of poor prognosis (Seidel et al., 2000). This correlation could reflect the tumor burden, but it may also relate to the need for a high level of syndecan-1 for growth, vascularization, and metastasis of these tumors *in vivo* (Khotskaya et al., 2009). Upregulation of syndecan-1 has also been reported in other hematological malignancies, including B-cell chronic lymphocytic leukemia (B-CLL) (Molica et al., 2006), acute lymphoblastic leukemia (ALL), acute myeloblastic leukemia (AML) (Seftalioglu and Karakus, 2003; Seftalioglu et al., 2003), and Hodgkin's lymphoma (Vassilakopoulos et al., 2005). A broad spectrum of lymphoid neoplasms are, however, CD138 negative (O'Connell et al., 2004).

High expression of syndecan-1 is often detected in hormone-related cancers, including ovarian cancer (Davies et al., 2004), breast cancer, and endometrial cancer. In breast cancer, high expression of syndecan-1 correlates to higher histological tumor grade, tumor size, and less favorable prognosis (Barbareschi et al., 2003), and syndecan-1 mediated signaling events are highly enriched in genome-wide association analysis (Menashe et al., 2010). Interestingly, there is a relationship between presence of syndecan-1 in high-grade tumors and absence of syndecan-4, these two syndecans being independent prognostic indicators in breast cancer (Lendorf et al., 2011). Increased expression of syndecan-1 has also been associated with occurrence and progression of endometrial cancer (Choi et al., 2007), which may relate to promotion of tumor growth and angiogenesis as shown in a xenograft model (Oh et al., 2010).

Conversely, in many other types of cancers, the malignant transformation is associated with downregulation of syndecan-1, and decreased levels have been shown in cell transformation models. For example, experimental inhibition of syndecan-1 transforms epithelial cells to a mesenchymal morphology (Kato et al., 1995), and an epithelioid phenotype can be restored when the cells are induced to reexpress syndecan-1 (Leppa et al., 1992). In the mouse epidermis, malignant transformation induced by ultraviolet irradiation decreases the amount of cell surface syndecan-1 (Inki et al., 1991). Syndecan-1 has also been associated with lung cancer; however, data are divergent in this context. Loss of syndecan-1 is strongly associated to aggressive phenotypes of cancers and poor prognosis. Thus syndecan-1 expression is downregulated in lung, colon, gastric, cervical, esophageal, oropharyngeal, and laryngeal cancers (Hirabayashi et al., 1998; Nakanishi et al., 1999; Rintala et al., 1999; Anttonen et al., 2001; Fujiya et al., 2001; Mikami et al., 2001; Numa et al., 2002; Shinyo et al., 2005; Jackson et al., 2007). In squamous cell carcinomas of the head and neck, and also uterine cervix, loss of syndecan-1 correlates with less differentiation (▶Figure 7.12) and worse clinical outcome (Inki et al., 1994; Orosz and Kopper, 2001), and similar association with increasing nuclear grade is shown in renal cell carcinoma (Gokden et al., 2006). In oropharyngeal cancers, the disappearance of cell surface reaction with intensive cytoplasmic positivity is characteristic, and loss of epithelial positivity together with stromal expression of syndecan-1 (▶Figure 7.12) correlates with poor prognosis (Mathe et al., 2006).

Downregulation of syndecan-1 has also been reported in some mesenchymal tumors including human fibrosarcoma (Orosz and Kopper, 2001), and a correlation between this expression and cell differentiation has been shown in malignant mesothelioma (Kumar-Singh et al., 1998; Dobra et al., 2000). The production of PGs and GAGs in malignant mesothelioma reflects the mesenchymal nature of the tumor; it produces a variety of macromolecules commonly found in the ECM of soft connective tissues (e.g. versicans, biglycan, decorin, perlecan, and hyaluronan) (Dobra et al., 2000). Malignant mesothelioma cells also synthesize a number of cell-associated PGs including syndecan-1, -2, and -4. Among them, syndecan-2 and -4 seem to dominate

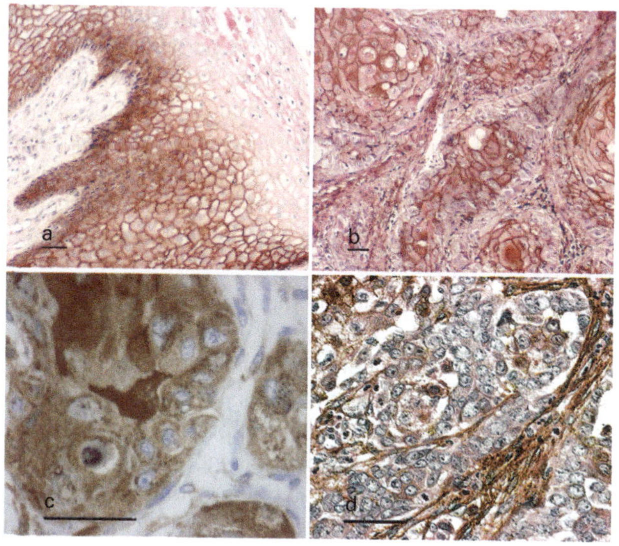

Figure 7.12 Immunocytochemical demonstration of syndecan-1 (CD138 epitope) in benign squamous epithelium and in squamous cell carcinoma. The reactivity of the benign epithelium (a) is mainly confined to the cell membranes of the less matured basal and parabasal cells. Reactivity in the carcinoma cell (b, c) is also seen in the cytoplasm and in a few nuclei. The amount of reactivity varies greatly between different tumor cells, indicating varying degrees of differentiation within the tumor tissue. Syndecan-1 may also accumulate in the tumor stroma (d), particularly in more aggressive tumors. Bar = 50 µm.

especially in the epithelioid subtype (Dobra et al., 2000). In these cells, the expression of syndecan-1 is relatively low in both phenotypes, while in another study, syndecan-1 expression is higher in the epithelioid cells, but reduced or lost in the sarcomatoid subtype.

Although immunohistochemical staining of syndecan-1 in fibrosarcoma sections was negative (Orosz and Kopper, 2001), some fibrosarcoma cell lines express syndecan-1 in culture. These fibrosarcoma cells migrate spontaneously on ECM components (Niggli et al., 2009), and shedding of syndecan-1 stimulates their migration (Endo et al., 2003). Overexpression of syndecan-2 results in increased migration, invasion, and anchorage-independent growth in HT1080 fibrosarcoma cells (Park et al., 2005). Fibrosarcoma cells overexpressing syndecan-1 have higher growth rates *in vitro* and *in vivo*, and tend to develop more lung metastases (Peterfia et al., 2006).

The expression level of syndecan-1 can vary considerably within a particular type of cancer, and most often it is related to the differentiation state and prognosis of the particular cancer (▶Figure 7.12). For example, several reports demonstrate that syndecan-1 is upregulated in prostate cancer, and elevated levels are here associated with a higher tumor grade and risk of recurrence (Zellweger et al., 2003, 2005; Chen et al., 2004). However, another report shows that syndecan-1 is almost absent in high-grade prostate carcinoma samples (Kiviniemi et al., 2004). Interestingly, in benign samples, syndecan-1 is expressed in basal and secretory epithelial cells with basolateral membrane distribution, whereas syndecan-2 is confined to the basal cells. In prostate cancer samples, the expression patterns of both syndecans change to granular-cytoplasmic localization (Ledezma et al., 2011). Also, loss of epithelial positivity together with stromal expression of syndecan-1 correlated with poor prognosis.

A recent study of CD138 reactivity in renal tumors demonstrates that both clear cell and papillary renal cell carcinomas show membranous staining, whereas in some of the other eosinophilic renal tumors there are varying degrees of cytoplasmic CD138 reactivity (Ozcan et al., 2011). In pancreatic cancer tissue samples, syndecan-1 is upregulated (Conejo et al., 2000); however, more than 70% of pancreatic carcinoma specimens have weak or no immunoreactivity in both epithelial and stromal compartments (Juuti et al., 2005).

The expression of syndecan-1, which is the major PG in the normal liver, varies when a hepatocellular a carcinoma develops (Zvibel et al., 2009). The expression of this PG increases in conditions like chronic hepatitis C infection and liver cirrhosis, both conditions having premalignant potentials. When hepatocellular carcinomas develop in noncirrhotic livers, the tumor cells express less syndecan-1, whereas the expression remains high in both the cirrhotic and noncirrhotic peritumoral liver parenchyma. Here the syndecan-1 expression is downregulated parallel with the histological dedifferentiation of the malignant hepatocyte, while on the condition that the hepatocellular carcinoma develops from a cirrhotic liver, the syndecan-1 expression is retained at a higher level. This indicates that either cirrhosis or the causative agent that induces liver cirrhosis is capable of upregulating the syndecan-1 expression, and this program is not effectively turned off by the malignant transformation. These observations imply that the assembly of PGs by particular tumors is determined by three independent factors: (1) the origin of the tumor tissue, (2) previous diseases influencing PG expression, and (3) biological features of tumors themselves.

7.5.8 Syndecan expression affects tumor cell behavior

Experimental studies on the role of syndecan-1 in malignancy have shown that syndecan-1 expression associates with the maintenance of epithelial morphology and inhibition of tumor cell growth and invasiveness. Several studies have also shown that modulation of syndecan-1 expression can affect tumor cell behavior.

Cell proliferation

Syndecan-1 overexpression influences tumor cell proliferation in a cell-type-dependent manner. In mouse squamous cell carcinoma (Hirabayashi et al., 1998), HT1080 fibrosarcoma cells (Peterfia et al., 2006), and human endometrial cancer cells (Choi et al., 2007) such overexpression increases cell proliferation. On the other hand, decreased cell growth has been seen when overexpressing full-length syndecan-1 in mouse mammary tumor cells (Leppa et al., 1992) and also in B6FS fibrosarcoma cell line and mesothelioma cells (Zong et al., 2010). This overexpression concomitantly decreases the expression of syndecan-2, rendering fibrosarcoma cells a more epithelial-like morphology (Zong et al., 2010). This duality in the role of syndecan-1 in cell proliferation indicates a complex regulatory mechanism, which is tissue and/or tumor type specific, and at least partly dependent upon serum conditions (Numa et al., 1995).

Studies have described the roles of distinct syndecan-1 domains in proliferation, mainly focusing on the ectodomain. Overexpression of various truncated syndecan-1 constructs that lack the C-terminal cytoplasmic tail induce hepatocyte (Cortes et al., 2007) and myeloma (Yang et al., 2002) proliferation. However, inhibitory effects have also been reported. For example, the entire mouse syndecan-1 ectodomain suppresses the growth of mouse mammary tumor cells (Mali et al., 1994), and a similar effect is obtained by transfection of these cells with minican, a shorter segment of the syndecan-1 ectodomain containing the distal GAG attachment sites (Viklund et al., 2002). Interestingly, a recent study shows differential roles for membrane-bound and soluble syndecan-1 in breast cancer progression (Nikolova et al., 2009). Overexpression of full-length syndecan-1 increases the proliferation of these cells, whereas shed syndecan-1 decreases it. Constitutive membrane-bound syndecan-1 inhibits invasiveness, whereas the soluble form promotes invasion of cells into Matrigel. Thus, the proteolytic conversion of syndecan-1 from a membrane-bound form into a soluble molecule marks a switch in these cells from a proliferative to an invasive phenotype.

Different syndecan-1 domains, however, seem to hamper the growth of mesothelioma and B6FS fibrosarcoma cells in two principally different ways. With the extracellular domain present, syndecan-1 leads to a longer S-phase; whereas the transmembrane/cytoplasmic domains result in a prolonged G0/G1 phase (Zong et al., 2010). At the same time, the growth rate of hepatoma cell lines is not affected by over expressing full-length syndecan-1, while a truncated construct, containing the intracellular and transmembrane domains together with the first four amino acids (DRKE) of the extracellular domain, inhibits cell proliferation and induces differentiation. Simultaneously, the ratio between cell surface and shed syndecan-1 decreases (Kovalszky et al., 2004). It seems that this hampered proliferation also correlates to the ratio of cell-surface-bound and shed extracellular syndecan-1, rather than to the absolute amounts of this PG, although the increased load of syndecan-1 cytoplasmic domain may also influence. Another effect of transfection with truncated syndecan-1 is a downregulation of ETS1 transcription factor – a key regulator of matrix metalloproteinase-7 (MMP7) – resulting in diminished activity of this enzyme. The role of ETS1 oncogene in hepato cellular carcinoma is well established, and it may regulate proliferation of the hepatoma cell lines (Ozaki et al., 2000).

The effect on proliferation varies between different fibrosarcoma cell lines. The HT1080 fibrosarcoma cells increase their proliferation rate after transfection with syndecan-1 contructs, whereas the B6FS cell line shows the opposite effect. As the truncated constructs are as effective as the full-length protein, it can be concluded that the extracellular domain is not needed to enhance the proliferation of these cells. These different cells seem to employ different signal transduction routes to influence cell behavior. The intracellular domain of syndecan-1, together with the first four extracellular amino acids, seems to be capable of generating signal transduction in hepatoma cells, resulting in decreased extracellular signal-regulated kinases 1 and 2 (ERK1/2) phosphorylation and downregulation of ETS1 transcription factor. As for the signaling through plasma membrane, preliminary results indicate the role of G proteins. The increased proliferation seen in the HT1080 cells goes together with enhanced expression of syndecan-2, and silencing of this PG hampers the effects. It therefore seems that syndecan-2 is a downstream effector; however, the cooperation between the two paralogs is important in certain cell types only and is not necessarily utilized by others (Park et al., 2005).

Cell adhesion, motility, and migration

Invasion is one of the key criteria of solid malignant tumors. An important factor in this process is the complex interplay between cell adhesion, motility, and migration. Tumor cells must adhere to the ECM and surfaces of adjacent cells as they invade. Cell motility and migration are dynamic processes that require continuous assembly and disassembly of such cell-cell and cell-matrix adhesions (Webb et al., 2002). These cell behaviors are mediated by the interactions between cell membrane receptors and the surrounding microenvironment, as well as the internal cytoskeleton. Among the many interacting molecules, syndecans are emerging as important regulators for these processes and thus crucial for tumor invasion.

Syndecan-1 decreases migration and motility and enhances adhesion of mesenchymal tumor cells in an expression-level-dependent manner (Zong et al., 2011). The distinct protein domains have different effects. Adhesion of tumor cells depends mainly on the extracellular domain with some contribution from the juxtamembrane DRKE motif. Effects on migration are more complex; the proadhesive effect of the extracellular domain is one factor hampering migration, but the transmembrane/cytoplasmic portions have more impact on cell motility. Gene microarray analysis identifies a number of differentially expressed genes in syndecan-1 overexpressing cells coding for proteins associated with cell adhesion and migration (Zong et al., 2011). Expression of syndecan-1 also inhibits the invasion of myeloma cells into type I collagen gel (Liebersbach and Sanderson, 1994) and an invasion regulatory domain within the ectodomain of syndecan-1 has been identified and proved to inhibit the invasion of these cells (Langford et al., 2005).

Overexpression of full-length syndecan-1, or a truncated variant that lacks the ectodomain, promotes, however, the formation of metastases of HT1080 fibrosarcoma in a mouse model (Peterfia et al., 2006). Downregulation of syndecan-1 expression by siRNA disrupts αVβ3-integrin-dependent cell spreading and migration of human mammary carcinoma cells (Beauvais et al., 2004). It has also recently been reported that overexpression of syndecan-2 enhances the migration and invasion of melanoma cells (Lee et al., 2009). Interestingly, syndecan-1 influences the expression profile of the other syndecan family members, in particular syndecan-2 and syndecan-4 (Zong et al., 2011), and it is possible that the effect on cell migration is modulated by an orchestrated interplay between different members of the syndecan family.

7.5.9 Potential for translation

There has been until now only a limited utility for syndecans in clinical practice. The CD138 epitope, located at the extracellular domain of syndecan-1, is characteristically present on plasma cells and therefore commonly used as an immunocytochemical diagnostic marker for malignant myeloma. Today, there are commercially available reagent kits to measure this epitope also in serum, allowing the use of syndecan-1 as a biomarker for estimating myeloma tumor burden (Seidel et al., 2000).

There are also other possible ways to use syndecan analyses in a diagnostic context. The different expression of syndecans-1 and -2 by cells of epithelial and mesenchymal origin can be detected immunocytochemically. It has thus been suggested as a tool to distinguish a malignant mesothelioma from an adenocarcinoma (Gulyas and Hjerpe, 2003), syndecan-2 reactivity being dominant in mesothelioma and syndecan- 1 in carcinomas. Like most other "mesothelioma-specific" immunocytochemical criteria, a typical reactivity will strongly support the diagnosis, however, not being entirely specific in itself. The expression of syndecans by tumors raises the possibility to analyze syndecans as biomarkers also for nonhematological tumors, although this is not yet established in clinical routine.

The rapidly expanding knowledge of syndecans and their importance for the tumor cell indicate possibilities for new therapeutic means and targeted therapies. Syndecans seem to influence the malignant cell in many different ways and have been described as "tuners of membrane signaling." It can therefore be speculated that targeting of these functions may have profound effects on the tumor cell behavior and provide a basis for subsequent targeted therapy in these tumors.

One such target could be a domain of the molecule with a specific function; either this function relates to interactions with the microenvironment, influences proliferation and migration, or regulates nuclear translocation. One interesting concept concerns possible interactions with specific GAG sequences, where specific interactions may be interfered with using soluble HS oligosaccharides to perturb the function of cell surface HSPGs. There are other possibilities to block functional domains, either by blocking antibodies or competitively by adding the free peptides in excess. Transfection of tumor cells with a construct expressing the syndecan NLS sequence (RMKKK) has profound effects on the tumor cell behavior (Zong et al., 2010, 2011), perhaps as a consequence of competition with the nuclear transport and/or binding of wild-type syndecan-1 in the nucleus.

HS and particularly syndecan-1 have been shown to enter the nucleus and to modulate crucial biological responses in target cells, including proliferation and migration (Fedarko et al., 1989; Roghani and Moscatelli, 1992; Cheng et al., 2001; Brockstedt et al., 2002; Zong et al., 2010, 2011). Moreover, xyloside priming modulates the biosynthesis and nuclear uptake of GAGs, resulting in secretion of an antiproliferative GAG, which specifically targets tumor cells in an autocrine fashion (Nilsson et al., 2010).

Accumulating evidence also suggests that many gene- and protein-delivery vehicles, including cell-penetrating peptides and nanoparticles, are largely dependent on negatively charged HSPGs for efficient intracellular delivery (Mislick and Baldeschwieler 1996; Belting, 2003; Belting et al., 2003; Sandgren et al., 2004; Poon and Gariepy 2007; Wittrup et al., 2007, 2009, 2011; Letoha et al., 2010). Such delivery of macromolecules across biological membranes offers promising possibilities of developing novel treatments (Belting et al., 2005; Raja et al., 2005; Belting and Wittrup, 2009a 2009b). The design of intracellular drug delivery vehicles, however, requires an increased understanding of the physiological ligands that mediate cellular communication and transport across the plasma membrane and also through the nuclear envelope.

These possibilities are still only ideas that need to be evaluated first in the laboratory. Before any of this can be tested in clinical settings, we need better knowledge about the precise mechanisms for these structure-function relationships, including specific protein and oligosaccharide sequences involved in this process. It is also necessary to better explain the apparent differences in syndecan-1 effects comparing tumor cells of different origin.

7.5.10 Take-home message

Syndecan-1 is involved in several functions of the tumor cell. The molecule may exert its function not only at the cell membrane but also in the cell nucleus. The different functions of this proteoglycan depend on its different domains and probably also on its location in the cell. The central role this proteoglycan plays in a malignant cell makes it an intriguing possible target in future tumor therapy.

References

Alexopoulou, A. N., Multhaupt, H. A., and Couchman, J. R. (2007). Syndecans in wound healing, inflammation and vascular biology. Int J Biochem Cell Biol *39* (3), 505–528.

Anttonen, A., Heikkila, P., Kajanti, M., Jalkanen, M., and Joensuu, H. (2001). High syndecan-1 expression is associated with favourable outcome in squamous cell lung carcinoma treated with radical surgery. Lung Cancer *32* (3), 297–305.

Au, Y. P., Dobrowolska, G., Morris, D. R., and Clowes, A.W. (1994). Heparin decreases activator protein-1 binding to DNA in part by posttranslational modification of Jun B. Circ Res *75* (1), 15–22.

Barbareschi, M., Maisonneuve, P., Aldovini, D., et al. (2003). High syndecan-1 expression in breast carcinoma is related to an aggressive phenotype and to poorer prognosis. Cancer *98* (3), 474–483.

Beauvais, D. M., Burbach, B. J., and Rapraeger, A. C. (2004). The syndecan-1 ectodomain regulates alphavbeta3 integrin activity in human mammary carcinoma cells. J Cell Biol *167* (1), 171–181.

Beauvais, D. M., Ell, B. J., McWhorter, A. R., and Rapraeger, A. C. (2009). Syndecan-1 regulates alphavbeta3 and alphavbeta5 integrin activation during angiogenesis and is blocked by synstatin, a novel peptide inhibitor. J Exp Med *206* (3), 691–705.

Belting, M. (2003). Heparan sulfate proteoglycan as a plasma membrane carrier. Trends Biochem Sci *28* (3), 145–151.

Belting, M., Mani, K., Jönsson, M., et al. (2003). Glypican-1 is a vehicle for polyamine uptake in mammalian cells: a pivotal role for nitrosothiol-derived nitric oxide. J Biol Chem *278* (47), 47181–47189.

Belting, M., Sandgren, S., Wittrup, A., et al. (2005). Nuclear delivery of macromolecules: barriers and carriers. Adv Drug Deliv Rev *57* (4), 505–527.

Belting, M., and Wittrup, A. (2009a). Developments in macromolecular drug delivery. Methods Mol Biol *480*, 1–10.

Belting, M., and Wittrup, A. (2009b). Macromolecular drug delivery: basic principles and therapeutic applications. Mol Biotechnol *43* (1), 89–94.

Bernfield, M., Götte, M., Park, P. W., et al. (1999). Functions of cell surface heparan sulfate proteoglycans. Annu Rev Biochem *68*, 729–777.

Bernfield, M., Kokenyesi, R., Kato, M., et al. (1992). Biology of the syndecans: a family of transmembrane heparan sulfate proteoglycans. Annu Rev Cell Biol *8*, 365–393.

Bobardt, M. D., Saphire, A. C., Hung, H. C., et al. (2003). Syndecan captures, protects, and transmits HIV to T lymphocytes. Immunity *18* (1), 27–39.

Börset, M., Hjertner, O., Yaccoby, S., Epstein, J., and Sanderson, R. D. (2000). Syndecan-1 is targeted to the uropods of polarized myeloma cells where it promotes adhesion and sequesters heparin-binding proteins. Blood *96* (7), 2528–2536.

Bourdon, M. A., and Ruoslahti, E. (1989). Tenascin mediates cell attachment through an RGD-dependent receptor. J Cell Biol *108* (3), 1149–1155.

Brockstedt, U., Dobra, K., Nurminen, M., and Hjerpe, A. (2002). Immunoreactivity to cell surface syndecans in cytoplasm and nucleus: tubulin-dependent rearrangements. Exp Cell Res *274* (2), 235–245.

Carey, D. J. (1997). Syndecans: multifunctional cell-surface co-receptors. Biochem J *327* (Pt 1), 1–16.

Carey, D. J., Evans, D. M., et al. (1992). Molecular cloning and characterization of N-syndecan, a novel transmembrane heparan sulfate proteoglycan. J Cell Biol *117* (1), 191–201.

Carey, D. J., Stahl, R. C., Stahl, R. C., et al. (1994). Aggregation-induced association of syndecan-1 with microfilaments mediated by the cytoplasmic domain. Exp Cell Res *214* (1), 12–21.

Chakravarti, R., Sapountzi, V., and Adams, J. C. (2005). Functional role of syndecan-1 cytoplasmic V region in lamellipodial spreading, actin bundling, and cell migration. Mol Biol Cell *16* (8), 3678–3691.

Chen, D., Adenekan, B., Chen, L., et al. (2004). Syndecan-1 expression in locally invasive and metastatic prostate cancer. Urology *63* (2), 402–407.

Chen, L., and Sanderson, R. D. (2009). Heparanase regulates levels of syndecan-1 in the nucleus. PLoS One *4* (3), e4947.

Cheng, F., Petersson, P., Arroyo-Yanguas, Y., and Westergren-Thorsson, G. (2001). Differences in the uptake and nuclear localization of anti-proliferative heparan sulfate between human lung fibroblasts and human lung carcinoma cells. J Cell Biochem *83* (4), 597–606.

Choi, D. S., Kim, J. H., Ryu, H. S., et al. (2007). Syndecan-1, a key regulator of cell viability in endometrial cancer. Int J Cancer *121* (4), 741–750.

Cizmeci-Smith, G., Stahl, R. C., Showalter, L. J., and Carey, D. J. (1993). Differential expression of transmembrane proteoglycans in vascular smooth muscle cells. J Biol Chem *268* (25), 18740–18747.

Conejo, J. R., Kleeff, J., Koliopanos, A., et al. (2000). Syndecan-1 expression is up-regulated in pancreatic but not in other gastrointestinal cancers. Int J Cancer *88* (1), 12–20.

Cook, D. M., Hinkes, M. T., Bernfield M., and Rauscher, F. J., 3rd. (1996). Transcriptional activation of the syndecan-1 promoter by the Wilms' tumor protein WT1. Oncogene *13* (8), 1789–1799.

Cortes, V., Amigo, L., Donoso, K., et al,. (2007). Adenovirus-mediated hepatic syndecan-1 overexpression induces hepatocyte proliferation and hyperlipidaemia in mice. Liver Int *27* (4), 569–581.

David, G., van der Schueren, B., Marynen, P., Cassiman, J. J., and van den Berghe, H. (1992). Molecular cloning of amphiglycan, a novel integral membrane heparan sulfate proteoglycan expressed by epithelial and fibroblastic cells. J Cell Biol *118* (4), 961–969.

Davies, E. J., Blackhall, F. H., Shanks, J. H., et al. (2004). Distribution and clinical significance of heparan sulfate proteoglycans in ovarian cancer. Clin Cancer Res *10* (15), 5178–5186.

Deepa, S. S., Yamada, S., Zako, M., Goldberger, O., and Sugahara, K. (2004). Chondroitin sulfate chains on syndecan-1 and syndecan-4 from normal murine mammary gland epithelial cells are structurally and functionally distinct and cooperate with heparan sulfate chains to bind growth factors. A novel function to control binding of midkine, pleiotrophin, and basic fibroblast growth factor. J Biol Chem *279* (36), 37368–37376.

Dingwall, C., Robbins, J., Dilworth, S. M., Roberts, B., and Richardson, W. D. (1988). The nucleoplasmin nuclear location sequence is larger and more complex than that of SV-40 large T antigen. J Cell Biol *107* (3), 841–849.

Dobra, K., Andang, M., Syrokou, A., Karamanos, N. K., and Hjerpe, A. (2000). Differentiation of mesothelioma cells is influenced by the expression of proteoglycans. Exp Cell Res *258* (1), 12–22.

Dobra, K., Nurminen, M., and Hjerpe, A. (2003). Growth factors regulate the expression profile of their syndecan co-receptors and the differentiation of mesothelioma cells. Anticancer Res *23* (3B), 2435–2444.

Dudás, J., Ramadori, G., Knittel, T., et al. (2000). Effect of heparin and liver heparan sulphate on interaction of HepG2-derived transcription factors and their cis-acting elements: altered potential of hepatocellular carcinoma heparan sulphate. Biochem J *350* (Pt 1), 245–251.

Edwards, I. J., Sun, H., Hu, Y., et al. (2008). In vivo and in vitro regulation of syndecan 1 in prostate cells by n-3 polyunsaturated fatty acids. J Biol Chem *283* (26), 18441–18449.

Elenius, K., Määttä, A., Salmivirta, M., and Jalkanen, M. (1992). Growth factors induce 3T3 cells to express bFGF-binding syndecan. J Biol Chem *267* (9), 6435–6441.

Elenius, K., Vainio, S., Laato, M., Salmivirta, M., Thesleff, I., and Jalkanen M. (1991). Induced expression of syndecan in healing wounds. J Cell Biol *114* (3), 585–595.

Endo, K., Takino, T., Miyamori, H., et al. (2003). Cleavage of syndecan-1 by membrane type matrix metalloproteinase-1 stimulates cell migration. J Biol Chem *278* (42), 40764–40770.

Esko, J. D., and Zhang, L. (1996). Influence of core protein sequence on glycosaminoglycan assembly. Curr Opin Struct Biol *6* (5), 663–670.

Fears, C. Y., and Woods, A. (2006). The role of syndecans in disease and wound healing. Matrix Biol *25* (7), 443–456.

Fedarko, N. S., Ishihara, M., and Conrad, H. E. (1989). Control of cell division in hepatoma cells by exogenous heparan sulfate proteoglycan. J Cell Physiol *139* (2), 287–294.

Fujiya, M., Watari, J., Ashida, T., et al. (2001). Reduced expression of syndecan-1 affects metastatic potential and clinical outcome in patients with colorectal cancer. Jpn J Cancer Res *92* (10), 1074–1081.

Gallo, R., Kim, C., Kokenyesi, R., Adzick, N. S., and Bernfield M. (1996). Syndecans-1 and -4 are induced during wound repair of neonatal but not fetal skin. J Invest Dermatol *107* (5), 676–683.

Gökden, N., Greene, G. F., Bayer-Garner, I. B., Spencer, H. J., Sanderson, R. D., and Gökden M. (2006). Expression of CD138 (Syndecan-1) in renal cell carcinoma is reduced with increasing nuclear grade. Appl Immunohistochem Mol Morphol *14* (2), 173–177.

Götte, M. (2003). Syndecans in inflammation. FASEB J *17* (6), 575–591.

Gould, S. E., Upholt, W. B., and Kosher, R. A. (1992). Syndecan 3: a member of the syndecan family of membrane-intercalated proteoglycans that is expressed in high amounts at the onset of chicken limb cartilage differentiation. Proc Natl Acad Sci U S A *89* (8), 3271–3275.

Gulyas, M., and Hjerpe, A. (2003). Proteoglycans and WT1 as markers for distinguishing adenocarcinoma, epithelioid mesothelioma, and benign mesothelium. J Pathol *199* (4), 479–487.

Habuchi, O. (2000). Diversity and functions of glycosaminoglycan sulfotransferases. Biochim Biophys Acta *1474* (2), 115–127.

Hayashi, K., Hayashi, M., Jalkanen, M., Firestone, J. H., Trelstad, R. L., and Bernfield, M. (1987). Immunocytochemistry of cell surface heparan sulfate proteoglycan in mouse tissues. A light and electron microscopic study. J Histochem Cytochem *35* (10), 1079–1088.

Hayashida, K., Stahl, P. D., and Park, P. W. (2008). Syndecan-1 ectodomain shedding is regulated by the small GTPase Rab5. J Biol Chem *283* (51), 35435–35444.

Hirabayashi, K., Numa, F., Suminami, Y., Murakami, A., Murakami, T., and Kato, H. (1998). Altered proliferative and metastatic potential associated with increased expression of syndecan-1. Tumour Biol *19* (6), 454–463.

Hooper, N. M., Karran, E. H., and Turner, A. J. (1997). Membrane protein secretases. Biochem J *321* (Pt 2), 265–279.

Hsia, E., Richardson, T. P., and Nugent, M. A. (2003). Nuclear localization of basic fibroblast growth factor is mediated by heparan sulfate proteoglycans through protein kinase C signaling. J Cell Biochem *88* (6), 1214–1225.

Ilan, N., Elkin, M., and Vlodavsky, I. (2006). Regulation, function and clinical significance of heparanase in cancer metastasis and angiogenesis. Int J Biochem Cell Biol *38* (12), 2018–2039.

Inki, P., Joensuu, H., Grénman, R., Klemi, P., and Jalkanen, M. (1994). Association between syndecan-1 expression and clinical outcome in squamous cell carcinoma of the head and neck. Br J Cancer *70* (2), 319–323.

Inki, P., Stenbäck, F., Talve, L., and Jalkanen, M. (1991). Immunohistochemical localization of syndecan in mouse skin tumors induced by UV irradiation. Loss of expression associated with malignant transformation. Am J Pathol *139* (6), 1333–1340.

Jaakkola, P., Kontusaari, S., Kauppi, T., Määtä, A., and Jalkanen, M. (1998). Wound reepithelialization activates a growth factor-responsive enhancer in migrating keratinocytes. FASEB J *12* (11), 959–969.

Jackson, L. L., Wade, Z., Hessler, R. B., Abdelsayed, R., Rogers, J. B., and Gourin, C. G. (2007). Quantitative analysis of syndecan-1 expression in dysplasia and squamous cell carcinoma of the oral cavity. Laryngoscope *117* (5), 868–871.

Joensuu, H., Anttonen, A., Eriksson, M., et al. (2002). Soluble syndecan-1 and serum basic fibroblast growth factor are new prognostic factors in lung cancer. Cancer Res *62* (18), 5210–5217.

Juuti, A., Nordling, S., Lundin, J., Louhimo, J., and Haglund, C. (2005). Syndecan-1 expression – a novel prognostic marker in pancreatic cancer. Oncology *68* (2–3), 97–106.

Kainulainen, V., Nelimarkka, L., Järveläinen, H., Laato, M., Jalkanen, M., and Elenius, K. (1996). Suppression of syndecan-1 expression in endothelial cells by tumor necrosis factor-alpha. J Biol Chem *271* (31), 18759–18766.

Kalderon, D., Roberts, B. L., Richardson, W. D., and Smith, A. E. (1984). A short amino acid sequence able to specify nuclear location. Cell *39* (3 Pt 2), 499–509.

Kato, M., Saunders, S., Nguyen, H., and Bernfield, M. (1995). Loss of cell surface syndecan-1 causes epithelia to transform into anchorage-independent mesenchyme-like cells. Mol Biol Cell *6* (5), 559–576.

Khotskaya, Y. B., Dai, Y., Ritchie, J. P., et al. (2009). Syndecan-1 is required for robust growth, vascularization, and metastasis of myeloma tumors in vivo. J Biol Chem *284* (38), 26085–26095.

Kim, C. W., Goldberger, O. A., Gallo, R. L., and Bernfield, M. (1994). Members of the syndecan family of heparan sulfate proteoglycans are expressed in distinct cell-, tissue-, and development-specific patterns. Mol Biol Cell *5* (7), 797–805.

Kiviniemi, J., Kallajoki, M., Kujala, I., et al. (2004). Altered expression of syndecan-1 in prostate cancer. APMIS *112* (2), 89–97.

Kjellen, L., and Lindahl, U. (1991). Proteoglycans: structures and interactions. Annu Rev Biochem *60*, 443–475.

Klemm, J. D., Schreiber, S. L., and Crabtree, G. R. (1998). Dimerization as a regulatory mechanism in signal transduction. Annu Rev Immunol *16*, 569–592.

Kojima, T., Shworak, N. W., and Rosenberg, R. D. (1992). Molecular cloning and expression of two distinct cDNA-encoding heparan sulfate proteoglycan core proteins from a rat endothelial cell line. J Biol Chem *267* (7), 4870–4877.

Kokenyesi, R., and Bernfield, M. (1994). Core protein structure and sequence determine the site and presence of heparan sulfate and chondroitin sulfate on syndecan-1. J Biol Chem *269* (16), 12304–12309.

Kovalszky, I., Dudás, J., Gallai, M., et al. (2004). [Proteoglycans in the liver]. Magy Onkol *48* (3), 207–213.

Kovalszky, I., Dudás, J., Oláh-Nagy, J., et al. (1998). Inhibition of DNA topoisomerase I activity by heparan sulfate and modulation by basic fibroblast growth factor. Mol Cell Biochem *183* (1–2), 11–23.

Kumar-Singh, S., Jacobs, W., Dhaene, K., et al. (1998). Syndecan-1 expression in malignant mesothelioma: correlation with cell differentiation, WT1 expression, and clinical outcome. J Pathol *186* (3), 300–305.

Lambaerts, K., Wilcox-Adelman, S. A., and Zimmermann, P. (2009). The signaling mechanisms of syndecan heparan sulfate proteoglycans. Curr Opin Cell Biol *21* (5), 662–669.

Langford, J. K., Yang, Y., Kieber-Emmons, T., and Sanderson, R. D. (2005). Identification of an invasion regulatory domain within the core protein of syndecan-1. J Biol Chem *280* (5), 3467–3473.

Leadbeater, W. E., Gonzalez, A. M., Logaras, N., Berry, M., Turnbull, J. E., and Logan, A. (2006). Intracellular trafficking in neurones and glia of fibroblast growth factor-2, fibroblast growth factor receptor 1 and heparan sulphate proteoglycans in the injured adult rat cerebral cortex. J Neurochem *96* (4), 1189–1200.

Ledezma, R., Cifuentes, F., Gallegos, I., et al. (2011). Altered expression patterns of syndecan-1 and -2 predict biochemical recurrence in prostate cancer. Asian J Androl *13* (3), 476–480.

Lee, J. H., Park, H., Chung, H., et al. (2009). Syndecan-2 regulates the migratory potential of melanoma cells. J Biol Chem *284* (40), 27167–27175.

Lendorf, M. E., Manon-Jensen, T., Kronqvist, P., Multhaupt, H. A., and Couchman, J. R. (2011). Syndecan-1 and syndecan-4 are independent indicators in breast carcinoma. J Histochem Cytochem *59* (6), 615–629.

Leppä, S., Mali, M., Miettinen, H. M., and Jalkanen, M. (1992). Syndecan expression regulates cell morphology and growth of mouse mammary epithelial tumor cells. Proc Natl Acad Sci U S A *89* (3), 932–936.

Letian, T., and Tianyu, Z. (2010). Cellular receptor binding and entry of human papillomavirus. Virol J *7*, 2.

Letoha, T., Keller-Pinter, A., Kusz, E., et al. (2010). Cell-penetrating peptide exploited syndecans. Biochim Biophys Acta *1798* (12), 2258–2265.

Liang, Y., Haring, M., Roughley, P. J., Margolis, R. K., and Margolis, R. U. (1997). Glypican and biglycan in the nuclei of neurons and glioma cells: presence of functional nuclear localization signals and dynamic changes in glypican during the cell cycle. J Cell Biol *139* (4), 851–864.

Liebersbach, B. F., and Sanderson, R. D. (1994). Expression of syndecan-1 inhibits cell invasion into type I collagen. J Biol Chem *269* (31), 20013–20019.

Lindahl, U., Kusche-Gullberg, M., and Kjellén, L. (1998). Regulated diversity of heparan sulfate. J Biol Chem *273* (39), 24979–24982.

Mali, M., Andtfolk, H., Miettinen, H. M., and Jalkanen, M. (1994). Suppression of tumor cell growth by syndecan-1 ectodomain. J Biol Chem *269* (45), 27795–27798.

Marynen, P., Zhang, J., Cassiman, J. J., Van den Berghe, H., and David, G. (1989). Partial primary structure of the 48- and 90-kilodalton core proteins of cell surface-associated heparan sulfate proteoglycans of lung fibroblasts. Prediction of an integral membrane domain and evidence for multiple distinct core proteins at the cell surface of human lung fibroblasts. J Biol Chem *264* (12), 7017–7024.

Mathe, M., Suba, Z., Németh, Z., et al. (2006). Stromal syndecan-1 expression is an adverse prognostic factor in oral carcinomas. Oral Oncol *42* (5), 493–500.

Menashe, I., Maeder, D., Garcia-Closas, M., et al. (2010). Pathway analysis of breast cancer genome-wide association study highlights three pathways and one canonical signaling cascade. Cancer Res *70* (11), 4453–4459.

Mikami, S., Ohashi, K., Usui, Y., et al. (2001). Loss of syndecan-1 and increased expression of heparanase in invasive esophageal carcinomas. Jpn J Cancer Res *92* (10), 1062–1073.

Mislick, K. A., and Baldeschwieler, J. D. (1996). Evidence for the role of proteoglycans in cation-mediated gene transfer. Proc Natl Acad Sci U S A *93* (22), 12349–12354.
Molica, S., Vitelli, G., Mirabelli, R., et al. (2006). Serum levels of syndecan-1 in B-cell chronic lymphocytic leukemia: correlation with the extent of angiogenesis and disease-progression risk in early disease. Leuk Lymphoma *47* (6), 1034–1040.
Nakanishi, K., Yoshioka, N., Oka, K., and Hakura, A. (1999). Reduction of syndecan-1 mRNA in cervical-carcinoma cells is involved with the 3′ untranslated region. Int J Cancer *80* (4), 527–532.
Niggli, V., Schlicht, D., and Affentranger, S. (2009). Specific roles of Rac1 and Rac2 in motile functions of HT1080 fibrosarcoma cells. Biochem Biophys Res Commun *386* (4), 688–692.
Nikolova, V., Koo, C. Y., Ibrahim, S. A., et al. (2009). Differential roles for membrane-bound and soluble syndecan-1 (CD138) in breast cancer progression. Carcinogenesis *30* (3), 397–407.
Nilsson, U., Johnsson, R., Fransson, L. Å., Ellervik, U., and Mani, K. (2010). Attenuation of tumor growth by formation of antiproliferative glycosaminoglycans correlates with low acetylation of histone H3. Cancer Res *70* (9), 3771–3779.
Nobuhisa, T., Naomoto, Y., Takaoka, M., et al. (2005). Emergence of nuclear heparanase induces differentiation of human mammary cancer cells. Biochem Biophys Res Commun *331* (1), 175–180.
Numa, F., Hirabayashi, K., Kawasaki, K., et al. (2002). Syndecan-1 expression in cancer of the uterine cervix: association with lymph node metastasis. Int J Oncol *20* (1), 39–43.
Numa, F., Hirabayashi, K., Tsunaga, N., et al. (1995). Elevated levels of syndecan-1 expression confer potent serum-dependent growth in human 293T cells. Cancer Res *55* (20), 4676–4680.
O'Connell, F. P., Pinkus, J. L., and Pinkus, G. S. (2004). CD138 (syndecan-1), a plasma cell marker immunohistochemical profile in hematopoietic and nonhematopoietic neoplasms. Am J Clin Pathol *121* (2), 254–263.
Oh, J. H., Lee, H. S., Park, S. H., Ryu, H. S., and Min, C. K. (2010). Syndecan-1 overexpression promotes tumor growth and angiogenesis in an endometrial cancer xenograft model. Int J Gynecol Cancer **20** (5), 751–756.
Ohkawa, T., Naomoto, Y., Takaoka, M., et al. (2004). Localization of heparanase in esophageal cancer cells: respective roles in prognosis and differentiation. Lab Invest *84* (10), 1289–1304.
Ojeh, N., Hiilesvuo, K., Wärri, A., Salmivirta, M., Henttinen, T., and Määttä, A. (2008). Ectopic expression of syndecan-1 in basal epidermis affects keratinocyte proliferation and wound re-epithelialization. J Invest Dermatol *128* (1), 26–34.
Okamoto, O., Bachy, S., Odenthal, U., et al. (2003). Normal human keratinocytes bind to the alpha3LG4/5 domain of unprocessed laminin-5 through the receptor syndecan-1. J Biol Chem *278* (45), 44168–44177.
Orosz, Z., and Kopper, L. (2001). Syndecan-1 expression in different soft tissue tumours. Anticancer Res *21* (1B), 733–737.
Ozaki, I., Mizuta, T., Zhao, G., et al. (2000). Involvement of the Ets-1 gene in overexpression of matrilysin in human hepatocellular carcinoma. Cancer Res *60* (22), 6519–6525.
Ozcan, A., Celik, E., Karslioğlu, Y., and Basal, S. (2011). CD138 expression in renal tumors and its usage in the differential diagnosis. Turk Patoloji Derg *27* (2), 110–115.
Park, H., Han, I., Kwon, H. J., and Oh, E. S. (2005). Focal adhesion kinase regulates syndecan-2-mediated tumorigenic activity of HT1080 fibrosarcoma cells. Cancer Res *65* (21), 9899–9905.
Perrimon, N., and Bernfield, M. (2001). Cellular functions of proteoglycans–an overview. Semin Cell Dev Biol *12* (2), 65–67.

Peterfia, B., Hollosi, P., Szilák, L., et al. (2006). [Role of syndecan-1 proteoglycan in the invasiveness of HT-1080 fibrosarcoma]. Magy Onkol *50* (2), 115–120.

Poon, G. M., and Gariepy, J. (2007). Cell-surface proteoglycans as molecular portals for cationic peptide and polymer entry into cells. Biochem Soc Trans *35* (Pt 4), 788–793.

Purushothaman, A., Chen, L., Yang, Y., and Sanderson, R. D. (2008). Heparanase stimulation of protease expression implicates it as a master regulator of the aggressive tumor phenotype in myeloma. J Biol Chem *283* (47), 32628–32636.

Purushothaman, A., Hurst, D. R., Pisano, C., Mizumoto, S., Sugahara, K., and Sanderson, R. D. (2011). Heparanase-mediated loss of nuclear syndecan-1 enhances histone acetyltransferase (HAT) activity to promote expression of genes that drive an aggressive tumor phenotype. J Biol Chem *286* (35), 30377–30383.

Raja, S. M., Metkar, S. S., Höning, S., et al. (2005). A novel mechanism for protein delivery: granzyme B undergoes electrostatic exchange from serglycin to target cells. J Biol Chem *280* (21), 20752–20761.

Rapraeger, A., Jalkanen, M., and Bernfield, M. (1986). Cell surface proteoglycan associates with the cytoskeleton at the basolateral cell surface of mouse mammary epithelial cells. J Cell Biol *103* (6 Pt 2), 2683–2696.

Rapraeger, A., Jalkanen, M., Endo, E., Koda, J., and Bernfield, M. (1985). The cell surface proteoglycan from mouse mammary epithelial cells bears chondroitin sulfate and heparan sulfate glycosaminoglycans. J Biol Chem *260* (20), 11046–11052.

Richardson, T. P., Trinkaus-Randall, V., and Nugent, M. A. (2001). Regulation of heparan sulfate proteoglycan nuclear localization by fibronectin. J Cell Sci *114* (Pt 9), 1613–1623.

Ridley, R. C., Xiao, H., Hata, H., Woodliff, J., Epstein, J., and Sanderson, R. D. (1993). Expression of syndecan regulates human myeloma plasma cell adhesion to type I collagen. Blood *81* (3), 767–774.

Rintala, M., Inki, P., Klemig, P., Jalkanen, M., and Grénman, S. (1999). Association of syndecan-1 with tumor grade and histology in primary invasive cervical carcinoma. Gynecol Oncol *75* (3), 372–378.

Robbins, J., Dilworth, S. M., Laskey, R. A., and Dingwall, C. (1991). Two interdependent basic domains in nucleoplasmin nuclear targeting sequence: identification of a class of bipartite nuclear targeting sequence. Cell *64* (3), 615–623.

Roghani, M., and Moscatelli, D. (1992). Basic fibroblast growth factor is internalized through both receptor-mediated and heparan sulfate-mediated mechanisms. J Biol Chem *267* (31), 22156–22162.

Sanderson, R. D., Lalor, P., and Bernfield, M. (1989). B lymphocytes express and lose syndecan at specific stages of differentiation. Cell Regul *1* (1), 27–35.

Sandgren, S., Wittrup, A., Cheng, F., et al. (2004). The human antimicrobial peptide LL-37 transfers extracellular DNA plasmid to the nuclear compartment of mammalian cells via lipid rafts and proteoglycan-dependent endocytosis. J Biol Chem *279* (17), 17951–17956.

Saunders, S., Jalkanen, M., O'Farrell, S., and Bernfield, M. (1989). Molecular cloning of syndecan, an integral membrane proteoglycan. J Cell Biol *108* (4), 1547–1556.

Schlessinger, J., Plotnikov, A. N., Ibrahimi, O. A., et al. (2000). Crystal structure of a ternary FGF-FGFR-heparin complex reveals a dual role for heparin in FGFR binding and dimerization. Mol Cell 6 (3), 743–750.

Schrage, Y. M., Hameetman, L., Szuhai, K., et al. (2009). Aberrant heparan sulfate proteoglycan localization, despite normal exostosin, in central chondrosarcoma. Am J Pathol *174* (3), 979–988.

Schubert, S. Y., Ilan, N., Shushy, M., Ben-Izhak, O., Vlodavsky, I., and Goldshmidt, O. (2004). Human heparanase nuclear localization and enzymatic activity. Lab Invest *84* (5), 535–544.

Seftalioglu, A., and Karakus, S. (2003). Syndecan-1/CD138 expression in normal myeloid, acute lymphoblastic and myeloblastic leukemia cells. Acta Histochem *105* (3), 213–221.

Seftalioglu, A., Karakus, S., Dundar, S., et al. (2003). Syndecan-1 (CD138) expression in acute myeloblastic leukemia cells–an immuno electron microscopic study. Acta Oncol *42* (1), 71–74.

Seidel, C., Sundan, A., Hjorth, M., et al. (2000). Serum syndecan-1: a new independent prognostic marker in multiple myeloma. Blood *95* (2), 388–392.

Shafti-Keramat, S., Handisurya, A., Kriehuber, E., Meneguzzi, G., Slupetzky, K., and Kirnbauer, R. (2003). Different heparan sulfate proteoglycans serve as cellular receptors for human papillomaviruses. J Virol *77* (24), 13125–13135.

Shimazu, A., Nah, H. D., Kirsch, T., et al. (1996). Syndecan-3 and the control of chondrocyte proliferation during endochondral ossification. Exp Cell Res *229* (1), 126–136.

Shinyo, Y., Kodama, J., Hasengaowa, Kusumoto, T., and Hiramatsu, Y. (2005). Loss of cell-surface heparan sulfate expression in both cervical intraepithelial neoplasm and invasive cervical cancer. Gynecol Oncol *96* (3), 776–783.

Shworak, N. W., Shirakawa, M., Mulligan, R. C., Rosenberg, R. D. (1994). Characterization of ryudocan glycosaminoglycan acceptor sites. J Biol Chem *269* (33), 21204–21214.

Silbert, J. E., and Sugumaran, G. (2002). Biosynthesis of chondroitin/dermatan sulfate. IUBMB Life *54* (4), 177–186.

Sun, H., Berquin, I. M., Owens, R. T., O'Flaherty, J. T., and Edwards, I. J. (2008). Peroxisome proliferator-activated receptor gamma-mediated up-regulation of syndecan-1 by n-3 fatty acids promotes apoptosis of human breast cancer cells. Cancer Res *68* (8), 2912–2919.

Tatrai, P., Dudas, J., Batmunkh, E., et al. (2006). Agrin, a novel basement membrane component in human and rat liver, accumulates in cirrhosis and hepatocellular carcinoma. Lab Invest *86* (11), 1149–1160.

Tatrai, P., Egedi, K., Somorácz, A., et al. (2010). Quantitative and qualitative alterations of heparan sulfate in fibrogenic liver diseases and hepatocellular cancer. J Histochem Cytochem *58* (5), 429–441.

Vassilakopoulos, T. P., Kyrtsonis, M. C., Papadogiannis, A., et al. (2005). Serum levels of soluble syndecan-1 in Hodgkin's lymphoma. Anticancer Res *25* (6C), 4743–4746.

Viklund, L., Loo, B. M., Hermonen, J., El-Darwish, K., Jalkanen, M., and Salmivirta, M. (2002). Expression and characterization of minican, a recombinant syndecan-1 with extensively truncated core protein. Biochem Biophys Res Commun *290* (1), 146–152.

Vlodavsky, I., Abboud-Jarrous, G., Elkin, M., et al. (2006). The impact of heparanese and heparin on cancer metastasis and angiogenesis. Pathophysiol Haemost Thromb *35* (1–2), 116–127.

Vlodavsky, I., Beckhove, P., Lerner, I., et al. (2011). Significance of heparanase in cancer and inflammation. Cancer Microenviron (Epub ahead of print).

Vlodavsky, I., Eldor, A., Bar-Ner, M., Fridman, R., Cohen, I. R., and Klagsbrun, M. (1988). Heparan sulfate degradation in tumor cell invasion and angiogenesis. Adv Exp Med Biol *233*, 201–210.

Webb, D. J., Parsons, J. T., and Horwitz, A. F. (2002). Adhesion assembly, disassembly and turnover in migrating cells – over and over and over again. Nat Cell Biol *4* (4), E97–E100.

Whiteford, J. R., Behrends, V., Kirby, H., Kusche-Gullberg, M., Muramatsu, T., and Couchman, J. R. (2007). Syndecans promote integrin-mediated adhesion of mesenchymal cells in two distinct pathways. Exp Cell Res *313* (18), 3902–3913.

Wittrup, A., Sandgren, S., Lilja, J., et al. (2007). Identification of proteins released by mammalian cells that mediate DNA internalization through proteoglycan-dependent macropinocytosis. J Biol Chem *282* (38), 27897–27904.

Wittrup, A., Zhang, S. H., and Belting, M. (2011). Studies of proteoglycan involvement in CPP-mediated delivery. Methods Mol Biol *683*, 99–115.

Wittrup, A., Zhang, S. H., ten Dam, G. B., et al. (2009). ScFv antibody-induced translocation of cell-surface heparan sulfate proteoglycan to endocytic vesicles: evidence for heparan sulfate epitope specificity and role of both syndecan and glypican. J Biol Chem *284* (47), 32959–32967.

Woods, A., and Couchman, J. R. (2001). Syndecan-4 and focal adhesion function. Curr Opin Cell Biol *13* (5), 578–583.

Woods, A., Longley, R. L., Tumova, S., and Couchman, J. R. (2000). Syndecan-4 binding to the high affinity heparin-binding domain of fibronectin drives focal adhesion formation in fibroblasts. Arch Biochem Biophys *374* (1), 66–72.

Worapamorn, W., Tam, S. P., Li, H., Haase, H. R., and Bartold, P. M. (2002). Cytokine regulation of syndecan-1 and -2 gene expression in human periodontal fibroblasts and osteoblasts. J Periodontal Res *37* (4), 273–278.

Xian, X., Gopal, S., and Couchman, J. R. (2010). Syndecans as receptors and organizers of the extracellular matrix. Cell Tissue Res *339* (1), 31–46.

Yamagata, M., Suzuki, S., et al. (1989). Regulation of cell-substrate adhesion by proteoglycans immobilized on extracellular substrates. J Biol Chem *264* (14), 8012–8018.

Yanagishita, M., and Hascall, V. C. (1992). Cell surface heparan sulfate proteoglycans. J Biol Chem *267* (14), 9451–9454.

Yang, Y., Macleod, V., et al. (2007). Heparanase enhances syndecan-1 shedding: a novel mechanism for stimulation of tumor growth and metastasis. J Biol Chem *282* (18), 13326–13333.

Yang, Y., Yaccoby, S., et al. (2002). Soluble syndecan-1 promotes growth of myeloma tumors in vivo. Blood *100* (2), 610–617.

Yoneda, A., and Couchman, J. R. (2003). Regulation of cytoskeletal organization by syndecan transmembrane proteoglycans. Matrix Biol *22* (1), 25–33.

Zellweger, T., Ninck, C., Bloch, M., et al. (2005). Expression patterns of potential therapeutic targets in prostate cancer. Int J Cancer *113* (4), 619–628.

Zellweger, T., Ninck, C., Mirlacher, M., et al. (2003). Tissue microarray analysis reveals prognostic significance of syndecan-1 expression in prostate cancer. Prostate *55* (1), 20–29.

Zhang, L., Sullivan, P., et al. (2010). Epidermal growth factor-induced heparanase nucleolar localization augments DNA topoisomerase I activity in brain metastatic breast cancer. Mol Cancer Res *8* (2), 278–290.

Zimmermann, P., and David, G. (1999). The syndecans, tuners of transmembrane signaling. FASEB J *13* (Suppl.), S91–S100.

Zimmermann, P., Zhang, Z., et al. (2005). Syndecan recycling [corrected] is controlled by syntenin-PIP2 interaction and Arf6. Dev Cell *9* (3), 377–388.

Zong, F., Fthenou, E., Castro, J., et al. (2010). Effect of syndecan-1 overexpression on mesenchymal tumour cell proliferation with focus on different functional domains. Cell Prolif *43* (1), 29–40.

Zong, F., Fthenou, E., Mundt, F., et al. (2011). Specific syndecan-1 domains regulate mesenchymal tumor cell adhesion, motility and migration. PLoS One *6* (6), e14816.

Zong, F., Fthenou, E., Wolmer, N., et al. (2009). Syndecan-1 and FGF-2, but not FGF receptor-1, share a common transport route and co-localize with heparanase in the nuclei of mesenchymal tumor cells. PLoS One *4* (10), e7346.

Zvibel, I., Halfon, P., Fishman, S., et al. (2009). Syndecan 1 (CD138) serum levels: a novel biomarker in predicting liver fibrosis stage in patients with hepatitis C. Liver Int *29* (2), 208–212.

7.6 Serglycin: a novel player in the terrain of neoplasia

Angeliki Korpetinou, Eleni Milia-Argeiti, Vassiliki Labropoulou, and Achilleas Theocharis

7.6.1 Introduction

Serglycin is an intracellular proteoglycan (PG) present in several different types of cells participating in a number of cellular processes reviewed by Schick in Chapter 3.8 in this book. It consists of a small core protein (158 amino acids in human) containing eight serine/glycine repeats, which serve as attachment sites for glycosaminoglycans (GAGs). Different types of GAGs are bound to the core protein and represent the main moieties that mediate the biological roles of the PG although core protein seems to be required for some biological functions. Heparin (HP) and heparan sulfate (HS) containing serglycin is stored inside connective tissue type mast cell granules and interacts with several granule components regulating their storage, secretion upon stimulation, activation, and transfer to target sites. It is also essential for granule formation and participates in inflammation and apoptosis. Serglycin-carrying chondroitin sulfate (CS) is found in other hematopoietic and nonhematopoietic cells, including cancer cells. The sulfation pattern seems to be important for the specific roles of serglycin among these cells. CS composed mainly of 4-sulfated disaccharides (CS-4S or CSA) is found in lymphocytes, natural killer (NK) cells, and monocytes. CS-4S is also found in platelets inside secretory vesicles, where it interacts with several components contributing to platelet aggregation during coagulation, atherosclerosis, and angiogenesis (Schick, 2010). However, hematopoietic cells including mucosal type mast cells, bone marrow-derived mast cells, and activated monocytes and macrophages express CS with a higher extent of sulfation containing higher amounts of disulfated units (CS-E, i.e., disulfated at C-4 and C-6; and CS-diB, i.e., disulfated at C-4 and C-2) (Kolset and Pejler, 2011). Both monocytes and macrophages constitutively secrete serglycin in complexes with partner molecules. Decreasing expression during granule maturation has been reported to neutrophils. Among nonhematopoietic cells, serglycin is expressed in pancreatic acinar cells, fibroblasts, endothelial, smooth muscle cells and is constitutively secreted in some cell types (Kolset and Tveit, 2008). Although the core protein does not contain a transmembrane domain, serglycin may also harbor at cell membrane possibly through binding to cell-surface molecules via ionic interactions.

7.6.2 Expression of serglycin in malignancies

Serglycin gene was cloned and sequenced for the first time from the rat yolk sac carcinoma cell line L2 (Bourdon et al., 1985). Since then serglycin transcript and/or core proteins carrying all possible types of GAG chains have been identified in several rat, murine, and human malignant cell lines and tissues. F9 teratocarcinoma cells (Grover et al., 1987) and numerous human hematopoietic tumor cell lines were found to express serglycin (Kolset and Gallagher, 1990; Stellrecht et al., 1991, 1993; Maillet et al., 1992; Schick and Senkowski-Richardson, 1992; Schick and Jacoby, 1995; Oynebråten et al., 2000). Interestingly, the megakaryocytic tumor cells synthesize a hybrid CS/HS serglycin (Schick and Jacoby, 1995). Moreover, in a variety of lymphoma, myeloma, mastocytoma, and thymoma cells, only serglycin containing CS-4S or CS-6S side chains has been identified to interact with cell surface cluster of differentiation (CD44) (Toyama-Sorimachi et al., 1997). Serglycin-bearing HS or HP was not capable for binding to CD44. In a recent study it has been shown that serglycin is highly expressed by leukemic blasts of patients with acute myeloid leukemia (Niemann et al., 2007). Serglycin plasma levels in acute myelogenous leukemia (AML) patients were significantly elevated compared to plasma levels in acute lymphocytic leukemia (ALL) patients, and the protein was localized specifically in immature blasts in the marrow of AML but not in bone marrow from patients with ALL. It is proposed that serglycin is a selective biomarker for acute myeloid leukemia compared to Philadelphia chromosome-negative chronic myeloproliferative disorders.

Serglycin distribution was examined in several multiple myeloma (MM) cell lines (Theocharis et al., 2006). It was found to represent the major PG to be synthesized by MM cells. Serglycin was constitutively secreted to the medium, but interestingly was also localized on the cell surface, attached through its GAG chains. Serglycin synthesized by MM cells carries CS chains, which are composed up to 93% by 4-sulfated disaccharides. Serglycin levels were elevated in the bone marrow aspirates of patients with newly diagnosed MM, suggesting a potent correlation of serglycin accumulation with disease progression and prognosis.

The expression of serglycin has been previously documented in numerous hematological malignancies and recently in carcinomas, where it associates with aggressive tumor cells' phenotype. High-throughput gene expression profiling analysis of nasopharyngeal carcinoma cell lines revealed the upregulation of serglycin in cells with high-metastatic potential in comparison to cells showing low-metastatic potential (Li et al., 2011). The clinical data revealed a potent prognostic importance of serglycin since it serves as an independent prognostic indicator for disease-free survival and distant metastasis-free survival of patients. In preliminary studies, we demonstrated high expression of serglycin and constitutive secretion in the culture medium of the high metastatic breast cancer MDA-MB-231 cells as compared to the low metastatic MDA-MB-468 and MCF-7 ones. Notably, immunohistochemical staining confirmed the expression of serglycin by cancer cells in breast carcinomas (Korpetinou et al., 2011). Further studies will help to elucidate whether the expression profile of this PG in other tumor types can serve as a prognostic marker.

7.6.3 Regulation of serglycin gene expression

Serglycin expression has been studied in several hematopoietic cell lines. Increased serglycin expression has been noticed during megakaryocytic (Stellrecht et al., 1993; Schick and Jacoby, 1995) and myeloblast differentiation where it coincides with granule biogenesis (Stellrecht et al., 1991). In contrast, serglycin expression is decreased during promyelocyte differentiation into mature neutrophils (Niemann et al., 2004). The expression of serglycin is differentially regulated upon treatment of leukemic cell lines with phorbol 12-myristate 13-acetate (PMA) (Schick et al., 2001b). The cell-specific manner response in serglycin expression upon treatment with PMA is associated with the presence of specific DNase-I hypersensitive sites within serglycin gene in the different cell types (Schick et al., 2001b). Serglycin expression is also modulated by viruses and transcription factors associated with tumor development. For example, serglycin is among other genes (vimentin, CD21, Epstein-Barr virus-induced genes 1 and 2 [EBI1 and EBI2], cathepsin H, annexin VI [p68], CD44, and the myristylated alanine-rich protein kinase C substrate [MARCKS]) induced by Epstein-Barr virus (EBV) infection of Burkitt's lymphoma cells *in vitro*. EBV is the cause of infectious mononucleosis, a benign proliferation of infected B lymphocytes, and can also cause acute and rapidly progressive B-lymphoproliferative disease in severely immunocompromised patients. This suggests an implication of serglycin in the EBV transformation of B lymphocytes (Birkenbach et al., 1993). Human serglycin gene contains a binding site for E-twenty-six-specific family of transcription factors (ETS) in the 5′ flanking region (–280 to –275 bp), which is also conserved in the mouse serglycin promoter region, suggesting that this site is important for transcriptional regulation. This site is functional and interacts with ETS1 and Friend leukemia integration 1 transcription factor proteins (FLI1). The expression of the human serglycin gene was shown to be upregulated in a number of human leukemic cell lines, ones that coincidentally have been shown to express high levels of ETS1 and FLI1 (Robinson et al., 1997). The ETS genes encode transcription factors that bind to specific DNA sequences and activate transcription of various cellular and viral genes. ETS transcription factors play important roles in hematopoiesis, angiogenesis, and organogenesis. The ETS genes can function as oncogenes and are overexpressed in human cancer. Conclusively, the fact that the expression of serglycin is regulated by tumor-promoting agents, viruses, and transcription factors implicated in tumor development may suggest a positive role for this PG in tumorigenesis.

7.6.4 Functional importance of serglycin in malignancies

Metastatic nasopharyngeal carcinoma cells highly express serglycin, which promotes motility, invasion, and metastasis via induction of vimentin, a marker of epithelial-to-mesenchymal transition (EMT), and participates in the regulation of migration (Li et al., 2011). Moreover, serglycin contributed to migration and invasion of the cells via yet unknown autocrine and paracrine mechanisms. The functions of serglycin depend on the fully glycosylated molecule. In nasopharyngeal cells serglycin with molecular weight of about 130 kDa is present intracellularly, but this proteoglycan with a molecular weight of about 300 kDa is also secreted in the culture medium.

Overexpression of serglycin by metastatic cells significantly increases only the amount of secreted serglycin suggesting that the secreted form of serglycin is mainly responsible for the metastatic potential of cancer cells. In addition, treatment of cancer cells with exogenously added glycosylated serglycin promotes cancer cell metastasis and invasion.

It is known that the overall structure of the intact serglycin is essential for high-affinity binding to other proteins. Proteins that bind to serglycin-carrying CS chains include matrix molecules like collagen and fibronectin. Serglycin present on the cell membrane of malignant cells may facilitate the adhesion of tumor cells to such matrix proteins. Intact serglycin also binds growth factors/cytokines and chemokines (platelet factor 4 [chemokine (CXC motif) ligand 4 (CXCL4)] and macrophage inflammatory protein-1α [MIP-1α, chemokine (C-C motif) ligand 3 (CCL3)]), lysozyme, and cell surface receptor CD44 (Schick, 2010). CD44 is the cell surface receptor for hyaluronan and has been recognized as a cancer stem cell marker for a variety of cancer types (Li et al., 2009; Casagrande et al., 2011; Liu et al., 2011; Ricardo et al., 2011; Su et al., 2011). CD44 is involved in cell-cell and cell-matrix interactions. It signals through several pathways regulating cancer cells' EMT, migration, metastasis, proliferation, apoptosis, and resistance (Zoller, 2011). Secreted serglycin may interact with CD44 present on cancer cell membrane and trigger CD44 signaling promoting cancer cell migration and invasion (▶Figure 7.13). It is known that other CS-bearing PGs such as versican interact with CD44 and hyaluronan on prostate cancer cells resulting in the remodeling of their pericellular environment and promotion of their motility (Ricciardelli et al., 2007). Furthermore, CS containing CS-E type of disulfated disaccharides present in the tumor microenvironment enhances CD44-dependent motility in tumor cells (Sugahara et al., 2008). Such CS chains may be synthesized and secreted in the tumor stroma by activated inflammatory cells as discussed previously.

Serglycin is also constitutively secreted in other malignancies such as MM (Theocharis et al., 2006). The secreted form of serglycin was found to influence the bone mineralization process through inhibition of the crystal growth rate of hydroxyapatite, providing a possible explanation for impaired bone formation and loss of bone mass commonly seen in MM patients (▶Figure 7.13). Moreover, secreted and cell surface associated serglycin demonstrates a role in protecting myeloma cells from complement system attack induced by antibody immunotherapy, therefore promoting the survival of malignant myeloma cells (▶Figure 7.13). Serglycin isolated from the culture medium of MM cell lines inhibits both the classical and the lectin pathway of complement system activation *in vitro* (Skliris et al., 2011). The inhibitory effect on the classical and lectin pathways was mediated by the interaction of serglycin with C1q and mannose-binding lectin (MBL), respectively. Both the core protein and the CS side chains are essential for the interaction between serglycin and MBL, whereas CS chains of serglycin are capable for interaction with C1q. The particular sulfation pattern of the CS-4S side chains bound on serglycin's core protein is essential for the inhibitory properties of the serglycin secreted from MM cells. Complement inhibition is a great limitation during antibody immunotherapy against several types of cancer. Tumor cells protect themselves by overexpressing membrane complement regulatory proteins and by binding soluble complement inhibitory factors from serum (Fishelson et al., 2003).

Drug resistance that is developed from tumor cells has become a great limitation of treatment of malignancies. Regarding the fact that several cell types are recruited from tumor cells and actively participate in tumor development and metastasis, the

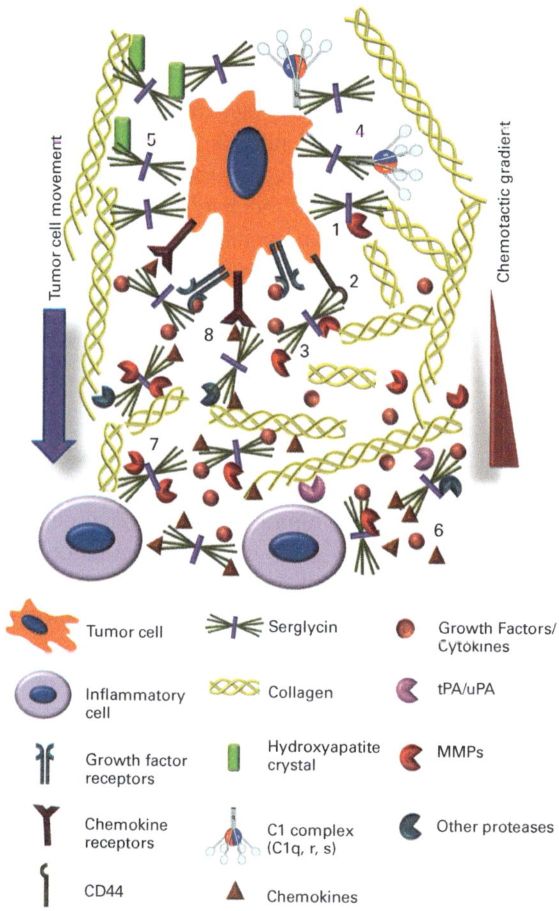

Figure 7.13 Functions of serglycin in malignancy. Aggressive tumor cells overexpress and secrete serglycin in the extracellular matrix, which promotes their migration, invasion, and formation of distant metastasis acting both in a paracrine and autocrine manner. Serglycin may be also secreted by tumor cells as a complex with matrix metalloproteinases (MMPs). This interaction modulates the functions of MMPs (1). Furthermore, serglycin binds to cell surface receptor CD44 and may trigger signaling events involved in tumor cell migration and invasion (2). The localization of serglycin on the cell membrane via a GAG-dependent manner may facilitate the anchorage of MMPs on tumor cell membrane enhancing pericellular matrix degradation (3). Cell surface associated and secreted serglycin inhibits both classical and lectin pathways of complement system through binding to C1q and mannose-binding lectin (MBL) suppressing immune system responses (4). The inhibition of classical pathway of complement system protects tumor cells by complement-dependent cytotoxicity induced by immunotherapy in malignancies. Secreted serglycin also binds to hydroxyapatite and inhibits bone formation in tumors grow at skeletal sites (5). Furthermore, inflammatory cells secrete serglycin complexes with various proteases, cytokines, growth factors, and chemokines creating an efficient chemotactic gradient promoting tumor cells' movement (6). Serglycin-bound proteases may rely on serglycin for optimal presentation of substrates since serglycin interacts with extracellular matrix molecules such as collagen (7). In addition, serglycin may facilitate the transport of growth factors and chemokines to tumor cells and assist in their presentation to receptors (8).

identification of factors that support this resistance is a wide area of interest. Beyer-Sehlmeyer et al. (1999) developed six tumor cell lines of hematopoietic origin that resist in doxorubicin, methotrexate, cisplatin, and vincristine treatment. Compared to the drug sensitive parental cell lines enhanced gene expression was identified for serglycin, sorcin, bone marrow proteoglycan gene, heat-shock protein 90, and prostate-tumor-inducing gene 1. The implication of serglycin produced from hematopoietic cells in drug resistance is of great interest and the mechanism of action is yet unknown.

Collectively, the expression of serglycin benefits tumor cells in multiple ways. Serglycin augments the invasion and metastasis of tumor cells, acts as a modulator of immune system in tumor microenvironment, and enriches tumor cells with resistance to various therapeutic agents.

7.6.5 Serglycin regulates the secretion of proteolytic enzymes

The release of proteolytic enzymes by tumor cells or stromal cells such as fibroblasts, endothelial cells, and inflammatory cells and the regulation of their activity in the tumor microenvironment are crucial for tumor progression (Gialeli et al., 2011). Serglycin interacts with matrix-degrading enzymes such as matrix metalloproteinases (MMPs). It has been shown that serglycin interacts with a fragment of C-terminal domain that comprises the hinge and hemopexin domain of MMP13 in chondrocytes. It is colocalized with MMP13 in cytoplasmic granules and may be implicated in the expression and regulation of MMP13 functions in this cellular system (Zhang et al., 2010). It has been demonstrated that tissue-type plasminogen activator (tPA) is colocalized with serglycin in human umbilical vein endothelial cells (HUVEC) (Schick et al., 2001a), whereas Madin-Darby canine kidney cells stably transfected with serglycin expressed elevated levels of MMP9 and urokinase plasminogen activator (uPA) both at mRNA and protein levels (Zernichow et al., 2006b). It is suggested that serglycin modulates the secretion level of proteases via as yet undefined mechanisms regulated in relation to the serglycin level. This relationship between the levels of serglycin and protease levels could accordingly be in support of cross-talk regulatory mechanisms.

In this line of evidences serglycin is one of the candidate proteoglycans synthesized and secreted by human acute monocytic leukemia cell line THP1 that forms complexes with proform of matrix metalloproteinase 9 (pro-MMP9) (Winberg et al., 2000). It binds to the hemopexin like domain of pro-MMP9 through its core protein forming a reduction sensitive heterodimer. The formation of heterodimers alters the mode of activation of pro-MMP9 and the interaction of the enzyme with its substrates (▶Figure 7.13) (Winberg et al., 2003; Malla et al., 2008). For example, pro-MMP9 in the heterodimer is activated in the presence of Ca^{2+}, although this cation stabilizes MMP9 without activating the proenzyme. The presence of this cation results in the cleavage of both the C-terminal hemopexin domain of the enzyme and the core protein of proteoglycans and the release of the activated enzyme. MM and solid tumors, which metastasize in the bones, induce bone destruction and release Ca^{2+}. The accumulation of serglycin within bone marrow in MM may be involved in the formation of heteromers with pro-MMP9 expressed by myeloma cells and to drive a calcium-induced activation of the enzyme.

Serglycin is also crucial for the storage of proteases in secretory granules in hematopoietic cells. Experiments in serglycin knockout mice revealed severely impaired storage of a number of granule-localized proteases (chymases, tryptases, and carboxypeptidase A) in mast cells (Åbrink et al., 2004; Braga et al., 2007). Similarly, the inactivation of serglycin gene resulted in the impaired storage of granzyme B (Grujic et al., 2005) in cytotoxic T lymphocytes (CTLs) and of elastase in the azurophilic granules of neutrophils (Niemann et al., 2007). Importantly, the lack of serglycin did not alter the levels of mRNA coding for these proteases, suggesting effects of serglycin on storage rather than on mRNA expression. The storage of granule proteases is critically depending on the high anionic charge imposed by the sulfation of GAG chains of serglycin. It has been shown that mast cell proteases that are dependent on serglycin can participate in the activation of pro-MMP2 (Lundequist et al., 2006).

The importance of different activation mechanisms for the different proMMPs will probably depend on tissue and the cell types involved. Serglycin is involved in the expression, secretion and activation of various proteolytic enzymes in different cell types and is most likely a regulatory molecule of proteolytic potential in tumor microenvironment (▶Figure 7.13).

7.6.6 Serglycin regulates the secretion and properties of inflammatory mediators

During tumor growth and progression, tumor cells degrade the surrounding basement membrane and invade to stroma, attach to the new site of tumor formation, and proliferate. If the developing tumor exceeds in size, the hypoxia caused by the lack of oxygen and nutrients will trigger angiogenesis (Yoshida et al., 2000). All these processes are regulated by mediators produced by and secreted from both tumor cells and the recruited cells from the tumor microenvironment. Fibroblasts, endothelial cells, and inflammatory cells such as macrophages, neutrophils, mast cells, and lymphocytes contribute to the formation of the tumor mass (Coussens and Werb, 2002).

Several cytokines and chemokines differentially regulated and distributed in the tumor microenvironment by these cell types are shown to interact with serglycin (▶Figure 7.13). Apart from the interaction of serglycin with several proteases in human hematopoietic and nonhematopoietic cells, this multifunctional PG has been referred to interact with several cytokines and chemokines inside the secretory vesicles of the cells. Moreover, it regulates the secretion and distribution of these molecules at the target sites. Infiltrating leukocytes as well as tumor cells express chemokine receptors, which permit them to respond to chemotactic gradients. The constitutive expression of chemokines by the metastatically affected organs can provide signaling cues for malignant cell homing (Labropoulou et al., 2011). Particularly, serglycin regulates the secretion of the chemokine growth-related oncogene-alpha (GRO-α/ CXCL1) to the apical side in HUVEC. Interestingly, the storage of GRO-α in the secretory vesicles was not depended on serglycin while upon stimulation with interleukin-1β (IL-1β), colocalization of the two molecules inside the vesicles was observed before secretion (Meen et al., 2011).

Serglycin is localized at the α granules of the platelets, but not at the dense granules, although it has been described that the presence of the PG can affect both of these structures and their functions. The α granules contain numerous proteins including growth factors, chemokines, cytokines and thrombogenic proteins, such as thrombospondin, fibronectin and von Willebrand factor. The absence of serglycin has multiple effects in platelets, including severely defective storage of CXCL4, β-thromboglobulin (CXCL7), and platelet-derived growth factor (PDGF) in α granules (Woulfe et al., 2008). Cytokines and chemokines secreted by activated platelets are important for the recruitment and activation of other cell types playing an important role in coagulation, inflammation, and angiogenesis (Schick, 2010). The lack of serglycin gene results in defective platelet aggregation and activation. Platelet function deficiencies are related to inadequate packaging and secretion of selected α-granule proteins and reduced secretion of dense granule contents critical for platelet activation. It has been suggested that platelets are involved in tumor metastasis (Gay and Felding-Habermann, 2011). Platelet serglycin may influence the release of growth factors, cytokines, and chemokines in the tumor microenvironment, which promote tumor cell growth and metastasis. For example platelet-tumor cell interactions are sufficient to prime tumor cells for subsequent metastasis. Platelet-derived transforming growth factor-beta (TGF-β) and direct platelet-tumor cell contacts synergistically activate the TGF-β/Smad and nuclear factor kappa-light-chain-enhancer of activated B cells (NF-κB) pathways in cancer cells, resulting in their transition to an invasive mesenchymal-like phenotype and enhanced metastasis *in vivo* (Labelle et al., 2011). Another possibility is that released serglycin either by tumor cells, for example nasopharyngeal carcinoma cells, or platelets is capable to interact with the surface of activated platelets and/or the surface of tumor cells. It is proposed that serglycin is bound on the surface of platelets (Schick, 2010), whereas serglycin has been identified on the surface of tumor cells (Bourdon et al., 1985; Theocharis et al., 2006). In this case, serglycin may participate in binding of tumor cells to the surface of activated platelets or by attachment to platelet microparticles and formation of hematogenous metastasis.

It has also been proposed that serglycin expressed in human monocytes participates in the regulation of autocrine cell growth (Kolset and Zernichow, 2008). Interestingly, inhibition of GAG chain polymerization led to rapid serglycin core protein degradation and inhibition of cell proliferation. Serglycin isolated from activated macrophages has been reported to interact, among others, with CCL3 and interleukin-1α (IL-1α) via its CS chains, but deletion of serglycin to mouse macrophages did not show any effect in the secretion of these molecules indicating a different contribution of serglycin (Zernichow et al., 2006a). It has also been shown that tumor necrosis factor-alpha (TNF-α) secretion is upregulated in serglycin null activated macrophages (Zernichow et al., 2006a).

Serglycin may act as vehicle for the extracellular delivery of inflammatory molecules with which it interacts within cells. After secretion, some of the serglycin-associated molecules will be released from serglycin due to the increased pH of the extracellular milieu, whereas others may remain attached after exocytosis. It is also possible that secreted serglycin may act as a scavenger that binds to and sequesters inflammatory molecules present in the extracellular microenvironment, thereby establishing chemotactic gradients (▶Figure 7.13). The association with serglycin could confer protection against proteolytic attack or aid the presentation of such molecules to their target cells (Kolset and Pejler, 2011). Thus, serglycin may affect the proliferation and motility of tumor cells as well as angiogenesis. It is therefore important to clarify the contribution of serglycin in the regulation of mediators involved in tumorigenesis.

7.6.7 Take-home message

Serglycin is a dominant proteoglycan expressed by inflammatory and hematological malignant cells. Recent data demonstrated that serglycin is synthesized and secreted by solid tumors as well. Overexpression of serglycin by tumor cells promotes the aggressive phenotype and confers resistance against drugs and immune system attack. It is now evident that serglycin is of great importance for the expression, secretion, and function of inflammatory mediators and proteolytic enzymes implicated in malignancy. Further studies to clarify the molecular mechanisms by which serglycin promotes tumorigenesis are needed.

References

Åbrink, M., Grujic, M., and Pejler, G. (2004). Serglycin is essential for maturation of mast cell secretory granule. J Biol Chem *279*, 40897–40905.

Beyer-Sehlmeyer, G., Hiddemann, W., Wörmann, B., and Bertram, J. (1999). Suppressive subtractive hybridisation reveals differential expression of serglycin, sorcin, bone marrow proteoglycan and prostate-tumour-inducing gene I (PTI-1) in drug-resistant and sensitive tumour cell lines of haematopoetic origin. Eur J Cancer *35* (12), 1735–1742.

Birkenbach, M., Josefsen, K., Yalamanchili, R., Lenoir, G., and Kieff, E. J. (1993). Epstein-Barr virus-induced genes: first lymphocyte-specific G protein-coupled peptide receptors. J Virol *67*, 2209–2220.

Bourdon, M. A., Oldberg, A., Pierschbacher, M., and Ruoslahti, E. (1985) Molecular cloning and sequence analysis of a chondroitin sulfate proteoglycan cDNA. Proc Natl Acad Sci U S A *82*, 1321–1325.

Braga, T., Grujic, M., Lukinius, A., Hellman, L., Abrink, M., and Pejler, G. (2007). Serglycin proteoglycan is required for secretory granule integrity in mucosal mast cells. Biochem J *403*, 49–57.

Casagrande, F., Cocco, E., Bellone, S., et al. (2011). Eradication of chemotherapy-resistant CD44+ human ovarian cancer stem cells in mice by intraperitoneal administration of Clostridium perfringens enterotoxin. Cancer *117*, 5519–5528.

Coussens, L. M., and Werb, Z. (2002). Inflammation and cancer. Nature *420*, 860–867.

Fishelson, Z., Donin, N., Zell, S., Schultz, S., and Kirschfink, M. (2003). Obstacles to cancer immunotherapy: expression of membrane complement regulatory proteins (mCRPs) in tumors. Mol Immunol *40*, 109–123.

Gay, L. J., and Felding-Habermann, B. (2011). Platelets alter tumor cell attributes to propel metastasis: programming in transit. Cancer Cell *20*, 553–554.

Gialeli, C., Theocharis, A. D., and Karamanos, N. K. (2011). Roles of matrix metalloproteinases in cancer progression and their pharmacological targeting. FEBS J *278*, 16–27.

Grover, A., Edwards, S. A., Bourdon, M., and Adamson, E. D. (1987) Proteoglycan-19, laminin and collagen type IV production is correlated with the levels of mRNA in F9 cell aggregates differentiating in the presence or absence of cyclic AMP. Differentiation *36*, 138–44.

Grujic, M., Braga, T., Lukinius, A., et al. (2005). Serglycin-deficient cytotoxic T lymphocytes display defective secretory granule maturation and granzyme B storage. J Biol Chem *280*, 33411–33418.

Kolset, S. O., and Gallagher, J. T. (1990). Proteoglycans in haemopoietic cells. Biochim Biophys Acta *1032*, 191–211.

Kolset, S. O., and Pejler, G. (2011). Serglycin: a structural and functional chameleon with wide impact on immune cells. J Immunol *187*, 4927–4933.

Kolset, S. O., and Tveit, H. (2008). Serglycin – structure and biology. Cell Mol Life Sci *65*, 1073–1085.

Kolset, S. O., and Zernichow, L. (2008). Serglycin and secretion in human monocytes. Glycoconj J *25*, 305–311.

Korpetinou, A., Triantaphyllidou, I. E., Giannopoulou, E., et al. (2011). Aggressive breast cancer cells secrete serglycin, a potent inhibitor of complement system. FEBS J *2787* (Suppl. 1), 419.

Labelle, M., Begum, S., and Hynes, R. O. (2011). Direct signaling between platelets and cancer cells induces an epithelial-mesenchymal-like transition and promotes metastasis. Cancer Cell *20*, 576–590.

Labropoulou, V. T., Theocharis, A. D., Symeonidis, A., Skandalis, S. S., Karamanos, N. K., and Kalofonos, H. P. (2011). Pathophysiology and pharmacological targeting of tumor-induced bone disease: current status and emerging therapeutic interventions. Curr Med Chem *18*, 1584–1598.

Li, C., Lee, C. J., and Simeone, D. M. (2009). Identification of human pancreatic cancer stem cells. Methods Mol Biol *568*, 161–173.

Li, X. J., Ong, C. K., Cao, Y., et al. (2011). Serglycin is a theranostic target in nasopharyngeal carcinoma that promotes metastasis. Cancer Res *71*, 3162–3172.

Liu, L. L., Fu, D., Ma, Y., and Shen, X. Z. (2011). The power and the promise of liver cancer stem cell markers. Stem Cells Dev *20*, 2023–2030.

Lundequist, A., Abrink, M., and Pejler, G. (2006) Mast cell dependent activation of pro matrix metalloprotease 2: a role for serglycin proteoglycan-dependent mast cell proteases. Biol Chem *387*, 1513–1519.

Maillet, P., Alliel, P. M., Mitjavila, M. T., Périn, J. P., Jollès, P., and Bonnet, F. (1992). Expression of the serglycin gene in human leukemic cell lines. Leukemia *11*, 1143–1147.

Malla, N., Berg, E., Uhlin-Hansen, L., and Winberg, J. O. (2008). Interaction of pro-matrix metalloproteinase-9/proteoglycan heteromer with gelatin and collagen. J Biol Chem *283*, 13652–13665.

Meen, A. J., Øynebråten, I., Reine, T. M., et al. (2011). Serglycin is a major proteoglycan in polarized human endothelial cells and is implicated in the secretion of the chemokine GROα/CXCL1. J Biol Chem *286*, 2636–2647.

Niemann, C. U., Abrink, M., Pejler, G., et al. (2007). Neutrophil elastase depends on serglycin proteoglycan for localization in granules. Blood *109*, 4478–4486.

Niemann, C. U., Cowland, J. B., Klausen, P., Askaa, J., Calafat, J., and Borregaard, N. (2004). Localization of serglycin in human neutrophil granulocytes and their precursors. J Leukoc Biol *76*, 406–415.

Niemann, C. U., Kjeldsen, L., Ralfkiaer, E., Jensen, M. K., and Borregaard. N. (2007). Serglycin proteoglycan in hematologic malignancies: a marker of acute myeloid leukemia. Leukemia *21*, 2406–2410.

Oynebråten, I., Hansen, B., Smedsrød, B., and Uhlin-Hansen, L. (2000). Serglycin secreted by leukocytes is efficiently eliminated from the circulation by sinusoidal scavenger endothelial cells in the liver. J Leukoc Biol *67*, 183–188.

Ricardo, S., Vieira, A. F., Gerhard, R., et al. (2011). Breast cancer stem cell markers CD44, CD24 and ALDH1: expression distribution within intrinsic molecular subtype. J Clin Pathol *64*, 937–946.

Ricciardelli, C., Russell, D. L., Ween, M. P., et al. (2007). Formation of hyaluronan- and versican-rich pericellular matrix by prostate cancer cells promotes cell motility. J Biol Chem *282*, 10814–10825.

Robinson, L., Panayiotakis, A., Papas, T. S., Kola, I., and Seth, A. (1997). ETS target genes: identification of egr1 as a target by RNA differential display and whole genome PCR techniques. Proc Natl Acad Sci U S A *94*, 7170–7175.

Schick, B. P. (2010). Serglycin proteoglycan deletion in mouse platelets: physiological effects and their implications for platelet contributions to thrombosis, inflammation, atherosclerosis, and metastasis. Prog Mol Biol Transl Sci *93*, 235–287.

Schick, B. P., Gradowski J. F., and San Antonio, J. D. (2001a). Synthesis, secretion, and subcellular localization of serglycin proteoglycan in human endothelial cells. Blood *97*, 449–458.

Schick, B. P., and Jacoby, J. A. (1995). Serglycin and betaglycan proteoglycans are expressed in the megakaryocytic cell line CHRF 288-11 and normal human megakaryocytes. J Cell Physiol *165*, 96–106.

Schick, B. P., Petrushina, I., Brodbeck, K. C., and Castronuevo, P. (2001b). Promoter regulatory elements and DNase I-hypersensitive sites involved in serglycin proteoglycan gene expression in human erythroleukemia, CHRF 288-11, and HL-60 cells. J Biol Chem *276*, 24726–24735.

Schick, B. P., and Senkowski-Richardson, S. (1992). Proteoglycan synthesis in human erythroleukaemia (HEL) cells. Biochem J *282*, 651–658.

Skliris, A., Happonen, K. E., Terpos, E., et al. (2011). Serglycin inhibits the classical and the lectin pathways of complement system via its glycosaminoglycan chains: implications for multiple myeloma. Eur J Immunol *41*, 1–14.

Stellrecht, C. M., Fraizer, G., Selvanayagam, C., Chao, L. Y., Lee, A., and Saunders, G. F. (1993). Transcriptional regulation of a hematopoietic proteoglycan core protein gene during hematopoiesis. J Biol Chem *268*, 4078–4084.

Stellrecht, C. M., Mars, W. M., Miwa, H., Beran, M., and Saunders, G. F. (1991). Expression pattern of a hematopoietic proteoglycan core protein gene during human hematopoiesis. Differentiation *48*, 127–135.

Su, Y. J., Lai, H. M., Chang, Y. W., Chen, G. Y., and Lee, J. L. (2011). Direct reprogramming of stem cell properties in colon cancer cells by CD44. EMBO J *30*, 3186–3199.

Sugahara, K. N., Hirata, T., Tanaka, T., et al. (2008). Chondroitin sulfate E fragments enhance CD44 cleavage and CD44-dependent motility in tumor cells. Cancer Res *68*, 7191–7199.

Theocharis, A. D., Seidel, C., Borset, M., et al. (2006). Serglycin constitutively secreted by myeloma plasma cells is a potent inhibitor of bone mineralization in vitro. J Biol Chem *281*, 35116–35128.

Toyama-Sorimachi, N., Kitamura, F., Habuchi, H., Tobita, Y., Kimata, K., and Miyasaka, M. (1997). Widespread expression of chondroitin sulfate-type serglycins with CD44 binding ability in hematopoietic cells. J Biol Chem *272*, 26714–26719.

Winberg, J. O., Berg, E., Kolset, S. O., and Uhlin-Hansen, L. (2003). Calcium-induced activation and truncation of promatrix metalloproteinase-9 linked to the core protein of chondroitin sulfate proteoglycans. Eur J Biochem *270*, 3996–4007.

Winberg, J. O., Kolset, S. O., Berg, E., and Uhlin-Hansen, L. (2000). Macrophages secrete matrix metalloproteinase 9 covalently linked to the core protein of chondroitin sulphate proteoglycans. J Mol Biol *304*, 669–680.

Woulfe, D. S., Lilliendahl, J. K., August, S., et al. (2008). Serglycin proteoglycan deletion induces defects in platelet aggregation and thrombus formation in mice. Blood *111*, 3458–3467.

Yoshida, B. A., Sokoloff, M. M., Welch, D. R., and Rinker-Schaefer, C. W. (2000) Metastasis-suppressor genes: a review and perspective. J Natl Cancer Inst *92*, 1717–1730.

Zernichow, L., Åbrink, M., Hallgren, J., Grujic, M., Pejler, G., and Kolset, S. O. (2006a). Serglycin is the major secreted proteoglycan in macrophages and has a role in the regulation of macrophage tumor necrosis factor-α secretion in response to lipopolysaccharide. J Biol Chem *281*, 26792–26801.

Zernichow, L., Dalen, K. T., Prydz, K., Winberg, J. O., and Kolset, S. O. (2006b). Secretion of proteases in serglycin transfected Madin-Darby canine kidney cells. FEBS J *273*, 536–547.

Zhang, L., Yang, M., Yang, D., Cavey, G., Davidson, P., and Gibson, G. (2010) Molecular interactions of MMP-13 C-terminal domain with chondrocyte proteins. Connect Tissue Res *51*, 230–239.

Zoller, M. (2011) CD44: can a cancer-initiating cell profit from an abundantly expressed molecule? Nat Rev Cancer *11*, 254–267.

7.7 Quantifying cell-ECM pathobiology in 3D

Joseph S. Maffei and Muhammad H. Zaman

7.7.1 Introduction

It has long been known that a relationship exists between the extracellular matrix (ECM) architecture and the progression of cancer. In opposition to the immune system, cancerous legions oftentimes are correlated with areas of inflammation as well as increased ECM degradation and reposition (Dvorak, 1986). As a result, cancerous tumors and their surrounding ECM are oftentimes stiffer than benign regions, and this stiffness may offer insight into the gradation of disease (Swaminathan et al., 2011). This has led to the indication that cancer may be brought on in part via the mechanical properties of the ECM (Ingber and Jamieson, 1985). Integrins, cell surface transmembrane proteins, are a key mediator in this process, acting both as adhesive receptors that adhere to ECM proteins, as well as a matrix "sensor" that transmits basal matrix tension through the cytoskeleton via intracellular attachment to actin fibers (Wang et al., 1993).

Recent studies have provided insight to the underlying consequences of a stiffer matrix, citing increased integrin-ECM interactions leading to mechanotransduction signaling within cells (Bershadsky et al., 2003). These mechanically transmitted forces are converted into a biochemical signal within the cell, guiding them toward certain phenotypes, and also acting as a positive feedback mechanism to those cells that continue to produce matrix stiffening ECM products as a result (Huang and Ingber, 2005). It is perhaps through this relationship that cells are able to sense their environment and extrude enzymes that cleave matrix proteins, enabling metastatic, malignant cells to escape from primary tumors and eventually lead to a secondary growth site. Matrix metalloproteinases (MMPs), enzymes produced by cells to cleave and remodel the nearby matrix, are key players in this regard and are often upregulated in cancerous tissue as well when cells come into contact with proteolytic targets such as collagen I (LaFleur et al., 2005 Sakai et al., 2011). It is therefore well established that studying the interaction between cells and the matrix can infer a great deal about both the underlying biochemical processes that may be occurring inside the cell, as well as a bulk classification of the tissue state as a whole.

7.7.2 Importance of three-dimensional culture systems

In cell-ECM adhesion

In search of investigating cell-ECM interactions, the dimensionality in which the study is performed cannot be overlooked. Traditionally, *in vitro* assays are done on two-dimensional (2D) tissue culture flasks, a vast simplification of how cells normally interact with their environment. While outstanding accomplishments have been conducted in a 2D environment highlighting the importance of substrate stiffness, ECM components, and integrin activity with regards to cancer aggressiveness; assays performed in 2D may undersell other vital cues involved in cellular behavior (Galvez et al., 2002; Yeung et al., 2005; Dydensborg et al., 2009). A major topic of both study and debate is one of focal adhesion sites, where integrin-ECM connections evolve into large multiprotein structures thought to enhance cell-ECM interactions and play a major role in migratory processes (Burridge et al., 1988, 1996). Whereas in 2D these structures are well defined, they have what seems to be a less prominent role when cells are cultured in three-dimensional (3D) lattices, differing in which proteins are present at the adhesion site. This is phenomenon can be observed via immunofluorescence of matrix adhesions of NIH-3T3 fibroblasts *in vitro* (▶Figure 7.14).

This suggests that 2D focal adhesions may represent an exaggerated case of what is actually visualized in 3D (Cukierman et al., 2001; Baumann, 2010). This is most likely due in part to the lack of dorsal cell-ECM interactions, forcing the cell to adopt an elongated structure rarely seen in 3D matrices. In addition, the disparity in spatial integrin-ligand binding may cause cells in 2D to differ genotypically and offer different signaling networks (Birgersdotter et al., 2005; Muthuswamy, 2011).

In angiogenesis

Another often overlooked factor when making the transition between 2D and 3D culture is the effect of dimensionality on the efficacy of angiogenesis in a growing tumor. Angiogenesis is the formation of new blood vessels, a process that is vital for growing

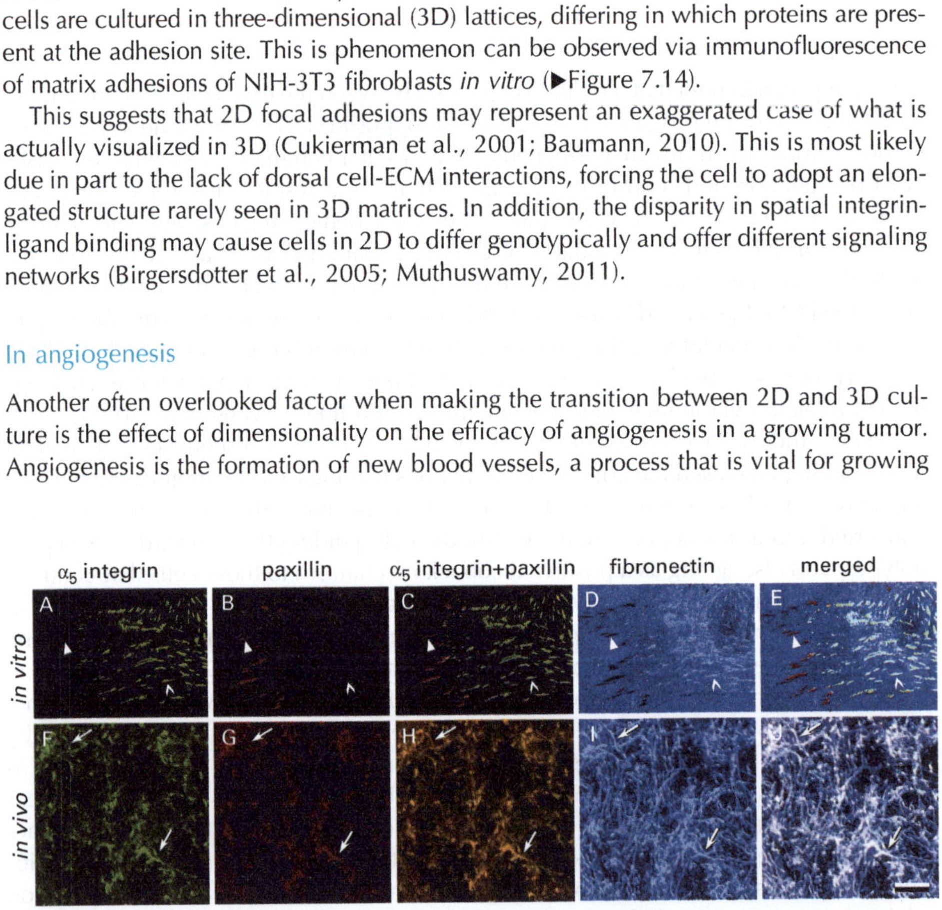

Figure 7.14 Matrix adhesions differ in two- and three-dimensional culture. Differences in adhesions can be qualitatively explored through immunofluorescence. NIH-3T3 fibroblasts are stained in 2D (A–E) or as cryostat sections of a mouse embryo, a native 3D morphology. (F–J) Only in 3D do α5 integrin and paxillin colocalize. Also, fibronectin shows a more fibril structure, showing a higher correlation to both integrin and paxillin in 3D compared to 2D. Focal adhesions (filled arrowheads) and fibrillar adhesions (open arrowheads) on 2D display how integrin and paxillin are separated spacially in 2D. Arrows in 3D represent areas of triple localization, which is absent in 2D. This display of key focal adhesion proteins lends insight to how matrix architecture can inform and alter how the cell organizes its morphology. Scale bar: 5 μm (Cukierman et al., 2001).

tumors. As tumors grow in size they require an enhanced system of vascularity in order to efficiently deliver oxygen to the regions on the tumor that are unable to absorb oxygen via simple diffusion through the tissue (Gupta and Massaque, 2006). Study of cells in monolayers in 2D ignores the spherical nature of tumorigenic growth and therefore does not fully recapitulate the hypoxic dilemma that normal cell masses encounter after a specified growth period (Zetter, 1998; Chaudary and Hill, 2007; Robinson et al., 2009). Culture in 3D has also shown to enhance the onset of angiogenesis through an increase in integrin engagement with the substrate as a result of spherical versus planar architecture (Stupack and Cheresh, 2004; Fischbach et al., 2009). This planar format of 2D cells also fails to capture angiogenic signals that are presented in the form of mechanical stresses due to flow or cell pulling (Yoshino et al., 2003). Finally, metastatic signals that enhance angiogenesis are often conveyed through simple cell density measures felt by a growing tumor, a characteristic that is not reproducible in 2D culture (Kuwano et al., 2004).

In cell migration and metastasis

A discussion regarding cancer and its progression would be egregiously incomplete without mention of what is highly regarded as the crux of the ongoing struggle to treat cancer, cell movement, and metastasis. While treating localized cancer legions has become a somewhat manageable process, after metastasis of the primary tumor the disease becomes inordinately more difficult to contain. Metastasis is a multistep process in where cells from the primary tumor first distance themselves, either genetically, phenotypically, or both from the overall cell mass. Reacting to either ECM or chemokine signals, these cells break away from the tumor mass and extravasate into the bloodstream, intravasate back through the vasculature at a new tissue site, and either undergo necrosis, lay in a quiescent state, or begin to proliferate into a new tumor (Kwong and Chin, 2009). Much of this relies upon the underlying state of the cell and the framework of the ECM where the cell relocates (Chambers et al., 2002).

Metastatic cells that release from the primary tumor oftentimes present a slightly different pheno- and/or genotype from the original cell mass (Yokota, 2000; Gupta and Massaque, 2006). Making matters worse, those cancer cells that do survive travel through the bloodstream or lymphatic system and begin to divide in another organ of the body are altered from their parent cells due to the new environment that they have adopted (Rofstad, 2000; Chambers et al., 2002). As a result, these new growth sites are often much more resistant to chemotherapy and complicate overall cancer treatment (Gu et al., 2004; Wei et al., 2009; Kim et al., 2010).

As highlighted through the enigmatic process of metastasis, a thorough understanding of cell migration at its basic level is critical toward capturing the entire picture of cancer progression. The migration of individual cells is a multistep process in which the cell polarizes, extends pseudopodia out from the leading edge, creates a traction force via actin and myosin contraction within the cell body, and retracts its rear end to release the previous adhesions and propel the cell forward (Sheetz et al., 1999). Including this dogma of cell movement, much has been revealed through analyzing cell migration on 2D surfaces (Lauffenburger and Horwitz, 1996; Ridley et al., 2003; Dobereiner et al., 2005). However, there are few, if any, physiologically relevant situations in which cells encounter a two-dimensional migration space.

This has led to interesting discoveries regarding important effectors of cell migration including the small Rho family of GTPases such as Rho, Rac, and Cdc42. Harnessing the capability to bind to and hydrolyze guanosine triphosphate (GTP) to guanosine diphosphate (GDP), these GTPases toggle between active and inactive states. As such, they are regarded as molecular switches in the cell due to their activation or suppression of various migratory pathways. Cells migrating in 3D matrices have been reported to exhibit smaller lamellipodia and reduced Rac activity, leading to less aggressive, but more directed cellular movements (Beningo et al., 2004; Pankov et al., 2005). In addition to these phenomenological changes in migration characteristics, 3D matrices tend to offer different ECM cues such as a more pliable, heterogeneous matrix leading to directed movement via durotaxis, and changes in integrin-ECM interactions (Cukierman et al., 2001; Lo et al., 2000). Three-dimensional environments also introduce physical barriers not seen in 2D environments. As a result, many cells exhibit proteolytic dependent migration requiring MMP degradation in order to fit through matrix pores, while also adapting to an amoeboid form, a method of migration completely absent in 2D (Wolf et al., 2003a; Lammermann and Sixt, 2009).

7.7.3 Advancements in 3D quantification

Dimensionality of cell culture *in vitro* effects, at least in some form, every underlying process associated with cancer development, growth, and malignancy. It is therefore evident that 3D culture systems be developed to best recreate native environments where cells are exposed to signals and connect with matrix fibers on all faces. Unfortunately, many outstanding tried-and-true protocols that are used in 2D experiments either do not translate well into 3D culture or were not designed to measure and address artifacts of 3D culture such as cell polarity and anisotropy, migration, and matrix degradation.

Matrix remodeling

Cancer progression is closely tied with local and global changes in the surrounding ECM (Stewart et al., 2004). MMPs have been shown to upregulate in cancerous tissue, likely a mechanism to degrade local collagenous fibers that perturb migratory patterns, or prohibit tumorigenic growth (Hotary et al., 2003; Lafleur et al., 2005). Quantifying this phenomenon in 3D has led to two general approaches: one in which fibers are imaged and analyzed in real time, and another in which cells are introduced into well-defined synthetic matrices.

Visualization of 3D collagen matrices can be achieved using confocal reflectance microscopy (CRM). This method uses low laser light to excite the matrix and takes advantage of the natural reflective properties of the fibers, collecting light in a narrow bandwidth centered on the excitation wavelength (Brightman et al., 2000). Combining this strategy with a confocal microscope results in a 3D reconstruction of the matrix that can be analyzed quantitatively using advanced image processing techniques. Using this system, it has been shown that cells lay down matrix proteins locally and also do so in a cell specific manner. Cell type also plays a large role in the degradation and depositing of fibers with each cell working to achieve a specific desired fibril density and matrix stiffness (Hartmann et al., 2006; Harjanto et al., 2011).

Matrix degradation can also be observed by incorporating into 3D gels a small percentage of fluorescently labeled fibers. Copolymerizing with the matrix and quenched until proteolytic cleavage, these fibers provide a means by which to assess matrix degradation by observing the resulting fluorescent signal in time. Such tactics have displayed the efficacy of MMP inhibitors to ablate collagen fiber degradation in 3D (Wolf et al., 2003b, 2007). Cells have also been shown to enhance their capability to degrading collagen fibers in the presence of the epidermal growth factor (EGF) (Kim et al., 2008).

Using a synthetic matrix to study cell induced remodeling allows for the probing of specific parameters in a controlled fashion. The well-known compound polyethylene glycol (PEG) can be used to create a fibrous matrix with defined pore sizes, and stiffness. In addition to having fine control over mechanical properties, PEG can be functionalized to contain a homogeneous pattern of integrin-binding sites based on the arginine, glycine, aspartic acid amino acid (RGD) motif recognized by integrins as well as proteolytic substrate regions (Lutolf et al., 2003; Raeber et al., 2005; Dikovsky et al., 2006). Cell culture within these gels provides the advantage of parsing out conflicting and overlapping characteristics of cell-matrix degradation such as pore size and integrin activity to assess fiber proteolysis in 3D and its subsequent effect on vital cellular processes such as migration (Stahl et al., 2010).

Cellular mechanical properties

Although only recently discovered to play a role in signaling and cell behavior, mechanical forces can drive complex processes such as gene expression and proliferation (Huang and Ingber, 1999). Integrins, the main mediator in the mechanical link between the ECM and intracellular signaling are intrinsically mechanosensitive, allowing for gradation of force transmission from the matrix to intracellular actin fibers and finally through a series of signaling networks (Katsumi et al., 2004). Acting upon these ECM cues, cells will utilize protein-based machinery to organize its actin network aiding in differentiation, migration, or optimal growth (Pelham and Wang, 1997; Discher et al., 2005; Engler et al., 2006). This remodeling, oftentimes representing itself in the form of cell contraction and actin reorganization, propagates as a change in the viscoelastic properties of the cytoplasm (Maniotis et al., 1997). It is therefore recommended to monitor cytoplasmic stiffness as a means to probe the state of the cell or assess how cells react to an external stimulus.

Measuring cell stiffness is a process originally devised to work in 2D, where the dorsal face of the cell was free to probe mechanically. Tactics such as atomic force microscopy (AFM) and magnetic bead twisting cytometry mechanically perturb cells plated on 2D surfaces and measure their stiffness using classical engineering and material science techniques (Haga et al., 2000; Wang et al., 2002; Lam et al., 2007). The lack of a free cell face presented in 3D culture is circumvented via particle tracking microrheology (▶Figure 7.15). Measuring cell stiffness is accomplished by inserting into the cell small fluorescent beads. These beads are tracked, using a high magnification, and high frame rates (~10 Hz) generating time dependent tracks. Using the mean square displacement from many bead tracks $G'(\omega)$ and $G''(\omega)$, the frequency dependent elastic and viscous components of the complex modulus can be computed (Wirtz, 2009; Fraley et al., 2011). These parameters can fully describe the viscoelastic nature of cells.

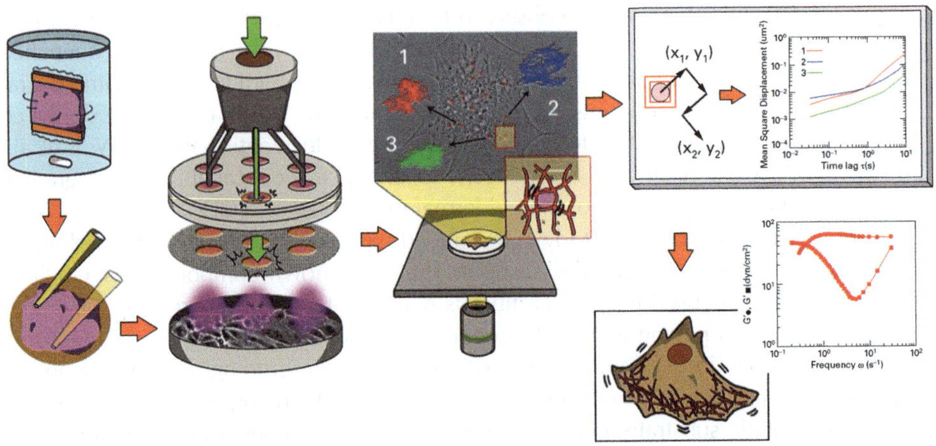

Figure 7.15 **Multiple particle tracking microrheology strategy.** Fluorescent micron-sized beads are dialyzed (A) and injected into cells using a ballistic system (B, C). Next, beads disperse within the cytoplasm of the cell, and the cell is imaged under a high-magnification objective (D). Using high spatiotemporal resolution, beads are tracked in real time, and their *x*- and *y*-coordinates are used to create tracks. From these tracks, the mean square displacement of each cell is calculated and plotted against the time lag of each track (E). Using mean square displacements, the local values of the frequency-dependent complex viscoelastic modulus are calculated (F). This allows for the stiffness to be probed intracellularly on a fine scale (Wirtz, 2009).

Microrheology has already uncovered distinct differences in the viscoelastic nature of cells on 2D plates compared to those within a 3D matrix resulting in a reduction of intracellular forces (Fraley et al., 2011). This change in stiffness and creep compliance has been demonstrated in prostate cancer to be dependent on integrin activity, suggesting that intracellular stiffness in 2D and 3D is in part controlled by the abundance of integrin-ECM interactions (Baker et al., 2009). It has also been shown that cells become more compliant when exposed to the vascular endothelial growth factor (VEGF), and this softening is Rho-kinase (ROCK) dependent (Panorchan et al., 2006). Microrheology measurements made on a series of cell lines ranging from slight malignancy to extreme invasiveness can uncover how cancer progression relates to intracellular stiffness. In fact, it has been shown that in 3D the overexpression of the human epidermal growth factor 2 (Her2 or ErbB2) in breast cancer cells causes them to stiffen in response to a stiffer ECM (Baker et al., 2010). The overexpression of Her2 is a hallmark of malignant breast cancer and this result details the acute interactions between cell and matrix that could potentially lead to a metastatic state.

Pushing and pulling: cell-matrix force interactions

Bead tracking microrheology, together with traction force microscopy (TFM) can be implemented to study matrix stiffness nearby migrating cells (Fraley et al., 2011). Measurements show that cells never push, but only pull on their matrix fibers, dependent on well-known proteins involved in cell contraction Rho-kinase and myosin II. Cells also induce asymmetric forces on the matrix during migration, deforming fibers on the migration front elastically, while inducing irreparable damage to the trail edge of migration channels (Bloom et al., 2008). Probing matrix deformations using bead microrheology can also provide a time dependent analysis of this phenomenon (Meshel et al., 2005). A more bulk representation of cell pulling is evaluated using 3D TFM. The concept of TFM was first developed for assessing cell-induced forces on polyacrylamide gels by analyzing the deformation fields of embedded fluorescent beads (Dembo and Wang, 1999). This process was done on 2D surfaces; however, the same principles have recently been applied to the three-dimensional realm. Calculating forces in 3D, cells are able to modulate and distort the matrix in and out of their plane of migration. Modes of migration, such as a rolling cell through 3D pores, has also been described, motions that are missed in 2D TFM (Franck et al., 2011).

Cell migration

Influenced by a plethora of factors provided by the ECM, cell migration is studied through assays founded upon matrix structure development and video microscopy techniques. Most migration studies are performed within 3D collagen matrices. Collagen solutions are polymerized at different concentrations and temperatures to achieve a fibrous network 100–300 nm in diameter, with pore sizes on the order of 1–10 μm (Elsdale and Bard, 1972; Turley et al., 1985). Cells may either be introduced to the collagen solution during polymerization, or seeded on top of a solid gel (Noble, 1987). Fluorescently staining embedded cells allows for time-lapse confocal fluorescent microscopy. Image tracks can then be analyzed to study migration speed and persistence, and directionality as a function of matrix properties or a chemotactic agent (Friedl et al., 1995). Cell migration has also been monitored using trans-well or Boyden chamber approaches. This strategy requires cells be seeded on top of ECM structures and sandwiched between heterogeneous aqueous media. The general setup introduces growth medium on the upper portion of the cells, while on the bottom, the media contains growth factors that presumably will attract the cells. End point z-dimension distances of cell travel within a given time are used to quantify cell invasiveness. While this method provides a meaningful way to quantify the aggressive nature of cell migration, it fails to capture cell movement on a single cell scale (Chen, 2005). These approaches represent the groundwork to probing cell movement as a function of ECM structure and composition, molecular cues, mechanical forces, and cell signaling on migration.

Monitoring cell movement in 3D presents a more *in vivo* like approach to studying the complex interplay between the major constituents that contribute to the migration mechanism. MMPs, integrins, ECM structure, and a vast network of signaling proteins all influence the direction, rate, and mode in which cells migrate. Proteolytic degradation is essential for many cells migrating through dense matrices where the pore size does not condone free movement (Friedl and Gilmour, 2009). Monitoring cells in 3D, collagen proteolysis can be imaged, and the importance of MMP degradation assessed by introducing active site MMP blockers and observing the resulting decrease in cell motility (Wolf et al., 2003a). In addition to simple steric hindrance stemming from fiber density, the ECM also provides cells with other cues. During a phenomenon known as durotaxis, cells tend to migrate toward stiffer surfaces (Harland et al., 2011). Other ECM cues such as the growth factor VEGF can also influence directional migration in 3D as well (Fiedler et al., 2005).

Migration in 3D matrices shows what seems to be both integrin dependent and independent movement. Common cells such as fibroblasts lose their capacity to migrate when incubated with anti-β1-integrin antibodies showing a spherical morphology indicative of a loss of adhesion sites with the matrix (Maaser et al., 1999). Despite the close ties integrins share with migration, examples of cells continuing to maneuver without integrin activity exist (Fassler and Meyer, 1995; Hirsch et al., 1996). How pervasive this migration mode is *in vivo* continues to be explored; however, 3D migration experiments have highlighted their usefulness by bringing to the forefront pertinent questions regarding the relationship between integrins and migration. One hypothesis currently in focus is the plasticity of the underlying migration mechanism in cells, where a loss in β1 integrin function is assuaged by other integrins or cell attachment mechanisms that begin to fill the void (Friedl and Wolf, 2010).

A growing interest with clinical implications stemming from motility is that of collective cell migration. Movement of single cells has been widely studied; however, what remains uncertain is the mechanism by which cells progress in tandem. Entwined within this challenge is the process of single cells breaking away from these collective cell masses, a vital underlying process of metastasis. Collective migration is very similar to its single cell counterpart; however, cells retain adherin junctions. Those cells that break away from tumor masses often undergo what is referred to the epithelial-to-mesenchymal transition (EMT) where they tend to "dedifferentiate" to a more stem cell like state, losing cell-cell adhesion proteins such as E- and N-cadherin (Friedl et al., 2004). Collective cell migration has also shown to be more dependent on proteolytic activity, and relies on integrin activity in a manner much different than single cell movement (Ilina and Friedl, 2009). While much remains to be explored on the front of migration, assays in 3D described earlier continue to provide the means to filling in the gaps in what presents a very complicated and interconnected problem.

Computational modeling

A perhaps underappreciated and overlooked aspect of studying cell-matrix interactions in 3D is computational modeling. Although frequently oversimplified, these computational approaches present methods to understand how cells behave in 3D as a function of many interdependent variables. They also provide a means by which to predict cellular behavior under specific conditions on both a single and multiscale level. Given the complex nature of cell migration and the multitude of reactions and forces occurring simultaneously during this process, computational models often differ greatly in how they approach the problem of reconstructing a framework to accurately portray systems *in vivo*.

Current various model systems are used to recreate cell migration and interaction with ECM fibers. Treating the system as a force-based model, cell migration speed and direction can be simulated. In this type of approach integrin-ligand reactions are translated into traction forces. The asymmetry of focal adhesions between the front and rear of the cell account for directional movement, counteracted by a viscous ECM (Zaman et al., 2005). Stochastic models base cell movement on random speed samples from a Gaussian distribution and can predict how persistent cells will migrate based on population (Parkhurst and Saltzman, 1992). Implementing Monte Carlo methods allow for re-creation of cell migration on a larger scale and present a more bulk analysis (Zaman et al., 2007).

Models can also be used to predict how cells will grow and interact with the ECM based on several cues such as nutrient composition. Visualizing the cell as a spheroid capable of shape change, a model has been created to simulate cell-cell interaction forces and forces present between cells and the ECM to depict a cell mass growing (Schaller and Meyer-Hermann, 2005). This has implications in tumor growth and also addresses how cells depend on metabolic factors diffusing in 3D. On the molecular scale, models have been able to quantitatively describe integrin clustering on the cell surface and describe their attachment to collagen fibers, providing insight into a system that is incredibly challenging to probe *in vitro* (Lepzelter and Zaman, 2010). Finally, using experimental kinetic parameters, ECM degradation can be quantitatively modeled (Karagiannis and Popel, 2004). While modeling cancer and its many effectors is still a field in its infancy, the potential for simulations to predict how cells react to specific cues on either a cellular or tumor basis is enticing. As models continue to develop and become more complex, they stand to play a larger role in clinical applications.

7.7.4 Future directions

In the past few decades outstanding strides have been made to incorporate into classical cancer pathobiology the complex interaction of cells and the ECM. While cell-matrix interactions continue to grow into a wider field of study, there remains much to be explored. Current 3D matrix solutions based on collagen fibers or polyacrylamide gels provide a convenient and inexpensive platform to perform experiments with a tight control on matrix composition (Friedl et al., 2004). As a result, model systems in 3D oftentimes uncover important phenomenological reactions missing in 2D (Cukierman et al., 2001). However, the gap between current 3D culture systems and *in vivo* biology remains steep.

Lacking in mainstream 3D model constructions is the incorporation of other ECM components such as fibronectin, laminin, proteoglycans, elastin, and several others (Alberts et al., 2002). Along with the mere presence of these constituents, no real research has set out to analyze the appropriate concentrations and organization of these elements, a complication that is most definitely tissue specific. Simple 3D matrices also neglect vital extracellular cues received from other cells types, including stromal cells that actively participate in cancer progression (De Weaver and Mareel, 2003).

Analyzing the forces between cell and matrix in 3D, model systems very rarely address the effects of external loading. Many tissues in the body react to stretch via special receptors or are exposed to cyclical strains and pulsatile flow (Morrow et al., 2007). These physiological processes cannot be ignored and their effect on cells, especially a growing tumor, needs to be characterized. Incorporating all these potential interactions is undoubtedly daunting, but represents the next stage in recreating *in vivo* like culture systems.

As culture systems continue to evolve, so must the methods used for quantification in these models. Current methods have displayed their excellence in probing cell speeds, stiffness, and proteolytic activity on a single cell level. Unfortunately, cancer is rarely ever a single cell event. Future emphasis needs to be placed on collective cell masses, with the capability of probing single cells within these tumors. Spatiotemporal analysis of these tumors can provide clues to how signals are propagated throughout the tumor, resulting in migration, proteolysis, and metastasis.

Perhaps the most neglected portion of quantification in 3D is the incorporation of computational modeling. Simulations in tandem with experimental techniques allows for a problem to analyzed in many different ways. The predictive nature of computational approaches may also uncover interesting avenues for future experimental research, paving the horizon for new discovery. Having tight control of specific variables and the ability to study numerically almost any desired reaction over a defined timescale also makes these works very alluring. The forefront of cancer pathobiology is extremely diverse and overwhelming. Only when all of these interactions and approaches are embraced and incorporated will a reliable system for complete characterization and quantification of cancer processes *in vitro* reach its full potential.

7.7.5 Take-home message

Study of cancer pathobiology *in vitro* demands robust platforms to quantitatively analyze how the ECM effects disease progression. A multitude of methods are beginning to emerge as the next wave of *in vitro* culture, highlighting the importance of taking experiments to the third dimension.

References

Alberts, B., Johnson, A., Lewis, J., Raff, M., Roberts, K., and Walter, P. (2002). Molecular Biology of the Cell, 4th edn. New York: Garland Science.

Baker, E. L., Bonnecaze, R. T., and Zaman, M. H. (2009). Extracellular matrix stiffness and architecture govern intracellular rheology in cancer. Biophys J *97*, 1013–1021.

Baker, E. L., Lu, J., Bonnecaze, R. T., and Zaman, M. H. (2010). Cancer cell stiffness: integrated roles of three-dimensional matrix stiffness and transforming potential. Biophys J *99*, 2048–2057.

Baumann, K. (2010). Cell migration: moving in 3D. Nat Rev Mol Cell Biol *11*, 465.

Beningo, K. A., Dembo, M., and Wang, Y. L. (2004). Responses of fibroblasts to anchorage of dorsal extracellular matrix receptors. Proc Natl Acad Sci U S A *101*, 18024–18029.

Bershadsky, A. D., Balaban, N. Q., and Geiger, B. (2003). Adhesion-dependent cell mechanosensitivity. Annu Rev Cell Dev Biol *19*, 677–695.

Birgersdotter, A., Sandberg, R., and Ernberg, I. (2005). Gene expression perturbation in vitro – a growing case for three dimensional (3D) culture systems. Semin Cancer Biol *15*, 405–412.

Bloom, R. J., George, J. P., Celedon, A., Sun, S. X., and Wirtz, D. (2008). Mapping local matrix remodeling induced by a migrating tumor cell using three-dimensional multiple-particle tracking. Biophys J *95*, 4077–4088.

Brightman A. O., Rajwa, B. P., Sturjis, J. E., McCallister, M. E., Robinson, J. P., and Voytik-Harbin, S. L. (2000). Time-lapse confocal reflection microscopy of collagen fibrillogenesis and extracellular matrix assembly in vitro. Biopolymers *54*, 222–234.

Burridge, K., and Chrzanowska-Wodnicka, M. (1996). Focal adhesions, contractility, and signaling. Annu Rev Cell Dev Biol *12*, 463–518.

Burridge, K., Fath, K., Kelly, T., Nuckolls, G., and Turner, C. (1988). Focal adhesions: transmembrane junctions between the extracellular matrix and the cytoskeleton. Annu Rev Cell Biol *4*, 487–525.

Chambers, A. F., Groom, A. C., and MacDonald, I. C. (2002). Dissemination and growth of cancer cells in metastatic sites. Nat Rev Cancer *2*, 563–572.

Chaudary, N., and Hill, R. P. (2007). Hypoxia and metastasis. Clin Cancer Res *13*, 1947–1949.

Chen, H. C. (2005). Boyden chamber assay. Methods Mol Biol *294*, 15–22.

Cukierman, E., Pankov, R., Stevens, D. R., and Yamada, K. M. (2001). Taking cell-matrix adhesions to the third dimension. Science *294*, 1708–1712.

Dembo, M., and Wang, Y. L. (1999). Stresses at the cell-to-substrate interface during locomotion of fibroblasts. Biophys J *76*, 2307–2316.

De Weaver, O., and Mareel, M. (2003). Role of tissue stroma in cancer cell invasion. J Pathol *200*, 429–447.

Dikovsky, D., Bianco-Peled, H., and Seliktar, D. (2006). The effect of structural alterations of PEG-fibrinogen hydrogel scaffolds on 3D cellular morphology and cellular migration. Biomaterials *27*, 1496–1506.

Discher, D. E., Janmey, P., and Wang, Y. L. (2005). Tissue cells feel and respond to the stiffness of their substrate. Science *310*, 1139–1143.

Dobereiner, H. G., Dubin-Thaler, B. J., Giannone, G., and Sheetz, M. P. (2005). Force sensing and generation in cell phases: analyses of complex functions. J Appl Physiol *98*, 1542–1546.

Dvorak, H. F. (1986). Tumors: wounds that do not heal: similarities between tumor stroma and wound healing. N Engl J Med *315*, 1650–1659.

Dydensborg, A. B., Teller, I. C., Groulx, J. F., et al. (2009). Integrin α6β4 inhibits colon cancer cell proliferation and c-Myc activity. BMC Cancer *9*, 223.

Elsdale, T., and Bard, J. (1972). Collagen substrata for studies on cell behavior. J Cell Biol *54*, 626–637.

Engler, A. J., Sen, S., Sweeney, H. L., and Discher, D. E. (2006). Matrix elasticity directs stem cell lineage specification. Cell *126*, 677–689.

Fassler, R., and Meyer, M. (1995). Consequences of lack of β1 integrin gene expression in mice. Genes Dev *9*, 1896–1908.

Fiedler, J., Leucht, F., Waltenberger, J., Dehio, C., and Brenner, R. E. (2005). VEGF-A and PlGF-1 stimulate chemotactic migration of human mesenchymal progenitor cells. Biochem Biophys Res Commun *334*, 561–568.

Fischbach, C., Kong, H. J., Hsiong, S. X., Evangelista, M. B., Yuen, W., and Mooney, D. J. (2009). Cancer cell angiogenic capability is regulated by 3D culture and integrin engagement. Proc Natl Acad Sci U S A *106*, 399–404.

Fraley, S. I., Hale, C. M., Bloom, R. J., Celedon, A., Lee, J. S. H., and Wirtz, D. (2011). Intra- and extracellular microrheology of endothelial cells in a 3D matrix. In Biophysical Regulation of Vascular Differentiation as Assembly, S. Gerecht, ed. New York: Springer, pp. 69–87.

Franck, C., Maskarinec, S. A., Tirrell, D. A., and Ravichandran, G. (2011). Three-dimensional traction force microscopy: a new tool for quantifying cell-matrix interactions. PLoS One *6*, e17833.

Friedl, P., and Gilmour, D. (2009). Collective cell migration in morphogenesis, regeneration and cancer. Nat Rev Mol Cell Biol *10*, 445–457.

Friedl, P., Hegerfeldt, Y., and Tusch, M. (2004). Collective cell migration in morphogenesis and cancer. Int J Dev Biol *48*, 441–449.

Friedl, P., Noble, P. B., and Zanker, K. S. (1995). T lymphocyte locomotion in a three-dimensional collagen matrix. Expression and function of cell adhesion molecules. J Immunol *154*, 4973–4985.

Friedl, P., and Wolf, K. (2010). Plasticity of cell migration: a multiscale tuning model. J Cell Biol *188*, 11–19.

Galvez, B. G., Matias-Roman, S., Yanez-Mo, M., Sanchez-Madrid, F., and Arroyo, A. G. (2002). ECM regulates MT1-MMP localization with β1 or αvβ3 integrins at distinct cell compartments modulating its internalization and activity on human endothelial cells. J Cell Biol *159*, 509–521.

Gu, B., Espana, L., Mendez, O., Torregrosa, A., and Sierra, A. (2004). Organ-selective chemoresistance in metastasis from human breast cancer cells: inhibition of apoptosis, genetic variability and microenvironment at the metastatic focus. Carcinogenesis *25*, 2293–2301.

Gupta, G. P., and Massaque, J. (2006). Cancer metastasis: building a framework. Cell *127*, 679–695.

Haga, H., Sasaki, S., Kawabata, K., Ito, E., Ushiki, T., and Sambongi, T. (2000). Elasticity mapping of living fibroblasts by AFM and immunofluorescence observation of the cytoskeleton. Ultramicroscopy *82*, 253–258.

Harjanto, D., Maffei, J. S., and Zaman, M. H. (2011). Quantitative analysis of the effect of cancer invasiveness and collagen concentration on 3D matrix remodeling. PLoS One *6* (9): e24891.

Harland, B., Walcott, S., and Sun, S. X. (2011). Adhesion dynamics and durotaxis in migrating cells. Phys Biol *8*, 015011.

Hartmann, A., Boukamp, P., and Friedl, P. (2006). Confocal reflection imaging of 3D fibrin fibers. Blood Cells Mol Dis *36*, 191–193.

Hirsch, E., Iglesias, A., Potocnik, A. J., Hartmann, U., and Fassler, R. (1996). Impaired migration but not differentiation of haematopoietic stem cells in the absence of β1 integrins. Nature *380*, 171–175.

Hotary, K. B., Allen, E. D., Brooks, P. C., Datta, N. S., Long, M. W., and Weiss, S. J. (2003). Membrane type 1 matrix metalloproteinase usurps tumor growth control imposed by the three-dimensional extracellular matrix. Cell *114*, 33–45.

Huang, S., and Ingber, D. E. (1999). The structural and mechanical complexity of cell-growth control. Nat Cell Biol *1*, E131–E138.

Huang, S., and Ingber, D. E. (2005). Cell tension, matrix mechanics, and cancer development. Cancer Cell *8*, 175–176.

Ilina, O., and Friedl, P. (2009). Mechanisms of collective cell migration at a glance. J Cell Sci *122*, 3203–3208.

Ingber D. E., and Jamieson, J. D. (1985). Cells as tensegrity structures: architectural regulation and histodifferentiation by physical forces transduced over basement membrane. In Gene Expression During Normal and Malignant Differentiation, P. Ekblom, ed. London: Academic Press, pp. 13–32.

Karagiannis, E. D., and Popel, A. S. (2004). A theoretical model of type I collagen proteolysis by matrix metalloproteinase (MMP) 2 and membrane type I MMP in the presence of tissue inhibitor of metalloproteinase 2. J Biol Chem *279*, 39105–39114.

Katsumi, A., Orr, A. W., Tzima, E., and Schwartz, M. A. (2004). Integrins in mechanotransduction. J Biol Chem *279*, 12001–12004.

Kim, H. D., Guo, T. W., Wu, A. P., Wells, A., Gertler, F. B., and Lauffenburger, D. A. (2008). Epidermal growth factor-induced enhancement of glioblastoma cell migration in 3D arises from an intrinsic increase in speed but an extrinsic matrix- and proteolysis-dependent increase in persistence. Mol Biol Cell *19*, 4249–4259.

Kim, M., Koh, Y. J., Kim, K. E., et al. (2010). CXCR4 signaling regulates metastasis of chemoresistant melanoma cells by a lymphatic metastatic niche. Cancer Res *70*, 10411–10421.

Kuwano, H., Miyazaki, T., Tsutsumi, S., et al. (2004). Cell density modulates the metastatic aggressiveness of a mouse colon cancer cell line, colon 26. Oncology *67*, 441–449.

Kwong, L. N., and Chin, L. (2009). The metastasis problem gets stickier. Cancer Cell *15*, 1–2.

Lafleur, M. A., Drew, A. F., de Sousa, E. L., et al. (2005). Upregulation of matrix metalloproteinases (MMPs) in breast cancer xenografts: a major induction of stromal MMP-13. Int J Cancer *20*, 544–554.

Lam, W. A., Rosenbluth, M. J., and Fletcher, D. A. (2007). Chemotherapy exposure increases leukemia cell stiffness. Blood *109*, 3505–3508.

Lammermann, T., and Sixt, M. (2009). Mechanical modes of "amoeboid" cell migration. Curr Opin Cell Biol *21*, 636–644.

Lauffenburger, D. A., and Horwitz, A. F. (1996). Cell migration: a physically integrated molecular process. Cell *84*, 359–369.

Lepzelter, D., and Zaman, M. H. (2010). Clustered diffusion of integrins. Biophys J *99*, 106–108.

Lo, C. M., Wang, H. B., Dembo, M., and Wang, Y. L. (2000). Cell movement is guided by the rigidity of the substrate. Biophys J *79*, 144–152.

Lutolf, M. P., Raeber, G. P., Zisch, A. H., Tirelli, N., and Hubbell, J. A. (2003). Cell-responsive synthetic hydrogels. Adv Mater *15*, 888–892.

Maaser, K., Wolf, K., Niggemann, B., Zanker, K. S., Brocker, E. B., and Friedl, P. (1999). Functional hierarchy of simultaneously expressed adhesion receptors: integrin $\alpha 2\beta 1$ but not CD44 mediates MV3 melanoma cell migration and matrix reorganization within three-dimensional hyaluronan-containing collagen matrices. Mol Biol Cell *10*, 3067–3079.

Maniotis, A. J., Chen, C. S., and Ingber, D. E. (1997). Demonstration of mechanical connections between integrins, cytoskeletal filaments, and nucleoplasm that stabilize nuclear structure. Proc Natl Acad Sci U S A *94*, 849–854.

Meshel, A. S., Wei, Q., Adelstein, R. S., and Sheetz, M. P. (2005). Basic mechanism of three-dimensional collagen fibre transport by fibroblasts. Nat Cell Biol *7*, 157–164.

Morrow, D., Sweeney, C., Birney, Y. A., et al. (2007). Biomechanical regulation of hedgehog signaling in vascular smooth muscle cells in vitro and in vivo. Am J Physiol Cell Physiol *292*, C488–C496.

Muthuswamy, S. K. (2011). 3D culture reveals a signaling network. Breast Cancer Res *13*, 103.

Noble, P. B. (1987). Extracellular matrix and cell migration: locomotory characteristics of MOS-11 cells within a three-dimensional hydrated collagen lattice. J Cell Sci *87*, 241–248.

Pankov, R., Endo, Y., Even-Ram, S., et al. (2005). A Rac switch regulates random versus directionally persistent cell migration. J Cell Biol *170*, 793–802.

Panorchan, P., Lee, J. S., Kole, T. P., Tseng, Y., and Wirtz, D. (2006). Microrheology and ROCK signaling of human endothelial cells embedded in a 3D matrix. Biophys J *91*, 3499–3507.

Parkhurst, M. R., and Saltzman, W. M. (1992). Quantification of human neutrophil motility in three-dimensional collagen gels. Effect of collagen concentration. Biophys J *61*, 306–315.

Pelham, R. J., Jr., and Wang, Y. (1997). Cell locomotion and focal adhesions are regulated by substrate flexibility. Proc Natl Acad Sci U S A *94*, 13661–13665.

Raeber, G. P., Lutolf, M. P., and Hubbell, J. A. (2005). Molecularly engineered PEG hydrogels: a novel model system for proteolytically mediated cell migration. Biophys J *89*, 1374–1388.

Ridley, A. J., Schwartz, M. A., Burridge, K., et al. (2003). Cell migration: integrating signals from front to back. Science *302*, 1704–1709.

Robinson, B. D., Sica, G. L., Liu, Y. F., et al. (2009). Tumor microenvironment of metastasis in human breast carcinoma: a potential prognostic marker linked to hematogenous dissemination. Clin Cancer Res *15*, 2433–2441.

Rofstad, E. K. (2000). Microenvironment-induced cancer metastasis. Int J Radiat Biol *76*, 589–605.

Sakai, K., Nakamura, T., Suzuki, Y., Imizu, T., and Matsumoto, K. (2011). 3D collagen-dependent cell surface expression of MT1-MMP and MMP-2 activation regardless of integrin-β1 function and matrix stiffness. Biochem Biophys Res Commun *19*, 98–103.

Schaller, G., and Meyer-Hermann, M. (2005). Multicellular tumor spheroid in an off-lattice Voronoi-Delaunay cell model. Phys Rev E *71*, 051910.

Sheetz, M. P., Felsenfeld, D., Galbraith, C. G., and Choquet, D. (1999). Cell migration as a five-step cycle. Biochem Soc Symp *65*, 233–243.

Stahl, P. J., Romano, N. H., Wirtz, D., and Yu, S. M. (2010). PEG-based hydrogels with collagen mimetic peptide-mediated and tunable physical cross-links. Biomacromolecules *11*, 2336–2344.

Stewart, D. A., Cooper, C. R., and Sikes, R. A. (2004). Changes in extracellular matrix (ECM) and ECM-associated proteins in the metastatic progression of prostate cancer. Reprod Biol Endocrinol *2*, 2.

Stupack, D. G., and Cheresh, D. A. (2004). Integrins and angiogenesis. Curr Top Dev Biol *64*, 201–238.

Swaminathan, V., Mythreye, K., O'Brien, T. E., Berchuck, A., Blobe, G. C., and Superfine, R. (2011). Mechanical stiffness grades metastatic potential in patient tumor cells and in cancer cell lines. Cancer Res *71*, 5075–5080.

Turley, E. A., Erickson, C. A., and Tucker, R. P. (1985). The retention and ultrastructural appearances of various extracellular matrix molecules incorporated into three-dimensional hydrated collagen lattices. Dev Biol *109*, 347–369.

Wang, N., Butler, J. P., and Ingber, D. E. (1993). Mechanotransduction across the cell surface and through the cytoskeleton. Science *260*, 1124–1127.

Wang, N., Tolic-Nørrelykke, I. M., Chen, J., et al. (2002). Cell prestress. I. Stiffness and prestress are closely associated in adherent contractile cells. Am J Physiol Cell Physiol *282* (3), C606–C616.

Wei, Y., Hu, G., and Kang, Y. (2009). Metadherin as a link between metastasis and chemoresistance. Cell Cycle *8*, 2132–2133.

Wirtz, D. (2009). Particle-tracking microrheology of living cells: principles and applications. Annu Rev Biophys *38*, 301–326.

Wolf, K., Mazo, I., Leung, H., et al. (2003a). Compensation mechanism in tumor cell migration: mesenchymal-amoeboid transition after blocking of pericellular proteolysis. J Cell Biol *160*, 267–277.

Wolf, K., Muller, R., Borgmann, S., Brocker E. B., and Friedl, P. (2003b). Amoeboid shape change and contact guidance: T-lyphocyte crawling through fibrillar collagen is independent of matrix remodeling by MMPs and other proteases. Blood *102*, 3262–3269.

Wolf, K., Wu, Y. I., Geiger, J., et al. (2007). Multi-step pericellular proteolysis controls the transition from individual to collective cancer cell invasion. Nat Cell Biol *9*, 893–904.

Yeung, T., Georges, P. C., Flanagan, L. A., et al. (2005). Cell Motil Cytoskeleton *60*, 24–34.

Yokota, J. (2000). Tumor progression and metastasis. Carcinogenesis *21*, 497–503.

Yoshino, H., Morita, I., Murota, S. I., and Ishikawa, I. (2003). Mechanical stress induces production of angiogenic regulators in cultured human gingival and periodontal ligament fibroblasts. J Periodontal Res *38*, 405–410.

Zaman, M. H., Kamm, R. D., Matsudaira, P., and Lauffenburger, D. A. (2005). Computational model for cell migration in three-dimensional matrices. Biophys J *89*, 1389–1397.

Zaman, M. H., Matsudairia, P., and Lauffenburger, D. A. (2007). Understanding effects of matrix protease and matrix organization on directional persistence and translational speed in three-dimensional cell migration. Ann Biomed Eng *35*, 91–100.

Zetter, B. R. (1998). Angiogenesis and tumor metastasis. Annu Rev Med *49*, 407–424.

7.8 Diabetic foot infections

Cecilia Rydén

7.8.1 Introduction

Human staphylococcal infection affects any organ, with skin infections and abscesses being the most common. Patients with diabetes mellitus are more susceptible to bacterial infection, due to the influence of metabolism on the immune defense, as well as the progressive neuropathy and angiopathy in later stages of diabetes. In patients suffering from diabetes mellitus, *Staphylococcus aureus* is the most frequently found microbe in skin and skeletal infections (Dinh et al., 2010). *S. aureus* is part of the normal flora of human skin and mucous membranes. Colonization with *S. aureus* is found in about one-third of the population at large, but persons with diabetes mellitus, skin disorders, or in frequent contact with medical care have a higher incidence of *S. aureus* colonization. On the other hand, everybody is colonized with coagulase negative staphylococci. However, these bacteria are commensals and symbionts, although often associated with infection of foreign material, and thus such infections are a fairly new problem to medicine (Campbell et al., 2008). Inserted devices are frequently used in medical care, catheters being most common, but also osteosynthesis material and prosthesis implants. Coagulase negative staphylococci are very efficient biofilm formers, and due to contamination during surgery from the patient's own flora or the surroundings, a concomitant implant may be colonized and cause a slowly progressing infection (Galdbart et al., 2000). *S. aureus* may also cause these types of infection of foreign, implanted material (Darouiche et al., 1997; Arciola et al., 2005; Campoccia et al., 2006; Otsuka et al., 2006).

Persons having frequent contact with hospital surroundings and medical care facilities may change their normal skin and mucous membrane microbial flora, their microbiome, and will carry antibiotic-resistant microbes including staphylococci of any kind, for example, MRSA (methicillin resistant *S. aureus*) and/or MRSE (*Staphylococcus epidermidis*, representing the coagulase negative staphylococci) (Eleftheriadou et al., 2010; Waninger et al., 2011). Staphylococci can survive on artificial surfaces such as turfs for at least 1 week (Vardakas et al., 2008). Thus, strictly performed hospital hygiene programs such as in Newcastle, UK, which are implemented and followed, are welcomed since they show that even high levels of MRSA in the hospital environment can decrease by 35% (Collins et al., 2011). In case of infection, MRSA will be much harder to treat since the number of effective antibiotics is limited for treatment of these bacteria. Even in the case of totally sensitive *S. aureus* infection, many different options for treatment are available, and no consensus on what and when to use the different kinds of antibiotics is presented, very few randomized trials have been performed, and regimens differ due to local prevalence of antibiotic resistant strains.

The regimens for soft tissue infection only versus osteitis should be different as for antimicrobial drug regimen and duration of treatment; thus optimizing diagnosis is essential for correct handling of the different types of infection (Mandal et al., 2002; Armstrong and Lipsky, 2004a, 2004b; Rich, 2005; Paydar et al., 2006; Andersen and Roukis, 2007; Matthews et al., 2007; Rao and Lipsky, 2007; Uzun et al., 2007; Eisenstein, 2008; Dinh et al., 2010). The specific location of staphylococci to skeletal structures involves several adhesion molecules, microbial surface components recognizing adhesive matrix molecules (MSCRAMMs) (Patti and Hook, 1994), including collagen and bone sialoprotein (BSP). We have studied the interaction between clinical isolates of *S. aureus* from patients suffering from osteomyelitis, septic arthritis and BSP and found a correlation with BSP binding in contrast to *S. aureus* strains isolated from patients with endocarditis and soft tissue infection (Ryden et al., 1987), and the MRSCRAMM was found to be a member of the Sdr family of surface proteins (Tung et al., 2000) that is specifically interacting with BSP (Ryden et al., 1989, 1997). Coagulase staphylococci from a patient with recurrent multifocal osteomyelitis were also found to bind BSP (Ryden et al., 1990a).

The most prevalent skeletal infections are foot infections frequently occurring in patients with diabetes mellitus. Soft tissue infection easily progresses into bone tissue causing osteitis in diabetic patients due to the vascular and neurophysiological changes occurring secondary to the metabolic disorder. *S. aureus* is almost always found in cultures of these wounds. Because these bacteria are primary pathogens, they could not be excluded as causative of the infection and thus can cause an underlying osteitis or, unless an osteitis is diagnosed, may contaminate cultures taken through the infected wound by puncture or biopsy of underlying bone tissue (Boulton et al., 2004; Ayello, 2005). While collecting patients for a study of diabetic foot infections, we performed wound cultures including an effort to get a control group of patients without any *S. aureus*–positive cultures. We almost failed since only 5 patients, which were previously treated with antibiotics, did not have cultures with *S. aureus* out of a total of 41 patients cultured (▶Table 7.1), although *S. aureus* was sometimes found together with other bacteria, though not primary pathogens for osteitis such as *Escherichia coli*. Anaerobes were often found in long-standing deeply penetrating wounds. Several studies report combined aerobic and anaerobic bacteria cultured from diabetic foot infections, and the frequency of cultures yielding *S. aureus* from 28%–76% (Goldstein et al.,

Table 7.1 Patients with immunoglobulin G (IgG) against BSP-binding protein (Bbp).

Diagnosis	*Patients with* S. aureus *infection*	*No* S. aureus *infection (controls)*
	Number of patients	*Number of controls*
	Total number of patients/number of patients with IgG against Bbp>cut off	
Diabetic foot osteitis	17/13	2/0
Soft tissue infection only	24/1	3/0
Respiratory tract infection	Not available	9/0
Healthy blood donor	Not available	29/1

1996; Tentolouris et al., 1999; Pellizzer et al., 2001; Kessler et al., 2006; Senneville et al., 2006, 2009). Combined neuropathic and angiopathic complications in diabetes patients increase the risk for severe outcome, such as infection postsurgery, as well as the mortality rate after infection (Sykes and Godsey, 1998; Schramm et al., 2006). In France, lower limb amputation rate in patients admitted due to diabetic foot problems is reported to be 40%. Management of diabetic angiopathy and osteitis involves surgical decisions both of vascular bypass and stent operations, as well as partial or total lower limb amputation (Frykberg, 2002, 2007; Zgonis et al., 2008; Game, 2010; Yudovsky et al., 2010). The symptoms of early neuropathy with dry feet due to decreased sweat secretion predisposes for ulcers at an early stage of neuropathic complication (Cavanagh et al., 2005; Cheer et al., 2009).

There is an association between hematogenous osteomyelitis and the growing skeleton. Since BSP is synthesized by osteoblasts, the interaction between staphylococci and BSP seems logical to study. BSP is abundant in the growing but also in regenerating bone, after fractures and insertion of joint prostheses. In countries with a low gross national product (GNP)/capita, skeletal infections commonly affect children, with an acute onset of symptoms in about 50% of the cases. However, in countries with a high GNP/capita, staphylococcal infections, and not least osteitis, are more prevalent in adults, the latter mainly contiguous skeletal infections. The cost for the patient as well as for society is increasing with increasing prevalence of diabetes. Unfortunately, we can foresee an increasing number of patients with diabetic complications since this diagnosis is becoming more prevalent in countries where it was formerly rare, and with increasing life expectancy, late complications will become manifest (Skyler et al., 2000).

Diagnosis of diabetic foot osteitis

Diagnosing bone tissue involvement in patients with either a local skeletal infection or a contiguous soft tissue infection invading the skeleton is often difficult. The probe to bone test has been debated; however, the results of a meta-analysis (Lavery et al., 2007) support using the test to exclude osteitis, with a negative predictive value of 98%. Thus, this cheap and always available test should be performed in every patient with diabetic ulcers (Dinh et al., 2010). Late changes on plain X-ray may often delay specific treatment of osteitis (O'Meara et al., 2006; Fard et al., 2007; Butalia et al., 2008; Hartemann-Heurtier et al., 2008). On the other hand, the decision to be made by the clinician in charge regarding type of therapy, and the duration of antibiotic treatment in any of these instances, is also very difficult, since radiographic changes of osteitis will be present even long after the infection is cured, which may lead to prolonged antibiotic treatment regimens (Bonham, 2001; Sheppard 2005; Berendt et al., 2008; Omar et al., 2008; Griffis et al., 2009). Guidelines for shorter treatment have not yet been well established, though this option is being considered more often since antibiotic resistance and selection of multidrug-resistant microbes may develop. However, scientific support is still lacking for such action. Prolonged antibiotic regimens may exert negative influences both on the individual patient and on the ecology of the environment.

The diagnosis of Charcot's disease, diabetic neuropathic osteoarthropathy, is challenging as a differential diagnosis from, for example, erysipelas or other early soft tissue infection, or even an osteitis without an ulcer. Patients with Charcot's are more prone to develop pressure wounds due to deformities of the skeletal structures of the feet, and thus prone to infectious osteitis. The diagnosis by radiographic methods may also be difficult in view of that both osteitis and Charcot's leads to decalcification of the skeleton. The handling of Charcot's foot, though, is of benefit in osteitis, that is, immediate relief of pressure, cast immobilization, and use of crutches, the important difference being no indication of antibiotic prescription in Charcot's foot. Such treatment should thus be avoided in this inflammatory situation, but it is crucial and should be started as soon as possible in infectious osteitis (Jude and Unsworth, 2004; Lipsky, 2004; Shank and Feibel, 2006; Kosinski and Joseph, 2007; Khanolkar et al., 2008; Lipsky, 2008; Miller and Henry, 2009; Kalish and Hamdan, 2010).

Microbiological diagnostic methods

Culture

The most convincing evidence of a bacterial correlation with a disease is a positive culture from a significant site of infection, but in case of diabetic foot infection, it is extremely hard to find patients with negative bacteriological cultures including cultures negative for *S. aureus*.

Positive blood cultures are more commonly found in patients with hematogenous infection than in patients with a contiguous infection such as localized osteomyelitis, a contained abscess, or a skin infection. Patients with osteitis secondary to a diabetic foot ulcer do not have positive blood cultures, unless sepsis has occurred. A microbial diagnosis by culture is mainly provided with a swab taken from the ulceration, after thorough cleaning to avoid contamination by the normal skin flora, and such cultures are accurate in 90% of cases for antimicrobial treatment decisions in wounds without bone penetration (Slater et al., 2004). The swab culture, though, is not accurate when the wounds penetrate into skeletal structures with only 65% accuracy (Slater et al., 2004). Most patients cultured this way have *S. aureus* isolated from their wounds, no difference be it a superficial soft tissue infection or an osteitis, but several other microbes are often found as well (Goldstein et al., 1996; Tentolouris et al., 1999; Pellizzer et al., 2001; Kessler et al., 2006; Senneville et al., 2006, 2009). Some of those bacteria constitute contamination by the patient's normal flora or other pathogens such as *Pseudomonas* or enterococci (Proctor et al., 2006). Aerobes are more frequently found in acute and superficial wounds, whereas anaerobes are cultured from deeply penetrating, long-standing wounds. A needle biopsy through the noninfected skin into an osteitis guided by X-ray reveals fewer bacteria than simultaneously taken swab cultures (Kessler et al., 2006), but when comparing needle puncture with percutaneous biopsy, the latter shows significantly more accuracy (Senneville et al., 2006) both in terms of diagnosing osteitis and also excluding patients from unnecessary antimicrobial treatment (Senneville et al., 2009). Swab cultures often yield twice as many bacterial species as bone biopsies (Senneville et al., 2006, 2009), and also more species than needle punctures (Kessler et al., 2006). It is very rare, though, that *S. aureus* is not found in diabetic wound infections (Dihn et al., 2010). Our study material included five patients with wounds without *S. aureus* in cultures totally, and those patients were all treated with antibiotics previously (Patti et al., 1993). *S. aureus* was the most frequently isolated pathogen in patients with only one pathogen found (Senneville et al., 2009).

The frequency of methicillin resistant *S. aureus*, MRSA, has been increasing and reflects the MRSA incidence of the society at large. In diabetic foot infections, MRSA was associated with prior antibiotic treatment. One should bear in mind, though, that a positive culture of the primary pathogen, *S. aureus*, in diabetic foot infection does not per se yield a correct diagnosis, nor does a negative culture exclude the possibility of infection, especially if the patient is already on antibiotic treatment (Nelson et al., 2006). Bone biopsy culture would be the preferential microbiological diagnostic procedure in osteitis, followed by needle puncture. If neither of these methods is possible, a swab culture could be performed, with the knowledge that not all microbes found on swab cultures are of significance for the infection.

PCR methods using 16S RNA

The method developed to find bacterial nucleic acid positivity has been tried in staphylococcal disease, but since it is very sensitive, the normal flora contamination, almost impossible to avoid, limits the use of this method. The 16S RNA may be used for normally sterile compartments (e.g. joint fluid, epidural abscesses) but is not practical in soft tissue infections or contiguous osteitis, which is the usual case in diabetic foot infection (Fuursted et al., 2008).

Diagnosing osteitis by radiography

To diagnose osteitis in the foot of a patient suffering from diabetes mellitus is a challenge both for radiologists and the clinician (Dihn et al., 2008, 2010; Yansouni et al., 2009). Thus, a normal radiography does not rule out an osteitis in the early stages of disease (reported sensitivity 54% and specificity 80% for osteitis diagnosis by plain X-ray; Dihn et al., 2010), and thus the patient has to be judged and treated according to the clinical picture and the physician's knowledge and experience of the disorder. Diabetes mellitus leads to osteopenia sometimes aggravated to fulfill the diagnosis of Charcot's disease. Osteopenia is bilateral; thus comparing an ulcerated foot with its normal counterpart may yield no added information but is usually performed. In case of severe osteitis/osteomyelitis, the mineralized matrix of the skeleton must decrease by 50% to be visualized on plain X-ray, which may take 4 weeks to occur. The use of computerized tomography (CT) scan may reveal changes in the skeleton as early as 1 week into the osteitis, though it usually takes 9–14 days. Sequestration of the skeletal structures may be visualized by CT scan. Magnetic resonance tomography imaging (MRI) shows early increase in fluid content and may add information mainly on edema and abscess formation, but it may also be helpful in differentiating osteitis from acute Charcot's. The cost of MRI, though, necessitates a modest use of this diagnostic tool, only when both other diagnostic methods and treatment have failed. Previously used radionuclide imaging, including the better but more expensive leukocyte scintigraphy, is expensive and not readily available and has been succeeded by CT. The combination of infectious osteitis and Charcot's disease in a patient makes diagnosis and judgment of treatment modality very difficult, not easily differentiated by either the clinician or the radiologist (Rosenthal et al., 1983; Gold et al., 1995; Newman, 1995; Santiago Restrepo et al., 2003; Pineda et al., 2006; Johnson et al., 2007; Rogers and Bevilacqua, 2008; Basu et al., 2009; Palestro and Love, 2009; Yudovsky et al., 2010).

Laboratory parameters in diabetic foot ulcers and osteitis

Clinical chemistry laboratory analysis is not specific for the soft tissue or skeletal damage found in diabetes mellitus, just like other soft tissue or skeletal infections. The wound-healing capacity has been studied in relation to laboratory parameters (Abu-Rumman et al., 2002) but, for diagnostic purposes, is of very limited benefit. C-reactive protein (CRP) is often increased but may also be normal in osteitis, and patients with diabetes tend to have a higher CRP than persons without diabetes (Vardakas et al., 2008). Since soft tissue infection leads to increase of CRP, it is often high in nontreated foot infection. Erythrocyte sedimentation rate (ESR) was increased in diabetes osteitis compared to cellulitis with 100% specificity in a small study of 29 patients and could thus support the diagnosis (Kaleta et al., 2001), whereas fever is often not present in these patients. Leukocyte counts, as well as neutrophils, may be normal or elevated and provide little help in diagnosing osteitis (Dinh et al., 2010). In conclusion, patients suffering from infection statistically have higher ESR, white blood cell counts (WBC), and procalcitonin (PCT), the latter being of little significance for diagnosis of diabetic foot infections for the time being (Uzun et al., 2007).

7.8.2 Serological diagnosis of osteitis in foot infection in diabetes mellitus

Bone sialoprotein binding by *S. aureus*

The staphylococcal adhesins of the family designated MSCRAMM have been shown to be involved in staphylococcal pathogenesis (Patti et al., 1992, 1993, 1994a, 1994b, 1995; Patti and Hook, 1994; Darouiche et al., 1997; Cassat et al., 2005; Campoccia et al., 2009). The intriguing selectivity in binding of BSP from bone matrix to clinical isolates from patients suffering from *S. aureus* osteitis or septic arthritis, compared to isolates from patients with endocarditis or soft tissue infection, prompted us to characterize the interaction, purify the staphylococcal-binding molecule, and use this molecule in clinical studies (Ryden et al., 1987, 1989, 1990a, 1997). Bone sialoprotein is mainly found in bone and calcified tissues (Franzen and Heinegard, 1985), most abundant in newly formed bone, thus in young persons and after damage to the skeleton (de Bri et al., 1996). The interaction between BSP and staphylococci was first described with *S. aureus* (Ryden et al., 1987), but later also with coagulase negative staphylococci including a strain from a patient suffering from chronic recurrent multifocal osteomyelitis (CRMO) (Ryden et al., 1990a). The interaction was significantly more frequent in clinical strains from patients suffering from osteitis and septic arthritis than from patients with sepsis only, endocarditis, and coagulase negative infection (Ryden et al., 1987). After characterizing the protein (Ryden et al., 1989), we purified the native BSP-binding protein (Bbp) from the clinical osteomyelitis *S. aureus* strain O24 and found a 97 kDa protein that after sequencing was found to belong to the Sdr family of cell-wall-associated proteins in gram-positive cocci (Ryden et al., 1997). A recombinant version of Bbp produced in *E. coli* was shown to interact with BSP, and it specifically inhibits binding of BSP to staphylococcal cells (Tung et al., 2000). Osteopontin, another glycoprotein found in bone tissue, as well as in many other tissues, does not show the selectivity in binding to clinical strains of *S. aureus* from different locations, nor do several other matrix molecules such as fibronectin, collagen, or fibrinogen (Ryden et al., 1987, 1989, 1990a, 1990b, 1997; Peacock et al., 2002; Dihn et al., 2010). Several Sdr family proteins of staphylococci have been characterized, including SdrC–G. SdrG interacts with fibrinogen, and SdrF of *S. epidermidis* binds collagen (Arrecubieta et al., 2007). SdrE is 76% homologous to Bbp in the A region and 96% in the B region of the protein but has not been shown to bind BSP (Davis et al., 2001). In fact, SdrE is negatively associated with skeletal infection, whereas a positive association exists to platelet activation. *S. aureus* Bbp gene expression is associated with expression of genes for methicillin resistance and Panton-Valentine leukocidin.

Bone-sialoprotein-binding protein of *S. aureus* as antigen in enzyme-linked immunosorbetnt assay (ELISA)

We evaluated antibody levels against Bbp in serum samples from patients with staphylococcal infection (Persson et al., 2009). Our hypothesis was that since BSP-binding correlated with skeletal involvement of *S. aureus*, such infections may evoke an immunoglobulin G (IgG) response in patients suffering from this type of infection, in contrast to other types of staphylococcal infection. The study we performed aimed at evaluating IgG antibodies to Bbp and also to investigate whether the correlation between osteitis and BSP interaction, previously reported, evoked an IgG response in serum from patients with different *S. aureus* infections. Patients included suffered from different types of *S. aureus* infection, including diabetic foot infection of soft tissue involvement or osteitis, as well as other types of infections. It was almost impossible to find wound infections that did not yield a positive culture for *S. aureus*; thus this group of patients was very limited. In addition to the two MSCRAMMs Bbp and SdrE, we used α toxin and teichoic acid as antigens in the ELISA and did serial dilutions of patient serum to determine the amount of IgG reacting with the respective antigen. The most convincing result from the study was that 100% of patients with diabetes and osteitis had increased levels of IgG against Bbp, whereas only a few of those with soft tissue infection showed any response with IgG against Bbp. There was no such striking difference in the results with the other antigens; SdrE did not evoke much of an IgG response in any of the patients tested.

The elevated IgG levels in serum at the first visit with the physician for presumed osteomyelitis probably reflects patients delay due to neuropathy rather than a very early IgG rise. Our results show that all patients with diabetic osteitis have significantly increased levels of IgG against Bbp (Persson et al., 2009), a finding that could aid in comparatively early diagnosis of osteomyelitis.

Several studies using serological assays have been performed and published trying to diagnose staphylococcal osteitis and soft tissue infection as well (Julander et al., 1983; Colque-Navarro et al., 2000, 2010; Elliott et al., 2000; Rafiq et al., 2000; Ryding et al., 2002; Casey et al., 2006; Jacobsson et al., 2010; Collins et al., 2011). Vaccines to evoke an immune response and thus prevent severe staphylococcal infection have been tried (Josefsson et al., 2001; Patti, 2004a, 2004b; Bloom et al., 2005; Capparelli et al., 2005; Clarke et al., 2006; Schaffer et al., 2006; Burian et al., 2012; Holtfreter et al., 2011; Joost et al., 2011) but are not yet clinically available. Vaccine antigen given intradermally to patients with diabetes mellitus does not yield a significant rise in antibodies in contrast to intramuscular injection, which induces an antibody response of the same range as in nondiabetic persons.

Our results show that absence of antibodies against Bbp strongly indicates that an ongoing staphylococcal infection does not affect bone tissue (Persson et al., 2009). Other studies support our finding of an association between the presence of Bbp in *S. aureus* and osteomyelitis. The conclusion from the study was that serum IgG directed against Bbp can serve as a marker of osteomyelitis, not least in diabetic foot infections.

7.8.3 Conclusion and summary

Bone sialoprotein interacts specifically with a cell-wall-anchored Sdr molecule of *S. aureus* and may be a major factor for localizing these bacteria to skeletal structures (Ryden et al., 1987, 1989, 1990b, 1997; Tung et al., 2000). An immunological response with increased IgG levels against Bbp is noted in patients with *S. aureus* osteitis, which may aid in diagnosis as well be used as a marker of diabetic foot osteitis versus soft tissue infection without osteitis of the diabetic foot (Persson et al., 2009).

The diagnosis of diabetic foot osteitis is very difficult, especially when soft tissue infection is deep and clinical examination does not rule out skeletal involvement. Microbiological and radiographic diagnostic measures are not specific and at an early stage of osteitis could not rule out such an infection. Radiological changes on plain X-rays are rare at an early stage of osteitis but could show up on MRI, a method that is not available 24 hours/day and certainly not at every clinic. If cultures turn out negative (i.e. no growth of microbes), the anti-Bbp ELISA could be helpful in deciding the treatment regimen, including the duration of antibiotic treatment. This may lead to prolonged antibiotic treatment regimens for patients, with possible side effects and environmental influence on antibiotic resistance.

The ELISA assay using Bbp as an antigen may be an answer to the question of what method to use for diagnosing patients with osteitis in diabetic foot infection, since all patients studied showed raised IgG levels against Bbp, in contrast to almost none of those with soft tissue infection only. We thus hope that a Bbp-based ELISA may be of help in diagnosing diabetic osteitis and made commonly available if these findings can be confirmed in larger studies and the outcome is positive.

7.8.4 Take-home message

The diagnosis of diabetic foot osteitis is often made on the clinical presentation and signs since laboratory parameters are not specific enough to discriminate soft tissue infection only from the deep engagement of skeletal structures. Radiographic changes appear at quite a late stage when an osteitis has been developing for weeks. Sophisticated methods such as MRI and CT scans are not available at every clinic, thus a screening test would be of great help for the clinician in diagnosing osteitis. IgG antibodies in serum against BSP-binding protein of *S. aureus* are increased in patients with diabetic foot osteitis compared to diabetes patients with soft tissue infection only and thus could be a helpful tool in diagnosing diabetic foot osteitis.

Acknowledgments

I am greatly indebted to Christian Johansson and Lena Persson, Uppsala University, Sweden, for performing the ELISA assays, as well as to professor Kristofer Rubin, Uppsala, Sweden, for valuable discussions.

References

Abu-Rumman, P. L., Armstrong, D. G., and Nixon, B. P. (2002). Use of clinical laboratory parameters to evaluate wound healing potential in diabetes mellitus. J Am Podiatr Med Assoc *92*, 38–47.

Andersen, C. A., and Roukis, T. S. (2007). The diabetic foot. Surg Clin North Am *87*, 1149–1177, x.

Arciola, C. R., Campoccia, D., Gamberini, S., Baldassarri, L., and Montanaro, L. (2005). Prevalence of cna, fnbA and fnbB adhesin genes among Staphylococcus aureus isolates from orthopedic infections associated to different types of implant. FEMS Microbiol Lett *246*, 81–86.

Armstrong, D. G., and Lipsky, B. A. (2004a). Advances in the treatment of diabetic foot infections. Diabetes Technol Ther *6*, 167–177.

Armstrong, D. G., and Lipsky, B. A. (2004b.) Diabetic foot infections: stepwise medical and surgical management. Int Wound J *1*, 123–132.

Arrecubieta, C., Lee, M. H., Macey, A., Foster, T. J., and Lowy, F. D. (2007). SdrF, a Staphylococcus epidermidis surface protein, binds type I collagen. J Biol Chem *282*, 18767–18776.

Ayello, E. A. (2005). What does the wound say? Why determining etiology is essential for appropriate wound care. Adv Skin Wound Care *18*, 98–109; quiz 110–111.

Basu, S., Chryssikos, T., Moghadam-Kia, S., Zhuang, H., Torigian, D. A., and Alavi, A. (2009). Positron emission tomography as a diagnostic tool in infection: present role and future possibilities. Semin Nucl Med *39*, 36–51.

Berendt, A. R., Peters, E. J., Bakker, K., et al. (2008). Diabetic foot osteomyelitis: a progress report on diagnosis and a systematic review of treatment. Diabetes Metab Res Rev *24* (Suppl. 1), S145–S161.

Bloom, B., Schelonka, R., Kueser, T., et al. (2005). Multicenter study to assess safety and efficacy of INH-A21, a donor-selected human staphylococcal immunoglobulin, for prevention of nosocomial infections in very low birth weight infants. Pediatr Infect Dis J *24*, 858–866.

Bonham, P. (2001). A critical review of the literature: Part II: antibiotic treatment of osteomyelitis in patients with diabetes and foot ulcers. J Wound Ostomy Continence Nurs *28*, 141–149.

Boulton, A. J., Kirsner, R. S., and Vileikyte, L. (2004). Clinical practice. Neuropathic diabetic foot ulcers. N Engl J Med *351*, 48–55.

Burian, M., Grumann, D., Holtfreter, S., Wolz, C., Goerke, C., and Bröker, B. M. (2012). Expression of staphylococcal superantigens during nasal colonization is not sufficient to induce a systemic neutralizing antibody response in humans. Eur J Clin Microbiol Infect Dis *31* (3), 251–256.

Butalia, S., Palda, V. A., Sargeant, R. J., Detsky, A. S., and Mourad, O. (2008). Does this patient with diabetes have osteomyelitis of the lower extremity? JAMA *299*, 806–813.

Campbell, S. J., Deshmukh, H. S., Nelson, C. L., et al. (2008). Genotypic characteristics of Staphylococcus aureus isolates from a multinational trial of complicated skin and skin structure infections. J Clin Microbiol *46*, 678–684.

Campoccia, D., Montanaro, L., and Arciola, C. R. (2006). The significance of infection related to orthopedic devices and issues of antibiotic resistance. Biomaterials *27*, 2331–2339.

Campoccia, D., Speziale, P., Ravaioli, S., et al. (2009). The presence of both bone sialoprotein-binding protein gene and collagen adhesin gene as a typical virulence trait of the major epidemic cluster in isolates from orthopedic implant infections. Biomaterials *30*, 6621–6628.

Capparelli, E. V., Bloom, B. T., Kueser, T. J., et al. (2005). Multicenter study to determine antibody concentrations and assess the safety of administration of INH-A21, a donor-selected human Staphylococcal immune globulin, in low-birth-weight infants. Antimicrob Agents Chemother *49*, 4121–4127.

Casey, A. L., Worthington, T., Bonser, R. S., Lambert, P. A., and Elliott, T. S. (2006). Rapid serodiagnosis of Staphylococcus aureus surgical site infection following median sternotomy. J Infect *52*, 276–281.

Cassat, J. E., Dunman, P. M., McAleese, F., Murphy, E., Projan, S. J., and Smeltzer, M. S. (2005). Comparative genomics of Staphylococcus aureus musculoskeletal isolates. J Bacteriol *187*, 576–592.

Cavanagh, P. R., Lipsky, B. A., Bradbury, A. W., and Botek, G. (2005). Treatment for diabetic foot ulcers. Lancet *366*, 1725–1735.

Cheer, K., Shearman, C., and Jude, E. B. (2009). Managing complications of the diabetic foot. BMJ *339*, b4905.

Clarke, S. R., Brummell, K. J., Horsburgh, M. J., et al. (2006). Identification of in vivo-expressed antigens of Staphylococcus aureus and their use in vaccinations for protection against nasal carriage. J Infect Dis *193*, 1098–1108.

Collins, J., Raza, M., Ford, M., Hall, L., Brydon, S., and Gould, F. K. (2011). Review of a three-year meticillin-resistant Staphylococcus aureus screening programme. J Hosp Infect *78*, 81–85.

Colque-Navarro, P., Jacobsson, G., Andersson, R., Flock, J. I., and Mollby, R. (2010). Levels of antibody against 11 Staphylococcus aureus antigens in a healthy population. Clin Vaccine Immunol *17*, 1117–1123.

Colque-Navarro, P., Palma, M., Soderquist, B., Flock, J. I., and Mollby, R. (2000). Antibody responses in patients with staphylococcal septicemia against two Staphylococcus aureus fibrinogen binding proteins: clumping factor and an extracellular fibrinogen binding protein. Clin Diagn Lab Immunol *7*, 14–20.

Darouiche, R. O., Landon, G. C., Patti, J. M., et al. (1997). Role of Staphylococcus aureus surface adhesins in orthopaedic device infections: are results model-dependent? J Med Microbiol *46*, 75–79.

Davis, S. L., Gurusiddappa, S., McCrea, K. W., Perkins, S., and Hook, M. (2001). SdrG, a fibrinogen-binding bacterial adhesin of the microbial surface components recognizing adhesive matrix molecules subfamily from Staphylococcus epidermidis, targets the thrombin cleavage site in the Bbeta chain. J Biol Chem *276*, 27799–27805.

de Bri, E., Reinholt, F. P., Heinegard, D., Mengarelli-Widholm, S., Norgard, M., and Svensson, O. (1996). Bone sialoprotein and osteopontin distribution at the osteocartilaginous interface. Clin Orthop Relat Res *330*, 251–260.

Dinh, M. T., Abad, C. L., and Safdar, N. (2008). Diagnostic accuracy of the physical examination and imaging tests for osteomyelitis underlying diabetic foot ulcers: meta-analysis. Clin Infect Dis *47*, 519–527.

Dinh, T., Snyder, G., and Veves, A. (2010). Current techniques to detect foot infection in the diabetic patient. Int J Low Extrem Wounds *9*, 24–30.

Eisenstein, B. I. (2008). Treatment challenges in the management of complicated skin and soft-tissue infections. Clin Microbiol Infect *14* (Suppl. 2), 17–25.

Eleftheriadou, I., Tentolouris, N., Argiana, V., Jude, E., and Boulton, A. J. (2010). Methicillin-resistant Staphylococcus aureus in diabetic foot infections. Drugs *70*, 1785–1797.

Elliott, T. S., Tebbs, S. E., Moss, H. A., et al. (2000). A novel serological test for the diagnosis of central venous catheter-associated sepsis. J Infect *40*, 262–266.

Fard, A. S., Esmaelzadeh, M., and Larijani, B. (2007). Assessment and treatment of diabetic foot ulcer. Int J Clin Pract *61*, 1931–1938.

Franzen, A., and Heinegard, D. (1985). Isolation and characterization of two sialoproteins present only in bone calcified matrix. Biochem J *232*, 715–724.

Frykberg, R. G. (2002). Diabetic foot ulcers: pathogenesis and management. Am Fam Physician *66*, 1655–1662.

Frykberg, R. G., Wittmayer, B., and Zgonis, T. (2007). Surgical management of diabetic foot infections and osteomyelitis. Clin Podiatr Med Surg *24*, 469–482, viii–ix.

Fuursted, K., Arpi, M., Lindblad, B. E., and Pedersen, L. N. (2008). Broad-range PCR as a supplement to culture for detection of bacterial pathogens in patients with a clinically diagnosed spinal infection. Scand J Infect Dis *40*, 772–777.

Galdbart, J. O., Allignet, J., Tung, H. S., Ryden, C., and El Solh, N. (2000). Screening for Staphylococcus epidermidis markers discriminating between skin-flora strains and those responsible for infections of joint prostheses. J Infect Dis *182*, 351–355.

Game, F. (2010). Management of osteomyelitis of the foot in diabetes mellitus. Nat Rev Endocrinol *6*, 43–47.

Gold, R. H., Tong, D. J., Crim, J. R., and Seeger, L. L. (1995). Imaging the diabetic foot. Skeletal Radiol *24*, 563–571.

Goldstein, E. J., Citron, D. M., and Nesbit, C. A. (1996). Diabetic foot infections. Bacteriology and activity of 10 oral antimicrobial agents against bacteria isolated from consecutive cases. Diabetes Care *19*, 638–641.

Griffis, C. D., Metcalfe, S., Bowling, F. L., Boulton, A. J., and Armstrong, D. G. (2009). The use of gentamycin-impregnated foam in the management of diabetic foot infections: a promising delivery system? Expert Opin Drug Deliv *6*, 639–642.

Hartemann-Heurtier, A., and Senneville, E. (2008). Diabetic foot osteomyelitis. Diabetes Metab *34*, 87–95.

Holtfreter, S., Jursa-Kulesza, J., Masiuk, H., et al. (2011). Antibody responses in furunculosis patients vaccinated with autologous formalin-killed Staphylococcus aureus. Eur J Clin Microbiol Infect Dis *30*, 707–717.

Jacobsson, G., Colque-Navarro, P., Gustafsson, E., Andersson, R., and Mollby, R. (2010). Antibody responses in patients with invasive Staphylococcus aureus infections. Eur J Clin Microbiol Infect Dis *29*, 715–725.

Johnson, P. T., Fayad, L. M., and Fishman, E. K. (2007). CT of the foot: selected inflammatory arthridites. J Comput Assist Tomogr *31*, 961–969.

Joost, I., Jacob, S., Utermöhlen, O., et al. (2011). Antibody response to the extracellular adherence protein (Eap) of Staphylococcus aureus in healthy and infected individuals. FEMS Immunol Med Microbiol *62* (1), 23–31.

Josefsson, E., Hartford, O., O'Brien, L., Patti, J. M., and Foster, T. (2001). Protection against experimental Staphylococcus aureus arthritis by vaccination with clumping factor A, a novel virulence determinant. J Infect Dis *184*, 1572–1580.

Jude, E. B., and Unsworth, P. F. (2004). Optimal treatment of infected diabetic foot ulcers. Drugs Aging *21*, 833–850.

Julander, I. G., Granstrom, M., Hedstrom, S. A., and Mollby, R. (1983). The role of antibodies against alpha-toxin and teichoic acid in the diagnosis of staphylococcal infections. Infection *11*, 77–83.

Kaleta, J. L., Fleischli, J. W., and Reilly, C. H. (2001). The diagnosis of osteomyelitis in diabetes using erythrocyte sedimentation rate: a pilot study. J Am Podiatr Med Assoc *91*, 445–450.

Kalish, J., and Hamdan, A. (2010). Management of diabetic foot problems. J Vasc Surg *51*, 476–486.

Kessler, L., Piemont, Y., Ortega, F., et al. (2006). Comparison of microbiological results of needle puncture vs. superficial swab in infected diabetic foot ulcer with osteomyelitis. Diabet Med *23*, 99–102.

Khanolkar, M. P., Bain, S. C., and Stephens, J. W. (2008). The diabetic foot. QJM *101*, 685–695.

Kosinski, M. A., and Joseph, W. S. (2007). Update on the treatment of diabetic foot infections. Clin Podiatr Med Surg *24*, 383–396, vii.

Lavery, L. A., Armstrong, D. G., Peters, E. J., and Lipsky, B. A. (2007). Probe-to-bone test for diagnosing diabetic foot osteomyelitis: reliable or relic? Diabetes Care *30*, 270–274.

Lipsky, B. A. (2004). A report from the international consensus on diagnosing and treating the infected diabetic foot. Diabetes Metab Res Rev *20* (Suppl. 1), S68–S77.

Lipsky, B. A. (2008). New developments in diagnosing and treating diabetic foot infections. Diabetes Metab Res Rev *24* (Suppl. 1), S66–S71.

Mandal, S., Berendt, A. R., and Peacock, S. J. (2002). Staphylococcus aureus bone and joint infection. J Infect *44*, 143–151.

Matthews, P. C., Berendt, A. R., and Lipsky B. A. (2007). Clinical management of diabetic foot infection: diagnostics, therapeutics and the future. Expert Rev Anti Infect Ther *5*, 117–127.

Miller, A. O., and Henry, M. (2009). Update in diagnosis and treatment of diabetic foot infections. Phys Med Rehabil Clin N Am *20*, 611–625.

Nelson, E. A., O'Meara, S., Craig, D., et al. (2006). A series of systematic reviews to inform a decision analysis for sampling and treating infected diabetic foot ulcers. Health Technol Assess *10*, iii–iv, ix–x, 1–221.

Newman, L. G. (1995). Imaging techniques in the diabetic foot. Clin Podiatr Med Surg *12*, 75–86.

Omar, N. S., El-Nahas, M. R., and Gray, J. (2008). Novel antibiotics for the management of diabetic foot infections. Int J Antimicrob Agents *31*, 411–419.

O'Meara, S., Nelson, E. A., Golder, S., Dalton, J. E., Craig, D., and Iglesias, C. (2006). Systematic review of methods to diagnose infection in foot ulcers in diabetes. Diabet Med *23*, 341–347.

Otsuka, T., Saito, K., Dohmae, S., et al. (2006). Key adhesin gene in community-acquired methicillin-resistant Staphylococcus aureus. Biochem Biophys Res Commun *346*, 1234–1244.

Palestro, C. J., and Love, C. (2009). Nuclear medicine and diabetic foot infections. Semin Nucl Med *39*, 52–65.

Patti, J. M. (2004a). A humanized monoclonal antibody targeting Staphylococcus aureus. Vaccine *22* (Suppl. 1), S39–S43.

Patti, J. M. (2004b). Immunotherapeutics for nosocomial infections. Expert Opin Investig Drugs *13*, 673–679.

Patti, J. M., Allen, B. L., McGavin, M. J., and Hook, M. (1994a). MSCRAMM-mediated adherence of microorganisms to host tissues. Annu Rev Microbiol *48*, 585–617.

Patti, J. M., Boles, J. O., and Hook, M. (1993). Identification and biochemical characterization of the ligand binding domain of the collagen adhesin from Staphylococcus aureus. Biochemistry *32*, 11428–11435.

Patti, J. M., Bremell, T., Krajewska-Pietrasik, D., et al. (1994b). The Staphylococcus aureus collagen adhesin is a virulence determinant in experimental septic arthritis. Infect Immun *62*, 152–161.

Patti, J. M., and Hook, M. (1994). Microbial adhesins recognizing extracellular matrix macromolecules. Curr Opin Cell Biol *6*, 752–758.

Patti, J. M., House-Pompeo, K., Boles, J. O., Garza, N., Gurusiddappa, S., and Hook, M. (1995). Critical residues in the ligand-binding site of the Staphylococcus aureus collagen-binding adhesin (MSCRAMM). J Biol Chem *270*, 12005–12011.

Patti, J. M., Jonsson, H., Guss, B., et al. (1992). Molecular characterization and expression of a gene encoding a Staphylococcus aureus collagen adhesin. J Biol Chem *267*, 4766–4772.

Paydar, K. Z., Hansen, S. L., Charlebois, E. D., Harris, H. W., and Young, D. M. (2006). Inappropriate antibiotic use in soft tissue infections. Arch Surg *141*, 850–854; discussion 855–856.

Peacock, S. J., Moore, C. E., Justice, A., et al. (2002). Virulent combinations of adhesin and toxin genes in natural populations of Staphylococcus aureus. Infect Immun *70*, 4987–4996.

Pellizzer, G., Strazzabosco, M., Presi, S., et al. (2001). Deep tissue biopsy vs. superficial swab culture monitoring in the microbiological assessment of limb-threatening diabetic foot infection. Diabet Med *18*, 822–827.

Persson, L., Johansson, C., and Ryden, C. (2009). Antibodies to Staphylococcus aureus bone sialoprotein-binding protein indicate infectious osteomyelitis. Clin Vaccine Immunol *16*, 949–952.

Pineda, C., Vargas, A., and Rodriguez, A. V. (2006). Imaging of osteomyelitis: current concepts. Infect Dis Clin North Am *20*, 789–825.

Proctor, R. A., von Eiff, C., Kahl, B. C., et al. (2006). Small colony variants: a pathogenic form of bacteria that facilitates persistent and recurrent infections. Nat Rev Microbiol *4*, 295–305.

Rafiq, M., Worthington, T., Tebbs, S. E., et al. (2000). Serological detection of Gram-positive bacterial infection around prostheses. J Bone Joint Surg Br *82*, 1156–1161.

Rao, N., and Lipsky, B. A. (2007). Optimising antimicrobial therapy in diabetic foot infections. Drugs *67*, 195–214.

Rich, P. (2005). Treatment of uncomplicated skin and skin structure infections in the diabetic patient. J Drugs Dermatol *4*, s26–29.

Rogers, L. C., and Bevilacqua, N. J. (2008). Imaging of the Charcot foot. Clin Podiatr Med Surg *25*, 263–274, vii.

Rosenthal, D. I., Mankin, H. J., and Bauman, R. A. (1983). Musculoskeletal applications for computed tomography. Bull Rheum Dis *33*, 1–4.

Ryden, C., Maxe, I., Franzen, A., Ljungh, A., Heinegard, D., and Rubin, K. (1987). Selective binding of bone matrix sialoprotein to Staphylococcus aureus in osteomyelitis. Lancet *2*, 515.

Ryden, C., Tung, H. S., Nikolaev, V., Engstrom, A., and Oldberg, A. (1997). Staphylococcus aureus causing osteomyelitis binds to a nonapeptide sequence in bone sialoprotein. Biochem J *327* (Pt 3), 825–829.

Ryden, C., Yacoub, A., Hirsch, G., Wendel, M., Oldberg, A., and Ljungh, A. (1990a). Binding of bone sialoprotein to Staphylococcus epidermidis isolated from a patient with chronic recurrent multifocal osteomyelitis. J Infect Dis *161*, 814–815.

Ryden, C., Yacoub, A., Kvarnstrom, A., et al. (1990b). Detection of Staphylococcus aureus infection by enzyme-linked immunosorbent assay and immunoblotting, using high molecular weight staphylococcal proteins. FEMS Microbiol Immunol *2*, 65–73.

Ryden, C., Yacoub, A. I., Maxe, I., et al. (1989). Specific binding of bone sialoprotein to Staphylococcus aureus isolated from patients with osteomyelitis. Eur J Biochem *184*, 331–336.

Ryding, U., Espersen, F., Soderquist, B., and Christensson, B. (2002). Evaluation of seven different enzyme-linked immunosorbent assays for serodiagnosis of Staphylococcus aureus bacteremia. Diagn Microbiol Infect Dis *42*, 9–15.

Santiago Restrepo, C., Gimenez, C. R., and McCarthy, K. (2003). Imaging of osteomyelitis and musculoskeletal soft tissue infections: current concepts. Rheum Dis Clin North Am *29*, 89–109.

Schaffer, A. C., Solinga, R. M., Cocchiaro, J., et al. (2006). Immunization with Staphylococcus aureus clumping factor B, a major determinant in nasal carriage, reduces nasal colonization in a murine model. Infect Immun *74*, 2145–2153.

Schramm, J. C., Dinh, T., and Veves, A. (2006). Microvascular changes in the diabetic foot. Int J Low Extrem Wounds *5*, 149–159.

Senneville, E., Melliez, H., Beltrand, E., et al. (2006). Culture of percutaneous bone biopsy specimens for diagnosis of diabetic foot osteomyelitis: concordance with ulcer swab cultures. Clin Infect Dis *42*, 57–62.

Senneville, E., Morant, H., Descamps, D., et al. (2009). Needle puncture and transcutaneous bone biopsy cultures are inconsistent in patients with diabetes and suspected osteomyelitis of the foot. Clin Infect Dis *48*, 888–893.

Shank, C. F., and Feibel, J. B. (2006). Osteomyelitis in the diabetic foot: diagnosis and management. Foot Ankle Clin *11*, 775–789.

Sheppard, S. J. (2005). Antibiotic treatment of infected diabetic foot ulcers. J Wound Care *14*, 260–263.

Skyler, J. S. (2000). The economic burden of diabetes and the benefits of improved glycemic control: the potential role of a continuous glucose monitoring system. Diabetes Technol Ther *2* (Suppl. 1), S7–S12.

Slater, R. A., Lazarovitch, T., Boldur, I., et al. (2004). Swab cultures accurately identify bacterial pathogens in diabetic foot wounds not involving bone. Diabet Med *21*, 705–709.

Sykes, M. T., and Godsey, J. B. (1998). Vascular evaluation of the problem diabetic foot. Clin Podiatr Med Surg *15*, 49–83.

Tentolouris, N., Jude, E. B., Smirnof, I., Knowles, E. A., and Boulton, A. J. (1999). Methicillin-resistant Staphylococcus aureus: an increasing problem in a diabetic foot clinic. Diabet Med *16*, 767–771.

Tung, H.-S., Guss, B., Hellman, U., Persson, L., Rubin, K., and Rydén, C. (2000). A bone sialoprotein-binding protein from *Staphylococcus aureus*: a member of the staphylococcal Sdr family. Biochem J *345*, 611–619.

Uzun, G., Solmazgul, E., Curuksulu, H., et al. (2007). Procalcitonin as a diagnostic aid in diabetic foot infections. Tohoku J Exp Med *213*, 305–312.

Vardakas, K. Z., Horianopoulou, M., and Falagas, M. E. (2008). Factors associated with treatment failure in patients with diabetic foot infections: an analysis of data from randomized controlled trials. Diabetes Res Clin Pract *80*, 344–351.

Waninger, K. N., Rooney, T. P., Miller, J. E., Berberian, J., Fujimoto, A., and Buttaro, B. A. (2011). Community-associated methicillin-resistant Staphylococcus aureus survival on artificial turf substrates. Med Sci Sports Exerc *43*, 779–784.

Yansouni, C. P., Mak, A., and Libman, M. D. (2009). Limitations of magnetic resonance imaging in the diagnosis of osteomyelitis underlying diabetic foot ulcers. Clin Infect Dis *48*, 135.

Yudovsky, D., Nouvong, A., and Pilon, L. (2010). Hyperspectral imaging in diabetic foot wound care. J Diabetes Sci Technol *4*, 1099–1113.

Zgonis, T., Stapleton, J. J., Girard-Powell, V. A., and Hagino, R. T. (2008). Surgical management of diabetic foot infections and amputations. AORN J *87*, 935–946; quiz 947–950.

8 Targeting tumor microenvironment at the ECM level

8.1 Introduction

Nikos K. Karamanos

Cancer is one of the leading causes of mortality and disease worldwide, and as a consequence, it has been the main project of several research groups throughout the world. During the past decade, cancer treatment research has focused on the cancer cell itself, identifying potential targets, such as oncogenes, intermediate cellular signal transducers, and cell surface receptors. The importance of the microenvironment in supporting malignant growth and the molecular events by which stromal cells may contribute to carcinogenesis have gained more attention (Polyak et al., 2009).

Numerous studies have demonstrated that the tumor microenvironment not only responds to and supports carcinogenesis, but also actively contributes to tumor initiation, progression, and metastasis. The underlying premise deals with the highly coordinated interplay between cancer cells and their surrounding microenvironment involving extracellular matrix (ECM), growth factors (GFs), and cytokines associated with ECM, as well as surrounding cells. The cells present in the tumor stroma are endothelial cells, fibroblasts, macrophages, mast cells, and neutrophils, as wells as pericytes and adipocytes (Liotta et al., 2001). It is well established that the cross talk between the tumor stroma and cancer cells is bidirectional. Tumor cells are able to shape their microenvironment and support the development of both nonmalignant cells and themselves. The biological importance of ECM macromolecules is highlighted throughout the previous thematic sections of this book. Cell migration, invasion, metastasis, and angiogenesis, all of significance in cancer progression, are dependent on the ECM properties of the tumor itself and the surrounding microenvironment, involving alterations in structure and/or gene expression, as well as activity modifications that take place during this process (see thematic minireview series by Iozzo and Karamanos, 2010; Hascall and Karamanos, 2011; Nagase and Karamanos, 2011). As consequence, pharmacological targeting of the tumor microenvironment unveils a promising tumor cell-directed therapy (Theocharis et al., 2010; Gialeli et al., 2011; Misra et al., 2011).

In an attempt to peel characteristic layers of this complex interplay, this thematic section provides the reader with a focused assessment of the different players of ECM considering their contribution in cancer progression and novel approaches with respect to their pharmacological targeting. These aspects are presented in ▶Figure 8.1.

Chapter 8.2 by Misra et al., entitled "Targeting the Tumor Microenvironment in Cancer Progression," introduces the reader to the tumor microenvironment, guiding them through important tumor characteristics from tumor interstitial fluid pressure to tumor acidosis and hypoxia, and neighbors like tumor-associated fibroblasts, macrophages, lymphocytes, and endothelial cells in correlation to ECM macromolecules. The importance of a cluster of differentiation (CD44) variant is also presented as a well-established example in prostate cancer targeting.

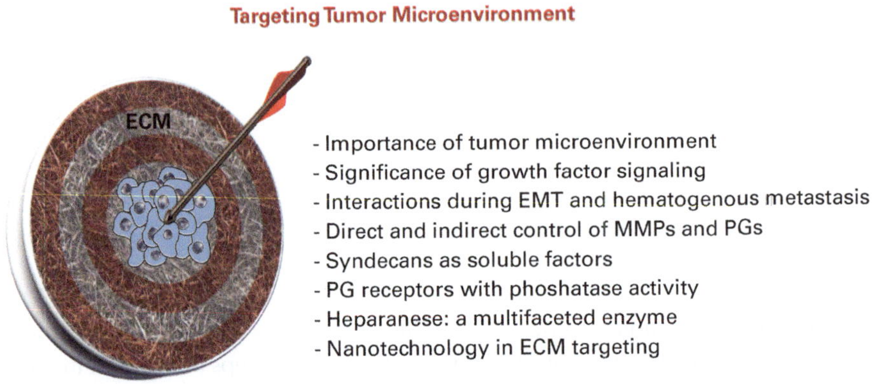

Figure 8.1 Tumor microenvironment supports tumor progression and metastasis. There is a highly coordinated interplay between cancer cells and their surrounding microenvironment. This cross talk between the tumor stroma and cancer cells plays a key role in shaping the microenvironment to support the tumor. Therefore, pharmacological targetingpharmacological targeting of the tumor microenvironment unveils a promising tumor cell-directed therapy. A focused assessment of the different players of ECM considering their contribution in cancer progression and novel approaches with respect to their pharmacological targeting is summarized in this diagram.

Chapter 8.3 by Nikitovic et al., entitled "Growth Factor Signaling and Extracellular Matrix," focuses on the signaling pathways activated by GFs, leading to the regulation of synthesis and degradation of ECM components, and on the reciprocal regulation of GF function by ECM components. Stromal and cancer cells may interact with each other through direct cell-cell contact or via paracrine signaling. Both cells secrete various ECM proteins, GFs that act as signal transducers through their cell surface receptors. GFs such as epidermal growth factor (EGF), platelet-derived growth factor (PDGF), fibroblast growth factor-2 (FGF2), and transforming growth factor-beta (TGF-β) are the main players of this network, and the disruption of their action may be a promising pharmacological approach in targeting cancer progression.

During cancer progression and metastasis, cancer cells undergo critical changes such as epithelial-to-mesenchymal transition (EMT) and hematogenous metastasis that occur to the primary tumor and the bloodstream, respectively. During EMT, epithelial cancer cells lose their polarity together with cell-cell contacts and then undergo a dramatic remodeling of the cytoskeleton and ECM. As cancer cells enter the bloodstream they form complexes with platelets and leucocytes in order to be protected by the host immune system. Once migrating to the suitable site, tumor cells reexpress E-cadherin and other epithelial markers via a process that is sometimes referred to as mesenchymal-to-epithelial transition (MET) (Kalluri and Weinberg, 2009). Chapter 8.4 by Pavao et al., entitled "Targeting Protein-Glycan Interactions at Cell Surface during EMT and Hematogenous Metastasis: Consequences on Tumor Invasion and Metastasis," underlines the role of ECM macromolecules and a possible intervention to these interactions with the use of heparin analogues, an important matrix macromolecule.

Considering the importance of ECM macromolecules encountered though this book, in Chapter 8.5 by Skandalis et al., entitled "Pharmacological Targeting of Proteoglycans and Metalloproteinases: An Emerging Aspect in Cancer Treatment," the novel roles of proteoglycans (PGs) and metalloproteinases are addressed in malignancy and their pharmacological targeting. The role of proteasome and its inhibition of action in ECM remodeling are also introduced.

The structure and function of the cell surface PGs, syndecans, are well characterized in three parts of this book. A new concept is introduced in Chapter 8.6 by Sanderson and Couchman, entitled "Targeting Syndecan Shedding in Cancer." Notably, syndecans can be shed by proteinases from the cell surface changing their localization and function as they are present as soluble factors to the extracellular milieu and the extent of shedding is related to disease progression (Choi et al., 2010; Manon-Jensen et al., 2010). Moreover, a promising approach of targeting the syndecan-1/heparanase axis for therapy in multiple myeloma is also presented in this chapter.

In Chapter 8.7 by Koutsioumpa and Papadimitriou, entitled "PG Receptors with Phosphatase Action in Cancer and Angiogenesis," fresh ideas on PG receptors with phosphatase activity are introduced to the reader. Pharmacological targeting of such phosphatase-active PG receptors may be a promising strategy for cancer prevention and treatment.

The fate of heparan sulfate (HS) PGs is regulated by an endoglycosidase-degrading HS enzyme, heparanase (Barash et., 2010). Chapter 8.8 by Vlodavsky et al., entitled "Heparanase, a Multifaceted Protein Involved in Cancer, Chronic Inflammation, and Kidney Dysfunction," summarizes the ongoing basic and translational research on the biology and clinical significance of the heparanase enzyme and its potential as a target.

Concluding this section, a novel approach in targeted therapy is presented using the recent ideas of nanotechnology. In Chapter 8.9, Ghatak et al. present the "Delivery Systems Targeting Cancer at the Level of ECM." This mainly focuses on the nonviral shRNA delivery systems, particularly CD44v6shRNA delivery to colon tumors as introduced in Chapter 8.2 of this part.

References

Barash, U., Cohen-Kaplan, V., Dowek, I., Sanderson, R. D., Ilan, N., and Vlodavsky, I. (2010). Proteoglycans in health and disease: new concepts for heparanase function in tumor progression and metastasis. FEBS J *277* (19), 3890–3903.

Choi, S., Lee, H., Choi, J. R., and Oh, E. S. (2010). Shedding; towards a new paradigm of syndecan function in cancer. BMB Rep *43* (5), 305–310.

Gialeli, Ch., Theocharis, A. D., and Karamanos, N. K. (2011). Matrix metalloproteinases in health and disease: roles in cancer progression and their pharmacological targeting. FEBS J *278*, 16–27.

Hascall, V., and Karamanos, N. Regulatory roles of hyaluronan in health and disease. (2011). FEBS J *278* (9), 1411.

Iozzo, R. V., and Karamanos, N. (2010). Proteoglycans in health and disease: emerging concepts and future directions. FEBS J *277* (19), 3863.

Kalluri, R., and Weinberg, R. A. (2009). The basics of epithelial-mesenchymal transition. J Clin Invest *119*, 1420–1428.

Liotta, L. A., and Kohn, E. C. (2001). The microenvironment of the tumour-host interface. Nature *411*, 375–379.

Manon-Jensen, T., Itoh, Y., and Couchman, J. R. (2010). Proteoglycans in health and disease: the multiple roles of syndecan shedding. FEBS J *277* (19), 3876–3889.

Misra, S., Heldin, P., Hascall, V. C., et al. (2011). Hyaluronan-CD44 interactions as potential targets for cancer therapy. FEBS J *278* (9), 1429–1443.

Nagase, H., and Karamanos, N. (2011). Metalloproteinases in health and disease: challenges and the future prospects. FEBS J *278* (1), 1.

Polyak, K., Haviv, I., and Campbell, G. I. (2009). Co-evolution of tumor cells and their microenvironment. Trends Genet *25*, 30–38.

Theocharis, A. D., Skandalis, S. S., Tzanakakis, G. N., and Karamanos, N. K. (2010). Proteoglycans in health and disease: novel roles for proteoglycans in malignancy and their pharmacological targeting. FEBS J *277*, 3904–3923.

8.2 Targeting the tumor microenvironment in cancer progression

Suniti Misra, Vincent C. Hascall, Nikos K. Karamanos, Roger R. Markwald, and Shibnath Ghatak

8.2.1 Targeting the tumor microenvironment

Interaction between tumor cells and the stromal compartment in the tumor microenvironment (TME) has a major role in growth and progression of solid tumors. The TME is the product of an emerging cross talk between various cell types. In epithelial tumors the cross talk is between tumor cells and stromal elements, including, the extracellular matrix (ECM) components, smooth muscle cells, tumor-associated endothelial cells (TAECs), immune and inflammatory cells (tumor-associated lymphocytes [TALs] and macrophages [TAMs]), tumor associated fibroblasts (TAFs) of various phenotypes, and blood and lymph vessels. Tumor cells residing in the stroma can modify stromal structure by altering the connective tissue sheath and modulating the anabolism and catabolism of resident cells and thus produce a stroma that is conducive for the tumor cell growth. Tumor stromal cells, including immune cells, mediate tumor progression by secreting growth factors/cytokines that promote angiogenesis, tumor cell proliferation, migration, and survival (Albini and Sporn, 2007) (▶Figure 8.2).

The interaction between the tumor cells and stromal cells gives rise to microenvironmental hallmark properties of solid tumor growth and progression. As shown in ▶Figure 8.3 these include the following: (1) increased self-proliferation by amplifying growth signals from autocrine and paracrine signals; (2) resistance to antiproliferative signals by inhibiting progression of G1 phase to S phase through c-Myc (cellular myelocytomatosis), or APC (adenomatous polyposis coli)/β-catenin overexpression; (3) evading apoptosis by increased cell survival through phosphatidylinositol 3-kinase (PI3K)/AKT pathway, loss of tumor suppressor phosphatase and tensin homolog (PTEN), and inhibition of caspases; (4) self-sufficiency in growth signals by overexpression of cell surface receptors, including receptor tyrosine kinases (RTKs) such as c-Met, epidermal growth factor receptor (EGFR), platelet-derived growth factor receptor (PDGFR), insulin-like growth factor 1 (IGF-1) receptor (IGF-IR), vascular endothelial growth factor receptor (VEGFR), and human epidermal growth factor receptor 2 (ErbB2), as well as adhesion receptors, including integrins and cluster of differentiation (CD44); (5) sustained angiogenesis by promoting angiogenic vascular endothelial growth factor (VEGF) and fibroblast growth factor (FGF) through upregulation of a Ras signaling pathway and amplification of growth factors and cytokines by matrix metalloproteinases (MMPs); (6) tumor invasion and metastasis through epithelial-mesenchymal transition (EMT), induction of MMPs, and loss of E-cadherins; and (7) avoidance of immune

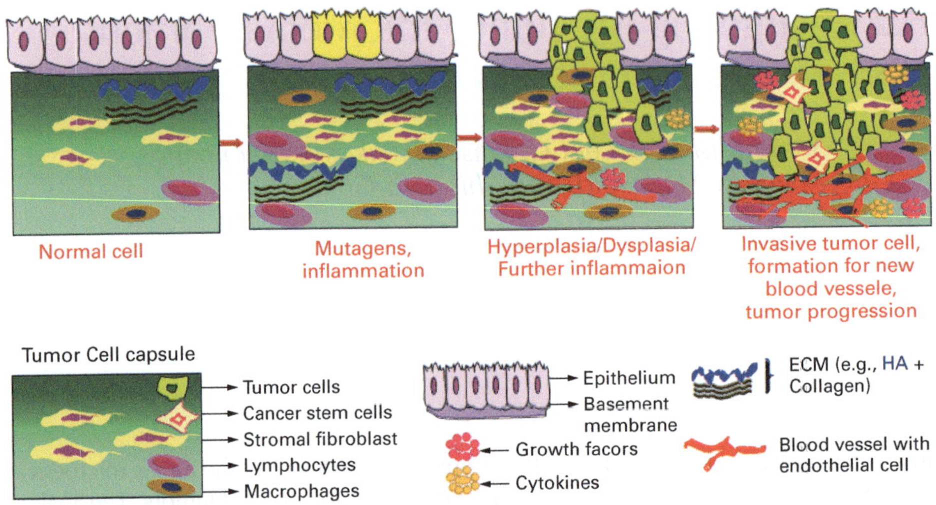

Figure 8.2 Multistep process of carcinogenesis. Normal epithelium is separated from the stromal compartment (fibroblasts, lymphocytes, macrophages, ECM [collagen, hyaluronan, and other ECM components]) by a basement membrane. During inflammation, the resident fibroblast cells become activated with the infiltration of macrophages and then lymphocytes in response to specific immune responses. This is followed by activation of endothelial cells and further activation of fibroblasts. During this process, ECM components such as HA and collagen are degraded with further increase of inflammation forming new blood vessels, and fibroblasts are converted to myofibroblasts with the release of growth factors and cytokines. Ultimately, all these processes lead to tumor progression.

surveillance by altering cell-induced immunity by tumors, which distorts tumor-associated endothelial cells to disrupt functions of natural killer (NK) cells, T cells, and macrophages, which promotes immune suppression.

The immune system eliminates cancer cells through cellular immunity. This further involves the following: (8) genomic instability in cancer cells, which generates random mutations including chromosomal rearrangements; (9) tumor-promoting inflammation where the inflammatory state of premalignant lesions is driven by cells of the immune system leading to tumor progression; and (10) Reprogramming of cellular energy metabolism to support continuous cell growth and proliferation, replacing the metabolic program of normal tissues. However, the majority (>80%) of cancers arise from epithelial tissue. The resulting epithelial cancer is composed of several distinct phases, including hyperplasia (excessive rate of cell division, leading to a larger than usual number of cells), mild dysplasia (precancerous lesion), severe dysplasia (advanced progression toward malignant phenotype), carcinoma in situ (uncontrolled growth of cells that remain in the original location), and invasive cancer (metastatic malignancy) (▶Figure 8.4).

This chapter focuses on (1) the contribution of the elements in the tumor microenvironment in cancer, with special reference to colon cancer, and (2) the potential molecular targets, in particular CD44v6, for colon cancer therapy.

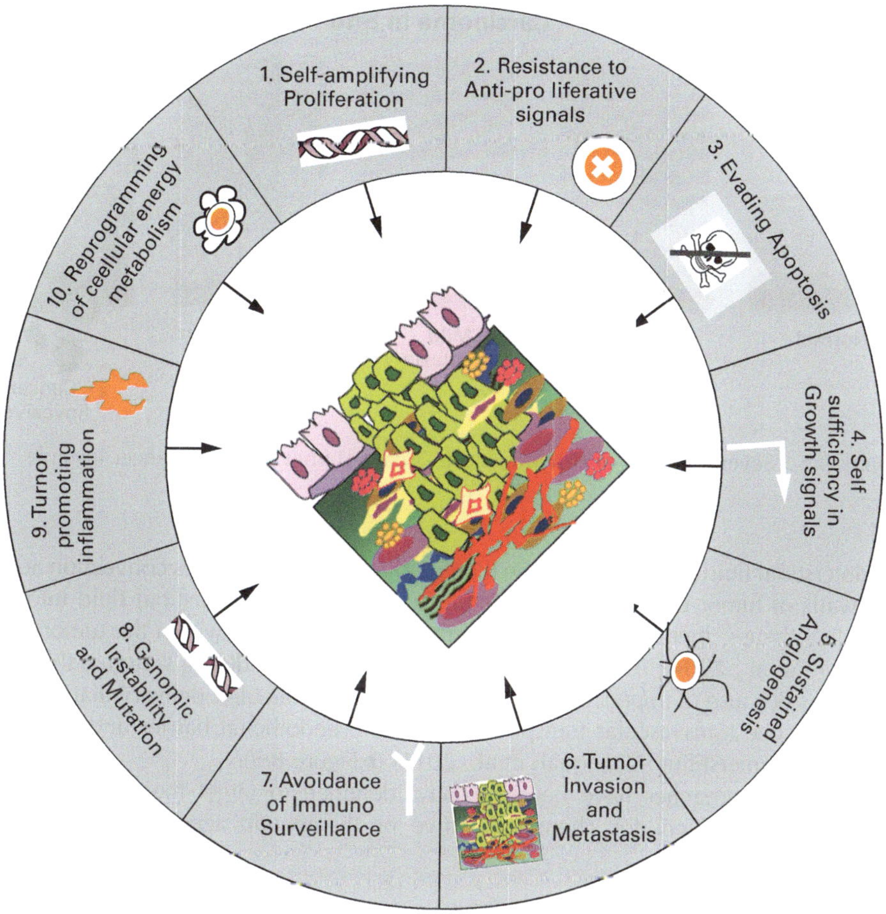

Figure 8.3 Acquired hallmark characteristics of a cancer cell.

Tumor interstitial fluid pressure

The interstitium of a tissue is the space between the cells and the vascular compartment, and it is essential for molecular transport from the blood vessel to the cells and back again. Normal interstitial pressure is close to atmospheric pressure. Solid tumors have elevated interstitial pressure because of interactions with the ECM components and stromal cells. At the onset of tumor development, this form of phenotype conversion arises because of the following: (1) the diminished function of the tumor blood vessels (through the secretion of VEGF or inhibition of the maturation of the tumor vasculature through transforming growth factor-beta [TGF-β] secretion); (2) the transport of the plasma macromolecules from the blood vessel into the tissue; (3) a high deposition of collagen/hyaluronan, leading to the formation of a dense network of ECM components in the tumor; (4) the TAFs-derived platelet-derived growth factor (PDGF) that promotes the interaction of integrins with the ECM; and (5) the acquisition of smooth muscle cell properties by TAFs that contract the tumor stroma and increase

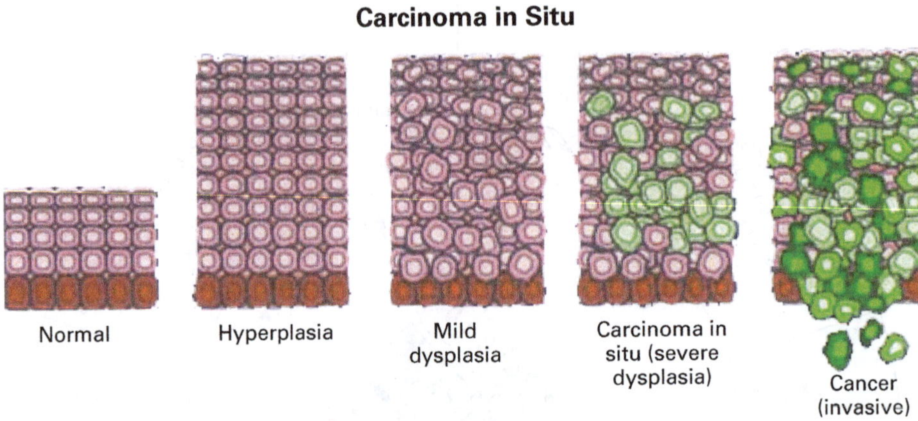

Figure 8.4 Several phases of carcinogenesis. Adapted from National Cancer Institute.

the interstitial fluid pressure (IFP). The increased IFP in tumor reduces convection across the walls of tumor blood vessels and reduces movement of interstitial fluid into surrounding tissues, thereby flushing out therapeutic agents from within the tumor (Desmouliere et al., 2004). As a result drug resistance develops during systemic therapies to tumors because of impeded drug (especially macromolecules) delivery that depends on convective transvascular transport to cross the endothelial barrier and migration through the interstitium (Danquah et al., 2011) (▶Figure 8.5).

Different approaches have been recently identified to target the components of the microenvironment for chemopreventive or therapeutic agents at all levels of carcinogenesis.

Targeting ECM

The interstitium (space between the cells and the vascular compartment) consists of a fibrillar collagen network embedded in a hydrophilic hyaluronan (HA)–based matrix (▶Figure 8.5). Intratumoral injection of collagenase or hyaluronidase reduced IFP and increased the uptake of drug macromolecules, in particular antibodies. Between collagenase and hyaluronidase, the degradation of the collagen network is more efficient than the degradation of the HA in response to antibody uptake. Notably, withdrawal of anti-VEGF therapy (Willett et al.,2004) followed by rapid revascularization of tumors increases the uptake of antibody and macromolecular drugs.

Targeting tumor-associated fibroblasts and macrophages

Clinical studies have shown that TAFs and TAMs can be targeted by imatinib (a small size inhibitor specific for the PDGF receptor) (Pietras, 2004) and Bevacizumab (an anti-VEGF monoclonal antibody) (Willett et al., 2004), which can lead to decreased IFP. Recent studies also indicate that HA/CD44 interaction can regulate PDGFR activity in fibroblasts (Li et al., 2007b) and tumor cells (Misra et al., 2006) (▶Figure 8.5).

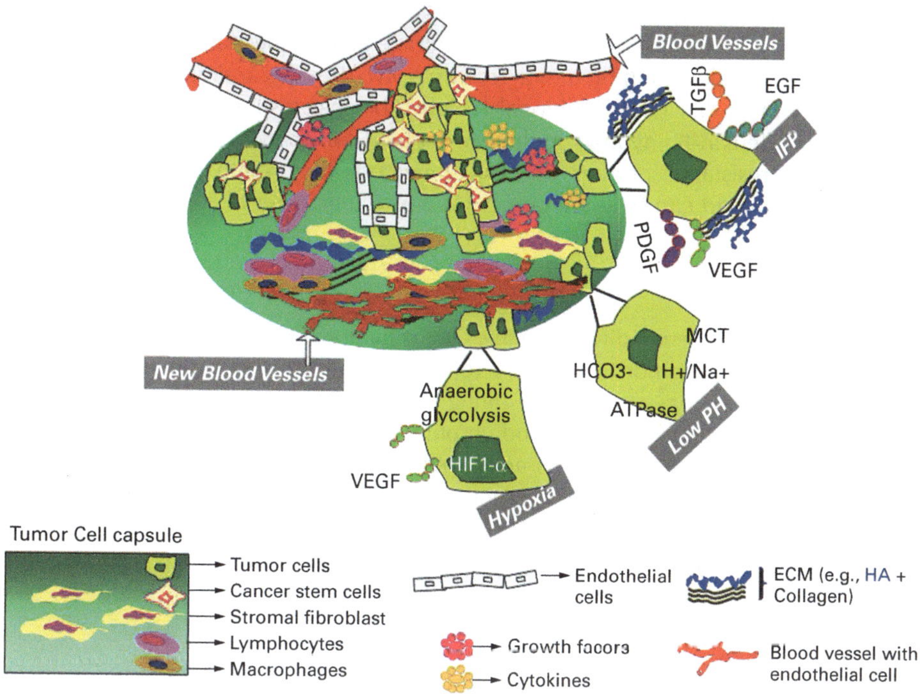

Figure 8.5 Microenvironmental phenotype changes present barriers to tumor therapy. Cancer cells are surrounded by cancer stem cells and embedded within the stromal cells (fibroblasts, macrophages, endothelial cells, lymphocytes, monocytes) in an ECM. Cancer cells are able to produce several angiogenic factors including prostaglandin E2 (PGE2), basic fibroblast growth factor (bFGF), vascular endothelial growth factor (VEGF), interleukin (IL)-8, transforming growth factor-β (TGF-β), which may in turn induce amplification of VEGF in stromal microenvironments. These secreted factors activate paracrine or autocrine loops in the stromal cells, which causes recruitment of stromal cells, proliferation and motility. The characteristic tumor microenvironmental features are hypoxia, interstitial fluid pressure (IFP), and low pH. Poor oxygen delivery by the defective vasculature and elevated oxygen consumption by the tumor cells result in hypoxia and activation of hypoxia-inducible factor 1-α (HIF1)-α. Tumor hypoxia increases glucose import and turns on anaerobic glycolysis and conversion to lactate. In this situation mitochondria oxygen consumption is downregulated, and endothelial cells secrete VEGF to form new blood vessels.

Tumor acidosis

The overall acidic nature of tumors was originally thought to be simply the conversion of glucose to lactic acid, and genetically modified model tumors have shown that impaired glycolysis results in lactate accumulation, which produces an acidic environment (Yamagata et al., 1998). The pH gradient across the plasma membrane is maintained by the monocarboxylate H^+ cotransporter, the vacuolar H^+ ATPase, the Na^+/H^+ exchanger, and the Na^+-dependent chloride/bicarbonate exchanger (Izumi et al., 2003) (▶Figure 8.5). It has been shown that pretreatment of cells with ATPase inhibiting drugs, such as omeprozole, sensitizes tumor cells to cisplatin, 5-fluorouracil, or vinblastine *in vitro*, and to cisplatin in tumors (Luciani et al., 2004) due to inhibition of intracellular acidification. This resulted in intracellular sequestration of the drugs and increased their uptake to inhibit tumor growth *in vivo*. Besides antitumor immunity, the low pH of tumors can have additional benefits for acid responsive gene targets, for example, interleukin-8, hypoxia inducible factor-1(HIF1) (Mekhail et al., 2004), a breast cancer resistance protein, and tumor necrosis factor-alpha, which contribute to tumor growth and can be good targets for novel anticancer therapy.

Tumor hypoxia

As the tumors grow, demand for oxygen exceeds the local supply of oxygen in tumors. Moreover, abnormal tumor vasculature reduces blood flow and inhibits delivery of oxygen in different areas of tumors resulting in regions of hypoxia, e.g. acute hypoxia arises due to inadequate perfusion and chronic hypoxia arises due to increased diffusion distances between tumor cells and blood vessels. Hypoxia causes activation of HIF1, a transcription factor that transactivates VEGF. HIF1 has been implicated in tumor progression and in drug resistance. The possibility of HIF1 blockade therefore, represents an interesting strategy for modifying the tumor physiology (Denko et al., 2003).

8.2.2 Cancer stem cells

Tumors consist of small populations of tumor initiating cells, also called cancer stem cells (CSCs) that are capable of initiating and maintaining tumor growth, and a large proportion of differentiated cells, which cannot maintain tumor growth. CSCs are found in most of the solid tumors, such as breast cancer (Ponti et al., 2005), pancreatic cancer (Li et al., 2007b), melanoma (Monzani et al., 2007), brain tumor (Singh et al., 2003), prostate cancer (Collins et al., 2005), liver cancer (Yang et al., 2008), ovarian cancer (Curley et al., 2009), and lung cancer (Eramo et al., 2008). The evidence for the existence of CSCs in human cancers *in vivo* was established by identifying: (1) specific cell surface markers, such as CD44, CD133, aldehyde dehydrogenase (ALDH), CD166, CD29, CD90, ABCG2, and ABCB5 (some of them being more specific to specific cancers than others), which should be able to initiate tumors and regulate their neoplastic proliferation *in vivo*; (2) the ability for self-renewal (i.e. the tumors that grow out from the CSCs should resemble the original malignancy); and (3) CSCs that are able to differentiate in marker-negative cells. CSCs are defined to have four functionally distinct characteristics: (1) retention of the capacity to initiate tumor CSCs, identified by a specific cell surface marker that can retain the ability to initiate tumors *in vivo* and control neoplastic proliferation; (2) the property of self-renewal; (3) differentiation, the asymmetrical self-renewing cell division producing one CSC and one more differentiated progenitor cell; and (4) state of equilibrium, the ability to attenuate differentiation and self-renewal according to epigenetic changes. Origin of the CSCs is still not clearly known, and they may indeed arise from tissue stem cells. In this traditional cancer model, metastases are considered to originate from clonal expansion of a subset of cancer cells with specific genotypic and epigenetic changes, and therefore are postulated to be substantially different from primary tumors (▶Figure 8.6A). CSCs may

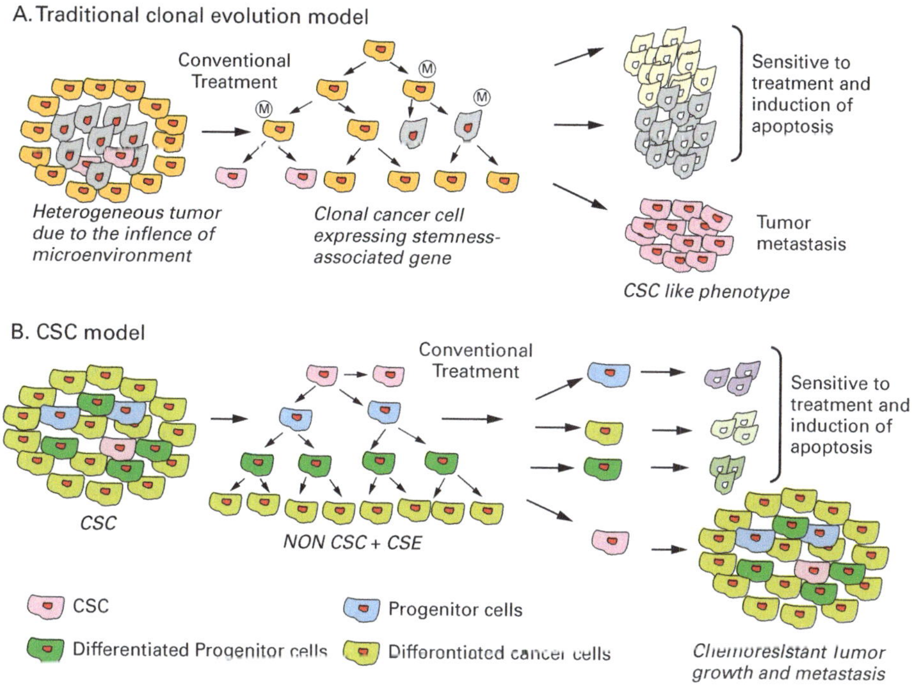

Figure 8.6 Cancer stem cell (CSC) model to describe metastases. (A) Traditional clonal evolution model explains how new mutation in cancer cells alters the cancers to become aggressive for metastatic tumor formation, which is postulated to be substantially different from primary tumors. (B) CSC model describes the heterogeneity of cancers and the hierarchy within a tumor by proposing that the metastaic aggressive tumor that grows out from the CSCs resembles the original malignancy.

arise from more differentiated progenitor cells as shown in ▶Figure 8.6B to attain the properties of CSCs.

CSC clinical perspectives

The most investigated CSC marker is CD44, originally identified as a leukocyte homing receptor and now shown to be the major receptor for HA (Aruffo et al., 1990). CSC markers used for colon are CD44 (Du et al., 2008), CD133 (Vermeulen et al., 2008) and CD166 [20], whereas expression of CD44/CD24 (Dalerba et al., 2007), CD29 (Vermeulen et al., 2008) and Lgr5 (leucine-rich repeat-containing G-protein coupled receptor 5) (Barker et al., 2009) are expressions for intestinal cancer. Recent studies demonstrate that intestinal/colon cancer niches, formed by intestinal subepithelial myofibroblasts located at the base of the crypt, regulate CSC fate, and maintain the correct balance between CSC self-renewal and differentiation. Recent evidence shows that CSCs are often resistant to commonly used chemotherapy and radiation because chemotherapeutics interfere with the rapidly growing cells to divide, while sparing the CSCs, which therefore may be responsible for the recurrence of cancer. Because tumor growth is believed to be mediated by CSCs, understanding CSC biology could lead to targeted therapies for the specific cancer types. However, the molecular mechanism(s) for the complex network that regulates CSC renewal and differentiation are not clear. Thus, targeting specific signaling pathways by specific anti-CSC drugs may be a better therapeutic option. For example sensitization by antibody against ABCB5 protein (Schatton et al., 2008) or silencing the ABCB5 gene (Huang et al., 2004), Cell surface receptor CD29, like CD24 and CD166, mediates cell to ECM adhesion. Since their functions in CSC biology are not clear, they are poor targets for CSC therapy. Knockdown of CD133 in colorectal cancer cells (CRC) did not influence the tumor formation capacity, specifically it did not suppress proliferation, migration, invasion and clonogenic cell division. This suggests that cancer stem cell marker CD133 has a high prognostic impact, but unknown functional relevance for the metastasis of human colon cancer (Horst et al., 2009). It might be possible to perturb CSC signaling as a relevant therapeutic approach for cancer therapy. Small-molecules that inhibit the Wnt pathway essential for maintaining stemness and crypt survival, and the Notch pathway, have better therapeutic options. Similarly, a new CSC-specific differentiation therapeutic agent, a potassium ionophore salinomycin, resulted in a significant decrease in tumor growth in a breast CSC mouse model by inhibiting potassium-positive channel-regulated migration and interfering with epithelial-mesenchymal transition and metastasis (Gupta et al., 2009). Recently, an immuno-therapeutic approach for immune clearance of the cancer-initiating cell population by blocking the immunoglobulin-like CD47 has been applied to CSCs from acute myeloid leukemia (AML) (Majeti et al., 2009) and human bladder (Chan et al., 2009).

One of the critical pathways of CSC angiogenesis is bone morphogenic protein (BMP) signaling that inhibits angiogenesis. BMP4 can activate a differentiation program and stimulate apoptosis in colon CSCs by reducing β-catenin activation through inhibition of the PI3K/AKT pathway (Shao et al., 2009). CD44 has also been used to detect CSCs in CRC (Dalerba et al., 2007), pancreatic cancer (Li et al., 2007a), prostate cancer (Collins et al., 2005) and head and neck cancers (Prince et al., 2007). Therefore, CD44 might have a functional role in the biology of CSCs that can be used for therapy. For example, knockdown of CD44 in cancer cell lines reduces anchorage independent growth *in vitro* and tumorgenicity *in vivo* (Ghatak et al., 2010; Misra et al., 2009). The transmembrane glycoprotein CD44 makes abundant use of alternative splicing to produce CD44variants (CD44v). Interestingly, upregulation of hyaluronan synthase-2 (HAS2) induces the v6 isoform of CD44 (CD44v6) that confers colon cells with metastatic potential by increasing activation of RTKs and cyclooxygenase-2 (COX2) activity in nonmetastatic Apc10.1 cells derived from Apc Min/+ mice adenomas (Misra et al., 2009). As a very important cell adhesion molecule, CD44v6 interacts with HA, and this interaction is critical for intestinal tumor cell proliferation, migration and growth (Misra et al., 2006, 2008a, 2008b, 2009). Additionally, CD44v6 is essential for activation of hepatocyte growth factor (HGF)–inducible c-Met (Orian-Rousseau et al., 2007), and activation of VEGFR2 in angiogenesis in human endothelial cells (Tremmel et al., 2009).

HGF is the ligand for c-Met and can restore the CSC phenotype in more differentiated colon cancer cells (Vermeulen et al., 2010), which may have a major role in maintaining a CSC state. Therefore, CD44 itself may have a role in regulating stemness in CSCs and thus may be a functional cell surface CSC marker. CD44 is known to have a role in normal myeloid differentiation, as this can be inhibited by CD44 blocking antibodies

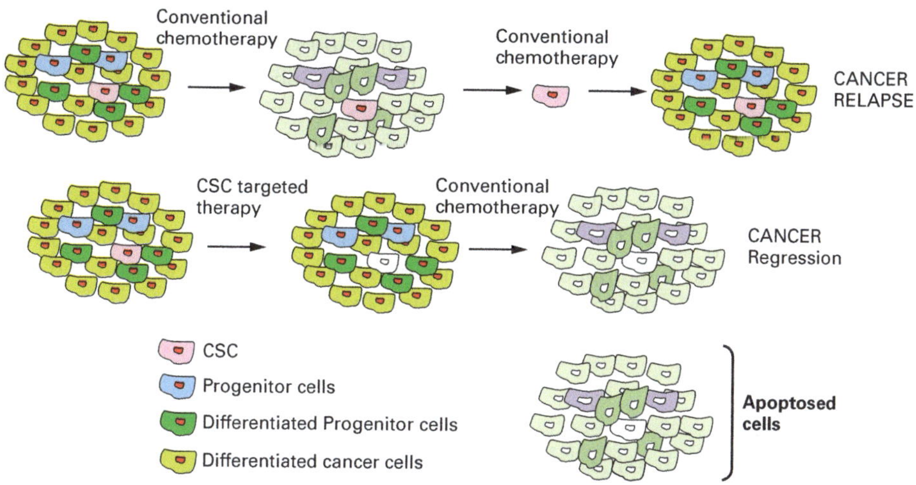

Figure 8.7 Model describing strategies for CSC therapy. Current chemotherapeutics do not affect CSCs and may cause relapse. Cell-specific CSC-targeted therapy including conventional cytotoxic chemotherapeutics, or immunotherapy, or differentiation-inducing agents, or signaling pathway inhibitors may regress the tumor completely.

(Miyake et al., 1990). CD44v6 has already been targeted for anticancer therapy in clinical trials. A clinical study used the CD44v6 monoclonal antibody (bivatuzumab) coupled with a cytotoxic drug mertansine. In this clinical phase-I study, when bivatuzumab/mertansine was injected weekly into squamous cell carcinoma patients for 3 weeks, the drug demonstrated a response of 10%, which was considered a success. The therapy was terminated because of death of one patient due to epidermal necrolysis (Riechelmann et al., 2008). In a recent study, we demonstrated that a novel intestine/colon-cancer-specific delivery of CD44v6shRNA in transferring coated nanoparticles blocks CD44v6 expression and diminishes adenoma growth by 40% after four intraperitoneal injections over ten days in Apc Min/+ mice (Misraet al., 2009, 2011). Thus, targeting CD44v6 with colon specific CD44v6shRNA delivery along with cytotoxic drugs could be a good strategy to attack CSCs in CRC patients. Strategies useful for sensitizing cancer stem cells are presented in ▶Figure 8.7.

8.2.3 Tumor angiogenesis: new concepts about the tumor microenvironment

During carcinogenesis and angiogenesis, stromal cells and their associated matrix as well as cells of the immune system, interact with the host tumor to create a unique environment that emerges in the course of tumor progression. Actually the host tumor always coordinates molecular and cellular events that take place in surrounding stromal cells (Bissell and Labarge, 2005). Since cancer is characterized by the uncontrolled division and proliferation of cells, a chronic inflammatory state with gene mutation or epigenetic events can promote cancer initiation. Additionally, nonsteroidal anti-inflammatory drugs (NS398, celecoxib) (NSAIDs) lower the risk significantly for cancer patients with certain chronic inflammatory conditions (Garcia-Rodriguez and Huerta-Alvarez, 2001). In normal tissue wounds, epithelial and stromal cell types engage in a reciprocal signaling dialogue that leads to the resolution of the wound. Once the wound is healed, the reciprocal signaling subsides. In, neoplasia, however, these inflammatory cells react paradoxically to the dysfunctional epithelial cells by promoting proteolytic enzymes, cytokines and chemokines, which are mitogenic for neoplastic cells, as well as for endothelial cells involved in inflammatory angiogenesis that potentiates tumor growth. The crucial role that blood vessels have in tumor growth and metastasis provides the opportunity to intervene with chemoprevention and chemotherapeutic strategies using therapeutic agents that reverse, inhibit, or prevent the tumor angiogenesis and tumor growth.

8.2.4 CD44 in tumor biology

CD44 interaction with hyaluronan and growth factors

CD44, the major receptor for the glycosaminoglycan HA, is a transmembrane protein encoded by a single gene. It is expressed on various cell types including epithelial cells, keratinocytes, leukocytes, fibroblasts, and endothelial cells. CD44 is subject to extensive alternative splicing, which is precisely regulated, and it occurs only in particular cell types and activation states. Due to alternative splicing, multiple forms of CD44 are generated that are further modified by *N*- and *O*-linked glycosylations. The smallest CD44 isoform that lacks variant exons, designated CD44s or CD44H (hematopoietic), is abundantly expressed by both normal and cancer cells, whereas the variant CD44 (CD44v) isoforms that contain a variable number of exon insertions (v1–v10) at the proximal plasma membrane external region are expressed primarily on the tumor cells.

The CD44 variant isoforms cover many malignant activities to promote cancer metastasis and poor patient outcome (Orian-Rousseau, 2010). First, CD44 communicates with the tumor microenvironment by binding to HA. CD44 contains a well-described lectin-like fold, link-homology HA-binding module (Banerji et al., 2007). Interaction with HA alters the CD44 conformation in such a way that its short cytoplasmic tail binds to ankyrin and ezrin-radixin-moesin (ERM) proteins, which guide CD44 to the leading edge of migrating cells (Lamontagne and Grandbois, 2008). The functional roles of CD44 are demonstrated from the studies where the CD44 facilitates binding to MMPs at the cell surface, and this aggregation of CD44 with MMPs (MMP2 and MMP9) is further stimulated by HA/CD44-interaction (Yu and Stamenkovic, 1999) (▶Figure 8.8). The extracellular domain of CD44 undergoes proteolysis by membrane-type MMPs (MT1-MMP), which releases the soluble extracellular domain of CD44 (Kajita et al., 2001). CD44 transmembrane cytoplasmic domains also undergo additional proteolytic cleavage to produce the CD44 intracellular fragment (CD44-ICD), which has been demonstrated to translocate to the nucleus and regulate various transcription of target genes, including CD44 itself, which provides an important regulatory mechanism for CD44 expression (Okamoto et al., 2001) (▶Figure 8.8). Proteolytic cleavage of CD44 results in an extracellular domain of CD44 (CD44-ECD), which is further cleaved and will not bind to HA because of the loss of binding domain (Cichy and Pure, 2003). In contrast, recombinant sol CD44 retains the HA binding

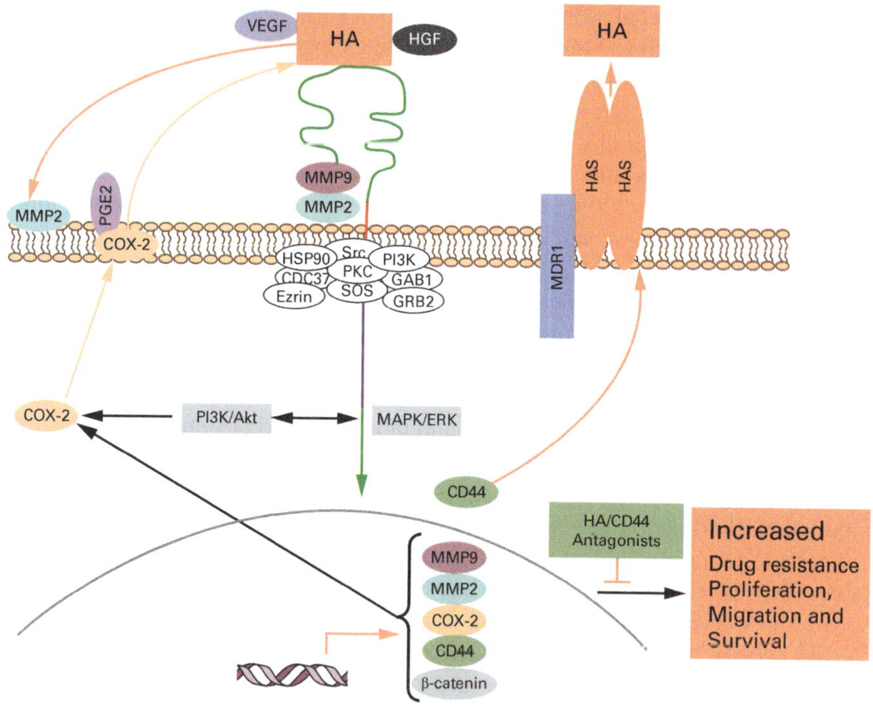

Figure 8.8 Molecular targets for prevention of colon tumor growth and angiogenesis. HGF and VEGF activate paracrine or autocrine loops in both epithelial cells and endothelial cells, resulting in neovascularization of the tumor through production of VEGF and MMP2 via PI3K regulated COX2 under the control of CD44v6 activation. By binding with HA/CD44variant, VEGF and HGF stimulate cell survival, drug resistance, migration, angiogenesis, and tumor growth. These activities were predominantly reinforced due to CD44v6 supporting coreceptor for VEGFR and c-Met, as well as by HA/CD44v6-associated highly organized downstream signaling as outlined in the figure. CD44v6shRNA inhibits all these activities in epithelial cells, endothelial cells, and inflammatory monocytes (Misra et. al., unpublished data).

domain and thus maintains the ability to serve as an antagonist to HA/CD44 interaction as shown previously (Yu and Stamenkovic, 2000). Cell surface CD44-associated MMP9 is also important for the cleavage of latent TGF-β, which promotes tumor growth, angiogenesis, and metastasis (Yu and Stamenkovic, 2000) (▶Figure 8.8). Second, although CD44 interacts with various components, such as cytokines and chemokines in the ECM, the interaction of CD44 with HA has a major impact on health and diseases (Ponta et al., 2003). HA/CD44 binding is upregulated by mitogenic stimuli, glycosylation of its extracellular domain and phosphorylation of specific residues of its cytoplasmic domain (Naor et al., 1997). CD44-HA binding has the following consequences: (1) CD44-assembled matrix strongly supports tumor cell adhesion and CD44-dependent tumor cell migration (Zoller, 2011); (2) leukocyte recruitment/rolling is via interaction of HA expressed by the endothelial cells with a CD44 isoform on leukocytes (Siegelman et al., 1999); (3) activation of various inflammatory cells, such as macrophages, is through CD44-HA-dependent signaling (Taylor and Gallo, 2006); and (4) CD44 initiates signal transduction via coupled RTKs (Misra et al., 2006), or via interaction of the cytoplasmic tail with non-RTK and linker proteins (Turley et al., 2002). On the other hand, increased HA induces metastatic potential in nonneoplastic cells, and CD44 coimmunoprecipitates with activated RTKs-ErbB2 family members (Ghatak et al., 2005; Misra et al., 2005, 2006). For example, the HA-CD44 interaction initiates ErbB2 phosphorylation, and stimulation of HA production induces assembly of a lipid raft integrated complex of ErbB2, CD44, ezrin, the chaperone molecule heat-shock protein 90 (HSP90), androgen receptor (AR), and cell division cycle control protein (CDC37) and PI3K, where activation of the PI3K/AKT pathway promotes increased cell survival activities and drug resistance (Misra et al., 2003, 2005) (▶Figure 8.8). Third, recent studies in pancreatic cells indicate that CD44v6 stimulates hyaluronan synthase transcription (Jung et al., 2011), and thus increased HA production is engaged in binding to a large range of cytokines and chemokines, including HGF, and VEGF thereby increasing the capacity of these ligands to interact with their receptors (▶Figure 8.8). Binding to these molecules, especially to VEGFA and HGF, is important for metastatic processes. HA may bind both MMP14 as well as its substrate (e.g. activatable MMP-2) (Couchman, 2010). Thus both HA and CD44v6 present themselves to c-Met and VEGFR, and the activation of these RTKs and their downstream signaling pathways require activation of CD44-associated phosphorylated ezrin, radixin, and moesin (ERM), which then induce Src (membrane-anchored tyrosine kinase protein), Rho (GTPase protein), and Rac1 (a subfamily of the Rho family of GTPases) signaling (Bourguignon et al., 2000; Gerritsen et al., 2003).

Combining VEGF-A with HGF enhances Rho and Rac1/mediated Ras/mitogen-activated protein kinase (MAPK) and the Ras-PI3K/AKT pathways probably through the more efficient recruitment of CD44v6 where CD44v6 exerts coreceptor function when it is activated by HGF/c-Met and VEGFR (Tremmel et al., 2009). Once CD44v6 is activated, these pathways are initiated without any additional stimuli as demonstrated in ▶Figure 8.8. Besides participating in the regulation of many genes required for cell growth, survival and invasion, HA/CD44v6 regulated PI3K/AKT is strongly coinflammatory upstream of COX2, which in turn regulates HA production (Misra et al., 2008a, 2008b). Although these effects have been extensively documented in tumor cells, it is now apparent that they are also exerted on components of the microenvironment.

Cellular and molecular targets of angiogenesis

The dysfunction of epithelial cells or their microenvironment is critical for carcinogenesis and metastasis. As discussed earlier, tumor cells are surrounded by stromal cells (nontumor cells) that are genetically stable compared to tumor cells, which undergo genetic mutation and develop drug resistance because of their intrinsic genetic instability. Thus, nontumor cells may prove to be good therapy targets in the tumor microenvironment provided these nontumor cells can keep normal physiological function while used as therapeutic targets. The macrophages of the dysfunctional tumor/microenvironment are primary sources of signaling molecules such as reactive oxygen and nitric oxide synthase, as well as prostaglandins and COX2 that regulate inflammation and angiogenesis (Coussens and Werb, 2002).

Loss of TGF-β signaling is a frequent occurrence in lymphocytes in many cancers suggesting that increased stromal TGF-β signaling can prevent the carcinogenesis (Suh et al., 2003). Recent studies indicate that inflammatory angiogenesis seems to be a central force in tumor growth, and inhibition of inflammation prevents angiogenesis. Inhibition of key signal activators of transcription factors such as signal transducer and activator of transcription (STATs), nuclear factor-kappa B (NF-κB), and HIF1 are good targets for inhibition of inflammation and angiogenesis (Albini and Sporn, 2007). Importantly, CD44v6shRNA inhibits malignant activities in CRC cells and endothelial cells, and prevents adenoma number and growth in Apc Min/+ mice indicating that CD44v6shRNA is a novel therapeutic agent for CRC (Misra et al., 2009, 2011).

CD44, a molecular target for therapy in the microenvironment

Since the detailed description of the expression of CD44v isoforms from less malignant to more advanced stages is beyond the scope of this review, we will highlight the relevance of CD44 variants in cancer, which seem to be suitable targets for anticancer therapy. Blocking of CD44 makes use of CD44 as a potential therapeutic target because of the following: (1) It communicates with the tumor surroundings (Ghatak et al., 2010). (2) The interaction of stromal-derived HA with the upregulated CD44v9 initiates signaling pathways that stabilize androgen receptor (AR) functions and induce antiapoptotic signaling (Ghatak et al., 2010). Colon cancer cells have the same mechanism but utilize CD44v6 (Misra et al., unpublished data). Silencing the appropriate CD44-variant inhibits tumor cell adhesion to tumor cell matrix and *in vitro* tumor cell invasion (Ghatak et al., 2010). (3) CD44 not only contributes to the assembly of the matrix surrounding the CSCs, but importantly, the feedback from the microenvironment to stem cells and CSCs also involves CD44 (Zoller, 2011). (4) It coordinates signal transduction through oncogene RTKs (Ghatak et al., 2005; Misra et al., 2005, 2006, 2008a, 2008b, 2009), and its stabilization of multidrug resistant (MDR) genes (Misra et al., 2003, 2005; Bourguignon et al., 2009). (5) Blocking of specific CD44v molecules, such as CD44v7 in AML, CD44v6 v10 in lymphoma, and CD44v6 in AML, blocks tumor growth (Avin et al., 2004; Liebisch et al., 2005). (6) Blocking of CD44 modulates the CD44-induced niche generation (Zoller, 2011). (7) Blocking the coreceptor function of CD44v6 for VEGFR2 and c-Met results in antiangiogenesis and poor cell survival (Tremmel et al., 2009). (8) Blocking CD44v6 showed promising antitumor effects and was considered successful in patients suffering from HNSCC with a 10% successful response rate in clinical trial (Riechelmann et al., 2008). (9) Tissue-specific delivery of CD44v6shRNA/nanoparticles prevents and inhibits *in vivo* tumor growth in Apc Min/+ mice (Misra et al., 2009, 2011). Thus, therapeutic approaches by targeting HA/CD44v interaction with CD44v-shRNA can target tumors at one or more of these levels: the microenvironment (stromal factors such as HGF and its inducers), receptor-based signals (select CD44v, Met/RTK), and signal transducers such as PI3K/AKT or MAPK (Misra et al., 2009, 2011; Ghatak et al., 2010).

It is now clear that CD44v6 isoforms and their binding to HA are crucial for progression and metastasis in various cancers including colon. CD44 has been defined as a CSC marker in many tumor entities, but it remains to be explored whether CSCs preferentially express CD44s, or CD44v6, or both. The mechanisms that relate to functions of CD44v6-overexpressing phenotypes in tumorigenicity, cancer progression, and therapeutic resistance have been identified. These results indicate that CD44v6 may contribute to the failure of existing therapies to consistently eradicate malignant tumors. Importantly, recent proof-of-principle studies have strengthened the rationale for developing CD44v6-targeted therapeutic modalities that might complement more conventional cancer therapies. Indeed, cell specific CD44v6shRNA/nanoparticle-targeted approaches have shown promise in preclinical models (Misra et al., 2009, 2011). Therefore, CD44v6 represent novel and translationally relevant targets for clinical cancer therapy.

8.2.5 Take-home message

The tumor microenvironment determines the fate of putative cancer cells and controls tumorigenesis, tumor progression, migration, invasion and eventually metastasis. In order to develop effective chemoprevention or therapy we need to understand the primary role of both the environment of the developing tumor and the secondary microenvironment site of metastatic tumor. Potential therapeutic target components of the tumor microenvironment include stromal cells (such as endothelial cells), tumor associated fibroblasts, macrophages, cell adhesion and regulatory molecules (such as CD44, integrins, growth factors), cytokines (such as, HGF, VEGF, IGF1, EGF), ECM molecules (such as thrombospondin and fibronectin), matrix-degrading proteases (MMPs), and tissue inhibitors of metalloproteinases (TIMPs). The potential for eliminating variants of the cell surface adhesion receptor CD44 to inhibit a tumorigenic phenotype has been shown in different experimental settings. For example it has been demonstrated that the suppression of CD44v6 isoform eliminates angiogenesis in endothelial cells, induces metastatic colon cancer epithelial cells to undergo apoptosis, and alters the transformed cells to normal epithelial cells. We have shown that tissue specific removal of CD44v6 from intestinal tumors in Apc Min/+ mice reverted tumorigenesis despite maintaining the standard CD44s. CD44v6 is present in colon cancer-associated epithelial, endothelial and monocyte cells. In addition, cancer stem cells are selected on the basis of CD44. Thus, our tissue specific delivery of CD44v6shRNA/nanoparticles may offer an exciting alternative to traditional tumor cell-directed therapy.

Acknowledgments

This work was supported, in whole or in part, by National Institutes of Health Grants P20RR021949 (to S. G.) and P20RR016434 (to S. M., S. G., and R. R. M.), 2P20RR16461-05A1 (to S. G., R. R. M., and L. P. C.) HL RO1 33756 and 1 P30AR050953 (to V. C. H.). This work was also supported by Mitral-07 CVD 04 (to R. R. M.), Medical University of South.

References

Albini, A., and Sporn, M. B. (2007). The tumour microenvironment as a target for chemoprevention. Nat Rev Cancer *7*, 139–147.

Aruffo, A., Stamenkovic, I., Melnick, M., Underhill, C. B., and Seed, B. (1990). CD44 is the principal cell surface receptor for hyaluronate. Cell *61*, 1303–1313.

Avin, E., Haimovich, J., and Hollander, N. (2004). Anti-idiotype x anti-CD44 bispecific antibodies inhibit invasion of lymphoid organs by B cell lymphoma. J Immunol *173*, 4736–4743.

Banerji, S., Wright, A. J., Noble, M., et al. (2007). Structures of the Cd44-hyaluronan complex provide insight into a fundamental carbohydrate-protein interaction. Nat Struct Mol Biol *14*, 234–239.

Barker, N., Ridgway, R. A., van Es, J. H., et al. (2009). Crypt stem cells as the cells-of-origin of intestinal cancer. Nature *457*, 608–611.

Bissell, M. J., and Labarge, M. A. (2005). Context, tissue plasticity, and cancer: are tumor stem cells also regulated by the microenvironment? Cancer Cell *7*, 17–23.

Bourguignon, L. Y., Spevak, C. C., Wong, G., Xia, W., and Gilad, E. (2009). Hyaluronan-CD44 interaction with protein kinase C (epsilon) promotes oncogenic signaling by the stem cell marker Nanog and the production of microRNA-21, leading to down-regulation of the tumor suppressor protein PDCD4, anti-apoptosis, and chemotherapy resistance in breast tumor cells. J Biol Chem *284*, 26533–26546.

Bourguignon, L. Y., Zhu, H., Shao, L., and Chen, Y. W. (2000). CD44 interaction with tiam1 promotes Rac1 signaling and hyaluronic acid- mediated breast tumor cell migration. J Biol Chem *275*, 1829–1838.

Chan, K. S., Espinosa, I., Chao, M., et al. (2009). Identification, molecular characterization, clinical prognosis, and therapeutic targeting of human bladder tumor-initiating cells. Proc Natl Acad Sci U S A *106*, 14016–14021.

Cichy, J., and Pure, E. (2003). The liberation of CD44. J Cell Biol *161*, 839–843.

Collins, A. T., Berry, P. A., Hyde, C., Stower, M. J., and Maitland, N. J. (2005). Prospective identification of tumorigenic prostate cancer stem cells. Cancer Res *65*, 10946–10951.

Couchman, J. R. (2010). Transmembrane signaling proteoglycans. Annu Rev Cell Dev Biol *26*, 89–114.

Coussens, L. M., and Werb, Z. (2002). Inflammation and cancer. Nature *420*, 860–867.

Curley, M. D., Therrien, V. A., Cummings, C. L., et al. (2009). CD133 expression defines a tumor initiating cell population in primary human ovarian cancer. Stem Cells *27*, 2875–2883.

Dalerba, P., Dylla, S. J., Park, I. K., et al. (2007). Phenotypic characterization of human colorectal cancer stem cells. Proc Natl Acad Sci U S A *104*, 10158–10163.

Danquah, M. K., Zhang, X. A., and Mahato, R. I. (2011). Extravasation of polymeric nanomedicines across tumor vasculature. Adv Drug Deliv Rev *63*, 623–639.

Denko, N. C., Fontana, L. A., Hudson, K. M., et al. (2003). Investigating hypoxic tumor physiology through gene expression patterns. Oncogene *22*, 5907–5914.

Desmouliere, A., Guyot, C., and Gabbiani, G. (2004). The stroma reaction myofibroblast: a key player in the control of tumor cell behavior. Int J Dev Biol *48*, 509–517.

Du, L., Wang, H., He, L., et al. (2008). CD44 is of functional importance for colorectal cancer stem cells. Clin Cancer Res *14*, 6751–6760.

Eramo, A., Lotti, F., Sette, G., et al. (2008). Identification and expansion of the tumorigenic lung cancer stem cell population. Cell Death Differ *15*, 504–514.

Garcia-Rodriguez, L. A., and Huerta-Alvarez, C. (2001). Reduced risk of colorectal cancer among long-term users of aspirin and nonaspirin nonsteroidal antiinflammatory drugs. Epidemiology *12*, 88–93.

Gerritsen, M. E., Tomlinson, J. E., Zlot, C., Ziman, M. and Hwang, S. (2003). Using gene expression profiling to identify the molecular basis of the synergistic actions of hepatocyte growth factor and vascular endothelial growth factor in human endothelial cells. Br J Pharmacol *140*, 595–610.

Ghatak, S., Hascall, V. C., Markwald, R. R., and Misra, S. (2010). Stromal hyaluronan interaction with epithelial CD44 variants promotes prostate cancer invasiveness by augmenting expression and function of hepatocyte growth factor and androgen receptor. J Biol Chem *285*, 19821–19832.

Ghatak, S., Misra, S., and Toole, B. P. (2005). Hyaluronan constitutively regulates ErbB2 phosphorylation and signaling complex formation in carcinoma cells. J Biol Chem *280*, 8875–8883.

Gupta, P. B., Onder, T. T., Jiang, G., et al. (2009). Identification of selective inhibitors of cancer stem cells by high-throughput screening. Cell *138*, 645–659.

Horst, D., Scheel, S. K., Liebmann, S., et al. (2009). The cancer stem cell marker CD133 has high prognostic impact but unknown functional relevance for the metastasis of human colon cancer. J Pathol *219*, 427–434.

Huang, Y., Anderle, P., Bussey, K. J., et al. (2004). Membrane transporters and channels: role of the transportome in cancer chemosensitivity and chemoresistance. Cancer Res *64*, 4294–4301.

Izumi, H., Torigoe, T., Ishiguchi, H., et al. (2003). Cellular pH regulators: potentially promising molecular targets for cancer chemotherapy. Cancer Treat Rev *29*, 541–549.

Jung, T., Gross, W., and Zoller, M. (2011). CD44v6 coordinates tumor matrix-triggered motility and apoptosis resistance. J Biol Chem *286*, 15862–15874.

Kajita, M., Itoh, Y., Chiba, T., et al. (2001). Membrane-type 1 matrix metalloproteinase cleaves CD44 and promotes cell migration. J Cell Biol *153*, 893–904.

Lamontagne, C. A., and Grandbois, M. (2008). PKC-induced stiffening of hyaluronan/CD44 linkage; local force measurements on glioma cells. Exp Cell Res *314*, 227–236.

Li, C., Heidt, D. G., Dalerba, P., et al. (2007a). Identification of pancreatic cancer stem cells. Cancer Res *67*, 1030–1037.

Li, L., Asteriou, T., Bernert, B., Heldin, C. H., and Heldin, P. (2007b). Growth factor regulation of hyaluronan synthesis and degradation in human dermal fibroblasts: importance of hyaluronan for the mitogenic response of PDGF-BB. Biochem J *404*, 327–336.

Liebisch, P., Eppinger, S., Schopflin, C., et al. (2005). CD44v6, a target for novel antibody treatment approaches, is frequently expressed in multiple myeloma and associated with deletion of chromosome arm 13q. Haematologica *90*, 489–493.

Luciani, F., Spada, M., De Milito, A., et al. (2004). Effect of proton pump inhibitor pretreatment on resistance of solid tumors to cytotoxic drugs. J Natl Cancer Inst *96*, 1702–1713.

Majeti, R., Chao, M. P., Alizadeh, A. A., et al. (2009). CD47 is an adverse prognostic factor and therapeutic antibody target on human acute myeloid leukemia stem cells. Cell *138*, 286–299.

Mekhail, K., Gunaratnam, L., Bonicalzi, M. E. and Lee, S. (2004). HIF activation by pH-dependent nucleolar sequestration of VHL. Nat Cell Biol *6*, 642–647.

Misra, S., Ghatak, S. and Toole, B. P. (2005). Regulation of MDR1 expression and drug resistance by a positive feedback loop involving hyaluronan, phosphoinositide 3-kinase, and ErbB2. J Biol Chem *280*, 20310–20315.

Misra, S., Ghatak, S., Zoltan-Jones, A., and Toole, B. P. (2003). Regulation of multidrug resistance in cancer cells by hyaluronan. J Biol Chem *278*, 25285–25288.

Misra, S., Hascall, V. C., Berger, F. G., Markwald, R. R., and Ghatak, S. (2008a). Hyaluronan, CD44, and cyclooxygenase-2 in colon cancer. Connect Tissue Res *49*, 219–224.

Misra, S., Hascall, V. C., De Giovanni, C., Markwald, R. R., and Ghatak, S. (2009). Delivery of CD44 shRNA/nanoparticles within cancer cells: perturbation of hyaluronan/CD44v6 interactions and reduction in adenoma growth in Apc Min/+ MICE. J Biol Chem *284*, 12432–12446.

Misra, S., Heldin, P., Hascall, V. C., et al. (2011). Hyaluronan-CD44 interactions as potential targets for cancer therapy. FEBS J *278* (9), 1429–1443.

Misra, S., Obeid, L. M., Hannun, Y. A., et al. (2008b). Hyaluronan constitutively regulates activation of COX-2-mediated cell survival activity in intestinal epithelial and colon carcinoma cells. J Biol Chem *283*, 14335–14344.

Misra, S., Toole, B. P., and Ghatak, S. (2006). Hyaluronan constitutively regulates activation of multiple receptor tyrosine kinases in epithelial and carcinoma cells. J Biol Chem *281*, 34936–34941.

Miyake, K., Underhill, C. B., Lesley, J., and Kincade, P. W. (1990). Hyaluronate can function as a cell adhesion molecule and CD44 participates in hyaluronate recognition. J Exp Med *172*, 69–75.

Monzani, E., Facchetti, F., Galmozzi, E., et al. (2007). Melanoma contains CD133 and ABCG2 positive cells with enhanced tumourigenic potential. Eur J Cancer *43*, 935–946.

Naor, D., Sionov, R. V., and Ish-Shalom, D. (1997). CD44: structure, function, and association with the malignant process. Adv Cancer Res *71*, 241–319.

Okamoto, I., Kawano, Y., Murakami, D., et al. (2001). Proteolytic release of CD44 intracellular domain and its role in the CD44 signaling pathway. J Cell Biol *155*, 755–762.

Orian-Rousseau, V. (2010). CD44, a therapeutic target for metastasising tumours. Eur J Cancer *46*, 1271–1277.

Orian-Rousseau, V., Morrison, H., Matzke, A., et al. (2007). Hepatocyte growth factor-induced Ras activation requires ERM proteins linked to both CD44v6 and F-actin. Mol Biol Cell *18*, 76–83.

Pietras, K. (2004). Increasing tumor uptake of anticancer drugs with imatinib. Semin Oncol *31*, 18–23.

Ponta, H., Sherman, L., and Herrlich, P. A. (2003). CD44: from adhesion molecules to signalling regulators. Nat Rev Mol Cell Biol *4*, 33–45.

Ponti, D., Costa, A., Zaffaroni, N., et al. (2005). Isolation and in vitro propagation of tumorigenic breast cancer cells with stem/progenitor cell properties. Cancer Res *65*, 5506–5511.

Prince, M. E., Sivanandan, R., Kaczorowski, A., et al. (2007). Identification of a subpopulation of cells with cancer stem cell properties in head and neck squamous cell carcinoma. Proc Natl Acad Sci U S A *104*, 973–978.

Riechelmann, H., Sauter, A., Golze, W., et al. (2008). Phase I trial with the CD44v6-targeting immunoconjugate bivatuzumab mertansine in head and neck squamous cell carcinoma. Oral Oncol *44*, 823–829.

Schatton, T., Murphy, G. F., Frank, N. Y., et al. (2008). Identification of cells initiating human melanomas. Nature *451*, 345–349.

Shao, E. S., Lin, L., Yao, Y., and Bostrom, K. I. (2009). Expression of vascular endothelial growth factor is coordinately regulated by the activin-like kinase receptors 1 and 5 in endothelial cells. Blood *114*, 2197–2206.

Siegelman, M. H., DeGrendele, H. C., and Estess, P. (1999). Activation and interaction of CD44 and hyaluronan in immunological systems. J Leukoc Biol *66*, 315–321.

Singh, S. K., Clarke, I. D., Terasaki, M., et al. (2003). Identification of a cancer stem cell in human brain tumors. Cancer Res *63*, 5821–5828.

Suh, N., Roberts, A. B., Birkey Reffey, S., et al. (2003). Synthetic triterpenoids enhance transforming growth factor beta/Smad signaling. Cancer Res *63*, 1371–1376.

Taylor, K. R., and Gallo, R. L. (2006). Glycosaminoglycans and their proteoglycans: host-associated molecular patterns for initiation and modulation of inflammation. FASEB J *20*, 9–22.

Tremmel, M., Matzke, A., Albrecht, I., et al. (2009). A CD44v6 peptide reveals a role of CD44 in VEGFR-2 signaling and angiogenesis. Blood *114*, 5236–5244.

Turley, E. A., Noble, P. W., and Bourguignon, L. Y. (2002). Signaling properties of hyaluronan receptors. J Biol Chem *277*, 4589–4592.

Vermeulen, L., De Sousa, E. M. F., van der Heijden, M., et al. (2010). Wnt activity defines colon cancer stem cells and is regulated by the microenvironment. Nat Cell Biol *12*, 468–476.

Vermeulen, L., Todaro, M., de Sousa Mello, F., et al. (2008). Single-cell cloning of colon cancer stem cells reveals a multi-lineage differentiation capacity. Proc Natl Acad Sci U S A *105*, 13427–13432.

Willett, C. G., Boucher, Y., di Tomaso, E., et al. (2004). Direct evidence that the VEGF-specific antibody bevacizumab has antivascular effects in human rectal cancer. Nat Med *10*, 145–147.

Yamagata, M., Hasuda, K., Stamato, T., and Tannock, I. F. (1998). The contribution of lactic acid to acidification of tumours: studies of variant cells lacking lactate dehydrogenase. Br J Cancer *77*, 1726–1731.

Yang, Z. F., Ho, D. W., Ng, M. N., et al. (2008). Significance of CD90+ cancer stem cells in human liver cancer. Cancer Cell *13*, 153–166.

Yu, Q., and Stamenkovic, I. (1999). Localization of matrix metalloproteinase 9 to the cell surface provides a mechanism for CD44-mediated tumor invasion. Genes Dev *13*, 35–48.

Yu, Q., and Stamenkovic, I. (2000). Cell surface-localized matrix metalloproteinase-9 proteolytically activates TGF-beta and promotes tumor invasion and angiogenesis. Genes Dev *14*, 163–176.

Zoller, M. (2011). CD44: can a cancer-initiating cell profit from an abundantly expressed molecule? Nat Rev Cancer *11*, 254–267.

8.3 Growth factor signaling and extracellular matrix

Dragana Nikitovic, Harris Pratsinis, Aikaterini Berdiaki, Chrisostomi Gialeli, Dimitris Kletsas, and George N. Tzanakakis

8.3.1 Introduction

Growth factors (GFs) are a group of biologically active polypeptides that can stimulate cellular division, growth, and differentiation. Through the regulation of these key biological functions GFs participate in the development and homeostasis of tissues throughout life (Goldring et al., 2006; Bobick et al., 2009) and are implicated in numerous pathological processes (Schmidt et al., 2006; Hayes and Ralphs, 2011). Extracellular matrices (ECMs), which are composed of a dynamic and complex array of macromolecules, provide cells with mechanical and structural support and are critically important for cell growth, survival, differentiation, and tissue morphogenesis. GFs and their respective receptors play important roles in all stages of matrix synthesis from early matrix deposition and subsequent accumulation to ongoing remodeling processes (Osada et al., 1996; Sawaji et al., 2008). This review focuses on the signaling pathways activated by GFs leading to the regulation of synthesis and degradation of ECM components, as well as, on the reciprocal regulation of GF-function by ECM components.

8.3.2 Interplay of growth factors and ECM

Extracellular matrices (ECM) composed of a dynamic and complex array of proteins constitute the cell microenvironment. Most important among these proteins are the collagens, elastin, laminins, fibronectin, glycosaminoglycans (GAGs), and proteoglycans (PGs). ECM provides the bulk, shape, and strength of many tissues *in vivo*. However, ECM provides much more than just mechanical and structural support, but is critically important for cell growth, survival, differentiation and morphogenesis. Thus, ECM imparts spatial context for signaling events by various cell surface growth factor receptors and adhesion molecules such as integrins (reviewed by Barkan et al., 2010; Rozario and DeSimone, 2010).

Signaling pathways

The interplay between GFs and ECM is being translated to specific signals transferred from the extracellular space to the cell nucleus through a complex system of signaling pathways, as depicted in ▶Figure 8.9. The starting points of this system are the growth factor receptors (GFRs), as well as, the ECM-receptors, e.g. the integrins or CD44.

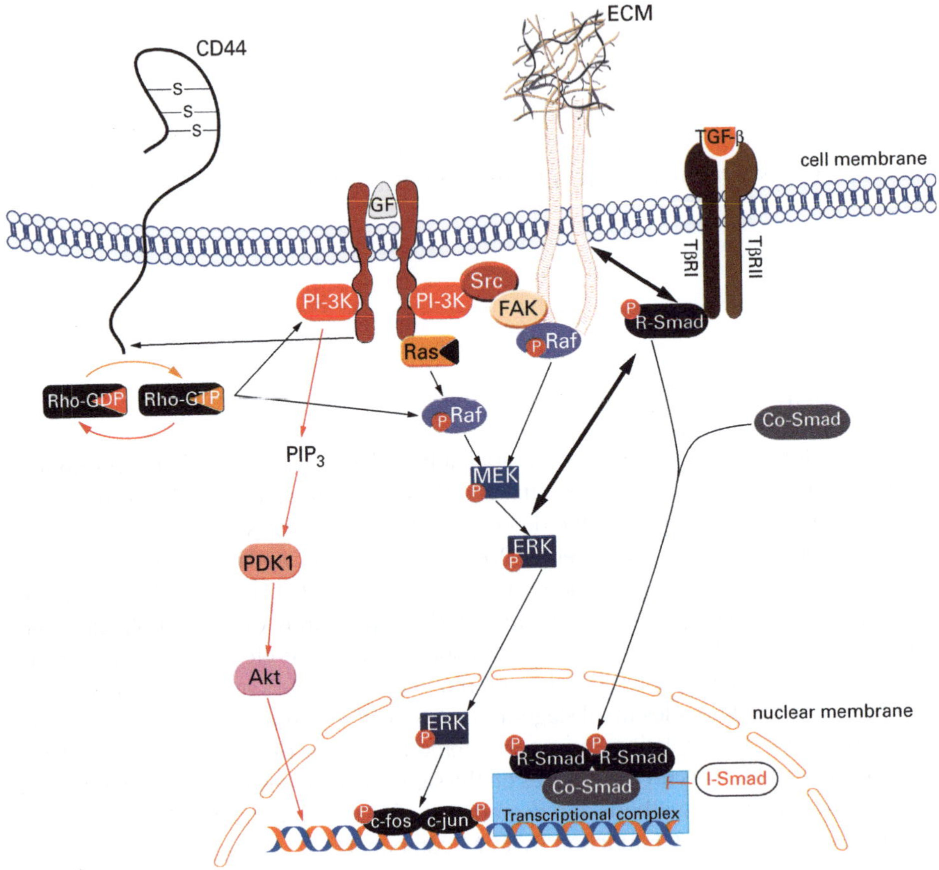

Figure 8.9 **Schematic presentation of major pathways involved in growth factor and/or extracellular matrix signal transduction from the cell membrane to the nucleus (see text).**

The GFRs are transmembrane proteins composed of an extracellular ligand-binding domain, a transmembrane section, and an intracellular catalytic domain capable of activation through autophosphorylation, usually at multiple tyrosine or serine and threonine residues (Malarkey et al., 1995; Shi and Massague, 2003). In many cases, homo- or heteropolymerization of more than one isoforms or receptor types, stabilized due to the extracellular presence of corresponding ligand dimers is required for the intracellular activation to occur (Heldin and Westermark, 1999; Moustakas and Heldin, 2009). Integrins are likewise heterodimeric transmembrane receptors composed of an alpha and a beta subunit noncovalently associated, with a large extracellular domain containing multiple motifs, which allows interaction with various extracellular molecules, a single membrane spanning part, and usually, a short cytoplasmic tail (Juliano and Haskill, 1993; Ross, 2004). The cluster of differentiation 44 (CD44), on the other hand, has a single large extracellular domain containing various GAG attachment sites, a transmembrane part and a relatively short cytoplasmic tail (Mythreye and Blobe, 2009). The GFR molecules usually show high specificity toward the ligands they bind, while integrins and CD44 interact with an array of different ECM molecules with overlapping binding affinities (Mythreye and Blobe, 2009; Kim et al., 2011).

The activation of GFRs (i.e. the phosphorylation of their intracellular domains) triggers a cascade of successive phosphorylations/activations of intracellular signaling substrates aiming at the conveyance of the extracellular signal to the cell nucleus, and hence at the regulation of gene expression. The most important among these substrates, at least regarding growth factor and ECM action include those discussed subsequently.

The highly conserved mitogen-activated protein kinase (MAPK) family, which in mammals is subdivided in at least four distinctly regulated groups (i.e. extracellular signal-regulated kinases 1 and 2 [ERK1/2], c-Jun-N-terminal kinases [JNKs], p38, and ERK5) (Chang and Karin, 2001). A common pattern of all MAPKs is that each one is activated by its specific upstream kinase, MAPK kinase (MAPKK) through phosphorylation of both threonine and tyrosine residues, while each MAPKK is similarly activated by a MAPKK kinase (MAPKKK) by phosphorylation on serine or threonine residues (Imajo et al., 2006). The archetypal signal transduction pathway of many growth factors, such as platelet-derived growth factor (PDGF), epidermal growth factor (EGF), and basic fibroblast growth factor (bFGF) to name only a few, includes the successive phosporylation of Ras, the MAPKKK c-Raf, the MAPKK MEK1/2 and the MAPK ERK1/2 (App et al., 1991; Friesel and Maciag, 1995; Heldin and Westermark, 1999). Besides GFRs, MAPKs also respond to chemical and physical stress (Chang and Karin, 2001). One of the major downstream targets of MAPKs is the transcription factor activator protein-1 (AP1), a heterodimeric protein complex assembled of different Jun (c-Jun, JunB, and JunD) and Fos (c-Fos, Fra-1, Fra-2, and FosB) subunits after phosphorylation by MAPKs (Whitmarsh and Davis, 1996), which regulates the transcription of many targets genes (e.g. matrix metalloproteinases, or MMPs) (Kajanne et al., 2007).

The phosphatidylinositol 3-kinase (PI3K) family, which comprises enzymes capable of phosphorylating phosphatidylinositol lipids at the D-3 position of their inositol ring after triggering by GFRs (Cantley, 2002). This generation of phosphatidylinositol (3,4,5)-trisphosphates recruits protein kinase B, also called AKT (PKB/AKT), at the cell membrane, where it undergoes phosphorylation in serine and threonine residues by various kinases, most notably phospholipid-dependent kinase-1 (PDK1) (Duronio et al., 1998). PKB/AKT in turn initiates further signaling cascades regulating a wide spectrum of cellular processes ranging from cell proliferation and apoptosis to cell migration or glucose metabolism (Brazil and Hemmings, 2001).

The Smad family of proteins, which are the main transducers of the signals triggered by ligands belonging to the transforming growth factor-beta (TGF-β) superfamily (Attisano and Lee-Hoeflich, 2001). Smad proteins are subdivided to three groups depending on their function: receptor-activated (R) Smads are directly phosphorylated by TGF-β (or activin) receptors (Smad2 and Smad3) or by bone morphogenetic protein (BMP) receptors (Smad1, Smad5, and Smad8), a fact allowing them to associate with the common-mediator (Co) Smad (Smad4) to form a heterotrimer, which is transferred in the nucleus to regulate gene expression, while the inhibitory (I) Smads (Smad6 and Smad7) negatively regulate signal strength and duration (Moustakas and Heldin, 2009). One of the special features of Smads is their capability for nucleocytoplasmic shuttling, allowing them to directly transfer signals from transmembrane receptors to DNA-binding transcriptional complexes (Massague et al., 2005).

In contrast to the GFRs, the cytoplasmic domains of integrins cannot be phosphorylated, but they affect their signaling activity by recruiting a wide range of accessory molecules (Loeser, 2002). In this way, the regions of the cell attaching to the ECM, the so-called focal adhesions (Schaller, 2001) are marked by integrin clustering (Lee and Juliano, 2004) and, intracellularly, by the redistribution of an astonishing amount of molecules (Miyamoto et al., 1995) that are proposed to form the "integrin adhesome" (Zaidel-Bar et al., 2007). The most significant among these molecules are the tyrosine kinases Src and focal adhesion kinase (FAK), since FAK autophosphorylation in Tyr397 is one of the earliest events following integrin ligation and clustering, which further induces Src binding and activation, while Src provokes supplementary FAK phosphorylation and stabilization of the complex (Kim et al., 2011). Moreover, activated FAK and Src recruit PI3K and Raf-1, thus providing a convergence point for ECM and growth factor signaling (Schaller, 2001; Slack-Davis et al., 2003; Edin and Juliano, 2005). Likewise, hyaluronan (HA) binding to CD44 activates Rho guanosine triphosphatases (GTPases) – a subclass of the Ras superfamily – most important among them being RhoA, Rac1 and Cdc42 (Wennerberg and Der, 2004), which are also GFR targets and activate downstream the MAPKs and/or the PI3K/AKT pathway (Bourguignon, 2008).

The abovementioned pathways have been shown to be interwoven in many instances, thus establishing a growth factor-ECM cross talk at the signaling level. For example, α2β1 integrin has been shown to activate the EGF receptor (Yu et al., 2000), which in turn can mediate the proliferative activity of fibronectin (Bill et al., 2004). Similarly, vitronectin enhances the proliferative effects of PDGF and TGF-β on human fibroblasts through the interaction of PDGF-β-receptor and TGF-β receptor II, respectively, with the integrin αvβ3 (Schneller et al., 1997; Scaffidi et al., 2004), while hyaluronan-induced CD44-Smad1 interaction in chondrocytes is necessary for the effects of BMP7 to occur (Peterson et al., 2004).

8.3.3 Growth factor signaling regulates ECM composition

GFs and their respective receptors play important roles in all stages of matrix synthesis from early matrix deposition and subsequent accumulation to ongoing remodeling processes (Osada et al., 1996; Sawaji et al., 2008) as depicted in ▶Figure 8.10.

Platelet-derived growth factor and ECM

PDGF is a potent mitogen for all cells of mesenchymal origin (Deuel, 1987; Heldin, 1992). This growth factor regulates other key cellular processes including chemotaxis, survival, apoptosis and transformation *in vitro* (Deuel et al., 1982; Doolittle et al., 1983; Senior et al., 1983; Barres et al., 1992; Kim et al., 1995). PDGF is produced by epithelial, endothelial, and other cell types that are in close apposition to PDGF receptors-expressing mesenchymal cells but also through mesenchymal cell autocrine function (Ataliotis and Mercola, 1997). Active PDGF is formed by heterodimerization or homodimerization of mainly two distinct but highly homologous polypeptides: A and B. Both A and B chains can be produced as long and short isoforms through alternative splicing. The short isoforms, bearing positively charged peptide sequences, are the most

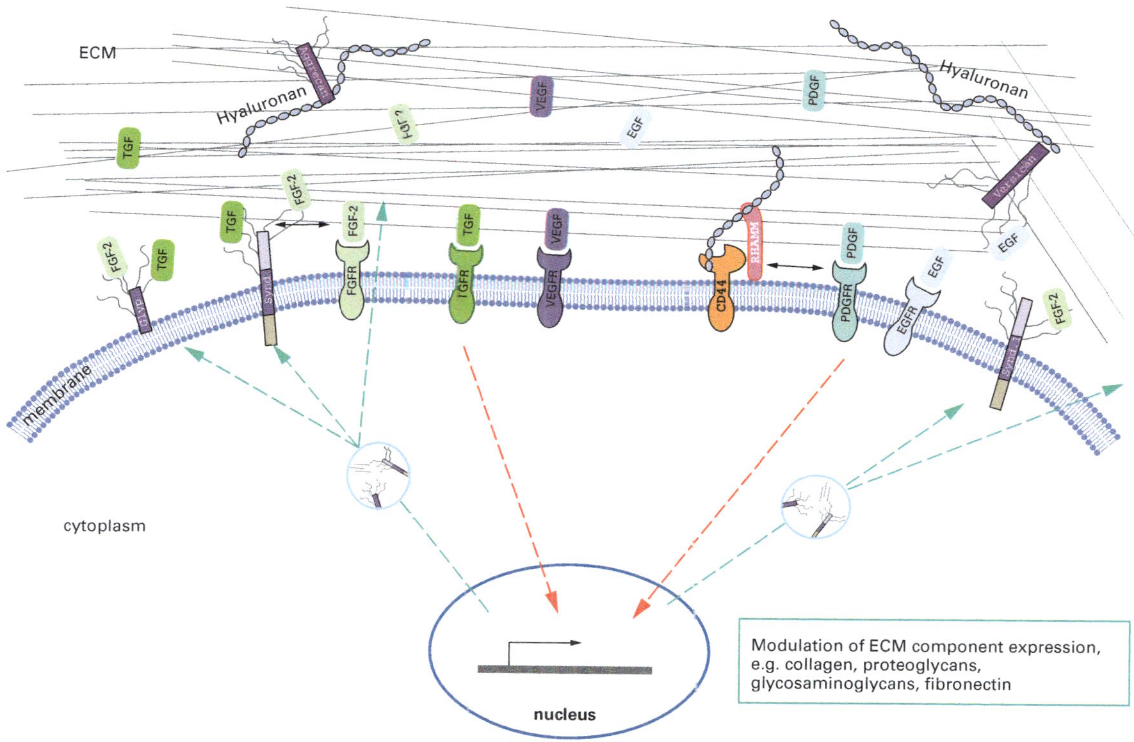

Figure 8.10 **Growth factors and their respective receptors signaling modulate changes in mRNA (red arrows) and protein expression of key matrix components (green arrows) including collagen, proteoglycans, glycosaminoglycans, fibronectin, and so forth.** Growth factors influence early matrix deposition and subsequent accumulation as well as its ongoing remodeling processes. (FGF2, fibroblast growth factor-2; TGF, transforming growth factor; synd, syndecan; glyp, glypican; VEGF, vascular endothelial growth factor; RHAMM, receptor for hyaluronan-mediated motility; PDGF, platelet-derived growth factor; EGF, epidermal growth factor; ECM, extracellular matrix; R, receptor.

prevalent in the ECM (Pollock and Richardson, 1992) whereas the long isoforms are mainly immobilized on the cell surface through the retention motif and perpetrate autocrine functions (Andersson et al., 1994). PDGF isoforms interact either through the retention motif, a carboxyterminal stretch of basic amino acids, or their positively charged peptide sequences in a highly specific manner with discriminate, highly negatively charged sequences found in GAGs (Feyzi et al., 1997; Garcia-Olivas et al., 2003). In both normal and transformed cells of mesenchymal origin, PDGF regulates the content and distribution of ECM components including GAGs (Heldin et al., 1989; Tzanakakis et al., 1995a, 1995b; Berdiaki et al., 2008), PGs (Syrokou et al., 1999), collagen (Canalis, 1981), fibronectin (Blatti et al., 1988), or MMP (Wagsater et al., 2009). These data establish a key role for PDGF in ECM accumulation and turnover. Furthermore, the previously discussed role correlates well with the proposed functional activities of PDGF *in vivo* during embryonic development, inflammation, and wound healing (Tzeng et al., 1985; Deuel, 1987; Molloy et al., 2003) as well as in different pathological conditions (Doolittle et al., 1983; Deuel, 1987; Ostendorf et al., 2012).

Epidermal growth factor and ECM

EGF is a small mitogenic protein, 53 amino acid residues long, that is thought to be involved in several physiological mechanisms such as normal cell growth, oncogenesis, and wound healing. This protein is the principal member of the EGF family that includes seven transmembrane growth factors: EGF itself, transforming growth factor-alpha (TGF-α), heparin-binding EGF-like growth factor (HB-EGF), amphiregulin (AREG), betacellulin (BTC), epiregulin (EREG), and epigen (EPGN) (Harris et al., 2003).

All members of the EGF family serve as agonists for human epidermal growth factor receptor (HER) family receptors. The HER family of receptor tyrosine kinases includes the epidermal growth factor receptor (EGFR/ErbB1/HER1), ErbB2/HER2/Neu, ErbB3/HER3, and ErbB4/HER4. Signaling by HER receptors is of principal importance in the control of cell fate, influencing proliferation, survival, or differentiation; hence, deregulated signaling by these receptors plays important roles in human malignancies (Gialeli et al., 2009; Yarden and Sliwkowski, 2001). Ligand binding causes the homo- and/or heterodimerization of HER receptors, leading to the activation of their intracellular tyrosine kinase domains providing a mechanism for coupling to downstream signaling cascades, including the Ras-Raf-ERK-MAPK, PI3K-AKT, protein kinase C (PKC), signal transducer and activator of transcription (STAT), and the Src kinase signal transduction pathways (Citri and Yarden, 2006).

EGF signaling is established to regulate ECM composition. Thus, EGF was found to stimulate mesothelioma matrix synthesis, specifically GAG/PG production, via receptor-growth factor complexes (Syrokou et al., 1999) and in a manner correlated to these cells differentiation state (Tzanakakis et al., 1996). The expression pattern of MMPs genes is transcriptionally coordinated by extracellular stimuli (Westermarck and Kahari, 1999) including EGF (Tian et al., 2007). Specifically, EGF induces matrix metalloproteinase-1 (MMP1), subsequent matrix remodeling and enhanced invasion in glioma and breast cancer cell lines via the MAPK pathway (Nut and Lunec, 1999; Anand et al., 2011; Park et al., 2011). Likewise, in premalignant keratinocytes, EGF and TGF-α stimulate the development of a collagenolytic ECM phenotype due to the upregulation of a number of MMPs, which enhances the epithelial-to-mesenchymal transition of these cells (Wilkins-Port and Higgins, 2007). Upregulation of membrane type-MMPs, especially MT1-MMP, is documented in glioma cells and squamous carcinoma cells upon treatment with EGF and subsequent increase in MMP2 activity (Sato et al., 1999; Van Meter et al., 2004). Similarly, EGF-mediated increase of MMP9 activity via MAPK and PI3K pathway is observed in breast cancer and human trophoblast cells, depicted in increased migration and invasive potential (Dilly et al., 2010).

Recent studies demonstrated a selective upregulation of α2 but not β1 integrin subunit in carcinoma and normal cells in response to EGF, associated with increased cell migration and invasion. Specifically, EGF is documented to increase the expression of α2β1 integrin as well as α6β1 integrin expression and localization (Chen et al., 1993; Yamanaka et al., 2003). It is worth mentioning that EGF induces the downregulation of E-cadherin, a principal adhesion molecule, via MAPK pathway and upregulation of snail (Cheng et al., 2010). Hyaluronan expression is also indirectly affected by EGF as its enzyme synthase hyaluronan synthase-2 (HAS2) is upregulated via PI3K and STAT pathway (Tammi et al., 2011).

Fibroblast growth factors and ECM

Fibroblast growth factors (FGFs) and their respective signaling pathway play significant roles in physiological processes including embryonic development, angiogenesis, neuronal differentiation and wound repair (Schonherr and Hausser, 2000) and have been involved in many pathological states (Wright et al., 1993).

Twenty two distinct FGFs have been identified in a variety of organisms from drosophila to human (Ornitz and Itoh, 2001). FGFs are differentially expressed in many if not all tissues, but the patterns and timing of their expression vary (Ornitz and Itoh, 2001). All the members of the FGF family share a conserved sequence of 120 amino acids and mediate their cellular responses by binding to and activating a family of four receptor tyrosine kinases (RTKs), fibroblast growth factor receptor-1 to -4 (FGFR1–4) (Lee et al., 1989; Givol and Yayon, 1992; Jaye et al., 1992; Eswarakumar et al., 2005). Through alternative splicing, structural RTK variants are generated that differ in their ligand-binding specificities and affinities (Schonherr and Hausser, 2000). In addition to these high-affinity signaling receptors, it is established that FGFs bind, albeit with lower affinity, heparan sulfate (HS) chains.

Fibroblast growth factor-2 (FGF2) has an outstanding role in the regulation of ECM synthesis of bone (Carinci et al., 2000), cardiovascular (Park et al., 2008) or cartilage tissues (Sawaji et al., 2008). Thus, modulation of ECM components by FGF2 in bone cells could explain the altered osteogenic process and account for pathological variations in cranial development (Bodo et al., 1999; Carinci et al., 2000). In malignant cells, an autocrine role of FGF2 in the regulation of ECM composition correlated to different osteosarcoma phenotypes has been postulated (Bodo et al., 2002). Moreover, FGF2 specifically regulates the organization of the osteosarcoma pericellular matrix, through increased HA biosynthesis and accumulation in the transcriptional upregulation of versican0/versican1 and HA synthase (HAS) genes respectively (Berdiaki et al., 2008). The importance of FGF2 signaling in heart development is highlighted by the finding that upon disruption of FGF2 signaling the developing outflow myocardium fails to produce ECM (Park et al., 2008). In human articular cartilage FGF2 has an important role in ECM remodeling (Sawaji et al., 2008).

Transforming growth factor-beta and ECM

TGF-β belongs to a large family of growth and differentiation factors (GDF), including BMPs, activins, and nodal (Moustakas and Heldin, 2009; Wharton and Derynck, 2009; Burch et al., 2011). Individual family members play crucial roles in fetal development and regulation of tissue homeostasis during adult life (Feng and Derynck, 2005; Massague and Gomis, 2006). The role of TGF-β in cell function appears to be more complex compared to other growth factors and is associated with a broad aspect of cellular processes and diverse disease states including cancer and atherosclerosis (Evanko et al., 1998; Padua and Massague, 2009; Burch et al., 2011).

TGF-β is synthesized as an integral part of a large molecule, the precursor-TGF-β containing the latency-associated proteins (LAPs) (ten Dijke and Arthur, 2007; Pohlers et al., 2009), which undergoes intracellular cleavage, subsequent secretion, and binding to specific ECM components (Ruiz-Ortega et al., 2007; Bernabeu et al., 2009; Rozario and DeSimone, 2010). Release of mature TGF-β from this complex can be achieved by various means including plasmin (Wan and Flavell, 2007), MMP2 and MMP9 (Leask and Abraham, 2004) or thrombospondin (Siegel and Massague, 2003; Pohlers et al., 2009) action. The mature TGF-β perpetrates signaling through multiple cell surface receptors (Massague et al., 2005; Wharton and Derynck, 2009; Burch et al., 2011). TGF-β cell surface receptors with intrinsic serine/threonine kinase activity also contain a cytoplasmic kinase domain with a weaker tyrosine kinase activity, classifying the receptors thus, as dual specificity kinases (Heldin et al., 1997). Five type II (TGF-βRII) and seven type I receptors (TGF-βRI) have been recognized in humans. TGF-β type III receptors (TGF-βRIII), betaglycan, and endoglin not only bind TGF-β, but they also modulate its signaling by through not fully identified mechanisms (Bernabeu et al., 2009). It is generally accepted that in pathological conditions like cancer, alterations of betaglycan and endoglin expression contribute to the deregulation of TGF-β signaling (Bernabeu et al., 2009).

The signaling pathways of the TGF-β superfamily members are required for the homeostasis and development of multicellular organisms (Wu and Hill, 2009) with important roles in tissue morphogenesis (Saika et al., 2001). Regarding cell proliferation, TGF-β shows a pleiotropic action by inhibiting epithelial, endothelial and hematopoietic cells, while it stimulates the growth of certain types of mesenchymal cells (Massague et al., 2000; Wenner and Yan, 2003). Interestingly, concerning human skin fibroblast proliferation, TGF-β-effects vary depending on the developmental stage of the donor (Pratsinis et al., 2004). In particular, TGF-β stimulates adult fibroblasts by inducing the synthesis and release of FGF2 and ultimately by the activation of the MEK-ERK pathway, while in fetal fibroblasts TGF-β inhibits proliferation via the activation of protein kinase-A (PKA) and the subsequent upregulation of the cyclin-dependent kinase inhibitors p21WAF1 and p15INK4 (Giannouli and Kletsas, 2006).

Furthermore, TGF-β has a key role in the processes of ECM structuring correlated to tumor dissemination (Chaudhury and Howe, 2009) and fibrosis (Bissel, 2001; Joseph et al., 2010). Thus, Tzanakakis et al. (1995a, 1997) have demonstrated the discriminate role of TGF-β2 in mesothelioma cell GAG synthesis on two mesothelioma cell sublines of fibroblast-like and epithelial morphology in a manner dependent on cell morphology (Tzanakakis et al., 1995a). Nikitovic et al. (2006) have demonstrated that TGF-β2 triggers the malignant phenotype pattern of versican and HA expression in human osteosarcoma cells suggesting that this growth factor may account for the metastatic potential of these cells (Nikitovic et al., 2006). Interestingly in fibrosarcoma cells, which likewise have mesenchymal origin, the effects of TGF-β2 on hyaluronan metabolism are well correlated to these cells metastatic phenotype (Berdiaki et al., 2008). Pancreatic stellate cells respond to TGF-β stimulation during the processes of fibrosis with enhanced collagen, fibronectin and proteoglycan synthesis and simultaneous blocking of protease secretion (reviewed by Menke and Adler, 2002). Notably, while TGF-β1 and TGF-β2 are promoting cutaneous scarring, the TGF-β3 isoform has been shown to act in an opposite direction (Shah et al., 1995).

Vascular endothelial growth factor and ECM

Members of vascular endothelial growth factor (VEGF) family of human origin are seven homodimeric, heparin-binding glycoproteins, encoded by genes located on different chromosomes (reviewed by Patil et al., 2012). Mammalian VEGFs bind to three types of RTKs, VEGFR1 (FMS-like tyrosine kinase [Flt]-1), VEGFR2 (KDR, Flk-1), and VEGFR3 (Flt-4) (Robinson et al., 2001). Ligand binding induces dimerization of VEGF receptors and their consequent tyrosine kinase activity. Signaling pathways activated by VEGFs mediate a plethora of biological processes in endothelial cells including cell proliferation, migration, survival, cel-cell communication, or differentiation, which ultimately control vessel formation (Shibuya and Claesson-Welsh, 2006). VEGF-signaling has been implicated in a variety of human diseases including tumor angiogenesis, tumor-dependent ascites formation, metastasis and inflammatory diseases (reviewed by Ferrara and Davis-Smyth, 1997; Karkkainen and Petrova, 2000).

ECM remodeling is one of the obligatory early steps in pathological angiogenesis (Chang et al., 2009). The initial step in this process is the generation of greatly enlarged "mother" vessels from preexisting normal venules by a process involving degradation of their rigid basement membranes. Thus, VEGF-A initiates "mother vessel" formation, in part, by inducing the expression of endothelial cell proteases such as a disintegrin and metalloproteinase with thrombospondin motifs-1 (ADAMTS1) and MMP15 that act in concert to degrade venular basement membrane versican (Fu et al., 2011). On the other hand, VEGF is likewise suggested to modulate ECM composition as it promotes collagen deposition during wound healing (Bao et al., 2009).

8.3.4 Effect of ECM on growth factor action

Through elegant feedback mechanisms, ECM components influence cellular processes by regulating the presentation and availability of growth factors and cytokines to cells (matricine signaling).Thus, PDGF-BB has been shown to bind ECM-associated HS, CS-A (chondroitin sulfate [CS] mainly sulfated at C-4 of galactosamine), dermatan sulfate (DS), and heparin, or it may be stored on the cell surface by binding to cell-associated HS chains (Garcia-Olivas et al., 2003; Iozzo, 2005). Furthermore, PDGF has been reported to interact with matricellular proteins (Nelson et al., 1997; Brekken and Sage, 2001) or collagens (Raines et al., 2000). These interactions are of utmost importance as the encounter of PDGF and mesenchymal cells typically takes place in a complex microenvironment rich in various ECM components, which participate in the regulation of PDGF signaling. Thus, binding of PDGF to matrix or cell associated GAGs can regulate its mitogenic function on normal cells. Specifically, in human smooth muscle cells, it has been shown that heparin (Lustig et al., 1996), HS, and DS (Fager et al., 1995) prevent the proliferation induced by PDGF. Moreover, it was established in fibroblasts that a stimulation of endogenous CS/DS production caused severe reduction of the PDGF induced proliferation (Fthenou et al., 2006). PDGF binding to secreted protein acidic and rich in cysteine (SPARC), a prototypical matricellular protein that functions to regulate cell-matrix interactions, modulates its downstream signaling (Brekken and Sage, 2001).

EGF signaling is partly regulated by cues originating in the ECM. In mesothelioma cells it is suggested that the interactions between GFs and GAGs, including those of EGF, regulate growth factor signaling correlated to these cells' phenotypic differentiation (Tzanakakis et al., 1995a). The synergistic effect of ECM-EGF complexes is for example depicted in the ability of vitronectin to associate with EGF, leading to increased cell motility of skin keratinocytes (Upton et al., 2008). Importantly, it is established that normal cells need to adhere to matrix in order to progress through the mid-G1 phase of the cell cycle after mitogen, including EGF treatment (Assoian, 1997) Therefore, cell adhesion to specific ECM molecules such as fibronectin activates integrin signaling essential for receptor response to EGF ligand, leading to control of transcriptional events and cell-cycle progression (Cabodi et al., 2004). It is established in literature that ECM interaction and composition determines the transcriptional EGF-mediated response. For example, adhesion of human embryonic cells to ECM molecules like laminin or fibronectin alters the intracellular signaling and how specific gene transcription responds to EGF stimulation (Yarwood and Woodgett, 2001). In continuation, joint integrin/EGFR signaling causes expression of cyclin D, activation of cyclin D-dependent kinases and degradation of cyclin-dependent kinase (CDK) inhibitor p27 to stimulate proliferation (Bill et al., 2004). Specifically, α2β1 integrin colocalizes with EGF receptor leading to activation of downstream signaling intermediates, like FAK or ERK, enhancing its activity (Yu et al., 2000). Another mechanism reported for EGF signaling regulation by integrins is the coordination of internalization and degradation of EGFR via α5β1 integrin interaction (Caswell et al., 2009). EGF-action is also regulated by certain ECM macromolecules like tenascin-C and laminin that bind directly with EGF receptor via EGF-like repeats in their structure (Tran et al., 2005). Furthermore it is suggested that the small leucine-rich proteoglycan (SLRP) member, decorin interacts with soluble EGF to regulate its downstream signaling ultimately resulting in modified cell-cycle progression (Mohan et al., 2011). Thus, EGF in concert with the contribution from the ECM components orchestrates complex genetic and biological responses to regulate key cellular functions.

FGF2 has been intensively investigated in light of its interaction with heparin, free HS chains and heparan sulfate proteoglycans (HSPGs). These interactions stabilize FGF2 to thermal denaturation, protect against proteolysis, limit its diffusion and release into interstitial spaces, present FGF2 to its receptors and facilitate FGFR dimerization necessary for signal transduction (Moscatelli, 1987; Flaumenhaft et al., 1990; Delehedde et al., 1996; Aviezer et al., 1997; Ornitz and Itoh, 2001). Numerous studies have established the importance of FGF2 and GAG interactions in the regulation of the growth rate, migration, adhesion, and differentiation of osteosarcoma, melanoma, fibrosarcoma, and colon carcinoma cells (Bodo et al., 2002; Berdiaki et al., 2008, 2009; Chatzinikolaou et al., 2008; Nikitovic et al., 2008; Chalkiadaki et al., 2009). Both cell-associated and extracellular HSPGs were shown to alter FGF2 signaling in melanoma cells and to increase their metastatic potential (Delehedde et al., 1996; Reiland et al., 2006). A rising amount of evidence has lately implicated CSPGs to participate in FGF2 signaling (Molteni, et al., 1999; Bao et al., 2004; Smith et al., 2007). Thus, Nikitovic et al. (2008) showed that CS/DS-containing proteoglycans, likely in cooperation with HS, participate in metastatic melanoma cell FGF2-induced mitogenic response.

The binding of TGF-β to ECM components is the main conduit for the regulation of TGF-β activity, the SLRPs being the key perpetrators (Schaefer and Iozzo, 2008). Thus, SLRP members like decorin, biglycan, and fibromodulin bind and regulate TGF-β by sequestering this growth factor within the ECM (Hildebrand et al., 1994). Both inhibition and stimulation of TGF-β activity after binding to SLRPs has been reported (Noble et al., 1992; Hausser et al., 1994; Takeuchi et al., 1994; Markmann et al., 2000). Recently, delayed wound closure in fibromodulin-deficient mice was reported to be associated with increased TGF-β3 signaling (Zheng et al., 2011). Likewise the SLRP, lumican through the modulation of the pericellular availability of TGF-β regulates its downstream intracellular signaling and thus affects osteoblastic cell functions (Nikitovic et al., 2011). The HSPG syndecan-2 is established to act as a coreceptor for TGF-β in conjunction with betaglycan. The shedding of syndecan-2 is proposed as an alternative pathway in the regulation of TGF-β signaling (Li et al., 2006).

Two key mechanisms suggested to regulate VEGF action are cues originating from the ECM and proteolytic processing (Ferrara, 2010). VEGF splice variants have a variable affinity to bind to heparin/HS, which results in membrane associated and ECM pools of this growth factor (Houck et al., 1992). The affinity of VEGF to bind heparin/HS discriminately regulates their bioavailability and activity (Park et al., 1993; Keyt et al., 1996). Indeed, the ECM HS-pool seems to be essential to spatially restrict VEGF to create the appropriate gradient to allow blood vessel branching to occur (Ruhrberg et al., 2002). Heparanase directly promotes angiogenesis by releasing heparin binding VEGF trapped within the ECM (Iozzo and San Antonio, 2001). HS proteoglycans can also reactivate oxidation-damaged forms of VEGF, a function that may be essential in hypoxic sites where new vessels are required (Gengrinovitch et al., 1999). Moreover, proteolytic processing of VEGF by plasmin and MMPs is also regulated by ECM cues as HSPGs potentiate VEGF cleavage by increasing the VEGF clearance time in tissues (Vempati et al., 2010). In conclusion, ECM-conveyed signals are essential cues for growth factor signaling in both physiological and transformed cells and represent potential therapeutic targets in the management of disease.

8.3.5 Pharmacological Interventions

The previously described features of the cells and their microenvironment (i.e. ECM molecules, growth factors, and their receptors, as well as the intracellular signaling cascades triggered by their interactions) offer the ground for a variety of pharmacological interventions in conditions and diseases related to abnormalities in ECM turnover, cell proliferation and/or survival, such as cancer, fibrosis, and systemic sclerosis, to name a few.

Such pharmacological interventions may target extracellularly the ligands triggering the pathological responses. Prominent example of this concept is the use of a monoclonal antibody against VEGF as a means to block neovascularization in tumors and in macular degeneration (Andriolo et al., 2009; Eichholz et al., 2010), and the attempts to reduce the levels of TGF-β – which is known to induce fibrosis – by administering neutralizing antibodies against this protein (Border and Noble, 1994; Denton et al., 2007). Furthermore, prevention of TGF-β-binding to the target cells by delivering soluble forms of its receptors is currently effective in preclinical experimental models (Prud'homme, 2007). Similar attempts to reduce TGF-β levels may target molecules that activate the latent form of this growth factor, such as MMP9 and CD44, which have been implicated in tumor invasion and angiogenesis through latent-TGF-β activation (Yu and Stamenkovic, 2000; for MMPs, see Chapter 8.5 by Skandalis et al. in the present section). On the other hand, in some cases, even the use of a specific ligand-isoform has been proposed for therapeutic applications (Shah et al., 1995; Ohno et al., 2011).

The above mentioned interventional strategies are not limited to the GFRs, since the anti-integrin monoclonal antibodies abciximab and natalizumab are already in clinical use, the former as antithrombotic agent and the latter for the treatment of multiple sclerosis and Crohn's disease (Byron et al., 2009).

A second level of pharmacological intervention, considered to be promising due to its specificity, is the blockade of a receptor by targeting either its extracellular ligand-binding domain or its intracellular kinase activity. There is a long list of pharmaceutical compounds that target the receptors of various growth factors, mainly with intended antitumor activity, either approved or in clinical trials. For example, cetuximab, matuzumab, and panitumumab are antibodies binding to the extracellular domain of EGFR, while gefitinib and erlotinib are small molecules inhibiting reversibly its tyrosine kinase activity, and canertinib and EKB-569 are irreversible EGFR-tyrosine kinase inhibitors (Ekman et al., 2007; Rocha-Lima et al., 2007; Höpfner et al., 2008a, 2008b). Furthermore, several inhibitors are used based on their wide target spectrum, such as sorafenib, which inhibits PDGFR, VEGFR, c-Raf, and B-Raf, while others are considered rather more specific, such as imatinib targeting mainly PDGFR, *c-kit*, and *bcl-abl* (Morabito et al., 2006; Homsi and Daud, 2007; Wilhelm et al., 2008).

Pharmacological inhibitors that act further downstream of the receptor level are targeting various kinases participating in the signal transduction process, such as rapamycin and its analogues sirolimus, everolimus, temsirolimus, and ridaforolimus, which have been used as immunosuppressants and anticancer agents and act by inhibiting the kinase mammalian target of rapamycin (mTOR) that phosphorylates AKT similarly to PI3K (Markman et al., 2010). More inhibitors targeting PI3K, and AKT, or dual inhibitors of PI3K and mTOR have also been designed and are currently under various stages of clinical evaluation (Markman et al., 2010). On the other hand, inhibitors targeting the MAPK pathway, such as PD98059, PD0325901, U0126, and CI-1040 (PD184352), have been shown to inhibit cancer cell proliferation *in vitro* and tumor growth in experimental animals (Wabnitz et al., 2004; Marampon et al., 2009; Henderson et al., 2010), but they were not so successful in clinical use until now (Pratilas and Solit, 2010).

A wide variety of other approaches are also under investigation, in order to block specific GF- and ECM-induced responses that account for various disorders and pathologies, making use of novel techniques, such as silencing of specific genes with antisense and ribozyme oligonucleotides or RNA interference (Liu et al., 2006). Promising results from the use of such strategies in experimental animal systems have to be complemented by more clinical trials, so that the current limitations can be overcome (Vachani et al., 2010).

8.3.6 Take-home message

GFs, GFRs and their interactions have crucial roles in ECM deposition, organization and remodeling while the ECM components themselves are essential cellular effectors that integrate and respond to the environmental stimulus through the regulation of GF signaling pathways. Correct integration of the GF-ECM signaling regulates key biological functions that participate in the development and homeostasis of tissues throughout life whereas, abnormalities in these interactions are implicated in numerous pathological processes. The described in this review intrinsic characteristics of the ECM molecules, GFs and their receptors, as well as the intracellular signaling cascades triggered by their interactions, offer the ground for targeted pharmacological interventions in conditions and diseases related to abnormalities in ECM turnover including cancer, fibrosis, and systemic sclerosis.

References

Anand, M., Van Meter, T. E., and Fillmore, H. L. (2011). Epidermal growth factor induces matrix metalloproteinase-1 (MMP-1) expression and invasion in glioma cell lines via the MAPK pathway. J Neurooncol *104* (3), 679–687.

Andersson, M., Ostman, A., Westermark, B., and Heldin, C. H. (1994). Characterization of the retention motif in the C-terminal part of the long splice form of platelet-derived growth factor A-chain. J Biol Chem *269*, 926–930.

Andriolo, R. B., Puga, M. E., Belfort Junior, R., and Atallah, A. N. (2009). Bevacizumab for ocular neovascular diseases: a systematic review. Sao Paulo Med J *127*, 84–91.

App, H., Hazan, R., Zilberstein, A., Ullrich, A., Schlessinger, J., and Rapp, U. (1991). Epidermal growth factor (EGF) stimulates association and kinase activity of raf-1 with the EGF receptor. Mol Cell Biol *11*, 913–919.

Assoian, R. K. (1997). Anchorage-dependent cell cycle progression. J Cell Biol *136*, 1–4.

Ataliotis, P., and Mercola, M. (1997). Distribution and functions of platelet-derived growth factors and their receptors during embryogenesis. Int Rev Cytol *172*, 95–127.

Attisano, L., and Lee-Hoeflich, S. T. (2001). The Smads. Genome Biol *2*, REVIEWS3010.

Aviezer, D., Iozzo, R. V., Noonan, D. M., and Yayon, A. (1997). Suppression of autocrine and paracrine functions of basic fibroblast growth factor by stable expression of perlecan antisense cDNA. Mol Cell Biol *17*, 1938–1946.

Bao, P., Kodra, A., Tomic-Canic, M., Golinko, M. S., Ehrlich, H. P., and Brem, H. (2009). The role of vascular endothelial growth factor in wound healing. J Surg Res *153*, 347–358.

Bao, X., Nishimura, S., Mikami, T., Yamada, S., Itoh, N., and Sugahara, K. (2004). Chondroitin sulfate/dermatan sulfate hybrid chains from embryonic pig brain, which contain a higher proportion of L-iduronic acid than those from adult pig brain, exhibit neuritogenic and growth factor binding activities. J Biol Chem *279*, 9765–9776.

Barkan, D., Green, J. E., and Chambers, A. F. (2010). Extracellular matrix: a gatekeeper in the transition from dormancy to metastatic growth. Eur J Cancer *46*, 1181–1188.

Barres, B. A., Hart, I. K., Coles, H. S., et al. (1992). Cell death in the oligodendrocyte lineage. J Neurobiol *23*, 1221–1230.

Berdiaki, A., Nikitovic, D., Tsatsakis, A., Katonis, P., Karamanos, N. K., and Tzanakakis, G. N. (2009). bFGF induces changes in hyaluronan synthase and hyaluronidase isoform expression and modulates the migration capacity of fibrosarcoma cells. Biochim Biophys Acta *1790*, 1258–1265.

Berdiaki, A., Zafiropoulos, A., Fthenou, E., et al. (2008). Regulation of hyaluronan and versican deposition by growth factors in fibrosarcoma cell lines. Biochim Biophys Acta *1780*, 194–202.

Bernabeu, C., Lopez-Novoa, J. M., and Quintanilla, M. (2009). The emerging role of TGF-beta superfamily coreceptors in cancer. Biochim Biophys Acta *1792*, 954–973.

Bill, H. M., Knudsen, B., Moores, S. L., et al. (2004). Epidermal growth factor receptor-dependent regulation of integrin-mediated signaling and cell cycle entry in epithelial cells. Mol Cell Biol *24*, 8586–8599.

Bissell, D. M. (2001). Chronic liver injury, TGF-beta, and cancer. Exp Mol Med *33*, 179–190.

Blatti, S. P., Foster, D. N., Ranganathan, G., Moses, H. L., and Getz, M. J. (1988). Induction of fibronectin gene transcription and mRNA is a primary response to growth-factor stimulation of AKR-2B cells. Proc Natl Acad Sci U S A *85*, 1119–1123.

Bobick, B. E., Chen, F. H., Le, A. M., and Tuan, R. S. (2009). Regulation of the chondrogenic phenotype in culture. Birth Defects Res C Embryo Today *87*, 351–371.

Bodo, M., Baroni, T., Carinci, F., et al. (1999). A regulatory role of fibroblast growth factor in the expression of decorin, biglycan, betaglycan and syndecan in osteoblasts from patients with Crouzon's syndrome. Eur J Cell Biol *78*, 323–330.

Bodo, M., Lilli, C., Bellucci, C., et al. (2002). Basic fibroblast growth factor autocrine loop controls human osteosarcoma phenotyping and differentiation. Mol Med *8*, 393–404.

Border, W. A., and Noble, N. A. (1994). Transforming growth factor beta in tissue fibrosis. N Engl J Med *331*, 1286–1292.

Bourguignon, L. Y. (2008). Hyaluronan-mediated CD44 activation of RhoGTPase signaling and cytoskeleton function promotes tumor progression. Semin Cancer Biol *18*, 251–259.

Brazil, D. P., and Hemmings, B. A. (2001). Ten years of protein kinase B signalling: a hard Akt to follow. Trends Biochem Sci *26*, 657–664.

Brekken, R. A., and Sage, E. H. (2001). SPARC, a matricellular protein: at the crossroads of cell-matrix communication. Matrix Biol *19*, 816–827.

Burch, M. L., Zheng, W., and Little, P. J. (2011). Smad linker region phosphorylation in the regulation of extracellular matrix synthesis. Cell Mol Life Sci *68*, 97–107.

Byron, A., Humphries, J. D., Askari, J. A., Craig, S. E., Mould, A. P., and Humphries, M. J. (2009). Anti-integrin monoclonal antibodies. J Cell Sci *122*, 4009–4011.

Cabodi, S., Moro, L., Bergatto, E., et al. (2004). Integrin regulation of epidermal growth factor (EGF) receptor and of EGF-dependent responses. Biochem Soc Trans *32*, 438–442.

Canalis, E. (1981). Effect of platelet-derived growth factor on DNA and protein synthesis in cultured rat calvaria. Metabolism *30*, 970–975.

Cantley, L. C. (2002). The phosphoinositide 3-kinase pathway. Science *296*, 1655–1657.

Carinci, P., Becchetti, E., and Bodo, M. (2000). Role of the extracellular matrix and growth factors in skull morphogenesis and in the pathogenesis of craniosynostosis. Int J Dev Biol *44*, 715–723.

Caswell, P. T., Vadrevu, S., and Norman, J. C. (2009). Integrins: masters and slaves of endocytic transport. Nat Rev Mol Cell Biol *10*, 843–853.

Chalkiadaki, G., Nikitovic, D., Berdiaki, A., et al. (2009). Fibroblast growth factor-2 modulates melanoma adhesion and migration through a syndecan-4-dependent mechanism. Int J Biochem Cell Biol *41*, 1323–1331.

Chang, L., and Karin, M. (2001). Mammalian map kinase signalling cascades. Nature *410*, 37–40.

Chang, S. H., Kanasaki, K., Gocheva, V., et al. (2009). VEGF-A induces angiogenesis by perturbing the cathepsin-cysteine protease inhibitor balance in venules, causing basement membrane degradation and mother vessel formation. Cancer Res *69*, 4537–4544.

Chatzinikolaou, G., Nikitovic, D., Asimakopoulou, A., Tsatsakis, A., Karamanos, N. K., and Tzanakakis, G. N. (2008). Heparin – a unique stimulator of human colon cancer cells' growth. IUBMB Life *60*, 333–340.

Chaudhury, A., and Howe, P. H. (2009). The tale of transforming growth factor-beta (TGFbeta) signaling: a soigné enigma. IUBMB Life *61*, 929–939.

Chen, J. D., Kim, J. P., Zhang, K., et al. (1993). Epidermal growth factor (EGF) promotes human keratinocyte locomotion on collagen by increasing the alpha 2 integrin subunit. Exp Cell Res *209*, 216–223.

Cheng, J. C., Klausen, C., and Leung, P. C. K. (2010). Hydrogen peroxide mediates EGF-induced down-regulation of E-cadherin expression via p38 MAPK and snail in human ovarian cancer cell. Mol Endocrinol *24* (8), 1569–1580.

Citri, A., and Yarden, Y. (2006). EGF-ERBB signalling: towards the systems level. Nat Rev Mol Cell Biol *7*, 505–516.

Delehedde, M., Deudon, E., Boilly, B., and Hondermarck, H. (1996). Heparan sulfate proteoglycans play a dual role in regulating fibroblast growth factor-2 mitogenic activity in human breast cancer cells. Exp Cell Res *229*, 398–406.

Denton, C. P., Merkel, P. A., Furst, D. E., et al. (2007). Recombinant human anti-transforming growth factor beta1 antibody therapy in systemic sclerosis: a multicenter, randomized, placebo-controlled phase I/II trial of CAT-192. Arthritis Rheum *56*, 323–333.

Deuel, T. F. (1987). Polypeptide growth factors: roles in normal and abnormal cell growth. Annu Rev Cell Biol *3*, 443–492.

Deuel, T. F., Senior, R. M., Huang, J. S., and Griffin, G. L. (1982). Chemotaxis of monocytes and neutrophils to platelet-derived growth factor. J Clin Invest *69*, 1046–1049.

Dilly, M., Hambruch, N., Haeger, J. D., and Pfarrer, C. (2010). Epidermal growth factor (EGF) induces motility and upregulates MMP-9 and TIMP-1 in bovine trophoblast cells. Mol Reprod Dev *77* (7), 622–629.

Doolittle, R. F., Hunkapiller, M. W., Hood, L. E., et al. (1983). Simian sarcoma virus onc gene, v-sis, is derived from the gene (or genes) encoding a platelet-derived growth factor. Science *221*, 275–277.

Duronio, V., Scheid, M. P., and Ettinger, S. (1998). Downstream signalling events regulated by phosphatidylinositol 3-kinase activity. Cell Signal *10*, 233–239.

Edin, M. L., and Juliano, R. L. (2005). Raf-1 serine 338 phosphorylation plays a key role in adhesion-dependent activation of extracellular signal-regulated kinase by epidermal growth factor. Mol Cell Biol *25*, 4466–4475.

Eichholz, A., Merchant, S., and Gaya, A. M. (2010). Anti-angiogenesis therapies: their potential in cancer management. Onco Targets Ther 3, 69–82.

Ekman, S., Bergqvist, M., Heldin, C. H., and Lennartsson, J. (2007). Activation of growth factor receptors in esophageal cancer – implications for therapy. Oncologist *12*, 1165–1177.

Eswarakumar, V. P., Lax, I., and Schlessinger, J. (2005). Cellular signaling by fibroblast growth factor receptors. Cytokine Growth Factor Rev *16*, 139–149.

Evanko, S. P., Raines, E. W., Ross, R., Gold, L. I., and Wight, T. N. (1998). Proteoglycan distribution in lesions of atherosclerosis depends on lesion severity, structural characteristics, and the proximity of platelet-derived growth factor and transforming growth factor-beta. Am J Pathol *152*, 533–546.

Fager, G., Camejo, G., Olsson, U., Ostergren-Lunden, G., Lustig, F., and Bondjers, G. (1995). Binding of platelet-derived growth factor and low density lipoproteins to glycosaminoglycan species produced by human arterial smooth muscle cells. J Cell Physiol *163*, 380–392.

Feng, X. H., and Derynck, R. (2005). Specificity and versatility in TGF-beta signaling through Smads. Annu Rev Cell Dev Biol *21*, 659–693.

Ferrara, N. (2010). Binding to the extracellular matrix and proteolytic processing: two key mechanisms regulating vascular endothelial growth factor action. Mol Biol Cell *21*, 687–690.

Ferrara, N., and Davis-Smyth, T. (1997). The biology of vascular endothelial growth factor. Endocr Rev *18*, 4–25.

Feyzi, E., Lustig, F., Fager, G., Spillmann, D., Lindahl, U., and Salmivirta, M. (1997). Characterization of heparin and heparan sulfate domains binding to the long splice variant of platelet-derived growth factor A chain. J Biol Chem *272*, 5518–5524.

Flaumenhaft, R., Moscatelli, D., and Rifkin, D. B. (1990). Heparin and heparan sulfate increase the radius of diffusion and action of basic fibroblast growth factor. J Cell Biol *111*, 1651–1659.

Friesel, R. E., and Maciag, T. (1995). Molecular mechanisms of angiogenesis: fibroblast growth factor signal transduction. FASEB J *9*, 919–925.

Fthenou, E., Zafiropoulos, A., Tsatsakis, A., Stathopoulos, A., Karamanos, N. K., and Tzanakakis, G. N. (2006). Chondroitin sulfate A chains enhance platelet derived growth factor-mediated signalling in fibrosarcoma cells. Int J Biochem Cell Biol *38*, 2141–2150.

Fu, Y., Nagy, J. A., Brown, L. F., et al. (2011). Proteolytic cleavage of versican and involvement of ADAMTS-1 in VEGF-A/VPF-induced pathological angiogenesis. J Histochem Cytochem *59*, 463–473.

Garcia-Olivas, R., Hoebeke, J., Castel, S., et al. (2003). Differential binding of platelet-derived growth factor isoforms to glycosaminoglycans. Histochem Cell Biol *120*, 371–382.

Gengrinovitch, S., Berman, B., David, G., Witte, L., Neufeld, G., and Ron, D. (1999). Glypican-1 is a VEGF165 binding proteoglycan that acts as an extracellular chaperone for VEGF165. J Biol Chem *274*, 10816–10822.

Gialeli, C., Kletsas, D., Mavroudis, D., Kalofonos, H. P., Tzanakakis, G. N., and Karamanos, N. K. (2009). Targeting epidermal growth factor receptor in solid tumors: critical evaluation of the biological importance of therapeutic monoclonal antibodies. Curr Med Chem *16*, 3797–3804.

Giannouli, C. C., and Kletsas, D. (2006). TGF-beta regulates differentially the proliferation of fetal and adult human skin fibroblasts via the activation of PKA and the autocrine action of FGF-2. Cell Signal *18*, 1417–1429.

Givol, D., and Yayon, A. (1992). Complexity of FGF receptors: genetic basis for structural diversity and functional specificity. FASEB J *6*, 3362–3369.

Goldring, M. B., Tsuchimochi, K., and Ijiri, K. (2006). The control of chondrogenesis. J Cell Biochem *97*, 33–44.

Harris, R. C., Chung, E., and Coffey, R. J. (2003). EGF receptor ligands. Exp Cell Res *284*, 2–13.

Hausser, H., Groning, A., Hasilik, A., Schonherr, E., and Kresse, H. (1994). Selective inactivity of TGF-beta/decorin complexes. FEBS Lett *353*, 243–245.

Hayes, A. J., and Ralphs, J. R. (2011). The response of foetal annulus fibrosus cells to growth factors: modulation of matrix synthesis by TGF-beta1 and IGF-1. Histochem Cell Biol *136* (2), 163–175.

Heldin, C. H. (1992). Structural and functional studies on platelet-derived growth factor. EMBO J *11*, 4251–4259.

Heldin, C. H., Ernlund, A., Rorsman, C., and Ronnstrand, L. (1989). Dimerization of B-type platelet-derived growth factor receptors occurs after ligand binding and is closely associated with receptor kinase activation. J Biol Chem *264*, 8905–8912.

Heldin, C. H., Miyazono, K., and ten Dijke, P. (1997). TGF-beta signalling from cell membrane to nucleus through SMAD proteins. Nature *390*, 465–471.

Heldin, C. H., and Westermark, B. (1999). Mechanism of action and *in vivo* role of platelet-derived growth factor. Physiol Rev *79*, 1283–1316.

Henderson, Y. C., Chen, Y., Frederick, M. J., Lai, S. Y., and Clayman, G. L. (2010). MEK inhibitor PD0325901 significantly reduces the growth of papillary thyroid carcinoma cells *in vitro* and *in vivo*. Mol Cancer Ther *9*, 1968–1976.

Hildebrand, A., Romaris, M., Rasmussen, L. M., et al. (1994). Interaction of the small interstitial proteoglycans biglycan, decorin and fibromodulin with transforming growth factor beta. Biochem J *302* (Pt 2), 527–534.

Homsi, J., and Daud, A. I. (2007). Spectrum of activity and mechanism of action of VEGF/PDGF inhibitors. Cancer Control *14*, 285–294.

Höpfner, M., Schuppan, D., and Scherübl, H. (2008a). Growth factor receptors and related signalling pathways as targets for novel treatment strategies of hepatocellular cancer. World J Gastroenterol *14*, 1–14.

Höpfner, M., Schuppan, D., and Scherübl, H. (2008b). Treatment of gastrointestinal neuroendocrine tumors with inhibitors of growth factor receptors and their signaling pathways: recent advances and future perspectives. World J Gastroenterol *14*, 2461–2473.

Houck, K. A., Leung, D. W., Rowland, A. M., Winer, J., and Ferrara, N. (1992). Dual regulation of vascular endothelial growth factor bioavailability by genetic and proteolytic mechanisms. J Biol Chem *267*, 26031–26037.

Imajo, M., Tsuchiya, Y., and Nishida, E. (2006). Regulatory mechanisms and functions of map kinase signaling pathways. IUBMB Life *58*, 312–317.

Iozzo, R. V. (2005). Basement membrane proteoglycans: from cellar to ceiling. Nat Rev Mol Cell Biol *6*, 646–656.

Iozzo, R. V., and San Antonio, J. D. (2001). Heparan sulfate proteoglycans: heavy hitters in the angiogenesis arena. J Clin Invest *108*, 349–355.

Jaye, M., Schlessinger, J., and Dionne, C. A. (1992). Fibroblast growth factor receptor tyrosine kinases: molecular analysis and signal transduction. Biochim Biophys Acta *1135*, 185–199.

Joseph, D. S., Malik, M., Nurudeen, S., and Catherino, W. H. (2010). Myometrial cells undergo fibrotic transformation under the influence of transforming growth factor beta-3. Fertil Steril *93*, 1500–1508.

Juliano, R. L., and Haskill, S. (1993). Signal transduction from the extracellular matrix. J Cell Biol *120*, 577–585.

Kajanne, R., Miettinen, P., Mehlem, A., et al. (2007). EGF-R regulates MMP function in fibroblasts through MAPK and AP-1 pathways. J Cell Physiol *212*, 489–497.

Karkkainen, M. J., and Petrova, T. V. (2000). Vascular endothelial growth factor receptors in the regulation of angiogenesis and lymphangiogenesis. Oncogene *19*, 5598–5605.

Keyt, B. A., Berleau, L. T., Nguyen, H. V., et al. (1996). The carboxyl-terminal domain (111–165) of vascular endothelial growth factor is critical for its mitogenic potency. J Biol Chem *271*, 7788–7795.

Kim, H. R., Upadhyay, S., Li, G., Palmer, K. C., and Deuel, T. F. (1995). Platelet-derived growth factor induces apoptosis in growth-arrested murine fibroblasts. Proc Natl Acad Sci U S A *92*, 9500–9504.

Kim, S. H., Turnbull, J., and Guimond, S. (2011). Extracellular matrix and cell signalling: the dynamic cooperation of integrin, proteoglycan and growth factor receptor. J Endocrinol *209*, 139–151.

Leask, A., and Abraham, D. J. (2004). TGF-beta signaling and the fibrotic response. FASEB J *18*, 816–827.

Lee, J. W., and Juliano, R. (2004). Mitogenic signal transduction by integrin- and growth factor receptor-mediated pathways. Mol Cells *17*, 188–202.

Lee, P. L., Johnson, D. E., Cousens, L. S., Fried, V. A., and Williams, L. T. (1989). Purification and complementary DNA cloning of a receptor for basic fibroblast growth factor. Science *245*, 57–60.

Li, M. O., Wan, Y. Y., Sanjabi, S., Robertson, A. K., and Flavell, R. A. (2006). Transforming growth factor-beta regulation of immune responses. Annu Rev Immunol *24*, 99–146.

Liu, X., Hu, H., and Yin, J. Q. (2006). Therapeutic strategies against TGF-beta signaling pathway in hepatic fibrosis. Liver Int *26*, 8–22.

Loeser, R. F. (2002). Integrins and cell signaling in chondrocytes. Biorheology *39*, 119–124.

Lustig, F., Hoebeke, J., Ostergren-Lunden, G., et al. (1996). Alternative splicing determines the binding of platelet-derived growth factor (PDGF-AA) to glycosaminoglycans. Biochemistry *35*, 12077–12085.

Malarkey, K., Belham, C. M., Paul, A., et al. (1995). The regulation of tyrosine kinase signalling pathways by growth factor and G-protein-coupled receptors. Biochem J *309*, 361–375.

Marampon, F., Bossi, G., Ciccarelli, C., et al. (2009). MEK/ERK inhibitor U0126 affects *in vitro* and *in vivo* growth of embryonal rhabdomyosarcoma. Mol Cancer Ther *8*, 543–551.

Markman, B., Dienstmann, R., and Tabernero, J. (2010). Targeting the PI3K/Akt/mTOR pathway – beyond rapalogs. Oncotarget *1*, 530–543.

Massague, J., and Gomis, R. R. (2006). The logic of TGFbeta signaling. FEBS Lett *580*, 2811–2820.

Markmann, A., Hausser, H., Schonherr, E., and Kresse, H. (2000). Influence of decorin expression on transforming growth factor-beta-mediated collagen gel retraction and biglycan induction. Matrix Biol *19*, 631–636.

Massague, J., Blain, S. W., and Lo, R. S. (2000). TGFbeta signaling in growth control, cancer, and heritable disorders. Cell *103*, 295–309.

Massague, J., Seoane, J., and Wotton, D. (2005). Smad transcription factors. Genes Dev *19*, 2783–2810.

Menke, A., and Adler, G. (2002). TGFbeta-induced fibrogenesis of the pancreas. Int J Gastrointest Cancer *31*, 41–46.

Miyamoto, S., Teramoto, H., Coso, O. A., et al. (1995). Integrin function: molecular hierarchies of cytoskeletal and signaling molecules. J Cell Biol *131*, 791–805.

Mohan, R. R., Tovey, J. C., Gupta, R., Sharma, A., and Tandon, A. (2011). Decorin biology, expression, function and therapy in the cornea. Curr Mol Med *11*, 110–128.

Molloy, T., Wang, Y., and Murrell, G. (2003). The roles of growth factors in tendon and ligament healing. Sports Med *33*, 381–394.

Molteni, A., Modrowski, D., Hott, M., and Marie, P. J. (1999). Alterations of matrix- and cell-associated proteoglycans inhibit osteogenesis and growth response to fibroblast growth factor-2 in cultured rat mandibular condyle and calvaria. Cell Tissue Res *295*, 523–536.

Morabito, A., De Maio, E., Di Maio, M., Normanno, N., and Perrone, F. (2006). Tyrosine kinase inhibitors of vascular endothelial growth factor receptors in clinical trials: current status and future directions. Oncologist *11*, 753–764.

Moscatelli, D. (1987). High and low affinity binding sites for basic fibroblast growth factor on cultured cells: absence of a role for low affinity binding in the stimulation of plasminogen activator production by bovine capillary endothelial cells. J Cell Physiol *131*, 123–130.

Moustakas, A., and Heldin, C. H. (2009). The regulation of TGFbeta signal transduction. Development *136*, 3699–3714.

Mythreye, K., and Blobe, G. C. (2009) Proteoglycan signaling co-receptors: roles in cell adhesion, migration and invasion. Cell Signal *21*, 1548–1558.

Nelson, P. R., Yamamura, S., and Kent, K. C. (1997). Platelet-derived growth factor and extracellular matrix proteins provide a synergistic stimulus for human vascular smooth muscle cell migration. J Vasc Surg *26*, 104–112.

Nikitovic, D., Assouti, M., Sifaki, M., et al. (2008). Chondroitin sulfate and heparan sulfate-containing proteoglycans are both partners and targets of basic fibroblast growth factor-mediated proliferation in human metastatic melanoma cell lines. Int J Biochem Cell Biol *40*, 72–83.

Nikitovic, D., Chalkiadaki, G., Berdiaki, A., et al. (2011). Lumican regulates osteosarcoma cell adhesion by modulating TGFbeta2 activity. Int J Biochem Cell Biol *43*, 928–935.

Nikitovic, D., Zafiropoulos, A., Katonis, P., et al. (2006). Transforming growth factor-beta as a key molecule triggering the expression of versican isoforms v0 and v1, hyaluronan synthase-2 and synthesis of hyaluronan in malignant osteosarcoma cells. IUBMB Life *58*, 47–53.

Noble, N. A., Harper, J. R., and Border, W. A. (1992). *In vivo* interactions of TGF-beta and extracellular matrix. Prog Growth Factor Res *4*, 369–382.

Nutt, J. E., and Lunec, J. (1999). Induction of metalloproteinase (MMP1) expression by epidermal growth factor (EGF) receptor stimulation and serum deprivation in human breast tumour cells. Eur J Cancer *32*, 2127–2135.

Ohno, S., Hirano, S., Kanemaru, S., et al. (2011). Prevention of buccal mucosa scarring with transforming growth factor beta3. Laryngoscope *121*, 1404–1409.

Ornitz, D. M., and Itoh, N. (2001). Fibroblast growth factors. Genome Biol *2*, REVIEWS3005.

Osada, R., Ohshima, H., Ishihara, H., et al. (1996). Autocrine/paracrine mechanism of insulin like growth factor-1 secretion, and the effect of insulin-like growth factor-1 on proteoglycan synthesis in bovine intervertebral discs. J Orthop Res *14*, 690–699.

Ostendorf, T., Eitner, F., and Floege, J. (2012). The PDGF family in renal fibrosis. Pediatr Nephrol *27* (7), 1041–1050.

Padua, D., and Massague, J. (2009). Roles of TGFbeta in metastasis. Cell Res *19*, 89–102.

Park, E. J., Watanabe, Y., Smyth, G., et al. (2008). An FGF autocrine loop initiated in second heart field mesoderm regulates morphogenesis at the arterial pole of the heart. Development *135*, 3599–3610.

Park, J. E., Keller, G. A., and Ferrara, N. (1993). The vascular endothelial growth factor (VEGF) isoforms: differential deposition into the subepithelial extracellular matrix and bioactivity of extracellular matrix-bound VEGF. Mol Biol Cell *4*, 1317–1326.

Park, S., Jung, H. H., Park, Y. H., Ahn, J. S., and Im, Y. H. (2011). ERK/MAPK pathways play critical roles in EGFR ligands-induced MMP1 expression. Biochem Biophys Res Commun *407* (4), 680–686.

Patil, A. S., Sable, R. B., and Kothari, R. M. (2012). Occurrence, biochemical profile of vascular endothelial growth factor (VEGF) isoforms and their functions in endochondral ossification. J Cell Physiol *227* (4), 1298–1308.

Peterson, R. S., Andhare, R. A., Rousche, K. T., et al. (2004). CD44 modulates Smad1 activation in the BMP-7 signaling pathway. J Cell Biol *166*, 1081–1091.

Pohlers, D., Brenmoehl, J., Loffler, I., et al. (2009). TGF-beta and fibrosis in different organs – molecular pathway imprints. Biochim Biophys Acta *1792*, 746–756.

Pollock, R. A., and Richardson, W. D. (1992). The alternative-splice isoforms of the PDGF A-chain differ in their ability to associate with the extracellular matrix and to bind heparin *in vitro*. Growth Factors *7*, 267–277.

Pratilas, C. A., and Solit, D. B. (2010). Targeting the mitogen-activated protein kinase pathway: physiological feedback and drug response. Clin Cancer Res *16*, 3329–3334.

Pratsinis, H., Giannouli, C. C., Zervolea, I., Psarras, S., Stathakos, D., and Kletsas, D. (2004). Differential proliferative response of fetal and adult human skin fibroblasts to transforming growth factor-beta. Wound Repair Regen *12*, 374–383.

Prud'homme, G. J. (2007). Pathobiology of transforming growth factor beta in cancer, fibrosis and immunologic disease, and therapeutic considerations. Lab Invest *87*, 1077–1091.

Raines, E. W., Koyama, H., and Carragher, N. O. (2000). The extracellular matrix dynamically regulates smooth muscle cell responsiveness to PDGF. Ann N Y Acad Sci *902*, 39–51; discussion 51–32.

Reiland, J., Kempf, D., Roy, M., Denkins, Y., and Marchetti, D. (2006). FGF2 binding, signaling, and angiogenesis are modulated by heparanase in metastatic melanoma cells. Neoplasia *8*, 596–606.

Robinson, C. J., and Stringer, S. E. (2001). The splice variants of vascular endothelial growth factor (VEGF) and their receptors. J Cell Sci *114*, 853–865.

Rocha-Lima, C. M., Soares, H. P., Raez, L. E., and Singal, R. (2007). EGFR targeting of solid tumors. Cancer Control *14*, 295–304.

Ross, R. S. (2004). Molecular and mechanical synergy: cross-talk between integrins and growth factor receptors. Cardiovasc Res *63*, 381–390.

Rozario, T., and DeSimone, D. W. (2010). The extracellular matrix in development and morphogenesis: a dynamic view. Dev Biol *341*, 126–140.

Ruhrberg, C., Gerhardt, H., Golding, M., et al. (2002). Spatially restricted patterning cues provided by heparin-binding VEGF-A control blood vessel branching morphogenesis. Genes Dev *16*, 2684–2698.

Ruiz-Ortega, M., Rodriguez-Vita, J., Sanchez-Lopez, E., Carvajal, G., and Egido, J. (2007). TGF-beta signaling in vascular fibrosis. Cardiovasc Res *74*, 196–206.

Saika, S., Saika, S., Liu, C. Y., et al. (2001). TGFbeta2 in corneal morphogenesis during mouse embryonic development. Dev Biol *240*, 419–432.

Sato, T., Iwai, M., Sakai, T., et al. (1999). Enhancement of membrane-type 1-matrix metalloproteinase (MT1-MMP) production and sequential activation of progelatinase A on human squamous carcinoma cells co-cultured with human dermal fibroblasts. Br J Cancer *80* (8), 1137–1143.

Sawaji, Y., Hynes, J., Vincent, T., and Saklatvala, J. (2008). Fibroblast growth factor 2 inhibits induction of aggrecanase activity in human articular cartilage. Arthritis Rheum *58*, 3498–3509.

Scaffidi, A. K., Petrovic, N., Moodley, Y. P., et al. (2004). Alpha(v)beta(3) integrin interacts with the transforming growth factor beta (TGFbeta) type II receptor to potentiate the proliferative effects of TGFbeta1 in living human lung fibroblasts. J Biol Chem *279*, 37726–37733.

Schaefer, L., and Iozzo, R. V. (2008). Biological functions of the small leucine-rich proteoglycans: from genetics to signal transduction. J Biol Chem *283*, 21305–21309.

Schaller, M.D. (2001). Biochemical signals and biological responses elicited by the focal adhesion kinase. Biochim Biophys Acta *1540*, 1–21.

Schmidt, M. B., Chen, E. H., and Lynch, S. E. (2006). A review of the effects of insulin-like growth factor and platelet derived growth factor on *in vivo* cartilage healing and repair. Osteoarthritis Cartilage *14*, 403–412.

Schneller, M., Vuori, K., and Ruoslahti, E. (1997). Alphavbeta3 integrin associates with activated insulin and PDGFbeta receptors and potentiates the biological activity of PDGF. EMBO J *16*, 5600–5607.

Schonherr, E., and Hausser, H. J. (2000). Extracellular matrix and cytokines: a functional unit. Dev Immunol *7*, 89–101.

Senior, R. M., Griffin, G. L., Huang, J. S., Walz, D. A., and Deuel, T. F. (1983). Chemotactic activity of platelet alpha granule proteins for fibroblasts. J Cell Biol *96*, 382–385.

Shah, M., Foreman, D. M., and Ferguson, M. W. (1995). Neutralisation of TGF-beta 1 and TGF-beta 2 or exogenous addition of TGF-beta 3 to cutaneous rat wounds reduces scarring. J Cell Sci *108*, 985–1002.

Shi, Y., and Massague, J. (2003). Mechanisms of TGF-beta signaling from cell membrane to the nucleus. Cell *113*, 685–700.

Shibuya, M., and Claesson-Welsh, L. (2006). Signal transduction by VEGF receptors in regulation of angiogenesis and lymphangiogenesis. Exp Cell Res *312*, 549–560.

Siegel, P. M., and Massague, J. (2003). Cytostatic and apoptotic actions of TGF-beta in homeostasis and cancer. Nat Rev Cancer *3*, 807–821.

Slack-Davis, J. K., Eblen, S. T., Zecevic, M., et al. (2003). PAK1 phosphorylation of MEK1 regulates fibronectin-stimulated MAPK activation. J Cell Biol *162*, 281–291.

Smith, S. M., West, L. A., Govindraj, P., Zhang, X., Ornitz, D. M., and Hassell, J. R. (2007). Heparan and chondroitin sulfate on growth plate perlecan mediate binding and delivery of FGF-2 to FGF receptors. Matrix Biol *26*, 175–184.

Syrokou, A., Tzanakakis, G. N., Hjerpe, A., and Karamanos, N. K. (1999). Proteoglycans in human malignant mesothelioma. Stimulation of their synthesis induced by epidermal, insulin and platelet-derived growth factors involves receptors with tyrosine kinase activity. Biochimie *81*, 733–744.

Takeuchi, Y., Kodama, Y., and Matsumoto, T. (1994). Bone matrix decorin binds transforming growth factor-beta and enhances its bioactivity. J Biol Chem *269*, 32634–32638.

Tammi, R. H., Passi, A. G., Rilla, K., et al. (2011). Transcriptional and post-translational regulation of hyaluronan synthesis. FEBS J *278* (9), 1419–1428.

ten Dijke, P., and Arthur, H. M. (2007). Extracellular control of TGFbeta signalling in vascular development and disease. Nat Rev Mol Cell Biol *8*, 857–869.

Tian, Y. C., Chen, Y. C., Chang, C. T., et al. (2007). Epidermal growth factor and transforming growth factor beta1 enhance HK-2 cell migration through a synergistic increase of matrix metalloproteinase and sustained activation of ERK signaling pathway. Exp Cell Res *313*, 2367–2377.

Tran, K. T., Lamb, P., and Deng, J. S. (2005). Matrikines and matricryptins: implications for cutaneous cancers and skin repair. J Dermatol Sci *40*, 11–20.

Tzanakakis, G. N., Hjerpe, A., and Karamanos, N. K. (1997). Proteoglycan synthesis induced by transforming and basic fibroblast growth factors in human malignant mesothelioma is mediated through specific receptors and the tyrosine kinase intracellular pathway. Biochimie *79*, 323–332.

Tzanakakis, G. N., Karamanos, N. K., and Hjerpe, A. (1995a). Effects on glycosaminoglycan synthesis in cultured human mesothelioma cells of transforming, epidermal, and fibroblast growth factors and their combinations with platelet-derived growth factor. Exp Cell Res *220*, 130–137.

Tzanakakis, G. N., Karamanos, N. K., Klominek, J., and Hjerpe, A. (1995b). Glycosaminoglycans from two human malignant mesothelioma cell lines: determination, distribution, and effect of platelet-derived growth factor on their synthesis. Biochem Cell Biol *73*, 59–66.

Tzanakakis, G. N., Karamanos, N. K., Syrokou, A., and Hjerpe, A. (1996). Effect of insulin and epidermal growth factors on the synthesis of glycosaminoglycans/proteoglycans in cultured human malignant mesothelioma cells of different phenotypic morphology. APMIS *104*, 718–728.

Tzeng, D. Y., Deuel, T. F., Huang, J. S., and Baehner, R. L. (1985). Platelet-derived growth factor promotes human peripheral monocyte activation. Blood *66*, 179–183.

Upton, Z., Cuttle, L., Noble, A., et al. (2008). Vitronectin: growth factor complexes hold potential as a wound therapy approach. J Invest Dermatol *128*, 1535–1544.

Vachani, A., Moon, E., Wakeam, E., and Albelda, S. M. (2010). Gene therapy for mesothelioma and lung cancer. Am J Respir Cell Mol Biol *42*, 385–393.

Van Meter, T. E., Broaddus, W. C., Rooprai, H. K., Pilkington, G. J., and Fillmore, H. L. (2004). Induction of membrane-type-1 matrix metalloproteinase by epidermal growth factor-mediated signaling in gliomas. Neuro Oncol *6* (3), 188–199.

Vempati, P., Mac Gabhann, F., and Popel, A. S. (2010). Quantifying the proteolytic release of extracellular matrix-sequestered VEGF with a computational model. PLoS One *5*, e11860.

Wabnitz, P. A., Mitchell, D., and Wabnitz, D. A. (2004). *In vitro* and *in vivo* metabolism of the anti-cancer agent CI-1040, a MEK inhibitor, in rat, monkey, and human. Pharm Res *21*, 1670–1679.

Wagsater, D., Zhu, C., Bjorck, H. M., and Eriksson, P. (2009). Effects of PDGF-C and PDGF-D on monocyte migration and MMP-2 and MMP-9 expression. Atherosclerosis *202*, 415–423.

Wan, Y. Y., and Flavell, R. A. (2007). "Yin-Yang" functions of transforming growth factor-beta and T regulatory cells in immune regulation. Immunol Rev *220*, 199–213.

Wenner, C. E., and Yan, S. (2003). Biphasic role of TGF-beta1 in signal transduction and crosstalk. J Cell Physiol *196*, 42–50.

Wennerberg, K., and Der, C. J. (2004). Rho-family GTPases: it's not only Rac and Rho (and I like it). J Cell Sci *117*, 1301–1312.

Westermarck, J., and Kahari, V. M. (1999). Regulation of matrix metalloproteinase expression in tumor invasion. FASEB J *13*, 781–792.

Wharton, K., and Derynck, R. (2009). TGFbeta family signaling: novel insights in development and disease. Development *136*, 3691–3697.

Whitmarsh, A. J., and Davis, R. J. (1996). Transcription factor ap-1 regulation by mitogen-activated protein kinase signal transduction pathways. J Mol Med (Berl) *74*, 589–607.

Wilhelm, S. M., Adnane, L., Newell, P., Villanueva, A., Llovet, J. M., and Lynch, M. (2008). Preclinical overview of sorafenib, a multikinase inhibitor that targets both raf and VEGF and PDGF receptor tyrosine kinase signaling. Mol Cancer Ther *7*, 3129–3140.

Wilkins-Port, C. E., and Higgins, P. J. (2007). Regulation of extracellular matrix remodeling following transforming growth factor-beta1/epidermal growth factor-stimulated epithelial-mesenchymal transition in human premalignant keratinocytes. Cells Tissues Organs *185*, 116–122.

Wright, J. A., Turley, E. A., and Greenberg, A. H. (1993). Transforming growth factor beta and fibroblast growth factor as promoters of tumor progression to malignancy. Crit Rev Oncog *4*, 473–492.

Wu, M. Y., and Hill, C. S. (2009). TGF-beta superfamily signaling in embryonic development and homeostasis. Dev Cell *16*, 329–343.

Yamanaka, I., Koizumi, M., Baba, T., Yamashita, S., Suzuki, T., and Kudo, R. (2003). Epidermal growth factor increased the expression of alpha2beta1 integrin and modulated integrin mediated signaling in human cervical adenocarcinoma cells. Exp Cell Res *286*, 165–74.

Yarden, Y., and Sliwkowski, M. X. (2001). Untangling the ErbB signalling network. Nat Rev Mol Cell Biol *2*, 127–37.

Yarwood, J. S, and Woodgett, R. J. (2001). Extracellular matrix composition determines the transcriptional response to epidermal growth factor receptor activation. Proc Natl Acad Sci U S A *98* (8), 4472–4477.

Yu, Q., and Stamenkovic, I. (2000). Cell surface-localized matrix metalloproteinase-9 proteolytically activates TGF-beta and promotes tumor invasion and angiogenesis. Genes Dev *14*, 163–176.

Yu, X., Miyamoto, S., and Mekada, E. (2000). Integrin alpha 2 beta 1-dependent EGF receptor activation at cell-cell contact sites. J Cell Sci *113*, 2139–2147.

Zaidel-Bar, R., Itzkovitz, S., Ma'ayan, A., Iyengar, R., and Geiger, B. (2007). Functional atlas of the integrin adhesome. Nat Cell Biol *9*, 858–867.

Zheng, Z., Nguyen, C., Zhang, X., et al. (2011). Delayed wound closure in fibromodulin-deficient mice is associated with increased TGF-beta3 signaling. J Invest Dermatol *131*, 769–778.

8.4 Targeting protein-glycan interactions at cell surface during EMT and hematogenous metastasis: consequences on tumor invasion and metastasis

Mauro S. G. Pavao, Eliene O. Kozlowski, and Felipe C.O.B. Teixeira

8.4.1 Introduction

Tumor cells that acquire the ability to migrate and colonize distant tissues are the most dangerous manifestation of cancer. Ninety percent of the cancer patients are affected by tumor growth at sites different from that where the primary tumor was first discovered (Chaffer and Weinberg, 2011). This behavior is known as metastatic disease and is the leading cause of death in these patients.

The ability of cancer cells to detach from the primary tumor and invade the adjacent stroma, enter the vascular system and travel in the blood and/or lymph, and extravasate and grow at distant tissues requires a substantial phenotypical change, which involves a multitude of cellular and biochemical events as a result of profound genetic rearrangement (Mani et al., 2008; Morel et al., 2008; Thiery et al., 2009). Informational molecules at the tumor microenvironment (stroma and/or extracellular matrix), synthesized by stroma cells or infiltrating cells recruited by the tumor, such as inflammatory cells and macrophages (Yang and Weinberg, 2008), start this complex process. After the binding of these molecules to their cognate cell surface receptors on tumor cells, the extracellular signal follows an intracellular pathway to the nucleus where different transcription factors modulate gene expression, producing specific phenotypical changes that capacitate tumor cells to colonize distant tissues (▶Figure 8.11) (Chaffer and Weinberg, 2011).

The great majority of cancers (~80%) with high risk of death are of epithelial origin (carcinomas). The carcinomas begin at the epithelial site of the basement membrane (BM). As the malignant disease progress, tumor cells acquire the ability to penetrate the BM, individually or in a group, and invade the adjacent stroma. Eventually, these cells reach the vascular system (blood and/or lymphatic vessels), gaining access to nutrients and oxygen carried by the blood. Frequently, however, tumor cells are able to stimulate angiogenesis at the stroma side of the BM by releasing angiogenic growth factors (Hanahan and Folkman, 1996; Bergers and Benjamin, 2003; Baeriswyl and Christofori, 2009). Once inside the vessels, tumor cells travel through the blood or lymph and colonize distant areas of the body.

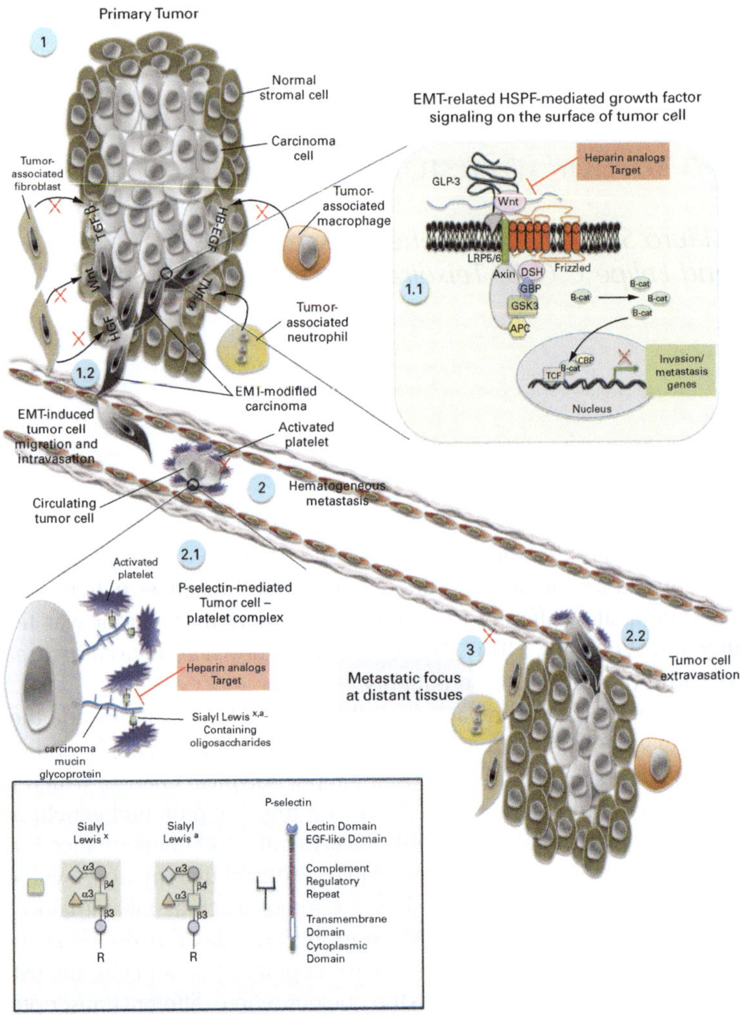

Figure 8.11 Cellular and molecular events during tumor invasion and metastasis: targeting glycan-protein interaction during tumor dissemination using heparin analogs. Tumor dissemination is a complex biological phenomenon, involving the escape of tumor cell from the primary tumor (1), migration to the adjacent stroma and intravasation in the vascular system (1.2), travel in the blood (2), extravasation from the vasculature (2.2), and growth at distant tissues (3). Two critical steps are required for the successful journey of metastatic tumor cells, named epithelial-to-mesenchymal transition (EMT), which occurs in the primary tumor (1), and the hematogenous metastasis, which occurs in the vascular system (2). Stromal growth factors and cell sulfate heparan sulfate proteoglycans (HSPGs) are key mediators of EMT (1), whereas tumor cell surface glycans and P-selectin on activated platelets are essential for hematogenous metastasis (2). Cell surface HSPGs, such as glypicans, act as coreceptors for the binding of several growth factors to their cognate tyrosine kinase receptors involved in EMT. For example, HSPG-mediated binding of Wnt proteins to Frizzled, its receptor on tumor cells, signals by the "canonical" pathway, which regulates the amount of cytoplasmic β-catenin. Intracellular accumulation of β-catenin results in its translocation to the nucleus,

8.4.2 Tumor invasion and metastasis

Two critical moments of carcinoma invasion and metastasis are the epithelial to mesenchymal transition and hematogenous metastasis. These two extraordinary biological phenomenon observed in cancer progression will be the subject of the present review.

Epithelial-to-mesenchymal transition

The first and crucial step of invasion and metastasis experienced by carcinoma cells in the primary tumor is the acquisition of a mesenchyme phenotype. This drastic change, named epithelial-to-mesenchymal transition (EMT) (Mani et al., 2008; Morel et al., 2008; Thiery et al., 2009), normally occurs at the invasive front of the tumor and involves the expression of mesenchyme proteins (vimentin, fibronectin, N-cadherin, and αvβ6 integrin) by the carcinoma cells that previously expressed epithelial proteins (cytokeratin and E-cadherin) (▶Figure 8.12A and B, C, and D) (Brabletz et al., 2001; Klymkowsky and Savagner, 2009; Polyak and Weinberg, 2009; Yilmaz and Christofori, 2009). As a result, cells assume a fibroblast-like form, losing their epithelial shape, obtaining the ability to migrate (▶Figure 8.12E and F). Additionally, upon EMT, the cells acquire the self-renewing trait associated with cancer stem cells (Mani et al., 2008; Morel et al., 2008).

In fact, EMT is a hidden morphogenetic program, normally observed during early embryogenesis (gastrulation – epiblast to mesoderm, somitogenesis – somite walls to sclerotome) (Barrallo-Gimeno and Nieto, 2005) and adult wound healing (Martin and Parkhurst, 2004; Nakamura and Tokura, 2011), and is pathologically activated during cancer progression. Several stroma-derived factors have been implicated in EMT (▶Figure 8.11) (Kalluri and Weinberg, 2009). Among them are the following: those produced by fibroblasts, including fibroblast growth factors (FGFs) (Strutz et al., 2002), hepatocyte growth factor (HGF) (Ding et al., 2010; Vergara et al., 2010; Ogunwobi and Liu, 2011; Toiyama et al., 2012), insulin-like growth factor-1 (IGF1) (Graham et al., 2008; Sivakumar et al., 2009; Lee et al., 2010; Natsuizaka et al., 2010; Walsh and

---

Figure 8.11 ***(Continued)***

where, upon binding to nuclear T-cell factor/lymphoid enhancing factor (TCF/LEF), it modulates the expression of a broad set of genes involved in tumor progression (1.1). During hematogenous metastasis (2), tumor cells form complexes with activated platelets (2.1). The interaction is mediated by tumor cell surface sialyl-Lewis^{x,a}-containing oligosaccharides and P-selectin on the surface of activated platelets (2.1). It has been proposed that this microemboli of tumor cells with platelets and other host cells is a mechanism that allows tumor cells to evade the immune defenses and eventually colonize distant organs, forming metastatic foci (3). The ability of heparin and heparin analogs to inhibit the binding of EMT-related growth factors to their receptors on tumor cells (red X in 1.1) and the interaction between tumor mucin-glycans and P-selectin on activated platelets (red X 2.2) illustrates a valuable example of how one drug can target two important steps of tumor invasion and metastasis. In fact, targeting EMT-signaling pathways both at the receptor and intracellular levels and the formation of tumor cell-platelet complex during hematogenous metastasis have been approached separately in several scientific investigations. However, no study has been conducted using heparin or heparin analogs to target both EMT-related signaling and hematogenous metastasis. ◇, *N*-acetylneuraminic acid; □, *N*-acetylglucosamine; O, galactose; Δ, fucose

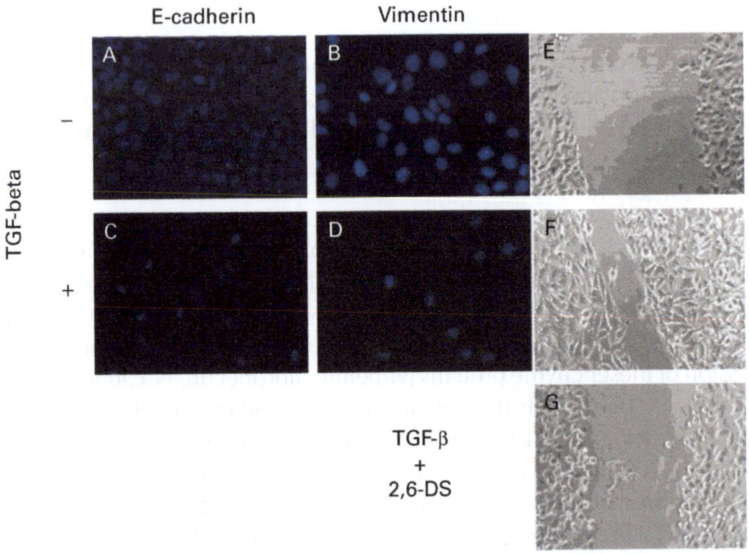

Figure 8.12 Cellular and functional features of EMT. The transition from an epithelial to a mesenchymal phenotype, observed upon EMT, involves the expression of mesenchymal proteins (vimentin) by the carcinoma cells that previously expressed epithelial protein (E-cadherin) (A–D). As a result, the cells assume a fibroblast-like form, losing their epithelial shape and acquiring the ability to migrate (E–G). In preliminary experiments from our laboratory, transforming growth factor-beta (TGF-β)–induced EMT in murine breast carcinoma M3 cells, as indicated by the loss of epithelial surface E-cadherin and the appearance of mesenchymal vimentin on the cells. (A–D) This results in acquisition of a mesenchymal phenotype and migration (E, F). 2,6-D-sulfated dermatan drastically attenuates cell migration, indicating that EMT was inhibited (G).

Damjanovski, 2011), and Wnts (Hosonaga et al., 2011; Howard et al., 2011; Ito et al., 2011; Jing et al., 2011; Okanami et al., 2011; Scheel et al., 2011); those produced by macrophages, including epidermal growth factor (EGF) (Dasgupta et al., 2010; Qian and Pollard, 2010; Wang and Xu, 2010; Xu et al., 2010; Alipio et al., 2011; Bocca et al., 2011; Richter et al., 2011; Yue et al., 2011; Zuo et al., 2011); those produced by inflammatory cells, including tumor necrosis factor-alpha (TNF-α) (Chuang et al., 2008; Takahashi et al., 2010; Asiedu et al., 2011); and those produced by myofibroblasts and epithelial cells, including transforming growth factor-beta (TGF-β) (Ikushima and Miyazono, 2010; Laberge et al., 2011; Shin et al., 2011).

Different from what is observed in cancer, EMT-related events involved in the transition of epithelial to mesenchymal cells during embryonic development are highly specific and coordinated. The sequence of steps initiates with the loss of apical-basal polarity resulted from the disarrangement of various types of cell-cell junctions (tight, adherens, and gap junctions) and degradation of the basement membrane (Peinado et al., 2004; Townsend et al., 2008; Micalizzi et al., 2010). Cell surface proteins involved in stable cell-cell and cell matrix interactions, such as E-cadherin and integrins are then substituted by N-cadherins and integrins with transient interaction characteristics (Micalizzi et al., 2010). This is followed by cytoskeleton changes where actin and cytokeratin are replaced by stress fibers and vimentin, respectively. Overall, these chances result in a spindle-like form cell with the ability to migrating throughout the extracellular matrix and resists to anoikis (Chaffer et al., 2007).

Cancer EMT does not follow a controlled morphogenetic program observed in embryonic EMT. In addition to nonanticipating genomic changes, tumor cells are influenced by an abnormal tumor environment, which can result in an incomplete EMT program. For example, in sarcomatoid tumor of the breast, the cells still express epithelial markers, such as keratin, despite a clear mesenchyme phenotype (Chaffer et al., 2007).

Heparin-binding growth factors in EMT

HGF/scatter factor

HGF/scatter factor (SF) was the first extracellular growth factor identified as an EMT inductor factor (Montesano et al., 1991a, 1991b). EMT can also be induced by TGF-β1, FGF, heparin-binding EGF (HB-EGF), and Wnt as mentioned earlier and reviewed in (Micalizzi et al., 2010), as well as by aberrant activation of nonreceptor protein kinases (PK), such as Src (Birchmeier et al., 2003; Avizienyte et al., 2004; Wang et al., 2008). Binding of HGF/SF and activation of its tyrosine kinase receptor, MET, modulates the expression of genes involved in invasion, protection from apoptosis and metastasis (Corso et al., 2005). EMT-mediated tumorigenesis induced by HGF/Met signaling is associated with the progression of several types of cancers and higher potential of tumor invasion and metastasis. In fact, the HGF/SF-MET signaling pathway is currently considered being a potential target in cancer therapy (Yap et al., 2011).

HGF/SF is a unique heparin-binding growth factor that binds to both heparan sulfate (Lyon et al., 1994, 2004; Rahmoune et al., 1998) and dermatan sulfate (Lyon et al., 1998, 2004; Bechard et al., 2001). Cell surface proteoglycans are required for the formation of active HGF/SF-MET signaling complexes. Several studies suggest that HGF/SF, the glycosaminoglycans, and MET form an active ternary complex in which sulfated glycosaminoglycan domains bring both ligand and receptor in proximity (Lietha et al., 2001; Lyon et al., 2002; Gherardi et al., 2003). Moreover, it is believed that proteoglycans have pivotal roles in localizing the signaling complex to the basolateral surface of cells restricting diffusion and protecting HGF/SF from proteolysis (Deakin and Lyon, 1999; Sergeant et al., 2000).

HB-EGF

HB-EGF is a heparin-binding member of the EGF family, which was first identified in the conditioned medium of human macrophages and is known as a potent mitogen and chemotactic factor for fibroblasts and smooth muscle cells (Higashiyama et al., 1991, 1993; Blotnick et al., 1994). Like others members of EGF family, HB-EGF binds and activates EGF receptors (erbBs) (Iwamoto and Mekada, 2000), stimulating the growth of several cells in an autocrine or paracrine manner and affecting stroma proliferation (Hisaka et al., 1999). HB-EGF is synthesized as a transmembrane protein, pro–HB-EGF, which is cleaved on the cell surface to yield a soluble mature growth factor (Goishi et al., 1995).

Recent studies have pointed to the involvement HB-EGF on tumor-related epithelial to mesenchymal transition. HB-EGF expression is altered in a number of cancer types including bladder (Adam et al., 2003a, 2003b; Ito et al., 2001a), prostate (Freeman et al., 1998), hepatic (Ito et al., 2001b), ovarian (Tanaka et al., 2005), and pancreatic (Ito et al., 2001). Although HB-EGF is typically expressed in epithelial-like cells in cancer, a robust expression of this growth factor has also been reported in the stroma (Freeman et al., 1998) and endothelium (Nolan-Stevaux et al., 2010) of organs, contributing with paracrine secretion to tumor cells. Little is known about the mechanisms of HB-EGF-induced EMT in carcinoma cells, but the role of HB-EGF in such transdifferentiation process is increasingly evident. Thus, HB-EGF might be a valuable target for tumor suppression. CRM197 (a mutated diphtheria toxin), a specific inhibitor of HB-EGF, which is part of clinical research studies in this field, was shown to prevent the properties associated with cell adhesion, invasion and angiogenesis and reduced the tumor burden in nude mice (Tanaka et al., 2005; Yotsumoto et al., 2008). New clinical studies are being performed currently to validate the therapeutic value of HB-EGF inhibitors in cancer patients. Since the effect of HB-EGF is dependent on ectodomain shedding by matrix metalloproteases, antimatrix metalloproteases might be also a relevant treatment targeting HB-EGF effects in cancer progression.

Fibroblast growth factor-2

Fibroblast growth factor-2 (FGF2), also referred to as basic FGF, is a member of the family of mitogen growth factors, which affects proliferation of mesenchymal or epithelial cells. FGFs, as FGF2 or FGF4, are key regulators of EMT during the development of mammalians or tumor progression (Strutz et al., 2002; Strutz and Neilson, 2003). These growth factors trigger cell signaling through binding to their cognate receptors (FGFRs). The binding and activation of FGFRs occurs in a heparan sulfate-dependent manner. FGFs control a multitude of cellular processes in different contexts, including proliferation, differentiation, survival and motility. FGF2 is relevant in angiogenesis and playing a critical role in cancer development (Dow and deVere White, 2000). Human melanomas express FGF2 and FGFRs simultaneously and their growth is stimulated in an autocrine manner (Yeoman, 1993). The inactivation of FGF2 by using antisense oligonucleotides or anti-FGF2 antibodies resulted in inhibition of melanoma cell proliferation *in vivo* and *in vitro* (Becker et al., 1989; Wang and Becker, 1997).

Wnts

Wnts are secreted lipid-modified proteins that act as signaling molecules involved in various biological processes during development and in cancer (Wodarz and Nusse, 1998). The Wnt signaling pathway is highly associated with tumor EMT-mediated invasion and metastasis (Scheel et al., 2011). Upon binding to the Frizzled family of receptors, Wnts signal by the "canonical" Wnt pathway, named β-catenin/Armadillo-dependent pathway, which regulates the amount of cytoplasmic β-catenin. Intracellular accumulation of β-catenin results in its translocation to the nucleus where, upon binding to nuclear T-cell factor/lymphoid enhancing factor (TCF/LEF), it modulates the expression of a broad set of genes involved in tumor progression (▶Figure 8.11, 1.1). In the absence of Wnt, β-catenin is phosphorylated by a destruction complex, leading to an ubiquitin-mediated proteolytic degradation. As a result, the prospective target genes are repressed. Wnt-β-catenin signaling is present in nearly all colorectal tumors (Huang and Du, 2008).

Wnt also signals via "noncanonical" pathways, the planar cell polarity (PCP) and Wnt/Ca^{2+} pathways (Veeman et al., 2003). In the Wnt/Ca^{2+} pathway, binding of Wnt to its receptor Frizzled, results in phospholipase C activation via G protein, leading to an increase in intracellular Ca^{+2} and activation of protein kinase C. Induction of this signaling pathway is involved in metastatic spread of melanoma cells via induction of EMT (Wang, 2009).

It has been shown that cell surface heparan sulfate proteoglycans (HSPGs) facilitate both the canonical and noncanonical Wnt signaling pathways (Capurro et al., 2005; O'Connell et al., 2009).

Cell surface heparan sulfate proteoglycans and growth factor signaling in cancer

Cell surface HSPGs act as coreceptors for the binding of several growth factors to their cognate tyrosine kinase receptors, involved in EMT and tumor dissemination. Two families of HSPGs carry the majority of the heparan sulfate on cells: glypicans, which are attached to the plasma membrane via glycosylphosphatidylinositol (GPI) anchors, and syndecans, which are transmembrane proteins (▶Figure 8.13) reviewed in Bishop et al. (2007). In mammalian cells, four syndecans with core protein masses varying from 31–45 kDa and five glypicans with core protein masses varying from 57–69 kDa, all encoded by separate genes, have been described to date (Bishop et al., 2007).

Glypicans

Glypican-1 (GLC1) is strongly expressed in human pancreatic cancer, occurring in both cancer cells and the adjacent fibroblasts. In pancreatic cancer cells, GLC1 is involved in the mitogenic responses of heparin-binding growth factors that are commonly overexpressed in pancreatic cancer, such as FGF2 and HB-EGF (Kleeff et al., 1998), as well as in responses mediated by members of the TGF-β family (Kayed et al., 2006). Enhanced expression of TGF-β and its type I and type II receptors (TGF-βRI and TGF-βRII) correlates

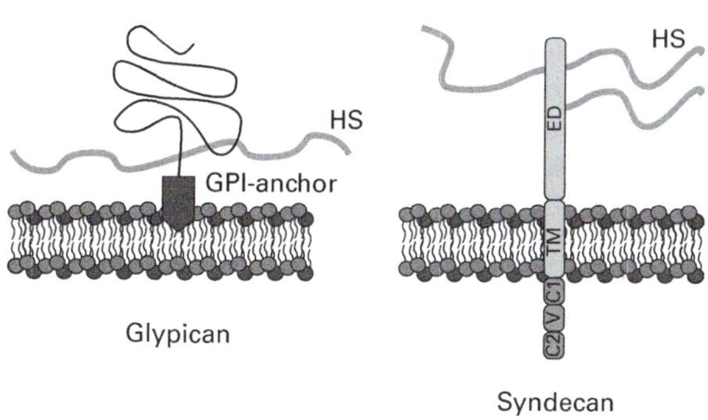

Figure 8.13 Cell surface heparan sulfate proteoglycans. Two families of HSPGs carry the majority of the heparan sulfate on cells: glypicans, which are attached to the plasma membrane via glycosylphosphatidylinositol anchors, and syndecans, which are transmembrane proteins. In mammalian cells, four syndecans with core protein masses varying from 31–45 kDa and five glypicans with core protein masses varying from 57–69 kDa, all encoded by separate genes, have been described to date.

with decreased survival and advanced stages in pancreatic ductal adenocarcinoma (Friess et al., 1993; Lu et al., 1997; Wagner et al., 1999). It has been shown that downregulation of GLC1 in pancreatic cancer results in an altered growth response mediated by TGF-β1, activin-A and bone morphogenetic protein-2 (BMP2) (Li et al., 2004; Kayed et al., 2006).

In addition to affecting the mitogenic effects of TGF-β1, GLC1 is also involved in the metastatic potential of human pancreatic tumor cell lines. Athymic mice lacking GLC1 exhibits decreased metastasis of PANC1 or T3M4 human pancreatic cancer cells and fewer pulmonary metastases following intravenous injection of murine B16-F10 melanoma cells (Aikawa et al., 2008). Moreover, GLC1 modulates the metastatic potential of human and mouse cancer cells. Overall, these data indicate that GLC1 is required for efficient TGF-β1 signaling in pancreatic cancer cells, involved in tumor growth and metastasis.

GLC1 also contributes for the progression of human breast cancer. It has been shown that GLC1 is strongly expressed in human breast cancer. Removal of GLC1 from MDA-MB-231 and MDA-MB-468 breast cancer cells by phosphoinositide-specific phospholipase-C attenuates the mitogenic response to HB-EGF and FGF2, indicating a role of GLC1 in the disease progression and malignancy (Matsuda et al., 2001).

Currently, there are no literature data implicating GLC1 in cancer EMT. However, the fact that it binds to several EMT-related heparin-binding growth factors, and that the lack of GLC1 in human breast cancer cell lines attenuates metastasis, strongly suggests its implication in signaling pathways during acquisition of a mesenchymal phenotype by carcinoma cells. In fact, a recent study showed that GLC1 is required for proper regulation of the canonical Wnt signaling, an important EMT-related factor during differentiation and organization of trigeminal placodal cells into ganglia (Shiau et al., 2010).

Glypican-3 (GLC3) is an important component of the Wnt signaling cell-surface complex that occur during embryonic development and in a few human cancers (Filmus, 2001; Filmus et al., 2008). In the Wnt signaling pathway, GLC3 facilitates the binding of Wnt to its cell-surface receptor, Frizzled, with the consequent increasing on the signal. GLC3 binds to both Wnt and to Frizzled, and the HS chains are partially required for the growth factor binding (▶Figure 8.11, 1.1) (Capurro et al., 2008).

Upregulation of GLC3 is observed in the majority of liver and colon cancers, but not in normal adult tissues (Filmus, 2001; Suzuki et al., 2010). In hepatocellular carcinoma (HCC), GLC3 is overexpressed and stimulates Wnt signaling, leading to tumor progression (Filmus and Capurro, 2008). A soluble form of GLC3 (sGLC3) is able to block Wnt signaling and inhibit the responses of Wnt-dependent events in HCC. In addition, the heparan sulfate chains of soluble GLC3 can also prevent the binding of other heparin-binding growth factors to their cognate receptors, attenuating tumor growth and invasion (Zittermann et al., 2010; Feng et al., 2011).

Recently, it has been reported that GLC3 is not only a diagnostic and prognostic marker in HCC, but can also be an ideal target for the therapy of HCC (Zou et al., 2010). In consonance with this report, an anti-GLC3 monoclonal antibody is currently undergoing clinical evaluation in patients with HCC. There is also recent evidence showing that soluble GLC3 may be a valuable serum biomarker for HCC (Ho and Kim, 2011). More recently, GLC3 expression was evaluated in 52 patients with clear cell carcinoma of the ovary. GLC3 overexpression was observed in about 42% of the patients and was associated with a low-level proliferation status of tumors and with a resistance to taxane-based chemotherapy (Umezu et al., 2010).

Glypican-5 (GLC5) has been shown to stimulate the proliferation of rhabdomyosarcoma cells (Williamson et al., 2007). The stimulation activity of GLC5 results from its ability to promote Hedgehog (Hh) signaling by increasing the binding of Sonic Hh to its receptor, patched-1, sonic hedgehog cell surface receptor (Ptc1). It has been demonstrated that GLC5 binds to both Hh and Ptc1 through its highly sulfated glycosaminoglycan chains. Therefore, it is proposed that GLC5 stimulates Hh signaling by facilitating/stabilizing the interaction between Hh and Ptc1 (Li et al., 2011).

Syndecans

Syndecan-1 (SDC1) is expressed predominantly in epithelial cells and binds specifically to FGF2 and its cognate receptor (FGFR1), forming the tertiary complex FGF2-SDC1-FGFR1 (Pellegrini et al., 2000; Schlessinger et al., 2000). After the binding of FGF and HS chains, FGFRs dimerize, enabling cytoplasmic kinase domains to autotransphosphorylate tyrosines on loop A, and become activated (Eswarakumar et al., 2005). In breast carcinoma cells, SDC1 regulates the formation of FGF2-FGFR1 complex formation, regulating the mitogen response of the growth factor during tumor progression (Mundhenke et al., 2002).

Genetic and biochemical studies indicate that SDC1 modulates Wnt signaling and is critical for Wnt1-mediated tumorigenesis of the mouse mammary gland (Alexander et al., 2000). This is in consonance with the fact that mammary tumor development is inhibited when *SDC1* gene is silenced (Liu et al., 2003). Mammary gland development requires syndecan-1 to create a beta-catenin/TCF-responsive mammary epithelial subpopulation. Additionally, it has been shown that Wnt signaling potentiates the stem cell activity of mouse mammary glands *in vivo*. Interestingly, this activity in drastically reduced in SDC1-deficient tumor-resistant mice mammary cells (Liu et al., 2004). Overall, these data suggest a possible linking between SDC1-mediated Wnt signaling and origin of cancer stem cell (CSC). In support of this hypothesis, recent studies demonstrated that characteristics of CSC could be acquired with the EMT program (Mani et al., 2008; Morel et al., 2008; Singh and Settleman, 2010), enabling cancer cells not only to propagate from the primary tumor but also to confer self-renewal capability required to clonal expansion at metastatic sites (Brabletz et al., 2005).

Increased levels of SDC1 and SDC4, associated with Wnt5A signaling have also been shown to correlate with increasing metastatic potential of melanoma cells. Accordingly, knockdown of SDC1 or 4 decreases cell invasion, indicating that these HSPGs are important components of the Wnt5A autocrine signaling loop and correlate to an increased metastatic potential of melanomas (O'Connell et al., 2009).

Hematogenous metastasis during tumor dissemination

The surface of carcinoma cells shows altered glycosylation patterns (Kim et al., 1996; Kim and Varki, 1997), expressing highly branched or sialylated oligosaccharides, especially fucosylated glycans containing sialyl-Lewis^x and sialyl-Lewis^a. The presence of these oligosaccharides in tumor cells is directly associated with a poor prognosis because of tumor progression and metastatic spread (Dennis and Laferte, 1987;

Hakomori, 1996; Kim and Varki, 1997). The sialyl-Lewis^{x}-oligosaccharides from the carcinoma cells are ligands of the three members of the selectin family of cell adhesion molecules. E-, P-, and L-selectins are vascular receptors for certain normal glycoproteins that contain sialyl-Lewis^{x,a} found on leucocytes and endothelium (Kansas, 1996). Therefore, by mediating the interactions of tumor cells with platelets and endothelium, selectins participate in the metastatic spread of tumor (Laubli and Borsig, 2010) (▶Figure 8.11, 2.2).

Hematogenous metastasis of tumor cells is a cascade of events involving the access of tumor cells into the bloodstream, evasion of innate immune surveillance, adhesion to vascular endothelium of distant organs with extravasation and colonization of tissues. During this process, tumor cells form complexes with platelets and leucocytes. It has been proposed that this microemboli of tumor cells with platelets and other host cells is a mechanism that allows tumor cells to evade the immune defenses and eventually colonize distant organs, forming metastatic foci (Borsig et al., 2001) (▶Figures 8.11 and ▶8.12). Several studies have shown that a few minutes after intravenous injection, tumor cells are detected in emboli inside pulmonary capillaries in association with platelets and fibrin (Karpatkin and Pearlstein, 1981; Gasic, 1984).

Studies from several groups have indicated that tumor metastasis in experimental animals can be inhibited by heparin (Zacharski et al., 1984; Engelberg, 1999; Hejna et al., 1999). Some clinical studies have also shown a beneficial effect of heparin in some types of human cancers (Fielding et al., 1992; Nitti et al., 1997). However, the antimetastatic effect of heparin is dissociated from its anticoagulant action (Zacharski et al., 1984; Zielinski and Hejna, 2000) and involved with the ability of heparin to inhibit the interaction of tumor cells with platelets. The glycosaminoglycan inhibits the binding of P- and L-selectin to their natural ligands (sialyl-Lewis^{x,a}-rich oligosaccharides) (Koenig et al., 1998). The inhibition of L- and P-selectin requires the presence of 6-*O*-sulfated glucosamine residues in the heparin molecule (Wang et al., 2002). Therefore, the heparin-mediated mechanism of metastasis attenuation involves the inhibition of the interaction of sialyl-Lewis^{x,a}-rich oligosaccharides on the surface of tumor cells and P-selectin on platelets (Borsig et al., 2001). In the presence of heparin, tumor cells loose the protection conferred by platelets becoming susceptible to the potentially cytotoxic action of blood innate immune cells, which leads to the inhibition of metastasis (▶Figure 8.11, 2.2). A unique intravascular injection of heparin promotes the immediate attenuation of the interaction of tumor cell-platelet, with a marked reduction of metastasis 6 weeks after the initial steps of the metastatic cascade (Borsig et al., 2001).

8.4.3 Unique glycosaminoglycans from marine invertebrates and their potential antitumor activity

Heparin-like analogs from marine invertebrates

A unique oversulfated chondroitin sulfate containing sulfated fucose branches and unusual 3-O-sulfated glucuronic acid (GlcA3S) residues has been described in the body wall of the holothurian *Holothuria grisea* (gray sea cucumber) (Vieira and Mourao, 1988; Vieira et al., 1991). The holothurian glycan has a chondroitin sulfate-E, [GlcA-GalNAc4S,6S – disaccharide units formed by glucuronic acid , beta 1-3 linked to 4-O- and 6-O- sulfated *N*-acetyl galactosamine] central core, but substituted at the 3-position of the glucuronic acid (GlcA) residues with sulfated fucose branches (Vieira and Mourao, 1988; Vieira et al., 1991). The sulfated fucose residues occur as 4-sulfate fucose, 2,4-disulfated fucose, or 3,4-disulfated fucose that are concentrated toward the nonreducing end of the polysaccharide chains (Mourao et al., 1996).

Dermatan sulfate polymers with high content of disulfated disaccharide units have been described in ascidians. A unique oversulfated dermatan sulfate consisting mainly of [2-sulfated Iduronic acid, beta 1,3 linked to 6-sulfated *N*-acetyl galactosamine (IdoA2S-GalNAc6S)] (80%) disaccharide units has been isolated from the internal organs of the *Phallusia nigra* (Pavao et al., 1995, 1998). Highly sulfated dermatan sulfates with the same core structure, [2-sulfated Iduronic acid, beta 1,3 linked to *N*-acetyl galactosamine (IdoA2-GalNAc)], but sulfated at carbon 4 of the GalNAc residues have also been described to occur in the extracellular matrix of different organs (intestine, heart, mantle, pharynx) of the ascidians *Styela plicata* (sea squirt) and *Halocynthia pyriformis* (Pavao et al., 1998; Hikino et al., 2003; Gandra et al., 2006).

The marine invertebrate glycosaminoglycans from holothurian and ascidians provide unique tools to determine specific sulfation patterns in these glycans required to bind EMT-related growth factors and adhesion molecules, such as selectins, that are involved in tumor invasion and metastasis. Additionally, the invertebrate glycans may become potential candidates of alternative drugs to be used in future clinical trials as an adjuvant antimetastatic therapy, since opposed to mammalian heparin, the invertebrate glycosaminoglycans possess low hemorrhagic effect, induces no toxicity, and their activities are retained even after oral administration (Pavao et al., 1998; Vicente et al., 2001, 2004; Cardilo-Reis et al., 2006; Fonseca and Mourao, 2006).

Possible effect of the invertebrate glycans on tumor invasion and metastasis

Two crucial moments of carcinoma invasion and metastasis are the EMT, which occurs in the primary tumor (▶Figure 8.11, 1), and the hematogenous metastasis, which occurs in the vascular system (▶Figures 8.11 and 8.12). Stroma growth factors and cell sulfate HSPG are key mediators of EMT, whereas tumor cell surface glycans and P-selectin on activated platelets are essential for hematogenous metastasis.

Previous work from our lab suggest that the unique glycosaminoglycans from marine invertebrates may attenuate the response of tumor cells to growth-factor-mediated EMT in the primary tumor and P-selectin-mediated formation of tumor cell-platelet complex during hematogenous metastasis. As a consequence, tumor cell dissemination can be drastically reduced.

Primary tumor

As discussed in the previous paragraphs, growth factors and other signaling molecules at the primary tumor microenvironment start the complex process of EMT. After the binding of EMT-related growth factors to their cell surface receptors on tumor cells, the extracellular signal follows its intracellular pathway to the nucleus where different transcription factors modulate gene expression, preparing the tumor cells to colonize distant tissues. Most of the EMT-related growth factors are heparin-binding factors and require cell surface HSPGs as coreceptors for the binding to their receptors on tumor cells. This fact raises the possibility that targeting growth factor–HS interaction at the tumor cell surface using nonanticoagulant glycosaminoglycans could reduce EMT-mediated invasion and migration of tumor cells from the primary tumor (▶Figure 8.11, 1.1).

In collaboration with John Gallagher and Malcon Lyon from Manchester (Catlow et al., 2008), we showed by surface plasmon resonance that the 2,6-sulfated dermatan sulfate obtained from the ascidian *P. nigra* binds to the EMT-related growth factor, HGF, with an affinity similar to that observed for mammalian heparan sulfate. Furthermore, in competition assay the *P. nigra* dermatan sulfate was shown to inhibit the binding of HGF to the MET receptor in MDCK (Madin Darby canine kidney cell line) cells (▶Table 8.1), suggesting that the invertebrate glycan can inhibit HGF-mediated EMT in tumor cells.

In a preliminary experiment to investigate the effect of 2,6-dermatan sulfate on TGF-β-mediated migration of tumor cell, we observed that the invertebrate glycan drastically attenuates migration of murine breast carcinoma M3 cells (▶Figure 8.12E–G). To confirm that TGF-β induced EMT we looked for the presence of epithelial E-cadherin and mesenchymal vimentin on the cells. TGF-β abolished the expression of E-cadherin and induced the expression of vimentin (▶Figure 8.12A–D). These results indicate that the 2,6-sulfated ascidian dermatan sulfate attenuates EMT-related growth factor signaling, suggesting that it could possibly be used as an inhibitor of EMT-mediated tumor cell migration.

Hematogenous metastasis

The fucosylated chondroitin sulfate from *H. grisea* was shown to inhibit the binding of P- and L-selectin to immobilized sialyl-Lewis^x, which is overexpressed in several tumor cells (Borsig et al., 2007). The glycan also inhibits the attachment of LS180 (intestinal human colon adenocarcinoma cell line) carcinoma cell to immobilized P- and L-selectins. As a result of its antiselectin effect, the holothurian glycan drastically attenuates experimental lung colonization by adenocarcinoma MC-38 (Murine colon adenocarcinoma cell line) cells. Interestingly, removal of the sulfated fucose branches on the glycan abolishes the inhibitory effect *in vitro* and *in vivo* (Borsig et al., 2007), indicating that the sulfated fucose branches are an absolute structural requirement for the antiselectin activity of the fucosylated chondroitin sulfate and its associated inhibitory effect on metastasis.

Table 8.1 Binding of soluble HGF/SF to surface immobilizes glycans. K_d values obtained by surface plasmon resonance. Inhibition of HGF-binding to MDCK cells.

HGF/SF-binding glycan	K_d *(nM)*	*Inhibition of the binding of HGF/SF to MDCK cells* IC_{50} *(M)*
Mammalian heparan sulfate	~0.2–3.0[a]	0.15[b]
Mammalian dermatan sulfate	~20[a]	0.31[b]
2,6-Disulfated dermatan sulfate from *P. nigra*	~1.0[a]	0.23[b]

[a] K_d values from Catlow et al. (2003).
[b] IC_{50} values from Catlow et al. (2008).

Table 8.2 Disaccharide composition and anti-P-selectin activity of dermatan sulfate polymers.

Dermatan sulfate	*Major disaccharide unit*	*Inhibition of tumor cell adhesion to P-selectin IC_{50} (μg/mL)*
Porcine dermatan sulfate	$[IdoA\text{-}GalNAc4S]_n$	>>100[a]
Ascidian 2,4-Disulfated dermatan	$[IdoA2S\text{-}GalNAc4S]_n$	13.51[a]
Ascidian 2,6-Disulfated dermatan	$[IdoA2S\text{-}GalNAc6S]_n$	12.19[a]
Oversulfated dermatan sulfate	$[IdoA\text{-}GalNAc4S,6S]_n$	12.56[a]

[a] IC_{50} values from Kozlowski et al. (2011b).

More recently, the dermatan sulfates with different sulfation patterns obtained from ascidians and a chemically oversulfated dermatan sulfate were used in experiments *in vitro* and *in vivo*, to determine if sulfation at a specific position in the disaccharide units of the glycosaminolgycan is necessary for its antiselectin activity and the associated biological effect on metastasis (Kozlowski et al., 2011b). The study revealed that dermatan sulfates containing mainly disulfated disaccharides units are potent inhibitors of P-selectin, regardless of the position of sulfation. Thus, if sulfation occurs at position 2 of the iduronic acid (IdoA) residues, the *N*-acetyl-D-galactosamine (GalNAc) needs to be sulfated either at position 4 or 6. However, if the IdoA residues are not sulfated, then the GalNAc needs to be sulfated at both positions 4 and 6. Dermatan sulfate containing one sulfate group at position 4 of the GalNAc residues has a very low antiselectin activity (▶Table 8.2). Therefore, dermatan sulfates containing disulfated disaccharide prevent the binding of tumor cells to immobilized P-selectins, as well as the formation of the tumor cell-platelet complex in the vasculature. As a consequence, metastasis is drastically attenuated by these glycans (Kozlowski et al., 2011b).

8.4.4 Challenges and future prospects

The ability of heparin-analogs to inhibit the binding of EMT-related growth factors to their receptors on tumor cells and the interaction between tumor mucin-glycans and P-selectin on activated platelets illustrates a valuable example on how one drug can target two important steps of tumor invasion and metastasis. In fact, targeting EMT-signaling pathways both at the receptor and intracellular levels (reviewed in Moon et al., 2004) and the formation of tumor cell-platelet complex during hematogenous metastasis have been approached separately in several scientific investigations (Borsig et al., 2001). However, no study has been conducted using heparin or heparin analogs to target both EMT-related signaling and hematogenous metastasis at the same time. Such a study can be conducted in an animal model of spontaneous metastasis using non-anticoagulant heparin analogs, where both the primary tumor and the hematogenous metastasis can be evaluated.

Based on the fact that heparin binds to different growth factors, cytokines and selectins, several parameters can be employed in the primary tumor to analyze growth-factor-mediated and cytokine-mediated responses, as well as P-selectin-mediated leukocyte recruitment. Examples include the presence of Wnt-mediated EMT (using epithelial and mesenchymal markers), vascular endothelial growth factor (VEGF)–mediated angiogenesis (using vascular markers), and P-selectin-mediated inflammatory cell infiltrate (using specific antibodies). In addition, evidence of hematogenous metastasis can be detected in the vasculature and in different organs (Borsig et al., 2007).

Although heparin is currently approved only as an anticoagulant and/or antithrombotic, its use as antimetastatic drug is only considered in clinical trials. In fact, a recent human clinical trial has been conducted to evaluate the effects of low-molecular-weight heparin on survival and disease progression in patients with hormone-refractory prostate cancer, locally advanced pancreatic cancer or non-small-cell lung carcinoma (ClinicalTrials.gov identifier: NCT00312013). Usually, drugs that alter blood coagulation are not considered a traditional antitumor agent.

Several heparin analogs have been described in different animal phyla (Kozlowski, 2011a). In addition to the glycans described in the present review, the structure, biological activity, and the mechanism of action of several other glycosaminoglycans from marine invertebrates have been extensively studied and evaluated in preclinical experiments in rodent animals with promising results (Kozlowski, 2011a). In general, these glycans can be isolated at reasonable yields, by procedures similar to those already employed in the preparation of pharmaceutical heparin. Several species of mollusks and sea cucumbers, including those containing high quantities of heparin analogs, have been successfully cultivated in different parts of the world. The cultivation employs developed aquaculture technologies capable of producing ton quantities of starting material (Bourne, 2000; Conand, 2004). Therefore, the critical conditions required to use marine invertebrates as a source of natural therapeutic compounds have already been established.

8.4.5 Take-home message

Strong evidence suggests that the marine glycosaminoglycans possess a significant therapeutic potential in cancer treatment, since they are able to act in two initial and fundamental steps of tumor dissemination, initially inhibiting EMT-related growth factor signaling in the primary tumor and later on the formation of microthrombus during hematogenous metastasis (▶Figure 8.11).

Acknowledgments

The work was supported by Conselho Nacional de Desenvolvimento Científico e Tecnológico (CNPq), Fundação de Amparo à Pesquisa do Estado do Rio de Janeiro (FAPERJ), and Mizutani Foundation for Glycoscience (to M. S. G. P.). M. S. G. P. is a research fellow from FAPERJ and CNPq.

References

Adam, R. M., Danciu, T., et al. (2003a). A nuclear form of the heparin-binding epidermal growth factor-like growth factor precursor is a feature of aggressive transitional cell carcinoma. Cancer Res *63* (2), 484–490.

Adam, R. M., Roth, J. A., et al. (2003b). Signaling through PI3K/Akt mediates stretch and PDGF-BB-dependent DNA synthesis in bladder smooth muscle cells. J Urol *169* (6), 2388–2393.

Aikawa, T., Whipple, C. A., et al. (2008). Glypican-1 modulates the angiogenic and metastatic potential of human and mouse cancer cells. J Clin Invest *118* (1), 89–99.

Alexander, C. M., Reichsman, F., et al. (2000). Syndecan-1 is required for Wnt-1-induced mammary tumorigenesis in mice. Nat Genet *25* (3), 329–332.

Alipio, Z. A., Jones, N., et al. (2011). Epithelial to mesenchymal transition (EMT) induced by bleomycin or TFG(b1)/EGF in murine induced pluripotent stem cell-derived alveolar type II-like cells. Differentiation *82* (2), 89–98.

Asiedu, M. K., Ingle, J. N., et al. (2011). TGFbeta/TNF(alpha)-mediated epithelial-mesenchymal transition generates breast cancer stem cells with a claudin-low phenotype. Cancer Res *71* (13), 4707–4719.

Avizienyte, E., Fincham, V. J., et al. (2004). Src SH3/2 domain-mediated peripheral accumulation of Src and phospho-myosin is linked to deregulation of E-cadherin and the epithelial-mesenchymal transition. Mol Biol Cell *15* (6), 2794–2803.

Baeriswyl, V., and Christofori, G. (2009). The angiogenic switch in carcinogenesis. Semin Cancer Biol *19* (5), 329–337.

Barrallo-Gimeno, A., and Nieto, M. A. (2005). The Snail genes as inducers of cell movement and survival: implications in development and cancer. Development *132* (14), 3151–3161.

Bechard, D., Gentina, T., et al. (2001). Endocan is a novel chondroitin sulfate/dermatan sulfate proteoglycan that promotes hepatocyte growth factor/scatter factor mitogenic activity. J Biol Chem *276* (51), 48341–48349.

Becker, D., Meier, C. B., et al. (1989). Proliferation of human malignant melanomas is inhibited by antisense oligodeoxynucleotides targeted against basic fibroblast growth factor. EMBO J *8* (12), 3685–3691.

Bergers, G., and Benjamin, L. E. (2003). Tumorigenesis and the angiogenic switch. Nat Rev Cancer *3* (6), 401–410.

Birchmeier, C., Birchmeier, W., et al. (2003). Met, metastasis, motility and more. Nat Rev Mol Cell Biol *4* (12), 915–925.

Bishop, J. R., Schuksz, M., et al. (2007). Heparan sulphate proteoglycans fine tune mammalian physiology. Nature *446* (7139), 1030–1037.

Blotnick, S., Peoples, G. E., et al. (1994). T lymphocytes synthesize and export heparin-binding epidermal growth factor-like growth factor and basic fibroblast growth factor, mitogens for vascular cells and fibroblasts: differential production and release by CD4+ and CD8+ T cells. Proc Natl Acad Sci U S A *91* (8), 2890–2894.

Bocca, C., Bozzo, F., et al. (2011). Celecoxib inactivates epithelial-mesenchymal transition stimulated by hypoxia and/or epidermal growth factor in colon cancer cells. Mol Carcinog. doi: 10.1002/mc.20846.

Borsig, L., Wang, L., et al. (2007). Selectin blocking activity of a fucosylated chondroitin sulfate glycosaminoglycan from sea cucumber. Effect on tumor metastasis and neutrophil recruitment. J Biol Chem *282* (20), 14984–14991.

Borsig, L., Wong, R., et al. (2001). Heparin and cancer revisited: mechanistic connections involving platelets, P-selectin, carcinoma mucins, and tumor metastasis. Proc Natl Acad Sci U S A *98* (6), 3352–3357.

Bourne, N. N. (2000). The potential for scallop culture – the next millennium. Aquacult Int *8* (2–3), 113–122.

Brabletz, T., Jung, A., et al. (2001). Variable beta-catenin expression in colorectal cancers indicates tumor progression driven by the tumor environment. Proc Natl Acad Sci U S A *98* (18), 10356–10361.

Brabletz, T., Jung, A., et al. (2005). Opinion: migrating cancer stem cells – an integrated concept of malignant tumour progression. Nat Rev Cancer *5* (9), 744–749.
Capurro, M. I., Xiang, Y. Y., et al. (2005). Glypican-3 promotes the growth of hepatocellular carcinoma by stimulating canonical Wnt signaling. Cancer Res *65* (14), 6245–6254.
Capurro, M. I., Xu, P., et al. (2008). Glypican-3 inhibits Hedgehog signaling during development by competing with patched for Hedgehog binding. Dev Cell *14* (5), 700–711.
Cardilo-Reis, L., Cavalcante, M. C., et al. (2006). In vivo antithrombotic properties of a heparin from the oocyte test cells of the sea squirt Styela plicata (Chordata-Tunicata). Braz J Med Biol Res *39* (11), 1409–1415.
Catlow, K., Deakin, J. A., et al. (2003). Hepatocyte growth factor/scatter factor and its interaction with heparan sulphate and dermatan sulphate. Biochem Soc Trans *31* (2), 352–353.
Catlow, K. R., Deakin, J. A., et al. (2008). Interactions of hepatocyte growth factor/scatter factor with various glycosaminoglycans reveal an important interplay between the presence of iduronate and sulfate density. J Biol Chem *283* (9), 5235–5248.
Chaffer, C. L., Thompson, E. W., et al. (2007). Mesenchymal to epithelial transition in development and disease. Cells Tissues Organs *185* (1–3), 7–19.
Chaffer, C. L., and Weinberg, R. A. (2011). A perspective on cancer cell metastasis. Science *331* (6024), 1559–1564.
Chuang, M. J., Sun, K. H., et al. (2008). Tumor-derived tumor necrosis factor-alpha promotes progression and epithelial-mesenchymal transition in renal cell carcinoma cells. Cancer Sci *99* (5), 905–913.
Conand, C. (2004). Present Status of World Sea Cucumber Resources and Utilisation: An International Overview. Rome: FAO.
Corso, S., Comoglio, P. M., et al. (2005). Cancer therapy: can the challenge be MET? Trends Mol Med *11* (6), 284–292.
Dasgupta, A., Alvarado, C. S., et al. (2010). Expression and functional role of inhibitor-of-apoptosis protein livin (BIRC7) in neuroblastoma. Biochem Biophys Res Commun *400* (1), 53–59.
Deakin, J. A., and Lyon, M. (1999). Differential regulation of hepatocyte growth factor/scatter factor by cell surface proteoglycans and free glycosaminoglycan chains. J Cell Sci *112* (Pt 12), 1999–2009.
Dennis, J. W., and Laferte, S. (1987). Tumor cell surface carbohydrate and the metastatic phenotype. Cancer Metastasis Rev *5* (3), 185–204.
Ding, W., You, H., et al. (2010). Epithelial-to-mesenchymal transition of murine liver tumor cells promotes invasion. Hepatology *52* (3), 945–953.
Dow, J. K., and deVere White, R. W. (2000). Fibroblast growth factor 2: its structure and property, paracrine function, tumor angiogenesis, and prostate-related mitogenic and oncogenic functions. Urology *55* (6), 800–806.
Engelberg, H. (1999). Actions of heparin that may affect the malignant process. Cancer *85* (2), 257–272.
Eswarakumar, V. P., Lax, I., et al. (2005). Cellular signaling by fibroblast growth factor receptors. Cytokine Growth Factor Rev *16* (2), 139–149.
Feng, M., Kim, H., et al. (2011). Recombinant soluble glypican 3 protein inhibits the growth of hepatocellular carcinoma in vitro. Int J Cancer *128* (9), 2246–2247.
Fielding, L. P., Hittinger, R., et al. (1992). Randomised controlled trial of adjuvant chemotherapy by portal-vein perfusion after curative resection for colorectal adenocarcinoma. Lancet *340* (8818), 502–506.
Filmus, J. (2001). Glypicans in growth control and cancer. Glycobiology *11* (3), 19R–23R.
Filmus, J., and Capurro, M. (2008). The role of glypican-3 in the regulation of body size and cancer. Cell Cycle *7* (18), 2787–2790.
Filmus, J., Capurro, M., et al. (2008). Glypicans. Genome Biol *9* (5), 224.

Fonseca, R. J., and Mourao, P. A. (2006). Fucosylated chondroitin sulfate as a new oral antithrombotic agent. Thromb Haemost *96* (6), 822–829.

Freeman, M. R., Paul, S., et al. (1998). Heparin-binding EGF-like growth factor in the human prostate: synthesis predominantly by interstitial and vascular smooth muscle cells and action as a carcinoma cell mitogen. J Cell Biochem *68* (3), 328–338.

Friess, H., Yamanaka, Y., et al. (1993). Enhanced expression of transforming growth factor beta isoforms in pancreatic cancer correlates with decreased survival. Gastroenterology *105* (6), 1846–1856.

Gandra, M., Kozlowski, E. O., et al. (2006). Collagen colocalizes with a protein containing a decorin-specific peptide in the tissues of the ascidian Styela plicata. Comp Biochem Physiol B Biochem Mol Biol *144* (2), 215–222.

Gasic, G. J. (1984). Role of plasma, platelets, and endothelial cells in tumor metastasis. Cancer Metastasis Rev *3* (2), 99–114.

Gherardi, E., Youles, M. E., et al. (2003). Functional map and domain structure of MET, the product of the c-met protooncogene and receptor for hepatocyte growth factor/scatter factor. Proc Natl Acad Sci U S A *100* (21), 12039–12044.

Goishi, K., Higashiyama, S., et al. (1995). Phorbol ester induces the rapid processing of cell surface heparin-binding EGF-like growth factor: conversion from juxtacrine to paracrine growth factor activity. Mol Biol Cell *6* (8), 967–980.

Graham, T. R., Zhau, H. E., et al. (2008). Insulin-like growth factor-I-dependent up-regulation of ZEB1 drives epithelial-to-mesenchymal transition in human prostate cancer cells. Cancer Res *68* (7), 2479–2488.

Hakomori, S. (1996). Tumor malignancy defined by aberrant glycosylation and sphingo (glyco)lipid metabolism. Cancer Res *56* (23), 5309–5318.

Hanahan, D., and Folkman, J. (1996). Patterns and emerging mechanisms of the angiogenic switch during tumorigenesis. Cell *86* (3), 353–364.

Hejna, M., Raderer, M., et al. (1999). Inhibition of metastases by anticoagulants. J Natl Cancer Inst *91* (1), 22–36.

Higashiyama, S., Abraham, J. A., et al. (1993). Heparin-binding EGF-like growth factor stimulation of smooth muscle cell migration: dependence on interactions with cell surface heparan sulfate. J Cell Biol *122* (4), 933–940.

Higashiyama, S., Abraham, J. A., et al. (1991). A heparin-binding growth factor secreted by macrophage-like cells that is related to EGF. Science *251* (4996), 936–939.

Hikino, M., Mikami, T., et al. (2003). Oversulfated dermatan sulfate exhibits neurite outgrowth-promoting activity toward embryonic mouse hippocampal neurons: implications of dermatan sulfate in neuritogenesis in the brain. J Biol Chem *278* (44), 43744–43754.

Hisaka, T., Yano, H., et al. (1999). Expressions of epidermal growth factor family and its receptor in hepatocellular carcinoma cell lines: relationship to cell proliferation. Int J Oncol *14* (3), 453–460.

Ho, M., and Kim, H. (2011). Glypican-3: a new target for cancer immunotherapy. Eur J Cancer *47* (3), 333–338.

Hosonaga, M., Ito, Y., et al. (2011). A lower dose of docetaxel at 60 mg/m(2) could be continued longer for controlling peripheral edema in patients with metastatic breast cancer. Breast Cancer. [Epub ahead of print]

Howard, S., Deroo, T., et al. (2011). A positive role of cadherin in Wnt/beta-catenin signalling during epithelial-mesenchymal transition. PLoS One *6* (8), e23899.

Huang, D., and Du, X. (2008). Crosstalk between tumor cells and microenvironment via Wnt pathway in colorectal cancer dissemination. World J Gastroenterol *14* (12), 1823–1827.

Ikushima, H., and Miyazono, K. (2010). TGFbeta signalling: a complex web in cancer progression. Nat Rev Cancer *10* (6), 415–424.

Ito, Y., Higashiyama, S., et al. (2001). Expression of heparin-binding epidermal growth factor-like growth factor in pancreatic adenocarcinoma. Int J Pancreatol *29* (1), 47–52.
Ito, Y., Iwase, T., et al. (2011). Eradication of breast cancer cells in patients with distant metastasis: the finishing touches? Breast Cancer. [Epub ahead of print]
Ito, Y., Takeda, T., et al. (2001a). Expression of heparin-binding epidermal growth factor-like growth factor in breast carcinoma. Breast Cancer Res Treat *67* (1), 81–85.
Ito, Y., Takeda, T., et al. (2001b). Expression of heparin binding epidermal growth factor-like growth factor in hepatocellular carcinoma: an immunohistochemical study. Oncol Rep *8* (4), 903–907.
Iwamoto, R., and Mekada, E. (2000). Heparin-binding EGF-like growth factor: a juxtacrine growth factor. Cytokine Growth Factor Rev *11* (4), 335–344.
Jing, Y., Han, Z., et al. (2011). Epithelial-mesenchymal transition in tumor microenvironment. Cell Biosci *1* (1), 29.
Kalluri, R., and Weinberg, R. A. (2009). The basics of epithelial-mesenchymal transition. J Clin Invest *119* (6), 1420–1428.
Kansas, G. S. (1996). Selectins and their ligands: current concepts and controversies. Blood *88* (9), 3259–3287.
Karpatkin, S., and Pearlstein, E. (1981). Role of platelets in tumor cell metastases. Ann Intern Med *95* (5), 636–641.
Kayed, H., Kleeff, J., et al. (2006). Correlation of glypican-1 expression with TGF-beta, BMP, and activin receptors in pancreatic ductal adenocarcinoma. Int J Oncol *29* (5), 1139–1148.
Kim, Y. J., and Varki, A. (1997). Perspectives on the significance of altered glycosylation of glycoproteins in cancer. Glycoconj J *14* (5), 569–576.
Kim, Y. S., Gum, J., Jr., et al. (1996). Mucin glycoproteins in neoplasia. Glycoconj J *13* (5), 693–707.
Kleeff, J., Ishiwata, T., et al. (1998). The cell-surface heparan sulfate proteoglycan glypican-1 regulates growth factor action in pancreatic carcinoma cells and is overexpressed in human pancreatic cancer. J Clin Invest *102* (9), 1662–1673.
Klymkowsky, M. W., and Savagner, P. (2009). Epithelial-mesenchymal transition: a cancer researcher's conceptual friend and foe. Am J Pathol *174* (5), 1588–1593.
Koenig, A., Norgard-Sumnicht, K., et al. (1998). Differential interactions of heparin and heparan sulfate glycosaminoglycans with the selectins. Implications for the use of unfractionated and low molecular weight heparins as therapeutic agents. J Clin Invest *101* (4), 877–889.
Kozlowski, E. O., Gomes, A. M., Sobral Silva, C., Sá Pereira, M., de Vilela Silva, A. C. E. S., and Pavão, M. S. G. (2011a). Structure and biological activities of glycosaminoglycan analogs from marine invertebrates: new therapeutic agents? In Glycans in Diseases and Therapeutics, M. S. G. Pavão, ed. Berlin: Springer, pp. 159–184.
Kozlowski, E. O., Pavao, M. S., et al. (2011b). Ascidian dermatan sulfates attenuate metastasis, inflammation and thrombosis by inhibition of P-selectin. J Thromb Haemost *9* (9), 1807–1815.
Laberge, R. M., Awad, P., et al. (2011). Epithelial-mesenchymal transition induced by senescent fibroblasts. Cancer Microenviron. *5* (1), 39–44.
Laubli, H., and Borsig, L. (2010). Selectins as mediators of lung metastasis. Cancer Microenviron *3* (1), 97–105.
Lee, J. J., Natsuizaka, M., et al. (2010). Hypoxia activates the cyclooxygenase-2-prostaglandin E synthase axis. Carcinogenesis *31* (3), 427–434.
Li, F., Shi, W., et al. (2011). Glypican-5 stimulates rhabdomyosarcoma cell proliferation by activating Hedgehog signaling. J Cell Biol *192* (4), 691–704.
Li, J., Kleeff, J., et al. (2004). Glypican-1 antisense transfection modulates TGF-beta-dependent signaling in Colo-357 pancreatic cancer cells. Biochem Biophys Res Commun *320* (4), 1148–1155.

Lietha, D., Chirgadze, D. Y., et al. (2001). Crystal structures of NK1-heparin complexes reveal the basis for NK1 activity and enable engineering of potent agonists of the MET receptor. EMBO J *20* (20), 5543–5555.

Liu, B. Y., Kim, Y. C., et al. (2003). Mammary gland development requires syndecan-1 to create a beta-catenin/TCF-responsive mammary epithelial subpopulation. Oncogene *22* (58), 9243–9253.

Liu, B. Y., McDermott, S. P., et al. (2004). The transforming activity of Wnt effectors correlates with their ability to induce the accumulation of mammary progenitor cells. Proc Natl Acad Sci U S A *101* (12), 4158–4163.

Lu, Z., Friess, H., et al. (1997). Presence of two signaling TGF-beta receptors in human pancreatic cancer correlates with advanced tumor stage. Dig Dis Sci *42* (10), 2054–2063.

Lyon, M., Deakin, J. A., and Gallagher, J. T. (2002). The mode of action of heparan and dermatan sulfates in the regulation of hepatocyte growth factor/scatter factor. J Biol Chem *277* (2), 1040–1046.

Lyon, M., Deakin, J. A. Lietha, D., Gherardi, E., and Gallagher, J. T. (2004). The interactions of hepatocyte growth factor/scatter factor and its NK1 and NK2 variants with glycosaminoglycans using a modified gel mobility shift assay. Elucidation of the minimal size of binding and activatory oligosaccharides. J Biol Chem *279* (42), 43560–43567.

Lyon, M., Deakin, J. A., Mizuno, K., Nakamura, T., and Gallagher, J. T. (1994). Interaction of hepatocyte growth factor with heparan sulfate. Elucidation of the major heparan sulfate structural determinants. J Biol Chem *269* (15), 11216–11223.

Lyon, M., Deakin, J. A., Rahmoune, H., Fernig, D. G., Nakamura, T., and Gallagher, J. T. (1998). Hepatocyte growth factor/scatter factor binds with high affinity to dermatan sulfate. J Biol Chem *273* (1), 271–278.

Mani, S. A., Guo, W., et al. (2008). The epithelial-mesenchymal transition generates cells with properties of stem cells. Cell *133* (4), 704–715.

Martin, P., and Parkhurst, S. M. (2004). Parallels between tissue repair and embryo morphogenesis. Development *131* (13), 3021–3034.

Matsuda, K., Maruyama, H., et al. (2001). Glypican-1 is overexpressed in human breast cancer and modulates the mitogenic effects of multiple heparin-binding growth factors in breast cancer cells. Cancer Res *61* (14), 5562–5569.

Micalizzi, D. S., Farabaugh, S. M., et al. (2010). Epithelial-mesenchymal transition in cancer: parallels between normal development and tumor progression. J Mammary Gland Biol Neoplasia *15* (2), 117–134.

Montesano, R., Matsumoto, K., et al. (1991a). Identification of a fibroblast-derived epithelial morphogen as hepatocyte growth factor. Cell *67* (5), 901–908.

Montesano, R., Schaller, G., et al. (1991b). Induction of epithelial tubular morphogenesis in vitro by fibroblast-derived soluble factors. Cell *66* (4), 697–711.

Moon, R. T., Kohn, A. D., et al. (2004). WNT and beta-catenin signalling: diseases and therapies. Nat Rev Genet *5* (9), 691–701.

Morel, A. P., Lievre, M., et al. (2008). Generation of breast cancer stem cells through epithelial-mesenchymal transition. PLoS One *3* (8), e2888.

Mourao, P. A., Pereira, M. S., et al. (1996). Structure and anticoagulant activity of a fucosylated chondroitin sulfate from echinoderm. Sulfated fucose branches on the polysaccharide account for its high anticoagulant action. J Biol Chem *271* (39), 23973–23984.

Mundhenke, C., Meyer, K., et al. (2002). Heparan sulfate proteoglycans as regulators of fibroblast growth factor-2 receptor binding in breast carcinomas. Am J Pathol *160* (1), 185–194.

Nakamura, M., and Tokura, Y. (2011). Epithelial-mesenchymal transition in the skin. J Dermatol Sci *61* (1), 7–13.

Natsuizaka, M., Ohashi, S., et al. (2010). Insulin-like growth factor-binding protein-3 promotes transforming growth factor-{beta}1-mediated epithelial-to-mesenchymal transition and motility in transformed human esophageal cells. Carcinogenesis *31* (8), 1344–1353.

Nitti, D., Wils, J., et al. (1997). Final results of a phase III clinical trial on adjuvant intraportal infusion with heparin and 5-fluorouracil (5-FU) in resectable colon cancer (EORTC GITCCG 1983–1987). European Organization for Research and Treatment of Cancer. Gastrointestinal Tract Cancer Cooperative Group. Eur J Cancer *33* (8), 1209–1215.

Nolan-Stevaux, O., Truitt, M. C., et al. (2010). Differential contribution to neuroendocrine tumorigenesis of parallel egfr signaling in cancer cells and pericytes. Genes Cancer *1* (2), 125–141.

O'Connell, M. P., Fiori, J. L., et al. (2009). Heparan sulfate proteoglycan modulation of Wnt5A signal transduction in metastatic melanoma cells. J Biol Chem *284* (42), 28704–28712.

Ogunwobi, O. O., and Liu, C. (2011). Hepatocyte growth factor upregulation promotes carcinogenesis and epithelial-mesenchymal transition in hepatocellular carcinoma via Akt and COX-2 pathways. Clin Exp Metastasis. *28* (8), 721–731.

Okanami, Y., Ito, Y., et al. (2011). Incidence of chemotherapy-induced amenorrhea in premenopausal patients with breast cancer following adjuvant anthracycline and taxane. Breast Cancer *18* (3), 182–188.

Pavao, M. S., Aiello, K. R., et al. (1998). Highly sulfated dermatan sulfates from Ascidians. Structure versus anticoagulant activity of these glycosaminoglycans. J Biol Chem *273* (43), 27848–27857.

Pavao, M. S., Mourao, P. A., et al. (1995). A unique dermatan sulfate-like glycosaminoglycan from ascidian. Its structure and the effect of its unusual sulfation pattern on anticoagulant activity. J Biol Chem *270* (52), 31027–31036.

Peinado, H., Portillo, F., et al. (2004). Transcriptional regulation of cadherins during development and carcinogenesis. Int J Dev Biol *48* (5–6), 365–375.

Pellegrini, L., Burke, D. F., et al. (2000). Crystal structure of fibroblast growth factor receptor ectodomain bound to ligand and heparin. Nature *407* (6807), 1029–1034.

Polyak, K., and Weinberg, R. A. (2009). Transitions between epithelial and mesenchymal states: acquisition of malignant and stem cell traits. Nat Rev Cancer *9* (4), 265–273.

Qian, B. Z., and Pollard, J. W. (2010). Macrophage diversity enhances tumor progression and metastasis. Cell *141* (1), 39–51.

Rahmoune, H., Rudland, P. S., et al. (1998). Hepatocyte growth factor/scatter factor has distinct classes of binding site in heparan sulfate from mammary cells. Biochemistry *37* (17), 6003–6008.

Richter, P., Umbreit, C., et al. (2011). EGF/TGFbeta1 co-stimulation of oral squamous cell carcinoma cells causes an epithelial-mesenchymal transition cell phenotype expressing laminin 332. J Oral Pathol Med *40* (1), 46–54.

Scheel, C., Eaton, E. N., et al. (2011). Paracrine and autocrine signals induce and maintain mesenchymal and stem cell states in the breast. Cell *145* (6), 926–940.

Schlessinger, J., Plotnikov, A. N., et al. (2000). Crystal structure of a ternary FGF-FGFR-heparin complex reveals a dual role for heparin in FGFR binding and dimerization. Mol Cell *6* (3), 743–750.

Sergeant, N., Lyon, M., et al. (2000). Stimulation of DNA synthesis and cell proliferation of human mammary myoepithelial-like cells by hepatocyte growth factor/scatter factor depends on heparan sulfate proteoglycans and sustained phosphorylation of mitogen-activated protein kinases p42/44. J Biol Chem *275* (22), 17094–17099.

Shiau, C. E., Hu, N., et al. (2010). Altering glypican-1 levels modulates canonical Wnt signaling during trigeminal placode development. Dev Biol *348* (1), 107–118.

Shin, J. A., Hong, O. K., et al. (2011). Transforming growth factor-beta induces epithelial to mesenchymal transition and suppresses the proliferation and transdifferentiation of cultured human pancreatic duct cells. J Cell Biochem *112* (1), 179–188.

Singh, A., and Settleman, J. (2010). EMT, cancer stem cells and drug resistance: an emerging axis of evil in the war on cancer. Oncogene *29* (34), 4741–4751.

Sivakumar, R., Koga, H., et al. (2009). Autocrine loop for IGF-I receptor signaling in SLUG-mediated epithelial-mesenchymal transition. Int J Oncol *34* (2), 329–338.

Strutz, F., and Neilson, E. G. (2003). New insights into mechanisms of fibrosis in immune renal injury. Springer Semin Immunopathol *24* (4), 459–476.

Strutz, F., Zeisberg, M., et al. (2002). Role of basic fibroblast growth factor-2 in epithelial-mesenchymal transformation. Kidney Int *61* (5), 1714–1728.

Suzuki, M., Sugimoto, K., et al. (2010). Up-regulation of glypican-3 in human hepatocellular carcinoma. Anticancer Res *30* (12), 5055–5061.

Takahashi, E., Nagano, O., et al. (2010). Tumor necrosis factor-alpha regulates transforming growth factor-beta-dependent epithelial-mesenchymal transition by promoting hyaluronan-CD44-moesin interaction. J Biol Chem *285* (6), 4060–4073.

Tanaka, Y., Miyamoto, S., et al. (2005). Clinical significance of heparin-binding epidermal growth factor-like growth factor and a disintegrin and metalloprotease 17 expression in human ovarian cancer. Clin Cancer Res *11* (13), 4783–4792.

Thiery, J. P., Acloque, H., et al. (2009). Epithelial-mesenchymal transitions in development and disease. Cell *139* (5), 871–890.

Toiyama, Y., Yasuda, H., et al. (2012). Co-expression of hepatocyte growth factor and c-Met predicts peritoneal dissemination established by autocrine hepatocyte growth factor/c-Met signaling in gastric cancer. Int J Cancer. *130* (12), 2912–2921.

Townsend, T. A., Wrana, J. L., et al. (2008). Transforming growth factor-beta-stimulated endocardial cell transformation is dependent on Par6c regulation of RhoA. J Biol Chem *283* (20), 13834–13841.

Umezu, T., Shibata, K., et al. (2010). Glypican-3 expression predicts poor clinical outcome of patients with early-stage clear cell carcinoma of the ovary. J Clin Pathol *63* (11), 962–966.

Veeman, M. T., Axelrod, J. D., et al. (2003). A second canon. Functions and mechanisms of beta-catenin-independent Wnt signaling. Dev Cell *5* (3), 367–377.

Vergara, D., Merlot, B., et al. (2010). Epithelial-mesenchymal transition in ovarian cancer. Cancer Lett *291* (1), 59–66.

Vicente, C. P., He, L., et al. (2004). Antithrombotic activity of dermatan sulfate in heparin cofactor II-deficient mice. Blood *104* (13), 3965–3970.

Vicente, C. P., Zancan, P., et al. (2001). Unbalanced effects of dermatan sulfates with different sulfation patterns on coagulation, thrombosis and bleeding. Thromb Haemost *86* (5), 1215–1220.

Vieira, R. P., and Mourao, P. A. (1988). Occurrence of a unique fucose-branched chondroitin sulfate in the body wall of a sea cucumber. J Biol Chem *263* (34), 18176–18183.

Vieira, R. P., Mulloy, B., et al. (1991). Structure of a fucose-branched chondroitin sulfate from sea cucumber. Evidence for the presence of 3-O-sulfo-beta-D-glucuronosyl residues. J Biol Chem *266* (21), 13530–13536.

Wagner, M., Kleeff, J., et al. (1999). Enhanced expression of the type II transforming growth factor-beta receptor is associated with decreased survival in human pancreatic cancer. Pancreas *19* (4), 370–376.

Walsh, L. A., and Damjanovski, S.(2011). IGF-1 increases invasive potential of MCF 7 breast cancer cells and induces activation of latent TGF-beta1 resulting in epithelial to mesenchymal transition. Cell Commun Signal *9* (1), 10.

Wang, L., Brown, J. R., et al. (2002). Heparin's anti-inflammatory effects require glucosamine 6-O-sulfation and are mediated by blockade of L- and P-selectins. J Clin Invest *110* (1), 127–136.

Wang, T., and Xu, Z. (2010). miR-27 promotes osteoblast differentiation by modulating Wnt signaling. Biochem Biophys Res Commun *402* (2), 186–189.

Wang, X., Nie, J., et al. (2008). Downregulation of Par-3 expression and disruption of Par complex integrity by TGF-beta during the process of epithelial to mesenchymal transition in rat proximal epithelial cells. Biochim Biophys Acta *1782* (1), 51–59.

Wang, Y. (2009). Wnt/Planar cell polarity signaling: a new paradigm for cancer therapy. Mol Cancer Ther *8* (8), 2103–2109.

Wang, Y., and Becker, D. (1997). Antisense targeting of basic fibroblast growth factor and fibroblast growth factor receptor-1 in human melanomas blocks intratumoral angiogenesis and tumor growth. Nat Med *3* (8), 887–893.

Williamson, D., Selfe, J., et al. (2007). Role for amplification and expression of glypican-5 in rhabdomyosarcoma. Cancer Res *67* (1), 57–65.

Wodarz, A., and Nusse, R. (1998). Mechanisms of Wnt signaling in development. Annu Rev Cell Dev Biol *14*, 59–88.

Xu, Z., Jiang, Y., et al. (2010). TGFbeta and EGF synergistically induce a more invasive phenotype of epithelial ovarian cancer cells. Biochem Biophys Res Commun *401* (3), 376–381.

Yang, J., and Weinberg, R. A. (2008). Epithelial-mesenchymal transition: at the crossroads of development and tumor metastasis. Dev Cell *14* (6), 818–829.

Yap, T. A., Sandhu, S. K., et al. (2011). HGF/c-MET targeted therapeutics: novel strategies for cancer medicine. Curr Drug Targets. *12* (14), 2045–2058.

Yeoman, L. C. (1993). An autocrine model for cell- and matrix-associated fibroblast growth factor. Oncol Res *5* (12), 489–499.

Yilmaz, M., and Christofori, G. (2009). EMT, the cytoskeleton, and cancer cell invasion. Cancer Metastasis Rev *28* (1–2), 15–33.

Yotsumoto, F., Yagi, H., et al. (2008). Validation of HB-EGF and amphiregulin as targets for human cancer therapy. Biochem Biophys Res Commun *365* (3), 555–561.

Yue, P., Zhang, X., et al. (2011). Hyperactive EGF receptor, Jaks and Stat3 signaling promote enhanced colony-forming ability, motility and migration of cisplatin-resistant ovarian cancer cells. Oncogene. *31* (18), 2309–2322.

Zacharski, L. R., Henderson, W. G., et al. (1984). Effect of warfarin anticoagulation on survival in carcinoma of the lung, colon, head and neck, and prostate. Final report of VA Cooperative Study #75. Cancer *53* (10), 2046–2052.

Zielinski, C. C., and Hejna, M. (2000). Warfarin for cancer prevention. N Engl J Med *342* (26), 1991–1993.

Zittermann, S. I., Capurro, M. I., et al. (2010). Soluble glypican 3 inhibits the growth of hepatocellular carcinoma in vitro and in vivo. Int J Cancer *126* (6), 1291–1301.

Zou, Z. Q., Ding, Y. P., et al. (2010). Gpc-3 is a notable diagnostic, prognostic and a latent targeted therapy marker in hepatocellular carcinoma. Hepatogastroenterology *57* (102–103), 1285–1290.

Zuo, J. H., Zhu, W., et al. (2011). Activation of EGFR promotes squamous carcinoma SCC10A cell migration and invasion via inducing EMT-like phenotype change and MMP-9-mediated degradation of E-cadherin. J Cell Biochem *112* (9), 2508–2517.

8.5 Pharmacological targeting of proteoglycans and metalloproteinases: an emerging aspect in cancer treatment

Spyros S. Skandalis, Chrisostomi Gialeli, Nikos Afratis, Alexios J. Aletras, Theodore Tsegenidis, Achilleas D. Theocharis, and Nikos K. Karamanos

8.5.1 Introduction

Tumors irrespective of their origin are heterogeneous cellular entities whose growth and progression greatly depend on reciprocal interactions between genetically altered (neoplastic) cells and their nonneoplastic microenvironment. Thus, microenvironmental factors promote many steps in carcinogenesis (e.g. proliferation, invasion, metastasis, angiogenesis, and chemoresistance). It is well established that cellular (myofibroblasts, vascular cells, and immune cells) and noncellular components, such as extracellular matrices (ECMs) proteins of the tumor microenvironment contribute to the antiapoptotic protection, survival and expansion of tumor cells. Tumor cells sense paracrine signals from the local microenvironment and communicate these signals with their stromal cells. In this way, they often alter the cellular and molecular composition of a particular tumor microenvironment to promote and maintain tumor progression. Hence, the notion of the tumor microenvironment as an integrated and essential part of the metastatic phenotype of carcinoma cells has been the subject of intense investigation. Proteoglycans (PGs) and matrix metalloproteinases (MMPs), which are involved in direct and indirect interactions between tumor cells and their microenvironment, have been identified as potential molecular targets that are hoped to advance the targeted cancer treatment in the future.

8.5.2 The importance of targeting at ECM level in tumor progression

Proteoglycans control numerous normal and pathological processes, among are morphogenesis, tissue repair, inflammation, vascularization, and cancer metastasis (Iozzo, 1998; Schaefer and Schaefer, 2010). During tumor development and growth, PG expression is markedly modified in the tumor microenvironment. Altered expression of PGs on tumor and stromal cell membranes affects cancer cell signaling, growth and survival, cell adhesion, migration and angiogenesis. The rapid increase in knowledge that PGs are among the key players in the tumor microenvironment and critical modulators of tumor progression, suggests their potential as pharmacological targets.

It has been recently suggested that pharmacological treatment may target PG metabolism, their utilization as targets for immunotherapy or their direct use as therapeutic agents (Theocharis et al., 2010).

MMPs are implicated in a variety of physiological processes, including wound healing, uterine involution, and organogenesis, as well as in pathological conditions, such as inflammation, vascular and autoimmune disorders, and carcinogenesis (Hadler-Olsen et al., 2011; Murphy and Nagase, 2011). MMPs are considered critical molecules in cancer cell migration, invasion and metastasis, and angiogenesis as they degrade various cell adhesion molecules, thereby modulating cell-cell and cell-ECMs interactions. The overexpression of MMPs in the tumor microenvironment depends not only on the cancer cells, but also on the neighboring stromal cells (endothelial cells, fibroblasts, macrophages, mast cells, neutrophils, pericytes, and adipocytes), which are induced by the cancer cells in a paracrine manner. Members of the MMP family exert different roles at different stages during cancer progression. In particular, they may promote or inhibit cancer development depending among other factors on the tumor stage, tumor site (primary, metastatic), enzyme localization (tumor cells, stroma), and substrate profile. As has been recently reviewed, targeting of MMPs by the development of effective and selective MMP inhibitors (MMPIs) is an emerging and promising area in the fight against cancer (Gialeli et al., 2011).

The proteasome pathway is the major mechanism of intracellular degradation of many proteins, implicated in several physiological cellular functions, including cell cycle control, transcription, cell signaling and apoptosis. Aberrant proteasome function is directly involved in the development and progression of malignancy (Navon and Ciechanover, 2009). The effect of proteasome blockade on the capacity of cells to selectively renew and degrade ECMs has sent encouraging signals that the development of specific proteasome inhibitors with regard to PGs and MMPs turnover could be a novel strategy to combat cancer (Skandalis et al., 2012). Currently, increasing efforts are being made to find proteasome inhibitors that target the degradation pathways of a single or a few proteins without affecting others.

Several therapeutic strategies based on the pharmacological targeting of PGs, MMPs, and the proteasome have been already elaborated. Some of them are either in preclinical and/or clinical studies. Even though these approaches turned out to be promising, the upcoming challenge will be to prove the efficacy of these strategies in improving treatment and prognosis of cancer patients.

8.5.3 Pharmacological targeting of proteoglycans

Extracellular matrix proteoglycans

Hyalectans

Versican – Versican regulates many cellular processes including adhesion, proliferation, apoptosis, migration, invasion, and ECMs assembly via the highly negatively charged chondroitin sulfate (CS)/dermatan sulfate (DS) side chains, and by interactions of the protein core G1 and G3 domains with other proteins (Wight, 2002; Theocharis, 2008). The significant role of versican in malignancy is discussed in Chapter 3.3, "The Pathobiology of Versican," by T. N. Wight. Versican biosynthesis is critically regulated by several matrix effectors, including platelet-derived growth factor (PDGF), transforming growth factor beta (TGF-β), interleukin-1β, epidermal growth factor (EGF), and insulin-like growth factor-1 (IGF-1) (Syrokou et al., 1999; Arslan et al., 2007; Berdiaki et al., 2008; Theocharis, 2008). Genetic and preclinical studies support the targeting of growth factor (PDGF, TGF-β, EGF, and vascular endothelial growth factor [VEGF]) signaling as a therapeutic strategy for combating cancer. In agreement with this concept, the effect of protein tyrosine kinase signaling pathways on versican synthesis can be reversed following treatment with specific tyrosine kinase inhibitors (TKIs) (Shimizu-Hirota et al., 2001). Accordingly, the TKI genistein can block versican expression induced by growth factors in malignant mesothelioma cell lines (Syrokou et al., 1999). However, there are no data to show that such approaches are effective in inhibiting the effects of versican in cancer cell models. In addition, versican synthesis is regulated by microRNAs (miRNA). miR-199a*, an onco-suppressor, leads to translational repression of versican gene. Several companies have developed miRNA-based therapeutics and a strategy that increases the natural dose of miR-199a*, by introducing a short double-stranded synthetic RNA that is loaded into RNA-induced silencing complex or by utilizing expression of the hairpin pre-miRNA in a viral vector expression system, may be useful for targeting versican among the other genes involved in tumorigenesis (Seto, 2010).

Manipulation of the versican catabolic pathways may also provide novel therapeutic targets for cancer invasion and metastasis. The use of an antibody against to the a disintegrin and metalloproteinase with thrombospondin motifs (ADAMTS)–specific versican cleavage site inhibits glioma cell migration in a dose-dependent manner, suggesting that the local accumulation of versican fragments may also promote cancer cell motility and invasion (Arslan et al., 2007). Notably, the broad-spectrum MMPI GM6001 (Galardin), which inhibits the activity of MMPs and ADAMTS proteases, has been shown to inhibit cancer cell invasion and metastasis in a transgenic breast cancer model (Almholt et al., 2008). The ability of MMPs and ADAMTS inhibitors to prevent versican catabolism and versican-induced motility and metastasis may be an interesting area of future study. Additionally, the use of hyaluronan oligomers is a potentially attractive agent to block the formation of large versican-hyaluronan aggregates and hyaluronan-cluster of differentiation 44 (CD44) interactions, as well as local tumor invasion. Specifically, disruption of the hyaluronan-CD44 interaction with hyaluronan oligomers has been shown to markedly inhibit the growth of B16F16 melanoma cells (Zeng et al., 1998). Moreover, hyaluronan oligomers inhibit the formation of receptor tyrosine kinases complexes and their phosphorylation in prostate, colon and breast carcinoma cells (reviewed in Misra et al., 2011).

Brevican – Upregulation and proteolytic cleavage of brevican increase the aggressiveness of glial tumors significantly (Nutt et al., 2001) and enhance cell adhesion and motility. The expression of a novel tumor-specific isoform of brevican that is localized on the cell membrane has been found in all high-grade gliomas and is suggested to play a significant role in glioma progression. In addition, the absence of brevican from benign gliomas prompts its use as a diagnostic marker to distinguish primary brain tumors of similar histology, but different pathological course (Viapiano et al., 2005). Inhibition of the expression of the tumor-specific isoform of brevican and inhibition of brevican cleavage may be a potential pharmacological target for the treatment of brain tumors.

Small leucine-rich proteoglycans

Decorin – Considerable effort has been recently applied to prove that decorin can be an antitumor therapeutic *in vivo*. Considering that chemotherapy is still the leading therapy for cancer patients and that decorin could be administered in concomitance with various compounds, it is relevant to understand their biological interaction. Decorin shows a synergistic effect with carboplatin in inhibiting ovarian cancer cell growth (Nash et al., 1999), whereas it antagonizes carboplatin and gemcitabine effects against pancreatic cancer cells (Koninger et al., 2004). Because of a lack of data and somewhat contrasting results, this area definitely needs more investigation.

Notably, adenoviral-mediated delivery of decorin slows the growth of lung, squamous and colon carcinoma tumor xenografts in immunocompromised mice (Tralhao et al., 2003), retards mammary adenocarcinoma growth and prevents metastatic spreading to the lungs reducing epidermal growth factor receptor 2 (ErbB2) receptor levels. Ectopic expression of decorin in a rat glioma model prolongs the survival of the animals and the size of the tumor is directly proportional to how early and how much decorin is expressed (Biglari et al., 2004). Administration of decorin protein core showed that it specifically localizes within the tumor, antagonizes epidermal growth factor receptor (EGFR) activity and induces apoptosis in the A431 squamous carcinoma model (Goldoni and Iozzo, 2008). The outcome is primary tumor growth inhibition because of the slower growth rate combined with apoptosis and impaired tumor metabolism. Decorin injected systemically can reduce breast tumor growth and metabolism and halt metastatic spread to the lungs (Goldoni et al., 2008). The finding that ectopic expression of decorin can revert the malignant phenotype in several cell lines of various histogenetic backgrounds (Santra et al., 1997) and can antagonize primary tumor growth and metastases *in vivo* further raises a hope for the postulated clinical application of decorin and related molecules. Decorin might be utilized in the near future as an adjunct "protein therapeutic" for solid tumors in which receptor tyrosine kinases play a key role.

Lumican – On the whole contrary to decorin, lumican is expressed by some tumor cell lines (Nikitovic et al., 2008), and its inhibition promotes cancer cell growth in some cases. Transfection of B16F1 mouse melanoma cells to express lumican or treatment with recombinant protein induces impaired anchorage-independent growth and the capacity to invade the ECMs (Vuillermoz et al., 2004). Importantly, lumican expression has been proposed as a prognostic factor in lymph-node-negative breast cancer (Troup et al., 2003). A recent study tested different recombinant and synthetic peptides and found an active site in the leucine-rich repeat 9 domain of the lumican core protein, which is able to inhibit melanoma cell migration (Zeltz et al., 2009). It will be interesting in future to test the effects of recombinant lumican on tumor growth and metastasis. Small leucine-rich proteoglycans (SLRPs) or peptides derived therefrom could be applied in the fight against cancer, because they represent a class of natural inhibitors of cancer growth.

Basement membrane proteoglycans

Basement membrane PGs are elongated molecules with a collage of domains that share structural and functional homology with numerous ECM proteins, growth factors and cell surface receptors. This class involves three main, well-characterized members:

perlecan, collagen type XVIII, and agrin, which are almost universally decorated with heparan sulfate (HS) side chains (Iozzo et al., 2009). Accumulating evidences indicate that heparan sulfate proteoglycans (HSPGs) act to inhibit cellular invasion by promoting tight cell-cell and cell-ECMs interactions, and by maintaining the structural integrity and self-assembly of ECMs. Notably, one of the characteristics of malignant transformation is downregulation of GAGs biosynthesis, especially of the HS chains.

Perlecan – The multiple functions of perlecan in development, cancer growth and angiogenesis are described by C. D. Willis, L. Schaefer, and R. V. Iozzo in Chapter 3.4, entitled "The Biology of Perlecan and its Bioactive Modules." Perlecan is a critical regulator of growth factor-mediated signaling and angiogenesis by binding growth factors via its HS side chains. The lack of HS in perlecan inhibits wound healing and fibroblast growth factor-2 (FGF2)–induced angiogenesis and tumor growth. Targeted knockdown of perlecan reduced the growth factor response, as revealed by decreased tumor growth and angiogenesis (Iozzo et al., 2009). In hepatoblastoma, xenografts treated with anti–vascular endothelial growth factor receptor (VEGFR) therapy, vessel recovery over time was associated with an increase in perlecan and heparanase expression around tumor vessels (Kadenhe-Chiweshe et al., 2008), suggesting a synergistic role of heparanase and/or proteases in the liberation of HS-bound VEGF and subsequent VEGFR2 activation. The C-terminal domain of perlecan, called endorepellin, blocks endothelial cell migration and capillary morphogenesis both *in vitro* and *in vivo* (Iozzo et al., 2009). Systemic delivery of human recombinant endorepellin to tumor xenograft-bearing mice causes a marked suppression of tumor growth and metabolic rate mediated by a sustained downregulation of the tumor angiogenic network (Bix et al., 2006). Utilization of endorepellin, a powerful angiostatic fragment of perlecan, as either a protein- or peptide-based pharmacological agent might represent a novel therapeutic rationale, especially when provided in combination with other tumor-suppressive compounds.

Collagen type XVIII – Endostatin, the C-terminal fragment of collagen type XVIII, is another potent antiangiogenic molecule that reduces tumor growth, choroidal neovascularization and wound healing (Iozzo et al., 2009). High levels of circulating endostatin reduce tumor burden, block the formation of pulmonary metastases and, notably, induce a total gene expression reprogramming, which ultimately disrupts endothelial cell migration. Endostatin downregulates several key components of the VEGF signaling cascade and, at the same time, stimulates the synthesis of thrombospondin, a powerful angiostatic protein and suppresses c-myc (Iozzo et al., 2009). Endostatin has been studied in phase I and II clinical trials for patients with metastatic cancer and has shown low efficacy; however, a new more stable version of endostatin has reentered the clinic and is now used in certain countries for the treatment of lung and gastric cancer (Sund and Kalluri, 2009). Recently, an immunoisolation device that contains endostatin-expressing cells was used effectively for the treatment of melanoma and Ehrlich tumors in mice (Rodrigues et al., 2010). This suggests that the macroencapsulation of engineered cells that produce endostatin may be innovate therapeutic strategy for treatment of malignancies.

Agrin – Agrin can be considered to be a marker of tumor angiogenesis in the liver because it is markedly deposited in proliferating bile ductules, in newly formed septal vessels in hepatic cirrhosis, in the angiogenic network of malignant hepatocellular carcinomas (HCCs) (Tatrai et al., 2006), and in cholangiocarcinoma (Batmunkh et al., 2007). The role of agrin in tumor angiogenesis is not yet established, but initial studies suggest that agrin's effect on tumor angiogenesis is likely context-dependent. It seems that the expression of agrin may be protective against disorganized angiogenesis in glioblastomas, whereas in hepatic malignancies it seems to support tumor angiogenesis, at least in the initial stages of tumor development. Similarly to perlecan, the utilization of the antiangiogenic C-terminal endorepellin-like fragment of agrin could represent a possible pharmacological agent to antagonize agrin in early stage tumor angiogenesis.

Cell surface proteoglycans

Glypicans – Glypicans influence tumor development and progression, and their expression is abnormal in various human tumors. Details on the structure, functions and pathobiology of glypicans are discussed in Chapter 3.7, "The Glypican Family," by J. Filmus and M. Cappuro. Notably, glypican-3 is a novel tumor marker for early stage melanoma and early stage HCC (Hippo et al., 2004; Nakatsura et al., 2004). It is not known, however, whether its elevated serum levels are important in tumor progression or are simply a reflection of aggressive tumors. The high levels of shed glypican may arise from increased expression of the PG or increased activity of tumor proteases. In this light, glypican-3 may be a candidate antigen for cancer immunotherapy in HCC. It is strongly expressed exclusively in HCC, is highly immunogenic and stimulates eradication by T cells of tumors expressing glypican-3 in mice and markedly inhibited growth of an established tumor. Glypican-3-derived peptide-pulsed vaccination is a novel strategy to prevent HCC and melanoma in patients and needs to be further developed as an anticancer therapy (Nakatsura et al., 2004). Interestingly, an anti-glypican-3 antibody (currently being tested in phase I clinical trials) induced antibody-dependent cellular cytotoxicity against glypican-3-positive HCC xenografts (Ishiguro et al., 2008; Zhu et al., 2011). Glypican-1 on target cells is recognized by some natural cytotoxicity receptors of natural killer cells and the suppression of glypican-1 in pancreatic cancer led to lower cytotoxic effects of natural killer cells (Bloushtain et al., 2004).

Syndecans – Syndecans are important mediators of cell spreading on ECMs and respond to growth factors and other biologically active polypeptides. The ectodomain of each syndecan is constitutively shed from many cultured cells, but is accelerated in response to wound healing and diverse pathophysiological events. Significant information regarding the biology of syndecans is provided in Chapter 3.6, "Structure and Function of Syndecans," by C. Pataki and J. R. Couchman, and Chapter 5.5, "Syndecans as Receptors for Pericellular Molecules," by X. Xian, J. R. Whiteford, and J. R. Couchman.

The observations that syndecan-1 is expressed at high levels on the surface of myeloma tumor cells and that shed syndecan-1 is an abundant component of the myeloma tumor environment revealed the potential of syndecan-1 as being a target for cancer treatment. The antitumor effect of murine/human chimeric syndecan-1-specific monoclonal antibody nBT062 conjugated with highly cytotoxic maytansinoid derivatives against multiple myeloma cells *in vitro* and *in vivo* was examined. Treatment significantly inhibited multiple myeloma tumor growth *in vivo* and prolonged host survival in both the xenograft mouse models of human multiple myeloma and the severe combined immunodeficiency (SCID)-hu mouse model. These results provide a preclinical framework supporting the evaluation of nBT062-maytansinoid derivatives in clinical trials to improve patient outcome in multiple myeloma (Ikeda et al., 2009). Another strategy for syndecan targeting in the treatment of malignancies is the inhibition of

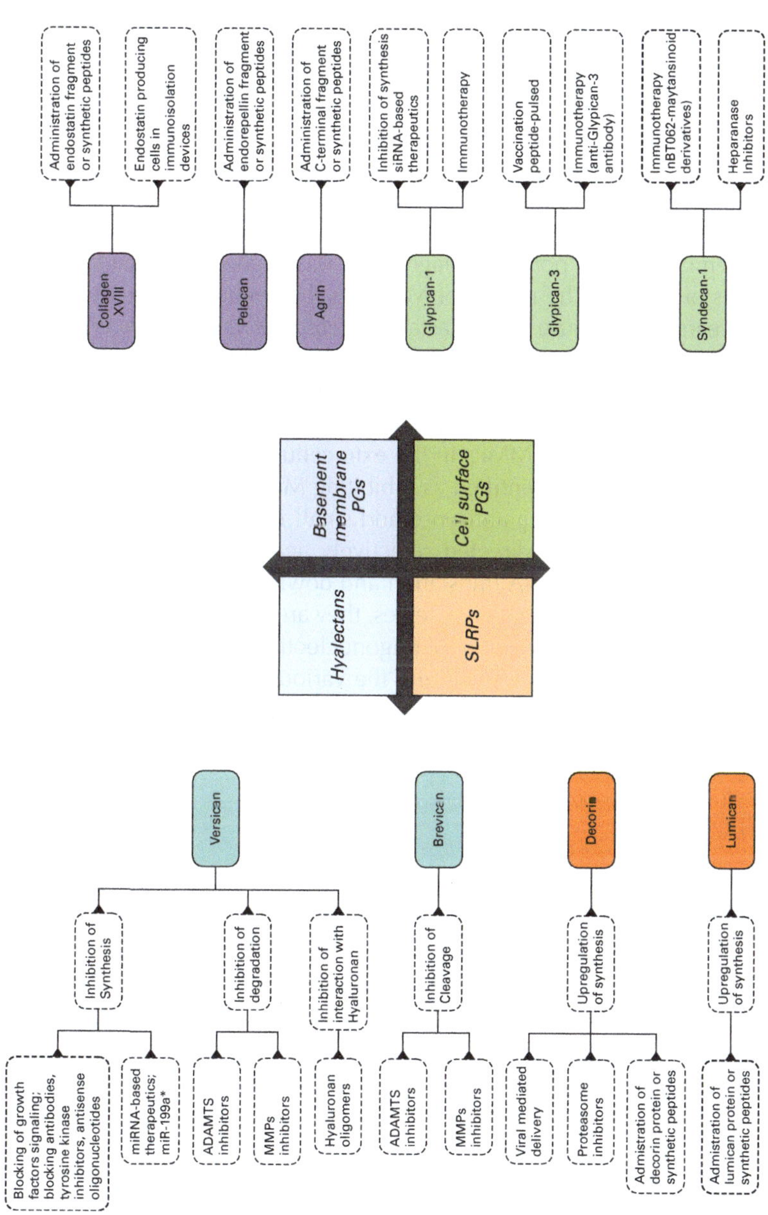

Figure 8.14 Levels of pharmacological targeting of PGs.

heparanase an enzyme that is known to modulate syndecans by cleaving the heparan sulfate side chains. The syndecan-1/heparanase axis affects multiple signaling pathways that drive tumor progression, thus providing an ideal therapeutic target, as described in Chapter 8.6, entitled, "Targeting Syndecan Shedding in Cancer," by R. D. Sanderson and J. R. Couchman and the review by Barash et al. (2010).

An overview of the level for pharmacological targeting of proteoglycans is presented in ▶Figure 8.14.

8.5.4 Pharmacological targeting of matrix metalloproteinases

MMPs have been considered potential diagnostic and prognostic biomarkers in many types and stages of cancer. The notion of MMPs as therapeutic targets of cancer was introduced 30 years ago because the metastatic potential of various cancers was correlated with the ability of cancer cells to degrade the basement membrane (Liotta et al., 1980). Several generations of synthetic MMPIs have been tested in phase III clinical trials in humans, including peptidomimetics, nonpeptidomimetics inhibitors and tetracycline derivatives, which target MMPs in the extracellular space. In addition, various natural compounds have been identified as inhibiting MMPs. Other strategies of MMP inhibition in development involve antisense and small interfering RNA (siRNA) technology. Antisense strategies are directed selectively against the mRNA of a specific MMP, resulting in decrease of RNA translation and downregulation of MMP synthesis. Despite the noted low toxicity of these strategies, they are still immature with respect to the effectiveness of the targeted delivery of oligonucleotides or siRNA to tumor cancer cells. The importance of MMPs targeting and the various strategies used has been recently reviewed by Gialeli et al. (2011). In this chapter, the major types of MMPIs are discussed, and their specificities and side effects are highlighted.

Synthetic inhibitors

Peptidomimetic MMPIs

Peptidomimetic MMPIs mimic the structure of collagen at the MMP cleavage site, functioning as competitive inhibitors, chelating the zinc ion present at the activation site (Betz et al., 1997). Based on the group that binds and chelates the zinc ion, peptidomimetics are subdivided into hydroxamates, carboxylates, hydrocarboxylates, sulfhydryls, and phosphoric acid derivatives. The earliest representative of this generation and the first MMPI that entered clinical trials is Batimastat (BB-94), a hydroxymate derivative with low water solubility and broad spectrum of inhibition (Macaulay et al., 1999). In order to overcome the solubility factor, Marimastat, another hydroxymate-based inhibitor, was introduced to be orally administered. However, it was also associated with musculoskeletal syndrome, probably due to the broad spectrum of inhibition (Steward and Thomas, 2000).

Nonpeptidomimetic MMPIs

Nonpeptidomimetic MMPIs were synthesized on the basis of the current knowledge of the three-dimensional conformation of the MMP active site in order to improve specificity and oral bioavailability. This generation comprises Tanomastat (BAY12–9566), Prinomastat (AG3340), BMS-275291, and CGS27023A (Brown, 2000). Musculoskeletal toxicity has been reported in clinical trials with CGS27023A, Prinomastat, and BMS-275291 (Gialeli et al., 2011).

Chemically modified tetracyclines

Another generation of MMPIs, tetracycline derivatives inhibit both the enzymatic activity and the synthesis of MMPs via blocking gene transcription. Chemically modified tetracyclines (CMTs), lacking antibiotic activities, may inhibit MMPs by binding to metal ions like zinc and calcium. This family of inhibitors, including Metastat (6-demethyl-6-deoxy-4-dedimethylamino tetracycline [COL-3]), minocycline, and doxycycline, cause limited systemic toxicity compared to regular tetracyclines. The CMT doxycycline is currently the only FDA-approved MMPI for prevention of periodontitis, whereas Metastat has entered phase II trials for Kaposi's sarcoma and brain tumors (Sapadin and Fleischmajer, 2006).

Novel mechanism-based inhibitors

A novel inhibitor, SB-3CT, was designed in order to selectively bind to the active site of gelatinases (MMP2 and MMP9) and reform the proenzyme structure. Specifically, the fundamental step in the inhibition of gelatinases by SB-3CT is enzyme-catalyzed ring opening of the thiirane, giving a stable zinc-thiolate species. It is reported to inhibit liver metastasis and increased survival in mouse models (Kruger et al., 2005).

Based on the importance of the a disintegrin and metalloproteinase (ADAM) family in cancer progression, small molecule inhibitors have been developed, such as INCB7839, which are currently tested in clinical trials (Moss and Bartsch, 2004).

Off-target inhibitors of MMPs

The off-target inhibitors of MMPs are molecules that influence MMPs and other ECM molecules indirectly. Such molecules are the bisphosphonates (BPs), analogs of PPi, which inhibit the function of the mevalonate pathway. Besides inhibition of osteoclast activity and bone resorption, BPs inhibit the enzymatic activity of various MMPs (Coxon et al., 2006). Another agent that has exerted inhibitory effects on MMPs is letrozole, a reversible nonsteroidal inhibitor of P450 aromatase. Gelatinases (MMP2 and 9) released by breast cancer cells, as well as functional invasion *in vitro*, are considerably suppressed by letrozole in a dose-dependent fashion, limiting in this way the metastatic potential of these cells (Mitropoulou et al., 2003).

It is worth mentioning that estrogen receptor-α suppression with siRNA in breast cancer cell lines abolishes the ability of estradiol to upregulate the expression of MMP9, highlighting the importance of the estrogen receptors signaling in the expression pattern of MMPs and therefore their potential pharmacological targeting (Kousidou et al., 2008).

Natural inhibitors of MMPs

Natural compounds with inhibitory effects on the activity of MMPs have been also used in order to avoid the negative results and toxicity issues raised by the use of synthetic

MMPIs. Oral administration of standardized extracts from shark cartilage, neovastat (AE-941), exerts antiangiogenic and antimetastatic activities, and these effects depend not only on the inhibition of MMPs enzymatic activity but also on the inhibition of VEGF (Falardeau et al., 2001). Genistein, another natural agent that has anticancer effects, is a soy isoflavonoid structurally similar to estradiol. Apart from its estrogenic and antiestrogenic properties, genistein confers tumor inhibition growth and invasion effects, interfering with the expression ratio and activity of several MMPs and tissue inhibitors of metalloproteinases (TIMPs) (Kousidou et al., 2005). TIMPs, the natural inhibitors of MMPs, have been also used to block MMPs activity. Although they have shown efficacy in experimental models, TIMPs may exert MMP-independent promoting effects (Brew and Nagase, 2010).

Antibody-based MMP inhibitors

As mentioned previously, most of the MMPIs demonstrated very limited clinical success, with severe toxicity and many side effects, due to their lack of specificity. Currently, efforts are being made toward the development of monoclonal antibodies (mAbs) against specific MMPs. Devy et al. (2009) presented DX-2400 as a selective antibody-based MMP inhibitor of MMP14. DX-2400 inhibited angiogenesis, in part, through inhibition of VEGF-driven cell invasion and pro-matrix metalloproteinase-2 (MMP2) activation. *In vivo* studies showed that DX-2400 markedly affected tumor growth of aggressive breast tumor cells (MDA-MB-231 and BT-474) when used as a single agent or in combination. In contrast, this compound did not alter the growth of less aggressive breast tumor cells (MCF7) (MMP14 negative) derived tumors, showing MMP14 dependency for DX-2400. In the MDA-MB-231 model, the antitumor effect of DX-2400 was associated with a strong decrease in tumor vascularization. In addition to its effects on primary tumor growth, DX-2400 also significantly reduced the number of metastatic foci *in vivo*. It seems that DX-2400 is a significant step toward the development of more specific MMP inhibitory drugs in the future. In another study, it has been reported that the use of monoclonal recombinant antibodies (SP1, SP2, SP3) specific to the catalytic domains of MMP1A, MMP2, and MMP3, respectively (Pfaffen et al., 2010). The full evaluation of their tumor-targeting potential requires additional studies.

It has been proposed that blockade of tumor cell receptor function with an antireceptor monoclonal antibody (mAb) might be an effective anticancer therapy. In line with this, murine mAb225 against EGF receptors led to a marked inhibition of cell growth in cultures of squamous cell carcinomas and in nude mouse xenografts. Notably, metastasis of xenografts was curtailed with mAb225 treatment, accompanied by a decrease in tumor production of MMP9 thus indicating, that using a monoclonal antibody against EGFR, one could successfully block indirectly the production of MMP9 (Mendelsohn and Baselga, 2000). These findings led to clinical trials of human/murine chimeric mAbC225 in combination with chemotherapy or radiotherapy showing quite promising results in advanced head and neck, colon, and pancreatic cancers. Another monoclonal antibody that affects the expression and activity of MMPs in an indirect mode is mAb4C5, which binds to the alpha and beta isoforms of heat-shock protein 90 (HSP90) and inhibits melanoma cell invasion and metastasis in murine models (Stellas et al., 2007). It has been shown that both isoforms of HSP90 interact with the inactive and active forms of MMP2 and MMP9 (Stellas et al., 2010). MAb4C5 additionally inhibits breast cancer cell invasion due to its ability to bind selectively to the extracellular pool of HSP90 and thus disrupt its association with the extracellular domain of human epidermal growth factor receptor 2 (HER2), which in turn leads to reduced HER2 phosphorylation (Sidera et al., 2008). With respect to MMPs, mAb 4C5 prevents MMP2 and MMP9 activation by disrupting their interaction with both isoforms of HSP90. Finally, mAb4C5 significantly reduced the metastatic dispositions of breast cancer cells into the lungs of SCID mice by blocking cell surface HSP90. This observation in line with the *in vitro* data led to the assumption that mAb4C5 exerts its activity by disrupting the interaction of extracellular HSP90 with MMP2 and MMP9, thus preventing activation of these MMPs, which, as is well documented, is necessary for *in vivo* cancer cell invasion. These data render mAb4C5 a strong candidate for further evaluation in clinical trials with cancer patients.

8.5.5 Pharmacological targeting of PGs/MMPs at the proteasome level

Recent success in the use of proteasome inhibitors (PIs) in the treatment of hematological malignancies validates the proteasome as a viable therapeutic target. The first PI that has come into clinical practice for the treatment of multiple myeloma and mantle cell lymphoma is bortezomib (Velcade, PS-341) (Richardson et al., 2003). Given that ECMs turnover and composition can be tightly regulated by proteasome activities, its pharmacological targeting may be considered a novel strategy to control the properties of tumor microenvironment. In line with this, it has been shown that decorin production was reestablished following treatment of ovarian cancer cells with the proteasome inhibitor MG-132, suggesting a regulatory mechanism for the production and accumulation of this tumor suppressive PG within the tumor stroma through inhibition of proteasome activities (Nash et al., 2002). Notably, other studies have shown that proteasome blockade by PIs resulted in a marked modification of multiple matrix macromolecules, such as collagen type-I, collagenase-1 (MMP 1) and stromelysin-1 (MMP 3), gene expression and activity by altering the activity of multiple transcription factors (Reunanen et al., 2002; Wu et al., 2002; Catalgol et al., 2009; Goffin et al., 2010).

In a forthcoming review, we highlight the possibility of enhancing or inhibiting the expression and bioactivity of certain matrix effectors, specifically PGs (i.e., decorin, glypican-1), GAGs (i.e., hyaluronan and heparan sulfate), MMPs (i.e., MMP 1, MMP 2, MMP 3), tissue inhibitors of metalloproteinases (TIMPs) (i.e., TIMP 1), and collagens (i.e., collagen type-I), with established roles in carcinogenesis and cancer progression, through the control of proteasome activities utilizing specific PIs (Skandalis et al., 2012). Importantly, controlled biosynthesis and turnover of key components of fibrosis matrices, which are known to increase risk to malignancy, such as collagen type-I, and/or their coordinated degradation by specific enzymes (i.e., the interstitial collagenase, MMP 1 and its inhibitor TIMP 1) utilizing selective proteasomal inhibition would provide a platform for the future design of therapies targeting the biomechanical properties of the cancer cell niche and microenvironment. PIs have been shown to possess the ability to overcome drug resistance and to synergize with a number of conventional therapies. The development of second-generation PIs that target the biosynthesis/

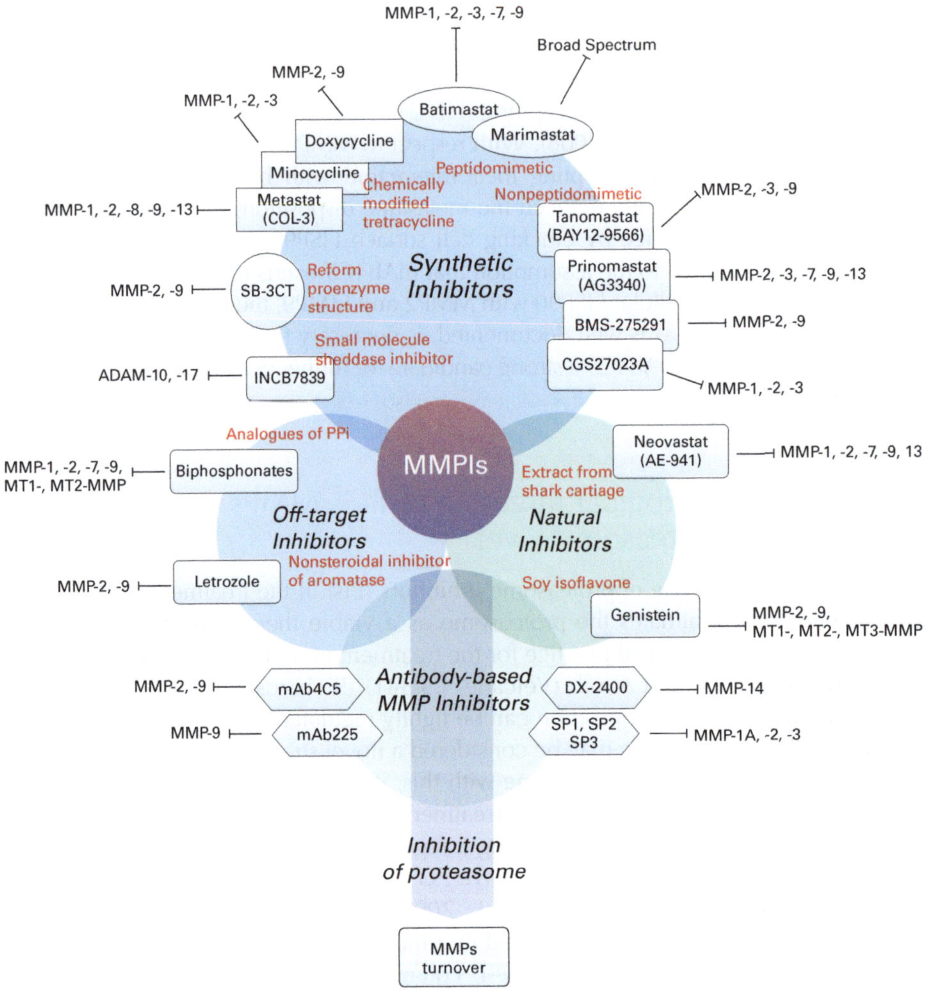

Figure 8.15 **An overview of matrix metalloproteinase inhibitors.**

degradation of specific PGs and MMPs in order to control tumor microenvironment is clearly required. An overview of the matrix metalloproteinase inhibitors is presented in ▶Figure 8.15.

8.5.6 Concluding remarks

PGs and MMPs constitute major extracellular regulators of the interactions between tumor cells and their microenvironment and, therefore, they have been identified as potential molecular targets that are expected to advance the targeted cancer treatment. Additionally, proteasome is a major intracellular component that controls the concentration and turnover of PGs and MMPs in ECMs and, consequently, in the tumor microenvironment. Pharmacological targeting of PGs, MMPs, and the proteasome shows encouraging results and may be a fruitful area for drug discovery and development in the future. An array of "protein therapeutics," functional PG-derived peptides, monoclonal antibodies, and synthetic and natural inhibitors of PGs, MMPs, and the proteolytic sites on the proteasome have been developed as therapeutic agents. Currently, there are several compounds either approved or in clinical trials for the treatment of multiple cancers and strokes. The development of a new generation of effective and selective compounds is an emerging and promising area of future research.

8.5.7 Take-home message

PGs and MMPs provide ideal therapeutic targets because of their multiple functions in regulating tumor-host cross talk. Their pharmacological targeting can be either direct or indirect by the use of specific proteasome inhibitors, since interfering with them could block their multiple effects on the properties of the tumor cell microenvironment that drive tumor progression.

Acknowledgments

This work has been cofinanced by the European Union (European Social Fund – ESF) and Greek national funds through the Operational Program "Education and Lifelong Learning" of the National Strategic Reference Framework (NSRF) – Research Funding Program: Thalis. Investing in knowledge society through the European Social Fund.

References

Almholt, K., Juncker-Jensen, A., Laerum, O. D., et al. (2008). Metastasis is strongly reduced by the matrix metalloproteinase inhibitor Galardin in the MMTV-PymT transgenic breast cancer model. Mol Cancer Ther *7*, 2758–2767.

Arslan, F., Bosserhoff, A. K., Nickl-Jockschat, T., Doerfelt, A., Bogdahn, U., and Hau, P. (2007). The role of versican isoforms V0/V1 in glioma migration mediated by transforming growth factor-beta2. Br J Cancer *96*, 1560–1568.

Barash, U., Cohen-Kapaln, V., Dowek, I., Sanderson, R. D., Ilan, N., and Vlodavsky, I. (2010). Proteoglycans in health and disease: new concepts for heparanase function in tumor progression and metastasis. FEBS J *277*, 3890–3903.

Batmunkh, E., Tatrai, P., Szabo, E., et al. (2007). Comparison of the expression of agrin, a basement membrane heparan sulfate proteoglycan, in cholangiocarcinoma and hepatocellular carcinoma. Hum Pathol *38*, 1508–1515.

Berdiaki, A., Zafiropoulos, A., Fthenou, E., et al. (2008). Regulation of hyaluronan and versican deposition by growth factors in fibrosarcoma cell lines. Biochim Biophys Acta *1780*, 194–202.

Betz, M., Huxley, P., Davies, S. J., et al. (1997). 1.8-Acrystal structure of the catalytic domain of human neutrophil collagenase (matrix metalloproteinase-8) complexed with a peptidomimetic hydroxamate primed-side inhibitor with a distinct selectivity profile. Eur J Biochem *247*, 356–363.

Biglari, A., Bataille, D., Naumann, U., et al. (2004). Effects of ectopic decorin in modulating intracranial glioma progression in vivo, in a rat syngeneic model. Cancer Gene Ther *11*, 721–732.

Bix, G., Castello, R., Burrows, M., et al. (2006). Endorepellin in vivo: targeting the tumor vasculature and retarding cancer growth and metabolism. J Natl Cancer Inst *98*, 1634–1646.

Bloushtain, N., Qimron, U., Bar-Ilan, A., et al. (2004). Membrane-associated heparan sulfate proteoglycans are involved in the recognition of cellular targets by NKp30 and NKp46. J Immunol *173*, 2392–2401.

Brew, K., and Nagase, H. (2010). The tissue inhibitors of metalloproteinases (TIMPs): an ancient family with structural and functional diversity. Biochim Biophys Acta *1803*, 55–71.

Brown, P. D. (2000). Ongoing trials with matrix metalloproteinase inhibitors. Expert Opin Investig Drugs *9*, 2167–2177.

Catalgol, B., Ziaja, I., Breusing, N., et al. (2009). The proteasome is an integral part of solar ultraviolet A radiation-induced gene expression. J Biol Chem *284*, 30076–30086.

Coxon, F. P., Thompson, K., and Rogers, M. J. (2006). Recent advances in understanding the mechanism of action of bisphosphonates. Curr Opin Pharmacol *6*, 307–312.

Devy, L., Huang, L., Naa, L., et al. (2009). Selective inhibition of matrix metalloproteinase-14 blocks tumor growth, invasion, and angiogenesis. Cancer Res *69*, 1517–1526.

Falardeau, P., Champagne, P., Poyet, P., Hariton, C., and Dupont, E. (2001). Neovastat, a naturally occuring multifunctional antiangiogenic drug, in phase III clinical trials. Semin Oncol *28*, 620–625.

Gialeli, C., Theocharis, A. D., and Karamanos, N. K. (2011). Roles of matrix metalloproteinases in cancer progression and their pharmacological targeting. FEBS J *278*, 16–27.

Goffin, L., Seguin-Estevez, Q., Alvarez, M., Reith, W., and Chizzolini, C. (2010). Transcriptional regulation of matrix metalloproteinase-1 and collagen 1A2 explains the anti-fibrotic effect exerted by proteasome inhibition in human dermal fibroblasts. Arthritis Res Ther *12*, R73.

Goldoni, S., and Iozzo, R. V. (2008). Tumor microenvironment: modulation by decorin and related molecules harboring leucine-rich tandem motifs. Int J Cancer *123*, 2473–2479.

Goldoni, S., Seidler, D. G., Heath, J., et al. (2008). An anti-metastatic role for decorin in breast cancer. Am J Pathol *173*, 844–855.

Hadler-Olsen, E., Fadness, B., Sylte, I., Uhlin-Hansen, L., and Winberg, J. O. (2011). Regulation of matrix metalloproteinase activity in health and disease. FEBS J *278*, 28–45.

Hippo, Y., Watanabe, K., Watanabe, A., et al. (2004). Identification of soluble NH2-terminal fragment of glypican-3 as a serological marker for early-stage hepatocellular carcinoma. Cancer Res *64*, 2418–2423.

Ikeda, H., Hideshima, T., Fulciniti, M., et al. (2009). The monoclonal antibody nBT062 conjugated to cytotoxic Maytansinoids has selective cytotoxicity against CD138-positive multiple myeloma cells in vitro and in vivo. Clin Cancer Res *15*, 4028–4037.

Iozzo, R. V. (1998). Matrix proteoglycans: from molecular design to cellular function. Annu Rev Biochem *67*, 609–652.

Iozzo, R. V., Zoeller, J. J., and Nystrom, A. (2009). Basement membrane proteoglycans: modulators par excellence of cancer growth and angiogenesis. Mol Cells *27*, 503–513.

Ishiguro, T., Sugimoto, M., Kinoshita, Y., et al. (2008). Anti-glypican-3 antibody as a potential antitumor agent for human liver cancer. Cancer Res *68*, 9832–9838.

Kadenhe-Chiweshe, A., Papa, J., McCrudden, K. W., et al. (2008). Sustained VEGF blockade results in microenvironmental sequestration of VEGF by tumors and persistent VEGF receptor-2 activation. Mol Cancer Res *6*, 1–9.

Koninger, J., Giese, N. A., di Mola, F. F., et al. (2004). Overexpressed decorin in pancreatic cancer: potential tumor growth inhibition and attenuation of chemotherapeutic action. Clin Cancer Res *10*, 4776–4783.

Kousidou, O. C., Mitropoulou, T. N., Roussidis, A. E., Kletsas, D., Theocharis, A. D., and Karamanos, N. K. (2005). Genistein suppresses the invasive potential of human breast cancer cells through transcriptional regulation of metalloproteinases and their tissue inhibitors. Int J Oncol *26*, 1101–1109.

Kousidou, O. Ch., Berdiaki, A., Kletsas, D., et al. (2008). Estradiol–estrogen receptor: a key interplay of the expression of syndecan-2 and metalloproteinase-9 in breast cancer cells. Mol Oncol *2*, 223–232.

Kruger, A., Arlt, M. J., Gerg, M., et al. (2005). Antimetastatic activity of a novel mechanism-based gelatinase inhibitor. Cancer Res *65*, 3523–3526.

Liotta, L. A., Tryggvason, K., Garbisa, S., Hart, I., Foltz, C. M., and Shafie, S. (1980). Metastatic potential correlates with enzymatic degradation of basement membrane collagen. Nature *284*, 67–68.

Macaulay, V. M., O'Byrne, K. J., Saunders, M. P., et al. (1999). Phase I study of intrapleural batimastat (BB-94), a matrix metalloproteinase inhibitor, in the treatment of malignant pleural effusions. Clin Cancer Res *5*, 513–520.

Mendelsohn, J., and Baselga, J. (2000) The EGF receptor family as targets for cancer therapy. Oncogene *19*, 6550–6565.

Misra, S., Heldin, P., Hascall, V. C., et al. (2011). Hyaluronan-CD44 interactions as potential targets for cancer therapy. FEBS J *278*, 1429–43.

Mitropoulou, T. N., Tzanakakis, G. N., Kletsas, D., Kalofonos, H. P., and Karamanos, N. K. (2003). Letrozole as a potent inhibitor of cell proliferation and expression of metalloproteinases (MMP-2 and MMP-9) by human epithelial breast cancer cells. Int J Cancer *104*, 155–160.

Moss, M. L., and Bartsch, J. W. (2004). Therapeutic benefits from targeting of ADAM family members. Biochemistry *43*, 7227–7235.

Murphy, G., and Nagase, H. (2011). Localising MMP activities in the pericellular environment. FEBS J *278*, 2–15.

Nakatsura, T., Kageshita, T., Ito, S., et al. (2004). Identification of glypican-3 as a novel tumor marker for melanoma. Clin Cancer Res *10*, 6612–6621.

Nash, M. A., Deavers, M. T., and Freedman, R. S. (2002). The expression of decorin in human ovarian tumors. Clin Cancer Res *8*, 1754–1760.

Nash, M. A., Loercher, A. E., and Freedman, R. S. (1999). In vitro growth inhibition of ovarian cancer cells by decorin: synergism of action between decorin and carboplatin. Cancer Res *59*, 6192–6196.

Navon, A., and Ciechanover, A. (2009). The 26 S proteasome: from basic mechanisms to drug targeting. J Biol Chem *284* (49), 33713–33718.

Nikitovic, D., Berdiaki, A., Zafiropoulos, A., et al. (2008). Lumican expression is positively correlated with the differentiation and negatively with the growth of human osteosarcoma cells. FEBS J *275*, 350–361.

Nutt, C. L., Zerillo, C. A., Kelly, G. M., and Hockfield, S. (2001). Brain enriched hyaluronan binding (BEHAB)/brevican increases aggressiveness of CNS-1 gliomas in Lewis rats. Cancer Res *61*, 7056–7059.

Pfaffen, S., Frey, K., Stutz, I., Roesli, C., and Neri, D. (2010) Tumour-targeting properties of antibodies specific to MMP-1A, MMP-2 and MMP-3. Eur J Nucl Med Mol Imaging *37* (8), 1559–1565.

Reunanen, N., Li, S. P., Ahonen, M., Foschi, M., Han, J., and Kahari, V. M. (2002). Activation of p38 alpha MAPK enhances collagenase-1 (matrix metalloproteinase (MMP)-1) and stromelysin-1 (MMP-3) expression by mRNA stabilization. J Biol Chem *277*, 32360–32368.

Richardson, P. G., Barlogie, B., Berenson, J., et al. (2003). A phase 2 study of bortezomib in relapsed, refractory myeloma. N Engl J Med *348*, 2609–2617.

Rodrigues, D. B., Chammas, R., Malavasi, N. V., et al. (2010). Anti-tumor therapy with macroencapsulated endostatin producer cells. BMC Biotechnol *10*, 19.

Santra, M., Mann, D. M., Mercer, E. W., Skorski, T., Calabretta, B., and Iozzo, R. V. (1997). Ectopic expression of decorin protein core causes a generalized growth suppression in neoplastic cells of various histogenetic origin and requires endogenous p21, an inhibitor of cyclin-dependent kinases. J Clin Invest *100*, 149–157.

Sapadin, A. N., and Fleischmajer, R. (2006). Tetracyclines: nonantibiotic properties and their clinical implications. J Am Acad Dermatol *54*, 258–265.

Schaefer, L., and Schaefer, R. M. (2010). Proteoglycans: from structural compounds to signaling molecules. Cell Tissue Res *339*, 237–246.

Seto, A. G. (2010). The road toward microRNA therapeutics. Int J Biochem Cell Biol *42*, 1298–1305.

Shimizu-Hirota, R., Sasamura, H., Mifune, M., et al. (2001). Regulation of vascular proteoglycan synthesis by angiotensin II type 1 and type 2 receptors. J Am Soc Nephrol *12*, 2609–2615.

Sidera, K., Gaitanou, M., Stellas, D., Matsas, R., and Patsavoudi, E. (2008). A critical role for HSP90 in cancer cell invasion involves interaction with the extracellular domain of HER-2. J Biol Chem *283*, 2031–2041.

Skandalis, S. S., Aletras, A. J., Gialeli, C., et al. (2012). Targeting the tumor proteasome as a mechanism to control the synthesis and bioactivity of matrix macromolecules. Curr Mol Med, in press.

Stellas, D., Hamidieh, A., and Patsavoudi, E. (2010). Monoclonal antibody 4C5 prevents activation of MMP2 and MMP9 by disrupting their interaction with extracellular HSP90 and inhibits formation of metastatic breast cancer cell deposits. BMC Cell Biol *11*, 51.

Stellas, D., Karameris, A., and Patsavoudi, E. (2007). Monoclonal antibody 4C5 immunostains human melanomas and inhibits melanoma cell invasion and metastasis. Clin Cancer Res *13*, 1831–1838.

Steward, W.,P., and Thomas, A.,L. (2000). Marimastat: the clinical development of a matrix metalloproteinase inhibitor. Expert Opin Investig Drugs *9*, 2913–2922.

Sund, M., and Kalluri, R. (2009). Tumor stroma derived biomarkers in cancer. Cancer Metastasis Rev *28*, 177–183.

Syrokou, A., Tzanakakis, G. N., Hjerpe, A., and Karamanos N. K. (1999). Proteoglycans in human malignant mesothelioma. Stimulation of their synthesis induced by epidermal, insulin and platelet-derived growth factors involves receptors with tyrosine kinase activity. Biochimie *81*, 733–744.

Tatrai, P., Dudas, J., Batmunkh, E., et al. (2006) Agrin, a novel basement membrane component in human rat and liver, accumulates in cirrhosis and hepatocellular carcinoma. Lab Invest *86*, 1149–1160.

Theocharis, A. D. (2008). Versican in health and disease. Connect Tissue Res *49*, 230–234.

Theocharis, A. D., Skandalis, S. S., Tzanakakis, G. N., and Karamanos, N. K. (2010). Proteoglycans in health and disease: novel roles for proteoglycan in malignancy and their pharmacological targeting. FEBS J *277*, 3904–3923.

Tralhao, J. G., Schaefer, L., Micegova, M., et al. (2003). In vivo selective and distant killing of cancer cells using adenovirus mediated decorin gene transfer. FASEB J *17*, 464–466.

Troup, S., Njue, C., Kliewer, E. V., et al. (2003). Reduced expression of the small leucine-rich proteoglycans, lumican, and decorin is associated with poor outcome in node-negative invasive breast cancer. Clin Cancer Res *9*, 207–214.

Viapiano, M. S., Bi, W. L., Piepmeier, J., Hockfield, S., and Matthews, R. T. (2005). Novel tumor-specific isoforms of BEHAB/brevican identified in human malignant gliomas. Cancer Res *65*, 6726–6733.

Vuillermoz, B., Khoruzhenko, A., D'Onofrio, M. F., et al. (2004). The small leucine-rich proteoglycan lumican inhibits melanoma progression. Exp Cell Res *296*, 294–306.

Wight, T. N. (2002). Versican: a versatile extracellular matrix proteoglycan in cell biology. Curr Opin Cell Biol *14*, 617–623.

Wu, H.-M., Chi, K.-H., and Lin, W.-W. (2002). Proteasome inhibitors stimulate activator protein-1 pathway via reactive oxygen species production. FEBS Lett *526*, 101–105.

Zeltz, C., Brezillon, S., Perreau, C., Ramont, L., Maquart, F.-X., and Wegrowski, Y. (2009). Lumcorin: a leucine-rich repeat 9-derived peptide from human lumican inhibiting melanoma cell migration. FEBS Lett *583*, 3027–3032.

Zeng, C., Toole, B. P., Kinney, S. D., Kuo, J. W., and Stamenkovic, I. (1998). Inhibition of tumor growth in vivo by hyaluronan oligomers. Int J Cancer *77*, 396–401.

Zhu, A. X., Gold, P. J., El-Khoueiry, A. B., Abrams, T. A., Morikawa, H., and Philip, P. A. (2011). A phase 1 study of GC33, a recombinant humanized antibody against glypican-3, in patients with advanced hepatocellular carcinoma. J Clin Oncol *29* (Suppl.), abstr 4085.

8.6 Targeting syndecan shedding in cancer

Ralph D. Sanderson and John R. Couchman

8.6.1 Introduction

Syndecans are type I membrane glycoproteins distributed widely on almost all vertebrate cell surfaces. The only notable exception is the erythrocyte. They possess glycosaminoglycan chains, usually heparan sulfate, but sometimes chondroitin or dermatan sulfate in addition. These chains, usually three or more per core protein, may be substantial in size, so that they are a dominant factor in the properties of the total molecule (Bass et al., 2009; Couchman, 2010). So far, there are no reports of syndecans occurring naturally on cell surfaces lacking glycosaminoglycan chains, and therefore they are considered full-time proteoglycans. Syndecans appear to be expressed as homodimers, since their transmembrane domains strongly favor self-association (Dews and Mackenzie, 2007). Therefore, a functional dimeric unit of syndecan may have six or more glycosaminoglycan chains. The structure of the syndecans, their glycosaminoglycans and their receptor and signaling functions are detailed elsewhere in this volume.

The ectodomains of the four mammalian syndecan core proteins have little sequence homology, except in proximity to the serine residues that are substrates for glycanation. Even between species there is variability, which initially led to the idea that the external core proteins of syndecans were little more than supports for the glycosaminoglycan chains. However, that view is changing in light of recent data showing that the ectodomains of syndecans-1, -2, and -4 can support integrin-mediated cell adhesion, in some cases independent of the carbohydrate chains (Whiteford et al., 2007; Beauvais et al., 2009). There are no data for syndecan-3 at the present time. Moreover, the syndecan core proteins share another property, they can all be subject to cleavage and release from the cell surface, a process known as shedding. Of the cleavage sites that have been mapped for a variety of proteinases, most are located in the membrane-proximal area. Therefore, the evidence suggests that syndecans have proteinase-sensitive domains that when targeted release a large portion of the ectodomain, complete with glycosaminoglycan chains.

This property of syndecans is by no means unique. Several families of cell surface receptors can be shed in this way. Proforms of growth factors are a prime example (Saftig and Reiss, 2011), and the importance of shedding in inflammatory disease is frequently emphasized. Others, such as the integrins, are by contrast, fairly proteinase resistant. For the syndecans, there may be a continual, constitutive shedding, at least in some cell lines (Wang et al., 2005), but the process can be considerably upregulated in cases of tissue damage or disease. Under these conditions, proteinases that shed syndecans become upregulated also, with the consequence of increased syndecan shedding (Choi et al., 2010; Manon-Jensen et al., 2010). This process can have several effects. First, signaling through the syndecan cytoplasmic domains will be affected. It has been assumed that signaling will terminate, and no evidence exists that the isolated transmembrane and cytoplasmic domains will continue to function. However, this cannot be ruled out without further experimentation. It is known that the transmembrane domains may be subject to further processing by the presenilin/γ-secretase complex (Schulz et al., 2003). In turn, this leads to decreased plasma membrane targeting of a calcium/calmodulin-associated serine kinase. Second, the ectodomain may be lost, and increased levels of shed syndecan in wound fluids, and the circulation of some patients, such as myeloma, are recorded (Kainulainen et al., 1998). Additionally, however, the shed ectodomain may be retained either at the same or adjacent cell surface through either heparan sulfate interactions or those of the core protein. There are precedents for both. The core proteins can trigger integrin-mediated responses either directly, in the case of syndecan-1, or indirectly, as indicated for syndecan-2 and -4 (Whiteford et al., 2007; Beauvais et al., 2009; Couchman, 2010). Therefore, new cell surface signaling events may supersede previous transmembrane signaling in response to syndecan shedding. In addition, shed syndecan in the pericellular domain can function as a competitive inhibitor of transmembrane proteoglycans for growth factors, cytokines, chemokines, or other heparan-sulfate-binding ligands (Elenius et al., 2004). Syndecan shedding, clearly, can have complex consequences.

8.6.2 Syndecan sheddases

Knowledge regarding when, where, and which proteinases are responsible for shedding is in its infancy. *In vitro* experiments have shown that several different classes of enzymes are capable of cleaving syndecan core proteins. Principal among them are the matrix metalloproteinases (MMPs). These belong to the metzincins, the zinc endopeptidases, and contain three major families, the serlysins, astacins, and a disintegrin and metalloproteinase (ADAMs/adamalysins). Several MMPs, including MMP2, MMP9, and ADAM17 tumor necrosis factor-α-converting enzyme (TACE) cleave syndecans-1 and -4 close to the plasma membrane (Brule et al., 2006; Preussmayer et al., 2010). MMP7, also known as matrilysin, and the transmembrane membrane type-1 matrix metalloproteinase (MT1-MMP [MMP14]) and membrane type-3 matrix metalloproteinase (MT3-MMP [MMP16]) cleave syndecan-1 (Endo et al., 2003; Li et al., 2002) and perhaps other syndecans, but this has yet to be confirmed. The importance of MMP7 and syndecan-1 shedding in lung inflammation and neutrophil response to chemokines has been elegantly demonstrated (Li et al., 2002). Other proteinases that cleave syndecan-4 include plasmin and thrombin, and the exact sites of cleavage have been identified in these cases (Schmidt et al., 2005). Two members of the a disintegrin and metalloproteinase with thrombospondin motifs (ADAMTS) family, these having thrombospondin motifs (TS motifs), ADAMTS1 and 4 can also cleave syndecan-4, but in this case close to the N terminus, between the first and second sites of heparan sulfate substitution (Rodríguez-Manzaneque et al., 2009). Cleavage here was reported to depress cell adhesion and increase migration. It remains, however, that most specific cleavage sites in syndecan core proteins have not been identified. In addition, it is likely that the list of proteinases with syndecan-shedding ability will continue to grow.

An interesting complication is that some of the enzymes that cleave syndecan core proteins additionally interact with heparan or chondroitin sulfate. The TS motifs of

ADAMTS proteinases have this capacity; it has been suggested that syndecan-4 heparan sulfate chains may regulate activation of ADAMTS5, with downstream impact on mitogen-activated protein kinase (MAPK)–dependent synthesis of MMP3. This may have relevance to arthritis, since ADAMTS4 and 5 are aggrecanases implicated in cartilage matrix degradation (Bondeson et al., 2008). Indeed the syndecan-4–null mouse has some resistance to arthritis in an experimental model system (Echtermeyer et al., 2009). Other enzymes that interact with heparan sulfate are MMP2, 7, 9, and 13, though the outcomes seem to vary. Heparan sulfate interaction with MMP2 inhibits activity, while the opposite is reported for MMP7 and MMP13 interactions with heparin (Yu and Woessner, 2001; Munesue et al., 2007).

8.6.3 Tissue inhibitors of metalloproteinases

Well known inhibitors of the MMPs are the tissue inhibitors of metalloproteinases (TIMPs), of which there are four in mammals. Early studies on sydnecan-1 shedding in murine mammary tumor cells showed that accelerated shedding in response to phorbol ester treatment, could be blocked only by TIMP3 (Fitzgerald et al., 2000). This member of the family is distinct in two ways. First, it has an ability to inhibit a wider range of proteinases that includes not only the MMPs, but also a range of ADAMs and ADAMTS enzymes (Nagase et al., 2006; Edwards et al., 2008). Notable among these are ADAM17 (TACE), ADAM10, ADAM12, and the aggrecanases ADAMTS4 and 5. Therefore, this had led to the idea that the accelerated shedding seen in the mammary tumor cells was ADAM-mediated, a hypothesis entirely consistent with the role of many ADAMs as receptor sheddases (Edwards et al., 2008). In addition, however, TIMP3 is unique in having the ability to bind glycosaminoglycans and synthetic analogs, including heparan, chondroitin, and dermatan sulfates; heparin; suramin; and pentosan (Yu et al., 2000). This potentially brings the inhibitor close to the substrate for some MMPs, and may trigger endocytosis of the proteinase. Consequently, while TIMP1, 2, and 4 may be diffusible, TIMP3 can be matrix-associated or cell surface localized (Nagase et al., 2006; Leco et al., 1994), by virtue of the widespread distribution of glycosaminoglycans in these tissue domains.

8.6.4 Syndecan shedding and cancer

Multiple myeloma

The initial idea that syndecan-1 shedding might be a target for cancer therapy arose from results of analysis of serum from multiple myeloma patients. In these patients, elevated levels of serum syndecan-1 correlated strongly with high tumor burden leading to speculation that high levels of serum syndecan-1 may be an indicator of poor progress (Dhodapkar et al., 1997). This was followed up by a study of 174 myeloma patients in which levels of serum syndecan-1 were elevated in 79% of patients when compared to normal controls (Seidel et al., 2000b). This study and others that followed confirmed that elevated syndecan-1 in the serum of myeloma patients is an independent predictor of poor prognosis.

Since syndecan-1 is expressed at high levels on the surface of myeloma tumor cells, initially it was not clear if shed syndecan-1 in patient serum simply identified an aggressive tumor phenotype, or if shed syndecan-1 was actually involved in promoting aggressive tumor behavior. To test this, cells overexpressing the shed form of syndecan-1 were implanted in mice. Tumors formed by these cells grew faster and were more metastatic than controls shedding normal levels of syndecan-1 (Yang et al., 2002). Also, when overexpressed, shed syndecan-1 accumulated not only in the serum of the mice, but also to concentrate within the interstitial matrix of the bone marrow, similar to what is observed in marrow of myeloma patients (Bayer-Garner et al., 2001; Yang et al., 2002; ▶Figure 8.16). This identified, shed syndecan-1 as an abundant component of the myeloma tumor microenvironment.

Several years later it was discovered that heparanase, an endoglycosidase that cleaves heparan sulfate chains, also stimulated growth and dissemination of myeloma tumors *in vivo*, similar to the effects seen with shed syndecan-1 (Yang et al., 2005). Further investigation revealed that heparanase upregulates expression and shedding of syndecan-1, suggesting that at least part of the ability of heparanase to promote an aggressive tumor phenotype comes from enhancing syndecan-1 shedding (Yang et al., 2007b). Mechanistically, increased shedding results from heparanase-mediated stimulation of expression of MMP9 and urokinase plasminogen activator and its receptor (uPA/uPAR), two proteases acting together to shed syndecan-1 from the cell surface (Purushothaman et al., 2008). Interestingly, heparanase-mediated increase in protease expression requires the presence of the enzymatically active form of heparanase indicating that changes in protease expression occur downstream of heparan sulfate degradation. Moreover, it was recently demonstrated that a reduction in the amount of syndecan heparan sulfate chains enhances core protein susceptibility to proteolytic cleavage by metalloproteinases, thus revealing that heparan sulfate regulates syndecan shedding (Ramani et al., 2012).

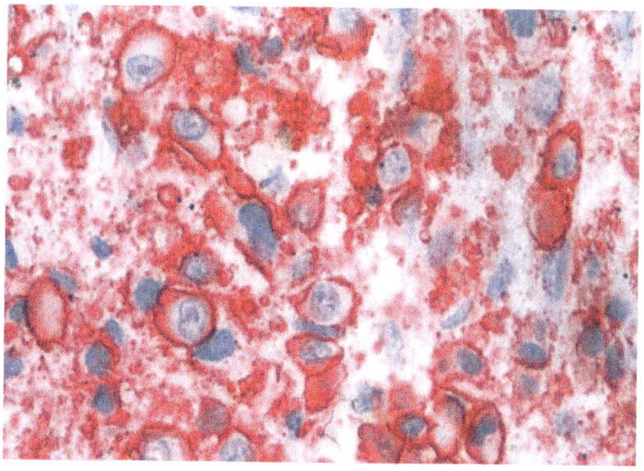

Figure 8.16 **Syndecan-1 staining in fibrotic bone marrow of a myeloma patient.** Shed syndecan-1 is localized within fibrotic regions of the bone marrow as well as on the surface of myeloma tumor cells. This concentration of shed syndecan-1 within the marrow likely helps catalyze tumor relapse following treatment by serving as a reservoir for growth factors and potentiating their effects in cell signaling. The myeloma patient bone marrow biopsy was immunostained with monoclonal antibody B-B4 against syndecan-1 followed by a biotinylated secondary antibody for 30 minutes, and the color generated using the DAKO Fast Red Substrate System. (Reprinted with permission from Bayer-Garner et al. (2001). Mod Pathol *14*, 1052–1058.)

Ongoing studies are revealing that the syndecan-1/heparanase axis plays a broad and dynamic regulatory role within the tumor microenvironment whereby it directs cross talk between tumor cells and host cells. A good example of this regulation of microenvironmental cross talk, and perhaps the most important contributor of the syndecan-1/heparanase axis to cancer progression, is the effect the axis has on stimulating tumor-mediated angiogenesis. Medium conditioned by myeloma cells expressing high levels of heparanase potently stimulate endothelial cell migration *in vitro* (Purushothaman et al., 2010). This effect was traced to elevated levels of secreted vascular endothelial growth factor (VEGF) and shed syndecan-1 in medium from heparanase-high cells compared to heparanase-low cells. Further, it was demonstrated that via its heparan sulfate chains, syndecan-1 binds to VEGF and anchors the complex to the extracellular matrix. The anchored complex then stimulates endothelial migration through activation of the VEGF receptor (Purushothaman et al., 2010). The core protein portion of syndecan-1 also plays a role in enhancing angiogenesis. This was demonstrated by showing that the endothelial migration induced by conditioned medium from heparanase-high cells could be blocked by synstatin, a peptide mimic of the syndecan-1 core protein (amino acids 92–119) (Purushothaman et al., 2010). This peptide interferes with the core-protein-mediated activation of $\alpha v\beta 3$ and $\alpha v\beta 5$ integrins on the endothelial surface, thereby blocking angiogenesis *in vitro* and *in vivo* (Beauvais et al., 2009). Therefore, shed syndecan-1 facilitates angiogenesis by performing multiple functions including binding to VEGF, anchoring the syndecan-1/VEGF complex to the extracellular matrix and activating integrins. Further supporting a role for syndecan-1 in myeloma angiogenesis is the previous finding of a correlation between high levels of shed syndecan-1 and high microvessel density in the bone marrow of myeloma patients (Andersen et al., 2005). In these patients, high microvessel density and high blood levels of interleukin(IL)-6, hepatocyte growth factor (HGF), and syndecan-1 were predictive of a shorter patient survival.

Another important player in the myeloma microenvironment is HGF, where mounting evidence indicates that this growth factor and its cell surface receptor c-Met help drive myeloma progression (Borset et al., 1996; Hov et al., 2009). Here the syndecan-1/heparanase axis also is involved because heparanase upregulates expression of HGF, and shed syndecan-1 binds to HGF in conditioned medium from heparanase-high myeloma cells (Ramani et al., 2011). Syndecan-1-facilitated c-Met signaling stimulates growth of myeloma cell lines, suggesting that enhanced levels of the syndecan-1/HGF complex could directly stimulate tumor cell proliferation (Derksen et al., 2002). Perhaps even more importantly, the syndecan-1/HGF complex stimulates paracrine signaling in osteoblasts leading to an increase in IL-11 signaling, resulting in increased osteoblast secretion of receptor activator of nuclear factor-kappa B ligand (RANKL) (Seidel et al., 2000a; Ramani et al., 2011). RANKL in turn drives osteoclastogenesis leading to enhanced bone destruction. This osteolysis fuels further tumor growth and causes devastating morbidity in myeloma patients (Mundy, 2002). HGF/c-Met paracrine signaling can also stimulate tumor-associated angiogenesis (Bussolino et al., 1992), consistent with the observation that in myeloma patients there is a positive correlation between high heparanase expression and high HGF expression and between high heparanase expression and high microvessel density within the bone marrow (Kelly et al., 2003; Ramani et al., 2011).

From these studies it has become clear that the shed syndecan-1 and heparanase promote an aggressive tumor phenotype in myeloma and that this is regulated by changes in gene expression (i.e. MMP9, VEGF, and HGF). Although the mechanism(s) whereby heparanase is controlling gene expression are not fully understood, there is evidence that syndecan-1 may play a role. Syndecan-1 is present in the nucleus of myeloma cells, but when heparanase expression is upregulated, levels of nuclear syndecan-1 drop dramatically (Chen and Sanderson, 2009). It has been demonstrated that heparan sulfate inhibits histone acetyltransferase, an enzyme associated with enhanced gene transcriptional activity (Buczek-Thomas et al., 2008), and thus a drop in nuclear syndecan-1 levels could lead to an increase in gene transcription. Consistent with this observation, addition of exogenous syndecan-1 to nuclear extracts from myeloma cells inhibited histone acetylation, and in cells expressing high heparanase and diminished nuclear syndecan-1, there is a readily detectable increase in nuclear histone acetylation (Purushothaman et al., 2011). Moreover, expression of MMP9 and VEGF in myeloma cells was blocked by chemical inhibitors of histone acetyltransferase. Together these findings suggest that when heparanase is increased in myeloma cells, levels of nuclear syndecan-1 are diminished resulting in increased activity of histone acetyltransferase and increased gene transcription.

Syndecans and solid tumors

The abundant evidence for a tumor-promoting role for shed syndecan-1 in myeloma is also paralleled by studies in solid tumors. For example, it has been suggested that syndecan-1 shed by pancreatic tumor cells acts to shuttle MMP7 into the stroma and away from tumor cells, perhaps protecting the protease from degradation (Ding et al., 2005). In some patients with infiltrating ductal carcinoma of the breast, high levels of shed syndecan-1 are found within stromal regions and can be indicative of poor prognosis, while stroma of normal breast tissue are devoid of syndecan-1 (Stanley et al., 1999; Tsanou et al., 2004; Lendorf et al., 2011). Studies utilizing mouse models of breast cancer revealed that syndecan-1 released by stromal cells enhances tumor growth and angiogenesis (Maeda et al., 2006; Su et al., 2007). In addition, as was found in myeloma cells, heparanase can stimulate shedding of syndecan-1 in breast cancer cell lines (Yang et al., 2007b). *In vivo*, this enhanced syndecan-1 shedding helps to promote a systemic osteolysis, thereby contributing to enhanced tumor morbidity and perhaps helping to establish the premetastatic niche (Kelly et al., 2010).

Targeting the syndecan-1/heparanase axis for therapy

Because of its multiple functions in regulating tumor-host cross talk, the syndecan-1/heparanase axis provides an ideal therapeutic target, since interfering with it could block multiple signaling pathways that drive tumor progression. Initial studies aimed at targeting heparan sulfate, heparanase or the syndecan-1 core protein may prove beneficial in attacking cancer. When animals bearing tumors established from cells harvested from myeloma patients were treated with a bacterial enzyme (heparinase III) that degrades heparan sulfate, growth of the tumors was dramatically inhibited (Yang et al., 2007a). It was also shown in this study that injection of tumor-bearing animals with fragments of heparan sulfate generated *ex vivo* using heparinase III inhibited tumor growth. This suggests that when exposed to certain lyases, there are cryptic regions within heparan sulfate that can inhibit tumor growth, perhaps by interfering with normal heparan sulfate function (Liu et al., 2002). Another approach has been to develop heparan sulfate mimics having low anticoagulant activity but retaining the ability to disrupt heparan sulfate-mediated signaling that drives cancer progression. One such compound recently developed is M402, a modified heparin that binds to fibroblast growth factor-2 (FGF2), VEGF, stromal cell-derived factor 1α (SDF1-α), and HGF and inhibits angiogenesis and metastatic growth of melanoma and colon carcinoma in animal models (Zhou et al., 2011). Interestingly, this compound also inhibits the activity of human heparanase, which may account for a portion of its antitumor activity. A similar strategy of developing modified heparins or heparin mimics as heparanase inhibitors has also shown success in animal models, and one, PI-88 has shown partial success in clinical trials, particularly in patients with hepatocellular carcinoma (Liu et al., 2009). Another heparanase inhibitor, SST0001, blocks growth and angiogenesis in animal models of myeloma and other cancers and will soon enter phase I clinical trials (Shafat et al., 2010; Ritchie et al., 2011). Preparation of SST0001 includes complete *N*-acetylation and controlled glycol-splitting of heparin, yielding a compound that is a potent heparanase inhibitor but with minimal anticoagulant activity (Naggi et al., 2005). All of these heparan sulfate mimics in addition to inhibiting heparanase also interfere with growth factor activities, similar to what was shown for M402. In addition, an added benefit of blocking heparanase activity is that this can reduce levels of shed syndecan-1 as was shown with SST0001 (Ritchie et al., 2011).

Yet another potential target for therapy is the syndecan-1 core protein. When downregulated in myeloma cells using shRNA, tumors formed by these cells grew much slower than tumors formed by cells expressing wild-type levels of the proteoglycan (Khotskaya et al., 2009). Knockdown of syndecan-1 also dramatically inhibited tumor angiogenesis and dissemination, consistent with the role of syndecan-1 as a key mediator of myeloma progression. Of importance, knockdown of syndecan-1 expression or inhibition of heparanase activity had no effect on growth of tumor cells *in vitro*; growth inhibitory effects were only seen *in vivo* (Khotskaya et al., 2009; Ritchie et al., 2011). This further supports the notion that the major role for the syndecan-1/heparanase axis in promoting tumor progression lies in regulating cross talk between tumor and host cells. Lastly, another approach to attack the syndecan-1 core protein is by blocking functions within the core protein. For example, as mentioned previously, the synstatin peptide that mimics the integrin activating region of syndecan-1 blocks angiogenesis and growth of breast tumors in mice (Beauvais et al., 2009).

8.6.5 Future prospects

Syndecans are emerging as important receptors in tumor progression in their own right. While long considered "coreceptors" always functioning in concert with other receptors such as growth factor receptors and adhesion receptors, it is clear that they can signal independently and, when shed, have multiple effects on the tumor environment. At least with respect to myeloma, there are encouraging prospects that targeting syndecans and their heparan sulfate chains may yield new approaches to treatment.

8.6.6 Take-home message

Syndecans are transmembrane proteoglycans; however, an important feature is their sensitivity to a range of proteinases that cleave the extracellular core protein. This may occur constitutively, but is markedly upregulated in several diseases. Shed ectodomains, which include the heparan sulfate chains, can retain biological properties that influence proliferation and cell behavior. The importance of this process is shown by syndecan-1, where shedding into the pericellular environment is now believed to be a key feature of myeloma. It may be that targeting the heparanase enzyme and syndecan-1 offer a novel path to treatment of this cancer.

References

Andersen, N. F., Standal, T., Nielsen, J. L., et al. (2005). Syndecan-1 and angiogenic cytokines in multiple myeloma: correlation with bone marrow angiogenesis and survival. Br J Haematol *128*, 210–217.

Bass, M. D., Morgan, M. R., and Humphries, M. J. (2009). Syndecans shed their reputation as inert molecules. Sci Signal *2*, pe18.

Bayer-Garner, I. B., Sanderson, R. D., Dhodapkar, M. V., Owens, R. B., and Wilson, C. S. (2001). Syndecan-1 (CD138) immunoreactivity in bone marrow biopsies of multiple myeloma: shed syndecan-1 accumulates in fibrotic regions. Mod Pathol *14*, 1052–1058.

Beauvais, D. M., Ell, B. J., McWhorter, A. R., and Rapraeger, A. C. (2009). Syndecan-1 regulates alphavbeta3 and alphavbeta5 integrin activation during angiogenesis and is blocked by synstatin, a novel peptide inhibitor. J Exp Med *206*, 691–705.

Bondeson, J., Wainwright, S., Hughes, C., and Caterson, B. (2008). The regulation of ADAMTS4 and ADAMTS5 aggrecanases in osteoarthritis: a review. Clin Exp Rheumatol *26*, 139–145.

Borset, M., Hjorth-Hansen, H., Seidel, C., Sundan, A., and Waage, A. (1996). Hepatocyte growth factor and its receptor c-met in multiple myeloma. Blood *88*, 3998–4004.

Brule, S., Charnaux, N., Sutton, A., et al. (2006). The shedding of syndecan-4 and syndecan-1 from HeLa cells and human primary macrophages is accelerated by SDF-1/CXCL12 and mediated by the matrix metalloproteinase-9. Glycobiology *16*, 488–501.

Buczek-Thomas, J. A., Hsia, E., Rich, C. B., Foster, J. A., and Nugent, M. A. (2008). Inhibition of histone acetyltransferase by glycosaminoglycans. J Cell Biochem *105*, 108–120.

Bussolino, F., Di Renzo, M. F., Ziche, M., et al. (1992). Hepatocyte growth factor is a potent angiogenic factor which stimulates endothelial cell motility and growth. J Cell Biol *119*, 629–641.

Chen, L., and Sanderson, R. D. (2009). Heparanase regulates levels of syndecan-1 in the nucleus. PLoS One *4*, e4947.

Choi, S., Lee, H., Choi, J. R., and Oh, E. S. (2010). Shedding; towards a new paradigm of syndecan function in cancer. BMB Rep *43*, 305–310.

Couchman, J. R. (2010). Transmembrane signalling proteoglycans. Annu Rev Cell Dev Biol *26*, 89–114.

Derksen, P. W., Keehnen, R. M., Evers, L. M., van Oers, M. H., Spaargaren, M., and Pals, S. T. (2002). Cell surface proteoglycan syndecan-1 mediates hepatocyte growth factor binding and promotes Met signalling in multiple myeloma. Blood *99*, 1405–1410.

Dews, I. C., and Mackenzie, K. R. (2007). Transmembrane domains of the syndecan family of growth factor coreceptors display a hierarchy of homotypic and heterotypic interactions. Proc Natl Acad Sci U S A *104*, 20782–20787.

Dhodapkar, M. V., Kelly, T., Theus, A., Athota, A. B., Barlogie, B., and Sanderson, R. D. (1997). Elevated levels of shed syndecan-1 correlate with tumour mass and decreased matrix metalloproteinase-9 activity in the serum of patients with multiple myeloma. Br J Haematol *99*, 368–371.

Ding, K., Lopez-Burks, M., Sanchez-Duran, J. A., Korc, M., and Lander, A. D. (2005). Growth factor-induced shedding of syndecan-1 confers glypican-1 dependence on mitogenic responses of cancer cells. J Cell Biol *171*, 729–738.

Echtermeyer, F., Bertrand, J., Dreier, R., et al. (2009). Syndecan-4 regulates ADAMTS-5 activation and cartilage breakdown in osteoarthritis. Nat Med *15*, 1072–1076.

Edwards, D. R., Handsley, M. M., and Pennington, C. J. (2008). The ADAM metalloproteinases. Mol Aspects Med *29*, 258–289.

Elenius, V., Götte, M., Reizes, O., Elenius, K., and Bernfield, M. (2004). Inhibition by the soluble syndecan-1 ectodomains delays wound repair in mice overexpressing syndecan-1. J Biol Chem *279*, 41928–41935.

Endo, K., Takino, T., Miyamori, H., et al. (2003). Cleavage of syndecan-1 by membrane type matrix metalloproteinase-1 stimulates cell migration. J Biol Chem *278*, 40764–40770.

Fitzgerald, M. L., Wang, Z., Park, P. W., Murphy, G., and Bernfield, M. (2000). Shedding of syndecan-1 and -4 ectodomains is regulated by multiple signalling pathways and mediated by a TIMP-3-sensitive metalloproteinase. J Cell Biol *148*, 811–824.

Hov, H., Tian, E., Holien, T., et al. (2009). c-Met signalling promotes IL-6-induced myeloma cell proliferation. Eur J Haematol *82*, 277–287.

Kainulainen, V., Wang, H., Schick, C., and Bernfield, M. (1998). Syndecans, heparan sulfate proteoglycans, maintain the proteolytic balance of acute wound fluids. J Biol Chem *273*, 11563–11569.

Kelly, T., Miao, H. Q., Yang, Y., et al. (2003). High heparanase activity in multiple myeloma is associated with elevated microvessel density. Cancer Res *63*, 8749–8756.

Kelly, T., Suva, L. J., Nicks, K. M., MacLeod, V., and Sanderson, R. D. (2010). Tumor-derived syndecan-1 mediates distal cross-talk with bone that enhances osteoclastogenesis. J Bone Miner Res *25*, 1295–1304.

Khotskaya, Y. B., Dai, Y., Ritchie, J. P., et al. (2009). Syndecan-1 is required for robust growth, vascularization and metastasis of myeloma tumors in vivo. J Biol Chem *284*, 26085–26095.

Leco, K. J., Khokha, R., Pavloff, N., Hawkes, S. P., and Edwards, D. R. (1994). Tissue inhibitor of metalloproteinases-3 (TIMP-3) is an extracellular matrix-associated protein with a distinctive pattern of expression in mouse cells and tissues. J Biol Chem *269*, 9352–9360.

Lendorf, M. E., Manon-Jensen, T., Kronqvist, P., Multhaupt, H. A. B., and Couchman, J. R. (2011). Syndecan-1 and syndecan-4 are independent indicators in breast carcinoma. J Histochem Cytochem *59*, 615–629.

Li, Q., Park, P. W., Wilson, C. L., and Parks, W. (2002). Matrilysin shedding of syndecan-1 regulates chemokine mobilization and transepithelial efflux of neutrophils in acute lung injury. Cell *111*, 635–646.

Liu, C. J., Lee, P. H., Lin, D. Y., et al. (2009). Heparanase inhibitor PI-88 as adjuvant therapy for hepatocellular carcinoma after curative resection: a randomized phase II trial for safety and optimal dosage. J Hepatol *50*, 958–968.

Liu, D., Shriver, Z., Venkataraman, G., El Shabrawi, Y., and Sasisekharan, R. (2002). Tumor cell surface heparan sulfate as cryptic promoters or inhibitors of tumour growth and metastasis. Proc Natl Acad Sci U S A *99*, 568–573.

Maeda, T., Desouky, J., and Friedl, A. (2006). Syndecan-1 expression by stromal fibroblasts promotes breast carcinoma growth in vivo and stimulates tumour angiogenesis. Oncogene *25*, 1408–1412.

Manon-Jensen, T., Itoh, Y., and Couchman, J. R. (2010). Proteoglycans in health and disease: the multiple roles of syndecan shedding. FEBS J *277*, 3876–3889.

Mundy, G. (2002). Metastasis to bone: causes, consequences and therapeutic opportunities. Nat Rev Cancer *2*, 584–593.

Munesue, S., Yoshitomi, Y., Kusano, Y., et al. (2007). A novel function of syndecan-2, suppression of matrix metalloproteinase-2 activation, which causes suppression of metastasis. J Biol Chem *282*, 28164–28174.

Nagase, H., Visse, R., and Murphy, G. (2006). Structure and function of matrix metalloproteinases and TIMPs. Cardiovasc Res *69*, 562–573.

Naggi, A., Casu, B., Perez, M., et al. (2005). Modulation of the heparanase-inhibiting activity of heparin through selective desulfation, graded N-acetylation, and glycol splitting. J Biol Chem *280*, 12103–12113.

Pruessmeyer, J., Martin, C., Hess, F. M., et al. (2010). A disintegrin and metalloproteinase 17 (ADAM17) mediates inflammation-induced shedding of syndecan-1 and -4 by lung epithelial cells. J Biol Chem *285*, 555–564.

Purushothaman, A., Chen, L., Yang, Y., and Sanderson, R. D. (2008). Heparanase stimulation of protease expression implicates it as a master regulator of the aggressive tumor phenotype in myeloma. J Biol Chem *283*, 32628–32636.

Purushothaman, A., Hurst, D. R., Pisano, C., Mizumoto, S., Sugahara, K., and Sanderson, R. D. (2011). Heparanase-mediated loss of nuclear syndecan-1 enhances histone acetyltransferase (HAT) activity to promote expression of genes that drive an aggressive tumor phenotype. J Biol Chem *286*, 30377–30383.

Purushothaman, A., Uyama, T., Kobayashi, F., et al. (2010). Heparanase-enhanced shedding of syndecan-1 by myeloma cells promotes endothelial invasion and angiogenesis. Blood *115*, 2449–2457.

Ramani, V. C., Yang, Y., Ren, Y., Nan, L., and Sanderson, R. D. (2011). Heparanase plays a dual role in driving hepatocyte growth factor (HGF) signaling by enhancing HGF expression and activity. J Biol Chem *286*, 6490–6499.

Ramani, V. C., Pruett, P. S., Thompson, C. A., DeLucas, L. D., and Sanderson, R. D. (2012). Heparan sulfate chains of syndecan-1 regulate ectodomain shedding. J Biol Chem *287*, 9952–9961.

Ritchie, J. P., Ramani, V. C., Ren, Y., et al. (2011). SST0001, a chemically modified heparin, inhibits myeloma growth and angiogenesis via disruption of the heparanase/syndecan-1 axis. Clin Cancer Res *17*, 1382–1393.

Rodríguez-Manzaneque, J. C., Carpizo, D., Plaza-Calonge Mdel, C., et al. (2009) Cleavage of syndecan-4 by ADAMTS1 provokes defects in adhesion. Int J Biochem Cell Biol *41*, 800–810.

Saftig, P., and Reiss, K. (2011). The "a disintegrin and metalloproteases" ADAM10 and ADAM17: novel drug targets with therapeutic potential? Eur J Cell Biol *90*, 527–535.

Schmidt, A., Echtermeyer, F., Alozie, A., Brands, K., and Buddecke, E. (2005). Plasmin and thrombin-accelerated shedding of syndecan-4 ectodomain generates cleavage sites at Lys(114)-Arg(115) and Lys(129)-Val(130) bonds. J Biol Chem *280*, 34441–34446.

Schulz, J. G., Annaert, W., Vandekerckhove, J., Zimmermann, P., De Strooper, B., and David, G. (2003). Syndecan 3 intramembrane proteolysis is presenilin/gamma-secretase-dependent and modulates cytosolic signalling. J Biol Chem *278*, 48651–48657.

Seidel, C., Borset, M., Hjertner, O., et al. (2000a). High levels of soluble syndecan-1 in myeloma-derived bone marrow: modulation of hepatocyte growth factor activity. Blood *96*, 3139–3146.

Seidel, C., Sundan, A., Hjorth, M., et al. (2000b). Serum syndecan-1: a new independent prognostic marker in multiple myeloma. Blood *95*, 388–392.

Shafat, I., Ben-Arush, M. W., Issakov, J., et al. (2010). Preclinical and clinical significance of heparanase in Ewing's sarcoma. J Cell Mol Med *15*, 1857–1864.

Stanley, M. J., Stanley, M. W., Sanderson, R. D., and Zera, R. (1999). Syndecan-1 expression is induced in the stroma of infiltrating breast carcinoma. Am J Clin Pathol *112*, 377–383.

Su, G., Blaine, S. A., Qiao, D., and Friedl, A. (2007). Shedding of syndecan-1 by stromal fibroblasts stimulates human breast cancer cell proliferation via FGF2 activation. J Biol Chem *282*, 14906–14915.

Tsanou, E., Ioachim, E., Briasoulis, E., et al. (2004). Clinicopathological study of the expression of syndecan-1 in invasive breast carcinomas. correlation with extracellular matrix components. J Exp Clin Cancer Res *23*, 641–650.

Wang, Z., Götte, M., Bernfield, M., and Reizes, O. (2005). Constitutive and accelerated shedding of murine syndecan-1 is mediated by cleavage of its core protein at a specific juxtamembrane site. Biochemistry *44*, 12355–12361.

Whiteford, J. R., Behrends, V., Kirby, H., Kusche-Gullberg, M., Muramatsu, T., and Couchman, J. R. (2007). Syndecans promote integrin-mediated adhesion of mesenchymal cells in two distinct pathways. Exp Cell Res *313*, 3902–3913.

Yang, Y., Macleod, V., Bendre, M., et al. (2005). Heparanase promotes the spontaneous metastasis of myeloma cells to bone. Blood *105*, 1303–1309.

Yang, Y., MacLeod, V., Dai, Y., et al. (2007a). The syndecan-1 heparan sulfate proteoglycan is a viable target for myeloma therapy. Blood *110*, 2041–2048.

Yang, Y., Macleod, V., Miao, H. Q., et al. (2007b). Heparanase enhances syndecan-1 shedding: a novel mechanism for stimulation of tumor growth and metastasis. J Biol Chem *282*, 13326–13333.

Yang, Y., Yaccoby, S., Liu, W., et al. (2002). Soluble syndecan-1 promotes growth of myeloma tumors in vivo. Blood *100*, 610–617.

Yu, W. H., and Woessner, J. F. (2001). Heparin-enhanced zymographic detection of matrilysin and collagenases. Anal Biochem *293*, 38–42.

Yu, W. H., Yu, S., Meng, Q., Brew, K., and Woessner, J. F. (2000). TIMP-3 binds to sulfated glycosaminoglycans of the extracellular matrix. J Biol Chem *275*, 31226–31232.

Zhou, H., Roy, S., Cochran, E., et al. (2011). M402, a novel heparan sulfate mimetic, targets multiple pathways implicated in tumor progression and metastasis. PLoS One *6*, e21106.

8.7 PG receptors with phosphatase action in cancer and angiogenesis

Marina Koutsioumpa and Evangelia Papadimitriou

8.7.1 Introduction

The regulation of protein phosphorylation, whether on serine (Ser), threonine (Thr), or tyrosine (Tyr) residues, plays a pivotal role in virtually all aspects of eukaryotic development. From the regulation of the cell cycle to cellular proliferation and differentiation, the delicate balance between the phosphorylation activity of kinases and the dephosphorylation activity of phosphatases controls the outcome of countless signal transduction cascades. The diversity of cellular contexts in which phosphorylation-dependent signaling mechanisms function and the conservation of similar signaling mechanisms in a wide variety of organisms make this a very exciting and dynamic frontier in cell biology. Although our insights into protein tyrosine phosphatase (PTP)–mediated signaling are not as good as into protein tyrosine kinase (PTK)–mediated signaling, it has been established definitively that several PTPs play crucial roles during development and in several pathophysiological conditions.

Following the purification of the first PTP, PTP1B, at least 75 distinct PTPs have been cloned on the basis of sequence homology in the catalytic domain. PTPs have been identified in many different species, ranging from bacteria to human. Random sequencing projects suggest that the human genome encodes approximately 500 PTPs. The PTP superfamily is structurally distinct from the Ser/Thr phosphatase family, consists of "classical" PTPs, dual specificity phosphatases (DSPs), and low-molecular-weight PTPs (LMW PTPs). All members have at least one Tyr phosphatase domain containing an 11-residue sequence motif, in which Arg and Cys residues are required for catalytic activity. The DSPs and LMW PTPs are cytoplasmically localized. The "classical" PTPs can be divided into two large groups, based on their overall structure, the cytoplasmic and the transmembrane PTPs, which are tentatively called receptor protein tyrosine phosphatases (RPTPs).

The RPTPs have been categorized based on their overall structure, and at least eight different types have been defined (▶Figure 8.17). Their extracellular domains are diverse, ranging from very short (e.g. RPTP-ε), to very long (e.g. LAR). Many different protein modules have been identified in the extracellular domains of RPTPs, including those reminiscent of cell adhesion molecules, such as fibronectin type III (FNIII)–like and immunoglobulin (Ig)–like domains. Other molecules that have been identified in the ectodomain of RPTPs are the meprin/A5-protein/PTPμ (MAM) domain, as well as a domain with high homology to carbonic anhydrases. Most RPTPs contain two tandem

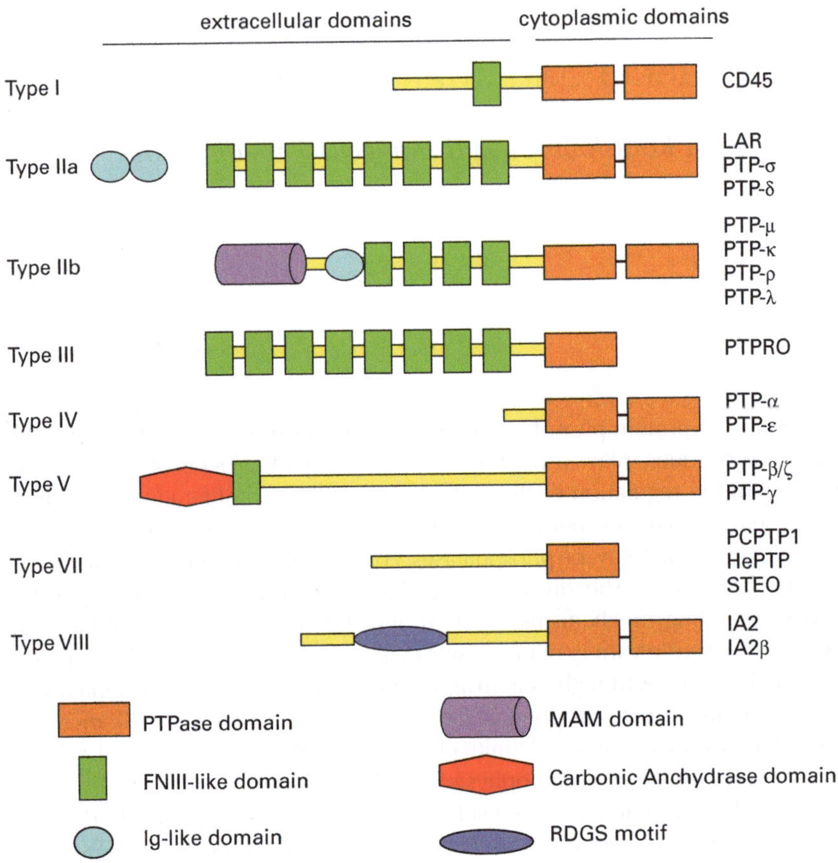

Figure 8.17 Schematic diagram of the eight RPTP subfamilies. Type I RPTPs contain a single FNIII-like domain extracellularly and two cytoplasmic PTPase domains. The type IIa RPTPs have large extracellular domains consisting of three NH2-terminal Ig-like and eight FNIII-like domains. Type IIb RPTPs have an extracellular MAM domain, a single Ig-like domain, and multiple FNIII-like domains. Type III RPTPs have a series of FNIII-like domains extracellularly but are unusual in that they have only one cytoplasmic PTPase domain. Type IV RPTPs have the shortest extracellular domains, which are often heavily glycosylated, while type V RPTP extracellular domains have a carbonic anhydrase domain, linked to a single FNIII-like domain. Type VII RPTPs have one cytoplasmic PTPase domain and a short extracellular domain, while type VIII RPTPs (thought to be catalytically inactive) contain an RDGS (Arg-Gly-Asp-Ser) adhesion recognition motif.

cytoplasmic PTPase domains, with the membrane proximal PTP domain containing the majority of the catalytic activity, which is essential for the function of several RPTPs. The diverse protein sequences flanking the highly conserved catalytic domains frequently serve a regulatory function, regulating the signaling molecules that interact with each receptor (reviewed in Johnson and Van Vactor, 2003).

8.7.2 Glycosylated transmembrane protein phosphatase receptors

Only few RPTPs are subject to extensive posttranslational glycosylation. It has been shown that leukocyte common antigen (cluster of differentiation 45 [CD45]), leukocyte common antigen-related protein (LAR), RPTP-α, RPTP-ε, RPTP-γ, RPTP-β/ζ, and density-enhanced phosphatase-1 (DEP1) are more or less glycosylated in their extracellular domain. CD45 is characterized as a glycoprotein, heavily *N*- and *O*-glycosylated (Barcley et al., 1987). Furthermore, it has been shown that LAR has multiple *N*-linked glycosylation sites (Krueger et al., 1990). RPTP-α and -ε contain extensive potential *N*-linked glycosylation sites (N-X-S or N-X-T) (Krueger et al., 1990), while RPTP-α also contains *O*-linked glycosylation sites (Daum et al., 1994). The extracellular domain of human RPTP-γ contains eight putative *N*-linked glycosylation sites, six of which are conserved in the murine homolog (Barnea et al., 1993). DEP1 has been proposed to be a chondroitin sulfate proteoglycan (Sörby, 2001), but among all RPTPs, only RPTP-β/ζ has been proved to be expressed in a proteoglycan form (Barnea et al., 1994). Proteoglycans bind many extracellular matrix components and growth factors through both their core protein and glycosaminoglycan portions and are thought to play important roles in the regulation of cell adhesion, growth, motility, and differentiation.

CD45

In mammals, many splice isoforms of CD45 (also known as RPTP-c) have been identified, all of which are expressed exclusively in the hematopoietic system. The ectodomain of CD45 does not resemble cell adhesion molecules, although a single FNIII-like domain and possibly spectrin-like repeats have been identified. CD45 dephosphorylates the Src-family members lck and fyn in T cells and has been implicated in antigen receptor-induced responses in both T and B cells (reviewed in Den Hertog et al., 1999). Moreover, CD45 was identified as the elusive Janus kinase (JAK) tyrosine phosphatase that negatively regulates cytokine receptor activation involved in the differentiation, proliferation, and antiviral immunity of hematopoietic cells (Irie-Sasaki et al., 2001). Based on the positive regulatory role on T cell receptor signaling, CD45 could thus be a useful treatment against autoimmunity diseases and transplant rejection. Indeed, a monoclonal antibody against CD45 has been shown to prevent renal allograft rejection in mice (Lazarovits et al., 1996). Modulation of CD45 splice variants provides a unique opportunity to design drugs that turn off or on antigen and cytokine receptor signaling in cancer, transplantation, or autoimmunity.

LAR

The mammalian LAR subfamily consists of three members, LAR (also known as PTPRf), RPTP-δ, and RPTP-σ, which have the ability to bind heparan sulfate proteoglycans (like agrin and collagen XVIII), through a perfectly conserved basic amino acid motif (KKXKK) in the first Ig-like domain (Aricescu et al., 2002). In epithelial cells, LAR associates with cadherin-catenin complex and dephosphorylates β-catenin, which correlates with its ability to inhibit cell migration and tumor formation (Muller et al., 1999). It was recently established that LAR reduces the basal c-Abl activity, thereby allowing for platelet-derived growth factor (PDGF) β-receptor kinase activation in mouse embryonic fibroblasts (Zheng et al., 2011). Moreover, LAR and Src act as death-associated protein kinase (DAPK) regulators through their reciprocal modification of DAPK Y491/492 residues, leading to tumor progression (Wang et al., 2007).

RPTP-α and -ε

RPTP-α and RPTP-ε differ from all other transmembrane PTPs, in that they display particularly short extracellular domains of only 123 and 27 residues, respectively. Their extracellular domains share no obvious sequence similarity, although both are rich in Ser and Thr (Krueger et al., 1990). RPTP-ε can support or inhibit mitogenic signaling in a context-dependent manner. Neu/RPTP-ε/Src signaling pathway exists in mammary tumor cells induced by Neu, in which Neu phosphorylates RPTP-ε, thereby driving the phosphatase to activate c-Src and contribute to the transformed phenotype of these cells (Berman-Golan and Elson, 2007). On the other hand, a clear role for RPTP-ε in downregulating signaling events was shown in the context of insulin receptor signaling (reviewed in Den Hertog et al., 1999), mitogen-activated protein kinase signaling (Toledano-Katchalski et al., 2003) and endothelial cell proliferation (Thompson et al., 2001).

RPTP-α is highly conserved in vertebrates and is mainly expressed in the developing brain of various species, especially in glia. Over the years, RPTP-α has been shown to be involved in a number of different signaling mechanisms. Previous reports indicated that RPTP-α plays a role in neuronal differentiation of P19 embryonal carcinoma cells and that its ectopic expression in fibroblasts leads to cell transformation and tumorigenesis. The majority of the roles played in these cellular processes involve the ability of RPTP-α to activate the proto-oncogenes Src and Fyn, by dephosphorylating their C-terminal inhibitory phosphotyrosine (reviewed in Den Hertog et al., 1999). Hence, inhibitors targeting RPTP-α might have therapeutic effects in tumors with elevated Src kinase activity.

RPTP-γ

RPTP-γ, together with RPTP-β/ζ, form a subfamily of enzymes characterized by the presence of a carbonic anhydrase-like and a FNIII-like domain in the N-terminal portion of the extracellular domain. The remainder of the RPTP-γ and RPTP-β/ζ extracellular domains does not share significant homology. RPTP-γ is thought to be a candidate tumor suppressor gene, whose functional loss is involved in the pathogenesis of kidney and lung tumors (LaForgia et al., 1991). In the same line, RPTP-γ expression is dramatically reduced in ovarian tumors (Van Niekerk and Poels, 1999). It was demonstrated that RPTP-γ is able to inhibit anchorage-independent growth of breast cancer cells in soft agar and reduce the estrogenic responses of MCF7 cell proliferation, suggesting that it may have a role in regulating the process of tumorigenesis in human breast (Liu et al., 2004).

DEP1

DEP1 (also known as CD148 or PTPRj) is expressed in several cell types, including endothelial and hematopoietic cells (reviewed in Matozaki et al., 2010). It comprises an extracellular domain containing eight FNIII-like motifs, a transmembrane domain, and a single intracellular catalytic domain. The expression levels of DEP1 were initially reported to increase with cell density, suggesting that it might work as a regulator of cell contact-mediated growth inhibition (Ostman et al., 1994). Later, DEP1 was identified as the gene associated with the mouse colon cancer susceptibility locus (*Scc1*) and was frequently found to be deleted or mutated in human cancers (Ruivenkamp et al., 2002). Furthermore, by generating a monoclonal antibody against DEP1 (Takahashi et al., 2006), it was suggested that DEP1 may be a valuable molecular target for antiangiogenesis therapy.

8.7.3 RPTP-β/ζ

Protein structure of RPTP-β/ζ

RPTP-β/ζ (also known as RPTP-ζ) is a member of type V RPTPs, initially isolated from neural tissue as a transmembrane protein-tyrosine-phosphatase that consists of a putative signal peptide, a very large extracellular domain containing an N-terminal sequence homologous to carbonic anhydrase, a transmembrane region and a cytoplasmic portion that contains two repeated PTPase domains. A shorter transmembrane and two secreted isoforms corresponding to the extracellular portions of the long and short transmembrane isoforms have been described, all considered splice variants of RPTP-β/ζ. Phosphacan and long RPTP-β/ζ receptor are heavily glycosylated chondroitin sulfate proteoglycans, whereas short isoform of phosphacan and short RPTP-β receptor isoforms are glycoproteins. The apparent molecular mass of the core and glycosylated forms of RPTP-β/ζ are approximately 250 and 300 kDa, respectively. However, the full-length form of RPTP-β/ζ has been also detected in a nonproteoglycan form in some tissues. Sulfation, carbohydrate composition and oligosaccharide structure of RPTP-β/ζ have been shown to be developmentally regulated and certain carbohydrates can alter its affinity for other proteins. Apart from the RPTP-β/ζ splicing variants that are normally expressed under physiological conditions, RPTP-β/ζ is cleaved by matrix metalloproteinase-9, tumor necrosis factor-α converting enzyme, presenilin/γ-secretase, and plasmin leading to secreted, transmembrane, or cytoplasmic forms of not yet fully identified biological significance (reviewed in Papadimitriou et al., 2009).

Molecules interacting with RPTP-β/ζ

Screening various extracellular matrix and cell adhesion molecules revealed that neutral cell adhesion molecule (N-CAM), neuron-glia cell adhesion molecule (Ng-CAM), and tenascin bind to phosphacan, through its complex-type *N*-linked oligosaccharides (Barnea et al., 1994; Milev et al., 1994). Amphoterin also colocalizes with phosphacan in the developing nervous system and clearly enhances binding of phosphacan to the neural cell adhesion molecule contactin (Milev et al., 1998). Maeda et al. (1996) firstly reported that phosphacan and RPTP-β/ζ are the functional receptors for the growth factor pleiotrophin (PTN), involved in neuronal migration. It has been shown that PTN binding to RPTP-β/ζ is mediated by both the protein core and the chondroitin sulfate chains of the receptor and this interaction is inhibited by exogenously supplied chondroitin sulfates. Moreover, the structure of chondroitin sulfate on phosphacan changes dynamically during development of the brain, concomitant with the changes in the affinity for PTN (Maeda et al., 2003). Finally, RPTP-β/ζ has been shown to be the putative F3 receptor on Swann cells (Thomaidou et al., 2001) and is also the receptor for Vac A, the toxin of the *Helicobacter pylori* (Fujikawa et al., 2003).

RPTP-β/ζ-mediated biological actions and signaling

It has been suggested that PTN binding to RPTP-β/ζ leads to dimerization of the receptor and inhibition of the PTPase activity, leading to increased tyrosine phosphorylation of β-catenin, disruption of the association of β-catenin with the cytoplasmic domain of the cadherins, destabilization of adherins junctions, and decreased homophilic cell-cell adhesion. PTN binding to RPTP-β/ζ activates protein kinase C (PKC), stimulates the PKC-catalyzed phosphorylation of β-adducin and the translocation of the latter to either nuclei or a membrane associated site, contributing to the disruption of cytoskeletal complexes. Fyn was found to interact with the intracellular domain of RPTP-β/ζ and possibly mediates PTN-stimulated cytoskeletal phenotype. Thus, PTN-RPTP-β/ζ signaling seems to regulate cytoskeletal stability, cell plasticity and cell-cell adhesion mechanisms. In addition, it was indicated that the PTN/RPTP-β/ζ signaling pathway is a critical regulator of the steady state levels of tyrosine phosphorylation and activation of anaplastic lymphoma kinase (ALK). Moreover, PTN binding to RPTP-β/ζ, increases the tyrosine phosphorylation of G protein coupled receptor kinase-interactor 1 and interaction with the PDZ domains containing membrane associated guanylate kinases, Magi1 and Magi3 (reviewed in Papadimitriou et al., 2009).

PTN binding to RPTP-β/ζ in endothelial cells leads to dephosphorylation and thus activation of c-Src, focal adhesion kinase, phosphatidylinositol-3-kinase and mitogen-activated protein kinases, all participating in PTN-induced endothelial cell migration and tube formation on matrigel. Recently, it was shown that in order for RPTP-β/ζ to induce cell migration, the presence of $\alpha_v\beta_3$ integrin is required. RPTP-β/ζ and $\alpha_v\beta_3$ form a functional complex on the surface of endothelial and glioma cell lines and RPTP-β/ζ seems to be responsible for β_3 tyrosine phosphorylation through the activation of c-Src (Mikelis et al., 2009, 2011).

Collectively, the PTN/RPTP-β/ζ signaling pathway is unique and important, as it coordinately regulates steady state levels of tyrosine phosphorylation of key proteins in different cellular systems and different signaling networks (reviewed in Papadimitriou et al., 2009).

Besides PTN, the other member of the same family of heparin-binding growth factors, midkine, is a soluble ligand for RPTP-β/ζ and it has been shown that RPTP-β/ζ and concerted signaling involving phosphoinositide-3 and mitogen-activated protein kinases, are required for midkine-induced migration (Qi et al., 2001), while Sakaguchi et al. (2003) suggested that the survival-promoting signal of midkine was received by a receptor complex containing RPTP-β/ζ.

RPTP-β/ζ and angiogenesis

Several reports indicate a positive correlation between RPTP-β/ζ and angiogenesis, a key step in the progress of many tumors. Initially, it was demonstrated that RPTP-β/ζ downregulation was shown to interrupt PTN signaling in human umbilical vein endothelial cells (HUVEC) and abolished its biological activity on cell migration and differentiation. More recently, it was suggested that $\alpha_v\beta_3$ integrin is activated by RPTP-β/ζ upon PTN binding, mediating the stimulatory effect of PTN in endothelial cell migration. Other studies indicated that PTN mediates the stimulatory effects of endothelial nitric oxide synthase/nitric oxide (eNOS/NO) and aprotinin on human endothelial cell migration in an RPTP-β/ζ-dependent manner (Koutsioumpa et al., 2009; Polytarchou et al., 2009), while RPTP-β/ζ may also act independently of PTN in aprotinin-induced endothelial cell migration (Koutsioumpa et al., 2009). Furthermore, expression of PTN and RPTP-β/ζ in human placenta suggests roles in trophoblast life cycle and angiogenesis (Ball et al., 2009). Moreover, the inhibitory effect of $PTN_{111-136}$ (a peptide corresponding to the C-terminal region of PTN) on prostate cancer (PC-3) tumor growth and angiogenesis *in vivo*, has been linked to a direct or indirect binding of this peptide to RPTP-β/ζ (Hamma-Kourbali et al., 2011).

RPTP-β/ζ and cancer

RPTP-β/ζ is expressed in human primary and metastatic melanomas (Goldmann et al., 2000), in late passage human osteosarcoma Saos-2 cells (Hausser and Brenner, 2005), in cancer stem-like cells (He et al., 2010), in meningiomas (Tong et al., 2006), glioblastomas and neuroblastomas (Goldmann et al., 2000). Chronic oxidative stress, which is frequently observed in many tumors, can induce genomic amplification of RPTP-β/ζ, leading to activation of β catenin pathways (Liu et al., 2007). Moreover, hypoxia inducible factor-2, which is activated in the hypoxic environment of many tumors, preferentially activates the RPTP-β/ζ gene in human embryonic kidney 293 T (HEK293T) cells (Wang et al., 2010). RPTP-β/ζ was shown to be localized not only in its normal association with the cell membrane, but also scattered in cytoplasm and in nuclei in different breast cancer cells and, in the case of infiltrating ductal carcinomas, its distribution changes as the breast cancer becomes more malignant (Perez-Pinera et al., 2007).

Many data declare the important role of RPTP-β/ζ in regulation of cancer cell motility, mostly studied in glioblastoma cells (Muller et al., 2003). Upregulation of PTN and RPTP-β/ζ in human astrocytic tumor cells can create an autocrine loop that is important for glioma cell migration (Ulbricht et al., 2003). Immobilized full-length PTN promotes haptotactic migration of U87MG, U373MG and LN229 glioma cells, in an RPTP-β/ζ-dependent fashion (Lu et al., 2005). In U373 glioma cells, PTN increases tyrosine phosphorylation of different RPTP-β/ζ substrates required for epithelial to mesenchymal transition (Perez-Pinera et al., 2006). Negative regulation of RPTP-β/ζ by leucine rich repeat containing 4 suppresses the invasion ability of gliomas cells (Wu et al., 2006). Ulbricht et al. (2006) reported suppression of glioblastoma growth *in vitro* and *in vivo* by RNA interference targeting of RPTP-β/ζ. Moreover, antibodies directed to RPTP-β/ζ and coupled to the cytotoxin saporin either directly or via a secondary antibody, kill glioma cells *in vitro* and significantly delay human glioma tumors in a mouse xenograft model (Foehr et al., 2006).

Besides glioblastomas, in an *in vivo* model of intestinal tumorigenesis, enterocytes showed decreased association between β-catenin and RPTP-β/ζ, which was consistent with increased levels of tyrosine-phosphorylated β-catenin and decreased cell migration (Carothers et al., 2001). Studies also revealed that PTN/RPTP-β/ζ signaling regulates menin-induced lung cancer cell migration through focal adhesion kinase, phosphatidylinositol 3-kinase and phosphorylated extracellular signal regulated kinase

1/2 (Feng et al., 2010). RPTP-β/ζ downregulation was finally shown to result in decreased migration, adhesion, and anchorage-independent growth of DU145 prostate cancer cells (Diamantopoulou et al., 2010).

8.7.4 Conclusions

Although the role initially attributed to RPTPs was limited to inhibition of tyrosine kinase pathways and was thus considered inhibitory, it is apparent that they are not always tumor suppressive and the potential of glycosylated RPTPs to promote angiogenesis and cancer is increasingly highlighted. RPTP-β/ζ is the only RPTP proven to exist in a proteoglycan form and seems to be involved in the progression of several tumor types, with the majority of the studies up to date being performed in glioblastoma cells/models. The studies cited previously, point to the need to better characterize the role of RPTP-β/ζ in diverse tumor types, as a potential therapeutic target in tumor growth and metastasis. Given the huge medical need and intense competition for innovative drugs to novel targets, targeting of RPTPs expression and/or signaling may be a promising strategy for cancer prevention and treatment.

8.7.5 Take-home message

The implication of glycosylated RPTPs in angiogenesis and cancer underlines the need to elucidate the signaling pathways that participate in their actions, ultimately contributing to development of possible new therapies for cancer or/and other angiogenesis-related pathologies.

References

Aricescu, A. R., McKinnell, I. W., Halfter, W., and Stoker, A. W. (2002). Heparan sulfate proteoglycans are ligands for receptor protein tyrosine phosphatase sigma. Mol Cell Biol *22*, 1881–1892.

Ball, M., Carmody, M., Wynne, F., et al. (2009). Expression of pleiotrophin and its receptors in human placenta suggests roles in trophoblast life cycle and angiogenesis. Placenta *30*, 649–653.

Barcley, A. N., Jackson, D. I. Willis, A. C., and Williams A. F. (1987). Lymphocyte specific heterogeneity in the rat leucocyte common antigen (T200) is due to differences in polypeptide sequences near the NH2-terminus. EMBO J *6*, 1259–1264.

Barnea, G., Grumet, M., Milev, P., et al. (1994). Receptor tyrosine phosphatase β is expressed in the form of proteoglycan and binds to the extracellular matrix protein tenascin. J Biol Chem *269*, 14349–14352.

Barnea, G., Silvennoinen, O., Shaanan, B., et al. (1993). Identification of a carbonic anhydrase-like domain in the extracellular region of RPTP gamma defines a new subfamily of receptor tyrosine phosphatases. Mol Cell Biol *13*, 1497–1506.

Berman-Golan, D., and Elson, A. (2007). Neu-mediated phosphorylation of protein tyrosine phosphatase epsilon is critical for activation of Src in mammary tumor cells. Oncogene *26*, 7028–7037.

Carothers, A. M., Melstrom, K. A., Jr., Mueller, J. D., Weyant, M. J., and Bertagnolli, M. M. (2001). Progressive changes in adherens junction structure during intestinal adenoma formation in Apc mutant mice. J Biol Chem *276*, 39094–39102.

Daum, G., Regenass, R., Sap, J., Schlessinger, J., and Fischer, E. H. (1994). Multiple forms of the human tyrosine phosphatase RPTPalpha. Isozymes and differences in glycosylation. J Biol Chem *269*, 10524–10528.

Den Hertog, J., Blanchetot, C., Buist, A., Overvoorde, J., van der Sar, A., and Tertoolen, L. G. (1999). Receptor protein-tyrosine phosphatase signalling in development. Int J Dev Biol *43*, 723–733.

Diamantopoulou, Z., Bermek, O., Polykratis, A., et al. (2010). A pleiotrophin C-terminus peptide induces anti-cancer effects through RPTPβ/ζ. Mol Cancer. *9*, 224.

Feng, Z. J., Gao, S. B., Wu, Y., Xu, X. F., Hua, X., and Jin, G. H. (2010). Lung cancer cell migration is regulated via repressing growth factor PTN/RPTP β/ζ signaling by menin. Oncogene *29*, 5416–5426.

Foehr, E. D., Lorente, G., Kuo, J., Ram, R., Nikolich, K., and Urfer, R. (2006). Targeting of the receptor protein tyrosine phosphatase beta with a monoclonal antibody delays tumor growth in a glioblastoma model. Cancer Res *66*, 2271–2278.

Fujikawa, A., Shirasaka, D., Yamamoto, S., et al. (2003). Mice deficient in protein tyrosine phosphatase receptor type Z are resistant to gastric ulcer induction by VacA of Helicobacter pylori. Nat Genet *33*, 375–381.

Goldmann, T., Otto, F., and Vollmer, E. (2000). A receptor-type protein tyrosine phosphatase PTPzeta is expressed in human cutaneous melanomas. Folia Histochem Cytobiol *38*, 19–20.

Hamma-Kourbali, Y., Bermek, O., Bernard-Pierrot, I., et al. (2011). The synthetic peptide P111–136 derived from the C-terminal domain of heparin affin regulatory peptide inhibits tumour growth of prostate cancer PC-3 cells. BMC Cancer *11*, 212.

Hausser, H. J., and Brenner, R. E. (2005). Phenotypic instability of Saos-2 cells in long-term culture. Biochem Biophys Res Commun *333*, 216–222.

He, J., Liu, Y., Xie, X., et al. (2010). Identification of cell surface glycoprotein markers for glioblastoma-derived stem-like cells using a lectin microarray and LC-MS/MS approach. J Proteome Res *9*, 2565–2572.

Irie-Sasaki, J., Sasaki, T., Matsumoto, W., et al. (2001). CD45 is a JAK phosphatase and negatively regulates cytokine receptor signalling. Nature *409*, 349–354.

Johnson, K. G., and Van Vactor, D. (2003). Receptor protein tyrosine phosphatases in nervous system development. Physiol Rev *83*, 1–24.

Koutsioumpa, M., Hatziapostolou, M., Mikelis, C., Koolwijk, P., and Papadimitriou, E. (2009). Aprotinin stimulates angiogenesis and human endothelial cell migration through the growth factor pleiotrophin and its receptor protein tyrosine phosphatase beta/zeta. Eur J Pharmacol *602*, 245–249.

Krueger, N. X., Streuli, M., and Saito, H. (1990). Structural diversity and evolution of human receptor-like protein tyrosine phosphatases. EMBO J *9*, 3241–3252.

LaForgia, S., Morse, B., Levy, J., et al. (1991). Receptor protein-tyrosine phosphatase gamma is a candidate tumor suppressor gene at human chromosome region 3p21. Proc Natl Acad Sci U S A *88*, 5036–5040.

Lazarovits, A. I., Poppema, S., Zhang, Z., et al. (1996). Therapy for mouse renal allograft rejection by monoclonal antibody to CD45RB. Transplant Proc *28*, 3208–3209.

Liu, S., Sugimoto, Y., Sorio, C., Tecchio, C., and Lin, Y. C. (2004). Function analysis of estrogenically regulated protein tyrosine phosphatase gamma (PTPgamma) in human breast cancer cell line MCF-7. Oncogene *23*, 1256–1262.

Liu, Y. T., Shang, D., Akatsuka, S., et al. (2007). Chronic oxidative stress causes amplification and overexpression of ptprz1 protein tyrosine phosphatase to activate beta-catenin pathway. Am J Pathol *171*, 1978–1988.

Lu, K. V., Jong, K. A., Kim, G. Y., et al. (2005). Differential induction of glioblastoma migration and growth by two forms of pleiotrophin. J Biol Chem *280*, 26953–26964.

Maeda, N., He, J., Yajima, Y., Mikami, T., Sugahara, K., and Yabe, T. (2003). Heterogeneity of the chondroitin sulfate portion of phosphacan/6B4 proteoglycan regulates its binding affinity for pleiotrophin/heparin binding growth-associated molecule. J Biol Chem *278*, 35805–35811.

Maeda, N., Nishiwaki, T., Shintani, T., Hamanaka, H., and Noda, M. (1996). 6B4 proteoglycan/phosphacan, an extracellular variant of receptor-like protein-tyrosine phosphatase ζ/RPTPβ, binds pleiotrophin/heparin-binding growth-associated molecule (HBGAM). J Biol Chem *271*, 21446–21452.

Matozaki, T., Murata, Y., Mori, M., Kotani, T., Okazawa, H., and Ohnishi, H. (2010). Expression, localization, and biological function of the R3 subtype of receptor-type protein tyrosine phosphatases in mammals. Cell Signal *22*, 1811–1817.

Mikelis, C., Lamprou, M., Koutsioumpa, M., et al. (2011). A peptide corresponding to the C-terminal region of pleiotrophin inhibits angiogenesis *in vivo* and *in vitro*. J Cell Biochem *112*, 1532–1543.

Mikelis, C., Sfaelou, E., Koutsioumpa, M., Kieffer, N., and Papadimitriou, E. (2009). Integrin alpha(v)beta(3) is a pleiotrophin receptor required for pleiotrophin-induced endothelial cell migration through receptor protein tyrosine phosphatase beta/zeta. FASEB J *23*, 1459–1469.

Milev, P., Friedlander, D. R., Sakurai, T., et al. (1994). Interactions of the chondroitin sulfate proteoglycan phosphacan, the extracellular domain of a receptor-type protein tyrosine phosphatase, with neurons, glia, and neural cell adhesion molecules. J Cell Biol *127*, 1703–1715.

Milev, P., Monnerie, H., Popp, S., Margolis, R. K., and Margolis, R. U. (1998). The core protein of the chondroitin sulfate proteoglycan phosphacan is a high-affinity ligand of fibroblast growth factor-2 and potentiates its mitogenic activity. J Biol Chem *273*, 21439–21442.

Muller, S., Kunkel, P., Lamszus, K., et al. (2003). A role for receptor tyrosine phosphatase zeta in glioma cell migration. Oncogene *22*, 6661–6668.

Muller, T., Choidas, A., Reichmann, E., and Ullrich, A. (1999). Phosphorylation and free pool of beta-catenin are regulated by tyrosine kinases and tyrosine phosphatases during epithelial cell migration. J Biol Chem *274*, 10173–10183.

Ostman, A., Yang, Q., and Tonks, N. (1994). Expression of DEP-1, a receptor like protein-tyrosine-phosphatase, is enhanced with increasing cell density. Proc Natl Acad Sci U S A *91*, 9680–9684.

Papadimitriou, E., Mikelis, C., Lampropoulou, E., et al. (2009). Roles of pleiotrophin in tumor growth and angiogenesis. Eur Cytokine Netw *20*, 180–190.

Perez-Pinera, P., Alcantara, S., Dimitrov, T., Vega, J. A., and Deuel, T. F. (2006). Pleiotrophin disrupts calcium-dependent homophilic cell-cell adhesion and initiates an epithelial-mesenchymal transition. Proc Natl Acad Sci U S A *103*, 17795–17800.

Perez-Pinera, P., Garcia-Suarez, O., Menendez-Rodriguez, P., et al. (2007). The receptor protein tyrosine phosphatase (RPTP) beta/zeta is expressed in different subtypes of human breast cancer. Biochem Biophys Res Commun *362*, 5–10.

Polytarchou, C., Hatziapostolou, M., Poimenidi, E., et al. (2009). Nitric oxide stimulates migration of human endothelial and prostate cancer cells through up-regulation of pleiotrophin expression and its receptor protein tyrosine phosphatase beta/zeta. Int J Cancer *124*, 1785–1793.

Qi, M., Ikematsu, S., Maeda, N., et al. (2001). Involvement of protein-tyrosine phosphatase z, mitogen-activates protein kinase, and phosphtidylinositol 3-kinase. J Biol Chem *276*, 15868–15875.

Ruivenkamp, C. A., van Wezel, L. T., Zanon, C., et al. (2002). Ptprj is a candidate for the mouse colon-cancer susceptibility locus Scc1 and is frequently deleted in human cancers. Nat Genet *31*, 295–300.

Sakaguchi, N., Muramatsu, H., Ichihara-Tanaka, K., et al. (2003). Receptor-type protein tyrosine phosphatase zeta as a component of the signaling receptor complex for midkine-dependent survival of embryonic neurons. Neurosci Res *45*, 219–224.

Sörby, M. (2001). Structural and Functional Studies of the Density Enhanced Receptor-Like Protein Tyrosine Phosphatase DEP-1. Acta Universitatis Upsaliensis. Comprehensive Summaries of Uppsala Dissertations from the Faculty of Medicine.

Takahashi, T., Takahashi, K., Mernaugh, R. L., Tsuboi, N., Liu, H., and Daniel, T. O. (2006). A monoclonal antibody against CD148, a receptor-like tyrosine phosphatase, inhibits endothelial-cell growth and angiogenesis. Blood *108*, 1234–1242.

Thomaidou, D., Coquillat, D., Meintanis, S., Noda, M., Rougon, G., and Matsas, R. (2001). Soluble forms of NCAM and F3 neuronal cell adhesion molecules promote Schwann cell migration: identification of protein tyrosine phosphatases zeta/beta as the putative F3 receptors on Schwann cells. J Neurochem *78*, 767–778.

Thompson, L. J., Jiang, J., Madamanchi, N., Runge, M. S., and Patterson, C. (2001). PTP-epsilon, a tyrosine phosphatase expressed in endothelium, negatively regulates endothelial cell proliferation. Am J Physiol Heart Circ Physiol *281*, 396–403.

Toledano-Katchalski, H., Kraut, J., Sines, T., et al. (2003). Protein tyrosine phosphatase epsilon inhibits signaling by mitogen-activated protein kinases. Mol Cancer Res *1*, 541–550.

Tong, Y., Mentlein, R., Buhl, R., et al. (2006). Overexpression of midkine contributes to anti-apoptotic effects in human meningiomas. J Neurochem *100*, 1097–1107.

Ulbricht, U., Brockmann, M. A., Aigner, A., et al. (2003). Expression and function of the receptor protein tyrosine phosphatase zeta and its ligand pleiotrophin in human astrocytomas. J Neuropathol Exp Neurol *62*, 1265–1275.

Ulbricht, U., Eckerich, C., Fillbrandt, R., Westphal, M., and Lamszus, K. (2006). RNA interference targeting protein tyrosine phosphatase ζ/receptor-type protein tyrosine phosphatase β suppresses glioblastoma growth *in vitro* and *in vivo*. J Neurochem *98*, 1497–1506.

Van Niekerk, C. C., and Poels, L. G. (1999). Reduced expression of protein tyrosine phosphatase gamma in lung and ovarian tumors. Cancer Lett *137*, 61–73.

Wang, V., Davis, D. A., Veeranna, R. P., Haque, M., and Yarchoan, R. (2010). Characterization of the activation of protein tyrosine phosphatase, receptor-type, Z polypeptide 1 (PTPRZ1) by hypoxia inducible factor-2 alpha. PLoS One *5*, e9641.

Wang, W. J., Kuo, J. C., Ku, W., et al. (2007). The tumor suppressor DAPK is reciprocally regulated by tyrosine kinase Src and phosphatase LAR. Mol Cell *27*, 701–716.

Wu, M., Gan, K., Huang, C., et al. (2006). LRRC4 controls *in vitro* invasion of glioblastoma cells through inhibiting RPTP-zeta expression. J Neurooncol *80*, 133–142.

Zheng, W., Lennartsson, J., Hendriks, W., Heldin, C. H., and Hellberg, C. (2011). The LAR protein tyrosine phosphatase enables PDGF β-receptor activation through attenuation of the c-Abl kinase activity. Cell Signal *23*, 1050–1056.

8.8 Heparanase, a multifaceted protein involved in cancer, chronic inflammation, and kidney dysfunction

Israel Vlodavsky, Michael Elkin, Benito Casu, Jin-Ping Li, Ralph D. Sanderson, and Neta Ilan

8.8.1 Introduction

Glycosaminoglycans (GAGs) are linear polysaccharides consisting of a repeating disaccharide generally of an acetylated amino sugar alternating with uronic acid. Notably, while 4 and 20 building blocks make nucleic acids and proteins, respectively, 32 disaccharide building blocks make up these complex, highly acidic and information dense biopolymers. The chemical heterogeneity and structural complexity of GAGs make investigations of these molecules most challenging, raising fundamental questions as to how topological positioning and function of cells and tissues are regulated by GAGs. The biosynthesis of heparan sulfate (HS) GAG takes place in the Golgi system and has been studied in great detail. Briefly, the polysaccharide chains are modified at various positions by sulfation, epimerization and *N*-acetylation, yielding clusters of sulfated disaccharides separated by low or nonsulfated regions (Kjellen and Lindahl, 1991; Iozzo and San Antonio, 2001). The sulfated saccharide domains provide numerous docking sites for a multitude of protein ligands, ensuring that a wide variety of bioactive molecules (i.e. cytokines, chemokines, growth factors, enzymes, protease inhibitors, ECM proteins) bind to the cell surface and the extracellular matrix (ECM) (Bernfield et al., 1999; Capila and Linhardt, 2002; Lindahl and Li, 2009) and thereby function in the control of normal and pathological processes, among which are morphogenesis, tissue repair, inflammation, vascularization, and cancer metastasis (Kjellen and Lindahl, 1991; Bernfield et al., 1999; Iozzo and San Antonio, 2001; Lindahl and Li, 2009). Unlike the well-resolved biosynthetic pathway, the mode of HS breakdown is less characterized. Enzymatic activity capable of cleaving glucuronidic linkages and releasing polysaccharide chains resistant to further degradation by the enzyme was first identified by Ogren and Lindahl (Ogren and Lindahl, 1975). The physiological function of this activity was initially implicated in degradation of macromolecular heparin to physiologically active fragments (Ogren and Lindahl, 1975; Thunberg et al., 1982). Subsequent studies revealed that the same enzyme (heparanase) is critically involved in various pathologies such as cancer progression (Parish et al., 2001; Vlodavsky and Friedmann, 2001), chronic inflammation (Li et al., 2008; Lerner et al., 2011) and kidney dysfunction (van den Hoven et al., 2006). As a direct result of these studies heparanase was advanced from being an obscure enzyme with a poorly understood function to a highly promising drug target, offering new treatment strategies for various cancers and other diseases. The present chapter summarizes our long-term and ongoing basic and translational research on the biology and clinical significance of the heparanase enzyme.

Heparan sulfate proteoglycans

From mice to worms, embryos that lack HS die during gastrulation (Kramer and Yost, 2003), suggesting a critical developmental role for heparan sulfate proteoglycans (HSPGs). HSPGs function is not limited to developmental processes but play key roles in numerous biological settings, including cytoskeleton organization, cell-cell and cell-ECM interactions (Iozzo and San Antonio, 2001; Simons and Horowitz, 2001; Sasisekharan et al., 2002). HSPGs exert their multiple functional repertoires via several distinct mechanisms that combine structural, biochemical and regulatory aspects. By interacting with other macromolecules such as laminin, fibronectin, and collagens I and IV, HSPGs contribute to the structural integrity, self-assembly, and insolubility of the ECM and basement membrane, thus intimately modulating cell-ECM interactions (Timpl and Brown, 1996; Bernfield et al., 1999; Hacker et al, 2005). Accumulating evidence indicates that HSPGs act to inhibit cellular invasion by promoting tight cell-cell and cell-ECM interactions, and by maintaining the structural integrity and self-assembly of the ECM (Sanderson, 2001; Timar et al., 2002). Notably, one of the characteristics of malignant transformation is down regulation of GAGs biosynthesis, especially of the HS chains (Sanderson, 2001; Timar et al., 2002). Low levels of cell surface HS also correlate with high metastatic capacity of many tumors. Biochemically, HSPGs often facilitate the biological activity of bound ligands by actively participating in receptor-ligand complex formation (Belting, 2003). In other cases, HSPGs mediate cellular uptake and catabolism of selected ligands (Belting, 2003), and/or sequester polypeptides to the ECM and cell surface, generally as an inactive reservoir (Folkman et al., 1988; Bashkin et al., 1989; Vlodavsky et al., 1991, 1996; Patel et al., 2007). Cleavage of HSPGs would ultimately release these proteins and convert them into bioactive mediators, ensuring rapid tissue response to local or systemic cues.

Mammalian heparanase

Heparanase is an endo-β-glucuronidase that cleaves HS side chains presumably at sites of low sulfation, releasing saccharide products with appreciable size (4–7 kDa) that can still associate with protein ligands and facilitate their biological potency. Mammalian cells express primarily a single dominant functional heparanase enzyme (heparanase-1) (Vlodavsky and Friedmann, 2001; Ilan et al., 2006; Barash et al., 2010b). A second heparanase (heparanase-2) has been cloned and sequenced but has not been shown to have HS degrading activity (McKenzie et al., 2000; Levy-Adam et al., 2010a). For simplification, throughout this chapter we will refer to heparanase-1 as heparanase. Enzymatic degradation of HS leads to disassembly of the ECM and is therefore involved in fundamental biological phenomena associated with tissue remodeling and cell migration, including inflammation, angiogenesis, and metastasis (Parish et al., 2001; Vlodavsky and Friedmann, 2001; Ilan et al., 2006; Barash et al., 2010b). Normally, heparanase is found mainly in platelets, mast cells, placental trophoblasts, keratinocytes, and leukocytes. Heparanase released from activated platelets and cells of the immune system facilitates extravasation of inflammatory and tumor cells (Vlodavsky et al., 1992). It also stimulates endothelial mitogenesis, primarily through release of HS-bound growth factors (i.e. fibroblast growth factor [FGF], hepatocyte growth factor [HGF], vascular endothelial growth factor [VEGF]) residing in the ECM (Elkin et al., 2001).

The heparanase mRNA encodes a 61.2 kDa protein with 543 amino acids. This proenzyme is posttranslationally cleaved into 8 and 50 kDa subunits that noncovalently associate to form the active heparanase (Vlodavsky and Friedmann, 2001; Levy-Adam et al., 2003; McKenzie et al., 2003). Heterodimer formation is essential for heparanase enzymatic activity (Levy-Adam et al., 2003; Nardella et al., 2004). The heparanase structure delineates a triosephosphate isomerase (TIM)-barrel fold harboring the enzyme's active site and a C-terminus domain that is critical for heparanase secretion and signaling function (Fux et al., 2009b). Site-directed mutagenesis revealed that similar to other glycosyl hydrolases, heparanase has a common catalytic mechanism that involves two conserved acidic residues, a putative proton donor at Glu225 and a nucleophile at Glu343 (Hulett et al., 2000). Cellular processing of the latent 65 kDa proheparanase into its active 8+50 kDa heterodimer involves removal of a 6 kDa linker segment and is inhibited by a cell permeable inhibitor of cathepsin L (Abboud-Jarrous et al., 2005). Moreover, multiple site-directed mutagenesis and cathepsin L gene silencing and knockout experiments indicate that cathepsin L is the predominant enzyme responsible for processing and activation of proheparanase (Abboud-Jarrous et al., 2008). Applying a structural model, it has been demonstrated that the linker segment, or even a small 1 kDa portion at its C terminus, render the active site inaccessible to the HS substrate and hence inhibits heparanase enzymatic activity (Abboud-Jarrous et al., 2008).

Cellular storage and secretion

While the latent 65 kDa heparanase protein is secreted by the conventional endoplasmic reticulum (ER)/Golgi pathway, the active 8+50 kDa heterodimer is not readily accessible to its extracellular substrate suggesting the existence of regulatory mechanism(s) by which intracellular heparanase is secreted in response to local or systemic cues. Several observations support the occurrence of such a scenario. For example, treatment of human microvascular endothelial cells with the proinflammatory cytokines tumor necrosis factor-alpha (TNF-α) and interleukin-1β (IL-1β) resulted in a marked increase of heparanase secretion (Chen et al., 2004). Secretion of heparanase in response to TNF-α was also noted in human peripheral T cells (Sotnikov et al., 2004). Interestingly, TNF-α and IL-1β had no effect on heparanase secretion from tumor-derived cells (our unpublished results), suggesting that effective stimuli may vary among cell types and biological settings. Instead, nucleotides, such as adenosine triphosphate (ATP), adenosine diphosphate (ADP), and adenosine were most effective in stimulating secretion of active heparanase by tumor cells (Shafat et al., 2006a). Regarded as a universal source of metabolic energy, extracellular ATP, as well as other nucleotides are capable of initiating signaling cascades through two classes of P2 receptors: P2X, which has an intrinsic activity of ion channel; and P2Y, a G-protein-coupled receptor (Communi et al., 2000). P2Y receptor activation is coupled to phospholipase C and adenylate cyclase, leading to protein kinase C (PKC) and protein kinase A (PKA) activation (Abbracchio and Burnstock, 1998; Communi et al., 2000;

van der Weyden et al., 2000). Remarkably, each and every cell line examined responded to nucleotides (ATP, ADP, and adenosine) by a stimulated secretion of active heparanase (Shafat et al., 2006a). Importantly, ATP exerted its maximal effect at a physiological concentration (1 μM), emphasizing the biological relevance of this mediator. Accordingly, heparanase secretion was inhibited by PKC inhibitors and P2Y receptor antagonists.

Several lines of evidence suggest that the secreted heparanase originated from intracellular pools, most likely late endosomes and lysosomes. The kinetics of heparanase secretion elicited by ATP resembled that of the lysosomal enzyme cathepsin D, supporting the notion that both enzymes were secreted from intracellular vesicles (Shafat et al., 2006a). Moreover, immunofluorescence staining revealed a clear transition in the localization of heparanase-positive vesicles toward the cell periphery, in response to stimulation with ATP (Shafat et al., 2006a). Thus, although not considered typical secretory vesicles, lysosomes may secrete their content under certain conditions and in response to the proper stimuli, concomitantly with elevated levels of secreted cathepsins found in several human malignancies (Turk et al., 2000, 2004). The occurrence of active heparanase extracellularly is supported by the establishment of an enzyme-linked immunosorbent assay (ELISA) method, which preferentially recognizes the active 50 kDa enzyme versus the latent 65 kDa protein (Shafat et al., 2006b). Applying this method, it was demonstrated that heparanase levels are elevated in the plasma of pediatric cancer patients, correlating with their response to anticancer treatments (Shafat et al., 2007). Likewise, elevated levels of heparanase were measured in the urine of bladder cancer (Shafat et al., 2008) and diabetic (Shafat et al., 2011b) patients, in correlation with disease progression.

Heparanase-protease cooperation and regulation of gene expression

Cross talk between heparanase and matrix metalloproteinases (MMPs) has been demonstrated. Enhanced expression of heparanase stimulates sustained extracellular signal-regulated kinase (ERK) phosphorylation that in turn drives MMP9 expression, while heparanase gene silencing resulted in reduced MMP9 activity (Purushothaman et al., 2008). Moreover, not only MMP9 but also urokinase plasminogen activator (uPA) and its receptor (uPAR), molecular determinants responsible for MMP9 activation, are upregulated by heparanase (Purushothaman et al., 2008). These findings provide evidence for cooperation between heparanase and MMPs in regulating HSPGs on the cell surface and likely in the ECM, and are supported by our generation and characterization of heparanase knockout mice. Despite the complete lack of heparanase gene expression and enzymatic activity, heparanase-null mice develop normally, are fertile, and exhibit no apparent anatomical or functional abnormalities (Zcharia et al., 2009). Notably, heparanase deficiency was accompanied by a marked elevation of MMP family members such as MMP2 and MMP14, in an organ-dependent manner, suggesting that MMPs provide tissue-specific compensation for heparanase deficiency (Zcharia et al., 2009). Coregulation of heparanase and MMPs was also noted by a marked decrease in MMP (primarily MMP2, 9, and 14) expression following overexpression of the heparanase gene in cultured human mammary carcinoma cells (Zcharia et al., 2009). These and other published (Chen and Sanderson, 2009) and unpublished results suggest that heparanase acts as a regulator of protease expression.

There is growing evidence that heparanase induces transcription of proangiogenic (i.e. VEGF-A, VEGF-C), prothrombotic (i.e. TF), mitogenic (HGF), osteolyic (receptor activator of nuclear factor-kappa B ligand, or RANKL) (Cohen-Kaplan et al., 2008b; Barash et al., 2010b; Yang et al., 2010; Ramani et al., 2011), and likely other effectors that condition the tumor microenvironment to promote an aggressive cancer phenotype (Cohen-Kaplan et al., 2008b; Purushothaman et al., 2008; Vlodavsky et al., 2011). Although HS has been associated with several functional roles within the nucleus, its inhibition of gene transcription via inhibition of topoisomerase I and histone H3 acetyltransferase (HAT) activity are particularly intriguing (Kovalszky et al., 1998; Buczek-Thomas et al., 2008). HATs regulate gene expression by catalyzing acetylation of the N-terminal region of histones, thereby modifying chromatin structure in a manner that facilitates transcriptional activation. Both heparin and HS can act as potent inhibitors of p300 and pCAF HAT activities (Buczek-Thomas et al., 2008). Notably, a marked reduction in the level of nuclear syndecan-1 was found following upregulation of heparanase in myeloma cells (Chen and Sanderson, 2009), possibly associated with nuclear localization of the enzyme (Ohkawa et al., 2004; Schubert et al., 2004; Kobayashi et al., 2006). This could lead to increased histone acetylation with an associated increase in gene transcription. In fact, heparanase-mediated loss of nuclear syndecan-1 enhances HAT activity to promote expression of genes that drive an aggressive tumor phenotype (Vlodavsky et al., 2011). Heparanase regulation of gene expression may thus be related to its ability to inhibit accumulation of HSPGs within the nucleus. Hence, strategies to enhance nuclear HS levels may prove effective in blocking at least some of the heparanase-mediated effects that promote tumor growth and metastasis. Notably, while nuclear localization of heparanase correlates with cell differentiation (Ohkawa et al., 2004; Doweck et al., 2006) and a favorable prognosis of cancer patients (Doweck et al., 2006), localization of heparanase to the cell cytoplasm predicts poor prognosis (Takaoka et al., 2003; Ohkawa et al., 2004; Doweck et al., 2006).

8.8.2 Involvement of heparanase in cancer progression

In early studies, heparanase activity was shown to be associated with the metastatic potential of tumor-derived cells such as B16 melanoma (Nakajima, 1983) and T lymphoma (Vlodavsky et al., 1983). These observations gained substantial support when specific molecular probes became available shortly after cloning of the heparanase gene (Hulett et al., 1999; Kussie et al., 1999; Toyoshima and Nakajima, 1999; Vlodavsky et al., 1999). Both overexpression (Cohen et al., 2006; Lerner et al., 2008) and silencing (Edovitsky et al., 2004; Lerner et al., 2008) of the heparanase gene clearly indicate that heparanase not only enhances cell dissemination, but also promotes the establishment of a vascular network that accelerates primary tumor growth and provides a gateway for invading metastatic cells (Ilan et al., 2006; Vreys and David, 2007). These studies enabled researchers to critically approve the notion that HS cleavage by heparanase is required for structural remodeling of the ECM underlying tumors and endothelial cells, thereby facilitating cell invasion (▶Figure 8.18), and providing a proof-of-concept for the prometastatic and proangiogenic capacity of heparanase.

The clinical significance of the enzyme in tumor progression emerged from a systematic evaluation of heparanase expression in primary human tumors. Immunohistochemistry, in situ hybridization, reverse transcriptase polymerase chain reaction (RT-PCR)

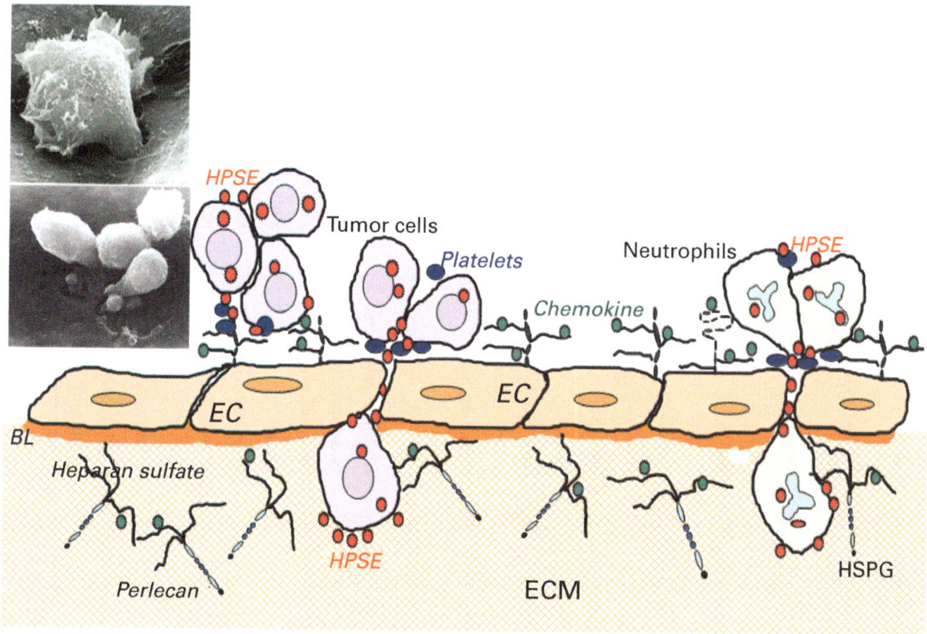

Figure 8.18 Heparanase-mediated extravasation of blood-borne cells. Heparanase expressed by tumor cells (left) and neutrophils (right) promotes cell invasion in between adjacent vascular endothelial cells (EC) and through their underlying basal lamina (BL) into the ECM. Shown schematically are also platelets (blue) contributing heparanase and facilitating cell invasion, heparanase (*HPSE*, red), HS-bound chemokines (green), and heparan sulfate (HS) proteoglycans (i.e. perlecan core protein and HS saccharide side chains). Left: scanning electron micrographs showing invasion of T-lymphoma cells, in the absence (top) or presence (bottom) of platelets, through a monolayer of cultured vascular EC.

and real time-PCR analyses revealed that heparanase is upregulated in essentially all major types of human cancer, namely carcinomas, sarcomas, and hematological malignancies (Ilan et al., 2006; Vreys and David, 2007; Fux et al., 2009b; Shafat et al., 2011a) (▶Table 8.3). Notably, increased heparanase levels were most often associated with reduced patient survival postoperation, increased tumor metastasis and higher microvessel density (Kelly et al., 2003; Ilan et al., 2006; Vreys and David, 2007) (▶Table 8.3), thus critically supporting the intimate involvement of heparanase in tumor progression and encouraging the development of heparanase inhibitors as anticancer therapeutics (Ferro et al., 2004; Miao et al., 2006; McKenzie, 2007; Vlodavsky et al., 2007; Casu et al., 2008).

Importantly, heparanase upregulation in human tumors (i.e. head and neck, tongue, hepatocellular, breast, and gastric carcinomas) is associated with large tumors (El-Assal et al., 2001; Maxhimer et al., 2002; Tang et al., 2002; Doweck et al., 2006; Nagler et al., 2007). Likewise, heparanase overexpression enhanced (Zetser et al., 2003; Yang et al., 2005; Cohen et al., 2006; Doviner et al., 2006; Barash et al., 2010a), while local delivery of antiheparanase siRNA inhibited (Lerner et al., 2008) the progression of tumor xenografts, altogether implying that heparanase function is not limited to

Table 8.3 **Correlation between heparanase expression and key clinical parameters in human cancer.**

Positive correlation with				*% Positive*	*Patient number*	*Carcinoma*	*Author (ref.)*
Survival	Metastasis		*MVD	85 (17/20)	40	Bladder	Gohji et al. (2001a)
	Metastasis			48 (32/67)	67	Bladder	Gohji et al. (2001b)
		Tumor size		36 (19/53)	53	Breast	Maxhimer et al. (2002)
Survival			MVD	49 (45/92)	92	Cervical	Shinyo et al. (2003)
				100	17	Colon	Friedman et al. (2000)
Survival	Metastasis			69 (37/54)	54	Colon	Nobuhisa et al. (2005)
Survival	Metastasis		MVD	25 (33/130)	130	Colorectal	Sato et al. (2004)
			MVD	50 (20/40)	40	Endometrial	Watanabe et al. (2003)
	Metastasis			49 (31/63)	63	Gastric	Endo et al. (2001)
Survival	Metastasis	Tumor size		83 (96/116)	116	Gastric	Tang et al. (2002)
Survival	Metastasis			80 (35/44)	44	Gastric	Takaoka et al. (2003)
Survival		Tumor size		86 (64/74)	74	Head and neck	Doweck et al. (2006)
		Tumor size	MVD	47 (26/55)	55	Hepatocellular	El-Assal et al. (2001)
Survival	Metastasis				76	Lung	Takahasi et al. (2004)
Survival	Metastsis			75 (85/114)	114	Lung	Cohen et al. (2008)
			MVD	86 (86/100)	100	Multiple myeloma	Kelly et al. (2003)
Survival				35 (16/46)	46	Nasopharyngeal	Bar-Sela et al. (2006)
Survival				62 (26/42)	42	Neuroblastoma	Zheng et al. (2009)
Survival				75 (25/33)	33	Pancreatic	Koliopanos et al. (2001)
Survival				78 (69/89)	89	Pancreatic	Kim et al. (2002)
Survival	Metastasis			76 (38/50)	50	Pancreatic	Rohloff et al. (2002)
Survival	Metastasis				70	Renal	Mikami et al. (2008)
Survival				70 (42/60)	60	**Salivary gland	Ben-Izhak et al. (2006)
		Tumor size		100	69	Ewing's sarcoma	Shafat et al. (2011a)
Survival		Tumor size		92 (55/60)	60	Tongue	Nagler et al. (2007)

% Positive = number of cases positively stained with antiheparanase antibodies.
*MVD = microvessel density.

tumor metastasis but is also engaged in accelerated vascularization and growth of the primary lesion (Cohen et al., 2006) (▶Table 8.3). Evidence indicates that heparanase not only assists in the breakdown of ECM but also is involved in regulating the bioavailability and activity of growth factors and cytokines. Briefly, as discussed previously, various heparin-binding growth factors are sequestered by HS in the ECM, providing a localized, readily accessible depot, protected from proteolytic degradation (Folkman et al., 1988; Vlodavsky et al., 1990) yet available to activate cells after being released by heparanase. Local release and activation of tissue-specific growth factors is clearly involved in creating a favorable "soil" for growth of the primary tumor as well as in dictating the organ selectivity of metastasis.

Signaling function of heparanase

Apart of their hydrolytic activity, ECM-degrading enzymes (i.e. cathepsins, plasminogen activators, MMPs, and heparanase) are also engaged in multiple signaling pathways, primarily by means of nonenzymatic activities that affect both the tumor cells and the tumor microenvironment (Kessenbrock et al., 2010). Compelling evidence indicate that heparanase facilitates the phosphorylation and activity of selected signaling molecules (i.e. Erk, Akt, p38, Src, epidermal growth factor receptor [EGFR], toll-like receptor [TLR]) and pathways. Signaling function requires heparanase secretion but not necessarily enzymatic activity (Zetser et al., 2006; Cohen-Kaplan et al., 2008a; Fux et al., 2009a). The concept of enzymatic activity-independent function of heparanase gained substantial support by identification of the heparanase C-terminus domain (C domain) as the molecular determinant behind its signaling capacity. The existence of a C domain emerged from a prediction of the three dimensional structure of a single chain heparanase enzyme (Fux et al., 2009a). In this protein variant, the linker segment was replaced by three glycine-serine repeats (GS3), resulting in a constitutively active enzyme (Nardella et al., 2004). The structure obtained clearly illustrates a TIM-barrel fold, in agreement with previous predictions (Hulett et al., 2000; Abboud-Jarrous et al., 2005). Notably, the structure also delineates a C-terminus fold positioned next to the TIM-barrel fold (Fux et al., 2009a). The predicted heparanase structure led to the premise that the seemingly distinct protein domains observed in the three dimensional model, namely the TIM-barrel and C-domain regions, mediate enzymatic and nonenzymatic functions of heparanase, respectively. Interestingly, cells transfected with the TIM-barrel construct (amino acids 36–417) failed to display heparanase enzymatic activity, suggesting that the C domain is required for the establishment of an active heparanase enzyme, possibly by stabilizing the TIM-barrel fold (Fux et al., 2009a). Deletion and site directed mutagenesis further indicated that the C domain plays a decisive role in heparanase enzymatic activity and secretion (Simizu et al., 2007; Birzele et al., 2008; Fux et al., 2009a;). Notably, Akt phosphorylation was stimulated by cells overexpressing the C domain (amino acids 413–543), while the TIM-barrel protein variant yielded no Akt activation compared with control, mock transfected cells (Fux et al., 2009a). These findings clearly indicate that the nonenzymatic signaling function of heparanase leading to activation of Akt is mediated by the C domain. The cellular consequences of C domain overexpression were best revealed by monitoring tumor xenograft development. Remarkably, tumor xenografts produced by C-domain-transfected glioma cells grew faster and appeared indistinguishable from those produced by cells transfected with the full-length heparanase in terms of tumor size and angiogenesis, yielding tumors sixfold larger than control. In contrast, progression of tumors produced by TIM-barrel-transfected cells appeared comparable with control mock transfected cells (Fux et al., 2009a). These results show, that in some tumor systems (i.e. glioma), heparanase facilitates primary tumor progression regardless of its enzymatic activity while in others (i.e. myeloma) heparanase enzymatic activity dominates. Enzymatic activity-independent function of heparanase is further supported by our identification of T5, a human splice variant of heparanase devoid of enzymatic activity (Barash et al., 2010a), and studies applying a C-domain-inducible system (unpublished data).

EGFR activation and downstream signaling

EGFR phosphorylation is markedly increased in cells (i.e. human glioma, meduloblastoma, prostate carcinoma, and breast carcinoma) overexpressing heparanase or in response to its exogenous addition, while heparanase gene silencing is accompanied by reduced EGFR and Src phosphorylation levels (Cohen-Kaplan et al., 2008a). Likewise, EGFR activation was observed following exogenous addition or overexpression of double mutated (DM; Glu225, Glu343), enzymatically inactive heparanase protein. Notably, enhanced EGFR phosphorylation by heparanase was restricted to selected tyrosine residues (i.e. 845, 1173) thought to be direct targets of Src rather than a result of receptor autophosphorylation (Haskell et al., 2001). Indeed, enhanced EGFR phosphorylation of tyrosine residues 845 and 1173 in response to heparanase was abrogated in cells treated with Src inhibitors or anti-Src siRNA (Cohen-Kaplan et al., 2008a). The functional significance of EGFR modulation by heparanase emerged by monitoring cell proliferation. Briefly, heparanase gene silencing was accompanied by a decrease in cell proliferation, while heparanase overexpression resulted in enhanced cell proliferation and formation of larger colonies in soft agar, in a Src- and EGFR-dependent manner (Cohen-Kaplan et al., 2008a). The clinical relevance of the heparanase-Src-EGFR pathway has been elucidated for head and neck carcinoma. Notably, heparanase expression in head and neck carcinomas correlated with phospho-EGFR immunostaining, and even more significant was the correlation between heparanase cellular localization (i.e. cytoplasmic vs nuclear), phospho-EGFR levels and clinical outcome (Cohen-Kaplan et al., 2008a).

Having found that in some cell lines (i.e. LNCaP prostate carcinoma) Akt phosphorylation levels were not elevated in response to EGFR activation, we assumed that signaling pathways other than Akt may be induced by EGFR. Signal transducer and activator of transcription (STAT) proteins can regulate several pathways important in oncogenesis, including cell-cycle progression, apoptosis, and tumor angiogenesis (Yu et al., 2009). Phosphorylation and nuclear translocation of STAT5b, but not STAT5a, was observed in response to heparanase overexpression or exogenous addition. Accordingly, reduced STAT5b phosphorylation and nuclear translocation were noted following heparanase gene silencing. Moreover, nuclear translocation of STAT5b following heparanase overexpression was efficiently attenuated in cells treated with Src or EGFR inhibitors, suggesting that Src-dependent EGFR activation by heparanase leads to activation of the STAT pathway (Cohen-Kaplan et al., 2012). Notably, enhanced proliferation of heparanase transfected cells was inhibited following STAT5b gene silencing, supporting a proliferative function of the heparanase-Src-EGFR axis.

Clinically, STAT3 phosphorylation was associated with head and neck cancer progression, EGFR phosphorylation, and heparanase expression and cellular localization. Importantly, cytoplasmic rather than nuclear phospho-STAT3 correlated with increased tumor size (T stage), number of infected neck lymph nodes, and reduced patient survival (Cohen Kaplan et al., 2012). Altogether, these studies provide a more realistic view of heparanase function in the course of tumor progression. Thus, while heparanase enzymatic activity has traditionally been implicated in tumor metastasis, the current view points to a multifaceted protein engaged in multiple aspects of tumor progression, combining enzymatic activity-dependent and independent functions of heparanase and affecting tumor metastasis, vascularization and EGFR activation.

Heparanase-inhibiting compounds

Compound PI-88, composed primarily of sulfated phosphomannopentaose and phosphomannotetraose oligosaccharide units, exhibits antiangiogenic, antimetastatic, and antirestenotic activities and is undergoing clinical trials for melanoma, myeloma, and hepatocellular carcinoma. This drug is thought to exert its biological effects by blocking the enzymatic activity of heparanase and by interfering with the action of HS-binding growth factors such as FGF1, FGF2, and VEGF (Khachigian and Parish, 2004). PI-88 interferes with a relatively broad range of protein-HS interactions, making interpretation regarding specificity and mode of action, questionable, similar to other polyanionic compounds. A highly promising compound is PG545, a synthetic, fully sulfated HS mimetic that has recently entered phase I trials for advanced cancer (Dredge et al., 2010, 2011). In contrast to other HS mimetics, which are mixtures, PG545 is a single molecular entity. It induces potent antiangiogenic activity *in vivo* and exhibits preclinical antitumor and antimetastatic efficacy with a pharmacokinetic profile that supports less frequent dosing compared with other HS mimetics (Dredge et al., 2011). Semisynthetic, metabolic stable sulfated trimannose C-C linked dimers (STMCs) endowed with heparanase and selectin inhibitory activity were recently developed and found to inhibit experimental metastasis (Borsig et al., 2011) and primary tumor growth (unpublished results). A number of heparanase-inhibiting small molecules were synthesized of which some yielded a partial response in models of tumor cell invasion and metastasis (Simizu et al., 2004). Neutralizing antiheparanase antibodies yielded a beneficial effect in experimental models of accelerated anti-glomerular basement membrane (GBM) disease (Levidiotis et al., 2005) and neointima formation (Myler et al., 2006). Ectopic miR-1258 has been recently demonstrated to negatively regulate heparanase expression and activity, resulting in inhibition of experimental breast cancer brain metastasis (Zhang et al., 2011). Although siRNA-mediated heparanase suppression has been previously reported (Edovitsky et al., 2004, 2006; Lerner et al., 2008) and applied in mice, it is challenging to maintain functional concentrations of siRNA *in vivo*. The development of miRNA-based approaches targeting heparanase may be of potential therapeutic value for the treatment of cancer patients with brain metastases (Zhang et al., 2011).

Heparin is one of the closest mimics of HS and is a natural choice as heparanase inhibitor. We have generated a novel chemically modified nonanticoagulant heparin (100% *N*-acetylated, 25% glycol-split heparin, termed HI-2 = SST0001; Sigma-Tau Research Switzerland, SA) that binds tightly to the heparanase active site and potently inhibits its enzymatic activity (Naggi et al., 2005; Vlodavsky et al., 2007; Casu et al.,

2008). Importantly, the modifications made to this heparin abolish its anticoagulant activity thereby enabling its use in high doses (Naggi et al., 2005; Vlodavsky et al., 2007; Casu et al., 2008). Moreover, glycol splitting introduces flexibility to the heparin molecule and confers resistance to cleavage by the enzyme, thereby strengthening the complex between the HS mimic and heparanase (Naggi et al., 2005; Vlodavsky et al., 2007; Casu et al., 2008). Compound SST0001 yielded impressive inhibition in preclinical models of multiple myeloma (Ritchie et al., 2011), Ewing's sarcoma (Shafat et al., 2011a), pancreatic carcinoma (Meirovitz et al., 2011) and diabetic nephropathy (Gil et al., 2012) (see Section 8.8.4). Notably, inhibition of cathepsin L by serpin was markedly augmented by heparin and HS (Higgins et al., 2010), suggesting that glycol-split heparin (i.e. SST0001) and related compounds will not only inhibit heparanase activity but also heparanase processing through cathepsin-L inhibition.

Multiple myeloma

Multiple myeloma is the second most prevalent hematologic malignancy. This B-lymphoid malignancy is characterized by tumor cell infiltration of the bone marrow, resulting in severe bone pain, fractures, and osteolytic bone disease. Heparanase expression and enzymatic activity was elevated in the bone marrow plasma of myeloma patients, correlating with poor prognosis (Kelly et al., 2003; Mahtouk et al., 2007). Heparanase overexpression in myeloma patients is associated with elevated microvessel density and syndecan-1 expression and shedding (Kelly et al., 2003; Mahtouk et al., 2007; Yang et al., 2007b; Purushothaman et al., 2010), determinants associated with aggressive myeloma progression. The heparanase-syndecan axis is therefore considered a viable target for myeloma therapy. Compound SST0001 diminishes heparanase-induced shedding of syndecan-1 and thereby disrupts the heparanase/syndecan-1 axis (Ritchie et al., 2011). Once inoculated subcutaneously into severe combined immunodeficient (SCID) mice, myeloma cells engineered to overexpress heparanase infiltrate the mouse femur compared with rare bone metastasis of control cells (Yang et al., 2005). In this model, heparanase transfected cells are metastatic well before the establishment of large tumors, indicating that metastasis is not simply a matter of tumor size. Heparanase transfected myeloma cells also metastasized to the liver, spleen, and lungs, while none of the mice bearing control cells had detectable evidence of micro- or macroscopic metastases (Yang et al., 2005). In a bone-to-bone metastasis model, cells are injected into the tibia and metastasize to the femurs contralateral to the injected tibia. Notably, myeloma cells were found in the contralateral femur of all mice injected with heparanase transfected cells compared with only 28% of bone-to-bone metastasis in mice injected with control cells (Yang et al., 2005). In both the subcutaneous and bone inoculation models, treatment with compound SST0001 profoundly inhibited the progression of tumor xenografts produced by myeloma cells (Yang et al., 2007a). Interestingly, inoculation of breast cancer cells overexpressing heparanase into the mammary fat pad promoted bone resorption in the absence of detectable bone metastases (Kelly et al., 2005), suggesting that heparanase stimulates the expression of a soluble osteolytic factor that function systemically. This notion is supported by the ability of medium conditioned by heparanase overexpressing breast cancer cells to stimulate a marked increase in osteoclastogenesis and concomitantly, bone resorption (Kelly et al., 2010). It was suggested that syndecan-1 shed by tumor cells exerts biological effects distal to the primary tumor and that it participates in driving osteoclastogenesis and the resulting bone destruction. Enhanced local and systemic osteolysis by heparanase has recently been demonstrated to be mediated, at least in part, by RANKL. The expression of this primary stimulator of osteoclastogenesis was increased two-fold in heparanase transfected myeloma cells, and the levels of RANKL correlated with heparanase expression in biopsies of bone metastases in myeloma patients (Yang et al., 2010). These results provide another indication for the ability of heparanase to modulate the transcription of genes (i.e. VEGF, TF, and HGF) involved in different aspects of tumorigenesis (Ilan et al., 2006; Fux et al., 2009b; Ramani et al., 2011).

Role of heparanase in radiation enhanced invasiveness of pancreatic carcinoma

Pancreatic cancer is one of the most aggressive neoplasm with an extremely low 5-year survival rate (Shaib et al., 2006; Borja-Cacho et al., 2008; Raimondi et al., 2009). Currently, pancreaticoduodenectomy is the only curative form of treatment; however, ~90% of pancreatic cancer patients miss the opportunity for complete surgical resection at the time of diagnosis (Borja-Cacho et al., 2008; Muller et al., 2008). Thus, radiotherapy remains a major component of treatment modalities for controlling pancreatic tumor progression. However, pancreatic cancer often shows resistance to ionizing radiation (IR), and randomized trials could not demonstrate benefit from radiation, revealing rather conflicting results (Neoptolemos et al., 2004; Cohen et al., 2005). Accumulating preclinical and clinical data suggest that IR may stimulate tumor aggressiveness (Camphausen et al., 2001; Neoptolemos et al., 2004; Ohuchida et al., 2004; Park et al., 2006), although the identity of downstream effectors acting at the cell or tissue levels and responsible for this effect, remains poorly investigated. Our recent results indicate that IR augments heparanase expression and thereby aggressiveness of pancreatic carcinoma both *in vitro* and *in vivo* (Meirovitz et al., 2011). Causal involvement of heparanase in pancreatic carcinoma progression is well-documented. There was a 30-fold increase in heparanase mRNA in pancreatic cancer tissue samples, in comparison to normal pancreatic tissue (Koliopanos et al., 2001). Moreover, elevated levels of the enzyme have been found in body fluids of patients with active pancreatic cancer disease (Quiros et al., 2006) as compared to healthy donors. Pancreatic cancer patients whose tumors exhibit high levels of the heparanase mRNA had a significantly shorter postoperative survival time than patients whose tumors contained relatively low levels of heparanase (Koliopanos et al., 2001; Rohloff et al., 2002). A recent finding that heparanase positivity is a highly significant independent variable for pancreatic adenocarcinoma dedifferentiation and lymph node metastasis further demonstrates a crucial role of the enzyme in the aggressiveness of pancreatic cancer (Hoffmann et al., 2008).

We have demonstrated that clinically relevant doses of IR augment invasive ability of pancreatic carcinoma cells *in vitro* and *in vivo* through upregulation of heparanase expression, and revealed that the molecular mechanism responsible for IR-induced heparanase transcription involves the early growth response 1 (Egr1) transcription factor (Meirovitz et al., 2011). As a transcriptional regulator, Egr1 can both induce and repress the expression of its target genes, including heparanase (de Mestre et al., 2003, 2005). Egr1 was previously shown to activate heparanase expression in T lymphocytes, prostate, breast, and colon carcinomas, but to inhibit its transcription in melanoma cells (de Mestre et al., 2005). Transactivation studies using Egr1 expression vector, cotransfected with a reporter construct encoding for luciferase (LUC) under the heparanase promoter, showed that in pancreatic carcinoma cells Egr1 acts to repress heparanase transcription (Meirovitz et al., 2011). Moreover, chromatin immunoprecipitation (ChIP) analysis revealed a marked decrease in occupancy of the heparanase promoter by Egr1 following IR treatment. We have found that pancreatic carcinoma cells express high basal levels of Egr1 and of the Egr1-binding protein NAB2, a transcriptional corepressor that directly interacts with Egr1 and represses activation of its target promoters (Srinivasan et al., 2006). It is therefore plausible that in the presence of both Egr1 and NAB2 the heparanase promoter is repressed in pancreatic cells. However, following IR-associated temporal decrease in Egr1, this repression is no longer operative, allowing for activation of the heparanase promoter.

Combination of radiotherapy with drugs that inhibit IR-induced tumor aggressiveness may be an attractive strategy to diminish adverse prometastatic action while retaining the therapeutic benefit of radiation, thus reducing resistance of pancreatic cancer to treatment. We have demonstrated that compound SST0001 (Casu et al., 2004, 2008; Naggi et al., 2005; Vlodavsky et al., 2007) attenuated radiation-induced invasiveness *in vitro* (Meirovitz et al., 2011). Moreover, spread of orthotopically growing pancreatic tumors was significantly reduced in mice treated with a combination of SST0001 and IR, as compared with either modality alone (Meirovitz et al., 2011). Taken together, our results support the combination of radiotherapy with a specific heparanase inhibitor as an effective strategy to prevent tumor resistance and progression, observed in many IR-treated pancreatic cancer patients. Notably, inducibility of heparanase by radiation appears not to be limited to pancreatic tissue, as it was recently demonstrated that heparanase is upregulated by IR in liver (Chung et al., 2010) and by ultraviolet B radiation in human skin (Iriyama et al., 2011).

8.8.3 Heparanase and inflammation

Inflammation plays an important role in host defense; however, when unrestrained, this process can lead to a number of debilitating diseases such as inflammatory bowel diseases (IBD), rheumatoid arthritis (RA), psoriasis, and asthma. Upregulation of heparanase was noted in various inflammatory conditions, often associated with degradation of HS and release of chemokines anchored within the ECM network and cell surfaces. Briefly, HS coordinates the inflammatory response by virtue of its interaction with different types of chemokines, cytokines, and selectins, largely due to the heterogeneous nature of the polysaccharide. Applying heparanase overexpressing mice we have recently demonstrated, for example, that endothelial HS cleavage impairs chemokine-mediated neutrophil trafficking, resulting in decreased ability to clear bacterial infections (Massena et al., 2011). Prior to cloning the heparanase gene, heparanase activity originating in activated cells of the immune system (T lymphocytes, neutrophils) has been found to contribute to their ability to penetrate blood vessels and accumulate in target organs (Vlodavsky et al., 1992). In subsequent studies we reported that degradation of HS in the subendothelial basement membrane resulted in vascular leakage, a hallmark of delayed type hypersensitivity skin reactions (Edovitsky et al., 2006). Notably, we have demonstrated that immunocytes are not the primary source of the enzyme in inflammation and that upregulation of heparanase, locally expressed (i.e. by vascular endothelium, skin keratinocytes, and colonic epithelium) at the site of inflammation, is an essential step in the inflammatory response.

The last decade critically revealed the decisive role of inflammatory responses in different stages of tumor development and metastasis (Grivennikov et al., 2010). The most frequently found immune cells within the tumor microenvironment are tumor-associated macrophages (Grivennikov et al., 2010; Mantovani and Sica, 2010). Notably, high density of these cells correlates with poor prognosis, while removal of macrophages almost completely ablated metastasis in a mouse of breast cancer (Pollard, 2004; Condeelis and Pollard, 2006; Dirkx et al., 2006; Mantovani and Sica, 2010). In a recent study we have demonstrated the importance of mucosal macrophages and heparanase in sustaining immune-epithelial cross talk underlying colitis-associated colon tumorigenesis (Lerner et al., 2011). Briefly, heparanase markedly affects the chronic phase of experimental colitis, as judged by activation of nuclear factor-kappa B (NF-κB) signaling, macrophage infiltration, expression of proinflammatory molecules, and enhanced angiogenesis (Lerner et al., 2011). Our studies revealed that TNF-α promotes continuous overexpression of heparanase by chronically inflamed colonic epithelium. Moreover, heparanase enzymatic activity profoundly stimulates macrophage activation by lipopolysaccharide (LPS) (Lerner et al., 2011), collectively suggesting the occurrence of heparanase-driven vicious cycle that drives colitis and the associated colon tumorigenesis (▶Figure 8.19). A similar scenario has been demonstrated in experimental psoriasis, a common inflammatory skin disease of unknown etiology (our unpublished results). Cross talk between immunocytes and keratinocytes, which results in the production of cytokines, chemokines, and growth factors, is thought to mediate the disease. These observations, the dramatic increase (>100-fold) in heparanase activity measured in synovial fluid of rheumatoid arthritis patients (Li et al., 2008), and the anti-inflammatory effect of newly generated heparanase-inhibiting molecules (Li and Vlodavsky, 2009) indicate that the enzyme is an attractive, yet under-investigated target for drug development in chronic inflammation and the associated tumorigenesis.

Heparanase powers a chronic inflammatory circuit that promotes colitis-associated tumorigenesis

The most feared long-term complication of IBD (in particular, ulcerative colitis [UC]) is colon carcinoma, as patients with UC have a risk of colorectal cancer that is an order of magnitude higher than the normal population (Clevers, 2006). In fact, colon carcinoma represents a paradigm for the association between inflammation and cancer (Karin and Greten, 2005). While significant progress has been made in deciphering the role of inflammatory cytokines (TNF-α, IL-1β, and IL-6) and their downstream transcription factors (NF-κB, STAT3) in tumor-stimulating cross talk between immune and epithelial cells (Greten et al., 2004; Karin and Greten, 2005; Popivanova et al., 2008), little is known about the role of ECM-degrading enzymes in this cross talk. Based on the preferential expression of heparanase in chronically inflamed colonic epithelium (Waterman et al., 2007) and increased incidence of colon cancer in colitis patients, we hypothesized that stimulation of heparanase expression plays an important role in the pathogenesis of UC, representing a mechanistic link between inflammation and cancer. Utilizing UC tissue specimens, along with a mouse model of colitis-associated cancer induced by the carcinogen azoxymethane (AOM), followed by the inflammatory agent dextran sodium sulfate (DSS) (Okayasu et al., 1996), we found that

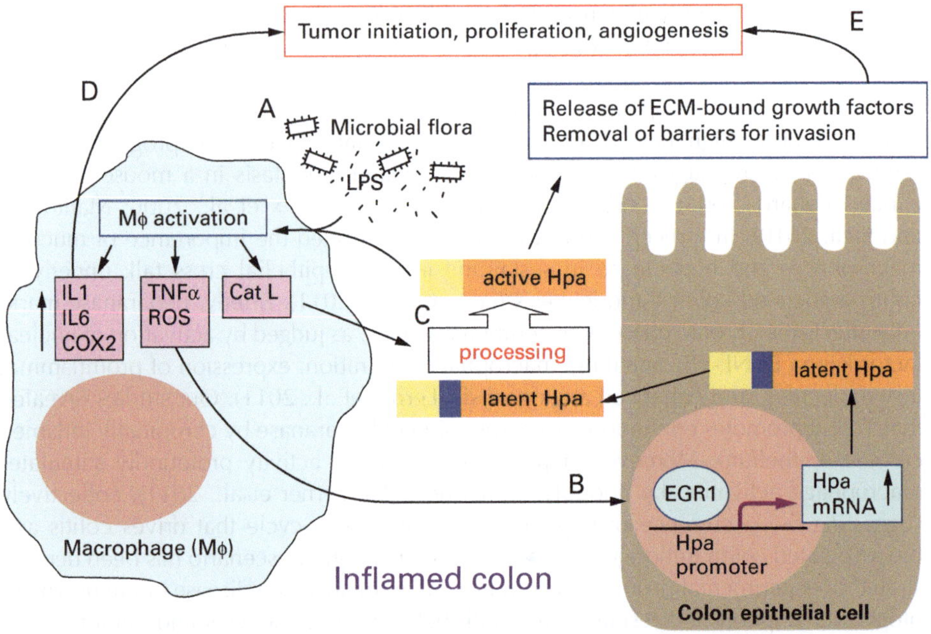

Figure 8.19 Heparanase-driven vicious cycle that powers chronic inflammation and the associated colon tumorigenesis. (A) Increased levels of TNF-α secreted by activated macrophages induce heparanase expression in the epithelial compartment via Egr1-dependent mechanism (B). (C) Secreted latent heparanase is processed into its active form by cathepsin (supplied by the macrophages) and, in turn, sensitizes macrophages to further activation (i.e. in the case of colitis by luminal flora), thus preventing inflammation resolution, switching macrophage responses to the chronic inflammation pattern, and creating a tumor-inducing inflammatory environment (D). In addition, heparanase promotes tumor progression via stimulation of angiogenesis, release of ECM-bound growth factors and bioactive HS fragments, and removal of extracellular barriers for invasion (E). (Reproduced from Lerner et al. (2011). J Clin Invest *121*, 1709–1721).

heparanase is constantly overexpressed by the colonic epithelium in UC and experimental colitis during both the acute and chronic phases of the disease. Moreover, heparanase overexpression preserves a chronic inflammatory conditions in DSS colitis and thus creates a tumor-promoting microenvironment characterized by enhanced NF-κB signaling (Greten et al., 2004; Karin, 2006), STAT 3 induction (Yu et al., 2009), and increased vascularization. Furthermore, we identified a novel biological mechanism contributing to chronic colitis and the associated colon tumorigenesis. This mechanism involves a self-sustained cycle through which heparanase of epithelial origin, acting synergistically with the local flora and cytokine milieu, facilitates abnormal recruitment and activation of innate immune cells (i.e. macrophages), which, in turn, stimulate further production of the enzyme by the colonic epithelium. Moreover, in chronic colitis activated macrophages represent a primary source of cathepsin L responsible for proteolytic activation of latent heparanase (Lerner et al., 2011). Analysis of the functional importance of macrophages in sustaining the pathogenesis of chronic colitis and the associated tumorigenesis is presented further on.

Heparanase-macrophage cross talk in chronic colitis

Macrophages are known to have a dual role in inflammation. In the scenario of inflammation resolution, macrophages perform phagocytosis and produce anti-inflammatory cytokines, thereby preventing inflammatory responses from lasting too long. However, if inflammation resolution is deregulated, macrophage response switches to the pattern of chronic inflammation. Recruitment and activation of macrophages within the intestinal mucosa play a key role in the pathogenesis of both human UC (Mahida, 2000; Sanchez-Munoz et al., 2008) and murine DSS colitis (Elson et al., 1995; Krieglstein et al., 2002). Moreover, activated macrophages are candidate cells linking between inflammation and cancer (Coussens et al., 2002; Pollard, 2004). Notably, the tumor promoting cytokines IL-1, IL-6, and TNF-α (Greten et al., 2004; Popivanova et al., 2008) are produced by activated macrophages and, along with macrophage-derived growth factors and reactive oxygen species, foster tumor initiation (Coussens et al., 2002; Sanchez-Munoz et al., 2008). Examining the involvement of macrophages in our system, we detected increased macrophage infiltration in the colon of heparanase transgenic (*Hpa*-tg) versus *wt* mice on day 80 of the chronic DSS colitis model (Lerner et al., 2011). Macrophages infiltrating the *Hpa*-tg colon manifested a striking proinflammatory profile, evidenced by increased number of TNF-α expressing macrophages and elevated NF-κB signaling. These findings led us to assume that heparanase overexpression directly affects macrophage activation. To recapitulate conditions occurring in UC (i.e. heparanase-rich environment and abundant microbial flora), we isolated mouse peritoneal macrophages and stimulated them with LPS in the absence or presence of recombinant active heparanase. Pretreatment with heparanase strongly sensitized macrophages to activation by LPS, as indicated by a marked increase in TNF-α, IL-6, and IL-12p35 (Lerner et al., 2011), all macrophage-derived cytokines known to be induced by TLR4 signaling and tightly involved in the pathogenesis of UC (Sanchez-Munoz et al., 2008). These findings are in agreement with previous reports showing that intact extracellular HS inhibits LPS-mediated TLR4 signaling and macrophage activation, and that its removal relieves this inhibition (Brunn et al., 2005). Given that one of the unique aspects of colorectal cancer development is the involvement of lumenal flora and TLR signaling (Fukata et al., 2007, 2009), the observed ability of heparanase to sensitize macrophages to LPS activation is of particular significance in light of the increased epithelial permeability to lumenal microbial products, characteristic of UC. Altogether, the previous data suggest that upregulated heparanase enables enhanced activation of macrophages, reprogramming their response from resolution of inflammation to unresolved chronic colitis. In addition, we found that activated macrophages are capable of inducing heparanase expression in colonic epithelial cells, most likely through TNF-α-mediated stimulation of Egr1, a powerful inducer of heparanase transcription in colonic tumor cells (de Mestre et al., 2005) and DSS colitis (Lerner et al., 2011). Notably, TNF-α was shown to stimulate Egr1 expression in IBD patients (Subbaramaiah et al., 2004). Moreover, due to their unique ability to secrete mature cathepsin L and allow extracellular accumulation of the active enzyme (Fiebiger et al., 2002), activated macrophages appear to be responsible for proteolytic activation of latent proheparanase in colitis. Thus, macrophages not only represent a cellular target for heparanase action, but also decisively upregulate heparanase in chronic colitis, both at the transcriptional and posttranslational levels. These results demonstrate and highlight the cooperation between two cellular compartments (i.e. colon epithelium and activated macrophages) in heparanase activation during inflammation-associated colon carcinogenesis. Altogether, our results (Lerner et al., 2011) indicate that heparanase generates a self-sustaining connection between chronic colitis and tumorigenesis (▶Figure 8.19). Briefly, macrophages activated by influx of the luminal flora secrete TNF-α and stimulate production of heparanase by the colon epithelium. The secreted 65 kDa latent heparanase is processed into its active form by cathepsin L, supplied by the activated macrophages. Enzymatically active heparanase sensitizes macrophages to further activation by microbial flora, thus preventing inflammation resolution, switching macrophage responses to the chronic inflammation pattern and creating tumor-inducing inflammatory environment (Lerner et al., 2011). In addition, high heparanase levels support tumor progression via stimulation of angiogenesis, release of ECM-bound growth factors and bioactive HS fragments and removal of extracellular barriers for invasion (▶Figure 8.19). Importantly, as presented in ▶Figure 8.18, the previously described self-sustaining tumor promoting inflammatory circuit may occur in other organs where inflammation plays a significant role in tumorigenesis and preferential expression of heparanase has been reported during cancer progression. The newly identified heparanase-powered vicious cycle may explain a yet poorly understood "multiplier effect" in IBD inflammation, in which even a small initial elevation in "initiating" inflammatory stimuli gives rise to large increases in downstream cytokines (Bouma and Strober, 2003). Thus, disruption of the heparanase-driven chronic inflammatory circuit is highly relevant to the design of therapeutic interventions in colitis and the associated cancer.

8.8.4 Heparanase and diabetic nephropathy

The kidneys represent primary targets of diabetes, and diabetic nephropathy (DN) is the leading cause of end stage renal disease in the western world (Ibrahim and Hostetter, 1997; Gilbertson et al., 2005; Atkins and Zimmet, 2010). DN is characterized by glomerular hyperfiltration, increased renal albumin permeability, and cellular and extracellular changes in the glomerular and tubulointerstitial compartments, collectively resulting in progression of proteinuria and renal failure. Studies linking heparanase to DN and other proteinuric disorders include elevated levels of heparanase in the kidneys of DN patients (Bitan et al., 2002; van den Hoven et al., 2006). Elevated levels of heparanase were also noted in the urine and plasma of type 1 (Katz et al., 2002) and type 2 (Shafat et al., 2011b) diabetic patients, patients with nondiabetic chronic kidney disease (Weinstein et al., unpublished), and in association with the development of proteinuria in transplanted kidneys (Weinstein et al., unpublished). Induction of glomerular heparanase expression was noted in murine models of streptozotocin (STZ)–induced diabetes (van den Hoven et al., 2006) and passive Heymann nephritis (Levidiotis et al., 2004), as well as *in vitro* studies demonstrating that hyperglycemic conditions enhance heparanase expression in rat and human glomerular epithelial cells (Maxhimer et al., 2005). Induction of glomerular heparanase in the course of diabetes may interfere with kidney function primarily through degradation of HS in the glomerular basement membrane (GBM). Indeed, GBM, along with fenestrated glomerular endothelium and podocyte foot processes/slit diaphragms, serves as the key functional component of the kidney filtration barrier, while HS represents a chief polysaccharide constituent of the GBM (Miner, 1999; Conde-Knape, 2001), playing a key space-filling and molecular-sieving role in the GBM. Findings supporting importance of HS integrity in permselectivity of the glomerular filtration (reviewed in Raats et al., 2000 include the following: (1) occurrence of massive proteinuria following administration of monoclonal anti-HS antibody to rats, (ii) increased GBM permeability as a result of HS removal, and (iii) decreased GBM HS content in human/experimental diabetes, which inversely correlates with the degree of proteinuria. Nevertheless, the impact of GBM-residing HS and its enzymatic degradation on DN development was questioned by several recent studies, showing no change in glomerular HS content/structure in early human and experimental DN (van den Born et al., 2006; Harvey and Miner, 2008). These reports challenged the notion that heparanase-mediated loss of HS in the GBM is the primary mechanism implicating the enzyme in DN.

Our generation of heparanase-null (*Hpse-KO*) mice (Zcharia et al., 2009) offered a unique opportunity to assess a causative role of heparanase in the pathogenesis of DN. Importantly, *Hpse-KO* mice failed to develop proteinuria in response to STZ-induced diabetes, and their urinary albumin excretion rate remained at the same level as before the onset of diabetes. In contrast, a greater than fivefold increase in urinary excretion rate of albumin was detected in *wt* mice following STZ-induced diabetes (Gil et al., 2012). In addition, heparanase deficiency resulted in amelioration of mesangial matrix expansion, macrophage infiltration, tubulointerstitial fibrosis and of , transforming growth factor-beta (TGF-β) overexpression in the diabetic kidneys of *Hpse-KO* versus *wt* mice. In further support of a causal involvement of heparanase in DN are our findings showing a lower degree of albuminuria in type 1 and type 2 diabetic mice treated with the heparanase inhibitor SST0001 versus mice treated with vehicle alone (Gil et al., 2012). Amelioration of albuminuria by SST0001 represents a proof of concept for heparanase inhibition as a relevant therapeutic approach in DN and warrants further studies aimed at identifying the most effective dose and administration schedule for SST0001 treatment, as well as at elucidating the molecular mechanism(s) underlying heparanase induction and pathogenic action in diabetic kidney. In agreement with the previously reported glucose responsiveness of Egr1 in other cell types (Josefsen et al., 1999), we have demonstrated dose-dependent induction of Egr1 expression in cultured kidney cells *in vitro*, and in diabetic kidney *in vivo*. ChIP analysis revealed increased occupancy of the heparanase promoter by Egr1 in the presence of elevated glucose levels, further confirming the role of Egr1 in glucose-dependent induction of heparanase in the kidney (Gil et al., 2012).

Apart of the straightforward heparanase mediated disruption of the GBM permselectivity function, it appears that heparanase-driven activation of macrophages may be critically important in DN, as well as in other kidney pathologies previously linked to heparanase. Important role of chronic inflammation and macrophages in DN (Tuttle, 2005; Nguyen et al., 2006; Navarro-Gonzalez and Mora-Fernandez, 2008; Tesch, 2010), taken together with the reduced number of macrophages infiltrating *Hpse-KO* diabetic kidneys and the recently revealed ability of heparanase to facilitate macrophage activation by LPS (Lerner et al., 2011) and by components of the diabetic milieu (our unpublished data), strongly suggest that under diabetic conditions, heparanase, induced in the kidney epithelium via Egr1-dependent mechanism sustains continuous activation of kidney-damaging macrophages, thus fostering DN development and

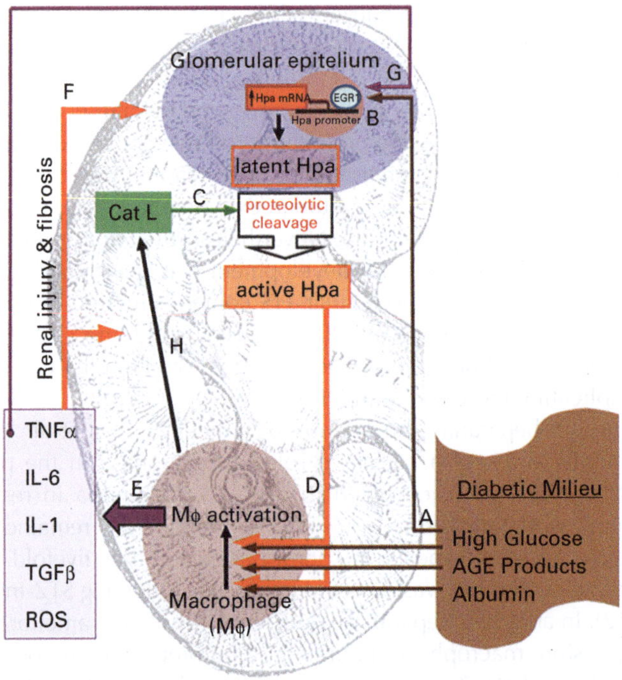

Figure 8.20 Heparanase-mediated cross talk that promotes inflammation and kidney injury in diabetic nephropathy. (A) Increased glucose levels induce heparanase expression in diabetic kidney via EGR1-dependent mechanism (B). (C) The overexpressed 65 kDa latent heparanase is processed into its enzymatically active (8 + 50 kDa) form by cathepsin L (Cat L), which is supplied, among other cells, by the macrophages activated by diabetic milieu components. (D) Active heparanase, in turn, sensitizes macrophages to further activation by diabetic milieu, thus preventing inflammation resolution, switching macrophage responses to the chronic inflammation pattern (E), and fostering macrophage-mediated renal damage. In addition, activated macrophages are capable of sustaining heparanase overexpression (via TNF-α-dependent mechanism) (G) and proteolytic activation (supplying CatL) (H).

progression (▶Figure 8.20). Studies are underway to accurately evaluate the pathophysiological mode of heparanase action in DN and thereby to better target patient populations in which future antiheparanase therapies could be particularly beneficial.

8.8.5 Challenges and future perspectives

Although much has been learned since the cloning of the human heparanase gene, the repertoire of heparanase functions in health and disease is only starting to emerge. Clearly, from activity implicated mainly in cell invasion associated with tumor metastasis, heparanase has turned into a multifaceted protein that appears to participate in essentially all major aspects of tumor progression, inflammation, and kidney dysfunction (Fux et al., 2009b; Barash et al., 2010b; Levy-Adam et al., 2010b). Heparanase knockout and overexpressing mice (Zcharia et al., 2004, 2009) are being utilized to examine the notion that heparanase is not only intrinsic to tumor cell invasion and angiogenesis, but also modifies, enzymatically and/or by virtue of its signaling and proadhesion functions (Levy-Adam et al., 2008, 2010b), the site of cell seeding and hence participates in establishing a "congenital soil" for successful homing and colonization of disseminated tumor cells.

While most attention was addressed to heparanase function in tumor biology, emerging evidence indicate that heparanase is also engaged in other pathologies such as chronic inflammation, atherosclerosis, and kidney dysfunction, often associated with degradation of HS, release of bioactive molecules anchored within the ECM network, and abnormal activation of innate immune cells (i.e. macrophages). We have demonstrated that activated cells of the immune system stimulate further production of the enzyme, thereby preserving chronic inflammation and creating a tumor-promoting microenvironment (Lerner et al., 2011). Mechanistic features of this vicious cycle are being investigated.

Accumulating evidence indicate that heparanase is an attractive tumor-associated antigen in cancer patients and that peptides derived from human heparanase can elicit a potent antitumor immune response, leading to lysis of heparanase-positive human tumor cells (Chen et al., 2008; Tang et al., 2010; Wang et al., 2011). Applying newly synthesized long peptides of heparanase that can be processed by dendritic cells, heparanase ranked among the most frequently recognized tumor antigens in patients with pancreatic, colorectal, or breast cancer (Sommerfeldt et al., 2006; Bonertz et al., 2009). Thus, heparanase provides a unique opportunity for a tumor-type independent, metastasis-selective tumor immunotherapy. While causing the generation of high frequencies of specific CD4 and CD8 memory T cells, heparanase, in contrast to most other tumor antigens, did not induce spontaneous regulatory T-cell responses (Bonertz et al., 2009). Antiheparanase immunotherapy is thus expected to be prolonged and more efficient due to the absence of T-suppressor cells. Moreover, the immune system appears capable of sensing ECM degradation and responding to HS degradation fragments released by heparanase from the ECM. Heparanase activity may thus directly activate antigen presenting cells and allow for tumor-antigen-specific T-cell responses in the absence of foreign inflammatory stimuli – a fundamental new mechanism for induction of antitumor immune responses.

Heparanase-inhibiting compounds are broad acting, targeting both the tumor and the tumor microenvironment (i.e. inflammation, angiogenesis). Novel heparanase inhibitors such as nonanticoagulant glycol-split heparin (SST0001) (Vlodavsky et al., 2007; Ritchie et al., 2011), heparin-derived HS mimetic (M402) (Zhou et al., 2011), STMCs (Borsig et al., 2011), PG545 (Dredge et al., 2011), and rationally designed structure-based compounds and site-specific neutralizing antibodies, are being developed for use in clinical settings to provide relief to patients with diabetes, colitis, and cancer.

8.8.6 Take-home message

Heparanase is a multifaceted protein endowed with enzymatic and nonenzymatic functions that appear to participate in major aspects of tumor progression, inflammation, and kidney dysfunction. Heparanase was advanced from being an obscure enzyme with a poorly understood function to a highly promising drug target, offering new treatment strategies for various cancers and other diseases (i.e. chronic colitis, diabetic nephropathy).

Acknowledgments

We thank Dr Claudio Pisano and Dr Allesandro Noseda (Sigma-Tau SpA, Pomezia, Italy) for their continuous support and active collaboration. This work was supported by National Institutes of Health (NIH) grants CA106456 (IV) and CA138535 (RDS), the United States-Israel Binational Science Foundation (BSF), the Ministry of Science & Technology of the State of Israel and the German Cancer Research Center (DKFZ), the Juvenile Diabetes Research Foundation (JDRF 1-2006-695 and 38-2009-635), and by a research contract from Sigma-Tau Research Switzerland S.A. I. Vlodavsky is a research professor of the ICRF. We gratefully acknowledge the contribution, motivation, and assistance of the research teams in the Hadassah-Hebrew University Medical Center (Jerusalem, Israel), the Cancer and Vascular Biology Research Center of the Rappaport Faculty of Medicine (Technion, Haifa), and the Ronzoni Institute for Chemical and Biochemical Research (Milan Italy).

References

Abboud-Jarrous, G., Atzmon, R., Peretz, T., et al. (2008). Cathepsin L is responsible for processing and activation of proheparanase through multiple cleavages of a linker segment. J Biol Chem *283*, 18167–18176.

Abboud-Jarrous, G., Rangini-Guetta, Z., Aingorn, H., et al. (2005). Site-directed mutagenesis, proteolytic cleavage, and activation of human proheparanase. J Biol Chem *280*, 13568–13575.

Abbracchio, M. P., and Burnstock, G. (1998). Purinergic signalling: pathophysiological roles. Jpn J Pharmacol *78*, 113–145.

Atkins, R. C., and Zimmet, P. (2010). Diabetic kidney disease: act now or pay later. J Am Soc Hypertens *4*, 3–6.

Bar-Sela, G., Kaplan-Cohen, V., Ilan, N., Vlodavsky, I., and Ben-Izhak, O. (2006). Heparanase expression in nasopharyngeal carcinoma inversely correlates with patient survival. Histopathology *49*, 188–193.

Barash, U., Cohen-Kaplan, V., Arvatz, G., et al. (2010a). A novel human heparanase splice variant, T5, endowed with protumorigenic characteristics. FASEB J *24*, 1239–1248.

Barash, U., Cohen-Kaplan, V., Dowek, I., Sanderson, R. D., Ilan, N., and Vlodavsky, I. (2010b). Proteoglycans in health and disease: new concepts for heparanase function in tumor progression and metastasis. FEBS J *277*, 3890–3903.

Bashkin, P., Doctrow, S., Klagsbrun, M., Svahn, C. M., Folkman, J., and Vlodavsky, I. (1989). Basic fibroblast growth factor binds to subendothelial extracellular matrix and is released by heparitinase and heparin-like molecules. Biochemistry *28*, 1737–1743.

Belting, M. (2003 0. Heparan sulfate proteoglycan as a plasma membrane carrier. Trends Biochem Sci *28*, 145–151.

Ben-Izhak, O., Kaplan-Cohen, V., Ilan, N., Gan, S., Vlodavsky, I., and Nagler, R. (2006). Heparanase expression in malignant salivary gland tumors inversely correlates with long-term survival. Neoplasia *8*, 879–884.

Bernfield, M., Gotte, M., Park, P. W., et al. (1999). Functions of cell surface heparan sulfate proteoglycans. Annu Rev Biochem *68*, 729–777.

Birzele, F., Csaba, G., and Zimmer, R. (2008). Alternative splicing and protein structure evolution. Nucleic Acids Res *36*, 550–558.

Bitan, M., Polliack, A., Zecchina, G., et al. (2002). Heparanase expression in human leukemias is restricted to acute myeloid leukemias. Exp Hematol *30*, 34–41.

Bonertz, A., Weitz, J., Pietsch, D. H., et al. (2009). Antigen-specific Tregs control T cell responses against a limited repertoire of tumor antigens in patients with colorectal carcinoma. J Clin Invest *119*, 3311–3321.

Borja-Cacho, D., Jensen, E. H., Saluja, A. K., Buchsbaum, D. J., and Vickers, S. M. (2008). Molecular targeted therapies for pancreatic cancer. Am J Surg *196*, 430–441.

Borsig, L., Vlodavsky, I., Ishai-Michaeli, R., Torri, G., and Vismara, E. (2011). Sulfated hexasaccharides attenuate metastasis by inhibition of P-selectin and heparanase. Neoplasia *13*, 445–452.

Bouma, G., and Strober, W. (2003). The immunological and genetic basis of inflammatory bowel disease. Nat Rev Immunol *3*, 521–533.

Brunn, G. J., Bungum, M. K., Johnson, G. B., and Platt, J. L. (2005). Conditional signaling by Toll-like receptor 4. FASEB J *19*, 872–874.

Buczek-Thomas, J. A., Hsia, E., Rich, C. B., Foster, J. A., and Nugent, M. A. (2008). Inhibition of histone acetyltransferase by glycosaminoglycans. J Cell Biochem *105*, 108–120.

Camphausen, K., Moses, M. A., Beecken, W.-D., Khan, M. K., Folkman, J., and O'Reilly, M. S. (2001). Radiation therapy to a primary tumor accelerates metastatic growth in mice. Cancer Res *61*, 2207–2211.

Capila, I., and Linhardt, R. J. (2002). Heparin-protein interactions. Angew Chem Int Ed Engl *41*, 391 412.

Casu, B., Guerrini, M., and Torri, G. (2004). Structural and conformational aspects of the anticoagulant and anti-thrombotic activity of heparin and dermatan sulfate. Curr Pharm Des *10*, 939–949.

Casu, B., Vlodavsky, I., and Sanderson, R. D. (2008). Non-anticoagulant heparins and inhibition of cancer. Pathophysiol Haemost Thromb *36*, 195–203.

Chen, G., Wang, D., Vikramadithyan, R., et al. (2004). Inflammatory cytokines and fatty acids regulate endothelial cell heparanase expression. Biochemistry *43*, 4971–4977.

Chen, L., and Sanderson, R. D. (2009). Heparanase regulates levels of syndecan-1 in the nucleus. PLoS One *4*, e4947.

Chen, T., Tang, X. D., Wan, Y., et al. (2008). HLA-A2-restricted cytotoxic T lymphocyte epitopes from human heparanase as novel targets for broad-spectrum tumor immunotherapy. Neoplasia *10*, 977–986.

Chung, S. I., Seong, J., Park, Y. N., Kim, W. W., Oh, H. J., and Han, K. H. (2010). Identification of proteins indicating radiation-induced hepatic toxicity in cirrhotic rats. J Radiat Res *51*, 643–650.

Clevers, H. (2006). Colon cancer–understanding how NSAIDs work. N Engl J Med *354*, 761–763.

Cohen, E., Doweck, I., Naroditsky, I., et al. (2008). Heparanase is overexpressed in lung cancer and correlates inversely with patient survival. Cancer *113*, 1004–1011.

Cohen, I., Pappo, O., Elkin, M., et al. (2006). Heparanase promotes growth, angiogenesis and survival of primary breast tumors. Int J Cancer *118*, 1609–1617.

Cohen, S. J., Dobelbower, R., Jr., Lipsitz, S., et al. (2005). A randomized phase III study of radiotherapy alone or with 5-fluorouracil and mitomycin-C in patients with locally advanced adenocarcinoma of the pancreas: Eastern Cooperative Oncology Group study E8282. Int J Radiat Oncol Biol Phys *62*, 1345–1350.

Cohen-Kaplan, V., Doweck, I., Naroditsky, I., Vlodavsky, I., and Ilan, N. (2008a). Heparanase augments epidermal growth factor receptor phosphorylation: correlation with head and neck tumor progression. Cancer Res *68*, 10077–10085.

Cohen-Kaplan, V., Jarbashian, J., Yanir, Y., et al. (2012). Heparanase induces STAT protein phosphorylation: pre-clinical and clinical significance in head and neck cancer. J Biol Chem *287*, 6668–6678.

Cohen-Kaplan, V., Naroditsky, I., Zetser, A., Ilan, N., Vlodavsky, I., and Doweck, I. (2008b). Heparanase induces VEGF C and facilitates tumor lymphangiogenesis. Int J Cancer *123*, 2566–2573.

Communi, D., Janssens, R., Suarez-Huerta, N., Robaye, B., and Boeynaems, J. M. (2000). Advances in signalling by extracellular nucleotides. the role and transduction mechanisms of P2Y receptors. Cell Signal *12*, 351–360.

Condeelis, J., and Pollard, J. W. (2006). Macrophages: obligate partners for tumor cell migration, invasion, and metastasis. Cell *124*, 263–266.

Conde-Knape, K. (2001). Heparan sulfate proteoglycans in experimental models of diabetes: a role for perlecan in diabetes complications. Diabetes Metab Res Rev *17*, 412–421.

Coussens, L. M., Fingleton, B., and Matrisian, L. M. (2002). Matrix metalloproteinase inhibitors and cancer: trials and tribulations. Science *295*, 2387–2392.

de Mestre, A. M., Khachigian, L. M., Santiago, F. S., Staykova, M. A., and Hulett, M. D. (2003). Regulation of inducible heparanase gene transcription in activated T cells by early growth response 1. J Biol Chem *278*, 50377–50385.

de Mestre, A. M., Rao, S., Hornby, J. R., Soe-Htwe, T., Khachigian, L. M., and Hulett, M. D. (2005). Early growth response gene 1 (EGR1) regulates heparanase gene transcription in tumor cells. J Biol Chem *280*, 35136–35147.

Dirkx, A. E., Oude Egbrink, M. G., Wagstaff, J., and Griffioen, A. W. (2006). Monocyte/macrophage infiltration in tumors: modulators of angiogenesis. J Leukoc Biol *80*, 1183–1196.

Doviner, V., Maly, B., Kaplan, V., et al. (2006). Spatial and temporal heparanase expression in colon mucosa throughout the adenoma-carcinoma sequence. Mod Pathol *19*, 878–888.

Doweck, I., Kaplan-Cohen, V., Naroditsky, I., Sabo, E., Ilan, N., and Vlodavsky, I. (2006). Heparanase localization and expression by head and neck cancer: correlation with tumor progression and patient survival. Neoplasia *8*, 1055–1061.

Dredge, K., Hammond, E., Davis, K., et al. (2010). The PG500 series: novel heparan sulfate mimetics as potent angiogenesis and heparanase inhibitors for cancer therapy. Invest New Drugs *28*, 276–283.

Dredge, K., Hammond, E., Handley, P., et al. (2011). PG545, a dual heparanase and angiogenesis inhibitor, induces potent anti-tumour and anti-metastatic efficacy in preclinical models. Br J Cancer *104*, 635–642.

Edovitsky, E., Elkin, M., Zcharia, E., Peretz, T., and Vlodavsky, I. (2004). Heparanase gene silencing, tumor invasiveness, angiogenesis, and metastasis. J Natl Cancer Inst *96*, 1219–1230.

Edovitsky, E., Lerner, I., Zcharia, E., Peretz, T., Vlodavsky, I., and Elkin, M. (2006). Role of endothelial heparanase in delayed-type hypersensitivity. Blood *107*, 3609–3616.

El-Assal, O. N., Yamanoi, A., Ono, T., Kohno, H., and Nagasue, N. (2001). The clinicopathological significance of heparanase and basic fibroblast growth factor expressions in hepatocellular carcinoma. Clin Cancer Res *7*, 1299–1305.

Elkin, M., Ilan, N., Ishai-Michaeli, R., et al. (2001). Heparanase as mediator of angiogenesis: mode of action. FASEB J *15*, 1661–1663.

Elson, C. O., Sartor, R. B., Tennyson, G. S., and Riddell, R. H. (1995). Experimental models of inflammatory bowel disease. Gastroenterology *109*, 1344–1367.

Endo, K., Maejara, U., Baba, H., et al. (2001). Heparanase gene expression and metastatic potential in human gastric cancer. Anticancer Res *21*, 3365–3369.

Ferro, V., Hammond, E., and Fairweather, J. K. (2004). The development of inhibitors of heparanase, a key enzyme involved in tumour metastasis, angiogenesis and inflammation. Mini Rev Med Chem *4*, 693–702.

Fiebiger, E., Maehr, R., Villadangos, J., et al. (2002). Invariant chain controls the activity of extracellular cathepsin L. J Exp Med *196*, 1263–1269.

Folkman, J., Klagsbrun, M., Sasse, J., Wadzinski, M., Ingber, D., and Vlodavsky, I. (1988). A heparin-binding angiogenic protein – basic fibroblast growth factor – is stored within basement membrane. Am J Pathol *130*, 393–400.

Friedmann, Y., Vlodavsky, I., Aingorn, H., et al. (2000). Expression of heparanase in normal, dysplastic, and neoplastic human colonic mucosa and stroma. Evidence for its role in colonic tumorigenesis. Am J Pathol *157*, 1167–1175.

Fukata, M., Chen, A., Vamadevan, A. S., et al. (2007). Toll-like receptor-4 promotes the development of colitis-associated colorectal tumors. Gastroenterology *133*, 1869–1881.

Fukata, M., Hernandez, Y., Conduah, D., et al. (2009). Innate immune signaling by Toll-like receptor-4 (TLR4) shapes the inflammatory microenvironment in colitis-associated tumors. Inflamm Bowel Dis *15*, 997–1006.

Fux, L., Feibish, N., Cohen-Kaplan, V., et al. (2009a). Structure-function approach identifies a COOH-terminal domain that mediates heparanase signaling. Cancer Res *69*, 1758–1767.

Fux, L., Ilan, N., Sanderson, R. D., and Vlodavsky, I. (2009b). Heparanase: busy at the cell surface. Trends Biochem Sci *34*, 511–519.

Gil, N., Goldberg, R., Neuman, T., et al. (2012). Heparanase is essential for the development of diabetic nephropathy in mice. Diabetes *61* (1), 208–216.

Gilbertson, D. T., Liu, J., Xue, J. L., et al. (2005). Projecting the number of patients with end-stage renal disease in the United States to the year 2015. J Am Soc Nephrol *16*, 3736–3741.

Gohji, K., Hirano, H., Okamoto, M., et al. (2001a). Expression of three extracellular matrix degradative enzymes in bladder cancer. Int J Cancer *95*, 295–301.

Gohji, K., Okamoto, M., Kitazawa, S., et al. (2001b). Heparanase protein and gene expression in bladder cancer. J Urol *166*, 1286–1290.

Greten, F. R., Eckmann, L., Greten, T. F., et al. (2004). IKKbeta links inflammation and tumorigenesis in a mouse model of colitis-associated cancer. Cell *118*, 285–296.

Grivennikov, S. I., Greten, F. R., and Karin, M. (2010). Immunity, inflammation, and cancer. Cell *140*, 883–899.

Hacker, U., Nybakken, K., and Perrimon, N. (2005). Heparan sulphate proteoglycans: the sweet side of development. Nat Rev Mol Cell Biol *6*, 530–541.

Harvey, S. J., and Miner, J. H. (2008). Revisiting the glomerular charge barrier in the molecular era. Curr Opin Nephrol Hypertens *17*, 393–398.

Haskell, M. D., Slack, J. K., Parsons, J. T., and Parsons, S. J. (2001). c-Src tyrosine phosphorylation of epidermal growth factor receptor, P190 RhoGAP, and focal adhesion kinase regulates diverse cellular processes. Chem Rev *101*, 2425–2440.

Higgins, W. J., Fox, D. M., Kowalski, P. S., Nielsen, J. E., and Worrall, D. M. (2010). Heparin enhances serpin inhibition of the cysteine protease cathepsin L. J Biol Chem *285*, 3722–3729.

Hoffmann, A. C., Mori, R., Vallbohmer, D., et al. (2008). High expression of heparanase is significantly associated with dedifferentiation and lymph node metastasis in patients with pancreatic ductal adenocarcinomas and correlated to PDGFA and via HIF1a to HB-EGF and bFGF. J Gastrointest Surg *12*, 1674–1681; discussion 1681–1682.

Hulett, M. D., Freeman, C., Hamdorf, B. J., Baker, R. T., Harris, M. J., and Parish, C. R. (1999). Cloning of mammalian heparanase, an important enzyme in tumor invasion and metastasis. Nat Med *5*, 803–809.

Hulett, M. D., Hornby, J. R., Ohms, S. J., et al. (2000). Identification of active-site residues of the pro-metastatic endoglycosidase heparanase. Biochemistry *39*, 15659–15667.

Ibrahim, H. N., and Hostetter, T. H. (1997). Diabetic nephropathy. J Am Soc Nephrol *8*, 487–493.

Ilan, N., Elkin, M., and Vlodavsky, I. (2006). Regulation, function and clinical significance of heparanase in cancer metastasis and angiogenesis. Int J Biochem Cell Biol *38*, 2018–2039.

Iozzo, R. V., and San Antonio, J. D. (2001). Heparan sulfate proteoglycans: heavy hitters in the angiogenesis arena. J Clin Invest *108*, 349–355.

Iriyama, S., Matsunaga, Y., Takahashi, K., Matsuzaki, K., Kumagai, N., and Amano, S. (2011). Activation of heparanase by ultraviolet B irradiation leads to functional loss of basement membrane at the dermal-epidermal junction in human skin. Arch Dermatol Res *303*, 253–261.

Josefsen, K., Sorensen, L. R., Buschard, K., and Birkenbach, M. (1999). Glucose induces early growth response gene (Egr-1) expression in pancreatic beta cells. Diabetologia *42*, 195–203.

Karin, M. (2006). Nuclear factor-kappaB in cancer development and progression. Nature *441*, 431–436.

Karin, M., and Greten, F. R. (2005). NF-kappaB: linking inflammation and immunity to cancer development and progression. Nat Rev Immunol *5*, 749–759.

Katz, A., Van-Dijk, D. J., Aingorn, H., et al. (2002). Involvement of human heparanase in the pathogenesis of diabetic nephropathy. Isr Med Assoc J *4*, 996–1002.

Kelly, T., Miao, H.-Q., Yang, Y., et al. (2003). High heparanase activity in multiple myeloma is associated with elevated microvessel density. Cancer Res *63*, 8749–8756.

Kelly, T., Suva, L. J., Huang, Y., et al. (2005). Expression of heparanase by primary breast tumors promotes bone resorption in the absence of detectable bone metastases. Cancer Res *65*, 5778–5784.

Kelly, T., Suva, L. J., Nicks, K. M., MacLeod, V., and Sanderson, R. D. (2010). Tumor-derived syndecan-1 mediates distal cross-talk with bone that enhances osteoclastogenesis. J Bone Miner Res *25*, 1295–1304.

Kessenbrock, K., Plaks, V., and Werb, Z. (2010). Matrix metalloproteinases: regulators of the tumor microenvironment. Cell *141*, 52–67.

Khachigian, L. M., and Parish, C. R. (2004). Phosphomannopentaose sulfate (PI-88): heparan sulfate mimetic with clinical potential in multiple vascular pathologies. Cardiovasc Drug Rev *22*, 1–6.

Kim, A. W., Xu, X., Hollinger, E. F., Gattuso, P., Godellas, C. V., and Prinz, R. A. (2002). Human heparanase-1 gene expression in pancreatic adenocarcinoma. J Gastrointest Surg *6*, 167–172.

Kjellen, L., and Lindahl, U. (1991). Proteoglycans: structures and interactions. Annu Rev Biochem *60*, 443–475.

Kobayashi, M., Naomoto, Y., Nobuhisa, T., et al. (2006). Heparanase regulates esophageal keratinocyte differentiation through nuclear translocation and heparan sulfate cleavage. Differentiation *74*, 235–243.

Koliopanos, A., Friess, H., Kleeff, J., et al. (2001). Heparanase expression in primary and metastatic pancreatic cancer. Cancer Res *61*, 4655–4659.

Kovalszky, I., Dudas, J., Olah-Nagy, J., et al. (1998). Inhibition of DNA topoisomerase I activity by heparan sulfate and modulation by basic fibroblast growth factor. Mol Cell Biochem *183*, 11–23.

Kramer, K. L., and Yost, H. J. (2003). Heparan sulfate core proteins in cell-cell signaling. Annu Rev Genet *37*, 461–484.

Krieglstein, C. F., Cerwinka, W. H., Sprague, A. G., et al. (2002). Collagen-binding integrin alpha1beta1 regulates intestinal inflammation in experimental colitis. J Clin Invest *110*, 1773–1782.

Kussie, P. H., Hulmes, J. D., Ludwig, D. L., et al. (1999). Cloning and functional expression of a human heparanase gene. Biochem Biophys Res Commun *261*, 183–187.

Lerner, I., Baraz, L., Pikarsky, E., et al. (2008). Function of heparanase in prostate tumorigenesis: potential for therapy. Clin Cancer Res *14*, 668–676.

Lerner, I., Hermano, E., Zcharia, E., et al. (2011). Heparanase powers a chronic inflammatory circuit that promotes colitis-associated tumorigenesis in mice. J Clin Invest *121*, 1709–1721.

Levidiotis, V., Freeman, C., Punler, M., et al. (2004). A synthetic heparanase inhibitor reduces proteinuria in passive Heymann nephritis. J Am Soc Nephrol *15*, 2882–2892.

Levidiotis, V., Freeman, C., Tikellis, C., Cooper, M. E., and Power, D. A. (2005). Heparanase inhibition reduces proteinuria in a model of accelerated anti-glomerular basement membrane antibody disease. Nephrology *10*, 167–173.

Levy-Adam, F., Feld, S., Cohen-Kaplan, V., et al. (2010a). Heparanase 2 interacts with heparan sulfate with high affinity and inhibits heparanase activity. J Biol Chem *285*, 28010–28009.

Levy-Adam, F., Feld, S., Suss-Toby, E., Vlodavsky, I., and Ilan, N. (2008). Heparanase facilitates cell adhesion and spreading by clustering of cell surface heparan sulfate proteoglycans. PLoS One *3*, e2319.

Levy-Adam, F., Ilan, N., and Vlodavsky, I. (2010b). Tumorigenic and adhesive properties of heparanase. Semin Cancer Biol *20*, 153–160.

Levy-Adam, F., Miao, H. Q., Heinrikson, R. L., Vlodavsky, I., and Ilan, N. (2003). Heterodimer formation is essential for heparanase enzymatic activity. Biochem Biophys Res Commun *308*, 885–891.

Li, J. P., and Vlodavsky, I. (2009). Heparin, heparan sulfate and heparanase in inflammatory reactions. Thromb Haemost *102*, 823–828.

Li, R. W., Freeman, C., Yu, D., et al. (2008). Dramatic regulation of heparanase activity and angiogenesis gene expression in synovium from patients with rheumatoid arthritis. Arthritis Rheum *58*, 1590–1600.

Lindahl, U., and Li, J. P. (2009). Interactions between heparan sulfate and proteins-design and functional implications. Int Rev Cell Mol Biol *276*, 105–159.

Mahida, Y. R. (2000). The key role of macrophages in the immunopathogenesis of inflammatory bowel disease. Inflamm Bowel Dis *6*, 21–33.

Mahtouk, K., Hose, D., Raynaud, P., et al. (2007). Heparanase influences expression and shedding of syndecan-1, and its expression by the bone marrow environment is a bad prognostic factor in multiple myeloma. Blood *109*, 4914–4923.

Mantovani, A., and Sica, A. (2010). Macrophages, innate immunity and cancer: balance, tolerance, and diversity. Curr Opin Immunol *22*, 231–237.

Massena, S., Christoffersson, G., Hjertstrom, E., et al. (2011). A chemotactic gradient sequestered on endothelial heparan sulfate induces directional intraluminal crawling of neutrophils. Blood *116*, 1924–1931.

Maxhimer, J. B., Quiros, R. M., Stewart, R., et al. (2002). Heparanase-1 expression is associated with the metastatic potential of breast cancer. Surgery *132*, 326–333.

Maxhimer, J. B., Somenek, M., Rao, G., et al. (2005). Heparanase-1 gene expression and regulation by high glucose in renal epithelial cells: a potential role in the pathogenesis of proteinuria in diabetic patients. Diabetes *54*, 2172–2178.

McKenzie, E., Tyson, K., Stamps, A., et al. (2000). Cloning and expression profiling of Hpa2, a novel mammalian heparanase family member. Biochem Biophys Res Commun *276*, 1170–1177.

McKenzie, E., Young, K., Hircock, M., et al. (2003). Biochemical characterization of the active heterodimer form of human heparanase (Hpa1) protein expressed in insect cells. Biochem J *373*, 423–435.

McKenzie, E. A. (2007). Heparanase: a target for drug discovery in cancer and inflammation. Br J Pharmacol *151*, 1–14.

Meirovitz, A., Hermano, E., Lerner, I., et al. (2011). Role of heparanase in radiation-enhanced invasiveness of pancreatic carcinoma. Cancer Res *71*, 2772–2780.

Miao, H. Q., Liu, H., Navarro, E., Kussie, P., and Zhu, Z. (2006). Development of heparanase inhibitors for anti-cancer therapy. Curr Med Chem *13*, 2101–2111.

Mikami, S., Oya, M., Shimoda, M., et al. (2008). Expression of heparanase in renal cell carcinomas: implications for tumor invasion and prognosis. Clin Cancer Res *14*, 6055–6061.

Miner, J. H. (1999). Renal basement membrane components. Kidney Int 56, 2016–2024.

Muller, M. W., Friess, H., Koninger, J., et al. (2008). Factors influencing survival after bypass procedures in patients with advanced pancreatic adenocarcinomas. Am J Surg *195*, 221–228.

Myler, H. A., Lipke, E. A., Rice, E. E., and West, J. L. (2006). Novel heparanase-inhibiting antibody reduces neointima formation. J Biochem (Tokyo) *139*, 339–345.

Naggi, A., Casu, B., Perez, M., et al. (2005). Modulation of the heparanase-inhibiting activity of heparin through selective desulfation, graded N-acetylation, and glycol splitting. J Biol Chem *280*, 12103–12113.

Nagler, R., Ben-Izhak, O., Cohen-Kaplan, V., et al. (2007). Heparanase up-regulation in tongue cancer: tissue and saliva analysis. Cancer *110*, 2732–2739.

Nakajima, M., Irimura, T., DiFerrante, D., DiFerrante, N., and Nicolson, G. L. (1983). Heparan sulfate degradation: relation to tumor invasion and metastatic properties of mouse B 16 melanoma sublines. Science (Wash DC) *220*, 611–613.

Nardella, C., Lahm, A., Pallaoro, M., Brunetti, M., Vannini, A., and Steinkuhler, C. (2004). Mechanism of activation of human heparanase investigated by protein engineering. Biochemistry *43*, 1862–1873.

Navarro-Gonzalez, J. F., and Mora-Fernandez, C. (2008). The role of inflammatory cytokines in diabetic nephropathy. J Am Soc Nephrol *19*, 433–442.

Neoptolemos, J. P., Stocken, D. D., Friess, H., et al. (2004). A randomized trial of chemoradiotherapy and chemotherapy after resection of pancreatic cancer. N Engl J Med *350*, 1200–1210.

Nguyen, D., Ping, F., Mu, W., Hill, P., Atkins, R. C., and Chadban, S. J. (2006). Macrophage accumulation in human progressive diabetic nephropathy. Nephrology *11*, 226–231.

Nobuhisa, T., Naomoto, Y., Ohkawa, T., et al. (2005). Heparanase expression correlates with malignant potential in human colon cancer. J Cancer Res Clin Oncol *131*, 229–237.

Ogren, S., and Lindahl, U. (1975). Cleavage of macromolecular heparin by an enzyme from mouse mastocytoma. J Biol Chem *250*, 2690–2697.

Ohkawa, T., Naomoto, Y., Takaoka, M., et al. (2004). Localization of heparanase in esophageal cancer cells: respective roles in prognosis and differentiation. Lab Invest *84*, 1289–1304.

Ohuchida, K., Mizumoto, K., Murakami, M., et al. (2004). Radiation to stromal fibroblasts increases invasiveness of pancreatic cancer cells through tumor-stromal interactions. Cancer Res *64*, 3215–3222.

Okayasu, I., Ohkusa, T., Kajiura, K., Kanno, J., and Sakamoto, S. (1996). Promotion of colorectal neoplasia in experimental murine ulcerative colitis. Gut *39*, 87–92.

Parish, C. R., Freeman, C., and Hulett, M. D. (2001). Heparanase: a key enzyme involved in cell invasion. Biochim Biophys Acta *1471*, M99–M108.

Park, C. M., Park, M. J., Kwak, H. J., et al. (2006). Ionizing radiation enhances matrix metalloproteinase-2 secretion and invasion of glioma cells through Src/epidermal growth factor receptor-mediated p38/Akt and phosphatidylinositol 3-kinase/Akt signaling pathways. Cancer Res *66*, 8511–8519.

Patel, V. N., Knox, S. M., Likar, K. M., et al. (2007). Heparanase cleavage of perlecan heparan sulfate modulates FGF10 activity during ex vivo submandibular gland branching morphogenesis. Development *134*, 4177–4186.

Pollard, J. W. (2004). Tumour-educated macrophages promote tumour progression and metastasis. Nat Rev Cancer *4*, 71–78.

Popivanova, B. K., Kitamura, K., Wu, Y., et al. (2008). Blocking TNF-alpha in mice reduces colorectal carcinogenesis associated with chronic colitis. J Clin Invest *118*, 560–570.

Purushothaman, A., Chen, L., Yang, Y., and Sanderson, R. D. (2008). Heparanase stimulation of protease expression implicates it as a master regulator of the aggressive tumor phenotype in myeloma. J Biol Chem *283*, 32628–32636.

Purushothaman, A., Uyama, T., Kobayashi, F., et al. (2010). Heparanase-enhanced shedding of syndecan-1 by myeloma cells promotes endothelial invasion and angiogenesis. Blood *115*, 2449–2457.

Quiros, R. M., Rao, G., Plate, J., et al. (2006). Elevated serum heparanase-1 levels in patients with pancreatic carcinoma are associated with poor survival. Cancer *106*, 532–540.

Raats, C. J., Van Den Born, J., and Berden, J. H. (2000). Glomerular heparan sulfate alterations: mechanisms and relevance for proteinuria. Kidney Int *57*, 385–400.

Raimondi, S., Maisonneuve, P., and Lowenfels, A. B. (2009). Epidemiology of pancreatic cancer: an overview. Nat Rev Gastroenterol Hepatol *6*, 699–708.

Ramani, V.C ., Yang, Y., Ren, Y., Nan, L., and Sanderson, R. D. (2011). Heparanase plays a dual role in driving hepatocyte growth factor (HGF) signaling by enhancing HGF expression and activity. J Biol Chem *286*, 6490–6499.

Ritchie, J. P., Ramani, V. C., Ren, Y., et al. (2011). SST0001, a chemically modified heparin, inhibits myeloma growth and angiogenesis via disruption of the heparanase/syndecan-1 axis. Clin Cancer Res *17*, 1382–1393.

Rohloff, J., Zinke, J., Schoppmeyer, K., et al. (2002). Heparanase expression is a prognostic indicator for postoperative survival in pancreatic adenocarcinoma. Br J Cancer *86*, 1270–1275.

Sanchez-Munoz, F., Dominguez-Lopez, A., and Yamamoto-Furusho, J. K. (2008). Role of cytokines in inflammatory bowel disease. World J Gastroenterol *14*, 4280–4288.

Sanderson, R. D. (2001). Heparan sulfate proteoglycans in invasion and metastasis. Semin Cell Dev Biol *12*, 89–98.

Sasisekharan, R., Shriver, Z., Venkataraman, G., and Narayanasami, U. (2002). Roles of heparan-sulphate glycosaminoglycans in cancer. Nat Rev Cancer *2*, 521–528.

Sato, T., Yamaguchi, A., Goi, T., et al. (2004). Heparanase expression in human colorectal cancer and its relationship to tumor angiogenesis, hematogenous metastasis, and prognosis. J Surg Oncol *87*, 174–181.

Schubert, S. Y., Ilan, N., Shushy, M., Ben-Izhak, O., Vlodavsky, I., and Goldshmidt, O. (2004). Human heparanase nuclear localization and enzymatic activity. Lab Invest *84*, 535–544.

Shafat, I., Ben-Arush, M. W., Issakov, J., et al. (2011a). Preclinical and clinical significance of heparanase in Ewing's sarcoma. J Cell Mol Med *15*, 1857–1864.

Shafat, I., Ben-Barak, A., Postovsky, S., et al. (2007). Heparanase levels are elevated in the plasma of pediatric cancer patients and correlate with response to anticancer treatment. Neoplasia *9*, 909–916.

Shafat, I., Ilan, N., Zoabi, S., Vlodavsky, I., and Nakhoul, F. (2011b.) Heparanase levels are elevated in the urine and plasma of type 2 diabetes patients and associate with blood glucose levels. PLoS One *6*, e17312.

Shafat, I., Pode, D., Peretz, T., Ilan, N., Vlodavsky, I., and Nisman, B. (2008). Clinical significance of urine heparanase in bladder cancer progression. Neoplasia *10*, 125–130.

Shafat, I., Vlodavsky, I., and Ilan, N. (2006a). Characterization of mechanisms involved in secretion of active heparanase. J Biol Chem *281*, 23804–23811.

Shafat, I., E. Zcharia, B. Nisman, et al. (2006b). An ELISA method for the detection and quantification of human heparanase. *Biochem* Biophys Res Commun *341*, 958–963.

Shaib, Y. H., Davila, J. A., and El-Serag, H. B. (2006). The epidemiology of pancreatic cancer in the United States: changes below the surface. Aliment Pharmacol Ther *24*, 87–94.

Shinyo, Y., Kodama, J., Hongo, A., Yoshinouchi, M., and Hiramatsu, Y. (2003). Heparanase expression is an independent prognostic factor in patients with invasive cervical cancer. Ann Oncol *14*, 1505–1510.

Simizu, S., Ishida, K., and Osada, H. (2004). Heparanase as a molecular target of cancer chemotherapy. Cancer Sci 95, 553–558.

Simizu, S., Suzuki, T., Muroi, M., et al. (2007). Involvement of disulfide bond formation in the activation of heparanase. Cancer Res *67*, 7841–7849.

Simons, M., and Horowitz, A. (2001). Syndecan-4-mediated signalling. Cell Signal *13*, 855–862.

Sommerfeldt, N., Beckhove, P., Ge, Y., et al. (2006). Heparanase: a new metastasis-associated antigen recognized in breast cancer patients by spontaneously induced memory T lymphocytes. Cancer Res *66*, 7716–7723.

Sotnikov, I., Hershkoviz, R., Grabovsky, V., et al. (2004). Enzymatically quiescent heparanase augments T cell interactions with VCAM-1 and extracellular matrix components under versatile dynamic contexts. J Immunol *172*, 5185–5193.

Srinivasan, R., Mager, G. M., Ward, R. M., Mayer, J., and Svaren, J. (2006). NAB2 represses transcription by interacting with the CHD4 subunit of the nucleosome remodeling and deacetylase (NuRD) complex. J Biol Chem *281*, 15129–15137.

Subbaramaiah, K., Yoshimatsu, K., Scherl, E., et al. (2004). Microsomal prostaglandin E synthase-1 is overexpressed in inflammatory bowel disease. Evidence for involvement of the transcription factor Egr-1. J Biol Chem *279*, 12647–12658.

Takahashi, H., Ebihara, S., Okazaki, T., et al. (2004). Clinical significance of heparanase activity in primary resected non-small cell lung cancer. Lung Cancer *45*, 207–214.

Takaoka, M., Naomoto, Y., Ohkawa, T., et al. (2003). Heparanase expression correlates with invasion and poor prognosis in gastric cancers. Lab Invest *83*, 613–622.

Tang, W., Nakamura, Y., Tsujimoto, M., et al. (2002). Heparanase: a key enzyme in invasion and metastasis of gastric carcinoma. Mod Pathol *15*, 593–598.

Tang, X. D., Liang, G. P., Li, C., et al. (2010). Cytotoxic T lymphocyte epitopes from human heparanase can elicit a potent anti-tumor immune response in mice. Cancer Immunol Immunother *59*, 1041–1047.

Tesch, G. H. (2010). Macrophages and diabetic nephropathy. Semin Nephrol *30*, 290–301.

Thunberg, L., Backstrom, G., Wasteson, A., Robinson, H. C., Ogren, S., and Lindahl, U. (1982). Enzymatic depolymerization of heparin-related polysaccharides. Substrate specificities of mouse mastocytoma and human platelet endo-beta-D-glucuronidases. J Biol Chem *257*, 10278–10282.

Timar, J., Lapis, K., Dudas, J., Sebestyen, A., Kopper, L., and Kovalszky, I. (2002). Proteoglycans and tumor progression: Janus-faced molecules with contradictory functions in cancer. Semin Cancer Biol *12*, 173–186.

Timpl, R., and Brown, J. C. (1996). Supramolecular assembly of basement membranes. Bioessays *18*, 123–132.

Toyoshima, M., and Nakajima, M. (1999). Human heparanase. Purification, characterization, cloning, and expression. J Biol Chem *274*, 24153–24160.

Turk, B., Turk, D., and Turk, V. (2000). Lysosomal cysteine proteases: more than scavengers. Biochim Biophys Acta *1477*, 98–111.

Turk, V., Kos, J., and Turk, B. (2004). Cysteine cathepsins (proteases) – on the main stage of cancer? Cancer Cell *5*, 409–410.

Tuttle, K. R. (2005). Linking metabolism and immunology: diabetic nephropathy is an inflammatory disease. J Am Soc Nephrol *16*, 1537–1538.

van den Born, J., Pisa, B., Bakker, M. A., et al. (2006). No change in glomerular heparan sulfate structure in early human and experimental diabetic nephropathy. J Biol Chem *281*, 29606–29613.

van den Hoven, M. J., Rops, A. L., Bakker, M. A., et al. (2006). Increased expression of heparanase in overt diabetic nephropathy. Kidney Int *70*, 2100–2108.

van der Weyden, L., Conigrave, A. D., and Morris, M. B. (2000). Signal transduction and white cell maturation via extracellular ATP and the P2Y11 receptor. Immunol Cell Biol *78*, 369–374.

Vlodavsky, I., Bar-Shavit, R., Ishai-Michaeli, R., Bashkin, P., and Fuks, Z. (1991). Extracellular sequestration and release of fibroblast growth factor: a regulatory mechanism? Trends Biochem Sci *16*, 268–271.

Vlodavsky, I., Beckhove, P., Lerner, I., et al. (2011). Significance of heparanase in cancer and inflammation. Cancer Microenviron, Aug. 3. PMID: 21811836.

Vlodavsky, I., Eldor, A., Haimovitz-Friedman, A., et al. (1992). Expression of heparanase by platelets and circulating cells of the immune system: possible involvement in diapedesis and extravasation. Invasion Metastasis *12*, 112–127.

Vlodavsky, I., and Friedmann, Y. (2001). Molecular properties and involvement of heparanase in cancer metastasis and angiogenesis. J Clin Invest *108*, 341–347.

Vlodavsky, I., Friedmann, Y., Elkin, M., et al. (1999). Mammalian heparanase: gene cloning, expression and function in tumor progression and metastasis. Nat Med *5*, 793–802.

Vlodavsky, I., Fuks, Z., Bar-Ner, M., Ariav, Y., and Schirrmacher, V. (1983). Lymphoma cells mediated degradation of sulfated proteoglycans in the subendothelial extracellular matrix: relation to tumor cell metastasis. Cancer Res *43*, 2704–2711.

Vlodavsky, I., Ilan, N., Naggi, A., and Casu, B. (2007). Heparanase: structure, biological functions, and inhibition by heparin-derived mimetics of heparan sulfate. Curr Pharm Des *13*, 2057–2073.

Vlodavsky, I., Korner, G., Ishai Michaeli, R., Bashkin, P., Bar-Shavit, R., and Fuks, Z. (1990). Extracellular matrix-resident growth factors and enzymes: possible involvement in tumor metastasis and angiogenesis. Cancer Metastasis Rev *9*, 203–226.

Vlodavsky, I., Miao, H. Q., Medalion, B., Danagher, P., and Ron, D. (1996). Involvement of heparan sulfate and related molecules in sequestration and growth promoting activity of fibroblast growth factor. Cancer Metastasis Rev *15*, 177–186.

Vreys, V., and David, G. (2007). Mammalian heparanase: what is the message? J Cell Mol Med *11*, 427–452.

Wang, G. Z., Tang, X. D., Lu, M. H., et al. (2011). Multiple antigenic peptides of human heparanase elicit a much more potent immune response against tumors. Cancer Prev Res *4*, 1285–1295.

Watanabe, M., Aoki, Y., Kase, H., and Tanaka, K. (2003). Heparanase expression and angiogenesis in endometrial cancer. Gynecol Obstet Invest *56*, 77–82.

Waterman, M., Ben-Izhak, O., Eliakim, R., Groisman, G., Vlodavsky, I., and Ilan, N. (2007). Heparanase upregulation by colonic epithelium in inflammatory bowel disease. Mod Pathol *20*, 8–14.

Yang, Y., Macleod, V., Bendre, M., et al. (2005). Heparanase promotes the spontaneous metastasis of myeloma cells to bone. Blood *105*, 1303–1309.

Yang, Y., MacLeod, V., Dai, Y., et al. (2007a). The syndecan-1 heparan sulfate proteoglycan is a viable target for myeloma therapy. Blood *110*, 2041–2048.

Yang, Y., Macleod, V., Miao, H. Q., et al. (2007b). Heparanase enhances syndecan-1 shedding: a novel mechanism for stimulation of tumor growth and metastasis. J Biol Chem *282*, 13326–13333.

Yang, Y., Ren, Y., Ramani, V. C., Nan, L., Suva, L. J., and Sanderson, R. D. (2010). Heparanase enhances local and systemic osteolysis in multiple myeloma by upregulating the expression and secretion of RANKL. Cancer Res *70*, 8329–8338.

Yu, H., Pardoll, D., and Jove, R. (2009). STATs in cancer inflammation and immunity: a leading role for STAT3. Nat Rev Cancer *9*, 798–809.

Zcharia, E., Jia, J., Zhang, X., et al. (2009). Newly generated heparanase knock-out mice unravel co-regulation of heparanase and matrix metalloproteinases. PLoS One *4*, e5181.

Zcharia, E., Metzger, S., Chajek-ShaulL, T., et al. (2004). Transgenic expression of mammalian heparanase uncovers physiological functions of heparan sulfate in tissue morphogenesis, vascularization, and feeding behavior. FASEB J *18*, 252–263.

Zetser, A., Bashenko, Y., Edovitsky, E., Levy-Adam, F., Vlodavsky, I., and Ilan, N. (2006). Heparanase induces vascular endothelial growth factor expression: correlation with p38 phosphorylation levels and Src activation. Cancer Res *66*, 1455–1463.

Zetser, A., Bashenko, Y., Miao, H.-Q., Vlodavsky, I., and Ilan, N. (2003). Heparanase affects adhesive and tumorigenic potential of human glioma cells. *Cancer Res.* 63, 7733–7741.

Zhang, L., Sullivan, P. S., Goodman, J. C., Gunaratne, P. H., and Marchetti, D. (2011). MicroRNA-1258 suppresses breast cancer brain metastasis by targeting heparanase. Cancer Res *71*, 645–654.

Zheng, L. D., Tong, Q. S., Tang, S. T., et al. (2009). Expression and clinical significance of heparanase in neuroblastoma. World J Pediatr *5*, 206–210.

Zhou, H., Roy, S., Cochran, E., et al. (2011). M402, a novel heparan sulfate mimetic, targets multiple pathways implicated in tumor progression and metastasis. PLoS One *6*, e21106.

8.9 Delivery systems targeting cancer at the level of ECM

Shibnath Ghatak, Vincent C. Hascall, Nikos K. Karamanos, Roger R. Markwald, and Suniti Misra

8.9.1 Introduction

Cancer is a complex group of diseases with many possible causes. The process of carcinogenesis can take years, if not decades. The question is: what causes a normal cell to be converted to a cancer cell? The human genome contains information in two forms: the genetic information provides the blueprint for the production of all the proteins necessary to create a life, while the epigenetic information provides instructions on how, where, and when the genetic information should be used. Epigenetic modification includes DNA methylation and alternate splicing of RNA. Evidence linking inflammation to epigenetics and cancer is rising – for example, proinflammatory prostaglandin-E2 and transforming growth factor-beta (TGF-β) as well as alternate spliced variant 6 of the cell surface hyaluronan (HA) receptor CD44 (CD44v6) that arises during inflammation/cancer are significantly linked to increased risk of colorectal cancer (CRC). Epigenetics provide indications that environmental factors linked with genetic changes occur in cancer, and that these epigenetic changes are reversible. Therefore, analyzing the alternate splicing patterns, or methylation patterns of certain genes that undergo these changes could provide the useful biomarkers for the precaution for people on the brink of developing cancer. For example, the reported increase in serum soluble CD44v6 could prove to be a valuable companion marker for assessing risk in patients with intestinal metaplasia. Arresting these reversible changes in genetic patterns before irreversible mutations occur is a very good way of preventing cancer.

Advances in the improvement of chemotherapy over the past three decades led to an improvement in patient survival, but there is still a need for the robust clinical outcome. The central problem in cancer chemotherapy is the severe toxic side effects of anticancer drugs on healthy tissues. Invariably the side effects impose dose reduction, treatment delay, or development of drug resistance to therapy (Park et al., 2002). Thus, one of the many challenges in chemotherapy is to deliver an effective dose of a cytotoxic agent to the tumor or at the site of the tumor to minimize unwanted detrimental side effects. The collection of genetic and epigenetic aberrations characterizing cancer cells provide new therapeutic targets such as microenvironment, blood vessels supporting tumor growth and more specific targeted therapeutics for cancer treatment and hopefully prevention.

Over the last two decades better understanding of signal pathways responsible for tumor survival along with the mapping of the human genome have changed cancer therapeutic potential from a mere randomly targeted chemotherapeutic approach to a highly selective targeted therapeutic strategy. Thus, the development of new strategies for the cure of cancer is a growing field of research, particularly with gene therapeutic approaches. Synthesizing interfering RNAs that mimic this process is an important method for transiently blocking protein expression that causes the disease. RNA interference (RNAi) is a potent and specific gene silencing event that can exploit an innate biological pathway and disrupt the flow of genetic information from RNA by degrading a target mRNA that has a complementary sequence to the siRNA. This is most commonly achieved by delivery of a synthetic double-stranded (ds) RNA targeted against the aberrant expression of a specific mRNA encoding the protein of interest (Xie et al., 2006). Inside the cell, RNAi occurs by loading the target mRNA with an RNA interference silencing complex (RISC) that attaches a short single stranded antisense RNA to a region that is complementary to the target mRNA (Shankar et al., 2005; Kumar et al., 2006). It is also possible to produce ds-RNA by delivery of a plasmid DNA carrying the RNA polymerase III promoter upstream of a short hairpin RNA (shRNA) sequence. Upon expression the plasmid produces a substrate for the endogenous endonuclease dicer that generates the 21-mer nucleotide siRNA. In mammalian cells transfection with an shRNA producing expression vector efficiently inhibits endogenous function of a specific gene by binding to and destroying its mRNA, which carries protein-building instructions from the DNA.

The inhibition of expression of the gene of interest occurs in a sequence-specific manner without induction of the interferon response. Thus, the siRNA or vectors that transcribe shRNA can be used to prevent the production of disease-causing proteins in the first place, rather than preventing their damage or neutralizing them once they are produced. However, the phenotypic changes induced by siRNAs only persist one week due to lack of transfer of siRNA into daughter cells or dilution of siRNA concentration after each cell division, which limits their utility for use in inhibiting tumor progression.

Targeting most cancers requires systemic delivery, which is more complex than delivering naked RNAi locally. The siRNAs without secondary modifications are unlikely to cause long lasting changes. However, systemic treatment of tumor-bearing animals with siRNAs is very limited because metastasis is the main cause of treatment failure and death from cancer. Accordingly, we prepared an expression vector driven by a human U6 promoter expressing shRNAs targeted to human CD44v6 and showed that it can inhibit expression of CD44v6 and key mediators of HA/CD44v6 signaling pathways in colon tumor cells. We then showed that intraperitoneal (IP) injection of such plasmids in a nanoparticle carrier into mice carrying adenoma tumors suppressed tumor growth in intestinal tumor tissues of *Apc* Min/+ mice (Misra et al., 2009).

There are excellent reviews on the extracellular matrix (ECM) and cell signaling with respect to the dynamic cooperation of the glycosaminoglycan hyaluronan (HA)/CD44, integrin, proteoglycan and growth factor receptor (Kim et al., 2011; Sebens and Schafer, 2011). In this chapter, we focus exclusively on various aspects of tumor growth and progression with respect to the involvement of HA and CD44v6 in CRC cell survival and on therapeutic strategies for CRC regression.

8.9.2 Targeting cancer

Barriers of anticancer treatment

- *In vitro* testing of targeted drug delivery technologies is done in cultured human cancer cells that express a specific surface marker. However, the IC_{50} (half maximal inhibitory concentration) values from the dose-response curves determined in drug delivery systems (DDSs) in the *in vitro* studies have been found to be difficult to predict therapeutic efficacy in clinical settings.
- *In vivo* testing of targeted drug delivery approaches are done using human cancer cell xenografts established in severe combined immunodeficiency (SCID) mice, or using mice having specific genetic alteration that leads to the onset of an oncogenic event. In most of these studies, however, tumors were not completely inhibited after treatment, and tumors reappear once the treatment has been stopped (Gao et al., 2011). Often these studies in xenograft animal models were not translated to potential *in vivo* outcomes in human studies (Jain and Stylianopoulos, 2010). Thus, translation of potential preclinical approaches to successful clinical trials have mostly not succeeded, and many times the translational approach depends on environmental characteristics and disease status at the time of treatment.
- The diverse perfusion and extravasations of abnormal tumor structure, and an unfriendly tumor microenvironment (see Chapter 8.2 by S. Misra and coworkers) can provide resistance to chemotherapy.
- The accumulation of ECM that consists of dense networks of collagen fibers and glycosaminoglycans creates a barrier for drug entry and blocks penetration of nanoparticles carrying drugs, leaving them concentrated in perivascular regions (Jain and Stylianopoulos, 2010; see Chapter 8.2 by S. Misra and coworkers).
- Cancer stem cells (CSCs) of the organs survive the chemotherapy because of their drug resistance capability and are able to replace healthy cells lost due to chemotherapy.

Targeted drug delivery

Targeted drug delivery refers to predominant drug accumulation within a desired target tissue that is independent of the method and route of drug administration (Torchilin, 2000).Targeting most cancers requires systemic delivery, which is more complex than delivering silencing genes locally. Targeting approaches include passive and active targeting.

Passive targeting

Passive targeting involves toxic chemotherapeutic drug accumulation in the areas of tumors with leaky vasculature and takes advantage of the permeability of tumor tissue.

Extended permeability and retention

Most nanoparticles exhibit the extended permeability and retention (EPR) effect (Maeda and Matsumura, 1989) (▶Figure 8.21).

This is a result of two issues: (1) increased permeability of the capillary endothelium in malignant tissues compared to that of normal tissues, and (2) the lack of tumor

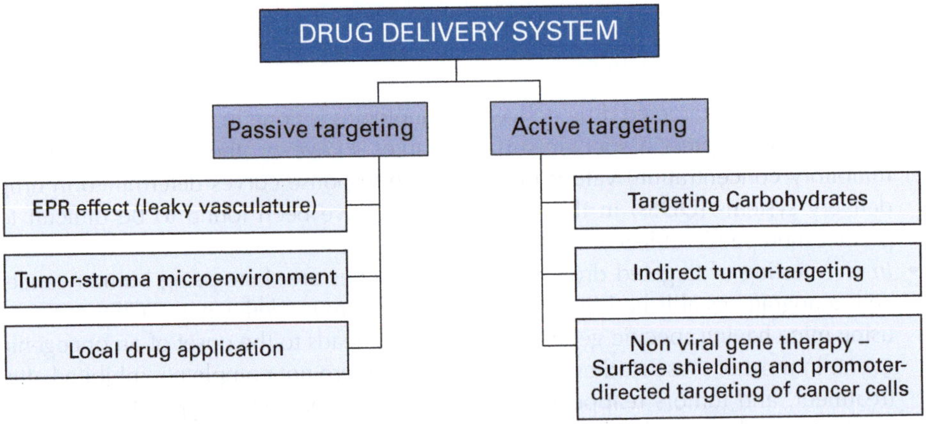

Figure 8.21 **Mechanism of DDS to tumor via passive and active targeting.**

lymphatic drainage within the tumor interstitium resulting in drug accumulation. If a therapeutic agent is coupled to a suitable biodegradable nanoparticle carrier, then such carriers have the potential of increasing the concentration of drug distribution in the tumor tissue due to the EPR effect.

Targeting the tumor microenvironment

The second passive targeting approach exploits the specific conditioning of the tumor micro environment to facilitate drug release from DDS. The conditions include a particular pH, certain enzymes or microflora in the organ or tumor, or the drug is conjugated to a tumor-specific molecule and is administered in an active state, which is known as tumor-activated prodrug therapy (Chari, 1998). The disadvantage is the development of severe organ cytotoxicity because this approach targets the whole organ, not the tumor alone.

Local drug application

In this approach the drug is delivered directly to tumor tissue avoiding systemic applications. For example, intratumoral application of mitomycin resulted in increased concentration of the drug with limited adverse drug effects on healthy organs as long as topically delivered drugs are retained in the tumor and do not enter the circulation. Disadvantage of this approach is for the cancers that are difficult to access for local drug delivery (for example lung cancer). Recently this was overcome by the use of anticancer aerosols (Gautam and Koshkina, 2003).

Active targeting

Active targeting is usually achieved by implementing a drug delivery system where a targeting moiety is conjugated with a nanoparticle, thereby allowing preferential accumulation of the drug in the individual cancer cells of the tumor itself, in intracellular organelles. The targets can be specific molecules in cancer cells or cell surface carbohydrates or receptors and/or antigens expressed on the cell membrane or elsewhere in malignant cells.

Targeting carbohydrates

Carbohydrate moieties can be used to target a DDS to lectins, and vice versa lectins can be used to target cell surface carbohydrates. These approaches are widely used in anti-cancer drug delivery systems for malignant colon cells, e.g. liposomes targeted to cell surface galectins have been used for the targeted delivery to malignant colon cells (Singh et al., 2001). The main disadvantage of such an approach is that the DDS is not cancer cell specific; it allows the drug to target whole organs due to widespread ligands for the lectins, and could thus be harmful to normal tissues with undesirable side effects.

Indirect tumor targeting

Ligands and antibodies, and their interactions, have the advantage of receptor-mediated endocytosis, which has specificity for cell surface receptors that are highly expressed in cancer cells. This markedly stimulates the internalization of the drug through bio-degradable nanoparticles that carry the optimal ligand/antibody. The best example is targeting of cytokines into the tumor microenvironment with immuno-cytokines in melanoma and colon cancer cells, which resulted in elimination of disease (Lode et al., 2000).

Nonviral gene therapy

Surface shielding and promoter-directed targeting of cancer cells

Nanoparticle properties: For possible advantageous cancer therapy over conventional medicine, DDS uses nanoparticles because they have the potential to facilitate better delivery of drugs to tumors owing to the EPR effect, and can deliver more than one therapeutic agent for combination therapy. Nanoparticles require intracellular delivery, and cellular internalization would depend on size, configuration and charge (Jain and Stylianopoulos, 2010). Long-term circulation in blood is important for targeted drug delivery and sustained release. Thus, nanoparticles with narrow size distribution (<500 nm) will be critical for optimizing desired bio-distribution parameters and receptor mediated endocytosis (Bae and Park, 2011; Jain and Stylianopoulos, 2010). Cationic nanoparticles preferentially target tumor endothelial cells compared with their neutral or anionic counterparts. On the other hand, neutral nanoparticles diffuse faster and distribute more homogeneously inside the tumor interstitial space than cationic and anionic nanoparticles. Cationic and anionic nanoparticles form aggregates with negatively charged (e.g. hyaluronan, proteoglycans) or positively charged (e.g. collagen) matrix molecules. With respect to shape, nanoparticles with linear, semiflexible configurations diffuse more efficiently in the interstitial matrix than do comparable sized, rigid spherical particles (Jain and Stylianopoulos, 2010). For higher circulation times and stabilization, nanoparticles containing polyethylene glycol or modified polyethylene glycol (PEG) are the most common approach to prevent destruction by serum proteins and phagocytosis by Kuppfer cells or hepatocytes (Jain and Stylianopoulos, 2010).

To limit the adverse side effects of cancer chemotherapy on healthy organs, we proposed a DDS with specific targeting ligands for cancer cells. The proposed DDS minimizes the uptake of the drug by normal cells and enhances the influx and retention of the drug in cancer cells. This delivery system includes three main components: (1) an apoptosis-inducing agent (anticancer drug), (2) a targeting moiety-penetration enhancer, and (3) a carrier. In the following sections, we describe one of the variants of such a system, which utilizes CD44v6shRNA as an apoptosis-inducing agent that perturbs HA-CD44 interaction and PEG and polyethylene imine (PEI) as a carrier (Misra et al., 2009).

8.9.3 CD44-HA in tumor biology

Structure of CD44

CD44, a major receptor (▶Figure 8.22) for HA, is a transmembrane protein encoded by a single gene, and is subject to extensive alternative splicing.

Due to alternative splicing, multiple forms of CD44 are generated that are further modified by *N*- and *O*-linked glycosylations. CD44 standard (CD44s) is the smallest CD44 isoform and lacks variant exons, CD44s contains an N-terminal signal sequence (exon 1), a link module that binds to HA (exons 2 and 3), a stem region (exons 4, 5, 16, and 17), a single-pass transmembrane domain (exon 18) and a cytoplasmic domain (exon 20). In all forms of CD44, exon 19 is spliced out so that the transmembrane domain with 73 amino acids encoded by exon 18 is followed by the cytoplasmic domain encoded by exon 20.

Influence of CD44 variant/HA in tumorigenesis

The affinity of CD44 for HA depends on posttranslational modification, such as modification in glycosylation (Skelton et al., 1998), and this function depends on cell type and growth condition. Isoforms that contain a variable number of exon insertions (v1–v10) at the proximal plasma membrane external region are expressed primarily on tumor cells (Ponta et al., 2003; Naor et al., 2008; Misra et al., 2009). CD44 in cancer cells interacts with the HA-rich microenvironments, which affects cell signaling pathways that trigger the ability of malignant cells to migrate, to invade basement membranes/ECM, and to lodge at distant sites of the tumor. Many of the downstream

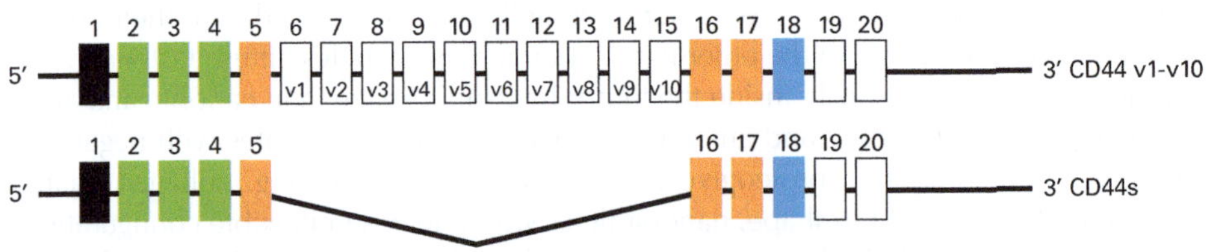

Figure 8.22 Alternative splicing in CD44 pre-mRNA. CD44 pre-mRNA is encoded by 20 exons. The common CD44s (hematopoietic) form contains no extra exons, and the protein may have serine motif encoded in exon 5 that initiates synthesis of a chondroitin sulfate or dermatan sulfate chain. Alternative splicing of CD44 predominantly involves variable insertion of 10 extra exons with combinations of exons 6–15 and spliced in v1–v10 into the stem region, of which v3 encodes a substitution site for a heparan sulfate chain. Variable numbers of the v exons can be spliced in epithelial cells, endothelial cells, and inflammatory monocytes, and also upregulated commonly on neoplastic transformation depending on the tissue.

pathways following CD44 activation become deregulated in cancer, leading to tumor growth, progression and metastasis (Misra et al., 2003, 2005, 2006, 2008a, 2008b, 2009; Bourguignon et al., 2007; Ghatak et al., 2010). Further, there is a connection between HA catabolism and energy generation most likely allowing HA to function as an alternate energy source to glucose for malignant cells (Smedsrod and Seljelid, 1991). HA alters the CD44 conformation in such a way that its short cytoplasmic tail binds to ankyrin and ezrin-radixin-moesin (ERM) proteins, which guide CD44 to the leading edge of migrating cells (Lamontagne and Grandbois, 2008) (▶Figure 8.23).

CD44 facilitates binding to matrix metalloproteinases (MMPs) at the cell surface, and this aggregation of CD44 with MMPs (MMP2/9) is further stimulated by HA/CD44-interaction (Yu and Stamenkovic, 1999; Jung et al., 2011). The extracellular domain of CD44 undergoes proteolysis (Kajita et al., 2001; Stamenkovic and Yu, 2009) (▶Figure 8.22). CD44 transmembrane and cytoplasmic domains can also undergo additional proteolytic cleavage followed by additional cleavage by γ-secretase. The CD44-β-peptide and CD44 intracellular domain (ICD) that are produced have been shown to be able to translocate to the nucleus and regulate various transcription of target genes,

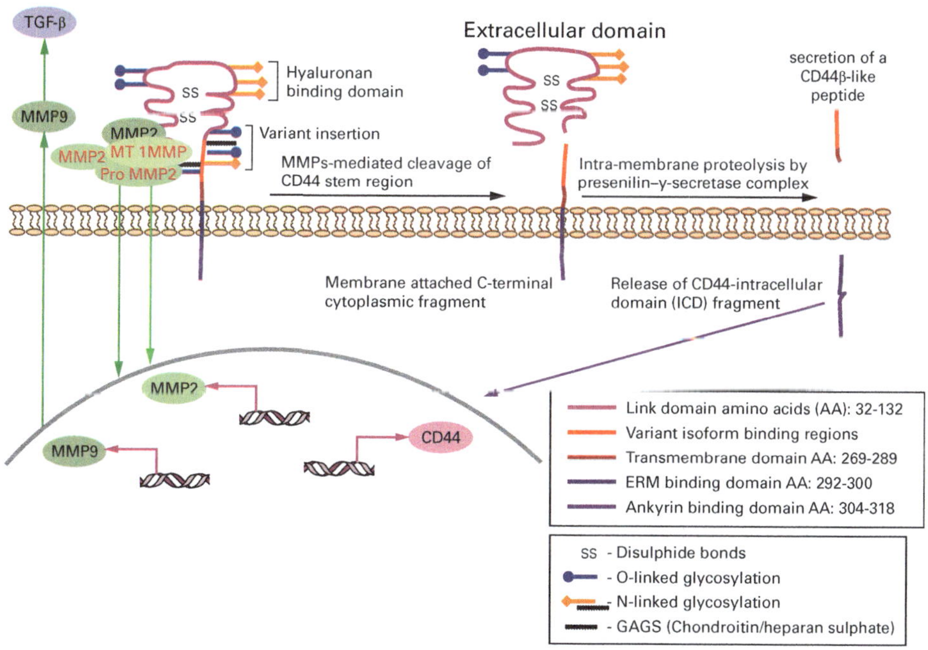

Figure 8.23 CD44 protein degradation and synthesis. The extracellular domain of CD44 has an amino terminal ~180 amino acids that has a HA-binding link module through which CD44 interacts with HA. This is followed by a mucin-like membrane proximal region for insertion of alternate splicing due to variant exons of CD44. CD44 anchors matrix metalloproteinases (MMPs) in the stem region (exons 4, 5, 16, and 17). MMPs can cleave the CD44 stem region and release the soluble extracellular domain of CD44 with a C-terminal region attached to the membrane. This activity is followed by intramembrane proteolysis by presenillin-γ-secretase, which releases CD44β-like peptide and CD44-intracellular domain (ICD), which in turn enters the nucleus for CD44 transcription. In this way, CD44 is recirculated to the system.

including CD44 itself, which provides an important regulatory mechanism for CD44 expression (Nagano and Saya, 2004) (▶Figure 8.22). Proteolytic cleavage of CD44 results in an extracellular domain of CD44 (sol CD44), which is further cleaved and will not bind to HA because of the loss of binding domain (Cichy and Pure, 2003). In contrast, recombinant sol CD44 retains the HA-binding domain and thus maintains the ability to serve as an antagonist to HA/CD44 interaction (Yu and Stamenkovic, 2000).

Besides creating a conducing environment for cell invasion, cell surface CD44-associated MMP9 is also important for the cleavage of latent TGF-β, which promotes angiogenesis, invasion and metastasis (Yu and Stamenkovic, 2000) (▶Figure 8.23). Combining vascular endothelial growth factor-A (VEGF-A) with hepatocyte growth factor (HGF) enhances the Rho and Rac1–mediated Ras-MAPK and the Ras-Phosphatidylinositol 3–kinase (PI3K)–Akt pathways, probably through the more efficient recruitment of CD44v6 where CD44v6 exerts coreceptor function when it is activated by HGF/c-Met and vascular endothelial growth factor receptor (VEGFR) (Tremmel et al., 2009). In addition to regulation of many genes required for cell growth, HA/CD44v6 regulated PI3K-Akt is strongly proinflammatory upstream of cyclooxygenase-2 (COX2), which in turn regulates HA production (Misra et al., 2008a, 2008b). Recent studies also indicated that niches are the physical environments that maintain CSCs in a variety of tissues, including human colon where they have been described as structures most likely formed by intestinal subepithelial myofibroblasts (ISEMFs) located at the base of the crypt. CD44 may also be actively engaged in remodeling a preexisting niche for CSCs (Zöller, 2011). Overexpression of the CD44v4–v7 and CD44v6–v9 isoforms in various carcinomas, including colorectal adenocarcinomas (Heider et al., 1993; Wielenga et al., 1993), has increased tumorigenicity and metastasis in multiple models (Heider et al., 1993). We and others have demonstrated that HA-oligomers, CD44siRNA and CD44 antibody that blocks binding to HA (Ghatak et al., 2002, 2005; Misra et al., 2003, 2005; Bourguignon et al., 2009) can uncouple constitutive HA/CD44 interaction and sensitize cancer cells to drugs, which partially inhibits tumor growth.

Importance of targeting CD44v6 in colon cancer

HA receptor CD44 has also been used to detect CSCs in CRC (Dalerba et al., 2007), pancreatic cancer (Li et al., 2007), prostate cancer (Collins et al., 2005) and head and neck cancers (Prince et al., 2007). Therefore, CD44 might have a functional role in the biology of CSCs that can be used for therapy (Zöller, 2011). For example, CSCs were screened for CD44 expression with antibodies recognizing all CD44 isoforms. CD44 isoforms are strongly overexpressed in dysplastic crypts and adenomas in humans and mice with mutant adenomatous polyposis coli (APC) (Wielenga et al., 1993). It remains to be explored whether CSCs preferentially express CD44s or CD44v6 or both. Since a small amount of CSCs have the capacity to bring cancer cell reappearance, we asked the question: could cancer cell recurrence be similar to the recovery process of normal healthy cells? Thus, we need to target the CSCs effectively. Complete cancer cell eradication may be possible using treatments that effectively target both cancer cells and their progenitor stem cells. As stem cells for various tissue cell types do not always reside in that tissue or organ (Visvader, 2011), the strategies to effectively target drugs to both tumor and nontumor sites may be what is required. In addition, tumor cell responses either *in vitro* or *in vivo* to cell-specific inhibition of HA/CD44 interaction at the transcription level remain largely unknown. Therefore, we developed tissue specific delivery of CD44v6shRNA/nanoparticle using a cre-Lox nonviral approach and asked whether the colon-specific promoter (Fabpl) driven cre plasmid and a floxed pSico-CD44v6shRNA plasmid packaged in transferrin (Tf)–coated nanoparticles as vehicles for delivery of the plasmids, can address the importance of CD44v6 in colon cancer (Misra et al., 2009). These approaches will be discussed in Section 8.9.4.

8.9.4 Strategies that target CD44 to perturb HA-CD44 interaction in tumors

HA-conjugated drug

Activated CD44 is overexpressed on solid tumors, but not on their nontumorigenic counterparts. CD44 can internalize HA (Culty et al., 1992), and drugs with low solubility can be successfully applied when conjugated with HA. HA-drug conjugates are then internalized via CD44, and the drug is released and activated mainly by intracellular enzymatic hydrolysis (Luo et al., 2000). Several preclinical studies have shown that HA chemically conjugated to cytotoxic agents improved anticancer properties of the agent *in vitro* (Akima et al., 1996; Coradini et al., 1999; Luo and Prestwich, 1999). For instance, the antimitotic chemotherapeutic agent paclitaxel has low water solubility. Upon conjugation to HA, its solubility and CD44 dependent cellular uptake increased *in vitro* (Luo et al., 2000).

HA-conjugated nanocarrier

HA when conjugated to a nanocarrier acts as a protective structural component and a targeting coating. Circulation time and biodistribution (pharmacokinetic properties) are influenced by incorporating the targeting and cell specific uptake properties of HA onto large carriers. Cargo liposomes or nanoparticles delivered to CD44 overexpressing cells include anticancer drugs: epirubicin (Akima et al., 1996), doxorubicin (Eliaz and Szoka, 2001), paclitaxel (Luo et al., 2000), and mitomycin c (Akima et al., 1996), as well as siRNA (Lee et al., 2007).

Targeting with anti-CD44 antibodies

Anti-CD44 antibodies against highly expressed variants can "actively" target drugs to CD44v, inhibit and disrupt CD44v matrix interactions, occupy CD44 and induce CD44 signaling, which can cause apoptosis (Song et al., 2004). Anti-CD44 antibodies targeting ligands for either radiolabels or anticancer chemotherapeutics partially stabilize some patients (Tijink et al., 2006; Sauter et al., 2007). CD44v6 expressed in breast, cervical and colon cancers, and in squamous cell carcinomas, was chosen as a model for therapy. A phase 1 clinical trial was done with an immunotoxin (humanized antibody coupled with a cytotoxic drug mertansine) against CD44v6 in 30 patients with incurable squamous cell carcinoma (Orian-Rousseau and Ponta, 2008). The trial had 10% success, which was considered to be a considerable success.

Tissue-specific deletion of CD44 variant signaling

The technique of using shRNA in an expression vector is an alternative strategy to stably suppress selected gene expression, which suggests that the use of shRNA expression vectors holds potential promise for therapeutic approaches for silencing disease-causing genes (Paul et al., 2002). There are two ways to deliver shRNA in cancer cells – either using a viral vector or a nonviral vector. Viral vectors have been used to achieve this proof of principle in animal models and, in selected cases, in human clinical trials (Raper et al., 2003).

Systemic targeting by viral vectors toward the desired tissue is difficult because the host immune responses activate viral clearance. Systemic administration of a large amount of adenovirus (e.g. into the liver) can be a serious health hazard that even caused the death of one patient (Raper et al., 2003). Nevertheless, there has been a considerable interest in developing nonviral vectors for gene therapy. In this regard, nonviral vectors, such as positively charged PEI-complexes shielded with PEG, can be used safely to avoid the nonspecific interactions with nontarget cells and blood components (Kursa et al., 2003). The model for the uptake of nanoparticles carrying multiple functional domains is illustrated in ▶Figure 8.24.

Nonviral vectors were once limited for their low gene transfer efficiency. However, the incorporation of various ligands, such as peptides, growth factors and proteins, or antibodies for targets highly expressed on cancer cells, circumvented this obstacle (Bellocq et al., 2003). Also, enhanced permeability due to aberrant vasculature in solid tumors and retention of ligand coated vectors around the receptors of tumor cells can increase chances for high probability of interaction with the cancer cells (Misra et al., 2009). Thus, the nonviral vectors can acquire high gene transfer efficiency (Misra et al., 2009). This concept has been tested by preparing nanoparticles with plasmids packed inside an outer PEG-PEI layer coated with Tf, an iron transporting protein (Bellocq et al., 2003; Misra et al., 2009) that binds with Tf-receptors (Tf-R) with high affinity (depicted in the model ▶Figure 8.24). The Tf-R is present at much higher levels on the tumor cells (Misra et al., 2009) than on phenotypically normal epithelial cells. Association of Tf with the nanoparticles significantly enhances transfection efficiency of shRNA generator-plasmids by promoting the internalization of nanoparticles in dividing and nondividing cells (Bellocq et al., 2003). After cellular association of nanoparticles to the target cells, particles are internalized by receptor-mediated endocytosis (Bellocq et al., 2003). This approach was tested for transfection of pSV-β-galactosidase/nanoparticles in cellular models (▶Figure 8.25).

Nanoparticles carrying multiple functional domains (surface shielding particles Tf-PEG-PEI, shRNA generator plasmids, and tissue specific promoter driven-Cre recombinase) can overcome the intracellular barriers for uptake, Therefore, we tested this approach in *Apc* Min/+ (mouse models for familial adenomatous polyposis [FAP]) mice (Misra et al., 2009) (▶Figure 8.26).

We successfully demonstrated that the CD44v6shRNA is localized into the colon tumor cells by an end point assay of CD44v6 expression, and by perturbation of HA-CD44v6 interaction as reflected in the reduction in number of tumors (Misra et al., 2009) (▶Figure 8.27). The tissue specific shRNA delivery was made possible by the use of Cre-recombinase produced in response to a colon tissue specific promoter, which deletes the interruption between the U6 promoter and the CD44v6shRNA

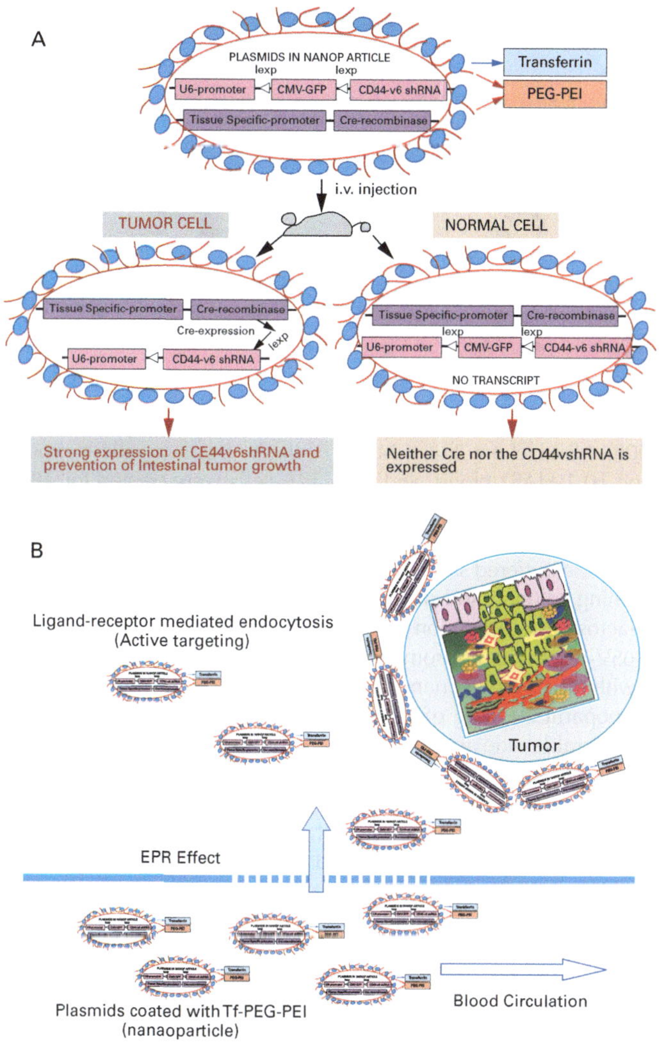

Figure 8.24 (A) Model for delivery of shRNA. Adapted from Misra et al. (2011). FEBS J *278*, 1429–43. This illustration depicts cellular uptake of plasmid Tf-PEG-PEI nanoparticles and the mechanism of action of shRNA. First, a pSico vector containing a U6 promoter-loxP-CMV-GFP-STOP signal-loxP-CD44vshRNA (gene of interest) is made. Second, an expression vector with the Cre recombinase gene controlled by the tissue specific promoter is created. Third, the two vectors are packaged in transferrin (Tf)–coated PEG-PEI nanoparticles that bind with Tf-receptors (Tf-R) present at high levels in the targeted tumor cells. Delivery of the vectors in normal and malignant cells from the targeted tissue results in deletion of the Stop signal and transcription of Cre recombinase driven by the tissue specific promoter. The target gene (CD44vshRNA) is then unlocked and transcribed through the strong U6 promoter for high expression. The normal tissue cells are not affected because they do not make the targeted CD44 variant. (B) Targeted delivery of plasmid/nanoparticles to the tumor. Tf-PEG-PEI nanoparticle coated plasmids (pSico-CD44v6shRNA/pFabpl-Cre) circulating in blood accumulate at tumor regions enhanced by the EPR effect. Endocytosis mediated by ligand-receptor interactions occurs because the nanoparticles are coated with the Tf ligand for the Tf-R receptor on the tumor cell surface.

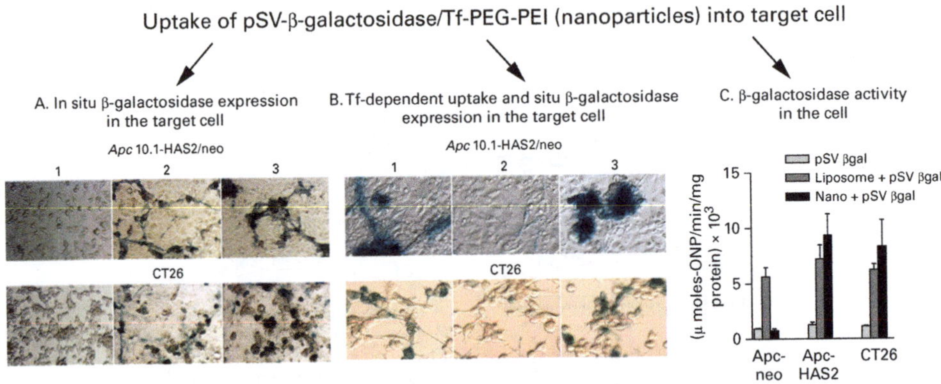

Figure 8.25 Delivery of pSV-β-galactosidase/nanoparticles in cell culture. Adapted from Misra et.al. (2009). J Biol Chem *284*, 12432. (A) In situ β-galactosidase expression in the target cell. The Apc 10.1-HAS2 clone, and the CT26 cells were treated for 48 h with the pSV-β-gal alone (panel 1), or with the pSV-β-gal with liposome (panel 2), or with the pSV-β-gal/nanoparticles (8 μg pSV-β-gal/mL) (panel 3). The transfected cells were fixed in 0.2% gluteraldehyde in phosphate buffered saline (PBS) and washed twice in PBS. The cells were treated with a β-Gal staining solution and digitally photographed. (B) Transferrin dependent uptake and in situ β-galactosidase expression in the target cell. The Apc 10.1-HAS2 cells were transfected with the pSV-β-gal with liposome (panel 1), or treated with Tf-R antibody and followed by transfection with the pSV-β-gal/nanoparticles (8 μg pSV-β-gal/mL) (panel 2), or treated with the pSV-β-gal/nanoparticles (8 μg pSV-β-gal/mL) alone (panel 3). The transfected cells were fixed in 0.2% gluteraldehyde in PBS and washed twice in PBS. The cells were treated with a β-Gal staining solution and digitally photographed. (C) β-galactosidase activity. Cell-free extracts of parallel cultures were prepared in 10 mM CHAP buffer and assayed for β-galactosidase activity using ortho-nitro-phenyl-β-galactoside (ONPG) as substrate. The results are expressed as μmoles of ONP formed/min/mg protein and represent +/– *SD* of triplicate assays from the untransfected, or liposome-transfected, or nanoparticle-treated cultures for each cell type.

oligonucleotide. This newly developed cell-specific shRNA delivery approach (Misra et al., 2009) confirmed that targeting the HA/CD44v6-induced signaling pathway inhibited distant colon tumor growth in Apc Min/+ mice (▶Figure 8.27).

Our recent *in vivo* studies with the C57Bl/6 (C57 black 6) mice provide evidence that systemic delivery of a mixture of the two plasmids in Tf/nanoparticles (pARR$_2$ Probasin-Cre/nanoparticles and the floxed pSico-CD44v9shRNA/nanoparticles) can target both localized and metastatic prostate cancer cells (Ghatak et al., unpublished data). This novel approach opens up new ways to combat cancer, and to understand tumorigenesis *in vivo* for the following reasons: (1) cell-specific shRNA is released by applying a tissue- specific promoter driven Cre-lox mechanism; (2) it silences the expression of the selected CD44 variant in the target tissue cancer cells; (3) it does not affect the normal target tissue cells, which do not express the targeted CD44 variant, and which rely on the CD44s (standard) form that is not affected by the plasmids; (4) the target CD44vshRNA will not be expressed in other types of cells because the tissue specific promoter only unlocks the Cre recombinase in the targeted tissue cells, thereby reducing potential side effects (Misra et al., 2009); (5) the nanoparticles that carry

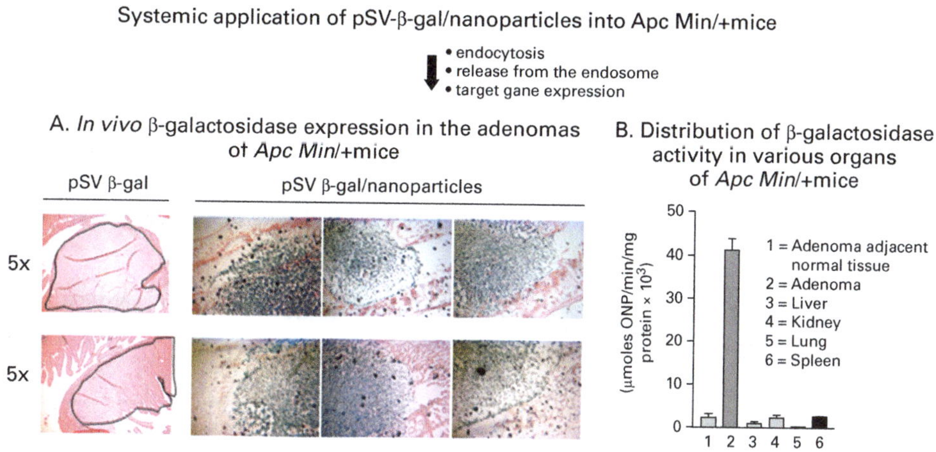

Figure 8.26 **Delivery of pSV-β-gal/nanoparticles in *Apc* Min/+ mice.** (A) Adapted from Misra et al. (2009). J Biol Chem *284*, 12432. *In vivo* β-galactosidase expression in the adenomas of Apc Min/+mice. Ten Apc Min/+ mice received pSV-β-gal plasmids (100 μg/100 μL, intraperitoneally [IP]) alone and 10 received pSV- β-gal/nanoparticles (100 μg/100 μL, IP), every other day. At day 10, mice were euthanized with isoflurane followed by cervical dislocation. Livers, spleens, kidneys and lungs were harvested and frozen immediately in dry ice as described in materials and methods. After counting the tumors, the flattened intestines from various treatment groups such as pSV-β-gal (Group1), pSV-β-gal/nanoparticles (Group 2) were rolled into "Swiss rolls" and fixed in OCT on dry ice. Frozen sections carrying the adenomas were stained with X-Gal (blue dye deposits around the tumor tissue at 5x magnification) and eosin, and the tumor bearing animals treated with pSV-β-gal showed no LacZ stain as outlined with a black line. (B) β-galactosidase activity in adenomas, adjacent normal tissue and other organs of *Apc* Min/+ mice by measuring the chromogen ortho-nitro-phenol released from the ortho-nitro phenyl glucoside.

plasmids are biodegradable and cleared from the system; (6) it addresses the pathophysiological role of HA-CD44v interactions in cancer; (7) it can establish diagnostic markers for the targeted cancer including CD44 variants, soluble CD44, and HA; and (8) it can establish CD44v/HA interactions as an innovative novel therapeutic target against cancer progression. Thus, the conditional suppression of gene expression by the use of an CD44vshRNA-expressing plasmid holds potential promise for therapeutic approaches for silencing HA/CD44 variant signaling and hence its downstream signaling that promotes disease-causing genes (Paul et al., 2002) (▶Figures 8.24–8.27).

Advantages of tumor-specific delivery of CD44 variant-shRNA versus other therapeutic strategies

First, this technique avoids multiple chemical steps to prepare HA-conjugated cytotoxic drugs and conjugation to nanocarriers. Second, it abolishes the CD44 variants in the cancer cells only. Third, a number of cell types in normal tissues that express CD44 standard or hematopoietic form will not be affected because these are not targeted. Fourth, inflammation- associated cancers accumulate activated immune cells that have upregulated transferrin receptors and CD44 variants. However, if they take up

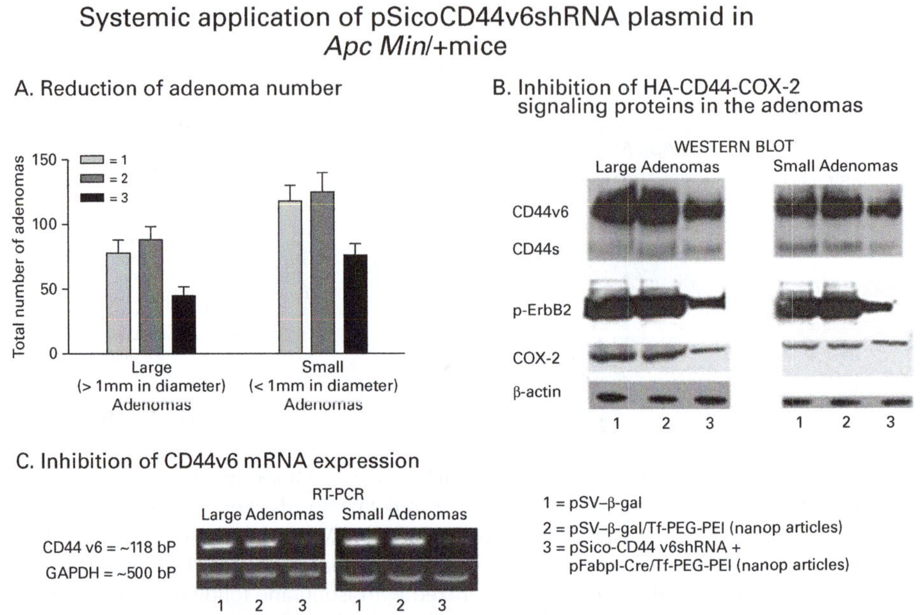

Figure 8.27 Effect of plasmid/nanoparticle treatment on protein expression, reverse transcription polymerase chain reaction (RT-PCR) analysis and number of adenomas in *Apc* Min/+ mice. Thirty *Apc* Min/+ mice were randomly divided into three groups. Group 1 received pSV-β-galactosidase (100 μg/100 μL, intraperitoneally [IP]) alone; Group 2 received pSV-β-gal nanoparticles (100 μg/100 μL, IP) targeted to the Tf-R, and Group 3 received pSico-CD44v6shRNA (75 μg) plus intestine/colon-specific pFabpl promoter-driven-Cre-recombinase (pFabpl-Cre) (25 μg)/nanoparticles ip every other day. Ten days after beginning treatment, the animals were sacrificed, and the large (>1 mm) and small (<1 mm) adenomas were counted (A). The tumor and adjacent normal tissues were subsequently processed for (B) Western blots for CD44, pErbB2 (phospho ErbB2), TErbB2 (total ErbB2), COX2, and β-actin. Total ErbB2 remained unchanged in all the treatment groups (data not shown). (C) RT-PCR analysis for CD44 variants from total RNA.

the nanoparticles, there will be no deletion of CD44 variants because the promoter is not lymphocyte specific (Misra et al., unpublished data). Fifth, it can also target colon rectal CSCs. To target activated lymphocytes, specific promoter driven-cre plasmids should be used. Sixth, accumulation of antibody in nontumor areas is a major limitation of anti-CD44 antibody therapy. Experiments so far do not produce any such effect in shRNA delivery.

8.9.5 Take-home message

For cancer therapies, the ideal targeted DDS is the one that delivers the drug only to the target tumor. Nonviral gene therapy approach is anticipated to lead to the development of novel, sophisticated, cell-specific, multifunctional applications that can recognize cancer cells, deliver drugs to target tissue, and at the same time help in reporting effects of therapy, and most importantly, monitor intracellular pathways to help prevent precancerous cells from becoming malignant.

Regardless of the increasing literature of studies conducted so far, the complete understanding of HA/CD44-induced signaling still remains elusive. However, both HA and CD44 variants appear to be vitally important from embryogenesis to morphogenesis, in inflammation, and in cancer, which accompanies over expression of CD44 and its splice variants and aberrant synthesis/turnover of HA.

Based on the functions of HA and its interaction with CD44 described previously, it seems likely that the impact of HA/CD44 and its variant-induced tumor growth is multifactorial. We are now focusing on how to better exploit this moderate amount of the drug (shRNA) delivered to the target tumor. As tumors may not be eradicated by just aiming at one target, we are focusing on developing strategies that deliver multiple drugs, and/or deliver the drug to multiple targets.

Acknowledgments

This work was supported, in whole or in part, by National Institutes of Health Grants P20RR021949 (to S. G.) and P20RR016434 (to S. M., S. G., and R. R. M.), 2P20RR16461-05A1 (to S. G., R. R. M., and L. P C.) HL RO1 33756 and 1 P30AR050953 (to V. C. H.). This work was also supported by Mitral-07 CVD 04 (to R. R. M.), Medical University of South Carolina University Research Council Project 2204000–24330 (to S. M.) and 2204000–24329 (to S. G.).

References

Akima, K., Ito, H., Iwata, Y., et al. (1996). Evaluation of antitumor activities of hyaluronate binding antitumor drugs: synthesis, characterization and antitumor activity. J Drug Target *4*, 1–8.

Bae, Y. H. and Park, K. (2011). Targeted drug delivery to tumors: myths, reality and possibility. J Control Release *153*, 198 205.

Bellocq, N. C., Pun, S. H., Jensen, G. S., and Davis, M. E. (2003). Transferrin-containing, cyclodextrin polymer-based particles for tumor-targeted gene delivery. Bioconjug Chem *14*, 1122–1132.

Bourguignon, L. Y., Spevak, C. C., Wong, G., Xia, W., and Gilad, E. (2009). Hyaluronan-CD44 interaction with protein kinase C(epsilon) promotes oncogenic signaling by the stem cell marker Nanog and the production of microRNA-21, leading to down-regulation of the tumor suppressor protein PDCD4, anti-apoptosis, and chemotherapy resistance in breast tumor cells. J Biol Chem *284*, 26533–26546.

Bourguignon, L. Y., Gilad, E., and Peyrollier, K. (2007). Heregulin-mediated ErbB2-ERK signaling activates hyaluronan synthases leading to CD44-dependent ovarian tumor cell growth and migration. *J Biol Chem 282*, 19426–19441.

Chari, R. V. (1998). Targeted delivery of chemotherapeutics: tumor-activated prodrug therapy. Adv Drug Deliv Rev *31*, 89–104.

Cichy, J., and Pure, E. (2003). The liberation of CD44. J Cell Biol *161*, 839–843.

Collins, A. T., Berry, P. A., Hyde, C., Stower, M. J., and Maitland, N. J. (2005). Prospective identification of tumorigenic prostate cancer stem cells. Cancer Res *65*, 10946–10951.

Coradini, D., Pellizzaro, C., Miglierini, G., Daidone, M. G., and Perbellini, A. (1999). Hyaluronic acid as drug delivery for sodium butyrate: improvement of the anti-proliferative activity on a breast-cancer cell line. Int J Cancer *81*, 411–416.

Culty, M., Nguyen, H. A., and Underhill, C. B. (1992). The hyaluronan receptor (CD44) participates in the uptake and degradation of hyaluronan. J Cell Biol *116*, 1055–1062.
Dalerba, P., Dylla, S. J., Park, I. K., et al. (2007). Phenotypic characterization of human colorectal cancer stem cells. Proc Natl Acad Sci U S A *104*, 10158–10163.
Eliaz, R. E., and Szoka, F. C., Jr. (2001). Liposome-encapsulated doxorubicin targeted to CD44: a strategy to kill CD44-overexpressing tumor cells. Cancer Res *61*, 2592–2601.
Gao, Z. G., Tian, L., Hu, J., Park, I. S., and Bae, Y. H. (2011). Prevention of metastasis in a 4T1 murine breast cancer model by doxorubicin carried by folate conjugated pH sensitive polymeric micelles. J Control Release *152*, 84–89.
Gautam, A., and Koshkina, N. (2003). Paclitaxel (taxol) and taxoid derivates for lung cancer treatment: potential for aerosol delivery. Curr Cancer Drug Targets *3*, 287–296.
Ghatak, S., Hascall, V. C., Markwald, R. R., and Misra, S. (2010). Stromal hyaluronan interaction with epithelial CD44 variants promotes prostate cancer invasiveness by augmenting expression and function of hepatocyte growth factor and androgen receptor. J Biol Chem *285*, 19821–19832.
Ghatak, S., Misra, S., and Toole, B. P. (2002). Hyaluronan oligosaccharides inhibit anchorage-independent growth of tumor cells by suppressing the phosphoinositide 3-kinase/Akt cell survival pathway. J Biol Chem *277*, 38013–38020.
Ghatak, S., Misra, S., and Toole, B. P. (2005). Hyaluronan constitutively regulates ErbB2 phosphorylation and signaling complex formation in carcinoma cells. J Biol Chem *280*, 8875–8883.
Heider, K. H., Hofmann, M., Hors, E., et al. (1993). A human homologue of the rat metastasis-associated variant of CD44 is expressed in colorectal carcinomas and adenomatous polyps. J Cell Biol *120*, 227–233.
Jain, R. K., and Stylianopoulos, T. (2010). Delivering nanomedicine to solid tumors. Nat Rev Clin Oncol *7*, 653–664.
Jung, T., Gross, W., and Zöller, M. (2011). CD44v6 coordinates tumor matrix-triggered motility and apoptosis resistance. J Biol Chem *286*, 15862–15874.
Kajita, M., Itoh, Y., Chiba, T., et al. (2001). Membrane-type 1 matrix metalloproteinase cleaves CD44 and promotes cell migration. J Cell Biol *153*, 893–904.
Kim, S. H., Turnbull, J., and Guimond, S. (2011). Extracellular matrix and cell signalling: the dynamic cooperation of integrin, proteoglycan and growth factor receptor. J Endocrinol *209*, 139–151.
Kumar, P., Lee, S. K., Shankar, P., and Manjunath, N. (2006). A single siRNA suppresses fatal encephalitis induced by two different flaviviruses. PLoS Med *3*, e96.
Kursa, M., Walker, G. F., Roessler, V., et al. (2003). Novel shielded transferrin-polyethylene glycol-polyethylenimine/DNA complexes for systemic tumor-targeted gene transfer. Bioconjug Chem *14*, 222–231.
Lamontagne, C. A., and Grandbois, M. (2008). PKC-induced stiffening of hyaluronan/CD44 linkage; local force measurements on glioma cells. Exp Cell Res *314*, 227–236.
Lee, H., Mok, H., Lee, S., Oh, Y. K., and Park, T. G. (2007). Target-specific intracellular delivery of siRNA using degradable hyaluronic acid nanogels. J Control Release *119*, 245–252.
Li, C., Heidt, D. G., Dalerba, P., et al. (2007). Identification of pancreatic cancer stem cells. Cancer Res *67*, 1030–1037.
Lode, H. N., Xiang, R., Perri, P., et al. (2000). What to do with targeted IL-2. Drugs Today (Barc) *36*, 321–336.
Luo, Y., and Prestwich, G. D. (1999). Synthesis and selective cytotoxicity of a hyaluronic acid-antitumor bioconjugate. Bioconjug Chem *10*, 755–763.
Luo, Y., Ziebell, M. R., and Prestwich, G. D. (2000). A hyaluronic acid-taxol antitumor bioconjugate targeted to cancer cells. Biomacromolecules *1*, 208–218.

Maeda, H., and Matsumura, Y. (1989). Tumoritropic and lymphotropic principles of macromolecular drugs. Crit Rev Ther Drug Carrier Syst *6*, 193–210.

Misra, S., Ghatak, S., and Toole, B. P. (2005). Regulation of MDR1 expression and drug resistance by a positive feedback loop involving hyaluronan, phosphoinositide 3-kinase, and ErbB2. J Biol Chem *280*, 20310–20315.

Misra, S., Ghatak, S., Zoltan-Jones, A., and Toole, B. P. (2003). Regulation of multidrug resistance in cancer cells by hyaluronan. J Biol Chem *278*, 25285–25288.

Misra, S., Hascall, V. C., Berger, F. G., Markwald, R. R., and Ghatak, S. (2008a). Hyaluronan, CD44, and cyclooxygenase-2 in colon cancer. Connect Tissue Res *49*, 219–224.

Misra, S., Hascall, V. C., De Giovanni, C., Markwald, R. R., and Ghatak, S. (2009). Delivery of CD44 shRNA/nanoparticles within cancer cells: perturbation of hyaluronan/CD44v6 interactions and reduction in adenoma growth in Apc Min/+ MICE. J Biol Chem *284*, 12432–12446.

Misra, S., Heldin, P., Hascall, V. C., et al. (2011). Hyaluronan-CD44 interactions as potential targets for cancer therapy. FEBS J *278*, 1429–1443.

Misra, S., Obeid, L. M., Hannun, Y. A., et al. (2008b). Hyaluronan constitutively regulates activation of COX-2-mediated cell survival activity in intestinal epithelial and colon carcinoma cells. J Biol Chem *283*, 14335–14344.

Misra, S., Toole, B. P., and Ghatak, S. (2006). Hyaluronan constitutively regulates activation of multiple receptor tyrosine kinases in epithelial and carcinoma cells. J Biol Chem *281*, 34936–34941.

Nagano, O., and Saya, H. (2004). Mechanism and biological significance of CD44 cleavage. Cancer Sci *95*, 930–935.

Naor, D., Wallach-Dayan, S. B., Zahalka, M. A., and Sionov, R. V. (2008). Involvement of CD44, a molecule with a thousand faces, in cancer dissemination. Semin Cancer Biol *18*, 260–267.

Orian-Rousseau, V., and Ponta, H. (2008). Adhesion proteins meet receptors: a common theme? Adv Cancer Res *101*, 63–92.

Park, J. W., Hong, K., Kirpotin, D. B., et al. (2002). Anti-HER2 immunoliposomes: enhanced efficacy attributable to targeted delivery. Clin Cancer Res *8*, 1172–1181.

Paul, C. P., Good, P. D., Winer, I., and Engelke, D. R. (2002). Effective expression of small interfering RNA in human cells. Nat Biotechnol *20*, 505–508.

Ponta, H., Sherman, L., and Herrlich, P. A. (2003). CD44: from adhesion molecules to signalling regulators. Nat Rev Mol Cell Biol *4*, 33–45.

Prince, M. E., Sivanandan, R., Kaczorowski, A., et al. (2007). Identification of a subpopulation of cells with cancer stem cell properties in head and neck squamous cell carcinoma. Proc Natl Acad Sci U S A *104*, 973–978.

Raper, S. E., Chirmule, N., Lee, F. S., et al. (2003). Fatal systemic inflammatory response syndrome in a ornithine transcarbamylase deficient patient following adenoviral gene transfer. Mol Genet Metab *80*, 148–158.

Sauter, A., Kloft, C., Gronau, S., et al. (2007). Pharmacokinetics, immunogenicity and safety of bivatuzumab mertansine, a novel CD44v6-targeting immunoconjugate, in patients with squamous cell carcinoma of the head and neck. Int J Oncol *30*, 927–935.

Sebens, S., and Schafer, H. (2011). The tumor stroma as mediator of drug resistance – a potential target to improve cancer therapy? Curr Pharm Biotechnol. [Epub ahead of print]

Shankar, P., Manjunath, N., and Lieberman, J. (2005). The prospect of silencing disease using RNA interference. JAMA *293*, 1367–1373.

Singh, M., Kisoon, N., and Ariatti, M. (2001). Receptor-mediated gene delivery to HepG2 cells by ternary assemblies containing cationic liposomes and cationized asialoorosomucoid. Drug Deliv *8*, 29–34.

Skelton, T. P., Zeng, C., Nocks, A., and Stamenkovic, I. (1998). Glycosylation provides both stimulatory and inhibitory effects on cell surface and soluble CD44 binding to hyaluronan. J Cell Biol *140*, 431–446.

Smedsrod, B., and Seljelid, R. (1991). Fate of intravenously injected aminated beta(1–3) polyglucose derivatized with 125I-tyraminyl cellobiose. Immunopharmacology *21*, 149–158.

Song, G., Liao, X., Zhou, L., Wu, L., Feng, Y., and Han, Z. C. (2004). HI44a, an anti-CD44 monoclonal antibody, induces differentiation and apoptosis of human acute myeloid leukemia cells. Leuk Res *28*, 1089–1096.

Stamenkovic, I., and Yu, Q. (2009). Shedding light on proteolytic cleavage of CD44: the responsible sheddase and functional significance of shedding. J Invest Dermatol *129*, 1321–1324.

Tijink, B. M., Buter, J., de Bree, R., et al. (2006). A phase I dose escalation study with anti-CD44v6 bivatuzumab mertansine in patients with incurable squamous cell carcinoma of the head and neck or esophagus. Clin Cancer Res *12*, 6064–6072.

Torchilin, V. P. (2000). Drug targeting. Eur J Pharm Sci *11* (Suppl. 2), S81–S91.

Tremmel, M., Matzke, A., Albrecht, I., et al. (2009). A CD44v6 peptide reveals a role of CD44 in VEGFR-2 signaling and angiogenesis. Blood *114*, 5236–5244.

Visvader, J. E. (2011). Cells of origin in cancer. Nature *469*, 314–322.

Wielenga, V. J., Heider, K. H., Offerhaus, G. J., et al. (1993). Expression of CD44 variant proteins in human colorectal cancer is related to tumor progression. Cancer Res *53*, 4754–4756.

Xie, F. Y., Woodle, M. C., and Lu, P. Y. (2006). Harnessing in vivo siRNA delivery for drug discovery and therapeutic development. Drug Discov Today *11*, 67–73.

Yu, Q., and Stamenkovic, I. (1999). Localization of matrix metalloproteinase 9 to the cell surface provides a mechanism for CD44-mediated tumor invasion. Genes Dev *13*, 35–48.

Yu, Q., and Stamenkovic, I. (2000). Cell surface-localized matrix metalloproteinase-9 proteolytically activates TGF-beta and promotes tumor invasion and angiogenesis. Genes Dev *14*, 163–176.

Zöller, M. (2011). CD44: can a cancer-initiating cell profit from an abundantly expressed molecule? Nat Rev Cancer *11*, 254–267.

Index

Printed and bound by CPI Group (UK) Ltd, Croydon, CR0 4YY
21/09/2026
14966558-0001